The Science and Engineering of Materials

The Science and Engineering of Materials

THIRD S.I. EDITION

Donald R. Askeland

S.I. Adaptation by
Frank Haddleton, Phil Green
and Howard Robertson

Stanley Thornes (Publishers) Ltd

First published in the USA by PWS Publishers 1984
Second edition 1989

SI edition first published in the UK by Chapman & Hall 1988
Second edition 1990
Third edition 1996

Reprinted in 1998 by:
Stanley Thornes (Publishers) Ltd
Ellenborough House
Wellington Street
CHELTENHAM
GL50 1YW
United Kingdom

98 99 00 01 02 / 10 9 8 7 6 5 4 3 2 1

A catalogue record for this book is available from the British Library

ISBN 0–7487–4083–X

Typeset by Logotechnics CPC Ltd, Sheffield
Printed and bound in China

Contents

Part 3

Engineering Materials

Part 4 Physical Properties of Engineering Materials

Preface

The Science and Engineering of Materials, Third Edition, continues the general theme of the earlier editions in providing an understanding of the relationship between structure, processing, and properties of materials. This text is intended for use by students of engineering rather than materials, at first degree level who have completed prerequisites in chemistry, physics, and mathematics. The author assumes these students will have had little or no exposure to engineering sciences such as statics, dynamics, and mechanics.

The material presented here admittedly cannot and should not be covered in a one-semester course. By selecting the appropriate topics, however, the instructor can emphasise metals, provide a general overview of materials, concentrate on mechanical behaviour, or focus on physical properties. Additionally, the text provides the student with a useful reference for accompanying courses in manufacturing, design, or materials selection.

In an introductory, survey text such as this, complex and comprehensive design problems cannot be realistically introduced because materials design and selection rely on many factors that come later in the student's curriculum. To introduce the student to elements of design, however, more than 100 examples dealing with materials selection and design considerations are included in this edition. These examples, which provide the student with the opportunity to apply the mechanical and physical properties of materials in the selection process, take three forms:

- **Design Examples** cover a range of design considerations, such as operating temperature, presence of corrosive media, economic considerations, recyclability, and environmental constraints, and their effect on materials and process selection, as in Example 13.2, 15.10, and 16.9.
- **Materials Selection Examples** specify a problem and essential design requirements. The information necessary to select the material and its processing route is provided in the solution, as in Example 13.10.
- **Open-Ended Examples** suggest that many solutions are possible, depending on considerations not described in the example, as in Examples 8.8 and 23.5.

Part I, which describes atomic bonding, atom arrangement, lattice imperfections, and diffusion, has been modified or supplemented with new topics in an effort to make this section of the text less metal-oriented. For example, to better explain slip in crystalline materials, including ceramics, we have introduced new illustrations of dislocations, along with an abbreviated version of the Peierls-Nabarro stress.

In Part II more examples and illustrations of mechanical testing of nonmetallics are included, such as impact properties of polymers and flexural strength and modulus of brittle ceramics and composites. In addition, treatment of fracture toughness and crack growth rates in Chapter 6 are greatly expanded. Chapter 8 has been expanded to include the solidification of polymers. Expanded coverage of intermetallic compounds is provided by Chapter 10, and the kinetics of phase transformations are introduced and related both to ceramics and polymers as well as to metals in Chapter 11.

In Part III the discussion of ferrous alloys in Chapter 12 is streamlined for easier understanding. Besides the addition of material on nonferrous alloys, Chapter 13 now includes a description of how metals are produced, a section on advanced aluminium alloys and processing techniques, examples of recycling and environmental problems, and a number of design and materials selection examples.

Chapter 14, on ceramics materials, includes an expanded discussion of deformation and mechanical failure, such as the importance of the inevitable presence of flaws in ceramics, and the necessity for using statistical treatments, such as Weibull statistics, to characterise the behaviour of ceramics. This expanded material also includes a discussion of the crucial role of processing on a ceramic's performance, including toughening mechanisms. Section 14.6 provides an example of this material. Although phase diagrams are still found in Chapter 14, they are now introduced throughout the chapter to illuminate the concepts being discussed. For instance, phase diagrams are used in Section 14.4 to illustrate the production of glasses. To expand the coverage of high temperature properties, such as creep in crystalline and glassy ceramics, Section 14.7 has been added.

Chapter 15, on polymers, has been rewritten and reorganised to include expanded coverage in Section 15.7 on deformation and failure. Section 15.3 provides a more comprehensive description of addition polymerisation; and a new section on adhesives illustrates this important application for polymer materials.

Chapter 16, on composites, has been rewritten to include additional treatment, in Sections 16.4 and 16.5, of the mechanical properties of fibre-reinforced composites. Section 16.10 now includes a more detailed description of honeycomb structures. The discussion of wood and concrete appears together in a separate and expanded Chapter 17, which includes a more detailed treatment of the behaviour of wood and the rationale for producing concrete mixes for specific applications.

In Part IV Chapter 18 covers the treatment of dielectric behaviour, thus allowing the student to understand the continuum of changes in the electrical behaviour of materials. Also introduced in this chapter is additional information on superconduction and conduction in polymers and composites. Chapter 20, on optical behaviour, has been modified; and magnetic properties and thermal behaviour are now discussed in Chapters 19 and 21, respectively.

In Part V the corrosion and degradation of polymers, including the use of biodegradable polymers, is treated in more detail in this edition.

A considerable number of the 'plug-and-crank' examples of the earlier editions have been replaced by the new selection and design examples, which show the student both how to perform the calculation and how the calculation can help to solve the applicable engineering problem.

Summaries found at the end of each chapter have been expanded and are more thorough than those in previous editions. Most of the problems, including suggestions for design and selection projects, are new to this edition.

In addition to the many content changes noted, this edition features a completely redrawn illustration program. These enhanced figures should further help the students to visualise the concepts being presented.

Supplements for instructors include:

- **an *Instructor's Solutions Manual*** that provides complete, worked-out solutions to all text problems and additional text items; and
- **a set of more than 100 full-colour photo micrographs** in slide format, classified by materials type, for in-class presentation.

Both are provided free to Askeland adopters.

Supplements for students include:

- ***Theorist notebooks for Science and Engineering of Materials*** (part of the *PWS Notebook Series*™), a collection of interactive problems supporting Askeland's chapters 2–18, for use with the *Student Edition of Theorist* symbolic algebra program for Macintosh (also published by PWS Publishing Company);
- **a *Materials Science and Engineering Lab Manual*,** by Sherif El Wakil (University of Massachusetts at Dartmouth), which contains a blend of classic materials science lab exercises for metals and for non-metallic materials including ceramics, composites and polymers;
- **a *CD-ROM for Materials Science*,** a collection of QuickTime® animated visualisations developed by John Russ (North Carolina State University), covering such important materials science concepts as crystal deformation, phase diagrams, shearing, and molecular bonding-distributed on CD-ROM disk for Macintosh and IBM Windows; and
- ***professional-quality crystallographic software for institutional site licence!*** *Software for the Science and Engineering of Materials,* developed by James T. Staley, Scientific Software Services, contains five programs, for IBM DOS and Macintosh computers, that let students visualise crystallographic planes and directions in cubic crystals and relate them to their Miller indices; visualise the atomic arrangements of atoms on the crystallographic planes of cubic and hexagonal metals and intermetallic compounds; study how types of incident X-radiation, crystal structure, atomic species and lattice parameters affect the spacing and intensities of spectral lines on a Debye-Scherer X-ray powder pattern; understand the lever law by graphically demonstrating its use for several two-component phase diagrams; and design structural components using interactive materials selection charts. Site licences (scaled to the number of students enrolled) are discounted 50% to schools adopting the Askeland text.

Acknowledgements

I am indebted to the many people who have provided the assistance, encouragement, and constructive criticism leading to the preparation of this third edition of the text.

My colleagues at UMR – Fred Kisslinger, Ron Kohser, Scott Miller, Chris Ramsay, Harry Weart, and Robert Wolf – have had the patience to use the text in our introductory courses and have provided invaluable suggestions.

Special thanks are extended to all of our students in MET 121 and AE 241; they are why I'm here.

I wish to acknowledge the many instructors who have read and used the text, and in particular the enormously helpful reviews of the manuscript offered by

C. M. Balik
North Carolina State University

James H. Edgar
Kansas State University

Michael W. Glasgow
Broome Community College

Philip J. Guichelaar
Western Michigan University

I. W. Hall
University of Delaware

D. Bruce Masson
Washington State University

Len Rabenberg
University of Texas at Austin

David A. Thomas
Lehigh University

James G. Vaughan
The University of Mississippi

Krishna Vedula
Iowa State University

Sherif D. El Wakil
University of Massachusetts/Dartmouth

Robert A. Wilke
The Ohio State University

Alan Wolfenden
Texas A & M University

Ernest G. Wolff
Oregon State University

Thanks most certainly go to everyone at PWS Publishing Company, Thompson Steele Book Production, and Vantage Art for their encouragement, advice, and particularly their patience. Thanks, Barbara, Sally and Elinor.

Finally, and most importantly, I am deeply indebted to my wife, Mary, and son, Per.

Donald R. Askeland
University of Missouri – Rolla

Preface

We classify materials into several major groups: metals, ceramics, polymers, semiconductors, and composites. The behaviour of the materials in each of these groups is determined by their structure. The electronic structure of an atom determines the nature of atomic bonding, which helps govern the mechanical and physical properties of a given material.

The arrangement of atoms into a crystalline or amorphous structure also influences a material's behaviour. Imperfections in atomic arrangement play a critical role in our understanding of deformation, failure, and mechanical properties.

Finally, the movement of atoms, known as diffusion, is important for many heat treatments and manufacturing processes, as well as for both physical and mechanical properties of materials.

In the following chapters, we introduce the structure-property-processing concept for controlling the behaviour of materials and examine the roles of atomic structure, atomic arrangement, imperfections, and atom movement. This examination lays the groundwork needed to understand the structure and behaviour of materials discussed later on.

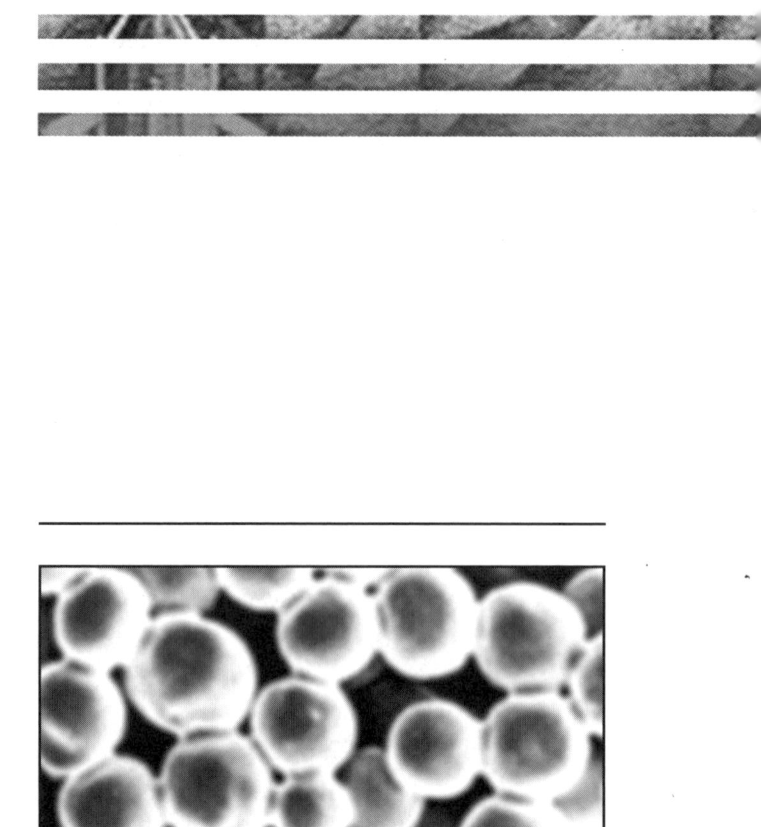

Powder consolidation of particles is a common method for manufacturing metal, ceramic, and composite materials. Diffusion of atoms to points of contact between the particles – in this case, spherical copper powders – during sintering causes the particles to become bonded. Continued sintering and diffusion eventually cause the pores between the particles to disappear. (*From* Metals Handbook, *Vol. 9, 9th Ed., ASM International, 1985*)

Part

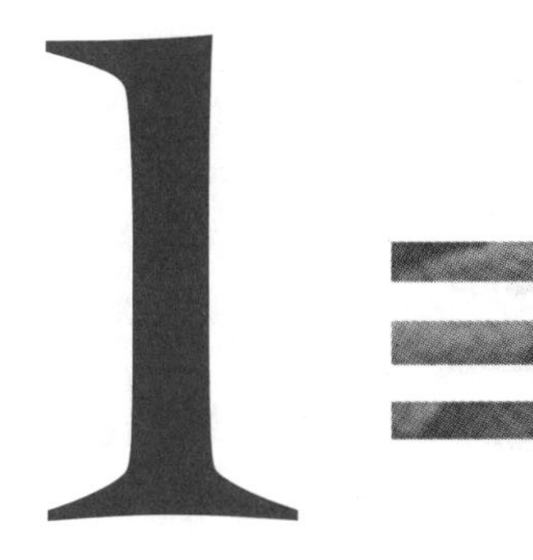

Atomic structure, Arrangement, and Movement

CHAPTER 1

Introduction to Materials

1.1 Introduction

All engineers are involved with materials on a daily basis in manufacturing and processing and in the design and construction of components or structures. They must select and use materials and analyse the failure of materials.

A number of important decisions must be made when selecting materials to be incorporated into a design, including whether the materials can consistently be formed into the correct shape and dimensional tolerances and maintain the correct shape during use, whether the required properties can be achieved and maintained during use, whether a material is compatible with and can be easily joined to other parts of an assembly, whether a material can be easily recycled, whether a material or its manufacture can cause environmental problems, and whether a material can be economically made into a useful part.

The intent of this text is to help the reader to become aware of the types of materials available, to understand their general behaviour and capabilities, and to recognise the effects of the environment and service conditions on their material's performance. This understanding is necessary for one to be able to participate in the design of reliable and economical components, systems, and processes that utilise the wide spectrum of materials.

1.2 Types of Materials

Materials are classified into five groups: metals, ceramics, polymers, semiconductors, and composite materials (Table 1.1). Materials in each of these groups possess different structures and properties. The differences in strength, which are compared in Figure 1.1, illustrate the wide range of properties available.

Metals Metal and alloys, including steel, aluminium, magnesium, zinc, cast iron, titanium, copper, and nickel, generally have the characteristics of good electrical and thermal conductivity, relatively high strength, high stiffness, ductility or formability, and shock resistance. They are particularly useful for structural or load-bearing applications. Although pure metals are occasionally used, combinations of metals called

Table 1.1 Representative examples, applications, and properties for each category of materials.

	Applications	Properties
Metals		
Copper	Electrical conductor wire	High electrical conductivity, good formability
Grey cast iron	Automobile engine blocks	Castable, machinable, vibration-damping, cheap
Alloy steels	Wrenches	Significantly strengthened by heat treatment
Ceramics		
SiO_2-Na_2O-CaO	Window glass	Optically transparent, thermal insulating
Al_2O_3, MgO, SiO_2	Refractories for containing molten metal	Thermal insulating, resistant to high temperature, relatively inert to molten metal
Barium titanate	Transducers for audio equipment	Converts sound to electricity (Piezoelectric behaviour)
Polymers		
Polyethylene	Food packaging	Easily formed into thin, flexible, airtight film
Epoxy	Encapsulation of integrated circuits	Electrically insulating and moisture-resistant
Phenolics	Adhesives for joining plies in plywood	Strong, moisture resistant
Semiconductors		
Silicon	Transistors and integrated circuits	Unique electrical behaviour
GaAs	Fibre-optic systems	Converts electrical signals to light
Composites		
Graphite-epoxy	Aircraft components	High strength-to-weight ratio
Tungsten carbide-cobalt	Carbide cutting tools for machining	High hardness, yet good shock resistance
Titanium-clad steel	Reactor vessels	Has the low cost and high strength of steel, with the corrosion resistance of titanium

alloys provide improvement in a particular desirable property or permit better combinations of properties. The section through a jet engine shown in Figure 1.2 illustrates the use of several metal alloys for a very critical application.

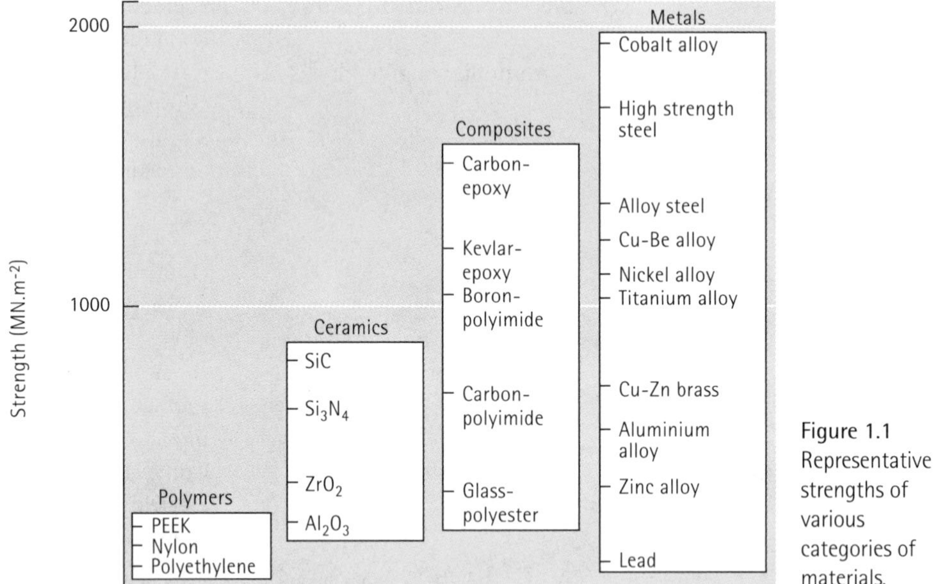

Figure 1.1
Representative strengths of various categories of materials.

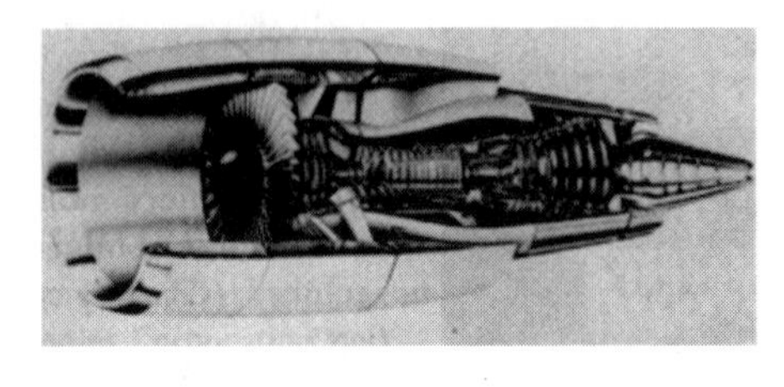

Figure 1.2
A section through a jet engine. The forward compression section operates at low to medium temperatures, and titanium parts are often used. The rear combustion section operates at high temperatures, and nickel-base superalloys are required. The outside shell experiences low temperatures, and aluminium and composites are satisfactory. *(Courtesy of GE Aircraft Engines.)*

Ceramics Brick, glass, tableware, refractories, and abrasives have low electrical and thermal conductivities and are often used as insulators. Ceramics are strong and hard, but also very brittle. New processing techniques make ceramics sufficiently resistant to fracture that they can be used in load-bearing applications, such as impellers in turbine engines (Figure 1.3).

Polymers Produced by creating large molecular structures from organic molecules in a process known as polymerisation, polymers include rubber, plastics, and many types of adhesives. Polymers have low electrical and thermal conductivities, are low in strength, and are not suitable for use at high temperatures. Thermoplastic polymers, in which the long molecular chains are not rigidly connected, have good formability; thermosetting polymers are stronger but more brittle because the molecular chains are tightly linked (Figure 1.4). Polymers are used in many applications including electronic devices (Figure 1.5).

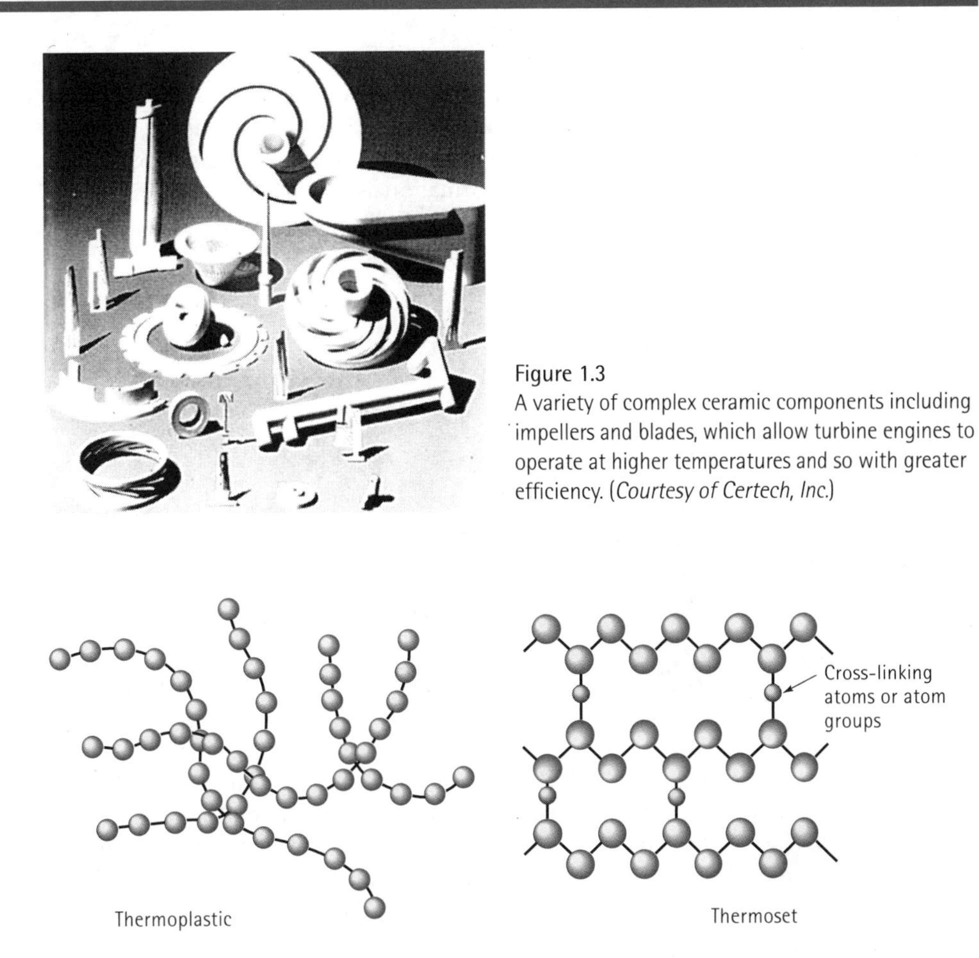

Figure 1.3
A variety of complex ceramic components including impellers and blades, which allow turbine engines to operate at higher temperatures and so with greater efficiency. (*Courtesy of Certech, Inc.*)

Figure 1.4
Polymerisation occurs when small molecules, represented by the circles, combine to produce larger molecules, or polymers. The polymer molecules can have a chainlike structure (thermoplastics) or can form three-dimensional networks (thermosets).

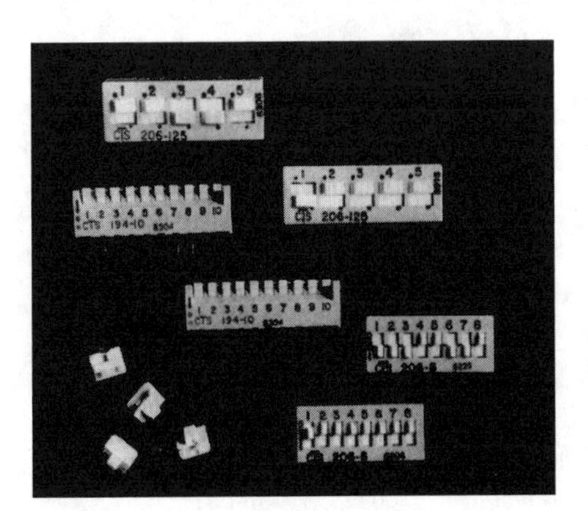

Figure 1.5
Polymers are used in a variety of electronic devices, including these computer dip switches, where moisture resistance and low conductivity are required. (*Courtesy of CTS Corporation.*)

Semiconductors Although silicon, germanium, and a number of compounds such as gallium arsenide (GaAs) are very brittle, they are essential for electronic, computer, and communication applications. The electrical conductivity of these materials can be controlled to enable their use in electronic devices such as transistors, diodes, and integrated circuits (Figure 1.6). Information is now being transmitted by light through fibre-optic systems; semiconductors, which convert electrical signals to light and vice versa, are essential components in these systems.

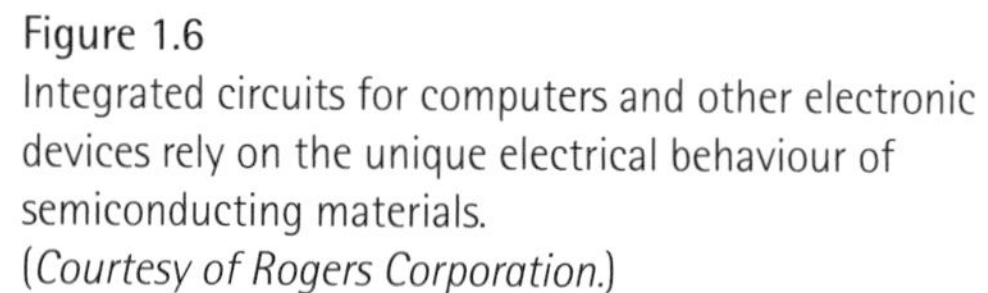

Figure 1.6
Integrated circuits for computers and other electronic devices rely on the unique electrical behaviour of semiconducting materials.
(*Courtesy of Rogers Corporation.*)

Composite Materials Composites are formed from two or more materials, producing properties not found in any single material. Concrete, plywood, and fibreglass are typical – although crude – examples of composite materials. With composites we can produce lightweight, very strong, very stiff, high temperature-resistant materials or we can produce hard, yet shock-resistant, cutting tools that would otherwise shatter. Advanced aircraft and aerospace vehicles rely heavily on composites such as carbon fibre reinforced polymers (Figure 1.7), and Kevlar reinforced polymers.

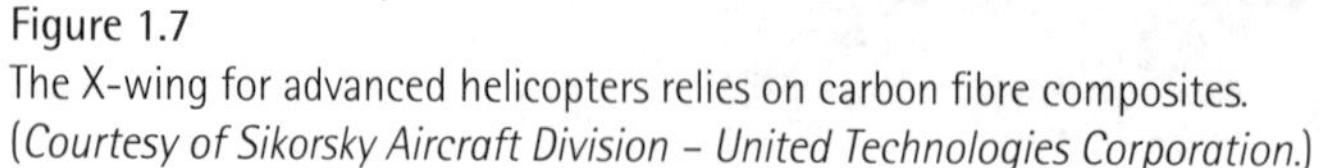

Figure 1.7
The X-wing for advanced helicopters relies on carbon fibre composites.
(*Courtesy of Sikorsky Aircraft Division – United Technologies Corporation.*)

EXAMPLE 1.1 **Materials Selection for an Electrical 'Black Box'**

Explain the selection of materials used in a cable to carry a current between two components of an electrical 'black box'.

SOLUTION

The material that actually carries the current must have high electrical conductivity. A metal such as copper aluminium, or gold is selected, but the metal wire must be insulated from the rest of the 'black box' to prevent short circuits or arcing. Although a ceramic coating provides excellent insulation, ceramics are brittle, and the wire cannot be bent without breaking the ceramic coating. Instead, we select a thermoplastic polymer or plastic coating with good insulating characteristics and good flexibility.

However, if the cable was to operate at elevated temperature e.g. near a furnace, then ceramic beading would be preferred to a plastic sheath.

EXAMPLE 1.2 **Materials Selection for a Coffee Cup**

Select a material from which a coffee cup might be produced. What particular property makes this material suitable?

SOLUTION

To avoid burning the user's hands, coffee cups should provide excellent thermal insulation. Both ceramics and polymers are appropriate due to their low thermal conductivity. Disposable expanded polystyrene cups are particularly effective because they contain many gas bubbles, which further improve insulation. This desirable physical property must, however, be weighed against the potential environmental damage caused by disposal of the polymer material. Ceramic cups can be reused and are of less danger to the environment. Metal cups would not be used, of course, because of their high thermal conductivity, and cause burnt fingers!

1.3 Structure-Property-Processing Relationship

A component must have the proper shape to perform its tasks for its expected lifetime. The materials engineer meets this requirement by taking advantage of a complex three-part relationship between the internal structure of the material, the processing of the material, and the final properties of the material (Figure 1.8).

When the materials engineer changes any one of these aspects of the relationship, either or both of the others also change.

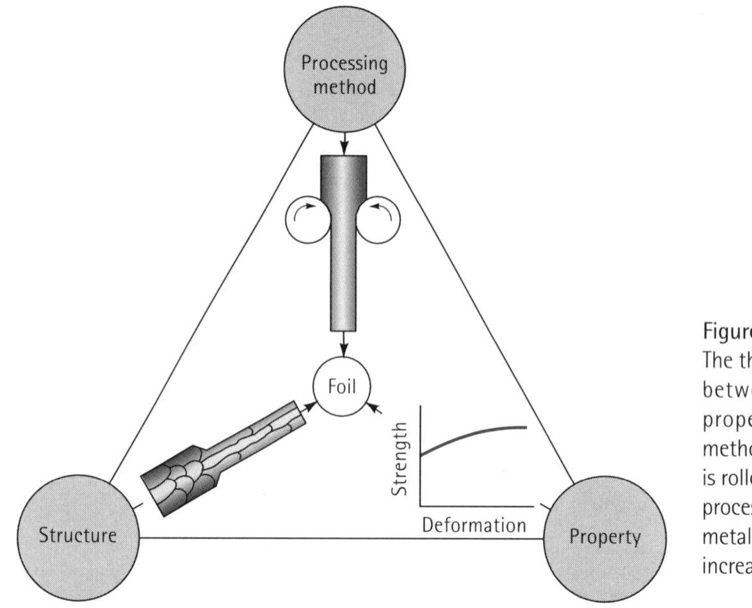

Figure 1.8
The three-part relationship between structure, properties, and processing method. When aluminium is rolled into foil, the rolling process changes the metal's structure and increases its strength.

Properties We can consider the properties of a material in two categories: mechanical and physical. Mechanical properties which describe how a material responds to an applied force, include strength and ductility. We are, however, often interested in how a material behaves when it is exposed to a sudden, intense blow (impact), continually cycled through an alternating force (fatigue), exposed to high temperatures (creep), or subjected to abrasive conditions (wear). Mechanical properties also determine the ease with which a material can be deformed into a useful shape. For example, a metal part formed by deep drawing must have high ductility to deform to the proper shape. A creep resistant alloy may be very strong and resist applied stress, but consequently is very difficult to deform and shape into a component. Small structural changes often have a profound effect on mechanical properties.

Physical properties. which include electrical, magnetic, optical, thermal, elastic, and chemical behaviour, depend on both structure and processing of a material. Even tiny changes in structure cause profound changes in the electrical conductivity of many semiconducting materials; for example high firing temperatures may greatly reduce the thermal insulation characteristics of ceramic brick.

EXAMPLE 1.3 Materials Selection for an Aircraft Wing

Describe some of the key mechanical and physical properties to consider in selecting a material for an aircraft wing.

SOLUTION

First, let's consider mechanical properties. The material must have strength to resist the forces acting on the wing. The wing is also exposed to a cyclical application of force during landing and takeoff, as well as vibration during flight, suggesting that fatigue properties are important. In supersonic flight, the wing may become very hot, so resistance to creep may be critical.

Physical properties are also important. Because the wing should be as light as possible, the material should have a low density. If the wing is exposed to a marine atmosphere, corrosion resistance is needed. In the event of a lightning strike, the electrical charge should be dissipated to avoid localised damage; consequently, the material must possess good conductivity.

Traditionally, aluminium alloys have been used to meet these requirements. Now, however, fibre-reinforced polymer matrix composites are being increasingly used in many advanced, high-performance aircraft because of their better strength-to-weight ratio and stiffness.

Structure The structure of a material can be considered on several levels (Figure 1.9). The arrangement of the electrons surrounding the nucleus of individual atoms affects electrical, magnetic, thermal, and optical behaviour. Furthermore, the electronic arrangement influences how the atoms are bonded to one another.

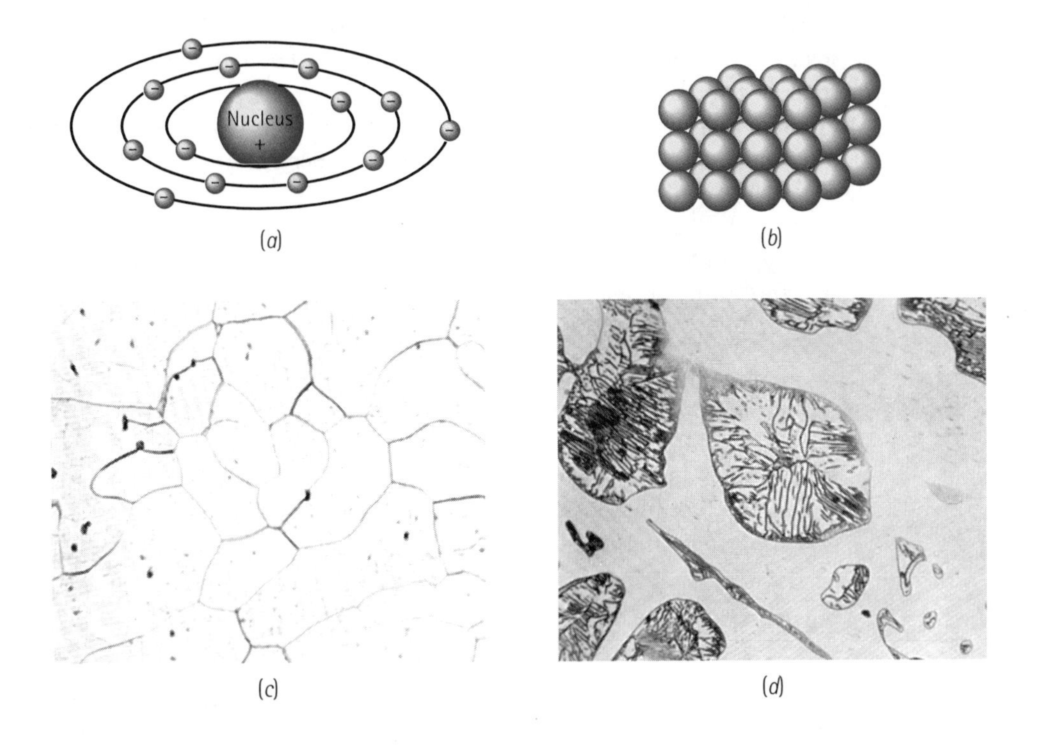

Figure 1.9
Four levels of structure in a material: (*a*) atomic structure, (*b*) crystal structure, (*c*) grain structure in iron (× 100), and (*d*) multiple-phase structure in white cast iron (× 200).

At the next level, the arrangement of the atoms is considered. Metals, semiconductors, many ceramics, and some polymers have a very regular atomic arrangement, or crystal structure. Other ceramic materials and many polymers have no orderly atomic arrangement. These amorphous, or glassy, materials behave very differently from crystalline materials. For instance, glassy polyethylene is transparent, whereas crystalline polyethylene is translucent. Imperfections in the atomic arrangement may be controlled to produce profound changes in properties.

A grain structure is found in most metals, semiconductors, and ceramics. The size and shape of the grains influence a material's behaviour. In some cases, as with silicon chips for integrated circuits or metals for jet engine parts, we wish to produce a material containing only one grain or a single crystal.

Finally, in most materials, more than one phase is present. A phase can be regarded as a distinctive part of a microstructure with each phase having its unique atomic arrangement and properties. Control of the type, size, distribution, and amount of these phases within the main body of a material provides an additional way to control properties.

Processing Materials processing produces the desired shape of a component from the initial formless material. Metals can be processed by pouring liquid metal into a mould (casting), joining individual pieces of metal (welding, brazing, soldering, adhesive bonding), forming the solid metal into useful shapes using high pressure (forging, drawing, extrusion, rolling, bending), compacting tiny metal powder particles into a solid mass (powder metallurgy), or removing excess material (machining). Similarly, ceramic materials can be formed into shapes by related processes such as casting, forming, extrusion, or compaction – often while wet – followed by heat treatment at high temperatures. Polymers are produced by injection of softened plastic into moulds (much like casting), drawing, and forming. Often a material is heat-treated at some temperature below its melting temperature to effect a desired change in structure. The type of processing we use depends, partly at least, on the properties – and thus the structure – of the material. It also depends on the shape of the component being produced.

EXAMPLE 1.4 Selection of a Process to Produce Tungsten Filaments

Tungsten has an unusually high melting temperature of 3410°C, making it difficult to process into useful shapes. Select a process by which small-diameter filaments can be produced from tungsten.

SOLUTION

Because of tungsten's high melting temperature, most casting processes cannot be used for it. A common method for producing tungsten is by a powder metallurgy process. Powder particles of tungsten oxide (WO_3), a ceramic, are heated in a hydrogen atmosphere; the subsequent reaction produces metallic tungsten particles and H_2O. The tungsten powder particles are consolidated at high temperatures and pressures into simple rods. The rods can then be wire-drawn (a forming process) to progressively smaller diameters until the correct size is produced (Figure 1.10).

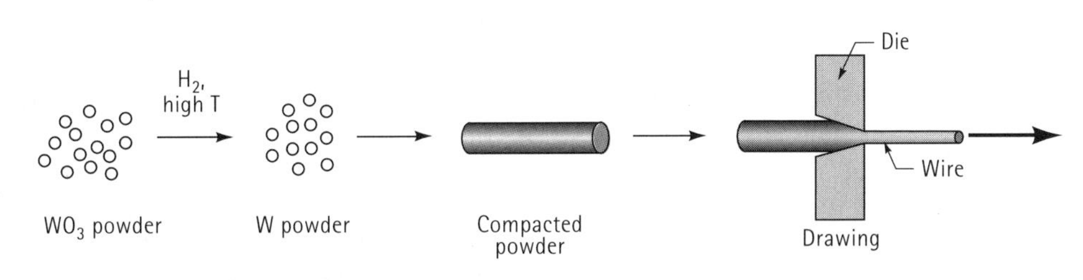

Figure 1.10
Tungsten oxide can be reduced to tungsten metal powder particles, which are compacted into a simple shape and drawn into wire.

1.4 Environmental Effects on Material Behaviour

The structure-property-processing relationship is influenced by the surroundings to which the material is subjected, including high temperature and corrosion.

Temperature Changes in temperature dramatically alter the properties of materials (Figure 1.11). Metals that have been strengthened by certain heat treatment or forming techniques may suddenly lose their strength when heated. High temperatures change the structure of ceramics and cause polymers to melt or char. (Very low temperatures, at the other extreme, may cause a metal or polymer to fail in a brittle manner, even though the applied loads are low.)

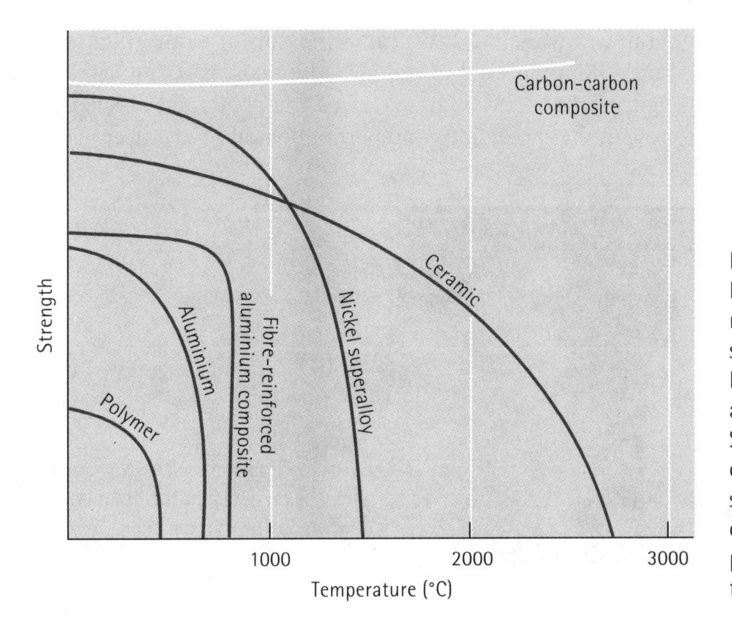

Figure 1.11
Increasing temperature normally reduces the strength of a material. Polymers are suitable only at low temperatures. Some composites, such as carbon-carbon composites, special alloys, and ceramics, have excellent properties at high temperatures.

The design of materials with improved resistance to temperature extremes is essential in many technologies, as illustrated by the increase in operating temperatures of aircraft and aerospace vehicles (Figure 1.12). As faster speeds are attained, more heating of the vehicle skin occurs because of friction with the air. At the same time, engines operate more efficiently at higher temperatures. So, in order to achieve higher speed and better fuel economy, new materials have gradually increased allowable skin and engine temperatures. But materials engineers are continually faced with new challenges. The USA 'national aerospace plane', (NASP) an advanced aircraft intended to carry passengers across the Pacific Ocean in less than three hours, will require the development of even more exotic materials and processing techniques in order to tolerate the higher temperatures that will be encountered. A similar challenge confronted the design engineers during the development of Concorde in the 1960s.

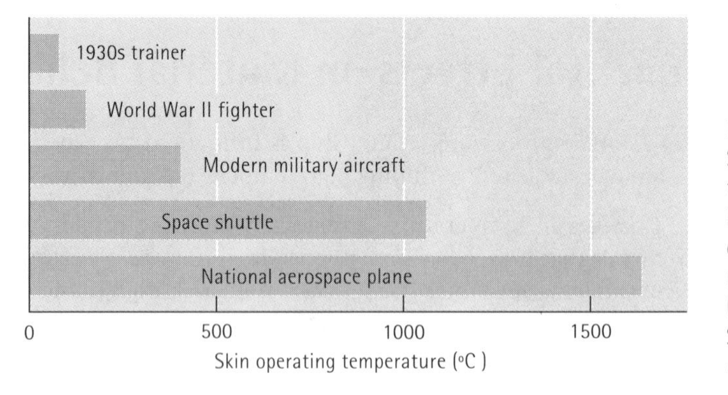

Figure 1.12
Skin operating temperatures for aircraft have increased with the development of improved materials.
(After M. Steinberg, Scientific American, October 1986.)

Corrosion Most metals and polymers react with oxygen or other gases, particularly at elevated temperatures. Metals and ceramics may disintegrate (Figure 1.13); polymers may become brittle. Materials are also attacked by corrosive liquids, leading to premature failure (Figure 1.14). The engineer faces the challenge of selecting materials or coatings that prevent these reactions and permit operation in extreme environments.

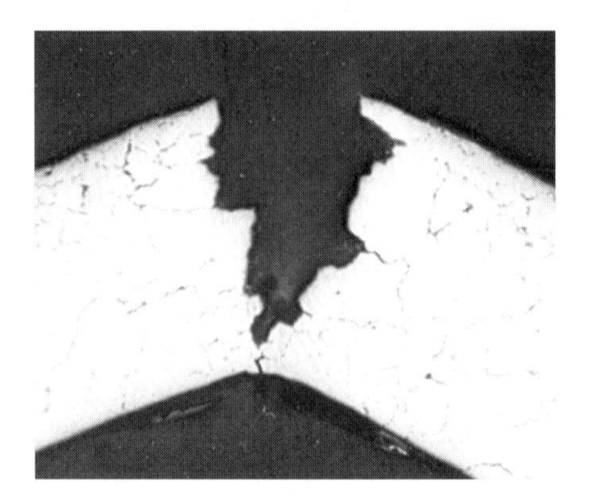

Figure 1.13
When hydrogen dissolves in tough pitch copper, steam is produced at the grain boundaries, creating thin voids. The metal is then weak and brittle and fails easily (× 50).

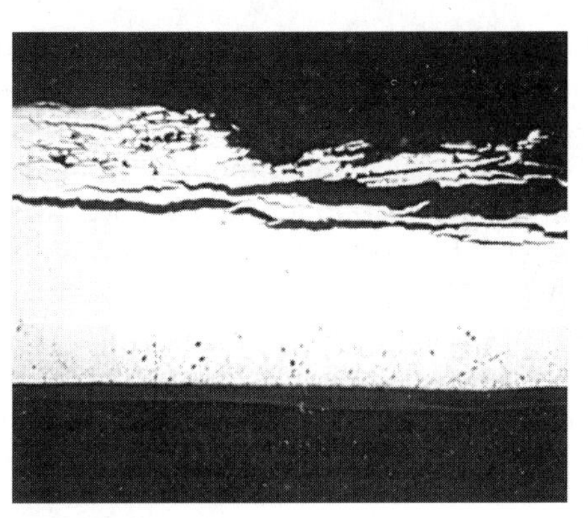

Figure 1.14
Attack on an aluminium fuel tank by bacteria in contaminated jet fuel causes severe corrosion, pitting, and eventually failure (× 10).

EXAMPLE 1.5 **Selection of a Process for Joining Titanium**

What precautions might have to be taken when joining titanium by a welding process?

SOLUTION

During welding, the titanium is heated to a high temperature. The high temperature may cause detrimental changes in the structure of the titanium, eliminating some of the strengthening mechanisms by which the properties of the metal were obtained. Furthermore, titanium reacts rapidly with oxygen, hydrogen, and other gases at high temperatures. A welding process must supply a minimum of heat while protecting the metal from the surrounding atmosphere. Special gases, such as argon, or even a vacuum may be needed to shield the molten weld from the air.

This also explains why an electric light bulb consists of a titanium wire enclosed in a glass bulb to prevent oxidation and failure of the wire during service.

1.5 Materials Design and Selection

When a material is designed for a given application, a number of factors must be considered. The material must acquire the desired physical and mechanical properties, must be capable of being processed or manufactured into the desired shape, and must provide an economical solution to the design problem. Satisfying these requirements in a manner that protects the environment – perhaps by encouraging recycling of the materials – is also essential. In meeting these design requirements, the engineer may have to make a number of trade-offs in order to produce a serviceable, yet marketable, product.

As an example, material cost is normally calculated on a cost per megagram basis. We must consider the density of the material, or its weight per unit volume, in our design and selection (Table 1.2). Aluminium may cost more per megagram than steel, but it is only one-third the weight of steel. Although parts made from aluminium may have to be thicker, the aluminium part may be less expensive than the one made from steel because of the weight difference.

In some instances, particularly in aerospace applications, weight is critical, since additional vehicle weight increases fuel consumption and reduces range. By using materials that are lightweight but very strong, aerospace vehicles can be designed to improve fuel utilisation. Many advanced aerospace vehicles use composite materials instead of aluminium. These composites, such as carbon-epoxy, are much more expensive than the traditional aluminium alloys; however, the fuel savings yielded by the higher strength-to-weight ratio of the composite (Table 1.2) may offset the higher initial cost of the aircraft.

Table 1.2 Strength-to-weight ratio of various materials.

Material	Strength $MN.m^{-2}$	Density $Mg.m^{-3}$	Strength-to-weight ratio $kg.m^2.s^{-2}$
Polyethylene	7	0.83	8
Pure aluminium	45	2.7	17
Al_2O_3	205	3.2	64
Epoxy	100	1.4	71
Heat-treated alloy steel	1650	7.8	212
Heat-treated aluminium alloy	590	2.7	219
Carbon-carbon composite	415	1.8	230
Kevlar-epoxy composite	1170	4.4	266
Carbon-epoxy composite	450	1.4	321
Heat-treated titanium alloy	550	1.4	393

SUMMARY

Selection of a material having the needed properties and the potential to be manufactured economically and safely into a useful product is a complicated process involving knowledge of the *structure-property-processing relationship*. A number of material categories must be considered:

- Metals have good strength, good ductility and formability, good electrical and thermal conductivity, and moderate temperature resistance.
- Ceramics are strong, serve as good electrical and thermal insulators, are often resistant to damage by high temperatures and corrosive environments, but are brittle.
- Polymers have relatively low strength, are not suitable for use at high temperatures, have good corrosion resistance, and – like ceramics – provide good electrical

and thermal insulation. Polymers may be either flexible or brittle, depending on whether a thermoplastic or thermosetting polymer is selected.

- Semiconductors possess unique electrical and optical properties that make them essential components in electronic and communication devices.
- Composites are mixtures of materials that provide unique combinations of mechanical and physical properties that cannot be found in any single material.

As we examine these materials, we come to understand the fundamentals of the structure of materials, how the structure affects the behaviour of the material, and the role that processing and the environment play in shaping the relationship between structure and properties. In particular, our discussion provides the basis for designing materials, components, and systems that are safe and economical.

GLOSSARY

Alloys
Combinations of metals that enhance the range of properties of the individual metals.

Ceramics
A group of materials characterised by good strength and high melting temperatures, but brittle and having poor electrical conductivity. Ceramic raw materials are typically compounds of metallic and nonmetallic elements.

Composites
A group of materials formed from mixtures of metals, ceramics, or polymers in such a manner that unusual combinations of properties are obtained.

Crystal structure
The arrangement of the atoms in a material into a regular repeatable pattern.

Density
Mass per unit volume of a material, usually expressed in units of $Mg.m^{-3}$. Note that $1\ Mg.m^{-3} \equiv 1\ g.cm^{-3}$

Mechanical properties
Properties of a material, such as strength, that describe how well a material withstands applied forces, including tensile or compressive forces, impact forces, cyclical or fatigue forces, or forces at high temperatures.

Metals
A group of materials having the general characteristics of good ductility, strength, and electrical conductivity.

Phase
A material having the same composition, structure, and properties everywhere under equilibrium conditions. A distinctive part of a microstructure.

Physical properties
Describe characteristics such as colour, elasticity, electrical or thermal conductivity, magnetism, and optical behaviour that generally are not influenced by forces acting on a material.

Polymerisation
The process by which small organic molecules are joined into giant molecules, or polymers.

Polymers
A group of materials normally obtained by joining organic molecules into giant molecular chains or networks. Polymers are characterised by low strengths, low melting temperatures, and poor electrical conductivity.

Semiconductors
A group of materials having intermediate electrical conductivity and other unusual physical properties.

Strength-to-weight ratio
The strength of a material divided by its density; materials with a high strength-to-weight ratio are strong but lightweight.

Thermoplastic
A special group of polymers that are easily formed into useful shapes. Normally, these polymers have a chainlike structure.

Thermosets
A special group of polymers that are normally quite brittle. These polymers typically have a three-dimensional network structure, and are less sensitive to temperature than are thermoplastics.

PROBLEMS

1.1 Iron is often coated with a thin layer of zinc if it is to be used outside. What characteristics do you think the zinc provides to this coated, or galvanised, steel? What precautions should be considered in producing this product? How will the recyclability of the product be affected?

1.2 We would like to produce a transparent canopy for an aircraft. If we were to use a ceramic (that is, traditional window glass) canopy, rocks or birds might cause it to shatter. Select a material that would minimise damage or at least keep the canopy from breaking into pieces.

1.3 Coiled springs ought to be very strong and stiff. Si_3N_4 is a strong, stiff material. Would you select this material for a spring? Explain.

1.4 Temperature indicators are sometimes produced from a coiled metal strip that uncoils a specific amount when the temperature increases. How does this work; from what kind of material would the indicator be made; and what are the important properties that the material in the indicator must possess?

1.5 You would like to design an aircraft that can be flown by human power nonstop for a distance of 30 km. What types of material properties would you recommend? What materials might be appropriate?

1.6 You would like to place a three-foot diameter microsatellite into orbit. The satellite will contain delicate electronic equipment that will send and receive radio signals from earth. Design the outer shell within which the electronic equipment is contained. What properties will be required, and what kind of materials might be considered?

1.7 What properties should the head of a carpenter's hammer possess? How would you manufacture a hammer head?

1.8 The hull of the space shuttle consists of ceramic tiles bonded to an aluminium skin. Discuss the design requirements of the shuttle hull that led to the use of this combination of materials. What problems in producing the hull might the designers and manufacturers have faced?

1.9 You would like to select a material for the electrical contacts in an electrical switching device which opens and closes frequently and forcefully. What properties should the contact material possess? What type of material might you recommend? Would Al_2O_3 be a good choice? Explain.

1.10 Aluminium has a density of 2.7 $Mg.m^{-3}$. Suppose you would like to produce a composite material based on aluminium having a density of 1.5 $Mg.m^{-3}$. Design a material that would have this density. Would introducing beads of polyethylene, with a density of 0.95 $Mg.m^{-3}$, into the aluminium be a likely possibility? Explain.

1.11 You would like to be able to identify different materials without resorting to chemical analysis or lengthy testing procedures. Describe some possible testing and sorting techniques you might be able to use based on the physical properties of materials.

1.12 You would like to be able to physically separate different materials in a scrap recycling plant. Describe some possible methods that might be used to separate materials such as polymers, aluminium alloys, and steels from one another.

1.13 Some pistons for automobile engines might be produced from a composite material containing small, hard silicon carbide particles in an aluminium alloy matrix. Explain what benefits each material in the composite may provide to the overall part. What problems might the different properties of the two materials cause in producing the part?

CHAPTER 2

Atomic Structure

2.1 Introduction

The structure of a material may be examined at four levels: atomic structure, atomic arrangement, microstructure, and macrostructure. Although the main thrust of this text is to understand and control the microstructure and macrostructure of various materials, we must first understand atomic and crystal structures.

Atomic structure influences how atoms are bonded together, an understanding of which in turn helps us to categorise materials as metals, semiconductors, ceramics, and polymers and permits us to draw some general conclusions concerning the mechanical properties and physical behaviour of these four classes of materials.

2.2 The Structure of the Atom

An atom is composed of a nucleus surrounded by electrons. The nucleus contains neutrons and positively charged protons and carries a net positive charge. The negatively charged electrons are held to the nucleus by an electrostatic attraction. The electrical charge q carried by each electron and proton is 1.60×10^{-19} coulomb (C). Because the number of electrons and protons in the atom are equal, the atom as a whole is electrically neutral.

The atomic number of an element is equal to the number of electrons or protons in each atom. Thus, an iron atom, which contains 26 electrons and 26 protons, has an atomic number of 26. Atomic number is the basis of the Periodic Table.

The atomic mass, M, of an atom is concentrated in the nucleus. The mass of a proton and a neutron is 1.67×10^{-24} g whilst the mass of an electron is much smaller at 9.11×10^{-28} g. Thus the mass of one atom would be a very small value in terms of grams or other practical units.

Atoms react with other atoms and so a practical concept is the mole; this is known as the Avogadro number (N_A) and equals 6.02×10^{23}. Consequently 32g of S will contain N_A number of atoms, 6.02×10^{23}. These weights are called relative atomic mass or relative atomic weight. The elements react together according to their relative atomic masses and valency.

One atomic mass unit (a.m.u.) is equal to 1/12 the mass of carbon 12. As an example, 1 mole of Fe contains 6.02×10^{23} and has a mass of 55.847g or 55.847 a.m.u.

To produce a mole of FeS, 55.85 g of Fe and 32.06 g of S would need to be heated together to react.

EXAMPLE 2.1

Calculate the number of atoms in 100 g of silver.

SOLUTION

The number of atoms can be calculated from the atomic mass and Avogadro's number. From Appendix A, the atomic mass, or weight, of silver is 107.868 $g.mol^{-1}$.

The number of atoms is:

$$\text{Number of Ag atoms} = \frac{(100\ \text{g})(6.02 \times 10^{23}\ \text{atoms.mol}^{-1})}{107.868\ \text{g.mol}^{-1}}$$

$$= 5.58 \times 10^{23}$$

2.3 The Electronic Structure of the Atom

Electrons occupy discrete energy levels within the atom. Each electron possesses a particular energy, with no more than two electrons in each atom having the same energy. This also implies that there is a discrete energy difference between any two energy levels.

Quantum Numbers The energy level to which each electron belongs is determined by four quantum numbers. The number of possible energy levels is determined by the first three quantum numbers.

1 The principal quantum number n is assigned integral values 1, 2, 3, 4, 5,... that refer to the quantum shell to which the electron belongs (Figure 2.1). Quantum shells are also assigned a letter; the shell for $n = 1$ is designated K, for $n = 2$ is L, for $n = 3$ is M, and so on.

2 The number of energy levels in each quantum shell is determined by the azimuthal quantum number l and the magnetic quantum number m_l. The azimuthal quantum numbers are also assigned numbers: $l = 0, 1, 2, ..., n - 1$. If $n = 2$, then there are also two azimuthal quantum numbers, $l = 0$ and $l = 1$. The azimuthal quantum numbers are designated by lower case letters:

s for $l = 0$ $\quad$ d for $l = 2$
p for $l = 1$ $\quad$ f for $l = 3$

1 Often the atomic mass is called the *atomic weight*.

The magnetic quantum number m_l gives the number of energy levels, or orbitals, for each azimuthal quantum number. The total number of magnetic quantum numbers for each l is $2l + 1$. The values for m_l are given by whole numbers between $-l$ and $+l$. For $l = 2$, there are $2(2) + 1 = 5$ magnetic quantum numbers, with values −2, −1, 0, +1, and +2.

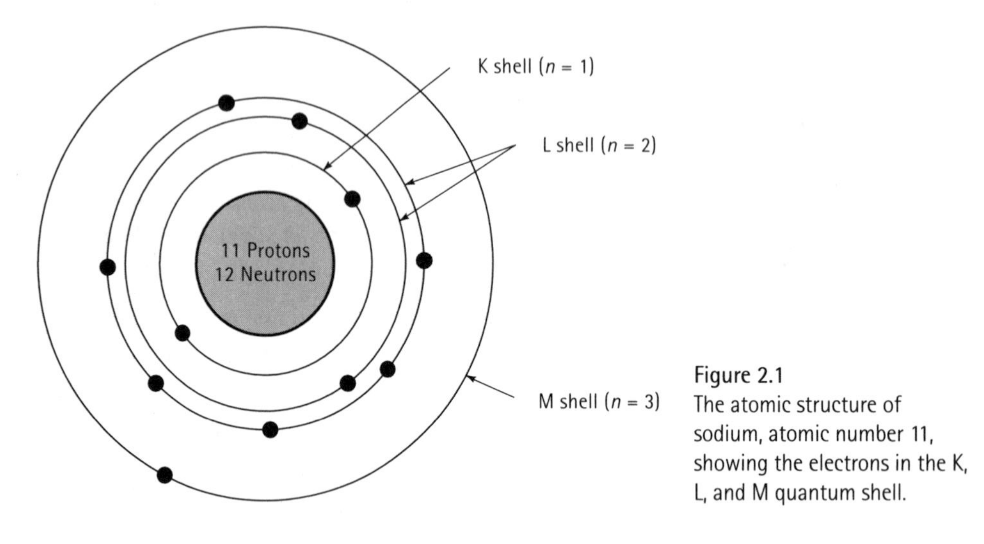

Figure 2.1
The atomic structure of sodium, atomic number 11, showing the electrons in the K, L, and M quantum shell.

3 The Pauli exclusion principle specifies that no more than two electrons, with opposing electronic spins, may be present in each orbital. The spin quantum number m_s, is assigned values +½ and −½ to reflect the different spins. Figure 2.2 shows the quantum numbers and energy levels for each electron in a sodium atom.

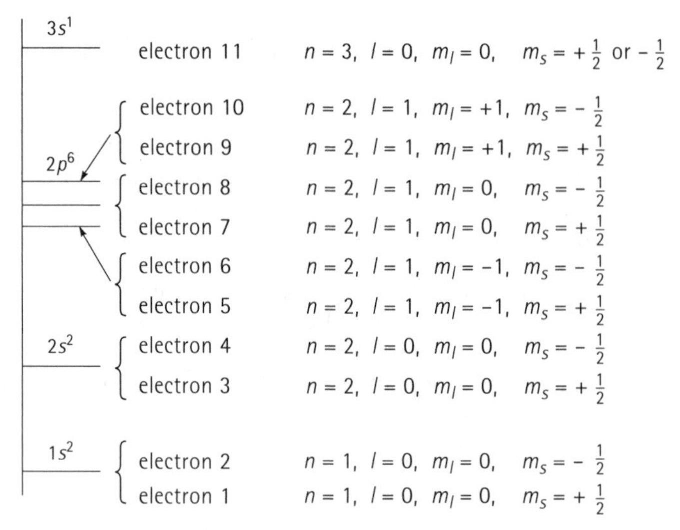

Figure 2.2
The complete set of quantum numbers for each of the 11 electrons in sodium.

The shorthand notation frequently used to denote the electronic structure of an atom combines the numerical value of the principal quantum number, the lower case letter notation for the azimuthal quantum number, and a superscript showing the number of electrons in each orbital. The shorthand notation for germanium, which has an atomic number of 32, is:

$1s^2 2s^2 2p^6 3s^2 3p^6 3d^{10} 4s^2 4p^2$.

The electronic configurations for the elements are summarised in Appendix C; the energy levels are summarised in Table 2.1.

Table 2.1 The pattern used to assign electrons to energy levels.

		l=0 (*s*)	l=1 (*p*)	l=2 (*d*)	l=3 (*f*)	l=4 (*g*)	l=5 (*h*)
$n = 1$	(K)	2					
$n = 2$	(L)	2	6				
$n = 3$	(M)	2	6	10			
$n = 4$	(N)	2	6	10	14		
$n = 5$	(O)	2	6	10	14	18	
$n = 6$	(P)	2	6	10	14	18	22

Note: 2, 6, 10, 14, ..., refer to the number of electrons in the energy level.

Deviations from Expected Electronic Structures The orderly building up of the electronic structure is not always followed, particularly when the atomic number is large and the *d* and *f* levels begin to fill. For example, we would expect the electronic structure of iron, atomic number 26, to be:

$1s^2 2s^2 2p^6 3s^2 3p^6 \boxed{3d^8}$

The actual structure, however, is:

$1s^2 2s^2 2p^6 3s^2 3p^6 \boxed{3d^6 4s^2}$

The unfilled 3*d* level causes the magnetic behaviour of iron, as shown in Chapter 19.

Valence The valence of an atom is related to the ability of the atom to enter into chemical combination with other elements and is often determined by the number of electrons in the outermost combined *sp* level. Examples of the valence are:

Mg: $1s^2 2s^2 2p^6 \boxed{3s^2}$ valence = 2

Al: $1s^2 2s^2 2p^6 \boxed{3s^2 3p^1}$ valence = 3

Ge: $1s^2 2s^2 2p^6 3s^2 3p^6 3d^{10} \boxed{4s^2 4p^2}$ valence = 4

Valence also depends on the nature of the chemical reaction. Phosphorus has a valence of five when it combines with oxygen. But the valence of phosphorus is only three – the electrons in the 3*p* level – when it reacts with hydrogen. Manganese may have a valence of 2, 3, 4, 6, or 7!

Atomic Stability and Electronegativity If an atom has a valence of zero, the element is inert. An example is argon, which has the electronic structure:

$1s^22s^22p^6 \boxed{3s^23p^6}$

Other atoms prefer to behave as if their outer *sp* levels are either completely full, with eight electrons, or completely empty. Aluminium has three electrons in its outer *sp* level. An aluminium atom readily gives up its outer three electrons to empty the 3*sp* level. The atomic bonding and the chemical behaviour of aluminium are determined by the mechanism through which these three electrons interact with surrounding atoms.

On the other hand, chlorine contains seven electrons in the outer 3*sp* level. The reactivity of chlorine is caused by its desire to fill its outer energy level by accepting an electron.

Electronegativity describes the tendency of an atom to gain an electron. Atoms with almost completely filled outer energy levels – such as chlorine – are strongly electronegative and readily accept electrons. However, atoms with nearly empty outer levels – such as sodium – readily give up electrons and are strongly electropositive. High atomic number elements also have a low electronegativity; because the outer electrons are at a greater distance from the positive nucleus, electrons are not as strongly attracted to the atom. Electronegativities for some elements are shown in Figure 2.3.

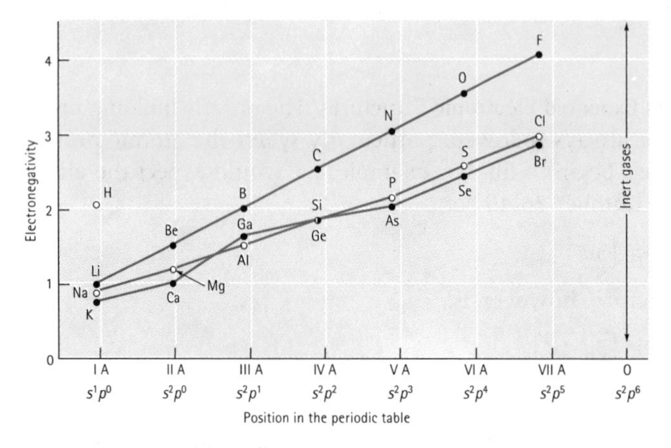

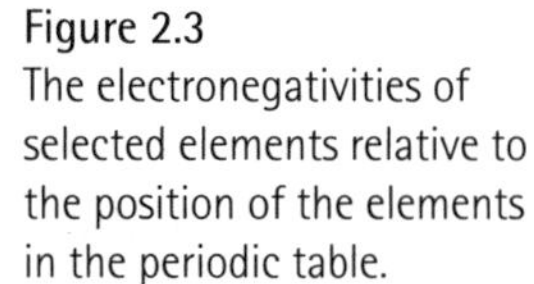
Figure 2.3
The electronegativities of selected elements relative to the position of the elements in the periodic table.

EXAMPLE 2.2

Using the electronic structures, compare the electronegativities of calcium and bromine.

SOLUTION

The electronic structures, obtained from Appendix C, are:

Ca: $1s^22s^22p^63s^23p^6 \boxed{4s^2}$

Br: $1s^2 2s^2 2p^6 3s^2 3p^6 3d^{10}$ $\boxed{4s^2 4p^5}$

Calcium has two electrons in its outer 4*s* orbital and bromine has seven electrons in its outer 4*s*4*p* orbital. Calcium, with an electronegativity of 1.0, tends to give up electrons and is strongly electropositive, but bromine, with an electronegativity of 2.8, tends to accept electrons and is strongly electronegative.

The familiar periodic table, shown inside the back cover of the book, is constructed in accordance with the electronic structure of the elements. Rows in the periodic table correspond to quantum shells, or principal quantum numbers. Columns typically refer to the number of electrons in the outermost *sp* energy level and correspond to the most common valence.

2.4 Atomic Bonding

There are four important mechanisms by which atoms are bonded in solids. In three of the four mechanisms, bonding is achieved when the atoms fill their outer *s* and *p* levels.

The Metallic Bond The metallic elements, which have a low electronegativity, give up their valence electrons to form a 'sea' of electrons surrounding the atoms (Figure 2.4). Aluminium, for example, gives up its three valence electrons, leaving behind a core consisting of the nucleus and inner electrons. Since three negatively charged

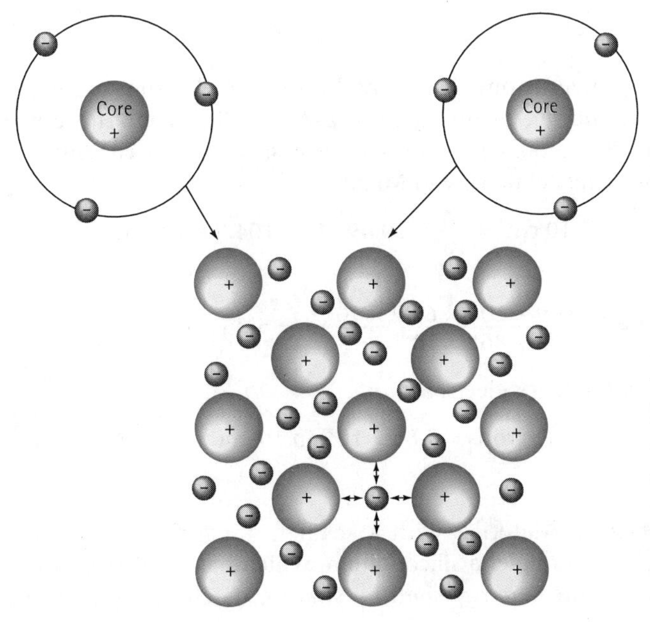

Figure 2.4
The metallic bond forms when atoms give up their valence electrons, which then form an electron sea. The positively charged atom cores are bonded by mutual attraction to the negatively charged electrons.

electrons are missing from this core, it has a positive charge of three. The valence electrons move freely within the electron sea and become associated with several atom cores. The positively charged atom cores are held together by mutual attraction to the nearby electron, thus producing a strong metallic bond.

Because their electrons are not fixed in any one position, metals are good electrical conductors. Under the influence of an applied voltage, the valence electrons move (Figure 2.5), causing a current to flow if the circuit is complete.

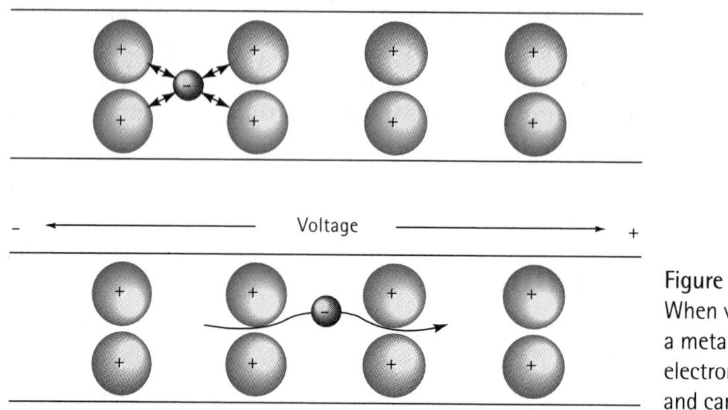

Figure 2.5
When voltage is applied to a metal, the electrons in the electron sea can easily move and carry a current.

EXAMPLE 2.3

Calculate the number of electrons capable of conducting an electrical charge in ten cubic centimetres of silver.

SOLUTION

The valence of silver is one, and only the valence electrons are expected to conduct the electrical charge. From Appendix A, we find that the density of silver is 10.49 $Mg.m^{-3}$. The atomic mass of silver is best written as 107.868×10^{-6} $Mg.mol^{-1}$ in order to match units of density in $Mg.m^{-3}$.

$$\text{Mass of silver is } 10\text{ cm}^3 = \frac{10}{10^6} \times 10.49\text{ Mg} = 104.9 \times 10^{-6}\text{ Mg}$$

$$\text{No. Atoms} = \frac{104.9 \times 10^{-6} \times 6.02 \times 10^{23}}{107.868 \times 10^{-6}} = 5.85 \times 10^{23}$$

and as there is 1 valence electron/atom,

Number valence electrons = 5.85×10^{23} in 10 cm^3 of silver.

The Covalent Bond Materials with a covalent bond share electrons among two or more atoms. For example, a silicon atom, which has a valence of four, obtains eight electrons in its outer energy shell by sharing its electrons with four surrounding silicon atoms (Figure 2.6). Each instance of sharing represents one covalent bond; thus, each silicon atom is bonded to four neighbouring atoms by four covalent bonds.

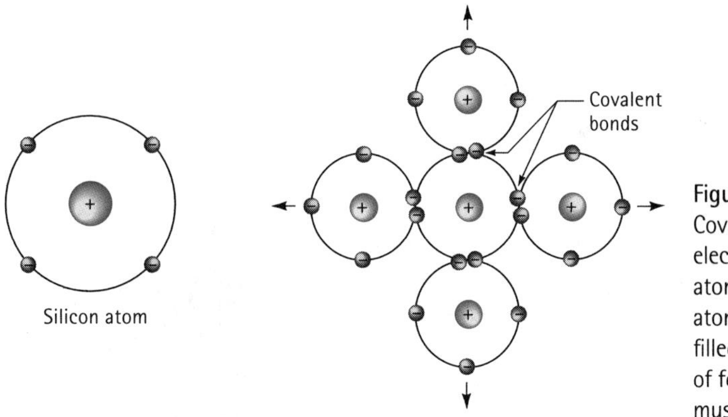

Figure 2.6
Covalent bonding requires that electrons be shared between atoms in such a way that each atom has its outer *sp* orbital filled. In silicon, with a valence of four, four covalent bonds must be formed.

In order for the covalent bonds to be formed, the silicon atoms must be arranged so the bonds have a fixed directional relationship with one another. In the case of silicon, this arrangement produces a tetrahedron, with angles of 109.5° between the covalent bonds (Figure 2.7).

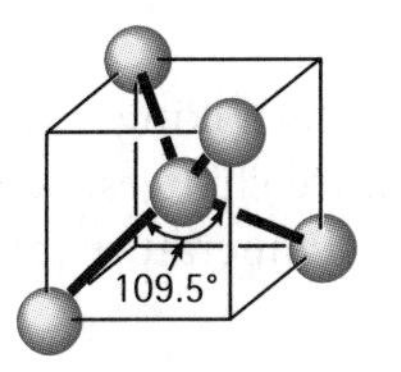

Figure 2.7
Covalent bonds are directional. In silicon, a tetrahedral structure is formed, with angles of 109.5° required between each covalent bond.

Although covalent bonds are very strong, materials bonded in this manner typically have poor ductility and poor electrical and thermal conductivity. For an electron to move and carry a current, the covalent bond must be broken, requiring high temperatures and voltage. Many ceramic, semiconductor, and polymer materials are fully or partly bonded by covalent bonds, explaining why glass shatters when dropped and why bricks are good insulating materials.

EXAMPLE 2.4

Describe how covalent bonding joins oxygen and silicon atoms in silica (SiO_2).

SOLUTION

Silicon has a valence of four and shares electrons, with four oxygen atoms, thus giving a total of eight electrons for each silicon atom. However, oxygen has a valence of six and shares electrons with two silicon atoms, giving oxygen a total of eight electrons.

Figure 2.8 shows one of the possible structures. As in silicon, a tetrahedral structure is produced.

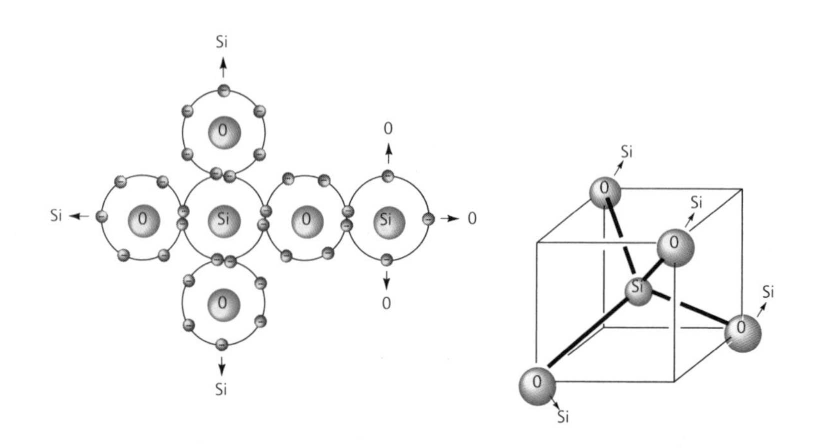

Figure 2.8
The tetrahedral structure of silica (SiO_2), which contains covalent bonds between silicon and oxygen atoms.

EXAMPLE 2.5 **Design of a Thermistor**

A thermistor is a device used to indicate temperature by taking advantage of the change in the electrical conductivity when the temperature changes. Select a material that might serve as a thermistor in the 500°C to 1 000°C temperature range.

SOLUTION

Two design requirements must be satisfied: First, a material with a high melting point must be selected; second, the electrical conductivity of the material must be sensitive to changes in temperature. Covalently bonded materials might be suitable. They often have high melting temperatures and, as more covalent bonds are broken when the temperature increases, increasing numbers of electrons become available to transfer electrical charge. The semiconductor silicon is one choice: Silicon melts at 1 410°C and is covalently bonded. A number of ceramic materials also have high melting points and behave as semiconducting materials. Polymers would *not* be suitable, even though the major bonding is covalent, because of their relatively low melting, or decomposition, temperatures.

The Ionic Bond When more than one type of atom is present in a material, one atom may donate its valence electrons to a different atom, filling the outer energy shell of the second atom. Both atoms now have filled (or empty) outer energy levels, but both have acquired an electrical charge and behave as ions. The atom that contributes the electrons is left with a net positive charge and is a cation, while the atom that accepts the electrons acquires a net negative charge and is an anion. The oppositely charged ions are then attracted to one another and produce the ionic bond. For example, attraction between sodium and chloride ions (Figure 2.9) produces sodium chloride (NaCl), or table salt. Electrical conductivity is poor; the electrical charge is transferred by the movement of entire ions (Figure 2.10), which, because of their size, do not move as easily as electrons.

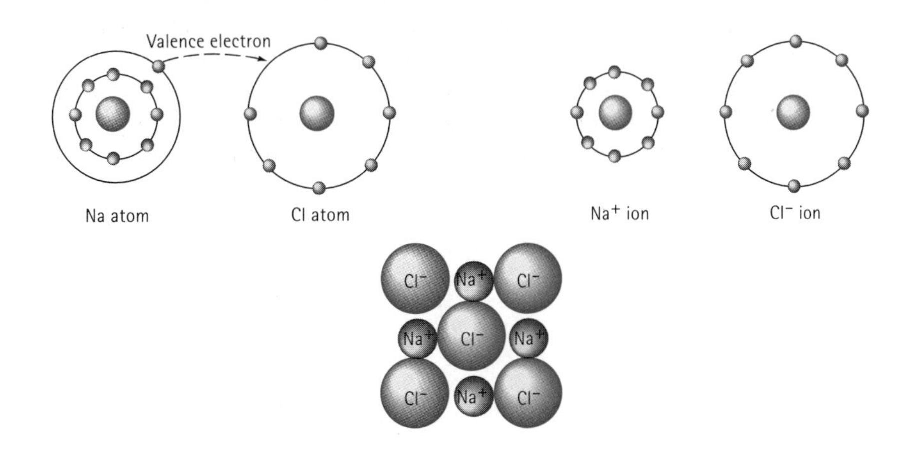

Figure 2.9
An ionic bond is created between two unlike atoms with different electronegativities. When sodium donates its valence electron to chlorine, each becomes an ion; attraction occurs, and the ionic bond is formed.

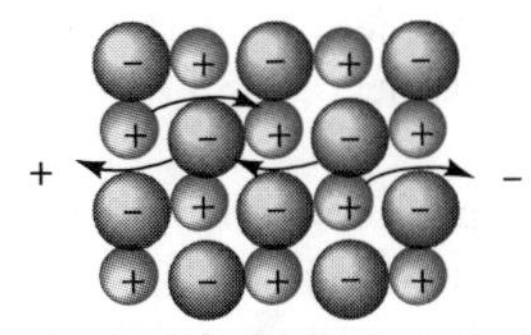

Figure 2.10
When voltage is applied to an ionic material, entire ions must move to cause a current to flow. Ion movement is slow and the electrical conductivity is poor.

EXAMPLE 2.6

Describe the ionic bonding between magnesium and chlorine.

SOLUTION

The electronic structures and valences are:

Mg: $1s^2 2s^2 2p^6 \boxed{3s^2}$ valence = 2

Cl: $1s^2 2s^2 2p^6 \boxed{3s^2 3p^5}$ valence = 7

Each magnesium atom gives up its two valence electrons, becoming a Mg^{2+} ion. Each chlorine atom accepts one electron, becoming a Cl^- ion. To satisfy the ionic bonding, there must be twice as many chloride ions as magnesium ions present, and a compound, $MgCl_2$, is formed.

Van der Waals Bonding Van der Waals bonds join molecules or groups of atoms by weak electrostatic attraction. Many plastics, ceramics, water, and other molecules are permanently polarised; that is, some portions of the molecule are positively charged, while other portions are negatively charged. The electrostatic attraction between the positively charged regions of one molecule and the negatively charged regions of a second molecule weakly bonds the two molecules together (Figure 2.11). This so-called *hydrogen bonding* occurs when hydrogen atoms represent one of the polarised regions.

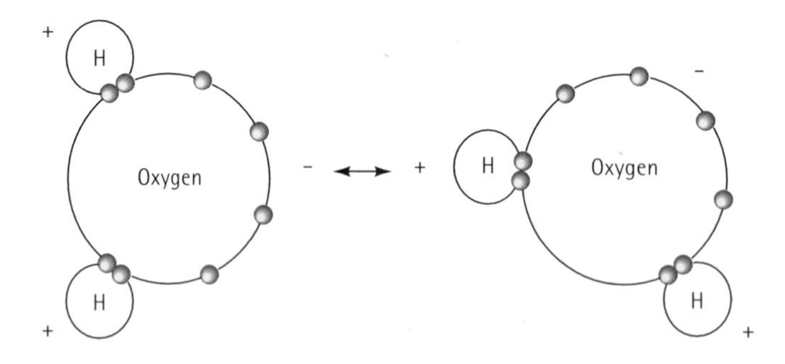

Figure 2.11
The Van der Waals bond is formed as a result of polarisation of molecules or groups of atoms. In water, electrons in the oxygen tend to concentrate away from the hydrogen. The resulting charge difference permits the molecule to be weakly bonded to other water molecules.

Van der Waals bonding is a secondary bond, but the atoms within the molecule or group of atoms are joined by strong covalent or ionic bonds. Heating water to the boiling point breaks the Van der Waals bond and changes water to steam, but much higher temperatures are required to break the covalent bonds joining oxygen and hydrogen atoms.

Van der Waals bonds can dramatically change the properties of materials. Since polymers normally have covalent bonds, we would expect polyvinyl chloride (PVC plastic) to be very brittle, but this material contains many long, chainlike molecules (Figure 2.12). Within each chain, bonding is covalent, but individual chains are bonded to one another by Van der Waals bonds. Polyvinyl chloride can be deformed by breaking the Van der Waals bonds, permitting the chains to slide past one another.

Mixed Bonding In most materials, bonding between atoms is a mixture of two or more types. Iron, for example, is bonded by a combination of metallic and covalent bonding which prevents atoms from packing as efficiently as we might expect.

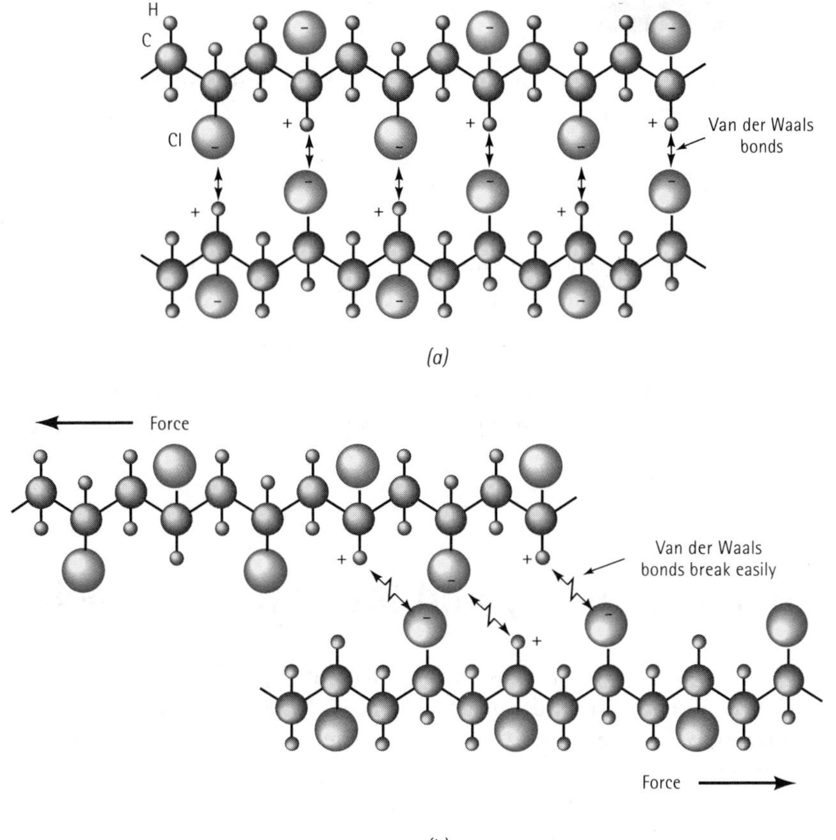

Figure 2.12
(*a*) In polyvinyl chloride, the chloride ions attached to the polymer chain have a negative charge and the hydrogen ions are positively charged. The chains are weakly bonded by Van der Waals bonds.
(*b*) When a force is applied to the polymer, the Van der Waals bonds are broken and the chains slide past one another.

Compounds formed from two or more metals (intermetallic compounds) may be bonded by a mixture of metallic and ionic bonds, particularly when there is a large difference in electronegativity between the elements. Because lithium has an electronegativity of 1.0 and aluminium has an electronegativity of 1.5, we would expect AlLi to have a combination of metallic and ionic bonding. On the other hand, because both aluminium and vanadium have electronegativities of 1.5, we would expect Al_3V to be bonded primarily by metallic bonds.

Many ceramic and semiconducting compounds, which are combinations of metallic and nonmetallic elements, have a mixture of covalent and ionic bonding. As the electronegativity difference between the atoms increases, the bonding becomes more ionic. The fraction of bonding that is covalent can be estimated from the equation

$$\text{Fraction covalent} = \exp(-0.25\Delta E^2) \tag{2.1}$$

where ΔE is the difference in electronegativities.

EXAMPLE 2.7

We used SiO_2 as an example of a covalently bonded material. What fraction of the bonding is covalent?

SOLUTION

From Figure 2.3, we estimate the electronegativity of silicon to be 1.8 and that of oxygen to be 3.5. The fraction of the bonding that is covalent is:

$$\text{Fraction covalent} = \exp\,[-0.25(3.5 - 1.8)^2] = \exp\,(-0.72) = 0.486$$

Although the covalent bonding represents only about half of the bonding, the directional nature of these bonds plays an important role in the eventual structure of SiO_2.

2.5 Binding Energy and Interatomic Spacing

Interatomic spacing, the equilibrium distance between atoms, is caused by a balance between repulsive and attractive forces. In the metallic bond, for example, the attraction between the electrons and the atom core is balanced by the repulsion between atom cores. Equilibrium separation occurs when the total energy of the pair of atoms is at a minimum or when no net force is acting to either attract or repulse the atoms (Figure 2.13).

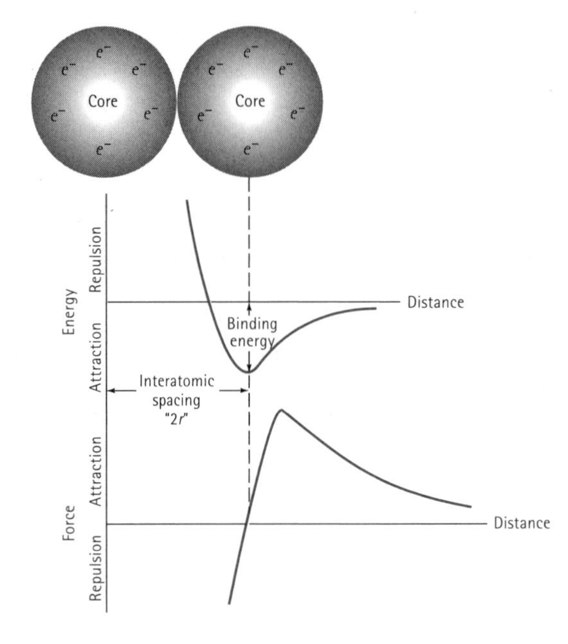

Figure 2.13
Atoms or ions are separated by an equilibrium spacing that corresponds to the minimum energy of the atoms or ions (or when zero force is acting to repel or attract the atoms or ions).

The interatomic spacing in a solid metal is equal to the atomic diameter, or twice the atomic radius r. We cannot use this approach for ionically bonded materials, however, since the spacing is the sum of the two different ionic radii. Atomic and ionic radii for the elements are listed in Appendix B and will be used in the next chapter.

The minimum energy in Figure 2.13 is the binding energy, or the energy required to create or break the bond. Consequently, materials having a high binding energy also have a high strength and a high melting temperature. Ionically bonded materials have a particularly large binding energy, because of the large difference in electronegativities between the ions (Table 2.2); metals have lower binding energies, because the electronegativities of the atoms are similar.

Table 2.2 Binding energies for the four bonding mechanisms.

Bond	Binding Energy ($kJ.mol^{-1}$)
Ionic	625 – 1550
Covalent	520 – 1250
Metallic	100 – 800
Van der Waals	<40

Other properties can be related to the force-distance and energy-distance expressions in Figure 2.13. For example, the modulus of elasticity of a material, which is the amount that a material will elastically stretch when a force is applied, is related to the slope of the force-distance curve (Figure 2.14). A steep slope, which correlates with a higher binding energy and a higher melting point, means that a greater force is required to stretch the bond; thus, the material has a high modulus of elasticity.

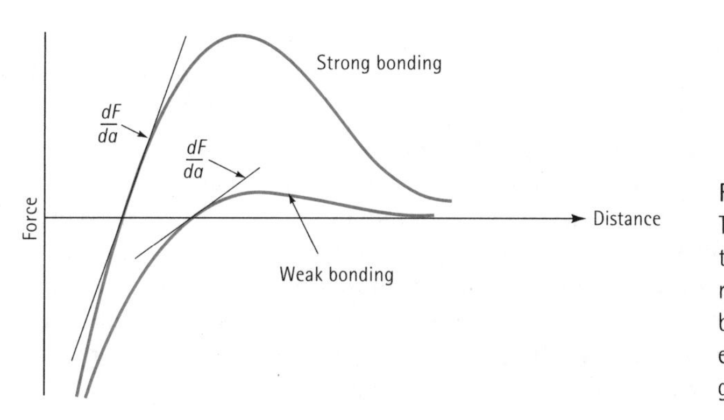

Figure 2.14
The force-distance curve for two materials, showing the relationship between atomic bonding and the modulus of elasticity. A steep dF/da slope gives a high modulus.

The coefficient of thermal expansion, which describes how much a material expands or contracts when its temperature is changed, is related to the strength of the atomic bonds. In order for the atoms to move from their equilibrium separation, energy must be introduced to the material. If a very deep energy trough caused by strong atomic bonding is characteristic of the material (Figure 2.15), the atoms separate to a lesser degree and have a low linear coefficient of thermal expansion. Materials with a low coefficient of thermal expansion maintain their dimensions more accurately when the temperature changes.

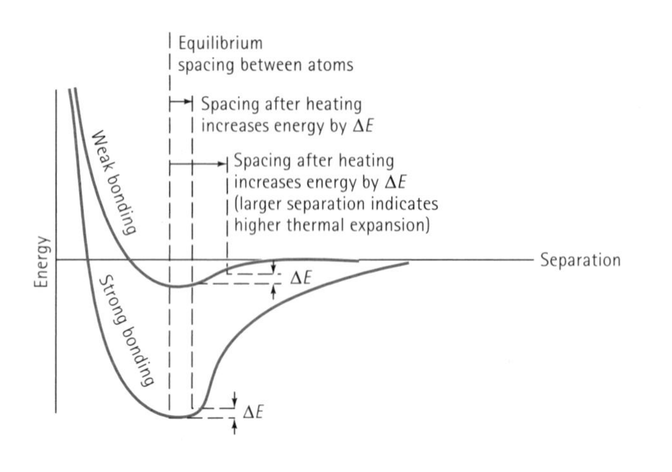

Figure 2.15
The energy-separation curve for two atoms. Materials that display a steep curve with a deep trough have low linear coefficients of thermal expansion.

EXAMPLE 2.8 Design of a Space Shuttle Arm

The space shuttle has a long manipulator arm that permits astronauts to launch and retrieve satellites. Select a suitable material for this device.

SOLUTION

Let's look at two of the many design considerations. First, the material should be stiff so that little bending occurs when a load is applied; this feature helps the operator manoeuvre the manipulator arm precisely. Generally, materials with strong bonding and high melting points also have a high modulus of elasticity, or stiffness. Second, the material should be light in weight to permit maximum payloads to be carried into orbit; a low density is thus desired.

Good stiffness is obtained from high melting-point metals (such as beryllium and tungsten), from ceramics, and from certain fibres (such as carbon). Tungsten, however, has a very high density, while ceramics are very brittle. Beryllium, which has a modulus of elasticity that is greater than that of steel and a density that is less than that of aluminium, might be an excellent candidate. The preferred material is a composite consisting of carbon fibres embedded in an epoxy matrix; the carbon fibres have an exceptionally high modulus of elasticity, while the combination of carbon and epoxy provide a very low-density material.

SUMMARY

The electronic structure of the atom, which is described by a set of four quantum numbers, helps determine the nature of atomic bonding and, hence, the physical and mechanical properties of materials. Atomic bonding is determined partly by how the valence electrons associated with each atom interact:

- For the metallic bond found in metals, the valence electrons are able to move easily. Consequently, metals tend to be ductile and to have good electrical and thermal conductivity.
- The covalent bond found in many ceramic, semiconductor, and polymer materials requires that atoms share valence electrons. The bonds are strong and highly directional, causing these materials to be brittle and to have poor electrical and thermal conductivity.
- The ionic bond found in many ceramics is produced when an electron is 'donated' from one atom to a different type of atom, creating positively charged cations and negatively charged anions. As in covalently bonded materials, these materials tend to be brittle and poor conductors in the solid state.
- The Van der Waals bonds are formed when atoms or groups of atoms have a nonsymmetrical electrical charge, permitting bonding by an electrostatic attraction. The Van der Waals bonds are particularly important in understanding the behaviour of thermoplastic polymers.
- The binding energy is related to the strength of the bonds and is particularly high in ionically and covalently bonded materials. Materials with a high binding energy often have a high melting temperature, a high modulus of elasticity, and a low coefficient of thermal expansion.

GLOSSARY

Anion
A negatively charged ion produced when an atom, usually of a nonmetal, accepts one or more electrons.

Atomic mass
The mass of the Avogadro number of atoms, $g.mol^{-1}$. Normally, this is the average number of protons and neutrons in the atom. Also called the atomic weight.

Atomic mass unit
The mass of an atom expressed in units based on $\frac{1}{12}$ of the mass of a carbon atom.

Atomic number
The number of protons or electrons in an atom.

Avogadro number
The number of atoms or molecules in a mole. The Avogadro number is 6.02×10^{23} per mole.

Binding energy

The energy required to separate two atoms from their equilibrium spacing to an infinite distance apart. Alternately, the binding energy is the strength of the bond between two atoms.

Cation

A positively charged ion produced when an atom, usually of a metal, gives up its valence electrons.

Coefficient of thermal expansion

The amount by which a material changes its dimensions when the temperature changes. A material with a low coefficient of thermal expansion tends to retain its dimensions when the temperature changes.

Covalent bond

The bond formed between two atoms when the atoms share their valence electrons.

Directional relationship

The bonds between atoms in covalently bonded materials form specific angles, depending on the material.

Electronegativity

The relative tendency of an atom to accept an electron and become an anion. Strongly electronegative atoms readily accept electrons.

Electropositivity

The relative tendency of an atom to give up an electron and become a cation. Strongly electropositive atoms readily give up electrons.

Interatomic spacing

The equilibrium spacing between the centres of two atoms. In solid elements, the interatomic spacing equals the apparent diameter of the atom.

Intermetallic compound

A compound such as Al_3V formed by two or more metallic atoms; bonding is typically a combination of metallic and ionic bonds.

Ionic bond

The bond formed between two different atom species when one atom (the cation) donates its valence electrons to the second atom (the anion). An electrostatic attraction binds the ions together.

Metallic bond

The electrostatic attraction between the valence electrons and the positively charged cores of the atoms.

Modulus of elasticity

The amount by which a material elastically stretches or deforms when a force, or stress, is applied. A material with a high modulus of elasticity experiences very little stretching, even for high forces.

Pauli exclusion principle

No more than two electrons in a material can have the same energy. The two electrons have opposite magnetic spins.

Polarised molecule
A molecule whose structure causes portions of the molecule to have a negative charge while other portions have a positive charge, leading to electrostatic attraction between the molecules (Van der Waals bond).

Quantum numbers
The numbers that assign electrons in an atom to discrete energy levels. The four quantum numbers are the principal quantum number n, the azimuthal quantum number l, the magnetic quantum number m_l, and the spin quantum number m_s.

Quantum shell
A set of fixed energy levels to which electrons belong. Each electron in the shell is designated by four quantum numbers.

Secondary bond
Relatively weak bonds, such as Van der Waals and hydrogen bonds, that typically join molecules to one another.

Valence
The number of electrons in an atom that participate in bonding or chemical reactions. Usually, the valence is the number of electrons in the outer *sp* energy level.

Van der Waals bond
A weak electrostatic attraction between polar molecules. The polar molecules have concentrations of positive and negative charges at different locations.

PROBLEMS

2.1 Aluminium foil used for storing food weighs about 500 g per square metre. How many atoms of aluminium are contained in this sample of foil?

2.2 Using the densities and atomic weights given in Appendix A, calculate and compare the number of atoms per cubic centimetre in
(a) lead and
(b) lithium.

2.3 Using data in Appendix A, calculate the number of iron atoms in one tonne. (A metric tonne = 1 Mg.)

2.4 Using data in Appendix A, calculate the volume in cubic centimetres occupied by one mole of boron.

2.5 In order to plate a steel part having a surface area of 0.13 m^2 with a 0.05 mm thick layer of nickel:
(a) How many atoms of nickel are required?
(b) How many moles of nickel are required?

2.6 Suppose an element has a valence of 2 and an atomic number of 27. Based only on the quantum numbers, how many electrons must be present in the 3*d* energy level?

2.7 Indium, which has an atomic number of 49, contains no electrons in its 4*f* energy levels. Based only on this information, what must be the valence of indium?

2.8 Without consulting Appendix C, describe the quantum numbers for each of the 18 electrons in the M shell of copper, using a format similar to that in Figure 2.2.

2.9 Electrical charge is transferred in metals by movement of valence electrons. How many potential charge carriers are there in an aluminium wire 1 mm in diameter and 100 m in length?

2.10 Increasing the temperature of a semiconductor breaks covalent bonds. For each broken bond, two electrons become free to move and transfer electrical charge. What fraction of the total valence electrons are free to move and what fraction of the covalent bonds must be broken in order that 5×10^{15} electrons conduct electrical charge in 50 g of silicon?

2.11 What fraction of the total silicon atoms must be replaced by arsenic atoms to obtain one million electrons that are free to move in 450g of silicon?

2.12 Methane (CH_4) has a tetrahedral structure similar to that of SiO_2 (Figure 2.8), with a carbon atom of radius 0.77×10^{-8} cm at the centre and hydrogen atoms of radius 0.46×10^{-8} cm at four of the eight corners. Calculate the size of the tetrahedral cube for methane.

2.13 The compound AlP is a compound semiconducting material having mixed ionic and covalent bonding. Calculate the fraction of the bonding that is ionic.

2.14 Calculate the fraction of bonding of MgO that is ionic.

2.15 Bonding in the intermetallic compound Ni_3Al is predominantly metallic. Explain why there will be little, if any, ionic bonding component. The electronegativity of nickel is about 1.8.

2.16 Plot the melting temperatures of elements in the IVB to VIIIB columns of the periodic table versus atomic number (that is, plot melting temperatures of Ti through Ni, Zr through Pd, and Hf through Pt). Discuss these relationships, based on atomic bonding and binding energy,

(a) as the atomic number increases in each row of the periodic table and

(b) as the atomic number increases in each column of the periodic table.

2.17 Plot the melting temperature of the elements in the IA column of the periodic table versus atomic number (that is, plot melting temperatures of Li through Cs). Discuss this relationship, based on atomic bonding and binding energy.

2.18 Beryllium and magnesium, both in the IIA column of the periodic table, are lightweight metals. Which would you expect to have the higher modulus of elasticity? Explain, considering binding energy and atomic radii and using appropriate sketches of force versus interatomic spacing.

2.19 Boron has a much lower coefficient of thermal expansion than aluminium, even though both are in the IIIA column of the periodic table. Explain, based on binding energy, atomic size, and the energy well, why this difference is expected.

2.20 Would you expect MgO or magnesium to have the higher modulus of elasticity? Explain.

2.21 Would you expect Al_2O_3 or aluminium to have the higher coefficient of thermal expansion? Explain.

2.22 Aluminium and silicon are side-by-side in the periodic table. Which would you expect to have the higher modulus of elasticity? Explain.

2.23 Explain why the modulus of elasticity of simple thermoplastic polymers, such as polyethylene and polystyrene, is expected to be very low compared with that of metals and ceramics.

2.24 Steel is coated with a thin layer of ceramic to help protect against corrosion. What do you expect to happen to the coating when the temperature of the steel is increased significantly? Explain.

Design Problems

2.25 You wish to introduce ceramic fibres into a metal matrix to produce a composite material, which is subjected to

(a) high forces and

(b) large temperature changes.

What design parameters might you consider to ensure that the fibres will remain intact and provide strength to the matrix? What problems might occur?

2.26 A turbine blade made of nickel may corrode in a jet engine. What design parameters would you consider selecting a coating that would protect the blade at the high operating temperatures, yet not break off when the blade cools to room temperature? What problems might occur? What types of materials might you select for the coating?

2.27 An extrinsic semiconductor can be produced by introducing 'impurity', or donor, elements into pure silicon. By doing so, additional electrons beyond those needed to participate in the bonding mechanism become part of the structure and can move. Design an alloy system that will cause this extrinsic semiconduction in silicon.

CHAPTER 3

Atomic Arrangement

3.1 Introduction

Atomic arrangement plays an important role in determining the microstructure and behaviour of a solid. For example, the atomic arrangement in aluminium provides good ductility, while that in iron provides good strength. Ceramic transducers capable of detecting tumours in the human body rely on an atomic arrangement that produces a permanent displacement of electrical charge within the material. Due to differing atomic arrangement, polyethylene is easily deformed, rubber can be elastically stretched, and epoxy is strong and brittle.

In this chapter, we describe typical atomic arrangements in perfect solid materials and develop the nomenclature used to characterise such arrangements. We will then be prepared to see how imperfections in the atomic arrangement permit us to understand both deformation and strengthening of many solid materials.

3.2 Short-Range Order Versus Long-Range Order

If we neglect imperfections in materials, there are three levels of atomic arrangement (Figure 3.1).

No Order In gases such as argon, the atoms have no order; argon atoms randomly fill up the space to which the gas is confined.

Short-Range Order A material displays short-range order if the special arrangement of the atoms extends only to the atom's nearest neighbours. Each water molecule in steam has a short-range order due to the covalent bonds between the hydrogen and oxygen atoms; that is, each oxygen atom is joined to two hydrogen atoms, forming an angle of 104.5° between the bonds. However, the water molecules have no special arrangement with one another.

A similar situation exists in ceramic glasses. In Example 2.4, we described the tetrahedral structure in silica that satisfies the requirements that four oxygen atoms be covalently bonded to each silicon atom. Because the oxygen atoms must form angles of 109.5° to satisfy the directionality requirements of the covalent bonds, a short-range order results. However, tetrahedral units may be joined together in a random manner.

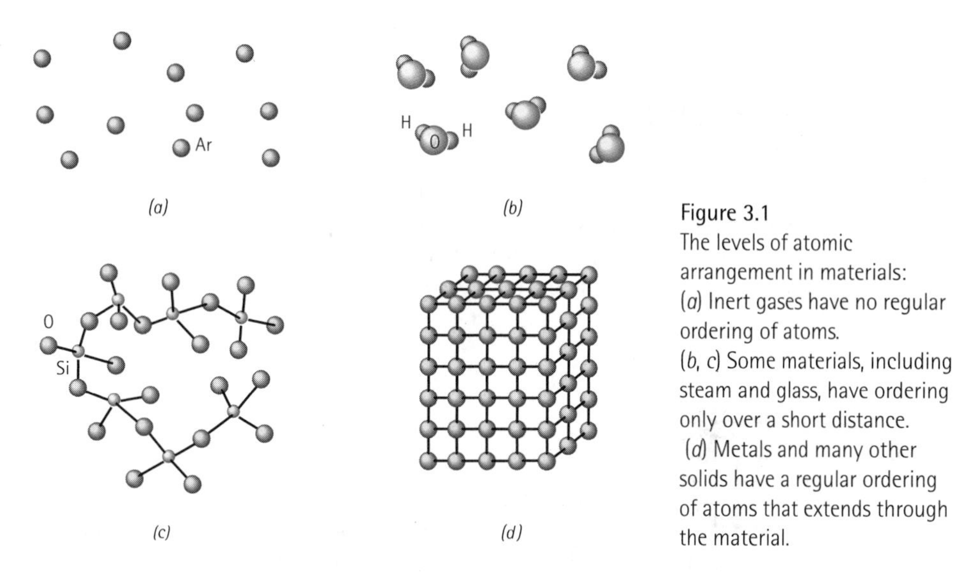

Figure 3.1
The levels of atomic arrangement in materials:
(*a*) Inert gases have no regular ordering of atoms.
(*b, c*) Some materials, including steam and glass, have ordering only over a short distance.
(*d*) Metals and many other solids have a regular ordering of atoms that extends through the material.

Polymers also display short-range atomic arrangements that closely resemble the silica glass structure. Polyethylene is composed of chains of carbon atoms, with two hydrogen atoms attached to each carbon. Because carbon has a valence of four and the carbon and hydrogen atoms are bonded covalently, a tetrahedral structure is again produced (Figure 3.2). Tetrahedral units can be joined in a random manner to produce polymer chains.

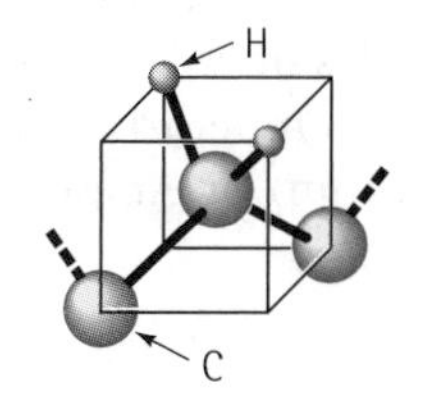

Figure 3.2
The tetrahedral structure in polyethylene.

Ceramics and polymers having only this short-range order are amorphous materials. **Glasses**, which form in both ceramic and polymer systems, are amorphous materials often having unique physical properties. A few specially prepared metals and semiconductors also possess only short-range order.

Long-Range Order Metals, semiconductors, many ceramics, and even some polymers have a crystalline structure in which the atoms display both short-range and long-range order; the special atomic arrangement extends throughout the entire material. The atoms form a regular repetitive, gridlike pattern, or lattice. The lattice is a collection of points, called lattice points which are arranged in a periodic pattern so that the surroundings of each point in the lattice are identical. One or more atoms are associated with each lattice point.

The lattice differs from material to material in both shape and size, depending on the size of the atoms and the type of bonding between the atoms. The crystal structure of a material refers to the size, shape, and atomic arrangement within the lattice.

3.3 Unit Cells

The unit cell is the subdivision of the crystalline lattice that still retains the overall characteristics of the entire lattice. In this text, we refer to the unit cell and the crystal structure interchangeably. A unit cell is shown in the lattice in Figure 3.3. By stacking identical unit cells, the entire lattice can be constructed.

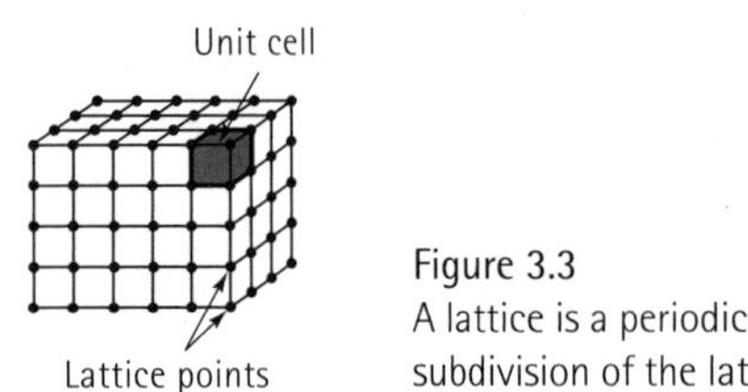

Figure 3.3
A lattice is a periodic array of points that define space. The unit cell (*shaded*) is a subdivision of the lattice that still retains the characteristics of the lattice.

We identify 14 types of unit cells, or Bravais lattices, grouped in seven crystal systems (Figure 3.4 and Table 3.1). Lattice points are located at the corners of the unit cells and, in some cases, at either faces or the centre of the unit cells. Let's look at some of the characteristics of a lattice or unit cell.

Lattice Parameter The lattice parameters which describe the size and shape of the unit cell, include the dimensions of the sides of the unit cell and the angles between the sides (Figure 3.5). In a cubic crystal system, only the length of one of the sides of the cube is necessary to completely describe the cell (angles of 90° are assumed unless otherwise specified). This length, measured at room temperature, is the lattice parameter a_0. The length is often given in nanometres (nm) or Angstrom (Å) units, where:

1 nanometre (nm) = 10^{-9} m

1 angstrom (Å) = 0.1 nm = 10^{-10} m

Several lattice parameters are required to define the size and shape of complex unit cells. For an orthorhombic unit cell, we must specify the dimensions of all three sides of the cell: a_0, b_0, and c_0. Hexagonal unit cells require two dimensions, a_0 and c_0, and the angle of 120° between the a_0 axes. The most complicated cell, the triclinic cell, is described by three lengths and three angles.

Number of Atoms per Unit Cell A specific number of lattice points define each of the unit cells. For example, the corners of the cells are easily identified, as are the body-centred (centre of the cell) and face-centred (centres of the six sides of the cell) positions (Figure 3.4). When counting the number of lattice points belonging to each unit cell, we must recognise that lattice points may be shared by more than one unit

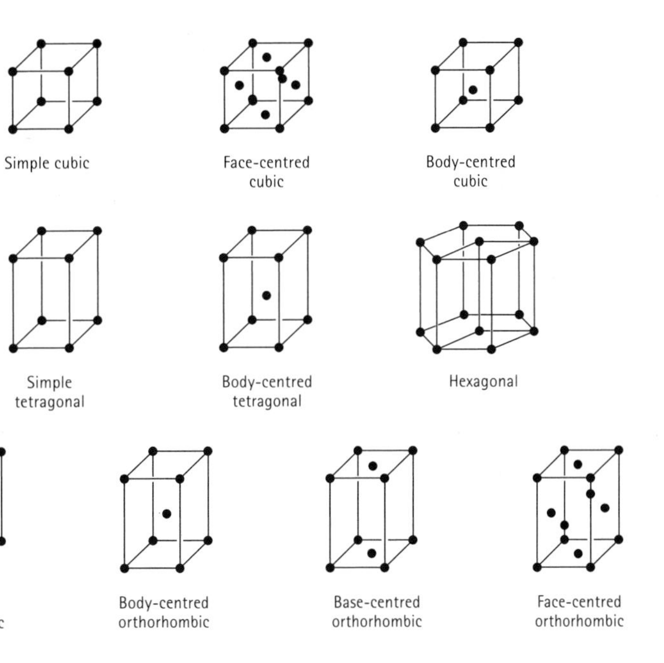

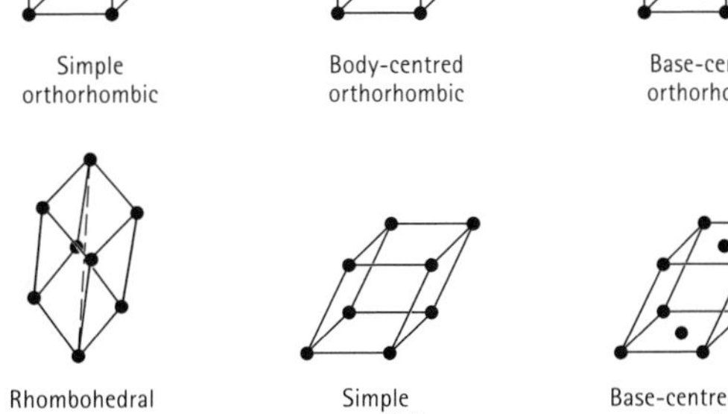

Figure 3.4
The fourteen types of unit cells, or Bravais lattices, grouped in seven crystal systems. Characteristics of the crystal systems are summarised in Table 3.1.

Table 3.1 Characteristics of the seven crystal systems.

Structure	Axes	Angles between Axes
Cubic	$a = b = c$	All angles equal 90°.
Tetragonal	$a = b \neq c$	All angles equal 90°.
Orthorhombic	$a \neq b \neq c$	All angles equal 90°.
Hexagonal	$a = b \neq c$	Two angles equal 90°. One angle equals 120°.
Rhombohedral	$a = b = c$	All angles are equal and none equals 90°.
Monoclinic	$a \neq b \neq c$	Two angles equal 90°. One angle (β) is not equal to 90°.
Triclinic	$a \neq b \neq c$	All angles are different and none equals 90°.

cell. A lattice point at a corner of one unit cell is shared by seven adjacent unit cells; only one-eighth of each corner belongs to one particular cell. Thus, the number of lattice points from all of the corner positions in one unit cell is:

$$\left(\frac{1}{8}\frac{\text{lattice point}}{\text{corner}}\right)\left(8\frac{\text{corners}}{\text{cell}}\right) = 1\frac{\text{lattice point}}{\text{unit cell}}$$

Corners contribute $\frac{1}{8}$ of a point, faces contribute $\frac{1}{2}$, and body-centred positions contribute a whole point.

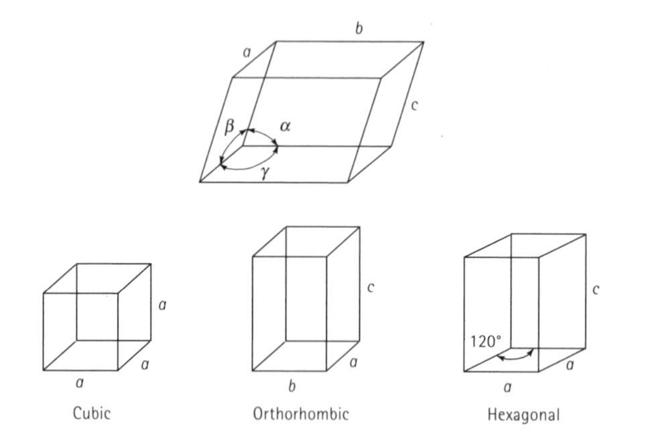

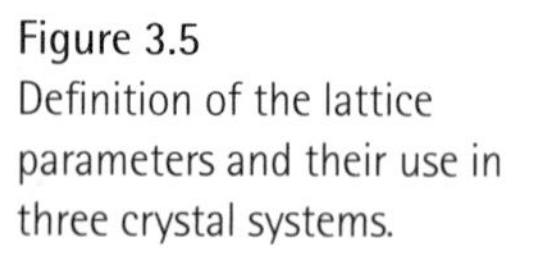
Figure 3.5
Definition of the lattice parameters and their use in three crystal systems.

The number of atoms per unit cell is the product of the number of atoms per lattice point and the number of lattice points per unit cell. In most metals, one atom is located at each lattice point, so the number of atoms is equal to the number of lattice points. The structures of simple cubic (SC), body-centred cubic (BCC), and face-centred cubic (FCC) unit cells, with one atom located at each lattice point, are shown in Figure 3.6. In more complicated structures, particularly polymer and ceramic materials, several or even hundreds of atoms may be associated with each lattice point, forming very complex unit cells.

Simple cubic

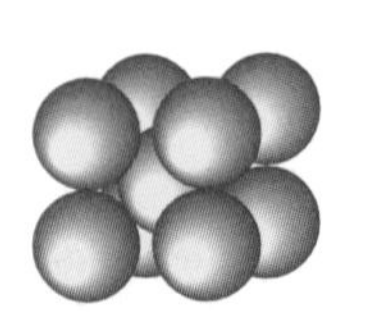
Body-centred cubic

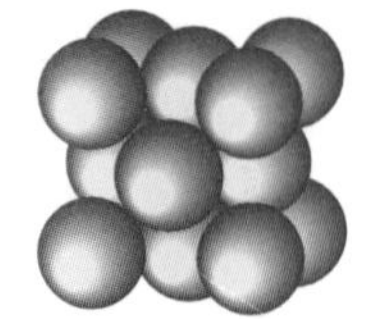
Face-centred cubic

Figure 3.6
The models for simple cubic (SC), body-centred cubic (BCC), and face-centred cubic (FCC) unit cells, assuming only one atom per lattice point.

EXAMPLE 3.1

Determine the number of lattice points per cell in the cubic crystal systems.

SOLUTION

In the SC unit cell, lattice points are located only at the corners of the cube:

$$\frac{\text{lattice points}}{\text{unit cell}} = (8 \text{ corners})\left(\frac{1}{8}\right) = 1$$

In BCC unit cells, lattice points are located at the corners and the centre of the cube:

$$\frac{\text{lattice points}}{\text{unit cell}} = (8 \text{ corners})\left(\frac{1}{8}\right) + (1 \text{ centre})(1) = 2$$

In FCC unit cells, lattice points are located at the corners and faces of the cube:

$$\frac{\text{lattice points}}{\text{unit cell}} = (8 \text{ corners})\left(\frac{1}{8}\right) + (6 \text{ faces})\left(\frac{1}{2}\right) = 4$$

Atomic Radius versus Lattice Parameter Directions in the unit cell along which atoms are in continuous contact are close-packed directions. In simple structures, particularly those with only one atom per lattice point, we use these directions to calculate the relationship between the apparent size of the atom and the size of the unit cell. By geometrically determining the length of the direction relative to the lattice parameters and then adding the number of atomic radii along this direction, we can determine the desired relationship.

EXAMPLE 3.2

Determine the relationship between the atomic radius and the lattice parameter in SC, BCC, and FCC structures when one atom is located at each lattice point.

SOLUTION

If we refer to Figure 3.7, we find that atoms touch along the edge of the cube in SC structures. The corner atoms are centred on the corners of the cube, so:

$$a_0 = 2r \qquad (3.1)$$

In BCC structures, atoms touch along the body diagonal, which is $\sqrt{3}a_0$ in length. There are two atomic radii from the centre atom and one atomic radius from each of the corner atoms on the body diagonal, so:

$$\sqrt{3}a_0 = 4r \quad \text{or} \quad a_0 = \frac{4r}{\sqrt{3}} \qquad (3.2)$$

In FCC structures, atoms touch along the face diagonal of the cube, which is $\sqrt{2}a_0$ in length. There are four atomic radii along this length – two radii from the face-centred atom and one radius from each corner, so:

$$\sqrt{2}a_0 = 4r \quad \text{or} \quad a_0 = \frac{4r}{\sqrt{2}} \tag{3.3}$$

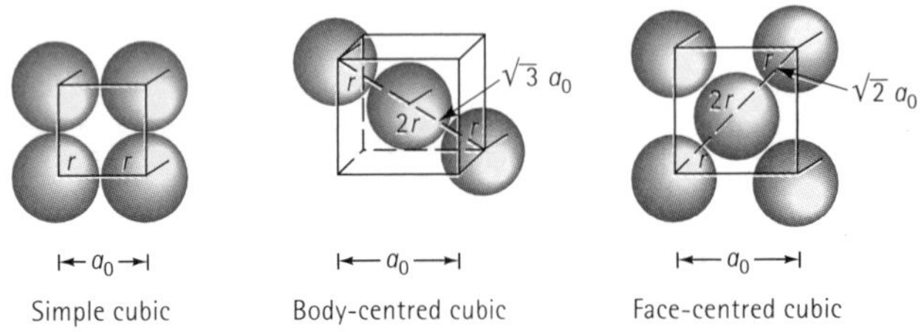

Simple cubic — Body-centred cubic — Face-centred cubic

Figure 3.7
The relationship between the atomic radius and the lattice parameter in cubic systems. (*See* Example 3.2.)

Coordination number The number of atoms touching a particular atom, or the number of nearest neighbours, is the coordination number and is one indication of how tightly and efficiently atoms are packed together. In cubic structures containing only one atom per lattice point, atoms have a coordination number related to the lattice structure. By inspecting the unit cells in Figure 3.8, we see that each atom in the SC structure has a coordination number of six, while each atom in the BCC structure has eight nearest neighbours. In Section 3.5, we will show that each atom in the FCC structure has a coordination number of 12, which is the maximum.

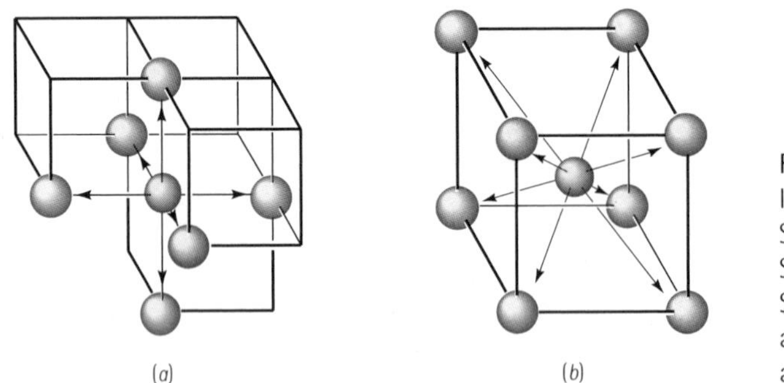

Figure 3.8
Illustration of coordination in SC (*a*) and BCC (*b*) unit cells. Six atoms touch each atom in SC, while the eight corner atoms touch the body-centred atom in BCC.

Packing Factor The packing factor is the fraction of space occupied by atoms, assuming that atoms are hard spheres. The general expression for the packing factor is:

$$\text{Packing factor} = \frac{(\text{number of atoms / cell})(\text{volume of each atom})}{\text{volume of unit cell}} \tag{3.4}$$

EXAMPLE 3.3

Calculate the packing factor for the FCC cell.

SOLUTION

There are four lattice points per cell; if there is one atom per lattice point, there are also four atoms per cell. The volume of one atom is $4\pi r^3/3$ and the volume of the unit cell is a_0^3.

$$\text{Packing factor} = \frac{(4\ \text{atoms}/\text{cell})(\frac{4}{3}\pi r^3)}{a_0^3}$$

Since, for FCC unit cells, $a_0 = 4r/\sqrt{2}$:

$$\text{Packing factor} = \frac{(4)\left(\frac{4}{3}\pi r^3\right)}{\left(4r/\sqrt{2}\right)^3} = 0.74$$

In metals, the packing factor of 0.74 in the FCC unit cell is the most efficient packing possible. BCC cells have a packing factor of 0.68 and SC cells have a packing factor of 0.52. Metals with only the metallic bond are packed as efficiently as possible. Metals with mixed bonding such as iron, may have unit cells with less than the maximum packing factor. No common engineering metals have the SC structure, although this structure is found in ceramic materials.

Density The theoretical density of a metal can be calculated using the properties of the crystal structure. The general formula is:

$$\text{Density}\ \rho = \frac{(\text{atoms}/\text{cell})(\text{atomic mass})}{(\text{volume of unit cell})(\text{Avogadro number})} \quad (3.5)$$

EXAMPLE 3.4

Determine the density of BCC iron.

SOLUTION

Atoms/cell = 2 (*refer to* Example 3.1)

From Appendix A, a_0 for BCC iron is 0.2866 nm = 2.866×10^{-10} m

Atomic mass is 55.847 $g.mol^{-1}$

Avogadro number $N_A = 6.02 \times 10^{23}$ $atoms.mol^{-1}$

Thus atomic mass of each atom of iron $= \dfrac{55.847}{6.02 \times 10^{23}}$ g

Volume of unit cell $= a_0^3 (2.866 \times 10^{-10}\ \text{m})^3 = 23.54 \times 10^{-30}\ \text{m}^3.\text{cell}^{-1}$

$$\text{Density} = \frac{2 \times 55.847}{6.02 \times 10^{23} \times 23.54 \times 10^{-30}}\ \text{g.m}^{-3} = 7.882 \times 10^6 = 7.882\ \text{Mg.m}^{-3}$$

The measured density is 7.870 Mg.m^{-3}. The slight discrepancy between the theoretical and measured densities is a consequence of defects in the lattice.

The Hexagonal Close-Packed Structure A special form of the hexagonal lattice, the hexagonal close-packed structure (HCP), is shown in Figure 3.9. The unit cell is the skewed prism, shown separately. The HCP structure has one lattice point per cell – one from each of the eight corners of the prism – but two atoms are associated with each lattice point. One atom is located at a corner, while the second is located within the unit cell.

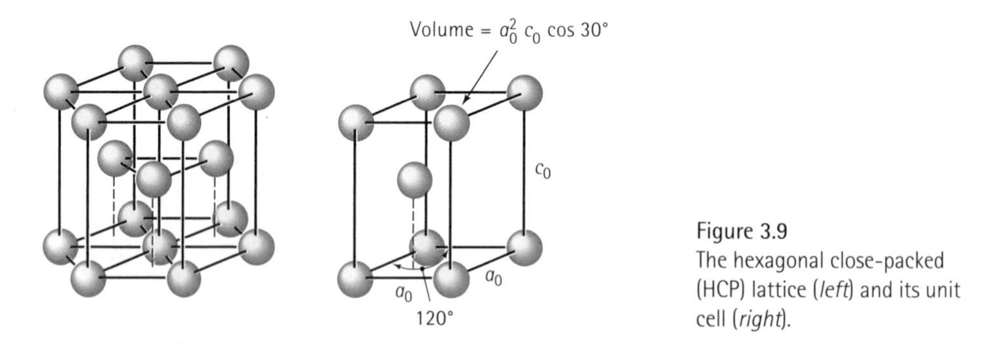

Figure 3.9
The hexagonal close-packed (HCP) lattice (*left*) and its unit cell (*right*).

In ideal HCP metals, the a_0 and c_0 axes are related by the ratio $c/a = 1.633$. Most HCP metals, however, have c/a ratios that differ slightly from the ideal value, because of mixed bonding. Because the HCP structure, like the FCC structure, has a very efficient packing factor of 0.74 and a coordination number of 12, a number of metals possess this structure. Table 3.2 summarises the characteristics of the most important crystal structures in metals.

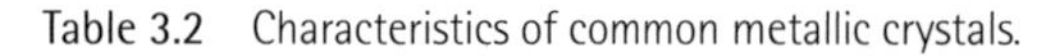

Table 3.2 Characteristics of common metallic crystals.

Structure	a_0 versus r	Atoms per Cell	Coordination Number	Packing Factor	Typical Metals
Simple cubic	$a_0 = 2r$	1	6	0.52	None
Body-centred cubic (BCC)	$a_0 = 4r/\sqrt{3}$	2	8	0.68	Fe, Ti, W, Mo, Nb, Ta, K, Na, V, Cr, Za
Face-centred cubic (FCC)	$a_0 = 4r/\sqrt{2}$	4	12	0.74	Fe, Cu, Al, Au, Ag, Pb, Ni, Pt
Hexagonal close-packed (HCP)	$a_0 = 2r$ $c_0 = 1.633a_0$	2	12	0.74	Ti, Mg, Zn, Be, Co, Zr, Cd

3.4 Allotropic or Polymorphic Transformations

Materials that can have more than one crystal structure are called allotropic or polymorphic. The term allotropy is normally reserved for this behaviour in pure elements, while polymorphism is a more general term. You may have noticed in Table 3.2 that some metals, such as iron and titanium, have more than one crystal structure. At low temperatures, iron has the BCC structure, but at higher temperatures, iron transforms to an FCC structure. These transformations provide the basis for the heat treatment of steel and titanium.

Many ceramic materials, such as silica (SiO_2), also are polymorphic. A volume change may accompany the transformation during heating or cooling; if not properly controlled, this volume change causes the material to crack and fail.

EXAMPLE 3.5 **Design of a Sensor to Measure a Volume Change**

To study how iron behaves at elevated temperatures, we would like to design an instrument that can detect – with a 1% accuracy – the change in volume of a 10 mm^3 iron cube when the iron is heated through its polymorphic transformation temperature. At 911°C, iron is BCC, with a lattice parameter of 0.2863 nm. At 913°C, iron is FCC, with a lattice parameter of 0.3591 nm. Determine the accuracy required of the measuring instrument.

SOLUTION

The volume change during the transformation can be calculated from crystallographic data. The volume of a unit cell of BCC iron before transforming is:

$$V_{\mathrm{BCC}} = a^3 = (0.2863 \text{ nm})^3 = 0.023467 \text{ nm}^3$$

This is the volume occupied by two iron atoms, since there are two atoms per unit cell in the BCC crystal structure.

The volume of the unit cell in FCC iron is:

$$V_{\mathrm{FCC}} = a_0^3 = (0.3591 \text{ nm})^3 = 0.046307 \text{ nm}^3$$

But this is the volume occupied by four iron atoms, as there are four atoms per FCC unit cell. Therefore we must compare two BCC cells (with a volume of 2(0.023467) = 0.046934 nm^3) with each FCC cell. The percent volume change during transformation is:

$$\text{Volume change} = \frac{(0.046307 - 0.046934)}{0.046934} \times 100 = -1.34\%$$

This indicates that the iron contracts upon heating.

The 10 mm side of the cube of iron contracts to 10 – 0.134 = 9.866 mm after transforming; therefore, to ensure 1% accuracy, the instrument must detect a change of:

$$(0.01)(0.134) = 0.00134 \text{ mm} \quad \text{or} \quad 1.34\ \mu\text{m}$$

3.5 Points, Directions, and Planes in the Unit Cell

Coordinates of Points We can locate certain points, such as atom positions, in the lattice or unit cell by constructing the right-handed coordinate system in Figure 3.10. Distance is measured in terms of the number of lattice parameters we must move in each of the *x*, *y*, and *z* coordinates to get from the origin to the point in question. The coordinates are written as the three distances, with commas separating the numbers.

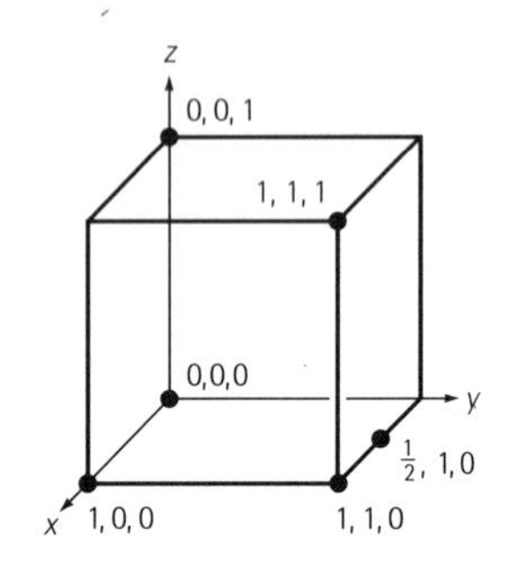

Figure 3.10
Coordinates of selected points in the unit cell. The numbers refer to the distance from the origin in terms of the numbers of lattice parameters.

Directions in the Unit Cell Certain directions in the unit cell are of particular importance. Metals deform, for example, in directions along which atoms are in closest contact. Miller indices for directions are the shorthand notation used to describe these directions. The procedure for finding the Miller indices for directions is as follows:

1 Using a right-handed coordinate system, determine the coordinates of two points that lie on the direction.

2 Subtract the coordinates of the 'tail' point from the coordinates of the 'head' point to obtain the number of lattice parameters travelled in the direction of each axis of the coordinate system.

3 Clear fractions and/or reduce the results obtained from the subtraction to lowest integers.

4 Enclose the numbers in square brackets [ ]. If a negative sign is produced, represent the negative sign with a bar over the number.

EXAMPLE 3.6

Determine the Miller indices of directions *A*, *B*, and *C* in Figure 3.11.

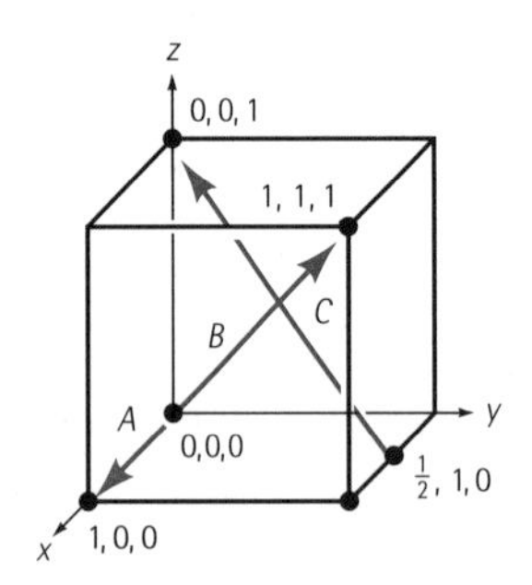

Figure 3.11
Crystallographic directions and coordinates required for Example 3.6.

SOLUTION

Direction A

1 Two points are 1, 0, 0, and 0, 0, 0
2 1, 0, 0 – 0, 0, 0 = 1, 0, 0
3 No fractions to clear or integers to reduce
4 [100]

Direction B

1 Two points are 1, 1, 1 and 0, 0, 0
2 1, 1, 1 – 0, 0, 0 = 1, 1, 1
3 No fractions to clear or integers to reduce
4 [111]

Direction C

1 Two points are 0, 0, 1 and $\frac{1}{2}$, 1, 0
2 0, 0, 1 – $\frac{1}{2}$, 1, 0 = –$\frac{1}{2}$, – 1, 1
3 2(–$\frac{1}{2}$, –1, 1) = –1, –2, 2
4 $[\bar{1}\bar{2}2]$

Several points should be noted about the use of Miller indices for directions:

1 Because directions are vectors, a direction and its negative are not identical; [100] is not equal to $[\bar{1}00]$. They represent the same line, but opposite directions.

2 A direction and its multiple are *identical*; [100] is the same direction as [200]. We just forgot to reduce to lowest integers.

3 Certain groups of directions are *equivalent*; they have their particular indices because of the way we construct the coordinates. For example, in a cubic system, a [100] direction is a [010] direction if we redefine the coordinate system as shown in Figure 3.12. We may refer to groups of equivalent directions as directions of a form. The special brackets < > are used to indicate this collection of directions. All of the directions of the form <110> are shown in Table 3.3. We would expect a material to have the same properties in each of these 12 directions of the form <110>.

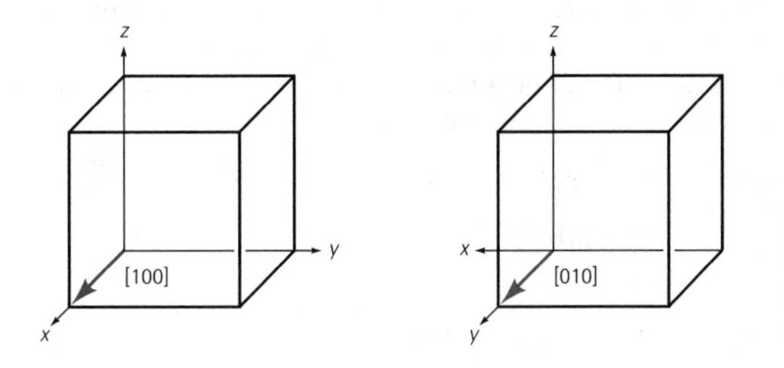

Figure 3.12
Equivalency of crystallographic directions of a form in cubic systems.

Table 3.3 Directions of the form <110> in cubic systems.

$$\langle 110 \rangle = \begin{cases} [110] & [\bar{1}\bar{1}0] \\ [101] & [\bar{1}0\bar{1}] \\ [011] & [0\bar{1}\bar{1}] \\ [1\bar{1}0] & [\bar{1}10] \\ [10\bar{1}] & [\bar{1}01] \\ [01\bar{1}] & [0\bar{1}1] \end{cases}$$

Another way of characterising equivalent directions is by the repeat distance, or the distance between lattice points along the direction. For example, we could examine the [110] direction in an FCC unit cell (Figure 3.13); if we start at the 0, 0, 0 location, the next lattice point is at the centre of a face, or a ½, ½, 0 site. The distance between lattice points is therefore one-half of the face diagonal, or $\frac{1}{2}\sqrt{2}a_0$. In copper, which has a lattice parameter of 0.36151 nm, the repeat distance is 0.2556 nm.

The linear density is the number of lattice points per unit length along the direction. In copper, there are two repeat distances along the [110] direction in each unit cell. This distance is $\sqrt{2}a_0 = 0.51125$ nm, so:

$$\text{Linear density} = \frac{2 \text{ repeat distances}}{0.51125 \text{ nm}} = 3.91 \text{ lattice points/nm}$$

Note that the linear density is also the reciprocal of the repeat distance.

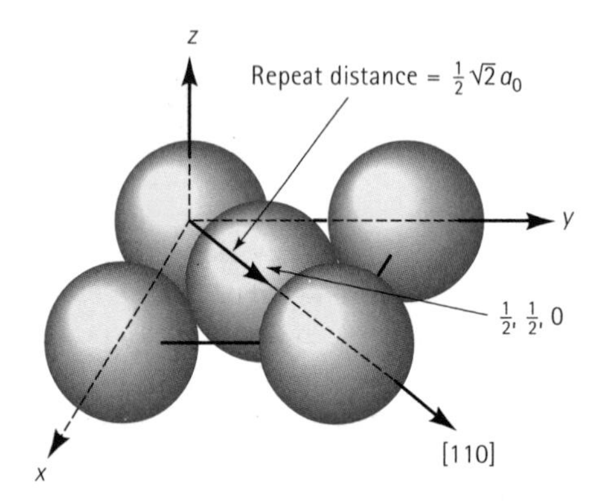

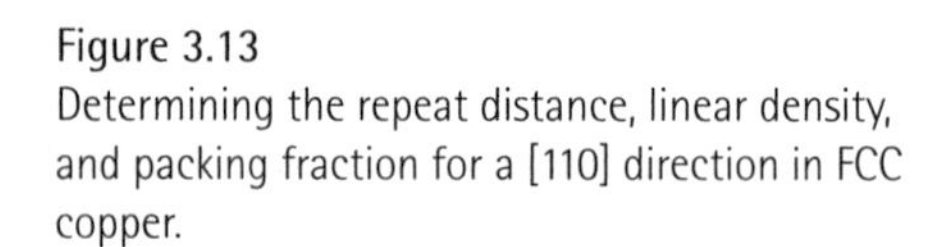
Figure 3.13
Determining the repeat distance, linear density, and packing fraction for a [110] direction in FCC copper.

Finally, we could compute the packing fraction of a particular direction, or the fraction actually covered by atoms. For copper, in which one atom is located at each lattice point, this fraction is equal to the product of the linear density and twice the atomic radius. For the [110] direction in FCC copper, the atomic radius $r = \sqrt{2}a_0/4 = 0.12781$ nm. Therefore, the packing fraction is:

$$\begin{aligned} \text{Packing fraction} &= (\text{linear density})(2r) \\ &= (3.91)(2)(0.12781) \\ &= 1.0 \end{aligned}$$

Atoms touch along the [110] direction, since the [110] direction is close-packed in FCC metals.

Planes in the Unit Cell Certain planes of atoms in a crystal are also significant; for example, metals deform along planes of atoms that are most tightly packed together. Miller indices are used as a shorthand notation to identify these important planes, as described in the following procedure.

1 Identify the points at which the plane intercepts the *x*, *y*, and *z* coordinates in terms of the number of lattice parameters. If the plane passes through the origin, the origin of the coordinate system must be moved!

2 Take reciprocals of these intercepts.

3 Clear fractions but do *not* reduce to lowest integers.

4 Enclose the resulting numbers in parentheses ( ). Again, negative numbers should be written with a bar over the number.

EXAMPLE 3.7

Determine the Miller indices of planes *A*, *B*, and *C* in Figure 3.14.

SOLUTION

Plane A

1 $x = 1, y = 1, z = 1$

2 $\frac{1}{x} = 1, \frac{1}{y} = 1, \frac{1}{z} = 1$

3 No fractions to clear.

4 [111]

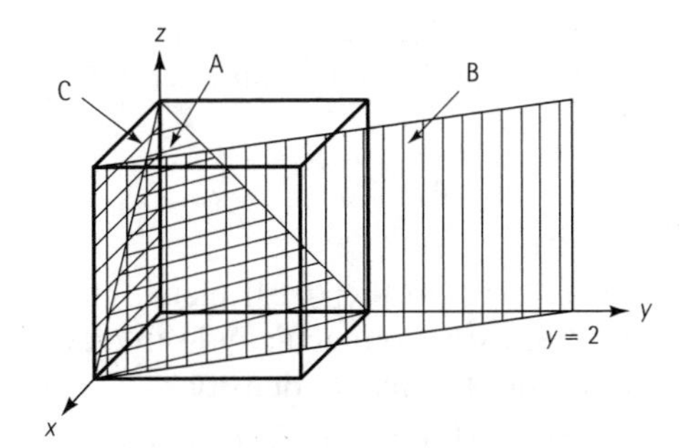

Figure 3.14
Crystallographic planes and intercepts for Example 3.7.

Plane B

1 The plane never intercepts the z axis, so $x = 1$, $y = 2$, and $z = \infty$.

2 $\frac{1}{x} = 1, \frac{1}{y} = \frac{1}{2}, \frac{1}{z} = 0$

3 Clear fractions: $\frac{1}{x} = 2, \frac{1}{y} = 1, \frac{1}{z} = 0$.

4 [210]

Plane C

1 We must move the origin, since the plane passes through 0, 0, 0. Let's move the origin one lattice parameter in the y-direction. Then, $x = \infty$, $y = -1$, and $z = \infty$.

2 $\frac{1}{x} = 0, \frac{1}{y} = -1, \frac{1}{z} = 0$

3 No fractions to clear.

4 $[0\bar{1}0]$

Several important aspects of the Miller indices for planes should be noted:

1 Planes and their negatives are identical (this was not the case for directions). Therefore, $(020) = (0\bar{2}0)$.

2 Planes and their multiples are not identical (again, this is the opposite of what we found for directions). We can show this by defining planar densities and planar packing fractions. The planar density is the number of atoms per unit area whose centres lie on the plane; the packing fraction is the fraction of the area of that plane actually covered by these atoms. Example 3.8 shows how these can be calculated.

EXAMPLE 3.8

Calculate the planar density and planar packing fraction for the (010) and (020) planes in simple cubic polonium, which has a lattice parameter of 0.334 nm.

SOLUTION

The two planes are drawn in Figure 3.15. On the (010) plane, atoms are centred at each corner of the cube face, with ¼ of each atom actually in the face of the unit cell. Thus, the total atoms on each face is one. The planar density is:

$$\text{Planar density (010)} = \frac{\text{atoms per face}}{\text{area of face}} = \frac{\text{1 atom per face}}{(0.334)^2}$$

$$= 8.96 \text{ atoms/nm}^2$$

The planar packing fraction is given by:

$$\text{Packing fraction (010)} = \frac{\text{area of atoms per face}}{\text{area of face}} = \frac{(1 \text{ atom})(\pi r^2)}{(a_0)^2}$$
$$= \frac{(\pi r^2)}{(2r)^2} = 0.79$$

However, no atoms are centred on the (020) planes. Therefore, the planar density and the planar packing fraction are both zero. The (010) and (020) planes are not equivalent!

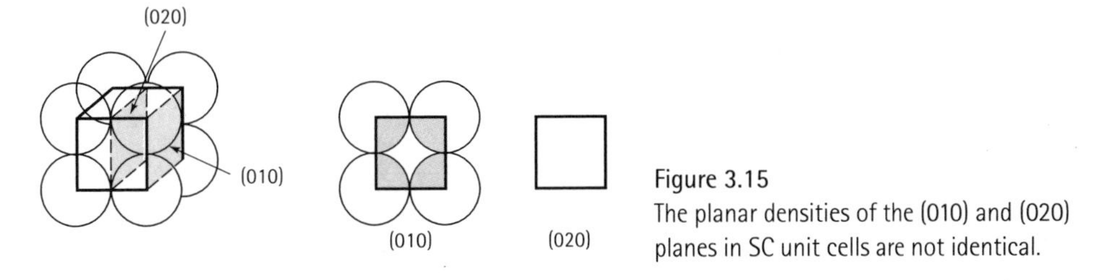

Figure 3.15
The planar densities of the (010) and (020) planes in SC unit cells are not identical.

3 In each unit cell, planes of a form represent groups of equivalent planes that have their particular indices because of the orientation of the coordinates. We represent these groups of similar planes with the notation { }. The planes of the form {110} in cubic systems are shown in Table 3.4.

4 In cubic systems, a direction that has the same indices as a plane is perpendicular to that plane.

Table 3.4 Planes of the form {110} in cubic systems.

Form	Planes
{110}	(110)
	(101)
	(011)
	$(1\bar{1}0)$
	$(10\bar{1})$
	$(01\bar{1})$

Note: The negatives of the planes are not unique planes.

Construction of Directions and Planes To construct a direction or plane in the unit cell, we simply work backwards. Example 3.9 shows how we might do this.

EXAMPLE 3.9

Draw (a) the $[1\bar{2}1]$ direction and (b) the $(\bar{2}10)$ plane in a cubic unit cell.

SOLUTION

a Because we know that we will need to move in the negative y-direction, let's locate the origin at 0, +1, 0. The 'tail' of the direction will be located at this new origin. A second point on the direction can be determined by moving +1 in the x-direction, –2 in the y-direction, and +1 in the z-direction [Figure 3.16(a)].

b To draw in the ($\bar{2}10$) plane, first take reciprocals of the indices to obtain the intercepts, that is:

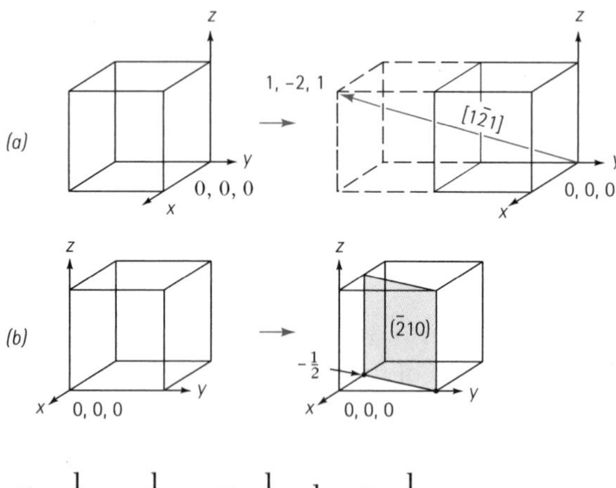

Figure 3.16
Construction of a direction (*a*) and plane (*b*) within a unit cell. (*See* Example 3.9.)

$$x = \frac{1}{-2} = -\frac{1}{2} \qquad y = \frac{1}{1} = 1 \qquad z = \frac{1}{0} = \infty$$

Since the x-intercept is in a negative direction, and we wish to draw the plane within the unit cell, let's move the origin +1 in the x-direction to 1, 0, 0. Then we can locate the x-intercept at –½ and the y-intercept at +1. The plane will be parallel to the z-axis [Figure 3.16(b)].

Miller Indices for Hexagonal Unit Cells A special set of Miller-Bravais indices has been devised for hexagonal unit cells because of the unique symmetry of the system (Figure 3.17). The coordinate system uses four axes instead of three, with the a_3 axis being redundant. The procedure for finding the indices of planes is exactly the same as before, but four intercepts are required, giving indices of the form (*hkil*). Because of the redundancy of the a_3 axis and the special geometry of the system, the first three integers in the designation, corresponding to the a_1, a_2, and a_3 intercepts, are related by $h + k = -i$.

Directions in HCP cells are denoted with either the three-axis or four-axis system. With the three-axis system, the procedure is the same as for conventional Miller indices; examples of this procedure are shown in Example 3.10. A more complicated procedure, by which the direction is broken up into four vectors, is needed for the four-axis system. We determine the number of lattice parameters we must move in each direction to get from the 'tail' to the 'head' of the direction, while for consistency still making sure that $h + k = -i$. This is illustrated in Figure 3.18, showing that the [010] direction is the same as the $[\bar{1}2\bar{1}0]$ direction.

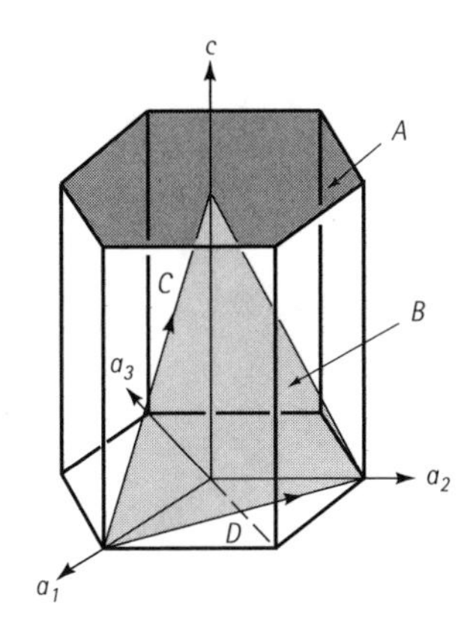

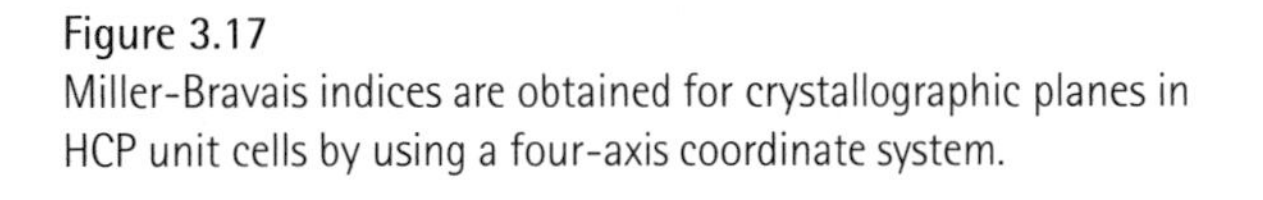

Figure 3.17
Miller-Bravais indices are obtained for crystallographic planes in HCP unit cells by using a four-axis coordinate system.

We can also convert the three-axis notation to the four-axis notation for directions by the following relationships, where h', k', and l' are the indices in the three-axis system:

$$\left.\begin{aligned} h &= \frac{1}{3}(2h' - k') \\ k &= \frac{1}{3}(2k' - h') \\ i &= -\frac{1}{3}(h' + k') \\ l &= l' \end{aligned}\right\} \quad (3.6)$$

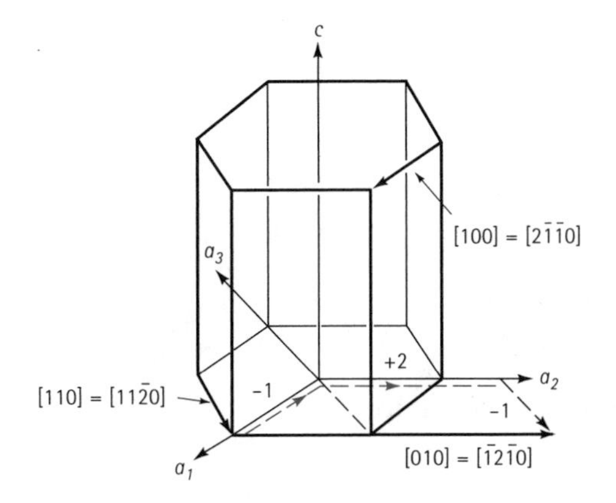

Figure 3.18
Typical directions in the HCP unit cell, using both three- and four-axis systems. The dashed lines show that the $[\bar{1}2\bar{1}0]$ direction is equivalent to a [010] direction.

After conversion, the values of h, k, i, and l may require clearing of fractions or reducing to lowest integers.

EXAMPLE 3.10

Determine the Miller-Bravais indices for planes A and B and directions C and D in Figure 3.17.

SOLUTION

Plane A

1 $a_1 = a_2 = a_3 = \infty$, $c = 1$

2 $\frac{1}{a_1} = \frac{1}{a_2} = \frac{1}{a_3} = 0$, $\frac{1}{c} = 1$

3 No fractions to clear.

4 (0001)

Plane B

1 $a_1 = 1$, $a_2 = 1$, $a_3 = -\frac{1}{2}$, $c = 1$

2 $\frac{1}{a_1} = 1$, $\frac{1}{a_2} = 1$, $\frac{1}{a_3} = -2$, $\frac{1}{c} = 1$

3 No fractions to clear.

4 $(11\bar{2}1)$

Direction C

1 Two points are 0, 0, 1 and 1, 0, 0.

2 0, 0, 1 – 1, 0, 0 = –1, 0, 1

3 No fractions to clear or integers to reduce.

4 $[\bar{1}01]$ or $[\bar{2}113]$

Direction D

1 Two points are 0, 1, 0 and 1, 0, 0.

2 0, 1, 0 – 1, 0, 0 = –1, 1, 0

3 No fractions to clear or integers to reduce.

4 $[\bar{1}10]$ or $[\bar{1}100]$

Close-Packed Planes and Directions In examining the relationship between atomic radius and lattice parameter, we looked for close-packed directions, where atoms are in continuous contact. We can now assign Miller indices to these close-packed directions, as shown in Table 3.5.

Table 3.5 Close-packed planes and directions.

Structure	Directions	Planes
SC	<100>	None
BCC	<111>	None
FCC	<110>	{111}
HCP	<100>, <110> or <11$\bar{2}$0>	(0001), (0002)

We can also examine FCC and HCP unit cells more closely and discover that there is at least one set of close-packed planes in each. Close-packed planes are shown in Figure 3.19. Notice that a hexagonal arrangement of atoms is produced in two dimensions. The close-packed planes are easy to find in the HCP unit cell; they are the (0001) and (0002) planes of the HCP structure and are given the special name basal planes. In fact, we can build up an HCP unit cell by stacking together close-packed planes in an ... *ABABAB* ... stacking sequence (Figure 3.19). Atoms on plane *B*, the (0002) plane, fit into the valleys between atoms on plane *A*, the bottom (0001) plane. If another plane identical in orientation to plane *A* is placed in the valleys of plane *B*, the HCP structure is created. Notice that all of the possible close-packed planes are parallel to one another. Only the basal planes – (0001) and (0002) – are close-packed.

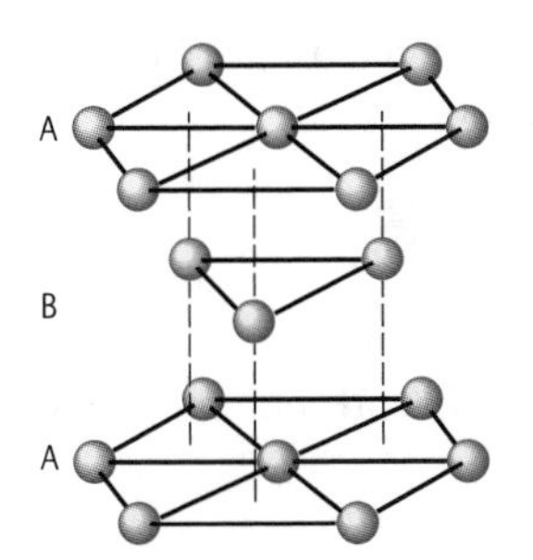

Figure 3.19
The *ABABAB* stacking sequence of close-packed planes produces the HCP structure.

From Figure 3.19, we find the coordination number of the atoms in the HCP structure. The centre atom in a basal plane is touched by six other atoms in the same plane. Three atoms in a lower plane and three atoms in an upper plane also touch the same atom. The coordination number is 12.

In the FCC structure, close-packed planes are of the form {111} (Figure 3.20). When parallel (111) planes are stacked, atoms in plane *B* fit over valleys in plane *A* and atoms in plane *C* fit over valleys in both planes *A* and *B*. The fourth plane fits directly over atoms in plane *A*. Consequently, a stacking sequence ... *ABCABCABC* ... is produced using the (111) plane. Again, we find that each atom has a coordination number of 12.

Unlike the HCP unit cell, there are four sets of nonparallel close-packed planes – (111), $(11\bar{1})$, $(1\bar{1}1)$, and $(\bar{1}11)$ – in the FCC cell. This difference between the FCC and HCP unit cells – the presence or absence of intersecting close-packed planes – affects the behaviour of metals with these structures.

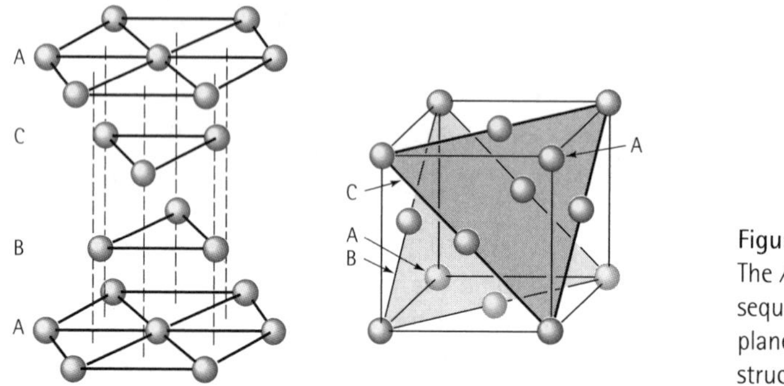

Figure 3.20
The *ABCABCABC* stacking sequence of close-packed planes produces the FCC structure.

Isotropic and Anisotropic Behaviour Because of differences in atomic arrangement in the planes and directions within a crystal, the properties also vary with direction. A material is anisotropic if its properties depend on the crystallographic direction along which the property is measured. For example, the modulus of elasticity of aluminium is 75.9×10^3 MN.m^{-2} in <111> directions, but only 63.4×10^3 MN.m^{-2} in <100> directions. If the properties are identical in all directions, the crystal is isotropic

Interplanar Spacing The distance between two adjacent parallel planes of atoms with the same Miller indices is called the interplanar spacing d_{hkl}. The interplanar spacing in cubic materials is given by the general equation

$$d_{hkl} = \frac{a_0}{\sqrt{h^2 + k^2 + l^2}} \tag{3.7}$$

where a_0 is the lattice parameter and *h*, *k*, and *l* represent the Miller indices of the adjacent planes being considered.

3.6 Interstitial Sites

In any of the crystal structures that have been described, there are small holes between the usual atoms into which smaller atoms may be placed. These locations are called interstitial sites.

An atom, when placed into an interstitial site, touches two or more atoms in the lattice. This interstitial atom has a coordination number equal to the number of atoms it touches. Figure 3.21 shows interstitial locations in the SC, BCC, and FCC structures. The cubic site, with a coordination number of eight, occurs in the SC structure. Octahedral sites give a coordination number of six, while tetrahedral sites give a coordination number of four. As an example, the octahedral sites in BCC unit cells are located at faces of the cube; a small atom placed in the octahedral site touches the four atoms at the corners of the face, the atom in the centre of the unit cell, plus another atom at the centre of the adjacent unit cell, giving a coordination number of six. In FCC unit cells, octahedral sites occur at the centre of each edge of the cube, as well as in the centre of the unit cell.

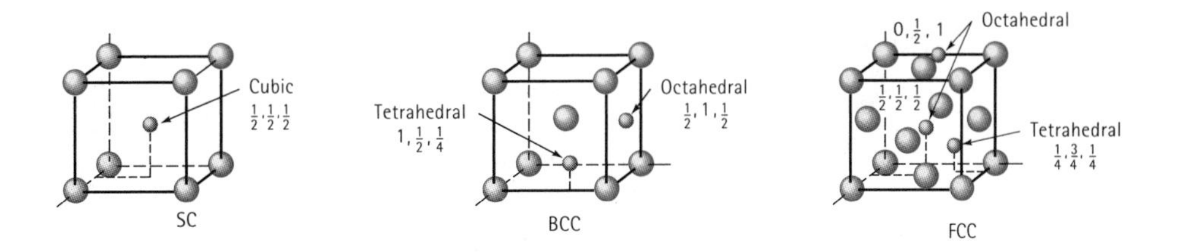

Figure 3.21
The location of the interstitial sites in cubic unit cells. Only representative sites are shown. *Also refer to* Table 3.6.

EXAMPLE 3.11

Calculate the number of octahedral sites that *uniquely* belong to one FCC unit cell.

SOLUTION

The octahedral sites include the 12 edges of the unit cell, with the coordinates

$$\frac{1}{2}, 0, 0 \quad \frac{1}{2}, 1, 0 \quad \frac{1}{2}, 0, 1 \quad \frac{1}{2}, 1, 1$$

$$0, \frac{1}{2}, 0 \quad 1, \frac{1}{2}, 0 \quad 1, \frac{1}{2}, 1 \quad 0, \frac{1}{2}, 1$$

$$0, 0, \frac{1}{2} \quad 1, 0, \frac{1}{2} \quad 1, 1, \frac{1}{2} \quad 0, 1, \frac{1}{2}$$

plus the centre position, ½, ½, ½. Each of the sites on the edge of the unit cell is shared between four unit cells, so only ¼ of each site belongs uniquely to each unit cell. Therefore, the number of sites belonging uniquely to each cell is:

$$(12 \text{ edges})(\tfrac{1}{4} \text{ per cell}) + 1 \text{ centre location} = 4 \text{ octahedral sites}$$

Interstitial atoms whose radii are slightly larger than the radius of the interstitial site may enter that site, pushing the surrounding atoms slightly apart. However, atoms whose radii are smaller than the radius of the hole are not allowed to fit into the interstitial site, because the ion would 'rattle' around in the site. If the interstitial atom becomes too large, it prefers to enter a site having a larger coordination number (Table 3.6). Therefore, an atom whose radius ratio is between 0.225 and 0.414 enters a tetrahedral site; if its radius is somewhat larger than 0.414, it enters an octahedral site instead. When atoms have the same size, as in pure metals, the radius ratio is one and the coordination number is 12, which is the case for metals with the FCC and HCP structures.

EXAMPLE 3.12 Design of a Radiation-absorbing Wall

We wish to produce a radiation-absorbing wall composed of 10 000 lead balls, each 30 mm in diameter, in a face-centred cubic arrangement. We decide that improved absorption will occur if we fill interstitial sites between the 30 mm balls with smaller balls. Design the size of the smaller lead balls and determine how many are needed.

Table 3.6 The coordination number and the radius ratio.

Coordination Number	Location of Interstitial	Radius Ratio	Representation
2	Linear	0–0.155	
3	Centre of triangle	0.155–0.225	
4	Centre of tetrahedron	0.225–0.414	
6	Centre of octahedron	0.414–0.732	
8	Centre of cube	0.732–1.000	

SOLUTION

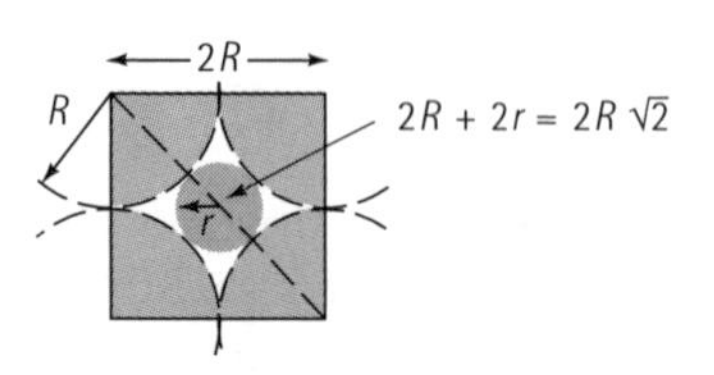

Figure 3.22
Calculation of octahedral interstitial site (*for* Example 3.12).

We can apply our knowledge of crystal structures to this design. For instance, we may decide to introduce small lead balls that just fit into all of the octahedral sites between the 30 mm balls. First, we can calculate the diameter of the octahedral sites located between the 30 mm diameter balls. Figure 3.22 shows the arrangement of the balls on a plane containing an octahedral site.

$$2R + 2r = 2R\sqrt{2}$$
$$r = \sqrt{2}R - R = (\sqrt{2} - 1)R$$
$$r / R = 0.414$$

This is consistent with Table 3.6. Since $r/R = 0.414$, the radius of the small lead balls is

$$r = 0.414R = (0.414)(30 \text{ mm}/2) = 6.21 \text{ mm}$$

From Example 3.11, we find that there are 4 octahedral sites in the FCC arrangement, which also has 4 lattice points. Therefore, we need the same number of small lead balls as large lead balls, or 10 000 small balls.

(As an exercise, the reader may wish to determine the change in packing factor due to the smaller balls; the reader may also compare tetrahedral sites to octahedral sites.)

3.7 Ionic Crystals

Many ceramic materials contain ionic bonds between the anions and the cations. These ionic materials must have crystal structures that ensure electrical neutrality, yet permit ions of different sizes to be packed efficiently.

Electrical Neutrality If the charges on the anion and the cation are identical, the ceramic compound has the formula AX, and the coordination number for each ion is identical to ensure a proper balance of charge. As an example, each cation may be surrounded by six anions, while each anion is, in turn, surrounded by six cations. However, if the valence of the cation is +2 and that of the anion is –1, then twice as many anions must be present, and the formula is of the form AX_2. The structure of the AX_2 compound must ensure that the coordination number of the cation is twice the coordination number of the anion. For example, each cation may have 8 anion nearest neighbours, while only 4 cations touch each anion.

Ionic Radii The crystal structures of the ionically bonded compounds often can be described by placing the cations at the normal lattice points of a unit cell, with the anions then located at one or more of the interstitial sites described in Section 3.6. The ratio of the sizes of the ionic radii of the anion and cation influences both the manner of packing and the coordination number (Table 3.6). A number of common structures in ceramic compounds are described below.

Caesium Chloride Structure Caesium chloride (CsCl) is simple cubic, with the 'cubic' interstitial site filled by the Cl anion (Figure 3.23). The ionic radius ratio, r_{Cs}/r_{Cl} =

0.167 nm/0.181 nm = 0.92, dictates that caesium chloride have a coordination number of eight. We can characterise the structure as a simple cubic structure with two ions – one Cs and one Cl – associated with each lattice point. This structure is possible when the anion and the cation have the same valence.

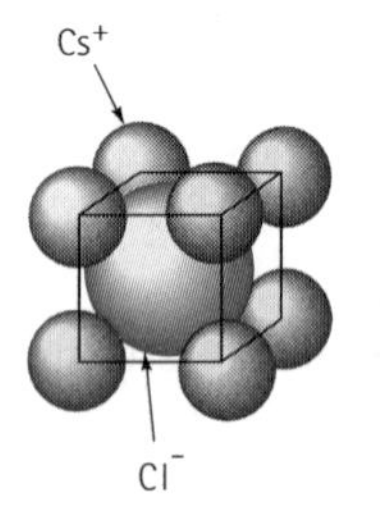

Figure 3.23
The caesium chloride structure, a SC unit cell with two ions (Cs^+ and Cl^-) per lattice point.

EXAMPLE 3.13

For KCl, (a) verify that the compound has the caesium chloride structure and (b) calculate the packing factor for the compound.

SOLUTION

a From Appendix B, $r_K = 0.133$ nm and $r_{Cl} = 0.181$ nm, so:

$$\frac{r_K}{r_{Cl}} = \frac{0.133}{0.181} = 0.735$$

Since $0.732 < 0.735 < 1.000$, the coordination number for each type of ion is eight and the CsCl structure is likely.

b The ions touch along the body diagonal of the unit cell, so:

$$\sqrt{3}a_0 = 2r_K + 2r_{Cl} = 2(0.133) + 2(0.181) = 0.628 \text{ nm}$$
$$a_0 = 0.363 \text{ nm}$$

$$\text{Packing factor} = \frac{\frac{4}{3}\pi r_K^3(1 \text{ K ion}) + \frac{4}{3}\pi r_{Cl}^3(1 \text{ Cl ion})}{a_0^3}$$

$$= \frac{\frac{4}{3}\pi(0.133)^3 + \frac{4}{3}\pi(0.181)^3}{(0.363)^3} = 0.725$$

Sodium Chloride Structure The radius ratio for sodium and chloride ions is r_{Na}/r_{Cl} = 0.097 nm/0.181 nm = 0.536; the sodium ion has a charge of +1; the chloride ion has a charge of –1. Therefore, based on the charge balance and radius ratio, each anion and cation must have a coordination number of six. The FCC structure, with Cl anions at FCC positions and Na cations at the four octahedral sites, satisfies these requirements (Figure 3.24). We can also consider this structure to be FCC with two ions – one Na and one Cl – associated with each lattice point. Many ceramics, including MgO, CaO, and FeO, have this structure.

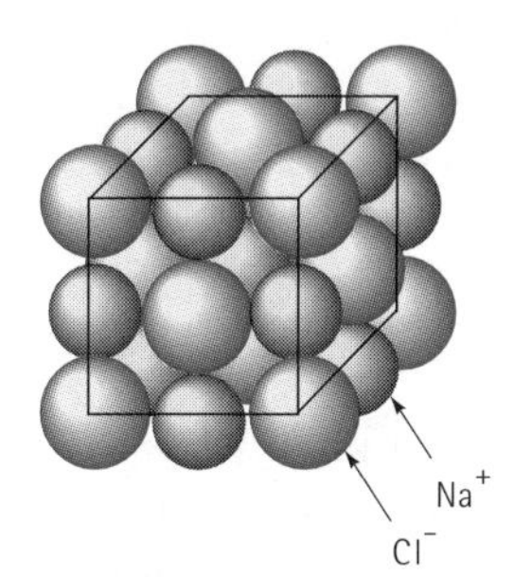

Figure 3.24
The sodium chloride structure, a FCC unit cell with two ions ($Na^+ + Cl^-$) per lattice point.

EXAMPLE 3.14

Show that MgO has the sodium chloride crystal structure and calculate the density of MgO.

SOLUTION

From Appendix B, $r_{Mg} = 0.066$ nm and $r_O = 0.132$ nm so:

$$\frac{r_{Mg}}{r_O} = \frac{0.066}{0.132} = 0.50$$

Since $0.414 < 0.50 < 0.732$, the coordination number for each ion is six, and the sodium chloride structure is possible.

The atomic weights are 24.312 and 16 $g.mol^{-1}$ for magnesium and oxygen, respectively. The ions touch along the edge of the cube, so:

$$a_0 = 2r_{Mg} + 2r_O = 2(0.066) + 2(0.132) = 0.396 \text{ nm} = 3.96 \times 10^{-10} \text{ m}$$

$$\rho = \frac{(4 \text{ Mg ions})(24.312) + (4 \text{ O ions})(16)}{(3.96 \times 10^{-10} \text{m})^3 (6.02 \times 10^{23})} = 4.31 \times 10^6 \text{g.m}^{-3} = 4.31 \text{ Mg.m}^{-3}$$

Zinc Blende Structure Although the Zn ions have a charge of +2 and S has a charge of –2, zinc blende (ZnS) cannot have the sodium chloride structure, because r_{Zn}/r_S = 0.074 nm/0.184 nm = 0.402. This radius ratio demands a coordination number of four, which in turn means that the sulphide ions enter tetrahedral sites in a unit cell, as indicated by the small 'cubelet' in the unit cell (Figure 3.25). The FCC structure, with Zn cations at the normal lattice points and S anions at half of the tetrahedral sites, can accommodate the restrictions of both charge balance and coordination number. A variety of materials, including the semiconductor GaAs, have this structure.

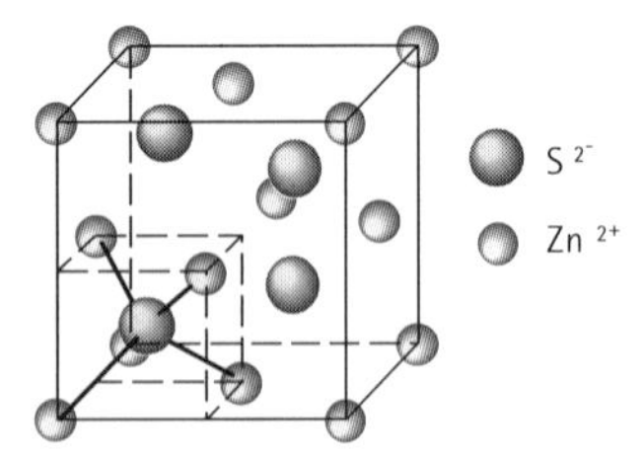

Figure 3.25
The zinc blende unit cell.

Fluorite Structure The fluorite structure is FCC, with anions located at all eight of the tetrahedral positions (Figure 3.26). Thus, there are four cations and eight anions per cell and the ceramic compound must have the formula AX_2, as in calcium fluorite, or CaF_2. The coordination number of the calcium ions is eight, but that of the fluoride ions is four, therefore assuring a balance of charge.

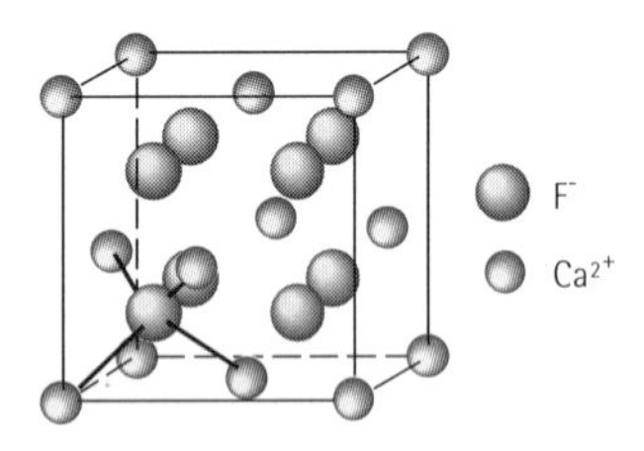

Figure 3.26
The fluorite unit cell.

3.8 Covalent Structures

Covalently bonded materials frequently must have complex structures in order to satisfy the directional restraints imposed by the bonding.

Diamond Cubic Structure Elements such as silicon, germanium, and carbon in its diamond form are bonded by four covalent bonds and produce a tetrahedron [Figure 3.27(a)]. The coordination number of each silicon atom is only four, because of the nature of the covalent bonding.

As these tetrahedral groups are combined, a large cube can be constructed [Figure 3.27(b)]. This large cube contains eight smaller cubes that are the size of the tetrahedral cube; however, only four of the cubes contain tetrahedra. The large cube is the diamond cubic, or DC, unit cell. The atoms on the corners of the tetrahedral cubes provide atoms at the regular FCC lattice points. However, four additional atoms are present within the DC unit cell from the atoms in the centre of the tetrahedral cubes. We can describe the DC lattice as an FCC lattice with two atoms associated with each lattice point. Therefore, there must be eight atoms per unit cell.

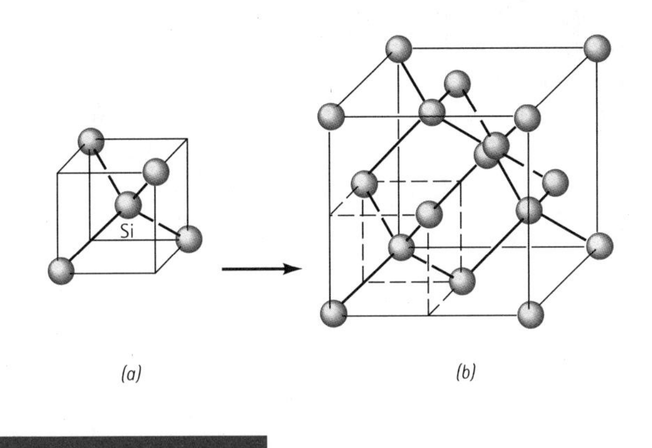

Figure 3.27
(a) Tetrahedron and (b) the diamond cubic (DC) unit cell. This open structure is produced because of the requirements of covalent bonding.

EXAMPLE 3.15

Determine the packing factor for diamond cubic silicon.

SOLUTION

We find that atoms touch along the body diagonal of the cell (Figure 3.28). Although atoms are not present at all locations along the body diagonal, there are voids that have the same diameter as atoms. Consequently:

$$\sqrt{3}a_0 = 8r$$

$$\text{Packing factor} = \frac{(8\ \text{atoms/cell})(\frac{4}{3}\pi r^3)}{a_0^3}$$

$$= \frac{(8)(\frac{4}{3}\pi r^3)}{(8r/\sqrt{3})^3}$$

$$= 0.34$$

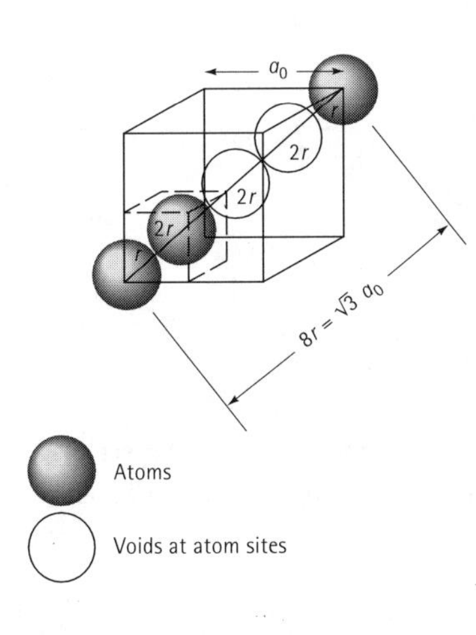

Figure 3.28
Determining the relationship between lattice parameter and atomic radius in a diamond cubic cell (*for* Example 3.15).

Crystalline Silica In a number of its forms, silica (or SiO_2) has a crystalline ceramic structure that is partly covalent and partly ionic. Figure 3.29 shows the crystal structure of one of the forms of silica, β-cristobalite, which is a complicated FCC structure. The ionic radii of silicon and oxygen are 0.042 nm and 0.132 nm, respectively, so the radius ratio is $r_{Si}/r_O = 0.318$ and the coordination number is four.

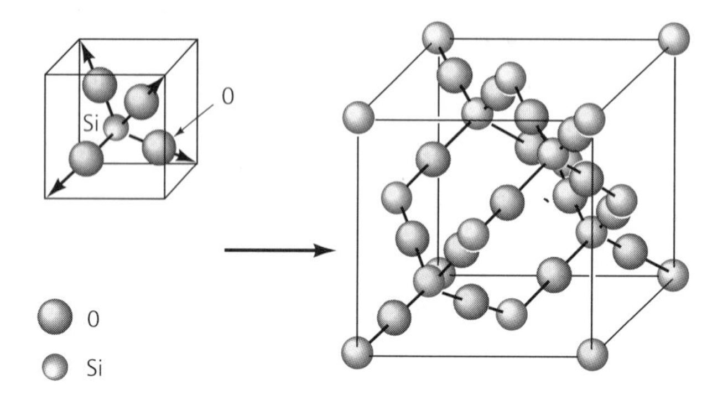

Figure 3.29
The silicon-oxygen tetrahedra and their combination to form the β-cristobalite form of silica.

Crystalline Polymers A number of polymers may form a crystalline structure. The dashed lines in Figure 3.30 outline the unit cell for the lattice of polyethylene. Polyethylene is obtained by joining C_2H_4 molecules to produce long polymer chains that form an orthorhombic unit cell. Some polymers, including nylon, can have several polymorphic forms.

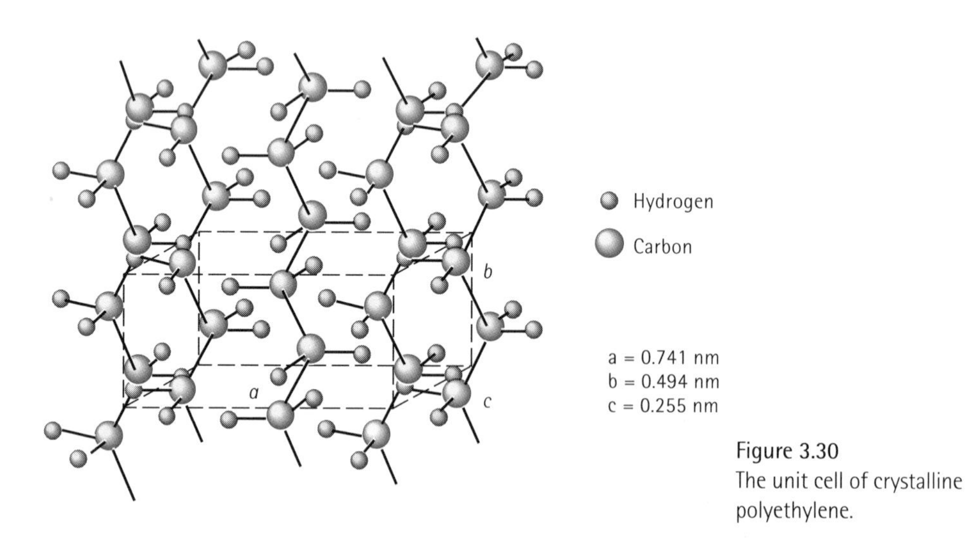

Figure 3.30
The unit cell of crystalline polyethylene.

EXAMPLE 3.16

How many carbon and hydrogen atoms are in each unit cell of crystalline polyethylene? There are twice as many hydrogen atoms as carbon atoms in the chain. The density of polyethylene is about 0.9972 $Mg.m^{-3}$.

SOLUTION

If we let x be the number of carbon atoms, then $2x$ is the number of hydrogen atoms. From the lattice parameters shown in Figure 3.30:

$$\rho = \frac{(x)(12\ \text{g.mol}^{-1}) + (2x)(1\ \text{g.mol}^{-1})}{(7.41\times10^{-10}\text{m})(4.94\times10^{-10}\text{m})(2.55\times10^{-10}\text{m})(6.02\times10^{23})}$$

$$0.9972 = \frac{14x\times10^{-6}}{562\times10^{-7}} = \text{Mg.m}^{-3}$$

$x = 4$ carbon atoms per cell, and

$2x = 8$ hydrogen atoms per cell

3.9 X-Ray Diffraction

Information about the crystal structure of a material can be obtained using X-ray diffraction. When a beam of X-rays having a single wavelength on the same order of magnitude as the atomic spacing in the material strikes that material, X-rays are scattered in all directions. Most of the radiation scattered from one atom cancels out radiation scattered from other atoms. However, X-rays that strike certain crystallographic planes at specific angles are reinforced rather than annihilated. This phenomenon is called diffraction. The X-rays are diffracted, or the beam is reinforced, when conditions satisfy Bragg's law

$$\sin\theta = \frac{\lambda}{2d_{hkl}} \tag{3.8}$$

where the angle θ is half the angle between the diffracted beam and the original beam direction, λ is the wavelength of the X-rays, and d_{hkl} is the interplanar spacing between the planes that cause constructive reinforcement of the beam (*see* Figure 3.31).

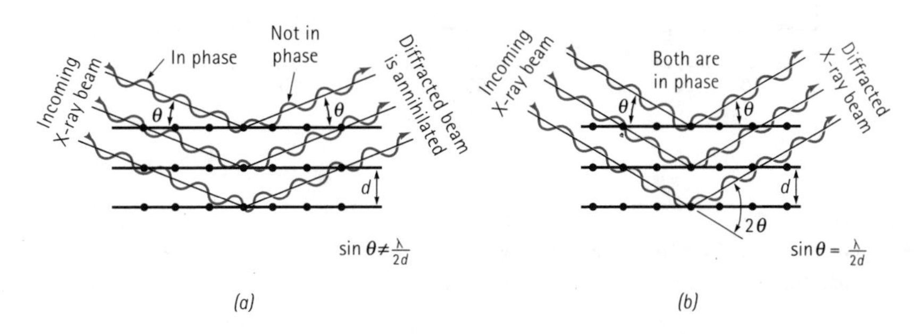

Figure 3.31
(*a*) Destructive and (*b*) reinforcing interactions between X-rays and the crystal structure of a material. Reinforcement occurs at angles that satisfy Bragg's law.

When the material is prepared in the form of a fine powder, there are always at least some powder particles whose (*hkl*) planes are oriented at the proper θ angle to satisfy Bragg's law. Therefore, a diffracted beam, making an angle of 2θ with the incident beam, is produced. In a diffractometer, a moving X-ray detector records the 2θ angles at which the beam is diffracted, giving a characteristic diffraction pattern (*see* Figure 3.32). If we know the wavelength of the X-rays, we can determine the interplanar spacings and, eventually, the identity of the planes that cause the diffraction.

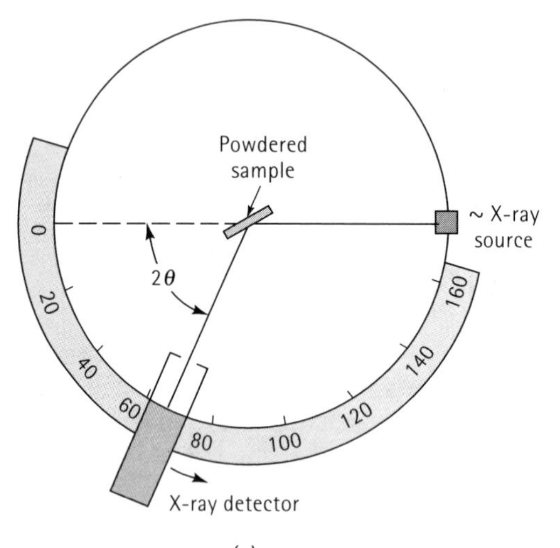

(a)

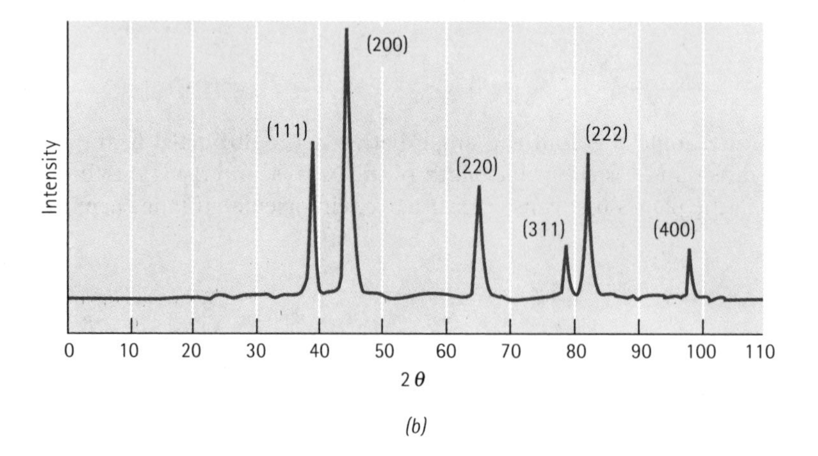

(b)

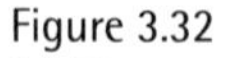

Figure 3.32
(*a*) Diagram of a diffractometer, showing the incident and diffracted beam, the sample in powdered form, and the X-ray detector. (*b*) The diffraction pattern obtained from a sample of gold powder.

To identify the crystal structure of a cubic material, we note the pattern of the diffracted lines – typically, by creating a table of $\sin^2 \theta$ values. By combining Equation 3.7 with Equation 3.8 for the interplanar spacing, we find that:

$$\sin^2\theta = \frac{\lambda^2}{4a_0^2}(h^2 + k^2 + l^2) \tag{3.9}$$

In simple cubic metals, all possible planes will diffract, giving an $h^2 + k^2 + l^2$ pattern of 1, 2, 3, 4, 5, 6, 8,... In body-centred cubic metals, diffraction occurs only from planes having an even $h^2 + k^2 + l^2$ sum of 2, 4, 6, 8, 10, 12, 14, 16, ... For face-centred cubic metals, more destructive interference occurs, and planes having $h^2 + k^2 + l^2$ of 3, 4, 8, 11, 12, 16, ... will diffract. By calculating the values of $\sin^2\theta$ and then finding the appropriate pattern, the crystal structure can be determined for metals having one of these simple structures, as illustrated in Example 3.17.

EXAMPLE 3.17

The results of an X-ray diffraction experiment using X-rays with $\lambda = 0.07107$ nm show that diffracted peaks occur at the following 2θ angles:

Peak	2θ	Peak	2θ
1	20.20	5	46.19
2	28.72	6	50.90
3	35.36	7	55.28
4	41.07	8	59.42

Determine the crystal structure, the indices of the plane producing each peak, and the lattice parameter of the material.

SOLUTION

We can first determine the $\sin^2\theta$ value for each peak, then divide through by the lowest denominator, 0.0308.

Peak	2θ	$\sin^2\theta$	$\sin^2\theta/0.0308$	$h^2 + k^2 + l^2$	$(h\,k\,l)$
1	20.20	0.0308	1	2	(110)
2	28.72	0.0615	2	4	(200)
3	35.36	0.0922	3	6	(211)
4	41.07	0.1230	4	8	(220)
5	46.19	0.1539	5	10	(310)
6	50.90	0.1847	6	12	(222)
7	55.28	0.2152	7	14	(321)
8	59.42	0.2456	8	16	(400)

When we do this, we find a pattern of $\sin^2\theta/0.0308$ values of 1, 2, 3, 4, 5, 6, 7, and 8. If the material were simple cubic, the 7 would not be present, because no planes have an $h^2 + k^2 + l^2$ value of 7. Therefore, the pattern must really be 2, 4, 6, 8, 10, 12, 14, 16,... and the material must be body-centred cubic. The (hkl) values listed give these required $h^2 + k^2 + l^2$ values.

We could then use 2θ values for any of the peaks to calculate the interplanar spacing and thus the lattice parameter. Picking peak 8:

$$2\theta = 59.42 \quad \text{or} \quad \theta = 29.71$$

$$d_{400} = \frac{\lambda}{2\sin\theta} = \frac{0.07107}{2\sin(29.71)} = 0.071699 \text{ nm}$$

$$a_0 = d_{400}\sqrt{h^2 + k^2 + l^2} = (0.071699)(4) = 0.2868 \text{ nm}$$

This is the lattice parameter for body-centred cubic iron.

SUMMARY

Atoms may be arranged in solid materials with either a short-range or long-range order. Amorphous materials, such as glasses and many polymers, have only a short-range order; crystalline materials, including metals and many ceramics, have both long- and short-range order. The long-range periodicity in these materials is described by the crystal structure:

- The atomic arrangement of crystalline materials is described by seven general crystal systems, which include 14 specific Bravais lattices. Examples include simple cubic, body-centred cubic, face-centred cubic, and hexagonal lattices.
- A crystal structure is characterised by the lattice parameters of the unit cell, which is the smallest subdivision of the crystal structure that still describes the overall structure of the lattice. Other characteristics include the number of lattice points and atoms per unit cell, the coordination number (or number of nearest neighbours) of the atoms in the unit cell, and the packing factor of the atoms in the unit cell.
- Allotropic, or polymorphic, materials have more than one possible crystal structure. Often, the structure and properties of materials having this characteristic can be controlled by special heat treatments.
- The atoms of metals having the face-centred cubic and hexagonal close-packed crystal structures are closely packed; atoms are arranged in a manner that occupies the greatest fraction of space. The FCC and HCP structures achieve this arrangement by different stacking sequences of close-packed planes of atoms.
- Points, directions, and planes within the crystal structure can be identified in a formal manner by the assignment of coordinates and Miller indices.

- Mechanical and physical properties may differ when measured along different directions or planes within a crystal; in this case, the crystal is said to be anisotropic. If the properties are identical in all directions, the crystal is isotropic.
- Interstitial sites, or holes between the normal atoms in a lattice, can be filled by other atoms or ions. The crystal structure of many ceramic materials can be understood by considering how these sites are occupied. Atoms or ions located in interstitial sites play an important role in strengthening materials, influencing the physical properties of materials, and controlling the processing of materials.

GLOSSARY

Allotropy
The characteristic of a material being able to exist in more than one crystal structure, depending on temperature and pressure.

Amorphous materials
Materials, including glasses, that have no long-range order, or crystal structure.

Anisotropy
Having different properties in different directions.

Atomic radius
The apparent radius of an atom, typically calculated from the dimensions of the unit cell, using close-packed directions.

Basal plane
The special name given to the close-packed plane in hexagonal close-packed unit cells.

Bragg's law
The relationship describing the angle at which a beam of X-rays of a particular wavelength diffracts from crystallographic planes of a given interplanar spacing.

Bravais lattices
The fourteen possible lattices that can be created using lattice points.

Close-packed directions
Directions in a crystal, along which atoms are in contact.

Coordination number
The number of nearest neighbours to an atom in its atomic arrangement.

Crystal structure
The arrangement of the atoms in a material into a regular repeatable lattice.

Cubic site
An interstitial position that has a coordination number of eight. An atom or ion in the cubic site touches eight other atoms or ions.

Density
Mass per unit volume of a material, usually in units of $Mg.m^{-3}$.

Diamond cubic
A special type of face-centred cubic crystal structure found in carbon, silicon, and other covalently bonded materials.

Diffraction
The constructive interference, or reinforcement, of a beam of X-rays interacting with a material. The diffracted beam provides useful information concerning the structure of the material.

Directions of a form
Crystallographic directions that all have the same characteristics, although their 'sense' is different. Denoted by < > brackets.

Glass
A solid noncrystalline material that has only short-range order between the atoms.

Interplanar spacing
Distance between two adjacent parallel planes with the same Miller indices.

Interstitial site
A location between the 'normal' atoms or ions in a crystal into which another – usually different – atom or ion is placed. Typically, the size of this interstitial location is smaller than the atom or ion that is to be introduced.

Isotropy
Having the same properties in all directions.

Lattice
A collection of points that divide space into smaller equally sized segments.

Lattice parameters
The lengths of the sides of the unit cell and the angles between those sides. The lattice parameters describe the size and shape of the unit cell.

Lattice points
Points that make up the lattice. The surroundings of each lattice point are identical anywhere in the material.

Linear density
The number of lattice points per unit length along a direction.

Long-range order
A regular repetitive arrangement of the atoms in a solid which extends over a very large distance.

Miller-Bravais indices
A special shorthand notation to describe the crystallographic planes in hexagonal close-packed unit cells.

Miller indices
A shorthand notation to describe certain crystallographic directions and planes in a material.

Octahedral site
An interstitial position that has a coordination number of six. An atom or ion in the octahedral site touches six other atoms or ions.

Packing factor
The fraction of space occupied by atoms.

Packing fraction
The fraction of a direction (linear packing fraction) or a plane (planar packing factor) that is actually covered by atoms or ions. When one atom is located at each lattice point, the linear packing fraction along a direction is the product of the linear density and twice the atomic radius.

Planar density
The number of atoms per unit area whose centres lie on the plane.

Planes of a form
Crystallographic planes that all have the same characteristics, although their orientations are different. Denoted by { } brackets.

Polymorphism
Allotropy, or having more than one crystal structure.

Repeat distance
The distance from one lattice point to the adjacent lattice point along a direction.

Short-range order
The regular and predictable arrangement of the atoms over a short distance – usually one or two atom spacings.

Stacking sequence
The sequence in which close-packed planes are stacked. If the sequence is *ABABAB*, a hexagonal close-packed unit cell is produced; if the sequence is *ABCABCABC*, a face-centred cubic structure is produced.

Tetrahedral site
An interstitial position that has a coordination number of four. An atom or ion in the tetrahedral site touches four other atoms or ions.

Tetrahedron
The structure produced when atoms are packed together with a fourfold coordination.

Unit cell
A subdivision of the lattice that still retains the overall characteristics of the entire lattice.

PROBLEMS

3.1 Calculate the atomic radius in cm for the following:

(a) BCC metal with a_0 = 0.3294 nm and one atom per lattice point

(b) FCC metal with a_0 = 0.4086 nm and one atom per lattice point

3.2 Determine the crystal structure for the following:

(a) a metal with a_0 = 0.4949 nm, r = 0.175 nm, and one atom per lattice point

(b) a metal with a_0 = 0.42906 nm, r = 0.1858 nm, and one atom per lattice point

3.3 The density of potassium, which has the BCC structure and one atom per lattice point, is 0.855 $Mg.m^{-3}$. The atomic weight of potassium is 39.09 $g.mol^{-1}$. Calculate

(a) the lattice parameter and

(b) the atomic radius of potassium.

3.4 The density of thorium, which has the FCC structure and one atom per lattice point, is 11.72 $Mg.m^{-3}$. The atomic weight of thorium is 232 $g.mol^{-1}$. Calculate

(a) the lattice parameter and

(b) the atomic radius of thorium.

3.5 A metal having a cubic structure has a density of 2.6 $Mg.m^{-3}$, an atomic weight of 87.62 $g.mol^{-1}$, and a lattice parameter of 0.60849 nm. One atom is associated with each lattice point. Determine the crystal structure of the metal.

3.6 A metal having a cubic structure has a density of 1.892 $Mg.m^{-3}$, an atomic weight of 132.91 $g.mol^{-1}$, and a lattice parameter of 0.613 nm. One atom is associated with each lattice point. Determine the crystal structure of the metal.

3.7 Indium has a tetragonal structure, with a_0 = 0.32517 nm and c_0 = 0.49459 nm. The density is 7.286 $Mg.m^{-3}$ and the atomic weight is 114.82 $g.mol^{-1}$. Does indium have the simple tetragonal or body-centred tetragonal structure?

3.8 Bismuth has a hexagonal structure, with a_0 = 0.4546 nm and c_0 = 1.186 nm. The density is 9.808 $Mg.m^{-3}$ and the atomic weight is 208.98 $g.mol^{-1}$. Determine

(a) the volume of the unit cell and

(b) how many atoms are in each unit cell.

3.9 Gallium has an orthorhombic structure, with a_0 = 0.45258 nm, b_0 = 0.45186 nm, and c_0 = 0.76570 nm. The atomic radius is 0.1218 nm. The density is 5.904 $Mg.m^{-3}$ and the atomic weight is 69.72 $g.mol^{-1}$. Determine

(a) the number of atoms in each unit cell and

(b) the packing factor in the unit cell.

3.10 Beryllium has a hexagonal crystal structure, with a_0 = 0.22858 nm and c_0 = 0.35842 nm. The atomic radius is 0.1143 nm, the density is 1.848 $Mg.m^{-3}$, and the atomic weight is 9.01 $g.mol^{-1}$. Determine

(a) the number of atoms in each unit cell and

(b) the packing factor in the unit cell.

3.11 Above 882°C, titanium has a BCC crystal structure, with a = 0.332 nm. Below this temperature, titanium has a HCP structure, with a = 0.2978 nm and c = 0.4735 nm. Determine the percent volume change when BCC titanium transforms to HCP titanium. Is this a contraction or expansion?

3.12 α-Mn has a cubic structure, with a_0 = 0.8931 nm and a density of 7.47 $Mg.m^{-3}$. β-Mn has a different cubic structure, with a_0 = 0.6326 nm and a density of 7.26 $Mg.m^{-3}$. The atomic weight of manganese is 54.938 $g.mol^{-1}$ and the atomic radius is 0.112 nm. Determine the percent volume change that would occur if α-Mn transforms to β-Mn.

3.13 A typical paper clip weighs 0.59 g and consists of BCC iron. Calculate

(a) the number of unit cells and

(b) the number of iron atoms in the paper clip. (*See* Appendix A for required data.)

3.14 Aluminium foil used to package food is approximately 25 μm thick. Assume that all of the unit cells of the aluminium are arranged so that a_0 is perpendicular to the foil surface. For a 100 mm × 100 mm square of the foil, determine

(a) the total number of unit cells in the foil and

(b) the thickness of the foil in number of unit cells. (*See* Appendix A.)

3.15 Determine the Miller indices for the directions in the cubic unit cell shown in Figure 3.33.

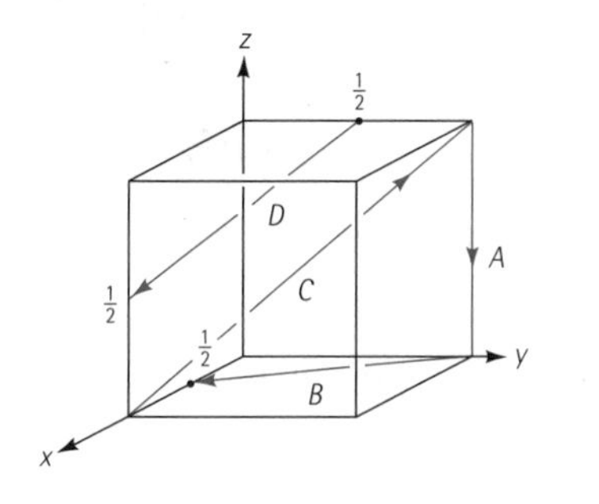

Figure 3.33 Directions in a cubic unit cell (*for* Problem 3.15).

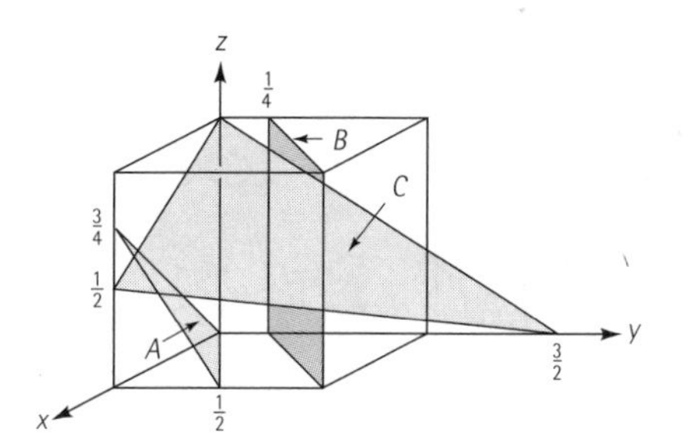

Figure 3.36 Planes in a cubic unit cell (*for* Problem 3.18).

3.16 Determine the indices for the directions in the cubic unit cell shown in Figure 3.34.

3.17 Determine the indices for the planes in the cubic unit cell shown in Figure 3.35.

3.18 Determine the indices for the planes in the cubic unit cell shown in Figure 3.36.

3.19 Determine the indices for the directions in the hexagonal lattice shown in Figure 3.37, using both the three-digit and four-digit systems.

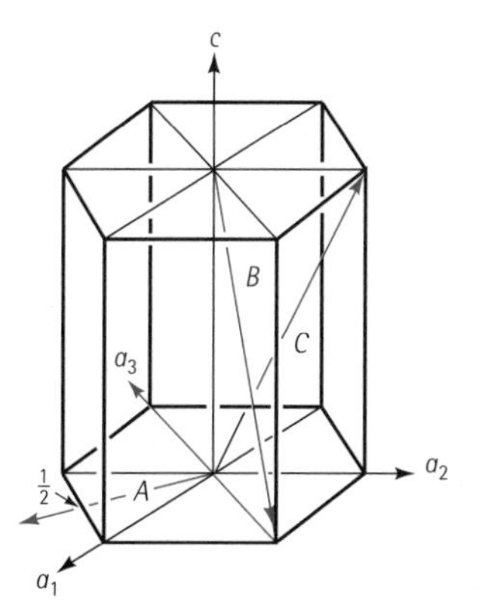

Figure 3.37 Directions in a hexagonal lattice (*for* Problem 3.19).

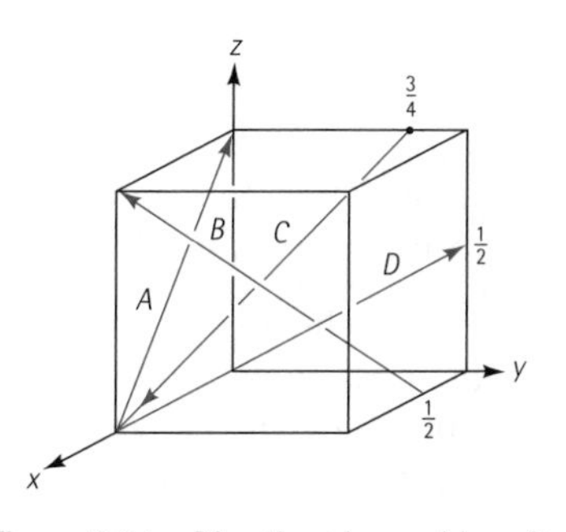

Figure 3.34 Directions in a cubic unit cell (*for* Problem 3.16).

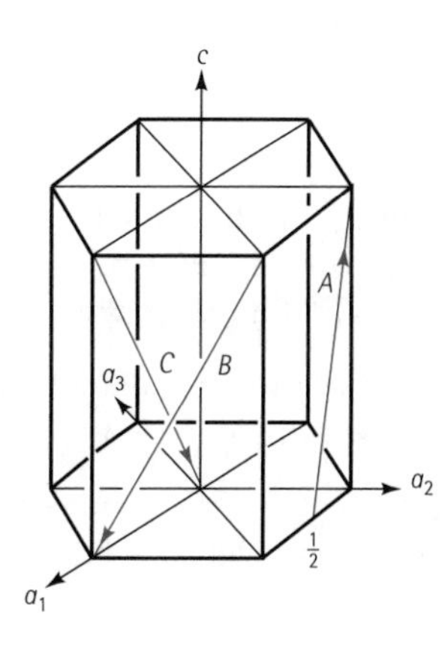

Figure 3.38 Directions in a hexagonal lattice (*for* Problem 3.20).

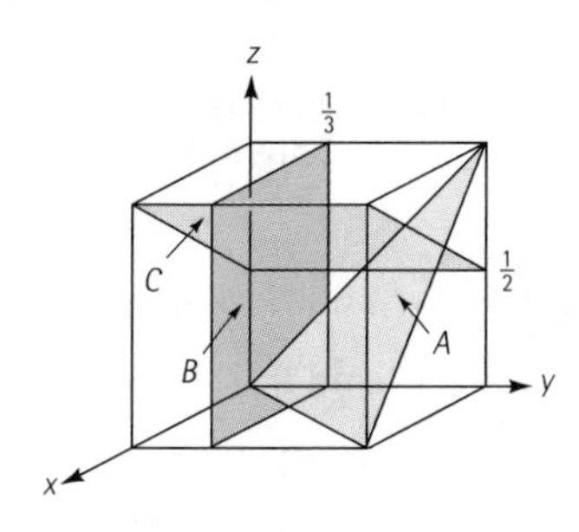

Figure 3.35 Planes in a cubic unit cell (*for* Problem 3.17).

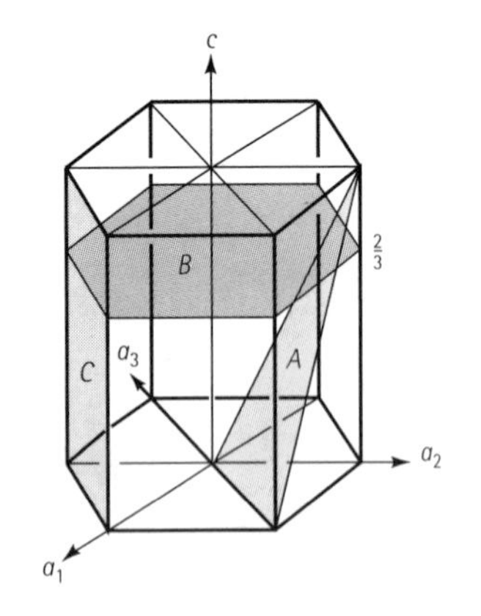

Figure 3.39 Planes in a hexagonal lattice (*for* Problem 3.21).

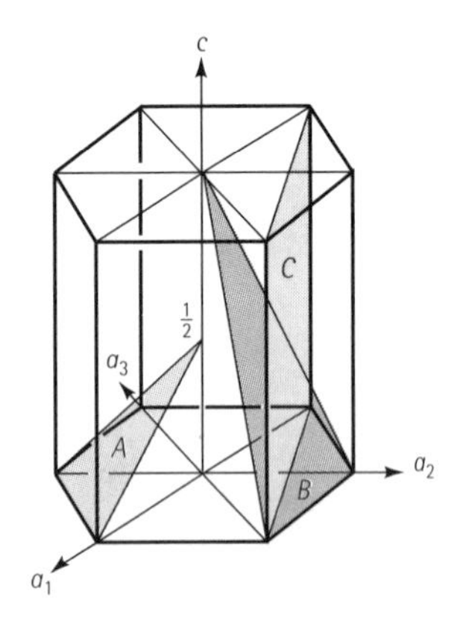

Figure 3.40 Planes in a hexagonal lattice (*for* Problem 3.22).

3.20 Determine the indices for the directions in the hexagonal lattice shown in Figure 3.38, using both the three-digit and four-digit systems.

3.21 Determine the indices for the planes in the hexagonal lattice shown in Figure 3.39.

3.22 Determine the indices for the planes in the hexagonal lattice shown in Figure 3.40.

3.23 Sketch the following planes and directions within a cubic unit cell:

(a) $[101]$ (b) $[0\bar{1}0]$ (c) $[12\bar{2}]$ (d) $[301]$

(e) $[\bar{2}01]$ (f) $[2\bar{1}3]$ (g) $(0\bar{1}\bar{1})$ (h) (102)

(i) (002) (j) $(1\bar{3}0)$ (k) $(\bar{2}12)$ (l) $(3\bar{1}\bar{2})$

3.24 Sketch the following planes and directions within a cubic unit cell:

(a) $[1\bar{1}0]$ (b) $[\bar{2}\bar{2}1]$ (c) $[410]$ (d) $[0\bar{1}2]$

(e) $[\bar{3}\bar{2}1]$ (f) $[1\bar{1}1]$ (g) $(11\bar{1})$ (h) $(01\bar{1})$

(i) (030) (j) $(1\bar{2}1)$ (k) $(11\bar{3})$ (l) $(0\bar{4}1)$

3.25 Sketch the following planes and directions within a hexagonal unit cell:

(a) $[01\bar{1}0]$ (b) $[11\bar{2}0]$ (c) $[\bar{1}011]$

(d) (0003) (e) $(\bar{1}010)$ (f) $(01\bar{1}1)$

3.26 Sketch the following planes and directions within a hexagonal unit cell:

(a) $[\bar{2}110]$ (b) $[11\bar{2}1]$ (c) $[10\bar{1}0]$

(d) $(1\bar{2}10)$ (e) $(\bar{1}\bar{1}22)$ (f) $(12\bar{3}0)$

3.27 What are the indices of the six directions of the form <110> that lie in the $(11\bar{1})$ plane of a cubic cell?

3.28 What are the indices of the four directions of the form <111> that lie in the $(\bar{1}01)$ plane of a cubic cell?

3.29 Determine the number of directions of the form <110> in a tetragonal unit cell and compare to the number of directions of the form <110> in an orthorhombic unit cell.

3.30 Determine the angle between the [110] direction and the (110) plane in a tetragonal unit cell; then determine the angle between the [011] direction and the (011) plane in a tetragonal cell. The lattice parameters are $a_0 = 0.4$ nm and $c_0 = 0.5$ nm. What is responsible for the difference?

3.31 Determine the Miller indices of the plane that passes through three points having the following coordinates:

(a)	0, 0, 1;	1, 0, 0;	and ½, ½, 0
(b)	½, 0, 1;	½, 0, 0;	and 0, 1, 0
(c)	1, 0, 0;	0,1, ½;	and 1, ½, ¼
(d)	1, 0, 0;	0, 0, ¼;	and ½, 1, 0

3.32 Determine the repeat distance, linear density, and packing fraction for FCC nickel, which has a lattice parameter of 0.35167 nm, in the [100], [110], and [111] directions. Which of these directions is close-packed?

3.33 Determine the repeat distance, linear density, and packing fraction for BCC lithium, which has a lattice parameter of 0.35089 nm, in the [100], [110], and [111] directions. Which of these directions is close-packed?

3.34 Determine the repeat distance, linear density, and packing fraction for HCP magnesium in the $[\bar{2}110]$ direction and the $[11\bar{2}0]$ direction. The lattice parameters for HCP magnesium are given in Appendix A.

3.35 Determine the planar density and packing fraction for FCC nickel in the (100), (110), and (111) planes. Which – if any – of these planes is close-packed?

3.36 Determine the planar density and packing fraction for BCC lithium in the (100), (110), and (111) planes. Which – if any – of these planes is close-packed?

3.37 Suppose that FCC rhodium is produced as a 1 mm-thick sheet, with the (111) plane parallel to the surface of the sheet. How many (111) interplanar spacings d_{111} thick is the sheet? See Appendix A for necessary data.

3.38 In a FCC unit cell, how many d_{111} are present between the 0, 0, 0 point and the 1, 1, 1 point?

3.39 Determine the minimum radius of an atom that will just fit into:

(a) the tetrahedral interstitial site in FCC nickel

(b) the octahedral interstitial site in BCC lithium

3.40 What is the radius of an atom that will just fit into the octahedral site in FCC copper without disturbing the lattice?

3.41 Using the ionic radii given in Appendix B, determine the coordination number expected for the following compounds:

(a) Y_2O_3 (b) UO_2 (c) BaO (d) Si_3N_4

(e) GeO_2 (f) MnO (g) MgS (h) KBr

3.42 Would you expect NiO to have the caesium chloride, sodium chloride, or zinc blende structure? Based on your answer, determine

((a) the lattice parameter,

(b) the density, and

(c) the packing factor for NiO.

3.43 Would you expect UO_2 to have the sodium chloride, zinc blende, or fluorite structure? Based on your answer, determine

(a) the lattice parameter,

(b) the density, and

(c) the packing factor for UO_2.

3.44 Would you expect BeO to have the sodium chloride, zinc blende, or fluorite structure? Based on your answer, determine

(a) the lattice parameter,

(b) the density, and

(c) the packing factor for BeO.

3.45 Would you expect CsBr to have the sodium chloride, zinc blende, fluorite, or caesium chloride structure? Based on your answer, determine

(a) the lattice parameter,

(b) the density, and

(c) the packing factor for CsBr.

3.46 Sketch the ion arrangement of the (110) plane of ZnS (with the zinc blende structure) and compare this arrangement to that on the (110) plane of CaF_2 (with the fluorite structure). Compare the planar packing fraction on the (110) planes for these two materials.

3.47 MgO, which has the sodium chloride structure, has a lattice parameter of 0.396 nm. Determine the planar density and the planar packing fraction for the (111) and (222) planes of MgO. What ions are present on each plane?

3.48 Polypropylene forms an orthorhombic unit cell with lattice parameters of a_0 = 1.450 nm, b_0 = 0.569 nm, and c0 = 0.740 nm. The chemical formula for the propylene molecule, from which the polymer is produced, is C_3H_6. The density of the polymer is about 0.90 $Mg.m^{-3}$. Determine the number of propylene molecules, the number of carbon atoms, and the number of hydrogen atoms in each unit cell.

3.49 The density of cristobalite is about 1.538 $Mg.m^{-3}$, and it has a lattice parameter of 0.8037 nm. Calculate the number of SiO_2 ions, the number of silicon ions, and the number of oxygen ions in each unit cell.

3.50 A diffracted X-ray beam is observed from the (220) planes of iron at a 2θ angle of 99.1° when X-rays of 0.15418 nm wavelength are used. Calculate the lattice parameter of the iron.

3.51 A diffracted X-ray beam is observed from the (311) planes of aluminium at a 2θ angle of 78.3° when X-rays of 0.15418 nm wavelength are used. Calculate the lattice parameter of the aluminium.

3.52 Figure 3.41 shows the results of an X-ray diffraction experiment in the form of the intensity of the diffracted peak versus the 2θ diffraction angle. If X-rays with a wavelength of 0.15418 nm are used, determine:

(a) the crystal structure of the metal,

(b) the indices of the planes that produce each of the peaks, and

(c) the lattice parameter of the metal.

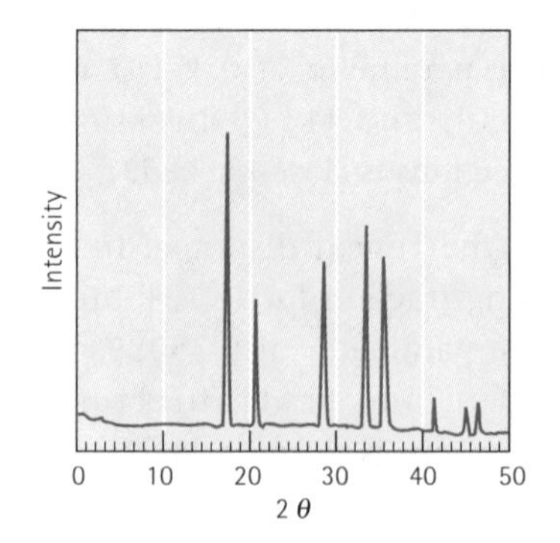

Figure 3.41 X-ray diffraction pattern (*for* Problem 3.52).

3.53 Figure 3.42 shows the results of an X-ray diffraction experiment in the form of the intensity of the diffracted peak versus the 2θ diffraction angle. If X-rays with a wavelength of 0.07107 nm are used, determine:

(a) the crystal structure of the metal,

(b) the indices of the planes that produce each of the peaks, and

(c) the lattice parameter of the metal.

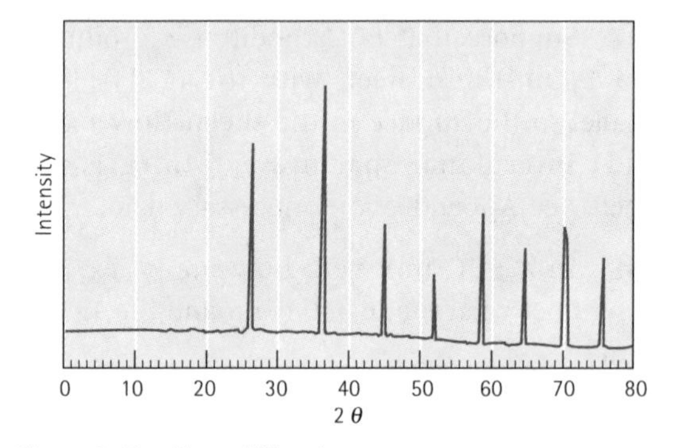

Figure 3.42 X-ray diffraction pattern (for Problem 3.53).

Design Problems

3.54 You would like to design a purification cell. The cell is to be composed of two sizes of spheres, with the smaller ones fitting into the holes between the larger ones. A wide variety of sizes are available, but the largest is 1 cm in diameter. A contaminated gas will flow through the cell, with the contaminants being absorbed at the surfaces of the spheres. We have found that we need a total surface area of 10 000 cm to accomplish the purification. Design such a cell.

3.55 You would like to sort iron specimens, some of which are FCC and others, BCC. Design an X-ray diffraction method by which this can be accomplished.

CHAPTER 4

Imperfections in the Atomic Arrangement

4.1 Introduction

The arrangement of the atoms in all materials contains imperfections which have a profound effect on the behaviour of the materials. By controlling lattice imperfections, we create stronger metals and alloys, more powerful magnets, improved transistors and solar cells, glassware of striking colours, and many other materials of practical importance.

In this chapter we introduce the three basic types of lattice imperfections: point defects, line defects (or dislocations), and surface defects. We must remember, however, that these imperfections only represent defects in the perfect atomic arrangement, not in the material itself. Indeed, these 'defects' may be intentionally added to produce a desired set of mechanical and physical properties. Later chapters will show how we control these defects through alloying, heat treatment, or processing techniques to produce improved engineering materials.

4.2 Dislocations

Dislocations are line imperfections in an otherwise perfect lattice. They are typically introduced into the lattice during solidification of the material or when the material is deformed. Although dislocations are present in all materials, including ceramics and polymers, they are particularly useful in explaining deformation and strengthening in metals. We can identify two types of dislocations: the screw dislocation and the edge dislocation.

Screw Dislocations The screw dislocation (Figure 4.1) can be illustrated by cutting partway through a perfect crystal, then skewing the crystal one atom spacing. If we follow a crystallographic plane one revolution around the axis on which the crystal was skewed, starting at point *x* and travelling equal atom spacings in each direction, we finish one atom spacing below our starting point (point *y*). The vector required to complete the loop and return us to our starting point is the Burgers vector **b**. If we continued our rotation, we would trace out a spiral path. The axis, or line around which we trace out this path, is the screw dislocation. The Burgers vector is parallel to the screw dislocation.

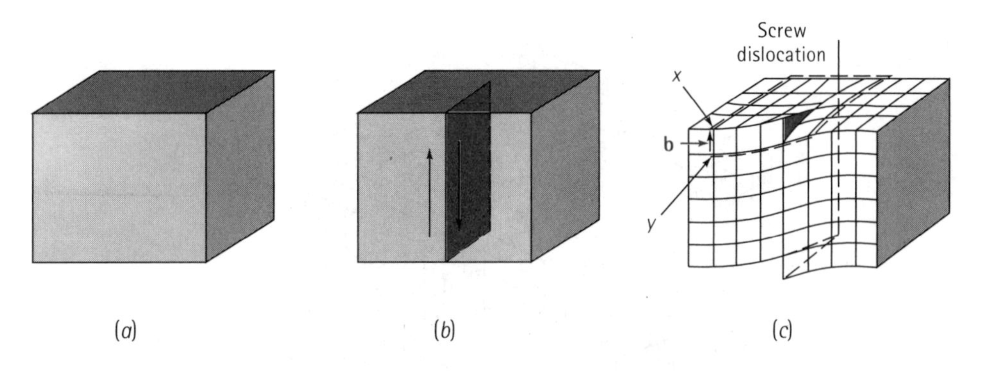

Figure 4.1
The perfect crystal (*a*) is cut and sheared one atom spacing, (*b*) and (*c*). The line along which shearing occurs is a screw dislocation. A Burgers vector **b** is required to close a loop of equal atom spacing around the screw dislocation.

Edge Dislocations An edge dislocation (Figure 4.2) can be illustrated by slicing partway through a perfect crystal, spreading the crystal apart, and partly filling the cut with an extra plane of atoms. The bottom edge of this inserted plane represents the edge dislocation, shown also in Figure 4.5. If we describe a clockwise loop around the edge dislocation, starting at point x and going an equal number of atoms spacings in each direction, we finish one atom spacing from starting point y. The vector required to complete the loop is, again, the Burgers vector. In this case, the Burgers vector is perpendicular to the dislocation. By introducing the dislocation, the atoms above the dislocation line are squeezed too closely together, while the atoms below the dislocation are stretched too far apart. The surrounding lattice has been disturbed by the presence of the dislocation.

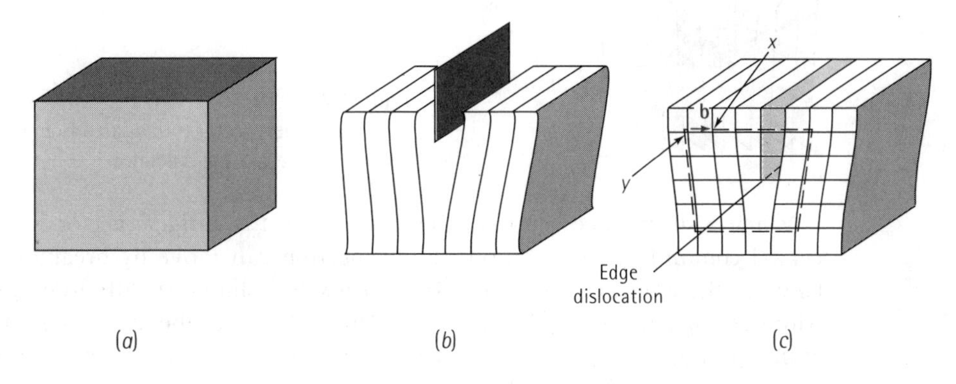

Figure 4.2
The perfect crystal in (*a*) is cut and an extra plane of atoms is inserted (*b*). The bottom edge of the extra plane is an edge dislocation (*c*). A Burgers vector **b** is required to close a loop of equal atom spacing around the edge dislocation.

Mixed dislocations As shown in Figure 4.3, mixed dislocations have both edge and screw components, with a transition region between them. The Burgers vector, however, remains the same for all portions of the mixed dislocation.

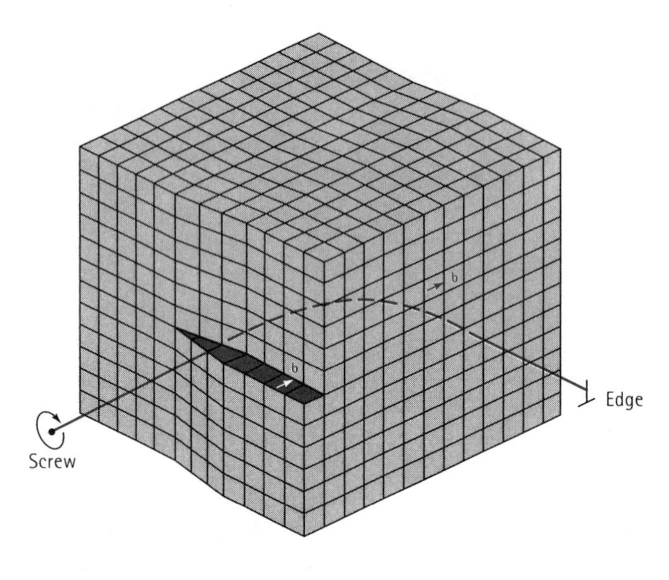

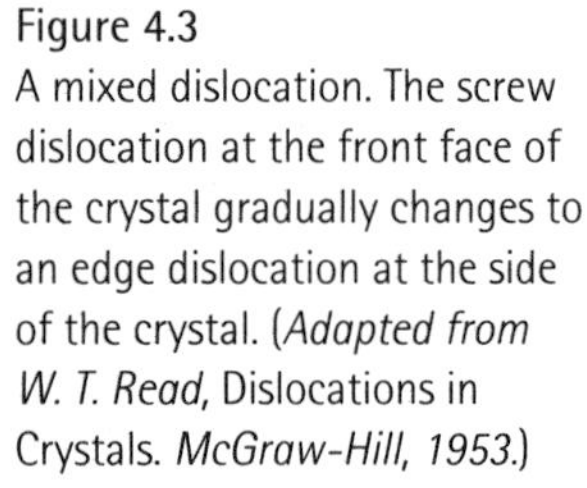
Figure 4.3
A mixed dislocation. The screw dislocation at the front face of the crystal gradually changes to an edge dislocation at the side of the crystal. (*Adapted from W. T. Read,* Dislocations in Crystals. *McGraw-Hill, 1953.*)

Slip We could translate the Burgers vector from the loop to the edge dislocation, as shown in Figure 4.4. After this translation, we find that the Burgers vector and the edge dislocation define a plane in the lattice. The Burgers vector and the plane are helpful in explaining how materials deform.

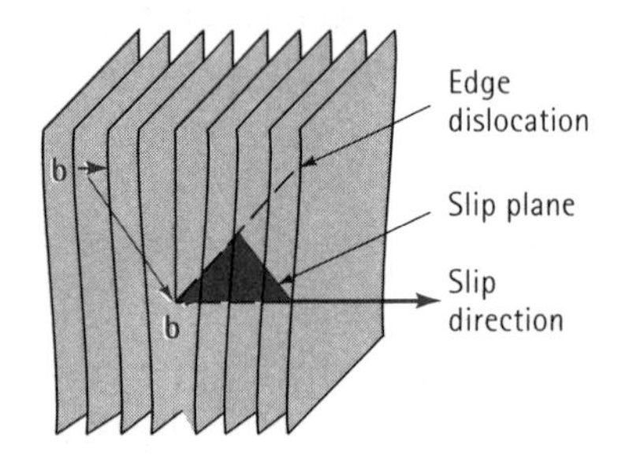

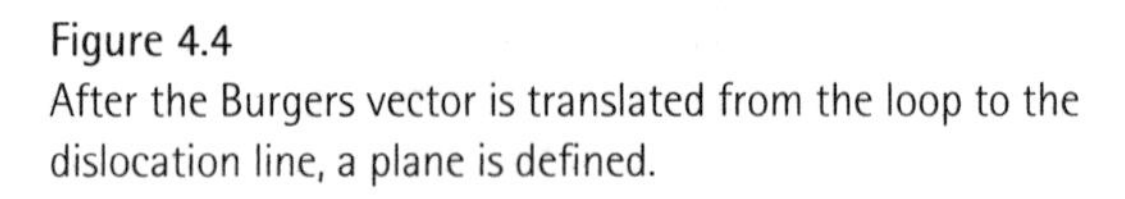
Figure 4.4
After the Burgers vector is translated from the loop to the dislocation line, a plane is defined.

When a shear force acting in the direction of the Burgers vector is applied to a crystal containing a dislocation, the dislocation can move by breaking the bonds between the atoms in one plane. The cut plane is shifted slightly to establish bonds with the original partial plane of atoms. This shift causes the dislocation to move one atom spacing to the side, as shown in Figure 4.5. If this process continues, the dislocation moves through the crystal until a step is produced on the exterior of the crystal; the crystal has then been deformed. If dislocations could be continually introduced into one side of the crystal and moved along the same path through the crystal, the crystal would eventually be cut in half.

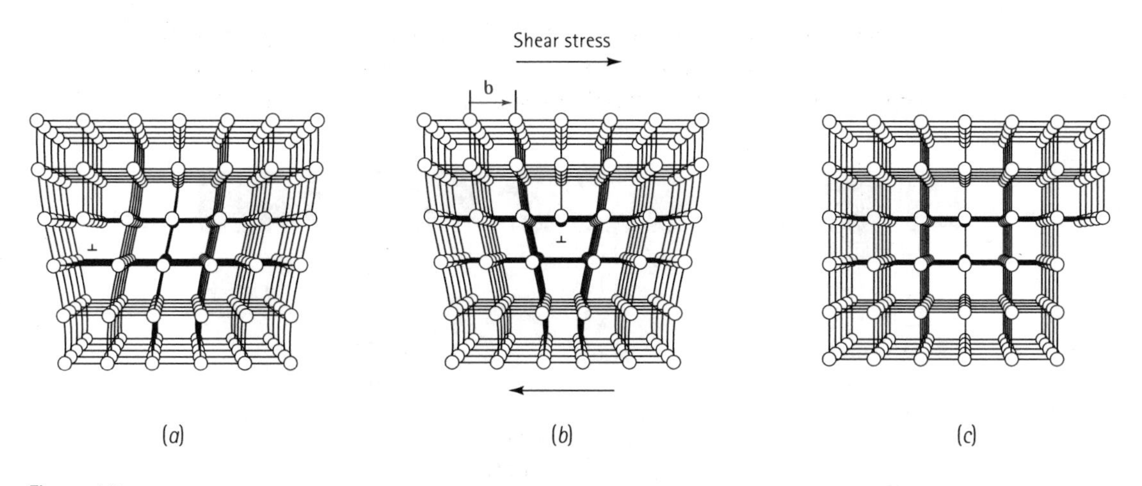

Figure 4.5
When a shear stress is applied to the dislocation in (*a*), the atoms are displaced, causing the dislocation to move one Burgers vector in the slip direction (*b*). Continued movement of the dislocation eventually creates a step (*c*), and the crystal is deformed. (*Adapted from A. G. Guy,* Essentials of Materials Science, *McGraw-Hill, 1976.*)

The process by which a dislocation moves and causes a material to deform is called slip. The direction in which the dislocation moves, the slip direction is the direction of the Burgers vector for edge dislocations. During slip, the edge dislocation sweeps out the plane formed by the Burgers vector and the dislocation; this plane is called the slip plane. The combination of slip direction and slip plane is the slip system. A screw dislocation produces the same result; the dislocation moves in a direction perpendicular to the Burgers vector although the crystal deforms in a direction parallel to the Burgers vector.

During slip, a dislocation moves from one set of surroundings to an identical set of surroundings. The Peierls-Nabarro stress (Equation 4.1) is required to move the dislocation from one equilibrium location to another,

$$\tau = c \exp(-kd/\mathbf{b}) \tag{4.1}$$

where τ is the shear stress required to move the dislocation, d is the interplanar spacing between adjacent slip planes, $\mathbf{b}$ is the Burgers vector, and both c and k are constants for the material. The dislocation moves in a slip system that requires the least expenditure of energy. Several important factors determine the most likely systems that will be active.

1. The stress required to cause the dislocation to move increases exponentially with the length of the Burgers vector. Thus, the slip direction should have a small repeat distance or high linear density. The close-packed directions in metals satisfy this criterion and are the usual slip directions.
2. The stress required to cause the dislocation to move decreases exponentially with the interplanar spacing of the slip planes. Slip occurs most easily between planes of atoms that are smooth (so there are smaller 'hills and valleys' on the surface) and between planes that are far apart (or have a relatively large interplanar spacing).

Planes with a high planar density fulfil this requirement. Therefore, the slip planes are typically close-packed planes or those as closely packed as possible. Common slip systems in several materials are summarised in Table 4.1.

Crystal Structure	Slip Plane	Slip Direction
BCC metals	$\{110\}$	$\langle 111 \rangle$
	$\{112\}$	
	$\{123\}$	
FCC metals	$\{111\}$	$\langle 110 \rangle$
HCP metals	$\{0001\}$	$\langle 100 \rangle$
	$\{11\bar{2}0\}$ (See note)	$\langle 110 \rangle$
	$\{10\bar{1}0\}$ (See note)	or $\langle 11\bar{2}0 \rangle$
	$\{10\bar{1}1\}$ (See note)	
MgO, NaCl (ionic)	$\{110\}$	$\langle 110 \rangle$
Silicon (covalent)	$\{111\}$	$\langle 110 \rangle$

Note: These planes are active in some metals and alloys or at elevated temperatures.

Table 4.1 Slip planes and directions in metallic structures.

3 Dislocations do not move easily in materials such as silicon or polymers, which have covalent bonds. Because of the strength and directionality of the bonds, the materials typically fail in a brittle manner before the force becomes high enough to cause appreciable slip.

4 Materials with an ionic bond, including many ceramics such as MgO, also are resistant to slip. Movement of a dislocation disrupts the charge balance around the anions and cations, requiring that bonds between anions and cations be broken. During slip, ions with a like charge must also pass close together, causing repulsion. Finally, the repeat distance along the slip direction, or the Burgers vector, is larger than in metals. Again, brittle failure of the material typically occurs before the dislocations move.

EXAMPLE 4.1

A sketch of a dislocation in MgO, which has the sodium chloride crystal structure and a lattice parameter of 0.396 nm, is shown in Figure 4.6. Determine the length of the Burgers vector.

SOLUTION

In Figure 4.6, we begin a clockwise loop around the dislocation at point *x*, then move equal atom spacings to finish at point *y*. The vector **b** is the Burgers vector. Because **b** is a [110] direction, it must be perpendicular to {110} planes. The length of **b** is the distance between two adjacent (110) planes. From Equation (3.7),

$$d_{110} = \frac{a_0}{\sqrt{h^2 + k^2 + l^2}} = \frac{0.396}{\sqrt{1^2 + 1^2 + 0^2}} = 0.280 \text{ nm}$$

The Burgers vector is a <110> direction that is 0.280 nm in length. Note, however, that two extra half-planes of atoms make up the dislocation – one composed of oxygen ions and one of magnesium ions.

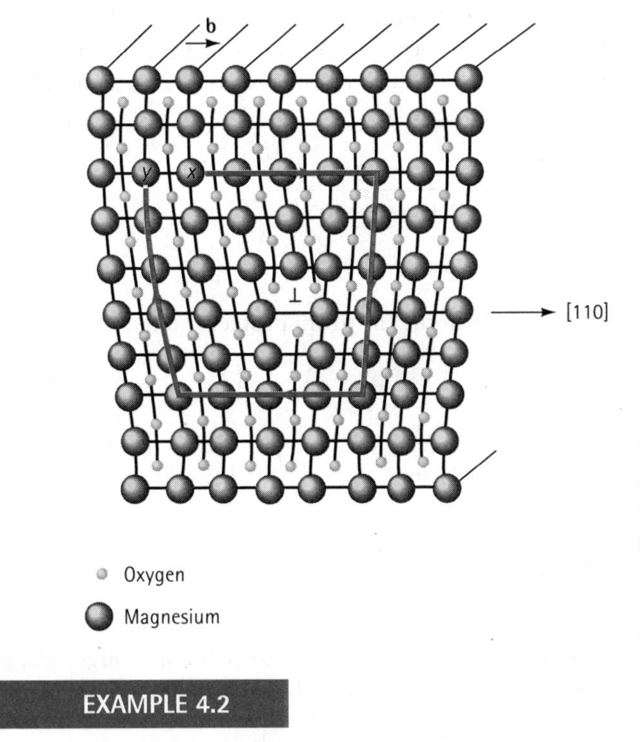

Figure 4.6
An edge dislocation in MgO showing the slip direction and Burgers vector (*for* Example 4.1). (*Adapted from W. D. Kingery, H. K. Bowen, and D. R. Uhlmann,* Introduction to Ceramics, *John Wiley, 1976.*)

EXAMPLE 4.2

Calculate the length of the Burgers vector in copper.

SOLUTION

Copper is FCC with a lattice parameter of 0.36151 nm. The close-packed directions, or the directions of the Burgers vector, are of the form <110>. The repeat distance along the <110> directions is half of the face diagonal, since lattice points are located at corners and centres of faces.

$$\text{Face diagonal} = \sqrt{2}a_0 = (\sqrt{2})(0.36151) = 0.51125 \text{ nm}$$

The length of the Burgers vector, or the repeat distance, is:

$$\mathbf{b} = \tfrac{1}{2}(0.51125 \text{ nm}) = 0.25563 \text{ nm}$$

EXAMPLE 4.3

The planar density of the (112) plane in BCC iron is 9.94×10^{18} atoms/m^2. Calculate (a) the planar density of the (110) plane and (b) the interplanar spacings for both the (112) and (110) planes. On which plane would slip normally occur?

SOLUTION

The lattice parameter of BCC iron is 0.2866 nm or 2.866×10^{-10} m. The (110) plane is shown in Figure 4.7, with the portion of the atoms lying within the unit cell being shaded. Note that a quarter of the four corner atoms plus the centre atom lie within an area of a_0 times $\sqrt{2}a_0$. The planar density is:

$$\text{Planar density (110)} = \frac{\text{atoms}}{\text{area}} = \frac{2}{(\sqrt{2})(2.866 \times 10^{-10} \text{ m})^2}$$

$$= 1.72 \times 10^{19} \text{ atoms/m}^2$$

Planar density (112) = 9.94×10^{18} atoms/m^2 (from problem statement)

The interplanar spacings are:

$$d_{110} = \frac{2.866 \times 10^{-10}}{\sqrt{1^2 + 1^2 + 0}} = 2.0266 \times 10^{-10} \text{ m}$$

$$d_{112} = \frac{2.866 \times 10^{-10}}{\sqrt{1^2 + 1^2 + 2^2}} = 1.17 \times 10^{-10} \text{ m}$$

The planar density and interplanar spacing of the (110) plane are larger than those for the (112) plane; therefore, the (110) plane would be the preferred slip plane.

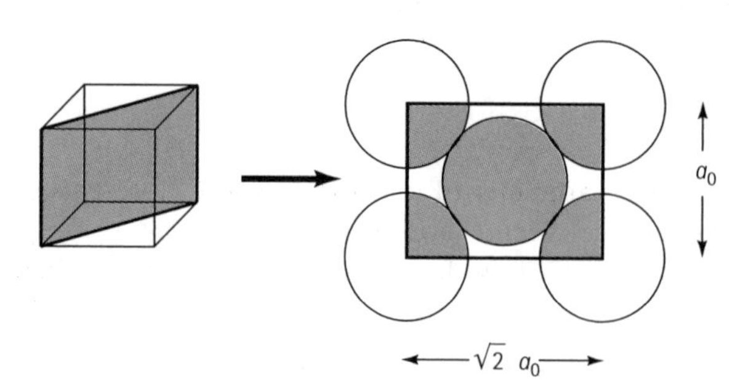

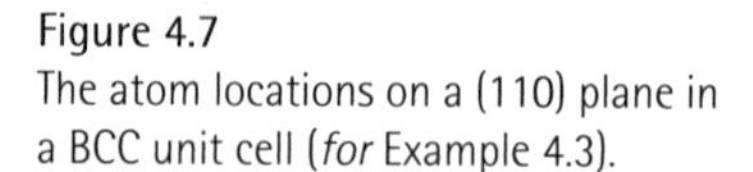

Figure 4.7
The atom locations on a (110) plane in a BCC unit cell (*for* Example 4.3).

4.3 Significance of Dislocations

Although slip can occur in some ceramics and polymers, the slip process is particularly helpful to us in understanding the mechanical behaviour of metals. First, slip explains why the strength of metals is much lower than the value predicted from the metallic bond. If slip occurs, only a tiny fraction of all of the metallic bonds across the interface need to be broken at any one time, and the force required to deform the metal is small.

Second, slip provides ductility in metals. If no dislocations were present, an iron bar would be brittle; metals could not be shaped by the various metal-working processes, such as forging, into useful shapes.

Third, we control the mechanical properties of a metal or alloy by interfering with the movement of dislocations. An obstacle introduced into the crystal prevents a dislocation from slipping unless we apply higher forces. If we must apply a higher force, then the metal must be stronger!

Enormous numbers of dislocations are found in materials. The dislocation density or total length of dislocations per unit volume, is usually used to represent the amount of dislocations present. Dislocation densities of 10 $m.mm^{-3}$ are typical of the softest metals, while densities up to 1 000 $km.mm^{-3}$ can be achieved by deforming the material.

The transmission electron microscope (TEM) is used to observe dislocations. In the TEM, electrons are focused on an extremely thin foil of the material; the beam of electrons interacts with imperfections in the material, causing differences in the fraction of the electrons that are transmitted. Consequently, different contrasts are observed when the transmitted beam is viewed on a fluorescent screen or a photographic plate. Figure 4.8 shows dislocations, which appear as dark lines, at very high magnifications.

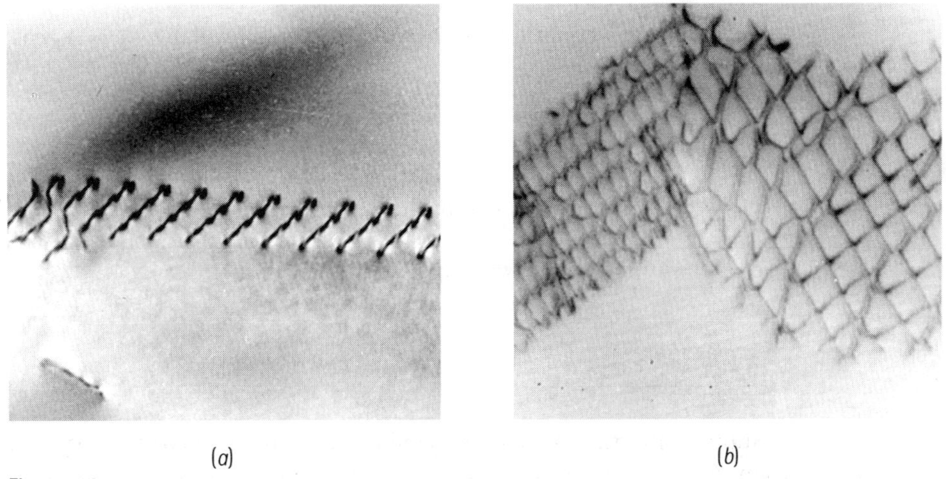

(a) (b)

Figure 4.8
Electron photomicrographs of dislocations in Ti_3Al: (*a*) dislocation pileups (× 36 500) and (*b*) dislocation network (× 15 750). (*Courtesy Gerald Feldewerth.*)

4.4 Schmid's Law

We can understand the differences in behaviour of metals that have different crystal structures by examining the force required to initiate the slip process. Suppose we apply a unidirectional force F to a cylinder of metal that is a single crystal (Figure 4.9). We can orient the slip plane and slip direction to the applied force by defining the angles λ and ϕ. λ is the angle between the slip direction and the applied force, and ϕ is the angle between the normal to the slip plane and the applied force.

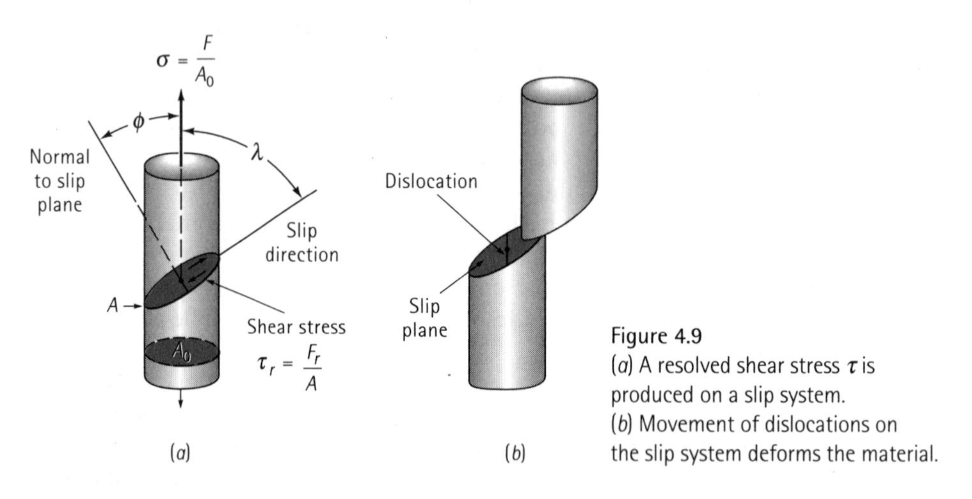

Figure 4.9
(*a*) A resolved shear stress τ is produced on a slip system.
(*b*) Movement of dislocations on the slip system deforms the material.

In order for the dislocation to move in its slip system, a shear force acting in the slip direction must be produced by the applied force. This resolved shear force F_r is given by:

$$Fr = F \cos \lambda$$

If we divide the equation by the area of the slip plane, $A = A_0/\cos \phi$, we obtain **Schmid's law**,

$$\tau_r = \sigma \cos \phi \cos \lambda \quad (4.2)$$

where:

$\tau_r = \dfrac{F_r}{A}$ = resolved shear *stress* in the slip direction

$\sigma = \dfrac{F}{A_0}$ = unidirectional *stress* applied to the cylinder

EXAMPLE 4.4

Suppose the slip plane is perpendicular to the applied stress σ, as in Figure 4.10. Then, $\phi = 0°$, $\lambda = 90°$, $\cos \lambda = 0$, and therefore $\tau_r = 0$. Even if the applied stress σ is enormous, no resolved shear stress develops along the slip direction and the dislocation cannot move. (You could perform a simple experiment to demonstrate this with a deck of cards. If you push on the deck at an angle, the cards slide over one another, as in the slip process. If you push perpendicular to the deck, however, the cards do not slide.) Slip cannot occur if the slip system is oriented so that either λ or ϕ is 90°.

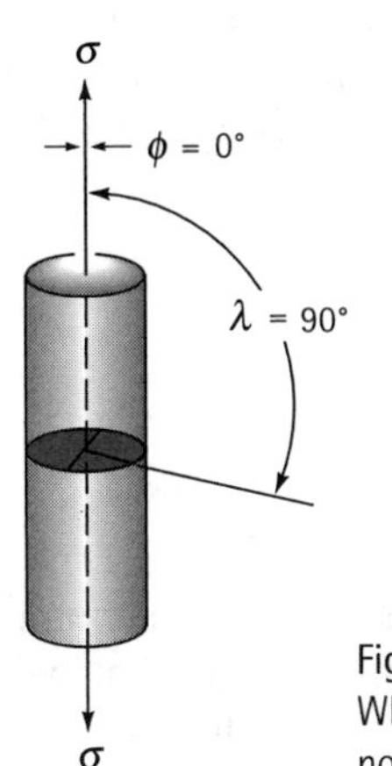

Figure 4.10
When the slip plane is perpendicular to the applied stress σ, the angle λ is 90° and no shear stress is resolved.

The critical resolved shear stress τ_{crss} is the shear stress required to break enough metallic bonds in order for slip to occur. Thus slip occurs, causing the metal to deform, when the *applied* stress produces a *resolved* shear stress that equals the *critical* resolved shear stress.

$$\tau_r = \tau_{crss} \tag{4.3}$$

EXAMPLE 4.5

We wish to produce a rod composed of a single crystal of pure aluminium, which has a crystal resolved shear stress of 1 $MN.m^{-2}$. We would like to orient the rod in such a manner that, when an axial stress of 3.5 $MN.m^{-2}$ is applied, the rod deforms by slip in a 45° direction to the axis of the rod and actuates a sensor that detects the overload. Evaluate the necessary angle of the slip plane and suggest how the rod might be produced.

SOLUTION

Dislocations begin to move when the resolved shear stress τ_r equals the critical resolved shear stress, 1 $MN.m^{-2}$. From Schmid's law:

$$\tau_r = \sigma \cos\lambda \cos\phi, \text{ or}$$

$$1 = (3.5) \cos\lambda \cos\phi$$

Because we wish slip to occur at a 45° angle to the axis of the rod, $\lambda = 45°$, and:

$$\cos\phi = \frac{1}{3.5\cos 45^\circ} = 0.404$$
$$\phi = 66.2^\circ$$

Therefore, we must produce a rod that is oriented such that $\lambda = 45°$ and $\phi = 66.2°$.

We might do this by a solidification process. We would orient a crystal of solid aluminium at the bottom of a mould. Liquid aluminium could be introduced into the mould. The liquid begins to solidify from the starting crystal and a single crystal rod of the proper orientation is produced.

4.5 Influence of Crystal Structure

We can use Schmid's law to compare the properties of metals having BCC, FCC, and HCP crystal structures. Table 4.2 lists three important factors that we can examine. We must be careful to note, however, that this discussion describes the behaviour of nearly perfect single crystals. Real engineering materials are seldom single crystals and always contain large numbers of defects.

Table 4.2 Summary of factors affecting slip in metallic structures.

Factor	FCC	BCC	HCP ($\frac{c}{a}$ >1.633)
Critical resolved shear stress ($MN.m^{-2}$)	0.35–0.7	35–70	0.35–0.7[a]
Number of slip systems	12	48	3[b]
Cross-slip	Can occur	Can occur	Cannot occur[b]
Summary of properties	Ductile	Strong	Relatively brittle

a For slip on basal planes.

b By alloying or heating to elevated temperatures, additional slip systems are active in HCP metals, permitting cross-slip to occur and thereby improving ductility.

Critical Resolved Shear Stress If the critical resolved shear stress in a metal is very high, the applied stress σ must also be high in order for τ_r to equal t_{crss}. If σ is large, the metal must have a high strength! In FCC metals, which have close-packed {111} planes, the critical resolved shear stress is low – about 0.35 to 0.7 $MN.m^{-2}$ in a perfect crystal; FCC metals tend to have low strengths. On the other hand, BCC crystal structures contain no close-packed planes and we must exceed a higher critical resolved shear stress – in the order of 70 $MN.m^{-2}$ in perfect crystals – before slip occurs; therefore BCC metals tend to have high strengths.

We would expect the HCP metals, because they contain close-packed basal planes, to have low critical resolved shear stresses. In fact, in HCP metals such as zinc that have a *c*/*a* ratio greater than or equal to the theoretical ratio of 1.633, the critical resolved shear stress is less than 0.7 $MN.m^{-2}$, just as in FCC metals. In HCP titanium, however, the *c*/*a* ratio is less than 1.633; the close-packed planes are spaced too closely together. Slip now occurs on planes such as $(10\bar{1}0)$, the 'prism' planes or faces of the hexagon, and the critical resolved shear stress is then as great as or greater than in BCC metals.

Number of Slip Systems If at least one slip system is oriented to give the angles λ and ϕ near 45°, then τ_r equals τ_{crss} at low applied stresses. Ideal HCP metals possess only one set of parallel close-packed planes, the (0001) planes, and three close-packed directions, giving three slip systems. Consequently, the probability of the close-packed planes and directions being oriented with λ and ϕ near 45° is very low. The HCP crystal may fail in a brittle manner without a significant amount of slip.

However, in HCP metals with a low *c*/*a* ratio, or when HCP metals are properly alloyed, or when the temperature is increased, other slip systems become active, making these metals less brittle than expected.

On the other hand, FCC metals contain four non-parallel close-packed planes of the form {111} and three close-packed directions of the form <110> within each plane, giving a total of 12 slip systems. At least one slip system is favourably oriented for slip to occur at low applied stresses, permitting FCC metals to have high ductilities.

Finally, BCC metals have as many as 48 slip systems that are nearly close-packed. Several slip systems are always properly oriented for slip to occur, allowing BCC metals to also have ductility.

Cross-Slip Suppose a screw dislocation moving on one slip plane encounters an obstacle and is blocked from further movement. The dislocation can shift to a second intersecting slip system, also properly oriented, and continue to move. This is called cross-slip. In many HCP metals, no cross-slip can occur because the slip planes are parallel, not intersecting. Therefore, the HCP metals tend to remain brittle. Fortunately, additional slip systems become active when HCP metals are alloyed or heated, thus improving ductility. Cross-slip is possible in both FCC and BCC metals because a number of intersecting slip systems are present. Cross-slip consequently helps maintain ductility in these metals.

4.6 Point Defects

Point defects are localised disruptions of the lattice involving one or, possibly, several atoms. These imperfections, shown in Figure 4.11, may be introduced by movement of the atoms when they gain energy by heating, during processing of the material, by introduction of impurities, or intentionally through alloying.

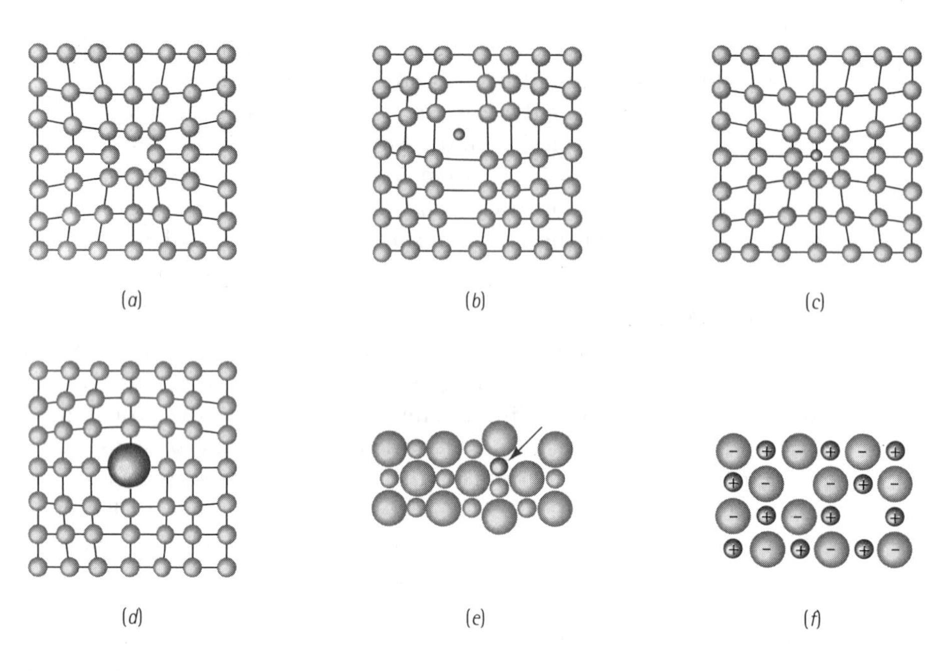

Figure 4.11
Point defects: (*a*) vacancy, (*b*) interstitial atom, (*c*) small substitutional atom, (*d*) large substitutional atom, (*e*) Frenkel defect, and (*f*) Schottky defect. All of these defects disrupt the perfect arrangement of the surrounding atoms.

Vacancies A vacancy is produced when an atom is missing from a normal site. Vacancies are introduced into the crystal during solidification, at high temperatures, or as a consequence of radiation damage. At room temperature, very few vacancies are present, but this number increases exponentially as we increase the temperature:

$$n_v = n \exp\left(\frac{-Q}{RT}\right) \tag{4.4}$$

where n_v is the number of vacancies per m^3; n is the number of lattice points per m^3; Q is the energy required to produce a vacancy, in $J.mol^{-1}$; R is the gas constant, 8.31 $J.mol^{-1}.K^{-1}$; and T is the temperature in degrees Kelvin. Due to the large thermal energy near the melting temperature, there may be as many as one vacancy per 1 000 lattice points.

EXAMPLE 4.6

Suggest a heat treatment that will provide 1 000 times more vacancies in copper than are normally present at room temperature. About 83 700 $J.mol^{-1}$ are required to produce a vacancy in copper.

SOLUTION

The lattice parameter of FCC copper is 0.36151 nm. The number of copper atoms, or lattice points, per m^3 is:

$$n = \frac{4 \text{ atoms / cell}}{(3.6151 \times 10^{-10} \text{ m})^3} = 8.47 \times 10^{28} \text{copper atoms/ m}^3$$

At room temperature, $T = 25 + 273 = 298$ K:

$$n_v = (8.47 \times 10^{28}) \exp [(-83\,700)/(8.31)(298)]$$

$$= 1.774 \times 10^{14} \text{ vacancies/m}^3$$

We wish to produce 1 000 times this number, or $n_v = 1.774 \times 10^{17}$ vacancies/m^3. We could do this by heating the copper to a temperature at which this number of vacancies forms:

$$n_v = 1.774 \times 10^{17} = (8.47 \times 10^{28}) \exp(-83\,700 / 8.31T)$$

$$\exp(-83\,700 / 8.31T) = \frac{1.774 \times 10^{17}}{8.47 \times 10^{28}} = 2.095 \times 10^{-12}$$

$$\frac{-83\,700}{8.31T} = \ln(2.095 \times 10^{-12}) = -26.89$$

$$T = \frac{83\,700}{(8.31)(26.89)} = 375 \text{ K} = 102°\text{C}$$

By heating the copper slightly above 100°C (perhaps even by placing the copper in boiling water), then rapidly cooling the copper back to room temperature, the number of vacancies trapped in the structure may be one thousand times greater than the equilibrium number of vacancies.

Interstitial Defects An interstitial defect is formed when an extra atom is inserted into the lattice structure at a normally unoccupied position. The interstitial sites were illustrated in Table 3.6. Interstitial atoms, although much smaller than the atoms located at the lattice points, are still larger than the interstitial sites that they occupy; consequently, the surrounding lattice is compressed and distorted. Interstitial atoms such as hydrogen are often present as impurities; carbon atoms are intentionally added to iron to produce steel. Once introduced, the number of interstitial atoms in the structure remains nearly constant, even when the temperature is changed.

Substitutional Defects A substitutional defect is introduced when one atom is replaced by a different type of atom. The substitutional atom remains at the original position. Substitutional atoms may either be larger than the normal atoms in the lattice (in which case the surrounding atoms are compressed) or smaller (causing the surrounding atoms to be in tension). In either case, the substitutional defect disturbs the surrounding lattice. Again, the substitutional defect can be introduced either as an impurity or as a deliberate alloying addition and, once introduced, the number of defects is relatively independent of temperature.

Other Point Defects An interstitialcy is created when an atom identical to those at the normal lattice points is located in an interstitial position. These defects are most likely to be found in lattices having a low packing factor.

A Frenkel defect is a vacancy-interstitial pair formed when an ion jumps from a normal lattice point to an interstitial site, leaving behind a vacancy. A Schottky defect is a pair of vacancies in an ionically bonded material; both an anion and a cation must be missing from the lattice if electrical neutrality is to be preserved in the crystal. These are common in ceramic materials with the ionic bond.

A final important point defect occurs when an ion of one charge replaces an ion of a different charge. This might be the case when an ion with a valence of +2 replaces an ion with a valence of +1 (Figure 4.12). In this case, an extra positive charge is introduced into the structure. To maintain a charge balance, a vacancy might be created where a +1 cation normally would be located. Again, this imperfection is observed in materials that have pronounced ionic bonding.

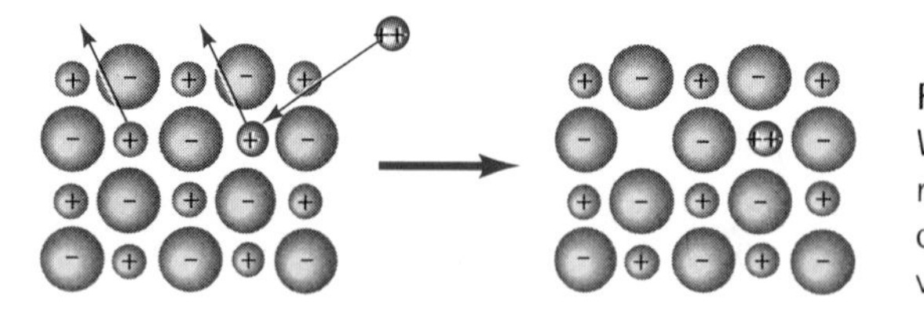

Figure 4.12
When a divalent cation replaces a monovalent cation, a second monovalent cation must also be removed, creating a vacancy.

Importance of Point Defects Point defects disturb the perfect arrangement of the surrounding atoms, distorting the lattice for perhaps hundreds of atom spacings from the actual point defect. A dislocation moving through the general vicinity of a point defect encounters a lattice in which the atoms are not at their equilibrium positions. This disruption requires that a higher stress be applied to force the dislocation past the defect, therefore increasing the strength of the material.

EXAMPLE 4.7

Determine the number of vacancies needed for a BCC iron lattice to have a density of 7.87 $Mg.m^{-3}$. The lattice parameter of the iron is 2.866×10^{-10} m.

SOLUTION

The expected theoretical density of iron can be calculated from the lattice parameter and the atomic mass. Since the iron is BCC, two iron atoms are present in each unit cell.

$$\rho = \frac{(2 \text{ atoms/cell})(55.847 \text{g.mol}^{-1})}{(2.866 \times 10^{-10} \text{m})^3 (6.02 \times 10^{23} \text{ atoms/mol})} = 7.8814 \text{ Mg.m}^{-3}$$

We would like to produce iron with a lower density. We could do this by intentionally introducing vacancies into the lattice. Let's calculate the number of iron atoms and vacancies that would be present in each unit cell for the required density of 7.87 $Mg.m^{-3}$:

$$\rho = \frac{(\text{atoms / cell})(55.847\,\text{g.mol}^{-1})}{(2.866 \times 10^{-10}\,\text{m})^3(6.02 \times 10^{23}\ \text{atoms / mol})} = 7.87\ \text{Mg.m}^{-3}$$

$$\text{Atoms / cell} = \frac{(7.87 \times 10^6)(2.866 \times 10^{-10})^3(6.02 \times 10^{23})}{55.847} = 1.9971$$

Or, there should be 0.0029 vacancies per unit cell. The number of vacancies per m^3 is:

$$\text{Vacancies / m}^3 = \frac{0.0029\ \text{vacancies / cell}}{(2.866 \times 10^{-10}\,\text{m})^3} = 1.23 \times 10^{26}$$

If additional information, such as the energy required to produce a vacancy in iron, were known, we might be able to establish a heat treatment (as we did in Example 4.6) to produce this concentration of vacancies.

EXAMPLE 4.8

In FCC iron, carbon atoms are located at octahedral sites at the centre of each edge of the unit cell (½, 0, 0) and at the centre of the unit cell (½, ½, ½). In BCC iron, carbon atoms enter tetrahedral sites, such as ¼, ½, 0. The lattice parameter is 0.3571 nm for FCC iron and 0.2866 nm for BCC iron. Carbon atoms have a radius of 0.071 nm. (1) Would we expect a greater distortion of the lattice by an interstitial carbon atom in FCC or BCC iron? (2) What would be the atomic percentage of carbon in each type of iron if all the interstitial sites were filled?

SOLUTION

1 We could calculate the size of the interstitial site at the ¼, ½, 0 location with the help of Figure 4.13. The radius R_{BCC} of the iron atom is:

$$R_{BCC} = \frac{\sqrt{3}a_0}{4} = \frac{\left(\sqrt{3}\right)(0.2866)}{4} = 0.1241\ \text{nm}$$

From Figure 4.13, we find that:

$$\left(\frac{1}{2}a_0\right)^2 + \left(\frac{1}{4}a_0\right)^2 = (r_{\text{interstitial}} + R_{BCC})^2$$

$$(r_{\text{interstitial}} + R_{BCC})^2 = 0.3125{a_0}^2 = (0.3125)(0.2866\ \text{nm})^2 = 0.02567$$

$$r_{\text{interstitial}} = \sqrt{0.02567} - 0.1241 = 0.0361\ \text{nm}$$

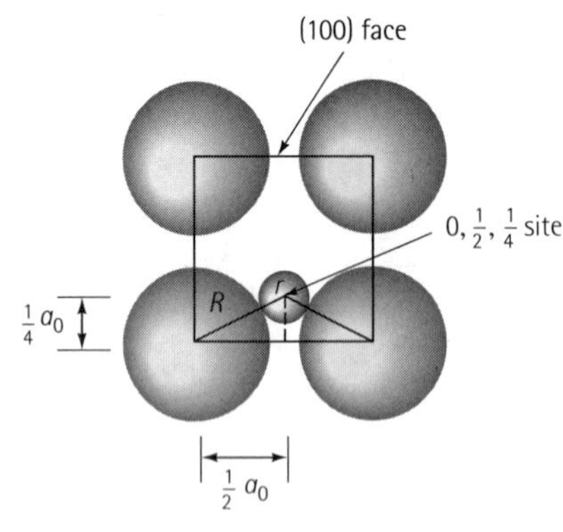

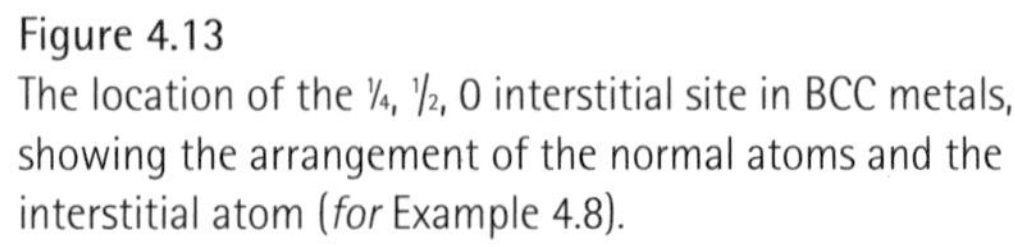
Figure 4.13
The location of the ¼, ½, 0 interstitial site in BCC metals, showing the arrangement of the normal atoms and the interstitial atom (*for* Example 4.8).

For FCC iron, the interstitial site such as the ½, 0, 0 lies along <100> directions. Thus, the radius of the iron atom and the radius of the interstitial site are:

$$R_{FCC} = \frac{\sqrt{2}a_0}{4} = \frac{(\sqrt{2})(0.3571)}{4} = 0.1263 \text{ nm}$$

$$2r_{\text{interstitial}} + 2R_{FCC} = a_0$$

$$r_{\text{interstitial}} = \frac{0.3571 - (2)(0.1263)}{2} = 0.0523 \text{ nm}$$

The interstitial site in the BCC iron is smaller than the interstitial site in the FCC iron. Although both are smaller than the carbon atom, carbon distorts the BCC lattice more than the FCC lattice. As a result, fewer carbon atoms are expected to enter interstitial positions in BCC iron than in FCC iron.

2 In BCC iron, two iron atoms are expected in each unit cell. We can find a total of 24 interstitial sites of the type ¼, ½, 0; however, since each site is located at a face of the unit cell, only half of each site belongs uniquely to a single cell. Thus:

(24 sites) (½) = 12 interstitial sites per unit cell

If all of the interstitial sites were filled, the atomic percentage of carbon contained in the iron would be:

$$\text{at\% C} = \frac{12 \text{ C atoms}}{12 \text{ C atoms} + 2 \text{ Fe atoms}} \times 100 = 86\%$$

In FCC iron, four iron atoms are expected in each unit cell, and the number of interstitial sites is:

12 edges (¼) + 1 centre = 4 interstitial sites per unit cell

Again, if all the interstitial sites were filled, the atomic percentage of carbon in the FCC iron would be:

$$\text{at\% C} = \frac{4 \text{ C atoms}}{4 \text{ C atoms} + 4 \text{ Fe atoms}} \times 100 = 50\%$$

As we will see in a later chapter, the maximum atomic percentage of carbon present in the two forms of iron under equilibrium conditions is:

BCC: 1.0%
FCC: 8.9%

Because of the strain imposed on the iron lattice by the interstitial atoms – particularly in the BCC iron – the fraction of the interstitial sites that can be occupied is quite small.

4.7 Surface Defects

Surface defects are the boundaries, or planes, that separate a material into regions, each region having the same crystal structure but different orientations.

Material Surface The exterior dimensions of the material represent surfaces at which the lattice abruptly ends. Each atom at the surface no longer has the proper coordination number and atomic bonding is disrupted. The exterior surface may also be very rough, may contain tiny notches, and may be much more reactive than the bulk of the material.

Grain Boundaries The microstructure of most materials consists of many grains. A grain is a portion of the material within which the arrangement of the atoms is identical. However, the orientation of the atom arrangement, or crystal structure, is different for each adjoining grain. Three grains are shown schematically in Figure 4.14; the lattice in each grain is identical but the lattices are oriented differently. A grain boundary, the surface that separates the individual grains, is a narrow zone in which the atoms are not properly spaced. That is to say, the atoms are so close at some locations in the grain boundary that they cause a region of compression, and in other areas they are so far apart that they cause a region of tension.

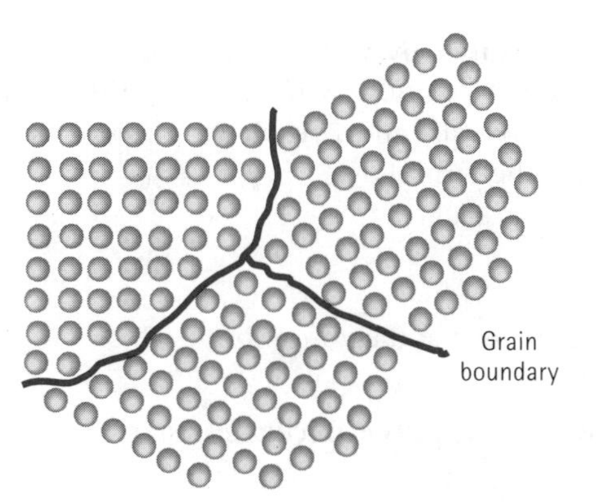

Figure 4.14
The atoms near the boundaries of the three grains do not have an equilibrium spacing or arrangement.

One method of controlling the properties of a material is by controlling the grain size. By reducing the grain size, we increase the number of grains and, hence, increase the amount of grain boundary. Any dislocation moves only a short distance before encountering a grain boundary, and the strength of the metal is increased. The Hall-Petch equation relates the grain size to the yield strength,

$$\sigma_y = \sigma_0 + Kd^{-1/2} \tag{4.5}$$

where σ_y is the yield strength or the stress at which the material permanently deforms, d is the average diameter of the grains, and σ_0 and K are constants for the metal. Figure 4.15 shows this relationship in steel. Later chapters will describe how the grain size can be controlled through solidification, alloying, and heat treatment.

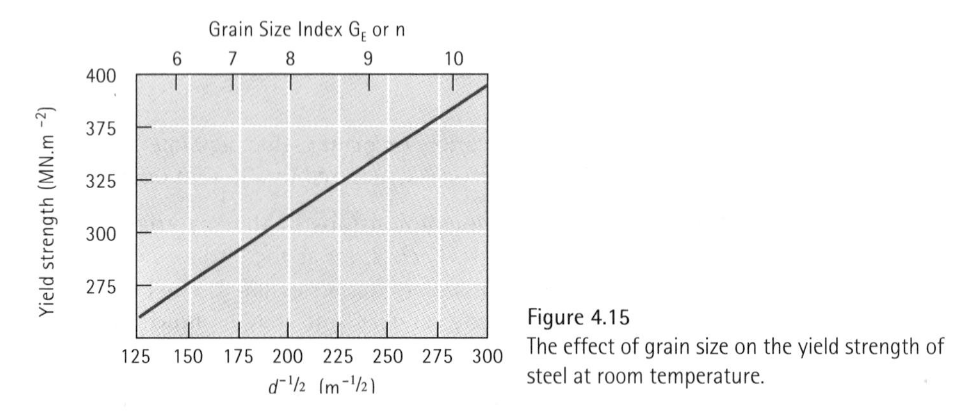

Figure 4.15
The effect of grain size on the yield strength of steel at room temperature.

Optical microscopy is used to reveal microstructural features such as grain boundaries that require less than about 2 000 magnification. The process of preparing a sample and observing or recording its microstructure is called metallography. A sample of the material is abraded and polished to a mirror-like finish. The surface is then exposed to chemical attack, or etching, with grain boundaries being attacked more aggressively than the remainder of the grain. Light from an optical microscope is reflected or scattered from the sample surface, depending on how the surface is etched. When more light is scattered from deeply etched features such as the grain boundaries, these features appear dark (Figure 4.16).

Specification of grain size One manner by which the grain size is defined is the designation of a grain size number or index which is specified in both **ASTM** (American Society for Testing & Materials) and **British Standards**. The ASTM grain size number **n** is determined from the formula:

$$\mathcal{N} = 2^{\mathbf{n}-1} \tag{4.6}$$

where

$\mathcal{N}$ = Number of grains per square inch at magnification ×100.
n = ASTM grain size number

The grain size index $\mathbf{G_E}$ (from BS 4490: 1989) is calculated using

$$\mathrm{m} = 2^{\mathbf{G_E}+3} \tag{4.7}$$

where

m = Number of grains per square millimetre at magnification × 1.
G_E = Grain size index.

For practical purposes the grain size number **n** and grain size index $\mathbf{G_E}$ are the same. A large grain size index or number indicates many grains, or a fine grain size, and correlates with high strengths. Table 4.3 gives the standard grain size designation and the corresponding grain relationships.

Table 4.3

Grain size index G_E or number n	Average grains per mm^2 at 1 ×	Average grains per $inch^2$ at 100 ×
0	8	0.5
1	16	1
2	32	2
3	64	4
4	128	8
5	256	16
6	512	32
7	1 024	64
8	2 048	128

From BS4490:1989.

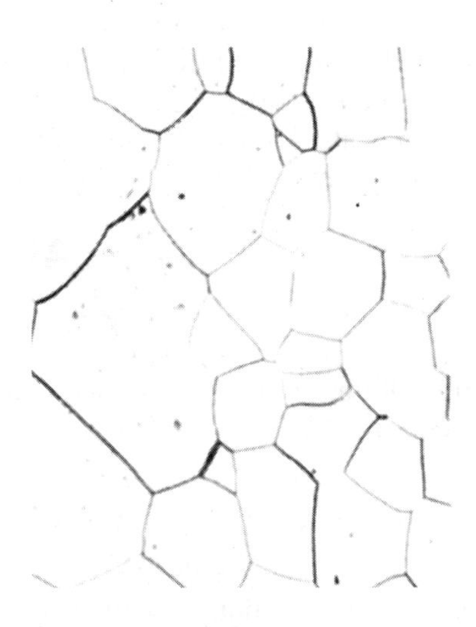

Figure 4.16
Microstructure of palladium, (× 100). (*From* Metals Handbook, *Vol. 9, 9th Ed., American Society for Metals, 1985.*)

EXAMPLE 4.9

We would like to produce a KCl ceramic part having a yield strength of 20 $MN.m^{-2}$. Based on previous data, we know that a grain size of 5 μm gives a strength of 28 $MN.m^{-2}$ and that a grain size of 100 μm gives a strength of 8 $MN.m^{-2}$. (1 μm is 10^{-6}m.)

SOLUTION

We can use the Hall-Petch equation to determine the required grain size in the ceramic. For a 5 μm grain size:

$$\sigma_y = \sigma_0 + Kd^{-1/2}$$
$$28 = \sigma_0 + K(5\times10^{-6})^{-1/2} = \sigma_0 + 447\ K$$
$$\sigma_0 = 28 - 447\ K$$

For a 100 μm grain size:

$$8 = \sigma_0 + K(100 \times 10^{-6})^{-1/2} = 28 - 447\ K + 100\ K$$
$$(447 - 100)K = 28 - 8$$
$$K = \frac{20}{347} = 0.058$$

and

$$\sigma_0 = 28 - (447)(0.058) = 2.07\ \text{MN.m}^{-2}$$

Therefore, the grain size required for a strength of 20 MN.m^{-2} is:

$$20 = 2.07 + 0.058d^{-1/2}$$
$$d^{-1/2} = \frac{20 - 2.07}{0.058}$$

So, $d = 10.47 \times 10^{-6}$m $= 10.47\ \mu$m

One method by which ceramic parts are produced is to initially crush the raw material (in this case KCl), pass the crushed material through a series of screens to separate out the particles that are too coarse and too fine, and then consolidate the particles into a part by pressing them into a die at a high temperature. In our case, we would want to use only particles that are approximately 10.47 μm in diameter.

EXAMPLE 4.10

Suppose we count 64 grains per square millimetre in a photomicrograph taken at magnification × 10. What is the grain size index, G_E?

SOLUTION

If we count 64 grains per square millimetre at magnification × 10, then at magnification × 1 we must have:

$$m = \left(\frac{10}{1}\right)^2 (64) = 640\ \text{grains}/\text{mm}^2 = 2^{G_E+3}$$
$$\log 640 = (G_E + 3)\log 2$$
$$2.806 = (G_E + 3)(0.301)$$
$$G_E = 6.32$$

Small Angle Grain Boundaries A small angle grain boundary is an array of dislocations that produces a small misorientation between the adjoining lattices (Figure 4.17). Because the energy of the surface is less than that at a regular grain boundary, the small angle grain boundaries are not as effective in blocking slip. Small angle boundaries formed by edge dislocations are called tilt boundaries, and those caused by screw dislocations are called twist boundaries

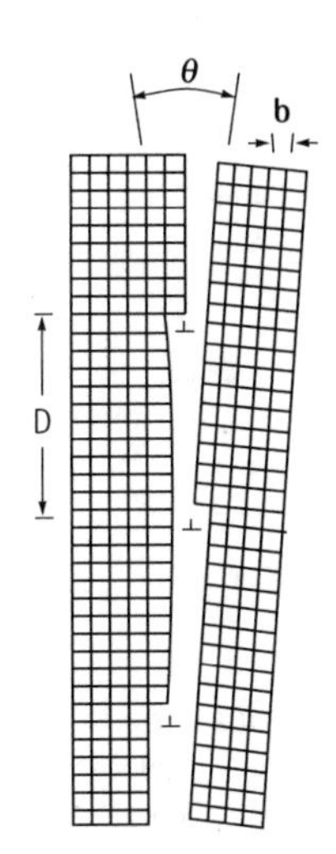

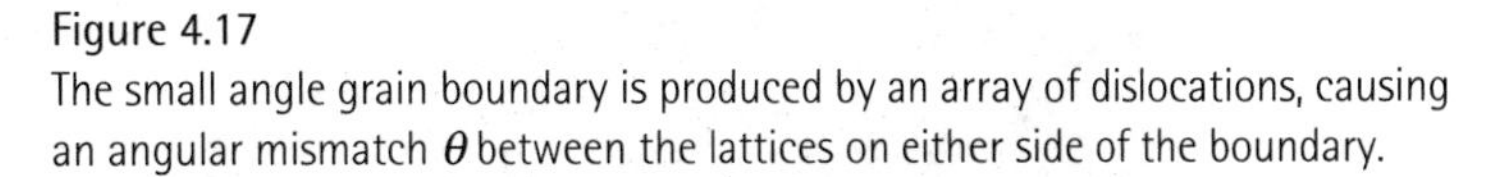

Figure 4.17
The small angle grain boundary is produced by an array of dislocations, causing an angular mismatch θ between the lattices on either side of the boundary.

Stacking Faults Stacking faults, which occur in FCC metals, represent an error in the stacking sequence of close-packed planes. Normally, a stacking sequence of *ABC ABC ABC* is produced in a perfect FCC lattice. But, suppose the following sequence is produced:

ABCABABCABC

In the portion of the sequence indicated, a type *A* plane is shown where a type *C* plane would normally be located. This small region, which has a HCP stacking sequence instead of the FCC stacking sequence, represents a stacking fault. Stacking faults interfere with the slip process.

Twin Boundaries A twin boundary is a plane across which there is a special mirror image misorientation of the lattice structure (Figure 4.18). Twins can be produced when a shear force, acting along the twin boundary, causes the atoms to shift out of position. Twinning occurs during deformation or heat treatment of certain metals. The twin boundaries interfere with the slip process and increase the strength of the metal. Movement of twin boundaries can also cause a metal to deform. Figure 4.18 shows that the formation of a twin has changed the shape of the metal.

The effectiveness of the surface defects in interfering with the slip process can be judged from the surface energies (Table 4.4). The high-energy grain boundaries are much more effective in blocking dislocations than either stacking faults or twin boundaries.

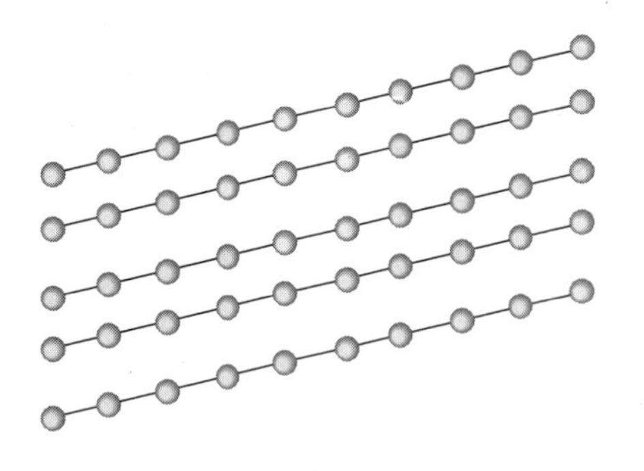

(a)

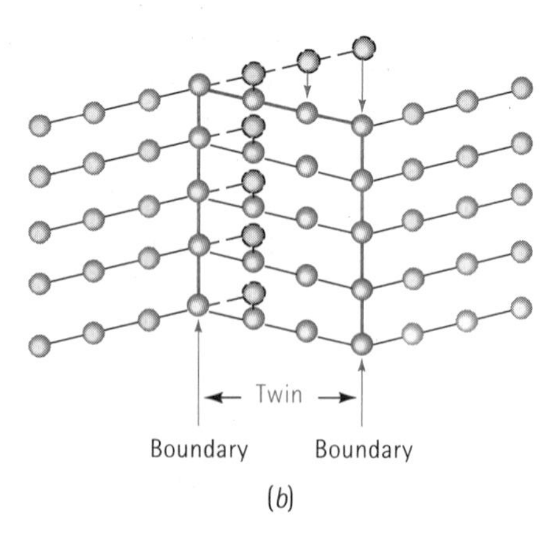

(b)

(c)

Figure 4.18
Application of a stress to the perfect crystal (a) may cause a displacement of the atoms (b), causing the formation of a twin. Note that the crystal has deformed a result of twinning. (c) A photomicrograph of twins within a grain of brass (× 250).

Table 4.4 Energies of surface imperfections in selected metals.

Surface imperfection ($J.m^{-2}\times10^{-3}$)	Al	Cu	Pt	Fe
Stacking fault energy	200	75	95	–
Twin boundary energy	120	45	195	190
Grain boundary energy	625	645	1 000	780

4.8 Control of the Slip Process

In a perfect crystal, the fixed, repeated arrangement of the atoms gives the lowest possible energy within the crystal. Any imperfection in the lattice raises the internal energy at the location of the imperfection. The local energy is increased because, near the imperfection, the atoms either are squeezed too closely together (compression) or are forced too far apart (tension).

One dislocation in an otherwise perfect lattice can move easily through the crystal if the resolved shear stress equals the critical resolved shear stress. However, if the dislocation encounters a region where the atoms are displaced from their usual positions, a higher stress is required to force the dislocation past the region of high local energy; thus, the material is stronger. We can, then, control the strength of a material by controlling the number and type of imperfections. Three common strengthening mechanisms are based on the three categories of lattice defects in crystals.

Strain Hardening Dislocations disrupt the perfection of the lattice. In Figure 4.19, the atoms below the dislocation line at point *B* are compressed, while the atoms above dislocation *B* are too far apart. If dislocation *A* moves to the right and passes near dislocation *B*, dislocation A encounters a region where the atoms are not properly arranged. Higher stresses are required to keep the second dislocation moving; consequently, the metal must be stronger. Increasing the number of dislocations further increases the potential of the material to strain harden. *Strain hardening* is discussed formally in Chapter 7.

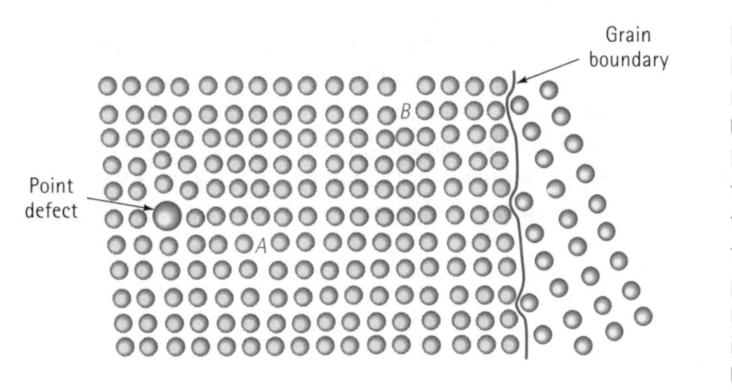

Figure 4.19
If the dislocation at point *A* moves to the left, it is blocked by the point defect. If the dislocation moves to the right, it interacts with the disturbed lattice near the second dislocation at point *B*. If the dislocation moves farther to the right, it is blocked by a grain boundary.

Solid Solution Strengthening Any of the point defects also disrupt the perfection of the lattice. If dislocation *A* moves to the left (Figure 4.19), it encounters a disturbed lattice caused by the point defect; higher stresses are needed to continue slip of the dislocation. By intentionally introducing substitutional or interstitial atoms, we cause *solid solution strengthening,* which is discussed in Chapter 9.

Grain Size Strengthening Finally, surface imperfections such as grain boundaries disturb the lattice. If dislocation *B* moves to the right (Figure 4.19), it encounters a grain boundary and is blocked. By increasing the number of grains or reducing the grain size, *grain size strengthening* is achieved. Control of grain size will be discussed in a number of later chapters.

EXAMPLE 4.11 **Materials Selection for a Stable Structure**

We would like to produce a bracket to hold ceramic bricks in place in a heat-treating furnace. The bracket should be strong, should possess some ductility so that it bends rather than fractures if overloaded, and should maintain most of its strength up to 600°C. Suggest a suitable material for this bracket, considering the various lattice imperfections as the strengthening mechanism.

SOLUTION

In order to serve up to 600°C, the bracket should *not* be produced from a polymer material. Instead, a metal or ceramic would be considered.

In order to have some ductility, dislocation must slip. Because slip in ceramics is difficult, the bracket should be produced from a metallic material. The metal should have a melting point well above 600°C; aluminium, with a melting point of 660°C, would not be suitable; iron, however, would be a reasonable choice.

We can introduce point, line, and surface imperfections into the iron to help produce strength, but we wish the imperfections to be stable as the service temperature increases. As shown in Chapter 5, grains can grow at elevated temperatures, reducing the number of grain boundaries and causing a decrease in strength. As indicated in Chapter 7, dislocations may be annihilated at elevated temperatures – again, reducing strength. The number of vacancies depends on temperature, so controlling these lattice defects may not produce stable properties.

The number of interstitial or substitutional atoms in the lattice does not, however, change with temperature. We might add carbon to the iron as interstitial atoms or substitute vanadium atoms for iron atoms at normal lattice points. These point defects continue to interfere with dislocation movement and help to keep the strength stable.

Of course, other design requirements may be important as well. For example, the steel bracket may deteriorate by oxidation or may react with the ceramic brick.

SUMMARY

- Imperfections, or defects, in the lattice of a crystalline material are of three general types: point defects, line defects or dislocations, and surface defects.
- Dislocations are line defects which, when a force is applied to the material, move and cause the material to deform.
  - The critical resolved shear stress is the stress required to move the dislocation.
  - The dislocation moves in a slip system, composed of a slip plane and a slip direction. The slip direction, or Burgers vector, is typically a close-packed direction. The slip plane is also normally close-packed or nearly close-packed.
  - In metallic crystals, the number and type of slip directions and slip planes influence the properties of the metal. In FCC metals, the critical resolved shear stress is low and an optimum number of slip planes are available; consequently, FCC metals tend to be ductile. In BCC metals, no close-packed planes are available and the critical resolved shear stress is high; thus, the BCC metals tend to be strong. The number of slip systems in HCP metals is limited, causing these metals to behave in a brittle manner.
- Point defects, which include vacancies, interstitial atoms, and substitutional atoms, introduce compressive or tensile strain fields that disturb the surrounding lattice.

As a result, dislocations cannot easily slip in the vicinity of point defects and the strength of the material is increased. The number of vacancies, or empty lattice points, depends on the temperature of the material; interstitial atoms (located in interstitial sites between the normal atoms) and substitutional atoms (which replace the host atom at lattice points) are often deliberately introduced and are typically unaffected by changes in temperature.

- Surface defects include grain boundaries. Producing a very small grain size increases the amount of grain boundary area; because dislocations cannot easily pass through a grain boundary, the material is strengthened.
- The number and type of lattice defects control the ease of movement of dislocations and, therefore, directly influence the mechanical properties of the material. Strain hardening is obtained by deforming the material causing dislocations to move until they become hindered by other dislocations or other lattice defects; solid solution strengthening involves the introduction of point defects; and grain size strengthening is obtained by producing a small grain size.

GLOSSARY

Burgers vector
The direction and distance that a dislocation moves in each step.

Critical resolved shear stress
The shear stress required to cause a dislocation to move and cause slip.

Cross-slip
A change in the slip system of a dislocation.

Dislocation
A line imperfection in the lattice of a crystalline material. Movement of dislocations helps explain how materials deform. Interference with the movement of dislocations helps explain how materials are strengthened.

Dislocation density
The total length of dislocation line per unit volume in a material.

Edge dislocation
A dislocation introduced into the lattice by adding an 'extra half plane' of atoms.

Frenkel defect
A pair of point defects produced when an ion moves to create an interstitial site, leaving behind a vacancy.

Grain
A portion of a solid material within which the lattice is identical and oriented in only one direction.

Grain boundary
A surface defect representing the boundary between two grains. The lattice has a different orientation on either side of the grain boundary.

Grain size index or number (G_E or n)

A measure of the size of the grains in a crystalline material obtained by counting the number of grains per unit area at magnification × 1 for G_E or ×100 for n.

Hall-Petch equation

The relationship between strength and grain size in a material – that is,

$$\sigma_y = \sigma_0 + Kd^{-1/2}.$$

Interstitial defect

A point defect produced when an atom is placed into the lattice at a site that is normally not a lattice point.

Interstitialcy

A point defect caused when a 'normal' atom occupies an interstitial site in the lattice.

Metallography

Preparation of a sample of a material by polishing and etching so that the structure can be examined using a microscope.

Mixed dislocation

A dislocation that contains partly edge components and partly screw components.

Peierls-Nabarro stress

The shear stress, which depends on the Burgers vector and the interplanar spacing, required to cause a dislocation to move – that is,

$$\tau = c \exp(-kd/b)$$

Point defects

Imperfections, such as vacancies, that are located at a single point in the lattice.

Schmid's law

The relationship between shear stress, the applied stress, and the orientation of the slip system – that is,

$$\tau = \sigma \cos \lambda \cos \phi.$$

Schottky defect

A pair of point defects in ionically bonded materials. In order to maintain a neutral charge, both a cation and an anion vacancy must form.

Screw dislocation

A dislocation produced by skewing a crystal so that one atomic plane produces a spiral ramp about the dislocation.

Slip

Deformation of a material by the movement of dislocations through the lattice.

Slip direction

The direction in the lattice in which the dislocation moves. The slip direction is the same as the direction of the Burgers vector.

Slip plane

The plane swept out by the dislocation line during slip. Normally, the slip plane is a close-packed plane, if one exists in the crystal structure.

Slip system

The combination of the slip plane and the slip direction.

Small angle grain boundary
An array of dislocations causing a small misorientation of the lattice across the surface of the imperfection.

Stacking fault
A surface defect in FCC metals caused by the improper stacking sequence of close-packed planes.

Substitutional defect
A point defect produced when an atom is removed from a regular lattice point and replaced with a different atom, usually of a different size.

Surface defects
Imperfections, such as grain boundaries, that form a two-dimensional plane within the lattice.

Tilt boundary
A small angle grain boundary composed of an array of edge dislocations.

Transmission electron microscope (TEM)
An instrument that, by passing an electron beam through a material, can detect microscopic structural features.

Twin boundary
A surface defect across which there is a mirror image misorientation of the lattice. Twin boundaries can also move and cause deformation of the material.

Twist boundary
A small angle grain boundary composed of an array of screw dislocations.

Vacancy
An atom missing from a lattice point.

PROBLEMS

4.1 What are the Miller indices of the slip directions:

(a) on the (111) plane in an FCC unit cell?

(b) on the (011) plane in a BCC unit cell?

4.2 What are the Miller indices of the slip planes in FCC unit cells that include the [101] slip direction?

4.3 What are the Miller indices of the {110} slip planes in BCC unit cells that include the [111] slip direction?

4.4 Calculate the length of the Burgers vector in the following materials:

(a) BCC niobium (b) FCC silver

(c) diamond cubic silicon

4.5 Determine the interplanar spacing and the length of the Burgers vector for slip on the expected slip systems in FCC aluminium. Repeat, assuming that the slip system is a (110) plane and a $[1\bar{1}1]$ direction. What is the ratio between the shear stresses required for slip for the two systems? Assume that $k = 2$ in Equation 4.1.

4.6 Determine the interplanar spacing and the length of the Burgers vector for slip on the $(110)/[1\bar{1}1]$ slip system in BCC tantalum. Repeat, assuming that the slip system is a $(111)/[1\bar{1}0]$ system. What is the ratio between the shear stresses required for slip for the two systems? Assume that $k = 2$ in Equation 4.1.

4.7 How many grams of aluminium, with a dislocation density of 10^{14} m.m^{-3}, are required to give a total dislocation length that would stretch from New York City to Los Angeles (4800 km)?

4.8 The distance from Earth to the Moon is 384 000 km. If this were the total length of dislocation in 1×10^{-6} cubic metres of material, what would be the dislocation density?

4.9 Suppose you would like to introduce an interstitial or large substitutional atom into the lattice near a dislocation. Would the atom fit more easily above or below the dislocation line shown in Figure 4.5(b)? Explain.

4.10 Compare the *c/a* ratios for the following HCP metals, determine the likely slip processes in each, and estimate the approximate critical resolved shear stress. Explain. (*See data in* Appendix A.)

(a) zinc (b) magnesium (c) titanium

(d) zirconium (e) rhenium (f) beryllium

4.11 A single crystal of an FCC metal is oriented so that the [001] direction is parallel to an applied stress of 35 MN.m^{-2}. Calculate the resolved shear stress acting on the (111) slip plane in the $[\bar{1}10]$, $[0\bar{1}1]$, and $[10\bar{1}]$ slip directions. Which slip system(s) will become active first?

4.12 A single crystal of a BCC metal is oriented so that the [001] direction is parallel to the applied stress. If the critical resolved shear stress required for slip is 80 MN.m^{-2}, calculate the magnitude of the applied stress required to cause slip to begin in the $[1\bar{1}1]$ direction on the (110), (011), and $(10\bar{1})$ slip planes.

4.13 Calculate the number of vacancies per m^3 expected in copper at 1085°C (just below the melting temperature). The energy for vacancy formation is 83 700 J.mol^{-1}.

4.14 The fraction of lattice points occupied by vacancies in solid aluminium at 660°C is 10^{-3}. What is the energy required to create vacancies in aluminium?

4.15 The density of a sample of FCC palladium is 11.98 Mg.m^3 and its lattice parameter is 0.38902 nm. Calculate

(a) the fraction of the lattice points that contain vacancies and

(b) the total number of vacancies in a cubic metre of Pd.

4.16 The density of a sample of HCP beryllium is 1.844 Mg.m^{-3} and the lattice parameters are a_0 = 0.22858 nm and c_0 = 0.35842 nm. Calculate

(a) the fraction of the lattice points that contain vacancies and

(b) the total number of vacancies in a cubic metre.

4.17 BCC lithium has a lattice parameter of 3.5089×10^{-10} m and contains one vacancy per 200 unit cells. Calculate

(a) the number of vacancies per cubic metre and

(b) the density of Li.

4.18 FCC lead has a lattice parameter of 0.4949 nm and contains one vacancy per 500 Pb atoms. Calculate

(a) the density and

(b) the number of vacancies per gram of Pb.

4.19 A niobium alloy is produced by introducing tungsten substitutional atoms in the BCC structure; eventually an alloy is produced that has a lattice parameter of 0.32554 nm and a density of 11.95 Mg.m^{-3}. Calculate the fraction of the atoms in the alloy that are tungsten.

4.20 Tin atoms are introduced into a FCC copper lattice, producing an alloy with a lattice parameter of 3.7589×10^{-10}m and a density of 8.772 Mg.m^{-3}. Calculate the atomic percentage tin present in the alloy.

4.21 We replace 7.5 percent of the chromium atoms in its BCC lattice with tantalum. X-ray diffraction shows that the lattice parameter is 0.29158 nm. Calculate the density of the alloy.

4.22 Suppose we introduce one carbon atom for every 100 iron atoms in an interstitial position in BCC iron, giving a lattice parameter of 0.2867 nm. For the Fe–C alloy, find

(a) the density and

(b) the packing factor.

4.23 The density of BCC iron is 7.882 $Mg.m^{-3}$ and the lattice parameter is 0.2866 nm when hydrogen atoms are introduced at interstitial positions. Calculate

(a) the atomic fraction of hydrogen atoms and

(b) number of unit cells on average that contain hydrogen atoms.

4.24 Suppose one Schottky defect is present in every tenth unit cell of MgO. MgO has the sodium chloride crystal structure and a lattice parameter of 0.396 nm. Calculate

(a) the number of anion vacancies per m^3 and

(b) the density of the ceramic.

4.25 ZnS has the zinc blende structure. If the density is 3.02 $Mg.m^{-3}$ and the lattice parameter is 0.59583 nm, determine the number of Schottky defects

(a) per unit cell and

(b) per cubic metre.

4.26 Suppose we introduce the following point defects. What other changes in each structure might be necessary to maintain a charge balance? Explain.

(a) Mg^{2+} ions substitute for yttrium atoms in Y_2O_3.

(b) Fe^{3+} ions substitute for magnesium ions in MgO.

(c) Li^{1+} ions substitute for magnesium ions in MgO.

(d) Fe^{2+} ions replace sodium ions in NaCl.

4.27 The strength of titanium is found to be 450 $MN.m^{-2}$ when the grain size is 17×10^{-6} m and 565 $MN.m^{-2}$ when the grain size is 0.8×10^{-6} m. Determine

(a) the constants in the Hall-Petch equation and

(b) the strength of the titanium when the grain size is reduced to 0.2×10^{-6} m.

4.28 A copper-zinc alloy has the following properties:

Grain diameter (mm)	Strength ($MN.m^{-2}$)
0.015	170
0.025	158
0.035	151
0.050	145

Determine

(a) the constants in the Hall-Petch equation and

(b) the grain size required to obtain a strength of 200 $MN.m^{-2}$.

4.29 For an SI grain size index of 8, calculate the number of grains per square millimetre.

4.30 Determine the SI grain size index if 20 grains/square millimetre are observed at a magnification of 40.

4.31 Determine the SI grain size index if 25 grains/square mm are observed at a magnification of 5.

4.32 Determine the ASTM grain size number for the materials in:

(a) Figure 4.16 (b) Figure 4.20 (c) Figure 1.9(c)

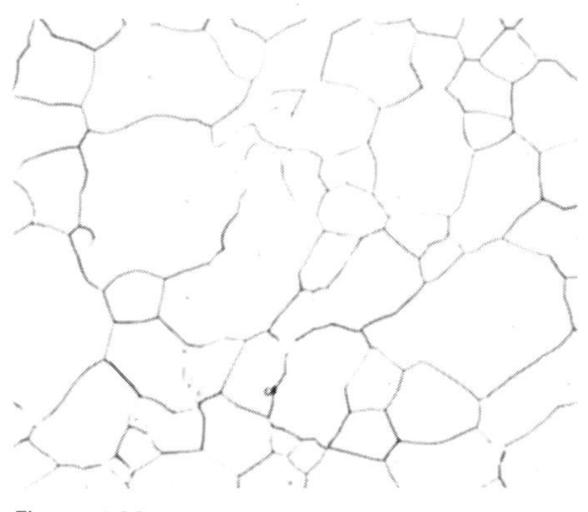

Figure 4.20
Microstructure of iron, *for* Problem 4.32 (× 500). (*From* Metals Handbook, *Vol. 9, 9th Ed., American Society for Metals, 1985.*)

4.33 The angle θ of a tilt boundary is given by $\sin(\theta/2) = b/2D$ (*See* Figure 4.17). Verify the correctness of this equation.

4.34 Calculate the angle θ of a small-angle grain boundary in FCC aluminium when the dislocations are 500 nm apart. (*See* Problem 4.33.)

4.35 For BCC iron, calculate average distance between dislocations in a small-angle grain boundary tilted 0.50°. (*See* Problem 4.33.)

4.36 Our discussion of Schmid's law dealt with single crystals of a metal. Discuss slip and Schmid's law in a polycrystalline material. What might happen as the grain size gets smaller and smaller?

4.37 The density of pure aluminium calculated from crystallographic data is expected to be 2.69955 $Mg.m^{-3}$. Suggest a suitable alloying element and calculate the atomic percentage of this element to produce an aluminium alloy that has a density of 2.6450 $Mg.m^{-3}$.

4.38 You would like a metal plate with good weldability. During the welding process, the metal next to the weld is heated almost to the melting temperature and depending on the welding parameters, may remain hot for some period of time. Design an alloy that will minimise the loss of strength in this 'heat-affected area' during the welding process.

CHAPTER 5

Atom Movement in Materials

5.1 Introduction

Diffusion is the movement of atoms within a material. Atoms move in a predictable fashion to eliminate concentration differences and produce a homogeneous, uniform composition. Movement of atoms is required for many of the treatments that we perform on materials. Diffusion is required for the heat treatment of metals, the manufacture of ceramics, the solidification of materials, the manufacture of transistors and solar cells, and the electrical conductivity of many ceramic materials. By understanding how mass is transferred by diffusion, we can design materials processing techniques, leak-proof devices, or even purification equipment.

In this chapter, we concentrate on understanding how diffusion occurs in solid materials. In addition, we discuss a number of examples of how diffusion is used to aid in the selection of materials and the design of manufacturing processes.

5.2 Stability of Atoms

Chapter 4 showed that imperfections could be introduced into the lattice of a crystal. However, these imperfections – and, indeed, even atoms in their normal lattice positions – are not stable or at rest. Instead, the atoms possess some thermal energy and they will move. For instance, an atom may move from a normal lattice point to occupy a nearby vacancy. An atom may move from one interstitial site to another. Atoms may jump across a grain boundary, causing the grain boundary to move.

The ability of atoms and imperfections to diffuse increases as the temperature, or thermal energy, possessed by the atoms increases. The rate of movement is related to temperature or thermal energy by the *Arrhenius* equation:

$$\text{Rate} = c_0 \exp\left(\frac{-Q}{RT}\right) \tag{5.1}$$

where c_0 is a constant, R is the gas constant ($8.314\ \text{J.mol}^{-1} \cdot \text{K}^{-1}$), T is the absolute temperature (K), and Q is the activation energy (J.mol^{-1}) required to cause the imperfection to move. This equation is derived from a statistical analysis of the probability that the atoms will have the extra energy Q needed to cause movement. The rate is related to the number of atoms that move.

We can rewrite the equation by taking natural logarithms of both sides:

$$\ln(\text{rate}) = \ln(c_0) - \frac{Q}{RT} \tag{5.2}$$

If we plot ln (rate) of some reaction versus $1/T$ (Figure 5.1), the slope of the curve will be $-Q/R$, and consequently Q can be calculated. The constant c_0 is the intercept when $1/T$ is zero.

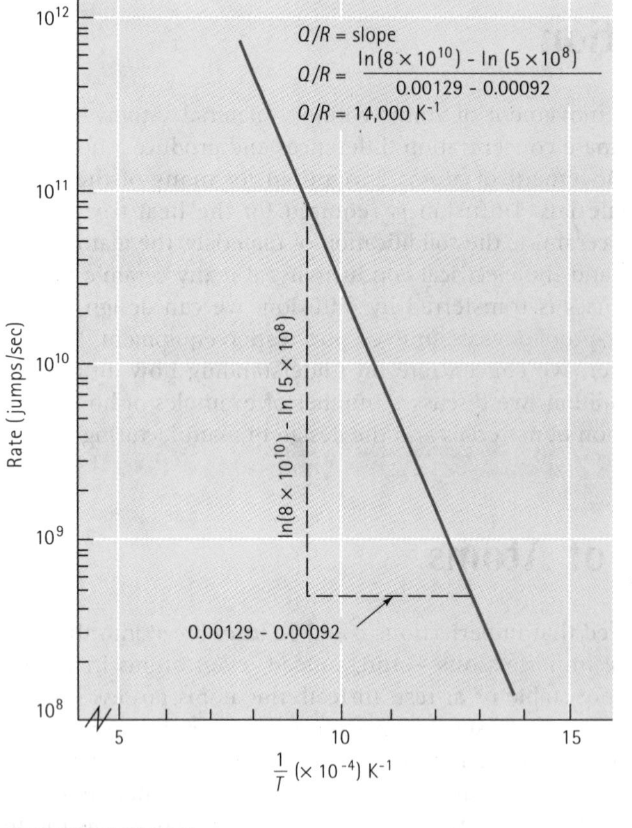

Figure 5.1
The Arrhenius plot of ln (rate) versus $1/T$ can be used to determine the activation energy required for a reaction.

EXAMPLE 5.1

Suppose that interstitial atoms are found to move from one site to another at the rates of 5×10^8 jumps/s at 500°C and 8×10^{10} jumps/s at 800°C. Calculate the activation energy Q for the process.

SOLUTION

Figure 5.1 represents the data on a ln (rate) versus $1/T$ plot; the slope of this curve, as calculated in the figure, gives Q/R = 14,000 K^{-1}, or Q = 116,400 $J.mol^{-1}$. Alternately, we could write two simultaneous equations:

$$5\times10^{8}=c_0\exp\left[\frac{-Q}{(8.314)(500+273)}\right]=c_0\exp(-0.000156Q)$$

$$8\times10^{10}=c_0\exp\left[\frac{-Q}{(8.314)(800+273)}\right]=c_0\exp(-0.000112Q)$$

Since

$$c_0=\frac{5\times10^{8}}{\exp(-0.000156Q)}$$

then

$$8\times10^{10}=\frac{(5\times10^{8})\exp(-0.000112Q)}{\exp(0.000156Q)}$$

$$\frac{8\times10^{10}}{5\times10^{8}}=160=\exp[(0.000156-0.000112)Q]=\exp(0.000044Q)$$

$$\ln(160)=5.075=0.000044Q$$

$$Q=\frac{5.075}{0.000044}=115340\ \text{J.mol}^{-1}$$

5.3 Diffusion Mechanisms

Even in absolutely pure solid materials, atoms move from one lattice position to another. This process, known as self-diffusion, can be detected by using radioactive tracers. Suppose we introduce a radioactive isotope of gold (Au^{198}) onto the surface of normal gold (Au^{197}). After a period of time, the radioactive atoms move into the regular gold, and eventually the radioactive atoms are uniformly distributed throughout the entire gold sample. Although self-diffusion occurs continually in all materials, the effect on the material's behaviour is not significant.

Diffusion of unlike atoms in materials also occurs (Figure 5.2). If a nickel sheet is bonded to a copper sheet, nickel atoms gradually diffuse into the copper and copper atoms migrate into the nickel. Again, the nickel and copper atoms eventually are uniformly distributed.

There are two important mechanisms by which atoms diffuse (Figure 5.3).

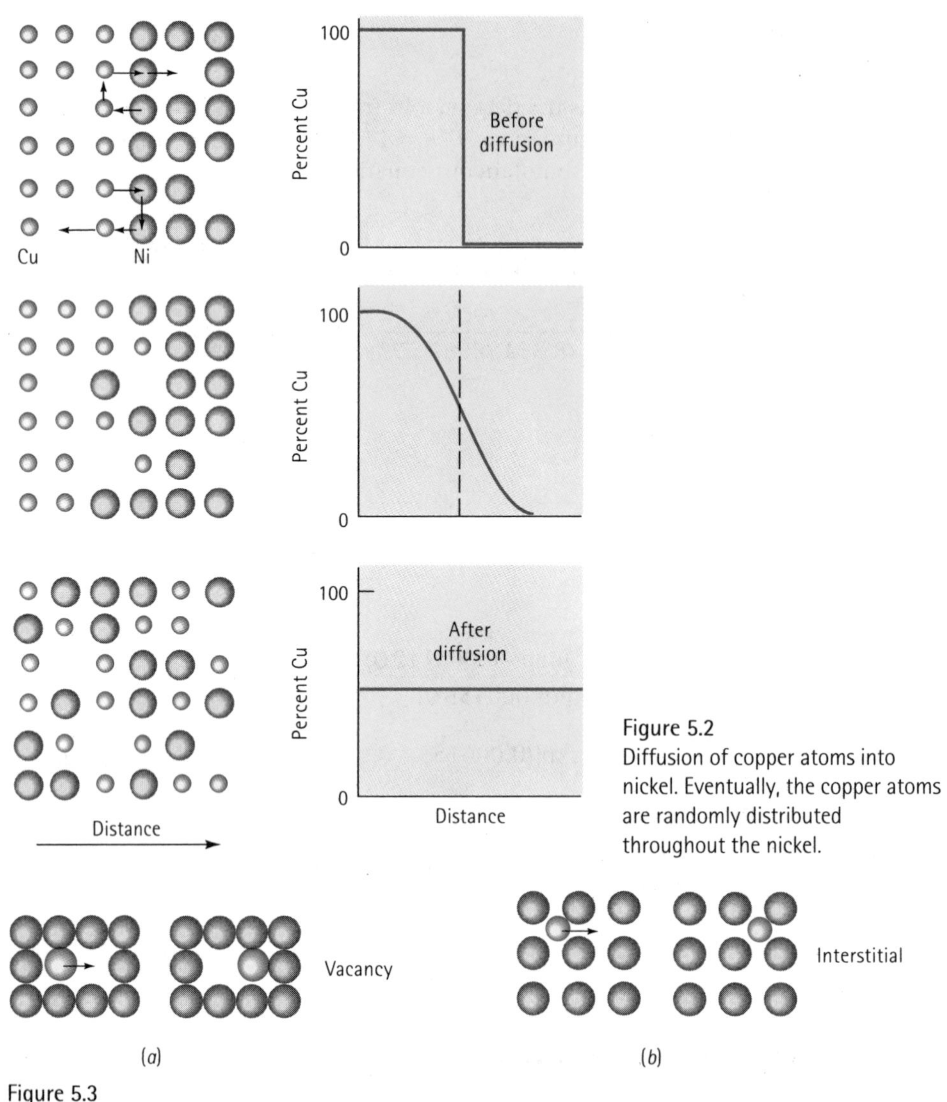

Figure 5.2
Diffusion of copper atoms into nickel. Eventually, the copper atoms are randomly distributed throughout the nickel.

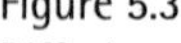

Figure 5.3
Diffusion mechanisms in materials: (*a*) vacancy or substitutional atom diffusion and (*b*) interstitial diffusion.

Vacancy Diffusion In self-diffusion and diffusion involving substitutional atoms, an atom leaves its lattice site to fill a nearby vacancy (thus creating a new vacancy at the original lattice site). As diffusion continues, we have a counter-current flow of atoms and vacancies, called vacancy diffusion. The number of vacancies, which increases as the temperature increases, helps determine the extent of both self-diffusion and diffusion of substitutional atoms.

Interstitial Diffusion When a small interstitial atom is present in the crystal structure, the atom moves from one interstitial site to another. No vacancies are required for this mechanism. Partly because there are many more interstitial sites than vacancies, interstitial diffusion is expected to be rapid.

5.4 Activation Energy for Diffusion

A diffusing atom must squeeze past the surrounding atoms to reach its new site. In order for this to happen, energy must be supplied to force the atom to its new position, as is shown schematically for vacancy and interstitial diffusion in Figure 5.4. The atom is originally in a low-energy, relatively stable location. In order to move to a new location, the atom must overcome an energy barrier. The energy barrier is the activation energy Q. Heat supplies the atom with the energy needed to exceed this barrier.

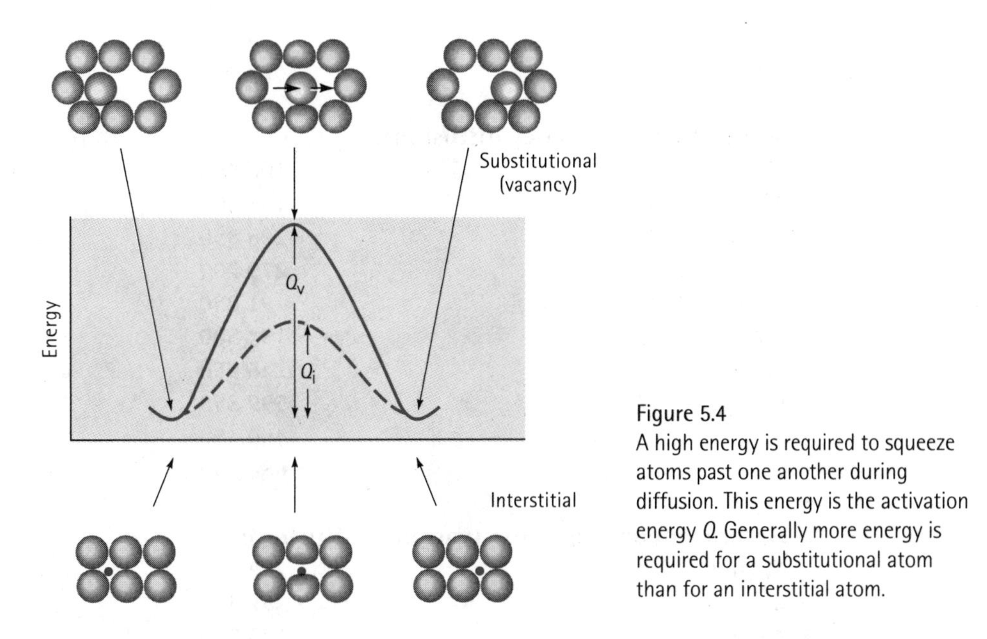

Figure 5.4
A high energy is required to squeeze atoms past one another during diffusion. This energy is the activation energy Q. Generally more energy is required for a substitutional atom than for an interstitial atom.

Normally less energy is required to squeeze an interstitial atom past the surrounding atoms; consequently, activation energies are lower for interstitial diffusion than for vacancy diffusion. Typical values for activation energies are shown in Table 5.1; a low activation energy indicates easy diffusion.

5.5 Rate of Diffusion (Fick's First Law)

The rate at which atoms diffuse in a material can be measured by the **flux** J, which is defined as the number of atoms passing through a plane of unit area per unit time (Figure 5.5). Fick's first law explains the net flux of atoms:

$$J = -D\frac{\Delta c}{\Delta x} \tag{5.3}$$

where J is the flux (atom.m^{-2}.s^{-1}), D is the diffusivity or diffusion coefficient (m^2.s^{-1}), and $\Delta c/\Delta x$ is the concentration gradient (atoms.m^{-3}.m^{-1}). Several factors affect the flux of atoms during diffusion.

Table 5.1 Diffusion data for selected materials.

Diffusion Couple	Q ($J.mol^{-1}$)	D_0 ($m^2.s^{-1}$)
Interstitial diffusion:		
C in FCC iron	137 700	2.3×10^{-5}
C in BCC iron	87 500	1.1×10^{-6}
N in FCC iron	144 900	3.4×10^{-7}
N in BCC iron	76 600	4.7×10^{-7}
H in FCC iron	43 100	6.3×10^{-7}
H in BCC iron	15 050	1.2×10^{-7}
Self-diffusion (vacancy diffusion):		
Pb in FCC Pb	108 400	1.27×10^{-4}
Al in FCC Al	134 800	1.0×10^{-5}
Cu in FCC Cu	206 350	3.6×10^{-5}
Fe in FCC Fe	279 200	6.5×10^{-5}
Zn in HCP Zn	91 250	1.0×10^{-3}
Mg in HCP Mg	134 800	1.0×10^{-4}
Fe in BCC Fe	246 550	4.1×10^{-4}
W in BCC W	599 850	1.88×10^{-4}
Si in Si (covalent)	460 450	0.18
C in C (covalent)	682 300	5.0×10^{-4}
Heterogeneous diffusion (vacancy diffusion):		
Ni in Cu	242 350	2.3×10^{-4}
Cu in Ni	257 450	6.5×10^{-5}
Zn in Cu	183 750	7.8×10^{-5}
Ni in FCC iron	267 900	4.1×10^{-4}
Au in Ag	190 450	2.6×10^{-5}
Ag in Au	168 300	7.2×10^{-6}
Al in Cu	165 350	4.5×10^{-6}
Al in Al_2O_3	477 200	2.8×10^{-3}
O in Al_2O_3	636 250	0.19
Mg in MgO	330 700	2.49×10^{-5}
O in MgO	343 650	4.3×10^{-9}

From several sources, including Y. Adda and J. Philibert, La Diffusion dans les Solides, *Vol.2, 1966.*

Concentration Gradient The concentration gradient shows how the composition of the material varies with distance: Δc is the difference in concentration over the distance Δx (Figure 5.6). The concentration gradient may be created when two materials of different composition are placed in contact, when a gas or liquid is in contact with a solid material, when nonequilibrium structures are produced in a material due to processing, and from a host of other sources.

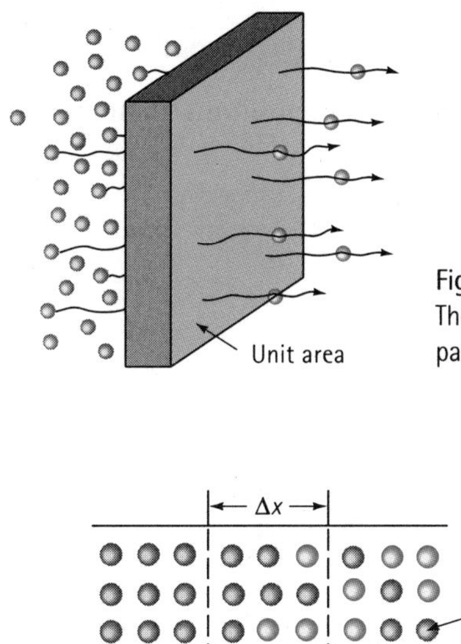

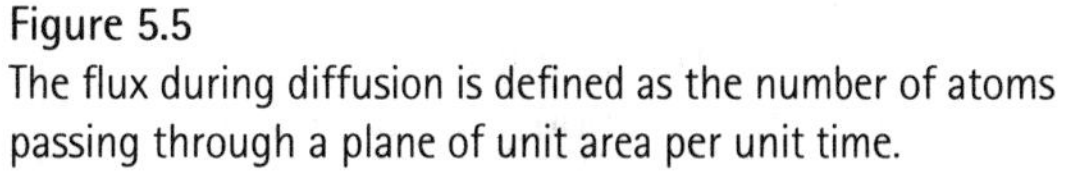

Figure 5.5
The flux during diffusion is defined as the number of atoms passing through a plane of unit area per unit time.

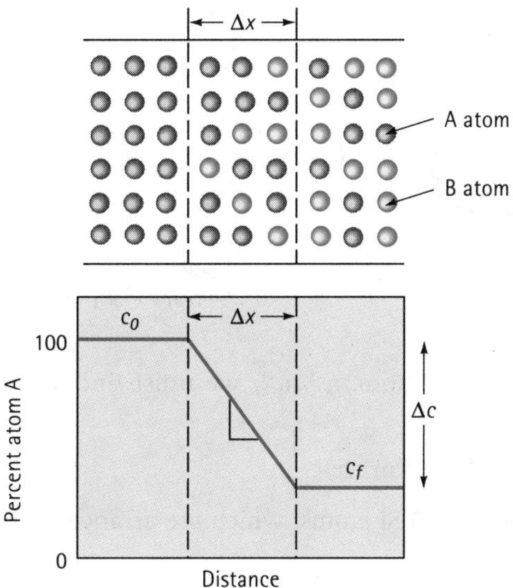

Figure 5.6
Illustration of the concentration gradient.

The flux at a particular temperature is constant only if the concentration gradient is also constant – that is, the compositions on each side of the plane in Figure 5.5 remain unchanged. In many practical cases, however, these compositions vary as atoms are redistributed, and thus the flux also changes. Often we find that the flux is initially high and then gradually decreases as the concentration gradient is reduced by diffusion.

EXAMPLE 5.2

One way to manufacture transistors, which amplify electrical signals, is to diffuse impurity atoms into a semiconductor material such as silicon. Suppose a silicon wafer 1 mm thick, which originally contains one phosphorus atom for every 10 000 000 Si atoms, is treated so that there are 400 P atoms for every 10 000 000 Si atoms at the surface (Figure 5.7). Calculate the concentration gradient (a) in atomic percent.m^{-1} and (b) in atoms.m^{-3}.m^{-1}. The lattice parameter of silicon is 0.54307 nm.

SOLUTION

(a) First, let's calculate the initial and surface compositions in atomic percent:

$$c_i = \frac{1 \text{ P atom}}{10\,000\,000 \text{ atoms}} \times 100 = 0.00001 \text{ at\% P}$$

$$c_s = \frac{400 \text{ P atoms}}{10\,000\,000 \text{ atoms}} \times 100 = 0.004 \text{ at\% P}$$

$$\frac{\Delta c}{\Delta x} = \frac{0.00001 - 0.004 \text{ at\% P}}{0.001 \text{ m}} = -3.99 \text{ at\% P.m}^{-1}$$

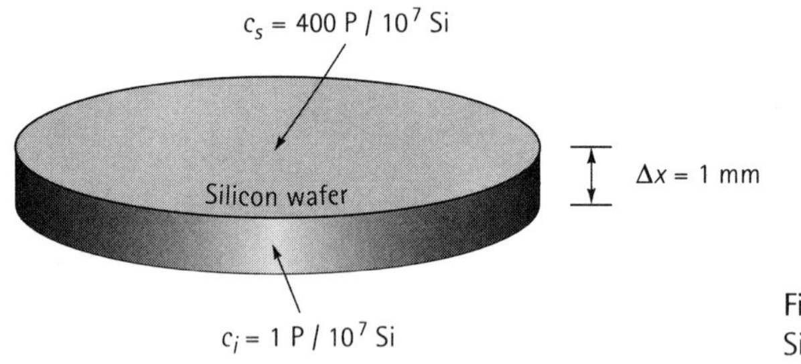

Figure 5.7
Silicon wafer (*for* Example 5.2).

(b) To find the gradient in terms of atom.m^{-3}.m^{-1}, we must find the volume of the unit cell:

$V_{cell} = (0.54307 \times 10^{-9}\text{m})^3 = 1.6 \times 10^{-28}\text{m}^3.\text{cell}^{-1}$ {note 1nm = 10^{-9} m}

The volume occupied by 10 000 000 Si atoms, which are arranged in a DC structure with 8 atoms/cell, is:

$$V = \frac{10\,000\,000 \text{ atoms}}{8 \text{ atoms/cell}}(1.6 \times 10^{-28}\text{m}^3.\text{cell}^{-1}) = 2 \times 10^{-22}\text{m}^3.\text{cell}^{-1}$$

The compositions in atoms.m^{-3} are:

$$c_i = \frac{1 \text{ P atom}}{2 \times 10^{-22}\text{m}^3} = 5 \times 10^{21} \text{ P atoms.m}^{-3}$$

$$c_s = \frac{400 \text{ P atoms}}{2 \times 10^{-22}\text{m}^3} = 2 \times 10^{24} \text{ P atoms.m}^{-3}$$

$$\frac{\Delta c}{\Delta x} = \frac{5 \times 10^{21} - 2000 \times 10^{21}}{10^{-3}}$$

$$= -1.995 \times 10^{27} \text{ atom. m}^{-3}.\text{m}^{-1}$$

EXAMPLE 5.3

A 0.5 mm layer of MgO is deposited between layers of nickel and tantalum to provide a diffusion barrier that prevents reactions between the two metals (Figure 5.8). At 1 400°C, nickel ions are created and diffuse through the MgO ceramic to the tantalum. Determine the number of nickel ions that pass through the MgO per second. The diffusion coefficient of nickel in MgO is 9×10^{-16} m^2.s^{-1}, and the lattice parameter of nickel at 1 400°C is 3.6×10^{-10} m.

SOLUTION

The composition of nickel at the Ni/MgO interface is 100% Ni, or

$$c_{\text{Ni/MgO}} = \frac{4\text{Ni atoms / unit cell}}{(3.6 \times 10^{-10}\,\text{m})^3} = 8.57 \times 10^{28}\ \text{atom.m}^{-3}$$

refer to Example 3.1 and Table 3.2.

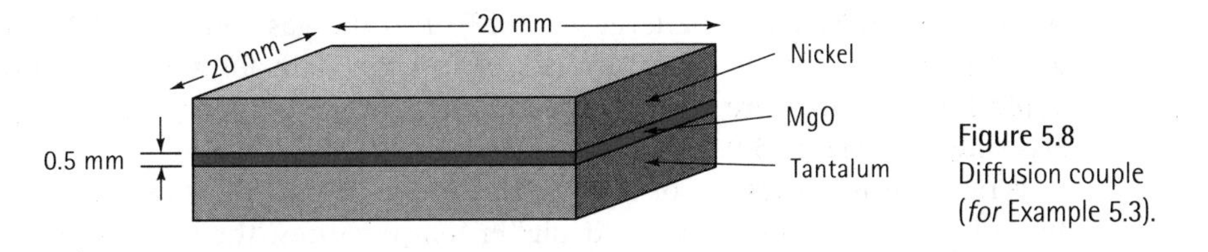

Figure 5.8
Diffusion couple (*for* Example 5.3).

The composition of nickel at the Ta/MgO interface is 0% Ni. Thus, the concentration gradient across the MgO is:

$$\Delta c / \Delta x = \frac{0 - 8.57 \times 10^{28}\ \text{atoms.m}^{-3}}{0.5 \times 10^{-3}\,\text{m}} = -1.71 \times 10^{32}\ \text{atoms.m}^{-3}.\text{m}^{-1}$$

The flux of nickel atoms through the MgO layer is:

$$J = -D\frac{\Delta c}{\Delta x} = -(9 \times 10^{-16}\,\text{m}^2.\text{s}^{-1})(-1.71 \times 10^{32}\ \text{atom.m}^{-3}.\text{m}^{-1})$$

$$J = 1.54 \times 10^{17}\,\text{Ni atoms.m}^{-2}.\text{s}^{-1}$$

The total number of nickel atoms crossing the 20 mm × 20 mm interface per second is:

$$\text{Total Ni atoms} = (J)(\text{Area}) = (1.54 \times 10^{17})(20 \times 10^{-3}\text{m})(20 \times 10^{-3}\text{m})$$
$$= 6.16 \times 10^{13}\ \text{Ni atoms.s}^{-1}$$

Although this appears to be very rapid, we would find that in one second, the volume of nickel atoms removed from the Ni/MgO interface is:

$$\frac{6.16\times10^{13}\,\text{Ni atoms.s}^{-1}}{8.57\times10^{28}\,\text{Ni atoms.m}^{-3}}=0.72\times10^{-15}\,\text{m}^3.\text{s}^{-1}$$

Or, the thickness by which the nickel layer is reduced each second is:

$$\frac{0.72\times10^{-15}\,\text{m}^3.\text{s}^{-1}}{(20\times10^{-3})^2}=1.8\times10^{-12}\,\text{m.s}^{-1}$$

For one micron (10^{-6} m) of nickel to be removed, the treatment requires

$$\frac{10^{-6}\,\text{m}}{1.8\times10^{-12}\,\text{m.s}^{-1}}=556\ 000\ \text{s}=154\ \text{h}$$

Temperature and the Diffusion Coefficient The diffusion coefficient D is related to temperature by an Arrhenius equation,

$$D=D_0\exp\left(\frac{-Q}{RT}\right) \tag{5.4}$$

where Q is the activation energy (J.mol^{-1}), R is the gas constant (8.314 J.mol^{-1}.K^{-1}), and T is the absolute temperature (K). D_0 is a constant for a given diffusion system. Typical values for D_0 are given in the Table 5.1, while the temperature dependence of D is shown in Figure 5.9 for several materials.

When the temperature of a material increases, the diffusion coefficient and the flux of atoms increase as well. At higher temperatures, the thermal energy supplied to the diffusing atoms permits the atoms to overcome the activation energy barrier and more easily move to new lattice sites. At low temperatures – often below about 0.4 times the absolute melting temperature of the material – diffusion is very slow and may not be significant. For this reason, the heat treatment of metals and the processing of ceramics are done at high temperatures, where atoms move rapidly to complete reactions or to reach equilibrium conditions.

EXAMPLE 5.4 Design of an Iron Membrane

A thick, impermeable pipe 30 mm in diameter and 100 mm long contains a gas that includes 0.5×10^{26} N atoms per m^3 and 0.5×10^{26} H atoms per m^3 on one side of an iron membrane (Figure 5.10). Gas is continuously introduced to the pipe to ensure a constant concentration of nitrogen and hydrogen. The gas on the other side of the membrane includes a constant 1×10^{24} atoms per m^3 and 1×10^{24} H atoms per m^3. The entire system is to operate at 700°C, where the iron has the BCC structure. Design an iron membrane that will allow no more than 1% of the nitrogen to be lost through the membrane each hour while allowing 90% of the hydrogen to pass through the membrane per hour.

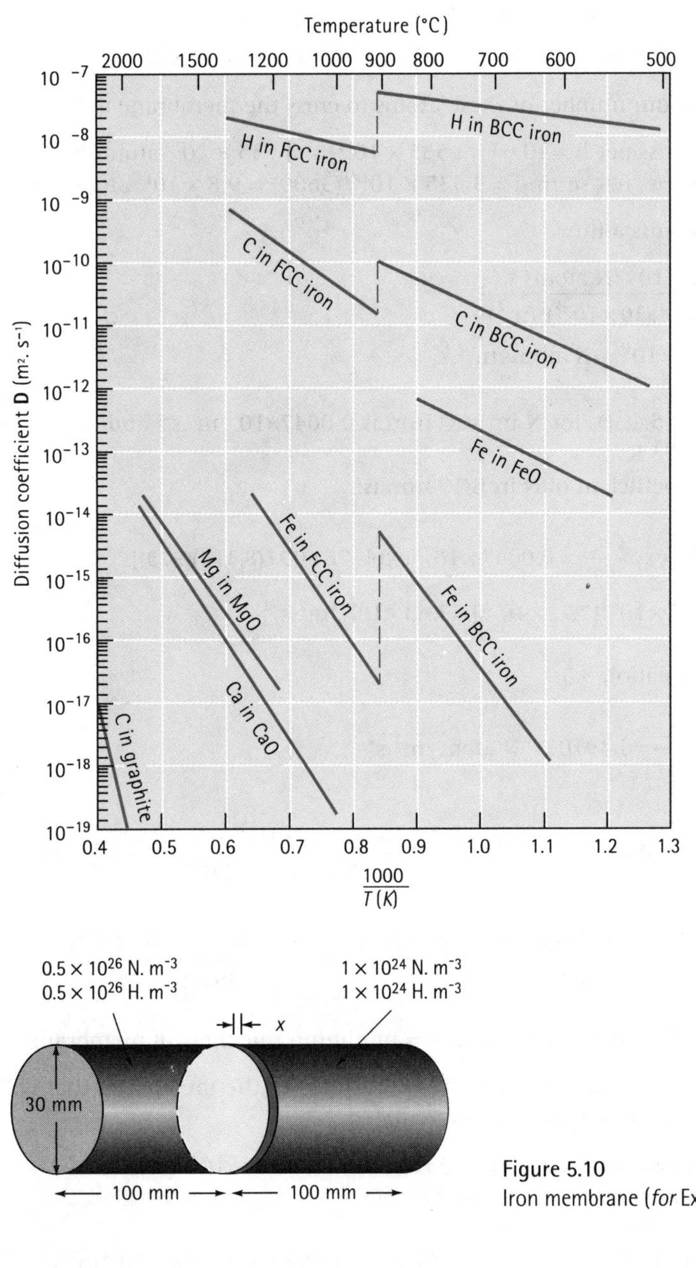

Figure 5.9
The diffusion coefficient D as a function of reciprocal temperature for several metals and ceramics. In this Arrhenius plot, D represents the rate of the diffusion process. A steep slope denotes a high activation energy.

Figure 5.10
Iron membrane (*for* Example 5.4).

SOLUTION

$$\text{Vol. of input cylinder} = (\pi/4)d^2.h = (\pi/4)(30 \times 10^{-3}\,\text{m})^2(100 \times 10^{-3}\,\text{m})$$
$$= 7.07 \times 10^{-5}\,\text{m}^3$$

The total number of nitrogen atoms

$(0.5 \times 10^{26}\ \text{N.m}^{-3})(7.07 \times 10^{-5}\text{m}^3) = 3.535 \times 10^{21}$ atoms

The maximum number of these atoms to cross the membrane is 1% of this total, or:

N atom loss per h = $(0.01)(3.535 \times 10^{21}) = 3.535 \times 10^{19}$ atoms.h^{-1}
N atom loss per second = $3.535 \times 10^{19}/(3600) = 9.8 \times 10^{15}$ atoms.s^{-1}

This represents a flux:

$$J = \frac{9.8 \times 10^{15}(\text{N atoms.s}^{-1})}{(\pi/4)(30 \times 10^{-3})^2(\text{m}^2)}$$

$$J = 1.39 \times 10^{19} \quad \text{N atoms.m}^{-2}.\text{s}^{-1}$$

From Table 5.1, D_0 for N in BCC iron is 0.0047×10^{-4} m^2.s^{-1} and Q is 76 600 J.mol^{-1}. At 700°C or 973 K,
Diffusion coefficient of N in BCC iron is:

$$D_N = D_0 \exp\frac{(-Q)}{RT} = 0.0047 \times 10^{-4} \exp[-76\,600/(8.314)(973)]$$

$$= 0.0047 \times 10^{-4}[7.72 \times 10^{-5}] = 3.63 \times 10^{-11}\text{m}^2.\text{s}^{-1}$$

From Equation 5.3:

$$J = D_N \frac{\Delta c}{\Delta x} = 1.391019 \text{ N atoms.m}^2.\text{s}^1$$

$$\text{rearranging } \Delta x = -D_N \Delta c / J = \frac{-3.63 \times 10^{-11}(1 \times 10^{24} - 0.5 \times 10^{26}\text{N.m}^{-3})}{1.39 \times 10^{19}}$$

$$\Delta x = \frac{-3.63 \times 10^{-11}(1 \times 10^{24} - 50 \times 10^{24})}{1.39 \times 10^{19}} = \frac{-3.63 \times 10^{-11}(-49 \times 10^{24})}{1.39 \times 10^{19}}$$

$\Delta x = 1.28 \times 10^{-4}$ m = 0.128 mm = minimum thickness of membrane

In a similar manner, the maximum thickness of the membrane that will permit 90% of the hydrogen to pass can be calculated:

H atom loss per h = $(0.90)(3.535 \times 10^{21}) = 3.18 \times 10^{21}$
H atom loss per s = 8.83×10^{17}
$J = 1.25 \times 10^{21}$ H atoms.m^{-2}.s^{-1}
$D_H = 0.0012 \times 10^{-4} \exp[-15050/(8.314)(9973)] = 1.87 \times 10^{-8}$ m^2.s^{-1}

$$\Delta x = \frac{-1.87 \times 10^{-8}(1 \times 10^{24} - 0.5 \times 10^{26})}{1.25 \times 10^{21}}$$

$= 7.33 \times 10^{-4} = 0.733$ mm maximum thickness of membrane

An iron membrane with a thickness between 0.128 and 0.733 mm will be satisfactory.

Factors Affecting Diffusion and the Activation Energy A small activation energy Q increases the diffusion coefficient and flux, because less thermal energy is required to overcome the smaller activation energy barrier. A number of factors influence the activation energy and, hence, the rate of diffusion. Interstitial diffusion, with a low activation energy, usually occurs much faster than vacancy, or substitutional, diffusion.

Activation energies are usually lower for atoms diffusing through open crystal structures than for close-packed crystal structures. Because the activation energy depends on the strength of atomic bonding, it is higher for diffusion of atoms in materials with a high melting temperature (Figure 5.11). Covalently bonded materials, such as carbon and silicon (Table 5.1), have unusually high activation energies, consistent with the high strength of their atomic bonds.

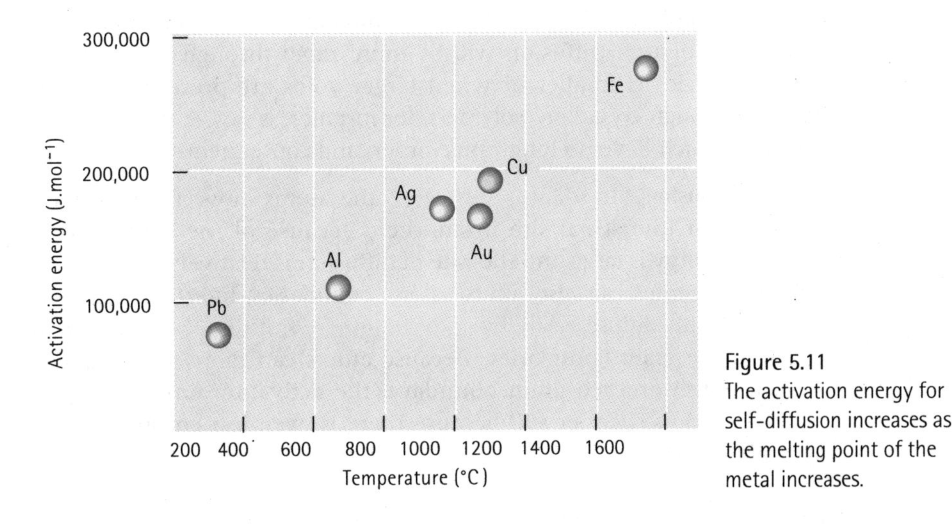

Figure 5.11
The activation energy for self-diffusion increases as the melting point of the metal increases.

In ionic materials, such as ceramics, a diffusing ion only enters a site having the same charge. In order to reach that site, the ion must physically squeeze past adjoining ions, pass by a region of opposite charge, and move a relatively long distance (Figure 5.12). Consequently, the activation energies are high and the rates of diffusion are lower for ionic materials than for metals.

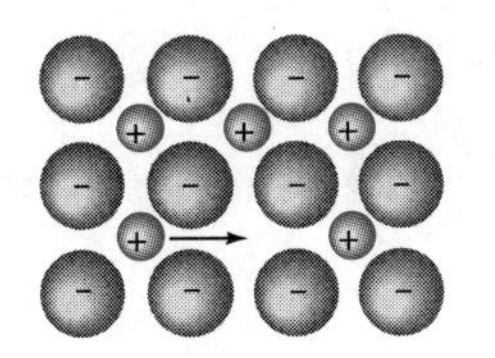

Figure 5.12
Diffusion in ionic compounds. Anions can only enter other anion sites.

We also find that, due to their smaller size, cations (with a positive charge) often have higher diffusion coefficients than anions (with a negative charge). In sodium chloride, for instance, the activation energy for diffusion of chloride ions is about twice that for diffusion of sodium ions.

Diffusion of the ions also provides a transfer of electrical charge; in fact, the electrical conductivity of ionically bonded ceramic materials is related to temperature by an Arrhenius equation. As the temperature increases, the ions diffuse more rapidly, electrical charge is transferred more quickly, and the electrical conductivity is increased.

In polymers, we are concerned with the diffusion of atoms or small molecules between the long polymer chains. For example, polymer films are typically used as packaging to store food. If air diffuses through the film, the food may spoil. If air diffuses through the rubber inner tube of an automobile tyre, the tyre will deflate. Diffusion of some more molecules into a polymer can cause swelling problems. On the other hand, diffusion is required to enable dyes to uniformly enter many of the synthetic polymer fabrics. Selective diffusion through polymer membranes is used to cause desalination of water; while water molecules pass through the polymer membrane, the ions in the salt are trapped.

In each of these examples, the diffusing atoms, ions, or molecules penetrate between the polymer chains rather than moving from one location to another within the chain structure. Diffusion will be more rapid through this structure when the diffusing species is smaller or when larger voids are present between the chains. Diffusion through crystalline polymers, for instance, is slower than through amorphous polymers, which have no long-range order, and consequently, have a lower density.

Types of Diffusion In volume diffusion, the atoms move through the crystal from one lattice or interstitial site to another. Because of the surrounding atoms, the activation energy is large and the rate of diffusion is relatively slow.

However, atoms can also diffuse along boundaries, interfaces, and surfaces in the material. Atoms diffuse easily by grain boundary diffusion because the atom packing is poor in the grain boundaries. Because atoms can more easily squeeze their way through the disordered grain boundary, the activation energy is low (Table 5.2). Surface diffusion is easier still because there is even less constraint on the diffusing atoms at the surface.

Table 5.2 The effect of the type of diffusion for thorium in tungsten and for self-diffusion in silver.

	Diffusion Coefficient ($m^2.s^{-1}$)	
Diffusion Type	Thorium in Tungsten	Silver in Silver
Surface	$0.47 \times 10^{-4} \exp(-277950/RT)$	$0.068 \times 10^{-4} \exp(-37250/RT)$
Grain boundary	$0.74 \times 10^{-4} \exp(-376750/RT)$	$0.24 \times 10^{-4} \exp(-95250/RT)$
Volume	$1.00 \times 10^{-4} \exp(-502300/RT)$	$0.99 \times 10^{-4} \exp(-190450/RT)$

Time Diffusion requires time; the units for flux are atoms.$m^{-2}.s^{-1}$! If a large number of atoms must diffuse to produce a uniform structure, long times may be required, even at high temperatures. Times for heat treatments may be reduced by using higher temperatures or by making the diffusion distances (related to Δx) as small as possible.

We find that some rather remarkable structures and properties are obtained if we prevent diffusion. Steels quenched rapidly from high temperatures to prevent diffusion form nonequilibrium structures that provide the basis for sophisticated heat treatments.

EXAMPLE 5.5

Consider a diffusion couple set up between pure tungsten and a tungsten-1 at% thorium alloy. After several minutes of exposure at 2 000°C, a transition zone of 0.1 mm thickness is established. What is the flux of thorium atoms at this time if diffusion is due to (a) volume diffusion, (b) grain boundary diffusion, and (c) surface diffusion? (*see* Table 5.2)

SOLUTION

The lattice parameter of BCC tungsten is about 0.3165 nm. Thus, the number of tungsten atom.m^{-3} is:

$$\frac{\text{W atoms}}{\text{m}^3} = \frac{2 \text{ atoms / cell}}{(3.165 \times 10^{-10})^3 \text{ m}^3 \text{ / cell}} = 6.3 \times 10^{28}$$

In the tungsten-1 at% thorium alloy, the number of thorium atoms is:

$$c_{\text{Th}} = (0.01)(6.3 \times 10^{28}) = 6.3 \times 10^{26} \text{ Th atoms.m}^{-3}$$

In the pure tungsten, the number of thorium atoms is zero. Thus, the concentration gradient is:

$$\frac{\Delta c}{\Delta x} = \frac{0 - 6.3 \times 10^{26}}{0.1 \times 10^{-3}\text{m}} = -6.3 \times 10^{30} \text{ Th atoms.m}^{-3}\text{.m}^{-1}$$

(a) Volume diffusion:

$$D = 1.0 \times 10^{-4} \exp\left(\frac{-502\,300}{(8.314)(2273)}\right)$$

$$= 2.86 \times 10^{-16} \text{m}^2\text{.s}^{-1}$$

$$J = -D\frac{\Delta c}{\Delta x} = -(2.86 \times 10^{-16})(-6.3 \times 10^{30})$$

$$= 18.0 \times 10^{14} \text{Th atoms.m}^{-2}\text{.s}^{-1}$$

(b) Grain boundary diffusion:

$$D = 0.74 \times 10^{-4} \exp\left(\frac{-37\,6750}{(8.314)(2\,273)}\right)$$

$$= 1.63 \times 10^{-13} \text{ m}^{-2}\text{.s}^{-1}$$

$$J = -(1.63 \times 10^{-13})(-6.3 \times 10^{30}) = 1.02 \times 10^{18} \text{ Th atoms.m}^{-2}\text{.s}^{-1}$$

(c) Surface diffusion:

$$D = 0.47 \times 10^{-4} \exp\left(\frac{277\,950}{8.314 \times 2273}\right)$$

$$= 1.93 \times 10^{-11} \mathrm{m^2.s^{-1}}$$

$$J = -(1.93 \times 10^{-11})(-6.3 \times 10^{30}) = 1.21 \times 10^{20} \text{ Th atoms.m}^{-2}.\mathrm{s}^{-1}$$

5.6 Composition Profile (Fick's Second Law)

Fick's second law, which describes the dynamic, or nonsteady state, diffusion of atoms, is the differential equation $dc/dt = D(d^2 c/dx^2)$, whose solution depends on the boundary conditions for a particular situation. One solution is

$$\frac{c_s - c_x}{c_s - c_0} = \operatorname{erf}\left(\frac{x}{2\sqrt{Dt}}\right) \tag{5.5}$$

where c_s is a constant concentration of the diffusing atoms at the surface of the material, c_0 is the initial uniform concentration of the diffusing atoms in the material, and c_x is the concentration of the diffusing atom at location x below the surface after time t. These concentrations are illustrated in Figure 5.13. The function erf is the error function and can be evaluated from Table 5.3.

The solution to Fick's second law permits us to calculate the concentration of one diffusing species near the surface of the material as a function of time and distance, provided that the diffusion coefficient D remains constant and the concentrations of the diffusing atom at the surface c_s and within the material c_0 remain unchanged. Fick's second law can also assist us in designing a variety of materials processing techniques, including the steel heat treatment described in Example 5.6.

$\frac{x}{2\sqrt{D\,t}}$	$\operatorname{erf} \frac{x}{2\sqrt{D\,t}}$
0	0
0.10	0.1125
0.20	0.2227
0.30	0.3286
0.40	0.4284
0.50	0.5205
0.60	0.6039
0.70	0.6778
0.80	0.7421
0.90	0.7970
1.00	0.8427
1.50	0.9661
2.00	0.9953

Table 5.3
Error function for Fick's second law.

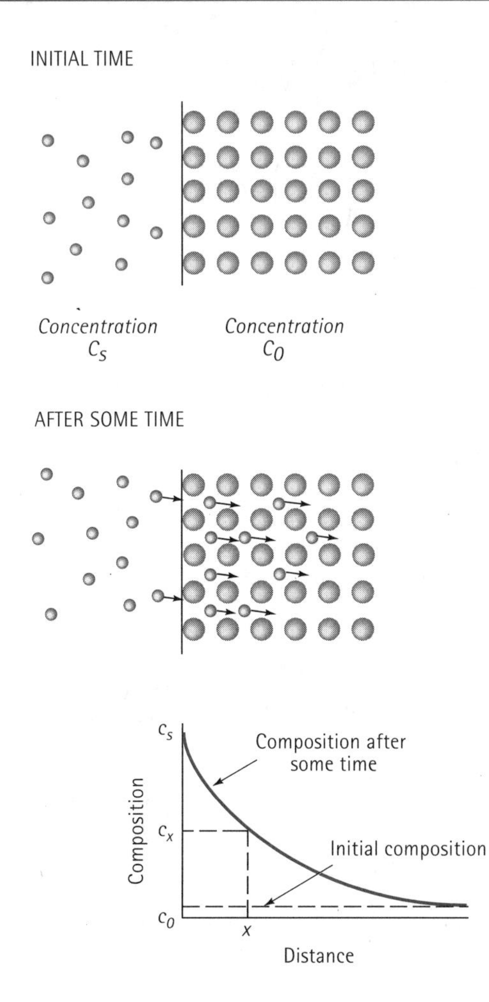

Figure 5.13 Diffusion of atoms into the surface of a material, illustrating the use of Fick's second law.

EXAMPLE 5.6 Design of a Carburising Treatment

The surface of a 0.1% C steel is to be strengthened by carburising. In carburising, the steel is placed in an atmosphere that provides 1.2% C at the surface of the steel at a high temperature. Carbon then diffuses from the surface into the steel. For optimum properties, the steel must contain 0.45% C at a depth of 2 mm below the surface. Design a carburising heat treatment that will produce these optimum properties. Assume that the temperature is high enough (at least 900°C) so that the iron has the FCC structure.

SOLUTION

We know that c_s = 1.2% C, c_0 = 0.1% C, c_x = 0.45% C, and x = 0.2 cm. From Fick's second law:

$$\frac{c_s - c_x}{c_s - c_0} = \frac{1.2 - 0.45}{1.2 - 0.1} = 0.68 = \text{erf}\left(\frac{2\times10^{-3}}{2\sqrt{Dt}}\right) = \text{erf}\left(\frac{10^{-3}}{\sqrt{Dt}}\right)$$

from Table 5.3, when $\text{erf}\left(\frac{x}{\sqrt{Dt}}\right) = 0.68, \left(\frac{x}{\sqrt{Dt}}\right) = 0.71$

squaring $\frac{10^{-6}}{Dt} = 0.5041; D = \frac{1.9837\times10^{-6}}{t}$

for C in FCC iron, D_0=0.23 × 10^{-4} $m^2.s^{-1}$ and Q = 137 700 $J.mol^{-1}$
and from equation

$$D = D_0 \exp\left(\frac{-Q}{RT}\right)$$

$$D = 0.23\times10^{-4} \exp\left(\frac{-137\,700}{8.314T}\right)$$

$$D = 0.23\times10^{-4} \exp\left(\frac{-16\,562}{T}\right)$$

equating equations in D, $\frac{1.9837\times10^{-6}}{t} = 0.23\times10^{-4} \exp\left(\frac{-16\,562}{T}\right)$

$$\frac{t}{1.9837\times10^{-6}} = \frac{1}{0.23\times10^{-4} \exp\left(\frac{-16\,562}{T}\right)}$$

$$t = \frac{0.0862}{\exp\left(\frac{-16\,562}{T}\right)}$$

There are many possible combinations of t and T, e.g:
when T = 900°C ≡ 1 173K, t = 116 806s = 32.4h
T = 1 000°C = 1 273K, t = 38 527s = 10.7h
T = 1 100°C = 1 372K, t = 15 068s = 4.2h
T = 1 200°C = 1 472K, t = 6 636s = 1.8h

The exact combination of temperature and time will depend on the maximum temperature that the heat treating furnace can reach, the rate at which parts must be produced, and the economics of the trade-offs between higher temperatures versus longer times.

Example 5.6 shows that one of the consequences of Fick's second law is that the same concentration profile can be obtained for different conditions, so long as the term Dt is constant. This permits us to determine the effect of temperature on the time required for a particular heat treatment to be accomplished.

EXAMPLE 5.7 **Design of a More Economical Heat Treatment**

We find that 10 h are required to successfully carburise a batch of 500 steel gears at 900°C, where the iron has the FCC structure. We find that it costs £1 000 per hour to operate the carburising furnace at 900°C and £1500 per hour to operate the furnace at 1 000°C. Is it economical to increase the carburising temperature to 1 000°C?

SOLUTION

The temperatures of interest are 900°C = 1 173 K and 1 000°C = 1 273 K. For carbon diffusing in FCC iron, the activation energy is 137 700 J.mol^{-1}. To achieve the same carburising treatment at 1 000°C as at 900°C:

$$D_{1\,273}t_{1\,273} = D_{1\,173}t_{1\,173}$$

$$t_{1\,273} = \frac{D_{1\,173}t_{1\,173}}{D_{1\,273}} = \frac{(10\text{h})\exp[-137\,700/(8.314)(1173)}{\exp[-137\,700/(8.314)(1\,273)]}$$

$$t_{1\,273} = \frac{10(7.377\times10^{-7})}{2.237\times10^{-6}}$$

$$t_{1\,273} = 3.298\text{h}$$

At 900°C, the cost per part is (£1 000/h)(10 h)/500 parts = £20/part
At 1 000°C, the cost per part is (£1 500/h)(3.298 h)/500 parts = £9.90/part

Considering only the cost of operating the furnace, increasing the temperature reduces the heat-treating cost of the gears and increases the production rate.

Equation 5.5 requires that there be a constant composition c_0 at the interface; this is the case in a process such as carburising of steel (Example 5.6), in which carbon is continuously supplied to the steel surface. In many cases, however, the surface concentration gradually changes during the process. In these cases, interdiffusion of atoms occurs, as shown in Figure 5.2, and Equation 5.5 is no longer valid.

Sometimes interdiffusion can cause difficulties. For example, when aluminium is bonded to gold at an elevated temperature, the aluminium atoms diffuse faster into the gold than gold atoms diffuse into the aluminium. Consequently, more total atoms eventually are on the original gold side of the interface than on the original aluminium side. This causes the physical location of the original interface to move towards the aluminium side of the diffusion couple. Any foreign particles originally trapped at the interface also move with the interface. This movement of the interface due to unequal diffusion rates is called the Kirkendall effect.

In certain cases, voids form at the interface as a result of the Kirkendall effect. In tiny integrated circuits, gold wire is welded to aluminium to provide an external lead for the circuit. During operation of the circuit, voids may form by coalescence of vacancies involved in the diffusion process; as the voids grow, the Au-Al connection is weakened and eventually may fail. Because the area around the connection discolours, this premature failure is called the purple plague. One technique for preventing this problem is to expose the welded joint to hydrogen. The hydrogen dissolves in the aluminium, fills the vacancies, and prevents self-diffusion of the aluminium atoms. This keeps the aluminium atoms from diffusing into the gold and embrittling the weld.

5.7 Diffusion and Materials Processing

Diffusional processes become very important when materials are used or processed at elevated temperatures. (Many important examples will be discussed in later chapters.) In this section, three cases in which diffusion is important will be considered.

Grain Growth A material composed of many grains contains a large number of grain boundaries, which represent a high-energy area because of the inefficient packing of the atoms. A lower overall energy is obtained in the material if the amount of grain boundary area is reduced by grain growth.

Grain growth involves the movement of grain boundaries, permitting some grains to grow at the expense of others. Diffusion of atoms across the grain boundary (Figure 5.14) is required, and, consequently, the growth of the grains is related to the activation energy needed for an atom to jump across the boundary. High temperatures or low activation energies increase the size of the grains. Many heat treatments of metals, which include holding the metal at high temperatures, must be carefully controlled to avoid excessive grain growth.

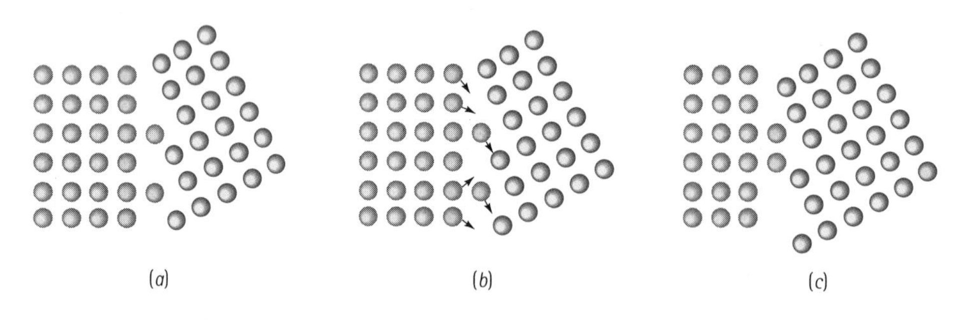

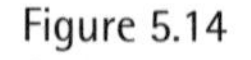

Figure 5.14
Grain growth occurs as atoms diffuse across the grain boundary from one grain to another.

Diffusion Bonding Diffusion bonding, a method used to join materials, occurs in three steps (Figure 5.15). The first step forces the two surfaces together at a high temperature and pressure, flattening the surface, fragmenting impurities, and producing a high atom-to-atom contact area. As the surfaces remain pressed together at high temperatures, atoms diffuse along grain boundaries to the remaining voids; the atoms condense and reduce the size of any voids in the interface. Because grain boundary diffusion is rapid, this second step may occur very quickly. Eventually, however, grain growth isolates the remaining voids from the grain boundaries. For the third step – final elimination of the voids – volume diffusion, which is comparatively slow, must occur.

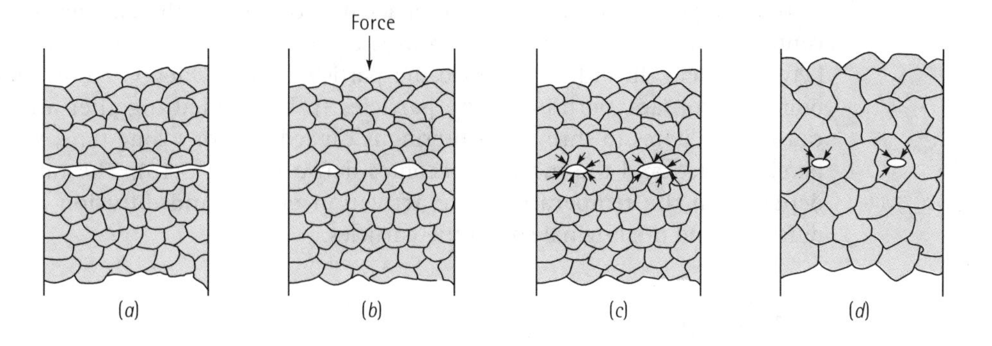

Figure 5.15
The steps in diffusion bonding: (*a*) Initially the contact area is small; (*b*) application of pressure deforms the surface, increasing the bonded area; (*c*) grain boundary diffusion permits voids to shrink; and (*d*) final elimination of the voids requires volume diffusion.

The diffusion bonding process is often used for joining reactive metals such as titanium, for joining dissimilar metals and materials, and for joining ceramics.

Sintering A number of materials are manufactured into useful shapes by a process that requires consolidation of small particles into a solid mass. Sintering is the high-temperature treatment that causes particles to join together and gradually reduces the volume of pore space between them. Sintering is a frequent step in the manufacture of ceramic components, as well as in the production of metallic parts by powder metallurgy. A variety of composite materials are produced using the same techniques.

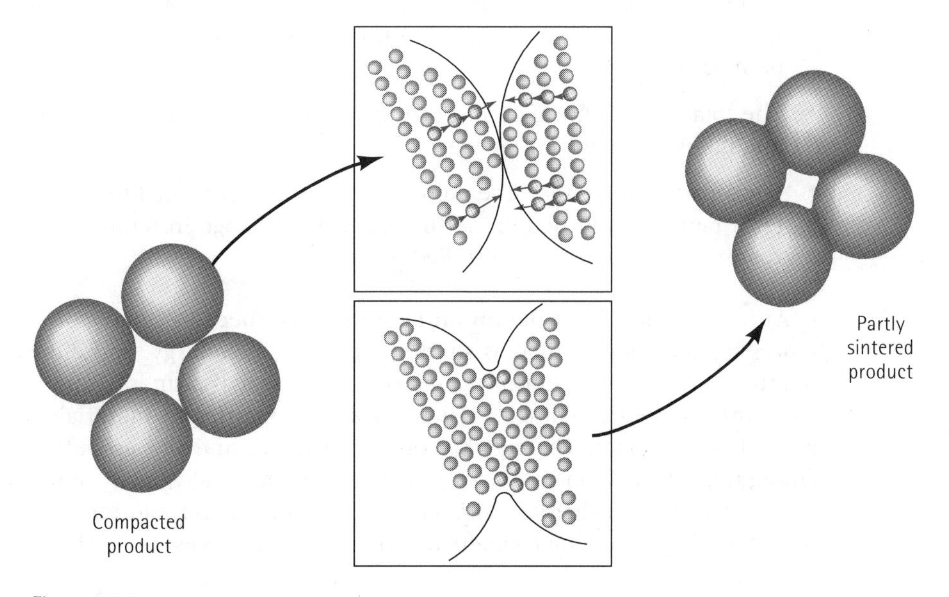

Figure 5.16
Diffusion processes during sintering and powder metallurgy. Atoms diffuse to points of contact, creating bridges and reducing the pore size.

When a powdered material is compacted into a shape, the powder particles are in contact with one another at numerous sites, with a significant amount of pore space between them. In order to reduce the particle surface energy, atoms diffuse to the points of contact, permitting the particles to be bonded together and eventually causing the pores to shrink. If sintering is carried out for a long time, the pores may be eliminated and the material becomes dense (Figure 5.16). The rate of sintering depends on the temperature, the activation energy and diffusion coefficient for diffusion, and the original size of the particles.

SUMMARY

Atoms move through a solid material – particularly at high temperatures when a concentration gradient is present – by diffusion mechanisms. Some key relationships involving diffusion include the following:

- The two important mechanisms for atom movement are vacancy diffusion and interstitial diffusion. Substitutional atoms in the lattice move by the vacancy mechanism.
- The rate of diffusion is governed by the Arrhenius relationship – that is, the rate increases exponentially with temperature. Diffusion is most significant at temperatures above about 0.4 times the melting temperature (in Kelvin) of the material.
- The activation energy Q describes the ease with which atoms diffuse, with rapid diffusion occurring for a low activation energy. A low activation energy and rapid diffusion rate are obtained for (1) interstitial diffusion compared with vacancy diffusion, (2) crystal structures with a low packing factor, (3) materials with a low melting temperature or weak atomic bonding, and (4) diffusion along grain boundaries or surfaces.
- The total movement of atoms, or flux, increases when the concentration gradient and diffusion coefficient increase.
- For a particular system, the amount of diffusion is related to the term Dt. The Dt term permits us to determine the effect of a change in temperature on the time required for a diffusion-controlled process.

Atom diffusion is of paramount importance, because many of the materials-processing techniques, such as sintering, powder metallurgy, and diffusion bonding, require diffusion. Furthermore, many of the heat treatments and strengthening mechanisms used to control structures and properties in materials are diffusion-controlled processes. The stability of the structure and the properties of materials during use at high temperatures depend on diffusion. Finally, many important materials are produced by deliberately preventing diffusion, thereby forming nonequilibrium structures. The following chapters describe many instances in which diffusion plays a significant role.

GLOSSARY

Activation energy

The energy required to cause a particular reaction to occur. In diffusion, the activation energy is related to the energy required to move an atom from one lattice site to another.

Concentration gradient

The rate of change of composition with distance in a nonuniform material, expressed as atoms.m^{-3}.m^{-1} or at% m^{-1}.

Diffusion

The movement of atoms within a material.

Diffusion bonding

A joining technique in which two surfaces are pressed together at high pressures and temperatures. Diffusion of atoms to the interface fills in voids and produces a strong bond.

Diffusion coefficient

A temperature-dependent coefficient related to the rate at which atoms diffuse. The diffusion coefficient depends on temperature and activation energy.

Diffusion distance

The maximum or desired distance that atoms must diffuse; often, the distance between the locations of the maximum and minimum concentrations of the diffusing atom.

Diffusivity

Another term for diffusion coefficient.

Fick's first law

The equation relating the flux of atoms by diffusion to the diffusion coefficient and the concentration gradient.

Fick's second law

The partial differential equation that describes the rate at which atoms are redistributed in a material by diffusion.

Flux

The number of atoms passing through a plane of unit area per unit time. This is related to the rate at which mass is transported by diffusion in a solid.

Grain boundary diffusion

Diffusion of atoms along grain boundaries. This is faster than volume diffusion, because the atoms are less closely packed in grain boundaries.

Grain growth

Movement of grain boundaries by diffusion in order to reduce the amount of grain boundary area. As a result, small grains shrink and disappear and other grains become larger.

Interdiffusion

Diffusion of different atoms in opposite directions. Interdiffusion may eventually produce an equilibrium concentration of atoms within the material.

Interstitial diffusion
Diffusion of small atoms from one interstitial position to another in the crystal structure.

Kirkendall effect
Physical movement of an interface due to unequal rates of diffusion of the atoms within the material.

Powder metallurgy
A method for producing metallic parts; tiny metal powders are compacted into a shape, which is then heated to allow diffusion and sintering to join the powders into a solid mass.

Purple plague
Formation of voids in gold-aluminium welds due to unequal rates of diffusion of the two atoms; eventually failure of the weld can occur.

Self-diffusion
The random movement of atoms within a pure material – that is, even when no concentration gradient is present.

Sintering
A high-temperature treatment used to join small particles. Diffusion of atoms to points of contact causes bridges to form between the particles. Further diffusion eventually fills in any remaining voids.

Surface diffusion
Diffusion of atoms along surfaces, such as cracks or particle surfaces.

Vacancy diffusion
Diffusion of atoms when an atom leaves a regular lattice position to fill a vacancy in the crystal. This process creates a new vacancy, and the process continues.

Volume diffusion
Diffusion of atoms through the interior of grains.

PROBLEMS

5.1 Atoms are found to move from one lattice position to another at the rate of 5×10^5 jumps per second at 400°C when the activation energy for their movement is 125 580 $J.mol^{-1}$. Calculate the jump rate at 750°C.

5.2 The number of vacancies in a material is related to temperature by an Arrhenius equation. If the fraction of lattice points containing vacancies is 8×10^{-5} at 600°C, determine the fraction of lattice points at 1 000°C.

5.3 The diffusion coefficient for Cr in Cr_2O_3 is 6×10^{-19} $m^2.s^{-1}$ at 727°C and is 1×10^{-13} $m^2.s^{-1}$ at 1 400°C. Calculate

(a) the activation energy and

(b) the constant D_0.

5.4 The diffusion coefficient for O in Cr_2O_3 is 4×10^{-19} $m^2.s^{-1}$ at 1 150°C, and 6×10^{-15} $m^2.s^{-1}$ at 1 715°C. Calculate

(a) the activation energy and

(b) the constant D_0.

5.5 A 0.2 mm thick wafer of silicon is treated so that a uniform concentration gradient of antimony is produced. One surface contains 1 Sb atom per 10^8 Si atoms and the other surface contains 500 Sb atoms per 10^8 Si atoms. The lattice parameter for Si is given in Appendix A. Calculate the concentration gradient in

(a) atomic percent Sb per m and

(b) Sb atoms.m^{-3}.m^{-1}.

5.6 When a Cu–Zn alloy solidifies, one portion of the structure contains 25 atomic percent zinc and another portion 0.025 mm away contains 20 atomic percent zinc. The lattice parameter for the FCC alloy is about 3.63×10^{-10} m. Determine the concentration gradient in

(a) atomic percent Zn per cm,

(b) weight percent Zn per cm, and

(c) Zn atoms.m^{-3}.m^{-1}.

5.7 A 0.025 mm BCC iron foil is used to separate a high hydrogen gas from a low hydrogen gas at 650°C. 5×10^{14} H atoms.m^{-3} are in equilibrium with the hot side of the foil and 2×10^{9} H atoms.m^{-3} are in equilibrium with the cold side. Determine

(a) the concentration gradient of hydrogen and

(b) the flux of hydrogen through the foil.

5.8 A 1 mm sheet of FCC iron is used to contain nitrogen in a heat exchanger at 1 200°C. The concentration of N at one surface is 0.04 atomic percent and the concentration at the second surface is 0.005 atomic percent. Determine the flux of nitrogen through the foil in N atoms.m^{-2}.s^{-1}.

5.9 A 4 cm-diameter, 0.5 mm-thick spherical container made of BCC iron holds nitrogen at 700°C. The concentration at the inner surface is 0.05 atomic percent and at the outer surface is 0.002 atomic percent. Calculate the number of grams of nitrogen that are lost from the container per hour.

5.10 A BCC iron structure is to be manufactured that will allow no more than 50 g of hydrogen to be lost per year through each square centimetre of the iron at 400°C. If the concentration of hydrogen at one surface is 0.005 H atom per unit cell and is 0.001 H atom per unit cell at the second surface, determine the minimum thickness of the iron.

5.11 Determine the maximum allowable temperature that will produce a flux of less than 2×10^{7} H atoms m^{-2}. s^{-1} through a BCC iron foil when the concentration gradient is -5×10^{24} atoms m^{-3}.m^{-1}.

5.12 The electrical conductivity of Mn_3O_4 is 8×10^{-16} ohm^{-1}.m^{-1} at 140°C and is 1×10^{-5} ohm^{-1}.m^{-1} at 400°C. Determine the activation energy that controls the temperature dependence of conductivity. Explain the process by which the temperature controls conductivity.

5.13 Compare the rate at which oxygen diffuses in Al_2O_3 with the rate at which aluminium diffuses in Al_2O_3 at 1 500°C. Explain the difference.

5.14 Compare the diffusion coefficients of carbon in BCC and FCC iron at the allotropic transformation temperature of 912°C and explain the difference.

5.15 Compare the diffusion coefficients for hydrogen and nitrogen in FCC iron at 1 000°C and explain the difference.

5.16 Explain why a polymer balloon filled with helium gas deflates over time.

5.17 A carburising process is carried out on a 0.10% C steel by introducing 1.0% C at the surface at 980°C, where the iron is FCC. Calculate the carbon content at 0.1 mm, 0.5 mm, and 1 mm beneath the surface after 1 h.

5.18 Iron containing 0.05% C is heated to 912°C in an atmosphere that produces 1.20% C at the surface and is held for 24 h. Calculate the carbon content at 0.5 mm beneath the surface if

(a) the iron is BCC and

(b) the iron is FCC.

Explain the difference.

5.19 What temperature is required to obtain 0.50% C at a distance of 0.5 mm beneath the surface of a 0.20% C steel in 2 h, when 1.10%

C is present at the surface? Assume that the iron is FCC.

5.20 A 0.15% C steel is to be carburised at 1 000°C, giving 0.35% C at a distance of 1 mm beneath the surface. If the surface composition is maintained at 0.90% C, what time is required?

5.21 A 0.02% C steel is to be carburised at 1 200°C in 4 h, with a point 0.6 mm beneath the surface reaching 0.45% C. Calculate the carbon content required at the surface of the steel.

5.22 A 1.2% C tool steel held at 1 150°C is exposed to oxygen for 48 h. The carbon content at the steel surface is zero. To what depth will the steel be decarburised to less than 0.20% C?

5.23 A 0.80% C steel must operate at 950°C in an oxidising environment where the carbon content at the steel surface is zero. Only the outermost 0.02 cm of the steel part can fall below 0.75% C. What is the maximum time that the steel part can operate?

5.24 A BCC steel containing 0.001% N is nitrided at 550°C for 5 h. If the nitrogen content at the steel surface is 0.08%, determine the nitrogen content at 0.25 mm from the surface.

5.25 What time is required to nitride a 0.002 N steel to obtain 0.12% N at a distance of 0.05 mm beneath the surface at 625°C? The nitrogen content at the surface is 0.15%.

5.26 We currently can successfully perform a carburising heat treatment at 1 200°C in 1 h. In an effort to reduce the cost of the brick lining in our furnace, we propose to reduce the carburising temperature to 950°C. What time will be required to give us a similar carburising treatment?

5.27 During freezing of a Cu-Zn alloy, we find that the composition is nonuniform. By heating the alloy to 600°C for 3 hours, diffusion of zinc helps to make the composition more uniform. What temperature would be required if we wished to perform this homogenisation treatment in 30 minutes?

5.28 A ceramic part made of MgO is sintered successfully at 1 700°C in 90 minutes. To minimise thermal stresses during the process, we plan to reduce the temperature to 1 500°C. Which will limit the rate at which sintering can be done: diffusion of magnesium ions or diffusion of oxygen ions? What time will be required at the lower temperature?

5.29 A Cu-Zn alloy has an initial grain diameter of 0.01 mm. The alloy is then heated to various temperatures, permitting grain growth to occur. The times required for the grains to grow to a diameter of 0.30 mm are:

Temperature (°C)	Time (minutes)
500	80 000
600	3 000
700	120
800	10
850	3

Determine the activation energy for grain growth. Does this correlate with the diffusion of zinc in copper? (**Hint:** Note that rate is the reciprocal of time.)

5.30 A sheet of gold is diffusion-bonded to a sheet of silver in 1 h at 700°C. At 500°C, 440 h are required to obtain the same degree of bonding, and at 300°C, bonding requires 1 530 years. What is the activation energy for the diffusion bonding process? Does it appear that diffusion of gold or diffusion of silver controls the bonding rate? (**Hint:** Note that rate is the reciprocal of time.)

Design Problems

5.31 Design a spherical tank, with a wall thickness of 20 mm, that will ensure that no more than 50 kg of hydrogen will be lost per year. The tank, which will operate at 500°C, can be made of nickel, aluminium, copper, and iron. The diffusion coefficient of hydrogen and the cost per kilogram for each available material is listed below:

	Diffusion Data		
Material	D_0 ($m^2.s^{-1}$)	Q $J.mol^{-1}$	Cost (£.kg^{-1})
Nickel	0.0055×10^{-4}	37 250	4.10
Aluminium	0.16×10^{-4}	43 280	0.60
Copper	0.011×10^{-4}	39 260	1.10
Iron	(*See* Table 5.1)		0.60

5.32 A steel gear initially containing 0.10% C is to be carburised so that the carbon content at a depth of 1.27 mm is 0.50% C. We can generate a carburising gas at the surface that contains anywhere from 0.95% C to 1.15% C. Design an appropriate carburising heat treatment.

5.33 When a valve casting containing copper and nickel solidifies under nonequilibrium conditions, we find that the composition of the alloy varies substantially over a distance of 0.05 mm. Usually we are able to eliminate this concentration difference by heating the alloy for 8 h at 1 200°C; however, sometimes this treatment causes the alloy to begin to melt, destroying the part. Design a heat treatment that will permit elimination of the nonuniformity without danger of melting. Assume that the cost of operating the surface per hour doubles for each 100°C increase in temperature.

In these chapters, we examine several methods used to control the structure and mechanical properties of materials. Three of these processes – grain size strengthening, solid solution strengthening, and strain hardening – rely on introducing and controlling the lattice imperfections discussed in Chapter 4.

We also obtain strengthening by creating multiple-phase materials, where each phase has a different composition or crystal structure. The interface between the phases provides strengthening by interfering with the deformation mechanism. Dispersion strengthening, age hardening, and a variety of phase transformations – which often rely on allotropic transformations – enable us to control the size, shape, and distribution of the phases in the material. The processing of the material, such as solidification processing, deformation processing, and heat treatment, is also employed to control microstructure and properties.

Before looking at strengthening mechanisms, we first briefly examine the mechanical testing of materials in order to understand the results of these tests, which are the mechanical properties of a material.

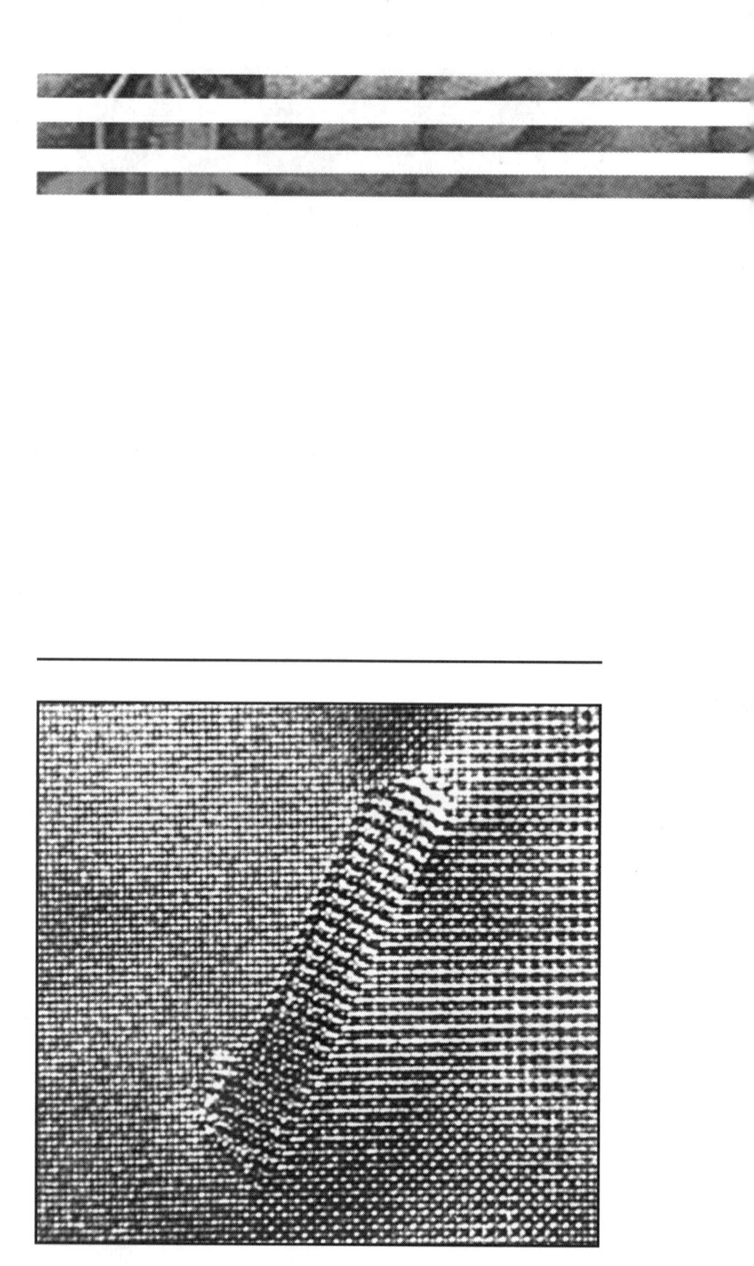

An Al_2MgCu precipitate is shown at the interface of an aluminium matrix (*upper left*) and an Al_3Li phase (*lower right*). The individual atoms in each phase are visible using atomic resolution microscopy. (*Courtesy of V. Radmilovic and G. J. Shiflet, University of Virginia.*)

Part

Controlling the Microstructure and Mechanical Properties of Materials

CHAPTER 6

Mechanical Testing and Properties

6.1 Introduction

We select a material by matching its mechanical properties to the service conditions required of the component. The first step in the selection process requires that we analyse the application to determine the most important characteristics that the material must possess. Should it be strong, or stiff, or ductile? Will it be subjected to repeated application of a high force, sudden intense force, high stress at elevated temperature, or abrasive conditions? Once we know the required properties, we can select the appropriate material using data listed in handbooks. We must, however, know how the properties listed in the handbook are obtained, know what the properties mean, and realise that the properties listed are obtained from idealised tests that may not apply exactly to real-life engineering applications.

In this chapter we study several tests that are used to measure how a material withstands an applied force. The results of these tests are the mechanical properties of the material.

6.2 The Tensile Test: Use of the Stress–Strain Diagram

The tensile test measures the resistance of a material to a static or slowly applied force. A test setup is shown in Figure 6.1; a typical specimen has a diameter of 12.5 mm and a gauge length of 50 mm. The specimen is placed in the testing machine and a force F, called the load, is applied. A strain gauge or extensometer is used to measure the amount that the specimen stretches between the gauge marks when the force is applied. Table 6.1 shows the effect of the load on the gauge length of an aluminium alloy test bar.

Engineering Stress and Strain The results of a single test apply to all sizes and shapes of specimens for a given material if we convert the force to stress and the distance between gauge marks to strain. Engineering stress and engineering strain are defined by the following equations,

$$\text{Engineering stress} = \sigma = \frac{F}{A_0} \tag{6.1}$$

$$\text{Engineering strain} = \varepsilon = \frac{l - l_0}{l_0} \tag{6.2}$$

where A_0 is the original cross-sectional area of the specimen before the test begins, l_0 is the original distance between the gauge marks, and l is the distance between the gauge marks after force F is applied. The conversions from load-gauge length to stress-strain are included in Table 6.1. The stress-strain curve (Figure 6.2) is used to record the results of a tensile test.

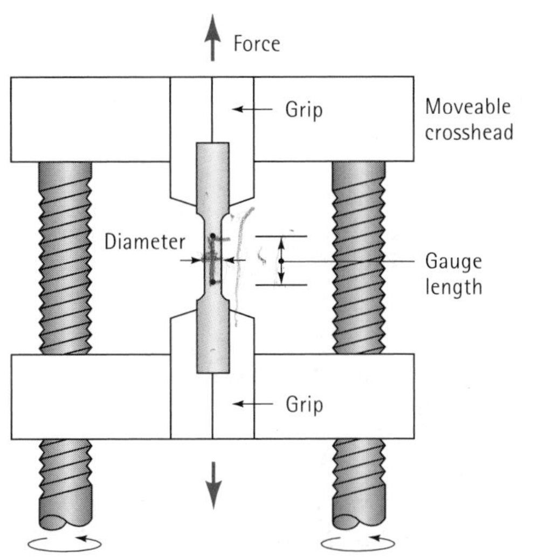

Figure 6.1
A unidirectional force is applied to a specimen in the tensile test by means of the moveable crosshead.

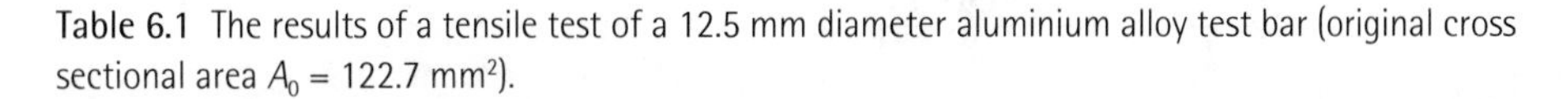

Table 6.1 The results of a tensile test of a 12.5 mm diameter aluminium alloy test bar (original cross sectional area A_0 = 122.7 mm²).

Measured		Calculated	
Load (kN)	Gauge Length (mm)	Engineering stress (MN.m⁻²)	Engineering strain (mm.mm⁻¹)
0	50.00	0	0
5	50.03	40.7	0.0006
10	50.06	81.5	0.0012
15	50.09	122.2	0.0018
20	50.12	163.0	0.0024
25	50.15	203.7	0.0030
30	50.185	244.5	0.0037
35.3	52.00	287.7	0.04
35.6 (Maximum)	53.00	290.1	0.06
33.8 (Fracture)	55.30	275.4	0.104

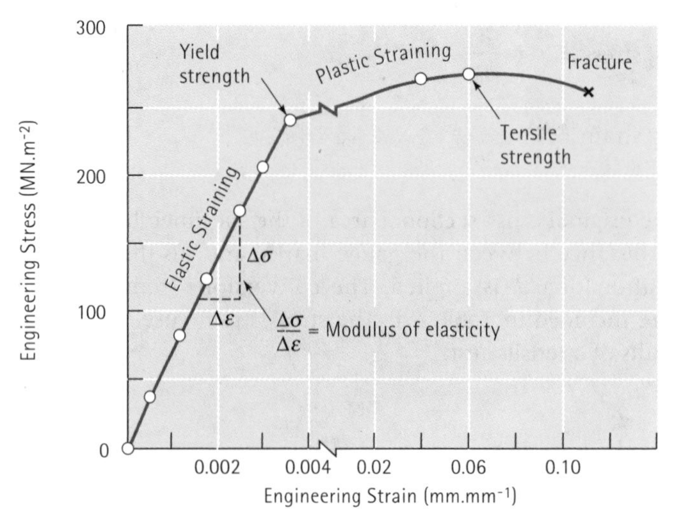

Figure 6.2
The stress-strain curve for an aluminium alloy plotted from data in Table 6.1.

EXAMPLE 6.1

Convert the load-gauge length data in Table 6.1 to engineering stress and strain and plot a stress-strain curve.

SOLUTION

For the 5 kN load:

$$\sigma = \frac{F}{A_0} = \frac{5\,000\text{ N}}{(\pi/4)(12.5\text{ mm})^2} = \frac{5\,000}{122.7} = 40.7\text{ N.mm}^{-2} \equiv 40.7\text{ MN.m}^{-2}$$

$$\varepsilon = \frac{l - l_0}{l_0} = \frac{50.03 - 50.00}{50.00} = 0.0006\text{ mm.mm}^{-1}$$

The results of similar calculations for each of the remaining loads are given in Table 6.1 and are plotted in Figure 6.2.

Units Many different units are used to report the results of the tensile test. The most common units for stress are $MN.m^{-2}$.

In fundamental S.I. units, stress = $N.m^{-2}$ and this stress unit is also called the Pascal (Pa); hence 1 Pa = $1N.m^{-2}$.

However, this is a very small value of stress in terms of engineering materials and hence $MN.m^{-2}$ or megapascals (MPa) are used; 1 MPa = 1 $MN.m^{-2}$.

The unit $MN.m^{-2}$ has found wide acceptance by engineers although MPa is also used. The important thing to remember is that they are equal.

Cross sectional areas in tensile tests are practically given in mm^2 rather than m^2.

Note that 1 $MN.m^{-2}$ = $\frac{1\text{ MN}}{1\text{ m}^2} = \frac{10^6\text{ N}}{10^6\text{ mm}^2} = 1\text{ N.mm}^{-2}$

Hence 1 $MN.m^{-2}$ = 1 N. mm^{-2} = 1 MPa

USA stress data is often found in pounds per square inch (psi). This can be converted by psi × 0.006895 → $MN.m^{-2}$.

The units for strain include millimetre/millimetre, metre/metre and inch/inch. Because strain is dimensionless (that is, the aforementioned units cancel each other out), no conversion factors are required to change the system of units. Indeed, the units are often not quoted for strain data.

EXAMPLE 6.2 **Design of a Suspension Rod**

An aluminium rod is to withstand an applied force of 200 kN (0.2 MN). To ensure a sufficient factor of safety, the maximum allowable stress on the rod is limited to 170 $MN.m^{-2}$. The rod must be at least 3.8 m long but must deform elastically no more than 6 mm when the force is applied. Design an appropriate rod.

SOLUTION

We can use the definition of engineering stress to calculate the required cross-sectional area of the rod:

$$A_0 = \frac{F}{\sigma} = \frac{0.2 \text{ MN}}{170 \text{ MN.m}^{-2}} = 1.18 \times 10^{-3} \text{ m}^{-2} = 1180 \text{ mm}^2$$

The rod could be produced in various shapes, provided that the cross-sectional area is 1180 mm^2. For a round cross-section, the minimum diameter to ensure that the stress is not too high is:

$$A_0 = \frac{\pi d^2}{4} = 1180 \text{ mm}^2 \text{or } d = 38.8 \text{ mm}$$

The maximum allowable elastic deformation is 6 mm. From the definition of engineering strain:

$$\varepsilon = \frac{l - l_0}{l_0} = \frac{\Delta l}{l_0} = \frac{6}{l_0}$$

From Figure 6.2, the strain expected for a stress of 170 $MN.m^{-2}$ is 0.0025 $mm.mm^{-1}$. If we use the cross-sectional area determined previously, the maximum length of the rod is:

$$0.0025 = \frac{\Delta l}{l_0} = \frac{6}{l_0} \quad \text{or } l_0 = 2400 \text{ mm} = 2.4 \text{ m}$$

However, the minimum length of the rod is specified as 3.8 m. To produce a longer rod, we might make the cross-sectional area of the rod larger. The minimum strain allowed for the 3.8 m rod is:

$$\varepsilon = \frac{\Delta l}{l_0} = \frac{6}{3800} = 0.00158 \text{ mm.mm}^{-1}$$

The stress, from Figure 6.2, is about 110 MN.m^{-2}, which is less than the maximum of 170 MN.m^{-2}. The minimum cross-sectional area then is:

$$A_0 = \frac{F}{\sigma} = \frac{0.2 \text{ MN}}{110 \text{ MN.m}^{-2}} = 1.82 \times 10^{-3} \text{ m} = 1820 \text{ mm}^2$$

In order to satisfy both the maximum stress and the minimum elongation requirements, the rod must be at least 1 820 mm^2, or a diameter of 48 mm.

6.3 Properties Obtained from the Tensile Test

Information concerning the strength, stiffness, and ductility of a material can be obtained from a tensile test.

Yield Strength The yield strength is the stress at which plastic deformation becomes noticeable. In metals, this is usually the stress required for dislocations to start to slip. The yield strength therefore is the strength that divides the *elastic* and *plastic* behaviour of the material. When designing a part that will not plastically deform in service, we must either select a material that has a high yield strength or make the component large so that the applied force produces a stress that is below the yield strength (*refer to* Example 6.2).

In many materials, the stress at which the material changes from elastic to plastic behaviour is not easily detected. In this case, we determine a proof strength or an offset yield strength [Figure 6.3(a)]. We construct a line parallel to the initial portion of the stress-strain curve but offset by 0.002 mm.mm^{-1} (0.2%) from the origin. The 0.2% proof strength is the stress at which our constructed line intersects the stress-strain curve. In Figure 6.3(a), the 0.2% proof strength for a solid solution aluminium alloy is 40 MN.m^{-2}.

The stress-strain curve for certain low-carbon steels displays a double yield point [Figure 6.3(b)]. The material is expected to plastically deform at stress σ_1. However, small interstitial atoms clustered around the dislocations interfere with slip and raise the yield point to σ_2. Only after we apply the higher stress σ_2 do the dislocations slip. After slip begins at σ_2, the dislocations move away from the clusters of small atoms and continue to move very rapidly at the lower stress σ_1. This is called the yield phenomenon of steel.

Tensile Strength The stress obtained at the highest applied force is the tensile strength, which is the maximum stress on the engineering stress-strain curve. In many ductile materials, deformation does not remain uniform. At some point, one region deforms more than others and a large local decrease in the cross-sectional area occurs (Figure 6.4). This locally deformed region is called a neck Because the cross-sectional area becomes smaller at this point, a lower force is required to continue its deformation, and the engineering stress, calculated from the original area A_0, decreases. The tensile strength is the stress at which necking begins in ductile materials.

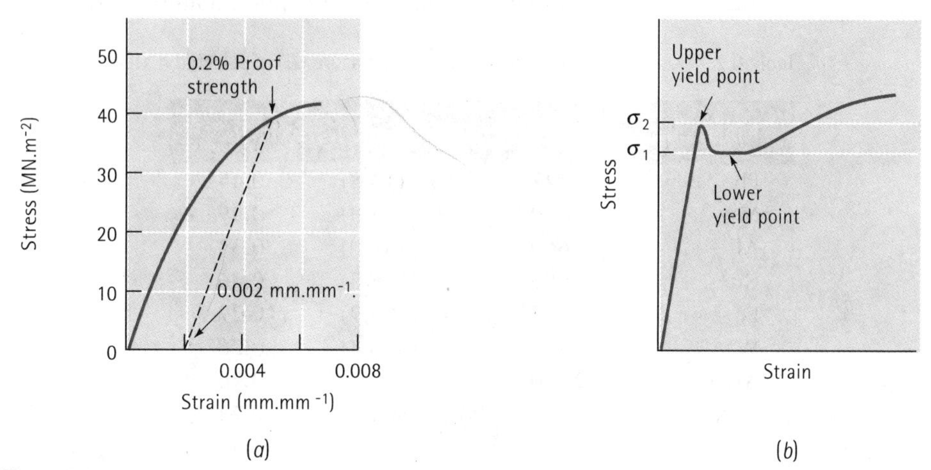

Figure 6.3
(*a*) Determining the 0.2% proof strength in an aluminium alloy and (*b*) upper and lower yield point behaviour in a low-carbon steel.

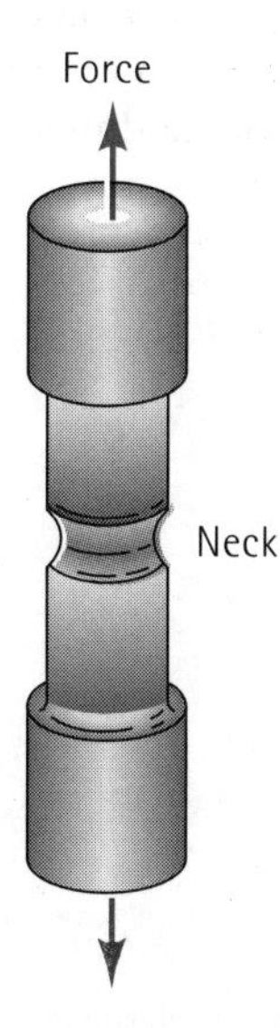

Figure 6.4
Localised deformation of a ductile material during a tensile test produces a necked region.

Elastic Properties The modulus of elasticity, or *Young's modulus, E,* is the slope of the stress-strain curve in the elastic region. This relationship is Hooke's law:

$$E = \frac{\sigma}{\varepsilon} \tag{6.2}$$

The modulus is closely related to the binding energies (Figure 2.14). A steep slope in the force-distance graph at the equilibrium spacing indicates that high forces are required to separate the atoms and cause the material to stretch elastically. Thus, the material has a high modulus of elasticity. Binding forces, and thus the modulus of elasticity, are typically higher for high melting point materials (Table 6.2).

Table 6.2 Elastic properties and melting temperature (T_m) of selected materials.

Material	T_m (°C)	E($GN.m^{-2}$)	μ
Pb	327	(13.8)	0.45
Mg	650	(44.8)	0.29
Al	660	(69.0)	0.33
Cu	1085	(124.8)	0.36
Fe	1538	(206.9)	0.27
W	3410	(408.3)	0.28
Al_2O_3	2020	(379.3)	0.26
Si_3N_4		(303.4)	0.24

The modulus is a measure of the stiffness of the material. A stiff material, with a high modulus of elasticity, maintains its size and shape even under an elastic load. Figure 6.5 compares the elastic behaviour of steel and aluminium. If a stress of 200 $MN.m^{-2}$ is applied to the shaft, the steel deforms elastically 0.001 $mm.mm^{-1}$; at the same stress, aluminium deforms 0.003 $mm.mm^{-1}$. Iron has a modulus of elasticity three times greater than that of aluminium. This units for modulus of elasticity are $GN.m^{-2}$ or GPa.

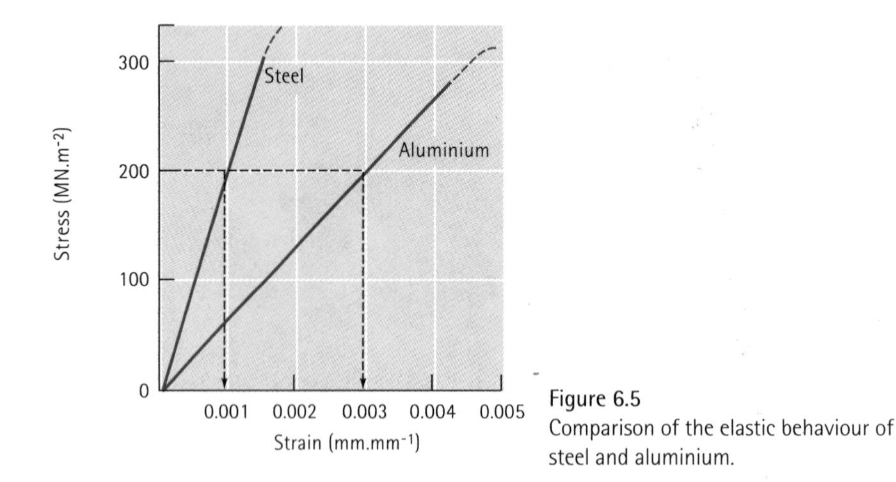

Figure 6.5
Comparison of the elastic behaviour of steel and aluminium.

The modulus of resilience (E_r), the area contained under the elastic portion of a stress-strain curve, is the elastic energy that a material absorbs during loading and subsequently releases when the load is removed. For linear elastic behaviour:

$$E_r = \frac{\text{yield strength}}{(2)(\text{strain at yielding })} \tag{6.4}$$

The ability of a spring or a golf ball to perform satisfactorily depends on a high modulus of resilience.

Poisson's ratio, μ, relates the longitudinal elastic deformation produced by a simple tensile or compressive stress to the lateral deformation that occurs simultaneously:

$$\mu = \frac{-\varepsilon_{\text{lateral}}}{\varepsilon_{\text{longitudinal}}} \tag{6.5}$$

Poisson's ratio is typically about 0.3 (Table 6.2).

EXAMPLE 6.3

From the data in Example 6.1, calculate the modulus of elasticity of the aluminium alloy. Use the modulus to determine the length of a 1.25 m bar to which a stress of 210 MN.m^{-2} is applied.

SOLUTION

From Figure 6.2:

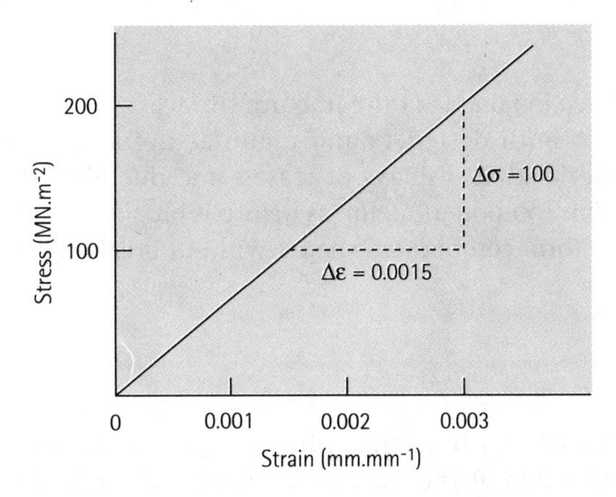

Modulus of elasticity = slope of linear part of the stress-strain curve.

$$E = \frac{\Delta\sigma}{\Delta\varepsilon} = \frac{100(\text{MN.m}^{-2})}{0.0015(\text{mm.mm}^{-1})} = 66\,700 \text{ MN.m}^{-2} = 66.7 \text{ GN.m}^{-2}$$

from the data, E = 66.7 GN.m^{-2}

rearranging Hooke's Law: $\varepsilon = \dfrac{\sigma}{E}$

when $\sigma = 210$ MN.m^{-2}, $\varepsilon = \dfrac{210}{66\,700} = 0.00315$ mm.mm^{-1}

The length of a 1.25 m bar, under a stress of 210 MN.m^{-2} is

1.25 + 1.25 × 0.00315 = 1.2539 m.

Note, the engineering strain is a ratio and the units e.g., mm.mm^{-1} are only included as a guide to how it was calculated. Thus when 1.25 was multiplied by 0.00336 it produced an answer in metres because the 1.25 was in metres.

Ductility Ductility measures the amount of plastic deformation that a material can withstand without breaking. We can measure the distance between the gauge marks on our specimen before and after the test. The % Elongation describes the extent to which the specimen stretches before fracture:

$$\% \text{ Elongation} = \frac{l_f - l_0}{l_0} \times 100 \qquad (6.6)$$

where l_f is the distance between gauge marks after the specimen breaks.

A second approach is to measure the percent change in cross-sectional area at the point of fracture before and after the test. The % Reduction in area describes the amount of thinning undergone by the specimen during the test:

$$\% \text{ Reduction in area} = \frac{A_0 - A_f}{A_0} \times 100 \qquad (6.7)$$

where A_f is the final cross-sectional area at the fracture surface.

Ductility is important to both designers and manufacturers. The designer of a component prefers a material that displays at least some ductility, so that, if the applied stress is too high, the component deforms before it breaks. Fabricators want a ductile material in order to form complicated shapes without breaking the material in the process.

EXAMPLE 6.4

The aluminium alloy in Example 6.1 has a final gauge length after failure of 55.02 mm and a final diameter of 9.85 mm at the fractured surface. Calculate the ductility of this alloy.

SOLUTION

$$\% \text{ Elongation} = \frac{l_f - l_0}{l_0} \times 100 = \frac{55.02 - 50}{50} \times 100 = 10.04\%$$

$$\% \text{ Reduction in area} = \frac{A_0 - A_f}{A_0} \times 100$$
$$= \frac{(\pi/4)(12.5)^2 - (\pi/4)(9.85)^2}{(\pi/4)(12.5)^2} \times 100$$
$$= 37.9\%$$

The final gauge length is less than 55.30 mm (*see* Table 6.1) because, after fracture, the elastic strain is recovered.

Effect of Temperature Tensile properties depend on temperature (Figure 6.6). Yield strength, tensile strength and modulus of elasticity decrease at higher temperatures, whereas ductility commonly increases. A materials fabricator may wish to deform a material at a high temperature (known as *hot working*) to take advantage of the higher ductility and lower required stress.

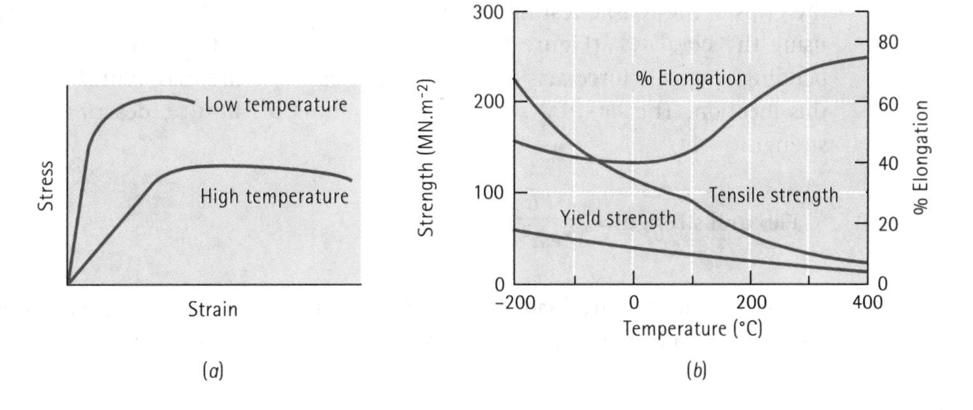

Figure 6.6
The effect of temperature (*a*) on the stress-strain curve and (*b*) on the tensile properties of an aluminium alloy.

6.4 The Bend Test for Brittle Materials

In ductile materials, the engineering stress-strain curve typically goes through a maximum; this maximum stress is the tensile strength of the material. Failure occurs at a lower stress after necking has reduced the cross-sectional area supporting the load. In more brittle materials, failure occurs at the maximum load, where the tensile strength and breaking strength are the same. In very brittle materials, including many ceramics, yield strength, tensile strength, and breaking strength are all the same (Figure 6.7).

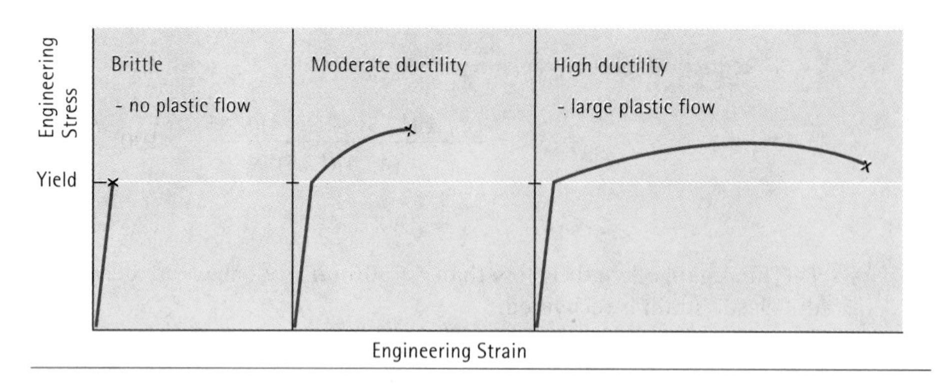

Note, all three hypothetical materials have nominally been given the same yield strength and modulus of elasticity values, but differ in their tensile strength and percentage elongation due to the varying amount of plastic flow available in the different materials.

Figure 6.7 The stress-strain behaviour of brittle materials compared with that of more ductile materials.

In many brittle materials, the normal tensile test cannot easily be performed because of the presence of flaws at the surface. Often, just placing a brittle material in the grips of the tensile testing machine causes cracking. These materials may be tested using the bend test (Figure 6.8). By applying the load at three points and causing bending, a tensile force acts on the material opposite the midpoint. Fracture begins at this location. The flexural strength, or modulus of rupture, describes the material's strength:

$$\text{Flexural strength} = \frac{3FL}{2wh^2} \tag{6.8}$$

where F is the fracture load, L is the distance between the two outer points, w is the width of the specimen, and h is the height of the specimen.

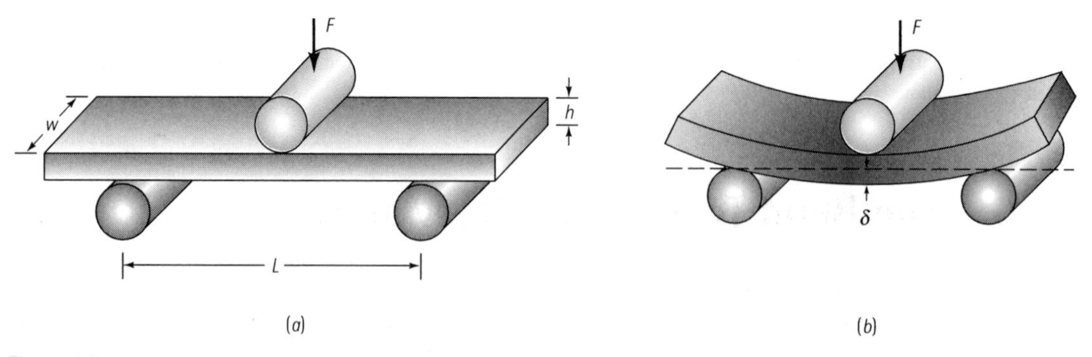

Figure 6.8
(*a*) The bend test often used for measuring the strength of brittle materials, and (*b*) the deflection δ obtained by bending.

The results of the bend test are similar to the stress-strain curves; however, the stress is plotted versus deflection rather than versus strain (Figure 6.9).

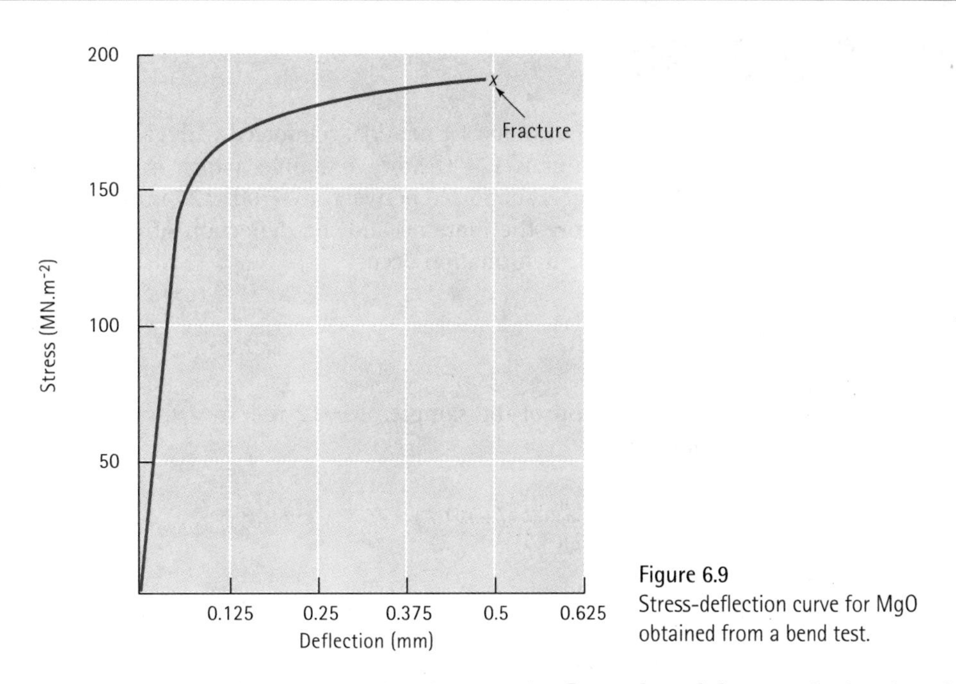

Figure 6.9
Stress-deflection curve for MgO obtained from a bend test.

The modulus of elasticity in bending, or the flexural modulus, is calculated in the elastic region of Figure 6.9:

$$\text{Flexural modulus} = \frac{L^3F}{4wh^3\delta} \tag{6.9}$$

where δ is the deflection of the beam when a force F is applied.

Since cracks and flaws tend to remain closed in compression, brittle materials are often designed so that only compressive stresses act on the part. Often, we find that brittle materials fail at much higher compressive stresses than tensile stresses (Table 6.3).

Table 6.3 Comparison of the tensile, compressive, and flexural strengths of selected ceramic and composite materials.

Material	Tensile Strength ($MN.m^{-2}$)	Compressive Strength ($MN.m^{-2}$)	Flexural Strength ($MN.m^{-2}$)
Polyester – 50% glass fibres	160	225	315
Polyester – 50% glass fibre fabric	260	190[a]	320
Al_2O_3 (99% pure)	210	2625	350
SiC (pressureless-sintered)	175	3920	560

a A number of composite materials are quite poor in compression.

EXAMPLE 6.5

The flexural strength of a composite material reinforced with glass fibres is 315 $MN.m^{-2}$ and the flexural modulus is 124 $GN.m^{-2}$ A sample, which is 12 mm wide, 9.5 mm high, and 200 mm long, is supported between two rods 125 mm apart. Determine the force required to fracture the material and the deflection of the sample at fracture, assuming that no plastic deformation occurs.

SOLUTION

Based on the description of the sample, w = 12 mm h = 9.5 mm and L = 125 mm. From Equation 6.8:

$$315 = \frac{3FL}{2wh^2} = \frac{(3)(F)(125)}{(2)(12)(9.5)^2} = 0.173F$$

$$F = \frac{315}{0.173} = 1819 \text{ N}$$

Therefore, the deflection, from Equation 6.9, is:

$$124 \times 10^3 = \frac{L^3F}{4wh^3\delta} = \frac{(125)^3(1819)}{(4)(12)(9.5)^3\delta}$$

$$\delta = 0.7 \text{ mm}$$

6.5 True Stress–True Strain

The decrease in engineering stress beyond the maximum point (*refer to* Figure 6.2) occurs because of our definition of engineering stress. We used the original area A_0 in our calculations, but this is not precise because the area continually decreases. We define true stress and true strain by the following equations:

$$\text{True stress} = \sigma_t = \frac{F}{A} \tag{6.10}$$

$$\text{True strain} = \int \frac{dl}{l} = \ln\left(\frac{l}{l_0}\right) = \ln\left(\frac{A_0}{A}\right) \tag{6.11}$$

where A is the actual or instantaneous area at which the load F is applied. The true stress-strain curve is compared with the engineering stress-strain curve in Figure 6.10. True stress continues to increase after necking because, although the load required decreases, the area decreases even more, so the stress is increasing up to fracture.

We often do not require true stress and true strain. When we exceed the yield strength, the material deforms. Our component has failed because it no longer has the original intended shape. Furthermore, a significant difference develops between the two curves only when necking begins. But when necking begins, our component is grossly deformed and no longer satisfies its intended use.

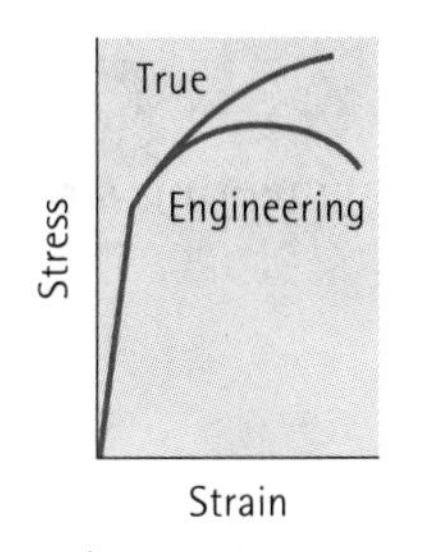

Figure 6.10
The relationship between the true stress–true strain diagram and the engineering stress–engineering strain diagram.

EXAMPLE 6.6

Compare engineering stress and strain with true stress and strain for the aluminium alloy in Example 6.1 at (a) the maximum load and (b) fracture. The diameter at maximum load is 12.3 mm and at fracture is 9.85 mm.

SOLUTION

a. At the tensile or maximum load:

$$\text{Engineering stress} = \frac{F}{A_0} = \frac{35.6\times10^3}{(\pi/4)(12.5)^2} = 290.1\ \text{MN.m}^{-2}$$

$$\text{True stress} = \frac{F}{A} = \frac{35.6\times10^3}{(\pi/4)(12.3)^2} = 299.6\ \text{MN.m}^{-2}$$

$$\text{Engineering strain} = \frac{l-l_0}{l_0} = \frac{53-50}{50} = 0.060\ \text{mm.mm}^{-1}$$

$$\text{True strain} = \ln\left(\frac{l}{l_0}\right) = \ln\left(\frac{53}{50}\right) = 0.058\ \text{mm.mm}^{-1}$$

b. At fracture:

$$\text{Engineering stress} = \frac{F}{A_0} = \frac{33.8\times10^3}{(\pi/4)(12.5)^2} = 275.4\ \text{MN.m}^{-2}$$

$$\text{True stress} = \frac{F}{A} = \frac{33.8\times10^3}{(\pi/4)(9.85)^2} = 443.6\ \text{MN.m}^{-2}$$

$$\text{Engineering strain} = \frac{l - l_0}{l_0} = \frac{55.2 - 50}{50} = 0.104 \text{ mm.mm}^{-1}$$

$$\text{True strain} = \ln\left(\frac{A_0}{A_f}\right) = \ln\left[\frac{(\pi/4)(12.5)^2}{(\pi/4)(9.85)^2}\right]$$

$$\ln(1.610) = 0.476 \text{ mm.mm}^{-1}$$

The true stress becomes much greater than the engineering stress only after necking begins.

6.6 The Hardness Test: Its Nature and Use

Hardness testing is one of the oldest mechanical testing methods. It is widely used in quality control because it is a relatively cheap test, rapid to perform and often non-destructive to the component tested (unlike the tensile test for example).

Hardness is the ability of a material to resist surface indentation or scratching. It therefore indicates resistance to abrasion or wear, however its main use is to check the quality of a product e.g., whether a heat treatment process has been carried out correctly so that the component's hardness meets that specified by the designer.

Hardness is not a fundamental property of a material, and so its value varies according to the test method. Hence the need to specify the test procedure; the Brinell Test (BS 240: 1986), the Vickers Test (BS 427: 1982) and the Rockwell Test (BS 891: 1989).

These are the most commonly used hardness tests in engineering and each has the same basic principle; the hardness is measured from an indentation produced in the component by applying a constant load on a specific indentor in contact with the surface of the component for a fixed time.

Brinell test An indentor comprising of a hardened steel or tungsten carbide ball is pressed into the surface for a standard time (10–15 secs) under a standard load. After removing the load, the circular indentation is then measured in two mutually perpendicular directions taking the average of the two readings [Figure 6.11(a)]. The Brinell Hardness Number (HB) is then calculated from

$$HB = \frac{\text{Applied load (kg)}}{\text{Surface area of impression (mm}^2)} \qquad (6.12)$$

$$HB = \frac{2F}{\pi D\{D - \sqrt{D^2 - d^2}\}}$$

where F = load in kilograms
D = ball diameter in mm
d = diameter of impression in mm

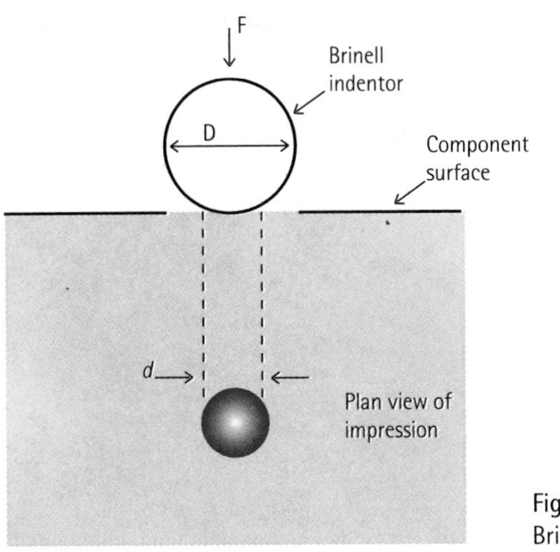

Figure 6.11(a)
Brinell indentor and impression.

With a soft material and a large load it would be possible to push the indentor in so far that it could only produce an impression $d = D$. In order to obtain accurate Brinell values the relationship $d = 0.25D$ to $0.50D$ must be maintained. Hence the ball diameter and load applied is specified for the material under test e.g.

for steels , $\dfrac{F}{D^2} = 30$

for copper alloys, $\dfrac{F}{D^2} = 10$

BS 240 contains F/D^2 values for a complete range of materials.

For example, the load used in a Brinell Test carried out on a steel component using a 10 mm diameter steel ball:

For steels, $\dfrac{F}{D^2} = 30$

when $D = 10$, $F = 30 \times 10^2 = 3\,000$ kg. Hence a 3 000 kg load is required.

The Brinell Test has the following limitations.

a) The impression is large (typically 2–4 mm in diameter) and this may act as a stress raiser in a component. It may also be unacceptable on grounds of appearance e.g., a car body panel, although acceptable on a car cylinder block.

b) The large depth of the impression precludes its use on plated or surface hardened components as the impression would also measure the underlying structure.

c) Very hard materials will deform the indentor, hence the Brinell Test is limited to materials of up to 450HB for a steel ball, and 600HB for a tungsten carbide ball.

Vickers test The Vickers indentor is a square based pyramid with an included angle of 136°, made from diamond. This indentor was designed to overcome the problems inherent in the Brinell Test of using a spherical indentor [Figures 6.11(b) and (c)]. The Vickers hardness (HV) is again a function of the applied load on the indentor and size of the resulting impression in the material being tested.

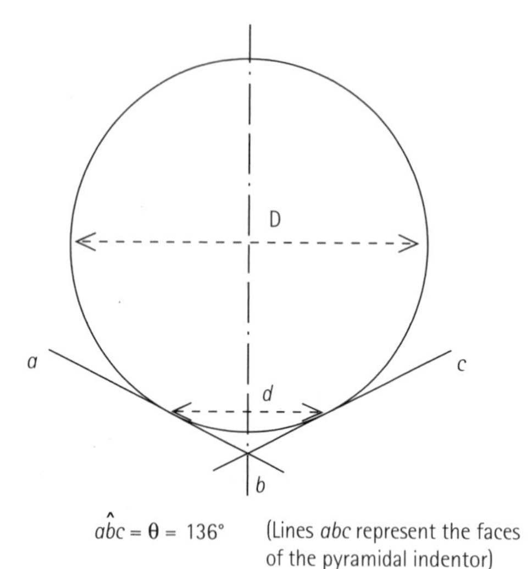

$a\hat{b}c = \theta = 136°$ (Lines *abc* represent the faces of the pyramidal indentor)

Figure 6.11(b)
Derivation of the Vickers diamond angle.

The advantage of $\theta = 136°$ is that HB ≈ HV, up to about 300. The Vickers Test has the following advantages over the Brinell Test.

a) Suitable for hard materials as well as soft materials.

b) There is no need to use the F/D^2 ratio for the material to be tested because all impressions are geometrically similar. The only criterion for load selection now is that the impression should be large enough to be read accurately. The Vickers hardness range is proportional, so a material of HV 400 is twice as hard as a material having a HV = 200.

Microhardness tests may also be made, with a Vickers indentor which is incorporated into a microscope so that very small indentations can be made under loads of 1–1 000 grams. The method is useful in determining the hardness of microconstituents e.g., phases in a microstructure.

The limitations of the Vickers Test.

a) The impression is small (difficult to see with the naked eye) and so the surface of the component must be polished flat with silicon carbide paper and the component surface must be secured perpendicular to the indentor during the test.

b) It takes a relatively long time to perform a Vickers Hardness Test.

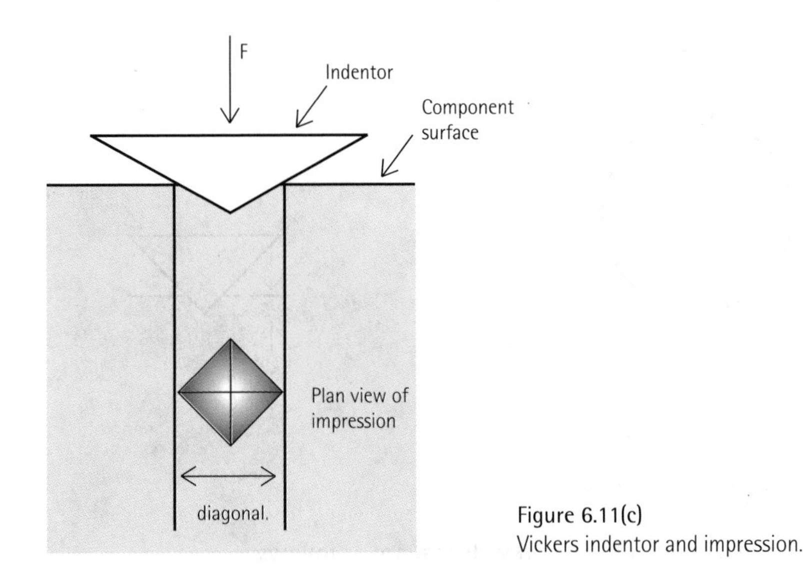

Figure 6.11(c)
Vickers indentor and impression.

Knoop test When the hardness of very thin layers is required the Knoop Test is preferred to the Vickers, because under the same load its rhombic diamond pyramid indentor penetrates less than the square diamond pyramid indentor of the Vickers Test [Figure 6.11(d)].

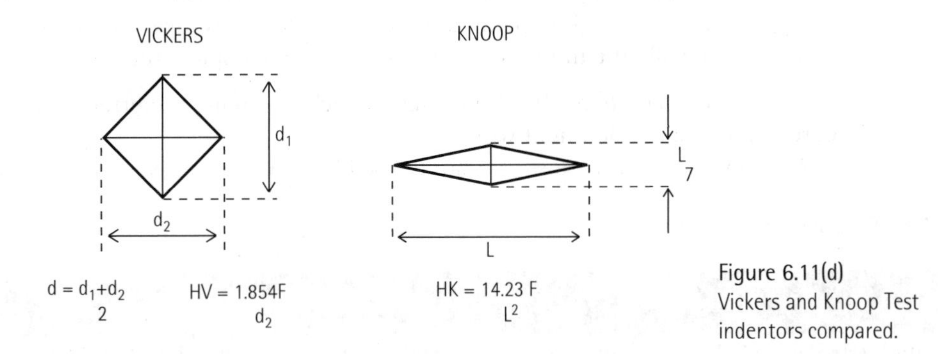

Figure 6.11(d)
Vickers and Knoop Test indentors compared.

Rockwell test The principle of the Rockwell test differs from that of the others in that the depth of the impression is related to the hardness rather than the diameter or diagonal of the impression [Figure 6.11(e)]. This greatly speeds up the measurement because the Rockwell machine is designed to record the depth of penetration of the indentor.

There are many Rockwell scales, but the most commonly used are the:

B-scale (1/16 inch diameter steel ball indentor; 100 kg load), used to measure the hardness (HRB) of non-ferrous metals.

C-scale (120° diamond cone indentor (called a BRALE); 150 kg load), used to measure the hardness (HRC) of steels.

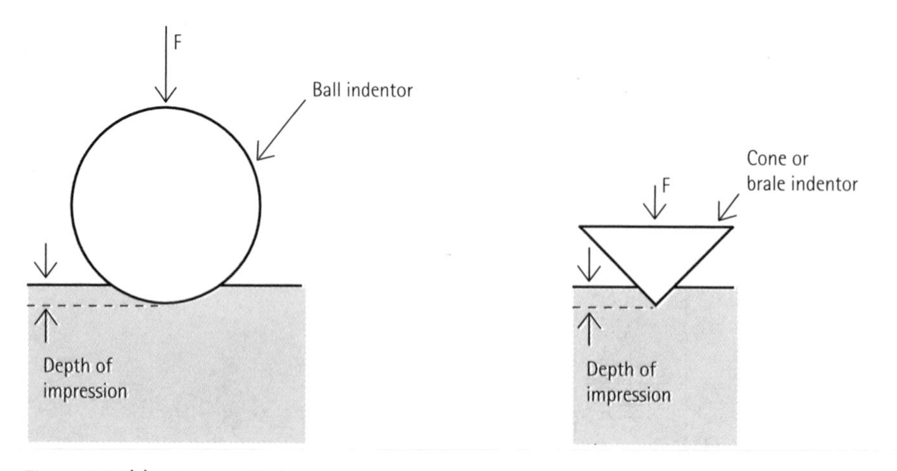

Figure 6.11(e) Rockwell indentor.

The advantages of the Rockwell Test are as follows:

a) It is a quickly made test and can be fitted into a production line, providing quality control on a line of components.
b) The impression size produced is between those of the Vickers and Brinell tests. Some surface irregularity can be accommodated, unlike the Vickers Test, because a minor load is applied initially to locate the indentor. A major load is then applied and the depth of penetration of the indentor is measured under this major loading. Finally the minor loading is removed to complete the test.

It is not as accurate as the Vickers test, which is usually preferred by technologists in research and development work.

A comparison of the most commonly used hardness tests is shown in Table 6.4.

Table 6.4 Comparison of hardness tests.

Test	Indentor	Load	Typical Applications
Brinell (HB)	1–10 mm diameter steel or tungsten carbide ball	Up to 3 000 kg for steel ball, depending upon F/D^2 ratio of material	Forged, rolled, cast components in ferrous and non-ferrous alloys
Vickers (HV)	Square based diamond pyramid	1–120 kg	All metal alloys and ceramics, needs surface preparation
Rockwell B-Scale (HRB)	1⁄16 inch diameter steel ball	10 kg Minor Load 100 kg Major Load	Low-strength steels and non-ferrous up to HV of 240
Rockwell C-Scale (HRC)	Diamond cone or Brale	10 kg Minor Load 150 kg Major Load	All metals with a machined surface finish or equivalent. High strength steels from HV 240–1 000

Hardness correlates well with wear resistance. A material used to crush or grind ore should be very hard to ensure that the material is not eroded or abraded by the hard feed materials. Similarly, gear teeth in the transmission or the drive system of a vehicle should be hard enough that the teeth do not wear out. Typically we find that polymer materials are exceptionally soft, metals have an intermediate hardness, and ceramics are exceptionally hard.

6.7 The Impact Test

When a material is subjected to a sudden, intense blow it often behaves in a more brittle manner than was observed in the tensile test. The main difference between these two tests, as experienced by the material, is the rate of loading. Whereas in the tensile test the tensile load is applied slowly and with gradual increase, in the impact test the full load is applied very rapidly. The strain rate during impact testing is much higher than during tensile testing. The impact test is widely used on steels because it can reveal brittleness under certain conditions that tensile testing fails to show.

Many impact testing methods have been devised, and for polymers are still being devised due to the inappropriateness of those developed mainly for metals. The well established impact tests include the *Charpy* and the *Izod* tests (BS 131) which are schematically shown in Figure 6.12.

Both tests are similar in principle, but differ in practical details and the results are not related. Hence quoted values of impact strength must indicate the test method to be of any value.

In principle, a heavy pendulum starting from rest at an elevation of h_0 swings down in an arc about a pivot reaching its maximum kinetic energy at a point vertically beneath the pivot. In a free swing it then rises, losing kinetic energy and eventually coming to rest at h_f. Note that without any interruption, $h_f = h_0$

When a specimen is placed in the path of the pendulum vertically beneath the pivot, its kinetic energy is absorbed in bending and possibly breaking the specimen. This loss of energy foreshortens the follow through, so $h_f < h_0$ and the difference between h_f and h_0 can be used to calculate the impact absorbed by the specimen. This is read from the direct scale reader and called the impact energy or impact strength. This energy is recorded in Joules, although Izod test values are also recorded in units of $J.m^{-1}$.

The ability of a material to withstand impact loading is often referred to as its toughness. Thus brittleness is revealed by a low impact energy value.

The specimens which may contain notches must be carefully made according to BS specification [Figure 6.12(b)]. The results must state the specimen used because any notch has a great effect on failure. The present tendency is to use the Charpy Test with a V-notch specimen. The advantage of the Charpy Test is that the specimen can be placed in its holder very quickly rather than having to be clamped as in the case in the Izod test. Consequently specimens at low temperatures can be easily tested in the Charpy and this is particularly important with steels testing.

One main feature of impact testing is that the results obtained are prone to scatter, hence several tests are required in order to assess the impact strength of a particular material.

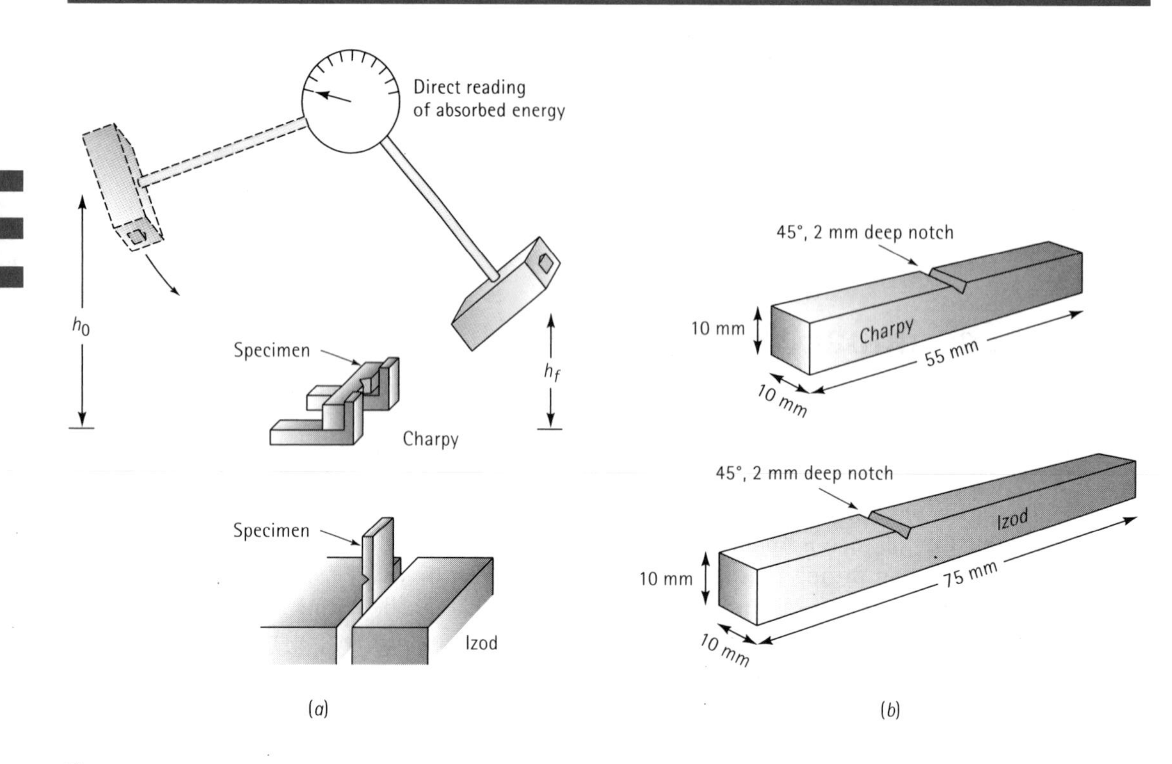

Figure 6.12
The impact test: (a) the Charpy and Izod tests, and (b) dimension of typical specimens.

6.8 Properties Obtained from the Impact Test

The results of a series of impact tests performed at various temperatures are shown in Figure 6.13 for a polymer.

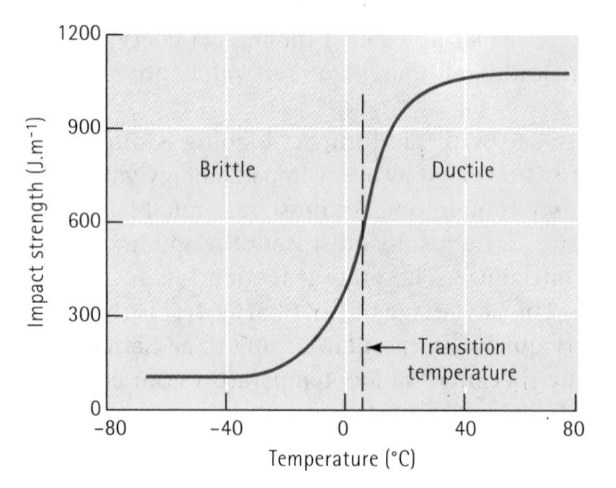

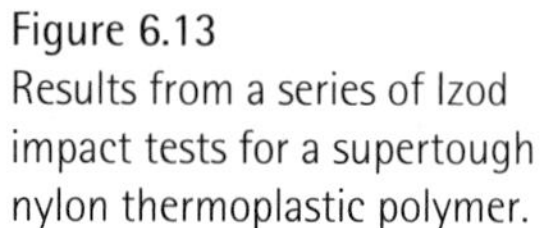

Figure 6.13
Results from a series of Izod impact tests for a supertough nylon thermoplastic polymer.

Transition Temperature The transition temperature is the temperature at which a material changes from ductile to brittle failure. This temperature may be defined by the average energy between the ductile and the brittle regions, at some specific absorbed energy, or by some characteristic fracture appearance. A material subjected to an impact blow during service should have a transition temperature *below* the temperature of the material's surroundings.

Not all materials have a distinct transition temperature (Figure 6.14). BCC metals have transition temperatures, but most FCC metals do not. FCC metals have high absorbed energies, with the energy decreasing gradually and sometimes even increasing as the temperature decreases.

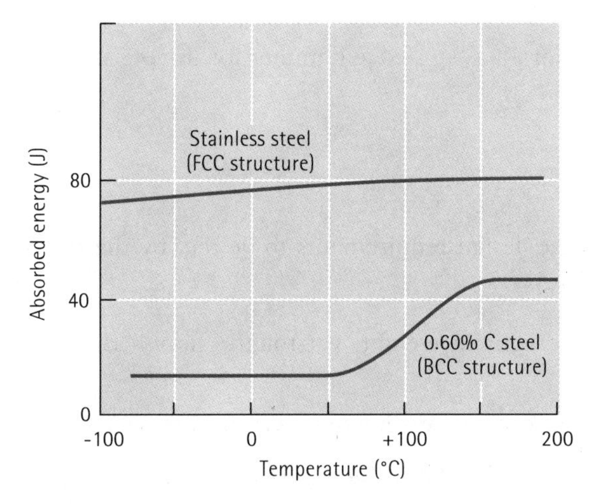

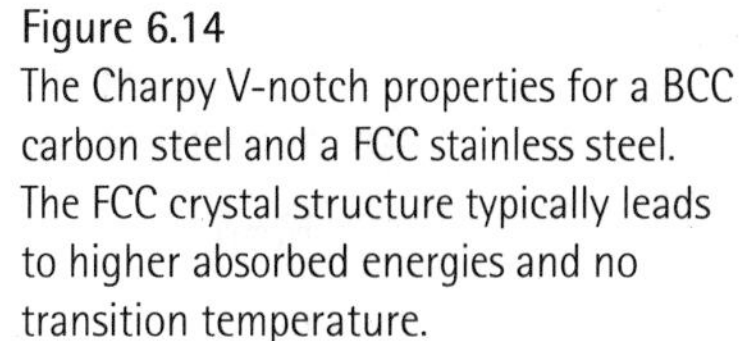
Figure 6.14
The Charpy V-notch properties for a BCC carbon steel and a FCC stainless steel. The FCC crystal structure typically leads to higher absorbed energies and no transition temperature.

Notch Sensitivity Notches caused by poor machining, fabrication, or design concentrate the stresses and reduce the toughness of materials. The notch sensitivity of a material can be evaluated by comparing the absorbed energies of notched versus unnotched specimens. The absorbed energies are much lower in notched specimens if the material is notch-sensitive.

Relationship to the Stress–Strain Diagram The energy required to break a material is related to the area contained within the true stress–true strain diagram (Figure 6.15). Metals with both high strength and high ductility have good toughness. Ceramics and many composites, on the other hand, have poor toughness, even though they have high strength, because they display virtually no ductility.

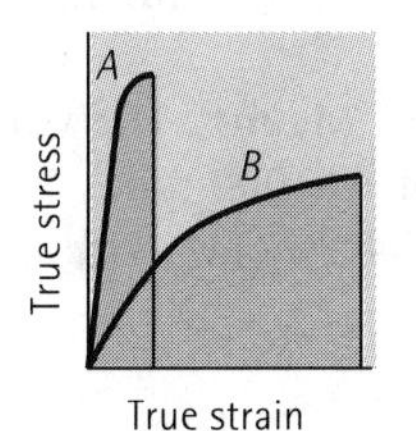

Figure 6.15
The area contained within the true stress–true strain curve is related to the impact energy. Although material *B* has a lower yield strength, it absorbs a greater energy than material *A*.

Use of Impact Properties Absorbed energy and transition temperature are very sensitive to loading conditions. For example, a higher rate of applying energy to the specimen reduces the absorbed energy and increases the transition temperature. The size of the specimen also affects the results; because it is more difficult for a thick material to deform, smaller energies are required to break thicker materials. Finally, the configuration of the notch affects the behaviour; a surface crack permits lower absorbed energies than does a V-notch. Because we often cannot predict or control all of these conditions, the impact test is best used for comparison and selection of materials.

EXAMPLE 6.7 Materials Selection for a Sledge Hammer

Select suitable materials for a 3.5 kg sledge hammer for driving steel fence posts into the ground.

SOLUTION

First we must consider the design requirements to be met by the sledge hammer. A partial list would include:

1. The handle should be light in weight, yet tough enough that it will not catastrophically break.
2. The head must not break or chip during use, even in subzero temperatures.
3. The head must not deform during continued use.
4. The sledge hammer should be inexpensive.

Although the handle could be a lightweight, tough composite material (such as a polymer reinforced with Kevlar fibres), a wooden handle about 750 mm long would be much less expensive and would still provide sufficient toughness. As shown in later chapter, wood can be categorised as a natural fibre-reinforced composite.

To produce the head, we prefer a material that has a low transition temperature, can absorb relatively high energy during impact, yet also has enough hardness to avoid deformation. The toughness requirement would rule out most ceramics. A face-centred cubic metal, such as FCC stainless steel or copper, might provide superior toughness even at low temperatures; however, these metals are relatively soft and expensive. An appropriate choice might be a normal BCC steel. Ordinary steel are inexpensive, have good hardness and strength, and have sufficient toughness at low temperatures.

In Appendix A, we find that the density of iron is 7.87 $Mg.m^{-3}$. The volume of steel required is $V = (3.5 \text{ kg})/(7\,870 \text{ kg.m}^{-3}) = 4.4\times10^{-4} \text{ m}^{-3}$ or 440 000 mm^3. To ensure that we will hit our target, the head might have a cylindrical shape, with a diameter of 65 mm. The length of the head would then be 133 mm.

6.9 Fracture Toughness

Fracture mechanics is the discipline concerned with the behaviour of materials containing cracks or other small flaws. All materials, of course, contain some flaws. What we wish to know is the maximum stress that a material can withstand if it contains flaws of a certain size and geometry. Fracture toughness measures the ability of a material containing a flaw to withstand an applied load. Unlike the results of an impact test, fracture toughness is a quantitative property of the material.

A typical fracture toughness test may be performed by applying a tensile stress to a specimen prepared with a flaw of known size and geometry (Figure 6.16). The stress applied to the material is intensified at the flaw, which acts as a *stress raiser*. For a simple test, the *stress intensity factor* K is

$$K = f\sigma\sqrt{\pi a} \tag{6.14}$$

where f is a geometry factor for the specimen and flaw, σ is the applied stress, and a is the flaw size (as defined in Figure 6.16). If the specimen is assumed to have an 'infinite' width then $f \cong 1.0$.

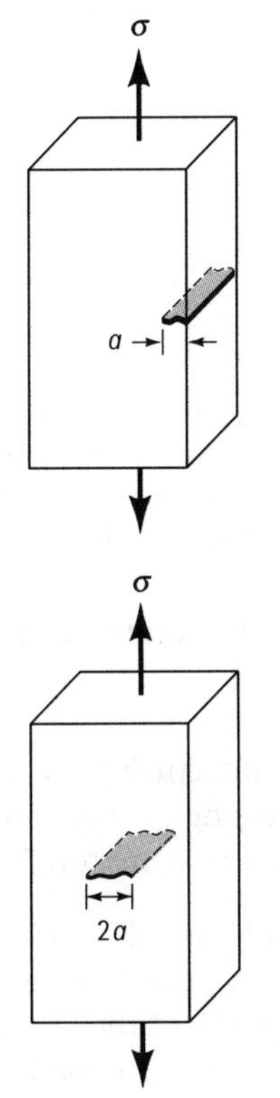

Figure 6.16
Schematic drawing of fracture toughness specimens with edge and internal flaws.

By performing a test on a specimen with a known flaw size, we can determine the value of K that causes the flaw to grow and cause failure. This critical stress intensity factor is defined as the *fracture toughness* K_c:

$$K_c = K \text{ required for a crack to propagate} \tag{6.15}$$

Fracture toughness depends on the thickness of the sample: as thickness increases, fracture toughness K_c decreases to a constant value (Figure 6.17). This constant is called the *plane strain fracture toughness* K_{Ic}. It is K_{Ic} that is normally reported as the property of a material. Table 6.5 compares the value of K_{Ic} to the yield strength of several materials. Units for fracture toughness are MPa.m$^{1/2}$ or MN.m$^{-3/2}$.

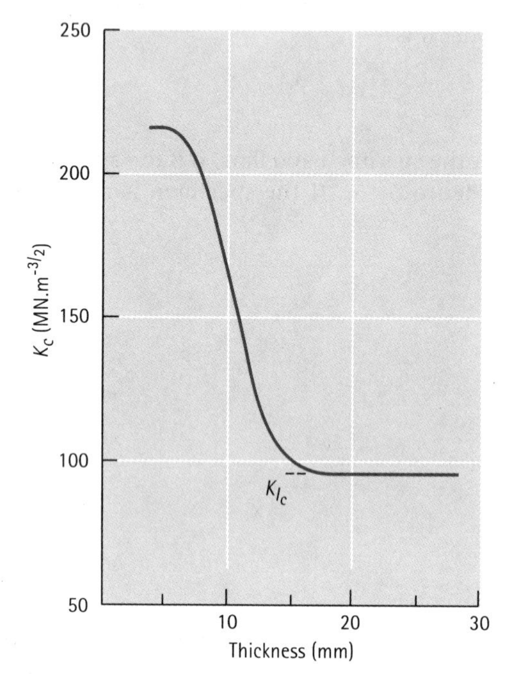

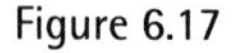

Figure 6.17

The fracture toughness K_c of a 2070 MN.m^{-2} yield strength steel decreases with increasing thickness, eventually levelling off at the plain strain fracture toughness K_{Ic}.

The ability of a material to resist the growth of a crack depends on a large number of factors:

1. Larger flaws reduce the permitted stress. Special manufacturing techniques, such as filtering impurities from liquid metals and hot pressing of particles to produce ceramic components, can reduce flaw size and improve fracture toughness.

2. The ability of a material to deform is critical. In ductile metals, the material near the tip of the flaw can deform, causing the tip of any crack to become blunt, reducing the stress intensity factor, and preventing growth of the crack. Increasing the strength of a given metal usually decreases ductility and gives a lower fracture toughness. (*See* Table 6.5.) Brittle materials such as ceramics and many polymers have much lower fracture toughness than metals.

3. Thicker, more rigid materials have a lower fracture toughness than thin materials.

Table 6.5 The plane strain fracture toughness K_{Ic} of selected materials.

Material	Fracture Toughness K_{Ic} $MN.m^{-3/2}$	Yield Strength $MN.m^{-2}$
Al-Cu alloy	24	455
	36	325
Ti-6% Al-4% V	55	900
	99	860
Ni-Cr steel	50	1640
	88	1420
Al_2O_3	1.76	210
Si_3N_4	5	550
Transformation toughened ZrO_2	11	415
Si_3N_4-SiC composite	56	825
Polymethyl methacrylate polymer	1	30
Polycarbonate polymer	3.3	60

4. Increasing the rate of application of the load, such as in an impact test, typically reduces the fracture toughness of the material.
5. Increasing the temperature normally increases the fracture toughness, just as in the impact test.
6. A small grain size normally improves fracture toughness, whereas more point defects and dislocations reduce fracture toughness. Thus, a fine-grained ceramic material may provide improved resistance to crack growth.

6.10 The Importance of Fracture Mechanics

The fracture mechanics approach allows us to design and select materials while taking into account the inevitable presence of flaws. There are three variables to consider: the property of the material (K_c or K_{Ic}), the stress σ that the material must withstand, and the size of the flaw a. If we know two of these variables, the third can be determined.

Selection of a Material If we know the maximum size a of flaws in the material and the magnitude of the applied stress, we can select a material that has a fracture toughness K_c or K_{Ic} large enough to prevent the flaw from growing.

Design of a Component If we know the maximum size of any flaw and the material (and therefore its K_c or K_{Ic}) has already been selected, we can calculate the maximum stress that the component can withstand. Then we can design the appropriate size of the part to ensure that the maximum stress is not exceeded.

Design of a Manufacturing or Testing Method If the material has been selected, the applied stress is known, and the size of the component is fixed, we can calculate the maximum size of a flaw that can be tolerated. A nondestructive testing technique that detects any flaw greater than this critical size can help ensure that the part will function safely. In addition, we find that, by selecting the correct manufacturing process, we can produce flaws that are smaller than this critical size.

EXAMPLE 6.8 **Selection of a Nondestructive Test**

A large steel plate used in a nuclear reactor has a plane strain toughness of 88 $MN.m^{-3/2}$ and is exposed to a stress of 310 $MN.m^{-2}$ during service. Select a testing or inspection procedure capable of detecting a crack at the surface of the plate before the crack is likely to grow at a catastrophic rate.

SOLUTION

We need to determine the minimum size of a crack that will propagate in the steel under these conditions. From Equation 6.14, assuming that $f = 1$:

$$K_{Ic} = f\sigma\sqrt{a\pi}$$
$$88 = (1)(310)\sqrt{a\pi}$$
$$a = 2.56 \times 10^{-2}\ \text{m} = 25.6\text{mm}$$

A 25.6 mm deep crack on the surface should be relatively easy to detect. Often, cracks of this size can be observed visually. A variety of other tests, such as dye penetrant inspection, magnetic particle inspection, and eddy current inspection, also detect cracks much smaller than this. If the growth rate of a crack is slow and inspection is performed on a regular basis, a crack should be discovered long before reaching this critical size. These tests are discussed in Chapter 23.

6.11 The Fatigue Test

A component is often subjected to the repeated application of a stress below the yield strength of the material. This cyclical stress may occur as a result of rotation, bending, or vibration. Even though the stress is below the yield strength, the material may fail after a large number of applications of the stress. This mode of failure is fatigue

Fatigue failures typically occur in three stages. First, a tiny crack initiates at the surface, often at a time well after loading begins. Next, the crack gradually propagates as the load continues to cycle. Finally, sudden fracture of the material occurs when the remaining cross-section of the material is too small to support the applied load.

A common method to measure a material's resistance to fatigue is the rotating cantilever beam test (Figure 6.18). One end of a machined, cylindrical specimen is mounted in a motor-driven chuck. A weight is suspended from the opposite end. The specimen initially has a tensile force acting on the top surface, while the bottom surface is compressed. After the specimen turns 90°, the locations that were originally in tension and compression have no stress acting on them. After a half revolution of 180°, the material that was originally in tension is now in compression. Thus, the stress at any one point goes through a complete sinusoidal cycle from maximum tensile stress to maximum compressive stress. The maximum stress acting on this type of specimen is given by

$$\sigma = \frac{10.18lF}{d^3} \tag{6.16}$$

where l is the length of the bar, F is the load, and d is the diameter.

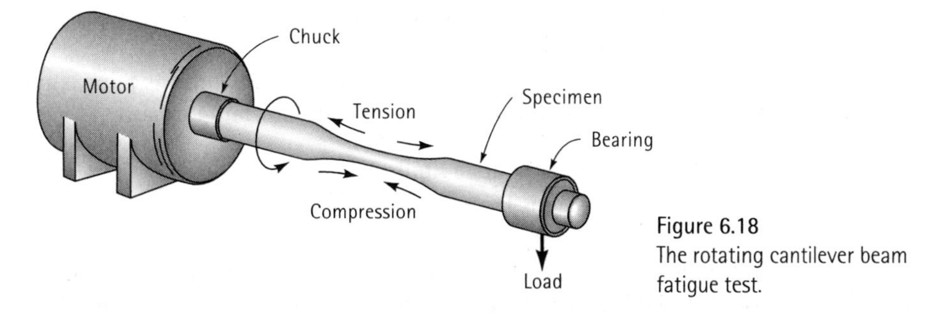

Figure 6.18
The rotating cantilever beam fatigue test.

After a sufficient number of cycles, the specimen may fail. Generally, a series of specimens are tested at different applied stresses. The results are presented as an S-N curve, with the stress (S) plotted versus the number of cycles (N) to failure (Figure 6.19).

Fatigue testing always produces results which show scatter and so statistical analysis is required before the S-N curve can be plotted as shown in Figure 6.19.

The large range of N-values means that to plot an S-N curve, a logarithmic scale is applied to the N-values. This affects the shape of the curve.

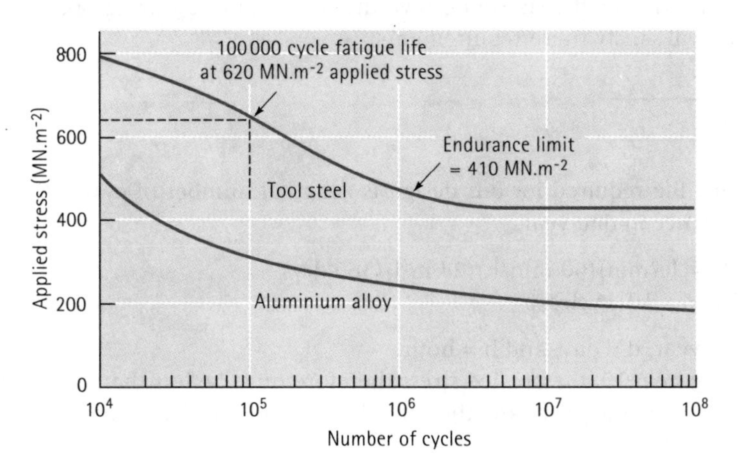

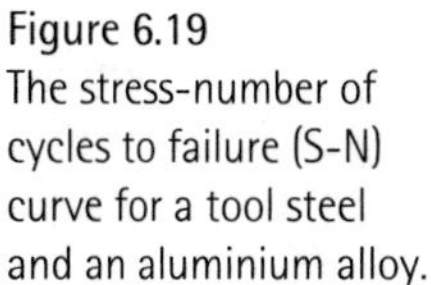

Figure 6.19
The stress-number of cycles to failure (S-N) curve for a tool steel and an aluminium alloy.

6.12 Results of the Fatigue Test

The fatigue test can tell us how long a part may survive or the maximum allowable loads that can be applied to prevent failure.

The endurance limit (also called the fatigue limit), which is the stress below which there is a 50% probability that failure by fatigue will never occur, is our preferred design criterion. To prevent a tool steel part from failing (Figure 6.19), we must be sure that the applied stress is below 410 $MN.m^{-2}$.

Fatigue life tells us how long a component survives at a particular stress. For example, if the tool steel (Figure 6.19) is cyclically subjected to an applied stress of 620 $MN.m^{-2}$, the fatigue life will be 100 000 cycles. Fatigue strength is the maximum stress for which fatigue will not occur within a particular number of cycles, such as 500 000 000. The fatigue strength is necessary for designing with aluminium and polymers, which have no endurance limit.

In some materials, including steels, the endurance limit is approximately half the tensile strength. The ratio is the endurance ratio, or fatigue ratio:

$$\text{Endurance ratio} = \frac{\text{endurance limit}}{\text{tensile strength}} \approx 0.5 \qquad (6.17)$$

The endurance ratio allows us to estimate fatigue properties from the tensile test.

Most materials are *notch sensitive*, with the fatigue properties particularly sensitive to flaws at the surface. Design or manufacturing defects concentrate stresses and reduce the endurance limit, fatigue strength, or fatigue life. Sometimes highly polished surfaces are prepared in order to minimise the likelihood of a fatigue failure.

EXAMPLE 6.9 **Design of a Rotating Shaft**

A solid shaft for a cement kiln produced from the tool steel in Figure 6.19 must be 2.44 m long and must survive continuous operation for one year with an applied load of 0.05 MN. The shaft makes one revolution per minute during operation. Design a shaft that will satisfy these requirements.

SOLUTION

The fatigue life required for our design is the total number of cycles N that the shaft will experience in one year:

$$N = (1 \text{ cycle/min})(60 \text{ min/h})(24 \text{ h/d})(365 \text{ d/y})$$
$$N = 5.256 \times 10^5 \text{ cycles/y}$$

where y = year, d = day, and h = hour.

From Figure 6.19, the applied stress therefore must be less than about 500 $MN.m^{-2}$. If Equation 6.16 is appropriate, then the diameter of the shaft must be:

$$\sigma = \frac{10.18lF}{d^3}$$

$$500\ \text{MN.m}^{-2} = \frac{(10.18)(2.44\ \text{m})(0.05\ \text{MN})}{d^3}$$

$$d = 0.136\ \text{m} = 136\ \text{mm}$$

A shaft with a diameter of 136 mm should operate for one year under these conditions. However, a significant margin of safety might be incorporated in the design. In addition, we might consider producing a shaft that would never fail. From Figure 6.19, the endurance limit is 410 MN.m^{-2}. The minimum diameter required to prevent failure would now be:

$$410 = \frac{(10.18)(2.44)(0.05)}{d^3}$$

$$d = 0.145\ \text{m} = 145\ \text{mm}$$

Selection of only a slightly larger shaft makes fatigue less likely to occur.

Other considerations might, of course, be important. High temperatures and corrosive conditions are inherent in producing cement. If the shaft is heated or attacked by the corrosive environment, fatigue is accelerated.

6.13 Application of Fatigue Testing

Material components are often subjected to loading conditions that do not give equal stresses in tension and compression (Figure 6.20). For example, the maximum stress during compression may be less than the maximum tensile stress. In other cases, the loading may be between a maximum and a minimum tensile stress; here the S-N curve is presented as the stress amplitude versus number of cycles to failure. *Stress amplitude* (σ_a) is defined as half of the difference between the maximum and minimum stresses; *mean stress* (σ_m) is defined as the average between the maximum and minimum stresses:

$$\sigma_a = \frac{\sigma_{max} - \sigma_{min}}{2} \tag{6.18}$$

$$\sigma_m = \frac{\sigma_{max} + \sigma_{min}}{2} \tag{6.19}$$

A compressive stress is conventionally written as a 'negative' stress. Thus, if the maximum tensile stress is 50 MN.m^{-2} and the minimum stress is a 10 MN.m^{-2} compressive stress, the stress amplitude is [50 – (–10)]/2 = 30 MN.m^{-2} and the mean stress is [50 + (–10)]/2 = 20 MN.m^{-2}.

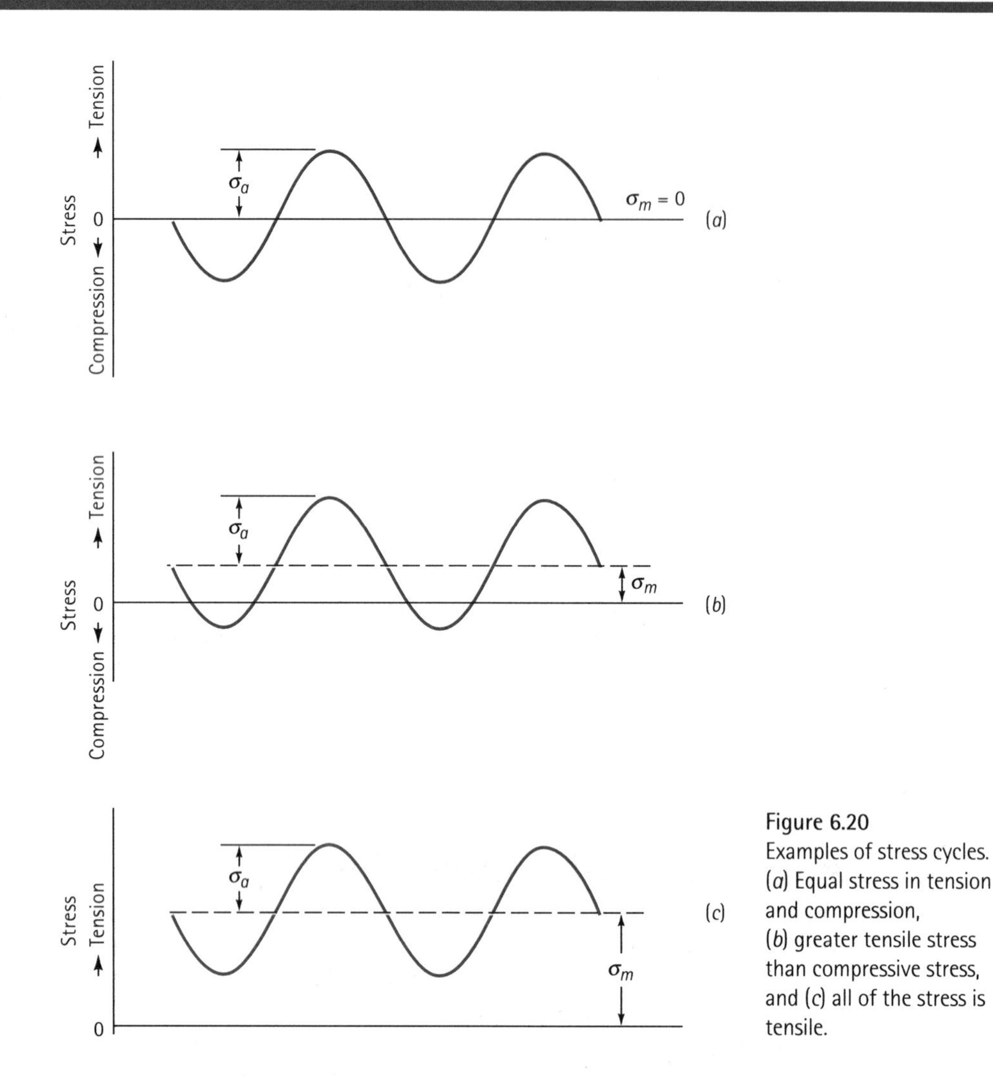

Figure 6.20
Examples of stress cycles. (a) Equal stress in tension and compression, (b) greater tensile stress than compressive stress, and (c) all of the stress is tensile.

As the mean stress increases, the stress amplitude must decrease to order for the material to withstand the applied stresses. This condition can be summarised by the Goodman relationship:

$$\sigma_a = \sigma_{fs}[1 - (\sigma_m/\sigma_T)] \qquad (6.20)$$

where σ_{fs} is the desired fatigue strength and σ_T is the tensile strength of the material. Therefore, in a typical rotating beam fatigue test, where the mean stress is zero, a relatively large stress amplitude can be tolerated without fatigue. If, however, an aircraft wing is loaded near its yield strength (high σ_m), vibrations of even a small amplitude may cause a fatigue crack to initiate and grow.

Crack Growth Rate In many cases, a component may not be in danger of failure even when a crack is present. To estimate when failure might occur, the rate of propagation of a crack becomes important. Figure 6.21 shows crack growth rate versus the range of stress intensity factor ΔK, which characterises crack geometry and the stress amplitude. Below a threshold ΔK, a crack does not grow; for somewhat higher stress-intensities, cracks grow slowly; and at still higher stress-intensities, a crack grows at a rate given by:

$$\frac{da}{dN} = C(\Delta K)^n \tag{6.21}$$

Finally, when ΔK is still higher, cracks grow in a rapid and unstable manner until fracture occurs.

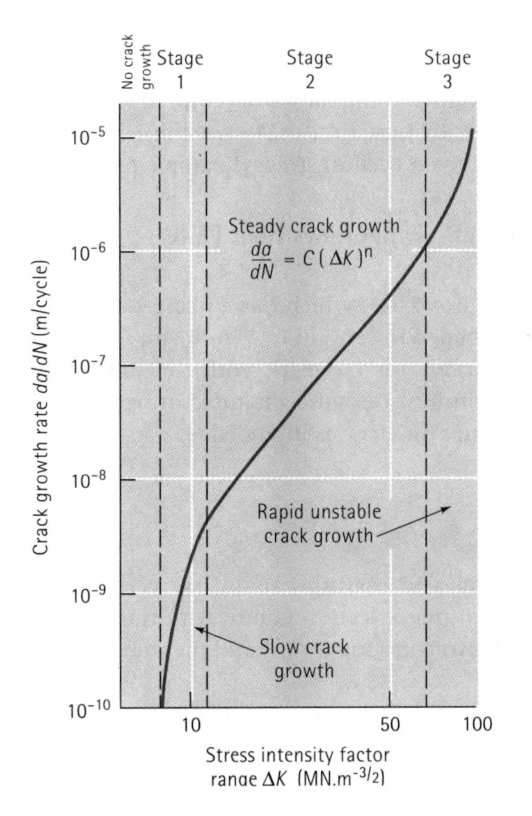

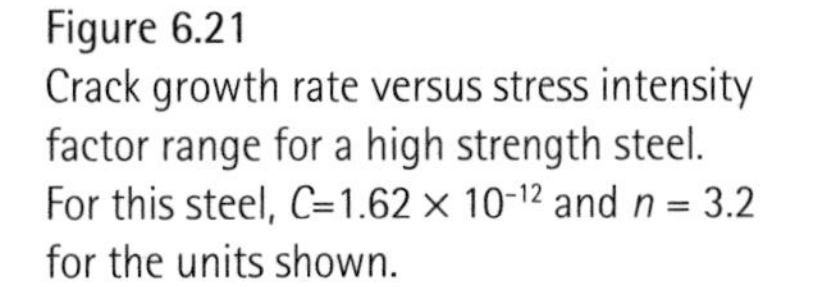

Figure 6.21
Crack growth rate versus stress intensity factor range for a high strength steel. For this steel, $C=1.62 \times 10^{-12}$ and $n = 3.2$ for the units shown.

The rate of crack growth increases as a crack increases in size, as predicted from stress intensity factor (Equation 6.14):

$$\Delta K = K_{max} - K_{min} = f\sigma_{max}\sqrt{\pi a} - f\sigma_{min}\sqrt{\pi a} = f\Delta\sigma\sqrt{\pi a} \tag{6.22}$$

If the cyclical stress $\Delta\sigma$ is not changed, than as crack length a increases, ΔK and the crack growth rate da/dN increase. In using this expression, however, one should note that a crack will not propagate during compression. Therefore, if σ_{min} is compressive or less than zero, then σ_{min} should be set equal to zero.

Knowledge of crack growth rate is of assistance in designing components and in nondestructive evaluation to determine if a crack poses imminent danger to the structure. One approach to this problem is to estimate the number of cycles required before failure occurs. By rearranging Equation 6.21 and substituting for ΔK:

$$dN = \frac{1}{Cf^n \Delta\sigma^n \pi^{n/2}} \frac{da}{a^{n/2}}$$

If we integrate this expression between the initial size of a crack and the crack size required for fracture to occur, we find that

$$N = \frac{2[(a_c)^{(2-n)/2} - (a_i)^{(2-n)/2}]}{(2-n)Cf^n \Delta\sigma^n \pi^{n/2}} \qquad (6.23)$$

where a_i is the initial flaw size and a_c is the flaw size required for fracture. If we know the material constants n and C in Equation 6.21, we can estimate the number of cycles required for failure for a given cyclical stress (Example 6.10).

EXAMPLE 6.10 Design of a Fatigue-Resistant Plate

A high-strength steel plate (Figure 6.21), which has a plane strain fracture toughness of 80 MN.m$^{-3/2}$, is alternately loaded in tension to 550 MN.m^{-2} and in compression to 60 MN.m^{-2}. The plate is to survive for 10 years, with the stress being applied at a frequency of once every 10 minutes. Design a manufacturing and testing procedure that ensures that the component will serve as intended.

SOLUTION

To design our manufacturing and testing capability, we must determine the maximum size of any flaws that might lead to failure within the 10-year period. The critical crack size (a_c), using the fracture toughness and the maximum stress, is:

$$K_{Ic} = f\sigma\sqrt{\pi a_c}$$

$$80 \text{ MN.m}^{-3/2} = (1)(550 \text{ MN.m}^{-2})\sqrt{\pi a_c}$$

$$a_c = 6.7 \times 10^{-3} \text{ m} = 6.7 \text{ mm}$$

The maximum stress is 550 MN.m^{-2}; however, the minimum stress is zero, not 60 MN.m^{-2} in compression, because cracks do not propagate in compression. Thus, $\Delta\sigma$ is:

$$\Delta\sigma = \sigma_{max} - \sigma_{min} = 550 - 0 = 550 \text{ MN.m}^{-2}$$

We need to determine the minimum number of cycles that the plate must withstand:

$$N = (1 \text{ cycle}/10 \text{ min})(60 \text{ min/h})(24 \text{ h/d})(365 \text{ d/y})(10 \text{ y})$$

$$N = 525\,600 \text{ cycles}$$

If we assume that $f = 1$ for all crack lengths and note that $C = 1.62 \times 10^{-12}$ and $n = 3.2$ in Equation 6.23, then;

$$525\,600 = \frac{2[(6.7\times10^{-3})^{(2-3.2)/2} - a_i^{(2-3.2)/2}]}{(2-3.2)(1.62\times10^{-12})(1)^{3.2}(550)^{3.2}\pi^{3.2/2}}$$

$$525\,600 = \frac{2[(0.0067)^{-0.6} - a_i^{-0.6}]}{(-1.2)(1.62\times10^{-12})(1)(550)^{3.2}\pi^{1.6}}$$

$$\frac{52560}{2} = \frac{[20.15 - a_i^{-0.6}]}{-7.13\times10^{-3}}$$

$$-187.47 = 20.15 - a_i^{-0.6}$$

$$a_i^{-0.6} = 207.6$$

$$a_i = 1.37\times10^{-4}\text{m}$$

$$a_i = 0.137 \text{ mm for surface flaws}$$

$$2a_i = 0.275 \text{ mm for internal flaws}$$

The manufacturing process must produce surface flaws smaller than 0.1307 mm in length. In addition, nondestructive tests must be available to ensure that cracks exceeding this length are not present.

Effect of Temperature As the material's temperature increases, both fatigue life and endurance limit decrease. Furthermore, a cyclical temperature change encourages failure by thermal fatigue; when the material heats in a nonuniform manner, some parts of the structure expand more than others. This nonuniform expansion introduces a stress within the material, and when the structure later cools and contracts, stresses of the opposite sign are imposed. As a consequence of the thermally induced stresses and strains, fatigue may eventually occur.

The frequency with which the stress is applied also influences fatigue behaviour. In particular, high frequency stresses may cause polymer materials to heat; at increased temperature, polymers fail more quickly.

6.14 The Creep Test

If we apply stress to a material at an elevated temperature, the material may stretch and eventually fail, even though the applied stress is *less* than the yield strength at that temperature. Plastic deformation at high temperatures is known as creep.

To determine the creep characteristics of a material, a constant stress is applied to a heated specimen in a creep test. As soon as the stress is applied, the specimen stretches elastically a small amount ε_0 (Figure 6.22), depending on the applied stress and the modulus of elasticity of the material at the high temperature.

Dislocation Climb High temperatures permit dislocations in a metal to climb. In climb, atoms move either to or from the dislocation line by diffusion, causing the dislocation to move in a direction that is perpendicular, not parallel, to the slip plane (Figure 6.23). The dislocation escapes from lattice imperfections, continues to slip, and causes additional deformation of the specimen even at low applied stresses.

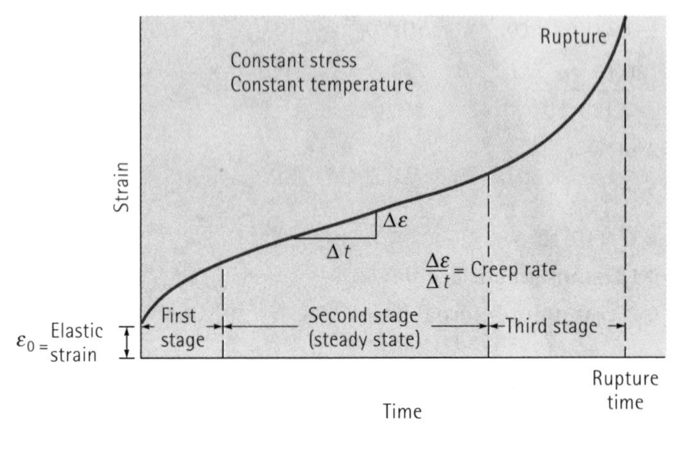

Figure 6.22
A typical creep curve showing the strain produced as a function of time for a constant stress and temperature.

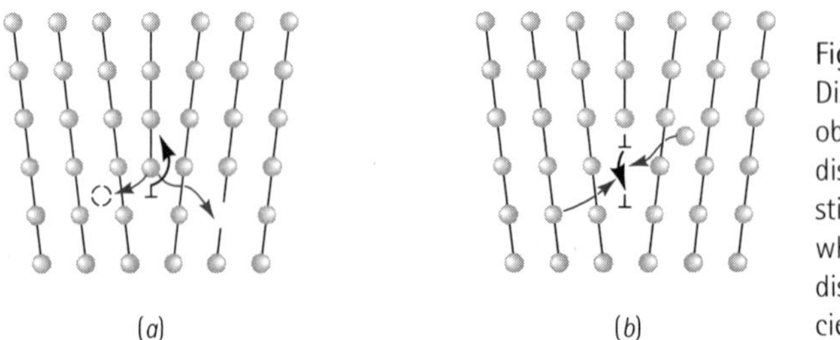

Figure 6.23
Dislocation can climb away from obstacles when atoms leave the dislocation line to create interstitials or to fill vacancies (*a*) or when atoms are attached to the dislocation line by creating vacancies or eliminating interstitials (*b*).

Creep Rate and Rupture Times During the creep test, strain or elongation is measured as a function of time and plotted to give the creep curve (Figure 6.22). In the first stage of creep of metals (primary creep), many dislocations climb away from obstacles slip, and contribute to deformation. Eventually, the rate at which dislocations climb away from obstacles equals the rate at which dislocations are blocked by other imperfections. This leads to second-stage, or steady-state, creep (secondary creep). The slope of the steady-state portion of the creep curve is the creep rate:

$$\text{Creep rate} = \frac{\Delta \text{ strain}}{\Delta \text{ time}} \tag{6.24}$$

Eventually, during third-stage creep (tertiary creep), necking begins, the stress increases, and the specimen deforms at an accelerated rate until failure occurs. The time required for failure to occur is the rupture time. Either a higher stress or a higher temperature reduces the rupture time and increases the creep rate (Figure 6.24).

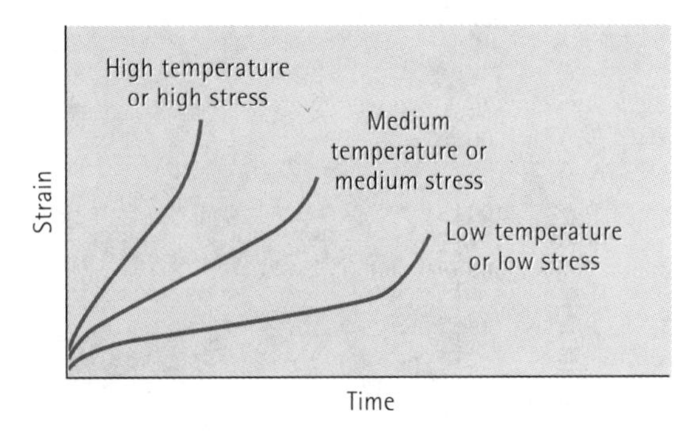

Figure 6.24
The effect of temperature or applied stress on the creep curve.

The combined influence of applied stress and temperature on the creep rate and rupture time (t_r) follows an Arrhenius relationship:

$$\text{Creep rate} = C\sigma^n \exp(-Q_c/RT) \tag{6.25}$$
$$t_r = K\sigma^m \exp(Q_m/RT) \tag{6.26}$$

where R is the gas constant, T is the temperature in Kelvin, C, K, n, and m are constants for the material. Q_c is the activation energy for creep, and Q_r is the activation energy for rupture. In particular, Q_c is related to the activation energy for self-diffusion when dislocation climb is important.

In crystalline ceramics, other factors – including grain boundary sliding and nucleation of microcracks – are particularly important. Often, a noncrystalline or glassy material is present at the grain boundaries; the activation energy required for a glass to deform is low, leading to high creep rates compared with completely crystalline ceramics. For the same reason, creep occurs at a rapid rate in ceramic glasses and amorphous polymers.

6.15 Use of Creep Data

The stress-rupture curves shown in Figure 6.25(a) permit us to estimate the expected lifetime of a component for a particular combination of stress and temperature. The Larson-Miller parameter, illustrated in Figure 6.25(b), is used to consolidate the stress-temperature-rupture time relationship into a single curve. The Larson-Miller parameter (L.M.) is:

$$\text{L.M.} = (T/1\,000)(A + B \ln t) \tag{6.27}$$

where T is in Kelvin, t is the time in hours, and A and B are constants for the material.

EXAMPLE 6.11 Design of Links for a Chain

Design an S.G. cast iron chain (Figure 6.26) to operate in a furnace used to fire ceramic bricks. The furnace is to operate without rupturing for 5 years at 600°C, with an applied load of 22 kN.

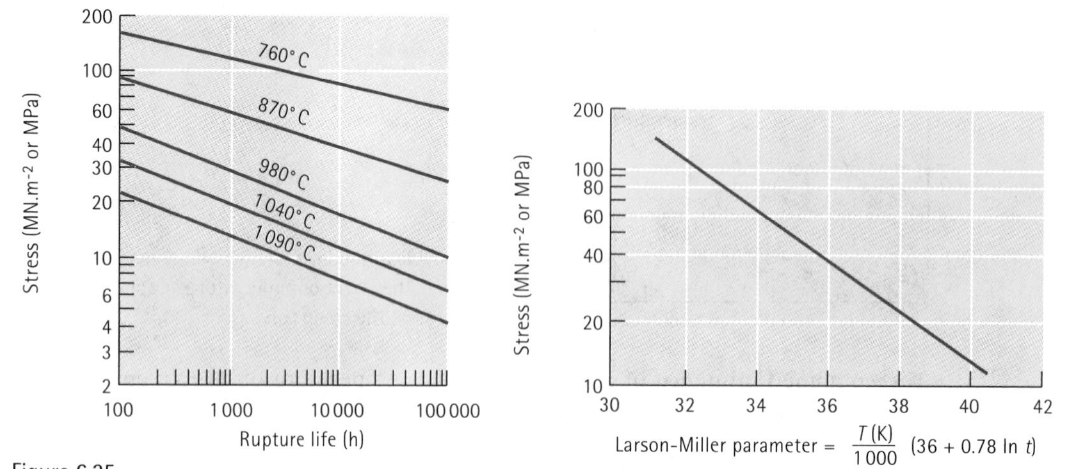

Figure 6.25
Results from a series of creep tests: (*a*) stress-rupture curves for an iron-chromium-nickel alloy and (*b*) the Larson-Miller parameter for ductile cast iron.

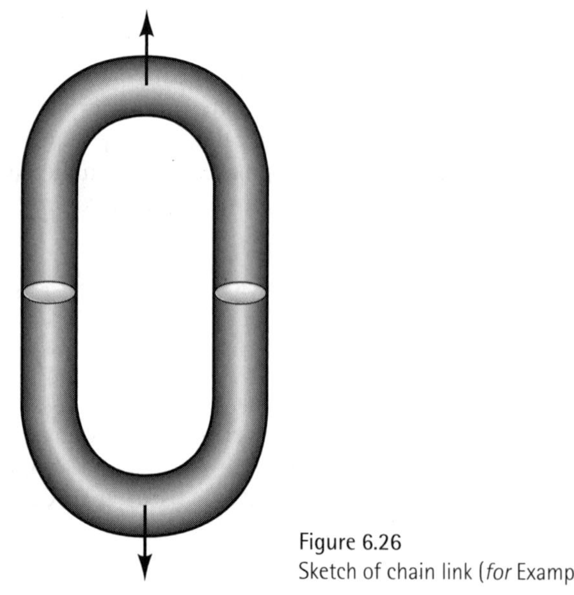

Figure 6.26
Sketch of chain link (*for* Example 6.11).

SOLUTION

The Larson-Miller parameter for S.G. cast iron is:

$$\text{L.M.} = \frac{T(36 + 0.78 \ln t)}{1000}$$

The chain is to survive 5 years, or:

$$t = (24\ \text{h/d})(365\ \text{d/y})(5\ \text{y}) = 43\,800\ \text{h}$$

$$\text{L.M.} = \frac{(600 + 273)[36 + 0.78\ \ln(43\,800)]}{1000} = 38.7$$

From Figure 6.25(b), the applied stress must be no more than 20 $MN.m^{-2}$.

The total cross-sectional area of the chain required to support the 22 kN (0.022 MN) load is:

$$A = F/\sigma = \frac{0.022\ \text{MN}}{20\ \text{MN.m}^{-2}} = 0.0011\ \text{m}^2 = 1100\ \text{mm}^2$$

The cross-sectional area of each 'half' of the iron link is then 550 mm^2 and, assuming a round cross-section:

$$d^2 = (4/\pi)A = (4/\pi)(550) = 700\ \text{mm}^2$$
$$d = 26.5\ \text{mm}$$

SUMMARY

The mechanical behaviour of materials is described by their mechanical properties, which are the results of idealised, simple tests. These tests are designed to represent different types of loading conditions. The properties of a material reported in various handbooks are the results of these tests. Consequently, we should always recall that handbook values are average results obtained from idealised tests and, must be used with some care.

- The tensile test describes the resistance of a material to a slowly applied stress. Important properties include yield strength (the stress at which the material begins to permanently deform), tensile strength (the stress corresponding to the maximum applied load), modulus of elasticity (the slope of the elastic portion of the stress-strain curve), and % Elongation and % Reduction in area (both, measures of the ductility of the material).
- The bend test is used to determine the tensile properties of brittle materials. A modulus of elasticity and a flexural strength (similar to a tensile strength) can be obtained.
- The hardness test measures the resistance of a material to penetration and provides a measure of the wear and abrasion resistance of the material. A number of hardness tests, including the Rockwell, Brinell and Vickers tests, are commonly used.
- The impact test describes the response of a material to a rapidly applied load. The Charpy and Izod tests are typical. The energy required to fracture the specimen is measured and can be used as the basis for comparison of various materials tested under the same conditions. In addition, a transition temperature above which the material fails in a ductile rather than a brittle manner can be determined.

- Fracture toughness describes how easily a crack or flaw in a material propagates. The plane strain fracture toughness K_{Ic} is a common result of these tests.
- The fatigue test permits us to understand how a material performs when a cyclical stress is applied. Important properties include fatigue or endurance limit (below which failure will not occur), fatigue strength (the maximum stress that will prevent failure within a given number of cycles), and fatigue life (the number of cycles a material will survive at a given stress). Knowledge of the rate of crack growth can help determine fatigue life.
- The creep test provides information on the load-carrying ability of a material at high temperatures. Creep rate and rupture time are important properties obtained from these tests.

GLOSSARY

Bend test
Application of a force to the centre of a bar that is supported on each end to determine the resistance of the material to a static or slowly applied load. Typically used for brittle materials.

Climb
Movement of a dislocation perpendicular to its slip plane by diffusion of atoms to or from the dislocation line.

Creep rate
The rate at which a material deforms when a stress is applied at a high temperature.

Creep test
Measures the resistance of a material to deformation and failure when subjected to a static load below the yield strength at an elevated temperature.

Ductility
The ability of a material to be permanently deformed without breaking when a force is applied.

Elastic deformation
Deformation of the material that is recovered when the applied load is removed.

% Elongation
The total percentage increase in the length of a specimen during a tensile test.

Endurance ratio
The endurance limit divided by the tensile strength of the material. The ratio is about 0.5 for many ferrous metals.

Engineering strain
The amount that a material deforms per unit length in a tensile test.

Engineering stress
The applied load, or force, divided by the original cross-sectional area of the material.

Fatigue life
The number of cycles permitted at a particular stress before a material fails by fatigue.

Fatigue limit or Endurance limit
The stress below which a material will not fail in a fatigue test.

Fatigue strength
The stress required to cause failure by fatigue in a given number of cycles, such as 500 million cycles.

Fatigue test
Measures the resistance of a material to failure when a stress below the yield strength is repeatedly applied.

Flexural modulus
The modulus of elasticity calculated from the results of a bend test, giving the slope of the stress-deflection curve.

Flexural strength
The stress required to fracture a specimen in a bend test. Also called the modulus of rupture.

Fracture mechanics
The study of a material's ability to withstand stress in the presence of a flaw.

Fracture toughness
The resistance of a material to failure in the presence of a flaw.

Hardness test
Measures the resistance of a material to penetration by a sharp object. Common hardness tests include the Brinell test, Rockwell test, Knoop test, and Vickers test.

Hooke's law
The relationship between stress and strain in the elastic portion of the stress-strain curve.

Impact energy
The energy required to fracture a standard specimen when the load is applied suddenly.

Impact test
Measures the ability of a material to absorb the sudden application of a load without breaking. The Charpy test is a commonly used impact test.

Larson–Miller parameter
A parameter used to relate the stress, temperature, and rupture time in creep.

Load
The force applied to a material during testing.

Modulus of elasticity
Young's modulus, or the slope of the stress-strain curve in the elastic region.

Modulus of resilience
The maximum elastic energy absorbed by a material when a load is applied.

Necking
Local deformation of a tensile specimen. Necking begins at the maximum stress in an engineering stress-strain curve.

Notch sensitivity

Measures the effect of a notch, scratch, or other imperfection on a material's properties, such as toughness or endurance limit.

Plastic deformation

Permanent deformation of a material when a load is applied, then removed.

Poisson's ratio

The ratio between the lateral and longitudinal strains in the elastic region.

Proof Strength

A yield strength obtained graphically that describes the stress that gives no more than a small specified amount of plastic deformation.

% Reduction in area

The total percentage decrease in the cross-sectional area of a specimen during the tensile test.

Rupture time

The time required for a specimen to fail by creep at a particular temperature and stress.

Stiffness

A qualitative measure of the elastic deformation produced in a material. A stiff material has a high modulus of elasticity.

Stress-rupture curve

A method of reporting the results of a series of creep tests by plotting the applied stress versus the rupture time.

Tensile strength

The stress that corresponds to the maximum load in a tensile test.

Tensile test

Measures the response of a material to a slowly applied uniaxial tensile force. The yield strength, tensile strength, modulus of elasticity, and ductility, are obtained.

Toughness

A qualitative measure of the impact properties of a material. A material that resists failure by impact is said to be tough.

Transition temperature

The temperature below which a material behaves in a brittle manner in an impact test.

True strain

The strain, given by $\varepsilon_t = \ln(l/l_0)$, produced in a material.

True stress

The load divided by the actual cross-sectional area of the specimen at that load.

Yield strength

The stress applied to a material that just causes permanent plastic deformation.

PROBLEMS

6.1 A 4 000 N force is applied to a 4 mm-diameter nickel wire having a yield strength of 300 $MN.m^{-2}$ and a tensile strength of 380 $MN.m^{-2}$. Determine

(a) whether the wire will plastically deform and

(b) whether the wire will experience necking.

6.2 A force of 100 000 N is applied to a 10 mm × 20 mm iron bar having a yield strength of 400 MPa and a tensile strength of 480 MPa. Determine

(a) whether the bar will plastically deform and

(b) whether the bar will experience necking.

6.3 Calculate the maximum force that a 5 mm-diameter rod of Al_2O_3, having a yield strength of 240 $MN.m^{-2}$, can withstand with no plastic deformation.

6.4 A force of 20 000 N will cause a 10 mm × 10 mm bar of magnesium to stretch from 100 mm to 100.45 mm. Calculate the modulus of elasticity.

6.5 A polymer bar's dimensions are 25 mm × 50 mm × 380 mm. The polymer has a modulus of elasticity of 4 $GN.m^{-2}$. What force is required to stretch the bar elastically to 387 mm?

6.6 An aluminium plate 5 mm thick is to withstand a force of 50 000 N with no permanent deformation. If the aluminum has a yield strength of 125 MPa, what is the minimum width of the plate?

6.7 A 75 mm-diameter rod of copper is to be reduced to a 50 mm-diameter rod by being pushed through an opening. To account for the elastic strain, what should be the diameter of the opening? The modulus of elasticity for the copper is 127 $GN.m^{-2}$ and the yield strength is 275 $MN.m^{-2}$.

6.8 A 1.5 mm-thick, 80 mm-wide sheet of magnesium that is originally 5 m long is to be stretched to a final length of 6.2 m. What should be the length of the sheet before the applied stress is released? The modulus of elasticity of magnesium is 45 $GN.m^{-2}$ and the yield strength is 200 $MN.m^{-2}$.

6.9 A steel cable 31 mm in diameter and 15.25 m long is to lift a 0.2 MN load. What is the length of the cable during lifting? The modulus of elasticity of the steel is 210 $GN.m^{-2}$.

6.10 The following data were collected from a 12 mm-diameter test specimen of magnesium:

Load (N)	Gauge length (mm)
0	30.0000
5 000	30.0296
10 000	30.0592
15 000	30.0888
20 000	30.15
25 000	30.51
26,500	30.90
27 000	31.50 (maximum load)
26,500	32.10
25 000	32.79 (fracture)

After fracture, the gauge length is 32.61 mm and the diameter is 11.74 mm. Plot the data and calculate

(a) the 0.2% proof strength,

(b) the tensile strength,

(c) the modulus of elasticity,

(d) the % elongation,

(e) the % reduction in area,

(f) the engineering stress at fracture,

(g) the true stress at fracture, and

(h) the modulus of resilience.

6.11 The following data were collected from a 20 mm-diameter test specimen of S.G. cast iron:

Load (N)	Gauge length (mm)
0	40.0000
25000	40.0185
50000	40.0370
75000	45.0555
90000	40.20
105000	40.60
120000	41.56
131000	44.00 (maximum load)
125000	47.52 (fracture)

After fracture, the gauge length is 47.42 mm and the diameter is 18.35 mm. Plot the data and calculate

(a) the 0.2% proof strength,

(b) the tensile strength,

(c) the modulus of elasticity,

(d) the % elongation,

(e) the % reduction in area,

(f) the engineering stress at fracture,

(g) the true stress at fracture, and

(h) the modulus of resilience.

6.12 A bar of Al_2O_3 that is 6 mm thick, 12 mm wide, and 225 mm long is tested in a three-point bending apparatus, with the supports located 150 mm apart. The deflection of the centre of the bar is measured as a function of the applied load. The data are shown below. Determine the flexural strength and the flexural modulus.

Force (N)	Deflection (mm)
64.5	0.065
128.5	0.13
193	0.195
257.5	0.26
382.5	0.38 (fracture)

6.13 A 10 mm-diameter, 305 mm-long titanium bar has a yield strength of 345 $MN.m^{-2}$, a modulus of elasticity of 110×103 $MN.m^{-2}$, and Poisson's ratio of 0.30. Determine the length and diameter of the bar when a 2.2 kN load is applied.

6.14 When a tensile load is applied to a 15 mm-diameter copper bar, the diameter is reduced to 14.98 mm. Determine the applied load, using the data in Table 6.3.

6.15 A three-point bend test is performed on a block of ZrO_2 that is 200 mm long, 12 mm wide, and 6 mm thick and is resting on two supports 100 mm apart. When a force of 1780 N is applied, the specimen deflects 0.94 mm and breaks. Calculate

(a) the flexural strength and

(b) the flexural modulus, assuming that no plastic deformation occurs.

6.16 A three-point bend test is performed on a block of silicon carbide that is 100 mm long, 15 mm wide, and 6 mm thick and is resting on two supports 75 mm apart. The sample breaks when a deflection of 0.09 mm is recorded. The flexural modulus for silicon carbide is 480 $GN.m^{-2}$. Assume that no plastic deformation occurs. Calculate

(a) the force that caused the fracture and

(b) the flexural strength.

6.17 A thermosetting polymer containing glass beads is required to deflects 0.5 mm when a force of 500 N is applied. The polymer part is 20 mm wide, 5 mm thick, and 100 mm long. If the flexural modulus is 6.9 $GN.m^{-2}$, determine the minimum distance between the supports. Will the polymer fracture if its flexural strength is 85 $MN.m^{-2}$? Assume that no plastic deformation occurs.

6.18 The flexural modulus of alumina is 350 $GN.m^{-2}$ and its flexural strength is 350 $MN.m^{-2}$. A bar of alumina 7.6 mm thick, 25 mm wide, and 250 mm long is placed on supports 175 mm apart. Determine the amount of deflection at the moment the bar breaks, assuming that no plastic deformation occurs.

6.19 A Brinell hardness measurement, using a 10 mm-diameter indenter and a 500 kg load, produces an indentation of 4.5 mm on an aluminium plate. Determine the Brinell hardness number (HB) of the metal.

6.20 The data below were obtained from a series of Charpy impact tests performed on four steels, each having a different manganese content. Plot the data and determine

(a) the transition temperature (defined by the the mean of the absorbed energies in the ductile and brittle regions) and

(b) the transition temperature (defined as the temperature that provides 50 J absorbed energy).

Plot the transition temperature versus manganese content and discuss the effect of manganese on the toughness of steel. What would be the minimum manganese allowed in the steel if a part is to be used at 0°C?

Test Temperature °C	Impact Energy (J)			
	0.30% Mn	0.39% Mn	1.01% Mn	1.55% Mn
−100	2	5	5	15
−75	2	5	7	25
−50	2	12	20	45
−25	10	25	40	70
0	30	55	75	110
25	60	100	110	135
50	105	125	130	140
75	130	135	135	140
100	130	135	135	140

6.21 The data below were obtained from a series of Charpy impact tests performed on four S.G. cast irons, each having a different silicon content. Plot the data and determine

(a) the transition temperature (defined by the mean of the absorbed energies in the ductile and brittle regions) and

(b) the transition temperature (defined as the temperature that provides 10 J absorbed energy).

Plot the transition temperature versus silicon content and discuss the effect of silicon on the toughness of the cast iron. What would be the maximum silicon allowed in the cast iron if a part is to be used at 25°C?

Test Temperature °C	Impact Energy (J)			
	2.55% Si	2.85% Si	3.25% Si	3.63% Si
−50	2.5	2.5	2	2
−25	3	2.5	2	2
0	6	5	3	2.5
25	13	10	7	4
50	17	14	12	8
75	19	16	16	13
100	19	16	16	16
125	19	16	16	16

6.22 FCC metals are often recommended for use at low temperatures, particularly when any sudden loading of the part is expected. Explain.

6.23 A steel part can be made by powder metallurgy (compacting iron powder particles and sintering to produce a solid) or by machining from a solid steel block. Which part is expected to have the higher toughness? Explain.

6.24 A number of aluminium-silicon alloys have a structure that includes sharp-edged plates of brittle silicon in the softer, more ductile aluminium matrix. Would you expect these alloys to be notch-sensitive in an impact test? Would you expect these alloys to have good toughness? Explain your answers.

6.25 Alumina Al_2O_3 is a brittle ceramic with low toughness. Suppose that fibres of silicon carbide SiC, another brittle ceramic with low toughness, could be embedded within the alumina. Would doing this affect the toughness of the ceramic matrix composite? Explain. (These materials are discussed in later chapters.)

6.26 A ceramic matrix composite contains internal flaws as large as 0.01 mm in length. The plane strain fracture toughness of the composite is 45 MPa$\sqrt{m}$ and the tensile strength is 550 MPa. Will the stress cause the composite to fail before the tensile strength is reached? Assume that $f = 1$.

6.27 An aluminium alloy that has a plane strain fracture toughness of 27 MN.m$^{-3/2}$ fails when a stress of 290 MN.m^{-2} is applied. Observation of the fracture surface indicates that fracture began at the surface of the part. Estimate the size of the flaw that initiated fracture. Assume that $f = 1.1$.

6.28 A polymer that contains internal flaws 1 mm in length fails at a stress of 25 MPa. Determine the plane strain fracture toughness of the polymer. Assume that $f = 1$.

6.29 A ceramic part for a jet engine has a yield strength of 520 MN.m^{-2} and a plane strain fracture toughness of 5.5 MN.m$^{-3/2}$. To be sure that the part does not fail, we plan to ensure that the maximum applied stress is only one-third the yield strength. We use a nondestructive test that will detect any internal flaws greater than 1.25 mm long. Assuming that $f = 1.4$, does our nondestructive test have the required sensitivity? Explain.

6.30 A cylindrical tool steel specimen that is 150 mm long and 6 mm in diameter rotates as a cantilever beam and is to be designed so that failure never occurs. Assuming that the maximum tensile and compressive stresses are equal, determine the maximum load that can be applied to the end of the beam. (*See* Figure 6.19.)

6.31 A 20 mm-diameter, 200 mm-long bar of an acetal polymer (Figure 6.27) is loaded on one end and is expected to survive one million cycles of loading, with equal maximum tensile and compressive stresses, during its lifetime. What is the maximum permissible load that can be applied?

6.32 A cyclical load of 6.67 kN is to be exerted at the end of a 250 mm-long aluminium beam (Figure 6.19). The bar must survive for at least 10^6 cycles. What is the minimum diameter of the bar?

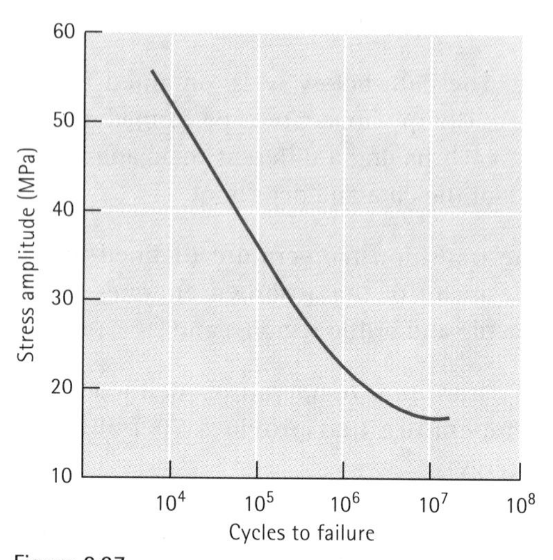

Figure 6.27

The S-N fatigue curve for an acetal polymer (*for* Problems 6.31, 6.33, *and* 6.34).

6.33 A cylindrical acetal polymer bar 200 mm long and 15 mm in diameter is subjected to a vibrational load at a frequency of 500 vibrations per minute, with a load of 50 N. How many hours will the part survive before breaking? (*See* Figure 6.27.)

6.34 Suppose that we would like a part produced from the acetal polymer shown in Figure 6.27 to survive for one million cycles under conditions that provide for equal compressive and tensile stresses. What is the fatigue strength, or maximum stress amplitude, required? What are the maximum stress, the minimum stress, and the mean stress on the part during its use? What effect would the frequency of the stress application have on your answers? Explain.

6.35 The high-strength steel in Figure 6.21 is subjected to a stress alternating at 200 revolutions per minute between 600 MPa and 200 MPa (both tension). Calculate the growth rate of a surface crack when it reaches a length of 0.2 mm in both m/cycle and m/s. Assume that $f = 1.2$.

6.36 The high-strength steel in Figure 6.21, which has a critical fracture toughness of 80 $MN.m^{-3/2}$, is subjected to an alternating stress varying from −900 $MN.m^{-3/2}$ (compression) to +900 $MN.m^{-2}$ (tension). It is to survive for 10^5 cycles before failure occurs. Assume that $f = 1$. Calculate

(a) the size of a surface crack required for failure to occur and

(b) the largest initial surface crack size that will permit this to happen.

6.37 The acrylic polymer from which Figure 6.28 was obtained has a critical fracture toughness of 2 $MPa\sqrt{m}$. It is subjected to a stress alternating between −10 and +10 MPa. Calculate the growth rate of a surface crack when it reaches a length of 5×10^{-6} m if $f = 1.3$.

6.38 Calculate the constants C and n in Equation 6.21 for the crack growth rate of an acrylic polymer. (*See* Figure 6.28.)

6.39 The acrylic polymer from which Figure 6.28 was obtained is subjected to an alternating stress between 15 MPa and 0 MPa. The largest surface cracks initially detected by nondestructive testing are 0.001 mm in length. If the critical fracture toughness of the polymer is 2 $MPa\sqrt{m}$, calculate the number of cycles required before failure occurs. Let $f = 1.2$. (**Hint:** use the results of Problem 6.38.)

6.40 Verify that integration of $da/dN = C(\Delta K)^n$ will give Equation 6.23.

6.41 The activation energy for self-diffusion in copper is 0.2 $MJ.mol^{-1}$. A copper specimen creeps at 0.002 $mm.mm^{-1}.h^{-1}$ when a stress of 100 $MN.m^{-2}$ is applied at 600°C. If the creep rate of copper is dependent on self-diffusion, determine the creep rate if the temperature is 800°C.

6.42 When a stress of 130 $MN.m^{-2}$ is applied to a material heated to 900°C, rupture occurs in 25 000 h. If the activation energy for rupture is 0.15 $MJ.mol^{-1}$, determine the rupture time if the temperature is reduced to 800°C.

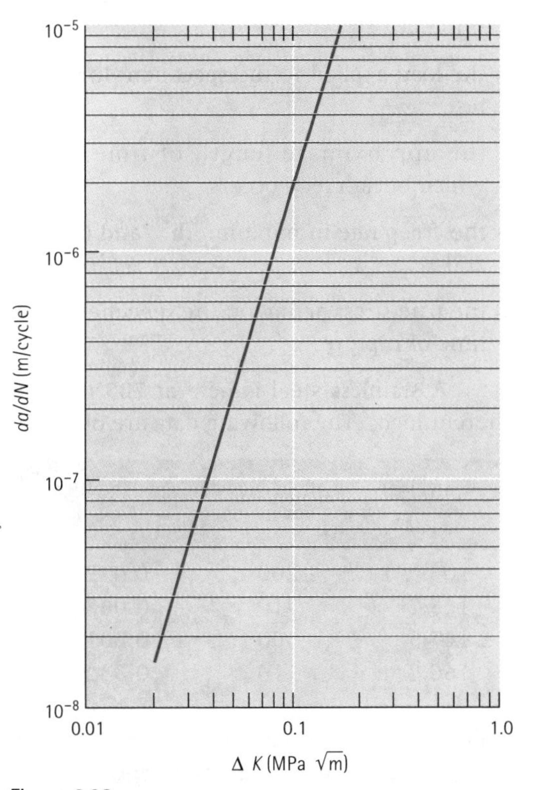

Figure 6.28
The crack growth rate for an acrylic polymer (*for* Problems 6.40, 6.41, *and* 6.42).

6.43 The following data were obtained from a creep test for a specimen having an initial gauge length of 50 mm and an initial diameter of 15 mm. The initial stress applied to the material is 67 $MN.m^{-2}$. The diameter of the specimen after fracture is 13.2 mm.

Length Between Gauge Marks (mm)	Time (h)
50.8	0
50.95	100
51.15	200
51.56	400
51.94	1 000
52.71	2 000
54.20	4 000
55.70	6 000
55.65	7 000
58.42	8 000 (fracture)

Determine

(a) the load applied to the specimen during the test,

(b) the approximate length of time during which linear creep occurs,

(c) the creep rate in $mm.mm^{-1}.h^{-1}$. and in $\%h^{-1}$, and

(d) the true stress acting on the specimen at the time of rupture.

6.44 A stainless steel is held at 705°C under different loads. The following data are obtained:

Applied Stress (MPa)	Rupture Time (h)	Creep Rate ($\%.h^{-1}$)
106.9	1 200	0.022
128.2	710	0.068
147.5	300	0.201
160.0	110	0.332

Determine the exponents n and m in Equations 6.25 and 6.26 that describe the dependence of creep rate and rupture time on applied stress.

6.45 Using the data in Figure 6.25(a) for an iron-chromium-nickel alloy, determine the activation energy Q_r and the constant m for rupture in the temperature range 980 to 1 090°C.

6.46 A 25 mm-diameter bar of an iron-chromium-nickel alloy is subjected to a load of 11 kN. How many days will the bar survive without rupturing at 980°C? (*See* Figure 6.25(a))

6.47 A 5 mm × 20 mm bar of an iron-chromium nickel alloy is to operate at 1 040°C for 10 years without rupturing. What is the maximum load that can be applied? (*See* Figure 6.25(a).)

6.48 An iron-chromium-nickel alloy is to withstand a load of 6670 N at 760°C for 6 years. Calculate the minimum diameter of the bar. (*See* Figure 6.25(a).)

6.49 A 30 mm-diameter bar of an iron-chromium-nickel alloy is to operate for 5 years under a load of 17.8 kN. What is the maximum operating temperature? (*See* Figure 6.25(a).)

6.50 A 25 mm × 50 mm S.G. cast iron bar must operate for 9 years at 650°C. What is the maximum load that can be applied? (*See* Figure 6.25(b).)

6.51 A S.G. cast iron bar is to operate at a stress of 40 MPa for 1 year. What is the maximum allowable temperature? (*See* Figure 6.25(b).)

Design Problems

6.52 A hook (Figure 6.29) for hoisting containers of ore in a mine is to be designed using a nonferrous (not based on iron) material. (A nonferrous material is used because iron and steel could cause a spark that would ignite explosive gases in the mine.) The hook must support a load of 111 kN, and a factor of safety of 50% should be used. We have determined that the cross-section labelled "?" is the most critical area; the rest of the device is already well over-designed. Determine the design requirements for this device and, based on the mechanical property data given in Chapters 13 and 14 and the metal prices given in Table 13.1, design the hook and select an economical material for the hook.

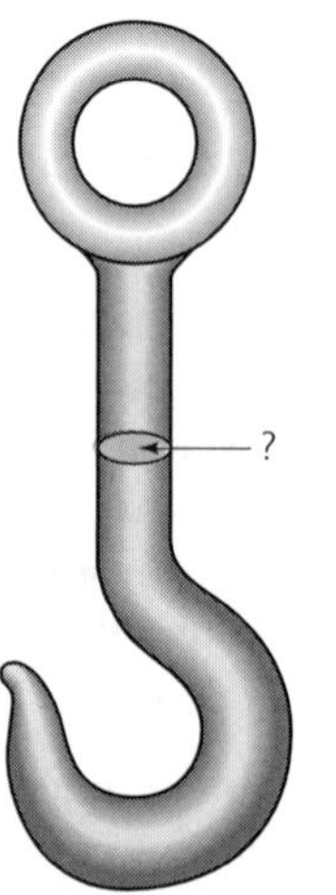

Figure 6.29
Hook (*for* Problem 6.52).

6.53 A support rod for the landing gear of a private aircraft is subjected to a tensile load during landing. The loads are predicted to be as high as 178 kN. Because the rod is crucial and failure could lead to a loss of life, the rod is to be designed with a factor of safety of 75% (that is, designed so that the rod is capable of supporting loads four times as great as expected). Operation of the system also produces loads that may induce cracks in the rod. Our non-destructive testing equipment can detect any crack greater than 0.5 mm deep. Based on the materials given in Table 6.5, design the support rod and the material, and justify your answer.

6.54 A lightweight rotating shaft for a pump on the National Aerospace plane is to be designed to support a cyclical load of 66.7 kN during service. The maximum stress is the same in both tension and compression. The endurance limits or fatigue strengths for several candidate materials are shown below. Design the shaft, including an appropriate material, and justify your solution.

Material	Endurance Limit/ Fatigue Strength ($MN.m^{-2}$)
Al-Mn alloy	110
Al-Mg-Zn alloy	225
Cu-Be alloy	295
Mg-Mn alloy	80
Be alloy	180
Tungsten alloy	320

6.55 A S.G. cast iron bar is to support a load of 178 kN in a heat-treating furnace used to make malleable cast iron. The bar is located in a spot that is continuously exposed to 500°C. Design the bar so that it can operate for at least 10 years without failing.

CHAPTER 7

Strain Hardening and Annealing

7.1 Introduction

In this chapter we will discuss three main topics: *cold working*, by which a metal is simultaneously deformed and strengthened; *hot working*, by which a metal is deformed at high temperatures without strengthening; and *annealing*, during which the effects of strengthening caused by cold working are eliminated or modified by heat treatment. The strengthening we obtain during cold working, which is brought about by slip becoming progressively more difficult, is called strain hardening or work hardening. By controlling these processes of deformation and heat treatment, we are able to process the material into a usable shape, yet still improve and control the properties.

The topics discussed in this chapter, pertain particularly to metals and alloys. Strain hardening, obtained by the movement, restriction and multiplication of dislocations, requires that the material have ductility. Brittle materials such as ceramics therefore do not respond well to strain hardening. We will show that deformation of thermoplastic polymers often produces a strengthening effect; however, the mechanism of deformation strengthening is completely different in polymers.

7.2 Relationship of Cold Working to the Stress-Strain Curve

A stress-strain curve for a ductile material is shown in Figure 7.1(a). If we apply a stress σ_1 that is greater than the yield strength, we cause a permanent deformation, or strain ε_1, when the stress is removed. If we remove a sample from the metal that had been stressed to σ_1 and retest that metal, we obtain the stress-strain curve in Figure 7.1(b). Our new test specimen would have a yield strength at σ_1 and would also have a higher tensile strength and a lower ductility. If we continue to apply a stress until we reach σ_2, then release the stress and again retest the metal, the new yield strength is σ_2. Each time we apply a higher stress, the yield strength and tensile strength increase and the ductility decreases. We eventually strengthen the metal until the yield, tensile, and breaking strengths are equal and there is no ductility [Figure 7.1(c)]. At this point, the metal can be plastically deformed no further.

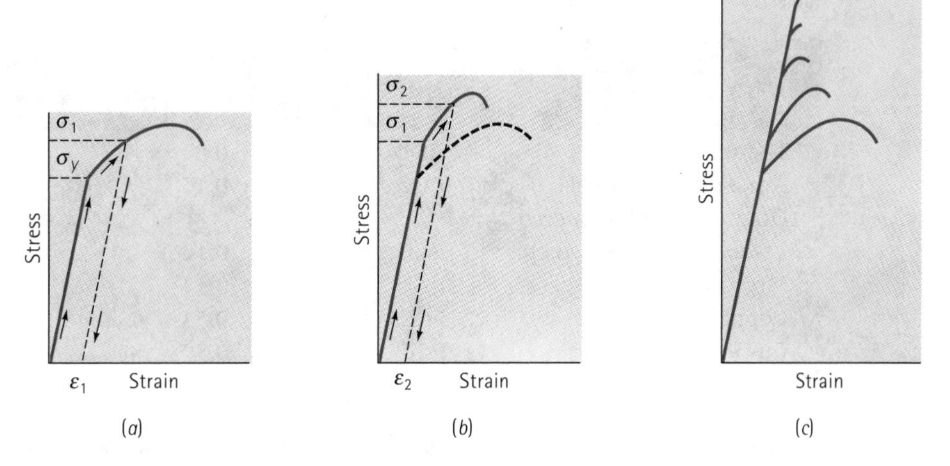

Figure 7.1
Development of strain hardening from the stress-strain diagram. (*a*) A specimen is stressed beyond the yield strength before the stress is removed. (*b*) Now the specimen has a higher yield strength and tensile strength, but lower ductility. (*c*) By repeating the procedure, the strength continues to increase and the ductility continues to decrease until the alloy becomes very brittle.

By applying a stress that exceeds the original yield strength of the metal, we have strain hardened or cold worked the metal, while simultaneously deforming it into a more useful shape.

Strain-Hardening Coefficient The response of the metal to cold working is given by the strain-hardening coefficient n, which is the slope of the plastic portion of the true stress–true strain curve in Figure 7.2 when a logarithmic scale is used:

$$\sigma_t = K\varepsilon_t^n \tag{7.1}$$

or

$$\ln \sigma_t = \ln K + n \ln \varepsilon_t$$

The constant K is equal to the stress when $\varepsilon_t = 1$.

The strain-hardening coefficient is relatively low for HCP metals but is higher for BCC and, particularly, for FCC metals (Table 7.1). Metals with a low strain-hardening coefficient respond poorly to cold working.

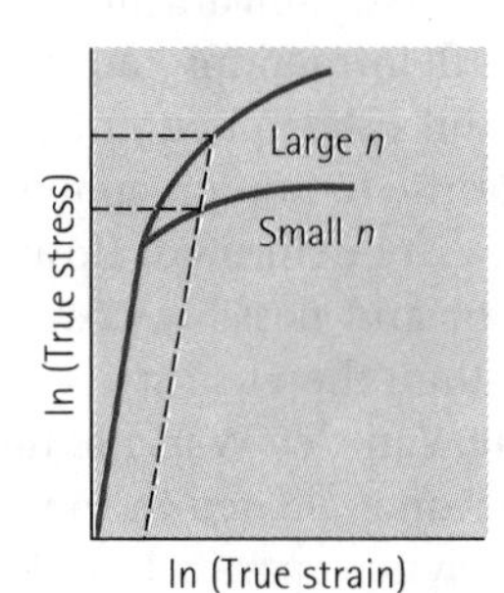

Figure 7.2
The true stress–true strain curves for metals with large and small strain-hardening coefficients. Larger degrees of strengthening are obtained for a given strain for the metal with the larger *n*.

Table 7.1 Strain-hardening coefficients of typical metals and alloys.

Metal	Crystal structure	n	K (MN.m^{-2})
Titanium	HCP	0.05	1200
Annealed alloy steel	BCC	0.15	640
Quenched and tempered medium-carbon steel	BCC	0.10	1570
Molybdenum	BCC	0.13	725
Copper	FCC	0.54	320
Cu-30% Zn	FCC	0.50	900
Austenitic stainless steel	FCC	0.52	1520

Adapted from G. Dieter, Mechanical Metallurgy, *McGraw-Hill, 1961, and other sources.*

7.3 Strain-Hardening Mechanisms

We increase the strengthening potential during deformation of a metal by increasing the number of dislocations. Before deformation, the dislocation density is about 10 m of dislocation line per cubic millimetre of metal – a relatively small number of dislocations.

When we apply a stress greater than the yield strength, dislocations begin to slip. Eventually, a dislocation moving on its slip plane encounters obstacles that pin the ends of the dislocation line. As we continue to apply the stress, the dislocation attempts to move by bowing in the centre. The dislocation may move so far that a loop is produced (Figure 7.3). When the dislocation loop finally touches itself, a new dislocation is created. The original dislocation is still pinned and can create additional dislocation loops. This mechanism for generating dislocations is called a Frank-Read source; Figure 7.3(e) shows an electron micrograph of a Frank-Read source.

The number of dislocations may increase to about 10^6 m of dislocation line per cubic millimetre of metal. We know that the more dislocations we have, the more likely they are to interfere with one another and the stronger the metal becomes.

Ceramics may contain dislocations and can even be strain-hardened to a small degree. However, ceramics are normally so brittle that significant deformation and strengthening are not possible at low temperatures; deformations can occur at high temperatures, but it is caused by sliding of grains and other phenomena. Likewise, covalently-bonded materials such as silicon are too brittle to harden appreciably.

Thermoplastic polymers will strengthen when they are deformed. However, this is *not* strain hardening but, instead, involves alignment and possibly crystallisation of the long, chainlike molecules. When a stress greater than the yield strength is applied to thermoplastic polymers such as polyethylene, the Van der Waals bonds between the chains are broken. The chains strengthen and become aligned in the direction of the applied stress (Figure 7.4). The strength of the polymer, particularly in the direction of the applied stress, increases as a result of the alignment.

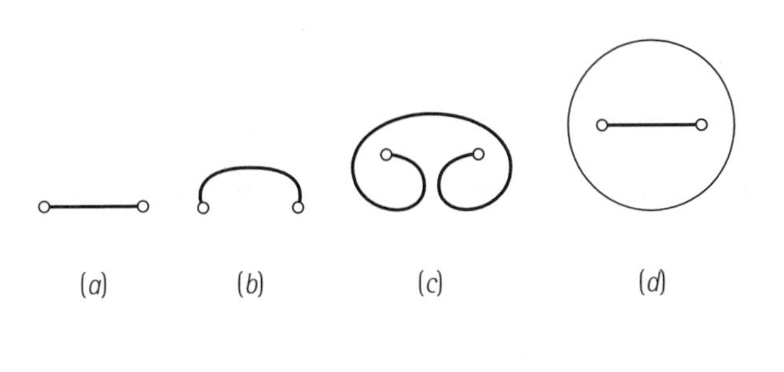

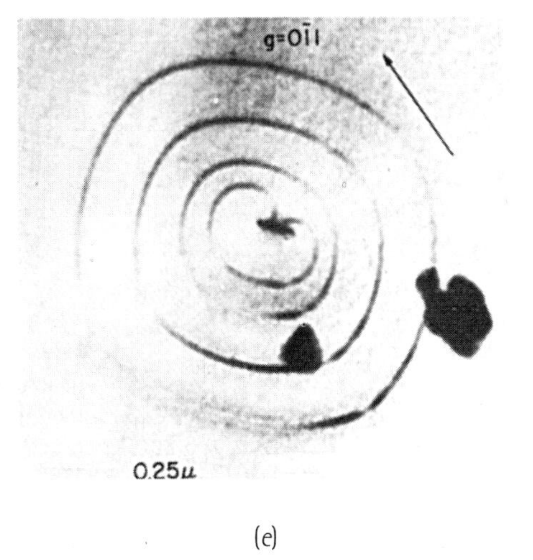

Figure 7.3
The Frank-Read source can generate dislocations. (*a*) A dislocation is pinned at its ends by lattice defects. (*b*) As the dislocation continues to move, the dislocation bows, eventually bending back on itself, (*c*). Finally the dislocation loop forms (*d*) and a new dislocation is created. (*e*) Electron micrograph of a Frank-Read source (× 30000). (*From J. Brittain, 'Climb Sources in Beta Prime-NiAl,'* Metallurgical Transactions, *Vol. 6A, April 1975.*)

7.4 Properties versus Percent Cold Work

Many techniques are used to simultaneously shape and strengthen a material by cold working (Figure 7.5). *Rolling* is used to produce metal plate, sheet, or foil. *Forging* deforms the metal into a die cavity, producing relatively complex shapes such as automotive crankshafts or connecting rods. In drawing, a metal or polymer rod is *pulled* through a die to produce a wire. In extrusion, a material is *pushed* through a die to form products of uniform cross-section, including rods, tubes, or aluminium trim for doors or windows. *Deep drawing* is used to form the body of aluminium beverage cans. *Stretch forming* and *bending* are used to shape sheet material. By controlling the amount of deformation, we control strain hardening. We normally measure the amount of deformation by defining the *percent cold work*:

$$\text{Percent cold work} = \frac{A_0 - A_f}{A_0} \times 100 \tag{7.2}$$

where A_0 is the original cross-sectional area of the metal and A_f is the final cross-sectional area after deformation.

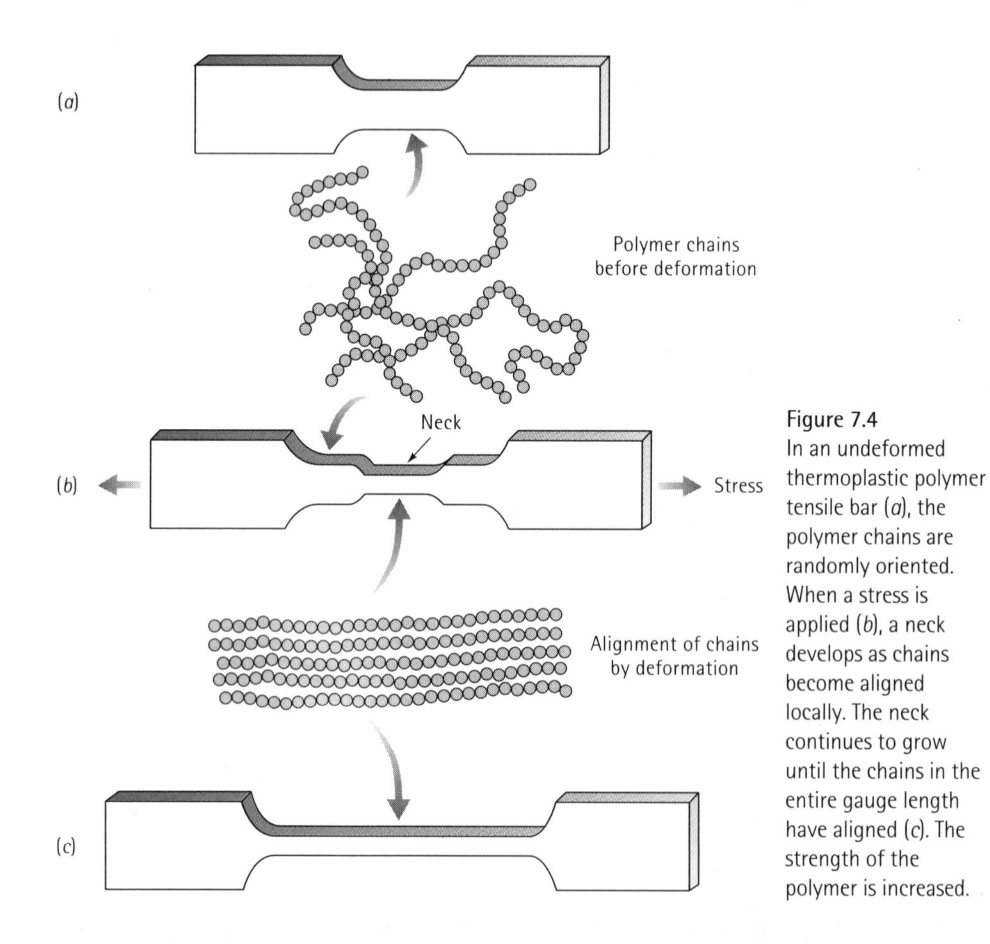

Figure 7.4
In an undeformed thermoplastic polymer tensile bar (*a*), the polymer chains are randomly oriented. When a stress is applied (*b*), a neck develops as chains become aligned locally. The neck continues to grow until the chains in the entire gauge length have aligned (*c*). The strength of the polymer is increased.

The effect of cold work on the mechanical properties of commercially pure copper is shown in Figure 7.6. As the cold work increases, both the yield and the tensile strength increase; however, the ductility decreases and approaches zero. The metal breaks if more cold work is attempted. Therefore, there is a maximum amount of cold work or deformation that we can do to a metal.

EXAMPLE 7.1

A 10 mm-thick copper plate is cold-reduced to 5 mm, and later further reduced to 1.6 mm. Determine the total percent cold work and the tensile strength of the 1.6 mm plate. (*See* Figure 7.7.)

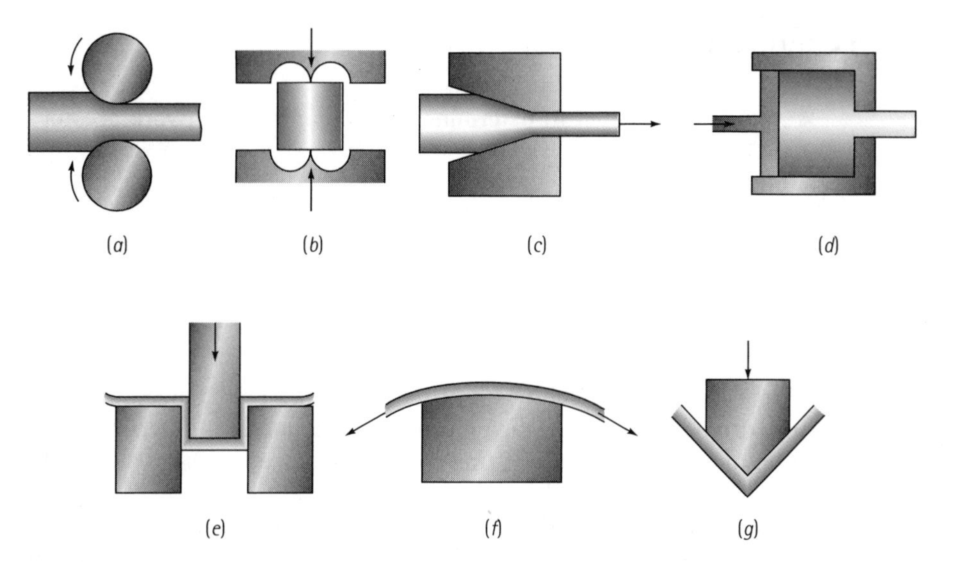

Figure 7.5
Schematic drawings of deformation processing techniques. (*a*) Rolling, (*b*) forging, (*c*) drawing, (*d*) extrusion, (*e*) deep drawing, (*f*) stretch forming, and (*g*) bending.

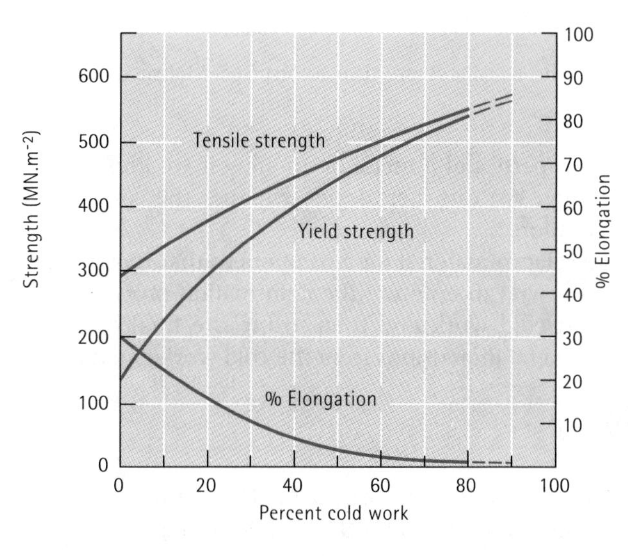

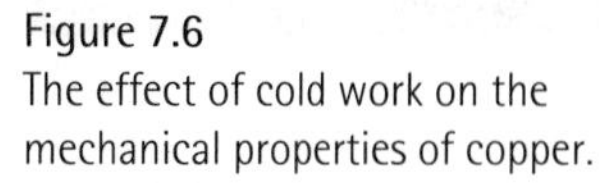

Figure 7.6
The effect of cold work on the mechanical properties of copper.

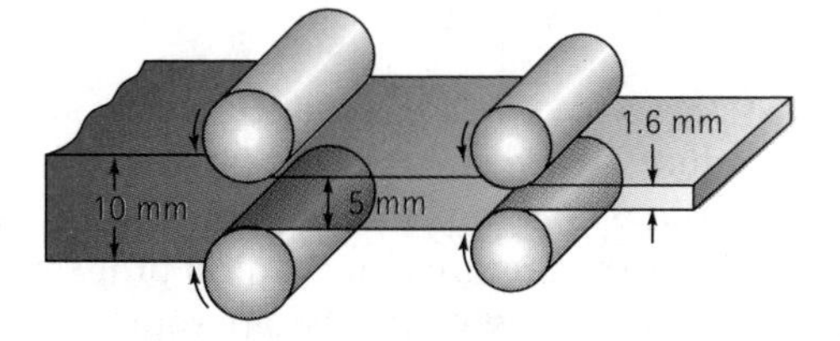

Figure 7.7
Diagram showing the rolling of a 10 mm plate to a 1.6 mm plate (Example 7.1).

SOLUTION

We might be tempted to determine the amount of cold work accomplished in each step, that is:

$$\% \text{ CW } = \frac{A_0 - A_f}{A_0} \times 100 = \frac{t_0 - t_f}{t_0} \times 100 = \frac{10 - 5}{10} \times 100 = 50\%$$

$$\% \text{ CW } = \frac{A_0 - A_f}{A_0} \times 100 = \frac{t_0 - t_f}{t_0} \times 100 = \frac{5 - 1.6}{5} \times 100 = 68\%$$

Note that, because the width of the plate does not change during rolling, the cold work can be expressed as the percentage reduction in the thickness t. We might then be tempted to combine the two cold work percentages (50% + 68% = 118%) to obtain the total cold work. *This would be incorrect.*

Our definition of cold work is the percentage change between the original and final cross-sectional areas; it makes no difference how many intermediate steps are involved. Thus, the total cold work is actually

$$\% \text{ CW } = \frac{t_0 - t_f}{t_0} \times 100 = \frac{10 - 1.6}{10} \times 100 = 84\%$$

and, from Figure 7.6, the tensile strength is about 565 $MN.m^{-2}$.

We can predict the properties of a metal or an alloy if we know the amount of cold work during processing. We can then decide whether the component has adequate strength at critical locations.

When we wish to select a material for a component that requires certain minimum mechanical properties, we can optimise the deformation process. We first determine the necessary percent cold work and then, using the final dimensions we desire, calculate the original metal dimensions from the cold work equation.

EXAMPLE 7.2

Establish a manufacturing process to produce a 1 mm-thick copper plate having at least 430 $MN.m^{-2}$ tensile strength, 400 $MN.m^{-2}$ yield strength, and 5% elongation.

SOLUTION

From Figure 7.6, we need at least 35% cold work to produce tensile strength of 430 $MN.m^{-2}$ and 40% cold work to produce a yield strength of 400 $MN.m^{-2}$, but we need less than 45% cold work to meet the 5% elongation requirement. Therefore, any cold work between 40% and 45% gives the required mechanical properties.

To produce the plate, a cold-*rolling* process would be appropriate. The original thickness of the copper plate prior to rolling can be calculated from Equation 7.2, assuming that the width of the plate does not change. Because there is a range of allowable cold work – between 40% and 45% – there is a range of initial plate thicknesses:

$$\% \ CW_{min} = 40 = \frac{t_{min} - 1}{t_{min}} \times 100 \therefore t_{min} = 1.67 \text{ mm}$$

$$\% \ CW_{max} = 45 = \frac{t_{max} - 1}{t_{max}} \times 100 \therefore t_{max} = 1.82 \text{ mm}$$

To produce the 1 mm copper plate, we begin with a 1.67 to 1.82 mm copper plate in the softest possible condition, then cold roll the plate 40% to 45% to achieve the 1 mm thickness.

7.5 Microstructure and Residual Stresses

During deformation, a fibrous microstructure is produced as the grains within the metal become elongated (Figure 7.8).

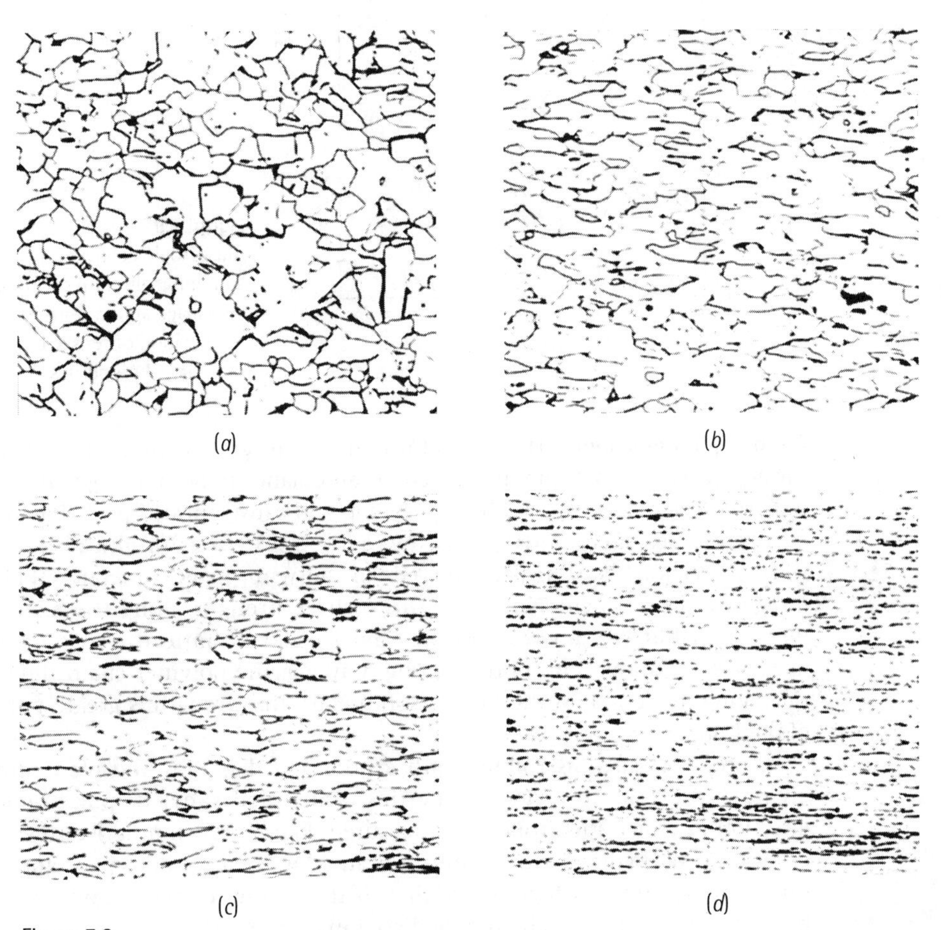

Figure 7.8
The fibrous grain structure of a low carbon steel produced by cold working: (*a*) 10% cold work, (*b*) 30% cold work, (*c*) 60% cold work, and (*d*) 90% cold work (× 250). (*From* Metals Handbook, *Vol. 9, 9th Ed. American Society For Metals, 1985.*)

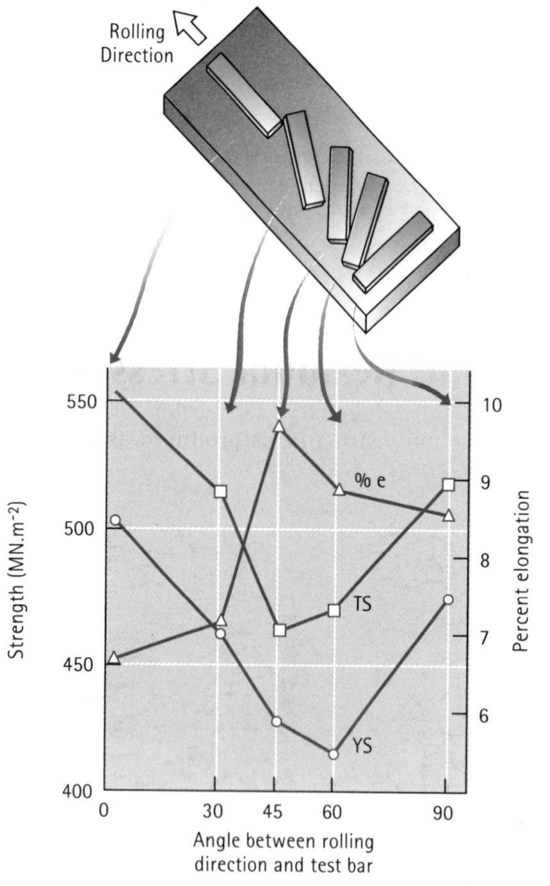

Figure 7.9
Anisotropic behaviour in a rolled aluminium-lithium sheet material used in aerospace applications. The sketch relates the position of tensile bars to the mechanical properties that are obtained.

Anisotropic Behaviour During deformation, the grains rotate as well as elongate, causing certain crystallographic directions and planes to become aligned. Consequently, preferred orientations, or textures, develop and cause anisotropic behaviour.

In processes such as wire drawing, a fibre texture is produced. In BCC metals, <110> directions line up with the axis of the wire. In FCC metals, <111> or <100> directions are aligned. This gives the highest strength along the axis of the wire, which is what we desire. A somewhat similar situation occurs when polymer materials are drawn into fibres; during drawing, the polymer chains line up side-by-side along the length of the fibre. As in metals, the strength is greatest along the axis of the fibre.

In processes such as rolling, both a preferred direction and plane are produced, giving a sheet texture. The properties of a rolled sheet or plate depend on the direction in which the property is measured. Figure 7.9 summarises the tensile properties of a cold worked aluminium-lithium alloy used for aerospace applications. For this alloy, strength is highest parallel to the rolling direction, whereas ductility is highest at a 45° angle to the rolling direction.

EXAMPLE 7.3

One method for producing fans for cooling automotive and truck engines is to stamp the blades from cold-rolled steel sheet, then attach the blades to a 'spider' that holds the blades in the proper position. A number of fan blades, all produced at the same time, have failed by the initiation and propagation of a fatigue crack transverse to the axis of the blade (Figure 7.10). All other fan blades perform satisfactorily. Provide an explanation for the failure of the blades and suggest a modification to the manufacturing process to prevent these failures.

SOLUTION

There may be several explanations for the failure of the blades – for example, the wrong steel may have been selected, the dies used to stamp the blades from the sheet may be worn, or the clearance between the parts of the dies may be incorrect, producing defects that initiate fatigue failure.

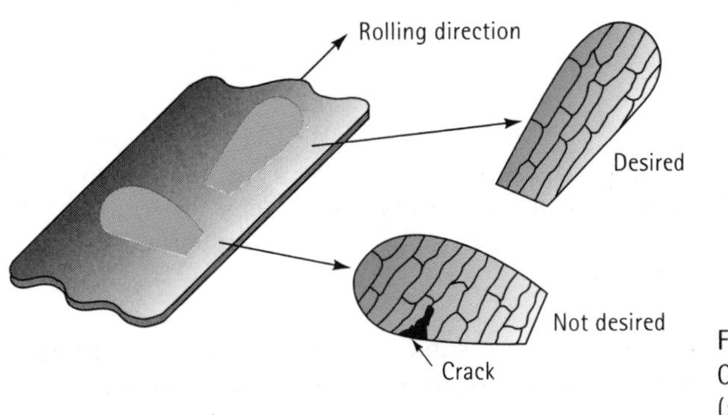

Figure 7.10
Orientations of samples (*for* Example 7.3).

The failures could also be related to the anisotropic behaviour of the steel sheet caused by rolling. To achieve the best performance from the blade, the axis of the blade should be aligned with the rolling direction of the steel sheet. This procedure produces high strength along the axis of the blade and, by assuring that the grains are aligned with the blade axis, reduces the number of grain boundaries along the leading edge of the blade that might help initiate a fatigue crack. Suppose your examination of the blade indicates that the steel sheet was aligned 90° from its usual position during stamping. Now the blade has a low strength in the critical direction and, in addition, fatigue cracks will more easily initiate and grow. This mistake in manufacturing has been the cause of failures and injuries to mechanics performing maintenance on road vehicles.

You might recommend that the manufacturing process be changed to ensure that the blades cannot be stamped from misoriented sheet. Perhaps special guides or locking devices on the die will ensure that the die is properly aligned with the sheet.

Residual Stresses Residual stresses develop during deformation. A small portion of the applied stress – perhaps about 10% – is stored within the structure as a tangled network of dislocations. The residual stresses increase the total energy of the structure.

The residual stresses are not uniform throughout the deformed metal. For example, high compressive residual stresses may be present at the surface of a rolled plate and tensile stresses may be stored in the centre. If we machine a small amount of metal from one surface of a cold-worked part, we remove metal that contains only compressive residual stresses. To restore the balance, the plate must distort.

Residual stresses also affect the ability of the part to carry a load (Figure 7.11). If a tensile stress is applied to a material that already contains tensile residual stresses, the total stress acting on the part is the sum of the applied and residual stresses. If, however, compressive stresses are stored at the surface of a metal part, an applied tensile stress must first balance the compressive residual stresses. Now the part may be capable of withstanding a larger than normal load.

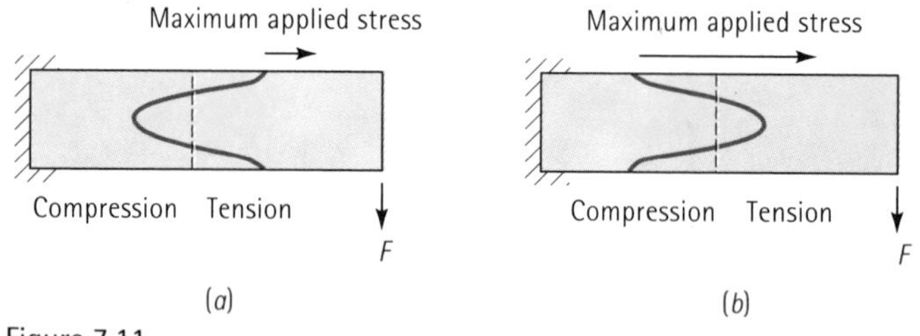

Figure 7.11
The compressive residual stresses can be harmful or beneficial. In (a), a bending force applies a tensile stress on the top of the beam. Since there are already tensile residual stresses at the top, the load-carrying characteristics are poor. In (b), the top contains compressive residual stresses. Now the load-carrying characteristics are very good.

Sometimes components that are subject to fatigue failure can be strengthened by shot peening. Bombarding the surface with steel shot propelled at a high velocity introduces compressive residual stresses at the surface that increase the resistance of the metal surface to fatigue failure.

EXAMPLE 7.4

Your company has produced several million shafts that have a fatigue strength of 140 MN.m^{-2}. The shafts are subjected to high bending loads during rotation. Your sales engineers report that the first few shafts placed into service failed in a short period of time by fatigue. Suggest a process by which the remaining shafts can be salvaged by improving their fatigue properties.

SOLUTION

Fatigue failures typically begin at the surface of a rotating part; thus, increasing the strength at the surface improves the fatigue life of the shaft. A variety of methods might be used to accomplish this.

If the shaft is made of steel, we could carburise the surface of the part. In carburising, carbon is diffused into the surface of the shaft. After an appropriate heat treatment, the higher carbon at the surface increases the strength of the surface and, perhaps more importantly, introduces *compressive* residual stresses at the surface.

We might consider cold working the shaft; cold working increases the yield strength of the metal and, if done properly, introduces compressive residual stresses. However, the cold work also reduces the diameter of the shaft and, because of the dimensional change, the shaft may not be able to perform its function.

Another alternative would be to shot peen the shaft. Shot peening introduces local compressive residual stresses at the surface without changing the dimensions of the part. If compressive stresses of 70 $MN.m^{-2}$ were introduced by shot peening, then the total applied stress that could be supported by the shaft may be $140 + 70 = 210$ $MN.m^{-2}$. This process, which is also inexpensive, might be sufficient to salvage the remaining shafts.

7.6 Characteristics of Cold Working

There are a number of advantages and limitations to strengthening a metal by cold working or strain hardening:

1. We can simultaneously strengthen the metal and produce the desired final shape.
2. We can obtain excellent dimensional tolerances and surface finishes by the cold-working process.
3. The cold-working process is an inexpensive method for producing large numbers of small parts, since high forces and expensive forming equipment are not needed.
4. Some metals, such as HCP magnesium, have a limited number of slip systems, and are rather brittle at room temperature; thus, only a small degree of cold working can be accomplished.
5. Ductility, electrical conductivity, and corrosion resistance are impaired by cold working. Because cold working reduces electrical conductivity less than other strengthening processes, such as introducing alloying elements (Figure 7.12), cold working is a satisfactory way to strengthen conductor materials, such as the copper wires used for transmission of electrical power.
6. Properly controlled residual stresses and anisotropic behaviour may be beneficial.
7. Some deformation processing techniques can be accomplished only if cold working occurs. For example, wire drawing requires that a rod be pulled through a die to produce a smaller cross-sectional area (Figure 7.13). For a given draw force F_d, a different stress is produced in the original and final wire. The stress on the initial wire must exceed the yield strength of the metal to cause deformation. The stress on the final wire must be less than its yield strength to prevent failure. This is accomplished only if the wire strain hardens during drawing.

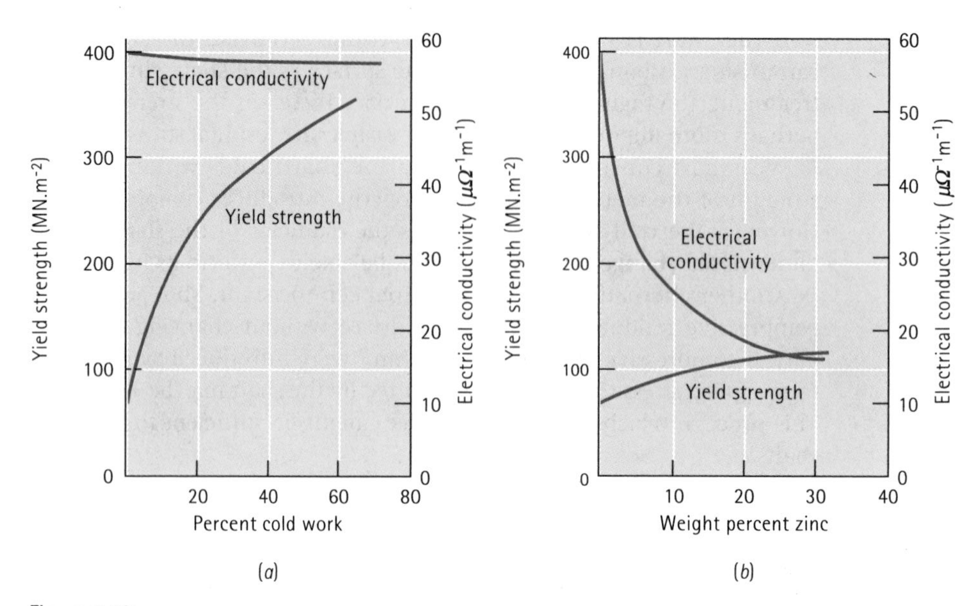

Figure 7.12
A comparison of strengthening copper by (*a*) cold working and (*b*) alloying with zinc. Note that cold working produces greater strengthening, yet has little effect on electrical conductivity.

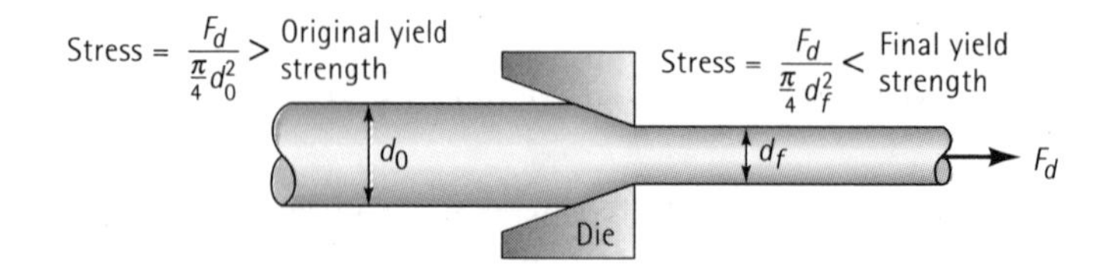

Figure 7.13
The wire-drawing process. The force F_d acts on both the original and final diameters. Thus, the stress produced in the final wire is greater than that in the original. If the wire did not strain harden during drawing, the final wire would break before the original wire was drawn through the die.

EXAMPLE 7.5

Establish a process to produce 5 mm-diameter copper wire.

SOLUTION

Wire drawing is the obvious manufacturing technique for this application. To produce the copper wire as efficiently as possible, we make the largest reduction in the diameter possible. Our design must ensure that the wire strain hardens sufficiently during drawing to prevent the drawn wire from breaking.

As an example calculation, let's assume that the starting diameter of the copper wire is 10 mm and that the wire is in the softest possible condition. The cold work is:

$$\% \text{ CW} = \frac{A_0 - A_f}{A_0} \times 100 = \frac{(\pi/4)d_0^2 - (\pi/4)d_f^2}{(\pi/4)d_0^2} \times 100$$

$$= \frac{(10)^2 - (5)^2}{(10)^2} \times 100 = 75\%$$

From Figure 7.6, the initial yield strength with 0% cold work is 140 MN.m^{-2}. The final yield strength with 75% cold work is about 535 MN.m^{-2} (with very little ductility). The draw force required to deform the initial wire is:

$F = \sigma_y A_0 = (140)(\pi/4)(10)^2 = 10\,996$ N

The stress acting on the wire after passing through the die is:

$$\sigma = \frac{F_d}{A_f} = \frac{10996}{(\pi/4)(5)^2} = 560 \text{MN.m}^{-2}$$

The applied stress of 560 MN.m^{-2} is greater than the 535 MN.m^{-2} yield strength of the drawn wire. The wire therefore breaks.

We can perform the same set of calculations for other initial diameters, with the results shown in Table 7.2 and Figure 7.14.

Table 7.2

d_0(mm)	% CW	Yield Strength of Drawn Wire (MN.m^{-2})	Draw Force (N)	Stress on Drawn Wire (MN.m^{-2})
6	31	350	3958	201
7	49	445	5388	274
8	61	480	7037	358
10	75	535	10996	560

The graph shows that the draw stress exceeds the yield strength of the drawn wire when the original diameter is about 9.8 mm. To produce the wire as efficiently as possible, the original diameter should be just under 9.8 mm.

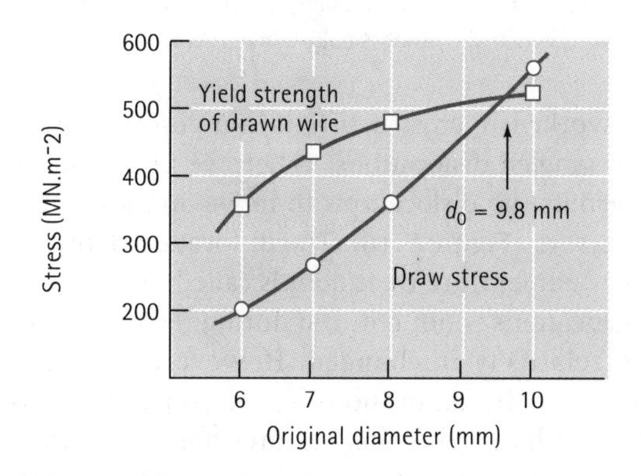

Figure 7.14
Yield strength and draw stress of wire (*for* Example 7.5).

7.7 The Three Stages of Annealing

Annealing is a heat treatment designed to eliminate the effects of cold working. It may be used to completely eliminate the strain hardening achieved during cold working; the final part is soft and ductile but still has a good surface finish and dimensional accuracy. Or, after annealing, additional cold work could be done, since the ductility is restored. By combining repeated cycles of cold working and annealing, large total deformations may be achieved. Finally, annealing at a low temperature may be used to eliminate the residual stresses produced during cold working without affecting the mechanical properties of the finished part. There are three stages in the annealing process; their effects on the properties of brass are shown in Figure 7.15:

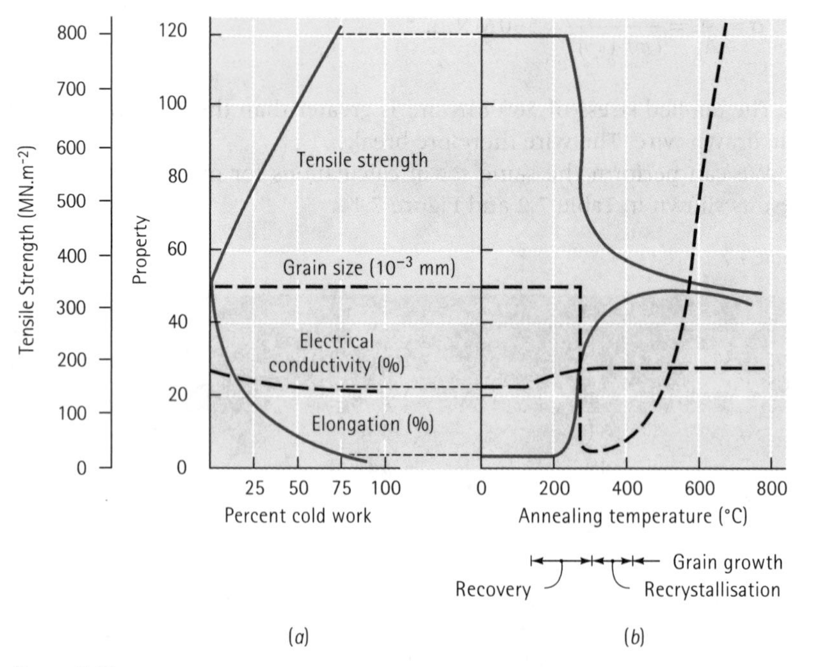

Figure 7.15
(*a*) The effect of cold work on the properties of a Cu-35% Zn alloy. (*b*) The effect of annealing temperature on the properties of a Cu-35% Zn alloy that is cold-worked 75%.

Recovery The original cold-worked microstructure is composed of deformed grains containing a large number of tangled dislocations. When we first heat the metal, the additional thermal energy permits the dislocations to move and form the boundaries of a polygonised subgrain structure (Figure 7.16). The dislocation density, however, is virtually unchanged. This low-temperature treatment is called recovery

Because the number of dislocations is not reduced during recovery, the mechanical properties of the metal are relatively unchanged. However, residual stresses are reduced or even eliminated when the dislocations are rearranged; recovery is often called a stress relief anneal. In addition, recovery restores high electrical conductivity to the metal, permitting us to produce copper or aluminium wire for transmission of electrical power that is strong yet still has high conductivity. Finally, recovery often improves the resistance of the material to corrosion.

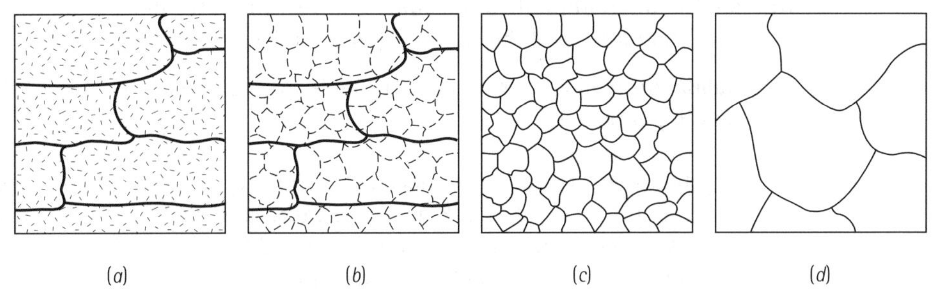

Figure 7.16
The effect of annealing temperature on the microstructure of cold-worked metals: (*a*) cold worked, (*b*) after recovery, (*c*) after recrystallisation, and (*d*) after grain growth.

Recrystallisation Recrystallisation occurs by the nucleation and growth of new grains containing few dislocations. When the metal is heated above the recrystallisation temperature, rapid recovery eliminates residual stresses and produces the polygonised dislocation structure. New small grains then nucleate at the cell boundaries of the polygonised structure, eliminating most of the dislocations (Figure 7.16). Because the number of dislocations is greatly reduced, the recrystallised metal has low strength but high ductility.

Grain Growth At still higher annealing temperatures, both recovery and recrystallisation occur rapidly, producing a fine recrystallised grain structure. The grains begin to grow, however, with favoured grains consuming the smaller grains (Figure 7.16). This phenomenon, called grain growth, was described in Chapter 5. Illustrated for a copper–zinc alloy in Figure 7.17, it is almost always undesirable.

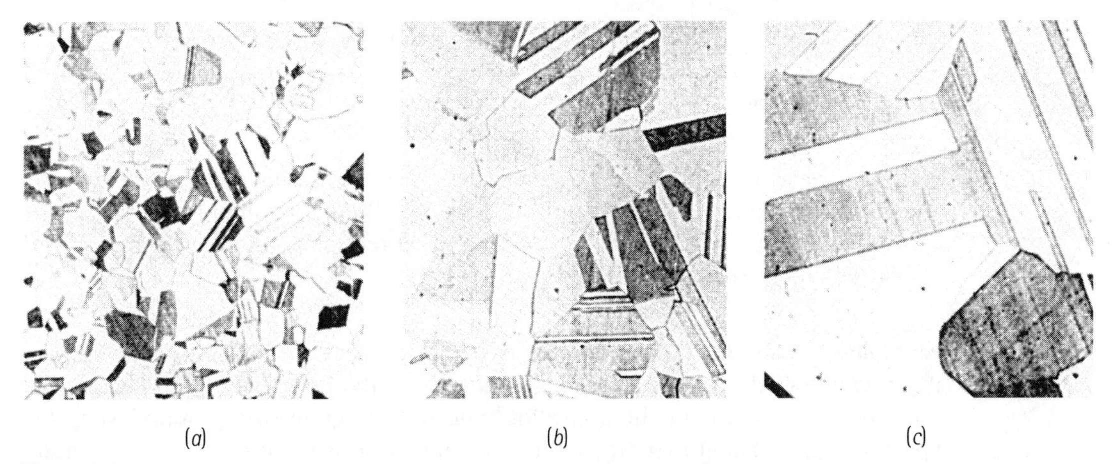

Figure 7.17
Photomicrographs showing the effect of annealing temperature on grain size in brass. Twin boundaries can also be observed in the structures. (*a*) Annealed at 400°C, (*b*) annealed at 650°C, and (*c*) annealed at 800°C (× 75). (*From R. Brick and A. Phillips,* The Structures and Properties of Alloys, *McGraw-Hill, 1949.*)

7.8 Control of Annealing

To establish an appropriate annealing heat treatment, we need to know the recrystallisation temperature and the size of the recrystallised grains.

Recrystallisation Temperature The recrystallisation temperature is affected by a variety of processing variables:

1. Recrystallisation temperature decreases when the amount of cold work increases. Greater amounts of cold work make the metal less stable and encourage nucleation of recrystallised grains. There is a minimum amount of cold work, about 30% to 40%, below which recrystallisation will not occur.
2. A small original cold-worked grain size reduces the recrystallisation temperature by providing more sites – the former grain boundaries – at which new grains can nucleate.
3. Pure metals recrystallise at lower temperatures than alloys.
4. Increasing the annealing time reduces the recrystallisation temperature (Figure 7.18), since more time is available for nucleation and growth of the new recrystallised grains.

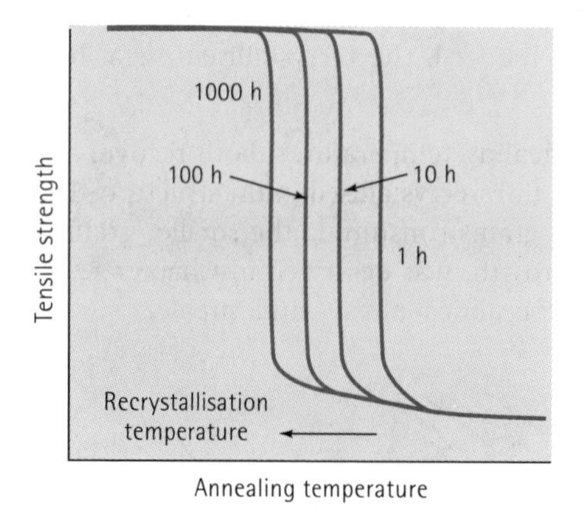

Figure 7.18
Longer annealing times reduce the recrystallisation temperature.

5. Higher melting point alloys have a higher recrystallisation temperature. Since recrystallisation is a diffusion-controlled process, the recrystallisation temperature is roughly proportional to $0.4T_m$ Kelvin. Typical recrystallisation temperatures for selected metals are shown in Table 7.3.

Recrystallised Grain Size A number of factors also influence the size of the recrystallised grains. Reducing the annealing temperature, the time required to heat to the annealing temperature, or the annealing time reduces grain size by minimising the opportunity for grain growth. Increasing the initial cold work also reduces final grain size by providing a greater number of nucleation sites for new grains. Finally, the presence of a second phase in the microstructure helps prevent grain growth and keeps the recrystallised grain size small.

Table 7.3
Typical recrystallisation temperatures for selected metals.

Metal	Melting Temperature (°C)	Recrystallisation Temperature (°C)
Sn	232	< Room temperature
Pb	327	< Room temperature
Zn	420	< Room temperature
Al	660	150
Mg	650	200
Ag	962	200
Cu	1 085	200
Fe	1 538	450
Ni	1 453	600
Mo	2 610	900
W	3 410	1 200

Adapted from R. Brick, A. Pense, and R. Gordon, Structure and Properties of Engineering Materials, *McGraw-Hill, 1977.*

7.9 Annealing and Materials Processing

The effects of recovery, recrystallisation, and grain growth are important in the processing and eventual use of a metal or an alloy.

Deformation Processing By taking advantage of the annealing heat treatment, we can increase the total amount of deformation we can accomplish. If we are required to reduce a 100 mm thick plate to a 1 mm thick sheet, we can do the maximum permissible cold work, anneal to restore the metal to its soft, ductile condition, then cold work again. We can repeat the cold work–anneal cycle until we approach the proper thickness. The final cold-working step can be designed to produce the final dimensions and properties required (Example 7.6).

EXAMPLE 7.6

We wish to produce a 1 mm-thick, 60 mm-wide copper strip having at least 400 MN.m^{-2} yield strength and at least 5% elongation. We are able to purchase 60 mm-wide strip only in thicknesses of 50 mm. Determine a process to produce the product we need.

SOLUTION

In Example 7.2, we found that the required properties can be obtained with a cold work of 40% to 45%. Therefore, the starting thickness must be between 1.67 mm and 1.82 mm, and this starting material must be as soft as possible – that is, in the annealed condition. Since we are able to purchase only 50 mm-thick stock, we must reduce the thickness of the 50 mm strip to between 1.67 and 1.82 mm, then anneal the strip prior to final cold working. But can we successfully cold work from 50 mm to 1.82 mm?

$$\%\ \mathrm{CW} = \frac{50 - 1.82}{50} \times 100 = 96.4\%$$

Based on figure 7.6, a maximum of about 90% cold work is permitted. Therefore we must do a series of cold work and anneal cycles. Although there are many possible combinations, one is as follows:

1. Cold work the 50 mm strip 80% to 10 mm:

$$80 = \frac{50 - t_i}{50} \times 100 \quad \text{or} \quad t_i = 10\,\mathrm{mm}$$

2. Anneal the 10 mm strip to restore the ductility. If we don't know the recrystallisation temperature, we can use the $0.4T_m$ relationship to provide an estimate. The melting point of copper is 1085°C:

   $T_r \cong (0.4)(1085 + 273) = 543\ \mathrm{K} = 270°\mathrm{C}$

3. Cold work the 10 mm-thick strip to 1.82 mm:

$$\%\ \mathrm{CW} = \frac{10 - 1.82}{10} \times 100 = 81.8\%$$

4. Again anneal the copper at 270°C to restore ductility.

5. Finally cold work 45%, from 1.82 mm to the final dimension of 1 mm. This process gives the correct final dimensions and properties.

High-Temperature Service Neither strain hardening nor grain size strengthening is appropriate for an alloy to be used at elevated temperatures, as in creep-resistant applications. When the cold-worked metal is placed into service at a high temperature, recrystallisation immediately causes a catastrophic decrease in strength. In addition, if the temperature is high enough, the strength continues to decrease because of growth of the newly recrystallised grains.

Joining Process When we join a cold-worked metal using a welding process, the metal adjacent to the weld heats above the recrystallisation and grain growth temperatures. This region is called heat-affected zone. The structure and properties in the heat-affected zone of a weld are shown in Figure 7.19. The properties are catastrophically reduced by the heat of the welding process.

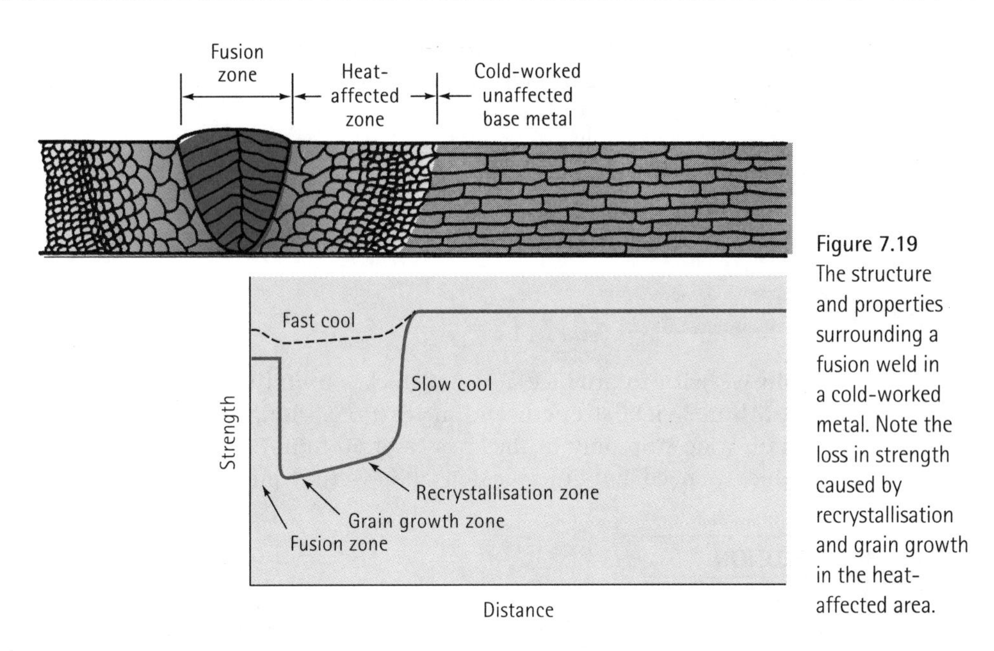

Figure 7.19
The structure and properties surrounding a fusion weld in a cold-worked metal. Note the loss in strength caused by recrystallisation and grain growth in the heat-affected area.

Welding processes such as electron-beam welding or laser welding, which provide high rates of heat input for brief times, minimise the exposure of the metal to temperatures above recrystallisation and minimise this type of damage.

7.10 Hot Working

We can deform a metal into a useful shape by hot working rather than cold working the metal. Hot working is defined as plastically deforming the metal at a temperature above the recrystallisation temperature. During hot working, the metal is continually recrystallised (Figure 7.20).

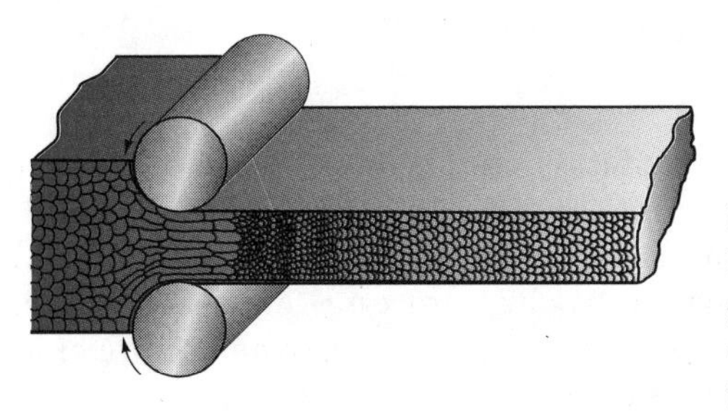

Figure 7.20
During hot working, the elongated, anisotropic grains immediately recrystallise. If the hot-working temperature is properly controlled, the final hot-worked grain size can be very fine.

Lack of Strengthening No strengthening occurs during deformation by hot working; consequently, the amount of plastic deformation is almost unlimited. A very thick plate can be reduced to a thin sheet in a continuous series of operations. The first steps in the process are carried out well above the recrystallisation temperature to take advantage to the lower strength of the metal. The last step is performed just above the recrystallisation temperature, using a large percent deformation in order to produce the finest possible grain size.

Hot working is well-suited for forming large parts, since the metal has a low yield strength and high ductility at elevated temperatures. In addition, HCP metals such as magnesium have more active slip systems at hot-working temperatures; the higher ductility permits larger deformations than are possible by cold working.

EXAMPLE 7.7

Again we wish to produce a 1 mm-thick, 60 mm-wide copper strip having at least 400 MN.m^{-2} yield strength and at least 5% elongation. We are able to purchase 60 mm-wide strip only in thicknesses of 50 mm. Establish a process to produce the product we need, but in fewer steps than were required in Example 7.6.

SOLUTION

In Example 7.6, we relied on a series of cold work–anneal cycles to obtain the required thickness. We could reduce the steps by hot rolling to the required intermediate thickness:

$$\%\ \text{HW} = \frac{50 - 1.82}{50} \times 100 = 96.4\%$$

$$\%\ \text{HW} = \frac{50 - 1.67}{50} \times 100 = 96.7\%$$

Because recrystallisation occurs simultaneously with hot working, we can obtain these large deformations and a separate annealing treatment is not required. Thus our process might be:

1. Hot work the 50 mm strip 96.4% to the intermediate thickness of 1.82 mm.
2. Cold work 45% from 1.82 mm to the final dimension of 1 mm. This sequence gives the correct dimensions and properties.

Elimination of Imperfections Some imperfections in the original metal may be eliminated or their effects minimised. Gas pores can be closed and welded shut during hot working – the internal lap formed when the pore is closed is eliminated by diffusion during the forming and cooling process. Composition differences in the metal can be reduced as hot working brings the surface and centre of the plate closer together, thereby reducing diffusion distances.

Anisotropic Behaviour The final properties in hot-worked parts are not isotropic. The forming rolls or dies, which are normally at a lower temperature than the metal, cool the surface more rapidly than the centre of the part. The surface then has a finer grain size than the centre. In addition, a fibrous structure is produced because inclusions and second-phase particles are elongated in the working direction.

Surface Finish and Dimensional Accuracy The surface finish is usually poorer than that obtained by cold working. Oxygen may react with the metal at the surface to form oxides, which are forced into the surface during forming. In some metals, such as tungsten and beryllium, hot working must be done in a protective atmosphere.

The dimensional accuracy is also more difficult to control during hot working. A greater elastic strain must be considered, since the modulus of elasticity is low at hot-working temperatures. In addition, the metal contracts as it cools from the hot-working temperature. The combination of elastic strain and thermal contraction requires that the part be made oversize during deformation; forming dies must be carefully designed, and precise temperature control is necessary if accurate dimensions are to be obtained.

7.11 Superplastic Forming

When specially heat-treated and processed, some materials can be uniformly deformed an exceptional amount – in some cases, more than 1 000%. This behaviour is called superplasticity. Often, superplastic forming can be coupled with diffusion bonding to produce complicated assemblies in a single step. Several conditions are required for a material to display superplastic behaviour:

1. The metal must have a very fine grain structure, with grain diameters less than about 0.005 mm.
2. The alloy must be deformed at a high temperature, often near 0.5 to 0.65 times the absolute melting point of the alloy.
3. A very slow rate of forming, or strain rate, must be employed. In addition, the stress required to deform the alloy must be very sensitive to the strain rate. If necking begins to occur, the necked region strains at a higher rate; the higher strain rate, in turn, strengthens the necked region and stops the necking, and the uniform deformation continues.
4. The grain boundaries in the alloy should allow grains to slide easily over one another and rotate when stress is applied. The proper temperature and a fine grain size are necessary for this to occur.

Superplastic forming is most commonly done for metals, including alloys such as Ti-6% Al-4% V, Cu-10% Al, and Zn-23% Al. Complex aerospace components are often produced using the superplastic titanium alloy. However, superplasticity is also found in materials that are normally considered to be brittle. These include a number of ceramic materials (Al_2O_3 and ZrO_2) and intermetallic compounds (Ni_3Si).

SUMMARY

The properties of metals can be controlled by combining plastic deformation and simple heat treatments. When a metal is deformed by cold working, strain hardening occurs as dislocations are 'locked' within the structure. Very large increases in strength may be obtained in this manner. Deformation of thermoplastic polymers provides strengthening by aligning the polymer chains (although this is not strain hardening). Strengthening in brittle materials, such as ceramics, is negligible. In metals:

  - Strain hardening, in addition to increasing strength and hardness, increases residual stresses, produces anisotropic behaviour, and reduces ductility, electrical conductivity, and corrosion resistance.
  - The amount of strain hardening is limited because of the simultaneous decrease in ductility; FCC metals typically have the best response to strengthening.
  - Strain hardening is not effective at elevated temperatures, where the effect of the cold work is eliminated by recrystallisation.

- Annealing is a heat treatment intended to eliminate all or a portion of the effects of strain hardening. The annealing process may involve as many as three steps.
  - Recovery occurs at low temperatures, eliminating residual stresses and restoring electrical conductivity without reducing the strength. A 'stress relief anneal' refers to recovery.
  - Recrystallisation occurs at higher temperatures and eliminates almost all of the effects of strain hardening. The dislocation density decreases dramatically during recrystallisation as new grains nucleate and grow.
  - Grain growth, which normally should be avoided, occurs at still higher temperatures.
- Hot working combines plastic deformation and annealing in a single step, permitting large amounts of plastic deformation without embrittling the material.
- Superplastic deformation provides unusually large amounts of deformation in some materials. Careful control of temperature, grain size, and strain rate are required for superplastic forming.

GLOSSARY

Annealing
A heat treatment used to eliminate part or all of the effects of cold working.

Cold working
Deformation of a metal below the recrystallisation temperature. During cold working, the dislocations become restricted, causing the metal to be strengthened as its shape is changed.

Drawing
A deformation processing technique by which a material is pulled through an opening in a die.

Extrusion
A deformation processing technique by which a material is pushed through an opening in a die.

Fibre texture
A preferred orientation of grains obtained during the drawing process. Certain crystallographic directions in each grain line up with the drawing direction, causing anisotropic behaviour.

Frank-Read source
A pinned dislocation which, under an applied stress, produces additional dislocations. This mechanism is at least partly responsible for strain hardening.

Heat-affected zone
The area adjacent to a weld that is heated during the welding process above some critical temperature at which a change in the structure, such as grain growth or recrystallisation, occurs.

Hot working
Deformation of a metal above the recrystallisation temperature. During hot working, only the shape of the metal changes; the strength remains relatively unchanged because no strain hardening occurs.

Polygonised structure
A subgrain structure produced in the early stages of annealing. The subgrain boundaries are a network of dislocations rearranged during heating.

Recovery
A low-temperature annealing heat treatment designed to eliminate residual stresses introduced during deformation without reducing the strength of the cold-worked material.

Recrystallisation
A medium-temperature annealing heat treatment designed to eliminate all of the effects of the strain hardening produced during cold working. Recrystallisation must be accomplished above the recrystallisation temperature.

Residual stresses
Stresses introduced in a material during processing which, rather than causing deformation of the material, remain stored in the structure. Later release of stresses as deformation can be a problem.

Sheet texture
A preferred orientation of grains obtained during the rolling process. Certain crystallographic directions line up with the rolling direction, and a preferred crystallographic plane becomes parallel to the sheet surface.

Shot peening
Introducing compressive residual stresses at the surface of a part by bombarding that surface with steel shot. The residual stresses may improve the overall performance of the material.

Strain hardening
Strengthening of a material by increasing the number of dislocations by deformation, or cold working. Also known as work hardening.

Strain-hardening coefficient
The effect that strain has on the resulting strength of the material. A material with a high strain-hardening coefficient obtains high strength with only small amounts of deformation or strain.

Strain rate
The rate at which a material is deformed. A material may behave much differently if it is slowly pressed into a shape rather than formed rapidly into a shape by an impact.

Stress relief anneal
The recovery stage of the annealing heat treatment, during which residual stresses are relieved without reducing the mechanical properties of the material.

Superplasticity
The ability of a material to deform uniformly by an exceptionally large amount. Careful control over temperature, grain size, and strain rate are required for a material to behave in a superplastic manner.

Work hardening
A term sometimes used instead of strain hardening or cold working to describe the effect of deformation on strengthening of materials.

PROBLEMS

7.1 A 12.5 mm-diameter metal bar with a 50 mm-gauge length is subjected to a tensile test. The following measurements are made:

Force (kN)	Gauge Length (mm)	Diameter (mm)
122	56.14	12.19
120	62.05	11.60
114	68.57	11.03

Determine the strain-hardening coefficient for the metal. Is the metal most likely to be FCC, BCC, or HCP? Explain.

7.2 A 15 mm-diameter metal bar with a 30 mm-gauge length is subjected to a tensile test. The following measurements are made:

Force (N)	Gauge Length (mm)	Diameter (mm)
16240	36.642	12.028
19066	44.754	10.884
19273	54.663	9.848

Determine the strain-hardening coefficient for the metal. Is the metal most likely to be FCC, BCC, or HCP? Explain.

7.3 A true stress–true strain curve is shown in Figure 7.21. Determine the strain-hardening coefficient for the metal.

7.4 A Cu-30% Zn alloy tensile bar has a strain-hardening coefficient of 0.50. The bar, which has an initial diameter of 10 mm and an initial gauge length of 30 mm, fails at an engineering stress of 120 $MN.m^{-2}$. After fracture, the gauge length is 35 mm and the diameter is 9.26 mm. No necking occurred. Calculate the true stress when the true strain is 0.05.

7.5 The Frank-Read source shown in Figure 7.3(e) has created four dislocation loops from the original dislocation line. Estimate the total dislocation line present in the photograph and determine the percent increase in the length of dislocations produced by the deformation.

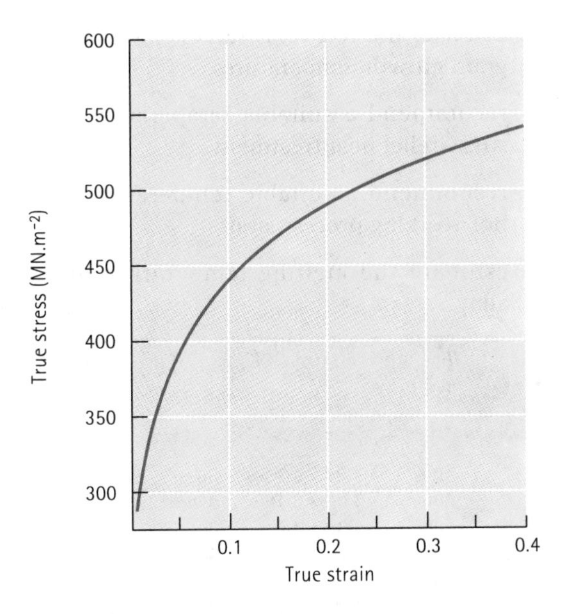

Figure 7.21

True stress-true strain curve (*for* Problem 7.3).

7.6 A 6 mm-thick copper plate is to be cold-worked 63%. Find the final thickness.

7.7 A 6 mm-diameter copper bar is to be cold-worked 63%. Find the final diameter.

7.8 A 50 mm-diameter copper rod is reduced to a 37.5 mm-diameter, then reduced again to a final diameter of 25 mm. In a second case, the 50 mm-diameter rod is reduced in one step from 50 mm to 25 mm diameter. Calculate the % CW for both cases.

7.9 A 3105 aluminium plate is reduced from 45 mm to 30 mm. Determine the final properties of the plate. (*See* Figure 7.22.)

7.10 A Çu-30% Zn brass bar is reduced from 25 mm diameter to a 10 mm diameter. Determine the final properties of the bar. (*See* Figure 7.23.)

7.11 A 3105 aluminium bar is reduced from a 25 mm diameter, to a 20 mm diameter, to a 15 mm diameter, to a final 10 mm diameter. Determine the % CW and the properties after each step of the process. Calculate the total percent cold work. (*See* Figure 7.22.)

7.12 We want a copper bar to have a tensile strength of at least 480 $MN.m^{-2}$ and a final diameter of 9.5 mm. What is the minimum diameter of the original bar? (*See* Figure 7.6.)

7.13 We want a Cu-30% Zn brass plate originally 30 mm thick to have a yield strength greater than 345 $MN.m^{-2}$ and a % elongation of at least 10%. What range of final thicknesses must be obtained? (*See* Figure 7.23.)

7.14 We want a copper sheet to have at least 345 $MN.m^{-2}$ yield strength and at least 10% elongation, with a final thickness of 30 mm. What range of original thickness must be used? (*See* Figure 7.6.)

7.15 A 3105 aluminium plate previously cold-worked 20% is 50 mm thick. It is then cold-worked further to 33 mm. Calculate the total percent cold work and determine the final properties of the plate. (*See* Figure 7.22.)

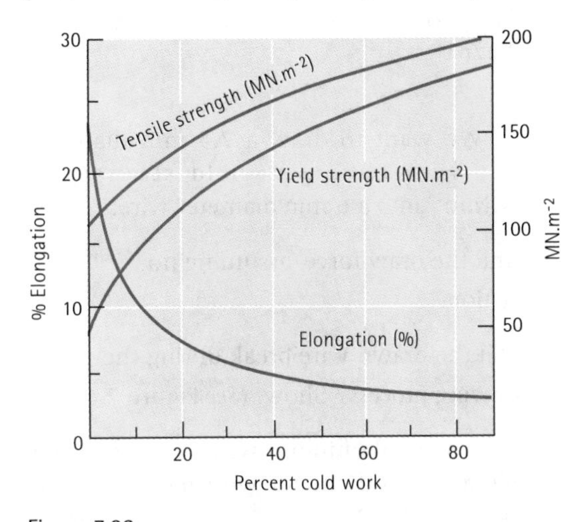

Figure 7.22

The effect of percent cold work on the properties of a 3105 aluminium alloy.

7.16 An aluminium-lithium strap 6 mm thick and 50 mm wide is to be cut from a rolled sheet, as described in Figure 7.9. The strap must be able to support a 150 kN load without plastic deformation. Determine the range of orientations from which the strap can be cut from the rolled sheet.

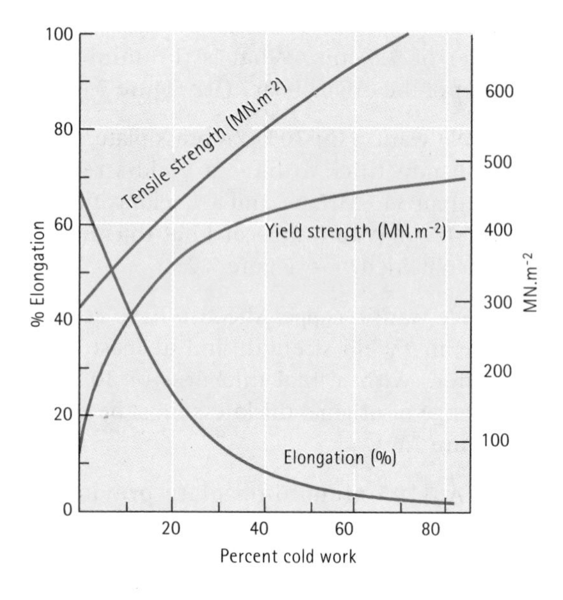

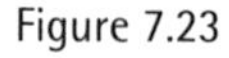

Figure 7.23

The effect of percent cold work on the properties of a Cu-30% Zn brass.

7.17 We want to draw a 7.5 mm-diameter copper wire having a yield strength of 140 MN.m^{-2} into a 6 mm diameter wire.

(a) Find the draw force, assuming no friction.

(b) Will the drawn wire break during the drawing process? Show. (*See* Figure 7.6.)

7.18 A 3105 aluminium wire is to be drawn to give a 1 mm-diameter-wire having a yield strength of 140 MN.m^{-2}.

(a) Find the original diameter of the wire,

(b) calculate the draw force required, and

(c) determine whether the as-drawn wire will break during the process. (*See* Figure 7.22.)

7.19 The following data were obtained when a cold-worked metal was annealed.

(a) Estimate the recovery, recrystallisation, and grain growth temperatures,

(b) recommend a suitable temperature for a stress-relief heat treatment,

(c) recommend a suitable temperature for a hot-working process, and

(d) estimate the melting temperature of the alloy.

Annealing Temperature (°C)	Electrical Conductivity (ohm^{-1}.m^{-1})	Yield Strength (MN.m^{-2})	Grain Size (mm)
400	3.04×10^7	86	0.10
500	3.05×10^7	85	0.10
600	3.36×10^7	84	0.10
700	3.45×10^7	83	0.098
800	3.46×10^7	52	0.030
900	3.46×10^7	47	0.031
1000	3.47×10^7	44	0.070
1100	3.47×10^7	42	0.120

7.20 The following data were obtained when a cold-worked metal was annealed.

(a) Estimate the recovery, recrystallisation, and grain growth temperatures,

(b) recommend a suitable temperature for obtaining a high-strength, high-electrical-conductivity wire,

(c) recommend a suitable temperature for a hot-working process, and

(d) estimate the melting temperature of the alloy.

Annealing Temperature (°C)	Residual Stresses (MN.m^{-2})	Tensile Strength (MN.m^{-2})	Grain Size (mm)
250	145	360	0.076
275	145	360	0.076
300	35	360	0.076
325	0	360	0.076
350	0	235	0.025
375	0	200	0.025
400	0	185	0.089
425	0	170	0.183

7.21 A titanium alloy contains a very fine dispersion of tiny Er_2O_3 particles. What will be the effect of these particles on the grain growth temperature and the size of the grains at any particular annealing temperature? Explain.

7.22 Determine the grain size Index G_E for each of the micrographs in Figure 7.17 and plot the grain size number versus the annealing temperature.

7.23 Using the data in Table 7.3, plot the recrystallisation temperature versus the melting temperature of each metal, using absolute temperatures (Kelvin). Measure the slope and compare with the expected relationship between these two temperatures. Is our approximation a good one?

7.24 We wish to produce a 7.5 mm-thick plate of 3105 aluminium having a tensile strength of at least 170 $MN.m^{-2}$ and a % elongation of at least 5%. The original thickness of the plate is 75 mm. The maximum cold work in each step is 80%. Describe the cold-working and annealing steps required to make this product. Compare this process with that you would recommend if you could do the initial deformation by hot working. (*See* Figure 7.22.)

7.25 We wish to produce a 5 mm diameter wire of copper having a minimum yield strength of 400 $MN.m^{-2}$ and a minimum % elongation of 5%. The original diameter of the rod is 50 mm and the maximum cold work in each step is 80%. Describe the cold-working and annealing steps required to make this product. Compare this process with that you would recommend if you could do the initial deformation by hot working. (*See* Figure 7.6.)

7.26 Determine, using one of the processes shown in Figure 7.5, a method to produce each of the following products. Should the process include hot working, cold working, annealing, or some combination of these? Explain your decisions.

(a) paper clips

(b) I-beams that will be welded to produce a portion of a bridge

(c) copper tubing that will connect a water tap to the main copper plumbing

(d) the steel tape in a tape measure

(e) a head for a carpenter's hammer formed from a round rod

7.27 We plan to join two sheets of cold-worked copper by soldering. (Soldering involves heating the metal to a high enough temperature that a filler material melts and is drawn into the joint.) Describe a soldering process that will not soften the copper. Explain. Could we use higher soldering temperatures if the sheet material were a Cu-30% Zn alloy? Explain.

7.28 We wish to produce a 1 mm-diameter copper wire having a minimum strength of 400 $MN.m^{-2}$ and a minimum % elongation of 5%. We start with a 20 mm-diameter rod. Design the process by which the wire can be drawn. Include all important details and explain.

CHAPTER 8

Principles of Solidification Strengthening and Processing

8.1 Introduction

In almost all metals and alloys, as well as in many semiconductors, composites, ceramics, and polymers, the material at one point in its processing is a liquid. The liquid solidifies as it cools below its freezing temperature. The material may be used in the as-solidified condition or may be further processed by mechanical working or heat treatment. The structures produced during solidification affect mechanical properties and influence the type of further processing needed. In particular, grain size and shape may be controlled by solidification.

During solidification, the atomic arrangement changes from, at best, a short-range order to a long-range order, or crystal structure. Solidification requires two steps: nucleation and growth. Nucleation occurs when a small piece of solid forms from the liquid. Growth of the solid occurs as atoms from the liquid are attached to the solid until no liquid remains.

In this chapter, the principles of solidification are introduced, concentrating on behaviour in pure materials. Subsequent chapters show how solidification differs in alloys and multiple-phase materials.

8.2 Nucleation

We expect a material to solidify when the liquid cools to just below its freezing (or melting) temperature, because the energy associated with the crystalline structure of the solid is then less than the energy of the liquid. The energy difference between liquid and solid is the volume free energy ΔG_v; as the solid grows in size, ΔG_v increases.

When the solid forms, however, an interface is created between it and the remaining liquid (Figure 8.1). A surface free energy σ is associated with this interface; the larger the solid, the greater the increase in surface energy. Thus, the total change in energy ΔG, shown in Figure 8.2, is:

$$\Delta G = \frac{4}{3}\pi r^3 \, \Delta G_v + 4\pi r^2 \sigma \qquad (8.1)$$

where $\frac{4}{3}\pi r^3$ is the volume of a spherical embryo of radius r, $4\pi r^2$ is the surface area of a spherical embryo, σ is the surface free energy, and ΔG_v is the volume free energy, which is a negative change.

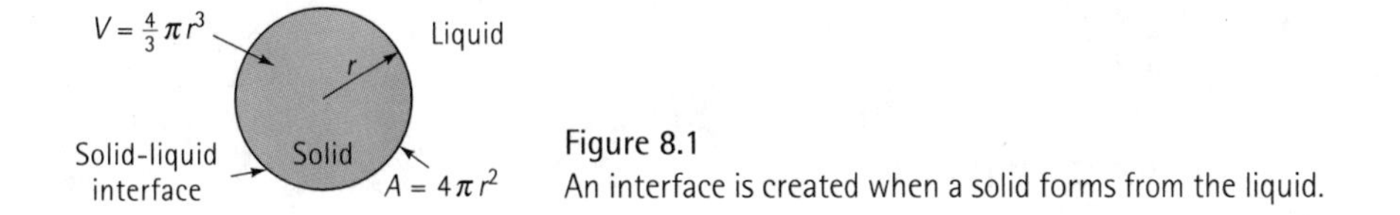

Figure 8.1
An interface is created when a solid forms from the liquid.

When the solid is very small (less than r^* in Figure 8.2), further growth causes the free energy to increase. Instead of growing, the solid prefers to remelt and cause the free energy to decrease; thus, the metal remains liquid. This small solid is called an embryo. Because the liquid is below the equilibrium freezing temperature, the liquid is undercooled. The undercooling ΔT is the equilibrium freezing temperature minus the actual temperature of the liquid.

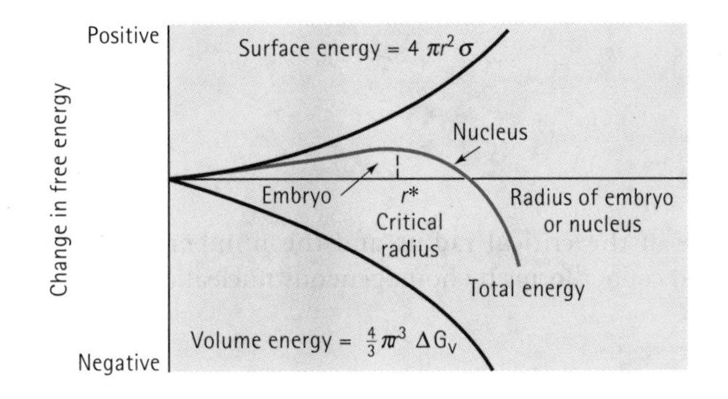

Figure 8.2
The total free energy of the solid-liquid system changes with the size of the solid. The solid is an embryo if its radius is less than the critical radius, and is a nucleus if its radius is greater than the critical radius.

But when the solid is larger than r^*, further growth causes the total energy to decrease. The solid that now forms is stable, nucleation has occurred, and growth of the solid particle – which is now called a nucleus – begins.

Homogeneous Nucleation As the liquid cools further below the equilibrium freezing temperature, two factors combine to favour nucleation. First, atoms cluster to form larger embryos. Second, the larger volume free energy difference between the liquid and the solid reduces the critical size of the nucleus. Homogeneous nucleation occurs when the undercooling becomes large enough to cause formation of a stable nucleus.

The size of the critical radius r^* is given by

$$r^* = \frac{2\sigma T_m}{\Delta H_f \Delta T} \tag{8.2}$$

where ΔH_f is the latent heat of fusion, T_m is the equilibrium solidification temperature in Kelvin, and $\Delta T = T_m - T$ is the undercooling when the liquid temperature is T. The latent heat of fusion represents the heat given up during the liquid-to-solid transformation. As the undercooling increases, the critical radius required for nucleation decreases. Table 8.1 presents values for σ, ΔH_f, and typical undercoolings observed experimentally for homogeneous nucleation.

Table 8.1
Values for freezing temperature, latent heat of fusion, surface energy, and maximum undercooling for selected materials.

Metal	Freezing Temperature (°C)	Latent Heat of Fusion $(J.m^{-3}) \times 10^6$	Surface Energy $(J.m^{-2}) \times 10^{-3}$	Typical Undercooling for Homogeneous Nucleation (°C)
Ga	30	488	56	76
Bi	271	543	54	90
Pb	327	237	33	80
Ag	962	965	126	250
Cu	1085	1628	177	236
Ni	1453	2756	255	480
Fe	1538	1737	204	420
NaCl	801			169
CsCl	645			152
H_2O	0			40

EXAMPLE 8.1

Calculate the size of the critical radius and the number of atoms in the critical nucleus when solid copper forms by homogeneous nucleation.

SOLUTION

From Table 8.1:

$$\Delta T = 236°C \qquad T_m = 1085 + 273 = 1358 \text{ K}$$

$$\Delta H_f = 1\,628 \times 10^6 \text{ J.m}^{-3}$$

$$\sigma = 177 \times 10^{-3} \text{ J.m}^{-2}$$

$$r^* = \frac{2\sigma T_m}{\Delta H_f \Delta T} = \frac{(2)(177 \times 10^{-3})(1\,358)}{(1\,628 \times 10^6)(236)} = 12.51 \times 10^{-10} \text{ m}$$

The lattice parameter for FCC copper is $a_0 = 0.3615$ nm $= 3.615 \times 10^{-10}$ m

$$V_{\text{unit cell}} = (a_0)^3 = (3.615 \times 10^{-10})^3 = 47.24 \times 10^{-30} \text{ m}^3$$

$$V_{r^*} = \tfrac{4}{3}\pi r^3 = (\tfrac{4}{3}\pi)(12.51 \times 10^{-10})^3 = 8200 \times 10^{-30} \text{ m}^3$$

The number of unit cells in the critical nucleus is

$$\frac{8\,200 \times 10^{-30}}{47.24 \times 10^{-30}} = 174 \text{ unit cells}$$

Since there are four atoms in each unit cell of FCC metals, the number of atoms in the critical nucleus must be:

(4 atom/cell)(174 cell/nucleus) = 696 atoms/nucleus

Heterogeneous Nucleation Except in unusual laboratory experiments, homogeneous nucleation never occurs in liquid metals. Instead, impurities in contact with the liquid, either suspended in the liquid or on the walls of the container that holds the liquid, provide a surface on which the solid can form (Figure 8.3). Now, a radius of curvature greater than the critical radius is achieved with very little total surface between the solid and liquid. Only a few atoms must cluster together to produce a solid particle that has the required radius of curvature. Much less undercooling is required to achieve the critical size, so nucleation occurs more readily. Nucleation on impurity surfaces is known as heterogeneous nucleation

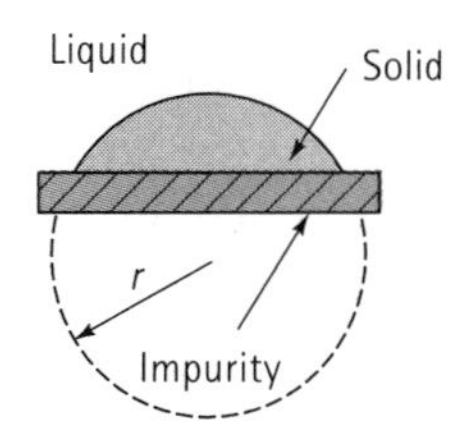

Figure 8.3
A solid forming on an impurity can assume the critical radius with a smaller increase in the surface energy. Thus, heterogeneous nucleation can occur with relatively low undercoolings.

Grain Size Strengthening Sometimes we intentionally introduce impurity particles into the liquid. Such practices are called grain refinement or inoculation. For example, a combination of 0.03% titanium and 0.01% boron is added to many liquid aluminium alloys. Tiny particles of Al_3Ti or TiB_2 form and serve as sites for heterogeneous nucleation. Grain refining or inoculation produces a large number of grains, each beginning to grow from one nucleus. The greater grain boundary surface area provides grain size strengthening in metals.

Glasses For rapid cooling rates, there may be insufficient time for nuclei to form and grow. When this happens, the liquid structure is locked into place and an amorphous – or glassy – solid forms. The complex crystal structure of many ceramic and polymer materials prevents nucleation of a solid crystalline structure even at slow cooling rates.

In metals, however, cooling rates of 10^6 °C/s or faster may be required to suppress nucleation of the crystal structure. The production of metallic glasses – as well as other unique structures – by rapid cooling has been termed rapid solidification processing. The high cooling rates are obtained using tiny metal powder particles or by forming continuous thin metallic ribbons about 0.04 mm in thickness.

Metallic glasses include complex iron-nickel-boron alloys containing chromium, phosphorus, cobalt, and other elements. Some metal glasses may obtain strengths in excess of 3500 $MN.m^{-2}$ while retaining fracture toughness of more than 11 $MN.m^{-3/2}$. Excellent corrosion resistance, magnetic properties, and other physical properties make these materials attractive for applications involved in electrical power, aircraft engines, tools and dies, and magnetism.

8.3 Growth

Once solid nuclei form, growth occurs as atoms are attached to the solid surface. The nature of the growth of the solid depends on how heat is removed from the system. Two types of heat must be removed: the specific heat of the liquid and the latent heat of fusion. The specific heat is the heat required to change the temperature of a unit weight of the material by one degree. The specific heat must be removed first, either by radiation into the surrounding atmosphere or by conduction into the surrounding mould, until the liquid cools to its freezing temperature. The latent heat of fusion must be removed from the solid-liquid interface before solidification is completed. The manner in which we remove the latent heat of fusion determines the growth mechanism and final structure.

Planar growth When a well-inoculated liquid cools under equilibrium conditions, the temperature of the liquid is greater than the freezing temperature and the temperature of the solid is at or below the freezing temperature. During solidification, the latent heat of fusion is removed by conduction from the solid-liquid interface through the solid to the surroundings. Any small protuberance that begins to grow on the interface is surrounded by liquid above the freezing temperature (Figure 8.4). The growth of the protuberance then stops until the remainder of the interface catches up. This growth mechanism, known as planar growth, occurs by the movement of a smooth solid-liquid interface into the liquid.

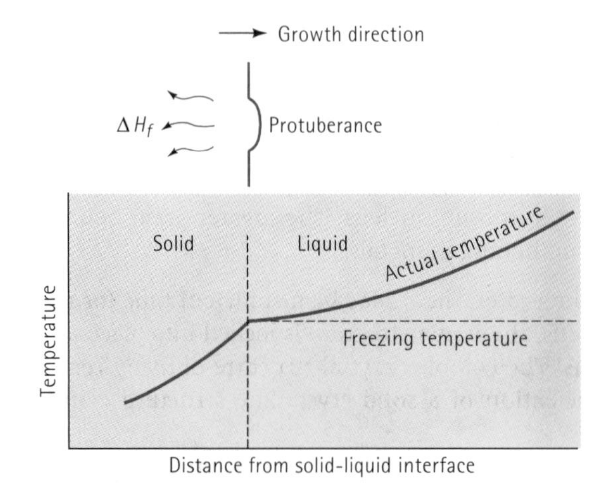

Figure 8.4
When the temperature of the liquid is above the freezing temperature, a protuberance on the solid-liquid interface will not grow, leading to maintenance of a planar interface. Latent heat is removed from the interface through the solid.

Dendritic Growth When nucleation is poor, the liquid undercools before the solid forms (Figure 8.5). Under these conditions, a small solid protuberance called a dendrite, which forms at the interface, is encouraged to grow. As the solid dendrite grows, the latent heat of fusion is conducted into the undercooled liquid, raising the temperature of the liquid toward the freezing temperature. Secondary and tertiary dendrite arms can also form on the primary stalks to speed the evolution of the latent heat. Dendritic growth continues until the undercooled liquid warms to the freezing temperature. Any remaining liquid then solidifies by planar growth. The difference between planar and dendritic growth arises because of the different sinks for the latent heat. The container or mould must absorb the heat in planar growth, but the undercooled liquid absorbs the heat in dendritic growth.

In pure metals, dendritic growth normally represents only a small fraction of the total growth:

$$\text{Dendritic fraction} = f = \frac{c\Delta T}{\Delta H_f} \tag{8.3}$$

where c is the specific heat of the liquid. The numerator represents the heat that the undercooled liquid can absorb, and the latent heat in the denominator represents the total heat that must be given up during solidification. As the undercooling ΔT increases, more dendritic growth occurs.

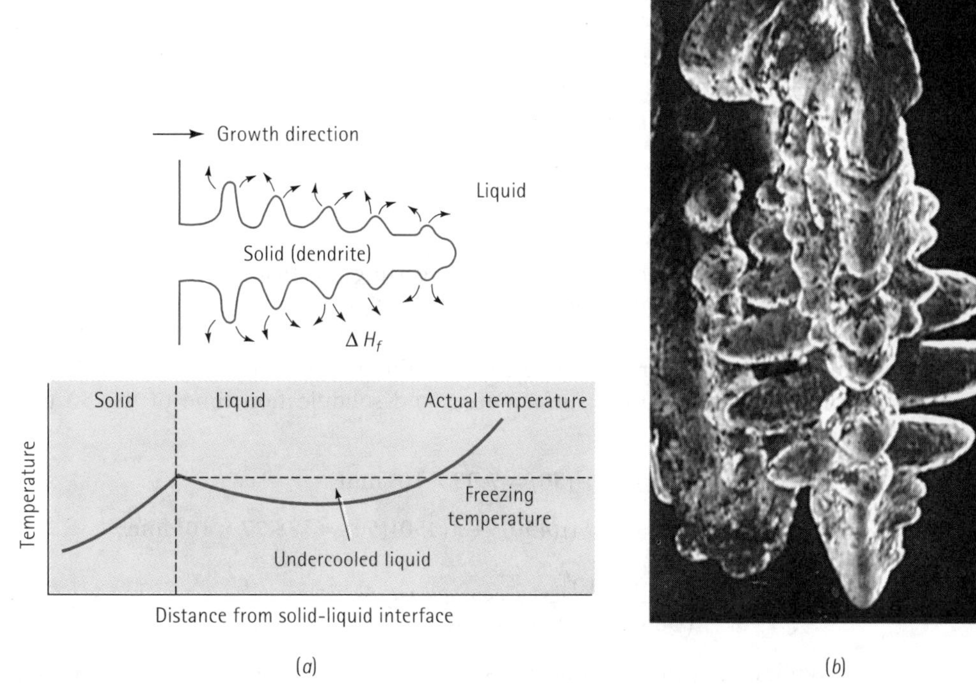

Figure 8.5
(*a*) If the liquid is undercooled, a protuberance on the solid-liquid interface can grow rapidly as a dendrite. The latent heat of fusion is removed by raising the temperature of the liquid back to the freezing temperature.
(*b*) Scanning electron micrograph of dendrites in steel (× 15).

8.4 Solidification Time and Dendrite Size

The rate at which growth of the solid occurs depends on the cooling rate, or rate of heat extraction. A fast cooling rate produces rapid solidification or short solidification times. The time t_s required for a simple casting to solidify completely can be calculated using *Chvorinov's rule*

$$t_s = B\left(\frac{V}{A}\right)^n \tag{8.4}$$

where V is the volume of the casting and represents the amount of heat that must be removed before freezing occurs, A is the surface area of the casting in contact with the mould and represents the surface from which heat can be transferred away from the casting, n is a constant (usually about 2), and B is the *mould constant.* The mould constant depends on the properties and initial temperatures of both the metal and the mould.

EXAMPLE 8.2 Casting for Improved Strength

Your company currently is producing a disc-shaped brass casting 50 mm thick and 450 mm in diameter. You believe that by making the casting solidify 25% faster, the improvement in the tensile properties of the casting will permit the casting to be made lighter in weight. Design the casting to permit this. Assume that the mould constant is 2 s.mm^{-2} for this process.

SOLUTION

One approach would be to use the same casting process, but reduce the thickness of the casting. The thinner casting would solidify more quickly and, because of the faster cooling, should have improved mechanical properties. Chvorinov's rule helps us calculate the required thickness. If d is the diameter and x is the thickness of the casting, then the volume, surface area, and solidification time of the 50 mm-thick casting are:

$$V = (\pi/4)d^2x = (\pi/4)(450)^2(50) = 7.95 \times 10^6 \text{ mm}^3$$

$$A = 2(\pi/4)d^2 + \pi dx = 2(\pi/4)(450)^2 + \pi(450)(50) = 388.77 \times 10^3 \text{ mm}^2$$

$$t = B\left(\frac{V}{A}\right)^2 = (2)\left(\frac{7.95\times10^6}{388.77\times10^3}\right)^2 = 837\,\text{s} = 13.95\,\text{min}$$

The solidification time of the redesigned casting should be 25% shorter than the current time, or $t_r = 0.75t$:

$$t_r = 0.75t = (0.75)(837) = 627.75 \text{ s}$$

Since the casting conditions have not changed, the mould constant B is unchanged. The V/A ratio of the new casting is:

$$t_r = B\left(\frac{V}{A}\right)^2 = (2)\left(\frac{V}{A}\right)^2 = 627.75$$

$$\left(\frac{V}{A}\right)^2 = 313.875 \quad \text{or} \quad \left(\frac{V}{A}\right) = 17.717$$

If x is the required thickness for our redesigned casting, then:

$$\frac{V_r}{A_r} = \frac{(\pi/4)d^2x}{2(\pi/4)d^2 + \pi dx} = \frac{(\pi/4)(450)^2(x)}{2(\pi/4)(450)^2 + \pi(450)x} = 17.717$$

$x = 42.06$ mm

This thickness provides the required solidification time, while reducing the overall weight of the casting by nearly 16%.

Solidification begins at the surface, where heat is dissipated into the surrounding mould material. The rate of solidification of a casting can be described by how rapidly the thickness d of the solidified skin grows:

$$d = k\sqrt{t} - c \qquad (8.5)$$

where t is the time after pouring, k is a constant for a given casting material and mould, and c is a constant related to the pouring temperature.

Effect on Structure and Properties The solidification time affects the size of the dendrites. Normally, dendrite size is characterised by measuring the distance between the secondary dendrite arms (Figure 8.6). The secondary dendrite arm spacing, or *SDAS*, is reduced when the casting freezes more rapidly. The finer, more extensive dendritic network serves as a more efficient conductor of the latent heat to the undercooled liquid. The *SDAS* is related to the solidification time by

$$SDAS = kt_s^m \qquad (8.6)$$

where m and k are constants depending on the composition of the metal. This relationship is shown in Figure 8.7 for several alloys. Small secondary dendrite arm spacings are associated with higher strengths and improved ductility (Figure 8.8).

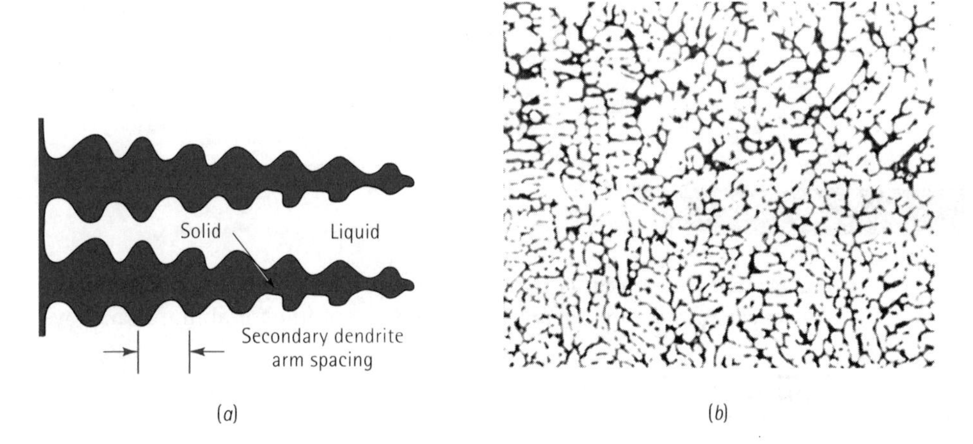

Figure 8.6
(*a*) The secondary dendrite arm spacing (*SDAS*). (*b*) Dendrites in an aluminium alloy (× 50). (*From* Metals Handbook, *Vol. 9, 9th Ed., American Society for Metals, 1985.*)

Rapid solidification processing is used to produce exceptionally fine secondary dendrite arm spacings; a common method is to produce very fine liquid droplets using special atomisation processes. The tiny droplets freeze at a rate of about 10^4 °C/s. This cooling rate is not rapid enough to form a metallic glass, but does produce a fine dendritic structure. By carefully consolidating the solid droplets by powder metallurgy processes, improved properties in the material can be obtained.

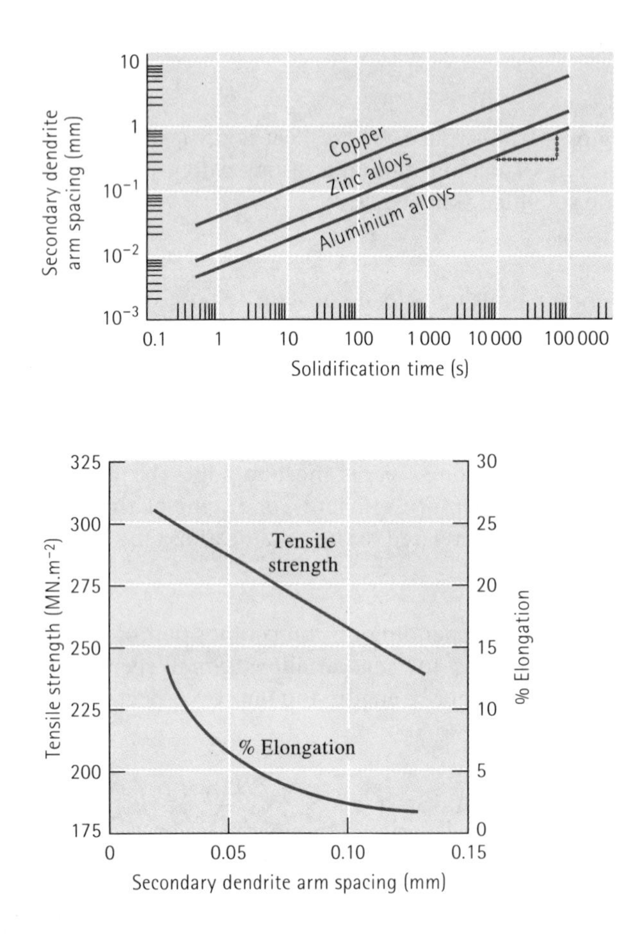

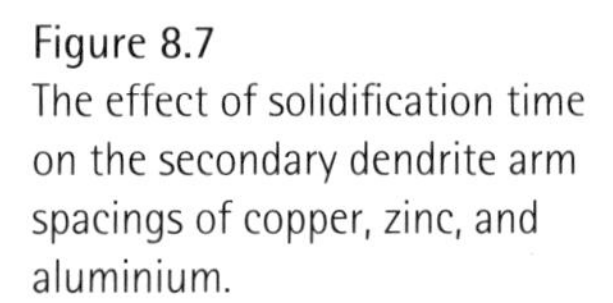

Figure 8.7
The effect of solidification time on the secondary dendrite arm spacings of copper, zinc, and aluminium.

Figure 8.8
The effect of the secondary dendrite arm spacing on the properties of an aluminium casting alloy.

EXAMPLE 8.3

Determine the constants in the equation that describes the relationship between secondary dendrite arm spacing and solidification time for aluminium alloys (Figure 8.7).

SOLUTION

We could obtain the value of *SDAS* at two times from the graph and calculate k and m using simultaneous equations. However, if the scales on the ordinate and abscissa are equal for powers of ten (as in Figure 8.7), we can obtain the slope m from the log-log plot by directly measuring the slope of the graph. In Figure 8.7, five equal units are marked on the vertical scale and 12 equal units are marked on the horizontal scale. The slope is:

$$m = \tfrac{5}{12} = 0.42$$

The constant k is the value of $SDAS$ when $t_s = 1$, since:

$$\log SDAS = \log k + m \log t_s$$

If $t_s = 1$, $m \log t_s = 0$, and $SDAS = k$, from Figure 8.7:

$$k = 8 \times 10^{-3} \text{ mm}$$

EXAMPLE 8.4

A 100 mm diameter aluminium bar solidifies to a depth of 12 mm beneath the surface in 5 minutes. After 20 minutes, the bar has solidified to a depth of 36 mm. How much time is required for the bar to solidify completely?

SOLUTION

From our measurements, we can determine the constants k and c in Equation 8.5:

$$12 \text{ mm} = k\sqrt{(5 \text{ min})} - c \quad or \quad c = k\sqrt{5} - 12$$
$$36 \text{ mm} = k\sqrt{(20 \text{ min})} - c = k\sqrt{20} - \left(k\sqrt{5} - 12\right)$$
$$36 = k\left(\sqrt{20} - \sqrt{5}\right) + 12$$
$$k = \frac{36 - 12}{4.472 - 2.236} = 10.733$$
$$c = (10.733)\sqrt{5} - 12 = 12$$

Solidification is complete when d = 50 mm (half the diameter, since freezing is occurring from all surfaces):

$$50 = 10.733\sqrt{t} - 12$$
$$\sqrt{t} = \frac{50 + 12}{10.733} = 5.78$$
$$t = 33.37 \text{min}$$

In actual practice, we would find that the total solidification time is somewhat longer than 33.37 min. As solidification continues, the mould becomes hotter and is less effective in removing heat from the casting.

EXAMPLE 8.5

Determine the thickness of an aluminium casting whose length is 300 mm and width is 200 mm, in order to produce a tensile strength of 275 $MN.m^{-2}$. The mould constant in Chvorinov's rule for aluminium alloys cast in a sand mould is 4 $s.mm^{-2}$.

SOLUTION

In order to obtain a tensile strength of 275 $MN.m^{-2}$, a secondary dendrite arm spacing of about 0.07 mm is required (*see* Figure 8.8). From Figure 8.7, we can determine that the solidification time required to obtain this spacing is about 300 s, or 5 min. From Chvorinov's rule:

$$t_s = B\left(\frac{V}{A}\right)^2$$

where $B = 4$ $s.mm^{-2}$ and x is the thickness of the casting. Since the length is 300 mm and the width is 200 mm:

$$V = (200)(300)(x) = 60 \times 10^3 x$$

$$A = (2)(200)(300) + (2)(x)(200) + (2)(x)(300) = 1\,000x + 120 \times 10^3$$

$$300 \text{ s} = 4 \text{ s.mm}^{-2}\left(\frac{60 \times 10^3 x}{1000x + 120 \times 10^3}\right)^2$$

$$\frac{60 \times 10^3 x}{1\,000x + 120 \times 10^3} = \sqrt{300/4} = 8.66$$

$$60 \times 10^3 x = 8\,660x + 1.0392 \times 10^6$$

$$x = 20.24 \text{ mm}$$

8.5 Cooling Curves

We can summarise our discussion to this point by examining a cooling curve, which shows how the temperature of a material changes with time (Figure 8.9). The liquid is poured into a mould at the pouring temperature. The difference between the pouring temperature and the freezing temperature is the superheat. The liquid cools as the specific heat is extracted by the mould until the liquid reaches the freezing temperature. The slope of the cooling curve before solidification begins is the *cooling rate* $\Delta T/\Delta t$.

If effective heterogeneous nuclei are present in the liquid, solidification begins at the freezing temperature. A thermal arrest, or plateau, is produced because of the evolution of the latent heat of fusion. The latent heat keeps the remaining liquid at the freezing temperature until all of the liquid has solidified and no more heat can be evolved. Growth under these conditions is planar. The total solidification time of the casting is the time required to remove both the specific heat of the superheated liquid and the latent heat of fusion. Measured from the time of pouring until solidification is complete, this time is given by Chvorinov's rule. The local solidification time is the time required to remove only the latent heat of fusion at a particular location in the casting; it is measured from when solidification begins until solidification is completed.

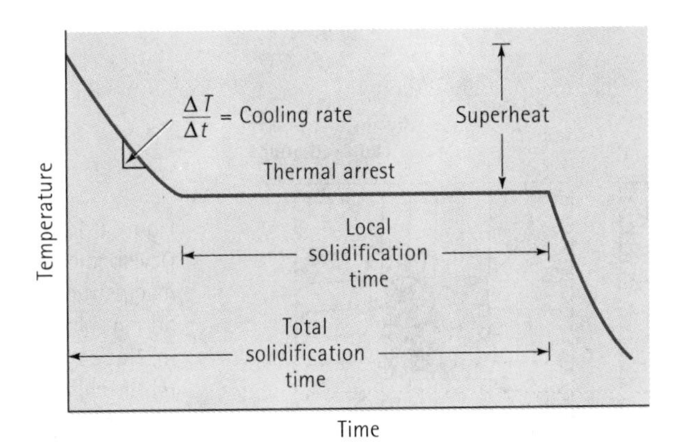

Figure 8.9
Cooling curve for the solidification of a pure material.

8.6 Casting or Ingot Structure

Molten metals are poured into moulds and permitted to solidify. Often the mould produces a finished shape, or casting. In other cases, the mould produces a simple shape, called an ingot, that requires extensive plastic deformation or machining before a finished product is created. A macrostructure, sometimes referred to as the ingot structure, is produced; it can consist of as many as three parts (Figure 8.10).

Chill Zone The chill zone is a narrow band of randomly oriented grains at the surface of the casting. The metal at the mould wall is the first to cool to the freezing temperature. The mould wall also provides many surfaces at which heterogeneous nucleation may take place.

Columnar Zone The columnar zone contains elongated grains oriented in a particular crystallographic direction. As heat is removed from the casting by the mould material, the grains in the chill zone grow in the direction opposite to the heat flow, or from the coldest toward the hottest areas of the casting. This tendency usually means that the grains grow perpendicular to the mould wall.

Grains grow fastest in certain crystallographic directions. In metals with a cubic crystal structure, grains in the chill zone that have a <100> direction perpendicular to the mould wall grow faster than other less favourably oriented grains (Figure 8.11). Eventually, the grains in the columnar zone have <100> directions that are parallel to one another, giving the columnar zone anisotropic properties.

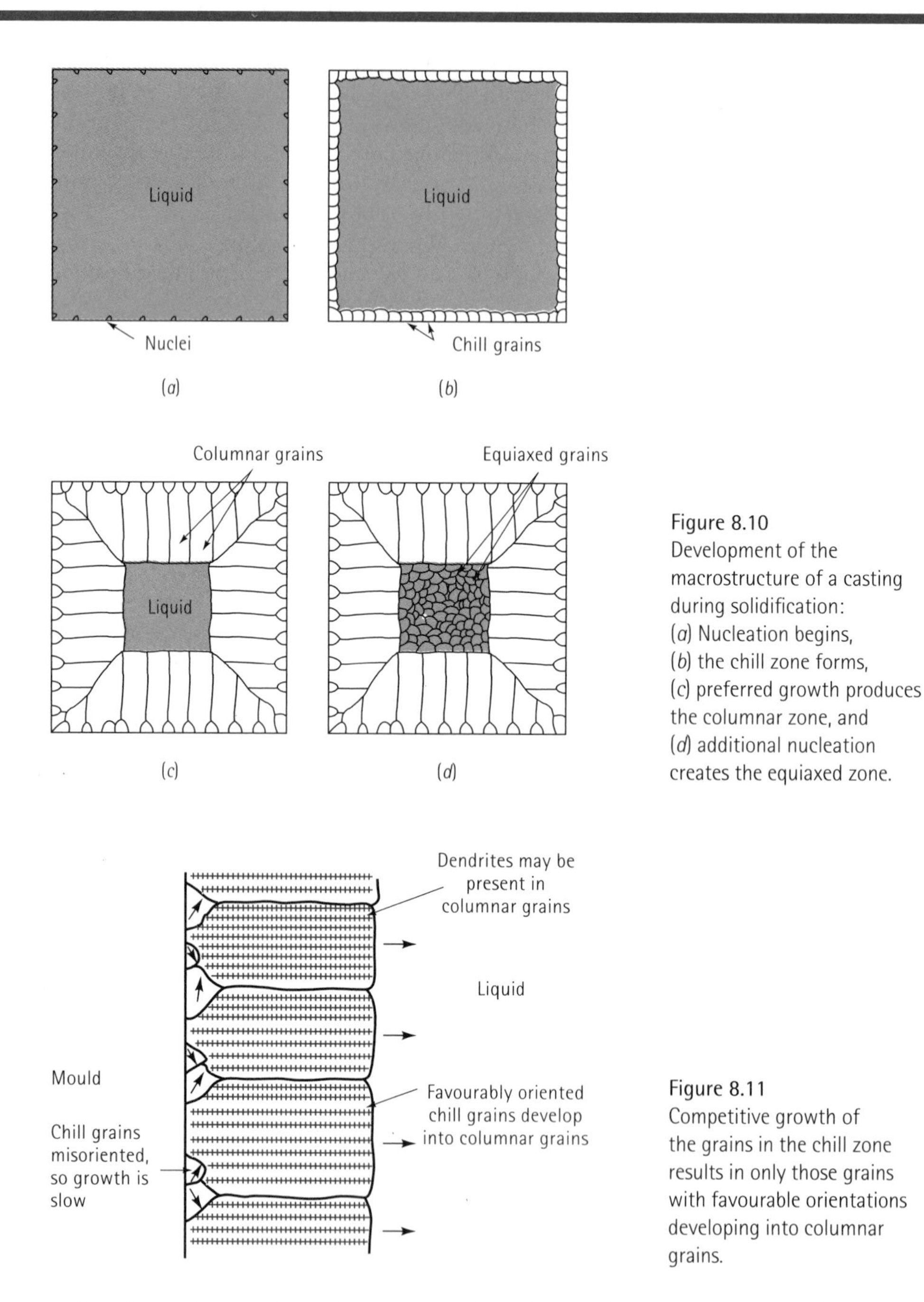

Figure 8.10
Development of the macrostructure of a casting during solidification:
(*a*) Nucleation begins,
(*b*) the chill zone forms,
(*c*) preferred growth produces the columnar zone, and
(*d*) additional nucleation creates the equiaxed zone.

Figure 8.11
Competitive growth of the grains in the chill zone results in only those grains with favourable orientations developing into columnar grains.

This formation of the columnar zone is influenced primarily by growth – rather than nucleation – phenomena. The grains may be composed of many dendrites if the liquid is originally undercooled. Or solidification may proceed by planar growth of the columnar grains if no undercooling occurs.

Equiaxed Zone Although the solid may continue to grow in a columnar manner until all of the liquid has solidified, an equiaxed zone frequently forms in the centre of the casting or ingot. The equiaxed zone contains new, randomly oriented grains, often caused by a low pouring temperature, alloying elements, or grain refining or inoculating agents. These grains grow as relatively round, or equiaxed, grains with a random orientation, and they stop the growth of the columnar grains. The formation of the equiaxed zone is a nucleation-controlled process and causes that portion of the casting to display isotropic behaviour.

8.7 Solidification of Polymers

Solidification of polymers is significantly different from that of metals, requiring long polymer chains to become closely aligned over relatively large distances. By doing so, the polymer grows as lamellar, or plate-like, crystals (Figure 8.12). The region between each lamella contains polymer chains arranged in an amorphous manner.

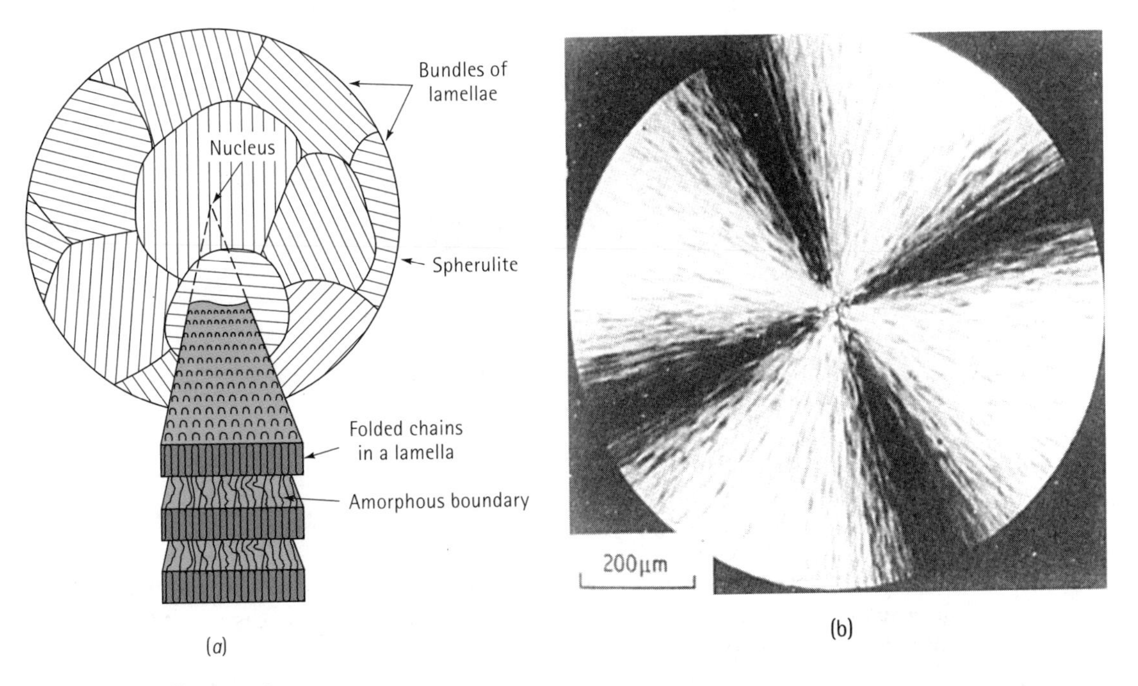

Figure 8.12
(a) Structure of a polymer spherulite: The spherulite consists of bundles of lamellae, each an individual crystal. The lamellae contain polymer chains formed into a crystalline structure. An amorphous boundary region separates the lamellae. *(b)* A spherulite in polystyrene (× 8000). (*From R. Young and P. Lovell,* Introduction to Polymers, *2nd. Ed., Chapman & Hall, 1991.*)

In addition, bundles of lamellae grow from a common nucleus, but the orientation of the lamellae differ from one bundle to another. As the bundles grow, they may produce a spheroidal shape called a spherulite. The spherulite is composed of many individual bundles of differently oriented lamellae.

Many polymers do not crystallise, or solidify, when cooled. The rate of nucleation of the solid may be too slow, or the complexity of the polymer chains may be so great that a crystalline solid does not form. Even when solidification does occur, crystallisation in polymers is never complete. Amorphous regions are present between the individual lamellae, between bundles of lamellae, and between individual spherulites.

8.8 Solidification Defects

Although there are many defects that potentially can be introduced during solidification, two deserve special mention.

Shrinkage Almost all materials are more dense in the solid state then in the liquid state. During solidification, the material contracts, or shrinks, as much as 7% (Table 8.2). Often, the bulk of the shrinkage occurs as cavities, if solidification begins at all surfaces of the casting, or pipes, if one surface solidifies more slowly then the others (Figure 8.13).

Table 8.2 Shrinkage during solidification for selected materials

Material	Shrinkage (%)
Al	7.0
Cu	5.1
Mg	4.0
Zn	3.7
Fe	3.4
Pb	2.7
Ga	+3.2 (expansion)
H_2O	+8.3 (expansion)

A common technique for controlling cavity and pipe shrinkage is to place a riser or an extra reservoir of metal, adjacent and connected to the casting. As the casting solidifies and shrinks, liquid metal flows from the riser into the casting to fill the shrinkage void. We need only assure that the riser solidifies after the casting and that there is an internal liquid channel that connects the liquid in the riser to the last liquid to solidify in the casting. Chvorinov's rule can be used to help design the size of the riser.

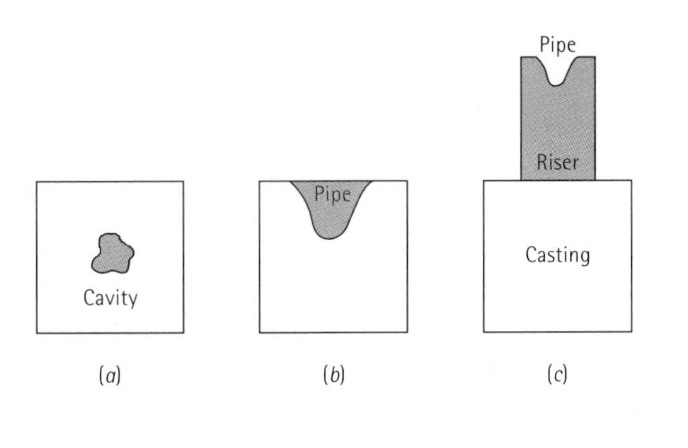

Figure 8.13
Several types of macroshrinkage can occur, including cavities and pipes. Risers can be used to help compensate for shrinkage.

8

EXAMPLE 8.6

Design a cylindrical riser, with a height equal to twice its diameter, that will compensate for shrinkage in a 20 mm × 80 mm × 160 mm casting (Figure 8.14).

SOLUTION

We know that the riser must freeze after the casting. To be conservative, however, we typically require that the riser take 25% longer to solidify than the casting. Therefore:

$$t_{\text{riser}} = 1.25 t_{\text{casting}} \quad \text{or} \quad B\left(\frac{V}{A}\right)_r^2 = 1.25B\left(\frac{V}{A}\right)_c^2$$

The mould constant B is the same for both casting and riser, so:

$$(V/A)_r = \sqrt{1.25}(V/A)_c$$
$$V_c = (20)(80)(160) = 256 \times 10^3 \text{ mm}^3$$
$$A_c = (2)(20)(80) + (2)(20)(160) + (2)(80)(160) = 35.2 \times 10^3 \text{ mm}^2$$

We can write equations for the volume and the area of a cylindrical riser in contact with the mould, noting that $H = 2D$:

$$V_r = (\pi/4)D^2 H = (\pi/4)D^2(2D) = (\pi/2)D^3$$
$$A_r = \pi DH = \pi D(2D) = (2\pi)D^2$$
$$\frac{V_r}{A_r} = \frac{(\pi/2)(D)^3}{(2\pi)(D)^2} = \frac{D}{4} > \sqrt{1.25}\left(\frac{256 \times 10^3}{32.5 \times 10^3}\right)$$
$$D = 35.2 \text{ mm} \quad H = 2D = 70.4 \text{ mm} \quad V_r = 68.51 \times 10^3 \text{ mm}^3$$

Although the volume of the riser is less then that of the casting, the riser solidifies slowly because of its compact shape.

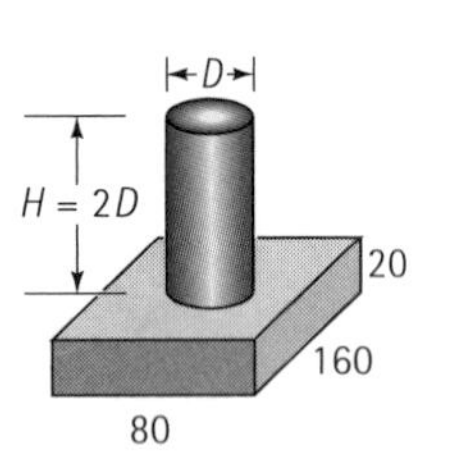

Figure 8.14
The geometry of the casting and riser (*for* Example 8.6).

Interdendritic shrinkage consists of small shrinkage pores between dendrites (Figure 8.15). This defect, also called *microshrinkage* or *shrinkage porosity,* is difficult to prevent by the use of risers. Fast cooling rates may reduce problems with interdendritic shrinkage; the dendrites may be shorter, permitting liquid to flow through the dendritic network to the solidifying solid interface. In addition, any shrinkage that remains may be finer and more uniformly distributed.

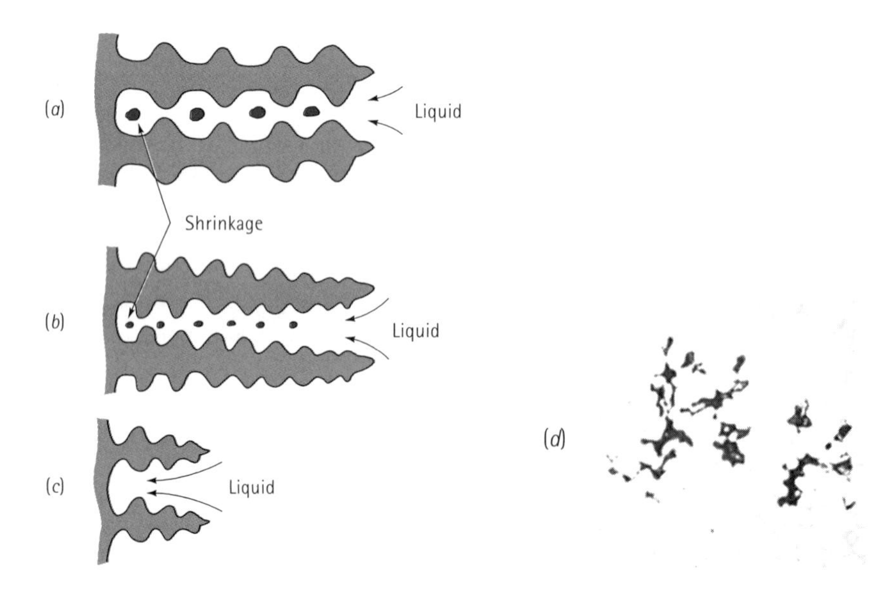

Figure 8.15
(*a*) Shrinkage can occur between the dendrite arms. Small secondary dendrite arm spacings result in smaller, more evenly distributed shrinkage porosity (*b*). Short primary arms can help avoid shrinkage (*c*). Interdendritic shrinkage in an aluminium alloy is shown in (*d*) (× 80).

Gas Porosity Many metals dissolve a large quantity of gas when they are liquid. Aluminium, for example, dissolves hydrogen. When the aluminium solidifies, however, the solid metal retains in its structure only a small fraction of the hydrogen (Figure 8.16). The excess hydrogen forms bubbles that may be trapped in the solid metal, producing gas porosity. The amount of gas that can be dissolved in molten metal is given by Sievert's law:

$$\text{Percent of gas} = K\sqrt{p_{\text{gas}}} \qquad (8.7)$$

where p_{gas} is the partial pressure of the gas in contact with the metal and K is a constant which, for a particular metal-gas system, increases with increasing temperature. We can minimise gas porosity in castings by keeping the liquid temperature low, by adding materials to the liquid to combine with the gas and form a solid, or by assuring that the partial pressure of the gas remains low. The latter may be achieved by placing the molten metal in a vacuum chamber or bubbling an invert gas through the metal. Because p_{gas} is low in the vacuum or inert gas, the gas leaves the metal, enters the vacuum or inert gas, and is carried away.

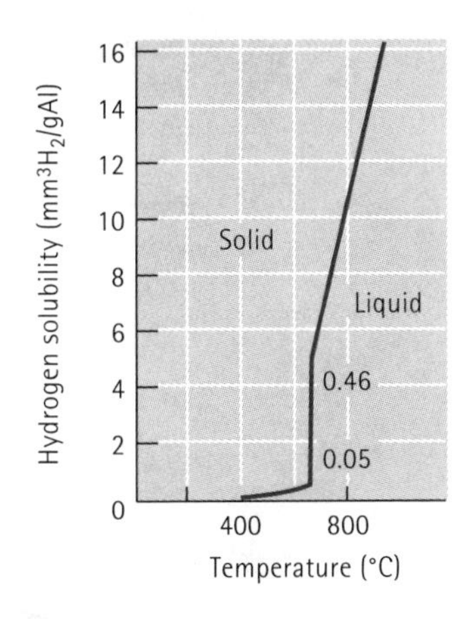

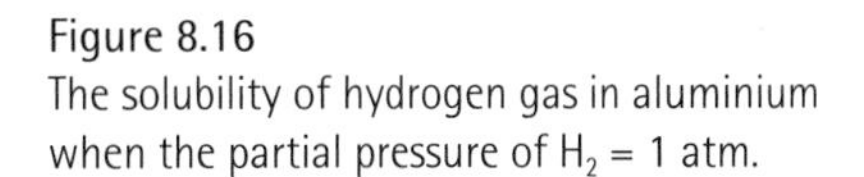
Figure 8.16
The solubility of hydrogen gas in aluminium when the partial pressure of H_2 = 1 atm.

EXAMPLE 8.7

After melting at atmospheric pressure, molten copper contains 0.01 weight percent oxygen. To assure that your castings will not be subject to gas porosity, you want to reduce the weight percent to less then 0.00001% prior to pouring. Suggest a degassing process for the copper.

SOLUTION

We can solve this problem in several ways. In one approach, the liquid copper is placed in a vacuum chamber; the oxygen is then drawn from the liquid and carried away into the vacuum. The vacuum required can be estimated from Sievert's law:

$$\frac{\%\mathrm{O}_{\mathrm{initial}}}{\%\mathrm{O}_{\mathrm{vacuum}}} = \frac{K\sqrt{p_{\mathrm{initial}}}}{K\sqrt{p_{\mathrm{vacuum}}}} = \sqrt{\left(\frac{1}{p_{\mathrm{vacuum}}}\right)}$$

$$\frac{0.01\,\%}{0.00001\,\%} = \sqrt{\left(\frac{1}{p_{\mathrm{vacuum}}}\right)}$$

$$\frac{1}{p_{\mathrm{vacuum}}} = (1000)^2 \quad \text{or} \quad p_{\mathrm{vacuum}} = 10^{-6}\,\mathrm{atm}$$

Another approach would be to introduce a copper-15% phosphorus alloy. The phosphorus reacts with oxygen to produce P_2O_5, which floats out of the liquid, by the reaction:

$$5O + 2P \rightarrow P_2O_5$$

Typically, about 0.01 to 0.02% P must be added to remove the oxygen.

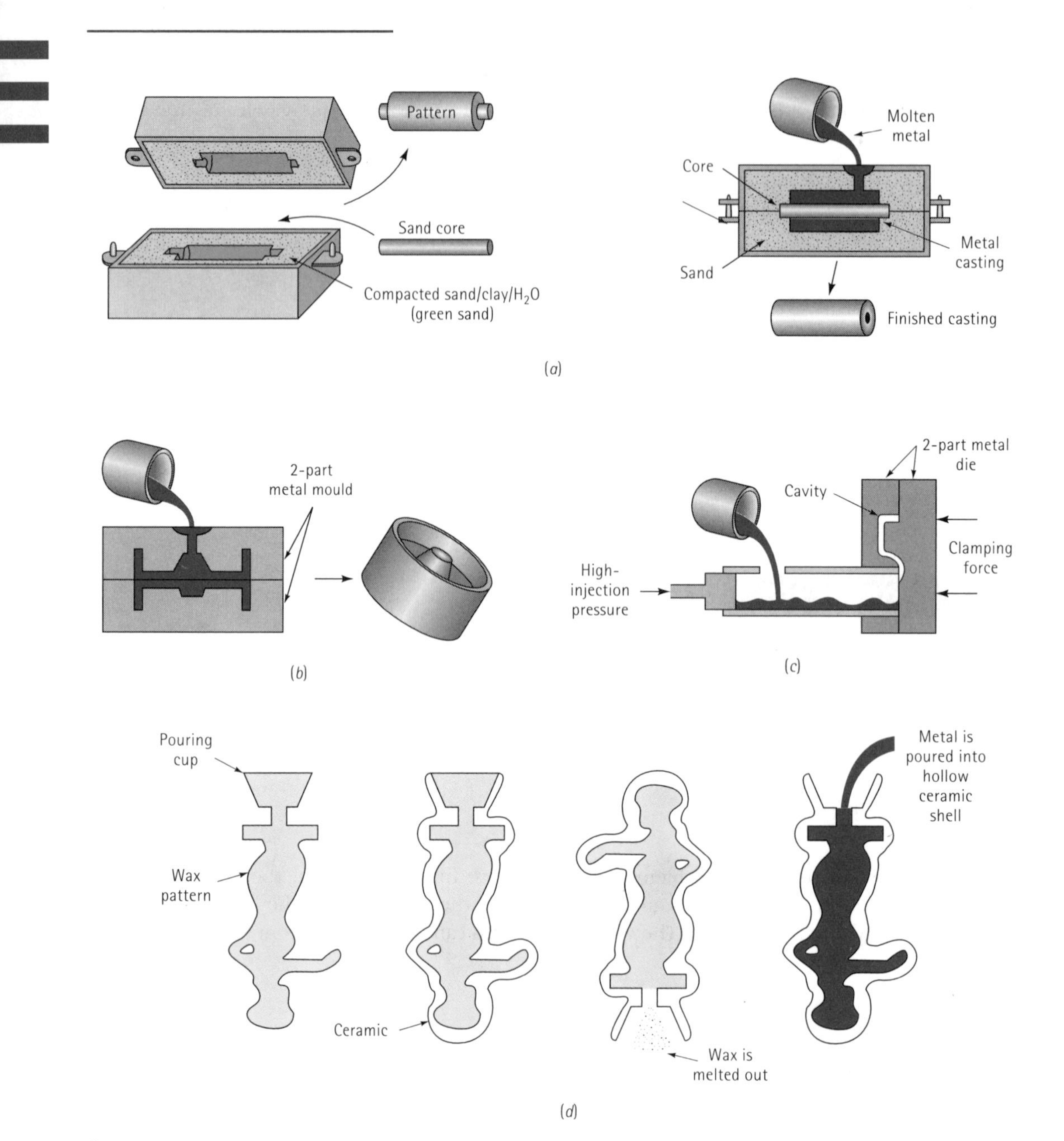

Figure 8.17
Four typical casting processes: (*a*) Green sand moulding, in which clay-bonded sand is packed around a pattern. Sand cores can produce internal cavities in the casting. (*b*) The permanent mould process, in which metal is poured into an iron or steel mould. (*c*) Die casting, in which metal is injected at high pressure into a steel die. (*d*) Investment casting, in which a wax pattern is surrounded by a ceramic; after the wax is melted and drained, metal is poured into the mould.

8.9 Casting Processes

Figure 8.17 summarises four of the dozens of commercial casting processes. Sand moulding processes include green sand moulding, for which silica (SiO_2) sand grains bonded with wet clay are packed around a removable pattern. Ceramic moulding processes use a fine-grained ceramic material as the mould; a slurry containing the ceramic may be poured around a reusable pattern, which is removed after the ceramic hardens. In investment casting, the ceramic slurry coats a wax pattern. After the ceramic hardens, the wax is melted and drained from the ceramic shell, leaving behind a cavity that is filled with molten metal.

In the permanent mould and die casting processes, a cavity is machined from metal. After liquid poured into the cavity solidifies, the mould is opened, the casting is removed, and the mould is reused. The processes using metal moulds tend to give highest strength castings because of the rapid solidification. Ceramic moulds, including those used in investment casting, are good insulators and give the slowest-cooling and lowest-strength castings.

Continuous Casting Continuous metal shapes can also be produced by the casting process. Figure 8.18 illustrates a common method for producing steel plate bars. The liquid metal is fed from a holding vessel (a tundish) into a water-cooled copper mould, which rapidly cools the surface of the steel. The partially solidified steel is withdrawn from the mould at the same rate that additional liquid steel is introduced into the mould. The centre of the steel casting finally solidifies well after the casting exits the mould. The continuously cast material is then cut into appropriate lengths by special cutting machines that travel with the moving steel strand. Similar processes are used to cast aluminium, copper, and even ceramic glasses.

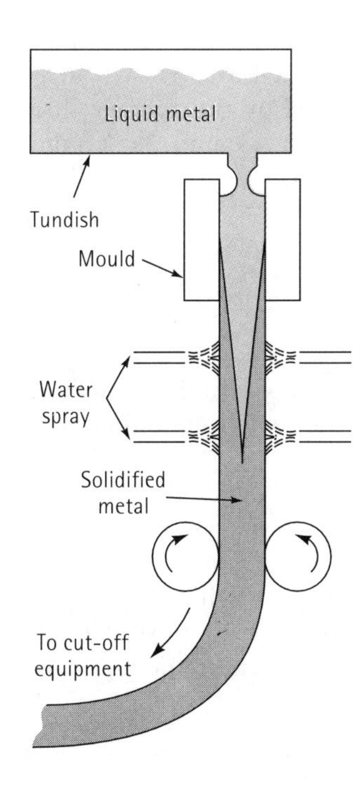

Figure 8.18
Vertical continuous casting, typical of that used in producing many steel products. Liquid metal contained in the tundish partially solidifies in a mould.

EXAMPLE 8.8

Figure 8.19 shows a method for continuous casting of 5 mm-thick, 1.2 m-wide aluminium plate that is subsequently rolled into aluminium foil. The liquid aluminium is introduced between two large steel rolls that slowly turn. We want the aluminium to be completely solidified by the rolls just as the plate emerges from the machine. The rolls act as a permanent mould with a mould constant B of about 0.5 s.mm^{-2} when the aluminium is poured at the proper superheat. Calculate the size of the rolls required for this process.

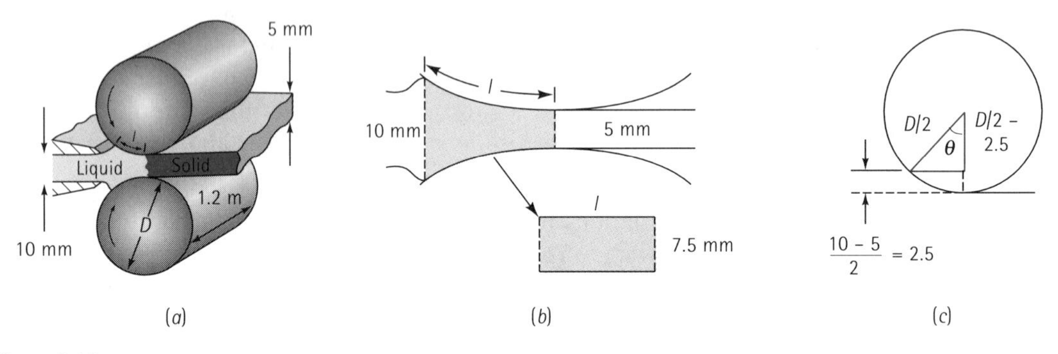

Figure 8.19
Horizontal continuous casting of aluminium (*for* Example 8.8).

SOLUTION

It would be helpful to simplify the geometry so that we can determine a solidification time for the casting. Let's assume that the shaded area shown in Figure 8.19(b) represents the 'casting' and can be approximated by the average thickness times a length and width. The average thickness is (10 mm + 5 mm)/2 = 7.5 mm. Then:

$$V = (\text{thickness})(\text{length})(\text{width}) = 7.5lw$$

$$A = 2(\text{length})(\text{width}) = 2lw$$

$$\frac{V}{A} = \frac{7.5lw}{2lw} = 3.75$$

Only the area directly in contact with the rolls is used in Chvorinov's rule, since little or no heat is transferred from other surfaces. The solidification time should be:

$$t_s = B\left(\frac{V}{A}\right)^2 = (0.5)(3.75)^2 = 7.0 \text{ seconds}$$

For the plate to remain in contact with the rolls for this period of time, the diameter of the rolls and the rate of rotation of the rolls must be established. Figure 8.19(c) shows that the angle θ between the points where the liquid enters and exits the rolls is:

$$\cos\theta = \frac{(D/2) - 2.5}{(D/2)} = \frac{D-5}{D}$$

The surface velocity of the rolls is the product of the circumference and the rate of rotation of the rolls, $v = \pi DR$, where R is in revolutions/minute. The velocity v is also the rate at which we can produce the aluminium plate. The time required for the rolls to travel the distance l must equal the required solidification time.

$$t = \frac{l}{v} = 7.0 \text{ s} = 0.116 \text{ min}$$

The length l is the fraction of the roll diameter that is in contact with the aluminium during freezing and can be given by

$$l = \frac{\pi D\theta}{360}$$

Then, by substituting for l and v in the equation for the time:

$$t = \frac{l}{v} = \frac{\pi D\theta}{360\pi DR} = \frac{\theta}{360R} = 0.116$$

$$R = \frac{\theta}{(360)(0.116)} = \theta/42$$

A number of combinations of D and R provide the required solidification rate. Let's calculate θ for several diameters and then find the required R.

D (mm)	$\theta°$	l(mm)	$R = \theta/42$ (R.P.M)	$v = \pi D R$ (mm.min^{-1})
600	7.402	38.76	0.176	331.75
900	6.042	47.45	0.144	407.15
1 200	5.232	54.79	0.125	471.24
1 500	4.679	61.25	0.111	523.08

As the diameter of the rolls increases, the contacts area (l) between the rolls and the metal increases. This, in turn, permits a more rapid surface velocity (v) of the rolls and increases the rate of production of the plate. However, the larger-diameter rolls do not need to rotate as rapidly to achieve these higher velocities.

In selecting our final design, we prefer to use the largest practical roll diameter to assure high production rates. As the rolls become more massive, however, they and their supporting equipment become more expensive.

In actual operation of such a continuous caster, faster speeds could be used, since the plate does not have to be completely solidified at the point where it emerges from the rolls.

Directional Solidification These are some applications for which a small equiaxed grain structure in the casting is not desired. Castings used for blades and vanes in turbine engines are an example (Figure 8.20). These castings are often made of cobalt or nickel superalloys by investment casting.

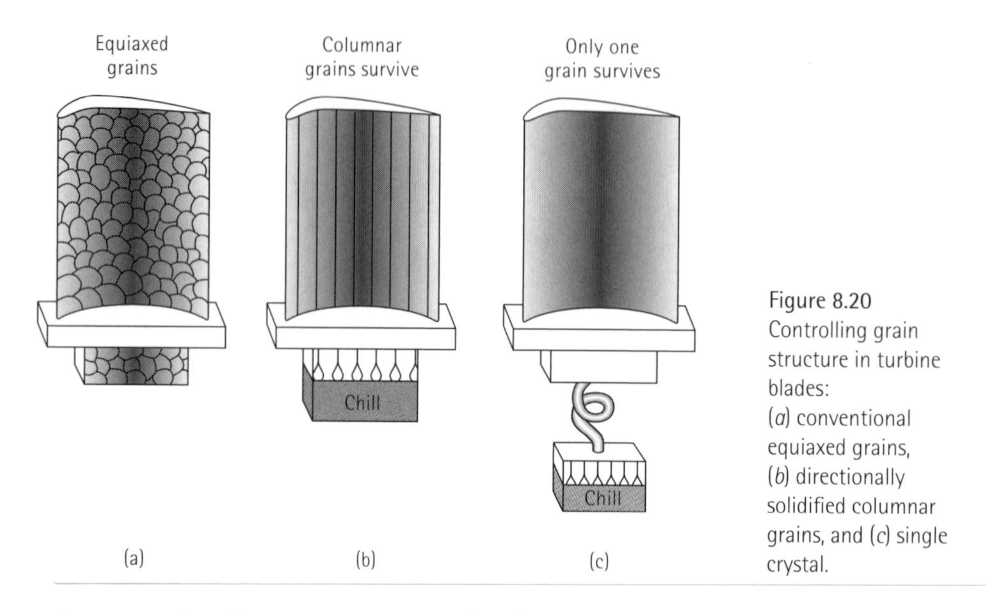

Figure 8.20 Controlling grain structure in turbine blades: (*a*) conventional equiaxed grains, (*b*) directionally solidified columnar grains, and (*c*) single crystal.

In conventionally cast parts, an equiaxed grain structure is produced. However, blades and vanes for turbine and jet engines fail along transverse grain boundaries. Better creep and fracture resistance are obtained using the directionally solidified (DS) technique. In the DS process, the mould is heated from one end and cooled from the other, producing a columnar microstructure with all of the grain boundaries running in the longitudinal direction of the part. No grain boundaries are present in the transverse direction [Figure 8.20(b)].

Still better properties are obtained by using a single crystal (SC) technique. Solidification of columnar grains again begins at a cold surface; however, due to the helical connection, only one columnar grain is able to grow to the main body of the casting [Figure 8.20(c)]. The single crystal casting has no grain boundaries at all and has its crystallographic planes and directions in an optimum orientation.

8.10 Solidification and Metals Joining

Solidification is also important in the joining of metals by fusion welding. In fusion-welding processes, a portion of the metals to be joined is melted and, in many instances, additional molten filler metal is added. The pool of liquid metal is called the fusion zone (Figure 8.21). When the fusion zone subsequently solidifies, the original pieces of metal are joined together.

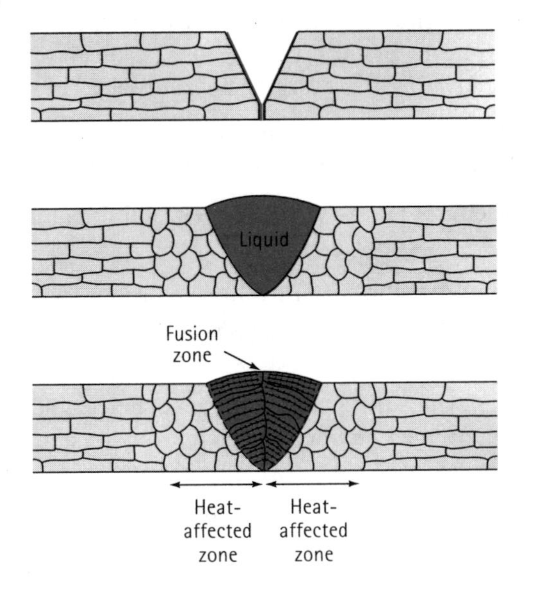

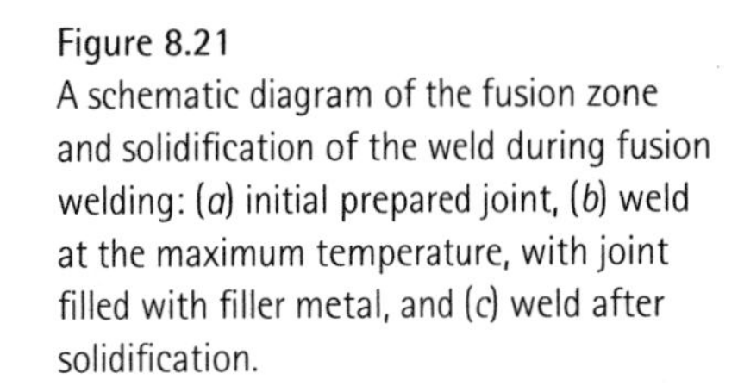
Figure 8.21
A schematic diagram of the fusion zone and solidification of the weld during fusion welding: (*a*) initial prepared joint, (*b*) weld at the maximum temperature, with joint filled with filler metal, and (*c*) weld after solidification.

During solidification of the fusion zone, nucleation is not required. The solid simply begins to grow from existing grains, frequently in a columnar manner. Growth of the solid grains in the fusion zone from the pre-existing grains is called epitaxial growth

The structure and properties in the fusion zone depend on many of the same variables as in a metal casting. Addition of inoculating agents to the fusion zone reduces the grain size. Fast cooling rates or short solidification times promote a finer microstructure and improved properties. Factors that increase the cooling rate include increased thickness of the metal, smaller fusion zones, low original metal temperatures, and certain types of welding processes. Oxyacetylene welding, for example, uses relatively low-intensity flames; consequently, welding times are long and the surrounding solid metal, which becomes very hot, is not an effective heat sink. Arc-welding processes provide a more intense heat source, thus minimising heating of the surrounding metal and providing faster cooling. Laser welding and electron-beam welding are exceptionally intense heat sources and produce very rapid cooling rates and potentially strong welds.

SUMMARY

One of the first opportunities to control the structure and mechanical properties of a material occurs during solidification. We can control the size and shape of the grains to improve overall properties, to obtain uniform properties, or – if we wish – to obtain anisotropic behaviour. We accomplish these ends by controlling nucleation and growth of the solid material from its liquid.

- Nucleation produces a critical-size solid particle from the liquid melt.
  - Homogeneous nucleation requires large undercoolings of the liquid and is not observed in normal solidification processing.

   - By introducing foreign particles into the liquid, nuclei are provided for heterogeneous nucleation. Done in practice by inoculation or grain refining, this process permits the grain size of the casting to be controlled.
   - Rapid cooling of the liquid can prevent nucleation and growth, giving amorphous solids, or glasses, with unusual mechanical and physical properties.
- Growth occurs as the nuclei grow into the liquid melt. Either planar or dendritic growth may be observed.
   - In planar growth, a smooth solid-liquid interface grows with little or no undercooling of the liquid. Special directional·solidification processes take advantage of planar growth.
   - Dendritic growth occurs when the liquid is undercooled. Rapid cooling, or a short solidification time, produces a finer dendritic structure and often leads to improved mechanical properties.
- Chvorinov's rule, $t_s = B(V/A)^n$, can be used to estimate the solidification time of a casting.

By controlling nucleation and growth, a casting may be given a columnar grain structure, an equiaxed grain structure, or a mixture of the two. Isotropic behaviour is typical of the equiaxed grains, whereas anisotropic behaviour is found in columnar grains.

In commercial solidification processing methods, defects in a casting (such as solidification shrinkage or gas porosity) can be controlled by proper design of the casting and riser system or by appropriate treatment of the liquid metal prior to casting.

GLOSSARY

Cavity shrinkage
A large void within a casting caused by the volume contraction that occurs during solidification.

Chill zone
A region of small, randomly oriented grains that forms at the surface of a casting as a result of heterogeneous nucleation.

Chvorinov's rule
The solidification time of a casting is directly proportional to the square of the volume-to-surface area ratio of the casting.

Columnar zone
A region of elongated grains having a preferred orientation that forms as a result of competitive growth during the solidification of a casting.

Critical radius r^*
The minimum size that must be formed by atoms clustering together in the liquid before the solid particle is stable and begins to grow.

Dendrite
The treelike structure of the solid that grows when an undercooled liquid nucleates.

Embryo
A tiny particle of solid that forms from the liquid as atoms cluster together. The embryo is too small to grow.

Epitaxial growth
Growth of a liquid onto an existing solid material without the need for nucleation.

Equiaxed zone
A region of randomly oriented grains in the centre of a casting produced as a result of widespread nucleation.

Fusion welding
Joining processes in which a portion of the materials must melt in order to achieve good bonding.

Fusion zone
The portion of a weld heated to produce all liquid during the welding process. Solidification of the fusion zone provides joining.

Gas porosity
Bubbles of gas trapped within a casting during solidification, caused by the lower solubility of the gas in the solid compared with that in the liquid.

Grain refinement
The addition of heterogeneous nuclei in a controlled manner to increase the number of grains in the casting.

Growth
The physical process by which a new phase increases in size. In the case of solidification, this refers to the formation of a stable solid as the liquid freezes.

Heterogeneous nucleation
Formation of a critically sized solid from the liquid on an impurity surface.

Homogeneous nucleation
Formation of a critically sized solid from the liquid by the clustering together of a large number of atoms at a high undercooling.

Ingot structure
The macrostructure of a casting, including the chill zone, columnar zone, and equiaxed zone.

Inoculation
The addition of heterogeneous nuclei in a controlled manner to increase the number of grains in a casting.

Interdendritic shrinkage
Small, frequently isolated pores between the dendrite arms formed by the shrinkage that accompanies solidification. Also known as microshrinkage or shrinkage porosity.

Latent heat of fusion ΔH_f
The heat evolved when a liquid solidifies. The latent heat of fusion is related to the energy difference between the solid and the liquid.

Local solidification time

The time required for a particular location in a casting to solidify once nucleation has begun.

Macrostructure

Features of a materials structure that typically can be observed by the naked eye.

Nucleation

The physical process by which a new phase is produced in a material. In the case of solidification, this refers to the formation of a tiny, stable solid in the liquid.

Nucleus

A tiny particle of solid that forms from the liquid as atoms cluster together. Because these particles are large enough to be stable, nucleation has occurred and growth of the solid can begin.

Pipe shrinkage

A large conical-shaped void at the surface of a casting caused by the volume contraction that occurs during solidification.

Planar growth

The growth of a smooth solid-liquid interface during solidification, when no undercooling of the liquid is present.

Pouring temperature

The temperature of the metal when it is poured into a mould during the casting process.

Rapid solidification processing

Producing unique material structures by promoting unusually high cooling rates during solidification.

Riser

An extra reservoir of liquid metal connected to a casting. If the riser freezes after the casting, the riser can provide liquid metal to compensate for shrinkage.

Secondary dendrite arm spacing (*SDAS*)

The distance between the centres of two adjacent secondary dendrite arms.

Sievert's law

The amount of a gas that dissolves in a metal is proportional to the partial pressure of the gas in the surroundings.

Specific heat

The heat required to change the temperature of a unit weight of the material one degree.

Spherulite

Spherical-shaped crystals produced when certain polymers solidify.

Superheat

The pouring temperature minus the freezing temperature.

Thermal arrest

A plateau on the cooling curve during the solidification of a material. The thermal arrest is caused by the evolution of the latent heat of fusion during solidification.

Total solidification time
The time required for the casting to solidify completely after the casting has been poured.

Undercooling
The temperature to which the liquid metal must cool below the equilibrium freezing temperature before nucleation occurs.

PROBLEMS

8.1 Suppose that liquid nickel is undercooled until homogeneous nucleation occurs. Calculate

(a) the critical radius of the nucleus required and

(b) the number of the nickel atoms in the nucleus.

Assume that the lattice parameter of the solid FCC nickel is 0.356 nm.

8.2 Suppose that liquid iron is undercooled until homogeneous nucleation occurs. Calculate

(a) the critical radius of the nucleus required and

(b) the number of iron atoms in the nucleus.

Assume that the lattice parameter of the solid BCC iron is 0.292 nm.

8.3 Suppose that the solid nickel was able to nucleate homogeneously with an undercooling of only 22°C. How many atoms would have to group together spontaneously for this to occur? Assume that the lattice parameter of the solid FCC nickel is 0.356 nm.

8.4 Suppose that the solid iron was able to nucleate homogeneously with an undercooling of only 15°C. How many atoms would have to group together spontaneously for this to occur? Assume that the lattice parameter of the solid BCC iron is 0.292 nm.

8.5 Calculate the fraction of solidification that occurs dendritically when iron nucleates

(a) at 10°C undercooling,

(b) at 100°C undercooling, and

(c) homogeneously.

The specific heat of iron is 5.78×10^6 J.m^{-3}.°C^{-1}.

8.6 Calculate the fraction of solidification that occurs dendritically when silver nucleates

(a) at 10°C undercooling,

(b) at 100°C undercooling, and

(c) homogeneously.

The specific heat of silver is 3.25 J.m^{-3}.°C^{-1}.

8.7 Analysis of a nickel casting suggests that 28% of the solidification process occurred in a dendritic manner. Calculate the temperature at which nucleation occurred. The specific heat of nickel is 4.1 J.m^{-3}.°C^{-1}.

8.8 A 50 mm cube solidifies in 4.6 min. Calculate

(a) the mould constant in Chvorinov's rule and

(b) the solidification time for a 12 mm × 12 mm × 150 mm bar cast under the same conditions.

Assume that $n = 2$.

8.9 A 50 mm diameter sphere solidifies in 1050 s. Calculate the solidification time for a 3 mm × 100 mm × 200 mm plate cast under the same conditions.

Assume that $n = 2$.

8.10 Find the constants B and n in Chvorinov's rule by plotting the following data on a log-log plot:

Casting Dimensions (mm)	Solidification Time (min)
12 × 200 × 300	1.03
50 × 75 × 250	4.53
60 cube	2.80
25 × 100 × 225	2.43

8.11 Find the constants B and n in Chvorinov's rule by plotting the following data on a log-log plot:

Casting Dimensions (mm)	Solidification Time (s)
10 × 10 × 60	28.58
20 × 40 × 40	98.30
40 × 40 × 40	155.89
80 × 60 × 50	306.15

8.12 A 75 mm-diameter casting was produced. The times required for the solid-liquid interface to reach different distances beneath the casting surface were measured and are shown in the following table:

Distance from surface (mm)	Time (s)
2.5	32.6
7.5	73.5
12.5	130.6
20	225.0
25	334.9

Determine

(a) the time at which solidification begins at the surface and

(b) the time at which the entire casting is expected to be solid.

(c) Suppose the centre of the casting actually solidified in 720 s. Explain why this time might differ from the time calculated in part (b).

8.13 Figure 8.6(b) shows a photograph of an aluminium alloy. Estimate

(a) the secondary dendrite arm spacing and

(b) the local solidification time for that area of the casting.

8.14 Figure 8.22 shows a photograph of FeO dendrites that have precipitated from a ceramic glass (an undercooled liquid). Estimate the secondary dendrite arm spacing.

8.15 Find the constants k and m relating the secondary dendrite arm spacing to the local solidification time by plotting the following data on a log-log plot:

Solidification Time (s)	*SDAS* (mm)
156	0.176
282	0.216
606	0.282
1356	0.374

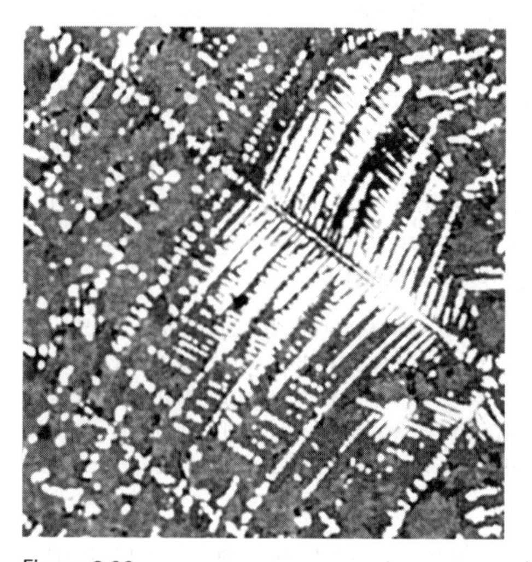

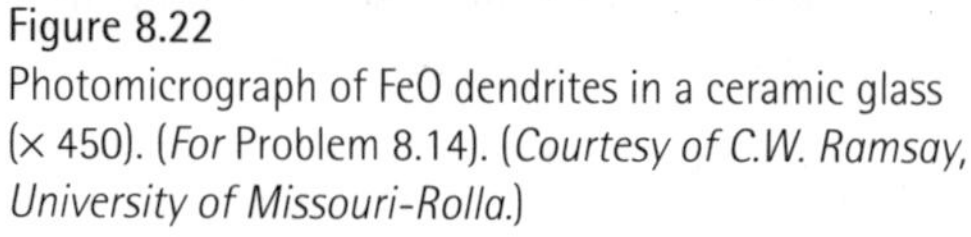

Figure 8.22
Photomicrograph of FeO dendrites in a ceramic glass (× 450). (*For* Problem 8.14). (*Courtesy of C.W. Ramsay, University of Missouri-Rolla.*)

8.16 Figure 8.23 shows dendrites in a titanium powder particle that has been rapidly solidified. Assuming that the size of the titanium dendrites is related to solidification time by the same relationship as in aluminium, estimate the solidification time of the powder particle.

Figure 8.23
Tiny dendrites exposed within a titanium powder particle produced by rapid solidification processing (× 2200). (*From J.D. Ayers and K. Moore, 'Formation of Metal Carbide Powder by Spark Machining of Reactive Metals,'* in Metallurgical Transactions, *Vol. 15A, June 1984, p.1120.*) (*For* Problem 8.16.)

8.17 The secondary dendrite arm spacing in an electron beam weld of copper is 9.5×10^{-3} mm. Estimate the solidification time of the weld.

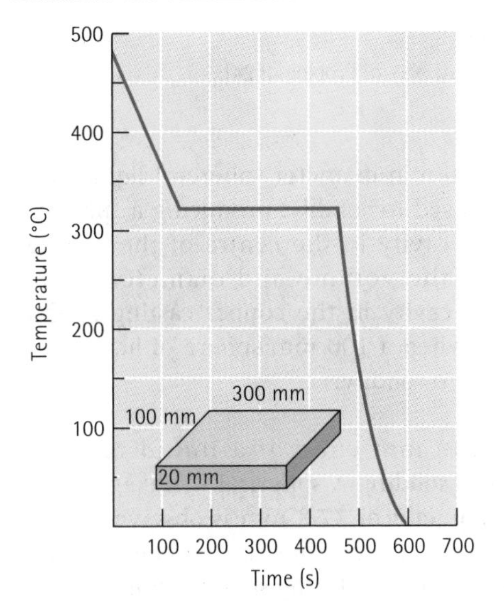

Figure 8.24
Cooling curve (*for* Problem 8.18).

8.18 A cooling curve is shown in Figure 8.24. Determine

(a) the pouring temperature,

(b) the solidification temperature,

(c) the superheat,

(d) the cooling rate, just before solidification begins,

(e) the total solidification time,

(f) the local solidification time, and

(g) the probable identity of the metal.

(h) If the cooling curve was obtained at the centre of the casting sketched in the figure, determine the mould constant, assuming that $n = 2$.

8.19 A cooling curve is shown in Figure 8.25. Determine

(a) the pouring temperature,

(b) the solidification temperature,

(c) the superheat,

(d) the cooling rate, just before solidification begins,

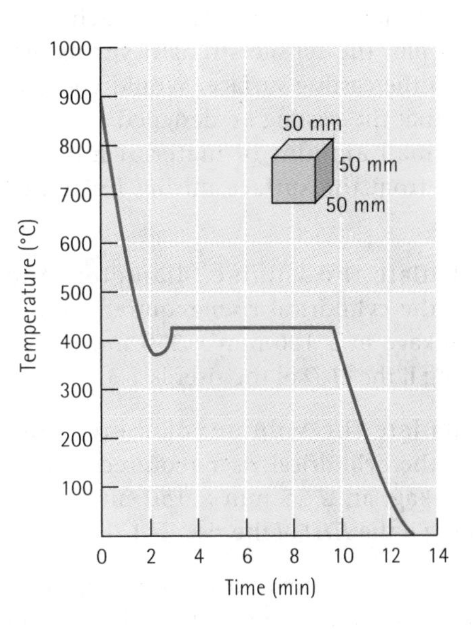

Figure 8.25
Cooling curve (*for* Problem 8.19).

(e) the total solidification time,

(f) the local solidification time,

(g) the undercooling, and

(h) the probable identity of the metal.

(i) If the cooling curve was obtained at the centre of the casting sketched in the figure, determine the mould constant, assuming that $n = 2$.

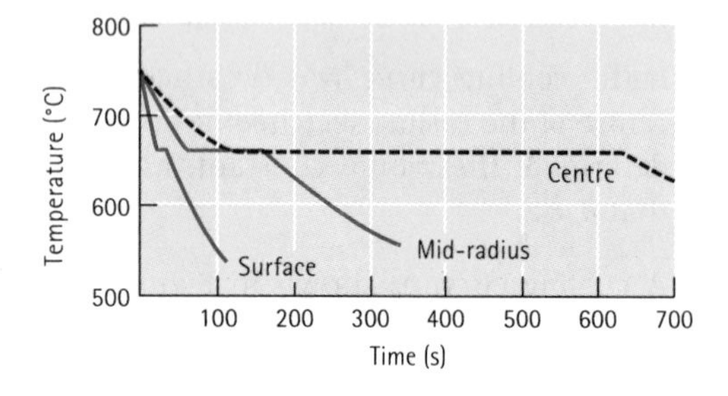

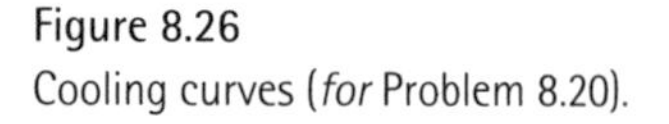

Figure 8.26
Cooling curves (*for* Problem 8.20).

8.20 Figure 8.26 shows the cooling curves obtained from several locations within a cylindrical aluminium casting. Determine the local solidification times and the *SDAS* at each location, then plot the tensile strength versus distance from the casting surface. Would you recommend that the casting be designed so that a large or small amount of material must be machined from the surface during finishing? Explain.

8.21 Calculate the volume, diameter, and height of the cylindrical riser required to prevent shrinkage in a 100 mm × 250 mm × 500 mm casting if the *H*/*D* of the riser is 1.5.

8.22 Calculate the volume, diameter, and height of the cylindrical riser required to prevent shrinkage in a 25 mm × 150 mm × 150 mm casting if the *H*/*D* of the riser is 1.0.

8.23 Figure 8.27 shows a cylindrical riser attached to a casting. Compare the solidification times for each casting section and the riser and determine whether the riser will be effective.

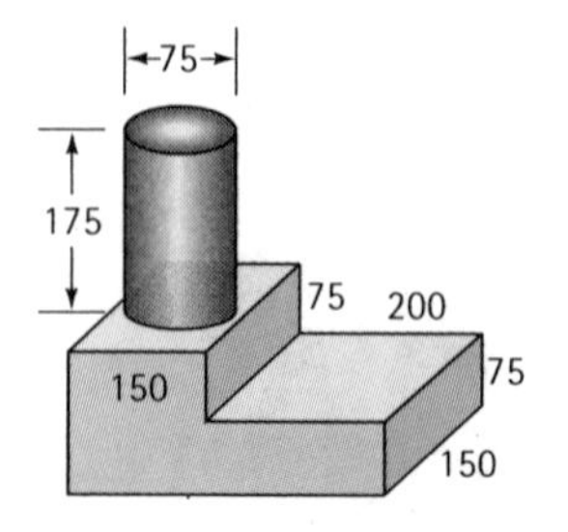

Figure 8.27
Step-block casting (*for* Problem 8.23).

8.24 Figure 8.28 shows a cylindrical riser attached to a casting. Compare the solidification times for each casting section and the riser and determine whether the riser will be effective.

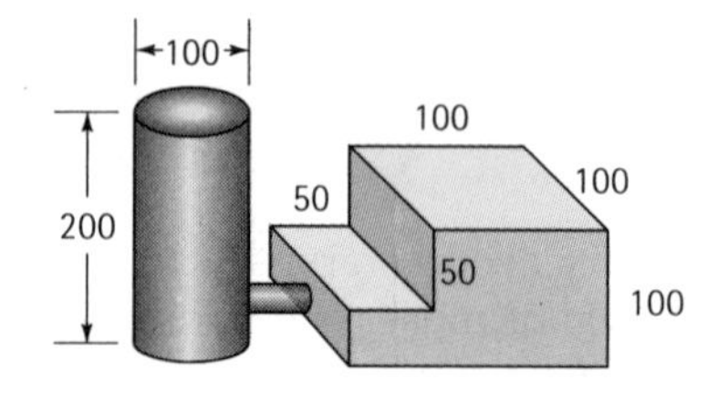

Figure 8.28
Step-block casting (*for* Problem 8.24).

8.25 A 100 mm-diameter sphere of liquid copper is allowed to solidify, producing a spherical shrinkage cavity in the centre of the casting. Compare the volume and diameter of the shrinkage cavity in the copper casting to that obtained when a 100 mm sphere of liquid iron is allowed to solidify.

8.26 A 100 mm cube of a liquid metal is allowed to solidify. A spherical shrinkage cavity with a diameter of 37.8 mm is observed in the solid casting. Determine the percent volume change that occurs during solidification.

8.27 A 20 mm × 40 mm × 60 mm magnesium casting is produced. After cooling to room temperature, the casting is found to weigh 80 g. Determine

(a) the volume of the shrinkage cavity at the centre of the casting and

(b) the percent shrinkage that must have occurred during solidification.

8.28 A 200 mm × 50 mm × 250 mm iron casting produced and, after cooling to room temperature, is found to weigh 19.9 kg. Determine

(a) the percent shrinkage that must have occurred during solidification and

(b) the number of shrinkage pores in the casting if all of the shrinkage occurs as pores with a diameter of 1.27 mm.

8.29 Liquid magnesium is poured into a 20 mm × 20 mm × 240 mm mould and, as a result of directional solidification, all of the solidification shrinkage occurs along the length of the casting. Determine the length of the casting immediately after solidification is completed.

8.30 A liquid cast iron has a density of 7.65 $Mg.m^{-3}$. Immediately after solidification, the density of the solid cast iron is found to be 7.71 $Mg.m^{-3}$. Determine the percent volume change that occurs during solidification. Does the cast iron expand or contract during solidification?

8.31 From Figure 8.16, find the solubility of hydrogen in liquid aluminium just before solidification begins when the partial pressure of hydrogen is 1 atm. Determine the solubility of hydrogen (in mm^3/g Al) at the same temperature if the partial pressure were reduced to 0.01 atm.

8.32 The solubility of hydrogen in liquid aluminium at 715°C is found to be 10 mm^3/g Al. If all of this hydrogen precipitated as gas bubbles during solidification and remained trapped in the casting, calculate the volume percent gas in the solid aluminium.

8.33 Aluminium is melted under conditions that give 0.6 mm^3 H_2 per g of aluminium. We have found that we must have no more than 0.02 mm^3 H_2 per g of aluminium in order to prevent the formation of hydrogen gas bubbles during solidification. Suggest a treatment process for the liquid aluminium that will assure that hydrogen porosity does not form.

8.34 When two 12 mm-thick copper plates are joined by using an arc-welding process, the fusion zone contains dendrites having a *SDAS* of 0.06 mm. However, this process produces large residual stresses in the weld. We have found that residual stresses are low when the welding conditions produce a *SDAS* of more then 0.2 mm Suggest a process by which we can accomplish low residual stresses. Justify your suggestion.

8.35 Design an efficient risering system for the casting shown in Figure 8.29. Be sure to include a sketch of the system, along with appropriate dimensions.

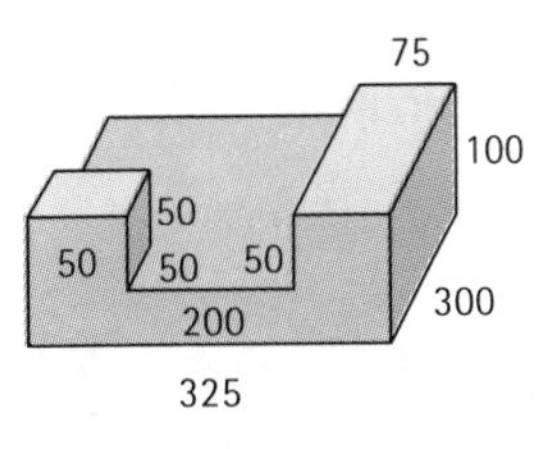

Figure 8.29
Casting to be risered (*for* Problem 8.35).

8.36 Establish a process that will produce a steel casting having uniform properties and high strength. Be sure to include the microstructure features you wish to control and explain how you would do so.

8.37 An aluminium casting is to be injected into a steel mould under pressure (die casting). The casting is essentially a 300 mm-long, 50 mm-diameter cylinder with a uniform wall thickness, and it must have a minimum tensile strength of 275 $MN.m^{-2}$. Based on the properties given in Figure 8.8, determine a suitable casting process.

CHAPTER 9

Solid Solution Strengthening and Phase Equilibrium

9.1 Introduction

The mechanical properties of materials can be controlled by addition of point defects such as substitutional and interstitial atoms. Particularly in metals, the point defects disturb the atomic arrangement in the lattice and interfere with the movement of dislocations, or slip. The point defects therefore cause the material to be solid solution strengthened.

In addition, the introduction of point defects changes the composition of the material and influences solidification behaviour. We examine this effect by introducing the equilibrium phase diagram. From the phase diagram, we can predict how a material will solidify under both equilibrium and nonequilibrium conditions.

9.2 Phases and the Unary Phase Diagram

Pure materials have many engineering applications, but frequently, alloys or mixtures of materials are used. There are two types of alloys: single-phase alloys and multiple-phase alloys. In this chapter we examine the behaviour of single-phase alloys. As a first step, let us define a phase and determine how the phase rule helps us to determine the state – solid, liquid, or gas – in which a pure material exists.

Phase A phase has the following characteristics; (1) a phase has the same structure or atomic arrangement throughout; (2) a phase has roughly the same composition and properties throughout; and (3) there is a definite interface between the phase and any surrounding or adjoining phases. For example, if we enclose a block of ice in a vacuum chamber [Figure 9.1(a)], the ice begins to melt and, in addition, some of the water vaporises. Under these conditions, we have three phases coexisting; solid H_2O, liquid H_2O and gaseous H_2O. Each of these forms of H_2O is a distinct phase; each has a unique atomic arrangement, unique properties, and a definite boundary between each form. In this case the phases have identical compositions, but that fact is not sufficient to permit us to call the entire system one phase.

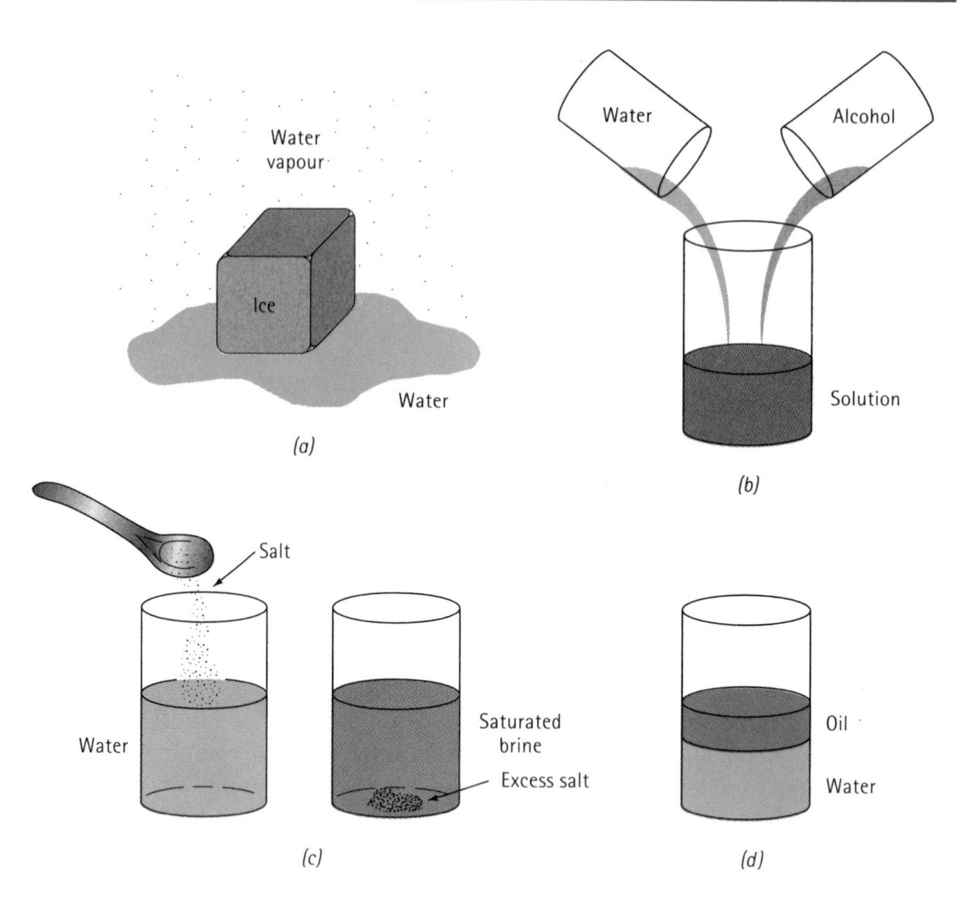

Figure 9.1
Illustration of phases and solubility: (*a*) The three forms of water – gas, liquid, and solid – are each a phase. (*b*) Water and alcohol have unlimited solubility. (*c*) Salt and water have limited solubility. (*d*) Oil and water have virtually no solubility.

Phase Rule The Gibbs phase rule describes the state of a material and has the general form:

$$F = C - P + 2 \tag{9.1}$$

In the phase rule, C is the number of components, usually elements or compounds, in the system; F is the number of degrees of freedom, or the number of variables, such as temperature, pressure, or composition, that are allowed to change independently without changing the number of phases in equilibrium; and P is the number of phases present. The constant 2 in the equation implies that both temperature and pressure are allowed to change.

As an example of the use of the phase rule, let us consider the case of pure magnesium. Figure 9.2 shows a unary (or one-component) phase diagram in which the lines divide the liquid, solid, and vapour phases. In the unary phase diagram, there is only one component – in this case, magnesium. Depending on the temperature and pressure, however, there may be one two, or even three phases present at any one time: solid magnesium, liquid magnesium and magnesium vapour. Note that at atmosphere pressure (one atmosphere ≅ 0.1 $MN.m^{-2}$), given by the dashed line, the intersection with the lines in the phase diagram give the usual melting and boiling temperatures for magnesium. At very low pressures the solid can *sublime*, or go directly to a vapour without melting when it is heated.

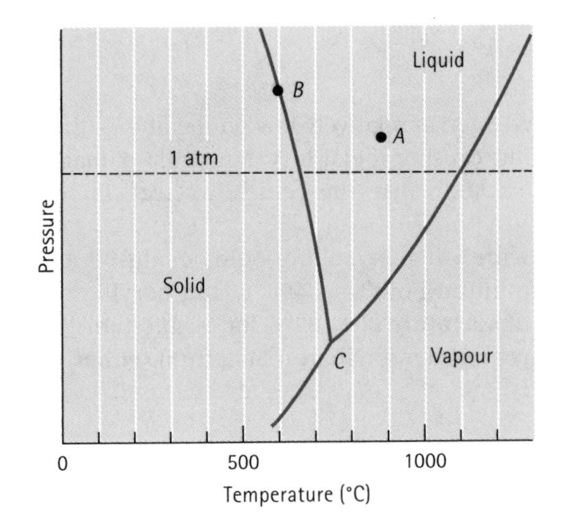

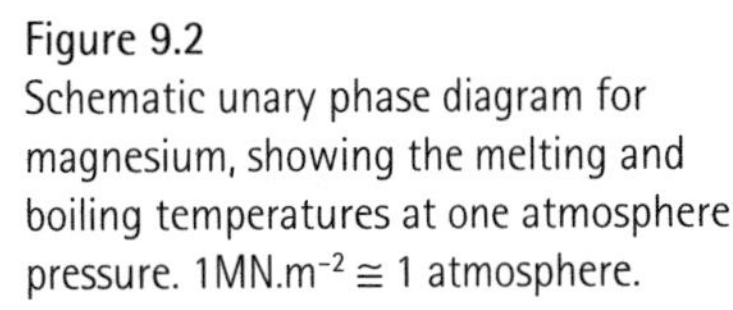

Figure 9.2
Schematic unary phase diagram for magnesium, showing the melting and boiling temperatures at one atmosphere pressure. 1$MN.m^{-2}$ ≅ 1 atmosphere.

Suppose we have a pressure and temperature that puts us at point A in the phase diagram, where the magnesium is all liquid. At this point, the number of components C is one (magnesium) and the number of phases is one (liquid), The phase rule tells us that

$$F = C - P + 2 = 1 - 1 + 2 = 2$$

or there are two degrees of freedom. Within limits, we can change either the pressure or the temperature, or both, and still be in an all-liquid portion of the diagram. Put another way, we must fix both the temperature and the pressure to know precisely where we are in the liquid portion of the diagram.

However, point B is at the boundary between the solid and liquid portions of the diagram. The number of components C is still one, but at point B the solid and liquid coexist, or the number of phases P is two. From the phase rule,

$$F = C - P + 2 = 1 - 2 + 2 = 1$$

or there is only one degree of freedom. For example, if we change the temperature, the pressure must also be adjusted if we are to stay on the boundary where the liquid and solid coexist. On the other hand, if we fix the pressure, the phase diagram tells us the temperature that we must have if solid and liquid are to coexist.

Finally, at point C, solid, liquid, and vapour coexist. While the number of components is still one, there are three phases. The number of degrees of freedom is:

$$F = C - P + 2 = 1 - 3 + 2 = 0$$

Now we have no degrees of freedom; all three phases coexist only if both the temperatures and the pressure are fixed. This point is the triple point.

EXAMPLE 9.1 Materials Selection for an Aerospace Component

Because magnesium is a very lightweight material, it has been suggested for use in an aerospace vehicle intended to enter the outer space environment. Is this a good design?

SOLUTION

In space, pressure is very low. Even at relatively low temperatures, the solid magnesium begins to change to a vapour, causing metal loss that could damage a space vehicle. In addition, solar radiation could cause the vehicle to heat, increasing the rate of magnesium loss.

A lightweight material with a higher boiling point might be a better choice. At atmospheric pressure, aluminium boils at 2 494°C and beryllium boils at 2 770°C compared with the boiling temperature of 1 107°C for magnesium. Although aluminium and beryllium are somewhat denser than magnesium, either might be a better selection.

9.3 Solubility and Solutions

When we begin to combine different materials, as when we add alloying elements to a metal, we produce solutions. We are interested in how much of each material we can combine without producing an additional phase. In other words, we are interested in the solubility of one material in another.

Unlimited Solubility Suppose we begin with a glass of water and a glass of alcohol. The water is one phase and the alcohol is a second phase. If we pour the water into the alcohol and stir, only one phase is produced [Figure 9.1(b)]. The glass contains a solution of water and alcohol that has unique structure, properties, and composition. Water and alcohol are soluble in each other. Furthermore, they display unlimited solubility: Regardless of the ratio of water and alcohol, only one phase is produced by mixing them together.

Similarly, if we were to mix any amounts of liquid copper and liquid nickel, only one liquid phase would be produced. The liquid alloy has the same composition, properties, and structure everywhere [Figure 9.3(a)], because nickel and copper have unlimited liquid solubility.

If the liquid copper-nickel alloy solidifies and cools to room temperature, only one solid phase is produced. After solidification, the copper and nickel atoms do not separate but, instead, are randomly located at the lattice points of an FCC lattice. Within the solid phase, the structure, properties and composition are uniform and no interface exists between the copper and nickel atoms. Therefore, copper and nickel also have unlimited solid solubility. The solid phase is a solid solution [Figure 9.3(b)].

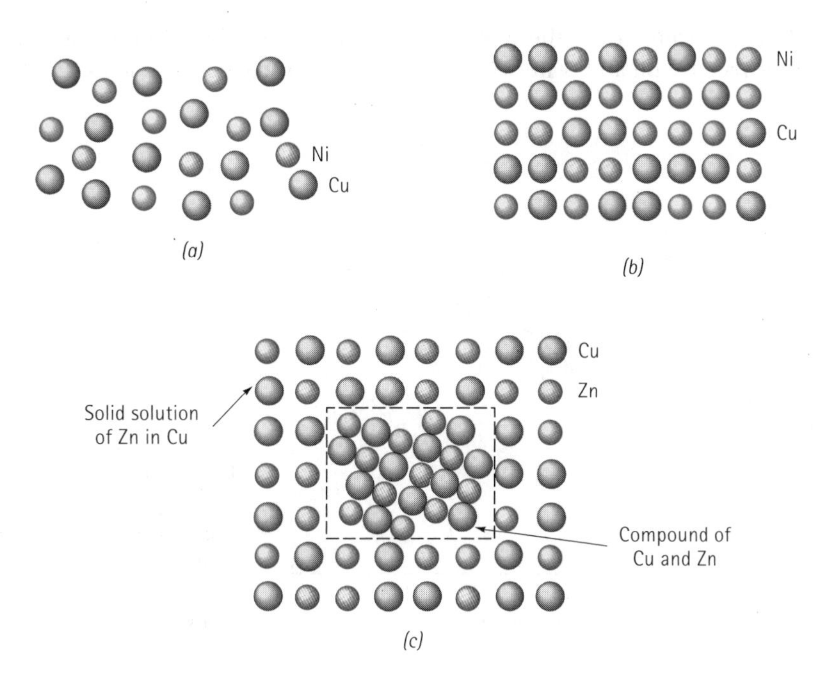

Figure 9.3
(*a*) Liquid copper and liquid nickel are completely soluble in each other. (*b*) Solid copper-nickel alloys display complete solid solubility, with copper and nickel atoms occupying random lattice sites. (*c*) In copper-zinc alloys containing more than 30% Zn, a second phase forms because of the limited solubility of zinc in copper.

A solid solution is not a mixture. A mixture contains more than one type of phase, whose characteristics are retained when the mixture is formed. But the components of a solid solution completely dissolve in one another and do not retain their individual characteristics.

Limited Solubility When we add a small quantity of salt (one phase) to a glass of water (a second phase) and stir, the salt dissolves completely in the water. Only one phase – salty water or brine – is found. However, if we add too much salt to the water, the excess salt sinks to the bottom of the glass [Figure 9.1(c)]. Now we have two phases – water that is saturated with salt plus excess solid salt. We find that salt has a limited solubility in water.

If we add a small amount of liquid zinc to liquid copper, a single liquid solution is produced. When that copper-zinc solution cools and solidifies, a single solid solution having an FCC structure results, with copper and zinc atoms randomly located at the normal lattice points, However, if the liquid solution contains more than about 30% Zn, some of the excess zinc atoms combine with some of the copper atoms to form a CuZn compound [Figure 9.3(c)]. Two solid phases now coexist: a solid solution of copper saturated with about 30% Zn plus a CuZn compound. The solubility of zinc in copper is limited. Figure 9.4 shows a portion of the Cu-Zn phase diagram illustrating the solubility of zinc in copper at low temperatures. The solubility increases with increasing temperature.

In the extreme case, there may be almost no solubility of one material in another. This is true for oil and water [Figure 9.1(d)] or for copper-lead alloys.

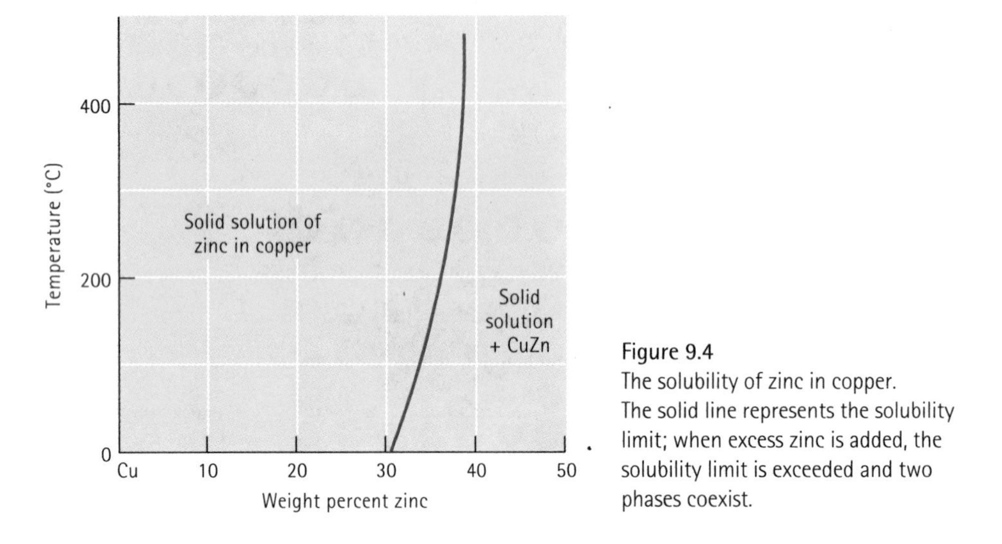

Figure 9.4
The solubility of zinc in copper. The solid line represents the solubility limit; when excess zinc is added, the solubility limit is exceeded and two phases coexist.

9.4 Conditions for Unlimited Solid Solubility

In order for an alloy system, such as copper-nickel, to have unlimited solid solubility, certain conditions must be satisfied. These conditions, the Hume-Rothery rules, are as follows:

1. *Size factor*: The atoms must be of similar size, with no more than a 15% difference in atomic radius, in order to minimise the lattice strain.
2. *Crystal structure*: The materials must have the same crystal structure; otherwise, there is some point at which a transition occurs from one phase to a second phase with a different structure.
3. *Valence*: The atoms must have the same valence; otherwise, the valence electron difference encourages the formation of compounds rather than solutions.
4. *Electronegativity*: The atoms must have approximately the same electronegativity. If the electronegativities differ significantly, compounds again form – as when sodium and chlorine combine to form sodium chloride.

Hume-Rothery's conditions must be met, but they are not necessarily sufficient, for two metals to have unlimited solid solubility.

Similar behaviour is observed between certain compounds, including ceramic materials. Figure 9.5 shows schematically the structure of MgO and NiO. But the Mg and Ni ions are similar in size and valence and, consequently, can replace one another in a sodium chloride type of lattice, forming a complete series of solid solutions of the form (Mg, Ni)O.

The solubility of interstitial atoms is always limited. Interstitial atoms are much smaller than the atoms of the host element, thereby violating the first of Hume Rothery's conditions.

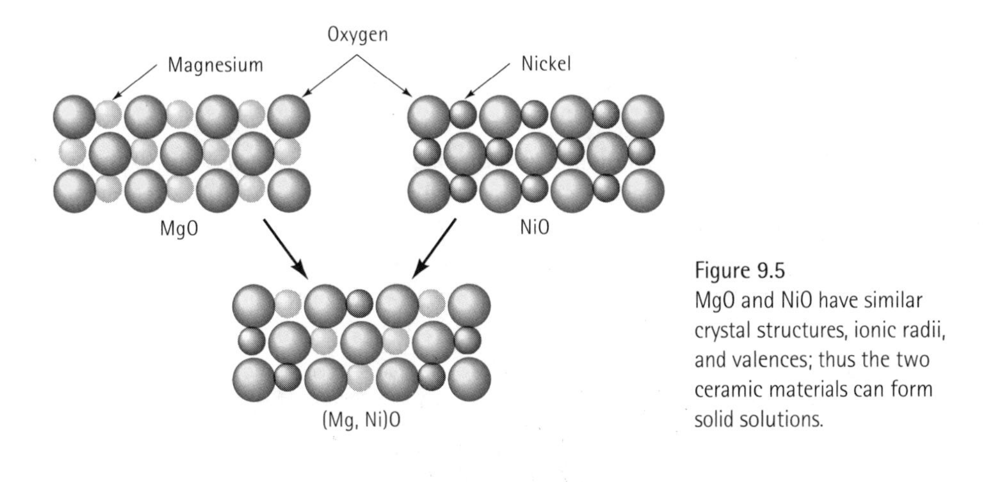

Figure 9.5
MgO and NiO have similar crystal structures, ionic radii, and valences; thus the two ceramic materials can form solid solutions.

EXAMPLE 9.2 Design of a Ceramic Solid Solution

While NiO can be added to MgO to produce a solid solution, NiO is relatively expensive. Design another ceramic system that can form a complete solid solution with MgO.

SOLUTION

In this case, we must consider oxide additives that have metal cations with the same valence and ionic radius as the magnesium cations. The valence of the magnesium ion is +2 and its ionic radius is 0.066 nm. From Appendix B, some other possibilities in which the cation has a valence of +2 include the following:

	r(nm)	r/r_{Mg}	Structure
Cd in CdO	$r_{Cd} = 0.097$	47%	NaCl
Ca in CaO	$r_{Ca} = 0.099$	50%	NaCl
Co in CoO	$r_{Co} = 0.072$	9%	NaCl
Fe in FeO	$r_{Fe} = 0.074$	12%	NaCl
Sr in SrO	$r_{Sr} = 0.112$	70%	NaCl
Zn in ZnO	$r_{Zn} = 0.074$	12%	NaCl

The percent difference in ionic radii and the crystal structures are also shown and suggest that the FeO-MgO system will display unlimited solid solubility. The CoO and ZnO systems also have appropriate radius ratios.

9.5 Solid Solution Strengthening

By producing solid solution alloys, we cause solid solution strengthening. In the copper-nickel system, we intentionally introduce a solid substitutional atom (say nickel) into the original lattice (say copper). The copper-nickel alloy has a strength greater than that of pure copper. Similarly, if less than 30% Zn is added to copper, the zinc behaves as a substitutional atom that strengthens the copper-zinc alloy, as compared with pure copper.

Degree of Solid Solution Strengthening The degree of solid solution strengthening depends on two factors. First, a large difference in atomic size between the original (or solvent) atom and the added (or solute) atom increases the strengthening effect. A larger size difference produces a greater disruption of the initial lattice, making slip more difficult (Figure 9.6).

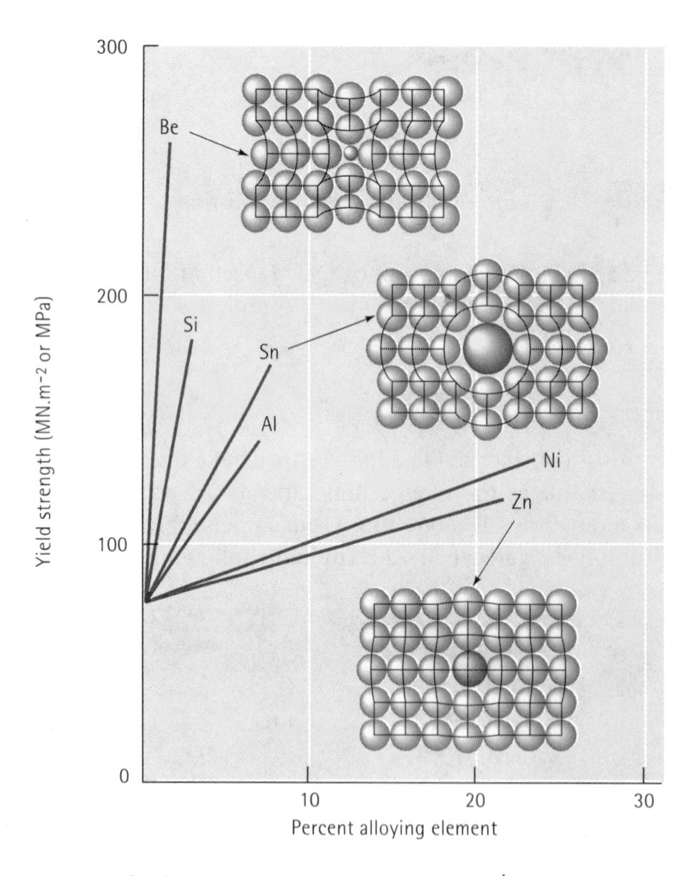

Figure 9.6
The effects of several alloying elements on the yield strength of copper. Nickel and zinc atoms are about the same size as copper atoms, but beryllium and tin atoms are much different from copper atoms. Increasing both atomic size difference and amount of alloying element increases solid solution strengthening.

Second, the greater the amount of alloying element added, the greater the strengthening effect (Figure 9.6). A Cu-20% Ni alloy is stronger than a Cu-10% Ni alloy. Of course, if too much of a large or small atom is added, the solubility limit may be exceeded and a different strengthening mechanism – dispersion strengthening – is produced. This mechanism is discussed in Chapter 10.

EXAMPLE 9.3

From the atomic radii, show whether the size difference between copper atoms and alloying atoms accurately predicts the amount of strengthening found in Figure 9.6.

SOLUTION

The atomic radii and percent size difference are shown below:

Metal	Radius ($m \times 10^{-10}$)	$\frac{r - r_{Cu}}{r_{Cu}} \times 100$
Cu	1.278	
Zn	1.332	+4.2%
Al	1.432	+12.1%
Sn	1.509	+18.1%
Ni	1.243	−2.7%
Si	1.176	−8.0%
Be	1.143	−10.6%

For atoms larger than copper – namely, zinc, aluminium, and tin – increasing the size difference increases the strengthening effect. Likewise for smaller atoms, increasing the size difference increases strengthening.

Effect of Solid Solution Strengthening on Properties The effects of solid solution strengthening on the properties of a material include the following (Figure 9.7):

1. The yield strength, tensile strength, and hardness of the alloy are greater than those of the pure metals.
2. Almost always, the ductility of the alloy is less than that of the pure metal. Only rarely, as in copper-zinc alloys, does solid solution strengthening increase both strength and ductility.
3. Electrical conductivity of the alloy is much lower than that of the pure metal. Solid solution strengthening of copper or aluminium wires used for transmission of electrical power is not recommended because of this pronounced effect.
4. The resistance to creep, or loss of strength at elevated temperatures, is improved by solid solution strengthening. High temperatures do not cause a catastrophic change in the properties of solid-solution-strengthened alloys. Many high-temperature alloys, such as those used for jet engines, rely partly on extensive solid solution strengthening.

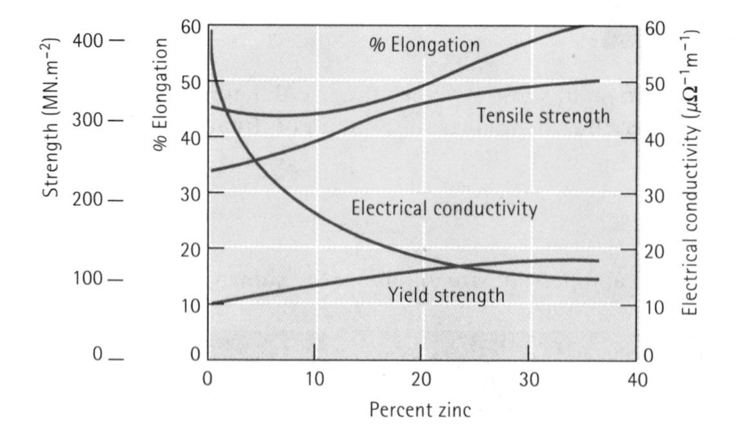

Figure 9.7
The effect of additions of zinc to copper on the properties of the solid solution-strengthened alloy. The increase in % elongation with increasing zinc content is not typical of solid solution strengthening.

9.6 Isomorphous Phase Diagrams

A phase diagram shows the phases and their compositions at any combination of temperature and alloy composition. When only two elements are present in the alloy, a binary phase diagram can be constructed. Isomorphous binary phase diagrams are found in a number of metallic and ceramic systems. In the isomorphous systems, which include the copper-nickel and NiO-MgO systems (Figure 9.8), only one solid phase forms; the two components in the system display complete solid solubility. There are several valuable pieces of information to be obtained from these phase diagrams, as follows.

Liquidus and Solidus Temperatures The upper curve on the diagram is the liquidus temperature for all of the copper-nickel alloys. We must heat a copper-nickel alloy above the liquidus to produce a completely liquid alloy that can then be cast into a useful shape. The liquid alloy begins to solidify when the temperature cools to the liquidus temperature. For the Cu-40% Ni alloy in Figure 9.8, the liquidus temperature is 1 280°C.

The solidus temperature for the copper-nickel alloys is the lower curve. A copper-nickel alloy is not completely solid until the metal cools below the solidus temperature. If we use a copper-nickel alloy at high temperatures, we must be sure that the service temperature is below the solidus so that no melting occurs. For the Cu-40% Ni alloy in Figure 9.8, the solidus temperature is 1 240°C.

Copper-nickel alloys melt and freeze over a range of temperatures, between the liquidus and the solidus. The temperature difference between the liquidus and the solidus is the freezing range of the alloy. Within the freezing range, two phases coexist; a liquid and a solid. The solid is a solution of copper and nickel atoms; solid phases are typically designated by a lower case Greek letter, such as α. For the Cu-40% Ni alloy in Figure 9.8, the freezing range is 1 280 – 1 240 = 40°C.

Phases Present Often we are interested in which phases are present in an alloy at a particular temperature. If we plan to make a casting, we must be sure that the metal is initially all liquid; if we plan to heat treat an alloy component, we must be sure

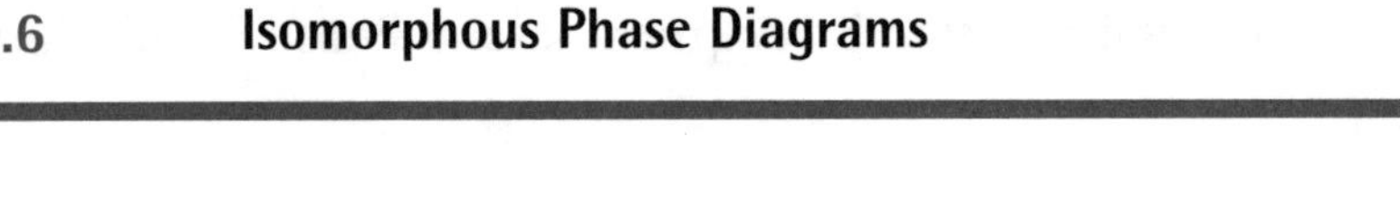

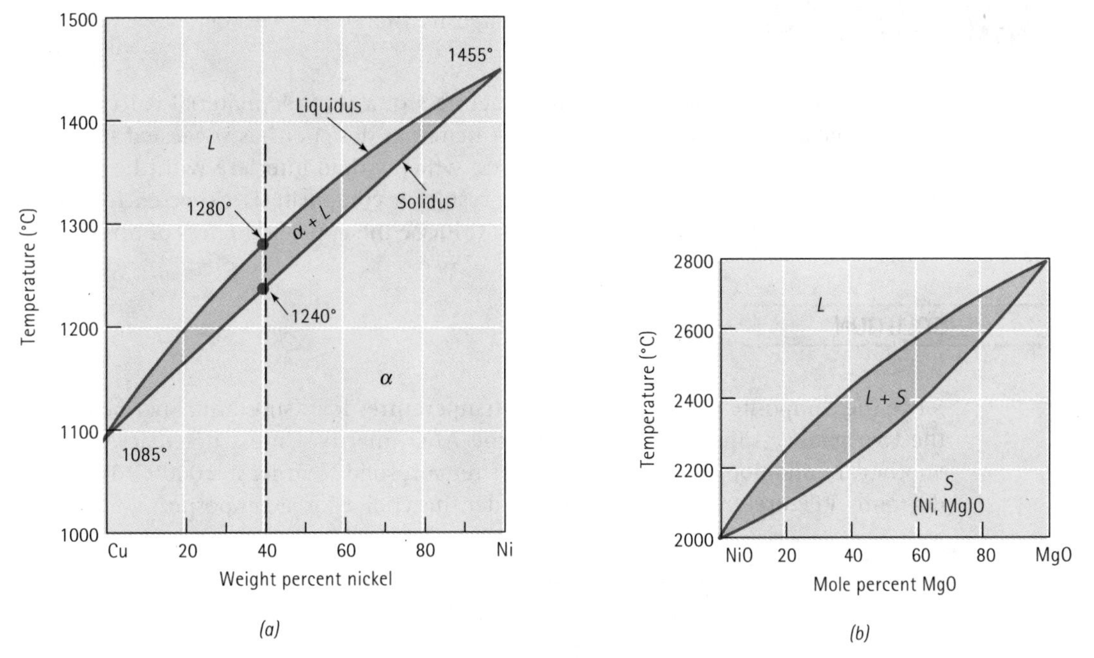

Figure 9.8
The equilibrium phase diagrams for the copper-nickel and NiO-MgO systems. The liquidus and solidus temperatures are shown for a Cu-40% Ni alloy.

that no liquid forms during the process. The phase diagram can be treated as a road map; if we know the coordinates – temperature and alloy composition – we can determine the phases present.

EXAMPLE 9.4 Design of a Refractory Brick

Select a NiO-MgO refractory material that can be melted and cast at 2 600°C but will not melt when placed into service at 2 300°C.

SOLUTION

We must have a material that has a liquidus temperature below 2 600°C but a solidus temperature above 2 300°C. The NiO-MgO phase diagram, Figure 9.8(*b*), permits us to design an appropriate composition for the refractory.

To produce a liquidus below 2 600°C, there must be less than about 65 mol% MgO in the refractory. To produce a solidus above 2 300°C, there must be at least about 50 mol% MgO present. Consequently, we can use any composition between 50 mol% MgO and 65% MgO. Our final decision will be based on other considerations, such as relative cost of the two oxides and compatibility with the environment of the refractory.

EXAMPLE 9.5 **Selection of a Composite Material**

One method to improve the fracture toughness of a ceramic material is to reinforce the ceramic matrix with ceramic fibres. A materials designer has suggested that Al_2O_3 could be reinforced with 25% Cr_2O_3 fibres, which would interfere with the propagation of any cracks in the alumina. The resulting composite is expected to operate under load at 2 000°C for several months. Criticise the appropriateness of this design.

SOLUTION

Since the composite will operate at high temperatures for a substantial period of time, the two phases – the Cr_2O_3 fibres and the Al_2O_3 matrix – must not react with one another. In addition, the composite must remain solid to at least 2 000°C. The phase diagram in Figure 9.9 permits us to consider this choice for a composite.

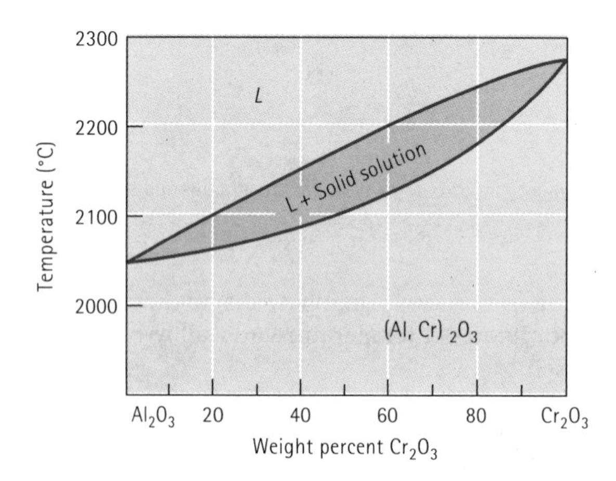

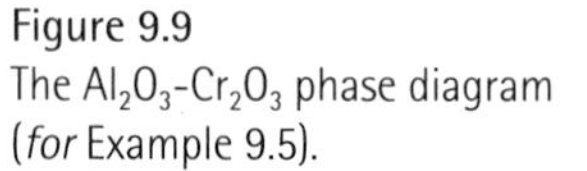
Figure 9.9
The Al_2O_3-Cr_2O_3 phase diagram (*for* Example 9.5).

Pure Cr_2O_3, pure Al_2O_3, and Al_2O_3-25% Cr_2O_3 have solidus temperatures above 2000°C; consequently, there is no danger of melting of any of the constituents. However, Cr_2O_3 and Al_2O_3 display unlimited solid solubility. At the high service temperature, 2 000°C, Al^{3+} ions will diffuse from the matrix into the fibre, replacing Cr^{3+} ions in the fibres. Simultaneously, Cr^{3+} ions will replace Al^{3+} ions in the matrix. Long before several months have elapsed, these diffusion processes cause the fibres to completely dissolve into the matrix. With no fibres remaining, the fracture toughness will again be poor.

Composition of Each Phase Each phase has a composition, expressed as the percentage of each element in the phase. Usually the composition is expressed in weight percent (wt%). When only one phase is present in the alloy, the composition of the phase equals the overall composition of the alloy. If the original composition of the alloy changes, then the composition of the phase must also change.

However, when two phases, such as liquid and solid, coexist, their compositions differ from one another and also differ from the original overall composition. In this case, if the original composition changes slightly, the composition of the two phases is unaffected, provided that the temperature remains constant.

This difference is explained by the Gibbs phase rule. In this case, unlike the example of pure magnesium described earlier, we keep the pressure fixed at one atmosphere (or about 0.1 $MN.m^{-2}$) which is normal for binary phase diagrams. The phase rule given by Equation 9.1 can be rewritten as

$$F = C - P + 1 \qquad \text{(for constant pressure)} \qquad (9.2)$$

where, again, C is the number of components, P is the number of phases, and F is the number of degrees of freedom. We now use a 1 instead of a 2 because we are holding the pressure constant at one atmosphere. In a binary system, the number of components C is two; the degrees of freedom that we have include changing the temperature and changing the composition of the phases present. We can apply this form of the phase rule to the Cu-Ni system, as shown in Example 9.6.

EXAMPLE 9.6

Determine the degrees of freedom in a Cu-40% Ni alloy at 1 300°C, 1 250°C, and 1 200°C.

SOLUTION

At 1 300°C, $P = 1$, since only one phase (liquid) is present; $C = 2$, since copper and nickel atoms are present. Thus:

$$F = 2 - 1 + 1 = 2$$

We must fix both the temperature and the composition of the liquid phase to completely describe the state of the copper-nickel alloy in the liquid region.

At 1 250°C, $P = 2$, since both liquid and solid are present; $C = 2$, since both copper and nickel atoms are present. Now:

$$F = 2 - 2 + 1 = 1$$

If we fix the temperature in the two-phase region, the compositions of the two phases are also fixed. Or, if the composition of one phase is fixed, the temperature and composition of the second phase are automatically fixed.

At 1 200°C, $P = 1$, since only one phase, solid is present; $C = 2$, since both copper and nickel atoms are present. Again,

$$F = 2 - 1 + 1 = 2$$

and we must fix both temperature and composition to completely describe the state of the solid.

Because there is only one degree of freedom in a two-phase region of a binary phase diagram, the compositions of the two phases are always fixed when we specify the temperature. This is true even if the overall composition of the alloy changes. We therefore can use a tie line to determine the composition of the two phases. A tie line is a horizontal line within a two-phase region drawn at the temperature of interest (Figure 9.10). Tie lines are not used in single-phase regions. In an isomorphous system, the tie line connects the liquidus and solidus points at the specified temperature. The ends of the tie line represent the compositions of the two phases in equilibrium.

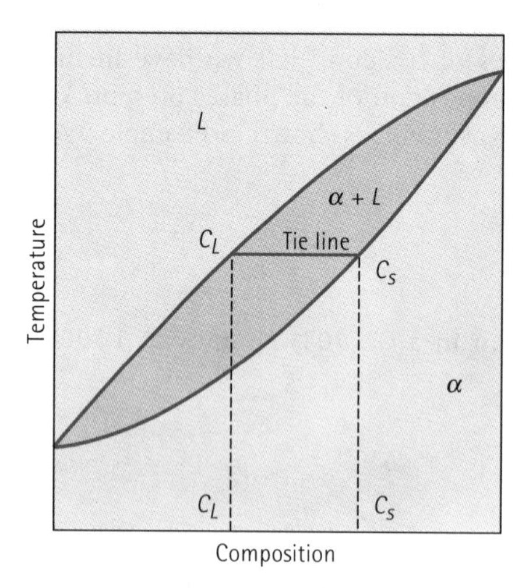

Figure 9.10
When an alloy is present in a two-phase region, a tie line at the temperature of interest fixes the composition of the two phases. This is a consequence of the Gibbs phase rule, which provides only one degree of freedom.

For any original composition lying between c_L and c_S, the composition of the liquid is c_L and the composition of the solid α is c_S.

EXAMPLE 9.7

Determine the composition of each phase in a Cu-40% Ni alloy at 1 300°C, 1 270°C, 1 250°C, and 1 200°C. (*See* Figure 9.11)

SOLUTION

The vertical line at 40% Ni represents the overall composition of the alloy:

1 300°C: Only liquid is present. The liquid must contain 40% Ni, the overall composition of the alloy.

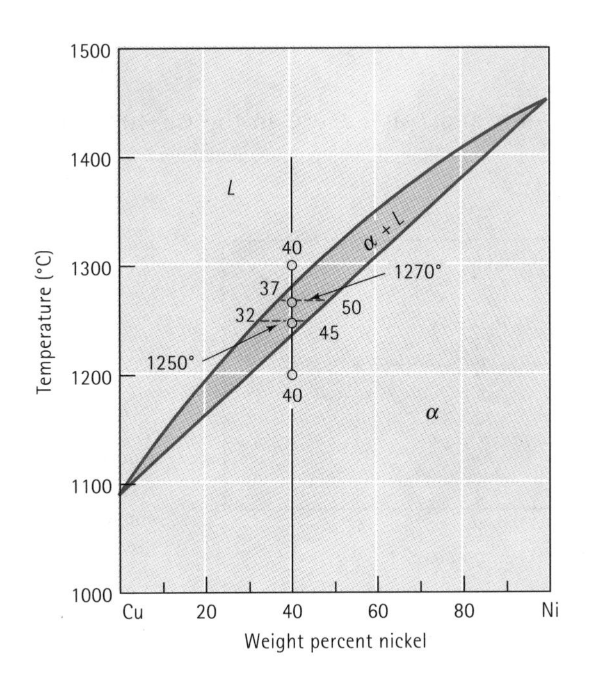

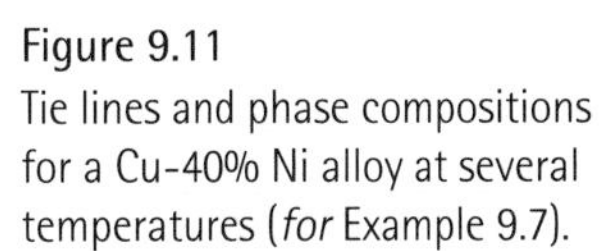
Figure 9.11
Tie lines and phase compositions for a Cu-40% Ni alloy at several temperatures (*for* Example 9.7).

1 270°C: Two phases are present. A horizontal line within the α + *L* field is drawn. The endpoint at the liquidus, which is in contact with the liquid region, is at 37% Ni. The endpoint at the solidus, which is in contact with the α region, is at 50% Ni. Therefore, the liquid contains 37% Ni and the solid contains 50% Ni.

1 250°C: Again two phases are present. The tie line drawn at this temperature shows that the liquid contains 32% Ni and the solid contains 45% Ni.

1 200°C: Only solid α is present, so the solid must contain 40% Ni.

In Example 9.7, we find that the solid *α* contains more nickel than the overall alloy and the liquid *L* contains more copper than the original alloy. Generally, the higher melting point element (in this case, nickel) is concentrated in the first solid that forms.

Amount of Each Phase (the Lever Law) Lastly, we are interested in the relative amounts of each phase present in the alloy. These amounts are normally expressed as weight percent (wt%).

In single-phase regions, the amount of the single phase is 100%. In two-phase regions, however, we must calculate the amount of each phase. One technique is to perform a materials balance, as shown in Example 9.8.

EXAMPLE 9.8

Calculate the amounts of α and L at 1 250°C in the Cu-40% Ni alloy shown in Figure 9.12.

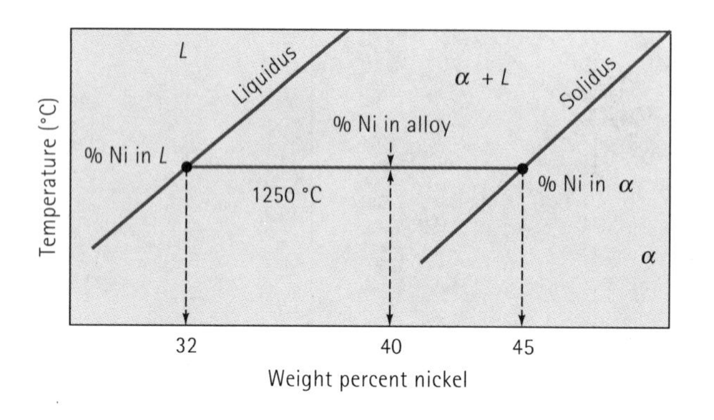

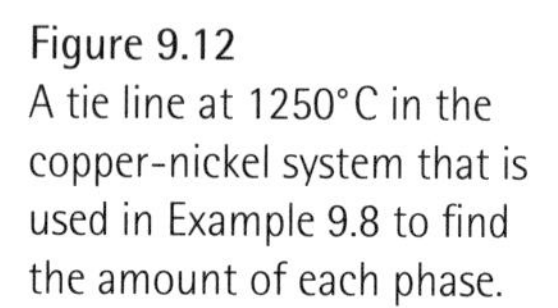

Figure 9.12
A tie line at 1250°C in the copper-nickel system that is used in Example 9.8 to find the amount of each phase.

SOLUTION

Let's say that x = fraction of the alloy that is solid, or α:

$$(\%\text{ Ni in }\alpha)(x) + (\%\text{ Ni in }L)(1-x) = (\%\text{ Ni in alloy})$$

By multiplying and rearranging:

$$x = \frac{(\%\text{ Ni in alloy}) - (\%\text{ Ni in }L)}{(\%\text{ Ni in }\alpha) - (\%\text{ Ni in }L)}$$

From the phase diagram at 1 250°C:

$$x = \frac{40-32}{45-32} = \frac{8}{13} = 0.62$$

If we convert from weight fraction to weight percent, the alloy at 1 250°C contains 62% α and 38% L.

To calculate the amounts of liquid and solid, we construct a lever on our tie line with the fulcrum of our lever being the original composition of the alloy. The leg of the lever *opposite* to the composition of the phase whose amount we are calculating is divided by the total length of the lever to give the amount of that phase. In Example 9.8, note that the denominator represents the total length of the tie line and the numerator is the portion of the lever that is *opposite* the composition of the solid we are trying to calculate.

The lever law in general can be written as:

$$\text{Phase percent} = \frac{\text{opposite arm of lever}}{\text{total length of tie line}} \times 100 \tag{9.3}$$

We can work the lever law in any two-phase region of a binary phase diagram. The lever law calculation is not used in single-phase regions because the answer is trivial (there is 100% of that phase present).

EXAMPLE 9.9

Determine the amount of each phase in the Cu-40% Ni alloy shown in Figure 9.11 at 1 300°C, 1 270°C, 1 250°C, and 1 200°C.

SOLUTION

1300°C: There is only one phase, so 100% L.

$$1270°\text{C}: \quad \%L = \frac{50-40}{50-37} \times 100 = 77\%$$

$$\%\alpha = \frac{40-37}{50-37} \times 100 = 23\%$$

$$1250°\text{C}: \quad \%L = \frac{45-40}{45-32} \times 100 = 38\%$$

$$\%\alpha = \frac{40-32}{45-32} \times 100 = 62\%$$

1200°C: There is only one phase, so 100% α.

Sometimes we wish to express composition as atomic percent (at%) rather than weight percent (wt%). For a Cu-Ni alloy, where M_{Cu} and M_{Ni} are the molecular weights, the following equations provide examples for making these conversions;

$$\text{at\% Ni} = \frac{\text{wt\% Ni} / M_{\text{Ni}}}{(\text{wt\% Ni} / M_{\text{Ni}}) + (\text{wt\% Cu} / M_{\text{Cu}})} \times 100 \tag{9.4}$$

$$\text{wt\% Ni} = \frac{(\text{at\% Ni})(M_{\text{Ni}})}{(\text{at\% Ni})(M_{\text{Ni}}) + (\text{at\% Cu})(M_{\text{Cu}})} \times 100 \tag{9.5}$$

9.7 Relationship Between Properties and the Phase Diagram

We have previously mentioned that a copper-nickel alloy may be stronger than either pure copper or pure nickel because of solid solution strengthening. The mechanical properties of a series of copper-nickel alloys are related to the phase diagram in Figure 9.13.

The strength of the copper increases by solid solution strengthening until about 60% Ni is added. Pure nickel is solid solution strengthened by the addition of copper until 40% Cu is added. The maximum strength is obtained for a Cu-60% Ni alloy, known as *Monel*. The maximum is closer to the pure nickel side of the phase diagram because pure nickel is stronger than pure copper.

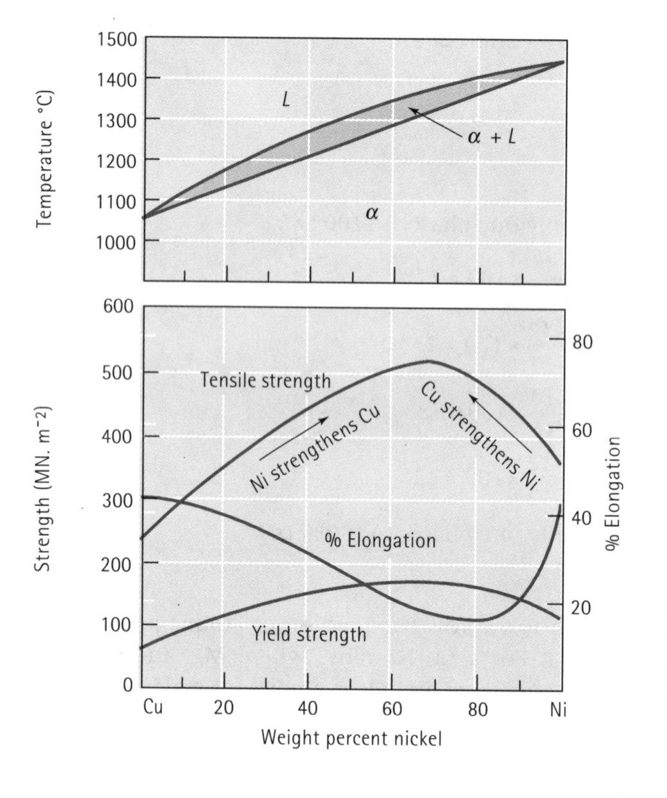

Figure 9.13
The mechanical properties of copper-nickel alloys. Copper is strengthened by up to 60% Ni and nickel is strengthened by up to 40% Cu.

EXAMPLE 9.10 Selection of an Alloy for a Casting

You need to produce a Cu-Ni alloy having a minimum yield strength of 140 MN.m^{-2}, a minimum tensile strength of 420 MN.m^{-2}, and a minimum % elongation of 20%.

You have in your inventory a Cu-20% Ni alloy and lots of pure nickel. Design a method for producing castings having the required properties.

SOLUTION

From Figure 9.13, we determine the required composition of the alloy. To meet the required yield strength, the alloy must contain between 30 and 90% Ni; for the tensile strength, 33 to 90% Ni is required. The required % elongation can be obtained for alloys containing less than 60% Ni or more than 90% Ni. To satisfy all of these conditions, we could use:

Cu-90% Ni or Cu-33% to 60% Ni

We prefer to select a low nickel content, since nickel is more expensive than copper. In addition, the lower nickel alloys have a lower liquidus, permitting castings to be made with less energy being expended. Therefore, a reasonable alloy might be Cu-35% Ni.

To produce this composition from the available melting stock, we must blend some of the pure nickel with the Cu-20% Ni ingot. If we wish to produce 10 kg of the alloy, then:

$$\text{Amount of Ni required } = (10 \text{ kg}) \frac{(35\%\,\text{Ni})}{100\%} = 3.5 \text{ kg}$$

Let y kg be the amount of Cu-20 ingot required,

then $(10 - y)$kg is the amount of pure Ni also required.

$$3.5 \text{ kg Ni } = y \times \frac{20}{100} + (10 - y) \times \frac{100}{100}$$

$$3.5 = 0.2x + 10 - y$$

$$0.8y = 6.5$$

$$y = 8.125 \text{ kg}$$

Therefore, we need to melt 8.125 kg of Cu-20% Ni with 1.875 kg of pure nickel to produce the required alloy. We would then heat the alloy above the liquidus temperature, which is 1 250°C for the Cu-35% Ni alloy, before pouring the liquid metal into an appropriate mould.

9.8 Solidification of a Solid Solution Alloy

When an alloy such as Cu-40% Ni is melted and cooled, solidification requires both nucleation and growth. Heterogeneous nucleation permits little or no undercooling, so solidification begins when the liquid reaches the liquidus temperature. The phase diagram (Figure 9.14), with a tie line drawn at the liquidus temperature, tells that the *first solid to form* has a composition of Cu-52% Ni.

Two conditions are required for growth of the solid α. First, growth requires that the latent heat of fusion, which evolves as the liquid solidifies, be removed from the solid-liquid interface. Second, and unlike the case of pure metals, diffusion must occur so that the compositions of the solid and liquid phases follow the solidus and liquidus curves during cooling. The latent heat of fusion is removed over a range of temperatures so that the cooling curve shows a change in slope, rather than a flat plateau (Figure 9.15).

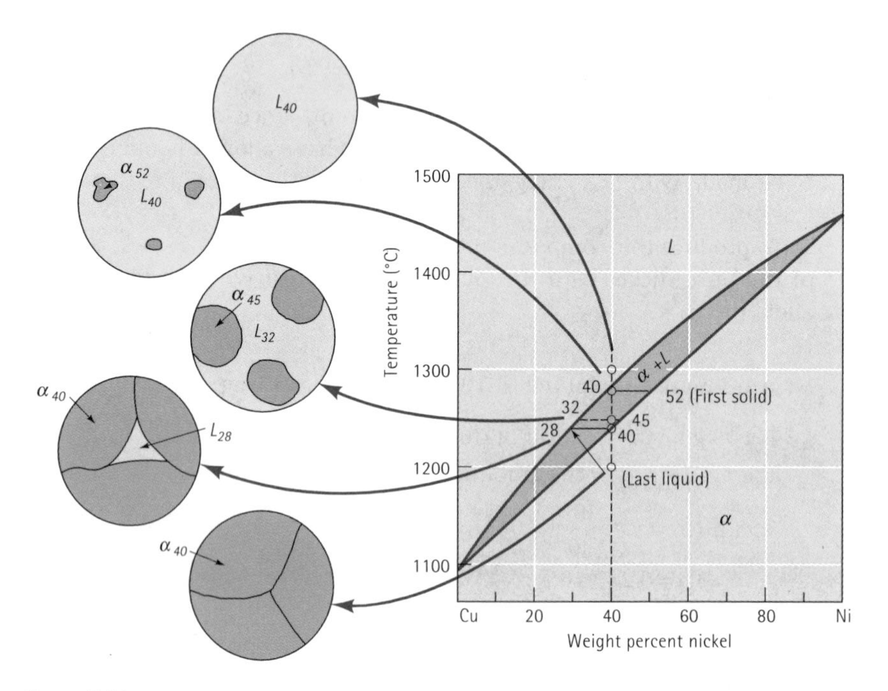

Figure 9.14
The change in structure of a Cu-40% Ni alloy during equilibrium solidification. The nickel and copper atoms must diffuse during cooling in order to satisfy the phase diagram and produce a uniform equilibrium structure.

At the start of freezing, the liquid contains Cu-40% Ni and the first solid contains Cu-52%. Nickel atoms must have diffused to and concentrated at the first solid to form. But after cooling to 1 250°C, solidification has advanced and the phase diagram tells us that now all of the liquid must contain 32% Ni and all of the solid must contain 45% Ni. On cooling from the liquidus to 1 250°C, some nickel atoms must diffuse from the first solid to the new solid, reducing the nickel in the first solid. Additional nickel atoms diffuse from the solidifying liquid to the new solid. Meanwhile, copper atoms have concentrated – by diffusion – into the remaining liquid.

This process must continue until we reach the solidus temperature, where the last liquid to freeze, which contains Cu-28% Ni, solidifies and forms a solid containing Cu-40% Ni. Just below the solidus, all of the solid must contain a uniform concentration of 40% Ni throughout.

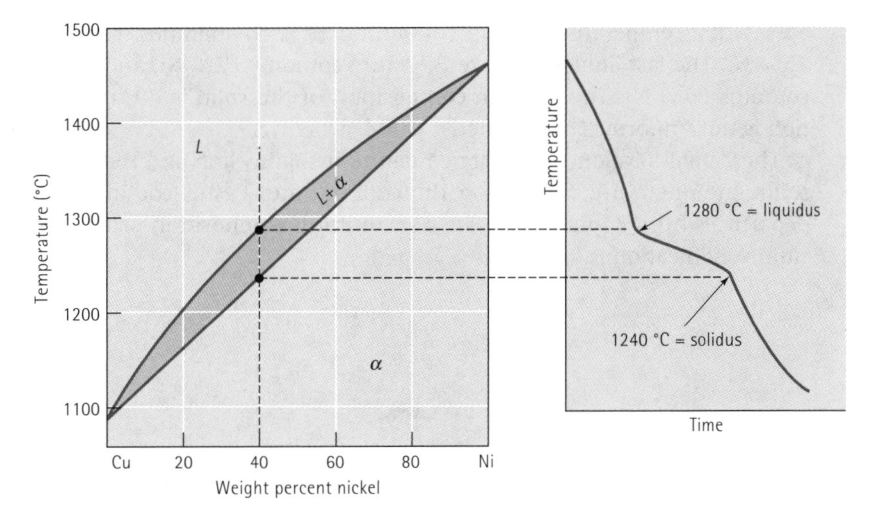

Figure 9.15
The cooling curve for an isomorphous alloy during solidification. The changes in slope of the cooling curve indicate the liquidus and solidus temperatures, in this case, for a Cu-40% Ni alloy.

In order to achieve this equilibrium final structure, the cooling rate must be extremely low. Sufficient time must be permitted for the copper and nickel atoms to diffuse and produce the compositions given by the phase diagram. In most practical casting situations, the cooling rate is too rapid to permit equilibrium.

9.9 Nonequilibrium Solidification and Segregation

When cooling is too rapid for atoms to diffuse and produce equilibrium conditions, nonequilibrium structures are produced in the casting. Let's see what happens to our Cu-40% Ni alloy on rapid cooling.

Again, the first solid, containing 52% Ni, forms on reaching the liquidus temperature (Figure 9.16). On cooling to 1 260°C, the tie line tells us that the liquid contains 34% Ni and the solid that forms at that temperature contains 46% Ni. Since diffusion occurs rapidly in liquids, we expect the tie line to predict the liquid composition accurately. However, diffusion in solids is comparatively slow. The first solid that forms still has about 52% Ni, but the new solid contains only 46% Ni. We might find that the average composition of the solid is 51% Ni. This gives a different nonequilibrium solidus than that given by the phase diagram. As solidification continues, the nonequilibrium solidus line continues to separate from the equilibrium solidus.

When the temperature reaches 1 240°C, the equilibrium solidus line, a significant amount of liquid remains. The liquid will not completely solidify until we cool to 1 180°C, where the nonequilibrium solidus intersects the original composition of 40% Ni. At that temperature, liquid containing 17% Ni solidifies, giving solid containing 25% Ni. The last liquid to freeze therefore contains 17% Ni, and the last solid to form contains 25% Ni. The average composition of the solid is 40% Ni, but the composition is not uniform. This is called a cored structure.

The actual location of the nonequilibrium solidus line and the final nonequilibrium solidus temperature depend on the cooling rate. Faster cooling rates cause greater departures from equilibrium. The non-uniform composition produced by nonequilibrium solidification is known as segregation.

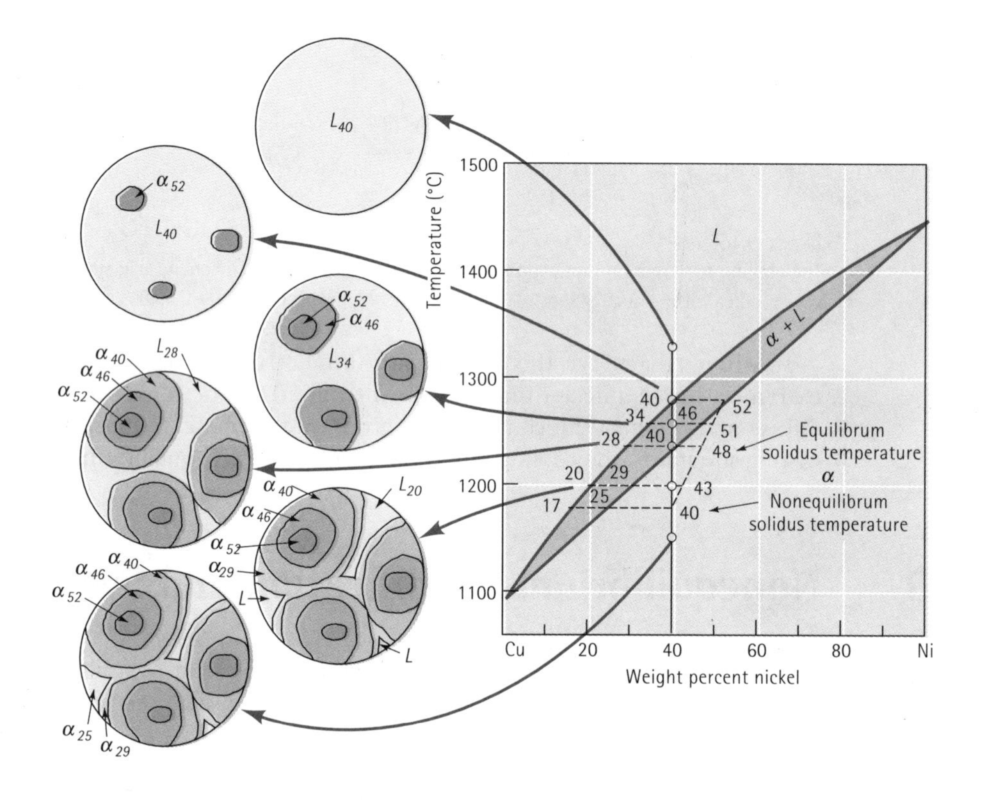

Figure 9.16
The change in structure of a Cu-40% Ni alloy during nonequilibrium solidification. Insufficient time for diffusion in the solid produces a segregated structure.

EXAMPLE 9.11

Calculate the composition and amount of each phase in a Cu-40% Ni alloy that is present under the nonequilibrium conditions shown in Figure 9.16 at 1 300°C, 1 280°C, 1 260°C, 1 240°C, 1 200°C, and 1 150°C. Compare with the equilibrium compositions and amounts of each phase.

SOLUTION

Temperature	Equilibrium		Nonequilibrium	
1 300°C	*L*: 40% Ni	100% *L*	*L*: 40% Ni	100% *L*
1 280°C	*L*: 40% Ni	100% *L*	*L*: 40% Ni	100% *L*
	α: 52% Ni	~0% α	α: 52% Ni	~0% α
1 260°C	*L*: 34% Ni	$\frac{46-40}{46-34} = 50\%\ L$	*L*: 34% Ni	$\frac{51-40}{51-34} = 65\%\ L$
	α: 46% Ni	$\frac{40-34}{46-34} = 50\%\ \alpha$	α: 51% Ni	$\frac{40-34}{51-34} = 35\%\ \alpha$
1 240°C	*L*: 28% Ni	~0% *L*	*L*: 28% Ni	$\frac{48-40}{48-28} = 40\%\ L$
	α: 40% Ni	100% α	α: 48% Ni	$\frac{40-28}{48-28} = 60\%\ \alpha$
1 200°C	α: 40% Ni	100% α	*L*: 20% Ni	$\frac{43-40}{43-20} = 13\%\ L$
			α: 43% Ni	$\frac{40-20}{43-20} = 87\%\ \alpha$
1 150°C	α: 40% Ni	100% α	α: 40% Ni	100% α

Microsegregation Microsegregation occurs over short distances, often between small dendrite arms. The centres of the dendrites, which represent the first solid to freeze, are rich in the higher melting point element in the alloy. The regions between the dendrites are rich in the lower melting point element, since these regions represent the last liquid to freeze. The composition and properties of *α* differ from one region to the next, and we expect the casting to have poorer properties as a result.

Microsegregation can cause hot shortness, or melting of the lower melting point interdendritic material at temperatures below the equilibrium solidus. When we heat the Cu-40% Ni alloy to 1 225°C, below the equilibrium solidus but above the nonequilibrium solidus, the low-nickel regions between the dendrites melt. Microsegregation can occur within a grain, and is called coring or between grains and is called interdendritic segregation

Homogenisation We can reduce the microsegregation and problems with hot shortness by means of a homogenisation heat treatment. If we heat the casting to a temperature below the nonequilibrium solidus, the nickel atoms in the centres of the dendrites diffuse to the interdendritic regions; copper atoms diffuse in the opposite direction (Figure 9.17). Since the diffusion distances are relatively short, only a few hours are required to eliminate most of the composition differences. The homogenisation time is related to

$$t = c(SDAS)^2/D_s \tag{9.6}$$

where $SDAS$ is the secondary dendrite arm spacing, D_s is the rate of diffusion of the solute in the matrix, and c is a constant. A small $SDAS$ reduces the diffusion distance and permits short homogenisation times.

If the ingot is not too embrittled by the coring, and it can be hot worked then this processing will accelerate the removal of the segregated structures. However, interdendritic segregation makes an ingot so brittle that hot working usually cracks it.

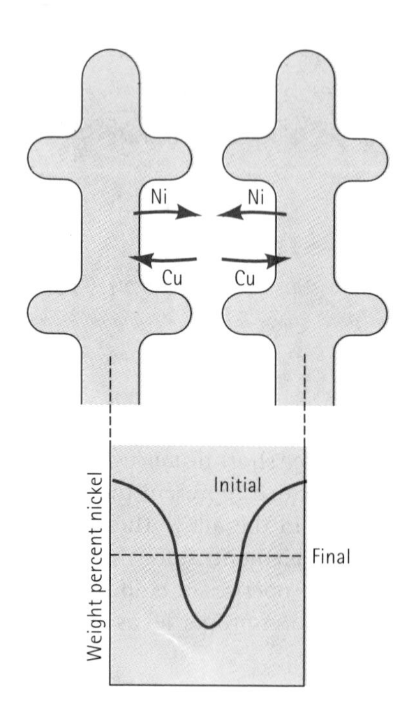

Figure 9.17
Microsegregation between dendrites can be reduced by a homogenisation heat treatment. Counterdiffusion of nickel and copper atoms may eventually eliminate the composition gradients and produce a homogeneous composition.

Macrosegregation Macrosegregation occurs over a large distance, for example between the surface and the centre of the casting, with the surface (which freezes first) containing slightly more than the average amount of the higher melting point metal. We cannot eliminate macrosegregation by a homogenisation treatment, because the diffusion distances are too great. Macrosegregation may also be reduced by *hot working*, which was discussed in Chapter 7.

SUMMARY

Solid solution strengthening is accomplished by the controlled addition of point defects, or alloying elements:

- The degree of solid solution strengthening increases when (1) the amount of alloying element increases and (2) the atomic size difference between the host material and the alloying element increases.
  - The amount of alloying element that we can add to produce solid solution strengthening is limited by the solubility of the alloying element in the host material. The solubility is limited when (1) the atomic difference is more than about 15%, (2) the alloying element has a different crystal structure than the host element, and (3) the valence and electronegativity of the alloying element are different from the host element.
  - In addition to increasing strength and hardness, solid solution strengthening typically decreases ductility and electrical conductivity. An important function of solid solution strengthening is to provide good high-temperature properties to the alloy.
- The addition of alloying elements to provide solid solution strengthening also changes the physical properties, including the melting temperature, of the alloy. The phase diagram helps explain these changes:
  - When complete solid solubility is obtained, an isomorphous phase diagram is produced.
  - As a result of solid solution strengthening solidification begins at the liquidus temperature and is completed at the solidus temperature; the temperature difference over which solidification occurs is the freezing range.
  - In two-phase regions of the phase diagram, the ends of a tie line fix the composition of each phase and the lever law permits the amount of each phase to be calculated.
- Segregation occurs during nonequilibrium solidification:
  - Microsegregation, or coring, occurs over small distances, often between dendrites. The centre of the dendrites are rich in the higher melting point element, whereas interdendritic regions, which solidify last, are rich in the lower melting point element. Homogenisation can reduce microsegregation. Hot working can also be effective.

- Macrosegregation describes differences in composition over long distances, such as between the surface and centre of a casting. Hot working may reduce macrosegregation.

GLOSSARY

Binary phase diagram
A phase diagram in which there are only two components.

Coring
Microsegregation within a grain.

Freezing range
The temperature difference between the liquidus and solidus temperatures.

Gibbs phase rule
Describes the number of degrees of freedom, or the number of variables that must be fixed to specify the temperature and composition of a phase.

Homogenisation heat treatment
The heat treatment used to reduce the microsegregation caused during nonequilibrium solidification.

Hot shortness
Melting of the lower melting point nonequilibrium material that forms due to segregation, even though the temperature is below the equilibrium solidus temperature.

Hume-Rothery rules
The conditions that an alloy system must meet if the system is to display unlimited solid solubility. Hume-Rothery's rules are necessary but are not sufficient to predict unlimited solubility.

Isomorphous phase diagram
A phase diagram that displays unlimited solid solubility.

Interdendritic Segregation
Microsegregation associated with the grain boundary region.

Lever law
A technique for calculating the amount of each phase in a two-phase system.

Limited solubility
When only a maximum amount of a solute material can be dissolved in a solvent material.

Liquidus
The temperature at which the first solid begins to form within the liquid during solidification.

Macrosegregation
The presence of composition differences in a material over large distances caused by nonequilibrium solidification.

Microsegregation

The presence of concentration differences in a material over short distances caused by nonequilibrium solidification. Various types of microsegregation include interdendritic segregation and coring.

Phase

A material having the same composition, structure, and properties everywhere under equilibrium conditions.

Phase diagram

A diagram showing the phases and their boundaries in terms of temperature and overall composition.

Segregation

The presence of nonequilibrium composition differences in a material, often caused by insufficient time for diffusion during solidification.

Solid solution

A solid phase that contains a mixture of more than one element, with the elements combining to give a uniform composition everywhere.

Solid solution strengthening

Increasing the strength of a material by introducing point defects into the structure in a deliberate and controlled manner.

Solidus

The temperature below which all liquid has completely solidified.

Solubility

The amount of one material that will completely dissolve in a second material without creating a second phase.

Tie line

A horizontal line drawn in a two-phase region of a phase diagram to assist in determining the compositions of the two phases.

Triple point

A pressure and temperature at which three phases of a single material are in equilibrium.

Unary phase diagram

A phase diagram in which there is only one component.

Unlimited solubility

When the amount of one material that will dissolve in a second material without creating a second phase is unlimited.

PROBLEMS

9.1 The triple point for water occurs at 709 Pa and 0.0075°C. Using this information and your knowledge of the behaviour of water at atmospheric pressure of 101 325 Pa, construct a schematic unary phase diagram.

9.2 The unary phase diagram for SiO_2 is shown in Figure 14.6. Locate the triple point where solid, liquid, and vapour coexist and give the temperature and the type of solid present. What do the other 'triple' points indicate?

9.3 Based on Hume-Rothery's conditions, which of the following systems would be expected to display unlimited solid solubility? Explain.

(a) Au-Ag (b) Al-Cu (c) Al-Au

(d) U-W (e) Mo-Ta (f) Nb-W

(g) Mg-Zn (h) Mg-Cd

9.4 Suppose 1 at% of the following elements is added to copper without exceeding the solubility limit. Which one would be expected to give the higher strength alloy? Is any of the alloying elements expected to have unlimited solid solubility in copper?

(a) Au (b) Mn (c) Sr

(d) Si (e) Co

9.5 Suppose 1 at% of the following elements is added to aluminium without exceeding the solubility limit. Which one would be expected to give the least reduction in electrical conductivity? Is any of the alloy elements expected to have unlimited solid solubility in aluminium?

(a) Li (b) Ba (c) Be

(d) Cd (e) Ga

9.6 Which of the following oxides is expected to have the largest solid solubility in Al_2O_3?

(a) Y_2O_3 (b) Cr_2O_3

(c) Fe_2O_3 (d) Ti_2O_3

9.7 Determine the liquidus temperature, solidus temperature, and freezing range for the following NiO-MgO ceramic compositions. (*See* Figure 9.8.)

(a) NiO-30 mol% MgO

(b) NiO-45 mol% MgO

(c) NiO-60 mol% MgO

(d) NiO-85 mol% MgO

9.8 Determine the liquidus temperature, solidus temperature, and freezing range for the following MgO-FeO ceramic compositions. (*See* Figure 9.18.)

(a) MgO-25% wt% FeO

(b) MgO-45% wt% FeO

(c) MgO-65% wt% FeO

(d) MgO-80% wt% FeO

9.9 Determine the phases present, the compositions of each phase, and the amount of each phase in mol% for the following NiO-MgO ceramics at 2400°C. (*See* Figure 9.8.)

(a) NiO-30 mol% MgO

(b) NiO-45 mol% MgO

(c) NiO-60 mol% MgO

(d) NiO-85 mol% MgO

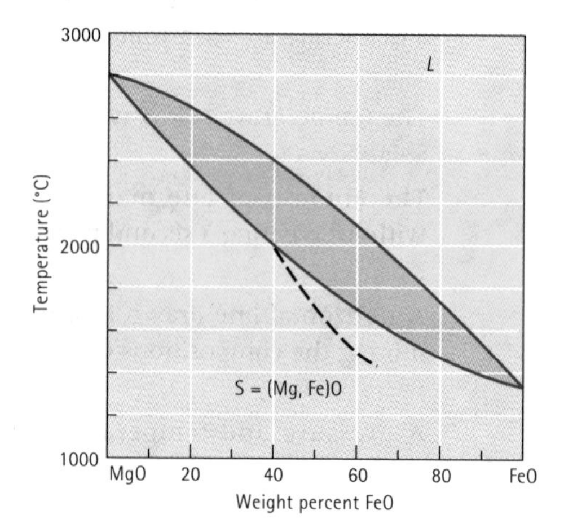

Figure 9.18

The equilibrium phase diagram for the MgO-FeO system.

9.10 Determine the phases present, the compositions of each phase and the amount of each phase in wt% for the following MgO-FeO ceramics at 2000°C. (*See* Figure 9.18.)

(a) MgO-25% wt% FeO

(b) MgO-45% wt% FeO

(c) MgO-60% wt% FeO

(d) MgO-80% wt% FeO

9.11 Consider an alloy of 65 wt% Cu and 35 wt% Al. Calculate the composition of the alloy in at%.

9.12 Consider a ceramic composed of 30 mol% MgO and 70 mol% FeO. Calculate the composition of the ceramic in wt%.

9.13 A NiO-20 mol% MgO ceramic is heated to 2 200°C. Determine

(a) the composition of the solid and liquid phases in both mol% and wt% and

(b) the amount of each phase in both mol% and wt%.

(c) Assuming that the density of the solid is 6.32 $Mg.m^{-3}$ and that of the liquid is 7.14 $Mg.m^{-3}$, determine the amount of each phase in vol%.

9.14 A Nb-60 wt% W alloy is heated to 2 800°C. Determine

(a) the composition of the solid and liquid phases in both wt% and at% and

(b) the amount of each phase in both wt% and at%.

(c) Assuming that the density of the solid is 16.05 $Mg.m^{-3}$ and that of the liquid is 13.91 $Mg.m^{-3}$, determine the amount of each phase in vol%. (*See* Figure 9.19.)

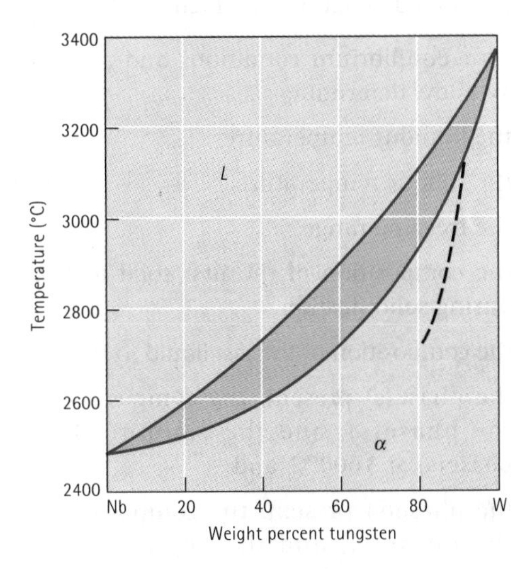

Figure 9.19
The equilibrium phase diagram for the Nb-W system.

9.15 How many grams of nickel must be added to 500 grams of copper to produce an alloy that has a liquidus temperature of 1 350°C? What is the ratio of the number of nickel atoms to copper atoms in this alloy?

9.16 How many grams of nickel must be added to 500 grams of copper to produce an alloy that contains 50 wt% α at 1 300°C?

9.17 How many grams of MgO must be added to 1 kg of NiO to produce a ceramic that has a solidus temperature of 2 200°C?

9.18 How many grams of MgO must be added to 1 kg of NiO to produce a ceramic that contains 25 mol% solid at 2 400°C?

9.19 We would like to produce a solid MgO-FeO ceramic that contains equal mol percentages of MgO and FeO at 1 200°C. Determine the wt% FeO in the ceramic. (*See* Figure 9.18.)

9.20 We would like to produce a MgO-FeO ceramic that is 30 wt% solid at 2 000°C. Determine the original composition of the ceramic in wt%. (*See* Figure 9.18.)

9.21 A Nb-W alloy held at 2 800°C is partly liquid and partly solid.

(a) If possible, determine the composition of each phase in the alloy.

(b) If possible, determine the amount of each phase in the alloy. (*See* Figure 9.19.)

9.22 A Nb-W alloy contains 55% α at 2 600°C. Determine

(a) the composition of each phase and

(b) the original composition of the alloy. (*See* Figure 9.19.)

9.23 Suppose a 0.5 tonne bath of a Nb-40 wt% W alloy is held at 2 800°C. How many grams of tungsten can be added to the bath before any solid forms? How many grams of tungsten must be added to cause the entire bath to be solid? (*See* Figure 9.19.) 1 tonne = 10^3 kg.

9.24 A fibre-reinforced composite material is produced, in which tungsten fibres are embedded in a Nb matrix. The composite is composed of 70 vol% tungsten.

(a) Calculate the wt% of tungsten fibres in the composite.

(b) Suppose the composite is heated to 2 600°C and held for several years. What happens to the fibres? Explain. (*See* Figure 9.19.)

9.25 Suppose a crucible made of pure nickel is used to contain 500 g of liquid copper at 1 150°C. Describe what happens to the system as it is held at this temperature for several hours. Explain.

9.26 Equal moles of MgO and FeO are combined and melted. Determine

(a) the liquidus temperature, the solidus temperature, and the freezing range of the ceramic and

(b) determine the phase(s) present, their composition(s), and their amount(s) at 1 800°C. (*See* Figure 9.18.)

9.27 Suppose 75 cm^3 of Nb and 45 cm^3 of W are combined and melted. Determine

(a) the liquidus temperature, the solidus temperature, and the freezing range of the alloy and

(b) determine the phase(s) present, their composition(s), and their amount(s) at 2 800°C. (*See* Figure 9.19.)

9.28 A NiO-60 mol% MgO ceramic is allowed to solidify. Determine

(a) the composition of the first solid to form and

(b) the composition of the last liquid to solidify under equilibrium conditions.

9.29 A Nb-35% W alloy is allowed to solidify. Determine

(a) the composition of the first solid to form and

(b) the composition of the last liquid to solidify under equilibrium conditions. (*See* Figure 9.19.)

9.30 For equilibrium conditions and a MgO-65 wt% FeO ceramic, determine

(a) the liquidus temperature,

(b) the solidus temperature,

(c) the freezing range,

(d) the composition of the first solid to form during solidification,

(e) the composition of the last liquid to solidify,

(f) the phase(s) present, the composition of the phase(s), and the amount of the phase(s) at 1 800°C, and

(g) the phase(s) present, the composition of the phase(s), and the amount of the phase(s) at 1 600°C. (*See* Figure 9.18.)

9.31 For the nonequilibrium conditions shown for the MgO-65 wt% FeO ceramic, determine

(a) the liquidus temperature,

(b) the nonequilibrium solidus temperature,

(c) the freezing range,

(d) the composition of the first solid to form during solidification,

(e) the composition of the last liquid to solidify,

(f) the phase(s) present, the composition of the phase(s), and the amount of the phase(s) at 1 800°C, and

(g) the phase(s) present, the composition of the phase(s), and the amount of the phase(s) at 1 600°C. (*See* Figure 9.18.)

9.32 For equilibrium conditions and a Nb-80 wt% W alloy, determine

(a) the liquidus temperature,

(b) the solidus temperature,

(c) the freezing range,

(d) the composition of the first solid to form during solidification,

(e) the composition of the last liquid to solidify,

(f) the phase(s) present, the composition of the phase(s), and the amount of the phase(s) at 3 000°C, and

(g) the phase(s) present, the composition of the phase(s), and the amount of the phase(s) at 2 800°C. (*See* Figure 9.19.)

9.33 For the nonequilibrium conditions shown for the Nb-80 wt% W alloy, determine

(a) the liquidus temperature,

(b) the nonequilibrium solidus temperature,

(c) the freezing range

(d) the composition of the first solid to form during solidification,

(e) the composition of the last liquid to solidify,

(f) the phase(s) present, the composition of the phase(s), and the amount of the phase(s) at 3000°C, and

(g) the phase(s) present, the composition of the phase(s), and the amount of the phase(s) at 2800°C. (*See* Figure 9.19.)

9.34 Figure 9.20 shows the cooling curve for a NiO-MgO ceramic. Determine

(a) the liquidus temperature,

(b) the solidus temperature,

(c) the freezing range,

(d) the pouring temperature,

(e) the superheat,

(f) the local solidification time,

(g) the total solidification time, and

(h) the composition of the ceramic.

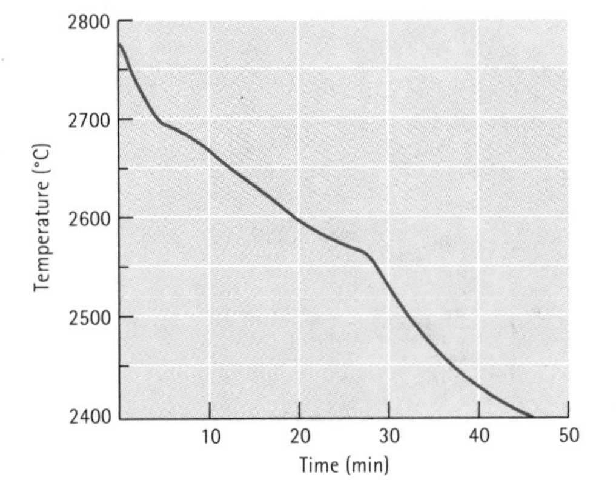

Figure 9.20
Cooling curve for a NiO-MgO ceramic (*for* Problem 9.34).

9.35 Figure 9.21 shows the cooling curve for an Nb-W alloy. Determine

(a) the liquidus temperature,

(b) the solidus temperature,

(c) the freezing range,

(d) the pouring temperature,

(e) the superheat,

(f) the local solidification time,

(g) the total solidification time,

(h) the composition of the alloy.

9.36 Cooling curves are shown in Figure 9.22 for several Mo-V alloys. Based on these curves, construct the Mo-V phase diagram.

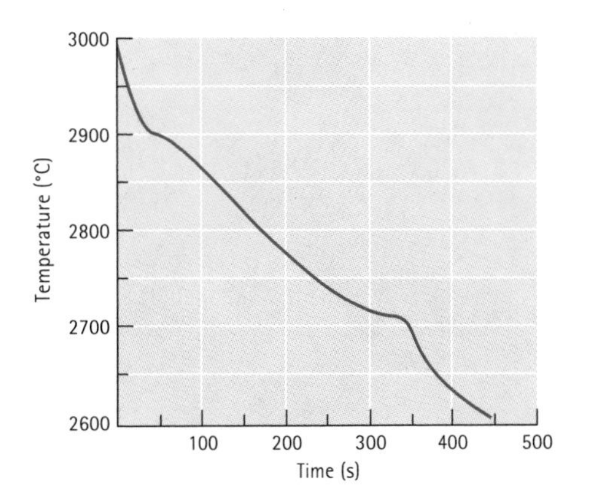

Figure 9.21
Cooling curve for a Nb-W alloy (*for* Problem 9.35).

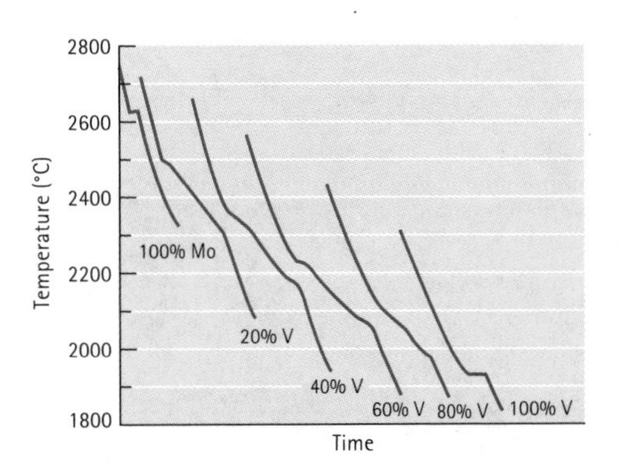

Figure 9.22
Cooling curves for a series of Mo-V alloys (*for* Problem 9.36).

Design Problems

9.37 Homogenisation of a slowly cooled Cu-Ni alloy having a secondary dendrite arm spacing of 0.025 cm requires 8 hours at 1 000°C. Design a process to produce a homogeneous structure in a more rapidly cooled Cu-Ni alloy having a *SDAS* of 0.005 cm.

9.38 Design a process to produce a NiO-60% MgO refractory whose structure is 40% glassy phase at room temperature. Include all relevant temperatures.

9.39 Design a method by which glass beads (having a density of 2.3 $Mg.m^{-3}$) can be uniformly mixed and distributed in a Cu-20% Ni alloy (density of 8.91 $Mg.m^{-3}$).

9.40 Suppose that MgO contains 5 mol% NiO. Design a solidification purification method that will reduce the NiO to less than 1 mol% in the MgO.

CHAPTER 10

Dispersion Strengthening by Solidification

10.1 Introduction

When the solubility of a material is exceeded by adding too much of an alloying element, a second phase forms and a two-phase alloy is produced. The boundary between the two phases is a surface at which the atomic arrangement is not perfect. In metals, this boundary interferes with the slip of dislocations and strengthens the material. The general term for strengthening by the introduction of a second phase is dispersion strengthening.

In this chapter we first discuss the fundamentals of dispersion strengthening to determine the structure we should aim to produce. Next, we examine the types of reactions that produce multiple-phase alloys. Finally, we look in some detail at methods to achieve dispersion strengthening by controlling the solidification process.

10.2 Principles of Dispersion Strengthening

More than one phase must be present in any dispersion-strengthened alloy. We call the continuous phase, which usually is present in larger amounts, the matrix. The second phase, usually present in smaller amounts, is the precipitate. In some cases, two phases form simultaneously. We define these structures differently, calling the intimate mixture of phases a microconstituent.

There are some general considerations for determining how the characteristics of the matrix and precipitate affect the overall properties of a metal alloy (Figure 10.1).

1. The matrix should be soft and ductile, although the precipitate should be strong. The precipitate interferes with slip, while the matrix provides at least some ductility to the overall alloy.
2. The hard precipitate should be discontinuous, while the soft, ductile matrix should be continuous. If the precipitate were continuous, cracks could propagate through the entire structure. However, cracks in the discontinuous, brittle precipitate are arrested by the precipitate-matrix interface.

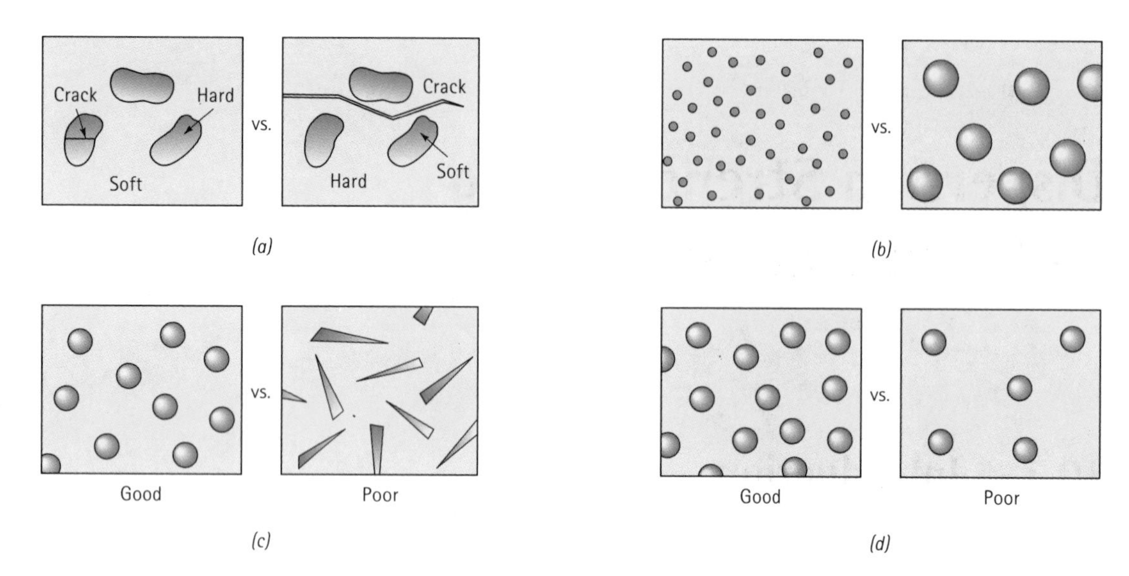

Figure 10.1
Considerations for effective dispersion strengthening: (*a*) The precipitate should be hard and discontinuous, (*b*) the precipitate particles should be small and numerous, (*c*) the precipitate particles should be round rather than needlelike, and (*d*) larger amounts of precipitate increase strengthening.

3. The precipitate particles should be small and numerous, increasing the likelihood that they interfere with the slip process.
4. The precipitate particles should be round, rather than needlelike or sharp-edged, because the rounded shape is less likely to initiate a crack or to act as a notch.
5. Large amounts of the precipitate increase the strength of the alloy.

Two-phase materials are also produced for reasons other than strengthening; in these cases, the above criteria may not apply. For example, the fracture toughness of materials may be improved by introducing a dispersed phase. Incorporating a ductile phase in a ceramic matrix or a rubber phase in a thermosetting polymer improves toughness; forming a dense network of needle-shaped precipitates in some titanium alloys helps impede the growth of cracks. Producing globules of very soft lead in copper improves machinability. Most of this chapter, however, concentrates on how precipitates improve mechanical properties, such as yield strength.

10.3 Intermetallic Compounds

Often dispersion-strengthened alloys contain an intermetallic compound. An intermetallic compound is made up of two or more elements, producing a new phase with its own composition, crystal structure, and properties. Intermetallic compounds are almost always very hard and brittle.

Stoichiometric intermetallic compounds have a fixed composition. Steels are strengthened by a stoichiometric compound, Fe_3C, that has a fixed ratio of three iron atoms to one carbon atom. Stoichiometric intermetallic compounds are represented in the phase diagram by a vertical line [Figure 10.2(a)].

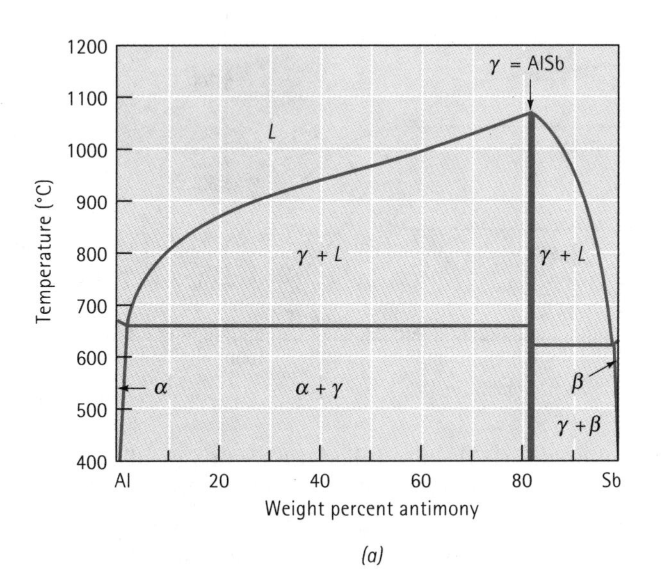

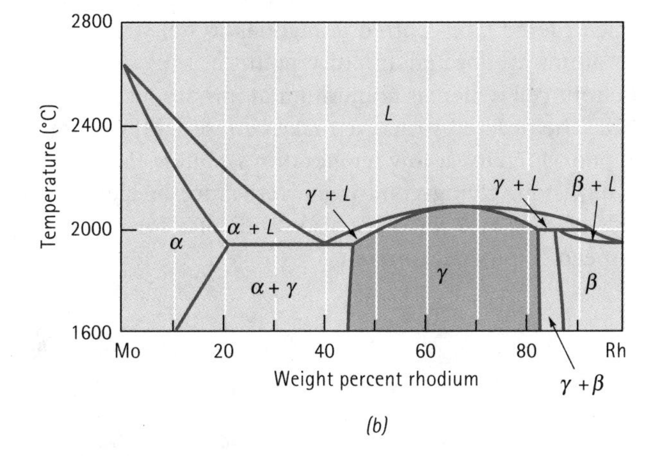

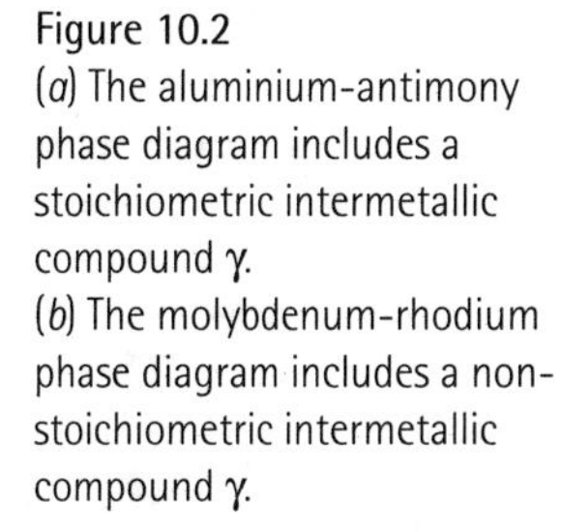
Figure 10.2
(*a*) The aluminium-antimony phase diagram includes a stoichiometric intermetallic compound γ.
(*b*) The molybdenum-rhodium phase diagram includes a non-stoichiometric intermetallic compound γ.

Nonstoichiometric intermetallic compounds have a range of compositions and are sometimes called intermediate solid solutions. In the molybdenum-rhodium system, the γ phase is an intermetallic compound [Figure 10.2(b)]. Because the molybdenum-rhodium atom ratio is not fixed, the γ can contain from 45 wt% to 83 wt% Rh at 1 600°C. Precipitation of the nonstoichiometric intermetallic $CuAl_2$ causes strengthening in a number of important aluminium alloys.

Intermetallic compounds are used to advantage by dispersing them into a softer, more ductile matrix, as we will see in this and subsequent chapters. However, there is considerable interest in using intermetallics by themselves, taking advantage of their high melting point, stiffness, and resistance to oxidation and creep. These new materials, which include Ti_3Al and Ni_3Al, maintain their strength and even develop usable ductility at elevated temperatures (Figure 10.3).

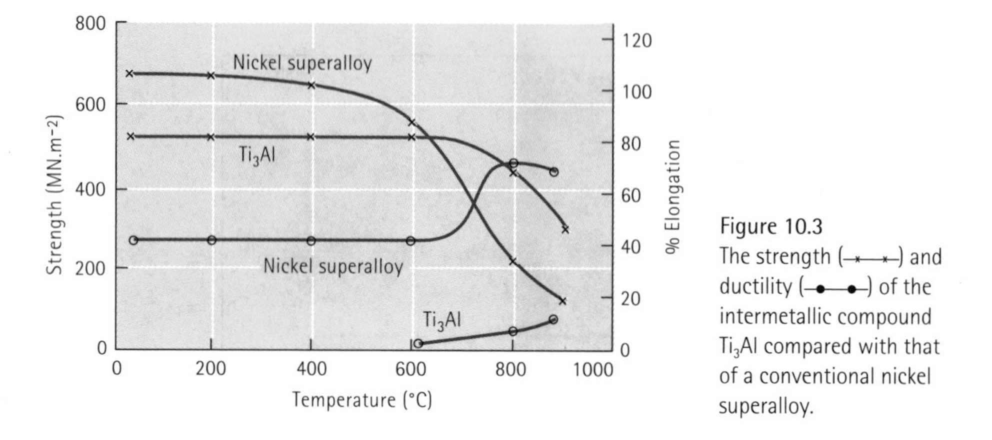

Figure 10.3
The strength (—×—×—) and ductility (—●—●—) of the intermetallic compound Ti_3Al compared with that of a conventional nickel superalloy.

The titanium aluminides, TiAl – also called the gamma (γ) alloy – and Ti_3Al – the α_2 alloy – are nonstoichiometric intermetallics considered for a variety of applications, including gas turbine engines and the U.S.A. national aerospace plane. Both have ordered crystal structures; in the ordered structure, the Ti and Al atoms occupy specific locations in the lattice, rather than random locations as in most solid solutions (Figure 10.4). In the ordered face-centred tetragonal crystal structure of TiAl (Figure 10.5), the titanium atoms are located at lattice points at the corners and the top and bottom faces of the unit cell, whereas aluminium atoms are located only at the other four faces of the cell. The ordered structure makes it more difficult for dislocations to move (resulting in poor ductility at low temperatures), but it also leads to a high activation energy for diffusion, giving good creep resistance at elevated temperatures. Ordered compounds of NiAl and Ni_3Al are also candidates for supersonic aircraft, jet engines, and high-speed commercial aircraft.

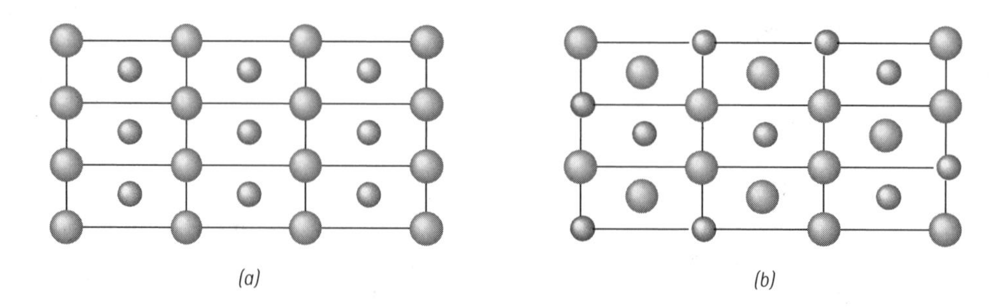

Figure 10.4
In an ordered lattice, the substituting atoms occupy lattice points (*a*), while in normal lattices, the substituting atoms are randomly located at the lattice points (*b*).

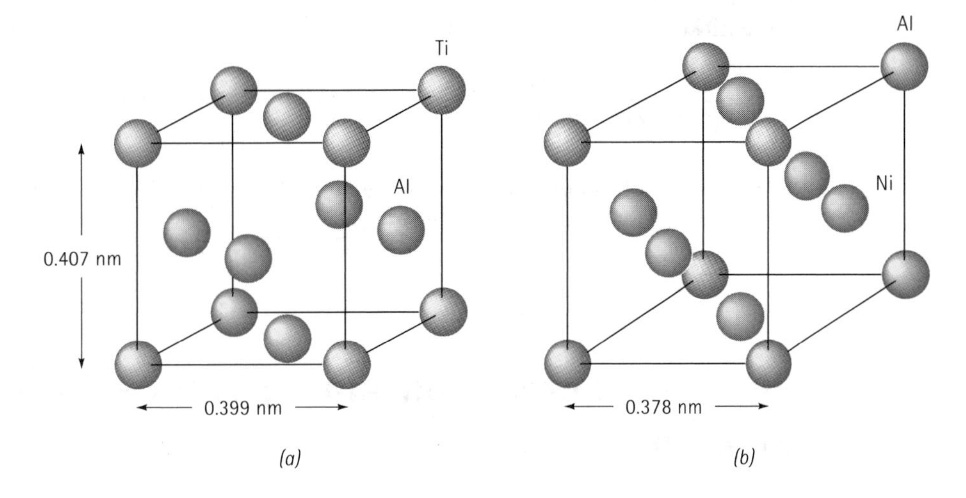

Figure 10.5
The unit cells of two intermetallic compounds: (*a*) TiAl has an ordered face-centred tetragonal cell and (*b*) Ni_3Al has an ordered face-centred cubic lattice.

EXAMPLE 10.1 Materials Selection for an Aerospace Vehicle

Select a material suitable for the parts of an aerospace vehicle that reach high temperatures during re-entry from Earth orbit.

SOLUTION

The material must withstand the high temperatures generated as the vehicle enters Earth's atmosphere, where the material is exposed to oxygen. Some ductility is needed to provide damage tolerance to the vehicle. Finally, the material should have a low density to minimise vehicle weight.

The requirement for ductility suggests a metallic alloy. High-temperature metals that might be considered include tungsten (density of 19.254 $Mg.m^{-3}$), nickel (density of 8.902 $Mg.m^{-3}$) and titanium (density of 4.507 $Mg.m^{-3}$). Tungsten and nickel are quite heavy and, as we will see in a later chapter, titanium has poor oxidation resistance at high temperatures. However, an intermetallic compound might be a solution. TiAl and Ni_3Al have good high-temperature properties and oxidation resistance and, at high temperatures, have at least some ductility. We can estimate their densities from their crystal structures.

The lattice parameters of the face-centred tetragonal TiAl unit are a_0 = 0.399 nm and c_0 = 0.407 nm. There are two titanium and two aluminium atoms in each unit cell. Therefore, using Appendix A:

$$\rho_{\mathrm{TiAl}} = \frac{(2\mathrm{Ti})(47.9)+(2\mathrm{Al})(29.981)}{(0.399\times10^{-9})^2(0.407\times10^{-9})(6.02\times10^{23})} = 3\,839\,414\ \mathrm{g.m^{-3}} = 3.84\ \mathrm{Mg.m^{-3}}$$

The lattice parameter for Ni_3Al is 0.378 nm = 0.378×10^{-9} m; each unit cell contains three nickel atoms and one aluminium atom:

$$\rho = \frac{(3\text{Ni})(58.71) + (1\text{Ti})(47.9)}{(0.378 \times 10^{-9})^3 (6.02 \times 10^{23})} = 6\,890\,233 \text{ g.m}^{-3} = 6.89 \text{ Mg.m}^{-3}$$

The TiAl is only about half the density of the Ni_3Al and, if other properties are comparable, TiAl might be our choice.

10.4 Phase Diagrams Containing Three-Phase Reactions

Many combinations of two elements produce more complicated phase diagrams than the isomorphous systems. These systems contain reactions that involve three separate phases, five of which are defined in Figure 10.6. Each of the reactions can be identified in a complex phase diagram by the following procedure:

1. Locate a horizontal line on the phase diagram. The horizontal line, which indicates the presence of a three-phase reaction, represents the temperature at which the reaction occurs under equilibrium conditions.

Eutectic	$L \rightarrow \alpha + \beta$	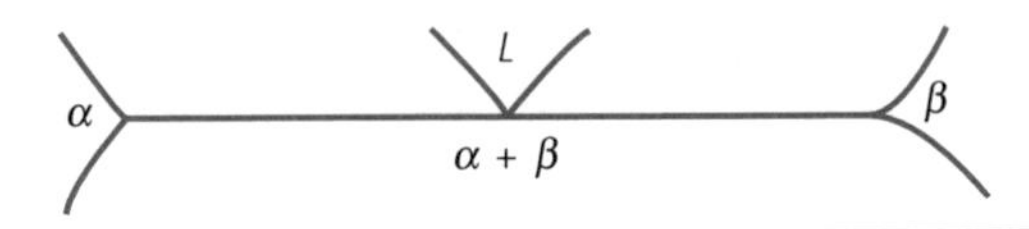 α; L; β; $\alpha + \beta$
Peritectic	$\alpha + L \rightarrow \beta$	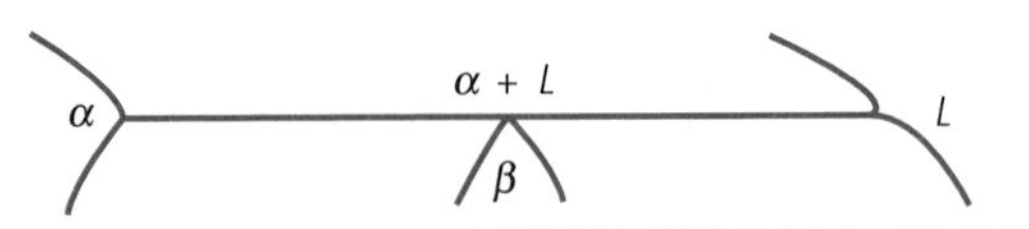 α; $\alpha + L$; L; β
Monotectic	$L_1 \rightarrow L_2 + \alpha$	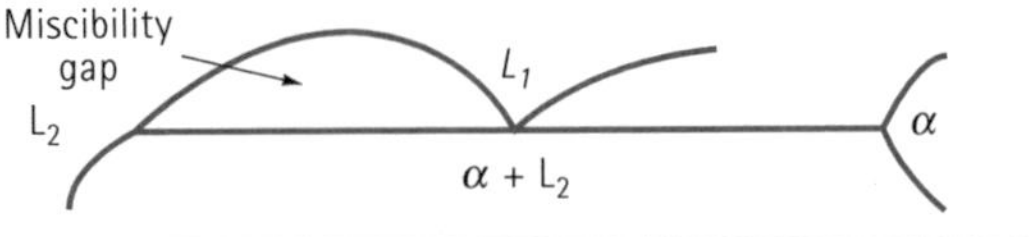 Miscibility gap; L_2; L_1; α; $\alpha + L_2$
Eutectoid	$\gamma \rightarrow \alpha + \beta$	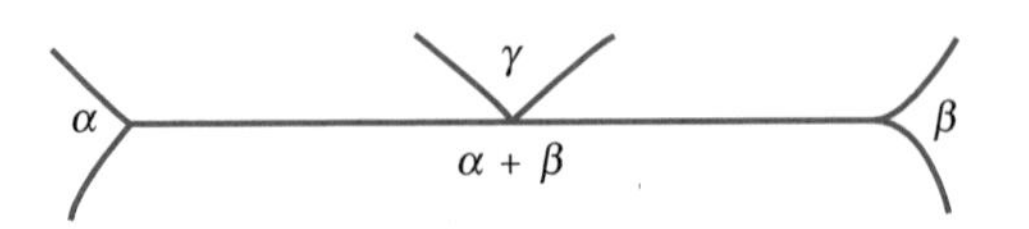 α; γ; β; $\alpha + \beta$
Peritectoid	$\alpha + \beta \rightarrow \gamma$	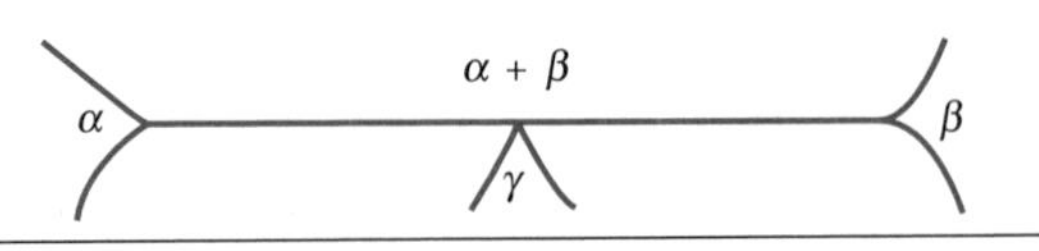 α; $\alpha + \beta$; β; γ

Figure 10.6
The five most important three-phase reactions in binary phase diagrams. L indicates a liquid phase.

2. Locate three distinct points on the horizontal line: the two endpoints plus a third point, often near the centre of the horizontal line. The centre point represents the composition at which the three-phase reaction occurs.
3. Look immediately above the centre point and identify the phase or phases present; look immediately below the centre point and identify the phase or phases present. Then write in reaction form the phase(s) above the centre point transforming to the phase(s) below the point. Compare the reaction with those in Figure 10.6 to identify the reaction.

EXAMPLE 10.2

Consider the phase diagram in Figure 10.7. Identify the three-phase reactions that occur.

SOLUTION

We find horizontal lines at 1 150°C, 920°C, 750°C, and 300°C:

1 150°C: The centre point is at 15% B. $\delta + L$ are present above the point, γ is present below. The reaction is:

$\delta + L \rightarrow \gamma$, or a *peritectic*

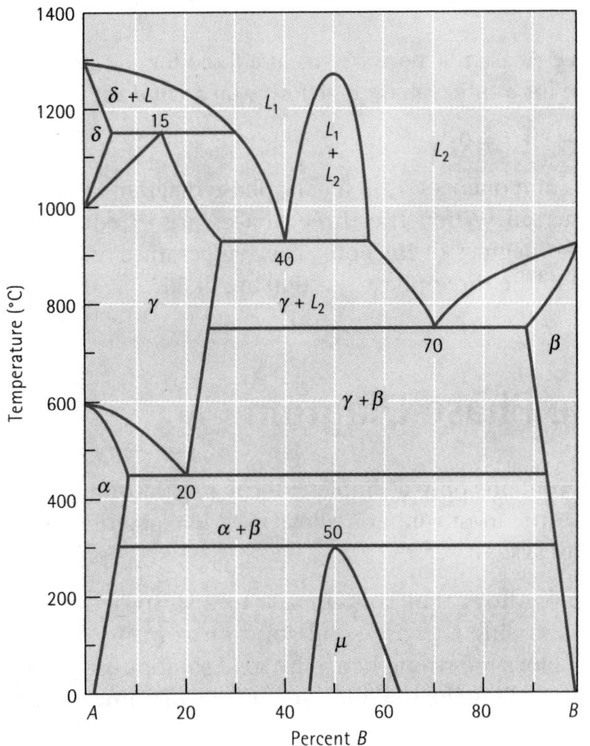

Figure 10.7
A hypothetical phase diagram (*for* Example 10.2).

920°C: This reaction occurs at 40% B:
$L_1 \rightarrow \gamma + L_2$, or a *monotectic*

750°C: This reaction occurs at 70% B:
$L \rightarrow \gamma + \beta$, or a *eutectic*

450°C: This reaction occurs at 20% B:
$\gamma \rightarrow \alpha + \beta$, or a *eutectoid*

300°C: This reaction occurs at 50% B:
$\alpha + \beta \rightarrow \mu$, or a *peritectoid*

The eutectic peritectic and monotectic reactions are part of the solidification process. Alloys used for casting or soldering are often based on the eutectic composition in order to take advantage of the low melting point of the eutectic reaction. The phase diagram of monotectic alloys contains a dome, or miscibility gap, in which two liquid phases coexist. In the copper-lead system, the monotectic reaction produces tiny globules of dispersed lead, which improves the machinability of the solidified copper alloy. Peritectic reactions lead to nonequilibrium solidification and segregation.

The eutectoid and peritectoid reactions are completely solid-state reactions. The eutectoid forms the basis for the heat treatment of several alloy systems, including steel. The peritectoid reaction is extremely slow, producing undesirable, nonequilibrium structures in alloys.

Each of these three-phase reactions occurs at a fixed temperature and composition. The Gibbs phase rule for a three-phase reaction is (at a constant pressure),

$$F = C - P + 1 = 2 - 3 + 1 = 0 \tag{10.1}$$

since there are two components C in a binary phase diagram and three phases P are involved in the reaction. When the three phases are in equilibrium during the reaction, there are no degrees of freedom. The temperature and the composition of each phase involved in the three-phase reaction are fixed.

10.5 The Eutectic Phase Diagram

The lead-tin system contains only a simple eutectic reaction (Figure 10.8). This alloy system is the basis for the most common alloys used for soldering. Let's examine four classes of alloys in this system.

Solid Solution Alloys Alloys that contain 0% to 2% Sn behave exactly like the copper-nickel alloys; a single phase solid solution α forms during solidification (Figure 10.9). These alloys are strengthened by solid solution strengthening, by strain hardening, and by controlling the solidification process to refine the grain structure.

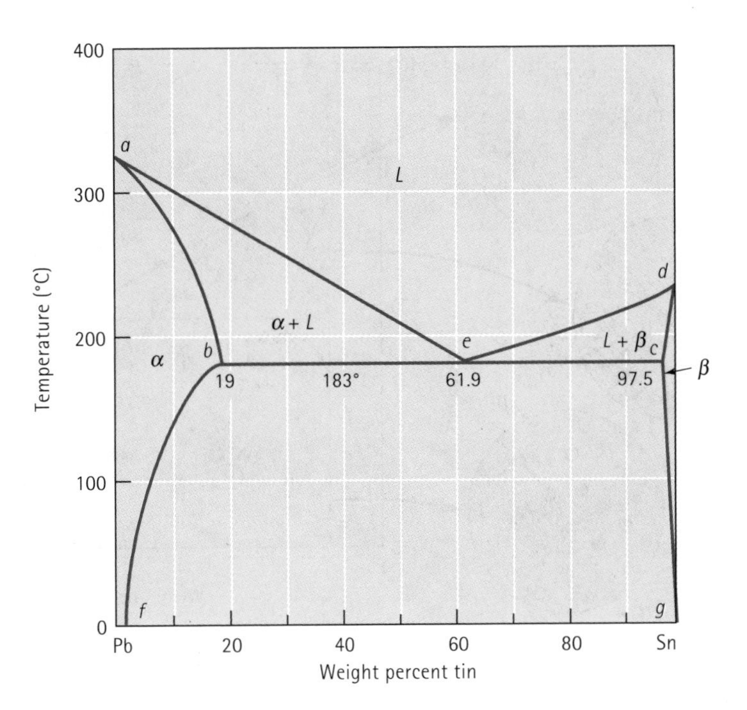

Figure 10.8
The lead-tin equilibrium phase diagram. Liquidus lines *ae*, *ed*. Solidus lines *ab*, *bc*, *cd*. Solvus lines *bf*, *cg*.

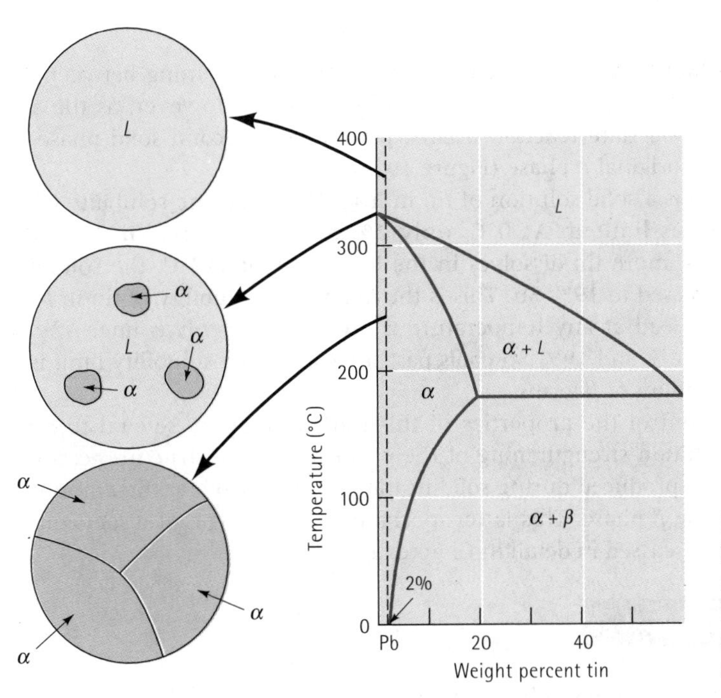

Figure 10.9
Solidification and microstructure of a Pb-2% Sn alloy. The alloy is a single-phase solid solution (α).

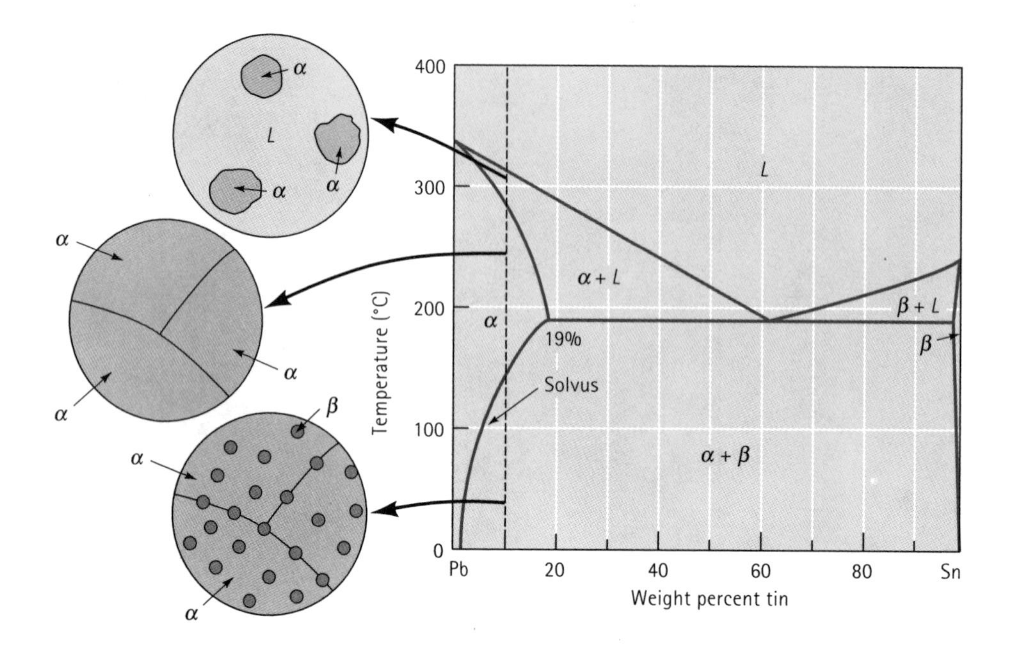

Figure 10.10
Solidification, precipitation, and microstructure of a Pb-10% Sn alloy. Some dispersion strengthening occurs as the β solid precipitates.

Alloys That Exceed the Solubility Limit Alloys containing between 2% and 19% Sn also solidify to produce a single solid solution α. However, as the alloy continues to cool, a solid state reaction occurs, permitting a second solid phase, β, to precipitate from the original α phase (Figure 10.10).

The α is a solid solution of tin in lead. However, the solubility of tin in the α solid solution is limited. At 0°C, only 2% Sn can dissolve in α. As the temperature increases, more tin dissolves in the lead until, at 183°C, the solubility of tin in lead has increased to 19% Sn. This is the maximum solubility of tin in lead. The solubility of tin in lead at any temperature is given by the solvus line. Any alloy containing between 2% and 19% Sn cools past the solvus, the solubility limit is exceeded, and a small amount of β forms.

We control the properties of this type of alloy by several techniques, including solid solution strengthening of the α portion of the structure, controlling the microstructure produced during solidification, and controlling the amount and characteristics of the β phase. This latter mechanism is one type of dispersion strengthening; it will be discussed in detail in Chapter 11.

EXAMPLE 10.3

Determine (a) the solubility of tin in solid lead at 100°C, (b) the maximum solubility of lead in solid tin, and (c) the amount of β that forms if a Pb-10% Sn alloy is cooled to 0°C.

SOLUTION

1. The 100°C temperature intersects the solvus line at 5% Sn. The solubility of tin in lead at 100°C therefore is 5% Sn.
2. The maximum solubility of lead in tin, which is found from the tin-rich side of the phase diagram, occurs at the eutectic temperature of 183°C and is 2.5% Pb (found from diagram by 100 – 97.5% Sn).
3. At 0°C, the 10% Sn alloy is in an $\alpha + \beta$ region of the phase diagram. By drawing a tie line at 0°C and working the lever law, we find that:

$$\%\beta = \frac{10-2}{100-2} \times 100 = 8.2\%$$

Eutectic Alloys The alloy containing 61.9% Sn has the eutectic composition (Figure 10.11). Above 183°C the alloy is all liquid and therefore must contain 61.9% Sn. After the liquid cools to 183°C, the eutectic reaction begins:

$$L_{61.9\%\ \mathrm{Sn}} \rightarrow \alpha_{19\%\ \mathrm{Sn}} + \beta_{97.5\%\ \mathrm{Sn}}$$

Two solid solutions – α and β – are formed during the eutectic reaction. The compositions of the two solid solutions are given by the ends of the eutectic line. The α is said to be lead-rich and the β is said to be tin-rich.

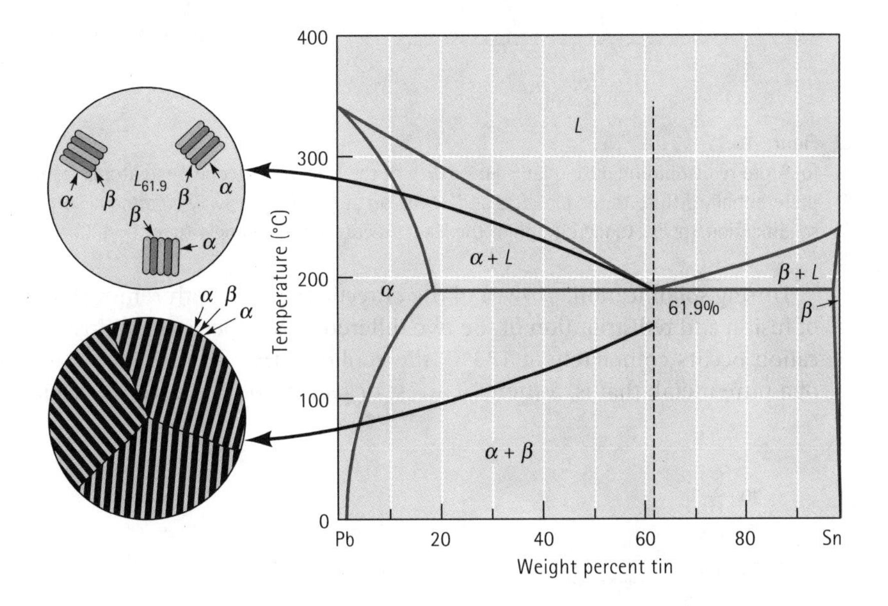

Figure 10.11
Solidification and microstructure of the eutectic alloy Pb-61.9% Sn.

As an α plate nucleates and grows in the eutectic liquid, it produces a lead-rich phase by a process of diffusion in which lead atoms move towards, and tin atoms move away from, this region (refer to Figure 10.12a). Consequently, when it reaches 19% Sn, 81% Pb it solidifies as an α platelet with an adjacent region of liquid rich in tin. This adjacent region continues to increase in tin content until at 97.5% Sn it solidifies as platelet of β. The process of eutectic solidification continues until all the liquid has solidified into lamellar structure of α and β, which is called the eutectic microconstituent. In the Pb-61.9% Sn alloy, the entire structure consists of eutectic microconstituent since all the liquid was at the eutectic composition and solidified at 183°C; the eutectic temperature (Figure 10.12b). There are many other eutectic systems which develop this lamellar structure consisting of two solid phases in intimate contact. Such a structure promotes strength and ductility; a combination required in many engineering material applications.

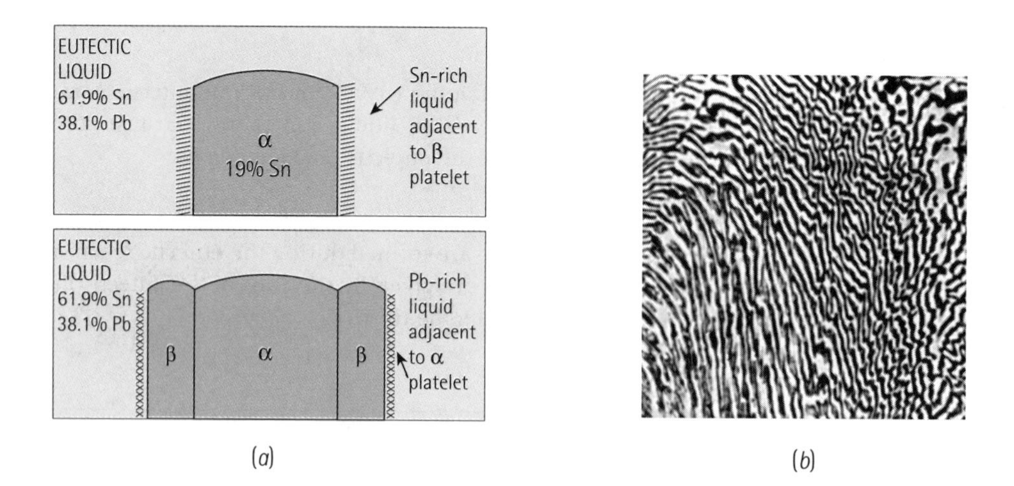

Figure 10.12
(a) Atom redistribution during lamellar growth of the lead-tin eutectic. Lead atoms from the liquid preferentially diffuse to the platelets and/or the tin atoms diffuse away creating an α platelet upon solidification. (b) Photomicrograph of the lead-tin eutectic microconstituent (× 400).

During solidification, growth of the eutectic requires both removal of the latent heat of fusion and redistribution of the two different atom species by diffusion. Since solidification occurs completely at 183°C, the cooling curve (Figure 10.13) is similar to that of a pure metal; that is, a thermal arrest or plateau occurs at the eutectic temperature.

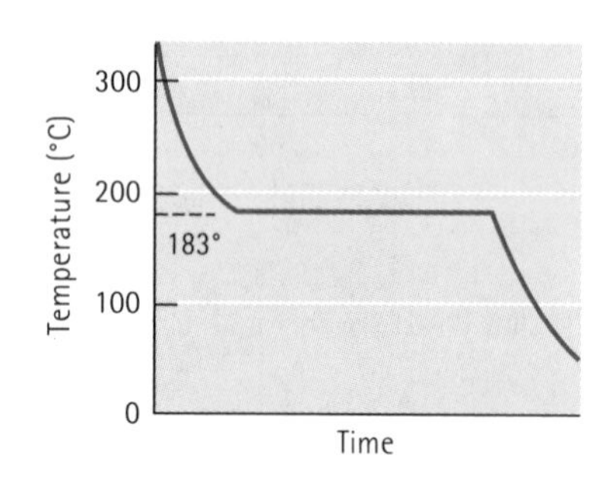

Figure 10.13
The cooling curve for a eutectic alloy is a simple thermal arrest, since eutectics freeze or melt at a single temperature.

EXAMPLE 10.4

Determine the amount and composition of each phase in the eutectic microconstituent in a lead-tin alloy.

SOLUTION

Using Figure 10.8, the eutectic microconstituent contains 61.9% Sn. We work the lever law at the eutectic temperature of 183°C, since that is the temperature at which the eutectic reaction is completed. The fulcrum of our lever is 61.9% Sn. The ends of the tie line coincide with the ends of the eutectic line.

$$\alpha\text{: Pb-19\% Sn} \quad \%\alpha = \frac{97.5-61.9}{97.5-19}\times 100 = 45\%$$

$$\beta\text{: Pb-97.5\% Sn} \quad \%\beta = \frac{61.9-19}{97.5-19}\times 100 = 55\%$$

Hypoeutectic and Hypereutectic Alloys As an alloy containing between 19% and 61.9% Sn cools, the liquid begins to solidify at the liquidus temperature. However, solidification is completed by going through the eutectic reaction (Figure 10.14). This solidification sequence occurs any time the vertical line corresponding to the original composition of the alloy crosses both the liquidus and the eutectic.

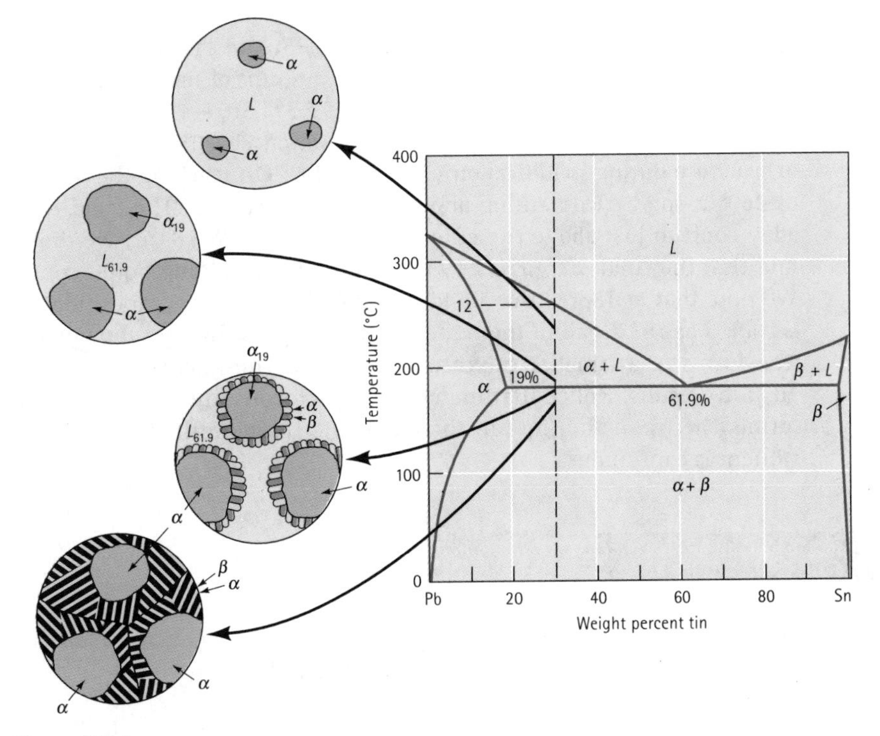

Figure 10.14
The solidification and microstructure of a hypoeutectic alloy (Pb-30% Sn).

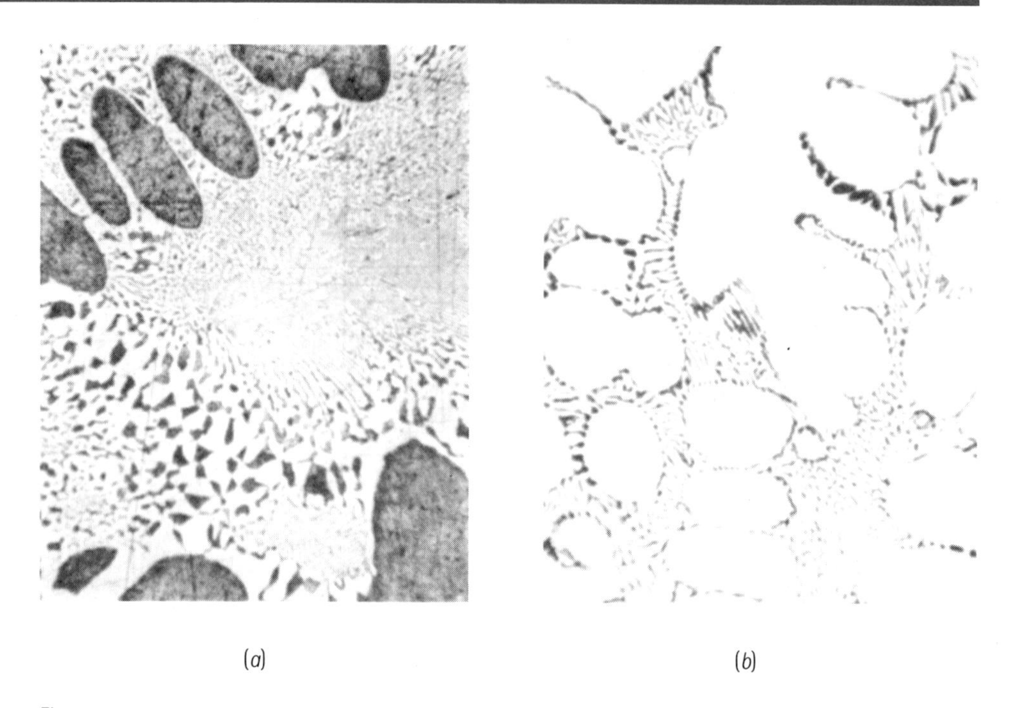

Figure 10.15
(a) A hypoeutectic lead-tin alloy. (b) A hypereutectic lead-tin alloy. The dark constituent is the lead-rich solid α, the light constituent is the tin-rich solid β, and the fine plate structure is the eutectic (× 400).

Alloys with compositions of 19% to 61.9% Sn are called hypoeutectic alloys, which are alloys containing less than the eutectic amount of tin. An alloy to the right of the eutectic composition, between 61.9% and 97.5% Sn, is hypereutectic.

Let's consider a hypoeutectic alloy containing Pb-30% Sn and follow the changes in structure during solidification (Figure 10.14). On reaching the liquidus temperature of 260°C, solid α containing about 12% Sn nucleates. The solid α grows until the alloy cools to just above the eutectic temperature. At 183°C, we draw a tie line and find that the solid α contains 19% Sn and the remaining liquid contains 61.9% Sn. We note that at 183°C, the liquid contains the eutectic composition! When the alloy is cooled at 183°C, all of the remaining liquid goes through the eutectic reaction and transforms to a lamellar mixture of α and β. The microstructure shown in Figure 10.15(a) results. Notice that the eutectic microconstituent surrounds the solid α that formed between the liquidus and eutectic temperatures. The eutectic microconstituent is continuous.

EXAMPLE 10.5

For a Pb-30% Sn alloy, determine the phases present, their amounts, and their compositions at 300°C, 200°C, 184°C, 182°C and 0°C.

SOLUTION

Temperature (°C)	Phases	Compositions	Amounts
300	L	L:30% Sn	$L = 100\%$
200	$\alpha + L$	L: 55% Sn	$L = \frac{30-18}{55-18} \times 100 = 32\%$
		α: 18% Sn	$\alpha = \frac{55-30}{55-18} \times 100 = 68\%$
184	$\alpha + L$	L: 61.9% Sn	$L = \frac{30-19}{61.9-19} \times 100 = 26\%$
		α: 19% Sn	$\alpha = \frac{61.9-30}{61.9-19} \times 100 = 74\%$
182	$\alpha + \beta$	α: 19% Sn	$\alpha = \frac{97.5-30}{97.5-19} \times 100 = 86\%$
		β: 97.5 Sn	$\beta = \frac{30-19}{97.5-19} \times 100 = 14\%$
0	$\alpha + \beta$	α: 2% Sn	$\alpha = \frac{100-30}{100-2} \times 100 = 71\%$
		β: 100% Sn	$\beta = \frac{30-2}{100-2} \times 100 = 29\%$

At just above the eutectic temperature, say 184°C, it can be seen that the structure consists of solid α in a dendritic form surrounded by a liquid of almost eutectic composition. During the eutectic reaction ($\alpha + \beta$) will be produced as a lamellar structure. Note the α phase is therefore present in both a dendritic form and a eutectic form. It is usual to call the dendritic form primary or proeutectic α to distinguish it from the eutectic form.

The microstructure shown in Figure 10.15(a) clearly shows this, and it is often more useful to calculate the proportions of primary α and eutectic ($\alpha + \beta$) rather than merely α and β at 182°C as was done in Example 10.5.

EXAMPLE 10.6

Determine the amounts and compositions of each microconstituent in a Pb-30% Sn alloy immediately after the eutectic reaction has been completed.

SOLUTION

The microconstituents are primary α and eutectic. We can determine their amounts and compositions if we look at how they form. The primary α is all of the solid α that forms before the alloy cools to the eutectic temperature; the eutectic microconstituent is all of the liquid that goes through the reaction. At the eutectic temperature of 183°C, the amounts and compositions of the two phases are:

$$\alpha\text{: } 19\%\text{ Sn} \quad \%\alpha = \frac{61.9-30}{61.9-19} \times 100 = 74\% = \%\text{ primary } \alpha$$

$$L\text{: } 61.9\%\text{ Sn} \quad \%L = \frac{30-19}{61.9-19} \times 100 = 26\% = \%\text{ eutectic}$$

The cooling curve for a hypoeutectic alloy is a composite of those for solid solution alloys and eutectic alloys (Figure 10.16). A change in slope occurs at the liquidus as primary α begins to form. Evolution of the latent heat of fusion slows the cooling rate as the solid α grows. When the alloy cools to the eutectic temperature, a thermal arrest is produced as the eutectic reaction proceeds at 183°C. The solidification sequence is similar in a hypereutectic alloy, giving the microstructure shown in Figure 10.15(b). Note that the primary phase is now the β phase rather than the α phase.

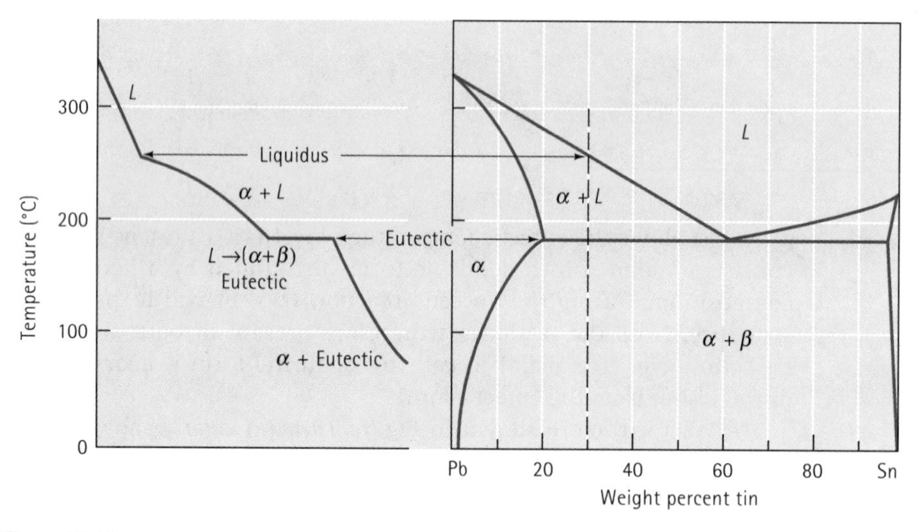

Figure 10.16
The cooling curve for a hypoeutectic Pb-30% Sn alloy.

10.6 Strength of Alloys Containing the Eutectic Phase

Each phase in the eutectic alloy is, to some degree, solid-solution-strengthened. In the lead-tin system, α, which is a solid solution of tin in lead, is stronger than pure lead. Some eutectic alloys can be strengthened by cold working. However, near eutectic alloys, being less ductile than single solid solution alloys, are often used to produce castings rather than being cold worked. We also control grain size by adding appropriate inoculants or grain refiners. Finally, we can influence the properties by controlling the amount and microstructure of the eutectic.

Eutectic Colony Size Eutectic colonies, or grains, each nucleate and grow independently. Within each colony, the orientation of the lamellae in the eutectic microconstituent is identical. The orientation changes on crossing a colony boundary [Figure 10.21(a)]. We can refine the eutectic colonies and improve the strength of the eutectic alloy by inoculation.

Interlamellar Spacing The interlamellar spacing of a eutectic is the distance from the centre of one α lamella to the centre of the next α lamella (Figure 10.17). A small interlamellar spacing indicates that the amount of α-β interface area is large. A small interlamellar spacing therefore increases the strength of the eutectic.

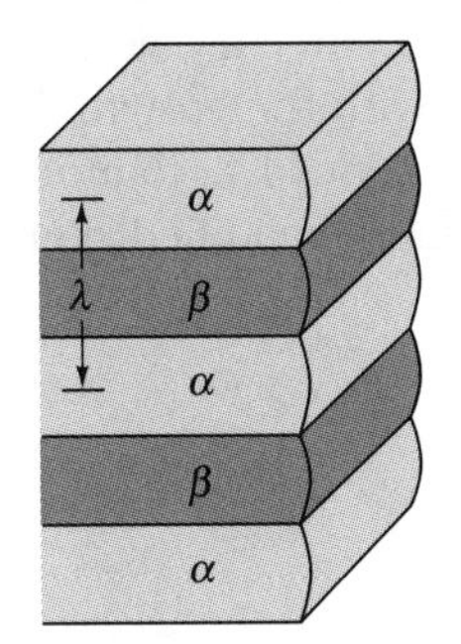

Figure 10.17
The interlamellar spacing in a eutectic microstructure.

The interlamellar spacing is determined primarily by the growth rate of the eutectic,

$$\lambda = cR^{-1/2} \qquad (10.2)$$

where R is the growth rate ($mm.s^{-1}$) and c is a constant. The interlamellar spacing for the lead-tin eutectic is shown in Figure 10.18. We can increase the growth rate R, and consequently reduce the interlamellar spacing, by increasing the cooling rate or reducing the solidification time.

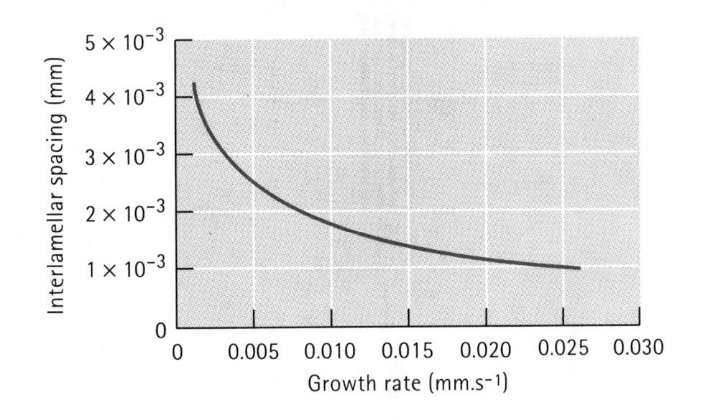

Figure 10.18
The effect of growth rate on the interlamellar spacing in the lead-tin eutectic.

EXAMPLE 10.7 Design of a Directional Solidification Process

Design a process to produce a single 'grain' of Pb-Sn eutectic microconstituent in which the interlamellar spacing is 0.0034 mm.

SOLUTION

We could use a directional solidification process to produce the single grain, while controlling the growth rate to assure that the correct interlamellar spacing is achieved. To obtain $\lambda = 0.0034$ mm, we need a growth rate of 0.0025 mm.s^{-1} (Figure 10.18).

Figure 10.19 shows how we might achieve this growth rate. The Pb-61.9% Sn alloy would be melted in a mould within a furnace. The mould would be withdrawn from the furnace at the rate of 0.0025 mm.s^{-1}, with the mould quenched with a water spray as it emerges from the furnace. If only one eutectic colony grows through the spiral, all of the lamellae are lined up in parallel with the growth direction. If the part to be made is 100 mm long, it would take 40 000 s, or 11 h, to produce the part.

This method has been used to produce directionally solidified high-temperature nickel-base eutectic parts for jet engines.

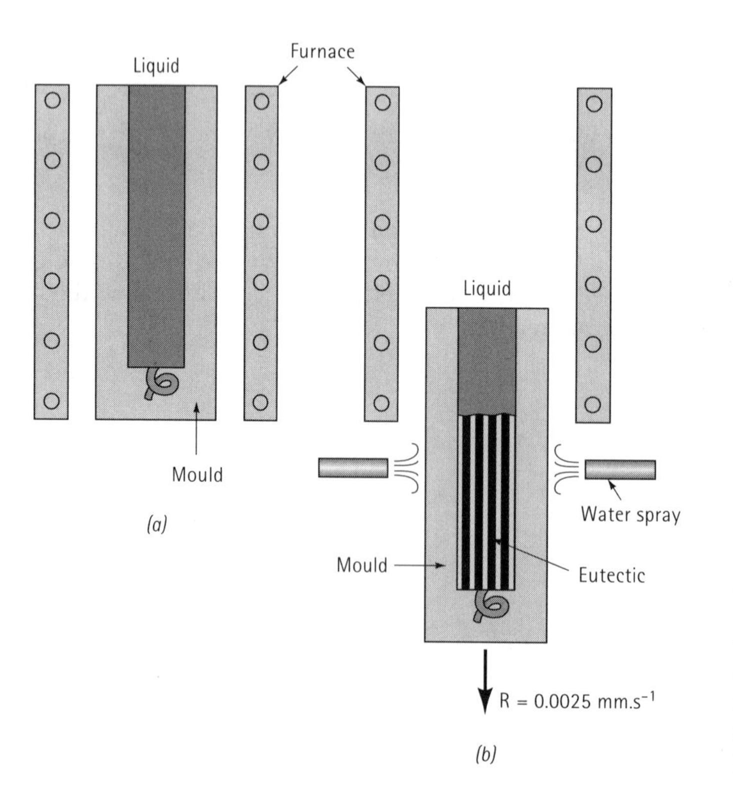

Figure 10.19
Directional solidification of a Pb-Sn eutectic alloy: (*a*) The metal is melted in the furnace, and (*b*) the mould is slowly withdrawn from the furnace and the casting is cooled (*for* Example 10.7).

Amount of Eutectic We also control the properties by the relative amounts of the primary microconstituent and the eutectic. In the lead-tin system, the amount of the eutectic microconstituent changes from 0% to 100% when the tin content increases from 19% to 61.9%. With increasing amounts of the stronger eutectic microconstituent, the strength of the alloy increases (Figure 10.20). Similarly, when we increase the lead added to tin from 2.5% to 38.1% Pb, the amount of primary β in the hypereutectic alloy decreases, the amount of the strong eutectic increases, and the strength increases. When both individual phases have about the same strength, the eutectic alloy is expected to have the highest strength due to effective dispersion strengthening.

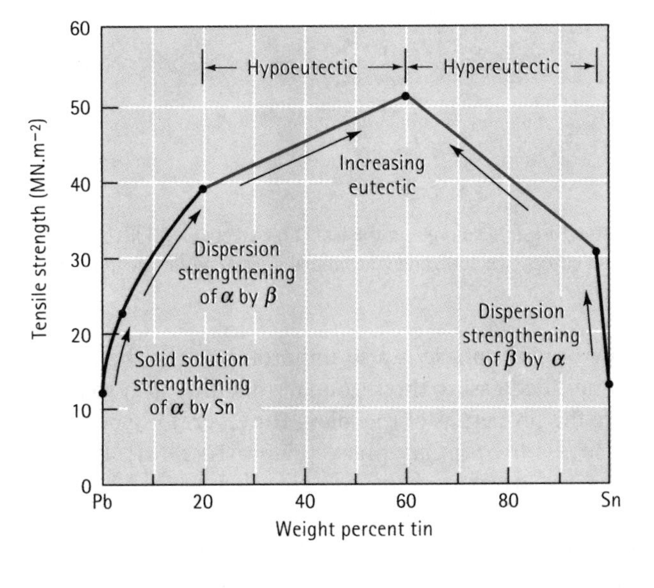

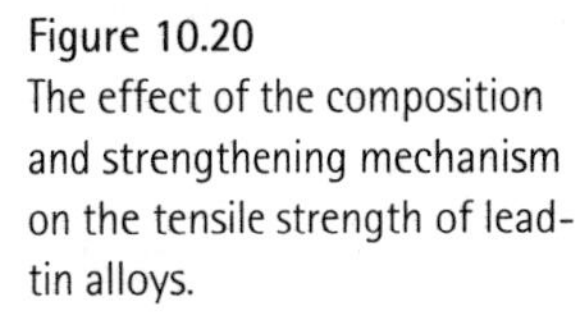
Figure 10.20
The effect of the composition and strengthening mechanism on the tensile strength of lead-tin alloys.

Microstructure of the Eutectic Not all eutectics give a lamellar structure. The shapes of the two phases in the microconstituent are influenced by the cooling rate, the presence of impurity elements, and the nature of the alloy (Figure 10.21).

The aluminium-silicon eutectic phase diagram (Figure 10.22) forms the basis for a number of important commercial alloys. However, the silicon portion of the eutectic grows as thin, flat plates that appear needlelike in a photomicrograph [Figure 10.21(b)]. The brittle silicon platelets concentrate stresses and reduce ductility and toughness.

The eutectic microstructure in aluminium-silicon alloys is altered by modification. Modification causes the silicon phase to grow as thin, interconnected rods between aluminium dendrites [Figure 10.21(c)], improving both tensile strength and % elongation. In two dimensions, the modified silicon appears to be composed of tiny, round particles. Rapidly cooled alloys, such as those in die casting, are naturally modified during solidification. At slower cooling rates, however, 0.02% Na or 0.01% Sr must be added to cause modification.

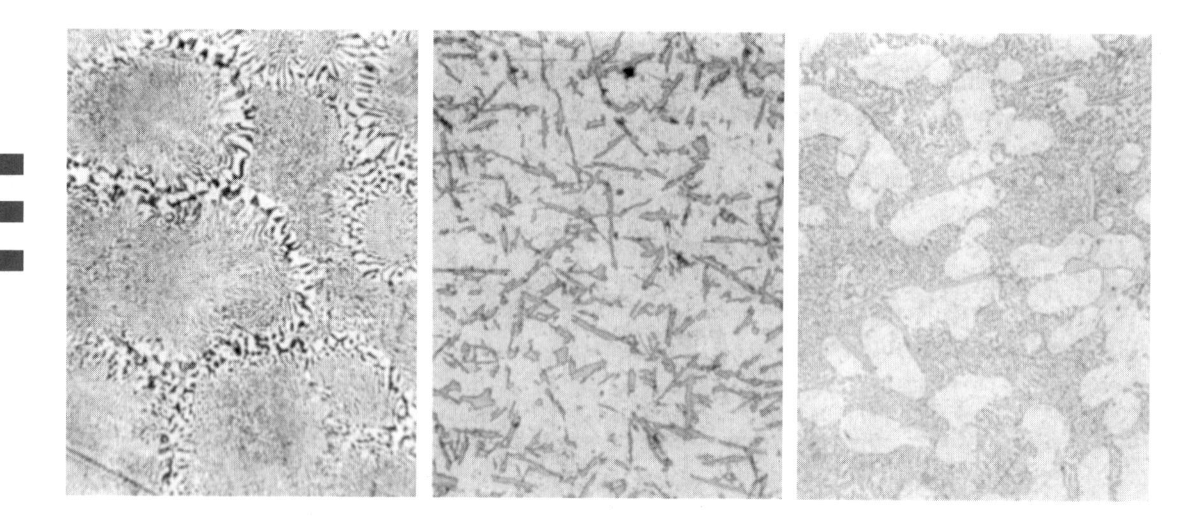

Figure 10.21
Typical eutectic microstructures: (*a*) colonies in the lead-tin eutectic (× 300), (*b*) needlelike silicon plates in the aluminium-silicon eutectic (× 100), and (*c*) rounded silicon rods in the modified aluminium-silicon eutectic (× 100).

The shape of the primary phase is also important. Often the primary phase grows in a dendritic manner; decreasing the secondary dendrite arm spacing of the primary phase may improve the properties of the alloy. However, in hypereutectic aluminium-silicon alloys, coarse β is the primary phase [Figure 10.23(a)]. Because β is hard, the hypereutectic alloys are wear-resistant and are used to produce automotive engine parts. But the coarse β causes poor machinability and gravity segregation (where the primary β floats to the surface of the casting during freezing). Addition of 0.05% P encourages nucleation of primary silicon, refines its size, and minimises its deleterious qualities [Figure 10.23(b)].

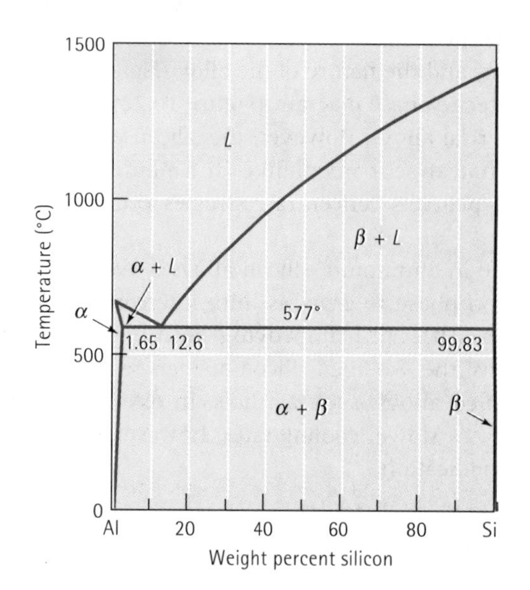

Figure 10.22
The aluminium-silicon phase diagram.

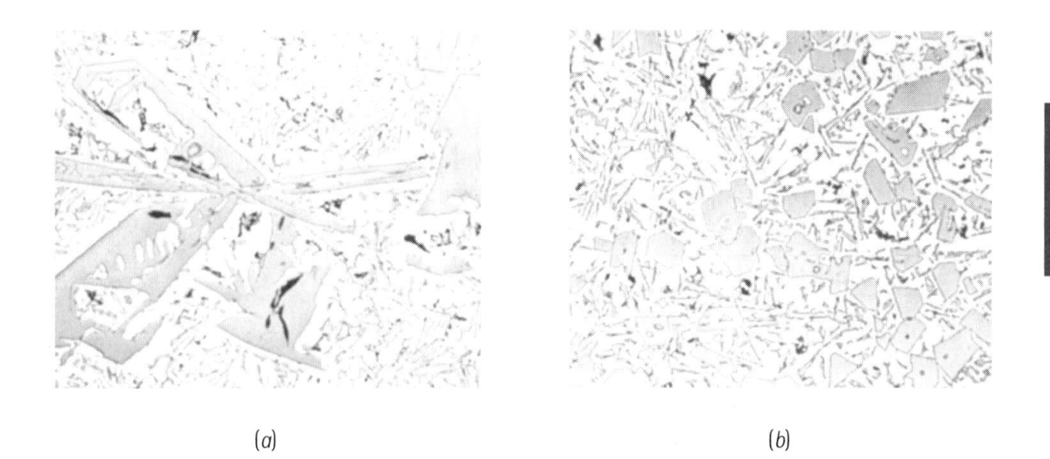

Figure 10.23
The effect of hardening with phosphorus on the microstructure of hypereutectic aluminium-silicon alloys: (*a*) coarse primary silicon and (*b*) fine primary silicon, as refined by phosphorus addition (× 75). (*From* Metals Handbook, *Vol.7, 8th Ed., American Society for Metals, 1972.*)

EXAMPLE 10.8 Selection of Materials for a Wiping Solder

One way to repair dents in a metal is to wipe a partly liquid-partly solid material into the dent, then allow this filler material to solidify. For our application, the wiping material should have the following specifications: (1) a melting temperature below 230°C, (2) a tensile strength in excess of 40 MN.m^{-2}, (3) be 60% to 70% liquid during application, and (4) the lowest possible cost. Select an alloy and repair procedure that will meet these specifications.

SOLUTION

Let's see if one of the Pb-Sn alloys will satisfy these conditions. First, the alloy must contain more than 40% Sn in order to have a melting temperature below 230°C (Figure 10.8). This low temperature will make it easier for the person doing the repairs to apply the filler.

Second, Figure 10.20 indicates that the tin content must lie between 20% and 80% to achieve the required 40 MN.m^{-2} tensile strength. In combination with the first requirement, any alloy containing between 40 and 80% Sn will be satisfactory.

Third, the cost of tin is about 10 times that of lead. Thus, an alloy of Pb-40% Sn might be the most economical choice.

Finally, the filler material must be at the correct temperature in order to be 60% to 70% liquid. As the calculations below show, the temperature must be between 200°C and 210°C:

$$\% L_{200} = \frac{40-18}{55-18} \times 100 = 60\% \qquad \% L_{210} = \frac{40-17}{50-17} \times 100 = 70\%$$

Our recommendation therefore is to use a Pb-40% Sn alloy applied at 205°C, a temperature at which there will be 65% liquid and 35% primary α.

EXAMPLE 10.9 **Design of a Wear-Resistant Cylinder Liner**

Design a lightweight cylindrical component that will provide excellent wear-resistance at the inner wall, yet still have reasonable ductility and toughness overall. Such a product might be used as a cylinder liner in an automotive engine.

SOLUTION

Many wear-resistant parts are produced from steels, which have a relatively high density, but the hypereutectic Al-Si alloys containing primary β may provide the wear-resistance that we wish at one-third the weight of the steel.

Since the part to be produced is cylindrical in shape, centrifugal casting (Figure 10.24) might be a unique method for producing it. In centrifugal casting, liquid metal is poured into a rotating mould and the centrifugal force produces a hollow shape. In addition, material that has a higher density than the liquid is spun to the outside wall of the casting, while material that has a lower density than the liquid migrates to the inner wall.

When we centrifugally cast a hypereutectic Al-Si alloy, primary β nucleates and grows. The density of the β is, according to Appendix A, 2.33 $Mg.m^{-3}$, compared with a density near 2.7 $Mg.m^{-3}$ for aluminium. As the primary β precipitates from the liquid, it spins to the inner surface. The result is a casting that is composed of eutectic microconstituent (with reasonable ductility) at the outer wall and a hypereutectic composition, containing large amounts of primary β, at the inner wall.

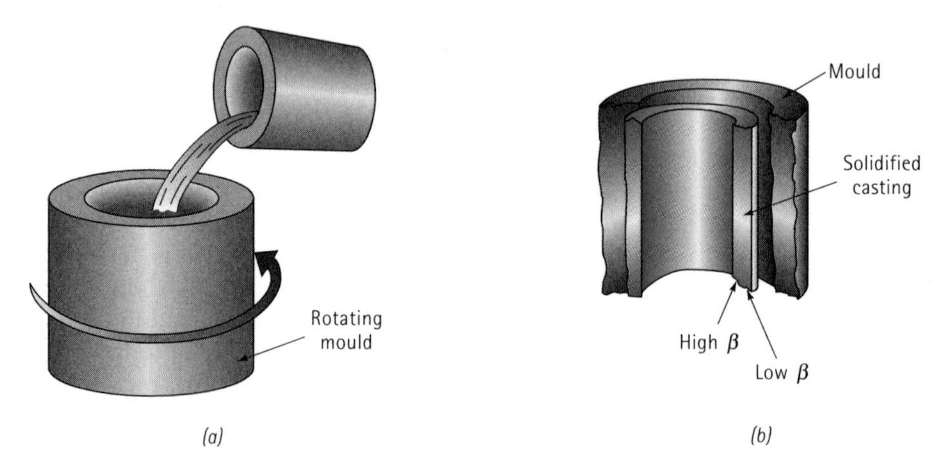

Figure 10.24
Centrifugal casting of a hypereutectic Al-Si alloy: (*a*) Liquid alloy is poured into a rotating mould, and (*b*) the solidified casting is hypereutectic at the inner diameter and eutectic at the outer diameter (*for* Example 10.9).

A typical alloy used to produce aluminium engine components is Al-17% Si. From Figure 10.22, the total amount of primary β that can form is calculated at 578°C, just above the eutectic temperature:

$$\% \text{ Primary } \beta = \frac{17-12.6}{99.83-12.6} \times 100 = 5.0\%$$

Although only 5.0% primary β is expected to form, the centrifugal action can double or triple the amount of β at the inner wall of the casting.

10.7 Eutectics and Materials Processing

Manufacturing processes take advantage of the low melting temperature associated with the eutectic reaction. The Pb-Sn alloys are the basis for a series of alloys used to produce filler materials for soldering. If, for example, we wish to join copper pipe, individual segments can be joined by introducing the low-melting-point eutectic Pb-Sn alloy into the joint (Figure 10.25). The copper is heated just above the eutectic temperature. The heated copper melts the Pb-Sn alloy, which is then drawn into the thin gap by capillary action. When the Pb-Sn alloy cools and solidifies, the copper is joined.

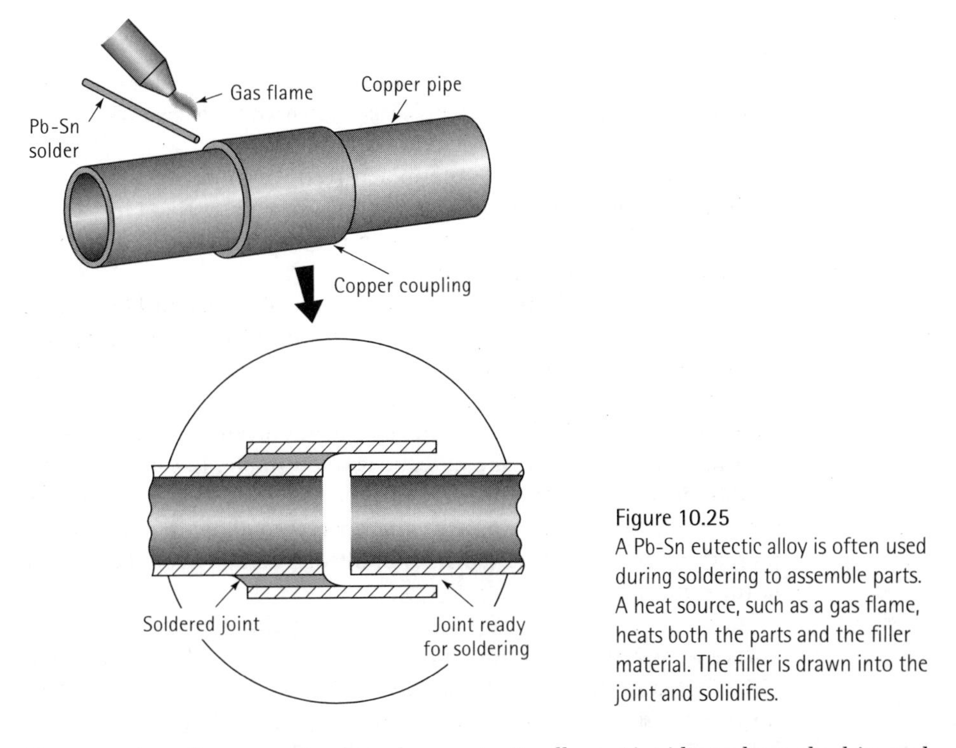

Figure 10.25
A Pb-Sn eutectic alloy is often used during soldering to assemble parts. A heat source, such as a gas flame, heats both the parts and the filler material. The filler is drawn into the joint and solidifies.

Many casting alloys are also based on eutectic alloys. Liquid can be melted in and poured into a mould at low temperatures, reducing energy costs involved in melting, minimising casting defects such as gas porosity, and preventing liquid metal-mould reactions. Cast iron and most aluminium casting alloys are eutectic alloys.

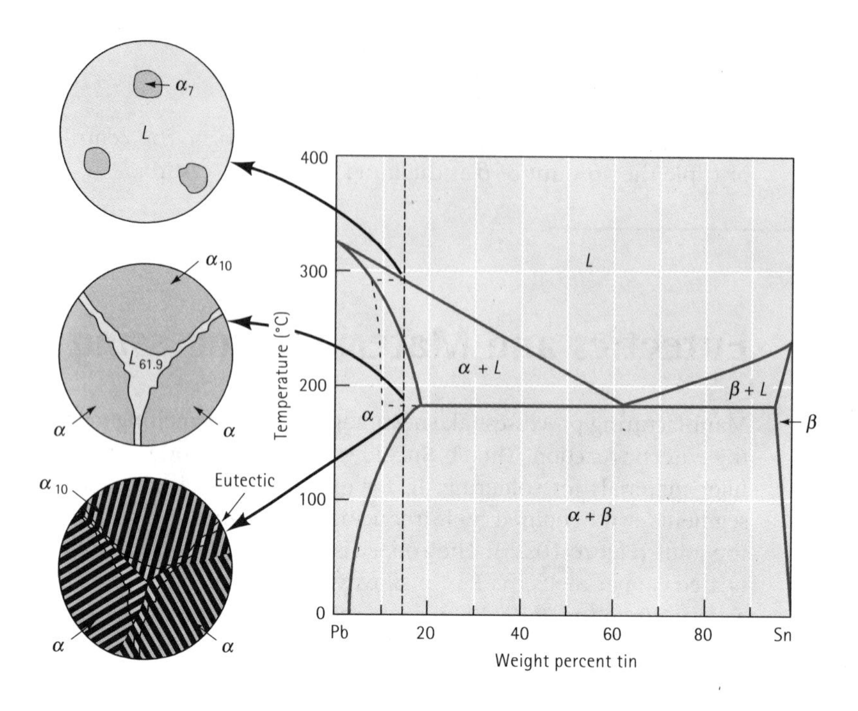

Figure 10.26
Nonequilibrium solidification and microstructure of a Pb-15% Sn alloy. A nonequilibrium eutectic microconstituent can form due to rapid solidification (— indicates the position of the equilibrium phase boundary lines).

The eutectic helps in the manufacture of ceramic glasses. Many common glasses are based on SiO_2, which melts at 1 710°C. By adding Na_2O to SiO_2, a eutectic reaction is introduced, with a eutectic temperature of about 790°C. The SiO_2-Na_2O glass can be produced at a low temperature.

The eutectic reaction may be used to speed diffusion bonding or to increase the rate of sintering of compacted powders for both metal and ceramic systems. In both cases, a liquid is produced to join dissimilar materials or powder particles, even though the temperature at which the processing is done is below the melting temperature of the individual constituents involved in the process.

Sometimes, however, the eutectic is undesirable. Because the eutectic is the last to solidify, it surrounds the primary phases. Eutectics that are brittle therefore embrittle the overall alloy, even if only a small percentage of the eutectic microconstituent is present in the structure. Deformation of such an alloy could cause failure through the brittle eutectic.

As another example, Al_2O_3 has a high melting point (2 020°C), which makes it attractive as a refractory for containing liquid steel. The melting temperature of CaO (2 570°C) is even higher. If an Al_2O_3 brick is placed into contact with a CaO brick, however, a series of eutectics is produced, giving a liquid with a melting temperature below the usual steel-making temperature. Thus, the refractory containing the liquid steel may fail. Also basic slags which are rich in CaO would attack an incorrectly chosen refractory lining.

10.8 Nonequilibrium Freezing in the Eutectic System

Suppose we have an alloy, such as Pb-15% Sn, that ordinarily solidifies as a solid solution alloy. The last liquid should freeze near 230°C, well above the eutectic. However, if the alloy cools too quickly, a nonequilibrium solidus curve is produced (Figure 10.26). The primary α continues to grow until, just above 183°C, the remaining nonequilibrium liquid contains 61.9% Sn. This liquid then transforms to the eutectic microconstituent, surrounding the primary α. For the conditions shown in Figure 10.26, the amount of nonequilibrium eutectic is:

$$\%\ \text{eutectic} = \frac{15-10}{61.9-10} \times 100 = 9.6\%$$

When heat treating an alloy such as Pb-15% Sn, we must keep the maximum temperature below the eutectic temperature of 183°C to prevent hot shortness.

10.9 Ternary Phase Diagrams

Many alloy systems are based on three or even more elements. When three elements are present, we have a ternary alloy. To describe the changes in structure with temperature, we must draw a three-dimensional phase diagram. Figure 10.27 shows a hypothetical ternary phase diagram made up of elements *A*, *B*, and *C*. Note that two binary eutectics are present on the two visible faces of the diagram; a third binary eutectic between elements *B* and *C* is hidden on the back of the plot.

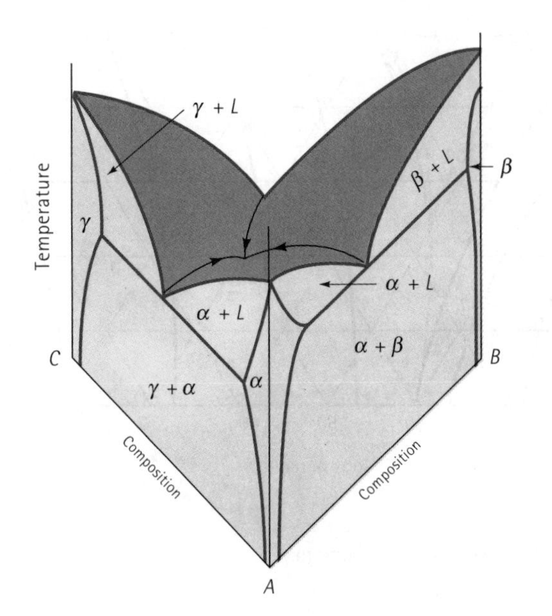

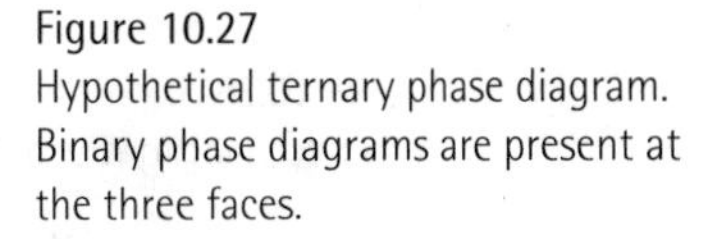
Figure 10.27
Hypothetical ternary phase diagram. Binary phase diagrams are present at the three faces.

It is difficult to use the three-dimensional ternary plot; however, we can present the information from the diagram in two dimensions by any of several methods, including the liquidus plot and the isothermal plot.

Liquidus Plot We note in Figure 10.27 that the temperature at which freezing begins is shaded. We could transfer these temperatures for each composition onto a triangular diagram, as in Figure 10.28, and plot the liquidus temperatures as isothermal contours. This presentation is helpful in predicting the freezing temperature of the material. The liquidus plot also gives the identity of the primary phase that forms during solidification for any given composition.

Isothermal Plot The isothermal plot shows the phases present in the material at a particular temperature. It is useful in predicting the phases and their amounts and compositions at that temperature. Figure 10.29 shows an isothermal plot from Figure 10.27 at room temperature.

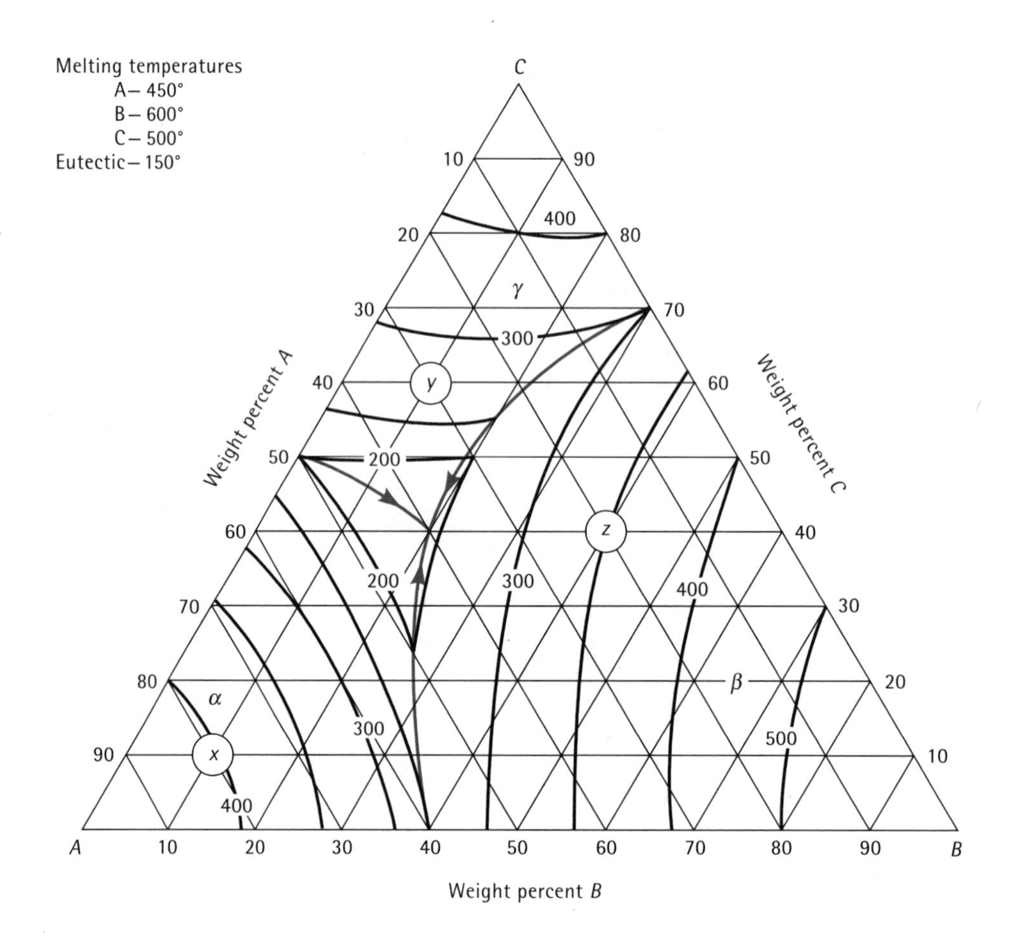

Figure 10.28
A liquidus plot for the hypothetical ternary phase diagram.

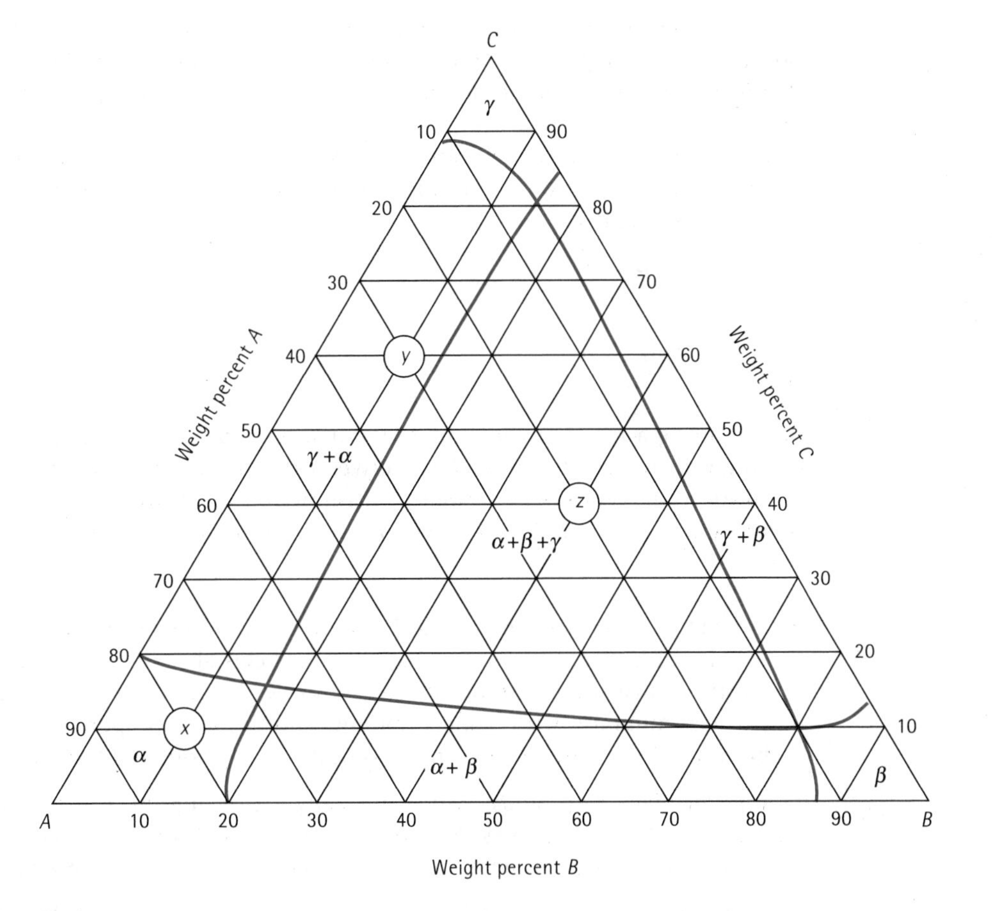

Figure 10.29
An isothermal plot at room temperature for the hypothetical ternary phase diagram.

EXAMPLE 10.10

Using the ternary plots in Figures 10.28 and 10.29, determine the liquidus temperature, the primary phase that forms during solidification, and the phases at room temperature for the following materials:

A-10% *B*-10% *C* *A*-10% *B*-60% *C* *A*-40% *B*-40% *C*

SOLUTION

The composition *A*-10% *B*-10% *C* is located at point *x* in the figures; from the isotherm in this region, the liquidus temperature is 400°C. The primary phase, as indicated in the diagram, is α. The final structure (Figure 10.29) is all α.

The composition A-10% B-60% C is located at point y; by interpolating the isotherms in this region, the liquidus temperature is about 270°C. The primary phase that forms in this section of the diagram is γ, and the room temperature phases are α and γ.

The composition A-40% B-40% C is located at point z; the liquidus temperature at this point is 350°C, and the point is in the primary β region. The room temperature phases are α, β and γ.

SUMMARY

By producing a material containing two or more phases, dispersion strengthening is obtained. In metals, the boundary between the phases impedes the movement of dislocations and improves strength. Introduction of multiple phases may provide other benefits, including improvement of the fracture toughness of ceramics and polymers.

- For optimum dispersion strengthening particularly in metals, a large number of small, hard, discontinuous precipitate particles should form in a soft, ductile matrix to provide the most effective obstacles to dislocations. Round precipitate particles minimise stress concentrations, and the final properties of the alloy can be controlled by the relative amounts of the precipitates and the matrix.
- Intermetallic compounds, which normally are strong but brittle, are frequently introduced as precipitates. Matrix structures based on intermetallic compounds have been introduced to take advantage of their high-temperature properties.
- Phase diagrams for materials containing multiple phases normally contain one or more three-phase reactions:
    - The eutectic reaction permits liquid to solidify as an intimate mixture of two solids. By controlling the solidification process, we can achieve a wide range of properties. Some of the factors that can be controlled include: the grain size or secondary dendrite arm spacings of primary microconstituents, the colony size of the eutectic microconstituent, the interlamellar spacing within the eutectic microconstituent, the microstructure, or shape, of the phases within the eutectic microconstituent, and the amount of the eutectic microconstituent that forms.
    - The eutectoid reaction causes a solid to transform to a mixture of two other solids. As shown in the next chapter, heat treatments to control the eutectoid reaction provide an excellent basis for dispersion strengthening.
    - In the peritectic and peritectoid reactions, two phases transform to a single phase on cooling. Dispersion strengthening does not occur, and severe segregation problems are often encountered.
    - The monotectic reaction produces a mixture of a solid and liquid. Although this reaction does not provide dispersion strengthening, it does provide other benefits, such as good machinability, to some alloys.

GLOSSARY

Dispersion strengthening
Increasing the strength of a material by mixing together more than one phase. By proper control of the size, shape, amount, and individual properties of the phases, excellent combinations of properties can be obtained.

Eutectic
A three-phase reaction in which one liquid phase solidifies to produce two solid phases.

Eutectoid
A three-phase reaction in which one solid phase transforms to two different solid phases.

Eutectic microconstituent
A characteristic mixture of two phases formed as a result of the eutectic reaction.

Hyper-
A prefix indicating that the composition of an alloy is more than the composition at which a three-phase reaction occurs.

Hypo-
A prefix indicating that the composition of an alloy is less than the composition at which a three-phase reaction occurs.

Hypereutectic alloys
Alloys containing more than the eutectic composition but containing at least some eutectic microconstituent.

Hypoeutectic alloys
Alloys containing less than the eutectic composition but containing at least some eutectic microconstituent.

Interlamellar spacing
The distance between the centre of a lamella or plate of one phase and the centre of the adjoining lamella or plate of the same phase.

Intermediate solid solution
A nonstoichiometric intermetallic compound displaying a range of compositions.

Intermetallic compound
A compound formed of two or more metals that has its own unique composition, structure, and properties.

Isothermal plot
A horizontal section through a ternary phase diagram showing the phases present at a particular temperature.

Lamella
A thin plate of a phase that forms during certain three-phase reactions, such as the eutectic or eutectoid.

Liquidus plot
A two-dimensional plot showing the temperature at which a three-component alloy system begins to solidify on cooling.

Matrix

Usually the continuous solid phase in a complex microstructure. Solid precipitates may form within the matrix.

Microconstituent

A phase or mixture of phases in an alloy that has a distinct appearance. Frequently, we describe a microstructure in terms of the microconstituents rather than the actual phases.

Miscibility gap

A region in a phase diagram in which two phases, with essentially the same structure, do not mix, or have no solubility in one another.

Modification

Addition of alloying elements, such as sodium or strontium, which change the microstructure of the eutectic microconstituent in aluminium-silicon alloys.

Monotectic

A three-phase reaction in which one liquid transforms to a solid and a second liquid on cooling.

Nonstoichiometric intermetallic compound

A phase formed by the combination of two components into a compound having a structure and properties different from either component. The nonstoichiometric compound has a variable ratio of the components present in the compound.

Ordered crystal structure

Solid solutions in which the different atoms occupy specific, rather than random, lattice sites.

Peritectic

A three-phase reaction in which a solid and a liquid combine to produce a second solid on cooling.

Peritectoid

A three-phase reaction in which two solids combine to form a third solid on cooling.

Precipitate

A solid phase that forms from the original matrix phase when the solubility limit is exceeded. In most cases, we try to control the formation of the precipitate to produce the optimum dispersion strengthening.

Primary microconstituent

The microconstituent that forms before the start of a three-phase reaction. This is also called a proeutectic phase.

Solvus

A solubility line that separates a single-solid phase region from a two-solid phase region in the phase diagram.

Stoichiometric intermetallic compound

A phase formed by the combination of two components into a compound having a structure and properties different from either component. The stoichiometric intermetallic compound has a fixed ratio of the components present in the compound. Also called an intermediate solid solution.

Ternary alloy
An alloy formed by combining three elements or components.

Ternary phase diagram
A phase diagram between three components showing the phases present and their compositions at various temperatures. This diagram requires a three-dimensional plot.

PROBLEMS

10.1 A hypothetical phase diagram is shown in Figure 10.30.

(a) Are any intermetallic compounds present? If so, identify them and determine whether they are stoichiometric or nonstoichiometric.

(b) Identify the solid solutions present in the system. Is either material *A* or *B* allotropic? Explain.

(c) Identify the three-phase reactions by writing down the temperature, the reaction in equation form, the composition of each phase in the reaction, and the name of the reaction.

10.2 The Cu-Zn phase diagram is shown in Figure 13.10.

(a) Are any intermetallic compounds present? If so, identify them and determine whether they are stoichiometric or nonstoichiometric.

(b) Identify the solid solutions present in the system.

(c) Identify the three-phase reactions by writing down the temperature, the reaction in equation form, and the name of the reaction.

10.3 A portion of the Al-Cu phase diagram is shown in Figure 11.5.

(a) Determine the formula for the θ compound.

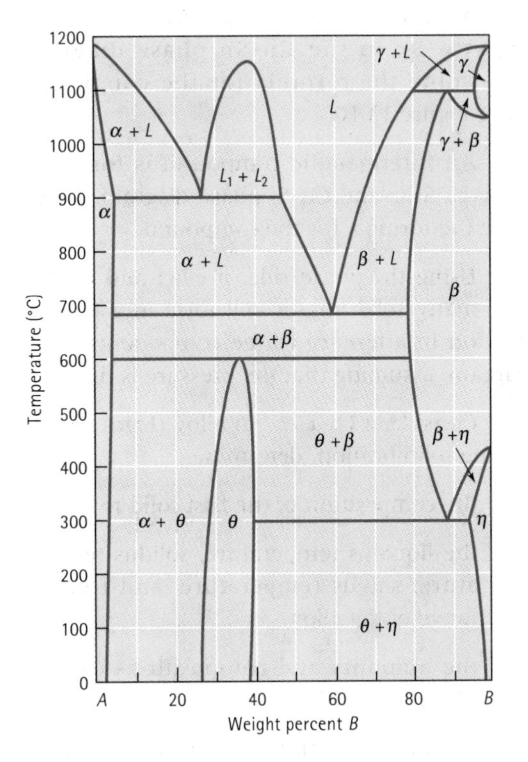

Figure 10.30
Hypothetical phase diagram (*for* Problem 10.1).

(b) Identify the three-phase reaction by writing down the temperature, the reaction in equation form, the composition of each phase in the reaction, and the name of the reaction.

10.4 The Al-Li phase diagram is shown in Figure 13.6.

(a) Are any intermetallic compounds present? If so, identify them and determine whether they are stoichiometric or nonstoichiometric. Determine the formula for each compound.

(b) Identify the three-phase reactions by writing down the temperature, the reaction in equation form, the composition of each phase in the reaction, and the name of the reaction.

10.5 An intermetallic compound is found for 38 wt% Sn in the Cu-Sn phase diagram. Determine the formula for the compound, using Figure 13.10.

10.6 An intermetallic compound is found for 10 wt% Si in the Cu-Si phase diagram. Determine the formula for the compound.

10.7 Using the phase rule, predict and explain how many solid phases will form in a eutectic reaction in a ternary (three-component) phase diagram, assuming that the pressure is fixed.

10.8 Consider a Pb-15% Sn alloy (Figure 10.8). During solidification, determine

(a) the composition of the first solid to form,

(b) the liquidus temperature, solidus temperature, solvus temperature, and freezing range of the alloy,

(c) the amounts and compositions of each phase at 260°C,

(d) the amounts and compositions of each phase at 183°C, and

(e) the amounts and compositions of each phase at 25°.

10.9 Consider an Al-12% Mg alloy (Figure 13.3). During solidification, determine

(a) the composition of the first solid to form,

(b) the liquidus temperature, solidus temperature, solvus temperature, and freezing range of the alloy,

(c) the amounts and compositions of each phase at 525°C,

(d) the amounts and compositions of each phase at 450°C, and

(e) the amounts and compositions of each phase at 25°C.

10.10 Consider a Pb-35% Sn alloy. Determine

(a) if the alloy is hypoeutectic or hypereutectic,

(b) the composition of the first solid to form during solidification,

(c) the amounts and compositions of each phase at 184°C,

(d) the amounts and compositions of each phase at 182°C,

(e) the amounts and compositions of each microconstituent at 182°C, and

(f) the amounts and compositions of each phase at 25°C.

10.11 Consider a Pb-70% Sn alloy. Determine

(a) if the alloy is hypoeutectic or hypereutectic,

(b) the composition of the first solid to form during solidification,

(c) the amounts and compositions of each phase at 184°C,

(d) the amounts and compositions of each phase at 182°C,

(e) the amounts and compositions of each microconstituent at 182°C, and

(f) the amounts and compositions of each phase at 25°C.

10.12 Calculate the total % β and the % eutectic microconstituent at room temperature for the following lead-tin alloys: 10% Sn, 20% Sn, 50% Sn, 60% Sn, 80% Sn and 95% Sn. Using Figure 10.20, plot the strength of the alloys versus the % β and the % eutectic and explain your graphs.

10.13 Consider an Al-4% Si alloy (Figure 10.22). Determine

(a) if the alloy is hypoeutectic or hypereutectic,

(b) the composition of the first solid to form during solidification,

(c) the amounts and compositions of each phase at 578°C,

(d) the amounts and compositions of each phase at 576°C,

(e) the amounts and compositions of each microconstituent at 576°C and

(f) the amounts and compositions of each phase at 25°C.

10.14 Consider an Al-25% Si alloy. Determine

(a) if the alloy is hypoeutectic or hypereutectic,

(b) the composition of the first solid to form during solidification,

(c) the amounts and compositions of each phase at 578°C,

(d) the amounts and compositions of each phase at 576°C,

(e) the amounts and compositions of each microconstituent at 576°C, and

(f) the amounts and compositions of each phase at 25°C.

10.15 A Pb-Sn alloy contains 45% α and 55% β at 100°C. Determine the composition of the alloy. Is the alloy hypoeutectic or hypereutectic?

10.16 An Al-Si alloy contains 85% α and 15% β at 500°C. Determine the composition of the alloy. Is the alloy hypoeutectic or hypereutectic?

10.17 A Pb-Sn alloy contains 23% primary α and 77% eutectic microconstituent. Determine the composition of the alloy.

10.18 An Al-Si alloy contains 15% primary β and 85% eutectic microconstituent. Determine the composition of the alloy.

10.19 Determine the maximum solubility for the following cases:

(a) lithium in aluminium (Figure 13.6)

(b) aluminium in magnesium (Figure 13.8)

(c) copper in zinc (Figure 13.10)

(d) carbon in γ-iron (Figure 11.13)

10.20 Determine the maximum solubility for the following cases:

(a) magnesium in aluminium (Figure 13.3)

(b) zinc in copper (Figure 13.10)

(c) beryllium in copper (Figure 13.10)

(d) Al_2O_3 in MgO (Figure 14.9)

10.21 Observation of a microstructure shows that there is 28% eutectic and 72% primary β in an Al-Li alloy (Figure 13.6).

(a) Determine the composition of the alloy and whether it is hypoeutectic or hypereutectic.

(b) How much α and β are in the eutectic microconstituent?

10.22 Write the eutectic reaction that occurs, including the compositions of the three phases in equilibrium, and calculate the amount of α and β in the eutectic microconstituent in the Mg-Al system (Figure 13.8).

10.23 Calculate the total amount of α and β and the amount of each microconstituent in a Pb-50% Sn alloy at 182°C. What fraction of the total α in the alloy is contained in the eutectic microconstituent?

10.24 Figure 10.31 shows a cooling curve for a Pb-Sn alloy. Determine

(a) the pouring temperature,

(b) the superheat,

(c) the liquidus temperature,

(d) the eutectic temperature,

(e) the freezing range,

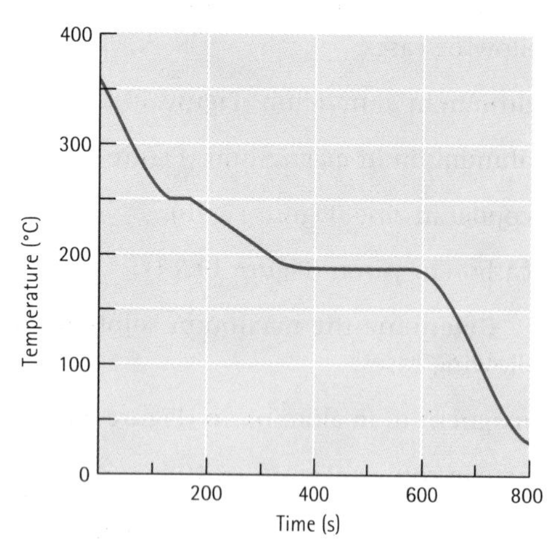

Figure 10.31
Cooling curve for a Pb-Sn alloy (*for* Problem 10.24).

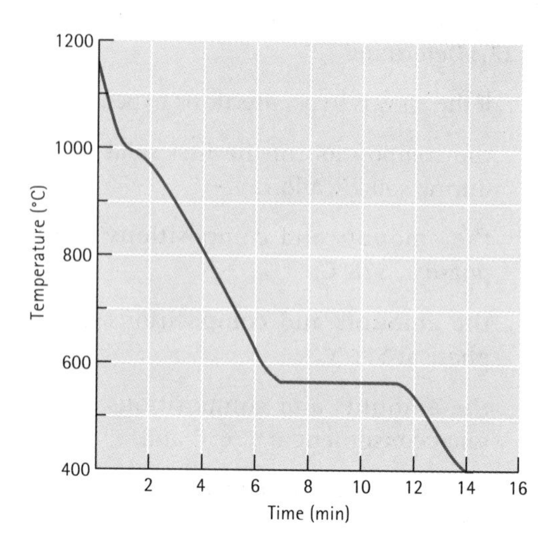

Figure 10.32
Cooling curve for an Al-Si alloy (*for* Problem 10.25).

(f) the local solidification time,

(g) the total solidification time, and

(h) the composition of the alloy.

10.25 Figure 10.32 shows a cooling curve for an Al-Si alloy. Determine

(a) the pouring temperature,

(b) the superheat,

(c) the liquidus temperature,

(d) the eutectic temperature,

(e) the freezing range,

(f) the local solidification time,

(g) the total solidification time, and

(h) the composition of the alloy.

10.26 Draw the cooling curves, including appropriate temperatures, expected for the following Al-Si alloys:

(a) Al-4% Si (b) Al-12.6% Si

(c) Al-25% Si (d) Al-65% Si

10.27 Based on the following observations, construct a phase diagram. Element *A* melts at 850°C and element *B* melts at 1 200°C. Element *B* has a maximum solubility of 5% in element *A*, and element *A* has a maximum solubility of 15% in element *B*. The number of degrees of freedom from the phase rule is zero when the temperature is 725°C and there is 35% *B* present. At room temperature, 1% *B* is soluble in *A* and 7% *A* is soluble in *B*.

10.28 Cooling curves are obtained for a series of Cu-Ag alloys (Figure 10.33). Use this data to produce the Cu-Ag phase diagram. The maximum solubility of Ag in Cu is 7.9% and the maximum solubility of Cu in Ag is 8.8%. The solubilities at room temperature are near zero.

10.29 The SiO-Al_2O_3 phase diagram is included in Figure 14.38. A refractory is required to contain molten metal at 1 900°C.

(a) Will pure Al_2O_3 be a potential candidate? Explain.

(b) Will Al_2O_3 contaminated with 1% SiO_2 be a candidate? Explain.

10.30 Consider the ternary phase diagram shown in Figures 10.28 and 10.29. Determine the liquidus temperature, the first solid to form, and the phases present at room temperature for the following compositions:

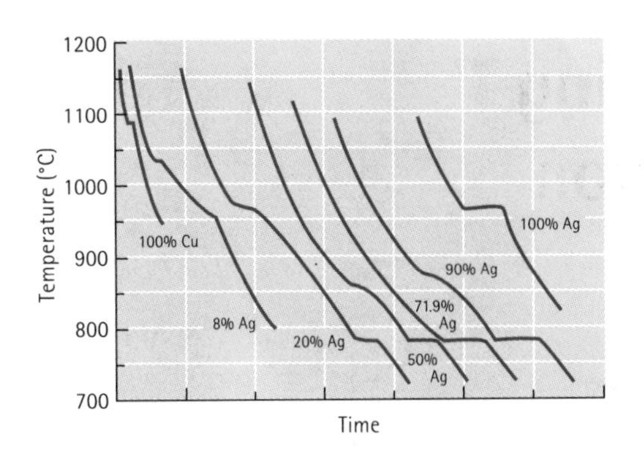

Figure 10.33
Cooling curves for a series of Cu-Ag alloys (*for* Problem 10.28).

(a) *A*-30% *B*-20% *C*

(b) *A*-10% *B*-25% *C*

(c) *A*-60% *B*-10% *C*

10.31 Consider the ternary phase diagram shown in Figures 10.28 and 10.29. Determine the liquidus temperature, the first solid to form, and the phases present at room temperature for the following compositions:

(a) *A*-5% *B*-80% *C*

(b) *A*-50% *B*-5% *C*

(c) *A*-30% *B*-35% *C*

10.32 Consider the liquidus plot in Figure 10.28.

(a) For a constant 20% *B*, draw a graph showing how the liquidus temperature changes from 80% *A*-20% *B*-0% *C* to 0% A-20% B-80% C.

(b) What is the composition of the ternary eutectic in this system?

(c) Estimate the temperature at which the ternary eutectic reaction occurs.

10.33 From the liquidus plot in Figure 10.28, prepare a graph of liquidus temperature versus percent *B* for a constant ratio of materials *A* and *C* (that is, from pure *B* to 50% *A*-50% *C* on the liquidus plot). Material *B* melts at 580°C.

Design Problems

10.34 Design a processing method that permits a Pb-15% Sn solidified under nonequilibrium conditions to be hot-worked.

10.35 Design a eutectic diffusion bonding process to join aluminium to silicon. Describe the changes in microstructure at the interface during the bonding process.

10.36 Design a directional solidification process that will give an interlamellar spacing of 0.0005 mm in a Pb-Sn eutectic alloy. You may need to determine the constant *c* in equation 10.2.

10.37 Design an Al-Si brazing alloy and process that will be successful in joining an Al-Mn alloy that has a liquidus of 659°C and a solidus of 656°C. Brazing, like soldering, involves introducing a liquid filler metal into a joint without melting the metals that are to be joined.

10.38 Your company would like to produce light-weight aluminium parts that have excellent hardness and wear-resistance. The parts must have a good combination of strength, ductility, and internal integrity. Design the process flow from the start of melting to the time that the liquid metal enters the mould cavity.

CHAPTER 11

Dispersion Strengthening by Phase Transformation and Heat Treatment

11.1 Introduction

In this chapter, we further discuss dispersion strengthening as we describe a variety of solid-state transformation processes, including age hardening and the eutectoid reaction. We also examine how nonequilibrium phase transformations – in particular, the martensitic reaction – provide strengthening. Each of these dispersion-strengthening techniques requires a heat treatment.

As we discuss these strengthening mechanisms, we must keep in mind the characteristics that produce the most desirable dispersion strengthening, as discussed in Chapter 10. The matrix should be relatively soft and ductile and the precipitate or second phase should be strong; the precipitate should be round and discontinuous; the precipitate particles should be small and numerous; and, in general, the more precipitate we have, the stronger the alloy will be. As in Chapter 10, we concentrate on how these reactions influence the strength of the materials, but these treatments influence other properties as well.

11.2 Nucleation and Growth in Solid-State Reactions

In order for a precipitate to form from a solid matrix, both nucleation and growth must occur. The total change in free energy required for nucleation of a spherical solid precipitate from the matrix is:

$$\Delta G = \frac{4}{3}\pi r^3 \Delta G_v + 4\pi r^2 \sigma + \frac{4}{3}\pi r^3 \varepsilon \tag{11.1}$$

The first two terms include the volume free energy change and the surface energy change, just as in solidification (Equation 8.1). However, the third term takes into account the strain energy ε introduced when the precipitate forms in a solid, rigid matrix. The precipitate does not occupy the same volume that is displaced, so additional energy is required to accommodate the precipitate in the matrix.

Nucleation As in solidification, nucleation occurs most easily on surfaces already present in the structure, thereby minimising the surface energy term. Thus, the precipitate nucleates most easily at grain boundaries or other lattice defects.

Growth Growth of the precipitate normally occurs by long-range diffusion and redistribution of atoms. Diffusing atoms must be detached from their original locations (perhaps at lattice points in a solid solution), move through the surrounding material to the nucleus, and be incorporated into the lattice of the precipitate. In some cases, the diffusing atoms might be so tightly bonded within an existing phase that the detachment process limits the rate of growth. In other cases, attaching the diffusing atoms to the precipitate – perhaps because of the lattice strain – limits growth. This result sometimes leads to the formation of precipitates that have a special relationship to the matrix structure that minimises the strain. In most cases, however, the controlling factor is the diffusion step.

Kinetics The overall rate, or *kinetics*, of the transformation process depends on both nucleation and growth. If more nuclei are present at a particular temperature, growth occurs from a larger number of sites and the phase transformation is completed in a shorter period of time. At higher temperatures, the diffusion coefficient is higher, growth rates are more rapid, and again we expect the transformation to be completed in a shorter time, assuming an equal number of nuclei.

The rate of transformation is given by Equation 11.2, with the fraction of the transformation f related to time t by

$$f = 1 - \exp(-ct^n), \quad (11.2)$$

where c and n are constants for a particular temperature. This Avrami relationship shown in Figure 11.1, produces a sigmoidal, or S-shaped, curve. An incubation time, t_0, during which no observable transformation occurs, is required for nucleation to occur. Initially, the transformation occurs slowly as nuclei form.

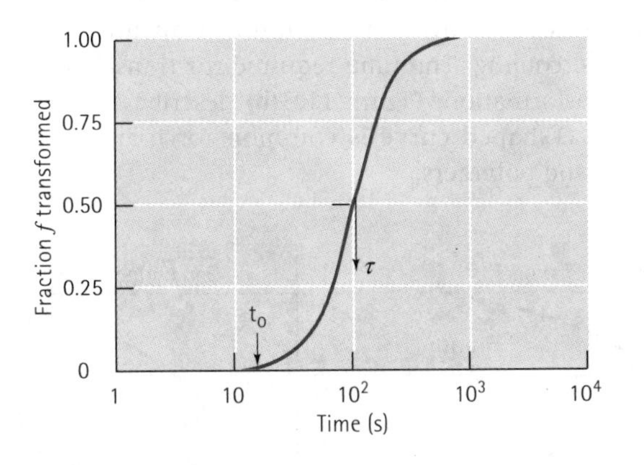

Figure 11.1
Sigmoidal curve showing the rate of transformation of FCC iron at a constant temperature. The incubation time t_0 and the time τ for 50% transformation are also shown.

Incubation is followed by rapid growth as atoms diffuse to the growing precipitate. Near the end of the transformation, the rate again slows as the source of atoms available to diffuse to the growing precipitate is depleted. The transformation is 50% complete in time τ; the rate of transformation is often given by the reciprocal of τ:

$$\text{Rate} = 1/\tau \quad (11.3)$$

Effect of Temperature In many phase transformations, the material undercools below the temperature at which the phase transformation occurs under equilibrium conditions. Because both nucleation and growth are temperature-dependent, the rate of transformation depends on the undercooling. The rate of nucleation is low for small undercoolings and increases for larger undercoolings (at least, up to a certain point). At the same time, the growth rate of the new phase decreases, because of slower diffusion, as the undercooling increases. The growth rate follows an Arrhenius relationship:

$$\text{Growth rate} = A \exp(-Q/RT) \tag{11.4}$$

where Q is the activation energy for the reaction, R is the gas constant, T is the temperature, and A is a constant.

Figure 11.2 shows sigmoidal curves at different temperatures for recrystallisation of copper; as the temperature increases, the rate of recrystallisation of copper *increases*, because growth is the most important factor for copper.

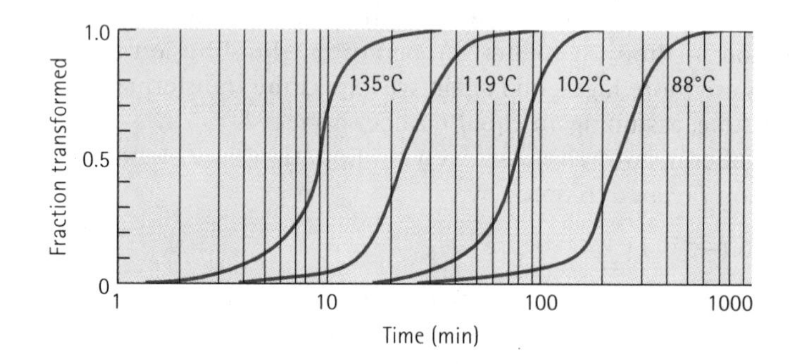

Figure 11.2
The effect of temperature on recrystallisation of cold-worked copper.

At any particular temperature, the overall rate of transformation is the product of the nucleation and growth rates. In Figure 11.3(a), the combined effect of the nucleation and growth rates is shown. A maximum transformation rate may be observed at a critical undercooling. The time required for transformation is inversely related to the rate of transformation; Figure 11.3(b) describes the time required for the transformation. This *C*-shaped curve is common for many transformations in metals, ceramics, glasses, and polymers.

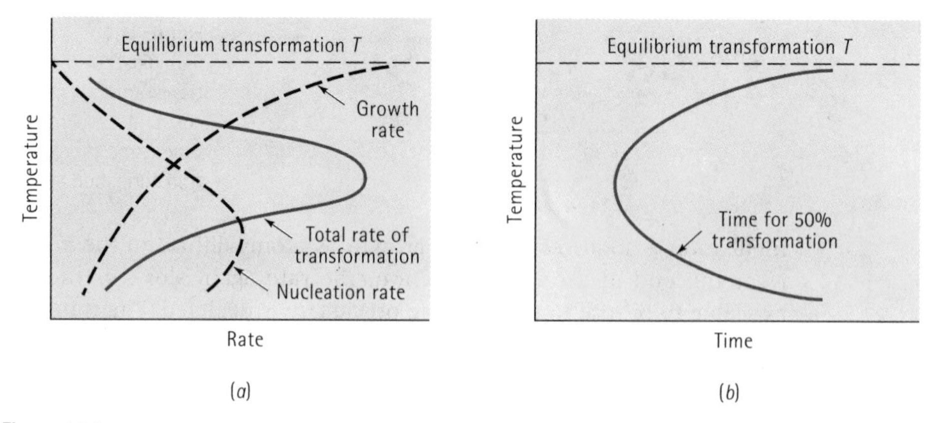

Figure 11.3
The effect of temperature on the rate of a phase transformation is the product of the growth rate and nucleation rate contributions, giving a maximum transformation rate at a critical temperature (*a*). Consequently, there is a minimum time required for the transformation, given by the "C-curve"(*b*).

In some processes, such as recrystallisation of a cold-worked metal, we find that the transformation rate continually decreases with decreasing temperature. In this case, nucleation occurs easily, and diffusion – or growth – predominates.

EXAMPLE 11.1

Determine the activation energy for the recrystallisation of copper from the sigmoidal curves in Figure 11.2.

SOLUTION

The rate of transformation is the reciprocal of the time τ required for half of the transformation to occur. From Figure 11.2, the times required for 50% transformation at several different temperatures can be calculated:

T (°C)	T (K)	τ (min)	Rate (min^{-1})
135	408	9	0.111
119	392	22	0.045
102	375	80	0.0125
88	361	250	0.0040

The rate of transformation is an Arrhenius equation, so a plot of ln (rate) versus $1/T$ (Figure 11.4) allows us to calculate the constants in the equation:

$$\text{slope} = \frac{-Q}{R} = \frac{\Delta \ln (\text{rate})}{\Delta(1/T)} = \frac{\ln (0.111) - \ln (0.004)}{1/408 - 1/361}$$

$$Q/R = 10\,414$$

$$Q = 86\,585 \text{ J.mol}^{-1}$$

$$0.111 = A \exp[-86\,585/(8.314)(408)]$$

$$A = 0.111/8.21 \times 10^{-12} = 1.35 \times 10^{10} \text{min}^{-1}$$

$$\text{rate} = 1.35 \times 10^{10} \exp(-86\,585/RT)$$

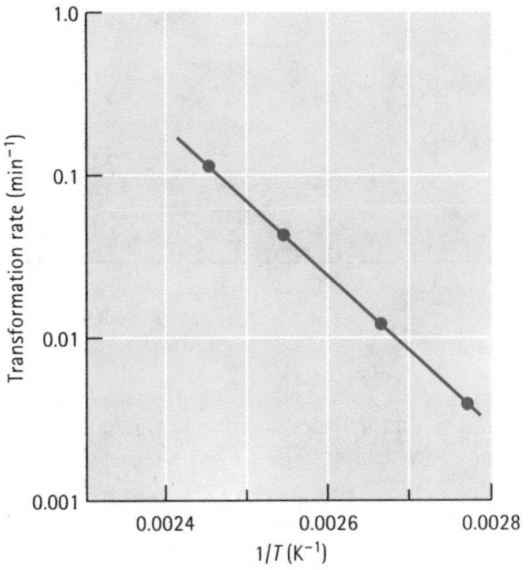

Figure 11.4
Arrhenius plot of transformation rate versus reciprocal temperature for recrystallisation of copper (*for* Example 11.1).

In this particular example, the rate at which the reaction occurs *increases* as the temperature increases, indicating that the reaction may be dominated by diffusion.

11.3 Alloys Strengthened by Exceeding the Solubility Limit

In Chapter 10 we found that lead-tin alloys containing about 2% to 19% Sn can be dispersion-strengthened because the solubility of tin in lead is exceeded.

A similar situation occurs in aluminium-copper alloys. For example, the Al-4% Cu alloy (shown in Figure 11.5) is completely α, or an aluminium solid solution, above 500°C. On cooling below the solvus temperature, a second phase, θ, precipitates. The θ phase, which is the hard, brittle intermetallic compound $CuAl_2$, provides dispersion strengthening. In a 4% Cu alloy, only about 7.5% of the final structure is θ. We must control the precipitation of the second phase to satisfy the requirements of good dispersion strengthening.

Widmanstatten Structure The second phase may grow so that certain planes and directions in the precipitate are parallel to preferred planes and directions in the matrix, creating a Widmanstatten structure. This growth mechanism minimises strain and surface energies and permits faster growth rates. Widmanstatten growth produces a characteristic appearance for the precipitate. When a needlelike shape is produced

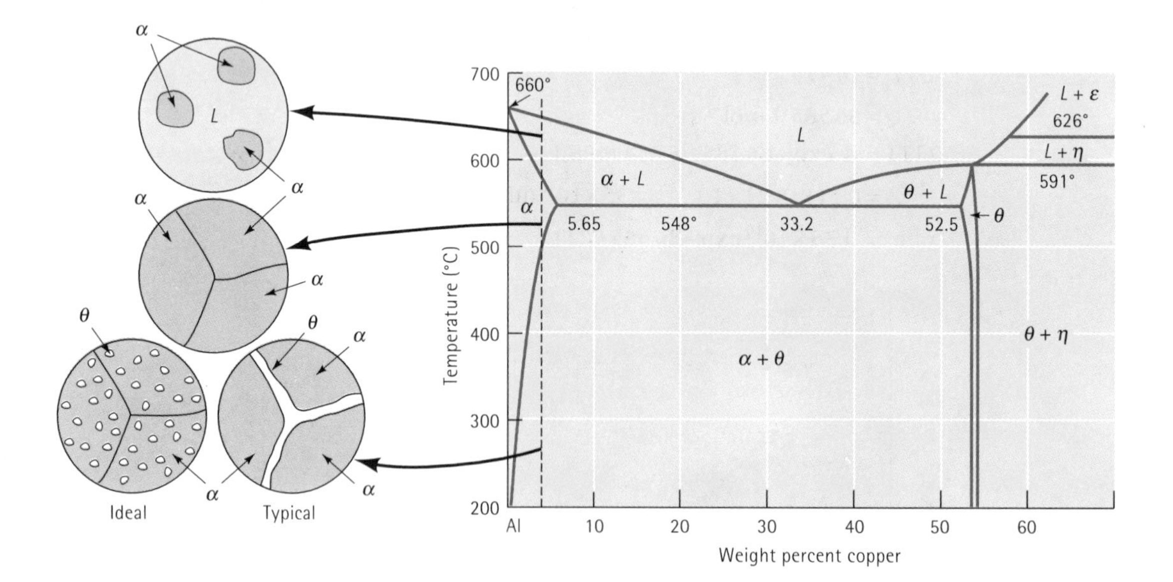

Figure 11.5
The aluminium-copper phase diagram and the microstructures that may develop during cooling of an Al-4% Cu alloy.

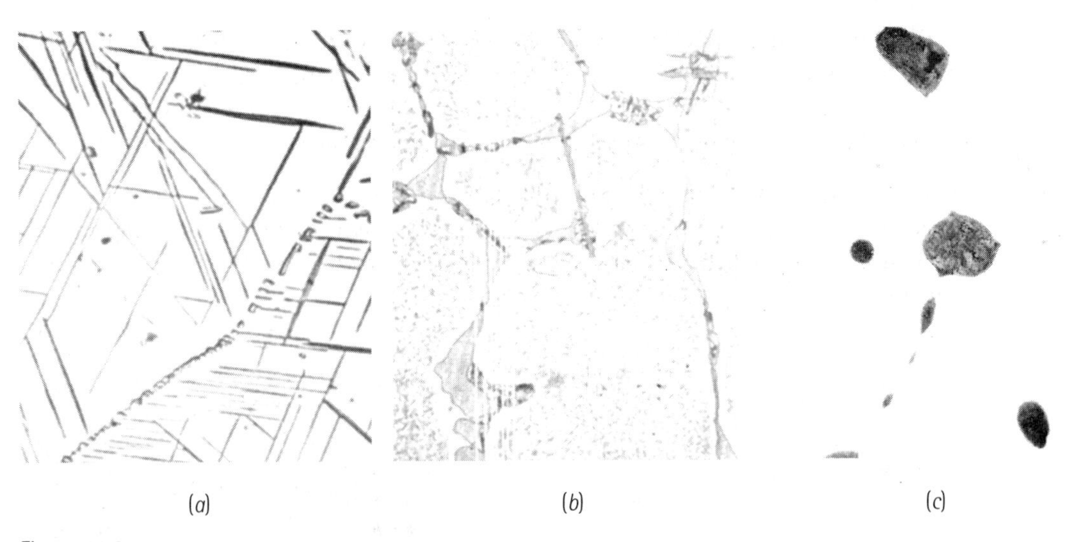

Figure 11.6
(*a*) Widmanstatten needles in a Cu-Ti alloy (× 420). (*From* Metals Handbook, *Vol. 9, 9th Ed., American Society for Metals, 1985.*) (*b*) Continuous θ precipitate in an Al-4% Cu alloy, caused by slow cooling (× 500). (*c*) Precipitates of lead at grain boundaries in copper (× 500).

[Figure 11.6(a)], the Widmanstatten precipitate may encourage the nucleation of cracks, thus reducing the ductility of the material. However, some of these structures make it more difficult for cracks, once formed, to propagate, therefore providing good fracture toughness. Certain titanium alloys and ceramics obtain toughness in this way.

Interfacial Energy Relationships We expect the precipitate to have a spherical shape in order to minimise surface energy. However, when the precipitate forms at an interface, the precipitate shape is also influenced by the interfacial energy associated with both the boundary between the matrix grains (γ_m) and the boundary between the matrix and the precipitate (γ_p). The interfacial surface energies fix a dihedral angle θ between the matrix-precipitate interface that, in turn, determines the shape of the precipitate (Figure 11.7). The relationship is:

$$\gamma_m = 2\gamma_p \cos\frac{\theta}{2} \qquad (11.5)$$

Figure 11.7
The effect of surface energy and the dihedral angle on the shape of a precipitate.

If the dihedral angle is small, the precipitate may be continuous. If the precipitate is also hard and brittle, the thin film that surrounds the matrix grains causes the alloy to be very brittle [Figure 11.6(b)]. On the other hand, discontinuous and even spherical precipitates form when the dihedral angle is large [Figure 11.6(c)].

Coherent Precipitate Even if we produce a uniform distribution of discontinuous precipitate, the precipitate may not significantly disrupt the surrounding matrix structure. Consequently, the precipitate blocks slip only if it lies directly in the path of the dislocation [Figure 11.8(a)].

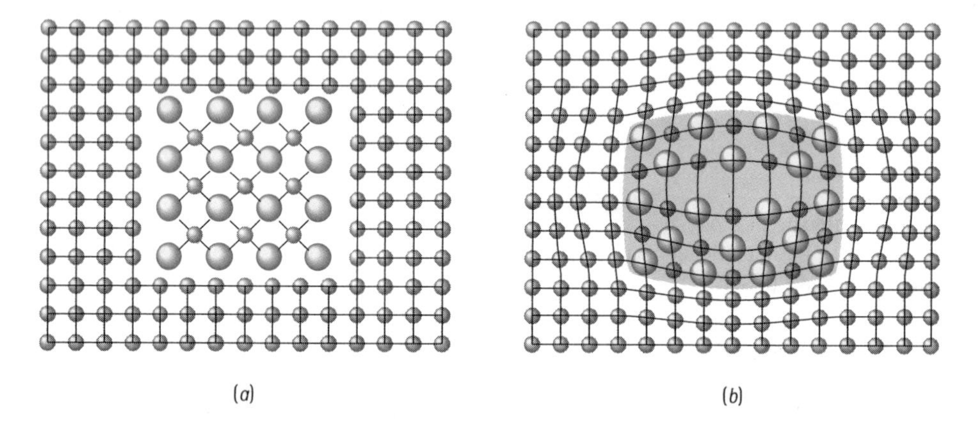

Figure 11.8
(a) A noncoherent precipitate has no relationship with the crystal structure of the surrounding matrix.
(b) A coherent precipitate forms so that there is a definite relationship between the precipitate's and the matrix's crystal structure.

But when a coherent precipitate forms, the planes of atoms in the lattice of the precipitate are related to – or even continuous with – the planes in the lattice of the matrix [Figure 11.8(b)]. Now a widespread disruption of the matrix lattice is created and the movement of a dislocation is impeded even if the dislocation merely passes near the coherent precipitate. A special heat treatment, such as age hardening, may produce the coherent precipitate.

11.4 Age Hardening or Precipitation Hardening

Age hardening, or precipitation hardening, produces a uniform dispersion of a fine, hard coherent precipitate in a softer, more ductile matrix. The Al-4% Cu alloy is a classical example of an age-hardenable alloy. There are three steps in the age-hardening heat treatment (Figure 11.9).

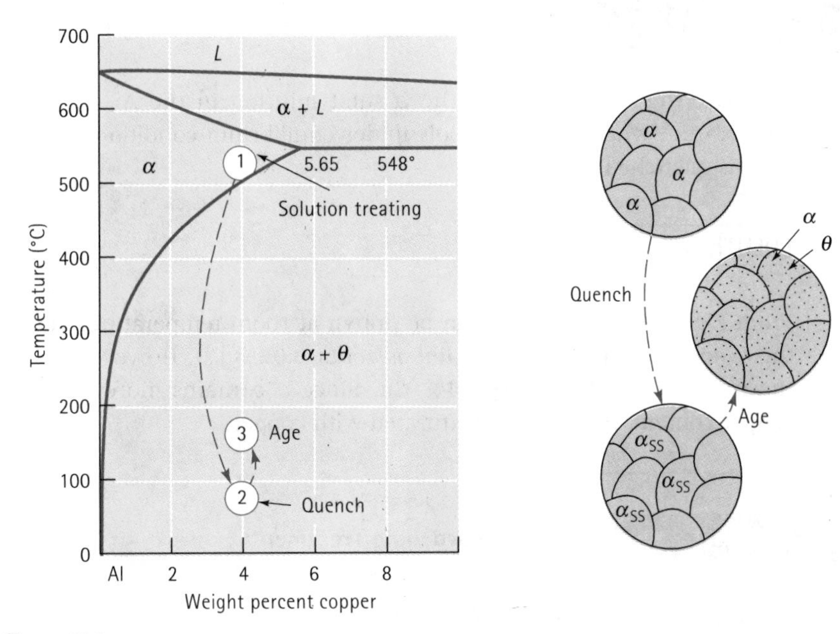

Figure 11.9
The aluminium-rich end of the aluminium-copper phase diagram showing the three steps in the age-hardening heat treatment and the microstructures that are produced.

Step 1: Solution Treatment In the solution treatment, the alloy is first heated above the solvus temperature and held until a homogeneous solid solution α is produced. This step dissolves the θ precipitate and reduces any segregation present in the original alloy.

We could heat the alloy to just below the solidus temperature and increase the rate of homogenisation. However, the presence of a nonequilibrium eutectic microconstituent may cause melting. Thus the Al-4% Cu alloy is solution-treated between 500°C and 548°C, that is, between the solvus and eutectic temperatures.

Step 2: Quench After solution treatment, the alloy, which contains only α in its structure, is rapidly cooled, or quenched. The atoms do not have time to diffuse to potential nucleation sites, so the θ does not form. After the quench, the structure still contains only α. The α is a supersaturated solid solution α_{ss} containing excess copper, and it is not an equilibrium structure.

Step 3: Age Finally, the supersaturated α is heated below the solvus temperature. At this *ageing* temperature, atoms diffuse only short distances. Because the supersaturated α is not stable, the extra copper atoms diffuse to numerous nucleation sites and precipitates grow. Eventually, if we hold the alloy for a sufficient time at the ageing temperature, the equilibrium $\alpha + \theta$ structure is produced. This, however, is not always desirable as higher strengths may be achieved by stopping the ageing process before an equilibrium structure is produced (*see* Nonequilibrium Precipitates *below*).

EXAMPLE 11.2

Compare the composition of the α solid solution in the Al-4% Cu alloy at room temperature when the alloy cools under equilibrium conditions with that when the alloy is quenched.

SOLUTION

From Figure 11.5, a tie line can be drawn at room temperature. The composition of the α determined from the tie line is about 0.02% Cu. However, the composition of the α after quenching is still 4% Cu. Since α contains more than the equilibrium copper content, the α is supersaturated with copper.

EXAMPLE 11.3 **Age-Hardening Treatment**

The magnesium-aluminium phase diagram is shown in Figure 13.3. Suppose a Mg-8% Al alloy is responsive to an age-hardening heat treatment. Establish a heat treatment for the alloy.

SOLUTION

Step 1: Solution-treat at a temperature between the solvus and the eutectic to avoid hot shortness. Thus, heat between 340°C and 451°C.

Step 2: Quench to room temperature fast enough to prevent the precipitate from forming.

Step 3: Age at a temperature below the solvus, that is, below 340°C.

Nonequilibrium Precipitates during Ageing During ageing of aluminium-copper alloys, a continuous series of precipitates forms before the equilibrium θ is produced. At the start of ageing, the copper atoms concentrate on {100} planes in the α matrix and produce very thin precipitates called Guinier-Preston (GP-I) zones. As ageing continues, more copper atoms diffuse to the precipitate and the GP-I zones thicken into thin disks, or GP-II zones. With continued diffusion, the precipitates develop a greater degree of order and are called θ'. Finally, the stable θ precipitate is produced.

The nonequilibrium precipitates – GP-I, GP-II, and θ' – are coherent precipitates. The strength of the alloy increases with ageing time as these coherent phases grow in size during the initial stages of the heat treatment. When these coherent precipitates are present, the alloy is in the aged condition. Figure 11.10 shows the structure of an aged Al-Ag alloy.

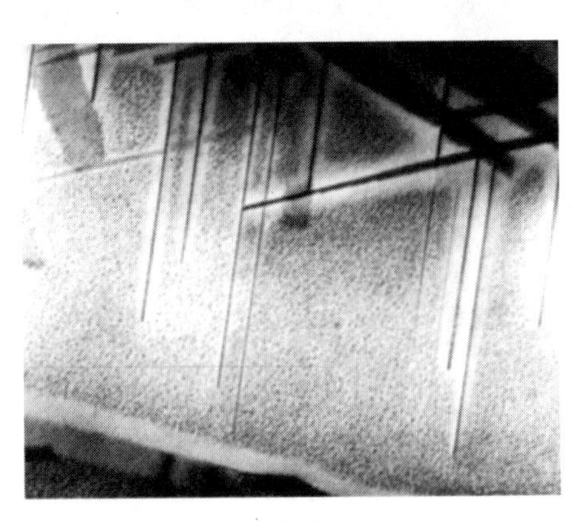

Figure 11.10
An electron micrograph of aged Al-15% Ag showing coherent γ' plates and round GP zones, × 40000. (*Courtesy, J. B. Clark.*)

When the stable noncoherent θ phase precipitates, the strength of the alloy begins to decrease. Now the alloy is in the overaged condition. The θ still provides some dispersion strengthening, but with increasing time, the θ grows larger and even the simple dispersion strengthening effect diminishes.

11.5 Effects of Ageing Temperature and Time

The properties of an age-hardenable alloy depend on both ageing temperature and ageing time (Figure 11.11). At 260°C, diffusion in the Al-4% Cu alloy is rapid, and precipitates quickly form. The strength reaches a maximum after less than 0.1 h exposure. Overageing occurs if the alloy is held for longer than 0.1 h.

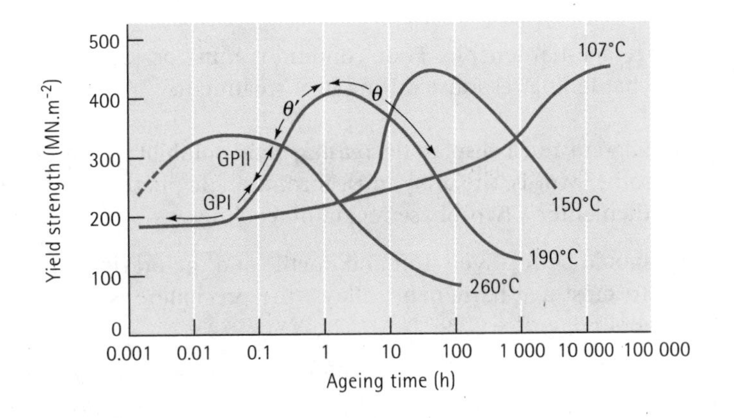

Figure 11.11
The effect of ageing temperature and time on the yield strength of an Al-4% Cu alloy.

At 190°C, which is a typical ageing temperature for many aluminium alloys, a longer time is required to produce the optimum strength. However, there are several benefits to using the lower temperature. First, the maximum strength increases as the ageing temperature decreases. Second, the alloy maintains its maximum strength over a long period of time. Third, the properties are more uniform. If the alloy is aged for only 10 min at 260°C, the surface of the part reaches the proper temperature and strengthens, but the centre remains cool and ages only slightly.

EXAMPLE 11.4

The operator of a furnace left for his hour lunch break without removing the Al-4% Cu alloy from the ageing furnace. Compare the effect on the yield strength of the extra hour of ageing for temperatures of 190°C and 260°C.

SOLUTION

At 190°C, the peak strength of 400 $MN.m^{-2}$ occurs at 2 h (Figure 11.11). After 3 h, the strength is essentially the same.

At 260°C, the peak strength of 340 $MN.m^{-2}$ occurs at 0.06 h. However, after 1.06 h, the strength decreases to 250 $MN.m^{-2}$.

Thus, the higher ageing temperature gives a lower peak strength and makes the strength more sensitive to ageing time.

Ageing at either 190°C or 260°C is called artificial ageing, because the alloy is heated to produce precipitation. Some solution-treated and quenched alloys age at room temperature; this is called natural ageing. Natural ageing requires long times – often several days – to reach maximum strength. However, the peak strength is higher than that obtained in artificial ageing, and no overageing occurs.

11.6 Requirements for Age Hardening

Not all alloys are age-hardenable. Four conditions must be satisfied for an alloy to have a true age-hardening response during heat treatment:

1. The phase diagram must display decreasing solid solubility with decreasing temperature. In other words, the alloy must form a single phase on heating above the solvus line, then enter a two-phase region on cooling.
2. The matrix should be relatively soft and ductile and the precipitate should be hard and brittle. In most age hardenable alloys, the precipitate is a hard brittle intermetallic compound.
3. The alloy must be quenchable. Some alloys cannot be cooled rapidly enough to suppress the formation of the precipitate. Quenching may, however, introduce residual stresses that cause distortion of the part. To minimise residual stresses, aluminium alloys are quenched in hot water, at about 80°C.
4. A coherent precipitate must form.

A number of important metal alloys, including certain stainless steels and alloys based on aluminium, magnesium, titanium, nickel, and copper, meet these conditions and are age-hardenable.

11.7 Use of Age Hardenable Alloys at High Temperatures

Based on our previous discussion, we would not select an age-hardened Al-4% Cu alloy for use at high temperatures. At service temperatures ranging from 100°C to 500°C, the alloy overages and loses it strength. Above 500°C, the second phase redissolves in the matrix and we do not even obtain dispersion strengthening. In general, the aluminium age-hardenable alloys are best suited for service near room temperature. However, some magnesium alloys may maintain their strength to about 250°C and certain nickel superalloys resist overageing at 1 000°C.

We also have problems when welding age-hardenable alloys (Figure 11.12). During welding, the metal adjacent to the weld is heated. The heat-affected area contains two principle zones. The lower temperature zone near the unaffected base metal is exposed to temperatures just below the solvus and may overage. The higher temperature zone is solution-treated, eliminating the effects of age-hardening. If the solution-treated zone cools slowly, stable θ may form at the grain boundaries, embrittling the weld area. Very fast welding processes such as electron-beam welding, complete reheat treatment of the area after welding, or welding the alloy in the solution-treated condition improve the quality of the weld.

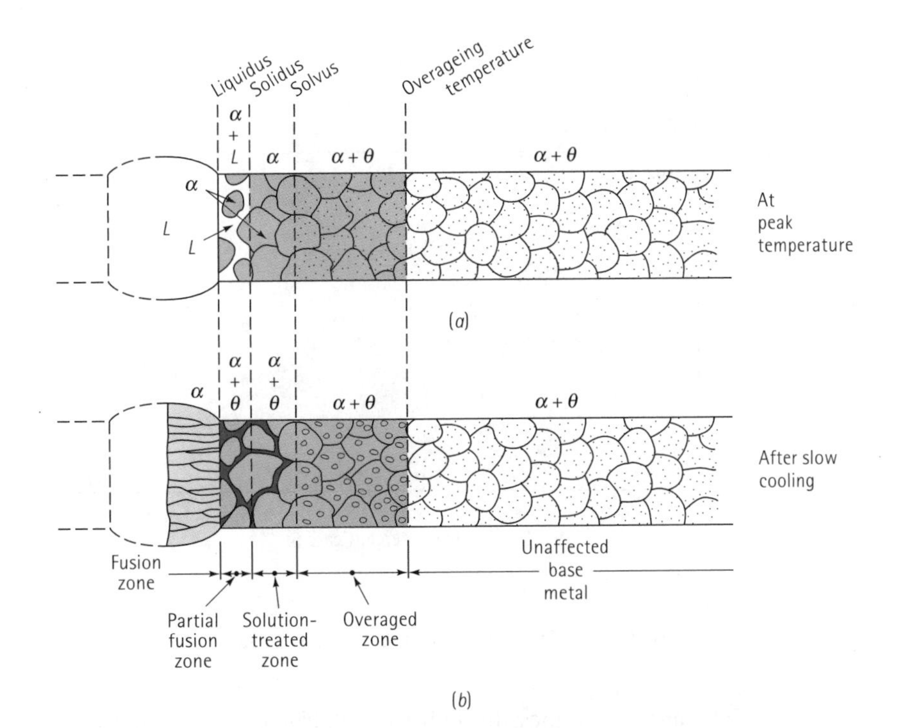

Figure 11.12
Microstructural changes that occur in age-hardened alloys during fusion welding: (*a*) microstructure in the weld at the peak temperature, and (*b*) microstructure in the weld after slowly cooling to room temperature.

11.8 The Eutectoid Reaction

In Chapter 10, we defined the eutectoid as a solid-state reaction in which one solid phase transforms to two other solid phases:

$$S_1 \rightarrow S_2 + S_3 \quad (11.6)$$

The formation of the two solid phases permits us to obtain dispersion strengthening. As an example of how we can use the eutectoid reaction to control the microstructure and properties of an alloy, let's examine the Fe-Fe_3C phase diagram (Figure 11.13), which is the basis for steels and cast irons. The following features should be noted.

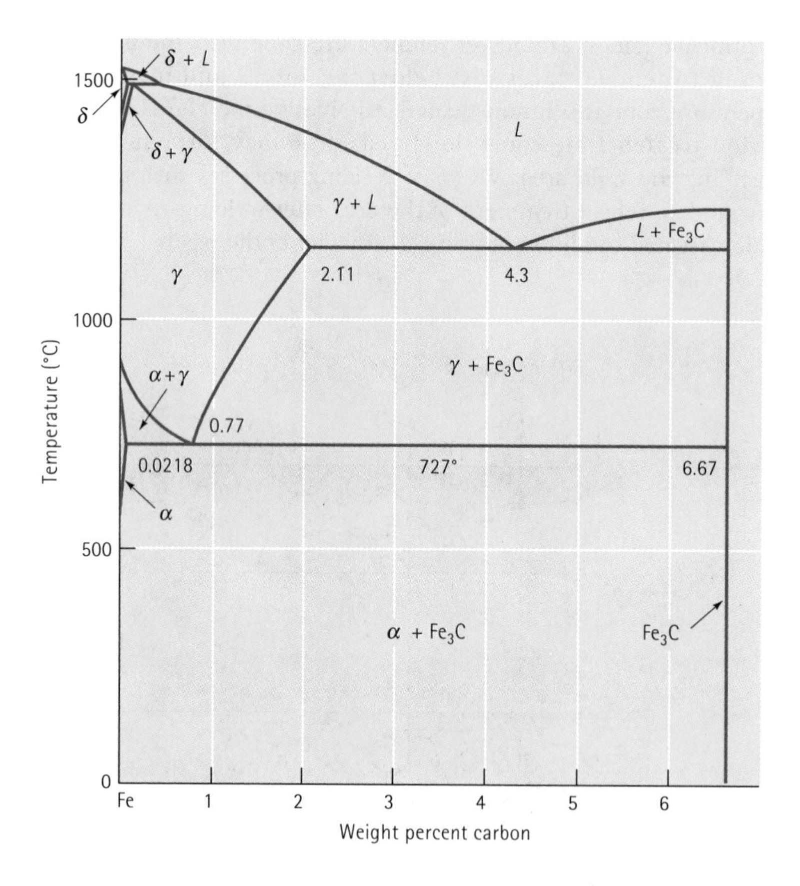

Figure 11.13
The Fe-Fe_3C phase diagram. The vertical line at 6.67% C is the stoichiometric compound Fe_3C.

Solid Solutions Iron goes through two allotropic transformations during heating or cooling. Immediately after solidification, iron forms a BCC structure called δ-ferrite. On further cooling, the iron transforms to a FCC structure called γ, or austenite. Finally, iron transforms back to the BCC structure at lower temperatures; this structure is called α, or ferrite. Both of the ferrites and the austenite are solid solutions of interstitial carbon atoms in iron (Example 4.8). Because interstitial holes in the FCC lattice are somewhat larger than holes in the BCC lattice, a greater number of carbon atoms can be accommodated in FCC iron. Thus, the maximum solubility of carbon in austenite is 2.11% C, whereas the maximum solubility of carbon in BCC iron is much lower – 0.0218% C in α and 0.09% C in δ. The solid solutions are relatively soft and ductile but are stronger than pure iron due to solid solution strengthening by the carbon.

Compounds A stoichiometric compound Fe_3C, or cementite, forms when the solubility of carbon in solid iron is exceeded. The Fe_3C contains 6.67% C, is extremely hard and brittle, and is present in all commercial steels. By properly controlling the amount, size, and shape of Fe_3C, we control the degree of dispersion strengthening and the properties of the steel.

The Eutectoid Reaction If we heat an alloy containing the eutectoid composition of 0.77% C above 727°C, we produce a structure containing only austenite grains. When austenite cools to 727°C, the eutectoid reaction begins:

$$\gamma_{0.77\%C} \rightarrow \alpha_{0.0218\%C} + Fe_3C_{6.67\%C} \quad (11.7)$$

As in the eutectic reaction, the two phases that form have different compositions, so atoms must diffuse during the reaction (Figure 11.14). Most of the carbon in the austenite diffuses to the Fe_3C and most of the iron atoms diffuse to α. This redistribution of atoms is easiest if the diffusion distances are short, which is the case when the α and Fe_3C grow as thin lamellae, or plates.

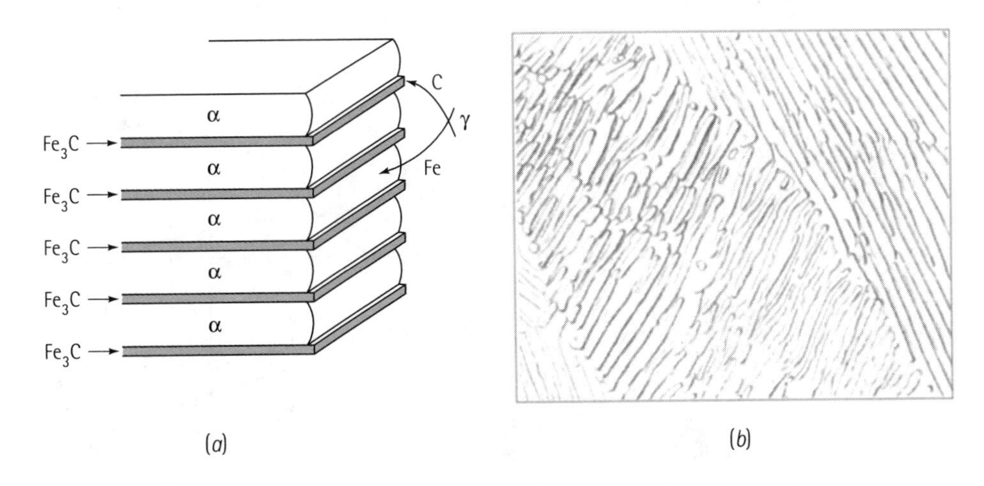

Figure 11.14
Growth and structure of pearlite: (*a*) redistribution of carbon and iron, and (*b*) photomicrograph of the pearlite lamellae (× 2000). (*From* Metals Handbook, *Vol. 7, 8th Ed., American Society for Metals, 1972.*)

Pearlite The lamellar structure of α and Fe_3C that develops in the iron-carbon system is called pearlite, which is a microconstituent in steel. The lamellae in pearlite are much finer than the lamellae in the lead-tin eutectic because the iron and carbon atoms must diffuse through solid austenite rather than through liquid.

EXAMPLE 11.5

Calculate the amounts of ferrite and cementite present in pearlite.

SOLUTION

Since pearlite must contain 0.77% C, using the lever law:

$$\%\ \alpha = \frac{6.67 - 0.77}{6.67 - 0.0218} \times 100 = 88.7\%$$

$$\%\ Fe_3C = \frac{0.77 - 0.0218}{6.67 - 0.0218} \times 100 = 11.3\%$$

In Example 11.5, we find that most of the pearlite is composed of ferrite. In fact, if we examine the pearlite closely, we find that the Fe_3C lamellae are surrounded by α. The pearlite structure therefore provides dispersion strengthening – the continuous ferrite phase is relatively soft and ductile and the hard brittle cementite is dispersed.

Primary Microconstituents Hypoeutectoid steels contain less than 0.77% C, and hypereutectoid steels contain more than 0.77% C. Ferrite is the primary or proeutectoid microconstituent in hypoeutectoid alloys, and cementite is the primary or proeutectoid microconstituent in hypereutectoid alloys. If we heat a hypoeutectoid steel containing 0.60% C above 750°C, only austenite remains in the microstructure.

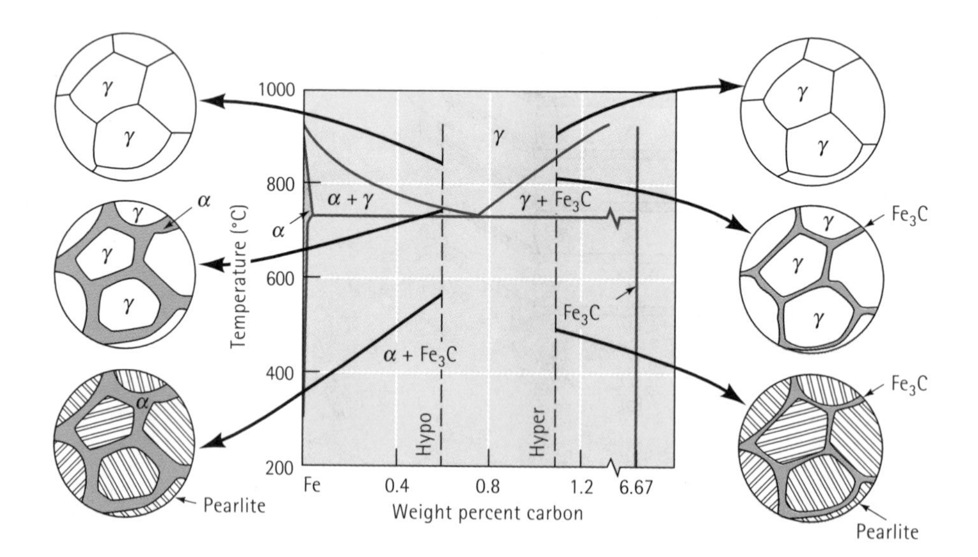

Figure 11.15
The evolution of the microstructure of hypoeutectoid and hypereutectoid steels during cooling, in relationship to the phase diagram.

Figure 11.15 shows what happens when the austenite cools. Just below 750°C, ferrite nucleates and grows, usually at the austenite grain boundaries. Primary ferrite continues to grow until the temperature falls to 727°C. The remaining austenite at that temperature is now surrounded by ferrite and has changed in composition from 0.60% C to 0.77% C. Subsequent cooling to below 727°C causes all of the remaining austenite to transform to pearlite by the eutectoid reaction. The final structure contains two phases – ferrite and cementite – arranged as two microconstituents – primary ferrite and pearlite.

The final microstructure contains islands of pearlite surrounded by the primary ferrite [Figure 11.16(a)]. This structure permits the alloy to be strong, due to the dispersion-strengthened pearlite, yet ductile, due to the continuous primary ferrite.

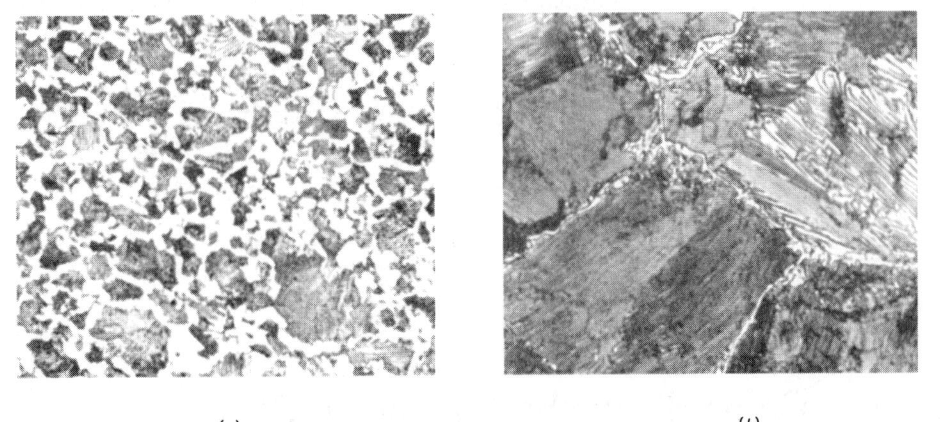

(*a*) (*b*)

Figure 11.16
(*a*) A hypoeutectoid steel showing primary α (white) and pearlite (× 400). (*b*) A hypereutectoid steel showing primary Fe_3C surrounding pearlite (× 800). (*From* Metals Handbook, *Vol. 7, 8th Ed., American Society for Metals, 1972.*)

In hypereutectoid alloys, however, the primary phase is Fe_3C, which again forms at the austenite grain boundaries. After the austenite cools through the eutectoid reaction, the steel contains hard, brittle cementite surrounding islands of pearlite [Figure 11.6(b)]. Now, because the hard, brittle microconstituent is continuous, the steel is also brittle. Fortunately, we can improve the microstructure and properties of the hypereutectoid steels by heat treatment.

EXAMPLE 11.6

Calculate the amounts and compositions of phases and microconstituents in a Fe-0.60% C alloy at 726°C.

SOLUTION

The phases are ferrite and cementite. Using a tie line and working the level law at 726°C, we find:

$$\alpha{:}0.0218\%\ \text{C} \qquad \%\ \alpha = \frac{6.67-0.60}{6.67-0.0218}\times 100 = 91.3\%$$

$$\text{Fe}_3\text{C}{:}6.67\%\ \text{C} \qquad \%\ \text{Fe}_3\text{C} = \frac{0.60-0.0218}{6.67-0.0218}\times 100 = 8.7\%$$

The microconstituents are primary ferrite and pearlite. If we construct a tie line just above 727°C, we can calculate the amounts and compositions of ferrite and austenite just before the eutectoid reaction starts. All of the austenite at that temperature will transform to pearlite; all of the ferrite will remain as primary ferrite:

$$\text{Primary } \alpha{:}\ 0.0218\%\ \text{C} \qquad \%\ \text{Primary } \alpha = \frac{0.77-0.60}{0.77-0.0218}\times 100 = 22.7\%$$

$$\text{Pearlite}{:}\ 0.77\%\ \text{C} \qquad \%\ \text{Pearlite} = \frac{0.60-0.0218}{0.77-0.0218}\times 100 = 77.3\%$$

11.9 Controlling the Eutectoid Reaction

We control dispersion strengthening in the eutectoid alloys in much the same way that we did in eutectic alloys.

Controlling the Amount of the Eutectoid By changing the composition of the alloy, we change the amount of the hard second phase. As the carbon content of a steel increases toward the eutectoid composition of 0.77% C, the amounts of Fe_3C and pearlite increase, thus increasing the strength. However, this strengthening effect eventually peaks and the properties level out or even decrease when the carbon content is too high (Table 11.1).

Controlling the Austenite Grain Size Pearlite grows as grains or *colonies*. Within each colony, the orientation of the lamellae is identical. The colonies nucleate most easily at the grain boundaries of the original austenite grains. We can increase the number of pearlite colonies by reducing the prior austenite grain size, usually by using low temperatures to produce the austenite. Typically, we can increase the strength of the alloy by reducing the grain size or increasing the number of colonies.

Controlling the Cooling Rate By increasing the cooling rate during the eutectoid reaction, we reduce the distance that the atoms are able to diffuse. Consequently, the lamellae produced during the reaction are finer or more closely spaced. By producing fine pearlite, we increase the strength of the alloy (Table 11.1 and Figure 11.17).

Table 11.1 The effect of carbon on the strength of steels.

	Slow cooling (Coarse Pearlite)			Fast cooling (Fine Pearlite)		
Carbon %	Yield Strength ($MN.m^{-2}$)	Tensile Strength ($MN.m^{-2}$)	% Elongation	Yield Strength ($MN.m^{-2}$)	Tensile Strength ($MN.m^{-2}$)	% Elongation
0.20	295	394	36.5	346	441	36.0
0.40	353	519	30.0	374	590	28.0
0.60	372	626	23.0	421	776	18.0
0.80	376	616	25.0	524	1010	11.0
0.95	379	657	13.0	500	1014	9.5

After Metals Progress Materials and Processing Databook, *1981.*

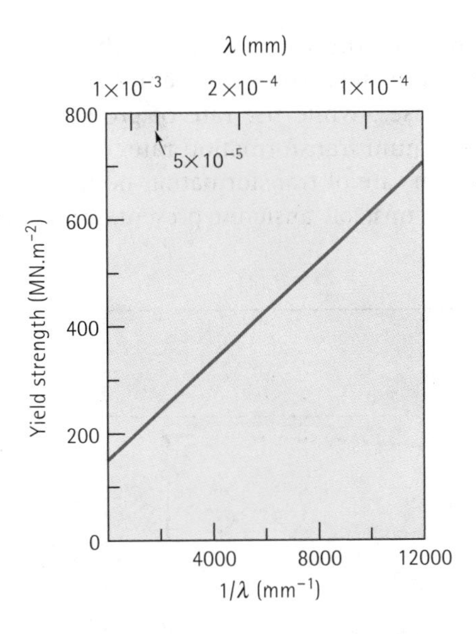

Figure 11.17
The effect of interlamellar spacing of pearlite on the yield strength of pearlite.

Controlling the Transformation Temperature The solid-state eutectoid reaction is rather slow, and the steel may cool below the equilibrium eutectoid temperature before the transformation begins. Lower transformation temperatures give a finer, stronger structure (Figure 11.18), influence the time required for transformation, and even alter the arrangement of the two phases. This information is contained in the time-temperature-transformation (TTT) diagram (Figure 11.19). This diagram, also called the isothermal transformation (IT) diagram or the C-curve, permits us to predict the structure, properties, and heat treatment required in steels.

The TTT diagram is a consequence of the kinetics of the eutectoid reaction. At any particular temperature, a sigmoidal curve describes the rate at which the austenite transforms to a mixture of ferrite and cementite (Figure 11.20). An incubation time is required for nucleation. The P_s line represents the time at which this transformation

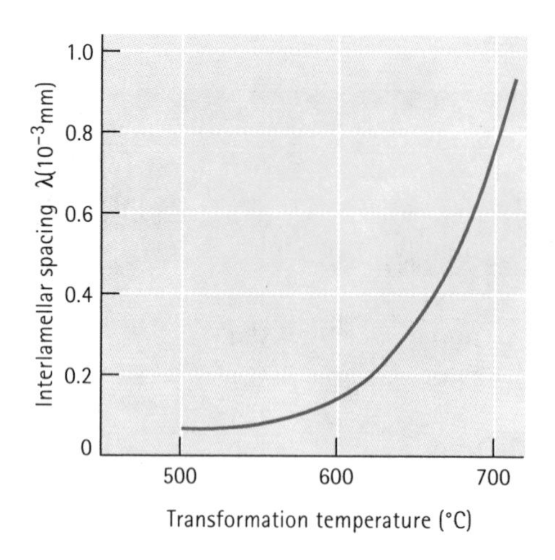

Figure 11.18
The effect of the austenite transformation temperature on the interlamellar spacing in pearlite.

starts. The sigmoidal curve also gives the time at which the transformation is completed; this time is given by the P_f line. When the temperature decreases from 727°C, the rate of nucleation increases, while the rate of growth of the eutectoid decreases. As in Figure 11.3, a maximum transformation rate, or minimum transformation time, is found; the maximum rate of transformation occurs near 550°C for a eutectoid steel (Figure 11.19). γ_u is the unstable austenite present during transformation.

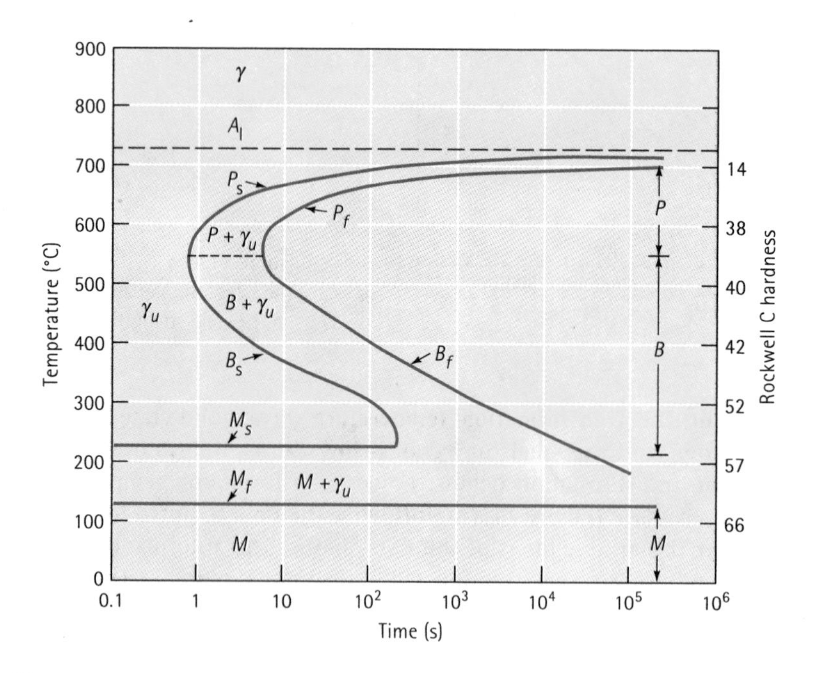

Figure 11.19
The time-temperature-transformation (TTT) diagram for a eutectoid steel.

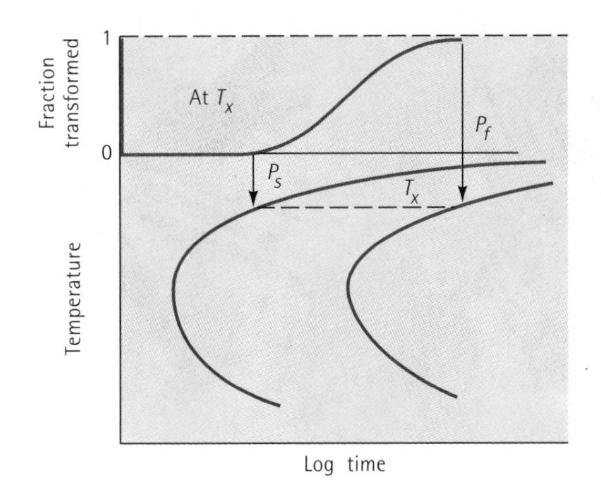

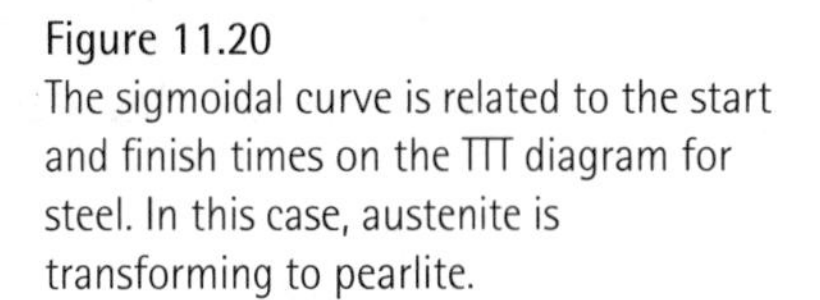
Figure 11.20
The sigmoidal curve is related to the start and finish times on the TTT diagram for steel. In this case, austenite is transforming to pearlite.

Two types of microconstituents are produced as a result of the transformation. Pearlite (P) forms above 550°C, bainite (B) forms at lower temperatures:

1. *Nucleation and growth of pearlite*: If we quench to just below the eutectoid temperature, the austenite is only slightly undercooled. Long times are required before stable nuclei for ferrite and cementite form. After pearlite begins to grow, atoms diffuse rapidly and *coarse* pearlite is produced; the transformation is complete at the pearlite finish (P_f) time. Austenite quenched to a lower temperature is more highly undercooled. Consequently, nucleation occurs more rapidly and the P_s is shorter. However, diffusion is also slower, so atoms diffuse only short distances and *fine* pearlite is produced. Even though growth rates are slower, the overall time required for the transformation is reduced because of the shorter incubation time. Finer pearlite forms in shorter times as we reduce the isothermal transformation temperature to about 550°C, which is the *nose*, or *knee*, of the TTT curve (Figure 11.19).

2. *Nucleation and growth of bainite*: At a temperature just below the nose of the TTT diagram, diffusion is very slow and total transformation times increase. In addition, we find a different microstructure! At low transformation temperatures, the lamellae in pearlite would have to be extremely thin and consequently, the boundary area between the ferrite and Fe_3C lamellae would be very large. Because of the energy associated with the ferrite-cementite interface, the total energy of the steel would have to be very high. The steel can reduce its internal energy by permitting the cementite to precipitate as discrete, rounded particles in a ferrite matrix. This new microconstituent, or arrangement of ferrite and cementite, is called bainite. Transformation begins at a bainite start (B_s) time and ends at a bainite finish (B_f) time.

The times required for austenite to begin and finish its transformation to bainite increase and the bainite becomes finer as the transformation temperature continues to decrease. The bainite that forms just below the nose of the curve is called course bainite, upper bainite, or feathery bainite. The bainite that forms at lower temperatures is called fine bainite, lower bainite, or acicular bainite. Figure 11.21 shows typical microstructures of bainite.

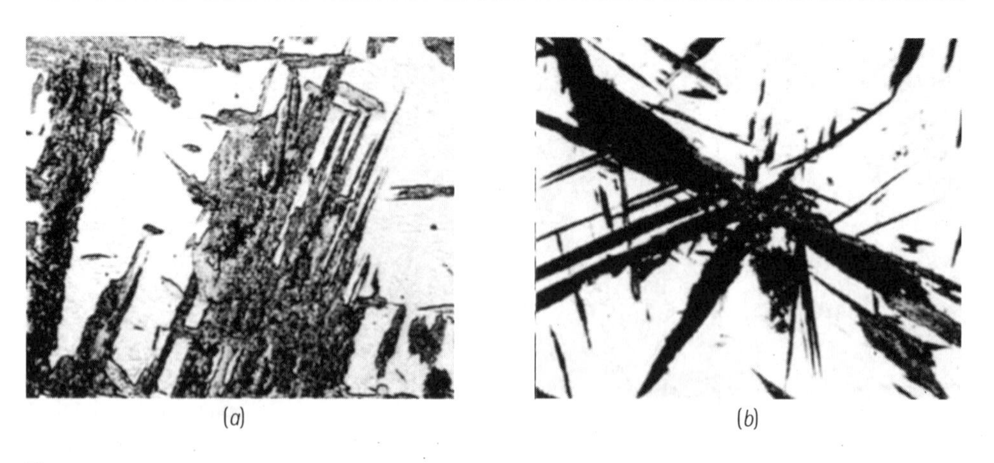

Figure 11.21
(*a*) Upper bainite (grey, feathery plates) (× 600). (*b*) Lower bainite (dark needles) (× 400). (*From* Metals Handbook, *Vol. 8, 8th Ed., American Society for Metals*, 1973.)

Figure 11.22 shows the effect of transformation temperature on the properties of a eutectoid steel. As the temperature decreases, there is a general trend toward higher strength and lower ductility due to the finer microstructure that is produced.

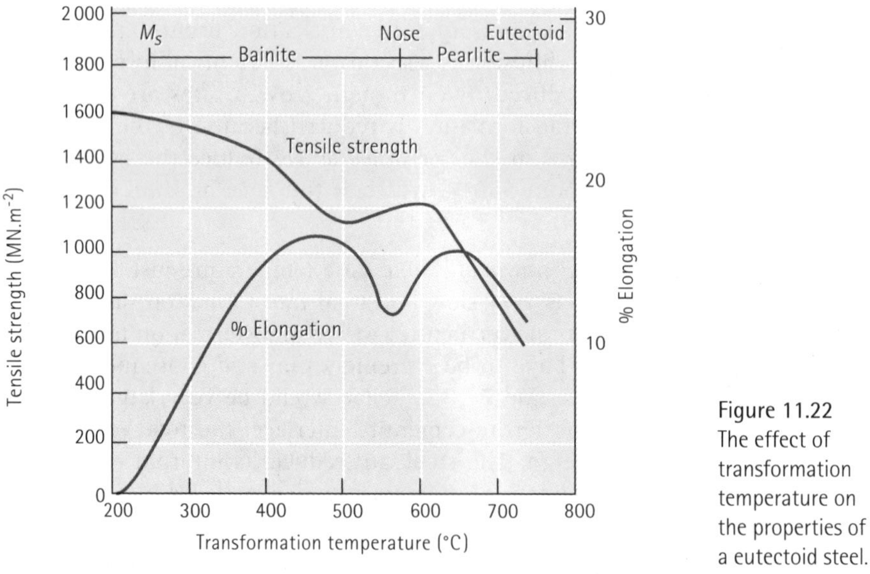

Figure 11.22
The effect of transformation temperature on the properties of a eutectoid steel.

EXAMPLE 11.7

Describe a heat treatment to produce the pearlite structure shown in Figure 11.14(b).

SOLUTION

First we need to determine the interlamellar spacing of the pearlite. If we count the number of lamellar spacings in the upper right of Figure 11.14(b), remembering that the interlamellar spacing is measured from one α plate to the next α plate, we find 14 spacings over a 20 mm distance. Due to the × 2 000 magnification, this 20 mm distance is actually 0.01 mm. Thus:

$$\lambda = \frac{0.01 \text{ mm}}{14 \text{ spacings}} = 0.714 \times 10^{-3} \text{mm}$$

If we assume that the pearlite formed by an isothermal transformation, we find from Figure 11.18 that the transformation temperature must have been approximately 700°C. From the TTT diagram (Figure 11.19), our heat treatment must have been:

1. Heat the steel to about 750°C and hold – perhaps for 1 h – to produce all austenite.
2. Quench to 700°C and hold for at least 10^5 s (the P_f time).
3. Cool to room temperature.

The steel should have a hardness of HRC 14 (Figure 11.19) and a yield strength of about 200 $MN.m^{-2}$ as shown in Figure 11.17.

EXAMPLE 11.8

Excellent combinations of hardness, strength, and toughness are obtained from bainite. One heat treater austenitised a eutectoid steel at 750°C, quenched and held the steel at 250°C for 15 min, and finally permitted the steel to cool to room temperature. Was the required bainitic structure produced?

SOLUTION

Let's examine the heat treatment using Figure 11.19. After heating at 750°C, the microstructure is 100% γ. After quenching to 250°C, unstable austenite remains for slightly more than 100 s, when fine bainite begins to grow. After 15 min, or 900 s, about 50% fine bainite has formed and the remainder of the steel still contains unstable austenite. As we will see later, the unstable austenite transforms to martensite when the steel is cooled to room temperature and the final structure is a mixture of bainite and hard, brittle martensite. The heat treatment was not successful! The heat treater should have held the steel at 250°C for at least 10^4 s, or about 3 h.

11.10 The Martensitic Reaction and Tempering

Martensite is a phase that forms as the result of a diffusionless solid-state transformation. Cobalt, for example, transforms from a FCC to a HCP crystal structure by a slight shift in the atom locations that alters the stacking sequence of close-packed planes. Because the reaction does not depend on diffusion, the martensite reaction is an athermal transformation – that is, the reaction depends only on the temperature, not on the time. The martensite reaction often proceeds rapidly, at speeds approaching the velocity of sound in the material.

Martensite in Steels In steels with less than about 0.2% C, the FCC austenite transforms to a supersaturated BCC martensite structure. In higher carbon steels, the martensite reaction occurs as FCC austenite transforms to BCT (body centred tetragonal) martensite. The relationship between the FCC austenite and the BCT martensite [Figure 11.23(a)] shows that carbon atoms in the 1/2,0,0 type of interstitial sites in the FCC cell can be trapped during the transformation to the body-centred structure, causing the tetragonal structure to be produced. As the carbon content of the steel increases, a greater number of carbon atoms are trapped in these sites, thereby increasing the difference between the *a*- and *c*-axes of the martensite [Figure 11.23(b)].

The steel must be quenched, or rapidly cooled, from the stable austenite region to prevent the formation of pearlite, bainite, or primary microconstituents. The martensite reaction begins in a eutectoid steel when austenite cools below 220°C,

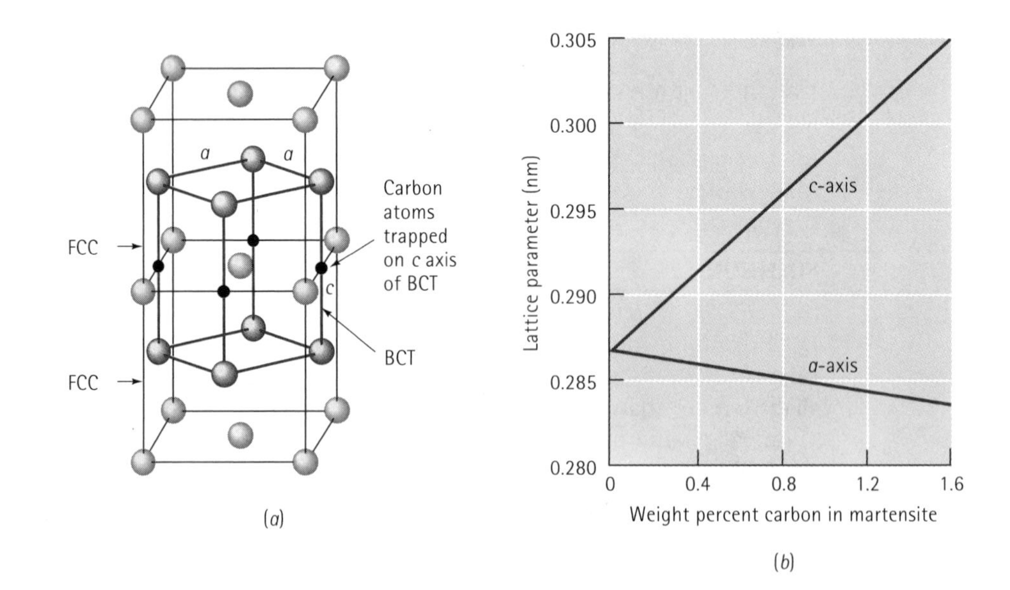

Figure 11.23
(*a*) The unit cell of BCT martensite is related to the FCC austenite unit cell. (*b*) As the percentage of carbon increases, more interstitial sites are filled by the carbon atoms and the tetragonal structure of the martensite becomes more pronounced.

the martensite start (M_s) temperature (Figure 11.19). The amount of martensite increases as the temperature decreases. When the temperature passes below the martensite finish temperature (M_f), the steel should contain 100% martensite. At any intermediate temperature, the amount of martensite does not change as the time at that temperature increases.

The composition of martensite must be the same as that of the austenite from which it forms. There is no long-range diffusion during the transformation that can change the composition. Thus, in iron-carbon alloys, the initial austenite composition and the final martensite composition are the same.

EXAMPLE 11.9 Heat Treatment for a Dual Phase Steel

Unusual combinations of properties can be obtained by producing a steel whose microstructure contains 50% ferrite and 50% martensite; the martensite provides strength and the ferrite provides ductility and toughness. Describe a heat treatment to produce a dual phase steel in which the composition of the martensite is 0.60% C.

SOLUTION

To obtain a mixture of ferrite and martensite, we need to heat-treat a hypoeutectoid steel into the $\alpha + \gamma$ region of the phase diagram. The steel is then quenched, permitting the γ portion of the structure to transform to martensite.

The heat treatment temperature is fixed by the requirement that the martensite contain 0.60% C. From the solubility line between the γ and the $\alpha + \gamma$ regions, we find that 0.60% C is obtained in austenite when the temperature is about 750°C. (Figure 12.2 might be used to provide a better estimate of this temperature.)

To produce 50% martensite, we need to select a steel that gives 50% austenite when the steel is held at 750°C. If the carbon content of the steel is x, then:

$$\%\gamma = \frac{x - 0.02}{0.60 - 0.02} \times 100 = 50 \quad \text{or} \quad x = 0.31\%\ \text{C}$$

Our final design is:

1. Select a hypoeutectoid steel containing 0.31% C.
2. Heat the steel to 750°C and hold (perhaps for 1 h, depending on the thickness of the part) to produce a structure containing 50% ferrite and 50% austenite, with 0.60% C in the austenite.
3. Quench the steel to room temperature. The austenite transforms to martensite, also containing 0.60% C.

Properties of Steel Martensite Martensite in steels is very hard and brittle. The BCT crystal structure has no close-packed slip planes in which dislocations can easily move. The martensite is highly supersaturated with carbon, since iron normally contains less than 0.0218% C at room temperature, and martensite contains the

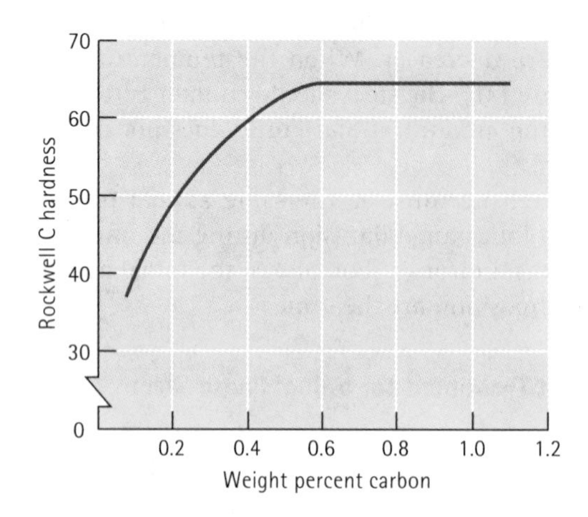

Figure 11.24
The effect of carbon content on the hardness of martensite in steels.

amount of carbon present in the steel. Finally, martensite has a fine grain size and an even finer substructure within the grains.

The structure and properties of steel martensites depend on the carbon content of the alloy (Figure 11.24). When the carbon content is low, the martensite grows in a 'lath' shape, composed of bundles of flat, narrow plates that grow side by side [Figure 11.25(a)]. This martensite is not very hard. At a higher carbon content, plate martensite grows, in which flat, narrow plates grow individually rather than as bundles [Figure 11.25(b)]. The hardness is much greater in the higher carbon, plate martensite structure, partly due to the greater distortion, or large *c*/*a* ratio, of the crystal structure.

Tempering of Steel Martensite Martensite is not an equilibrium structure. When martensite in a steel is heated below the eutectoid temperature, the stable α and Fe_3C precipitate. This process is called tempering. The decomposition of martensite in steels causes the strength and hardness of the martensite to decrease while the ductility and impact properties are improved (Figure 11.26).

Figure 11.25
(*a*) Lath martensite in low-carbon steel (× 80). (*b*) Plate martensite in high-carbon steel (× 400). (*From* Metals Handbook, *Vol. 8, 8th Ed., American Society for Metals,* 1973.)

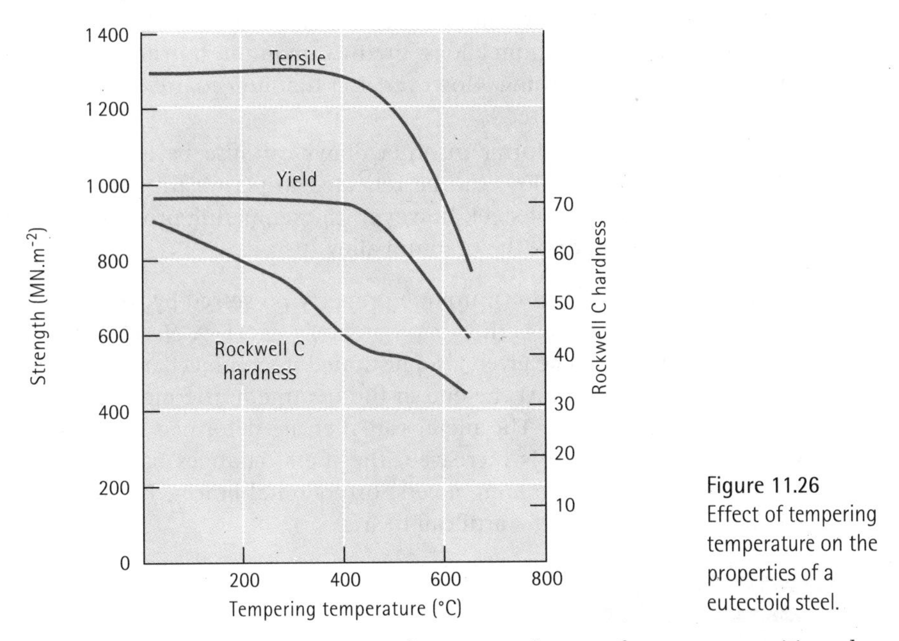

Figure 11.26
Effect of tempering temperature on the properties of a eutectoid steel.

At low tempering temperatures, the martensite may form two transition phases – a lower carbon martensite and a very fine nonequilibrium ε-carbide, or $Fe_{2.4}C$. The steel is still strong, brittle, and perhaps even harder than before tempering. At higher temperatures, the stable α and Fe_3C form and the steel becomes softer and more ductile. If the steel is tempered just below the eutectoid temperature, the Fe_3C becomes very coarse and the dispersion-strengthening effect is greatly reduced. By selecting the appropriate tempering temperature, a wide range of properties can be obtained. The product of the tempering process is a microconstituent called tempered martensite (Figure 11.27).

Martensite in Other Systems The characteristics of the martensite reaction are different in other alloy systems. For example, martensite can form in ironbase alloys that contain little or no carbon by a transformation of the FCC crystal structure to a BCC crystal structure. In certain high-manganese steels and stainless steels, the FCC structure changes to a HCP crystal structure during the martensite transformation. In addition, the martensite reaction may occur during the transformation of many polymorphic ceramic materials, including ZrO_2, and even in some crystalline polymers.

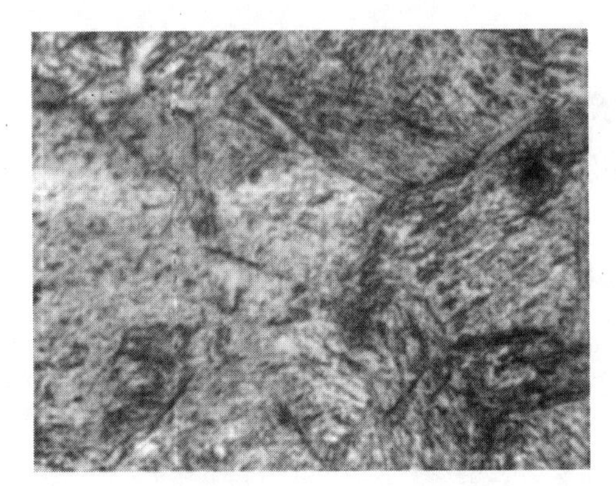

Figure 11.27
Tempered martensite in steel (× 500). (*From* Metals Handbook, *Vol. 9, 9th Ed., American Society for Metals*, 1985.)

The properties of martensite in other alloys are also different from the properties of steel martensite. In titanium alloys, the BCC titanium transforms to a HCP martensite structure during quenching. However, the titanium martensite is softer and weaker than the original structure.

The martensite that forms in other alloys can also be tempered. The martensite produced in titanium alloys can be reheated to permit the precipitation of a second phase. Unlike the case of steel, however, the tempering process *increases,* rather than decreases, the strength of the titanium alloy.

The Shape-Memory Effect A unique property possessed by some alloys that undergo the martensitic reaction is the shape-memory effect. A Ni-50% Ti alloy and several copper-base alloys can be given a sophisticated thermomechanical treatment to produce a martensitic structure. At the end of the treatment, the metal has been deformed to a predetermined shape. The metal can then be deformed into a second shape; but when the temperature is increased, the metal changes back to its original shape! Applications include actuating levers, orthodontal braces, blood clot filters, engines, and perhaps – eventually – artificial hearts.

EXAMPLE 11.10

At times, you need to join titanium tubing in the field. Devise a method for doing this quickly.

SOLUTION

Titanium is quite reactive and, unless special welding processes are used, may be contaminated. In the field, we may not have access to these processes. Therefore, we wish to make the joint without resorting to high-temperature processes.

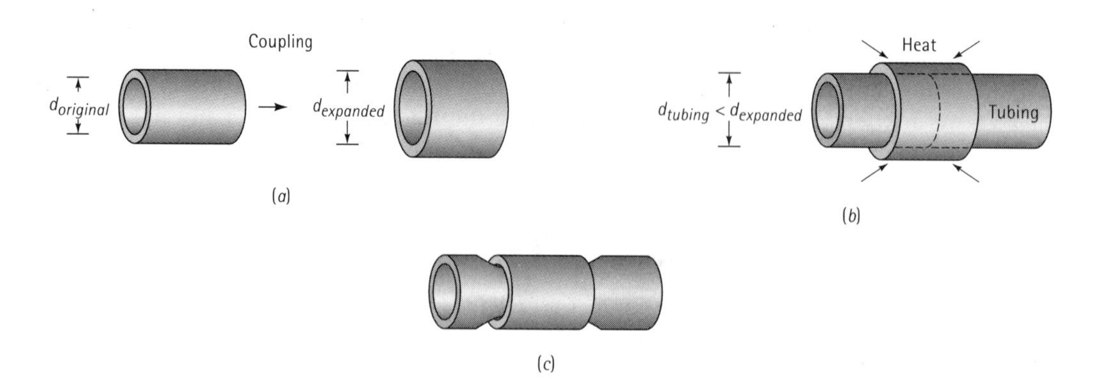

Figure 11.28
Use of memory alloys for coupling tubing: A memory alloy coupling is expanded (*a*) so it fits over the tubing (*b*). When the coupling is reheated, it shrinks back to its original diameter (*c*), squeezing the tubing for a tight fit (*for* Example 11.10).

We can take advantage of the shape-memory effect for this application (Figure 11.28). Ahead of time, we can set a Ni-Ti coupling into a small diameter, then deform it into a larger diameter. In the field, the coupling is slipped over the tubing and heated (at a low enough temperature so that the titanium tubing is not contaminated). The coupling contracts back to its predetermined shape, producing a strong mechanical bond to join the tubes.

SUMMARY

Solid-state phase transformations, which have a profound effect on the structure and properties of a material, can often be controlled by proper heat treatments. These heat treatments are designed to provide an optimum distribution of two or more phases in the microstructure. Dispersion strengthening permits a wide variety of structures and properties to be obtained.

- These transformations typically require both nucleation and growth of new phases from the original structure. The kinetics of the phase transformation help us understand the mechanisms that control the reaction and the rate at which the reaction occurs, enabling us to apply a suitable heat treatment to produce the desired microstructure. Reference to appropriate phase diagrams also helps us select the necessary compositions and temperatures.
- Age hardening, or precipitation hardening, is one powerful method for controlling the optimum dispersion strengthening in many metallic alloys. In age hardening, a very fine widely dispersed coherent precipitate is allowed to precipitate by a heat treatment that includes (a) solution treating to produce a single-phase solid solution, (b) quenching to retain that single phase, and (c) ageing to permit a precipitate to form. In order for age hardening to occur, the phase diagram must show decreasing solubility of the solute in the solvent as the temperature decreases.
- The eutectoid reaction can be controlled to permit one type of solid to transform to two different types of solid. The kinetics of the reaction depend on nucleation of the new solid phases and diffusion of the different atoms in the material to permit the growth of the new phases. The most widely used eutectoid reaction occurs in producing steels from iron-carbon alloys:
    - Either pearlite or bainite can be produced as a result of the eutectoid reaction in steel. In addition, primary ferrite or primary cementite may be present, depending on the carbon content of the alloy.
    - Factors that influence the mechanical properties of the microconstituent produced by the eutectoid reaction include (a) the composition of the alloy (amount of eutectoid microconstituent), (b) the grain size of the original solid, the eutectoid microconstituent, and any primary microconstituents (c) the fineness of the structure within the eutectoid microconstituent (interlamellar

spacing), (d) the cooling rate during the phase transformation, and (e) the temperature at which the transformation occurs (the amount of undercooling).

- A martensitic reaction occurs with no long-range diffusion. Again, the best known example occurs in steels:
    - The amount of martensite that forms depends on the temperature of the transformation (an athermal reaction).
    - Martensite in steels is very hard and brittle, with the hardness determined primarily by the carbon content.
    - The amount and composition of the martensite are the same as the austenite from which it forms.
- Martensite can be tempered. During tempering, a dispersion-strengthened structure is produced. In steels, tempering reduces the strength and hardness but improves the ductility and toughness of the microstructure.
- Since optimum properties are obtained through heat treatment, we must remember that the structure and properties may change when the material is used at or exposed to elevated temperatures. Overageing or overtempering can occur as a natural extension of the phenomena governing these transformations when the material is placed into service.

GLOSSARY

Age hardening
A special dispersion-strengthening heat treatment. By solution treatment, quenching, and ageing, a coherent precipitate forms that provides a substantial strengthening effect. Also known as precipitation hardening.

Artificial ageing
Reheating a solution-treated and quenched alloy to a temperature below the solvus in order to provide the thermal energy required for a precipitate to form.

Athermal transformation
When the amount of the transformation depends only on the temperature, not on the time.

Austenite
The name given to the FCC crystal structure of iron.

Avrami relationship
Describes the fraction of a transformation that occurs as a function of time.

Bainite
A two-phase microconstituent, containing ferrite and cementite, that forms in steels that are isothermally transformed at relatively low temperatures.

Cementite
The hard, brittle intermetallic compound Fe_3C that, when properly dispersed, provides the strengthening in steels.

Coherent precipitate
A precipitate whose crystal structure and atomic arrangement have a continuous relationship with the matrix from which the precipitate formed. The coherent precipitate provides excellent disruption of the atomic arrangement in the matrix and provides excellent strengthening.

Dihedral angle
The angle that defines the shape of a precipitate particle in the matrix. The dihedral angle is determined by the relative surface energies.

Ferrite
The name given to the BCC crystal structure of iron.

Guinier-Preston zones
Tiny clusters of atoms that precipitate from the matrix in the early stages of the age-hardening process. Although the GP zones are coherent with the matrix, they are too small to provide optimum strengthening.

Interfacial energy
The energy associated with the boundary between two phases.

Isothermal transformation
When the amount of a transformation at a particular temperature depends on the time permitted for the transformation.

Martensite
A metastable phase formed in steel and other materials by a diffusionless, athermal transformation.

Natural ageing
When a coherent precipitate forms from a solution-treated and quenched age hardenable alloy at room temperature, providing optimum strengthening.

Pearlite
A two-phase lamellar microconstituent, containing ferrite and cementite, that forms in steels cooled in a normal fashion or isothermally transformed at relatively high temperatures.

Shape-memory effect
The ability of certain materials to develop microstructures that, after being deformed, can return the material to its initial shape when heated.

Solution treatment
The first step in the age-hardening heat treatment. The alloy is heated above the solvus temperature to dissolve any second phase and to produce a homogeneous single-phase structure.

Strain energy
The energy required to permit a precipitate to fit into the surrounding matrix during nucleation and growth of the precipitate.

Supersaturated solid solution
The solid solution formed when a material is rapidly cooled from a high-temperature single-phase region to a low-temperature two-phase region without the second phase precipitating. Because the quenched phase contains more alloying elements than the solubility limit, it is supersaturated in that element.

Tempering
A low-temperature heat treatment used to reduce the hardness of martensite in steels by permitting the martensite to begin to decompose to the equilibrium phases.

TTT diagram
The time-temperature-transformation diagram describes the time required at any temperature for a phase transformation to begin and end. The TTT diagram assumes that the temperature is constant during the transformation.

Widmanstatten structure
The precipitation of a second phase from the matrix when there is a fixed crystallographic relationship between the precipitate and matrix crystal structures. Often needlelike or platelike structures form in the Widmanstatten structure.

PROBLEMS

11.1 Determine the constants c and n in Equation 11.2 that describe the rate of crystallisation of polypropylene at 140°C. (*See* Figure 11.29.)

11.2 Determine the constants c and n in Equation 11.2 that describe the rate of recrystallisation of copper at 135°C. (*See* Figure 11.2.)

11.3 Determine the activation energy for crystallisation of polypropylene, using the curves in Figure 11.29.

11.4 (a) Recommend an artificial age-hardening heat treatment for a Cu-1.2% Be alloy. (*see* Figure 13.10.) Include appropriate temperatures.

(b) Compare the amount of the γ_2 precipitate that forms by artificial ageing at 400°C with the amount of the precipitate that forms by natural ageing.

11.5 Suppose that age hardening is possible in the Al-Mg system. (*See* Figure 13.3.)

(a) Recommend an artificial age-hardening heat treatment for each of the following alloys and

(b) compare the amount of the β precipitate that forms from your treatment of each alloy.

(c) Testing of the alloys after the heat treatment reveals that little strengthening occurs as a result of the heat treatment. Which of the requirements for age hardening is likely not satisfied?

(a) Al-4% Mg (b) Al-6% Mg (c) Al-12% Mg

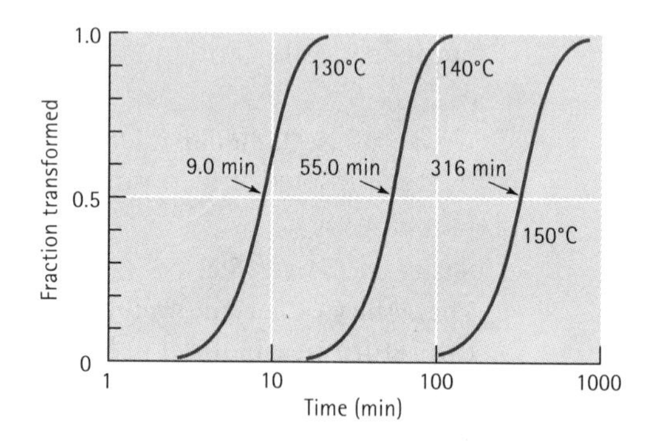

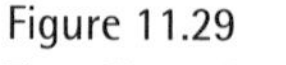
Figure 11.29
The effect of the temperature on the crystallisation of polypropylene (*for* Problems 11.1 *and* 11.3).

11.6 An Al-2.5% Cu alloy is solution-treated, quenched, and overaged at 230°C to produce a stable microstructure. If the spheroidal θ precipitates that form have a diameter of 9 000 nm and a density of 4.26 $Mg.m^{-3}$, determine the number of precipitate particles per mm^3.

11.7 Figure 11.30 shows a hypothetical phase diagram. Determine whether each of the following alloys might be good candidates for age hardening, and explain your answer. For those alloys that might be good candidates, describe the heat treatment required, including recommended temperatures.

(a) A-10% B (b) A-20% B

(c) A-55% B (d) A-87% B

(e) A-95% B

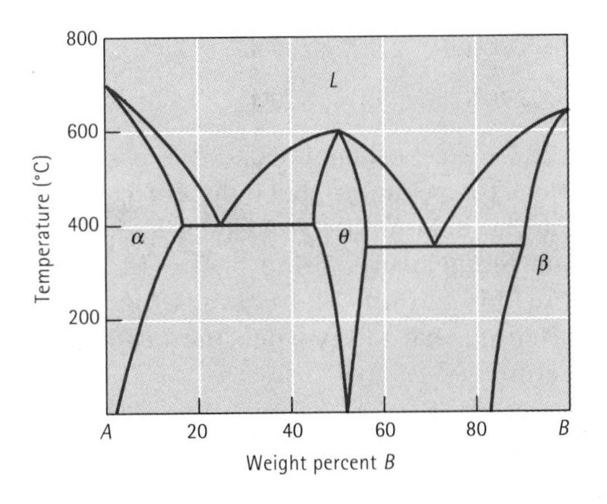

Figure 11.30
Hypothetical phase diagram (*for* Problem 11.7).

11.8 Figure 11.1 shows the sigmoidal curve for the transformation of austenite. Determine the constants c and n in Equation 11.2 for this reaction. By comparing this figure with the TTT diagram (Figure 11.19), estimate the temperature at which this transformation occurred.

11.9 For an Fe-0.35% C alloy, determine

(a) the temperature at which austenite first begins to transform on cooling,

(b) the primary microconstituent that forms,

(c) the composition and amount of each phase present at 728°C,

(d) the composition and amount of each phase present at 726°C, and

(e) the composition and amount of each microconstituent present at 726°C.

11.10 For an Fe-1.15% Cu alloy, determine

(a) the temperature at which austenite first begins to transform on cooling,

(b) the primary microconstituent that forms,

(c) the composition and amount of each phase present at 728°C,

(d) the composition and amount of each phase present at 726°C, and

(e) the composition and amount of each microconstituent present at 726°C.

11.11 A steel contains 8% cementite and 92% ferrite at room temperature. Estimate the carbon content of the steel. Is the steel hypoeutectoid or hypereutectoid?

11.12 A steel contains 18% cementite and 82% ferrite at room temperature. Estimate the carbon content of the steel. Is the steel hypoeutectoid or hypereutectoid?

11.13 A steel contains 18% pearlite and 82% primary ferrite at room temperature. Estimate the carbon content of the steel. Is the steel hypoeutectoid or hypereutectoid?

11.14 A steel contains 94% pearlite and 6% primary cementite at room temperature. Estimate the carbon content of the steel. Is the steel hypoeutectoid or hypereutectoid?

11.15 A steel contains 55% α and 45% γ at 750°C. Estimate the carbon content of the steel.

11.16 A steel contains 96% γ and 4% Fe_3C at 800°C. Estimate the carbon content of the steel.

11.17 A steel is heated until 40% austenite, with a carbon content of 0.5% forms. Estimate the temperature and the overall carbon content of the steel.

11.18 A steel is heated until 85% austenite, with a carbon content of 1.05%, forms. Estimate the temperature and the overall carbon content of the steel.

11.19 Determine the eutectoid temperature, the composition of each phase in the eutectoid reaction, and the amount of each phase present in the eutectoid microconstituent for the following systems. Comment on whether you expect the eutectoid microconstituent to be ductile or brittle.

(a) ZrO_2-CaO (*see* Figure 14.23)

(b) Cu-Al at 11.8% Al (*see* Figure 13.10)

(c) Cu-Zn at 47% Zn (*see* Figure 13.10)

(d) Cu-Be (*see* Figure 13.10)

11.20 Compare the interlamellar spacing and the yield strength when a eutectoid steel is isothermally transformed to pearlite at

(a) 700°C and

(b) 600°C.

11.21 An isothermally transformed eutectoid steel is found to have a yield strength of 410 $MN.m^{-2}$. Estimate

(a) the transformation temperature and

(b) the interlamellar spacing in the pearlite.

11.22 Determine the required transformation temperature and microconstituent if a eutectoid steel is to have the following hardnesses:

(a) HRC 38 (b) HRC 42

(c) HRC 48 (d) HRC 52

11.23 Describe the hardness and microstructure in a eutectoid steel that has been heated to 800°C for 1 h, quenched to 350°C and held for 750 s, and finally quenched to room temperature.

11.24 Describe the hardness and microstructure in a eutectoid steel that has been heated to 800°C, quenched to 650°C and held for 500 s, and finally quenched to room temperature.

11.25 Describe the hardness and microstructure in a eutectoid steel that has been heated to 800°C, quenched to 300°C and held for 10 s, and finally quenched to room temperature.

11.26 Describe the hardness and microstructure in a eutectoid steel that has been heated to 800°C, quenched to 300°C and held for 10 s, quenched to room temperature, and then reheated to 400°C before finally cooling to room temperature again.

11.27 A steel containing 0.3% C is heated to various temperatures above the eutectoid temperature, held for 1 h and then quenched to room temperature. Using Figure 12.2, determine the amount, composition, and hardness of any martensite that forms when the heating temperature is:

(a) 728°C (b) 750°C

(c) 790°C (d) 850°C

11.28 A steel containing 0.95% C is heated to various temperatures above the eutectoid temperature, held for 1 h, and then quenched to room temperature. Using Figure 12.2 determine the amount and composition of any martensite that forms when the heating temperature is:

(a) 728°C (b) 750°C

(c) 780°C (d) 850°C

11.29 A steel microstructure contains 75% martensite and 25% ferrite; the composition of the martensite is 0.6% C. Using Figure 12.2, determine

(a) the temperature from which the steel was quenched and

(b) the carbon content of the steel.

11.30 A steel microstructure contains 92% martensite and 8% Fe_3C; the composition of the martensite is 1.10% C. Using Figure 12.2, determine

(a) the temperature from which the steel was quenched and

(b) the carbon content of the steel.

11.31 A steel containing 0.8% C is quenched to produce all martensite. Estimate the volume change that occurs, assuming that the lattice parameter of the austenite is 0.36 nm. Does the steel expand or contract during quenching?

11.32 Describe the complete heat treatment required to produce a quenched and tempered eutectoid steel having a tensile strength of at least 860 $MN.m^{-2}$. Include appropriate temperatures.

11.33 Describe the complete heat treatment required to produce a quenched and tempered eutectoid steel having a HRC hardness of less than 50. Include appropriate temperatures.

11.34 In eutectic alloys, the eutectic microconstituent is generally the continuous one, but in the eutectoid structures, the primary microconstituent is normally continuous. By describing the changes that occur with decreasing temperature in each reaction, explain why this difference is expected.

11.35 Describe how the memory metals might be useful as plates to be surgically placed around broken bones to provide more rapid healing.

11.36 You wish to attach aluminium sheet to the frame on the 24th floor of a building. You plan to use rivets made of an age-hardenable aluminium, but the rivets must be soft and ductile in order to close. After the sheets are attached, the rivets must be very strong. Describe a method for producing, using, and strengthening the rivets.

11.37 Describe a process to produce a polypropylene polymer with a structure that is 75% crystalline. Figure 11.29 will provide appropriate data.

11.38 An age-hardened Al-Cu bracket is used to hold a heavy electrical sensing device on the outside of a steel-making furnace. Temperatures may exceed 200°C. Is this a good choice? Explain. If it is not, suggest an appropriate material for the bracket and explain why your choice is acceptable.

11.39 You use an arc-welding process to join a eutectoid steel. Cooling rates may be very high following the joining process. Describe what happens in the heat-affected area of the weld and discuss the problems that might occur. Suggest a joining process that may minimise these problems.

The mechanical properties of each can be predicted and controlled by understanding the atomic bonding, atomic arrangement, and strengthening mechanisms discussed in the previous sections. This fact is particularly evident in Chapters 12 and 13, in which the ideas of solid solution strengthening, strain hardening, and dispersion strengthening are applied to ferrous and nonferrous alloys.

The discussion of ceramics and polymers in Chapters 14 and 15 emphasises the importance of atomic bonding and atomic arrangement.

The mechanical properties of ceramics and polymers are explained in these chapters by mechanisms that do not involve dislocation movement.

Composite materials are even more difficult to categorise because of the many types and intended uses of the materials, as pointed out in Chapter 16. Many composites, are designed to provide special characteristics that go beyond conventional methods for controlling the structure-property relationship. Construction materials such as wood and concrete, described in Chapter 17, are special types of 'composite' materials.

Frequently we find that complex parts and structures are composed of materials from several or even all of these groups. Each group has its own unique set of properties that best suit the individual application.

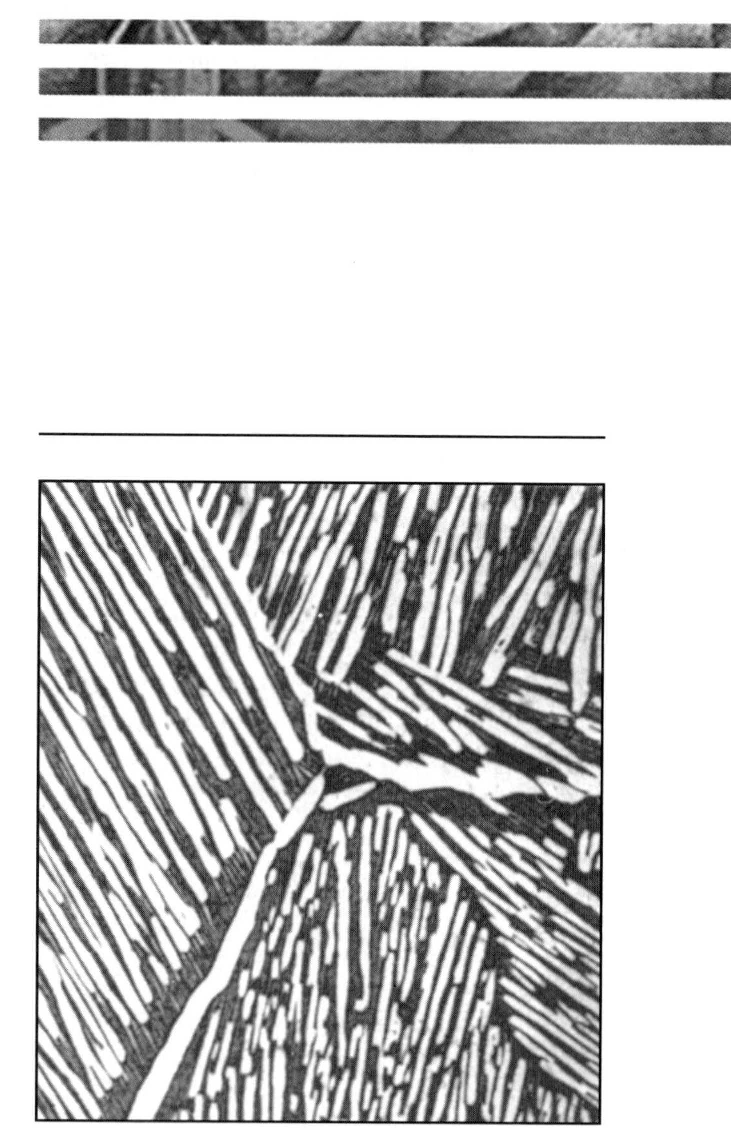

Appropriate design of heat treatments enables the engineer to control the microstructure and mechanical properties of metal alloys. In this example, the polymorphic behaviour of a nonferrous titanium alloy permits α plates of titanium to form in a matrix of β titanium. The plate-like structure interferes with the growth of cracks, therefore improving the fracture toughness of the alloy. (*From* Metals Handbook, *Vol. 2, 10th Ed., ASM International, 1990.*)

Part

Engineering Materials

CHAPTER 12

Ferrous Alloys

12.1 Introduction

Ferrous alloys, which are based on iron-carbon alloys, include plain-carbon steels, alloy and tool steels, stainless steels, and cast irons. Steels are typically produced in two ways: by refining iron ore or by recycling scrap steel (Figure 12.1).

In producing primary steel, iron ore (iron oxide) is heated in a *blast furnace* in the presence of coke (carbon) and oxygen. The carbon reduces the iron oxide to liquid pig iron, with carbon monoxide and carbon dioxide produced as by-products. Limestone, added to help remove impurities, melts and produces a liquid slag. Because the liquid pig iron contains very large amounts of carbon, oxygen is blown into the *basic oxygen furnace* to eliminate the excess carbon and produce liquid steel.

Steel is also produced by recycling steel scrap. The scrap is often melted in an *arc furnace*, in which the heat of the arc melts the scrap. Many alloys and special steels are also produced using electric melting.

Liquid steel is sometimes directly poured into moulds to produce finished steel castings; it is also allowed to solidify into shapes that are later processed by a metal-forming technique such as rolling or forging. In the latter case, the steel is either poured into large ingot moulds or is continuously cast into regular shapes (as described in Figure 8.18).

All of the strengthening mechanisms apply to at least some of the ferrous alloys. In this chapter we discuss how to use the eutectoid reaction to control the structure and properties of steels through heat treatment and alloying. We also examine two special classes of ferrous alloys: stainless steels and cast irons.

12.2 Designations for Steels

The Fe-Fe_3C phase diagram provides the basis for understanding the treatment and properties of steels. The phase diagram, phases, and microconstituents in steels were discussed in Chapter 11. The dividing point between steels and cast irons is 2.11% C (Figure 11.13), where the eutectic reaction becomes possible. For steels, we concentrate on the eutectoid portion of the diagram (Figure 12.2), in which the solubility lines and the eutectoid isotherm are specially identified. The A_3 shows the temperature at which ferrite starts to form on cooling; the A_{cm} shows the temperature at which cementite starts to form; and the A_1 is the eutectoid temperature.

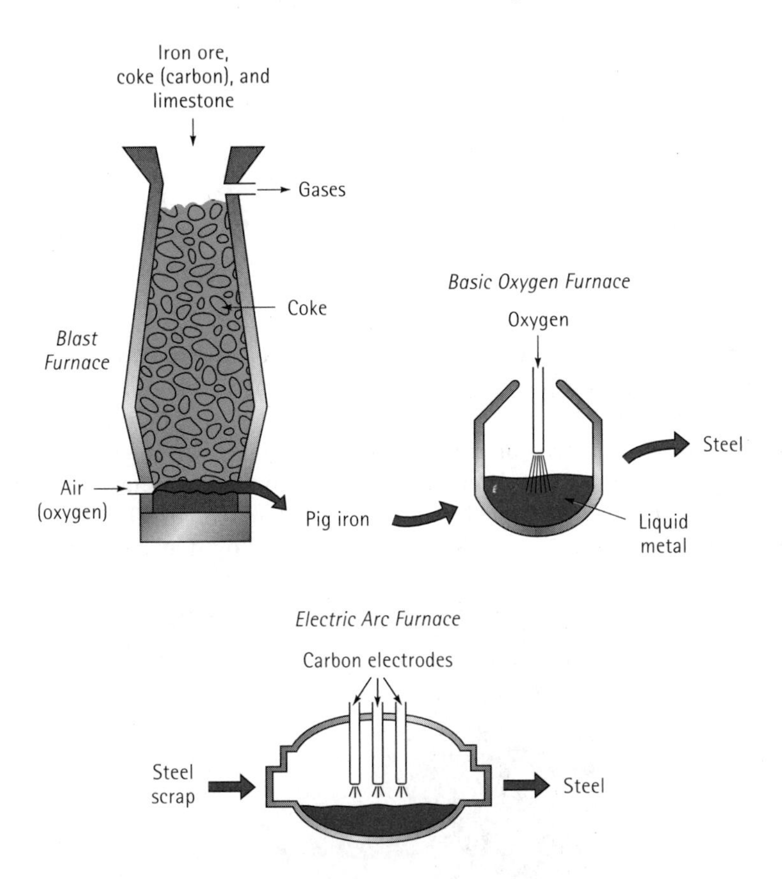

Figure 12.1
In a blast furnace, iron ore is reduced using coke (carbon) and air to produce liquid pig iron. The high carbon content in the pig iron is reduced by introducing oxygen into the basic oxygen furnace to produce liquid steel. An electric arc furnace can be used to produce liquid steel by melting scrap.

Almost all of the heat treatments of a steel are directed toward producing the mixture of ferrite and cementite that gives the proper combination of properties. Figure 12.3 shows the three important microconstituents, or arrangements of ferrite and cementite, that are usually sought. Pearlite is a lamellar mixture of ferrite and cementite. In bainite, which is obtained by transformation of austenite at a large undercooling, the cementite is more rounded than in pearlite. Tempered martensite, a mixture of very fine and nearly round cementite in ferrite, forms when martensite is reheated following its formation by quenching.

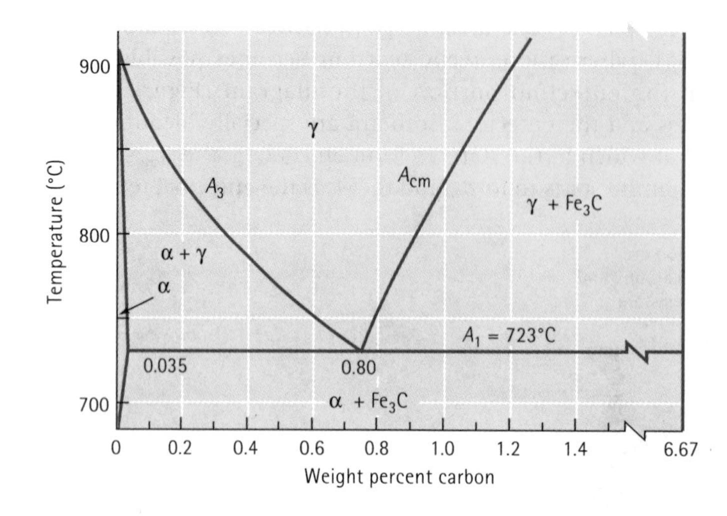

Figure 12.2
The eutectoid portion of the Fe-Fe_3C phase diagram.

Designations The AISI (American Iron and Steel Institute) and SAE (Society Of Automotive Engineers) provide designation systems (Table 12.1) that use a four- or five-digit number. The first two numbers refer to the major alloying elements present, and the last two or three numbers refer to the major alloying elements present, and the last two or three numbers refer to the percentage of carbon. An AISI 1040 steel is a plain-carbon steel with 0.40% C. An SAE 10120 steel is a plain-carbon steel containing 1.20% C. An AISI 4340 steel is an alloy steel containing 0.40% C.

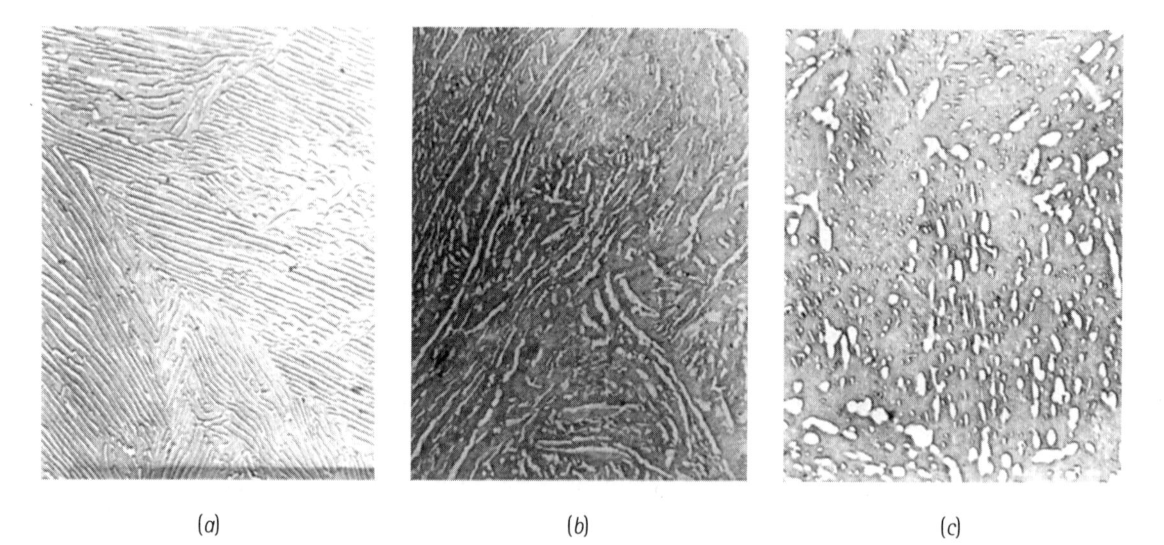

(*a*) (*b*) (*c*)

Figure 12.3
Electron micrographs of (*a*) pearlite, (*b*) bainite, and (*c*) tempered martensite, illustrating the differences in cementite size and shape among these three microconstituents (×7 500). (*From* The Making, Shaping, and Treating of Steel, *10th Ed., Courtesy Association of Iron and Steel Engineers.*)

Table 12.1 Compositions of selected BS and AISI-SAE steels.

AISI-SAE Number	Typical BS Grade	% C	% Mn	%Si	%Ni	%Cr	Others
1020	040A20	0.18–0.23	0.30–0.60				
1040	080A40	0.37–0.44	0.60–0.90				
1060	080A62	0.55–0.65	0.60–0.90				
1080	070A78	0.75–0.88	0.60–0.90				
1095	060A96	0.90–1.03	0.30–0.50				
1140	212M44	0.37–0.44	0.70–1.00				0.08–0.13% S
4140	708M40	0.38–0.43	0.75–1.00	0.15–0.30		0.80–1.10	0.15–0.25% Mo
4340	817M40	0.38–0.43	0.60–0.80	0.15–0.30	1.65–2.00	0.70–0.90	0.20–0.30% Mo
4620	665H20	0.17–0.22	0.45–0.65	0.15–0.30	1.65–2.00		0.20–0.30% Mo
52100	534A99	0.98–1.10	0.25–0.45	0.15–0.30		1.30–1.60	
8620	805H20	0.18–0.23	0.70–0.90	0.15–0.30	0.40–0.70	0.40–0.60	0.15–0.25% V
9260	250A58	0.56–0.64	0.75–1.00	1.80–2.20			

The British Standards Institute (BSI) also provide a specification system for wrought steels (Table 12.1) which is comparable to the AISI system. BS970 provides a six-digit code for each grade of steel. The first three digits denote the family of steels to which the alloy belongs (000 – 199 for plain carbon steels, 300 – 499 for stainless and heat resisting steels, etc.). The fourth character shows whether the steel is being ordered for its chemical composition limits (A), mechanical property limits (M), hardenability limits (H) or as a stainless steel (S). The final two digits represent 100 × the carbon content of the steel (so 40 denotes 0.40% C content).

EXAMPLE 12.1 **Selection of a Method to Determine AISI Number**

An unalloyed steel tool used for machining aluminium automobile wheels has been found to work well, but the purchase records have been lost and you do not know the steel's composition. The microstructure of the steel is tempered martensite, and you cannot estimate the composition of the steel from the structure. Select a treatment that may help determine the steel's carbon content.

SOLUTION

You have no access to equipment that would permit you to analyse the chemical composition directly. Since the entire structure of the steel is a very fine tempered martensite, you can do a simple heat treatment to produce a structure that can be analysed more easily. This can be done in two different ways.

The first way is to heat the steel to a temperature just below the A_1 temperature and hold for a long time. The steel overtempers and large Fe_3C spheres form in a ferrite matrix. We then estimate the amount of ferrite and cementite and calculate the carbon content using the lever law. If we measure 16% Fe_3C using this method, the carbon content is:

$$\%\ Fe_3C = \frac{x - 0.035}{6.67 - 0.035} \times 100 = 16 \quad \text{or} \quad x = 1.1\%\ C$$

A better approach, however, is to heat the steel above the A_{cm} to produce all austenite. If the steel then cools slowly, it transforms to pearlite and a primary microconstituent. If, when we do this, we estimate that the structure contains 95% pearlite and 5% primary Fe_3C then:

$$\%\ \text{Pearlite} = \frac{6.67 - x}{6.67 - 0.8} \times 100 = 95 \quad \text{or} \quad x = 1.09\%\ C$$

The carbon content is on the order of 1.09 to 1.1%, consistent with a 10110 AISI-SAE steel.

In this procedure, we assume that the weight and volume percentages of the microconstituents are the same, which is nearly the case in steels.

12.3 Simple Heat Treatments

Four simple heat treatments – process annealing, annealing, normalising, and spheroidising – are commonly used for steels (Figure 12.4). These heat treatments are used to accomplish one of three purposes: (1) eliminating cold work, (2) controlling dispersion strengthening, or (3) improving machinability.

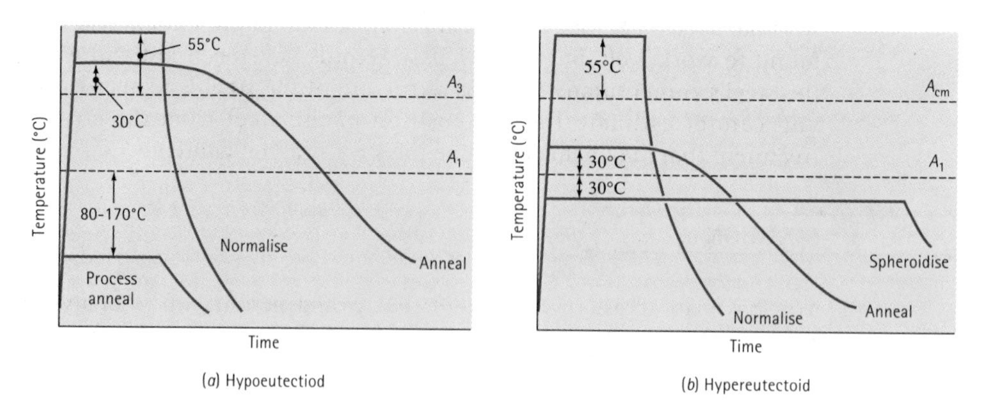

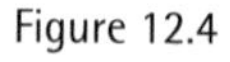

Figure 12.4
Schematic summary of the simple heat treatments for (*a*) hypoeutectoid steels and (*b*) hypereutectoid steels.

Process Annealing-Eliminating Cold Work The recrystallisation heat treatment used to eliminate the effect of cold working in steels with less than about 0.25% C is called a process anneal. The process anneal is done 80°C to 170°C below the A_1 temperature. This is above the recrystallisation temperature of ferrite.

Annealing and Normalising – Dispersion Strengthening Steels can be dispersion-strengthened by controlling the fineness of pearlite. The steel is initially heated to produce homogeneous austenite, a step called austenitising Annealing, or a full anneal, allows the steel to cool slowly in a furnace, producing coarse pearlite. Normalising allows the steel to cool more rapidly, in air, producing fine pearlite. Figure 12.5 shows the typical properties obtained by annealing and normalising plain-carbon steels.

For annealing, austenitising of hypoeutectoid steels is done at about 30°C above the A_3, producing 100% γ. However, austenitising of a hypereutectoid steel is done at about 30°C above the A_1, producing austenite and Fe_3C; this process prevents the formation of a brittle, continuous film of Fe_3C at the grain boundaries that occurs on slow cooling from the 100% γ region. In both cases, the slow furnace cool and coarse pearlite provide relatively low strength and good ductility.

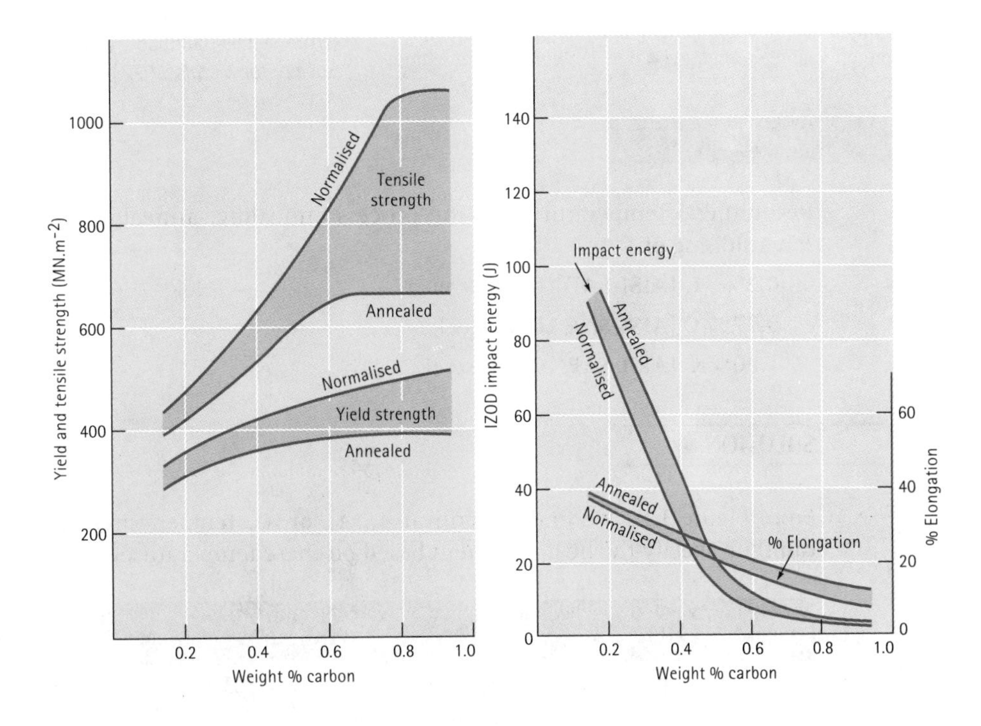

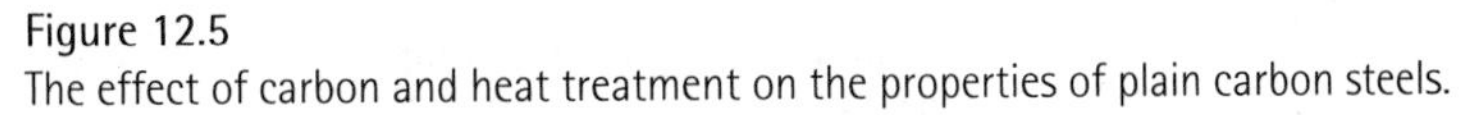
Figure 12.5
The effect of carbon and heat treatment on the properties of plain carbon steels.

For normalising, austenitising is done at about 55°C above the A_3 or A_{cm}; the steel is then removed form the furnace and cooled in air. The faster cooling gives fine pearlite and provides higher strength.

Spheroidising-Improving Machinability High-carbon steels, which contain a large amount of Fe_3C, have poor machining characteristics. During the *spheroidising* treatment, which requires several hours at about 30°c below the A_1, the Fe_3C changes shape into large, spherical particles in order to reduce boundary area. The microstructure, known as spheroidised carbide, has a continuous matrix of soft, machinable ferrite (Figure 12.6). After machining, the steel is given a more sophisticated heat treatment to produce the required properties. A similar structure occurs when martensite is tempered just below the A_1 for long periods of time.

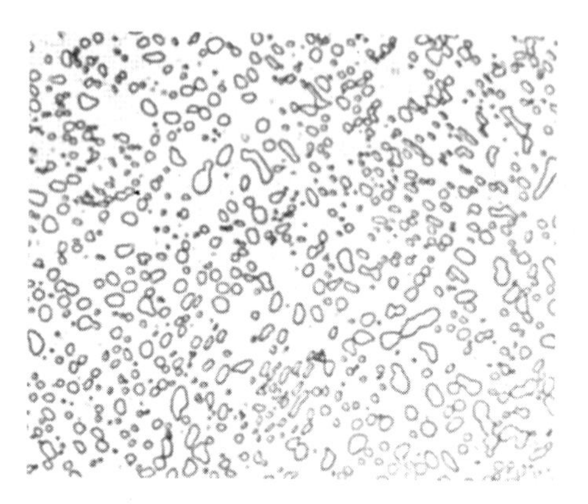

Figure 12.6
The microstructure of spheroidised carbide, with Fe_3C particles dispersed in a ferrite matrix (× 850). (*From* Metals Handbook, *Vol. 7, 8th Ed., American Society for Metals, 1972.*)

EXAMPLE 12.2

Recommend temperatures for the process annealing, annealing, normalising, and spheroidising of:

0.2% C [AISI-SAE 1020] steel;

0.77% C [AISI-SAE 1077] steel;

1.20% C [AISI-SAE 10120] steel.

SOLUTION

From Figure 12.2 we find the critical A_1, A_3, or A_{cm} temperatures for each steel. We can then specify the heat treatment based on these temperatures.

	1020	1077	10120
Critical temperatures	A_1 = 723°C A_3 = 830°C	A_1 = 723°C	A_1 = 723°C A_{cm} = 895°C
Process annealing	723 – (80 to 170) = 553°C to 643°C	Not done	Not done
Annealing	830 + 30 = 860°C	723 + 30 = 753°C	723 + 30 = 753°C
Normalising	830 + 55 = 885°C	723 + 55 = 778°C	895 + 55 = 950°C
Spheroidising	Not done	723 – 30 = 693°C	723 – 30 = 693°C

12.4 Isothermal Heat Treatments

The effect of transformation temperature on the properties of a eutectoid steel, e.g. AISI-SAE 1080 or B.S. 070A78, was discussed in Chapter 11. As the isothermal transformation temperature decreases, pearlite becomes progressively finer before bainite begins to form instead. At very low temperatures, martensite is obtained.

Austempering and Isothermal Annealing The isothermal transformation heat treatment used to produce bainite, called austempering, simply involves austenitising the steel, quenching to some temperature below the nose of the TTT curve, and holding at that temperature until all of the austenite transforms to bainite (Figure 12.7).

Annealing and normalising are usually used to control the fineness of pearlite. However, pearlite formed by an isothermal anneal (Figure 12.7) may give more uniform properties, since the cooling rates and microstructure obtained during continuous cooling, annealing, and normalising vary across the cross section of the steel.

Effect of Carbon on the TTT Diagram In either a hypoeutectoid or a hypereutectoid steel, the TTT diagram must reflect the possible formation of a primary phase. The isothermal transformation diagrams for a 0.5% C and a 1.1% C steel are shown in Figure 12.8. The most remarkable change is the presence of a 'wing' which begins at the nose of the curve and becomes asymptotic to the A_3 or A_{cm} temperature. The wing represents the ferrite start (F_s) time in hypoeutectoid steels or the cementite start (C_s) time hypereutectoid steels.

When a 0.5% C steel is austenitised, quenched, and held between the A_1 and the A_3, primary ferrite nucleates and grows; eventually an equilibrium amount of ferrite and austenite result. Similarly, primary cementite nucleates and grows to its equilibrium amount in a 1.1% C steel held between the A_{cm} and A_1 temperatures.

If an austenitised 0.5% C steel is quenched to a temperature between the nose and the A_1 temperatures, primary ferrite again nucleates and grows until reaching the equilibrium amount. The remainder of the austenite then transforms to pearlite. A similar situation, producing primary cementite and pearlite, is found for the hypereutectoid steel.

If we quench below the nose of the curve, only bainite forms, regardless of the carbon content of the steel.

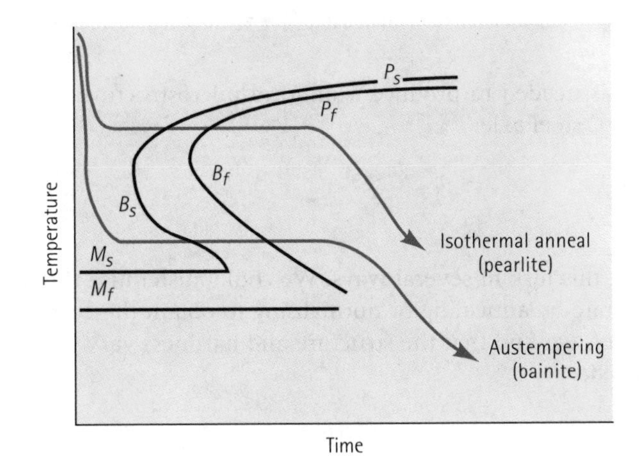

Figure 12.7
The austempering and isothermal anneal heat treatments in a eutectoid steel (0.8% C steel e.g. AISE-SAE 1080).

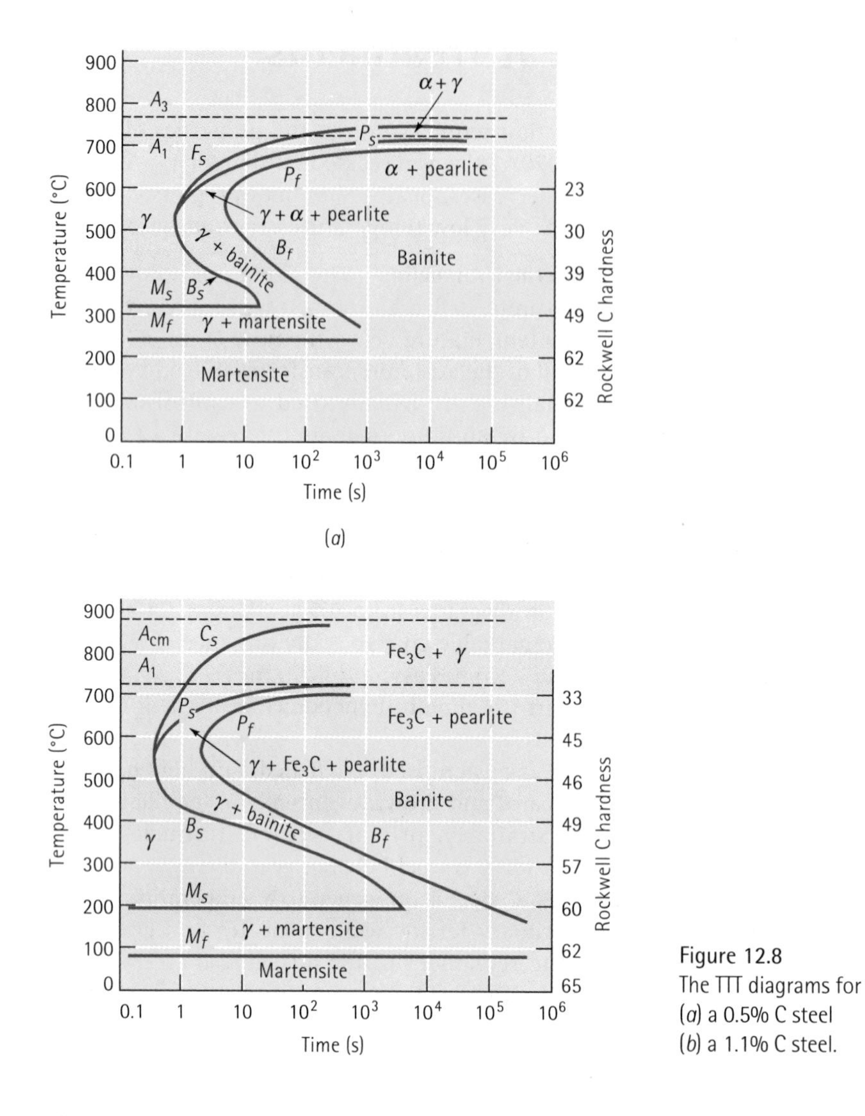

Figure 12.8
The TTT diagrams for
(*a*) a 0.5% C steel
(*b*) a 1.1% C steel.

EXAMPLE 12.3 Specification of a Heat Treatment for an Axle

A heat treatment is needed to produce a uniform microstructure and a hardness of HRC 23 in a 0.5% C steel axle.

SOLUTION

We might attempt this task in several ways. We could austenitise the steel, then cool at an appropriate rate by annealing or normalising to obtain the correct hardness. By doing this, however, we find that the structure and hardness vary from the surface to the centre of the axle.

A better approach is to use an isothermal heat treatment. From Figure 12.8, we find that a hardness of HRC 23 is obtained transforming austenite to a mixture of ferrite and pearlite at 600°C. From Figure 12.2, we find that the A_3 temperature is 770°C. Therefore, our heat treatment is:

1. Austenitise the steel at 770 + (30 to 55) = 805°C to 825°C, holding for, perhaps, 1 h and obtaining 100% γ.
2. Quench the steel to 600°C and hold for a minimum of 10 s. Primary ferrite begins to precipitate from the unstable austenite after about 1.0 s. After 1.5 s, pearlite begins to grow, and the austenite is completely transformed to ferrite and pearlite after about 10 s. After this treatment, the microconstituents present are:

   from Figure 12.2

$$\text{Primary } \alpha = \frac{0.8-0.5}{0.8-0.035} \times 100 = 39\%$$

$$\text{Pearlite} = \frac{0.5-0.035}{0.8-0.035} \times 100 = 61\%$$

3. Cool in air to room temperature, preserving the equilibrium amounts of primary ferrite and pearlite. The microstructure and hardness are uniform because of the isothermal anneal.

Interrupting the Isothermal Transformation Complicated microstructures are produced by interrupting the isothermal heat treatment. For example, we could austenitise the 0.5% C steel (Figure 12.9) at 800°C, quench to 650°C and hold for 10 s (permitting some ferrite and pearlite to form), then quench to 350°C and hold for 1 h (3 600 s). Whatever unstable austenite remained before quenching to 350°C transforms to bainite. The final structure is ferrite, pearlite, and bainite.

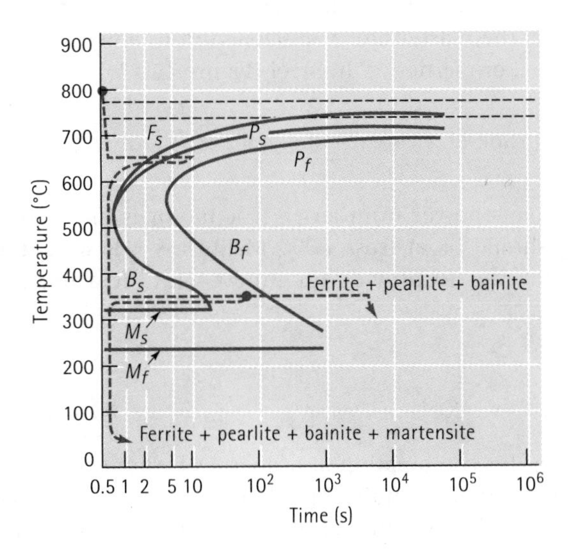

Figure 12.9
Producing complicated structures by interrupting the isothermal heat treatment of a 0.5% C steel (AISI-SAE 1050).

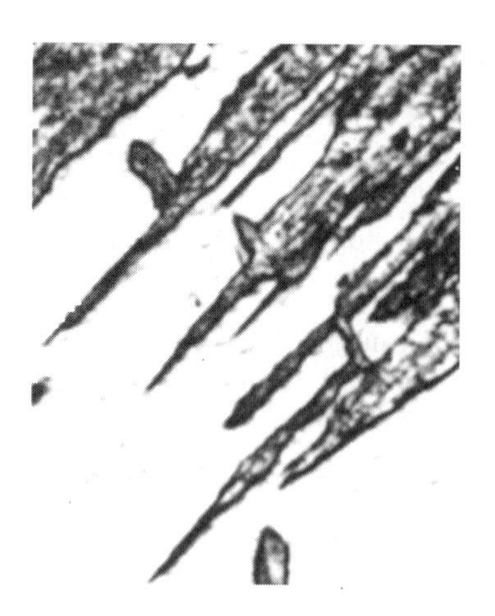

Figure 12.10
Dark feathers of bainite surrounded by light martensite, obtained by interrupting the isothermal transformation process (× 1 500). (*From* Metals Handbook, *Vol. 9, 9th Ed., American Society for Metals, 1985.*)

We could complicate the treatment further by interrupting the treatment at 350°C after 1 min (60 s) and quenching. Any austenite remaining after 1 min at 350°C forms martensite. The final structure now contains ferrite, pearlite, bainite, and martensite. Note that each time we change the temperature, we start at zero time!

Figure 12.10 shows the structure obtained by interrupting the transformation to bainite of a 0.5% C steel by quenching the remaining austenite to martensite. Because such complicated mixtures of microconstituents produce unpredictable properties, these structures are seldom produced intentionally.

12.5 Quench and Temper Heat Treatments

We can obtain an even finer dispersion of Fe_3C if we first quench the austenite to produce martensite, then temper. During tempering, an intimate mixture of ferrite and cementite forms from the martensite, as discussed in Chapter 11. The tempering treatment controls the final properties of the steel (Figure 12.11).

EXAMPLE 12.4 Outline of a Quench and Temper Treatment

A rotating shaft that delivers power from an electric motor is made from a 0.5% C steel. Its yield strength should be at least 1 000 $MN.m^{-2}$, yet it should also have at least 15% elongation in order to provide toughness. Outline a heat treatment to produce this part.

SOLUTION

We are not able to obtain this combination of properties by annealing or normalising (Figure 12.5). However a quench and temper heat treatment produces a microstructure that can provide both strength and toughness. Figure 12.11 shows that the

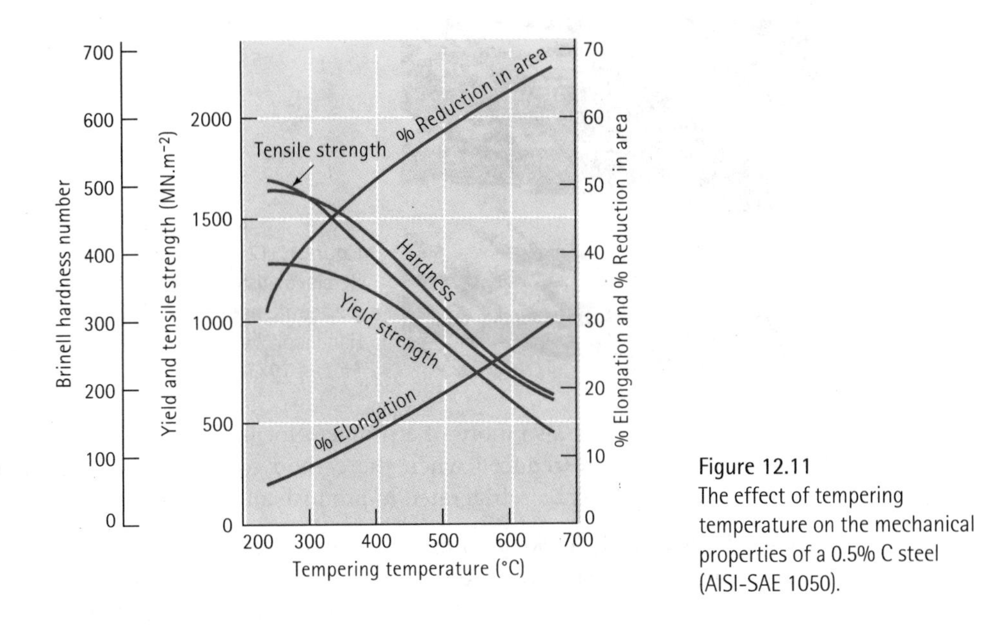

Figure 12.11
The effect of tempering temperature on the mechanical properties of a 0.5% C steel (AISI-SAE 1050).

yield strength exceeds 1 000 Mn.m^{-2} if the steel is tempered below 460°C, whereas the elongation exceeds 15% if tempering is done above 425°C. The A_3 temperature for the steel is 770°C. A possible heat treatment is:

1. Austenitise above the A_3 temperature of 770°C for 1 h. An appropriate temperature may be 770 + 55 = 825°C.
2. Quench rapidly to room temperature. Since the M_f is about 250°C, martensite will form.
3. Temper by heating the steel to 440°C. Normally, 1 h will be sufficient if the steel is not too thick.
4. Cool to room temperature.

Retained Austenite There is a large volume expansion when martensite forms from austenite. As the martensite plates form during quenching, they surround and isolate small pools of austenite (Figure 12.12), which deform to accommodate the lower-density martensite. However, for the remaining pools of austenite to transform, the surrounding martensite must deform. Because the strong martensite resists the transformation, either the existing martensite cracks or the austenite remains trapped in the structure as retained austenite

Retained austenite can be a serious problem. Martensite softens and becomes more ductile during tempering. After tempering, the retained austenite cools below the M_s and M_f temperatures and transforms to martensite, since the surrounding tempered martensite can deform. But now the steel contains more hard, brittle martensite! A second tempering step may be needed to eliminate the martensite formed from the retained austenite.

Figure 12.12
Retained austenite (white) trapped between martensite needles (black) (× 1000). (*From* Metals Handbook, *Vol. 8, 8th Ed., American Society for Metals, 1973*).

Retained austenite is also more of a problem for high-carbon steels. The start and finish temperatures are reduced when the carbon content increases (Figure 12.13). High-carbon steels must be refrigerated to produce all martensite.

Residual Stresses and Cracking Residual stresses are also produced because of the volume change. The surface of the quenched steel cools rapidly and transforms to martensite. When the austenite in the centre later transforms, the hard surface is placed in tension, while the centre is compressed. If the residual stresses exceed the yield strength, quench cracks form at the surface (Figure 12.14). However, if we first cool to just above the M_s and hold until the temperature equalises in the steel, subsequent quenching permits all of the steel to transform to martensite at about the same time. This heat treatment is called martempering or marquenching (Figure 12.15).

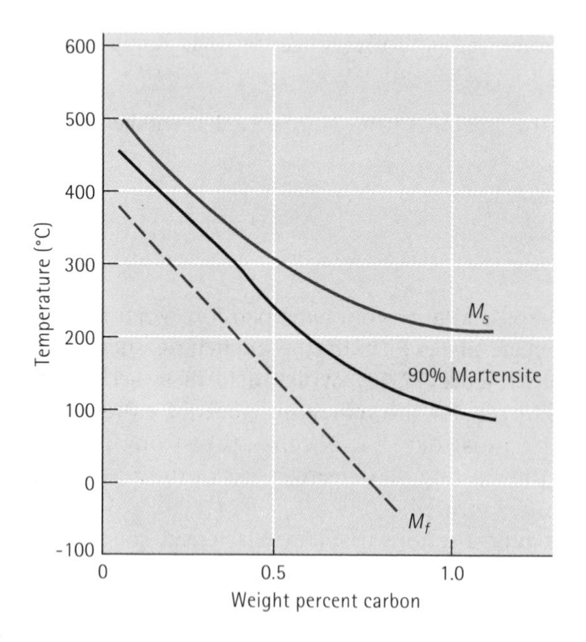

Figure 12.13
Increasing carbon reduces the M_s and M_f temperatures in plain-carbon steels.

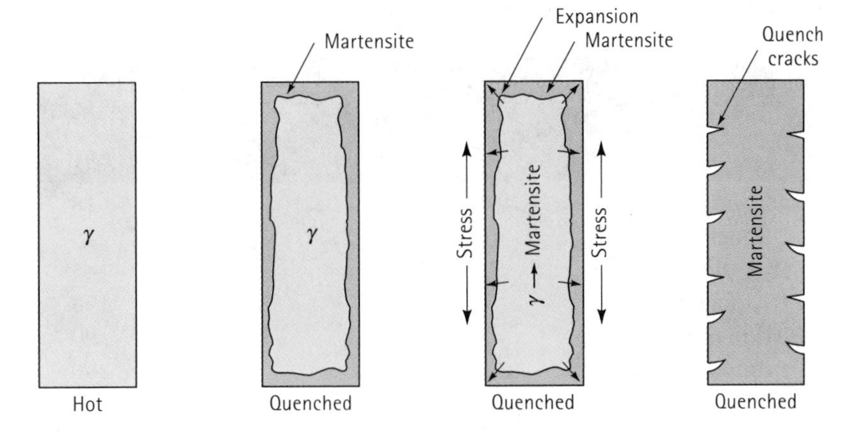

Figure 12.14
Formation of quench cracks caused by residual stresses produced during quenching. The figure illustrates the development of stresses as the austenite transforms to martensite during cooling.

Quench Rate In using the TTT diagram, we assumed that we could cool from the austenitising temperature to the transformation temperature instantly. Because this is not true, undesired microconstituents may form during quenching. For example, pearlite may form as the steel cools past the nose of the curve, particularly because the time of the nose is less than one second in plain-carbon steels.

The rate at which the steel cools during quenching depends on several factors. First, the surface always cools faster than the centre of the part. In addition, as the size of the part increases, the cooling rate at any location is slower. Finally, the cooling rate depends on the temperature and heat transfer characteristics of the quenching medium (Table 12.2). Quenching in oil, for example, produces a lower H coefficient, or slower cooling rate, than quenching in water or brine.

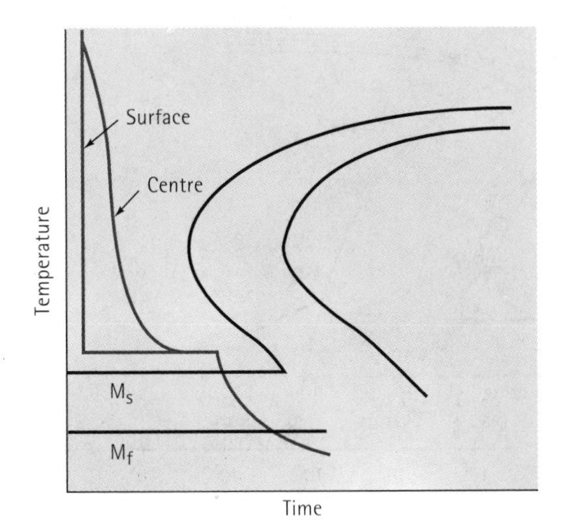

Figure 12.15
The martempering heat treatment, designed to reduce residual stresses and quench cracking.

Table 12.2 The H coefficient, or severity of the quench, for several quenching media.

Medium	H Coefficient	Cooling Rate at the Centre of a 25 mm Bar (°C/s)
Oil (no agitation)	0.25	18
Oil (agitation)	1.0	45
H_2O (no agitation)	1.0	45
H_2O (agitation)	4.0	190
Brine (no agitation)	2.0	90
Brine (agitation)	5.0	230

Continuous Cooling Transformation Diagrams We can develop a continuous cooling transformation (CCT) diagram by determining the microstructures produced in a steel at various rates of cooling. The CCT curve for a 1080 steel is shown in Figure 12.16. The CCT diagram differs from the TTT diagram (Figure 11.19) in that longer times are required for transformations to begin and no bainite region is observed.

If we cool a eutectoid steel at 5°C/s, the CCT diagram tells us that we obtain coarse pearlite; we have annealed the steel. Cooling at 35°C/s gives fine pearlite, and is a normalising heat treatment. Cooling at 100°C/s permits pearlite to start forming, but the reaction is incomplete and the remaining austenite changes to martensite. We obtain 100% martensite and thus are able to perform a quench and temper heat treatment, only if we cool faster than 140°C/s. Other steels, such as the low-carbon steel in Figure 12.17 have more complicated CCT diagrams.

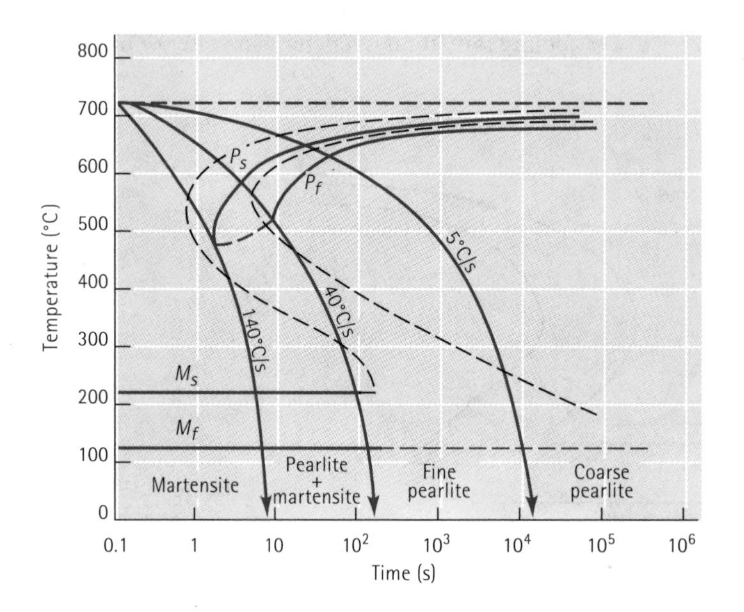

Figure 12.16
The CCT diagram (solid lines) for a eutectoid steel [AISI-SAE 1080] compared with the TTT diagram (dashed lines).

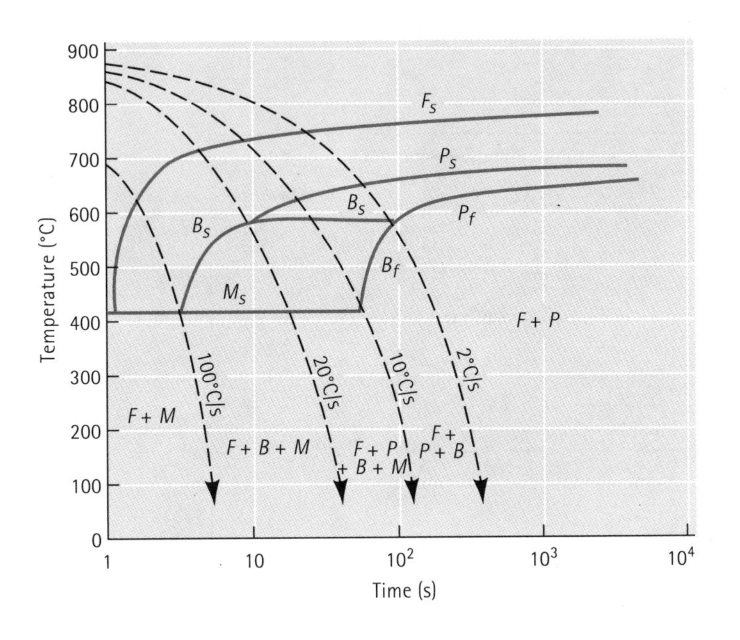

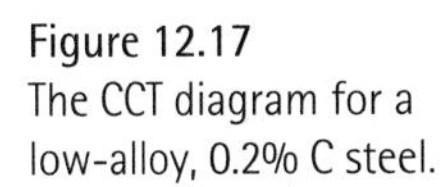
Figure 12.17
The CCT diagram for a low-alloy, 0.2% C steel.

12.6 Effect of Alloying Elements

Alloying elements are added to steels to (a) provide solid solution strengthening of ferrite, (b) cause the precipitation of alloy carbides rather than Fe_3C, (c) improve corrosion resistance and other special characteristics of the steel, and (d) improve hardenability. Improving hardenability is most important in alloy and tool steels.

Hardenability In plain-carbon steels, the nose of the TTT and CCT curves occurs at very short times; hence, very fast cooling rates are required to produce all martensite. In thin sections of steel, the rapid quench produces distortion and cracking. In thick steels, we are unable to produce martensite. All common alloying elements in steel shift the TTT and CCT diagrams to longer times, permitting us to obtain all martensite even in thick sections at slow cooling rates. Figure 12.18 shows the TTT and CCT curves for AISI-SAE 4340 (or BS 817M40).

Hardenability refers to the ease with which martensite forms. Plain-carbon steels have low hardenability – only very high cooling rates produce all martensite. Alloyed steels have high hardenability – even cooling in air may produce martensite. Hardenability does not refer to the hardness of the steel. A low-carbon, high-alloy steel may easily form martensite but, because of the low carbon content, the martensite is not hard.

Effect on the Phase Diagram When alloying elements are added to steel, the binary Fe-Fe_3C phase diagram is altered (Figure 12.19). Alloying elements reduce the carbon content at which the eutectoid reaction occurs and change the A_1, A_3, and A_{cm} temperatures. A steel containing only 0.6% C is hypoeutectoid and would operate at 700°C without forming austenite; the same steel containing 6% Mn is hypereutectoid and austenite forms at 700°C.

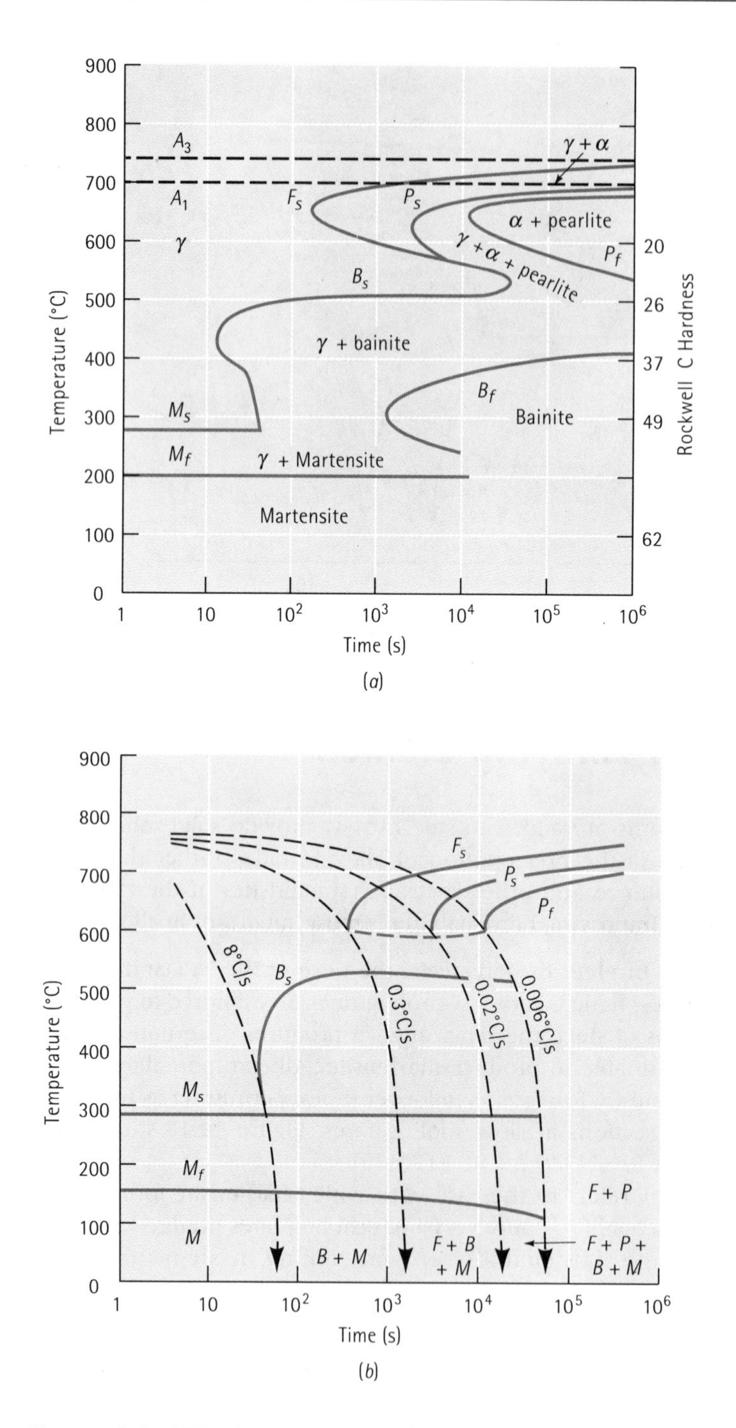

Figure 12.18
(a) TTT and (b) CCT curves for a low alloy Ni-Cr-Mo steel with 0.4% C [AISI-SAE 4340, BS 817M40].

Shape of the TTT Diagram Alloying elements may introduce a 'bay' region into the TTT diagram, as in the case of the 4340 steel (Figure 12.18). The bay region is used as the basis for thermomechanical heat treatment known as ausforming. A steel can be

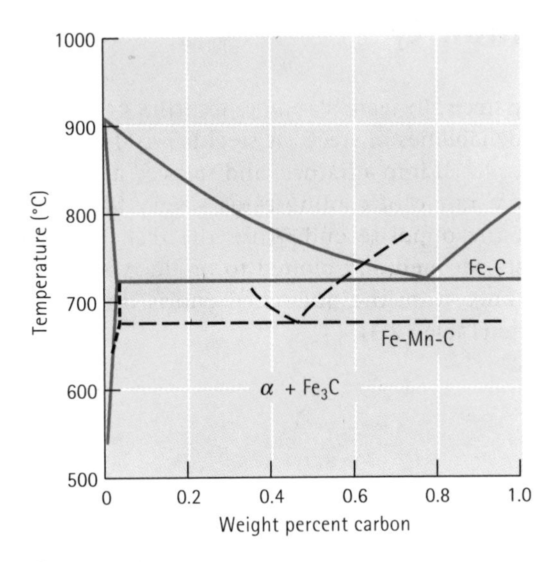

Figure 12.19
The effect of 6% manganese on the eutectoid portion of the Fe-Fe_3C phase diagram.

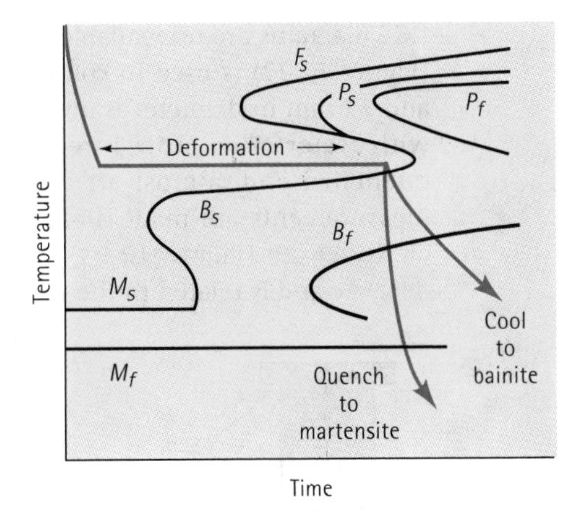

Figure 12.20
When alloying elements introduce a bay region into the TTT diagram, the steel can be ausformed.

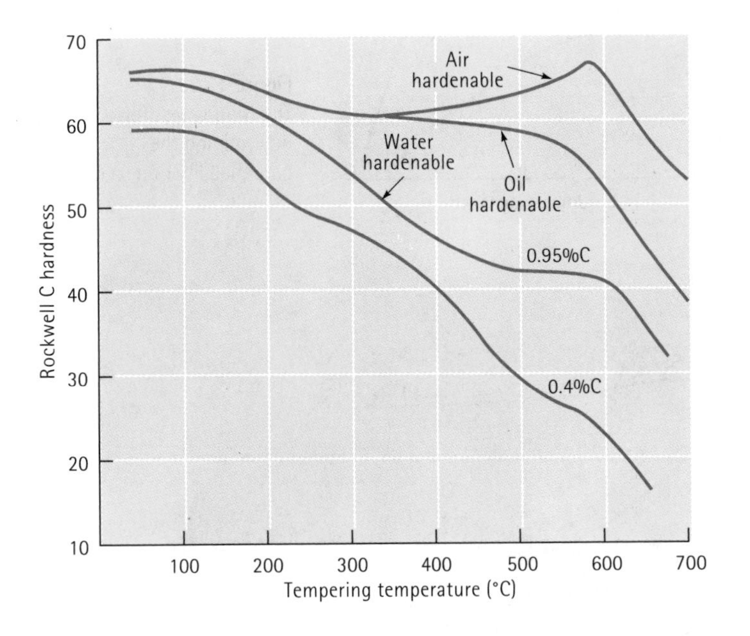

Figure 12.21
The effect of alloying elements on the tempering curves of steels. The air-hardenable steel shows a secondary hardening peak.

austenitised, quenched to the bay region, plastically deformed, and finally quenched to produce martensite (Figure 12.20).

Tempering Alloying elements reduce the rate of tempering compared with that of a plain-carbon steel (Figure 12.21). This effect may permit the alloy steels to operate more successfully at higher temperatures than plain-carbon steels.

12.7 Application of Hardenability

CCT diagrams are unavailable for many steels. Instead, a Jominy test (B.S.4437:1987) (Figure 12.22) is used to compare hardenabilities of steels. A steel bar 100 mm long and 25 mm in diameter is austenitised, placed into a fixture, and sprayed at one end with water. This procedure produces a range of cooling rates – very fast at the quenched end, almost air cooling at the opposite end. After the test, hardness measurements are made along the test specimen and plotted to produce a hardenability curve (Figure 12.23). The distance from the quenched end is the Jominy distance and is related to the cooling rate (Table 12.3).

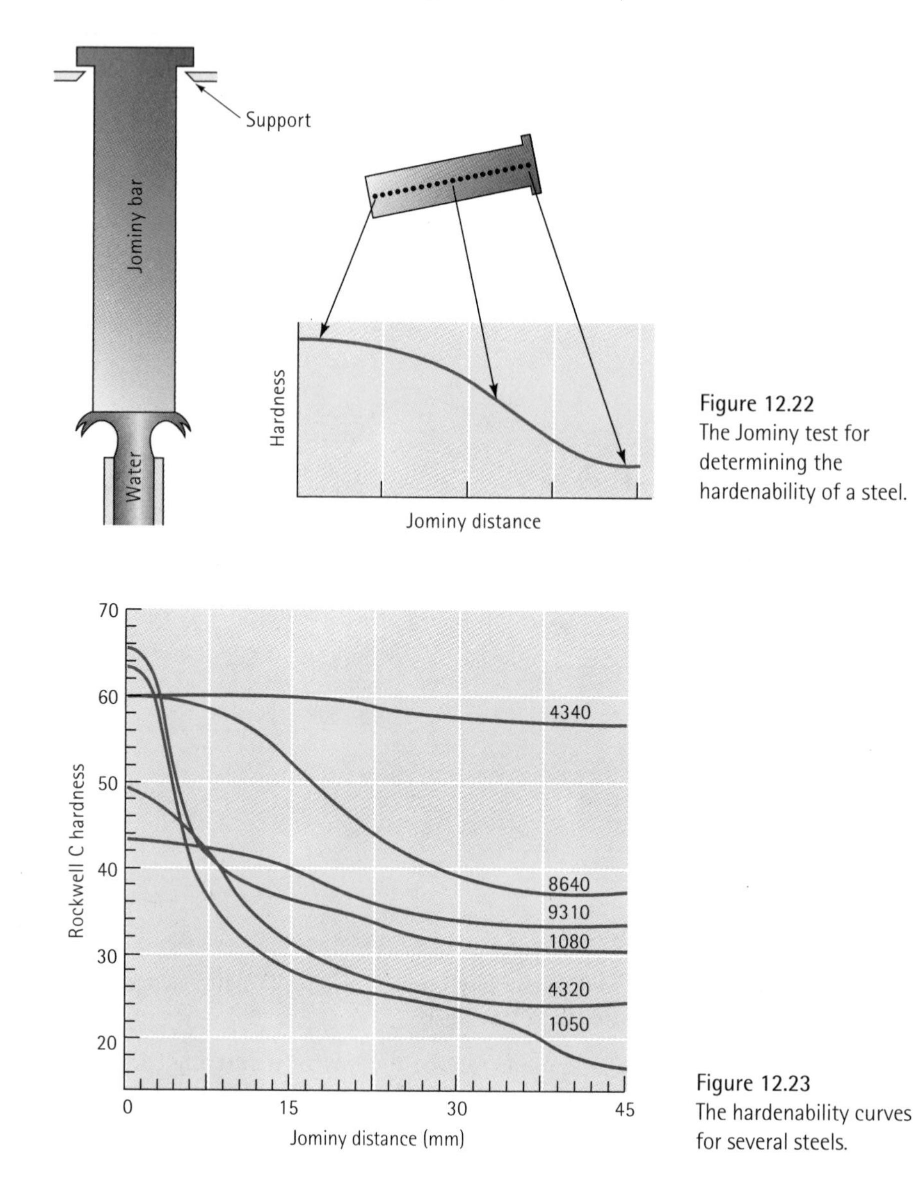

Figure 12.22
The Jominy test for determining the hardenability of a steel.

Figure 12.23
The hardenability curves for several steels.

Table 12.3 The relationship between cooling rate and Jominy distance.

Jominy Distance (mm)	Cooling Rate (°C.s^{-1})
1.5	300
3	100
5	60
7	35
9	25
11	17.5
13	15
15	12
20	7.5
25	5.5
30	4.5
35	3.8
40	3.2
45	3.0
50	2.8
55	2.5
60	2.2

Virtually any steel transforms to martensite at the quenched end. Thus, the hardness at zero Jominy distance is determined solely by the carbon content of the steel. At larger Jominy distances, there is a greater likelihood that bainite or pearlite will form instead of martensite. An alloy steel with a high hardenability (such as AISI-SAE 4340) maintains a rather flat hardenability curve; a plain-carbon steel (such as AISI-SAE 1050) has a curve that drops off quickly. The hardenability is determined primarily by the alloy content of the steel.

We can use hardenability curves in selecting or replacing steels in practical applications. The fact that two different steels cool at the same rate if quenched under identical conditions helps in this selection process.

EXAMPLE 12.5 Specification of a Wear-Resistant Gear

A gear made from AISI-SAE 9310 steel, which has an as-quenched hardness at a critical location of HRC 40, wears at an excessive rate. Tests have shown that an as-quenched hardness of at least HRC 50 is required at that critical location. Specify a steel that would be appropriate.

SOLUTION

We know that if different steels of the same size are quenched under identical conditions, their cooling rates or Jominy distances are the same. From Figure 12.23, a hardness of HRC 40 in a AISI-SAE 9310 steel corresponds to a Jominy distance of 15 mm ($10°C.s^{-1}$). If we assume the same Jominy distance, the other steels shown in Figure 12.23 have the following hardnesses at the critical location:

AISI-SAE 1050	HRC 28
AISI-SAE 1080	HRC 36
AISI-SAE 4320	HRC 31
AISI-SAE 8640	HRC 52
AISI-SAE 4340	HRC 60

Both the 8640 and 4340 steels are appropriate. The 4320 steel has too low a carbon content ever to reach HRC 50; the 1050 and 1080 have enough carbon, but the hardenability is too low. In Table 12.1, we find that the 86xx steels contain less alloying elements than the 43xx steels; thus the 8640 steel is probably less expensive than the 4340 steel and might be our best choice.

In another simple technique, we utilise the severity of the quench and the Grossman chart (Figure 12.24) to determine the hardness at the *centre* of a round bar. The bar diameter and H coefficient, or severity of the quench in Table 12.2, give the Jominy distance at the centre of the bar. We can then determine the hardness from the hardenability curve of the steel. (*See* Example 12.6.)

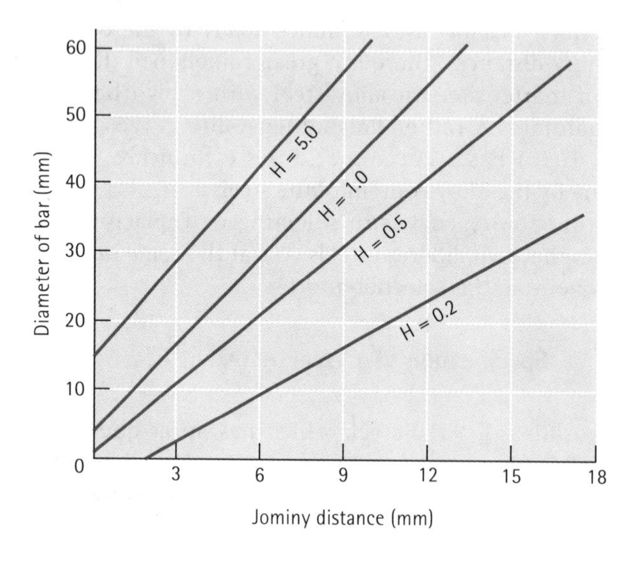

Figure 12.24
The Grossman chart used to determine the hardenability at the centre of a steel bar for different quenchants.

EXAMPLE 12.6 **Design of a Quenching Process**

Design a quenching process to produce a minimum hardness of HRC 40 at the centre of a 30 mm diameter AISI-SAE 4320 steel bar.

SOLUTION

Several quenching media are listed in Table 12.2. We can find an approximate H coefficient for each of the quenching media, then use Figure 12.24 to estimate the Jominy distance in a 30 mm diameter bar for each media. Finally, we can use the hardenability curve (Figure 12.23) to find the hardness in the 4320 steel. The results are listed below.

	H coefficient	Jominy distance (mm)	HRC
Oil (no agitation)	0.25	13.5	33
Oil (agitation)	1.00	6.5	44
H_2O (no agitation)	1.00	6.5	44
H_2O (agitation)	4.00	4	46
Brine (no agitation)	2.00	5	45
Brine (agitation)	5.00	3.5	47

The last three methods, based on brine or agitated water, are satisfactory. Using an unagitated brine quenchant might be least expensive, since no extra equipment is needed to agitate the quenching bath. However, H_2O is less corrosive than the brine quenchant.

12.8 Special Steels

There are many special categories of steels, including tool steels, high-strength – low-alloy steels, microalloyed steels, dual-phase steels, and maraging steels.

Tool steels are usually high-carbon steels that obtain high hardnesses by a quench and temper heat treatment. Their applications include cutting tools in machining operations, dies for die casting, forming dies, and other uses in which a combination of high strength, hardness, toughness, and temperature resistance is needed.

Alloying elements improved the hardenability and high-temperature stability of the tool steels. The water-hardenable steels such as AISI-SAE 1095 (0.95% C) must be quenched rapidly to produce martensite and also soften rapidly even at relatively low temperatures; oil-hardenable steels form martensite more easily, temper more slowly, but still soften at high temperatures. The air-hardenable and special tool steels may harden to martensite while cooling in air; in addition, these steels may not soften until near the A_1 temperature. In fact, the highly alloyed tool steels may pass through a secondary hardening peak near 500°C as the normal cementite dissolves and hard alloys carbides precipitate (Figure 12.21). The alloy carbides are particularly stable, resist growth or spheroidisation, and are important in establishing the high-temperature resistance of these steels.

High-strength–low-alloy (HSLA) steels and microalloyed steels are low-carbon steels containing small amounts of alloying elements. The HSLA steels are specified on the basis of yield strength, with grades up to 550 MN.m^{-2}; the steels contain the least amount of alloying element that still provides the proper yield strength without heat treatment. In microalloyed steels, careful processing permits precipitation of carbides and nitrides of Cb, V, Ti, or Zr, which provide dispersion strengthening and a fine grain size.

Dual-phase steels contain a uniform distribution of ferrite and martensite, with the dispersed martensite providing yield strengths of 400 to 1 000 MN.m^{-2} These low-carbon steels do not contain enough alloying elements to have good hardenability using normal quenching processes. But when the steel is heated into the ferrite-plus-austenite portion of the phase diagram, the austenite phase becomes enriched in carbon, which provides the needed hardenability. During quenching, only the austenite portion transforms to martensite (Figure 12.25).

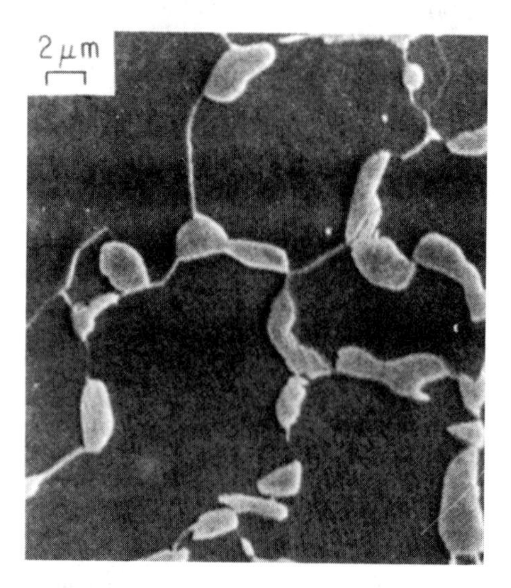

Figure 12.25
Microstructure of a dual-phase steel, showing islands of light martensite in a ferrite matrix (× 2 500). (*From G. Speich, 'Physical Metallurgy of Dual-Phase Steels,'* Fundamentals of Dual-Phase Steels, *The Metallurgical Society of AIME, 1981.*)

Maraging steels are low-carbon, highly alloyed steels. The steels are austenitised and quenched to produce a soft martensite that contains less than 0.3% C. When the martensite is aged at about 500°C, intermetallic compounds such as Ni_3Ti, Fe_2Mo, and Ni_3Mo precipitate.

Many steels are also coated, usually to provide good corrosion protection. *Galvanised* steel is coated with thin layer of zinc, *terne* steel is coated with lead, and other steels are coated with aluminium or tin.

12.9 Surface Treatments

We can, by proper heat treatment, produce a structure that is hard and strong at the surface, so that excellent wear and fatigue resistance are obtained, but at the same times gives a soft, ductile, tough core that provides good resistance to impact failure.

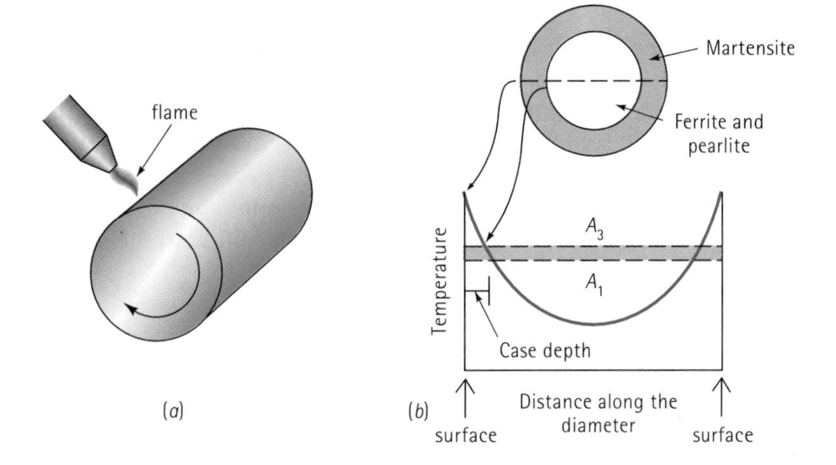

Figure 12.26
(*a*) Surface hardening by localised heating. (*b*) Only the surface heats above the A_1 temperature and is quenched to martensite.

Selectively Heating the Surface We could begin by rapidly heating the surface of a medium-carbon steel above the A_3 temperature (the centre remains below the A_1). After the steel is quenched, the centre is still a soft mixture of ferrite and pearlite, while the surface is martensite (Figure 12.26). The depth of the martensite layer is the case depth. Tempering produces the desired hardness at the surface. We can provide local heating of the surface by using a gas flame, an induction coil, a laser beam, or an electron beam. We can, if we wish, harden only selected areas of the surface that are most subject to failure by fatigue or wear.

Carburising and Nitriding For even better toughness, we start with a low-carbon steel. In carburising, carbon is diffused into the surface of the steel at a temperature above the A_3 (Figure 12.27). A high carbon content is produced at the surface due to rapid diffusion and the high solubility of carbon in austenite. When the steel is then quenched and tempered, the surface becomes a high-carbon tempered martensite, while the ferritic centre remains soft and ductile. The thickness of the hardened surface, again called the case depth, is much smaller in carburised steels than in flame- or induction-hardened steels.

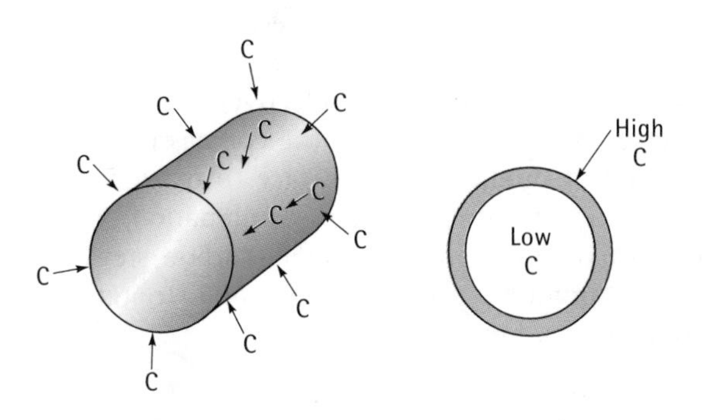

Figure 12.27
Carburising of a low-carbon steel to produce a high-carbon, wear-resistant surface.

Nitrogen provides a hardening effect similar to that of carbon. In cyaniding the steel is immersed in a liquid cyanide bath which permits both carbon and nitrogen to diffuse into the steel. In carbonitriding, a gas containing carbon monoxide and ammonia is generated and both carbon and nitrogen diffuse into the steel. Finally, only nitrogen diffuses into the surface from a gas in nitriding. Nitriding is carried out below the A_1 temperature.

In each of these processes, compressive residual stresses are introduced at the surface, providing excellent fatigue resistance in addition to the good combination of hardness, strength, and toughness.

EXAMPLE 12.7 Selection of Surface-Hardening Treatments for a Drive Train

Select the materials and heat treatments for an automobile axle and drive gear (Figure 12.28).

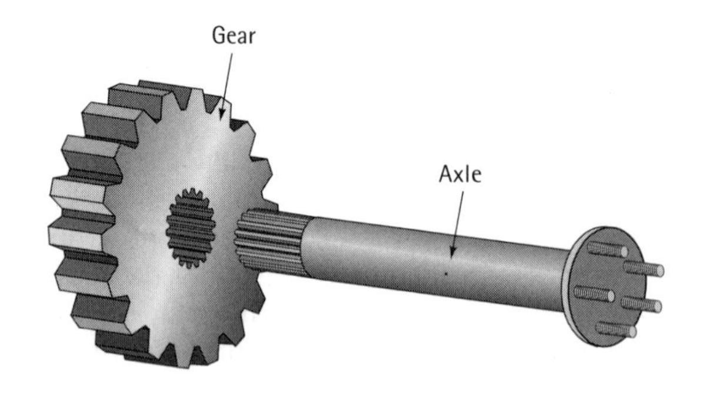

Figure 12.28
Axle and gear (*for* Example 12.7).

SOLUTION

Both parts require good fatigue resistance. The gear also should have a good hardness to avoid wear, and the axle should have good overall strength to withstand bending and torsional loads. Both parts should have good toughness. Finally, since millions of these parts will be made, they should be inexpensive.

Quenched and tempered alloy steels might provide the required combination of strength and toughness; however, the alloy steels are expensive. An alternative approach for each part is described below.

The axle might be made from a forged 0.5% C steel [e.g. AISI-SAE 1050] containing a matrix of ferrite and pearlite. The axle could be surface-hardened, perhaps by moving the axle through an induction coil to selectively heat the surface of the steel above the A_3 temperature (about 770°C). After the coil passes any particular location of the axle, the cold interior quenches the surface to martensite. Tempering then softens the martensite to improve ductility. This combination of carbon content and heat treatment meets our requirements. The plain carbon steel is inexpensive; the core of ferrite and pearlite produces good toughness and strength; and the hardened surface provides good fatigue and wear resistance.

The gear is subject to more severe loading conditions, for which the 0.5% C steel does not provide sufficient toughness, hardness, and wear resistance. Instead, we might carburise a 1.0% C steel for the gear. The original steel contains mostly ferrite, providing good ductility and toughness. By performing a gas carburising process above the A_3 temperature (about 860°C), we introduce about 1.0% C in a very thin case at the surface of the gear teeth. This high carbon case, which transforms to martensite during quenching, is tempered to control the hardness. Now we obtain toughness due to the low-carbon ferrite core, wear resistance due to the high carbon surface, and fatigue resistance due to the high-strength surface containing compressive residual stresses introduced during carburising. In addition, the plain carbon 0.1% C steel is an inexpensive starting material that is easily forged into a near-net shape prior to heat treatment.

12.10 Weldability of Steel

During welding, the metal nearest the weld heats above the A_1 temperature and austenite forms (Figure 12.29). During cooling, the austenite in this heat-affected zone transforms to a new structure, dependent on the cooling rate and the CCT diagram for the steel. Plain low-carbon steels have such a low hardenability that normal cooling rates seldom produce martensite. However, an alloy steel may have to be preheated to slow down the cooling rate or postheated to temper any martensite that forms.

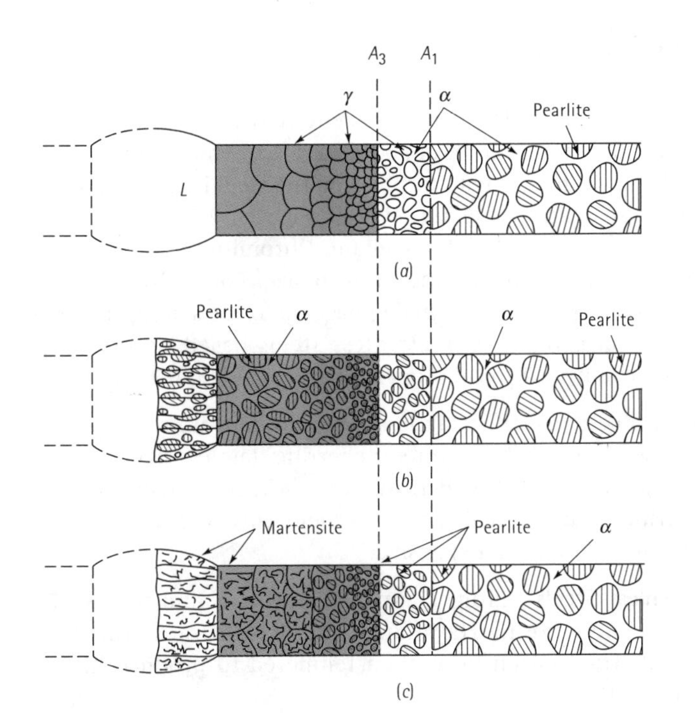

Figure 12.29
The development of the heat-affected zone in a weld: (a) the structure at the maximum temperature, (b) the structure after cooling in a steel of low hardenability, and (c) the structure after cooling in a steel of high hardenability.

A steel that is originally quenched and tempered has two problems during welding. First, the portion of the heat-affected zone that heats above the A_1 may form martensite after cooling. Second, a portion of the heat-affected zone below the A_1 may overtemper. Normally, we should not weld a steel in the quenched and tempered condition.

EXAMPLE 12.8

Compare the structures in the heat-affected zones of welds in 1080 and 4340 steels if the cooling rate in the heat-affected zone is 5°C.s^{-1}.

SOLUTION

From the CCT diagrams, Figures 12.16 and 12.18, the cooling rate in the weld produces the following structures:

1080: 100% pearlite

4340: Bainite and martensite

The high hardenability of the alloy steel reduces the weldability, permitting martensite to form and embrittle the weld.

12.11 Stainless Steels

Stainless steels are selected for their excellent resistance to corrosion. All true stainless steels contain a minimum of about 12% Cr, which permits a thin, protective surface layer of chromium oxide to form when the steel is exposed to oxygen.

Chromium is also a *ferrite stabilising element.* Figure 12.30(a) illustrates the effect of chromium on the iron-carbon phase diagram. Chromium causes the austenite region to shrink, while the ferrite region increases in size. For high-chromium, low-carbon compositions, ferrite is present as a single phase up to the solidus temperature.

There are several categories of stainless steels based on crystal structure and strengthening mechanism. Typical properties are included in Table 12.4.

Ferritic Stainless Steels Ferritic stainless steels contain up to 30% Cr and less than 0.12% C. Because of the BCC structure, the ferritic stainless steels have good strengths and moderate ductilities derived from solid solution strengthening and strain hardening. Ferritic stainless steels have excellent corrosion resistance and moderate formability and are relatively inexpensive.

Martensitic Stainless Steels From Figure 12.30(a), we find that a 17% Cr-0.5%-C alloy heated to 1 200°C forms 100% austenite, which transforms to martensite on quenching in oil. The martensite is then tempered to produce high strengths and hardnesses [Figure 12.31(a)].

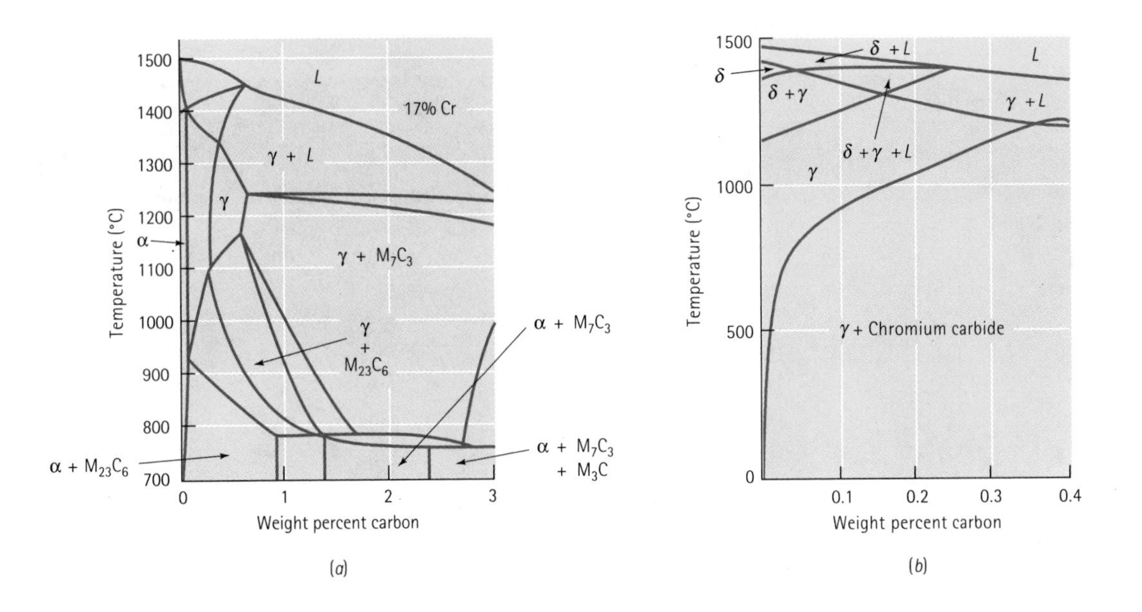

Figure 12.30
(*a*) The effect of 17% chromium on the iron-carbon phase diagram. At low carbon contents, ferrite is stable at all temperatures. (*b*) A section of the iron-chromium-nickel-carbon phase diagram at a constant 18% Cr-8% Ni. At low carbon contents, austenite is stable at room temperature.

The chromium content is usually less than 17% Cr; otherwise, the austenite field becomes so small that very stringent control over both austenitising temperature and carbon content is required. Lower chromium contents also permit the carbon content to vary from about 0.1% to 1.0%, allowing martensites of different hardnesses to be produced. The combination of hardness, strength, and corrosion resistance makes the alloys attractive for applications such as high quality knives, ball bearings, and valves.

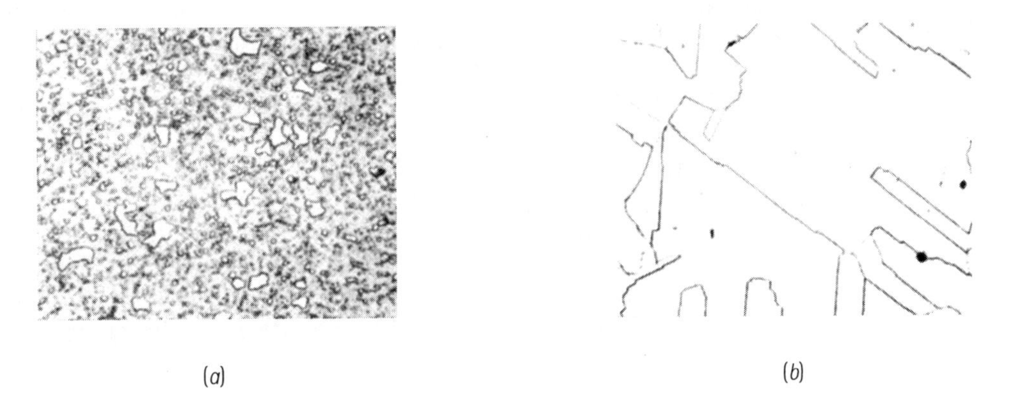

Figure 12.31
(*a*) Martensitic stainless steel containing large primary carbides and small carbides formed during tempering (× 350). (*b*) Austenitic stainless steel (× 500). (*From* Metals Handbook, *Vols. 7 and 8, 8th Ed., American Society for Metals, 1972, 1973.*)

Table 12.4 Typical compositions and properties of stainless steels.

AISI-SAE Grade	Typical BS Grade	% C	% Cr	% Ni	Others	Tensile Strength (MN.m^{-2})	Yield Strength (MN.m^{-2})	% Elongation	Condition
Austenitic:									
201		0.15	17	5	6.5% Mn	650	310	40	Annealed
304	304S15	0.08	19	10		520	205	30	Annealed
						1275	965	9	Cold-worked
304L	304S12	0.03	19	10		520	205	30	Annealed
316	316S16	0.08	17	12	2.5% Mo	520	205	30	Annealed
321	321S12	0.08	18	10	0.4% Ti	585	240	55	Annealed
347	347S17	0.08	18	11	0.8% Nb	620	240	50	Annealed
Ferritic:									
430	430S15	0.12	17			450	205	22	Annealed
442	442S19	0.12	20			520	275	20	Annealed
Martensitic:									
416	416S21	0.15	13		0.6% Mo	1240	965	18	Quenched and tempered
431	431S29	0.20	16	2		1380	1035	16	Quenched and tempered
440C		1.10	17		0.7 Mo	1965	1895	2	Quenched and tempered
Precipitation hardening:									
17-4		0.07	17	4	0.4% Nb	1310	1170	10	Age-hardened
17-7		0.09	17	7	1.0% Al	1650	1585	6	Age-hardened

Austenitic Stainless Steels Nickel, which is an austenite stabilising element, increases the size of the austenite field, while nearly eliminating ferrite from the iron-chromium-carbon alloys [Figure 12.30(b)]. If the carbon content is below about 0.03%, the carbides do not form and the steel is virtually all austenite at room temperature [Figure 12.31(b)].

The FCC austenitic stainless steels have excellent ductility, formability, and corrosion resistance. Strength is obtained by extensive solid solution strengthening, and the austenitic stainless steels may be cold-worked to higher strengths than the ferritic stainless steels. The steels have excellent low-temperature impact properties, since they have no transition temperature. Furthermore, the austenitic stainless steels are not ferromagnetic. Unfortunately, the high nickel and chromium contents make the alloys expensive.

Precipitation-Hardening (PH) Stainless Steels The precipitation-hardening (or PH) stainless steels contain Al, Nb, or Ta and derive their properties from solid-solution strengthening, strain hardening, age hardening, and the martensitic reaction. The steel is first heated and quenched to permit the austenite to transform to martensite.

Reheating permits precipitates such as Ni_3Al to form from the martensite. High mechanical properties are obtained even with low carbon contents.

Duplex Stainless Steels In some cases, mixtures of phases are deliberately introduced into the stainless steel structure. By appropriate control of the composition and heat treatment, a duplex stainless steel containing approximately 50% ferrite and 50% austenite can be produced. This combination provides a set of mechanical properties, corrosion resistance, formability, and weldability not obtained in any one of the usual stainless steels.

EXAMPLE 12.9 Selection of a Test to Separate Stainless Steels

In order to efficiently recycle stainless steel scrap, we wish to separate the high-nickel stainless steel from the low-nickel stainless steel. Describe a method for doing this.

SOLUTION

Performing a chemical analysis on each piece of scrap is tedious and expensive. Sorting based on hardness might be less expensive; however, because of the different types of treatments – such as annealing, cold working, or quench and tempering – the hardness may not be related to the steel composition.

The high-nickel stainless steels are ordinarily austenitic, whereas the low-nickel alloys are ferritic or martensitic. An ordinary magnet will be attracted to the low-nickel ferritic and martensitic steels, but will not be attracted to the high-nickel austenitic steel. We might specify this simple and inexpensive magnetic test for our separation process.

12.12 Phase Transformations in Cast Irons

Cast irons are iron-carbon-silicon alloys, typically containing 2% to 4% C and 0.5% to 3% Si, that pass through the eutectic reaction during solidification.

The microstructures of the five important types of cast irons are shown schematically in figure 12.32. Grey cast iron contains small, interconnected graphite flakes that cause low strength and ductility. White cast iron is a hard, brittle alloy containing massive amounts of Fe_3C. Malleable cast iron, formed by the heat treatment of white iron, produces rounded clumps of graphite. Spheroidal Graphite (S. G. iron) or nodular cast iron contains spheroidal graphite particles obtained during solidification. Compacted graphite cast iron contains rounded but interconnected (vermicular) graphite also produced during solidification.

To understand the origin of these cast irons, we must examine the phase diagram, solidification, and phase transformations of the alloys.

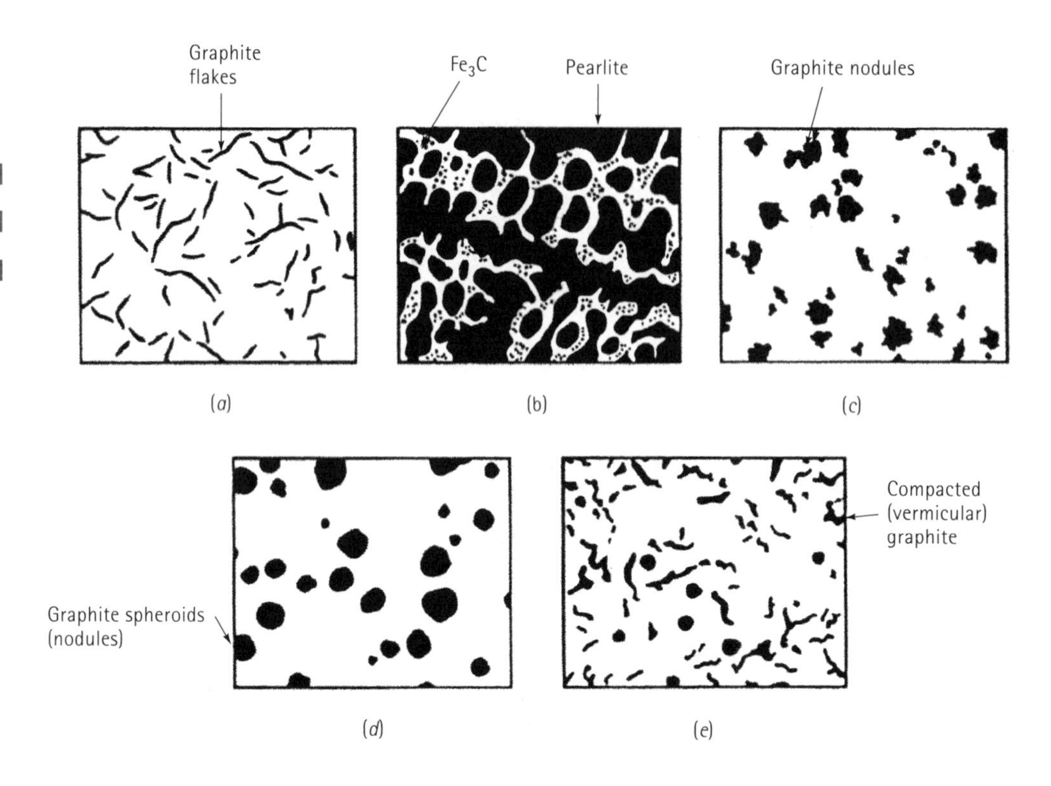

Figure 12.32
Schematic drawings of the five types of cast iron: (*a*) grey iron, (*b*) white iron, (*c*) malleable iron, (*d*) S. G. iron, and (*e*) compacted graphite iron.

The Eutectic Reaction in Cast Irons As shown in Figure 12.32, the carbon in a cast iron may be either in the form of graphite or Fe_3C. The iron-carbon phase diagram, Figure 12.33 therefore contains the two systems; the solid lines show the equilibrium reactions between iron and graphite and the dotted lines show the metastable, iron-Fe_3C reactions.

Based on the Fe-Fe_3C phase diagram (dashed lines in Figure 12.33), the eutectic reaction that occurs in cast irons at 1140°C is:

$$L \rightarrow \gamma + Fe_3C \tag{12.1}$$

If we produce a cast iron using only iron-carbon alloys, this reaction produces *white cast iron,* with a microstructure composed of Fe_3C and pearlite. This system, however, is really a metastable phase diagram. Under truly equilibrium conditions, the eutectic reaction is:

$$L \rightarrow \gamma + \text{graphite} \tag{12.2}$$

The Fe-C phase diagram is shown as solid lines in Figure 12.33. When the stable $L \rightarrow \gamma +$ graphite eutectic reaction occurs at 1 146°C, grey, ductile, or compacted graphite cast iron forms.

In Fe-C alloys, the liquid easily undercools 6°C (the temperature difference between the stable and metastable eutectic temperatures) and white iron forms.

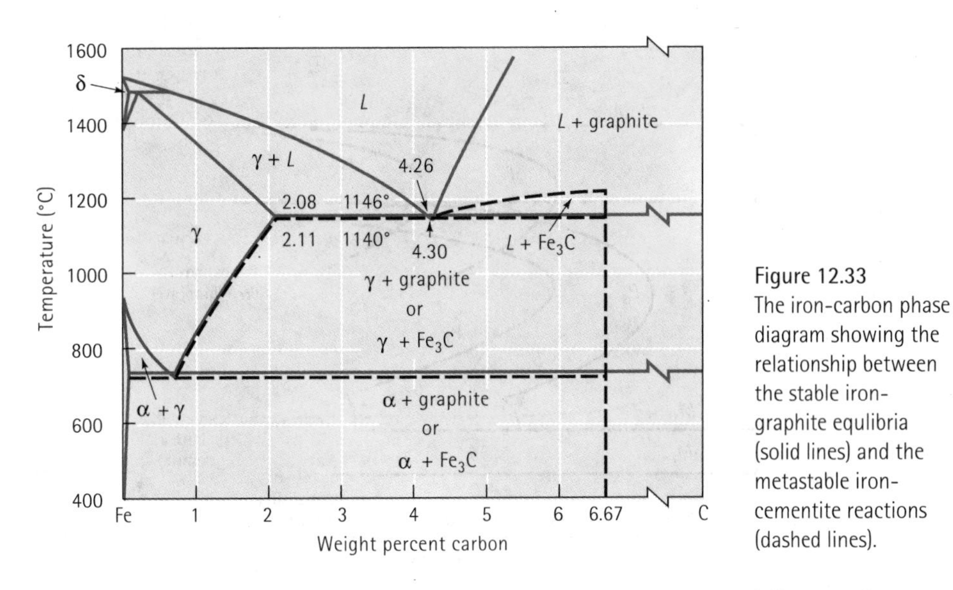

Figure 12.33
The iron-carbon phase diagram showing the relationship between the stable iron-graphite equlibria (solid lines) and the metastable iron-cementite reactions (dashed lines).

Adding about 2% silicon to the iron increases the temperature difference between the eutectics, permitting larger undercoolings to be tolerated and more time for the stable graphite eutectic to nucleate and grow. Silicon therefore is a graphite stabilising element. Elements such as chromium and bismuth have the opposite effect and encourage white cast iron.

We can also introduce inoculants, such as FeSi alloys, to encourage the nucleation of graphite, or we can reduce the cooling rate of the casting to provide more time for growth of graphite.

Silicon also reduces the amount of the carbon contained in the eutectic. We can take this effect into account by defining the carbon equivalent (CE):

$$CE = \%\ C + \frac{1}{3}\ \%\ Si \tag{12.3}$$

The eutectic composition is always near 4.3% CE. A high-carbon equivalent encourages the growth of the graphite eutectic.

The Eutectoid Reaction in Cast Irons The matrix structure and properties of each type of cast iron are determined by how the austenite transforms during the eutectoid reaction. In the Fe-Fe_3C phase diagram used for steels, the austenite transformed to ferrite and cementite, often in the form of pearlite. However, silicon also encourages the *stable* eutectoid reaction:

$$\gamma \rightarrow \alpha + \text{graphite} \tag{12.4}$$

Under equilibrium conditions, carbon atoms diffuse from the austenite to existing graphite particles, leaving behind the low-carbon ferrite.

The transformation diagram (Figure 12.34) describes how the austenite might transform during heat treatment. Annealing (or furnace cooling) of cast iron gives a soft ferritic matrix. Normalising, or air cooling, gives a pearlitic matrix. The cast irons can also be austempered to produce bainite, or can be quenched to martensite and tempered. Austempered S. G. iron, with strengths of up to 1 400 MN.m^{-2}, is used for high-performance gears.

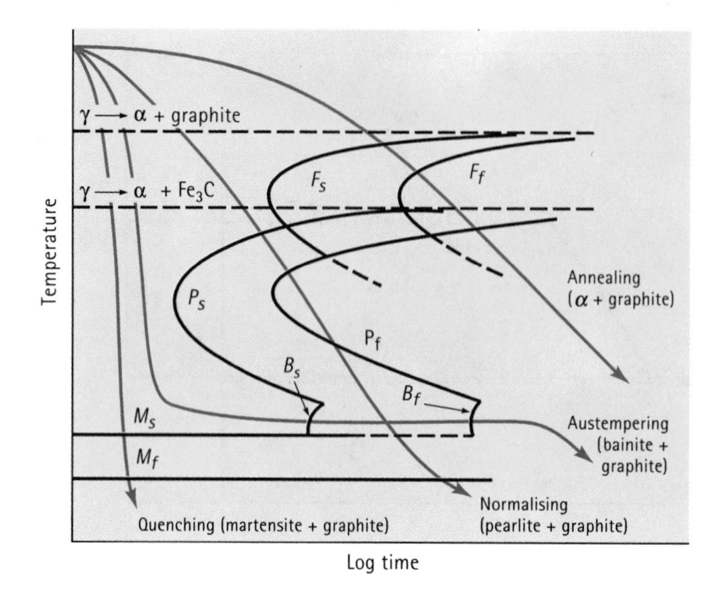

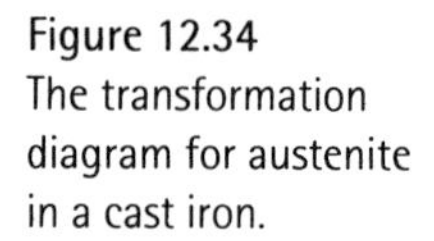
Figure 12.34
The transformation diagram for austenite in a cast iron.

12.13 Characteristics and Production of the Cast Irons

In order to produce the desired type of cast iron, with the correct mechanical properties, it is important to control the eutectic reaction and the eutectoid reaction. The eutectic reaction is affected by the way the cast iron is solidified and the eutectoid reaction is affected by subsequent cooling rate and by any heat treatment process.

Grey cast iron Grey iron is the most common of the cast irons and is a relatively cheap engineering material. Solidification produces interconnected graphite flakes, which resemble a number of potato crisps glued together at a single location (Figure 12.35). The point at which the flakes are connected is the original graphite nucleus. Grey cast iron contains many clusters, or eutectic cells, of graphite flakes with each cell representing 1 nucleation event of

liquid → graphite + austenite

Inoculation, produced by adding small amounts of ferro-silicon alloy or rapid cooling rates, helps to produce finer eutectic cells thus improving strength of the casting. The graphite flakes have a low mechanical strength and are brittle so that they are considered equivalent to tiny cracks within the pearlite matrix. Hence grey iron has an elongation of only about 1%.

The grey irons are specified in B.S. by a grade number that gives the minimum tensile strength of the iron (Table 12.5) in a standard test piece machined from a 30 mm diameter cast test bar. A grade 150 iron would therefore have minimum tensile strength of 150 $MN.m^{-2}$. Figure 12.36 illustrates that for any grade of iron the tensile strength can vary with cooling rate and/or casting size and cross section. Too rapid cooling can cause chilling particularly at the surface of a casting. Carbides form and a white iron structure is produced making any subsequent machining very difficult.

Table 12.5 Typical compositions and properties of grey irons. (BS. 1452: 1990).

Grade (MN.m^{-2})	Approx. C.E.	Brinell Hardness for 40-80mm section	% Elongation	un-notched impact (J)
150	4.5	170–100	0.6	8–13
200	4.2	190–120	0.4	8–16
250	3.85	220–145	0.5	13–23
300	3.65	240–165	0.5	16–31
350	3.5	260–185	0.5	24–47

In cast irons with lower carbon equivalents, the nominal tensile strength is increased, since a smaller amount of graphite is produced during the solidification. Even higher strengths can be produced by alloy additions or heat treatment.

In spite of its low tensile strength and its brittle behaviour grey iron has a number of attractive properties for an engineer. The flakes do not act as stress raisers under compressive loading. So with careful design grey iron can bear large loads. The machinability of grey iron is excellent since the graphite flakes act as chip breakers. Resistance to sliding wear is good; the porous graphite flakes absorb and hold lubricant and because graphite is soft and slippery, may even provide some self-lubrication. Very good thermal conductivity is obtained due to interconnected graphite flakes. Vibration damping characteristics are exceptional particularly with coarse graphite flakes. The properties have made grey cast iron a successful material used in engine blocks and machine tool bases. Grey iron is not recommended for making bells however!

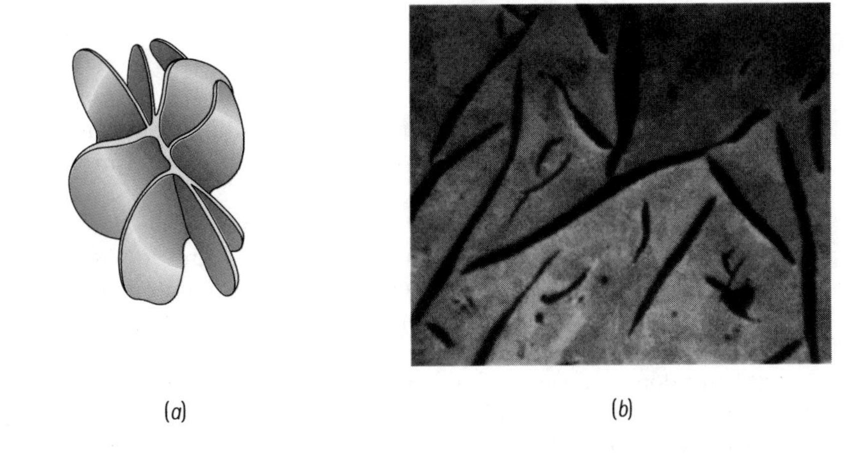

(*a*) (*b*)

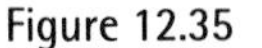

Figure 12.35
(*a*) Sketch and (*b*) photomicrograph of the flake graphite in grey cast iron (× 100).

White Cast Iron Low carbon equivalent cast iron containing about 2.5% Carbon and 1.5% Silicon solidifies without graphite formation and produces a very hard material which is not machinable. This however is the base material used in the malleabilising heat treatment process (discussed in the next section). Its microstructure is shown in Figure 12.38(a).

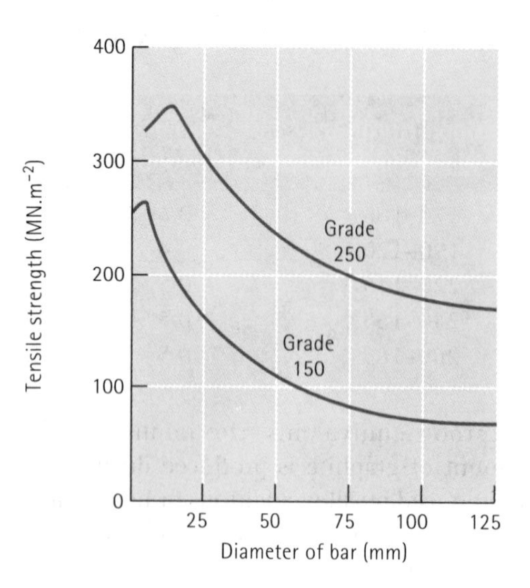

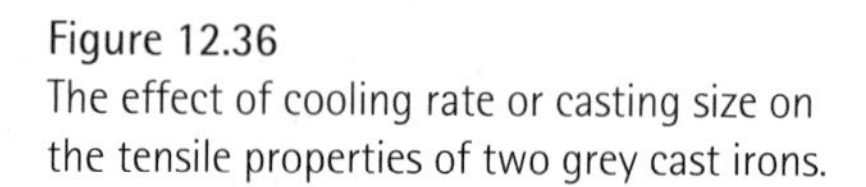
Figure 12.36
The effect of cooling rate or casting size on the tensile properties of two grey cast irons.

Alloy additions Chromium, Nickel and Molybdenum are made to the white iron in order to make it even harder and more wear resistant. Such an alloy solidifies as martensite making it ideal for parts in excavators and other wear resistant applications.

EXAMPLE 12.10 Materials Selection for an Inexpensive Pair of Scissors

Design a pair of inexpensive scissors with a hard edge for cutting paper.

SOLUTION

High-quality scissors, such as those used in cutting fabric, are often produced from a hardened stainless steel. However, we are interested in producing a much less expensive product.

If we consider the design of the blades of scissors, we find that the cross-section of the blades is a wedge. If the carbon equivalent of a grey iron is correctly adjusted, the tip of the wedge cools rapidly enough to produce white iron, while the remainder of the blade forms grey iron. The white iron, at the cutting surface, is hard and can be sharpened.

But don't try to use these inexpensive scissors as a lever. Both the white and grey portions of the blade are very brittle and the blade will break, rather than bend, when any unusual force is applied.

Malleable Cast Iron Malleable iron is produced by heat treating unalloyed 3% carbon equivalent (2.5% C, 1.5% Si) white iron. During the malleabilising heat treatment, the cementite formed during solidification is decomposed and graphite clumps, or nodules, are produced. The nodules, or temper carbon, often resemble popcorn. The rounded graphite shape permits a good combination of strength and ductility.

Table 12.6 Grades of Malleable Cast Iron (B.S. 6681: 1986).

Grade/ Designation	Minimum Tensile Strength ($MN.m^{-2}$)	Minimum 0.2% Proof Stress ($MN.m^{-2}$)	Minimum Elongation %	Maximum Brinell Hardness Number
Blackheart				
B 35 – 12	350	200	12	150
B 32 – 10	320	190	10	150
Whiteheart				
W 35 – 04	360	–	3	230
W 40 – 05	420	230	4	220
Pearlitic				
P 45 – 06	450	270	6	200
P 55 – 04	550	340	4	230

The production of malleable iron requires several steps (Figure 12.37). Graphite nodules nucleate as the white iron is slowly heated. During first stage graphitisation (FSG), cementite decomposes to the stable austenite and graphite phases as carbon in Fe_3C diffuses to the graphite nuclei. Following FSG, the austenite transforms during cooling.

Figure 12.38 shows the microstructures of the original white iron and the two types of malleable iron that can be produced. To make *ferritic malleable iron*, the casting is cooled slowly through the eutectoid temperature range to cause second stage graphitisation (SSG). The ferritic malleable iron has good toughness compared with that of other irons because its low carbon equivalent reduces the transition temperature below room temperature. It also contains the ductile phase, ferrite.

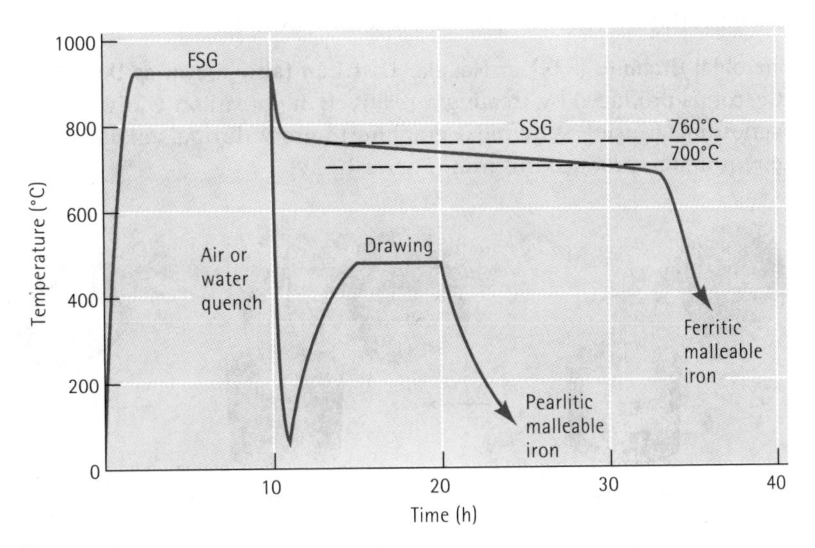

Figure 12.37
The heat treatments for ferritic and pearlitic malleable irons.

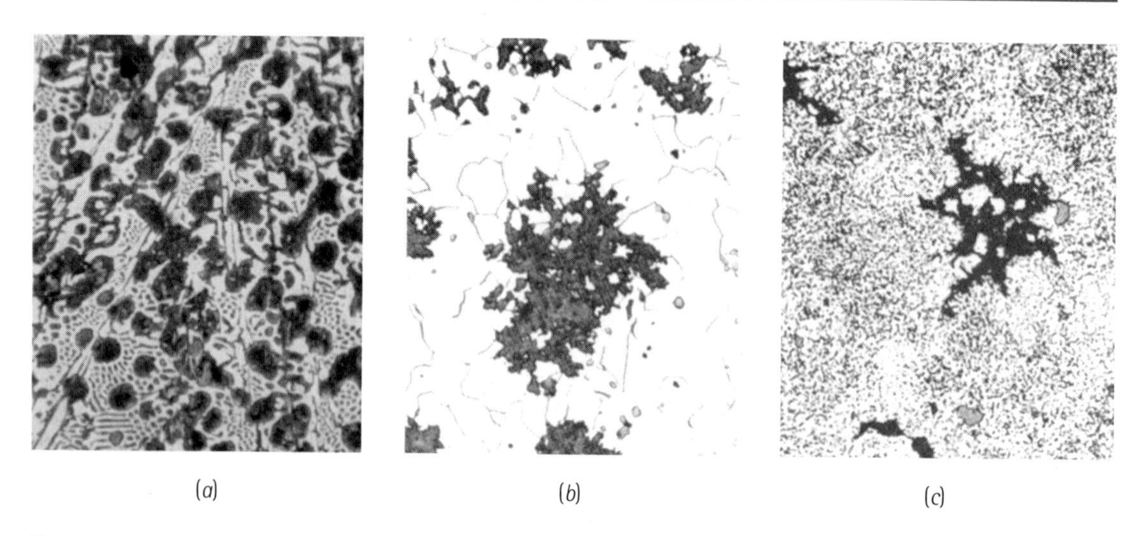

Figure 12.38
(*a*) White cast iron prior to heat treatment (× 100). (*b*) Ferritic malleable iron with graphite nodules and small MnS inclusions in a ferrite matrix (× 200). (*c*) Pearlite malleable iron drawn to produce a tempered martensite matrix (× 500). (Images (*b*) and (*c*) are from Metals Handbook, *Vols. 7 and 8, 8th Ed., American Society for Metals, 1972, 1973.*)

Pearlitic malleable iron is obtained when austenite is cooled in air or oil to form pearlite or martensite. In either case, the matrix is hard and brittle. The iron is then drawn at a temperature below the eutectoid. Drawing is a heat treatment process that tempers the martensite or spheroidises the pearlite. A higher drawing temperature decreases strength and increases ductility and toughness.

The Blackheart and Whiteheart matrix depends upon the section size of the casting and the process details. The Blackheart matrix is ferrite and is usually best achieved in thin cross sections. The Whiteheart matrix is a duplex structure being ferrite near its surface and pearlite near the centre of larger cross sections.

The pearlitic malleable iron has a pearlitic matrix.

Spheroidal Graphite (S.G.) or Nodular Cast Iron (also known as Ductile Iron in the USA)
S.G. iron is produced by treating a relatively high-carbon equivalent liquid iron with magnesium, causing spheroidal graphite to grow during solidification. Several steps are required to produce this iron (Figure 12.39).

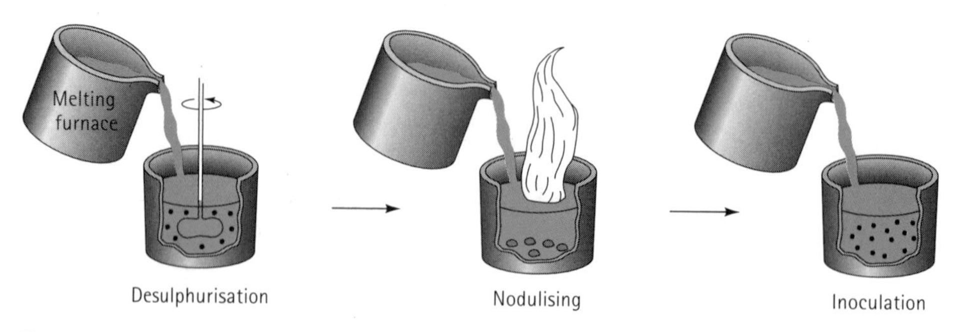

Figure 12.39
Schematic diagram of the treatment of ductile iron.

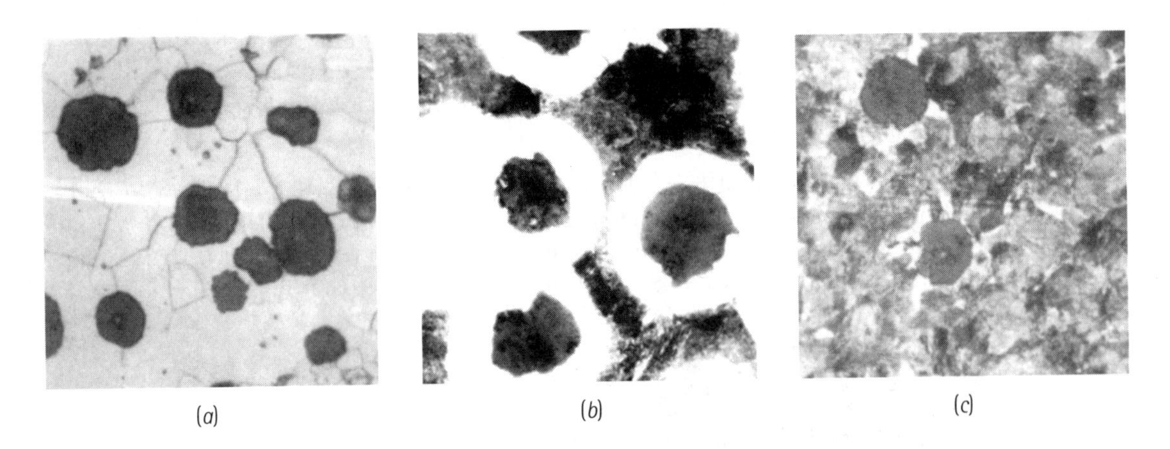

Figure 12.40
(*a*) Annealing ductile iron with a ferrite matrix (× 250). (*b*) As-cast ductile iron with a matrix of ferrite (white) and pearlite (× 250). (*c*) Normalised ductile iron with pearlite matrix (× 250).

1. *Desulphurisation*. Sulphur causes graphite to grow as flakes. Low-sulphur irons are obtained by melting low-sulphur charge materials; by melting in furnaces that remove sulphur from iron during melting; or by mixing the iron with a desulphurising agent such as calcium carbide.
2. *Nodulising*. Magnesium, added in the nodulising step, removes any sulphur and oxygen still in the liquid metal and provides a residual of 0.03% Mg, which causes growth of the graphite to be spheroidal. The magnesium is added near 1 500°C. Unfortunately, magnesium vaporises near 1150°C, so many nodulising alloys contain magnesium diluted with ferrosilicon to reduce the violence of the reaction and permit higher magnesium recoveries are used.

   Fading, or gradual, nonviolent vapourisation or oxidation of magnesium, must also be controlled. If the iron is not poured within a few minutes after nodulising, the iron reverts to grey iron.
3. *Inoculation*. Magnesium by itself is an effective carbide stabiliser, and it causes white iron to form during solidification! Consequently, we must inoculate the iron with FeSi alloys after nodulising. The inoculating effect also fades with time.

Table 12.7 Grades of S.G. Cast Iron (B.S. 2789: 1985).

Grade	Minimum Tensile Strength ($MN.m^{-2}$)	Minimum 0.2% Proof Stress ($MN.m^{-2}$)	Elongation %	Matrix	Heat Treatment
350/22	350	220	22	Ferrite	Annealed
420/12	420	270	12	Mainly Ferrite	Annealed
500/7	500	320	7	Ferrite & Pearlite	Annealed
800/2	800	480	2	Tempered Martensite	Quenched & Tempered

Compared with grey iron, S.G. cast iron has excellent strength, ductility, and toughness. Ductility and strength are also higher than in malleable irons, but due to the higher silicon content in ductile iron, the toughness is lower. Typical structures of ductile irons are shown in Figure 12.40.

Compacted Graphite Cast Iron The graphite shape in compacted graphite cast iron is intermediate between flakes and spheres, with numerous rounded rods of graphite that are interconnected to the nucleus of the eutectic cell (Figure 12.41). This graphite, sometimes called vermicular graphite, also forms when S. G. iron fades.

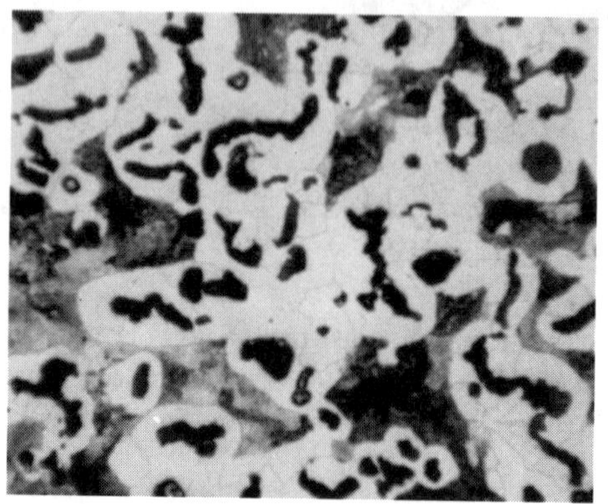

Figure 12.41
The structure of compacted graphite cast iron, with a matrix of ferrite (white) and pearlite (grey) (× 250).

The compacted graphite permits strengths and ductilities that exceed those of grey cast iron, but allows the iron to retain good thermal conductivity and vibration damping properties. The treatment for the compacted graphite iron is similar to that for ductile iron; however, only about 0.015% Mg is introduced during nodulising.

EXAMPLE 12.11 Materials Selection for a Cast Iron Key

Select an appropriate grade of cast iron for the key of your favourite castle door.

SOLUTION

High stresses may be applied when the key is used to turn the rusty lock in your castle door. To ensure that the key does not break off in the lock it should be strong ductile. This rules out grey and white cast irons.

The key likely has a small diameter and, consequently, will solidify very quickly. Even with effective inoculation: we may not be able to encourage the stable graphite eutectic reaction, therefore, S. G. iron may be difficult to produce. However, the rapid cooling rate will easily produce white cast iron, which can then be heat-treated to produce malleable iron. Perhaps a P45–06 or P55–04 grade would be appropriate.

EXAMPLE 12.12 **Materials Selection for a Bottle-Making Mould**

Select an appropriate grade of cast iron for a mould into which hot glass can be introduced and, with air pressure, be blown into the shape of a bottle.

SOLUTION

The mould will be made in two parts by pouring the liquid iron into a sand mould. At least some machining will be required to produce a smooth surface against which the glass can be formed.

We wish to produce the glass bottles as quickly as possible to maximise the production rate. There is, however, an optimum mould temperature for forming the bottles. Each time we introduce the hot glass, the mould also heats. This heating slows the rate of cooling of the bottle, which increases the time before the bottle can be removed from the mould. In addition, we have to cool the mould back to the optimum mould temperature. To minimise the time required for each cycle, we would like to remove the heat from the mould-glass interface as quickly as possible.

The mould does not heat uniformly during each cycle; the surface next to the glass heats to the highest temperature, causing more expansion of the mould at the surface. This expansion can lead to cyclical stresses caused by the manufacture of the glass which, in turn, can cause thermal fatigue and cracking of mould.

Grey cast iron may be the best choice for this application. Machinability of grey iron is excellent, reducing the finishing costs of the mould. The graphite flakes distribute the thermal stresses, providing resistance to thermal fatigue. Finally, the interconnected graphite flakes provide excellent thermal conductivity, enabling heat to be removed quickly from the mould surface.

SUMMARY

The properties of steels, determined by dispersion strengthening, depend on the amount, size, shape, and distribution of cementite. These factors are controlled by alloying and heat treatment.

- A process anneal recrystallises cold-worked steels.
- Spheroidising produces large, spheroidal Fe_3C and good machinability in high-carbon steels.
- Annealing, involving a slow furnace cool after austenitising, gives a coarse pearlitic structure containing lamellar Fe_3C.
- Normalising, involving an air cool after austenitising, gives a fine pearlitic structure and higher strength compared with annealing.

- In isothermal annealing, pearlite with a uniform interlamellar spacing is obtained by transforming the austenite at a constant temperature.
- Austempering is used to produce bainite, containing rounded Fe_3C, by an isothermal transformation.
- Quench and temper heat treatments require the formation and decomposition of martensite, providing exceptionally fine dispersions of round Fe_3C.

We can better understand the mechanics of the heat treatments by use of TTT diagram, CCT diagrams, and hardenability curves.

- The TTT diagrams describe how austenite transforms to pearlite and bainite at a constant temperature.
- The CCT diagrams describe how austenite transforms during continuous cooling. These diagrams give the cooling rates needed to obtain martensite in quench and temper treatments.
- The hardenability curves compare the ease with which different steels transform to martensite.
- Alloying elements increase the times required for transformations in the TTT diagrams, reduce the cooling rates necessary to produce martensite in the CCT diagrams, and improve the hardenability of the steel.

Special steels and heat treatments provide unique properties or combinations of properties. Of particular importance are surface-hardening treatments, such as carburising, that produce an excellent combination of fatigue and impact resistance. Stainless steels, which contain a minimum of 12% Cr, have excellent corrosion resistance.

Cast irons, by definition, undergo the eutectic reaction during solidification. Depending on the composition and treatment, either γ and Fe_3C or γ and graphite form during freezing:

- White cast iron, with good wear resistance, is obtained when Fe_3C forms during the eutectic reaction.
- Malleable cast iron, with good strength, ductility, and toughness, is produced by heat-treating white cast iron to form rounded graphite.
- By producing graphite directly solidification, grey cast iron, S. G. iron, and compacted graphite cast iron are produced. Because graphite flakes form in grey iron, its strength and ductility are limited. The graphite spheres that form in S. G. iron as a result of the addition of magnesium permit good strength and ductility. Compacted graphite iron has intermediate properties.

GLOSSARY

Annealing (cast iron)
A heat treatment used to produce a ferrite matrix in a cast iron by austenitising, then furnace cooling.

Annealing (steel)

A heat treatment used to produce a soft, coarse pearlite in a steel by austenitising, then furnace cooling.

Ausforming

The thermomechanical heat treatment in which austenite is plastically deformed below the A_1 temperature, then permitted to transform to bainite or martensite.

Austempering

The isothermal heat treatment by which austenite transforms to bainite.

Austenitising

Heating a steel or cast iron to a temperature where homogeneous austenite can form. Austenitising is the first step in most of the heat treatments for steel and cast irons.

Carbon equivalent

Carbon plus one-third of the silicon in a cast iron.

Carbonitriding

Hardening the surface of a steel with carbon and nitrogen obtained from a special gas atmosphere.

Carburising

A group of surface-hardening techniques by which carbon diffuses into steel.

Case depth

The depth below the surface of a steel to which hardening occurs by surface hardening and carburising processes.

Cast iron

Ferrous alloys containing sufficient carbon so that eutectic reaction occurs during solidification.

Compacted graphite cast iron

A cast iron treated with small amounts of magnesium and titanium to cause graphite to grow during solidification as an interconnected, coral-shaped precipitate, giving properties midway between grey and S.G. iron.

Cyaniding

Hardening the surface of a steel with carbon and nitrogen obtained from a bath of liquid cyanide solution.

Drawing

Reheating a malleable iron in order to reduce the amount of carbon combined as cementite by spheroidising pearlite, tempering martensite, or graphitising both.

Dual-phase steels

Special steels treated to produce martensite dispersed in a ferrite matrix.

Duplex stainless steel

A special class of stainless steels containing a microstructure of ferrite and austenite.

Eutectic cell

A cluster of graphite flakes and austenite produced during solidification that are all interconnected to a common nucleus.

Fading

The loss of the nodulising or inoculating effect in cast irons as a function of time, permitting undesirable changes in microstructure and properties.

First stage graphitisation

The first step in the heat treatment of a malleable iron, during which the massive carbides formed during solidification are decomposed to graphite and austenite.

Grey cast iron

Cast iron which, during solidification, permits graphite flakes to grow, causing low strength and poor ductility.

Hardenability

The ease with which a steel can be quenched to form martensite. Steels with high hardenability form martensite even on slow cooling.

Hardenability curves

Graphs showing the effect of cooling rate on the hardness of an as-quenched steel.

Inoculation

The addition of an agent to the molten cast iron that provides nucleation sites at which graphite precipitates during solidification.

Isothermal annealing

Heat treatment of a steel by austenitising, cooling, rapidly to a temperature between the A_1 and the nose of the TTT curve, and holding until the austenite transforms to pearlite.

Jominy distance

The distance from the quenched end of a Jominy bar. The Jominy distance is related to the cooling rate.

Jominy test

The test used to evaluate hardenability. An austenitised steel bar is quenched at one end only, thus producing a range of cooling rates along the bar.

Malleable cast iron

Cast iron obtained by a lengthy heat treatment, during which cementite decomposes to produce rounded clumps of graphite. Good strength and ductility are obtained as a result of this structure.

Maraging steels

A special class of alloy steels that obtain high strengths by a combination of martensite and age-hardening reactions.

Martempering

Quenching austenite to a temperature just above the M_s and holding until the temperature is equalised throughout the steel before further cooling to produce martensite. This process reduces residual stresses and quench cracking. Also known as marquenching.

Nitriding

Hardening the surface of a steel with nitrogen obtained from a special gas atmosphere.

Nodulising

The addition of magnesium to molten cast iron to cause the graphite to precipitate as spheres rather than as flakes during solidification.

Normalising
A simple heat treatment obtained by austenitising and air cooling to produce a fine pearlite structure. This can be done for steels and cast irons.

Process anneal
A low-temperature heat treatment used to eliminate all or part of the effect of cold working in steels.

Quench cracks
Cracks that form at the surface of a steel during quenching due to tensile residual stresses that are produced because of the volume change that accompanies the austenite-to-martensite transformation.

Retained austenite
Austenite that is unable to transform into martensite during quenching because of the volume expansion associated with the reaction.

Second stage graphitisation
The second step in the heat treatment of malleable irons that are to have a ferritic matrix. The iron is cooled slowly from the first stage graphitisation temperature so that austenite transforms to ferrite and graphite rather than to pearlite.

Secondary hardening peak
Unusually high hardness in a steel tempered at a high temperature caused by the precipitation of alloy carbides.

S.G. cast iron
Cast iron treated with magnesium to cause graphite to precipitate during solidification as spheres, permitting excellent strength and ductility. Also known as ductile cast iron and nodular cast iron.

Spheroidised carbide
A microconstituent containing coarse spheroidal cementite particles in a matrix of ferrite, permitting excellent machining characteristics in high-carbon steels. Refer to Figure 12.6.

Stainless steels
A group of ferrous alloys that contain at least 12% Cr, providing extraordinary corrosion resistance.

Tempered martensite
The microconstituent of ferrite and cementite formed when martensite is tempered.

Tool steels
A group of high-carbon steels that provide combinations of high hardness, toughness, or resistance to elevated temperatures.

Vermicular graphite
The rounded, interconnected graphite that forms during the solidification of cast iron. This is the intended shape in compacted graphite iron, but it is a defective shape in S.G. iron.

White cast iron
Cast iron that produces cementite rather than graphite during solidification. The white irons are very hard and brittle.

PROBLEMS

12.1 Calculate the amounts of ferrite, cementite, primary microconstituent, and pearlite in the following plain carbon steels:

(a) 0.15% C (b) 0.35% C

(c) 0.95% C (d) 1.3% C

12.2 Estimate the % carbon for steels having the following microstructures:

(a) 38% pearlite-62% primary ferrite

(b) 93% pearlite-7% primary cementite

(c) 97% ferrite-3% cementite

(d) 86% ferrite-14% cementite

12.3 Complete the following table:

	0.35% C steel	1.15% C steel
A_1 temperature		
A_3 or A_{cm} temperature		
Full annealing temperature		
Normalising temperature		
Process annealing temperature		
Spheroidising temperature		

12.4 The pearlite in a eutectoid steel (0.8% C) consists of cementite platelets which are 4 × 10^{-4} mm thick, and the ferrite platelets which are 14 × 10^{-4} mm thick. In a spheroidised 0.8% C steel, the cementite spheres are 4 × 10^{-2} mm in diameter. Estimate the total interface area between the ferrite and cementite in a cubic centimetre of each steel. Determine the percentage reduction in surface area when the pearlitic steel is spheroidised. The density of ferrite is 7.87 $Mg.m^{-3}$ and that of cementite is 7.66 $Mg.m^{-3}$.

12.5 Describe the microstructure present in a 0.5% C steel after each step in the following heat treatments:

(a) heat at 820°C, quench to 650°C and hold for 90 s, and quench to 25°C,

(b) heat at 820°C, quench to 450°C and hold for 90 s, and quench to 25°C,

(c) heat at 820°C, and quench to 25°C,

(d) heat at 820°C, quench to 720°C and hold for 100 s, and quench to 25°C,

(e) heat at 820°C, quench to 720°C and hold for 100 s, quench to 400°C and hold 500 s, and quench to 25°C,

(f) heat at 820°C, quench to 720°C and hold for 100 s, quench to 400°C and hold for 10 s, and quench to 25°C,

(g) heat at 820°C, quench to 25°C, heat to 500°C and hold for 10^3 s, and air cool to 25°C.

12.6 Describe the microstructure present in a 1.1% C steel after each step in the following heat treatments:

(a) heat to 900°C, quench to 400°C and hold for 10^3 s, and quench to 25°C,

(b) heat to 900°C, quench to 600°C and hold for 50 s, and quench to 25°C,

(c) heat to 900°C, and quench to 25°C,

(d) heat to 900°C, quench to 300°C and hold for 200 s, and quench to 25°C,

(e) heat to 900°C, quench to 675°C and hold for 1 s, and quench to 25°C,

(f) heat to 900°C, quench to 675°C and hold for 1 s, quench to 400°C and hold for 900 s, and slowly cool to 25°C,

(g) heat to 900°C, quench to 675°C and hold for 1 s, quench to 300°C and hold for 103 s, and air cool to 25°C,

(h) heat to 900°C, quench to 300°C and hold for 100 s, quench to 25°C, heat to 450°C for 3600 s, and cool to 25°C.

12.7 Recommend appropriate isothermal heat treatment to obtain the following, including appropriate temperatures and times:

(a) an isothermally annealed 0.5%C steel with HRC 23,

(b) an isothermally annealed 1.1% C steel with HRC 40,

(c) an isothermally annealed 0.8% C steel with HRC 38,

(d) an austempered 0.5% C steel with HRC 40,

(e) an austempered 1.1% C steel with HRC 55,

(f) an austempered 0.8% C steel with HRC 50.

12.8 Compare the minimum times required to isothermally anneal the following steels at 600°C. Discuss the effect of the carbon content of the steel on the kinetics of nucleation and growth during the heat treatment.

(a) 0.5%C (b) 0.8%C (c) 1.1%C

12.9 We wish to produce a 0.5%C steel that has a Brinell hardness of at least 330 and an elongation of at least 15%.

(a) Recommend a heat treatment, including appropriate temperatures, that permits this to be achieved. Determine the yield strength and tensile strength that are obtained by this heat treatment.

(b) What yield and tensile strength would be obtained in a 0.8% C steel by the same heat treatment?

(c) What yield strength, tensile strength, and Elongation would be obtained in the 0.5% C steel if it were normalised?

12.10 We wish to produce a 0.5% C steel that has a tensile strength of at least 1200 $Mn.m^{-2}$ and a % reduction in area of at least 50%.

(a) Recommend a heat treatment, including appropriate temperatures, that permits this to be achieved. Determine the Brinell hardness number, % Elongation, and yield strength that are obtained by this heat treatment.

(b) What yield strength and tensile strength would be obtained in a 0.8% C steel by the same heat treatment?

(c) What yield strength, tensile strength, and elongation would be obtained in the 0.5% C steel if it were annealed?

12.11 A 0.3.% C steel is given an improper quench and temper heat treatment, producing a final structure composed of 60% martensite and 40% ferrite. Estimate the carbon content of the martensite and the austenitising temperature that was used. What austenitising temperature would you recommend?

12.12 A 0.5% C steel should be austenitised at 820°C, quenched in oil to 25°C, and tempered at 400°C for an appropriate time.

(a) What yield strength, hardness, and % Elongation would you expect to obtain from this heat treatment?

(b) Suppose the actual yield strength of the steel is found to be 860 $Mn.m^{-2}$. What might have gone wrong in the heat treatment to cause this low strength?

(c) Suppose the hardness is found to be HB 525. What might have gone wrong in the heat treatment to cause this high hardness?

12.13 A part produced from a low-alloy, 0.2% C steel (Figure 12.17) has a microstructure containing ferrite, pearlite, bainite, and martensite after quenching. What microstructure would be obtained if we used a 0.8% steel? What microstructure would be obtained if we used an AISI–SAE 4340 steel?

12.14 Fine pearlite and a small amount of martensite are found in a quenched 1080 steel. What microstructure would be expected if we used a low-alloy, 0.2% C steel? What microstructure would be obtained if we used an AISI-SAE 4340 steel?

12.15 We have found that an AISI-SAE 1070 steel, when austenitised at 750°C, forms a structure containing pearlite and a small amount of grain boundary ferrite that gives acceptable strength and ductility. What changes in the microstructure, if any, would be expected if the AISI-SAE 1070 steel contained an alloying element such as Mo or Cr? Explain.

12.16 Using the TTT diagrams, compare the hardenabilities of 4340 and 1050 steels by determining the times required for isothermal transformation of ferrite and pearlite (F_s, P_s, and P_f) to occur at 650°C.

12.17 We would like to obtain a hardness of HRC 38 to 40 in a quenched steel. What range of cooling rates would we have to obtain for the following steels? Are some steels inappropriate?

(a) 4340 (b) 8640 (c) 9310

(d) 4320 (e) 1050 (f) 1080

12.18 A steel part must have an as-quenched hardness of HRC 35 in order to avoid excessive wear rates during use. When the part is made from 4320 steel, the hardness is only HRC 32. Determine the hardness if the part were made under identical conditions, but with the following steels. Which, if any, of these steels would be better choices than 4320?

(a) 4340 (b) 8640 (c) 9310

(d) 1050 (e) 1080

12.19 A part produced from a 4320 steel has a hardness of HRC 35 at a critical location after quenching. Determine

(a) the cooling rate at that location and

(b) the microstructure and hardness that would be obtained if the part were made of a 1080 steel.

12.20 A 1080 steel is cooled at the fastest possible rate that still permits all pearlite to form. What cooling rate, Jominy distance, and hardness are expected for this cooling rate?

12.21 Determine the hardness and microstructure at the centre of a 38 mm diameter 1080 steel bar produced by quenching in

(a) unagitated oil,

(b) unagitated water, and

(c) agitated brine.

12.22 A 50 mm-diameter bar of 4320 steel is to have a hardness of at least HRC 35. What is the minimum severity of the quench (H coefficient)? What type of quenching medium would you recommend to produce the desired hardness with the least chance of quench cracking?

12.23 A steel bar is to be quenched in agitated water. Determine the maximum diameter of the bar that will produce a minimum hardness of HRC 40 if the bar is:

(a) 1050 (b) 1080 (c) 4320

(d) 8640 (e) 4340

12.24 The centre of a 25mm-diameter bar of 4320 steel has a hardness of HRC 40. Determine the hardness and microstructure at the centre of a 50 mm bar of 1050 steel quenched in the same medium.

12.25 A 1010 steel is to be carburised using a gas atmosphere that produces 1.0% C at the surface of the steel. The case depth is defined as the distance below the surface that contains at least 0.5% C. If carburising is done at 1 000°C, determine the time required to produce a case depth of 0.25 mm. (*See* Chapter 5 for review.)

12.26 A 0.15% C steel is to be carburised at 1 050°C for 2 h using a gas atmosphere that produces 1.2% C at the surface of the steel. Plot the percent carbon versus the distance from the surface of the steel. If the steel is slowly cooled after carburising, determine the amount of each phase and microconstituent at 0.05 mm intervals from the surface. (*See* Chapter 5.)

12.27 A 0.5% C steel is welded. After cooling, hardnesses in the heat-affected zone are obtained at various locations from the edge of the fusion zone. Determine the hardnesses expected at each point if a 0.8% C steel were welded under the same conditions. Predict the microstructure at each location in the as-welded 0.8% C steel.

Distance from Edge of Fusion Zone	Hardness in 0.5% steel weld
0.05 mm	HRC 50
0.10 mm	HRC 40
0.15 mm	HRC 32
0.20 mm	HRC 28

12.28 We wish to produce a martensitic stainless steel containing 17% Cr. Recommend a carbon content and austenitising temperature that would permit us to obtain 100% martensite during the quench. What microstructure would be produced if the martensite were then tempered until the equilibrium phases formed?

12.29 Occasionally, when an austenitic stainless steel is welded, the weld deposit may be slightly magnetic. Based on the Fe-Cr-Ni-C phase diagram [Figure 12.30(b)], what phase would you expect is causing the magnetic behaviour? Why might this phase have formed? what could you do to restore the nonmagnetic behaviour?

12.30 A tensile bar of a grey iron casting is found to have a tensile strength of 345 $MN.m^{-2}$. Why is the tensile strength greater than that given by the class number? (B.S. gives strength as 275 $MN.m^{-2}$). What do you think is the diameter of the test bar?

12.31 You would like to produce a grey iron casting that freezes with no primary austenite or graphite. If the carbon content in the iron is 3.5%, what percentage of silicon must you add?

12.32 We find that first stage graphitisation during the production of a 25 mm-thick malleable iron casting can be accomplished in 6 h if the white iron casting is slowly heated to the FSG temperature. What will be the effect of the following changes on the FSG time? Explain.

(a) increasing the rate at which the casting is heated to the FSG temperature,

(b) producing malleable iron from a 12 mm-thick white iron casting,

(c) increasing the silicon content of the white iron by 0.25%.

12.33 When the thickness of an S. G. iron casting increases, the number of graphite nodules normally decreases.

(a) What effect will this result have on the amount of ferrite present in the matrix? Explain.

(b) Suppose that you observed the opposite effect of thickness on the amount of ferrite. How would you explain this phenomenon?

12.34 We would like to produce a 420/12 grade of S. G. iron without heat treating.

(a) What major phase should be present in the matrix?

(b) Would increasing the number of the graphite nodules produced during solidification help or hinder efforts to produce this matrix? Explain. Suggest a method by which the number of graphite nodules might be changed.

(c) What changes in the composition of the iron might help produce the desired matrix?

12.35 Compare the expected hardenabilities of a plain carbon steel, a malleable cast iron, and an S. G. cast iron. Explain why you expect different hardenabilities.

12.36 A B35/12 malleable cast iron is produced, requiring both FSG and SSG treatments. What went wrong with the treatment if:

(a) The final matrix includes ferrite with 10% pearlite

(c) the final structure includes ferrite with 15% massive Fe_3C

Design Problems

12.37 We would like to produce a 50 mm-thick steel wear-plate for a rock-crushing unit. To avoid frequent replacement of the wear plate, the hardness should exceed HRC 38 within 6 mm of the steel surface. However, the centre of the plate should have a hardness of no more than HRC 32 to ensure some toughness. We have only a water quench available to us. Design the plate, assuming that we only have the steels given in Figure 12.23 available to us.

12.38 A quenched and tempered 1.1% C steel is found to have surface cracks that cause the heat treated part to be rejected by the customer. Why did the cracks form? Design a heat treatment, including appropriate temperatures and times, that will minimise these problems.

12.39 Design a corrosion-resistant steel to use for a pump that transports liquid helium at 4 K in a superconducting magnet.

12.40 Design a heat treatment for a hook made of 25 mm-diameter steel rod having a microstructure containing a mixture of ferrite, bainite, and martensite after quenching. Estimate the mechanical properties of your hook.

12.41 Design a annealing treatment for a 0.5% C steel. Be sure to include details of temperatures, cooling rates, microstructures, and properties.

12.42 Design a process to produce a 5 mm-diameter steel shaft having excellent toughness, yet excellent wear and fatigue resistance. The surface hardness should be at least HRC 60, and the hardness 0.1 mm beneath the surface should be approximately HRC 50. Describe the process, including details of the heat-treating atmosphere, the composition of the steel, temperatures, and times.

CHAPTER 13

Nonferrous Alloys

13.1 Introduction

The ferrous alloys – even stainless steels and cast irons – use similar methods for controlling microstructures and properties. However, the structures and behaviour of the different groups of nonferrous alloys have enormous differences. Melting temperatures, for example, vary from near room temperature for gallium to over 3 000°C for tungsten. Strengths vary from 5 $MN.m^{-2}$ to over 1 500 $MN.m^{-2}$. Aluminium, magnesium, and beryllium (the 'light metals') have very low densities, whereas lead and tungsten have exceptionally high densities.

In many applications, weight is a critical factor. To relate the strength of the material to its weight, a specific strength, or strength-to-weight ratio, is defined:

$$\text{Specific strength} = \frac{\text{strength}}{\text{density}} \qquad (13.1)$$

Table 13.1 compares the specific strength of some high-strength nonferrous alloys.

Another factor in designing with nonferrous metals is their cost, which also varies considerably. Table 13.1 gives the appropriate price of metals in 1992. One should note, however, that the price of the metal is only a small portion of the cost of a part. Fabrication and finishing, not to mention marketing and distribution, often contribute much more to the overall cost of a part.

13.2 Aluminium Alloys

Aluminium is the second most plentiful metal on earth, but, until the late 1800s, was expensive and difficult to produce. The 2.75 kg cap installed on the top of the Washington Monument in 1884 was one of the largest aluminium parts made up to that time. Development of electrical power and the Hall–Heroult process for electrolytically reducing Al_2O_3 to liquid metal (Figure 13.1) allowed aluminium to become one of the most widely used and inexpensive engineering materials. Applications number in the millions, including beverage cans, household appliances, chemical processing equipment, electrical power transmission equipment, automotive components, and aerospace parts and assemblies.

Table 13.1 Specific strength and cost of nonferrous alloys.

Metal	Density ($Mg.m^{-3}$)	Tensile* Strength ($MN.m^{-2}$)	Specific Tensile Strength ($m^2.s^{-2}$)	Cost per kg (US$)
Aluminium	2.70	570	211 000	1.30
Beryllium	1.85	380	205 000	660.00
Copper	8.93	1300	146 000	2.45
Lead	11.36	70	6 000	0.80
Magnesium	1.74	380	218 000	3.00
Nickel	8.90	1360	153 000	9.00
Titanium	4.51	1350	299 000	12.15
Tungsten	19.25	1030	54 000	22.00
Zinc	7.13	520	73 000	1.25
Iron	7.87	2070	263 000	0.22

*Figures represent the highest strengths available from commercial alloys.

General Properties of Aluminium Aluminium has a density of 2.70 $Mg.m^{-3}$, or one-third the density of steel, and a modulus of elasticity of 70 $GN.m^{-2}$. Although aluminium alloys have low tensile properties compared with those of steel, their specific strength (or strength-to-weight ratio) is excellent. Aluminium is often used when weight is an important factor, as in aircraft and automotive applications.

Aluminium also responds readily to strengthening mechanisms. Table 13.2 compares the strength of pure annealed aluminium with that of alloys strengthened by various techniques. The alloys may be 30 times stronger than pure aluminium.

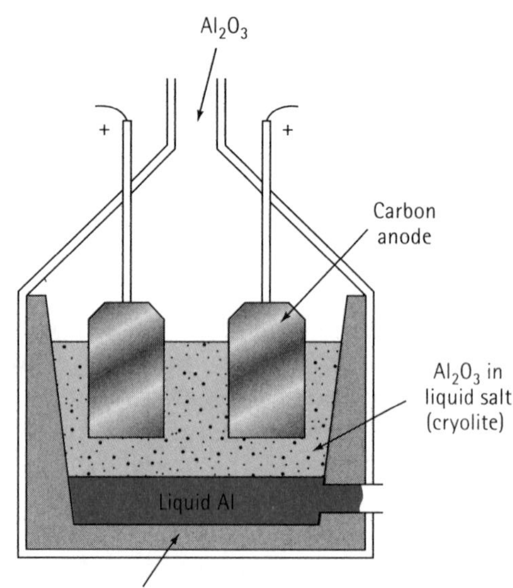

Figure 13.1
Production of aluminium in an electrolytic cell.

Table 13.2 The effect of strengthening mechanisms in aluminium and aluminium alloys.

Material	Tensile Strength ($MN.m^{-2}$)	Yield Strength ($MN.m^{-2}$)	% Elongation	Yield Strength (alloy) / Yield Strength (pure)
Pure Al (99.999% Al)	45	17	60	
Commercially pure Al (99% Al)	90	35	45	2.0
Solid-solution strengthened Al alloy (Al, 1.2% Mn)	110	41	35	2.4
75% cold-worked Al (99% Al)	165	152	15	8.8
Dispersion-strengthened Al alloy (Al, 5% Mg)	290	152	35	8.8
Age-hardened Al alloy (Al, 5.6% Zn, 2.5% Mg)	572	503	11	29.2

Aluminium's beneficial physical properties include high electrical and thermal conductivity, nonmagnetic (paramagnetic) behaviour, and excellent resistance to oxidation and corrosion. Aluminium reacts with oxygen, even at room temperature, to produce an extremely thin aluminium oxide (Al_2O_3) layer that protects the underlying metal from many corrosive environments.

Aluminium does not, however, display a high fatigue limit, so failure by fatigue may eventually occur even at low stresses. Because of its low melting temperature, aluminium does not perform well at elevated temperatures. Finally, aluminium alloys have a low hardness, leading to poor wear resistance.

EXAMPLE 13.1

A steel cable 12.5 mm in diameter has a yield strength of 310 $MN.m^{-2}$. The density of steel is about 7.87 $Mg.m^{-3}$. Based on the data in Table 13.5, determine (a) the maximum load that the steel cable can support, (b) the diameter of a cold-worked aluminium-manganese alloy (3004-H18) required to support the same load as the steel, and (c) the weight per metre of the steel cable versus the aluminium alloy cable.

SOLUTION

a. Load $= F = \sigma_Y A = 310 \left(\frac{\pi}{4}\right)(0.0125)^2 = 0.038$ MN = 38 kN

b. The yield strength of the aluminium alloy is 250 $MN.m^{-2}$. Thus:

$$A = \frac{\pi}{4} d^2 = \frac{F}{\sigma_Y} = \frac{0.038}{250} = 1.52 \times 10^{-4} \text{ m}^2$$

$$d = 0.0139 \text{ m} = 13.9 \text{ mm}$$

c. Density of steel $= \rho = 7.87$ Mg.m^{-3}

Density of aluminium $= \rho = 2.70$ Mg.m^{-3}

Weight of 1 m of steel $= Al\rho$ $= \frac{\pi}{4}(0.0125)^2(1)(7.87) = 9.66 \times 10^{-4}$ Mg $= 0.966$ kg

Weight of 1 m of aluminium $= Al\rho = \frac{\pi}{4}(0.0139)^2(1)(2.70) = 4 \times 10^{-4}$ Mg $= 0.410$ kg

Although the yield strength of the aluminium is lower than that of the steel and the cable must be larger in diameter, the aluminium cable weighs less than half as much as the steel cable.

Designation Aluminium alloys can be divided into two major groups: wrought and casting alloys, depending on their method of fabrication. Wrought alloys, which are shaped by plastic deformation (hot and/or cold working), have compositions and microstructures significantly different from casting alloys, reflecting the different requirements of the manufacturing process. Within each major group we can divide the alloys into two subgroups: heat-treatable and nonheat-treatable alloys.

Aluminium alloys are designated by the numbering system shown in Table 13.3. The first number specifies the principle alloying elements, and the remaining numbers refer to the specific composition of the alloy. This IADS (International Alloy Designation System) numbering system has been adopted by most countries.

The degree of strengthening is given by the temper designation T or H, depending on whether the alloy is heat-treated or strain-hardened (Table 13.4). Other designations indicate whether the alloy is annealed (O), solution-treated (W), or used in the as-fabricated condition (F). The numbers following the T or H indicate the amount of strain hardening, the exact type of heat treatment, or other special aspects of the processing of the alloy. Typical alloys and their properties are included in Table 13.5.

Wrought Alloys The 1xxx, 3xxx, 5xxx, and most of the 4xxx wrought alloys are not age-hardenable. The 1xxx and 3xxx alloys are single-phase alloys except for the presence of small amounts of inclusions or intermetallic compounds (Figure 13.2). Their properties are controlled by strain hardening, solid solution strengthening, and grain-size control. However, because the solubilities of the alloying elements in aluminium are small at room temperature, the degree of solid solution strengthening is limited.

The 5xxx alloys contain two phases at room temperature – α, a solid solution of magnesium in aluminium, and Mg_2Al_3, a hard, brittle intermetallic compound (Figure 13.3). The aluminium-magnesium alloys are strengthened by a fine dispersion of Mg_2Al_3 as well as by strain hardening, solid solution strengthening, and grain-size control. However, because Mg_2Al_3 is not coherent, age-hardening treatments are not possible.

Table 13.3 IADS (International Alloy Designation System) designation system for aluminium alloys.

Designation	Alloy	Heat treatment
Wrought alloys:		
1xxx	Commercially pure Al (>99% Al)	Not age-hardenable
2xxx	Al-Cu and Al-Cu-Li	Age-hardenable
3xxx	Al-Mn	Not age-hardenable
4xxx	Al-Si and Al-Mg-Si	Age-hardenable if magnesium is present
5xxx	Al-Mg	Not age-hardenable
6xxx	Al-Mg-Si	Age-hardenable
7xxx	Al-Mg-Zn	Age-hardenable
8xxx	Al-Li, Sn, Zr, B, Fe or Cr	Mostly age-hardenable
Casting alloys:		
1xx	Commercially pure Al	Not age-hardenable
2xx	Al-Cu	Age-hardenable
3xx	Al-Si-Cu or Al-Mg-Si	Some are age-hardenable
4xx	Al-Si	Not age-hardenable
5xx	Al-Mg	Not age-hardenable
7xx	Al-Mg-Zn	Age-hardenable
8xx	Al-Sn	Age-hardenable

Table 13.4 Temper designations for aluminium alloys.

F As-fabricated (hot-worked, forged, cast, etc.)

O Annealed (in the softest possible condition)

H Cold-worked

- H1x – cold worked only. (x refers to the amount of cold work and strengthening.)
  - H12 – cold work that gives a tensile strength midway between the O and H14 tempers.
  - H14 – cold work that gives a tensile strength midway between the O and H18 tempers.
  - H16 – cold work that gives a tensile strength midway between the H14 and H18 tempers.
  - H18 – cold work that gives about 75% reduction.
  - H19 – cold work that gives a tensile strength at least 15 $MN.m^{-2}$ greater than that obtained by the H18 temper.
- H2x – cold-worked and partly annealed.
- H3x – cold-worked and stabilised at a low temperature to prevent age-hardening of the structure.

W Solution-treated

T Age-hardened

- T1 – cooled from the fabrication temperature and naturally aged.
- T2 – cooled from the fabrication temperature, cold-worked, and naturally aged.
- T3 – solution treated, cold-worked, and naturally aged.
- T4 – solution treated and naturally aged.
- T5 – cooled from the fabrication temperature and artificially aged.
- T6 – solution-treated and artificially aged.
- T7 – solution-treated and stabilised by overageing.
- T8 – solution-treated, cold-worked, and artificially aged.
- T9 – solution-treated, artificially aged, and cold-worked.
- T10 – cooled from the fabrication temperature, cold-worked, and artificially aged.

Table 13.5 Properties of typical aluminium alloys.

Alloy		Tensile Strength ($MN.m^{-2}$)	Yield Strength ($MN.m^{-2}$)	% Elongation	Applications
Nonheat-treatable wrought alloys:					
1100-O	>99% Al	90	35	40	Electrical components, foil, food processing
1100-H18		165	150	10	
3004-O	1.2% Mn-1.0% Mg	180	70	25	Beverage can bodies, architectural uses
3004-H18		285	250	9	
4043-O	5.2% Si	145	70	22	Filler metal for welding
4043-H18		285	270	1	
5182-O	4.5% Mg	290	130	25	Beverage can tops, marine components
5182-H19		420	395	4	
Heat-treatable wrought alloys:					
2024-T4	4.4% Cu	470	325	20	Truck wheels
2090-T6	2.4% Li-2.7% Cu	550	517	6	Aircraft skins
4032-T6	12% Si-1% Mg	380	320	9	Pistons
6061-T6	1% Mg-0.6% Si	310	275	15	Canoes, railway carriages
7075-T6	5.6% Zn-2.5% Mg	570	505	11	Aircraft frames
Casting alloys:					
201-T6	4.5% Cu	485	435	7	Transmission housings
319-F	6% Si-3.5% Cu	185	125	2	General purpose castings
356-T6	7% Si-0.3% Mg	230	165	3	Aircraft fittings
380-F	8.5% Si-3.5% Cu	315	160	3	Motor housings
390-F	17% Si-4.5% Cu	285	240	1	Automotive engines
443-F	5.2% Si (sand cast)	130	55	8	Food handling equipment, marine fittings
	(permanent mould)	160	60	10	
	(die cast)	230	110	9	

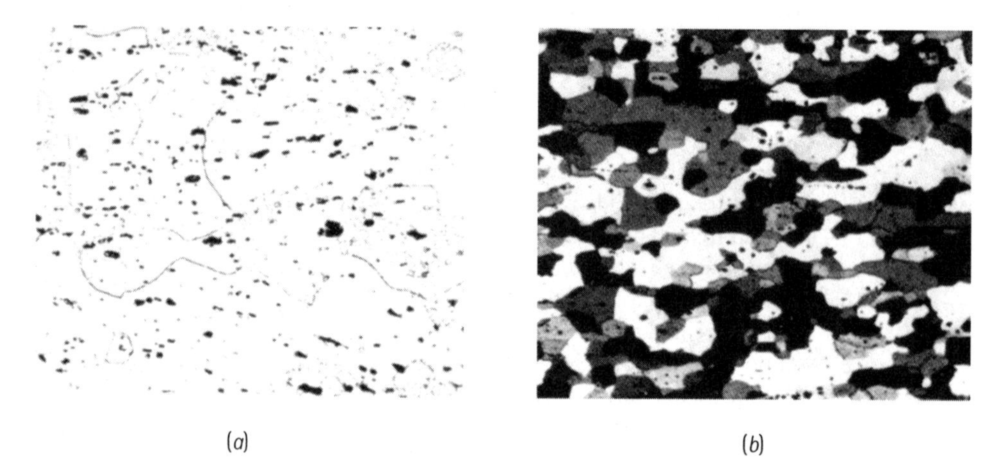

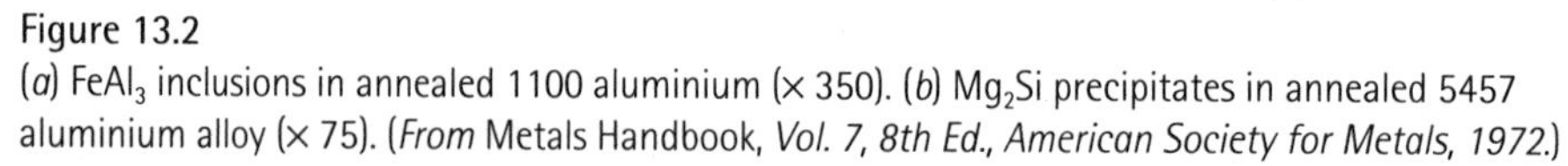

Figure 13.2
(*a*) $FeAl_3$ inclusions in annealed 1100 aluminium (× 350). (*b*) Mg_2Si precipitates in annealed 5457 aluminium alloy (× 75). (*From* Metals Handbook, *Vol. 7, 8th Ed., American Society for Metals, 1972.*)

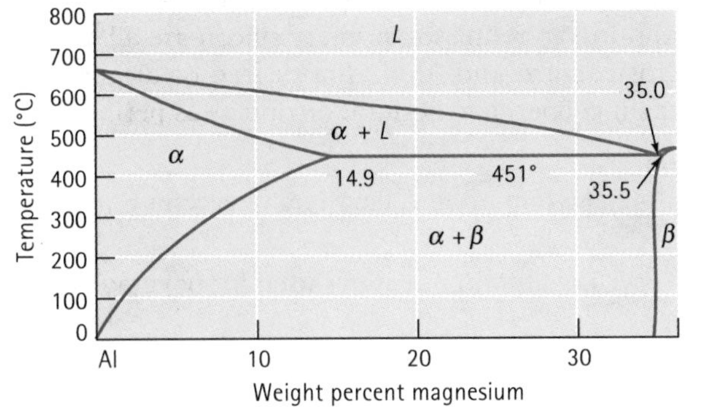

Figure 13.3
Portion of the aluminium-magnesium phase diagram.

The 4xxx series alloys also contain two phases, α and nearly pure silicon, β (Figure 10.22). Alloys that contain both silicon and magnesium can be age-hardened by permitting Mg_2Si to precipitate.

The 2xxx, 6xxx, and 7xxx alloys are age-hardenable alloys. Although excellent specific strengths are obtained for these alloys, the amount of precipitate that can form is limited. In addition, they cannot be used at temperatures above approximately 175°C in the aged condition.

Casting Alloys Many of the common aluminium casting alloys shown in Table 13.5 contain enough silicon to cause the eutectic reaction, giving the alloys low melting points, good fluidity, and good castability. Fluidity is the ability of the liquid metal to flow through a mould without prematurely solidifying, and castability refers to the ease with which a good casting can be made from the alloy.

The properties of the aluminium-silicon alloys are controlled by solid solution strengthening of the α aluminium matrix, dispersion strengthening by the β phase, and solidification, which controls the primary grain size and shape as well as the nature of the eutectic microconstituent. Fast cooling obtained in die casting or permanent mould casting increases strength by refining grain size and the eutectic microconstituent (Figure 13.4). Grain refinement using boron and titanium additions,

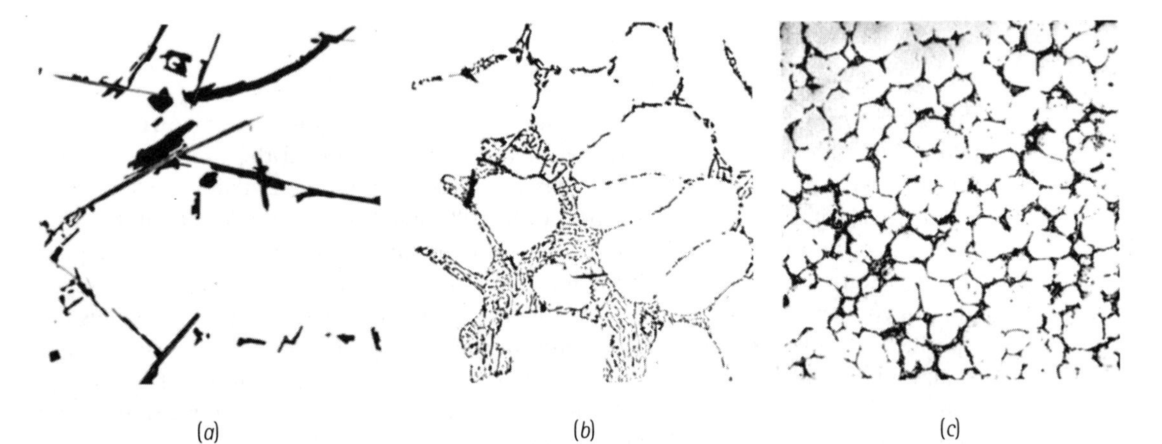

Figure 13.4
(a) Sand-cast 443 aluminium alloy containing coarse silicon and inclusions. (b) Permanent-mould 443 alloy containing fine dendrite cells and fine silicon due to faster cooling. (c) Die-cast 443 alloy with a still finer microstructure (× 350). (*From* Metals Handbook, *Vol. 7, 8th Ed., American Society for Metals, 1972.*)

modification using sodium or strontium to change the eutectic structure, and hardening with phosphorus to refine the primary silicon are all done in certain alloys to improve the microstructure and, thus, the degree of dispersion strengthening. Many alloys also contain copper, magnesium, or zinc, thus permitting age-hardening.

EXAMPLE 13.2 Selection of an Aluminium Recycling Process

Select a method for recycling aluminium alloys used for beverage cans.

SOLUTION

Recycling of aluminium is advantageous because only a fraction of the energy required to produce aluminium from Al_2O_3 is required. However, recycling of beverage cans does present several difficulties.

The beverage cans are made of two aluminium alloys (3004 for the main body, and 5182 for the lids) having different compositions (Table 13.5). The 3004 alloy has the exceptional formability needed to perform the deep drawing process; the 5182 alloy is harder and permits the pull-tops to function properly. When the cans are remelted, the resulting alloy contains both Mg and Mn and is not suitable for either application.

One approach to recycling the cans is to separate the two alloys from the cans. The cans are shredded, then heated to remove the lacquer that helps protect the cans during use. We could then further shred the material at a temperature where the 5182 alloy begins to melt. The 5182 alloy has a wider freezing range than the 3004 alloy and breaks into very small pieces; the more ductile 3004 alloy remains in larger pieces. The small pieces of 5182 can therefore be separated by passing the material through a screen. The two separated alloys can then be melted, cast, and rolled into new can stock.

An alternative method would be to simply remelt the cans. Once the cans have been remelted, we could bubble chlorine gas through the liquid alloy. The chlorine reacts selectively with the magnesium, removing it as a chloride. The remaining liquid can then be adjusted to the proper composition and be recycled as 3004 alloy.

Advanced Aluminium Alloys A number of improvements over conventional aluminium alloys and manufacturing methods have extended the usefulness of this metal. Alloys containing lithium have been introduced, particularly for the aerospace industry. Lithium has a density of 0.534 $Mg.m^{-3}$; consequently, the density of the Al-Li alloys may be up to 10% less than that of conventional aluminium alloys (Figure 13.5). The modulus of elasticity is increased, and the strength can equal or exceed that of conventional alloys (*see* alloy 2090 in Table 13.5). The low density makes the specific strength excellent and the improvement in specific stiffness is even greater making these alloys very desirable for aerospace structural applications. These alloys have a slow fatigue crack growth rate, resulting in improving fatigue resistance, and have good toughness at cryogenic temperatures. The alloys can also be super-plastically formed into complex shapes. The Al-Li alloys find applications for floors, skins, and frames in military and commercial aircraft.

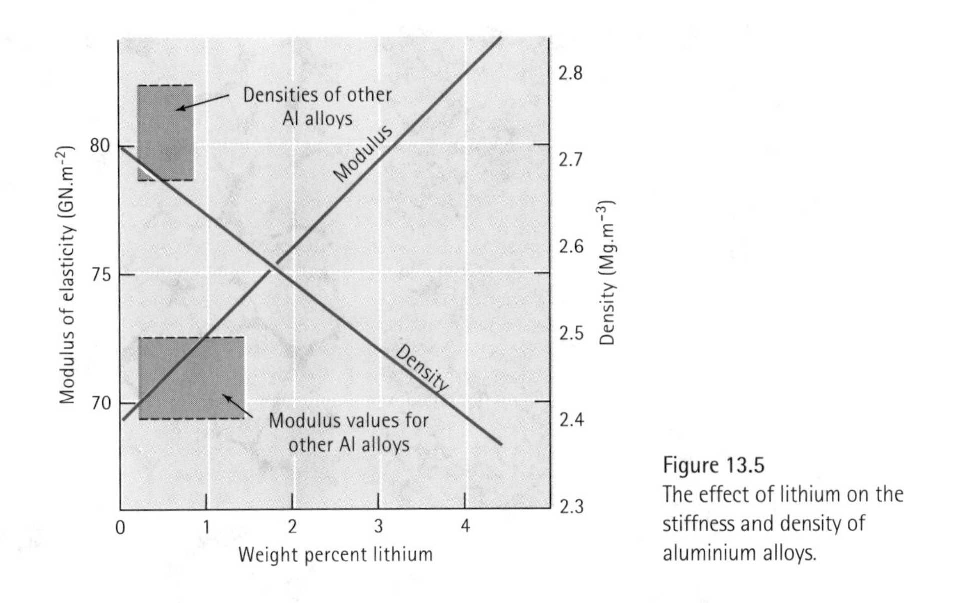

Figure 13.5
The effect of lithium on the stiffness and density of aluminium alloys.

The high strength of the Al-Li alloys is a result of age-hardening (Figure 13.6). Alloys containing up to 2.5% Li can be heat-treated by conventional methods. Additional Li (up to 4%) can be introduced by rapid solidification processing, further enhancing both the light weight and maximum strength.

Modern manufacturing methods help to improve the strength of aluminium alloys, particularly at elevated temperatures. *Rapid solidification processing,* in which the liquid alloy is broken into tiny drops that quickly solidify, is combined with powder metallurgy in the production and fabrication of new alloys. A group of aluminium alloys containing transition elements such as iron and chromium contain tiny intermetallic compounds (dispersoids) such as Al_6Fe. Although the room temperature properties of these advanced alloys are similar to those of conventional alloys, the dispersoids are stable at higher temperatures, providing good properties at temperatures

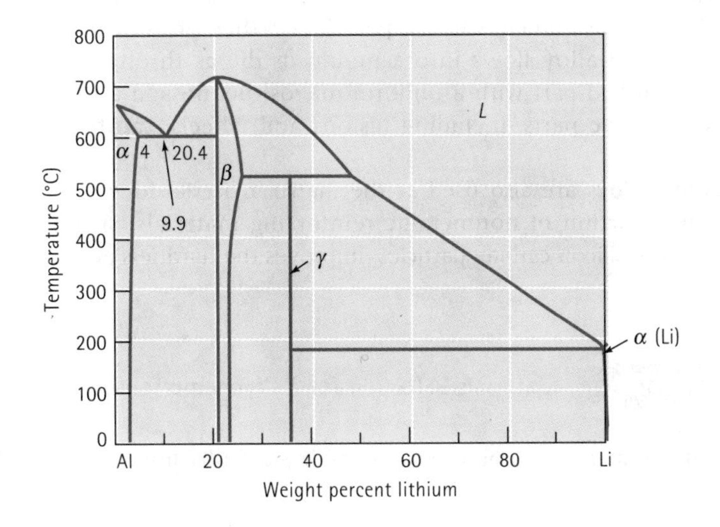

Figure 13.6
The aluminium-lithium phase diagram.

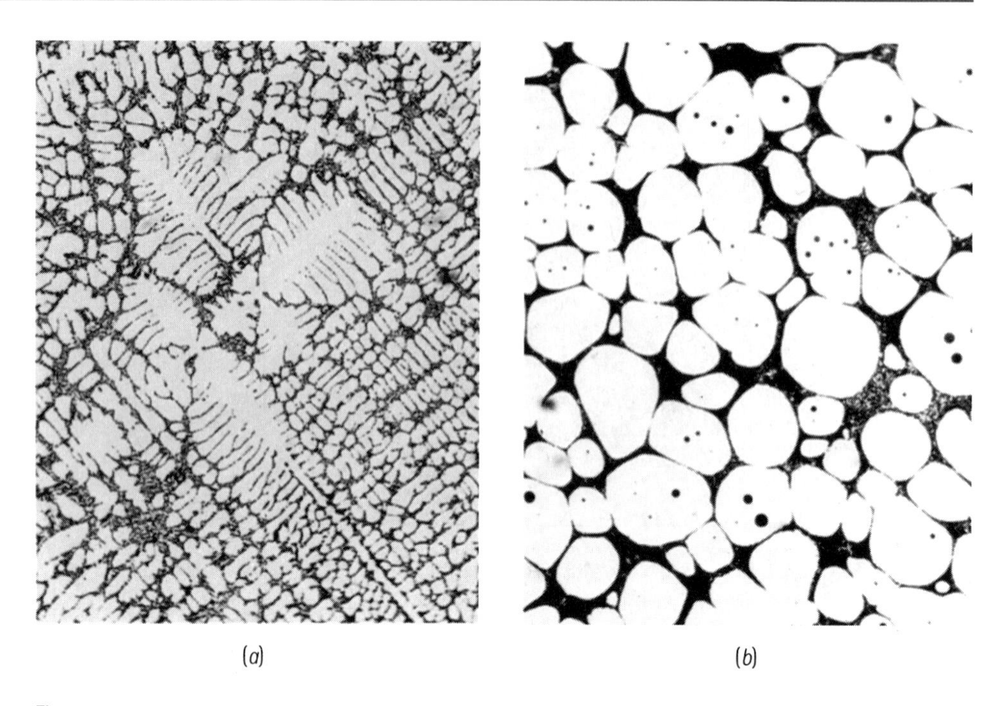

Figure 13.7
The normal dendritic structure in a hypoeutectic Al-Si casting alloy (*a*) can be broken up by stirring action in thixocasting (*b*). The primary aluminium phase forms a round, equiaxed structure in the thixocast material (× 200). (*From* Metals Handbook, *Vol. 2, 10th Ed., ASM International, 1990.*)

where recrystallisation or overageing would otherwise occur. Potential applications include a variety of structural aerospace components.

Aluminium alloys are also processed by thixocasting (also known as rheocasting); the aluminium alloy is vigorously stirred during solidification to break the dendritic structure into small, rounded primary aluminium grains surrounded by a eutectic microconstituent (Figure 13.7). The billet produced during this process is later reheated between the liquidus and eutectic temperatures. When pressure is applied, the partly liquid-partly solid alloy flows into a mould or die as through it were all liquid, producing a finished part with a uniform microstructure and a minimum of casting defects. Automotive parts, including pistons and wheels, can be produced by this method.

Aluminium alloys are also used as the matrix material for *metal-matrix composites* (MMCs). Introduction of nonmetallic reinforcing materials, such as boron fibres, alumina fibres or silicon carbide particles, improves the hardness and high-temperature properties.

EXAMPLE 13.3 Materials Selection for a Cryogenic Tank

Select a suitable material to be used to contain liquid hydrogen fuel for the national aerospace plane (NASP).

SOLUTION

Liquid hydrogen is stored below –253°C; therefore, our tank must have good cryogenic properties. The tank is subject to high stresses, particularly when the plane is inserted into orbit, and it should have good fracture toughness to minimise the chances of catastrophic failure. Finally, it should be light in weight to permit higher payloads or less fuel consumption.

Lightweight aluminium would appear to be a good choice. Figure 6.6 shows that, at very low temperatures, both strength and ductility increase in aluminium. Because of its good ductility, we expect aluminium to also have good fracture toughness, particularly when the alloy is in the annealed condition.

One of the most common cryogenic aluminium alloys is 5083-O. Aluminium-lithium alloys are also being considered for low-temperature applications to take advantage of their even lower density. Alloy 2090-T4 would therefore be another possible choice.

EXAMPLE 13.4 Selection of a Casting Process for Wheels

Select a suitable casting process to produce automotive wheels having reduced weight and consistent and uniform properties.

SOLUTION

Many automotive wheels are produced by permanent mould casting of 356 aluminium alloy. In permanent mould casting, liquid aluminium is introduced into a heated cast iron mould and solidifies. Risers may be needed to ensure that shrinkage voids are not created. This need may require that the wheel be designed to ensure good casting characteristics, rather than to minimise weight. The casting may also cool at very different rates, producing differences in microstructure, such as secondary dendrite arm spacing, and properties throughout the wheel.

An alternative process might be to use the thixocasting process. We would select an alloy with a wide freezing range so that a significant portion of the solidification process occurs by growth of dendrites. A hypoeutectic aluminium-silicon alloy might be appropriate. In the thixocasting process, the dendrites are broken up by stirring during solidification. The billet is later reheated to cause melting of just the eutectic portion of the alloy, and it is then forced into the mould in its semisolid condition at a temperature below the liquidus temperature. When the alloy again solidifies, the primary aluminium phase will be uniform, round grains (rather than dendrites) surrounded by a continuous matrix of eutectic. Because approximately half of the alloy is already solid at the time of injection, the total amount of shrinkage is small, reducing the possibility of internal defects. This result also reduces the requirement for risers, which in turn provides more freedom in designing the wheel for its eventual function rather than for its ease of manufacture.

13.3 Magnesium Alloys

Magnesium, which is often extracted electrolytically from concentrated magnesium chloride in seawater, is lighter than aluminium, with a density of 1.74 Mg.m^{-3}, and it melts at a slightly lower temperature than aluminium (650°C). In many environments, the corrosion resistance of magnesium approaches that of aluminium; however, exposure to salts, such as that near a marine environment, causes rapid deterioration. Although magnesium alloys are not as strong as aluminium alloys, their specific strengths are comparable. Consequently, magnesium alloys are used in aerospace applications, high-speed machinery, and transportation and materials handling equipment.

Magnesium, however, has a low modulus of elasticity (45 GN.m^{-2}) and poor resistance to fatigue, creep, and wear. Magnesium also poses a hazard during casting and machining, since it combines easily with oxygen and burns. Finally, the response of magnesium to strengthening mechanisms is relatively poor.

Structure and Properties Pure magnesium, which has an HCP structure, is less ductile than aluminium. However, magnesium alloys do have some ductility because alloying increases the number of active slip planes. Some deformation and strain hardening can be accomplished at room temperature, and the alloys can be readily deformed at elevated temperatures. Strain hardening produces a relatively small effect in pure magnesium because of the low strain-hardening coefficient.

As in aluminium alloys, the solubility of alloying elements in magnesium at room temperature is limited, causing only a small degree of solid solution strengthening. However, the solubility of many alloying elements increases with temperature, as shown in the Mg-Al phase diagram (Figure 13.8). Therefore, alloys may be strengthened by either dispersion strengthening or age hardening. Some age-hardened magnesium alloys, such as those containing Zr, Th, Ag, or Ce, have good resistance to overageing at temperatures as high as 300°C. Alloys containing up to 9% Li have

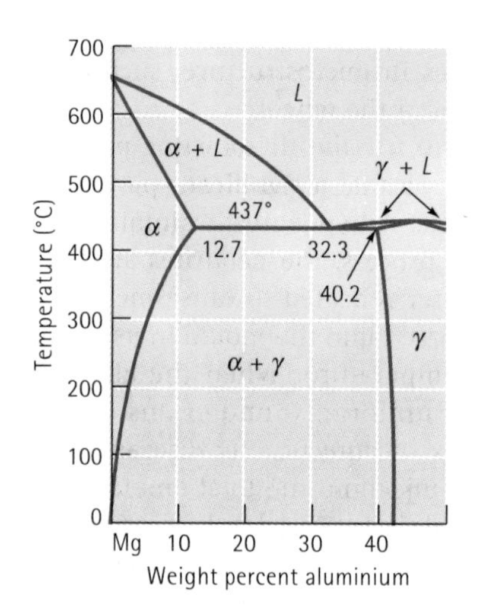

Figure 13.8
The magnesium-aluminium phase diagram.

Table 13.6 Properties of typical magnesium alloys.

Alloy (ASTM Designation)	Composition	Tensile Strength ($MN.m^{-2}$)	Yield Strength ($MN.m^{-2}$)	% Elongation
Pure Mg:				
Annealed		160	90	3–15
Cold-worked		180	115	2–10
Casting alloys:				
AM100-T6	10% Al-0.1% Mn	275	150	1
AZ81A-T4	7.6% Al-0.7% Zn	275	85	15
ZK61A-T6	6% Zn-0.7% Zr	310	195	10
Wrought alloys:				
AZ80A-T5	8.5% Al-0.5% Zn	380	275	7
ZK40A-T5	4% Zn-0.45% Zr	275	255	4
HK31A-H24	3% Th-0.6% Zr	260	205	8

exceptionally light weight. Properties of typical magnesium alloys are listed in Table 13.6. The numbering system developed by the American Society for Testing Materials (ASTM) has been adopted by most countries.

Advanced magnesium alloys include those with very low levels of impurities and those containing large amounts (>5%) of cerium and other rare earths. These alloys form a protective MgO film that improves corrosion resistance. Rapid solidification processing permits larger amounts of alloying elements to be dissolved in the magnesium, further improving corrosion resistance. Improvements in strength, particularly at high temperatures, can be obtained by introducing ceramic particles or fibres such as silicon carbide into the metal.

EXAMPLE 13.5 Materials Selection for a Printer Carrier Unit

Select a suitable material for a carrier unit that operates at high speeds in a printer for a computer.

SOLUTION

The carrier unit moves the printing head back and forth at high speeds. In order for it to operate most efficiently and rapidly, the inertial forces should be minimised. We can do this by making the device as light as possible. Selecting a magnesium alloy would be an economical choice. Its light weight minimises inertial forces, yet reasonable strengths can be obtained by using die casting (fast solidification rates) or an age-hardening heat treatment.

EXAMPLE 13.6 Materials Selection for a Gear Box for a Naval Helicopter

Select a suitable material for a gear box for the rotors on a naval helicopter.

SOLUTION

The gear box must have reasonable strength, and we would like to use a lightweight material so that the helicopter can carry more fuel. But the gear box will be exposed to a saltwater, marine environment. Aluminium alloys might typically be used for such an application, but magnesium is still lighter. Unfortunately, conventional magnesium alloys are subject to corrosion in the marine environment.

One design possibility would be to select a high-purity magnesium alloy. Minimising impurities improves the corrosion resistance, even in a saltwater environment. The magnesium gear box may be as much as 30% lighter than a comparable aluminium part.

13.4 Beryllium

Beryllium is lighter than aluminium, with a density of 1.848 $Mg.m^{-3}$, yet it is stiffer than steel, with a modulus of elasticity of 290 $GN.m^{-2}$. Beryllium alloys, which have yield strengths of 200 to 350 $MN.m^{-2}$, have high specific strengths and maintain both strength and stiffness to high temperatures (Figure 13.9). Instrument grade beryllium is used in inertial guidance systems where the elastic deformation must be minimal; structural grades are used in aerospace applications; and nuclear applications take advantage of the transparency of beryllium to electromagnetic radiation.

Unfortunately, beryllium is expensive, brittle, reactive, and toxic. Its production is quite complicated, requiring that a beryllium sulphate be obtained from the ore, converted to beryllium hydroxide, dissolved and precipitated as a fluoride, and finally reacted with magnesium to produce the metal. The limited availability of ores and the expensive processing make the cost of the beryllium metal high. Beryllium has an HCP crystal structure and so limited ductility at room temperature, but when exposed to the atmosphere at elevated temperatures, it rapidly oxidises to form BeO. These problems require the use of sophisticated manufacturing techniques, such as vacuum casting, vacuum forging, and powder metallurgy, further adding to the expense. Finally, BeO is a carcinogenic material for some people, and special care and equipment are used when beryllium is processed.

EXAMPLE 13.7 Materials Selection for a Mirror for Satellite Signalling

Select a suitable material for a mirror for a satellite imaging system.

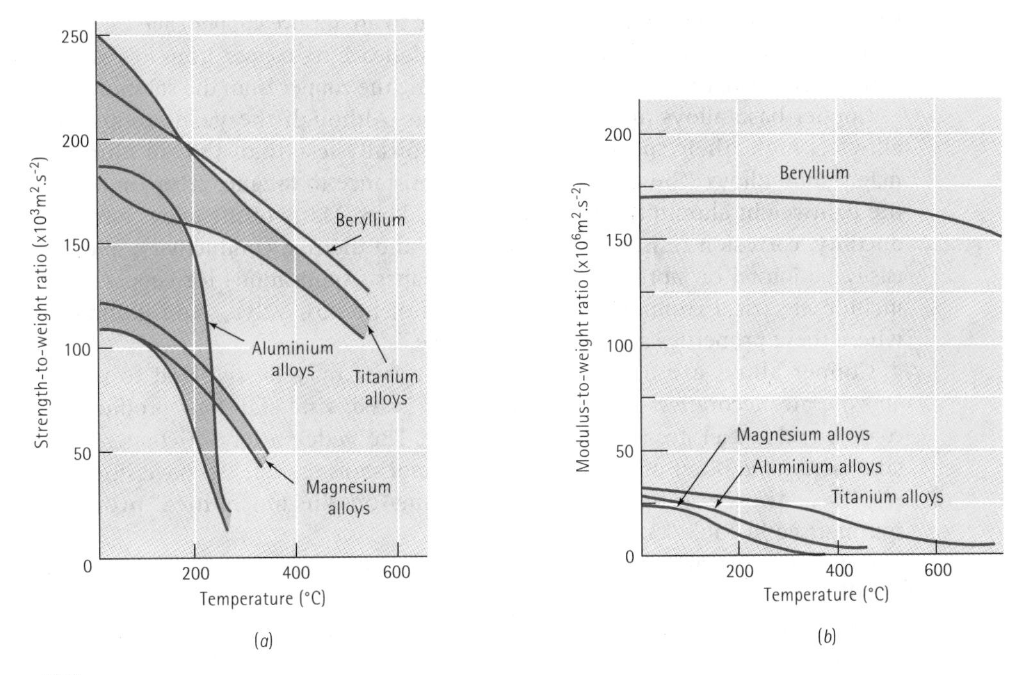

Figure 13.9
A comparison of the strength-to-weight (*a*) and modulus of elasticity-to-weight (*b*) ratios of beryllium and other nonferrous alloys.

SOLUTION

A large mirror is required to help transmit signals to and from satellites. The mirror must maintain precise dimensions and geometries if it is to transmit the signals most efficiently. The mirror should also be as light as possible so that it can be easily transported into orbit. Finally, the mirror should not degenerate during use.

The light metals – aluminium, magnesium, and beryllium – satisfy the weight requirement. Because of its low boiling temperature, however, magnesium may evaporate in the vacuum of space. In order to maintain the most precise size and shape, we need to select a material that undergoes the least amount of deformation – even elastic deformation – when used. Our design might specify beryllium, for its modulus of elasticity is more than four times that of aluminium.

13.5 Copper Alloys

Copper is typically produced by a pyrometallurgical (high-temperature) process. The copper ore containing a high sulphur content is concentrated, then converted into a copper sulphide-iron sulphide matte. Oxygen introduced to the matte converts the iron sulphide to iron oxide and the copper sulphide to an impure copper called blister copper, which is then purified. Other methods include leaching copper from low-sulphur ores with a weak acid, then electrolytically extracting the copper from the solution.

Copper-base alloys are heavier than iron. Although the yield strength of some alloys is high, their specific strength is typically less than that of aluminium or magnesium alloys. The alloys have better resistance to fatigue, creep, and wear than the lightweight aluminium and magnesium alloys. Many of the alloys have excellent ductility, corrosion resistance, and electrical and thermal conductivity, and most can easily be joined or fabricated into useful shapes. Applications for copper-base alloys include electrical components (such as wire), pumps, valves, and plumbing parts, where these properties are used to advantage.

Copper alloys are also unusual in that they may be selected to produce an appropriate decorative colour. Pure copper is red; zinc additions produce a yellow colour, and nickel produces a silver colour. The wide variety of copper-base alloys take advantage of all of the strengthening mechanisms that we have discussed. The effects of these strengthening mechanisms on the mechanical properties are summarised in Table 13.7.

Coppers containing less than 1% impurities are used for electrical applications. Small amounts of cadmium, silver and Al_2O_3 improve their hardness without significantly impairing conductivity. The single-phase copper alloys are strengthened by cold-working. Examples of this effect are shown in Table 13.7. The FCC copper has excellent ductility and a high strain-hardening coefficient.

Table 13.7 Properties of typical copper alloys obtained by different strengthening mechanisms.

Material	Tensile Strength (MN.m^{-2})	Yield Strength (MN.m^{-2})	% Elongation	Strengthening Mechanism
Pure Cu, annealed	210	35	60	
Commercially pure Cu, annealed to coarse grain size	220	70	55	
Commercially pure Cu, annealed to fine grain size	235	75	55	Grain size
Commercially pure Cu, cold-worked 70%	395	365	4	Strain hardening
Annealed Cu-35% Zn	325	105	62	Solid solution
Annealed Cu-10% Sn	455	195	68	Solid solution
Cold-worked Cu-35% Zn	675	435	3	Solid solution + strain hardening
Age-hardened Cu-2% Be	1310	1205	4	Age-hardening
Quenched and tempered Cu-Al	760	415	5	Martensitic reaction
Cast manganese bronze	490	195	30	Eutectoid reaction

Solid-Solution-Strengthened Alloys A number of copper-base alloys contain large quantities of alloying elements, yet remain single phase. Important binary phase diagrams are shown in Figure 13.10.

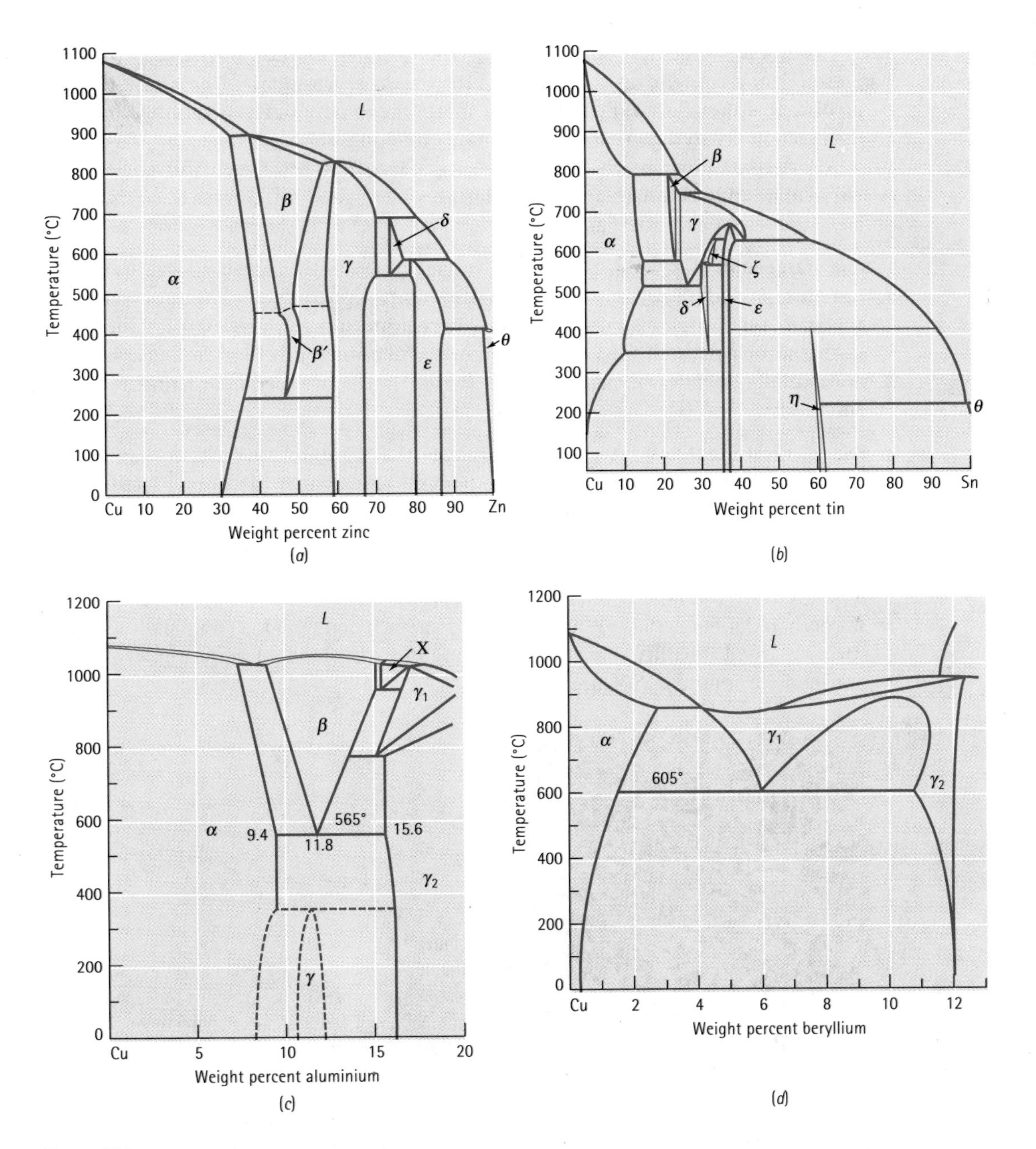

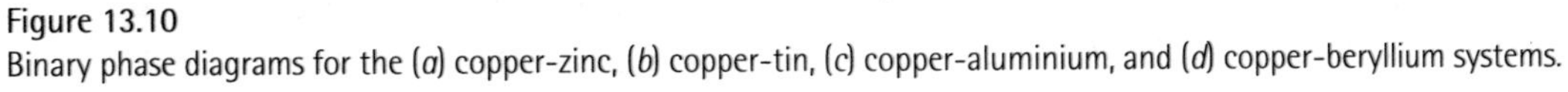
Figure 13.10
Binary phase diagrams for the (*a*) copper-zinc, (*b*) copper-tin, (*c*) copper-aluminium, and (*d*) copper-beryllium systems.

The copper-zinc, or brass, alloys with less than 40% Zn form single-phase solid solutions of zinc in copper. The mechanical properties – even elongation – increase as the zinc content increases. These alloys can be cold-formed into rather complicated yet corrosion-resistant components. Manganese bronze is a particularly high-strength alloy containing manganese as well as zinc for solid solution strengthening.

Tin bronzes, called phosphor bronzes when a small amount of phosphorus is also present, may contain up to 10% Sn and remain single phase. The phase diagram predicts that the alloy will contain the Cu_3Sn (ε) compound. However, the kinetics of the reaction are so slow that the precipitate often does not form.

Alloys containing less than about 9% Al or less than 3% Si are also single phase. These aluminium bronzes and silicon bronzes have good forming characteristics and are often selected for their good strength and excellent toughness.

Age-Hardenable Alloys A number of copper-base alloys display an age-hardening response, including zirconium copper, chromium copper, and beryllium copper. The copper-beryllium alloys, also known as beryllium bronzes, are used for their high strength, their high stiffness (making them useful as high precision springs), and their nonsparking qualities (making them useful for tools to be used near flammable gases and fluids).

Phase Transformations Aluminium bronzes that contain over 9% Al can form β phase on heating above 565°C, the eutectoid temperature [Figure 13.10(c)]. On subsequent cooling, the eutectoid reaction produces a lamellar structure, or pearlite, that contains a brittle γ_2 compound. The low-temperature peritectoid reaction $\alpha + \gamma_2 \rightarrow \gamma$, normally does not occur. The eutectoid product is relatively weak and brittle, but we can rapidly quench the β to produce martensite, or β', which has high strength and low ductility. When β' is subsequently tempered, a combination of high strength, good ductility, and excellent toughness is obtained as fine platelets of α precipitate from the β' (Figure 13.11).

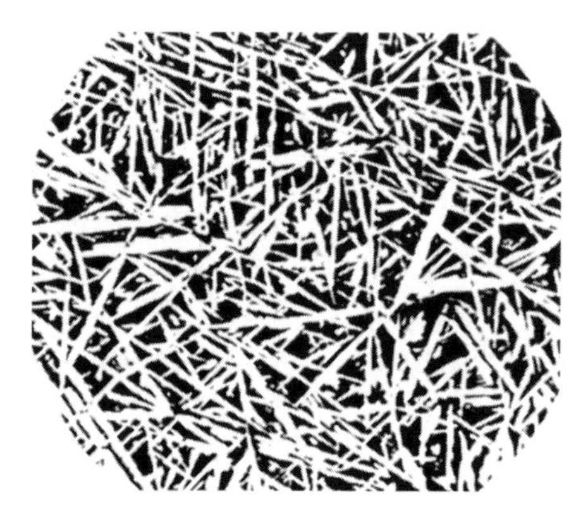

Figure 13.11
The microstructure of a quenched and tempered aluminium bronze containing alpha plates in a beta matrix (× 150). (*From* Metals Handbook, *Vol. 7, 8th Ed., American Society for Metals, 1972.*)

Leaded Copper Alloys Virtually any of the wrought copper alloys may contain up to 4.5% Pb. The lead forms a monotectic reaction with copper and produces tiny lead spheres as the last liquid to solidify. The lead improves machining characteristics.

Even larger amounts of lead are used for copper casting alloys. It helps provide lubrication and embeddability, by which hard particles or grit are embedded in the soft lead spheres, and therefore helps to minimise wear.

Use of leaded copper alloys, however, has a major environmental impact. Lead is known to provide a variety of health problems. Use of lead in plumbing fixtures, for example, may lead to high levels in drinking water. Alternative methods, such as introducing phosphorus to the copper, may provide the good machinability needed to produce copper parts economically.

EXAMPLE 13.8 **Materials Selection for an Electrical Switch**

Select a suitable material for the contacts for a switch or relay that opens and closes a high-current electrical circuit.

SOLUTION

When the switch or relay opens and closes, contact between the conductive surfaces can cause wear and result in poor contact and arcing. A high hardness would minimise wear, but the contact materials must allow the high current to pass through the connection without overheating or arching.

Therefore, our design must provide for both good electrical conductivity and good wear resistance. A relatively pure copper alloy dispersion strengthened with a hard phase that does not disturb the copper lattice would, perhaps, be ideal. In a Cu-Al_2O_3 alloy, the hard ceramic oxide particles provide wear resistance but do not interfere with the electrical conductivity of the copper matrix.

EXAMPLE 13.9 **Selection of a Heat Treatment for a Cu-Al Alloy Gear**

Select the heat treatment required to produce a high-strength aluminium bronze gear containing 10% Al.

SOLUTION

The aluminium bronze can be strengthened by a quench and temper heat treatment. We must heat above 900°C to obtain 100% β for a Cu-10% Al alloy [Figure 13.10(c)]. The eutectoid temperature for the alloy is 565°C. Therefore, our recommended heat treatment is:

1. Heat the alloy to 950°C and hold to produce 100% β.
2. Quench the alloy to room temperature to cause β to transform to martensite, β' which is supersaturated in copper.

3. Temper below 565°C; a temperature of 400°C might be suitable. During tempering, the martensite transforms to α and γ_2. The amount of the γ_2 that forms at 400°C is:

$$\%\ \gamma_2 = \frac{10-9.4}{15.6-9.4} \times 100 = 9.7\%$$

4. Cool rapidly to room temperature so that the equilibrium γ does not form.

Note that if tempering were carried out below about 370°C, γ would form rather than γ_2.

13.6 Nickel and Cobalt

Nickel and cobalt alloys are used for corrosion protection and for high-temperature resistance, taking advantage of their high melting points and high strengths. Nickel is FCC and has good formability; cobalt is an allotropic metal, with an FCC structure above 417°C and an HCP structure at lower temperatures. Special cobalt alloys are used for exceptional wear resistance and, because of resistance to human body fluids, for prosthetic devices. Typical alloys and their applications are listed in Table 13.8.

Nickel and Monel Nickel and its alloys have excellent corrosion resistance and forming characteristics. When copper is added to nickel, the maximum strength is obtained near 60% Ni. A number of alloys, called Monels, with approximately this composition are used for their strength and corrosion resistance in salt water and at

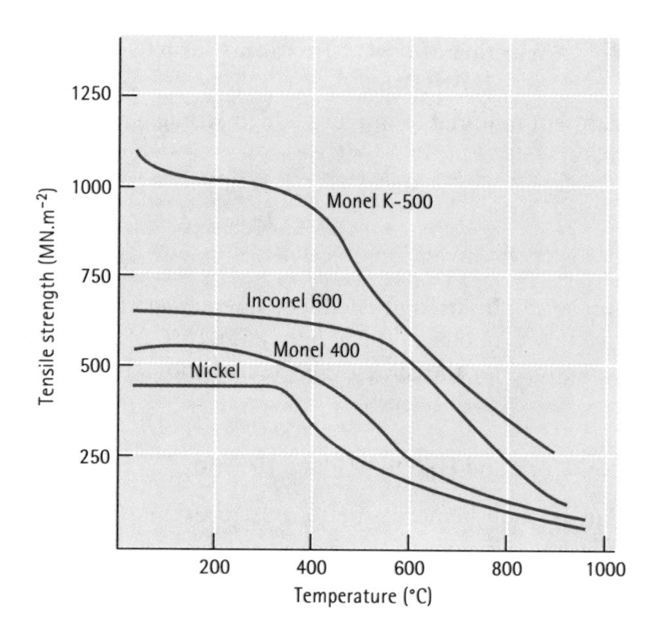

Figure 13.12
The effect of temperature on the tensile strength of several nickel-base alloys.

Table 13.8 Compositions, properties, and applications for selected nickel and cobalt alloys.

Material	Tensile Strength ($MN.m^{-2}$)	Yield Strength ($MN.m^{-2}$)	% Elongation	Strengthening Mechanism	Applications
Pure Ni (99.9% Ni)	345	110	45	Annealed	Corrosion resistance
	655	620	4	Cold-worked	Corrosion resistance
Ni-Cu alloys:					
Monel 400 (Ni-31.5% Cu)	540	270	37	Annealed	Valves, pumps, heat exchangers
Monel K-500 (Ni-29.5% Cu-2.7% Al-0.6% Ti)	1 030	760	30	Aged	Shafts, springs, impellers
Ni superalloys:					
Inconel 600 (Ni-15.5% Cr-8% Fe)	620	200	49	Carbides	Heat-treatment equipment
Hastelloy B-2 (Ni-28% Mo)	900	415	61	Carbides	Corrosion resistance
DS-Ni (Ni-2% ThO_2)	490	330	14	Dispersion	Gas turbines
Fe-Ni superalloys:					
Incoloy 800 (Ni-46% Fe-21% Cr)	615	258	37	Carbides	Heat exchangers
Co superalloys:					
Stellite 6B (60% Co-30% Cr-4.5% W)	1 220	710	4	Carbides	Abrasive wear resistance

elevated temperatures. Some of the Monels contain small amounts of aluminium and titanium. These alloys show an age-hardening response by the precipitation of γ', a coherent Ni_3Al or Ni_3Ti precipitate which nearly doubles the tensile properties. The precipitates resist overageing at temperatures up to 425°C (Figure 13.12).

Several special properties can be obtained in nickel alloys. Nickel can be used to produce permanent magnets by virtue of its ferromagnetic behaviour. A Ni-50% Ti alloy has the shape-memory effect discussed in Chapter 11. A Ni-36% Fe alloy (Invar) displays practically no expansion during heating; this effect is exploited in producing bimetallic composite materials.

Superalloys Superalloys are nickel, iron-nickel, and cobalt alloys that contain large amounts of alloying elements intended to produce a combination of high strength at elevated temperatures, resistance to creep at temperatures up to 1 000°C, and resistance to corrosion. Yet these excellent high-temperature properties are obtained even though the melting temperatures of the alloys are about the same as that for steels. Typical applications include vanes and blades for turbine and jet engines, heat exchangers, chemical reaction vessel components, and heat-treating equipment.

To obtain high strength and creep resistance, the alloying elements must produce a strong, stable microstructure at high temperatures. Solid solution strengthening, dispersion strengthening, and precipitation hardening are generally employed.

Solid Solution Strengthening Large additions of chromium, molybdenum, and tungsten and smaller additions of tantalum, zirconium, niobium, and boron provide solid solution strengthening. The effects of solid solution strengthening are stable and, consequently, make the alloy resistant to creep, particularly when large atoms such as molybdenum and tungsten (which diffuse slowly) are used.

Carbide Dispersion Strengthening All of the alloys contain a small amount of carbon which, by combining with other alloying elements, produces a network of fine, stable carbide particles. The carbide network interferes with dislocation movement and prevents grain boundary sliding. The carbides include TiC, BC, ZrC, TaC, Cr_7C_3, $Cr_{23}C_6$, Mo_6C, and W_6C, although often they are more complex and contain several alloying elements. Stellite 6B (Table 13.8), a cobalt-base superalloy, has unusually good wear resistance at high temperatures due to these carbides.

Precipitation Hardening Many of the nickel and nickel-iron superalloys that contain aluminium and titanium form the coherent precipitate γ' (Ni_3Al or Ni_3Ti) during ageing. The γ' particles (Figure 13.13) have a crystal structure and lattice parameter similar to that of the nickel matrix; this similarity leads to a low surface energy and minimises overageing of the alloys, providing good strength and creep resistance even at high temperatures.

By varying the ageing temperature, precipitates of various sizes can be produced. Small precipitates, formed at low ageing temperatures, can grow between the larger precipitates produced at higher temperatures, therefore increasing the volume percentage of the γ' and further increasing the strength [Figure 13.13(b)].

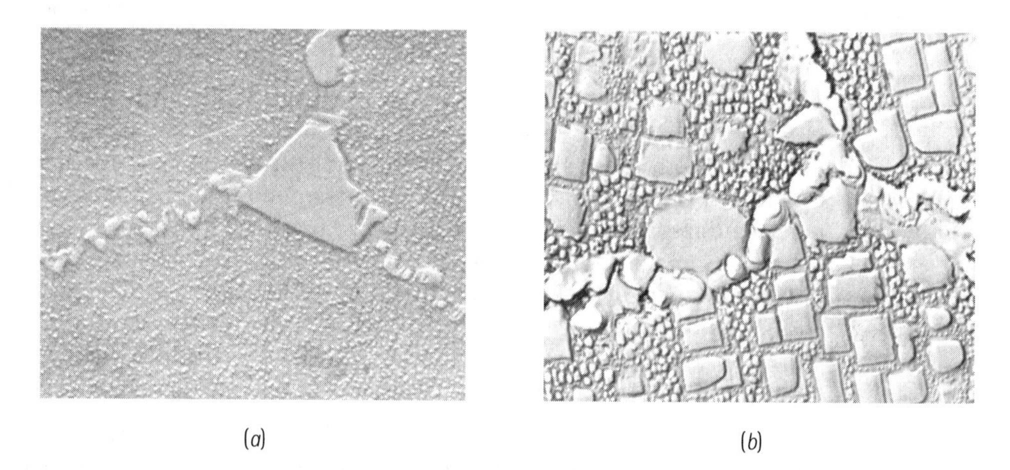

(a) (b)

Figure 13.13
(a) Microstructure of a superalloy, with carbides at the grain boundaries and γ' precipitates in the matrix (× 15000). (b) Microstructure of a superalloy aged at two temperatures, producing both large and small cubic γ' precipitates (× 10000). (*From* Metals Handbook, *Vol. 9, 9th Ed., American Society for Metals, 1985.*)

The high-temperature use of the superalloys can be improved when a ceramic or intermetallic compound coating is used. One method for doing this is to first coat the superalloy with a metallic bond coat composed of a complex NiCoCrAlY alloy, then apply an outer coating of a ZrO_2-base ceramic. The coating helps reduce oxidation of the superalloy and permits jet engines to operate at higher temperatures and with greater efficiency.

EXAMPLE 13.10 Materials Selection for a High-Performance Jet Engine

Select a nickel-base superalloy for producing turbine blades for a gas turbine aircraft engine that will have a particularly long creep rupture time at temperatures approaching 1 100°C.

SOLUTION

First, we need a very stable microstructure. Addition of aluminium or titanium permits the precipitation of up to 60 vol% of the γ' phase during heat treatment and may permit the alloy to operate at temperatures approaching 0.85 times the absolute melting temperature. Addition of carbon and alloying elements such as tantalum and hafnium permits the precipitation of alloy carbides that prevent grain boundaries from sliding at high temperatures. Other alloying elements, including molybdenum and tungsten, provide solid solution strengthening.

Second, we might produce a directionally solidified or even single-crystal turbine blade (Chapter 10). In directional solidification, only columnar grains form during

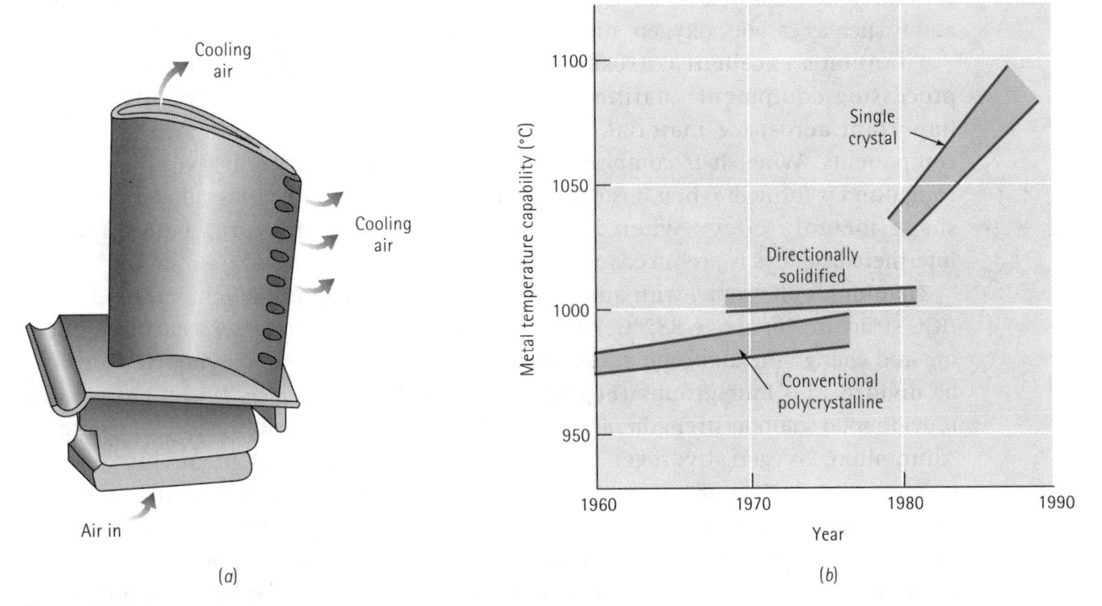

Figure 13.14
(*a*) A turbine blade designed for active cooling by a gas. (*b*) The high-temperature capability of superalloys has increased with improvements in manufacturing methods. (*For* Example 13.10.)

freezing, eliminating transverse grain boundaries that might nucleate cracks. In single-crystal solidification, no grain boundaries are present. Alternatively, we might use the investment casting process, being sure to pass the liquid superalloy through a filter to trap any tiny inclusions before the metal enters the ceramic investment mould.

We would then heat-treat the casting to ensure that the carbides and γ' precipitate with the correct size and distribution. Multiple ageing temperatures might be used to ensure that the largest possible volume percent γ' is formed.

Finally, the blade might contain small cooling channels along its length. Air for combustion in the engine can pass through these channels, providing active cooling to the blade, before reacting with fuel in the combustion chamber.

Figure 13.14 shows the improvements in performance that can be obtained using these design methods.

13.7 Titanium Alloys

Titanium is produced from TiO_2 by the Kroll process. The TiO_2 is converted to $TiCl_4$, which is subsequently reduced to titanium metal by sodium or magnesium. Due to its high affinity for oxygen, all melting and casting processes have to be carried out under vacuum. Titanium provides excellent corrosion resistance, high specific strength, and good high-temperature properties. Strengths up to 1 400 MN.m^{-2}, coupled with a density of 4.505 Mg.m^{-3}, provide the excellent specific mechanical properties. An adherent, protective TiO_2 film provides excellent resistance to corrosion and contamination below 535°C. Above 535°C, the oxide film breaks down and small atoms such as carbon, oxygen, nitrogen, and hydrogen embrittle the titanium.

Titanium's excellent corrosion resistance provides applications in chemical processing equipment, marine components, and biomedical implants. It is an important aerospace material, finding applications as airframe and jet engine components. When it is combined with niobium, a superconductive intermetallic compound is formed; when it is combined with nickel, the resulting alloy displays the shape-memory effect; when it is combined with aluminium, a new class of intermetallic alloys is produced, as discussed in Chapter 10.

Titanium is allotropic, with an HCP crystal structure (α) at low temperatures and a BCC structure (β) above 882°C. Alloying elements provide solid solution strengthening and change the allotropic transformation temperature. The alloying elements can be divided into four groups (Figure 13.15). Additions such as tin and zirconium provide solid solution strengthening without affecting the transformation temperature. Aluminium, oxygen, hydrogen, and other alpha stabilising elements increase the temperature at which α transforms to β. Beta stabilisers such as vanadium, tantalum, molybdenum, and niobium lower the transformation temperature, even causing β to be stable at room temperature. Finally, manganese, chromium, and iron produce a eutectoid reaction, reducing the temperature at which the α–β transformation occurs and producing a two-phase structure at room temperature. Several categories of titanium and its alloys are listed in Table 13.9.

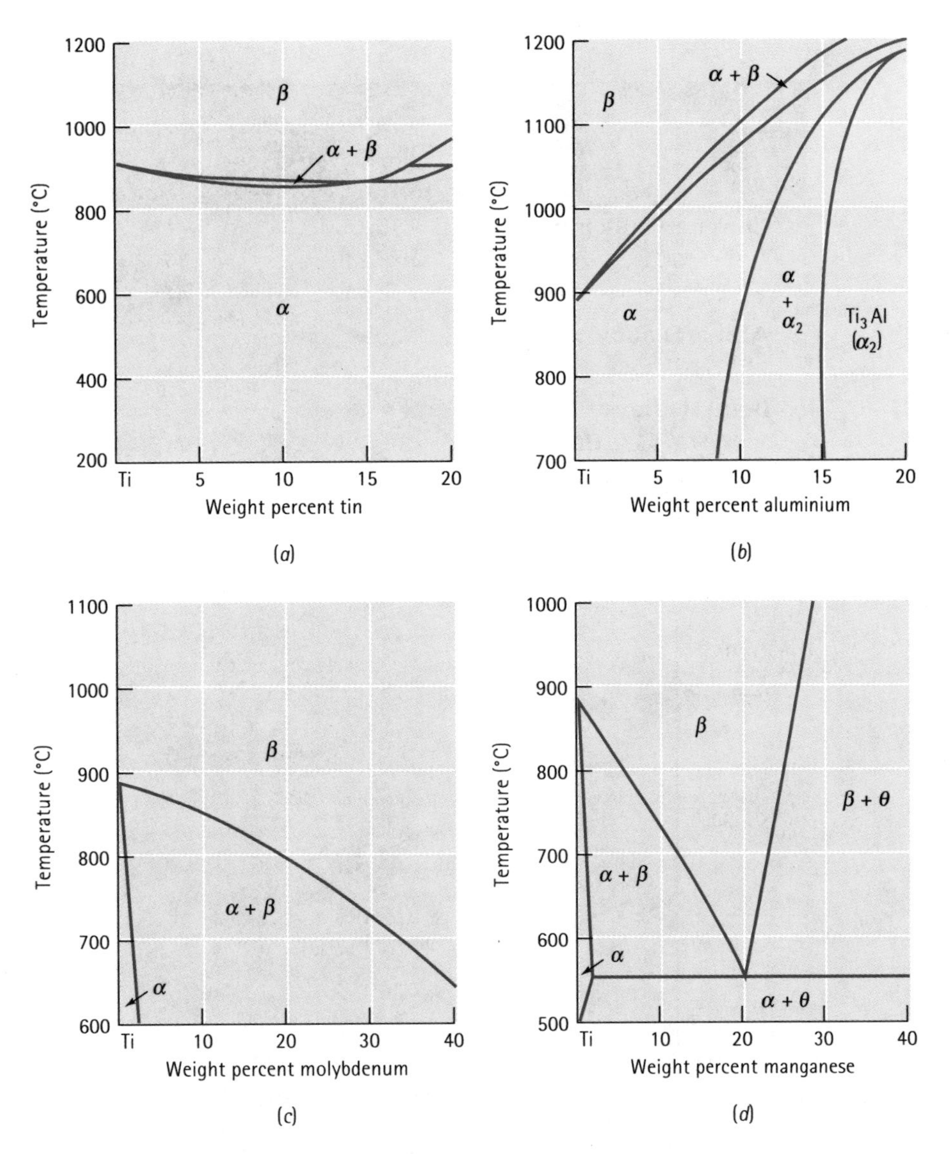

Figure 13.15
Portions of the phase diagrams for (*a*) titanium-tin, (*b*) titanium-aluminium, (*c*) titanium-molybdenum, and (*d*) titanium-manganese.

Commercially Pure Titanium Unalloyed titanium is used for its superior corrosion resistance. Impurities, such as oxygen, increase the strength of the titanium (Figure 13.16) but reduce corrosion resistance. Applications include heat exchangers, piping, reactors, pumps, and valves for the chemical and petrochemical industries.

Alpha Titanium Alloys The most common of all-alpha alloys contains 5% Al and 2.5% Sn, which provide solid solution strengthening to the HCP alpha. The alpha alloys are annealed at high temperatures in the β region. Rapid cooling gives an

Table 13.9 Properties of selected titanium alloys.

Material	Tensile Strength (MN.m^{-2})	Yield Strength (MN.m^{-2})	% Elongation
Commercially pure Ti:			
99.5% Ti	240	170	24
99.0% Ti	550	485	15
Alpha-Ti alloys:			
5% Al-2.5% Sn	860	780	15
Beta-Ti alloys:			
13% V-11% Cr-3% Al	1290	1210	5
Alpha-beta Ti alloys:			
6% Al-4% V	1030	970	8

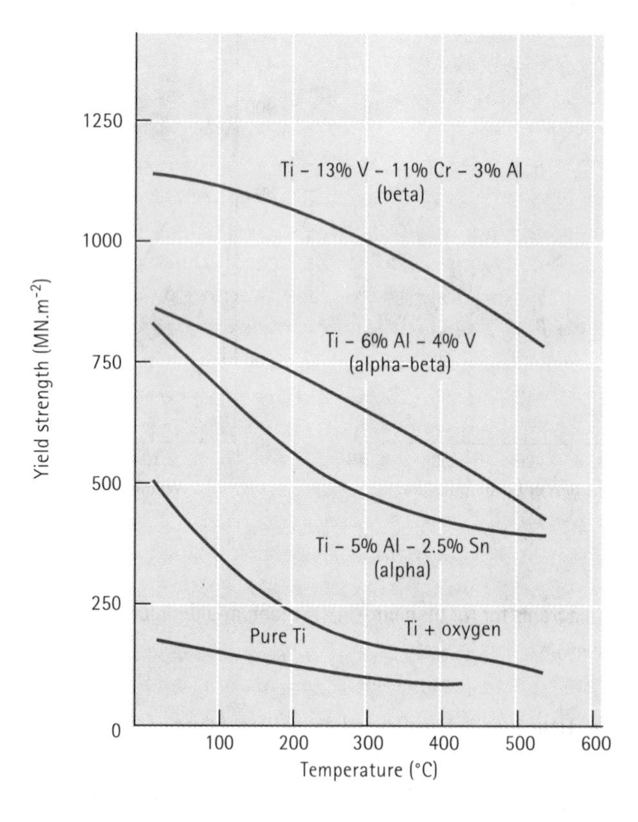

Figure 13.16
The effect of temperature on the yield strength of selected titanium alloys.

acicular, or Widmanstatten, α grain structure [Figure 13.17 (b)] that provides good resistance to fatigue; furnace cooling gives a more platelike α structure that provides better creep resistance.

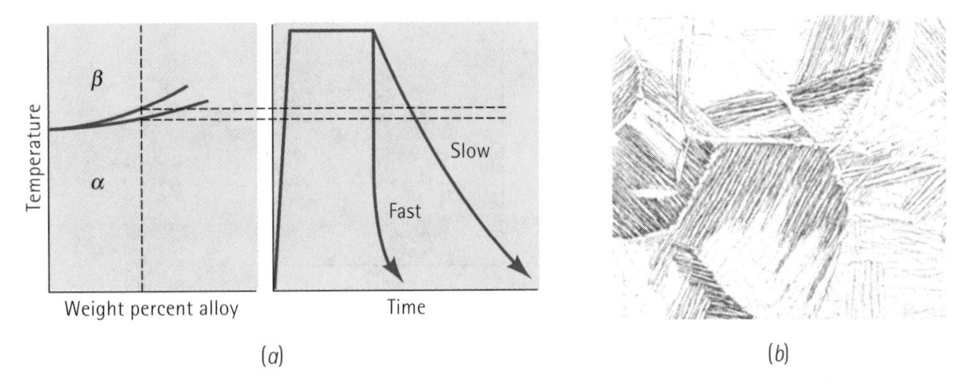

Figure 13.17
(*a*) Annealing and (*b*) microstructure of rapidly cooled alpha titanium (× 100). Both the grain boundary precipitate and the Widmanstatten plates are alpha. (*From* Metals Handbook, *Vol. 7, 8th Ed., American Society for Metals, 1972.*)

Beta Titanium Alloys Although large additions of vanadium or molybdenum produce an entirely β structure at room temperature, none of the so-called beta alloys are actually alloyed to that extent. Instead, they are rich in β stabilisers, so that rapid cooling produces a metastable structure composed of all β. Strengthening is obtained both from the large amount of solid-solution-strengthening alloying elements and by ageing the metastable β structure to permit α to precipitate. Applications include high-strength fasteners, beams, and other fittings for aerospace applications.

Alpha-Beta Titanium Alloys With proper balancing of the α and β stabilisers, a mixture of α and β is produced at room temperature. Ti-6% Al-4% V, an example of this approach, is by far the most common of all the titanium alloys. Because the alloys contain two phases, heat treatments can be used to control the microstructure and properties.

Annealing provides a combination of high ductility, uniform properties, and good strength. The alloy is heated just below the β-transus temperature, permitting a small amount of α to remain and prevent grain growth (Figure 13.18). Slow cooling causes equiaxed α grains to form; the equiaxed structure provides good ductility and formability while making it difficult for fatigue cracks to nucleate. Faster cooling, particularly from above the α–β transus temperature, produces an acicular – or 'basketweave' – alpha phase [Figure 13.18 (c)]. Although fatigue cracks may nucleate more easily in this structure, cracks must follow a tortuous path along the boundaries between the α and β. This condition results in a low fatigue crack growth rate, good fracture toughness, and good resistance to creep.

Two possible microstructures can be produced when the β phase is quenched from a high temperature. The phase diagram in Figure 13.19 includes a dashed martensite start line, which provides the basis for a quench and temper treatment. The β transforms to titanium martensite (α') in an alloy that crosses the M_S line on cooling. The titanium martensite is a relatively soft supersaturated phase. When α' is reheated, tempering occurs by the precipitation of β from the supersaturated α':

$\alpha' \rightarrow \alpha + \beta$ precipitates

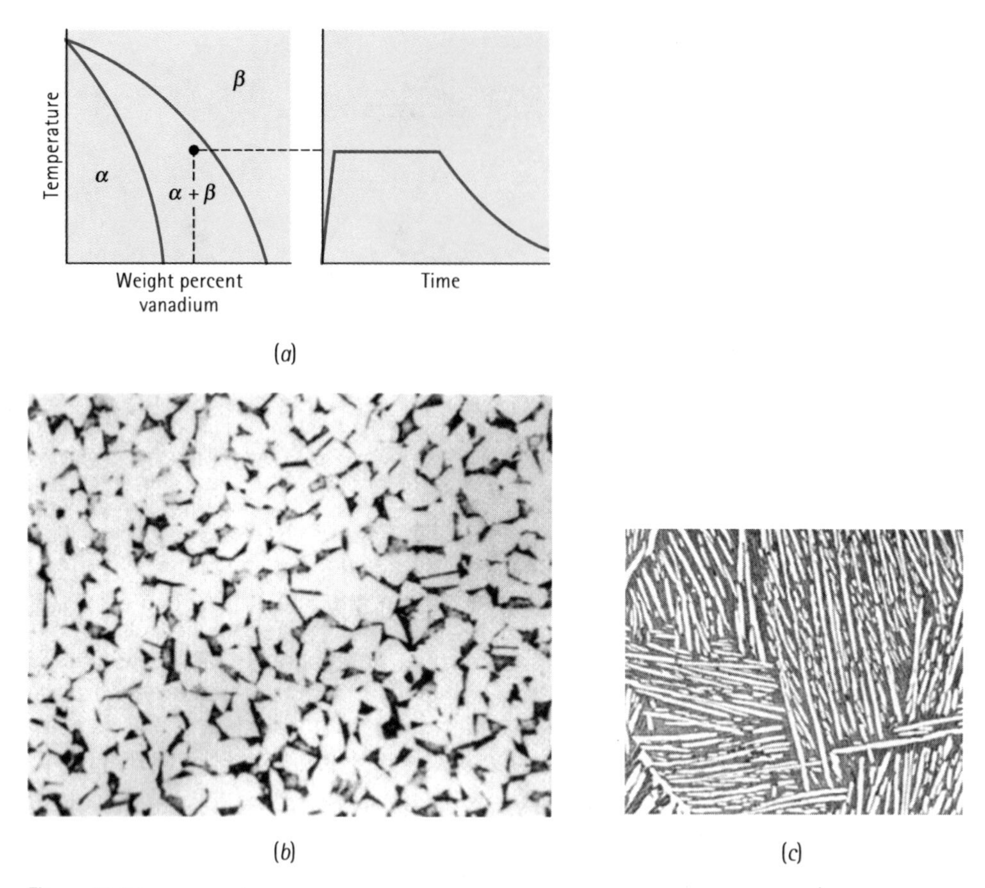

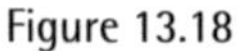
Figure 13.18
Annealing of an alpha-beta titanium alloy. (*a*) Annealing is done just below the α–β transus temperature, (*b*) slow cooling gives equiaxed α grains (× 250), and (*c*) rapid cooling yields acicular α grains (× 2 500). (*From* Metals Handbook, *Vol. 7, 8th Ed., American Society for Metals, 1972.*)

Fine β precipitates initially increase the strength compared with the α', opposite to what is found when a steel martensite is tempered. However, softening occurs when tempering is done at too high a temperature.

More highly alloyed α–β alloys are age-hardened. When the β in these alloys is quenched, β_{ss}, which is supersaturated in titanium, remains. When β_{ss} is aged, α precipitates in a Widmanstatten structure, improving the strength and fracture toughness (Figure 13.19):

$$\beta_{ss} \rightarrow \beta + \alpha \text{ precipitates}$$

Components for airframes, rockets, jet engines, and landing gear are typical applications for the heat-treated alpha-beta alloys. Some alloys, including the Ti-6% Al-4% V alloy, are superplastic and can be deformed as much as 1 000%.

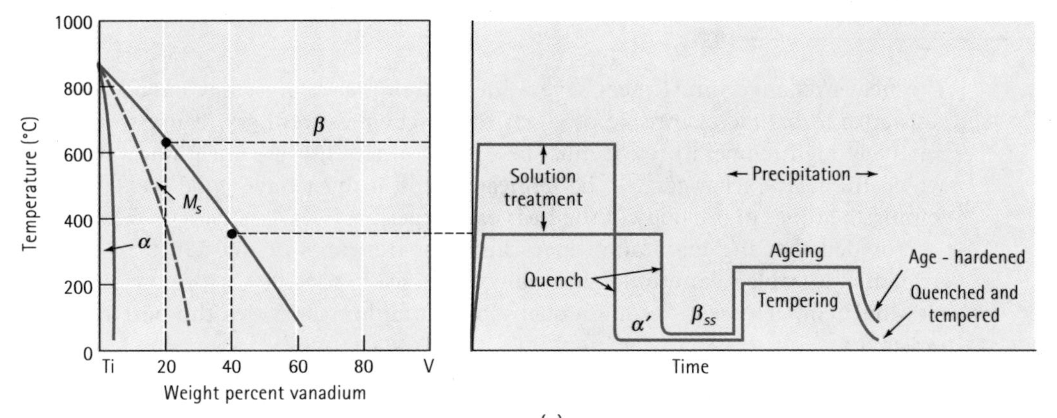

(*a*)

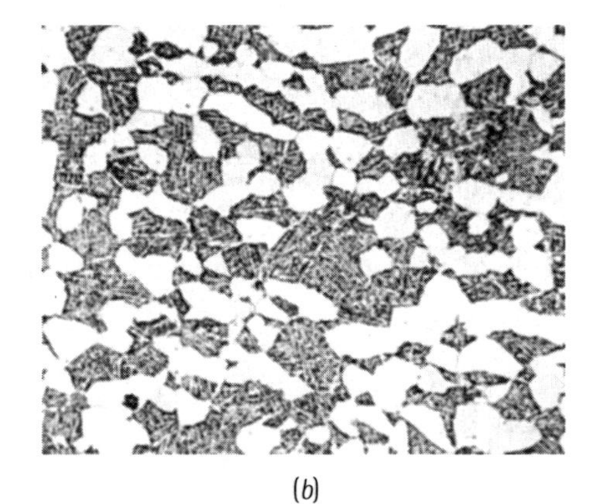

(*b*)

Figure 13.19
(*a*) Heat treatment and (*b*) microstructure of the alpha-beta titanium alloys. The structure contains primary α (large white grains) and a dark β matrix with needles of α formed during ageing (× 250). (*From* Metals Handbook, *Vol. 7, 8th Ed., American Society for Metals, 1972.*)

EXAMPLE 13.11 Materials Selection for a Heat Exchanger

Select a suitable material for a 1.5 m-diameter, 10 m-long heat exchanger for the petrochemical industry (Figure 13.20).

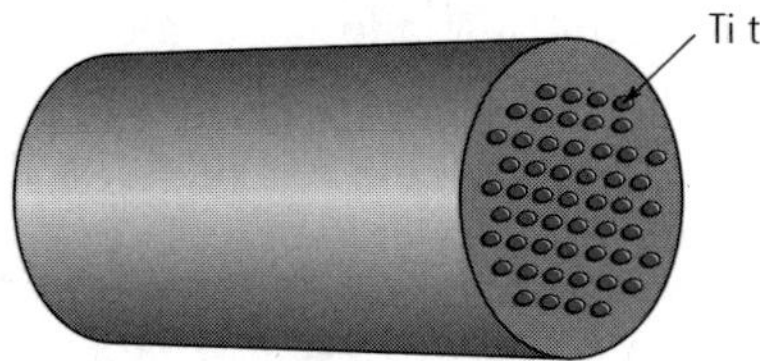

Figure 13.20
Sketch of a heat exchanger using titanium tubes (*for* Example 13.11).

SOLUTION

The heat exchanger must meet several design criteria. It must have good corrosion resistance to handle aggressive products of the chemical refinery; it must operate at relatively high temperatures; it must be easily formed into the sheet and tubes from which the heat exchanger will be fabricated; and it must have good weldability for joining the tubes to the body of the heat exchanger.

Provided that the maximum operating temperature is below 535°C so that the oxide film is stable, titanium might be a good choice to provide corrosion resistance at elevated temperatures. A commercially pure titanium provides the best corrosion resistance.

Pure titanium also provides superior forming and welding characteristics and would, therefore, be our most logical selection. If pure titanium does not provide sufficient strength, an alternative is an alpha-titanium alloy, still providing good corrosion resistance, forming characteristics, and weldability but also somewhat improved strength.

EXAMPLE 13.12 Materials Selection for a Connecting Rod

Select a suitable material for a high-performance connecting rod for the engine of a racing automobile (Figure 13.21).

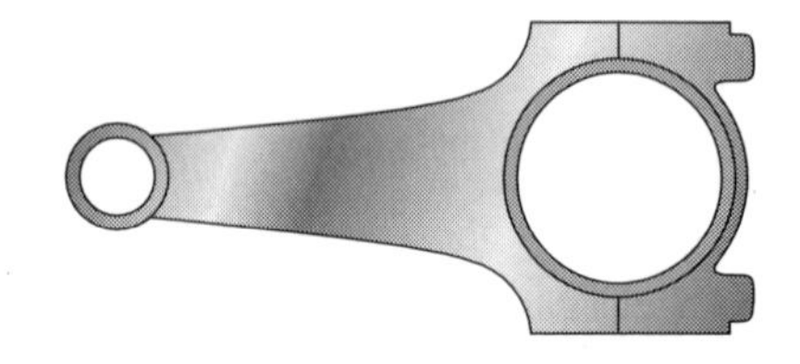

Figure 13.21
Sketch of connecting rod (*for* Example 13.12).

SOLUTION

A high-performance racing engine requires materials that can operate at high temperatures and stresses while minimising the weight of the engine. In normal automobiles, the connecting rods are often a forged steel or a malleable cast iron. We might be able to save considerable weight by replacing these parts with titanium.

To achieve high strengths, we might consider an alpha-beta titanium alloy. Because of its availability, the Ti-6% Al-4% V alloy is a good choice. The alloy is heated to about 1 065°C, which is in the all-β portion of the phase diagram. On quenching, a titanium martensite forms; subsequent tempering produces a microstructure containing β precipitates in an α matrix.

When the heat treatment is performed in the all-β region, the martensite has an acicular structure, which reduces the rate of growth of any fatigue cracks that might develop.

13.8 Refractory Metals

The refractory metals, which include tungsten, molybdenum, tantalum, and niobium (or columbium), have exceptionally high melting temperatures (above 1 925°C) and, consequently, have the potential for high-temperature service. Applications include filaments for light bulbs, rocket nozzles, nuclear power generators, electronic capacitors, and chemical processing equipment. The metals, however, have a high density, limiting their specific strengths (Table 13.10).

Table 13.10 Properties of refractory metals.

Metal	Melting Temperature (°C)	Density ($Mg.m^{-3}$)	T = 1 000°C Tensile Strength ($MN.m^{-2}$)	T = 1 000°C Yield Strength ($MN.m^{-2}$)	Transition Temperature (°C)
Nb	2468	8.57	117	55	−140
Mo	2610	10.22	345	207	30
Ta	2996	16.6	186	165	−270
W	3410	19.25	455	103	300

Oxidation The refractory metals begin to oxidise between 200°C and 425°C and are rapidly contaminated or embrittled. Consequently, special precautions are required during casting, hot working, welding, or powder metallurgy. The metals must also be protected during service at elevated temperatures. For example, the tungsten filament in a light bulb is protected by a vacuum.

For some applications, the metal may be coated with silicide or aluminide coating. The coating must (a) have a high melting temperature, (b) be compatible with the refractory metal, (c) provide a diffusion barrier to prevent contaminants from reaching the underlying metal, and (d) have a coefficient of thermal expansion similar to that of the refractory metal. Coatings are available that protect the metal to about 1 650°C.

Forming Characteristics The refractory metals, which have a BCC crystal structure, display a ductile-to-brittle transition temperature. Because the transition temperatures for niobium and tantalum are below room temperature, these two metals can readily be formed. However, annealed molybdenum and tungsten normally have a transition temperature above room temperature, causing them to be brittle. Fortunately, if these metals are hot-worked to produce a fibrous microstructure, the transition temperature is lowered and the forming characteristics are improved.

Alloys Large increases in both room temperature and high-temperature mechanical properties are obtained by alloying. Tungsten alloyed with hafnium, rhenium, and carbon can operate up to 2 100°C. These alloys typically are solid-solution-strengthened; in fact, tungsten and molybdenum form a complete series of solid solutions, much like copper and nickel. Some alloys, such as W-2% ThO_2, are dispersion-strengthened by oxide particles during their manufacture by powder metallurgy processes, and are subsequently used for applications such as lightbulb filaments. Composite materials, such as niobium reinforced with tungsten fibres, may also improve high-temperature properties.

SUMMARY

Nonferrous alloys, or those based on metals other than iron, have a tremendous range of physical and mechanical properties. Extremes in density, specific strength, melting temperature, and corrosion resistance are found from one group of alloys to the next. As a group, however, the nonferrous alloys still take advantage of all of the strengthening mechanisms discussed in previous chapters, and much of their behaviour can be explained in these terms.

- The 'light metals' include low-density alloys based on aluminium, magnesium, titanium, and beryllium. These alloys have a high specific strength due to their low density and, as a result, find many aerospace applications. Excellent corrosion resistance and electrical conductivity of aluminium also provide for a vast number of applications. The most important and powerful strengthening mechanism in these alloys is age hardening. Aluminium and magnesium are limited to use at low temperatures because of the loss of their mechanical properties as a result of overageing or recrystallisation. Beryllium, on the other hand, has an exceptional strength-to-weight ratio, maintains its strength at high temperatures, and is unusually stiff. Titanium alloys have intermediate densities and temperature resistance, along with excellent corrosion resistance, leading to applications in aerospace and chemical processing. These alloys show a powerful response to strengthening by age hardening and quench and temper heat treatments.
- Nickel and cobalt alloys, including superalloys, provide good properties at even higher temperatures. Combined with their good corrosion resistance, these alloys find many applications in aircraft engines and chemical processing equipment. Strengthening even at high temperatures is usually obtained by age hardening, solid solution strengthening, and dispersion strengthening due to alloy carbides.
- Refractory metals are able to operate at the highest temperatures, although they may have to be protected from oxidation by appropriate atmospheres or coatings.
- Copper alloys can be strengthened by all of the strengthening mechanisms discussed in the text; many applications are found, particularly in electronics, power generation and transmission, and processing of chemicals.

GLOSSARY

Blister copper
An impure form of copper obtained during the refining process.

Brass
A group of copper-base alloys, normally containing zinc as the major alloying element.

Castability
The ease with which a metal can be poured into a mould to make a casting without producing defects or requiring unusual or expensive techniques to prevent casting problems.

Fluidity
The ability of liquid metal to fill a mould cavity without prematurely freezing.

Hall–Heroult process
An electrolytic process by which aluminium is extracted from its ore.

Monel
A copper-nickel alloy, containing approximately 60% Ni, that gives the maximum strength in the binary alloy system.

Nonferrous alloy
An alloy based on some metal other than iron.

Refractory metals
Metals having a very high melting temperature, above 2 000°C.

Specific strength
The ratio of strength to density. Also called the strength-to-weight ratio.

Superalloys
A group of nickel, iron-nickel, and cobalt alloys that have exceptional heat resistance, creep resistance, and corrosion resistance.

Temper designation
A shorthand notation using letters and numbers to describe the processing of an alloy. H tempers refer to cold-worked alloys; T tempers refer to age-hardening treatments.

Thixocasting (Rheocasting)
A process by which a material is stirred during solidification, producing a partly liquid, partly solid structure that behaves as a solid when no external force is applied, yet flows as a liquid under pressure.

Wrought alloys
Alloys that are shaped by a deformation process.

PROBLEMS

13.1 In some cases, we may be more interested in cost per unit volume than in cost per unit weight. Rework Table 13.1 to show the cost of each metal in terms of US $\$/m^3$. Does this change alter the relationship between the different metals?

13.2 Assuming that the density remains unchanged, compare the specific strength of the 2090-T6 aluminium alloy to that of a die-cast 443-F aluminium alloy (Table 13.5). If you considered the actual density, do you think the difference between the specific strengths would increase or become smaller? Explain.

13.3 Explain why aluminium alloys containing more than about 15% Mg are not used.

13.4 Calculate the modulus of elasticity-to-density ratio (also called the specific modulus) of an Al-3% Li alloy and compare with the ratio for pure aluminium.

13.5 Estimate the secondary dendrite arm spacing for each structure in Figure 13.4 and, from Figure 8.7, estimate the solidification time obtained by each of the three casting processes. Do you expect higher strengths for die casting, permanent mould casting, or sand casting? Explain.

13.6 Would you expect a 2024-T9 aluminium alloy to be stronger or weaker than a 2024-T6 alloy? Explain.

13.7 Based on the data in Figure 7.22, estimate the mechanical properties if the 3105 aluminium alloy is in the H18 condition.

13.8 Estimate the tensile strength expected for the following aluminium alloys:

(a) 1100-H14 (b) 5182-H12

(c) 3004-H16

13.9 Suppose, by rapid solidification from the liquid state, that a supersaturated Al-7% Li alloy can be produced and subsequently aged. Compare the amount of β that will form in this alloy with that formed in a 2090 alloy.

13.10 Determine the amount of Mg_2Al_3 (β) expected to form in a 5182-O aluminium alloy.

13.11 Based on the phase diagrams, which of the following alloys would be most suited for thixocasting? Explain your answer. (*See* Figures 10.22, 11.5, *and* 13.3.)

(a) Al-12% Si (b) Al-1% Cu

(c) Al-10% Mg

13.12 From the data in Table 13.6, estimate the ratio by which the yield strength of magnesium can be increased by alloying and heat treatment and compare with that of aluminium alloys.

13.13 Suppose a 600 mm-long round bar is to support a load of 1.8 kN without any permanent deformation. Calculate the minimum diameter of the bar if it is made of

(a) AZ80A-T5 magnesium alloy and

(b) 6061-T6 aluminium alloy.

Calculate the weight of the bar and the approximate cost (based on pure Al and Mg) in each case.

13.14 A 10 m rod 5 mm in diameter must elongate no more than 2 mm under load. What is the maximum force that can be applied if the rod is made of:

(a) aluminium

(b) magnesium

(c) beryllium

13.15 A Cu-20% Sn alloy is found, after cooling from the liquid to room temperature, to have a microstructure containing 50% α, 30% β, and 20% γ. What microstructure would you predict under equilibrium conditions? Explain why the observed microstructure is not unexpected.

13.16 We say that copper can contain up to 40% Zn or 9% Al and still be single phase. How do we explain this statement in view of the phase diagrams in Figure 13.10?

13.17 Compare the percentage increase in the yield strength of commercially pure annealed aluminium, magnesium, and copper by strain hardening. Explain the differences observed.

13.18 We would like to produce a quenched and tempered aluminium bronze containing 13% Al. Recommend a heat treatment, including appropriate temperatures. Calculate the amount of each phase after each step of the treatment.

13.19 A number of casting alloys have very high lead contents; however, the Pb content in wrought alloys is comparatively low. Why isn't more lead added to the wrought alloys? What precautions must be taken when a leaded wrought alloy is hot-worked or heat-treated?

13.20 Would you expect the fracture toughness of quenched and tempered aluminium bronze to be high or low? Would there be a difference in the resistance of the alloy to crack nucleation compared with crack growth? Explain.

13.21 Based on the photomicrograph in Figure 13.13(a), would you expect the γ' precipitate or the carbides to provide a greater strengthening effect in superalloys at low temperatures? Explain.

13.22 The density of Ni_3Al is 6.56 $Mg.m^{-3}$. Suppose a Ni-5 wt% Al alloy is heat-treated so that all of the aluminium reacts with nickel to produce Ni_3Al. Determine the volume percentage of the Ni_3Al precipitate in the nickel matrix.

13.23 Figure 13.13(b) shows a nickel superalloy containing two sizes of γ' precipitates. Which precipitate likely formed first? Which precipitate formed at the higher temperature? What does our ability to perform this treatment suggest concerning the effect of temperature on the solubility of Al and Ti in nickel? Explain.

13.24 When steel is joined using arc welding, only the liquid fusion zone must be protected by a gas or flux. However, when titanium is welded, both the front and back sides of the welded metal must be protected. Why must these extra precautions be taken when joining titanium?

13.25 Both a Ti-15% V alloy and a Ti-35% V alloy are heated to a temperature at which all β just forms. They are then quenched and reheated to 300°C. Describe the changes in microstructure during the heat treatment for each alloy, including the amount of each phase. What is the matrix and what is the precipitate in each case? Which is an age-hardening process? Which is a quench and temper process? [*See* Figure 13.19(a).]

13.26 The θ phase in the Ti-Mn phase diagram has the formula MnTi. Calculate the amount of α and θ in the eutectoid microconstituent. [*See* Figure 13.15(d).]

13.27 The temperature of a coated tungsten part is increased. What happens when the protective coating on a tungsten part expands more than the tungsten? What happens when the protective coating on a tungsten part expands less than the tungsten?

13.28 Determine the ratio of the yield strengths of the strongest Al, Mg, Cu, Ti, and Ni alloy to the yield strength of the pure metal. Compare the alloy systems and rank them in order of their response to strengthening mechanisms. Try to explain their order.

13.29 Determine the specific yield strength of the strongest Al, Mg, Cu, Ti, and Ni alloys. Use the densities of the pure metals, in $Mg.m^{-3}$, in your calculations. Try to explain their order.

13.30 Based on the phase diagrams, estimate the solubilities of Ni, Zn, Al, Sn, and Be in copper at room temperature. Are these solubilities expected in view of Hume-Rothery's conditions for solid solubility? Explain.

Design problems

13.31 A part for an engine mount for a private aircraft must occupy a volume of 0.06 litres with a minimum thickness of 5 mm and a minimum width of 40 mm. The load on the part during service may be as much as 75 000 N. The part is expected to remain below 100°C during service. Select a suitable material and its treatment that will perform satisfactorily in this application.

13.32 You wish to design the rung on a ladder. The ladder should be light in weight so that it can easily be transported and used. The rungs on the ladder should be 6 mm × 25 mm and are 300 mm long. Select a suitable material and its processing for the rungs.

13.33 We have determined that we need an alloy having a density of 2.3 – 0.05 $Mg.m^{-3}$ that must be strong, yet still have some ductility. Design a material and its processing that might meet these requirements.

13.34 We wish to design a mounting device that will position and aim a laser for precision cutting of a composite material. What design requirements might be important? Select a suitable material and its processing that might meet these requirements.

13.35 Select a nickel-titanium alloy that will produce 60 volume percent Ni_3Ti precipitate in a pure nickel matrix.

13.36 An actuating lever in an electrical device must open and close almost instantly and carry a high current when closed. What design requirements would be important for this application? Select a suitable material and its processing to meet these requirements.

13.37 A fan blade in a chemical plant must operate at temperatures as high as 400°C under rather corrosive conditions. Occasionally, solid material is ingested and impacts the fan. What design requirements would be important? Select a suitable material and its processing for this application.

CHAPTER 14

Ceramic Materials

14.1 Introduction

Ceramic materials are complex chemical compounds and solutions containing both metallic and nonmetallic elements. Alumina (Al_2O_3), for example, is a ceramic composed of metallic (aluminium) and nonmetallic (oxygen) atoms. Ceramic materials have a wide range of mechanical and physical properties. Applications vary from pottery, brick, tile, cooking ware, and soil pipe to glass, refractories, magnets, electrical devices, fibres, and abrasives. The tiles that protect the space shuttle are silica, a ceramic material. For most of these applications, there is one particular property or combination of properties of the ceramic that is essential, cannot be obtained by any other material, and therefore forms the basis for its selection.

Because of their ionic or covalent bonding, ceramics are usually hard, brittle, high-melting-point materials with low electrical and thermal conductivity, good chemical and thermal stability, and high compressive strengths. However, ceramics are somewhat of an enigma. Although they are, indeed, brittle, some ceramic-matrix composites (such as Si_3N_4-SiC) obtain fracture toughness values greater than some metals (such as age-hardened aluminium alloys) and some are even superplastic. Although most ceramics are good electrical and thermal insulators, SiC and AlN have thermal conductivities near that of metals! Ceramics such as FeO and ZnO are semi-conductors and, in addition, ceramic superconducting materials such as $YBa_2Cu_3O_{7-x}$ have been discovered.

In this chapter, we discuss the mechanisms by which these materials deform when a load is applied. Of critical importance is the observation that ceramic materials are brittle, that flaws inevitably present in the structure may cause the ceramic to fail in a brittle manner, that the size and number of flaws differ in each individual ceramic part, and that the mechanical properties can only be described in a statistical manner. This behaviour makes the mechanical behaviour of ceramics less predictable than that of metals; it is this unpredictability that limits the use of ceramics in high-strength, critical applications.

In later chapters, the electrical, magnetic, thermal, and optical properties of ceramics will be discussed and contrasted with those of other materials.

14.2 The Structure of Crystalline Ceramics

In Chapter 3, several crystal structures for ionically bonded materials were introduced. In these structures, the ions in the unit cells fit into lattice sites that provide the proper coordination and ensure that a proper charge balance is obtained. A large number of ceramics – including CaO, MgO, MnS, NiO, MnO, FeO, and HfN – possess the sodium chloride structure. The zinc blende structure is typical of ZnS, BeO, and SiC, whereas a number of ceramics, including CaF_2, ThO_2, CeO_2, UO_2, ZrO_2, and HfO_2, have the fluorite structure. Most ceramics, however, have more complicated crystal structures, including those described in Figure 14.1.

Perovskite Structure The perovskite unit cell [Figure 14.1(a)] is found in several important electrical ceramics, such as $BaTiO_3$ and $SrTiO_3$. Three types of ions are present in the cell. If barium ions are located at the corners of a cube, oxygen ions fill face-centred sites and titanium ions occupy body-centred sites. Distortion of the unit cell produces an electrical signal, permitting some titanates to serve as transducers.

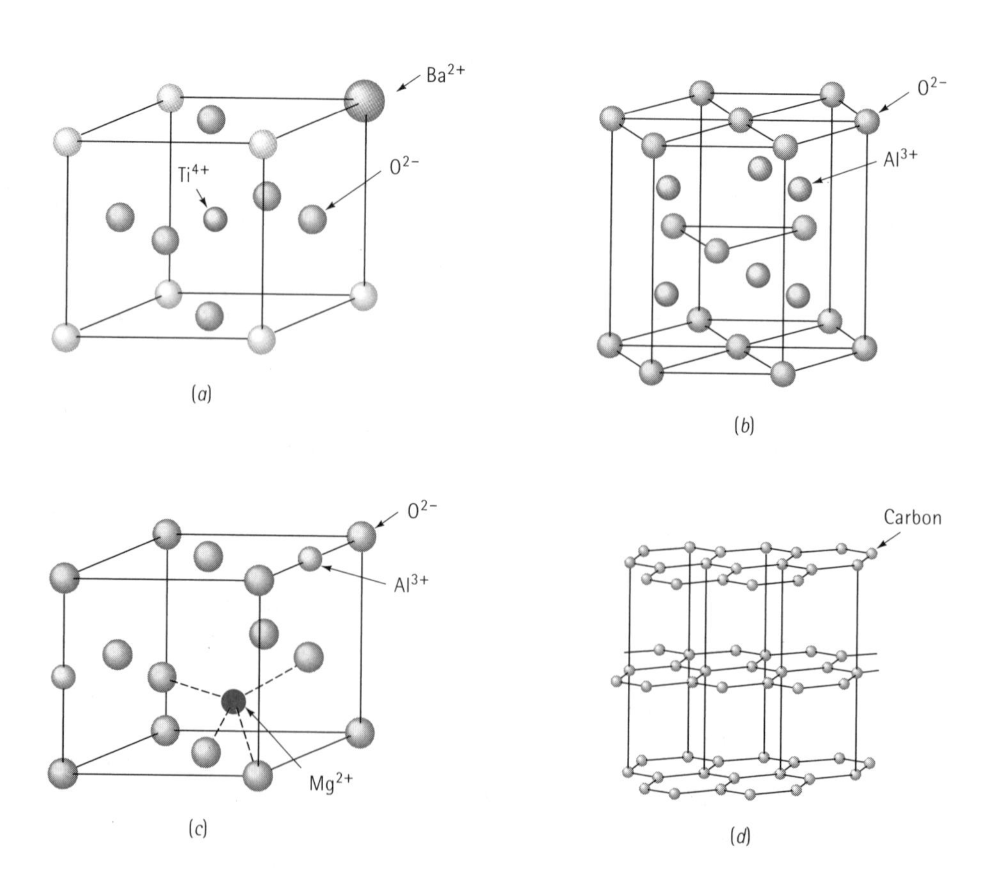

Figure 14.1
Complex ceramic crystal structures: (a) perovskite, (b) a portion of the corundum cell (two-thirds of the Al^{3+} sites are actually occupied), (c) a portion of the spinel cell, and (d) graphite.

Corundum Structure One of the forms of alumina, Al_2O_3, has the corundum crystal structure, which is similar to a hexagonal close-packed structure; however, 12 aluminium ions and 18 oxygen ions are associated with each unit cell [Figure 14.1(b)]. Alumina is a common refractory, electrical insulator, and abrasive material. A number of other ceramics, including Cr_2O_3 and Fe_2O_3, have this structure.

Spinel structure The spinel structure [Figure 14.1(c)], typical of $MgAl_2O_4$, has a cubic unit cell that can be viewed as consisting of eight smaller cubes. In each of the smaller cubes, the oxygen ions are located in normal face-centred cubic positions. Within the smaller cubes are four octahedral interstitial sites and eight tetrahedral interstitial sites; the cations fit into three of these 12 available sites. In *normal* spinels, the divalent ions (such as Mg^{2+}) fit into tetrahedral sites and the trivalent ions (like Al^{3+}) fit into octahedral sites. In *inverse* spinels, the divalent ion and half of the trivalent ions are located at octahedral sites. Many important electrical and magnetic ceramics, including Fe_3O_4, have this structure.

Graphite Graphite, a crystalline form of carbon, is sometimes considered a ceramic material, although carbon is an element rather than a combination of metal and nonmetal atoms. Graphite has a hexagonal layered structure [Figure 14.1(d)] and is used as a refractory material, a lubricant, and a fibre.

EXAMPLE 14.1

Corundum or Al_2O_3, has a hexagonal unit cell [Figure 14.1 (b)]. The lattice parameters of alumina are $a_0 = 0.475$ nm and $c_0 = 1.299$ nm and the density is about 3.98 $Mg.m^{-3}$. How many Al_2O_3 groups, Al^{3+} ions, and O^{2-} ions are present in a hexagonal prism having these dimensions?

SOLUTION

The molecular weight of alumina is $2(26.98) + 3(16) = 101.96\ g.mol^{-1} = 101.96 \times 10^6\ Mg.mol^{-1}$. The volume of the hexagonal prism is :

$$V = a_0^2 c_0 \cos 30 = (0.475)^2(1.299)(\cos 30) = 0.25382\ nm^3$$

$$= 0.25382 \times 10^{-27}\ m^3/prism$$

If x is the number of Al_2O_3 groups in the prism, then:

$$3.98 = \frac{101.96 \times 10^{-6} x}{(0.25382 \times 10^{-27})(6.02 \times 10^{23})}$$

$$x = \frac{(3.98)(0.25382 \times 10^{-27})(6.02 \times 10^{23})}{101.96 \times 10^{-6}}$$

$$x = 6$$

Thus, a hexagonal prism having the dimensions given in the problem contains 6 Al_2O_3 groups, with 12 aluminium ions and 18 oxygen ions.

14.3 The Structure of Crystalline Silicates

Some ceramic materials contain covalent bonds. An example is the cristobalite form of SiO_2, or silica, which is an important raw material for ceramics (Figure 14.2). The arrangement of the atoms in the unit cell provides the proper coordination, balances the charge, and in addition ensures that the directionality of the covalent bonds is not violated.

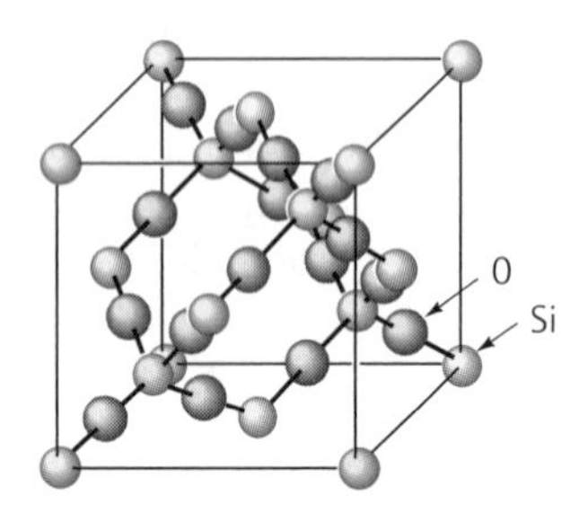

Figure 14.2
The crystal structure of cristobalite, one of the forms of SiO_2.

In silica, covalent bonding requires that the silicon atoms have four nearest neighbours (four oxygen atoms), thus creating a tetrahedral structure. The silicon-oxygen tetrahedra are the basic building blocks for silica, more complicated crystalline structures such as clays, and glassy silicate structures. Silica tetrahedra, SiO_4^{4-}, behave as ionic groups; the oxygen ions at the corners of the tetrahedra are attached to other ions, or one or more of the oxygen ions can be shared by two tetrahedral groups to satisfy the charge balance. Figure 14.3 summarises these structures.

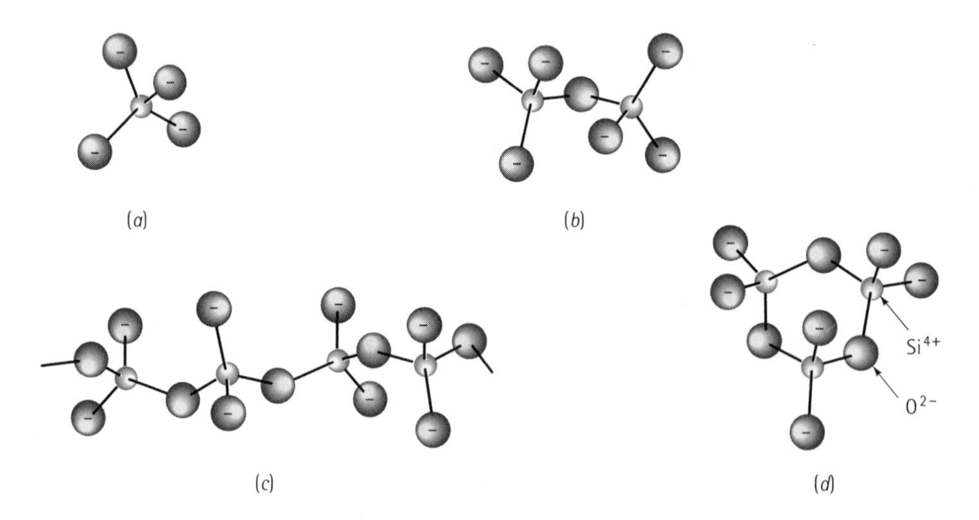

Figure 14.3
Arrangement of silica tetrahedra: (*a*) orthosilicate island, (*b*) pyrosilicate island, (*c*) chain, and (*d*) ring. Positive ions are attached to the silicate groups.

Silicate Compounds When two Mg^{2+} ions are available to combine with one tetrahedron, a compound Mg_2SiO_4, or forsterite, is produced. The two Mg^{2+} ions satisfy the charge requirements and balance the SiO_4^{4-} ions. The Mg_2SiO_4 groups, in turn, produce a three-dimensional crystalline structure. Similarly, Fe^{2+} ions can combine with silica tetrahedra to produce Fe_2SiO_4. Mg_2SiO_4 and Fe_2SiO_4 form a series of solid solutions known as *olivines* or orthosilicates.

Two silicate tetrahedra can combine by sharing one corner to produce a double tetrahedron, or a $Si_2O_7^{6-}$ ion. This ionic group can then combine with other ions to produce pyrosilicate, or double tetrahedron, compounds.

Ring and Chain Structures When two corners of the tetrahedron are shared with other tetrahedral groups, rings and chains with the formula $(SiO_3)_n^{2n-}$ form, where n gives the number of SiO_3^{2-} groups in the ring or chain. A large number of ceramic materials have this metasilicate structure. Wollastonite ($CaSiO_3$) is built from Si_3O_9 rings; beryl ($Be_3Al_2Si_6O_{18}$) contains larger Si_6O_{18} rings; and enstatite ($MgSiO_3$) has a chain structure.

Sheet structures (Clays) When the O : Si ratio gives the formula Si_2O_5, the tetrahedra combine to form sheets (Figure 14.4). Ideally, three of the oxygen atoms in each tetrahedron are located in a single plane, forming a hexagonal pattern. The silicon atoms in the tetrahedra form a second plane, also with a hexagonal pattern. The fourth oxygen atom in each tetrahedron is present in a third plane. The oxygen atoms in this third plane are ionically bonded to other groups of atoms, forming material such as clay, mica, and talc.

Kaolinite, a common clay, is composed of the silicate sheet ionically bonded to a sheet composed of Al and OH groups, producing thin, hexagonal-shaped platelets of

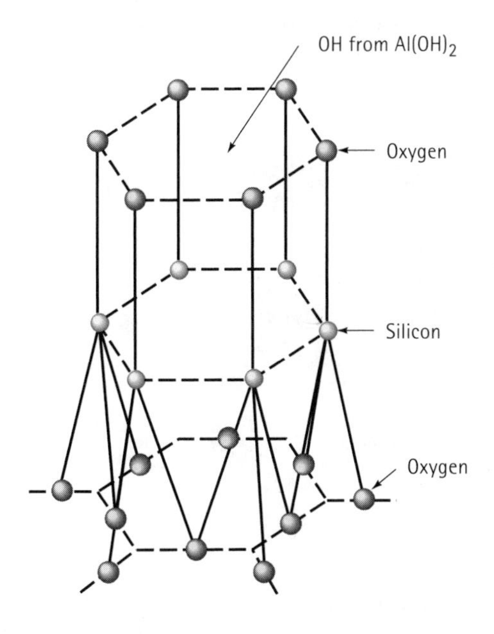

Figure 14.4
The silica tetrahedra produces hexagonal-shaped sheet structures, forming the basis for clays and other minerals. Each silicon atom in the middle plane is covalently bonded to four oxygen atoms. (Note that the drawing is distorted in the c-axis to better show the arrangement of the ions.)

clay with the formula $Al_2Si_2O_5(OH)_4$ [Figure 14.5(a)]. Montmorillonite, or $Al_2(Si_2O_5)_2(OH)_2$ contains two silicate sheets sandwiched around a central Al plus (OH) layer [Figure 14.5(b)]. Bonding within each of the clay platelets is achieved by a combination of covalent and ionic bonding; the separate platelets are bonded to one another by weak Van der Waals bonds. Clays are important components of many ceramic materials.

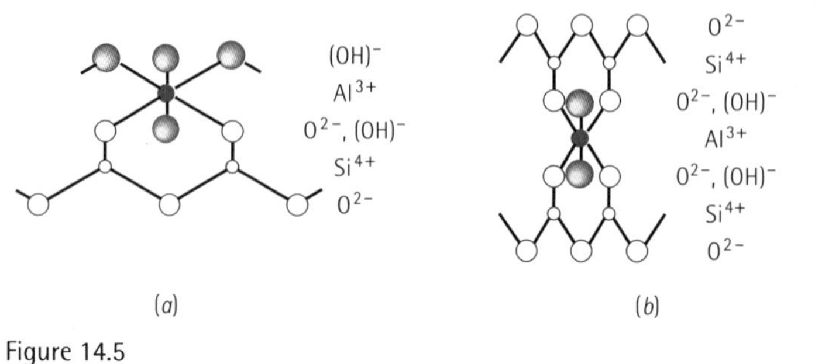

Figure 14.5
The silicate sheet structure that forms the basis for clays. (*a*) Kaolinite clay and (*b*) montmorillonite clay.

Silica Finally, when all four corners of the tetrahedra are shared with other silica tetrahedra, silica (SiO_2) is produced, as in cristobalite. Silica can exist in several allotropic forms. As the temperature increases, silica changes from α-quartz to β-quartz to β-tridymite to β-cristobalite to liquid. The pressure-temperature equilibrium diagram in Figure 14.6 shows the stable forms of silica. An abrupt change in the dimensions of silica accompanies the transformation of α-quartz to β-quartz. These changes are shown in Figure 14.7 for quartz. High stresses and even cracking accompany the volume change.

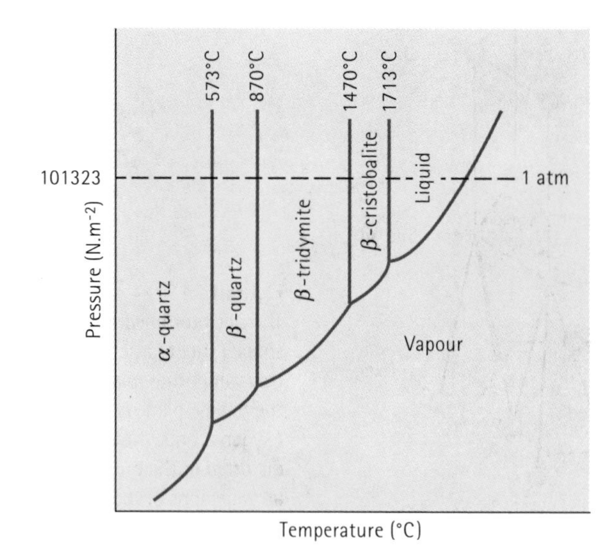

Figure 14.6
The pressure-temperature phase diagram for SiO_2.

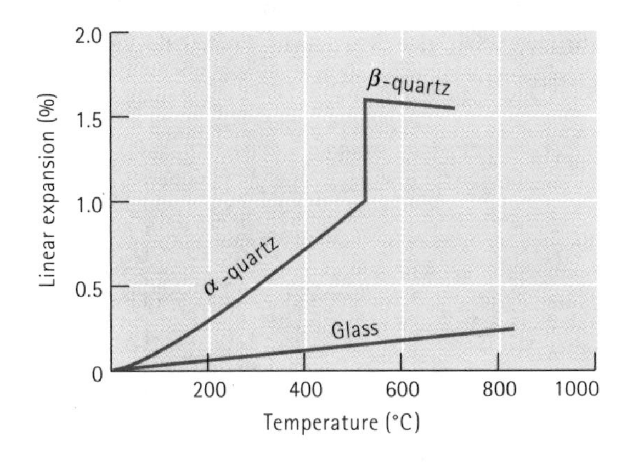

Figure 14.7
The expansion of quartz. In addition to the regular – almost linear – expansion, a large, abrupt expansion accompanies the α to β-quartz transformation. However, glasses expand uniformly.

EXAMPLE 14.2

Determine the type of silicate structure expected for each of the following complex ceramics:

$CaO \cdot MnO \cdot 2SiO_2$ $\quad Na_2O \cdot 2SiO_2$ $\quad Sc_2O_3 \cdot 2SiO_2$ $\quad 3FeO \cdot Al_2O_3 \cdot 3SiO_2$

SOLUTION

If we rearrange the chemical formulas of the ceramics, we can isolate the ratios of Si and O in the structures:

$$Cao \cdot MnO \cdot 2SiO_2 = CaMn(SiO_3) \text{ or metasilicate}$$

$$Na_2O \cdot 2SiO_2 = Na_2(Si_2O_5) \text{ or sheet structure}$$

$$Sc_2O_3 \cdot 2SiO_2 = Sc_2(Si_2O_7) \text{ or pyrosilicate}$$

$$3FeO \cdot Al_2O_3 \cdot 3SiO_2 = Fe_3Al_2(SiO_4)_3 \text{ or orthosilicate}$$

14.4 Imperfections in Crystalline Ceramic Structures

As in metals, the structures of ceramic materials contain a variety of imperfections. Point defects are particularly important for physical properties, such as electrical conductivity. Mechanical properties are influenced by surfaces, including grain boundaries, particle surfaces, and pores.

Point Defects Substitutional and interstitial solid solutions form in ceramic materials. the NiO-MgO system (Figure 9.8), Al_2O_3-Cr_2O_3 system (Figure 9.9), and the MgO-FeO system (Figure 9.18) display a complete series of substitutional solid solutions and have isomorphous phase diagrams. Likewise, the olivines, $(Mg, Fe)_2SiO_4$, display a complete range of solubility, with the Mg^{2+} and Fe^{2+} ions replacing one another completely in the silicate structure (Figure 14.8).

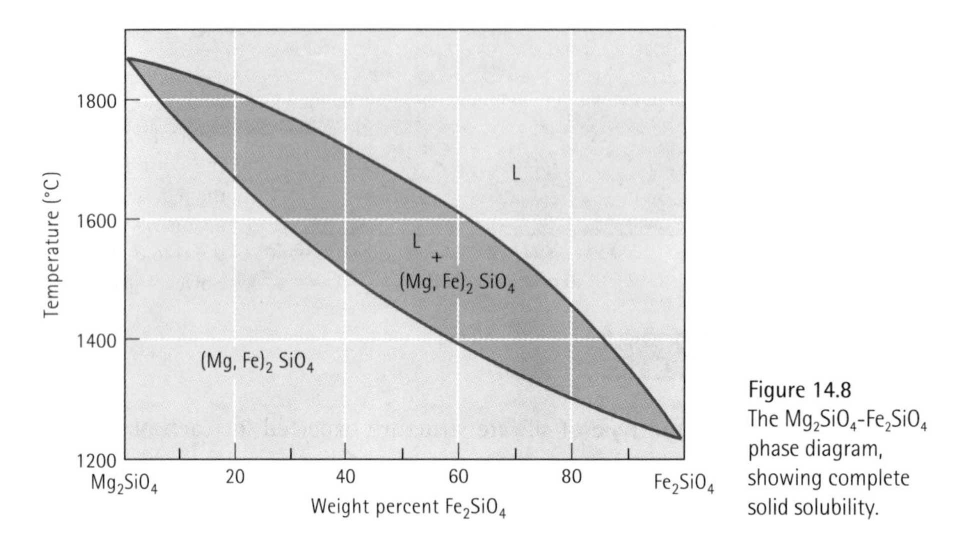

Figure 14.8
The Mg_2SiO_4-Fe_2SiO_4 phase diagram, showing complete solid solubility.

Solid solutions may provide unusual physical properties; for example, adding Cr_2O_3 to Al_2O_3 produces ruby, which can serve as a laser.

Often, the solid solubility of one phase in another is limited. In the MgO-Al_2O_3 system (Figure 14.9), some Al_2O_3 is soluble in MgO above about 1 600°C, whereas virtually no MgO is soluble in Al_2O_3 at any temperature. This system also includes an intermediate solid solution, $MgAl_2O_4$, or spinel. In other systems, such as the SiO_2-MgO system (Figure 14.10), no solid solubility is found, and two stoichiometric compounds, $MgSiO_3$ (enstatite) and Mg_2SiO_4 (forsterite), are observed.

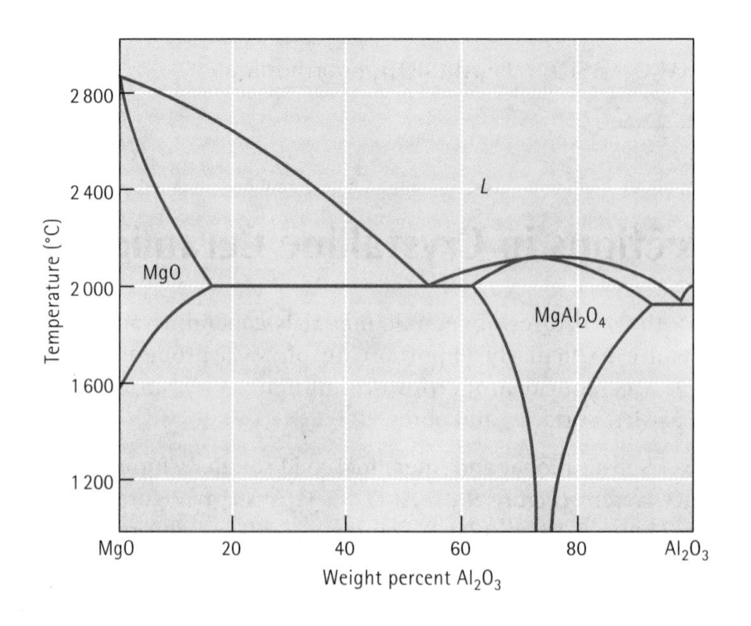

Figure 14.9
The MgO-Al_2O_3 phase diagram, showing limited solid solubility and the presence of $MgAl_2O_4$, or spinel.

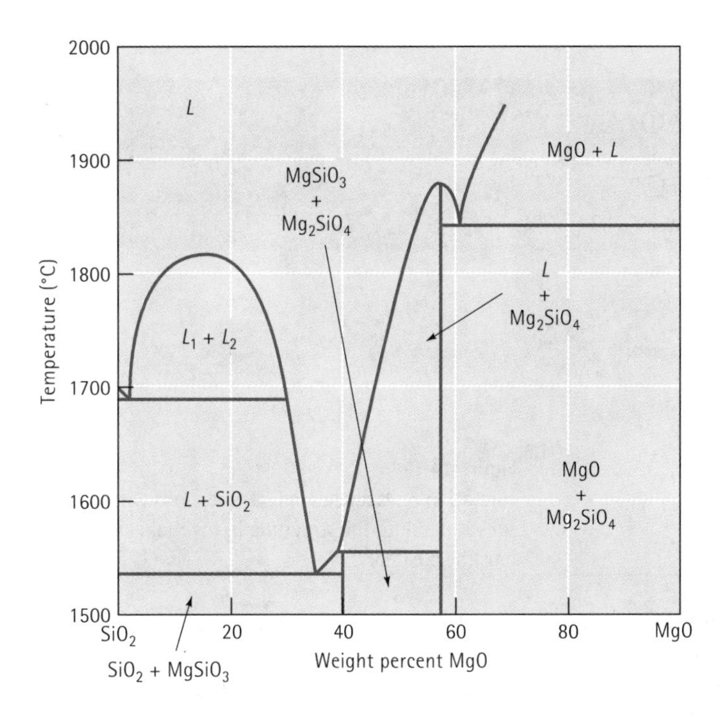

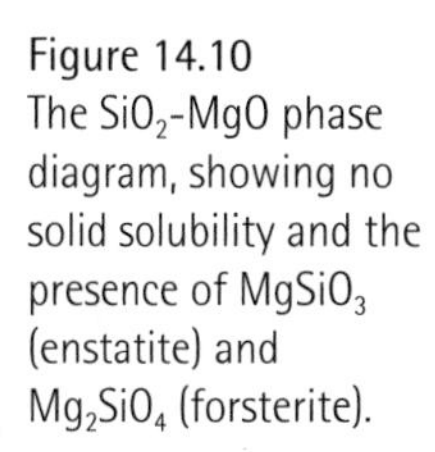

Figure 14.10
The SiO_2-MgO phase diagram, showing no solid solubility and the presence of $MgSiO_3$ (enstatite) and Mg_2SiO_4 (forsterite).

Maintaining a balanced charge distribution is difficult when solid solution ions are introduced. However, charge deficiencies or excesses can be accommodated in ceramic materials in several ways. For example, if an Al^{3+} ion in the centre of a montmorillonite clay platelet is replaced by a Mg^{2+} ion, the clay platelet has an extra negative charge. To equalise the charge, a positively charged ion, such as sodium or calcium, is adsorbed onto the surface of the clay platelet (Figure 14.11). The type and number of adsorbed ions affect the surface chemistry of the clay platelets; this, in turn, influences the formability and strength of ceramic products based on clays.

A second way of accommodating the unbalanced charge is to create vacancies (as described in Chapter 4). We might expect FeO to contain equal numbers of Fe^{2+} and O^{2-} ions in the sodium chloride structure. However, FeO is always a nonstoichiometric structure formed when two Fe^{3+} ions substitute for three Fe^{2+} ions. This creates a vacancy where an iron ion normally would be located (Figure 14.12). When Fe^{3+} ions are present, fewer Fe ions than oxygen ions are present in the structure and the

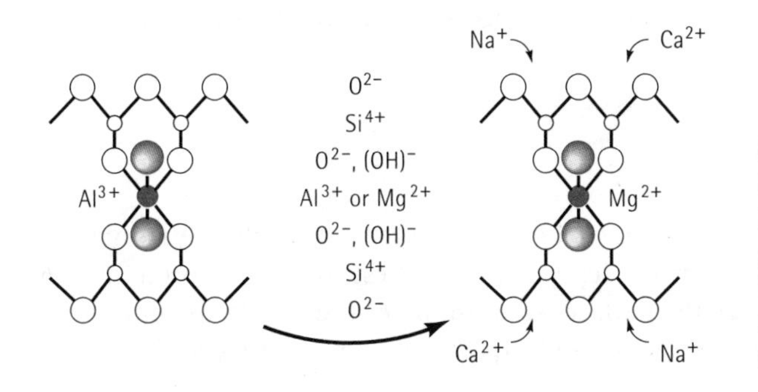

Figure 14.11
Replacement of an Al^{3+} ion by a Mg^{2+} ion in a montmorillonite clay platelet produces a charge imbalance which permits cations, such as sodium or calcium, to be attracted to the clay.

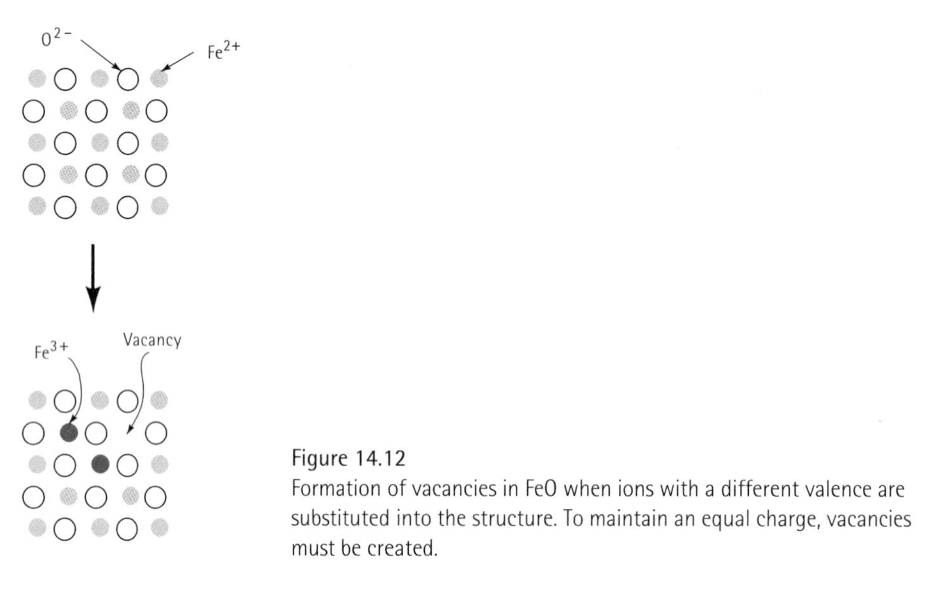

Figure 14.12
Formation of vacancies in FeO when ions with a different valence are substituted into the structure. To maintain an equal charge, vacancies must be created.

stoichiometric ratio of 1 Fe to 1 O is not obtained. The iron oxide is represented by the formula $Fe_{1-x}O$ to indicate its nonstoichiometric nature. Defects such as this occur frequently in ceramic materials and often lead to special properties. For example, FeO behaves as a semiconductor.

We could also substitute more than one type of ion. For example, we can introduce a Li^+ ion and a Fe^{3+} ion to substitute for two Mg^{2+} ions in MgO. In this mechanism, vacancies do not need to be created. Vacancies may also be present as Frenkel defects or Schottky defects (Figure 4.11). The Frenkel defect occurs when an ion leaves its normal position and a vacancy remains. The Schottky defect is a pair of vacancies – a cation vacancy and an anion vacancy.

EXAMPLE 14.3 Selection of a Method to Change the Composition of Clay

When water is added to montmorillonite (or bentonite) clay, the clay acts as a binder to hold sand grains together, producing green sand moulds for making metal castings. Clay mined in the southern United States contains 1% Ca, which is adsorbed onto the surface of the clay platelets. Better mould properties (such as less thermal expansion and improved recyclability) would be obtained if Na rather than Ca were adsorbed on the clay. Describe a process to convert the calcium bentonite to sodium bentonite.

SOLUTION

We could simply purchase sodium bentonite, which already contains adsorbed Na, but shipping charges from deposits in the West might be prohibitive. An alternative method is to change the surface composition of the Southern bentonite by mixing the clay with sodium carbonate, Na_2CO_3. The Na ions and Ca ions replace one another, with the Na ions becoming attached to the clay platelets and the Ca ions forming $CaCO_3$. This 'activated' clay has properties that resemble those of Western bentonite.

Montmorillonite's formula is $Al_2(Si_2O_5)_2(OH)_2$, and it has a molecular weight of 360.28 $g.mol^{-1}$. In one kg of clay, there are 1 000/360.28 = 2.776 moles of montmorillonite.

If 1 wt% of the clay is calcium (M = 40.08 $g.mol^{-1}$), then in one kg of clay there are approximately 10 g of Ca, or 10/40.08 = 0.25 moles of calcium. If all of the Ca^{2+} is replaced by all of the Na^+ ions in the sodium carbonate, then to maintain a charge balance we need to introduce 0.50 moles of sodium (M = 22.99 $g.mol^{-1}$), or 22.99 × 0.50 = 11.50 g Na. The molecular weight of Na_2CO_3 is 105.98$g.mol^{-1}$. We need one mole of sodium carbonate for two moles of Na, or we need 0.25 moles of sodium carbonate to provide sufficient Na to replace the Ca:

$$\text{g of } Na_2CO_3 = (0.25)(105.98) = 26.50 \text{ g}$$

Under ideal conditions, assuming that all of the sodium carbonate is consumed, our process design is to produce an activated clay by mixing 26.50 g of sodium carbonate with 1000g of the Southern bentonite.

EXAMPLE 14.4

Ceramic compounds containing lattice imperfections may act as semiconductors. In FeO, each vacancy introduces one carrier of electrical charge. Specify a FeO ceramic that provides 5.7×10^{21} vacancies (and therefore 5.7×10^{21} charge carriers) per cm^3. FeO has the sodium chloride structure and a lattice parameter of 0.412 nm.

SOLUTION

Because of the sodium chloride structure we know that there should be 4 Fe^{2+} sites and 4 O^{2-} sites. To obtain 5.7×10^{21} charge carriers, or Fe^{2+} vacancy sites, we need:

$$5.7 \times 10^{21} \times 10^6 \text{ vac.m}^{-3})(4.12 \times 10^{-8} \times 10^2\text{m})^3 = 0.4 \text{ vacancies.cell}^{-1}$$

To produce Fe^{2+} vacancies in FeO, we must replace a fraction of the ions with Fe^{3+} ions. When two Fe^{3+} ions are introduced, three Fe^{2+} ions must be removed, creating one vacancy. Let's assume that we have 25 unit cells of FeO, or 100 oxygen ions. There are also 100 iron ion sites. The number of vacancies required is (0.4 vacancies/cell)(25 cells) = 10 vacancies. To obtain 10 vacancies, we need to add 20 Fe^{3+} ions and remove 30 Fe^{2+} ions. Therefore we expect to have:

100 oxygen ions
70 Fe^{2+} ions
20 Fe^{3+} ions
10 vacancies

These ions constitute a 'formula' of $Fe_{0.9}O$, which permits the FeO to behave as a semiconductor and to have the required number of charge carriers.

The atomic percent oxygen in this structure is:

$$\text{at\% O} = \frac{100 \text{ oxygen atoms}}{100 \text{ O} + 70 \text{ Fe}^{2+} + 20 \text{ Fe}^{3+}} \times 100 = 52.6\%$$

The weight percent oxygen in this structure is:

$$\text{wt\% O} = \frac{(52.6)(16 \text{ g.mol}^{-1})}{(52.6)(16 \text{ g.mol}^{-1}) + (47.4)(55.847 \text{ g.mol}^{-1})} \times 100 = 24.1\%$$

Dislocations Dislocations are observed in some ceramic materials, including LiF, sapphire (Al_2O_3), and MgO. However, the dislocations do not easily move, a consequence of the large Burgers vector, the presence of relatively few slip systems, and the necessity to break strong ionic bonds and then force ions past oppositely charged ions during the slip process. Because slip is not likely to occur, a crack is not blunted by deformation of the material ahead of that crack, which thus continues to propagate. The ceramic is brittle.

Although dislocations move more easily at high temperatures, deformation is more likely to occur by mechanisms such as grain boundary sliding or viscous flow of glassy phases.

Surface Defects Grain boundaries (Figure 14.13) and particle surfaces are important surface defects in ceramics. Typically, ceramics with a small grain size are stronger than coarse-grained ceramics. Finer grain sizes help reduce stresses that develop at grain boundaries due to anisotropic expansion and contraction. Normally, a fine grain size is produced by beginning with finer ceramic raw materials.

Particle surfaces, which represent planes of broken, unsatisfied covalent or ionic bonds, are reactive. Gaseous molecules, for example, may be adsorbed onto the surface to reduce the surface energy. In clay deposits, foreign ions may be attracted to the platelet surface (Figure 14.14), altering the composition, properties, and formability of clay and clay products.

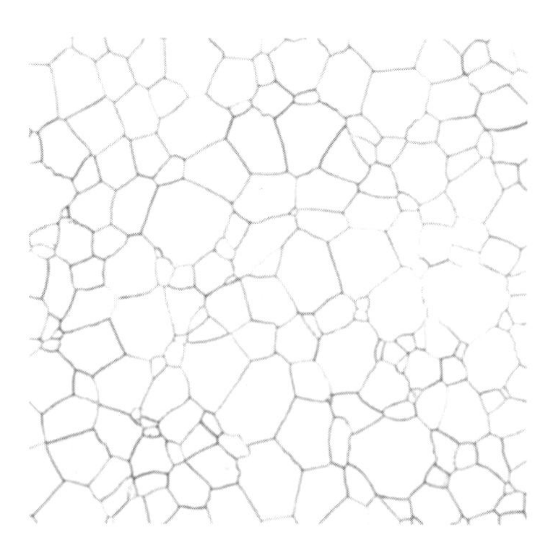

Figure 14.13
Grain structure in PLZT, a lead-lanthanum-zirconium titanate used as a ceramic sensor material (× 600). (*Courtesy of G. Haertling.*)

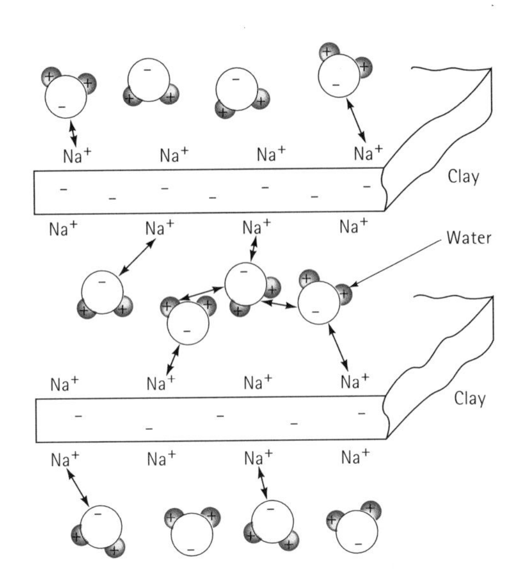

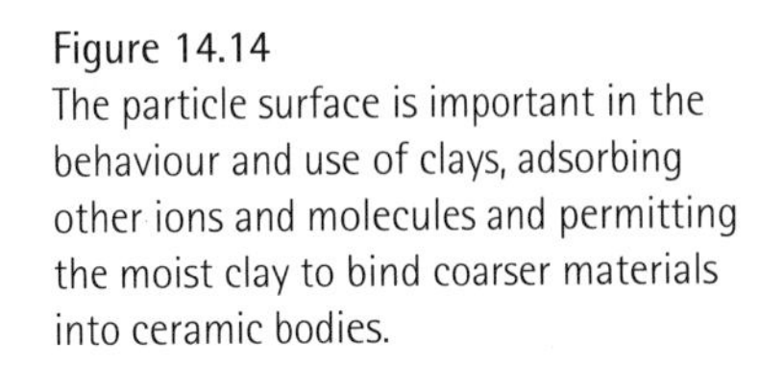
Figure 14.14
The particle surface is important in the behaviour and use of clays, adsorbing other ions and molecules and permitting the moist clay to bind coarser materials into ceramic bodies.

Porosity Pores can be considered a special type of surface defect. The pores in a ceramic may be either interconnected or closed. The apparent porosity measures the interconnected pores and determines the permeability, or the ease with which gases and fluids seep through the ceramic component. The apparent porosity is determined by weighing the dry ceramic (W_d), then reweighing the ceramic both when it is suspended in water (W_s) and after it is removed from the water (W_w). Using units of grams and cm^3:

$$\text{Apparent porosity} = \frac{W_w - W_d}{W_w - W_s} \times 100 \tag{14.1}$$

The true porosity includes both interconnected and closed pores. The true porosity, which better correlates with the properties of the ceramic, is:

$$\text{True porosity} = \frac{\rho - B}{\rho} \times 100 \tag{14.2}$$

where

$$B = \frac{W_d}{W_w - W_s} \tag{14.3}$$

B is the bulk density and ρ is the true density or specific gravity of the ceramic. The bulk density is the weight of the ceramic divided by its volume.

EXAMPLE 14.5

Silicon carbide particles are compact and fired at a high temperature to produce a strong ceramic shape. The specific gravity of SiC is 3.2 Mg.m^{-3}. The ceramic shape subsequently is weighed when dry (360 g), after soaking in water (385 g), and while suspended in water (224 g). Calculate the apparent porosity, the true porosity, and the fraction of the pore volume that is closed.

SOLUTION

$$\text{Apparent porosity} = \frac{W_w - W_d}{W_w - W_s} \times 100 = \frac{385 - 360}{385 - 224} \times 100 = 15.5\%$$

$$\text{Bulk density} = B = \frac{W_d}{W_w - W_s} = \frac{360}{385 - 224} = 2.24$$

$$\text{True Porosity} = \frac{\rho - B}{\rho} \times 100 = \frac{3.2 - 2.24}{3.2} \times 100 = 30\%$$

The closed pore percentage is the true porosity minus the apparent porosity, or 30 – 15.5 = 14.5%. Thus:

$$\text{Fraction closed pores} = \frac{14.5}{30} = 0.483$$

14.5 The Structure of Ceramic Glasses

The most important of the noncrystalline ceramic materials are glasses. A glass is a solid material that has hardened and become rigid without crystallising. A glass in some ways resembles an undercooled liquid. However, below the glass transition temperature (Figure 14.15), the rate of volume contraction on cooling is reduced and the material can be considered a glass rather than an undercooled liquid. The glassy structures are produced by joining silica tetrahedra or other ionic groups to produce a solid but noncrystalline framework structure (Figure 14.16).

We can also find noncrystalline structures in exceptionally fine powders such as gels or colloids. In these materials, particles sizes may be 10 nm or less. These amorphous materials, which include some cements and adhesives, are produced by condensation of vapours, electrodeposition, or chemical reactions.

Silicate Glasses The silicate glasses are the most widely used. *Fused silica*, formed from pure SiO_2, has a high melting point and the dimensional changes during heating and cooling are small (Figure 14.7). Generally, however the silicate glasses contain additional oxides (Table 14.1). While oxides such as silica behave as glass formers, an intermediate oxide (such as lead or aluminium oxide) does not form a glass by itself but is incorporated into the network structure of the glass formers. A third group of oxides, the modifiers, break up the network structure and eventually cause the glass to devitrify, or crystallise.

Modified Silicate Glasses Modifiers break up the silica network if the oxygen-to-silicon ratio increases significantly. When Na_2O is added, for example, the sodium ions enter holes within the network rather than becoming part of the network.

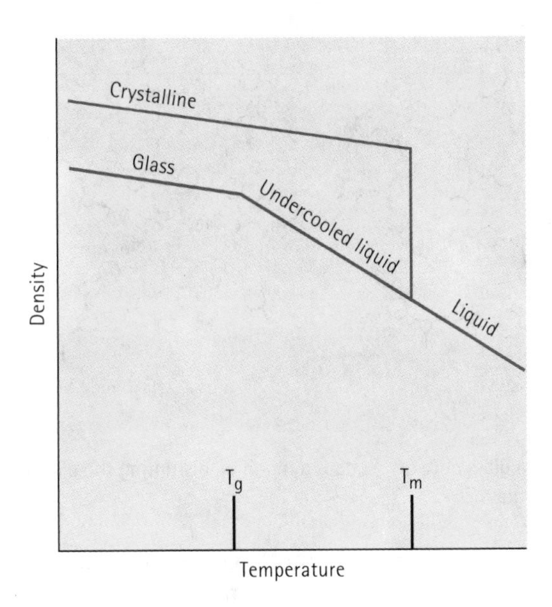

Figure 14.15
When silica crystallises on cooling, an abrupt change in the density is observed. For glassy silica, however, the change in slope at the glass transition temperature indicates the formation of a glass from the undercooled liquid.

Table 14.1 Division of the oxides into glass formers, intermediates, and modifiers.

Glass Formers	Intermediates	Modifiers
B_2O_3	TiO_2	Y_2O_3
SiO_2	ZnO	MgO
GeO_2	PbO	CaO
P_2O_5	Al_2O_3	PbO_2
V_2O_3	BeO	Na_2O

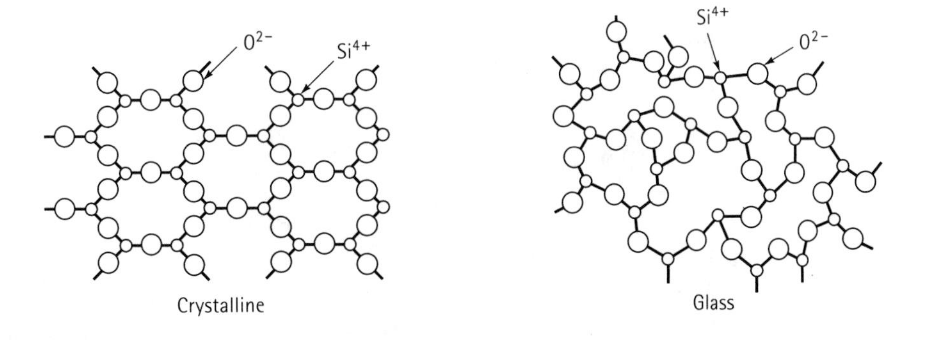

Figure 14.16
Crystalline and glassy silicate structures. Both structures have a short-range order, but only the crystalline structure has a long-range order.

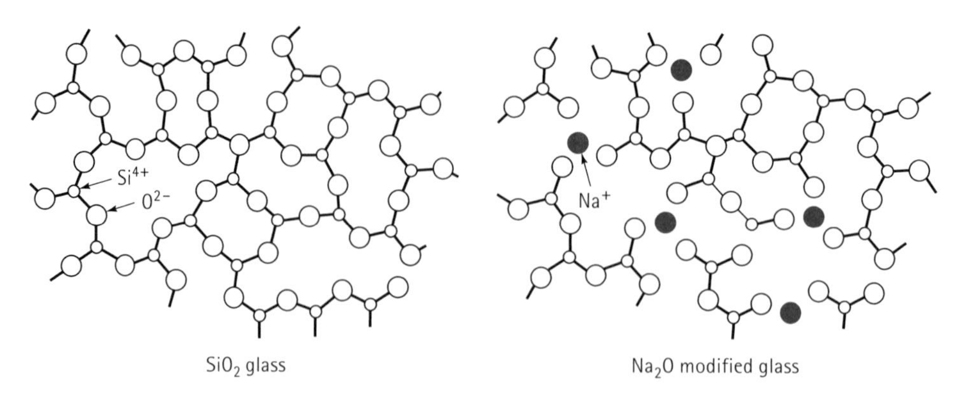

Figure 14.17
The effect of Na_2O on the silica glass network. Soda is a modifier, disrupting the glassy network and reducing the ability to form a glass.

However, the oxygen ion that enters with the Na_2O does become part of the network (Figure 14.17). When this, happens, there aren't enough silicon ions to combine with the extra oxygen ions and keep the network intact. Eventually, a high O : Si ratio causes the remaining silica tetrahedra to form chains, rings, or compounds, and the silica no longer transforms to a glass. When the O : Si ratio is above about 2.5, silica glasses are difficult to form; above a ratio of three, a glass forms only when special precautions are taken, such as use of rapid cooling rates.

Modification also lowers the melting point and viscosity of silica, making it possible to produce glass at lower temperatures. The effect of Na_2O additions to silica is shown in Figure 14.18. These glasses are further modified by adding CaO, which reduces solubility the of the glass in water.

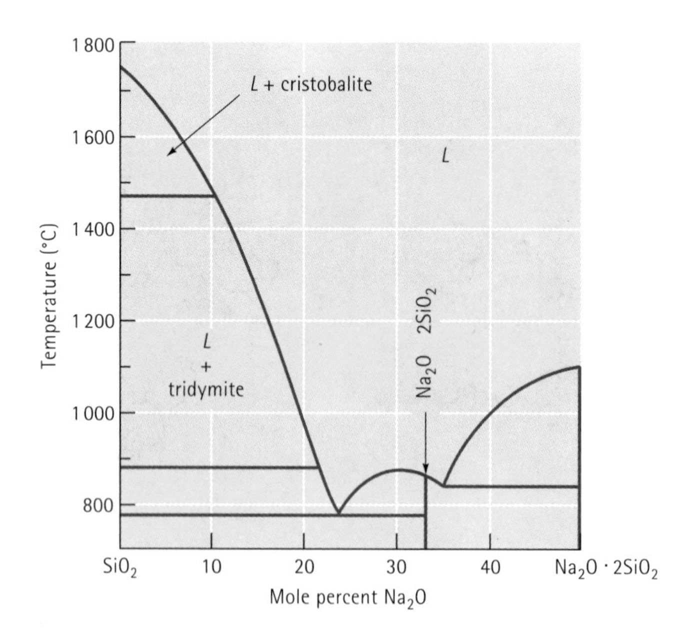

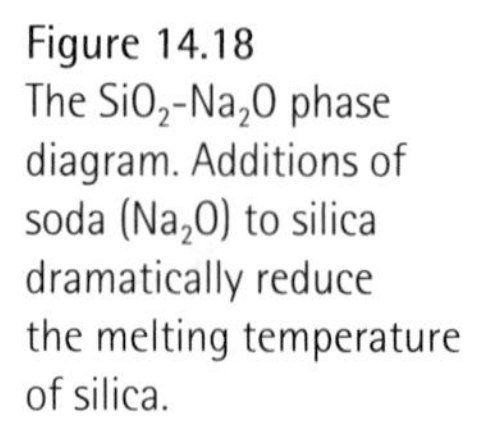
Figure 14.18
The SiO_2-Na_2O phase diagram. Additions of soda (Na_2O) to silica dramatically reduce the melting temperature of silica.

EXAMPLE 14.6 Specification of a Glass

We produce good chemical resistance in a glass when we introduce B_2O_3 into silica. to ensure that we have good glass-forming tendencies, we wish the O : Si ratio to be no more than 2.5, but we also want the glassware to have a low melting temperature to make the glass-forming process easier and more economical. Specify such a glass.

SOLUTION

Because B_2O_3 reduces the melting temperature of silica, we would like to add as much as possible. We also, however, want to ensure that the O : Si ratio is no more than 2.5, so the amount of B_2O_3 is limited. As an example, let's determine the amount of B_2O_3 we must add to obtain exactly an O : Si ratio of 2.5. Let f_B be the mole fraction of B_2O_3 added to the glass, and $1 - f_B$ be the mole fraction of SiO_2:

$$\frac{O}{Si} = \frac{\left(3\frac{\text{O ions}}{B_2O_3}\right)(f_B) + \left(2\frac{\text{O ions}}{SiO_2}\right)(1-f_B)}{\left(1\frac{\text{Si ion}}{SiO_2}\right)(1-f_B)}$$

$$3f_B + 2 - 2f_B = 2.5 - 2.5f_B \quad \text{or} \quad f_B = 0.143$$

Therefore, we must produce a glass containing no more than 14.3 mol% B_2O_3. In weight percent:

$$\text{wt\%}B_2O_3 = \frac{(f_B)(69.62 \text{ g.mol}^{-1})}{(f_B)(69.62 \text{ g.mol}^{-1}) + (1-f_B)(60.8\text{g.mol}^{-1})} \times 100$$

$$\text{wt\%}B_2O_3 = \frac{(0.143)(69.62)}{(0.143)(69.62\ ) + (0.857)(60.08)} \times 100 = 16.2$$

14.6 Mechanical Failure of Ceramics

Ceramic materials, both crystalline and noncrystalline, are very brittle, particularly at low temperatures. Problems with brittle failure of ceramics are heightened by the presence of flaws such as small cracks, porosity, foreign inclusions, glassy phases, or a large grain size, typically introduced during the manufacturing process. Flaws vary in size, shape, and orientation, both within a single part and from part to part.

Brittle Fracture Any crack or imperfection limits the ability of a ceramic to withstand a tensile stress. This is because a crack (sometimes called a Griffith flaw) concentrates and magnifies the applied stress. Figure 14.19 shows a crack of length *a* at the surface of a brittle material. the radius r of the crack is also shown. When a tensile stress σ is applied, the actual stress at the crack tip is:

$$\sigma_{\text{actual}} \cong 2\sigma\sqrt{a/r} \tag{14.4}$$

For very thin cracks (small r) or long cracks (large a), the ratio $\sigma_{\text{actual}}/\sigma$ becomes large, or the stress is magnified. If the magnified stress exceeds the yield strength, the crack grows and eventually causes failure, even though the actual applied stress σ is small.

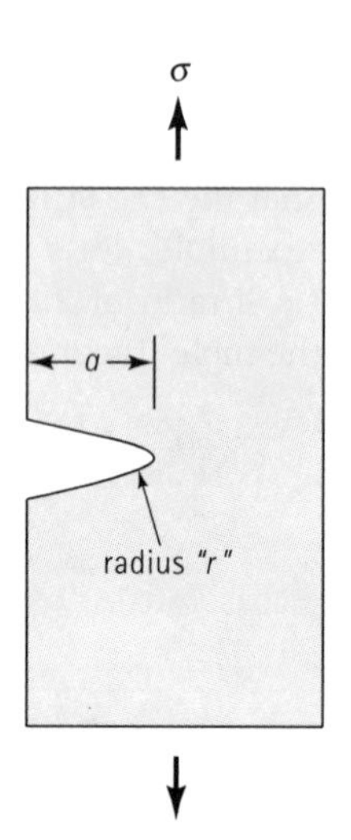

Figure 14.19
Schematic diagram of the Griffith flaw in a ceramic.

In a different approach, we recognise that an applied stress causes an elastic strain, related to the modulus of elasticity E, within the material. When a crack propagates, this strain energy is released, reducing the overall energy. At the same time, however, two new surfaces are created by the extension of the crack; this increases the energy associated with the surface. By balancing the strain energy and the surface energy, we find that the critical stress required to propagate the crack is given by the Griffith equation,

$$\sigma_{\text{critical}} = \sqrt{2E\gamma/\pi a} \tag{14.5}$$

where a is the length of a surface crack (or half the length of an internal crack) and γ is the surface energy (per unit area). Again this equation shows that even small flaws severely limit the strength of the ceramic.

We might also note that if we rearrange Equation 6.14, which described the stress concentration factor K, we obtain:

$$\sigma = \frac{K}{f\sqrt{\pi a}} \tag{14.6}$$

This equation is similar in form to Equation 14.5. Each of these equations points out the dependence of the mechanical properties on the size of flaws present in the ceramic. Development of manufacturing processes to minimise the flaw size becomes crucial in improving the strength of ceramics.

The flaws are most important when tensile stresses act on the material. Compressive stresses try to close rather than open a crack; consequently, ceramics often have very good compressive strengths.

EXAMPLE 14.7

An advanced ceramic, sialon, has a tensile strength of 400 MN.m^{-2}. A thin crack 0.25 mm deep is observed before a sialon part is tested. The part unexpectedly fails at a stress of 3 MN.m^{-2} by propagation of the crack. Estimate the radius of the crack tip.

SOLUTION

The failure occurred because the 3 MN.m^{-2} applied stress, magnified by the stress concentration at the tip of the crack, produced an actual stress equal to the tensile strength. From Equation 14.4:

$$\sigma_{\text{actual}} = 2\sigma\sqrt{a/r}$$

$$400\,\text{MN.m}^{-2} = (2)(3\,\text{MN.m}^{-2})\ \sqrt{0.25\times 10^{-3}\,\text{m}/r} \quad \text{where } r \text{ is in metres}$$

$$66.7 = \sqrt{0.25\times 10^{-3}/r}$$

$$0.25\times 10^{-3}/r = 4444.4$$

$$r = \frac{0.25\times 10^{-3}}{4444.4} = 5.625\times 10^{-8}\,\text{m}$$

$$r = 56.25\times 10^{-9}\,\text{m} = 56.25\,\text{nm}$$

The likelihood of our being able to measure a radius of curvature of this size by any method of nondestructive testing is virtually nil. Therefore, although Equation 14.4 may help illustrate the factors that influence how a crack propagates in a brittle material, it does not help in predicting the strength of actual ceramic parts.

EXAMPLE 14.8 **Design of a Ceramic Support**

Design a supporting 75 mm wide plate made of sialon, which has a fracture toughness of 10 NM.m$^{-3/2}$, that will withstand a tensile load of 175 kN. The part is to be nondestructively tested to ensure that no flaws are present that might cause failure.

SOLUTION

Let's assume that we have three nondestructive testing methods available to us: X-ray radiography can detect flaws larger than 0.5 mm, gamma-ray radiography can detect flaws larger than 0.2 mm, and ultrasonic inspection can detect flaws larger than 0.1 mm. For these flaw sizes, we must now calculate the minimum thickness of the plate that will ensure that flaws of these sizes will not propagate.

From our fracture toughness equation, assuming that $f = 1$;

$$\text{Stress} = \sigma_{\max} = \frac{K_{Ic}}{\sqrt{\pi a}} = \frac{F}{A}$$

$$A = \frac{F\sqrt{\pi a}}{K_{Ic}} = \frac{(0.175\,\text{MN})(\sqrt{\pi})(\sqrt{a})}{10\,\text{MN.m}^{-3/2}}$$

$$\text{Area} = 3.1\times10^{-2}\sqrt{a}(\text{metre}^2)$$

$$\text{Thickness} = \frac{\text{Area}}{\text{Width}} = \frac{3.1\times10^{-2}\sqrt{a}}{75\times10^{-3}} = 0.413\sqrt{a}\ (\text{metre})$$

Substituting for various 'a' values according to N.D.T. method:

NDT Method	Smallest detectable crack 'a' (m)	Minimum area (m^2)	(mm^2)	Minimum thickness (m)	(mm)	Maximum stress ($MN.m^{-2}$)
X-radiography	0.5×10^{-3}	6.9×10^{-4}	690	9.2×10^{-3}	9.2	253
γ-radiography	0.2×10^{-3}	4.4×10^{-4}	440	5.8×10^{-3}	5.8	398
Ultrasonic	0.1×10^{-3}	3.1×10^{-4}	310	4.1×10^{-3}	4.1	565

Our ability to detect flaws, coupled with our ability to produce the ceramic with flaws smaller than our detection limit, significantly affects the maximum stress than can be tolerated and, hence, the size of the part. In this example, the part can be smaller if ultrasonic inspection is available.

The fracture toughness is also important. Had we used Si_3N_4, with a fracture toughness of 3 $MN.m^{-3/2}$ instead of the sialon, we could repeat the calculations and show that, for ultrasonic testing, the minimum thickness is 13.8 mm and the maximum stress is only 170 $MN.m^{-2}$.

Statistical Treatment of Fracture Because the tensile properties of ceramics depend so critically on the size and geometry of the ever-present flaws, there is considerable scatter in the values for the strength determined from a tensile, bending, or fatigue test. Ceramic parts produced from identical materials by identical methods fail at very different applied loads. In order to design structural parts using ceramics, the probability that a flaw is present that will cause failure to occur at any given stress must be known. The Weibull distribution and Weibull modulus provide one statistical approach to designing with ceramics.

The Weibull distribution shown in Figure 14.20(a) describes the fraction of samples that fail at different applied stresses. At low stresses, a small fraction of samples contain flaws enough to cause fracture; most fail at an intermediate applied stress; and a few contain only small flaws and do not fail until large stresses are applied. To provide predictability, we prefer a very narrow distribution.

The probability of failure can be related to the failure stress by

$$\ln\left[\ln\left(\frac{1}{1-P}\right)\right]=m\ln(\sigma_f) \qquad (14.7)$$

where P is the cumulative probability of failure, σ_f is the stress at which failure occurs, and m is the Weibull modulus. The probability of failure is shown in a cumulative manner in Figure 14.20(b) for Al_2O_3 prepared by two different processes. When the applied stress is high, there is a very high probability that any sample will fail. As the stress decreases the probability that any sample will fail also decreases. Even for low applied stresses, there is a finite possibility that a that a sample contains a flaw large enough to propagate. It is this small probability, even at low stresses, that limits the use of ceramic materials for critical applications.

The Weibull modulus m is the slope of the cumulative probability curve. For design of critical, load-bearing ceramic parts, the Weibull modulus should be large; a high slope represents a ceramic with a narrow range of flaw sizes and helps us design more reliable parts. In Figure 14.20(b), the Weibull modulus for conventionally processed alumina is only about half that of alumina prepared using exceptionally small powder particles. The advanced alumina is both stronger and more reliable than the conventional material. A Weibull modulus of 10 to 20 is typical of the advanced ceramic materials. Unfortunately, in order to produce the desired Weibull modulus, high-purity raw materials and complex processing are normally required, making reliable ceramic parts expensive.

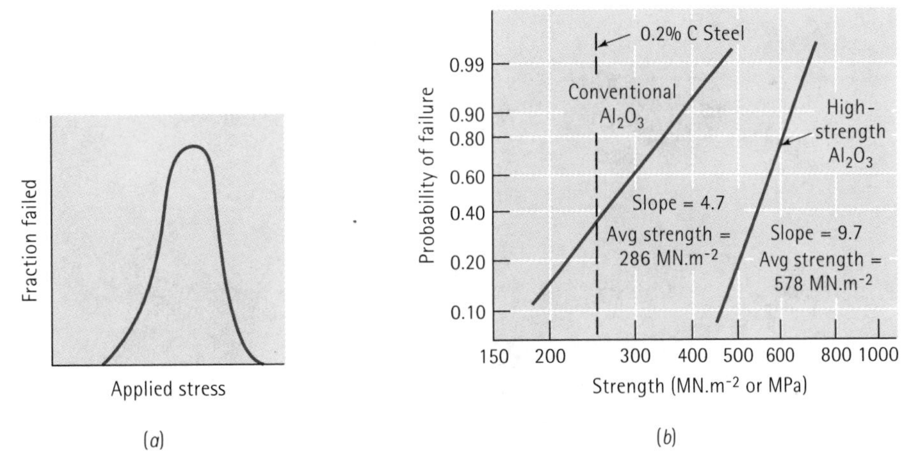

Figure 14.20
(*a*) The Weibull distribution describes the fraction of samples that fail at any given applied stress.
(*b*) A cumulative plot (using special graph paper) of the probability that a sample will fail at any given stress yields the Weibull modulus, or slope. Alumina, produced by two different methods, is compared with low-carbon steel. Good reliability in design is obtained for a high Weibull modulus.

EXAMPLE 14.9

Seven silicon carbide specimens were tested and the following fracture strengths were obtained: 23 MN.m^{-2}, 49 MN.m^{-2}, 34 MN.m^{-2}, 30 MN.m^{-2}, 55MN.m^{-2}, 43 MN.m^{-2}, and 40 MN.m^{-2}. Estimate the Weibull modulus for the data and discuss the reliability of the ceramic.

SOLUTION

One simple, though not completely accurate, method for determining the behaviour of the ceramic is to assign a numerical rank (1 to 7) to the specimens, with the specimen having the lowest fracture strength assigned the value 1. The total number of specimens is n (in our case, 7). The cumulative probability P is then the numerical rank divided by $n + 1$ (in our case, 8). We can then plot ln [ln (1/1 – P)] versus ln σ_f. The following table and Figure 14.21 show the results of these calculations.

i^{th} Specimen	σ_f (MN.m^{-2})	P	$\ln\left[\ln\left(\frac{1}{1-P}\right)\right]$
1	23	1/8 = 0.125	–2.013
2	30	2/8 = 0.250	–1.246
3	34	3/8 = 0.375	–0.755
4	40	4/8 = 0.500	–0.367
5	43	5/8 = 0.625	–0.019
6	49	6/8 = 0.750	+0.327
7	55	7/8 = 0.875	+0.732

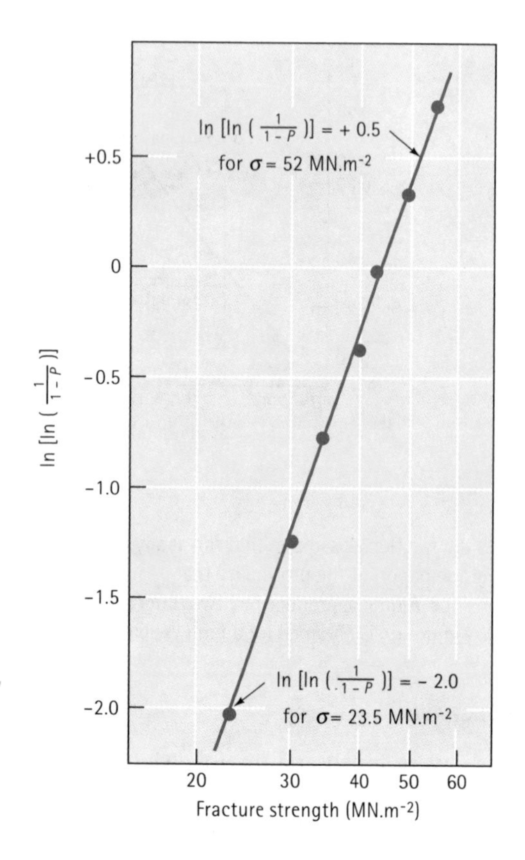

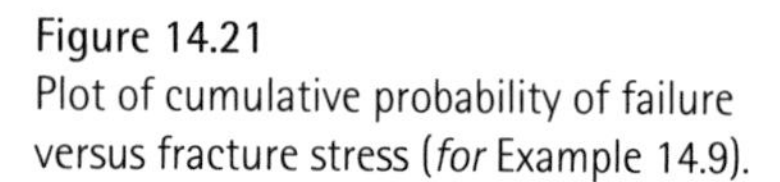

Figure 14.21
Plot of cumulative probability of failure versus fracture stress (*for* Example 14.9).

The slope of the curve, or the Weibull modulus m, is (using the two points indicated on the curve):

$$m = \frac{0.5-(-2.0)}{\ln(52)-\ln(23.5)} = \frac{2.5}{3.951-3.157} = 3.15$$

This low Weibull modulus suggests that the ceramic has a highly variable fracture strength, making it difficult to use reliably in high-load-bearing applications.

Toughening Methods Several methods are used to improve fracture toughness, which in turn leads to higher fracture strengths and higher service stresses. One traditional method for improving toughness is to surround the brittle ceramic particles with a softer, tougher matrix material. This is done in producing cermet cutting tools and abrasives, which are really composite materials [Figure 14.22(a)]. As an example, very hard tungsten carbide (WC) particles are embedded in a matrix of cobalt metal. The composite part remains the high hardness and cutting ability of WC, but the softer, more ductile cobalt deforms and absorbs energy. Other cermets,

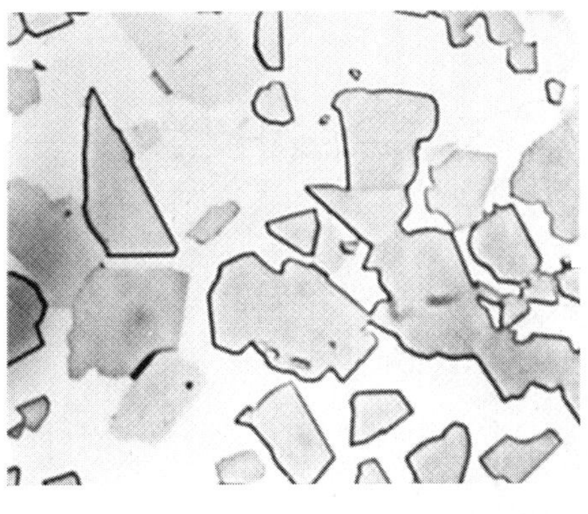

(a)

Figure 14.22
Microstructures of toughened ceramics:
(a) a tungsten carbide-cobalt matrix cermet (× 1 500),
(b) a glass-ceramic composite reinforced with SiC fibres (× 100), and
(c) partially stabilised ZrO_2, containing plates of the tetragonal phase in a monoclinic matrix (× 15 000).
(*From* Metals Handbook, *Vol. 9, 9th Ed., American Society for Metals, 1985.*)

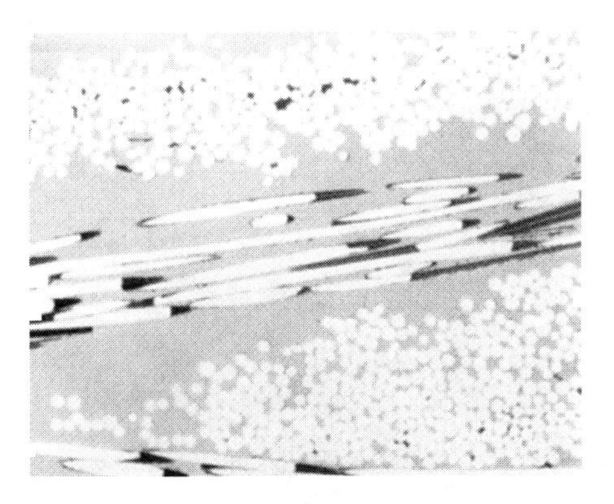

(b)

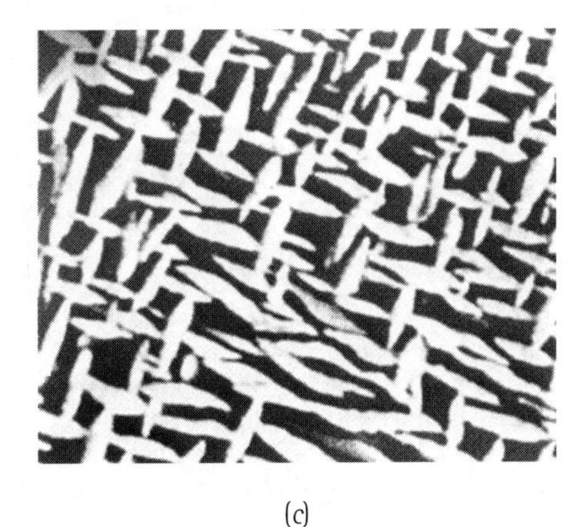

(c)

such as TiC particles in a nickel matrix or TiB in a cobalt matrix, provide good high-temperature strength and corrosion resistance for applications such as rocket motors, and UO_2 in an aluminium matrix serves as a nuclear fuel.

Another approach is to create ceramic matrix composites (CMCs) by introducing ceramic fibres or agglomerates into the ceramic matrix. When a crack tries to propagate in the matrix, it encounters the interface between the matrix and the ceramic fibre; the interface helps block the propagation of the crack [Figure 14.22(b)].

Cracks or stress concentrations may be introduced by phase transformations when a ceramic is heated or cooled. For example, zirconia (ZrO_2) transforms from a tetragonal structure to a monoclinic structure on cooling (Figure 14.23); this process leads to a large volume change. Because the resulting stresses cannot be relieved by plastic deformation, they initiate or propagate cracks in the part. Adding CaO, MgO, or other materials to the zirconia, however, forms a cubic solid solution that is stable at all temperatures. Forming the cubic solid solution, called *stabilised zirconia*, eliminates the phase transformation and makes it possible to use the material as a refractory. However, the fracture toughness is still only about 2 $MN.m^{-3/2}$.

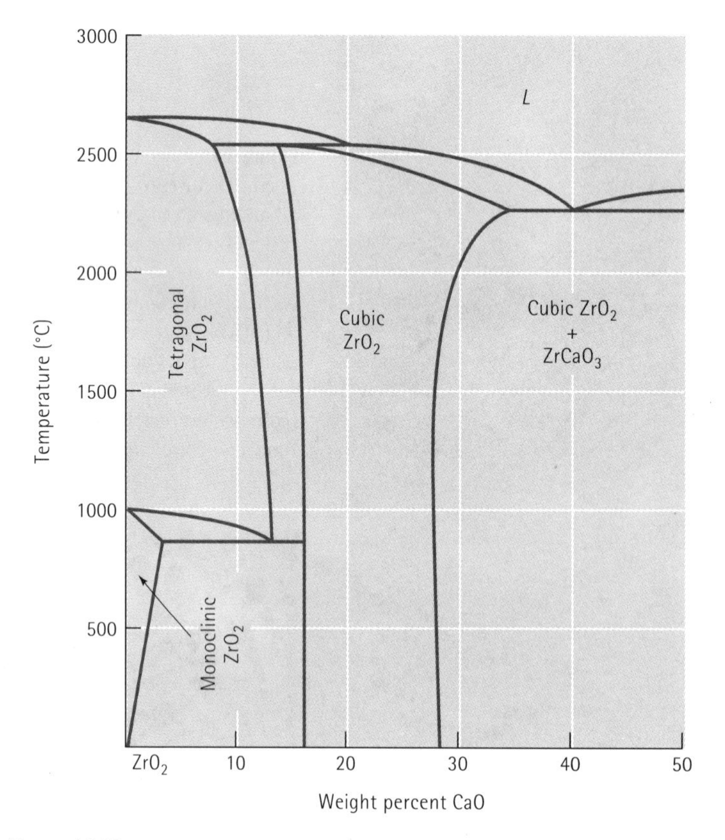

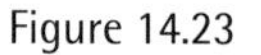

Figure 14.23
The ZrO_2-CaO phase diagram. A polymorphic phase transformation occurs for pure ZrO_2. Adding 16 to 26% CaO produces a single cubic zirconia phase at all temperatures.

But certain ceramic materials can be transformation-toughened. In zirconia, for instance, the energy of a crack can be absorbed by a metastable phase present in the original structure. This absorption of the crack energy, which effectively blunts the crack growth, permits the metastable phase to transform into the more stable form and simultaneously helps to close the crack. In *partially stabilised zirconia* (PSZ), only a small amount of the stabilising oxide is added. A matrix of the monoclinic phase is toughened with plates of the tetragonal phase formed by a martensitic transformation [Figure 14.22 (c)], giving fracture toughness values as high as 8 $MN.m^{-3/2}$.

The processing of the ceramic is also critical in improving toughness. Processing techniques that produce exceptionally fine-grained, high-purity, completely dense ceramics improve strength and toughness. Another processing approach is to deliberately introduce many tiny microcracks that are too small to propagate on their own; however, these microcracks can help blunt other, larger cracks that may try to grow.

14.7 Deformation of Ceramics at High Temperatures

Because dislocations are not mobile in ceramics at low temperatures, no significant plastic deformation is observed. At higher temperatures, viscous flow and grain boundary sliding become important deformation mechanisms. Viscous flow occurs in glasses and ceramics containing a mixture of glassy and crystalline phases; grain boundary sliding occurs in ceramics that are primarily crystalline.

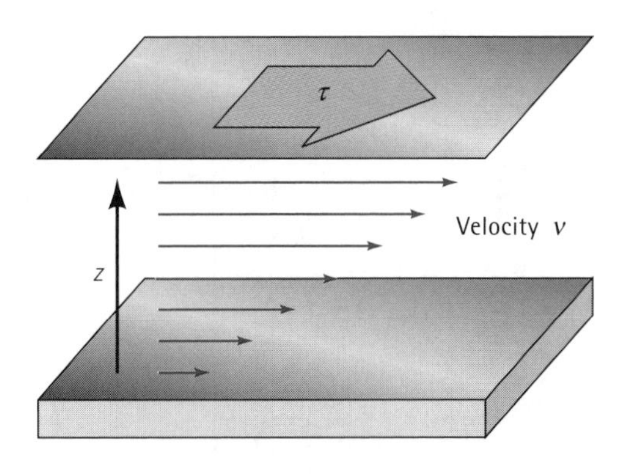

Figure 14.24
Viscosity is related to the velocity gradient produced in a liquid when a shear stress τ is applied.

Viscous Flow of Glass A glass deforms by viscous flow if the temperature is sufficiently high. The application of a shear stress τ causes a liquid to flow: the velocity at which the liquid flows, in turn, varies with location. Near the point at which the shear stress is applied, the liquid flows rapidly; further away, the liquid moves more slowly (Figure 14.24). Consequently, a velocity gradient dv/dz is created. The viscosity η is defined as:

$$\eta = \frac{\tau}{dv/dz} \tag{14.8}$$

The S.I. units for viscosity are $Ns.m^{-2}$ or Pa.s. In water, even a low shear stress causes the liquid to flow at a high velocity; the viscosity η is low, approximately 0.001 $Ns.m^{-2}$ at 20°C. In a thicker liquid such as glycerine, a higher shear stress is required to produce the same rate of flow and the viscosity is higher (about 1.5 $Ns.m^{-2}$ for glycerine).

In a ceramic glass, groups of atoms, such as silicate islands, rings, or chains, move past one another in response to stress, permitting deformation. However, the attraction between these groups of atoms offers resistance to the applied shear stress. At high temperatures – say above the melting point of the glass – this resistance is very low and the liquid glass is pourable; that is, it deforms and flows under its own weight. The viscosity of a liquid glass is typically less than about 50 $Ns.m^{-2}$ – more viscous than water, but still capable of flowing easily.

The viscosity of glass, however, is dependent on temperature:

$$\eta = \eta_0 \exp\frac{Q_\eta}{RT} \tag{14.9}$$

As the temperature decreases, the viscosity increases, and glass becomes more difficult to deform. The activation energy Q_η is related to the ease with which the atom groups move past one another. The addition of modifiers, such as Na_2O, breaks up the network structure, permits the atom groups to move more easily, reduces Q_η, and reduces the viscosity of glass (Figure 14.25).

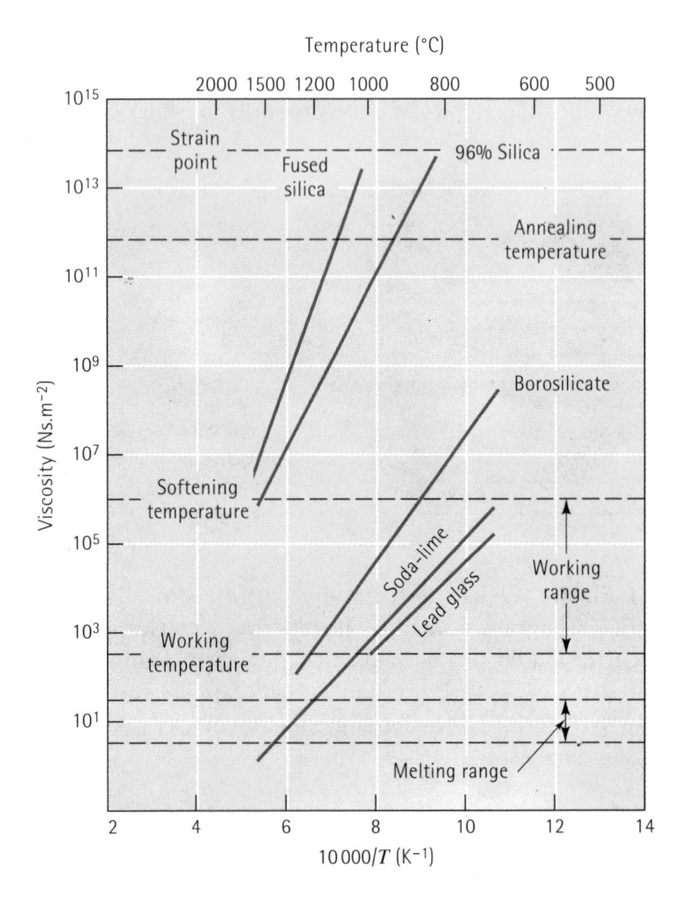

Figure 14.25
The effect of temperature and composition on the viscosity of glass.

Several critical processing temperatures are related to the viscosity of glass. The *melting range,* in which glass is fluid, occurs when the viscosity is very low – about 5 to 50 Ns.m^{-2}. The viscosities in the *working range* vary from 10^3 to 10^6 Ns.m^{-2}; in the working range, glass can be deformed into useful shapes. At still lower temperatures, the *annealing point* occurs; here the viscosity is approximately 10^{12} Ns.m^{-2} and there may be just enough mobility of the glassy chains so that residual stresses can be reduced. At even lower temperatures, below the *strain point,* glass appears completely rigid.

Creep in Ceramics Because ceramics are often designed for use at high temperatures, creep resistance is an important property. Crystalline ceramics have good resistance to creep due to their high melting points and high activation energies for diffusion. Figure 14.26 compares the flexural strength of several ceramic materials with a Ni-Cr superalloy; the ceramics tend to maintain their strength – sometimes, above 1 200°C.

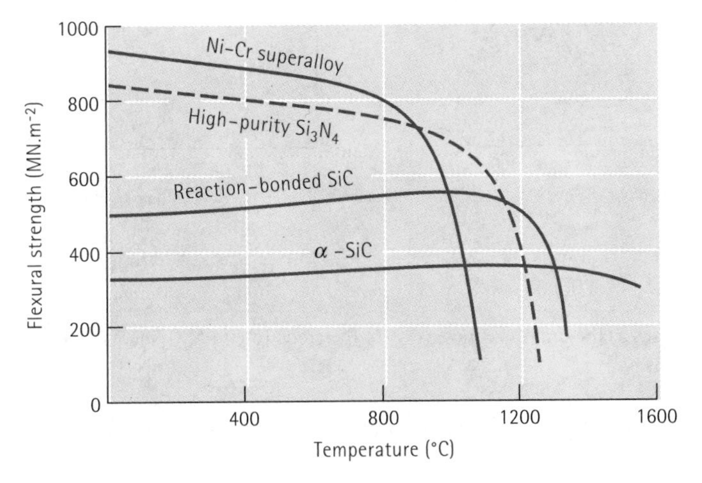

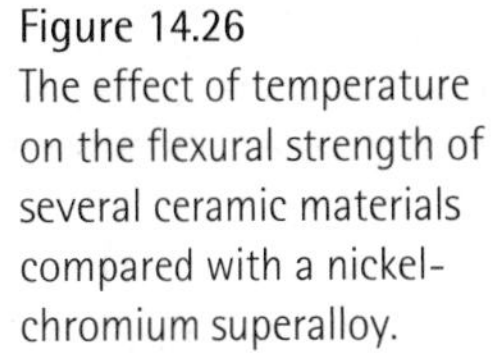

Figure 14.26
The effect of temperature on the flexural strength of several ceramic materials compared with a nickel-chromium superalloy.

Creep in crystalline ceramics often occurs as a result of grain boundary sliding. As the grains slide past one another, cracks may initiate and eventually cause a failure. Several factors make grain boundary sliding easier and, consequently, reduce the creep resistance:

1. *Grain size.* Smaller grain size increase the creep rate; more grain boundaries are present and grain boundary sliding is easier [Figure 14.27(a)].
2. *Porosity.* Increasing porosity in the ceramic reduces the cross-sectional area and increases the stress acting on the ceramic for a given load; pores also may make grain boundary sliding easier. Consequently, the creep rate increases [Figure 14.27(b)].
3. *Impurities.* Various impurities may lead to the formation of glassy phases at the grain boundaries, permitting creep by viscous flow.
4. *Temperature.* High temperatures reduce the strength of the grain boundaries, increase the rate of diffusion, and encourage the formation of glassy phases.

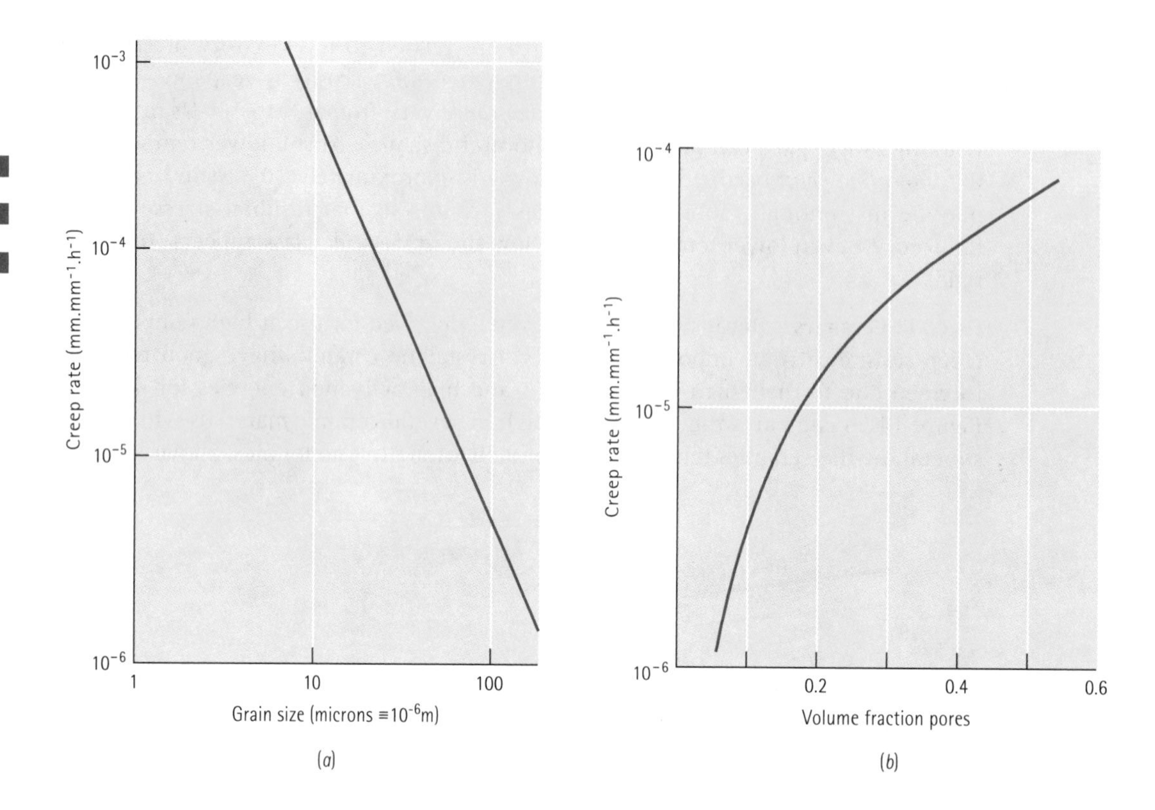

Figure 14.27
(*a*) The creep rate in MgO decreases as the grain size increases. (*b*) The creep rate in Al_2O_3 increases when the ceramic contains a larger amount of porosity.

The creep rate of glasses is related closely to viscosity. In some glasses, the creep rate $d\varepsilon/dt$ is given by

$$\frac{d\varepsilon}{dt} = \frac{\sigma}{\eta} \tag{14.10}$$

where σ is the applied stress. Thus, the creep rate increases exponentially with increasing temperature (decreasing viscosity). In silicate glasses, the best resistance to creep is obtained for pure silica; as modifying oxides such as MgO, SrO, and PbO are added, viscosity and, therefore, resistance to creep decrease.

Many ceramics contain a mixture of both glassy and crystalline phases. In the production of many ceramics, a glassy phase is formed between crystalline particles; the glassy phase bonds the crystalline particles together to produce the required ceramic shape. When the ceramic is heated, however, viscous flow of the glassy phase encourages grain boundary sliding and reduces the high-temperature strength and creep resistance. On the other hand, if a crystalline phase is permitted to precipitate within the glassy phase, the viscosity of the glassy phase increases and creep resistance improves. This characteristic is also used to advantage in glass ceramics; crystalline precipitates in what is originally an all-glass ceramic increase the high-temperature properties of the part.

EXAMPLE 14.10

A stress of 100 $MN.m^{-2}$ is applied to a 1 m-tall sheet of the soda-lime glass used as a viewing port into an oven. The temperature of the glass is 150°C. What is the strain rate in the glass? How much will it stretch in 20 years of continuous use?

SOLUTION

Figure 14.25 does not include the viscosity at 25°C; let's use that figure to determine the constants in Equation 14.9. From the graph for the soda-lime glass, we find that the viscosity is 10^1 when $10\,000/T = 6.1$, or $T = 1\,639$ K. Similarly, $\eta = 10^5$ when $10\,000/T = 9.9$, or $T = 1\,010$ K. Using Equation 14.9 and producing two simultaneous equations:

$$\ln 10^1 = 2.3026 = \ln \eta_0 + \frac{Q_\eta}{8.314(1639)} = \ln \eta_0 + 0.000073 Q_\eta$$

$$\ln 10^5 = 11.513 = \ln \eta_0 + \frac{Q_\eta}{8.314(1010)} = \ln \eta_0 + 0.000119 Q_\eta$$

Thus,

$$Q_\eta = \frac{9.2104}{0.000046} = 200226 \text{ J.mol}^{-1}$$

and

$$\ln \eta_0 = 11.513 - (0.000119)(200226) = -12.314$$

$$\eta_0 = 4.49 \times 10^{-6} \text{ Ns.m}^{-2}$$

At 150°C, or 423 K, the viscosity is:

$$\eta = 4.49 \times 10^{-6} \exp \frac{200226}{(8.314)(423)} = 2.389 \times 10^{19} \text{ Ns.m}^{-2}$$

From Equation 14.10, the strain rate is:

$$\frac{d\varepsilon}{dt} = \frac{\sigma}{\eta} = \frac{100 \times 10^6 \text{ N.m}^{-2}}{2.389 \times 10^{19} \text{ Ns.m}^{-2}}$$

$$\frac{d\varepsilon}{dt} = 4.186 \times 10^{-12} \text{ s}$$

There are 6.3×10^8 seconds in 20 years. The total strain is:

$$\Delta\varepsilon = (4.186 \times 10^{-12})(6.3 \times 10^8) = 0.0026 \text{ m.m}^{-1}$$

The 1 m-length of glass will stretch 2.6 mm in its 20-year service.

14.8 Processing and Applications of Ceramic Glasses

Glasses are manufactured into useful articles at a high temperature, with viscosity controlled so that the glass can be shaped without breaking. Figure 14.25 helps us understand the processing in terms of the viscosity ranges.

1. *Liquid range.* Sheet and plate glass are produced when the glass is in the molten state. Techniques include rolling the molten glass through water-cooled rolls or floating the molten glass over a pool of liquid tin (Figure 14.28). The liquid tin process produces an exceptionally smooth surface on the glass.

   Some glass shapes, including large optical lenses, are produced by casting the molten glass into a mould, then assuring that cooling is as slow as possible to minimise residual stresses and avoid cracking of the glass part. Glass fibres may be produced by drawing the liquid glass through small openings in a platinum die [Figure 14.29(c)]. Typically, many fibres are produced simultaneously for a single die.

2. *Working range.* Shapes such as those of containers or light bulbs can be formed by pressing, drawing, or blowing glass into moulds (Figure 14.29), A hot *gob* of liquid glass may be preformed into a crude shape (parison), then pressed or blown into a heated die to produce the final shape. The glass is heated to the working range so that the glass is formable but not 'runny.'

3. *Annealing range.* Some ceramic parts may be annealed to reduce residual stresses introduced during forming. Large glass castings, for example, are often annealed and slowly cooled to prevent cracking. Some glasses may be heat-treated to cause devitrification, or precipitation of a crystalline phase from the glass.

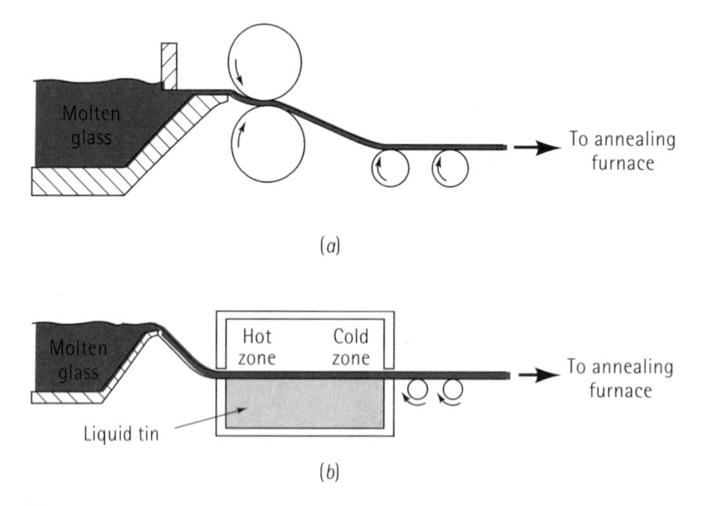

Figure 14.28
Techniques for manufacturing sheet and plate glass: (*a*) rolling and (*b*) floating the glass on molten tin.

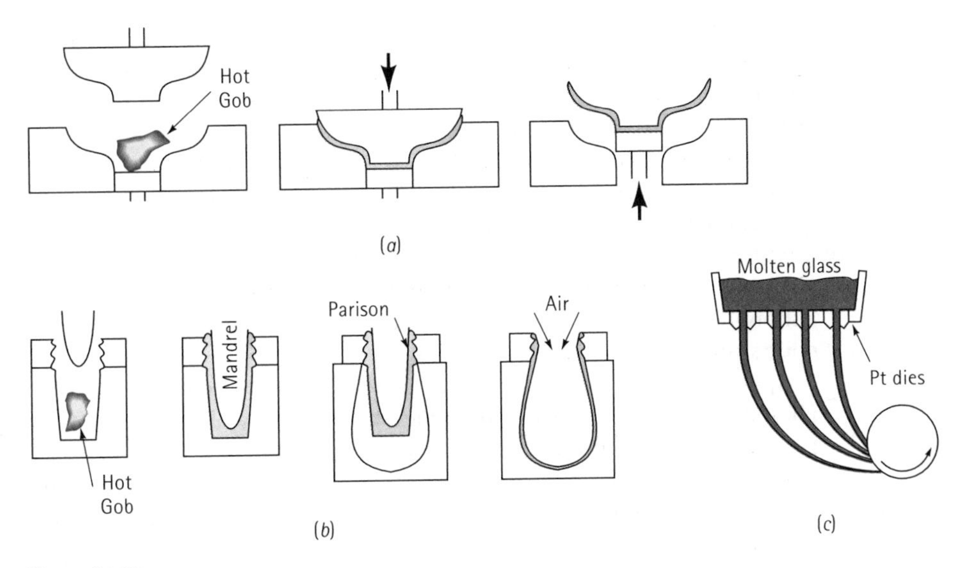

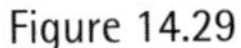

Figure 14.29
Techniques for forming glass products: (*a*) pressing, (*b*) press and blow process, and (*c*) drawing of fibres.

Tempered glass is produced by quenching the surface of plate glass with air, causing the surface layers to cool and contract. When the centre cools, its contraction is restrained by the already rigid surface, which is placed in compression (Figure 14.30). Prestressed glass is capable of withstanding much higher tensile stresses and impact blows than untempered glass.

Glass Compositions and Phase Diagrams Pure SiO_2 must be heated to very high temperatures to obtain viscosities that permit economical forming. Most commercial glasses are based on silica; modifiers such as soda (Na_2O) are added to break down the network structure and reduce the melting point, whereas lime (CaO) is added to reduce the solubility of the glass in water. The most common commercial glass contains approximately 75% SiO_2, 15% Na_2O, and 10% CaO. Figure 14.31 is a liquidus plot of the ternary phase diagram showing the effect of both soda and lime on silica; the square shows that the SiO_2-15% Na_2O-10% CaO composition has a liquidus temperature of about 1 100°C. Although inexpensive to produce, the soda-lime glasses have poor resistance to chemical attack and thermal stresses. Table 14.2 compares the compositions of several typical glasses.

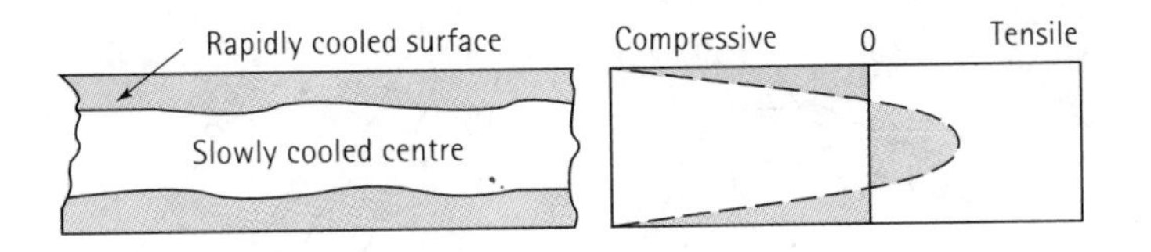

Figure 14.30
Tempered glass is cooled rapidly to produce compressive residual stresses at the surface.

Borosilicate glasses, which contain about 15% B_2O_3, have excellent chemical and dimensional stability. Their uses include laboratory glassware (Pyrex) and containers for the disposal of high-level radioactive nuclear waste. Calcium aluminoborosilicate glass – or E-glass – is used as a general purpose fibre for composite materials, such as fibreglass. Aluminosilicate glass, with 20% Al_2O_3 and 12% MgO, and high-silica glasses, with 3% B_2O_3, are excellent for high-temperature resistance and for protection against heat or thermal shock. S-glass, a magnesium aluminosilicate, is used to produce high-strength fibres for composite materials. Fused silica, or virtually pure SiO_2, has the best resistance to high temperature, thermal shock, and chemical attack, although it is also expensive. Particularly high-quality fused silica is used for fibre-optic systems.

Special optical qualities can also be obtained, including sensitivity to light. Photochromic glass, which is darkened by the ultraviolet portion of sunlight, is used for sunglasses. Photosensitive glass darkens permanently when exposed to ultraviolet light; if only selected portions of the glass are exposed and then immersed in hydrofluoric acid, etchings can be produced. Polychromatic glasses are sensitive to all light, not just ultraviolet radiation.

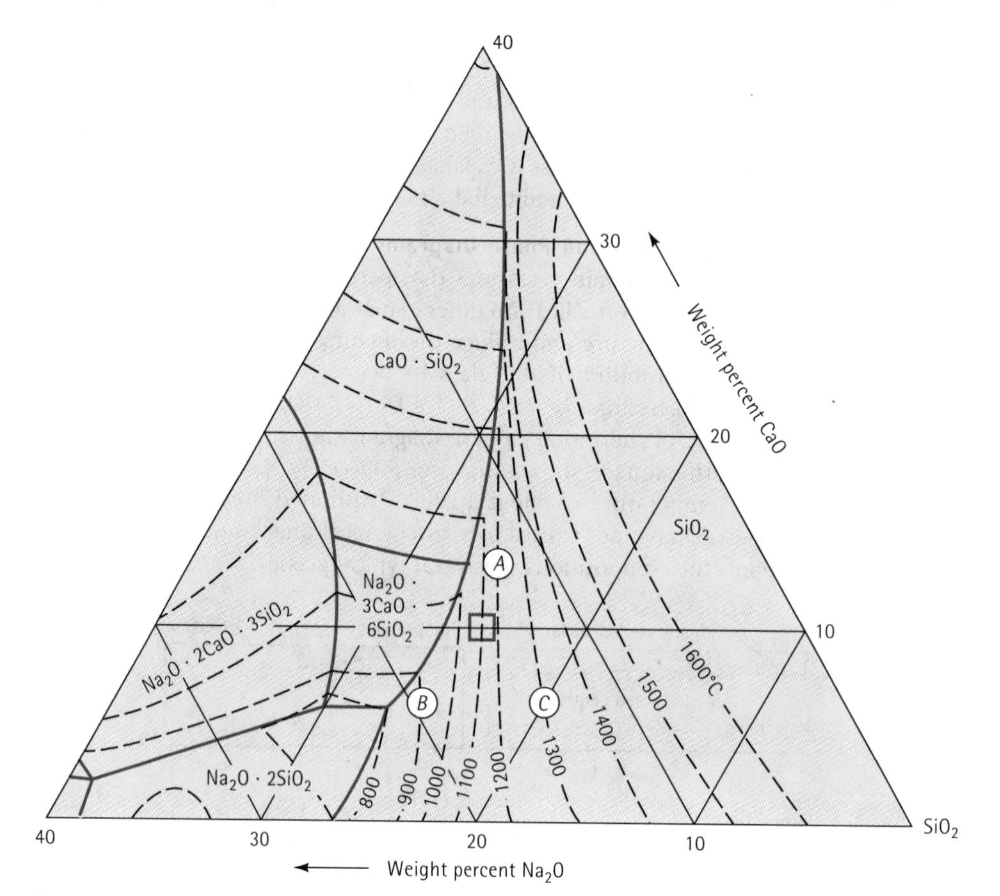

Figure 14.31
The liquidus plot for the SiO_2-CaO-Na_2O phase diagram.

Table 14.2 Compositions of typical glasses (in weight percent).

Glass	SiO_2	Al_2O_3	CaO	Na_2O	B_2O_3	MgO	PbO	Others
Fused silica	99							
Vycor	96				4			
Pyrex	81	2		4	12			
Glass jars	74	1	5	15		4		
Window glass	72	1	10	14		2		
Plate glass	73	1	13	13				
Light bulbs	74	1	5	16		4		
Fibres	54	14	16		10	4		
Thermometer	73	6		10	10			
Lead glass	67			6			17	10% K_2O
Optical flint	50			1			19	13% BaO, 8% K_2O, ZnO
Optical crown	70			8	10			2% BaO, 8% K_2O
E-glass fibres	55	15	20		10			
S-glass fibres	65	25				10		

EXAMPLE 14.11 Specification of a Soda-Lime Glass

Specify a soda-lime capable of being cast at a temperature of 1 000°C.

SOLUTION

For casting, the glass should be heated above its liquidus temperature. Its viscosity should be low enough so that it flows easily into the mould. Therefore, if we want to pour the liquid at 1 000°C, we might select a glass composition that has a lower liquidus – say 900°C. Suppose we used the following compositions (indicated on the silica-soda-lime phase diagram, Figure 14.31):

Glass *A*: 74% SiO_2–13% CaO–13% Na_2O

Glass *B*: 74% SiO_2–6% CaO–20% Na_2O

Glass *C*: 80% SiO_2–7% CaO–13% Na_2O

From the liquidus plot, we find that glass *A* has a liquidus of 1 200°C, glass *B* has a liquidus of 900°C, and glass *C* has a liquidus of 1 300°C. Of these three glasses, glass *B* is our obvious choice.

Of course, other compositions might also have a liquidus of 900°C. Increasing the CaO slightly still gives with the required liquidus but processing will be more difficult; decreasing the CaO also gives a glass with the required liquidus, but the solubility of the glass in water is higher.

14.9 Processing and Applications of Glass-Ceramics

Glass-ceramics are partly crystalline, partly glassy ceramics. A product that contains very low porosity can be obtained by producing a shape with conventional glass-forming techniques, such as pressing or blowing. Glass, however, has poor creep resistance. Its high-temperature properties can be improved by the precipitation of a crystalline phase after the part has been fabricated.

The first step in producing a glass-ceramic is to ensure that crystallisation does not occur during cooling from the forming temperature. A continuous cooling transformation diagram, much like the CCT diagram for steels, applies to glass [Figure 14.32(a)].

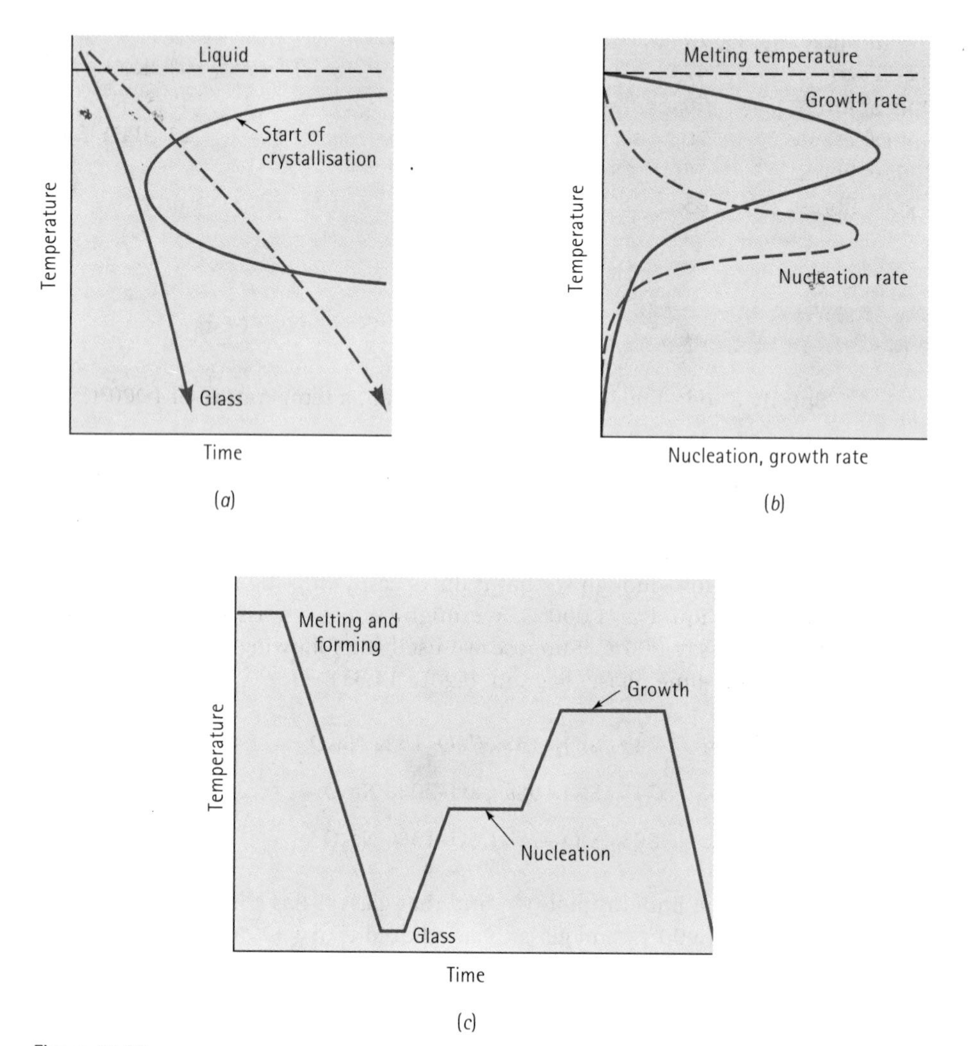

Figure 14.32
Producing a glass-ceramic: (*a*) Cooling must be rapid to avoid the start of crystallisation. (*b*) The rate of nucleation of precipitates is high at low temperatures, whereas the rate of growth of the precipitates is high at higher temperatures. (*c*) A typical process includes several steps to ensure that the correct structure is produced in a short overall time.

If glass cools too slowly, a transformation line is crossed; nucleation and growth of the crystals begin, but in an uncontrolled manner. Addition of modifying oxides to glass, much like addition of alloying elements to steel, shifts the transformation curve to longer times and prevents devitrification even at slow cooling rates.

Nucleation of the crystalline phase is controlled in two ways. First, the glass contains agents, such as TiO_2, that provide nucleation sites. Second, a heat treatment designed to provide the appropriate number of nuclei; the temperature should be relatively low in order to maximise the rate of nucleation [Figure 14.32(b)]. However, the overall rate of crystallisation depends on the growth rate of the crystals once nucleation occurs; higher temperatures are required to maximise the growth rate. Consequently, a heat treatment schedule similar to that shown in Figure 14.32(c) can be used. The low temperature step provides nucleation sites, and the high-temperature step speeds the rate of growth of the crystals; as much as 90% of the part may crystallise.

This special structure can provide good mechanical strength and toughness, often with a low coefficient of thermal expansion and high-temperature corrosion resistance. Perhaps the most important glass-ceramic is based on the Li_2O-Al_2O_3–SiO_2 system. These materials are used for cooking utensils (Corning Ware®) and ceramic tops for stoves. Other glass-ceramics are used in communication and computer applications.

14.10 Processing and Applications of Clay Products

Crystalline ceramics are often manufactured into useful articles by preparing a shape, or compact, composed of the raw materials in a fine powder form. The powders are then bonded by chemical reaction, partial or complete vitrification (melting), or sintering.

Clay products form a group of traditional ceramics used for producing pipe, brick, cooking ware, and other common products. Clay, such as kaolinite, and water serve as the initial binder for the ceramic powders, which are typically silica. Other materials, such as feldspar $[(K, Na)_2O \cdot Al_2O_3 \cdot 6SiO_2]$, serve as fluxing (glass-forming) agents during later heat treatment.

Forming Techniques for Clay Products The powders, clay, flux, and water are mixed and formed into a shape(Figure 14.33). Dry or semidry mixtures are mechanically pressed into 'green' (unbaked) shapes of sufficient strength to be handled. For more uniform compaction of complex shapes, isostatic pressing may be done; the powders are placed into a rubber mould and subjected to high pressures through a gas or liquid medium. Higher moisture contents permit the powders to be more plastic or formable. Hydroplastic forming processes, including extrusion, jiggering, and hand working, can be applied to these plastic mixes. Ceramic slurries containing large amounts of organic plasticisers, rather than water, can be injected into moulds.

Still higher moisture contents permit the formation of a slip, or pourable slurry, containing fine ceramic powder. The slip is poured into a porous mould. The water in the slip nearest to the mould wall is drawn into the mould, leaving behind a soft solid which has a low moisture content. When enough water has been drawn from the slip

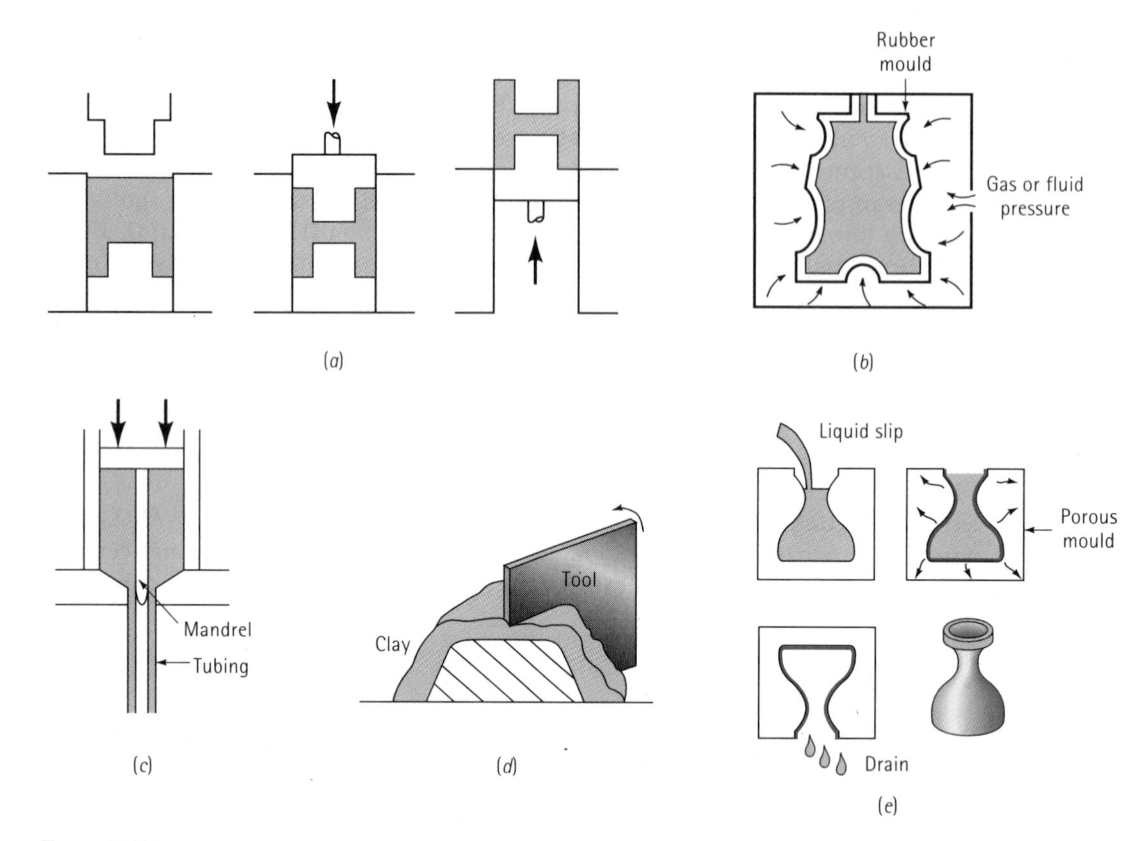

Figure 14.33
Processes for shaping crystalline ceramics: (*a*) pressing, (*b*) isostatic pressing, (*c*) extrusion, (*d*) jiggering, and (*e*) slip casting.

to produce a desired thickness of solid, the remaining liquid slip is poured from the mould, leaving behind a hollow shell. Slip casting is used in manufacturing wash basins and other commercial products.

After forming, the ceramic bodies – or greenware – which are still weak, contain water or other lubricants, and are porous, and subsequent drying and firing are required.

Drying and Firing of Clay Products During drying, excess moisture is removed and large dimensional changes occur (Figure 14.34). Initially, the water between the clay platelets – the interparticle water – evaporates and provides most of the shrinkage. Relatively little dimensional change occurs as the remaining water between the pores evaporates. The temperature and humidity are carefully controlled to provide uniform drying throughout the part, thus minimising stresses, distortion, and cracking.

The rigidity and strength of a ceramic part are obtained during firing. During heating, the clay dehydrates, eliminating the hydrated water that is part of the kaolinite crystal structure, and vitrification, or melting, begins (Figure 14.35).

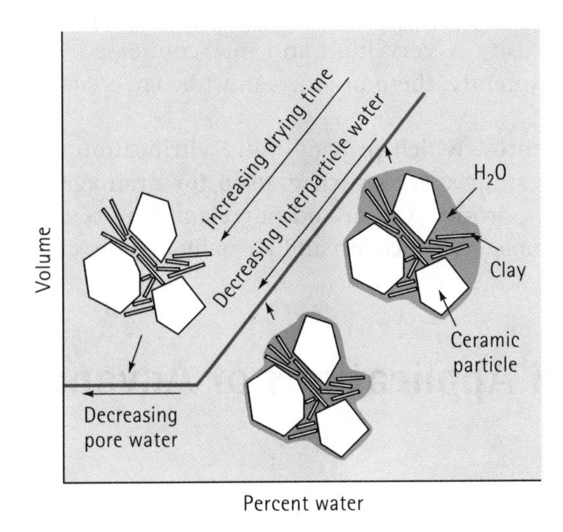

Figure 14.34
The change in the volume of a ceramic body as moisture is removed during drying. Dimensional changes cease after the interparticle water is gone.

Impurities and the fluxing agent react with the ceramic particles (SiO_2) and clay, producing a low-melting-point liquid phase at the grain surfaces. The liquid helps eliminate porosity and, after cooling, changes to a rigid glass that binds the ceramic particles. This glassy phase provides a ceramic bond, but it also causes additional shrinkage of the entire ceramic body.

The grain size of the final part is determined primarily by the size of the original powder particles. Furthermore, as the amount of flux increases, the melting temperature decreases; more glass forms, and the pores become rounder and smaller. A smaller initial grain size accelerates this process by providing more surface area at which vitrification can occur.

Applications of Clay Products Many structural clay products and whitewares are produced using these processes. Brick and tile used for construction are pressed or extruded into shape, dried, and fired to produce the ceramic bond. Higher firing temperatures or finer original particle sizes produce more vitrification, less porosity, and higher density. The higher density improves mechanical properties but reduces the insulating qualities of the brick or tile.

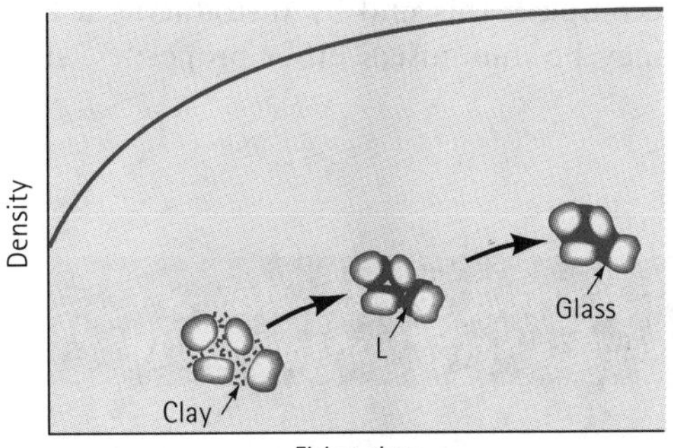

Figure 14.35
During firing, clay and other fluxing materials react with coarser particles to produce a glassy bond and reduce porosity.

Earthenware are porous clay bodies fired at relatively low temperatures. Little vitrification occurs; the porosity is very high and interconnected, and earthenware ceramics may leak. Consequently, these products must be covered with an impermeable glaze.

Higher firing temperatures, which provide more vitrification and less porosity, produce stoneware. The stoneware, which is used for drainage and sewer pipe, contains only 2% to 4% porosity. China and porcelain require even higher firing temperatures to cause complete vitrification and virtually no porosity.

14.11 Processing and Applications of Advanced Ceramics

Advanced structural ceramics are designed to optimise mechanical properties at elevated temperatures. In order to achieve these properties, exceptional control of purity, processing, and microstructure are required, in comparison with traditional ceramics. Raw materials are often synthesised by complex methods to obtain the proper powder purity. Special techniques are also used to form these materials into useful products. As a comparison, the effect of processing on silicon nitride shown in Table 14.3.

Pressing and Sintering Many of the more advanced ceramics begin in a powder form, are mixed with a lubricant to improve compaction, and are pressed into a shape. The pressed shape is sintered to develop the required microstructure and properties. Vitrification is not desired; instead, diffusion achieves the desired strength.

During sintering, ions first diffuse along grain boundaries and surfaces to the points of contact between particles, providing bridging and connection of the individual grains (Figure 14.36). Further grain boundary diffusion shrinks the pores and increases the density, while the pores become more rounded. Finer initial particle sizes and higher temperatures accelerate the rate of pore shrinkage.

Even after long sintering times, however, porosity may still remain in the ceramic part and the probability of failure may be prohibitively high. Sintering aids may be added to the raw materials to make it easier to develop maximum density. Sintering aids, however, typically accomplish this end by introducing a low melting glassy phase. Although porosity may be minimised, other properties, such as creep resistance, are impaired.

Table 14.3 Properties of Si_3N_4.

Process	Compressive Strength ($MN.m^{-2}$)	Flexural Strength ($MN.m^{-2}$)
Slip casting	140	70
Reaction bonding	1 035	240
Hot pressing	3 450	895

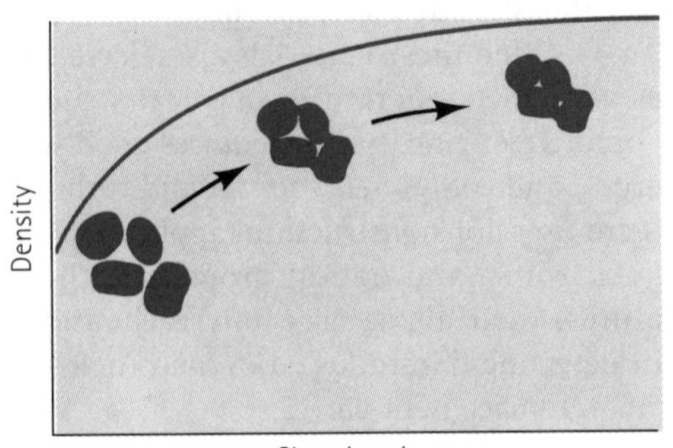

Figure 14.36
During sintering, diffusion produces bridges between the particles and eventually causes the pores to be filled in.

In some cases – particularly for the advanced ceramics – forming of the powder compacts is done at high temperatures by hot pressing or hot isostatic pressing (HIP). In the HIP process, the powders are sealed in metal or glass containers and are then simultaneously heated and compacted in a pressurised inert gas container. This process permits use of less lubricant and provides for at least some simultaneous sintering, resulting in parts with low porosity and desirable mechanical properties.

An important difference between advanced ceramics and typical metals is that once sintering is finished and the ceramic part is manufactured, the microstructure of the ceramic is fixed.

Reaction bonding Some ceramics, such as Si_3N_4, are produced by reaction bonding. Silicon is formed into a shape and then reacted with nitrogen to form the nitride. Reaction bonding, which can be done at lower temperatures, provides better dimensional control compared with hot pressing; however, lower densities and mechanical properties are obtained.

Sol Gel Processing The sol gel process is used to produce and consolidate exceptionally fine, pure ceramic powders. A liquid colloidal solution containing dissolved metallic ions is prepared. Hydrolysis reactions form an organometallic solution, or sol, composed of polymer-like chains containing metallic ions and oxygen. Amorphous oxide particles form from the solution, producing a rigid gel. The gel is dried and fired to provide sintering and densification of a final ceramic part; the sintering temperatures are low due to the highly reactive powders. Higher firing temperatures permit the production of glasses and glass-ceramics. The sol gel process can be used to produce UO_2 for nuclear reactor fuels, perovskite structures such as barium titanate for electronic devices, ultra-fine grained alumina for high-strength structural applications, and a wide variety of other ceramics.

Joining and Assembly of Ceramic Components Ceramics are often made as monolithic components rather than assemblies of numerous components. When two ceramic parts are placed in contact under a load, stress concentrations at the brittle surface are created, leading to an increased probability of failure.

In addition, methods for joining ceramic parts into a larger assembly are limited. The brittle ceramics cannot be joined by fusion welding or deformation bonding processes. At low temperatures, adhesive bonding using polymer materials may be accomplished; ceramic cements may be used at higher temperatures. Diffusion bonding is used to join ceramics and also to join ceramics to metals.

Advanced Materials and Applications Advanced ceramics include carbides, borides, nitrides, and oxides (Table 14.4). Often these materials are selected for both mechanical and physical properties at high temperatures. Typical structural applications include the 'all ceramic' automotive engine, which requires good wear resistance and elevated temperature properties, and components for jet and turbine engines. A large group of advanced ceramics are used for nonstructural applications, taking advantage of their unique magnetic, electronic, and optical properties, their good corrosion resistance at high temperatures, their ability to catalyse chemical reactions, their ability to serve as sensors for detecting hazardous gases, and their suitability for prosthetic devices and other 'human replacement parts'.

Alumina (Al_2O_3) is used to contain molten metal or to operate at high temperatures where good strength is required. Alumina is also used as an insulating substrate to support integrated circuits. One classical application is for insulators in spark plugs. Some unique applications are also being found in dental and medical use, including restoration of teeth, bone filler, and orthopaedic implants. Chromium-doped alumina is a laser.

Aluminium nitride (AlN) provides good electrical insulation but has a high thermal conductivity. Because its coefficient of thermal expansion is similar to that of silicon, AlN is a suitable replacement for Al_2O_3 as a substrate material for integrated circuits. Cracking is minimised and electrical insulation is obtained, yet heat generated by the electronic circuit can be quickly removed. It is also better suited than many competing materials for use in electrical circuits operating at a high frequency.

Boron carbide (B_4C) is very hard yet unusually lightweight. In addition to its use as nuclear shielding, it finds uses in applications requiring excellent abrasion resistance and as a portion of bulletproof armour plate, although it has rather poor properties at high temperatures.

Silicon carbide (SiC) provides outstanding oxidation resistance at temperatures even above the melting point of steel. SiC is often used as a coating for metals, carbon-carbon composites, and other ceramics to provide protection at these extreme temperatures. SiC is also used as a particulate and fibrous reinforcement in both metal matrix and ceramic matrix composites (Figure 14.37).

Table 14.4 Mechanical properties of selected advanced ceramics.

Material	Density $Mg.m^{-3}$	Tensile Strength $MN.m^{-2}$	Flexural Strength $MN.m^{-2}$	Compressive Strength $MN.m^{-2}$	Young's Modulus $(MN.m^{-2} \times 10^3)$	Fracture Toughness $MN.m^{-3/2}$
Al_2O_3	3.98	210	550	2760	386	5.5
SiC (sintered)	3.1	175	550	3860	414	4
Si_3N_4 (reaction bonded)	2.5	140	240	1035	207	3
Si_3N_4 (hot pressed)	3.2	550	895	3450	310	5.5
Sialon	3.24	415	965	3450	310	10
ZrO_2 (partially stabilised)	5.8	450	690	1860	207	11
ZrO_2 (transformation toughened)	5.8	345	790	1725	200	12

(a) (b)

Figure 14.37
Silicon carbide reinforcement materials: (a) SiC whiskers and (b) SiC single crystal platelets. (*Courtesy of American Matrix, Inc.*)

Silicon nitride (Si_3N_4) has properties similar to those of SiC, although its oxidation resistance and high-temperature strength are somewhat lower. Both silicon nitride and silicon carbide are likely candidates for components for automotive and gas turbine engines, permitting higher operating temperatures and better fuel efficiencies with less weight than traditional metals and alloys.

Sialon is formed when aluminium and oxygen partially substitute for silicon and nitrogen in silicon nitride. The general form of the material is $Si_{6-z}Al_zO_zN_{8-z}$; when $z = 3$, the formula is $Si_3Al_3O_3N_5$. The sialon crystals are typically embedded in a glassy phase based on Y_2O_3. The glassy phase is then allowed to devitrify by a heat treatment to improve the creep resistance. The result is a ceramic that is relatively lightweight, with a low coefficient of thermal expansion, good fracture toughness, and a higher strength than many of the other common advanced ceramics. Sialon may find applications in engine components and other applications involving both high temperatures and demanding wear conditions.

Titanium boride (TiB_2) is a good conductor of both electricity and heat. In addition, it provides excellent toughness. TiB_2, along with boron carbide, silicon carbide, and alumina, finds application in producing armour.

Urania (UO_2) is widely used as a nuclear reactor fuel. This material has exceptional dimensional stability because its crystal structure can accommodate the products of the fission process.

EXAMPLE 14.12 Design of a Ceramic Connecting Rod

Design a ceramic connecting rod that connects the piston to the crankshaft in an automotive engine. Metal connecting rods typically have yield strengths of about 550 $MN.m^{-2}$.

SOLUTION

To replace the metal part, we might consider an advanced ceramic with appropriate properties. Table 14.4 shows that Al_2O_3 and sintered SiC have flexural strengths that are near 550 $MN.m^{-2}$, whereas hot-pressed Si_3N_4, sialon, and ZrO_2 have flexural strengths well above 550 $MN.m^{-2}$.

However, Table 14.4 does not include any measure of the reliability of the ceramics, such as the Weibull modulus. Although the advanced ceramics are capable of good strengths and high-temperature resistance, there is a greater probability that they will also contain or develop critically sized flaws. While the stresses acting on a connecting rod are generally compressive in nature, and the ceramics all have good compressive strengths, any problems that might produce tensile stresses – particularly in a part containing a flaw – may have catastrophic results.

We might also have problems attaching the bearing cap to the remainder of the connecting rod. Appropriate fasteners might have to be designed. When the connecting rod is assembled, we might have points of contact at which stress concentrations develop, resulting in locations at which a critical flaw might easily develop.

At this time, we likely would recommend that ceramics not be considered for this application.

14.12 Refractories

Refractory materials are important components of the equipment used in the production, refining, and handling of metals and glasses, for constructing heat treating furnaces, and for other high-temperature processing equipment. The refractories must survive at high temperatures without being corroded or weakened by chemical attack by the surrounding environment. Typical refractories are composed of coarse oxide particles bonded by a finer refractory material. The finer material melts during firing, providing bonding. In some cases, refractory bricks contain about 20% to 25% apparent porosity to provide improved thermal insulation.

Refractories are often divided into three groups – acid, basic, and neutral – based on their chemical behaviour (Table 14.5).

Acid Refractories Common acidic refractories include silica, alumina, and fireclay (an impure kaolinite). Pure silica is sometimes used to contain molten metal. In some applications, the silica may be bonded with small amounts of boron oxide, which melts and produces the ceramic bond. When a small amount of alumina is added to silica, the refractory contains a very low-melting-point eutectic microconstituent (Figure 14.38) and is not suited for refractory applications at temperatures above about 1600°C, a temperature often required for steelmaking. However, when larger amounts of alumina are added, the microstructure contains increasing amounts of mullite, $3Al_2O_3 \cdot 2SiO_2$, which has a high melting temperature. These fireclay refractories are generally relatively weak, but they are inexpensive. Alumina contents above about 50% constitute the high-alumina refractories.

Table 14.5 Compositions of typical refractories (weight percents).

Refractory	SiO_2	Al_2O_3	MgO	Fe_2O_3	Cr_2O_3
Acidic					
Silica	95–97				
Superduty fire brick	51–53	43–44			
High-alumina fire brick	10–45	50–80			
Basic					
Magnesite			83–93	2–7	
Olivine	43		57		
Neutral					
Chromite	3–13	12–30	10–20	12–25	30–50
Chromite-magnesite	2–8	20–24	30–39	9–12	30–50

From Ceramic Data Book, *Cahners Publishing Co., 1982*

Basic Refractories A number of refractories are based on MgO (magnesia, or periclase). Pure MgO has a high melting point, good refractoriness, and good resistance to attack by the basic environments often found in steel-making processes. Olivine refractories contain forsterite, or Mg_2SiO_4, and also have high melting points (Figure 14.8). Other magnesia refractories may include CaO or carbon. Typically, the basic refractories are more expensive than the acid refractories.

Neutral Refractories These refractories, which include chromite and chromite-magnesite, might be used to separate acid and basic refractories, preventing them from attacking one another.

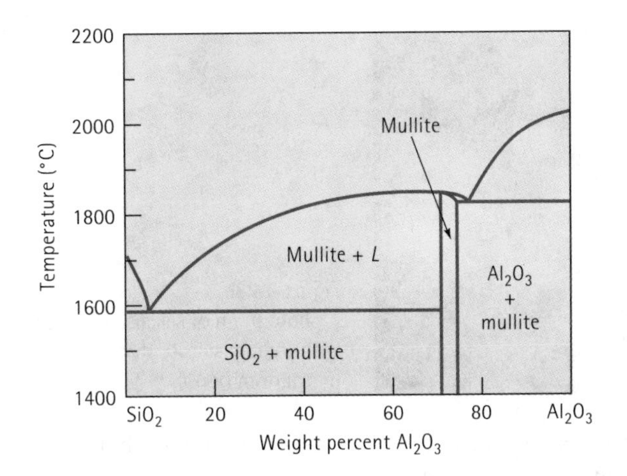

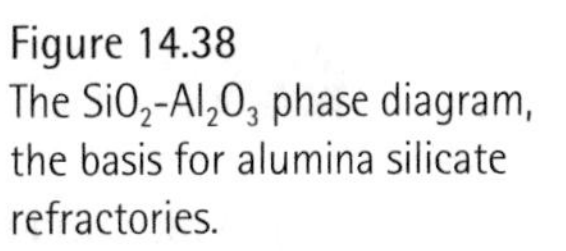

Figure 14.38
The SiO_2-Al_2O_3 phase diagram, the basis for alumina silicate refractories.

Special Refractories Carbon, or graphite, is used in many refractory applications, particularly when oxygen is not readily available. Other refractory materials include zirconia (ZrO_2), zircon ($ZrO_2 \cdot SiO_2$) and a variety of nitrides, carbides, and borides.

Most of the carbides, such TiC and ZrC, do not resist oxidation well, and their high-temperature applications are best suited to reducing conditions. However, silicon carbide is an exception; when SiC is oxidised at high temperatures, a thin layer of SiO_2 forms at the surface, protecting the SiC from further oxidation up to about 1 500°C. Nitrides and borides also have high melting temperatures and are less susceptible to oxidation. Some of the oxides and nitrides are candidates for use in jet engines.

14.13 Other Ceramic Materials and Applications

In addition to their use in producing construction materials, appliances, structural materials and refractories, ceramics find a host of other applications, including the following.

Cements Ceramic raw materials are joined using a binder that does not require firing or sintering in a process called cementation. A chemical reaction converts a liquid resin to a solid that joins the particles. In the case of sodium silicate solution into a glassy material:

$$x\mathrm{Na_2O} \cdot y\mathrm{SiO_2} \cdot \mathrm{H_2O} + \mathrm{CO_2} \rightarrow \text{glass}$$

Figure 14.39 shows silica sand grains used to produce moulds for metal casting. The liquid sodium silicate coats the sand grains and provides bridges between the sand grains. Introduction of the CO_2 converts the bridges to a solid, joining the sand grains.

Fine alumina powder solutions catalysed with phosphoric acid produce an aluminium phosphate cement:

$$\mathrm{Al_2O_3} + 2\mathrm{H_3PO_4} \rightarrow 2\mathrm{AlPO_4} + 3\mathrm{H_2O}$$

Figure 14.39
A photograph of silica sand grains bonded with sodium silicate through the cementation mechanism (× 60).

When alumina particles are bonded with the aluminium phosphate cement, refractories capable of operating at temperatures as high as 1 650°C are produced.

Plaster of paris, or gypsum, is another material that is hardened by a cementation reaction:

$$CaSO_4 \cdot \frac{1}{2}H_2O + \frac{3}{2}H_2O \rightarrow CaSO_4 \cdot 2H_2O$$

When the liquid slurry reacts, interlocking solid crystals of gypsum ($CaSO_4 \cdot 2H_2O$) grow, with very small pores between the crystals. Larger amounts of water in the original slurry provide more porosity, but they also decrease the strength of the final plaster. One of the important uses of this material is for construction of walls in buildings.

The most common and important of the cementation reactions occurs in Portland cement, which is used to produce concrete. These cements are discussed in Chapter 17.

Coatings Ceramics are often used to provide protective coatings to other materials. Common commercial coatings include glazes and enamels. Glazes are applied to the surface of a ceramic material to seal a permeable clay body, to provide protection and decoration, or for special purposes. Enamels are applied to metal surfaces. The enamels and glazes are clay products that vitrify easily during firing. A common composition is $CaO \cdot Al_2O_3 \cdot 2SiO_2$.

Special colours can be produced in glazes and enamels by the addition of other minerals. Zirconium silicate gives a white glaze, cobalt oxide makes the glaze blue, chromium oxide produces green, lead oxide gives a yellow colour, and a red glaze may be produced by adding a mixture of selenium and cadmium sulphides.

One of the problems encountered with a glaze or enamel is surface cracking, or crazing, which occurs when the glaze has a coefficient of thermal expansion different than that of the underlying material. This is frequently the most important factor in determining the composition of the coating.

Special coatings are used for advanced ceramics and high-service temperature metals. SiC coatings are applied to carbon-carbon composite materials to improve their oxidation resistance. Zirconia coatings may be applied to nickel-base superalloys to provide thermal barriers that protect the metal from melting or adverse reactions.

Fibres Fibres are produced from ceramic materials for several uses: as a reinforcement in composite materials, for weaving into fabrics, or for use in fibre-optic systems. Borosilicate glass fibres, the most commonly produced fibres, provide strength and stiffness in fibreglass. Fibres can be produced from a variety of other ceramics, including alumina, silicon carbide, and boron carbide.

A special type of fibrous material is the silica tile used to provide the thermal protection system for the space shuttle. Silica fibres are bonded with a silica powder to produce an exceptionally lightweight tile with densities as low as $0.144\ Mg.m^{-3}$; the tile is coated with special high-emissivity glazes to permit protection up to 1 300°C.

SUMMARY

Ceramics are combinations of metallic and nonmetallic elements that form hard, brittle, high-melting-point compounds. Typical ceramics are electrical and thermal insulators with good chemical stability and good strength in compression.

Ceramics, however, inevitably contain flaws. Because the brittle ceramic cannot plastically deform, these flaws limit the ability of the material to withstand a tensile load. Because the nature of the flaws varies from part to part, design of ceramic parts for critical applications is more important in determining the sensitivity of the material to flaws; fracture toughness provides another measure of the safety of ceramic materials.

Three basic microstructures are found in ceramics: crystalline, glassy, and crystalline-glassy mixture.

- Crystalline ceramics have good high-temperature properties; plastic deformation under load occurs only at high temperatures due to grain boundary sliding. Some crystalline ceramics can be toughened by a number of mechanisms, including fibre reinforcement, embedding ceramic particles in a metallic matrix, transformation toughening, appropriate manufacturing and fabrication methods, and careful control of purity and grain size.
- Glassy ceramics have poor high-temperature properties. Plastic deformation occurs by viscous flow of the glass, often permitting high creep rates.
- Ceramics containing a mixture of crystalline and glassy phases are common. Glassy phases are promoted during firing or sintering to improve bonding between crystalline particles; in glass-ceramics, crystalline phases are allowed to precipitate in a glassy matrix to improve strength. In both cases, however, viscous flow may occur at high temperatures and limit creep resistance.
  The mechanical behaviour of ceramics is highly dependant on the method of manufacture. Ceramic properties are very sensitive to the purity of the raw materials and control of the microstructure, including grain boundary phases and porosity, which are consequences of the production method. Sophisticated production of powders and processing of the powders into monolithic parts are required, particularly for advanced structural ceramics.

GLOSSARY

Apparent porosity
The percentage of a ceramic body that is composed of interconnected porosity.

Bulk density
The mass of a ceramic body per unit volume, including closed and interconnected porosity.

Cementation
Bonding ceramic raw materials into a useful product, using binders that form a glass or gel without firing at high temperatures.

Ceramic bond
Bonding ceramic materials by permitting a glassy product to form at high firing temperatures.

Cermet
A composite containing ceramic particles in a metal matrix, providing a good combination of hardness with other properties such as toughness.

Devitrification
The precipitation of a crystalline product from a glassy product, usually at high temperatures.

Firing
Heating a ceramic body at a high temperature to cause a ceramic bond to form.

Flux
Additions to ceramic raw materials that reduce the melting temperature.

Glass-ceramics
Ceramic shapes formed in the glassy state and later allowed to crystallise during heat treatment to achieve improved strength and toughness.

Glass formers
Oxides with a high bond strength that easily produce a glass during processing.

Glass transition temperature
The temperature at which an undercooled liquid becomes a glass.

Griffith flaw
A crack or other imperfection in a brittle material that concentrates and magnifies an applied stress.

Hydroplastic forming
A number of processes by which a moist ceramic clay body is formed into a useful shape.

Intermediates
Oxides that, when added to a glass, help to extend the glassy network, although the oxides normally do not form a glass themselves.

Metasilicates
A group of silicate structures having a ring or chain structure.

Modifiers
Oxides that, when added to a glass, disrupt the glassy network, eventually causing crystallisation.

Orthosilicates
A group of silicate structures based on a single silicate tetrahedral unit. Also known as olivines.

Parison
A crude glassy shape that serves as an intermediate step in the production of glass ware. The parison is later formed into a finished product.

Pyrosilicates
A group of silcate structures based on a pair of silicate tetrahedral units.

Reaction bonding
A ceramic processing technique by which a shape is made using one material that is later converted to a ceramic material by reaction with a gas.

Refractories
A group of ceramic materials capable of withstanding high temperatures for prolonged periods of time.

Slip

A liquid slurry that is poured into a mould. When the slurry begins to harden at the mould surface, the remaining liquid slurry is decanted, leaving behind a hollow ceramic casting.

Slip casting

Forming a hollow ceramic part by introducing a pourable slurry into a mould. The water in the slurry is extracted into the porous mould, leaving behind a drier surface. Excess slurry can then be decanted.

Sol gel process

A method for producing ceramic materials. A polymer-like solution (sol) containing metal ions and oxygen is prepared; the solid oxide (gel) precipitates from the solution and is subsequently fired.

Tempered glass

Glass that is prestressed during cooling to improve its strength.

Transformation toughening

Improving the toughness of ceramic materials by taking advantage of volume changes that accompany a polymorphic transformation induced by a crack.

True porosity

The percentage of a ceramic body that is composed of both closed and interconnected porosity.

Viscous flow

Deformation of a glassy material at high temperatures.

Vitrification

Melting, or formation of a glass.

Weibull distribution

A plot of the frequency of failure versus the applied stress.

Weibull modulus

A measure of the reliability of a ceramic, obtained as the slope of a graph of cumulative probability for failure versus strength.

PROBLEMS

14.1 Calculate the lattice parameter, packing factor, and density expected for $BaTiO_3$ [Figure 14.1(a)], using the data in the appendices.

14.2 Calculate the packing factor and density expected for $MgAl_2O_4$ [Figure 14.1(c)] if the lattice parameter is 0.808 nm.

14.3 Quartz (SiO_2) has a hexagonal crystal structure, with lattice parameters of $a_0 = 0.4913$ nm and $c_0 = 0.5405$ nm and a density of 2.65 $Mg.m^{-3}$. Determine

(a) the number of SiO_2 groups in quartz and

(b) the packing factor of the quartz unit cell.

14.4 Tungsten carbide (WC) has a hexagonal structure, with lattice parameters of $a_0 = 0.291$ nm and $c_0 = 0.284$ nm. If the density of WC is

15.77 $Mg.m^{-3}$, determine the number of tungsten and carbon atoms per cell.

14.5 Determine whether the following are orthosilicate, pyrosilicate, metasilicate, or sheet types of ceramics.

(a) $FeO \cdot SiO_2$

(b) $3BeO \cdot Al_2O_3 \cdot 6SiO_2$

(c) $Li_2O \cdot Al_2O_3 \cdot 4SiO_2$

(d) $CaO \cdot Al_2O_3 \cdot 2SiO_2$

(e) $2CaO \cdot MgO \cdot 2SiO_2$

(f) $Al_2O_3 \cdot 2SiO_2$

14.6 The density of orthorhombic forsterite (Mg_2SiO_4) is 3.21 $Mg.m^{-3}$ and the lattice parameters are a_0 = 0.476 nm, b_0 = 1.020 nm, and c_0 = 0.599 nm. Calculate the number of Mg^{2+} ions and the number of SiO_4^{4-} ionic groups in each unit cell.

14.7 Suppose 10% of the Al^{3+} ions in montmorillonite are replaced by Mg^{2+} ions. How many grams of Na^+ ions will be attracted to the clay per kg of clay?

14.8 Show that Mg_2SiO_4 and Fe_2SiO_4 are expected to display complete solid solubility.

14.9 A typical composition for FeO (wustite) is 52 at% O. Calculate the number of Fe^{3+} ions and the number of vacancies per cm^3 expected in this typical composition. FeO has the sodium chloride crystal structure.

14.10 Each vacancy in FeO provides one charge carrier that will contribute to electrical conductivity in a ceramic. If the ratio between Fe^{3+} ions to Fe^{2+} ions is 1 to 25, calculate the number of charge carriers per cubic centimetre. This is one way of producing a semiconducting ceramic material.

14.11 Using the $MgO \cdot Al_2O_3$ phase diagram, determine the weight percent Al_2O_3 if the spinel had the stoichiometric composition.

(a) Is the spinel nonstoichiometric on the MgO-rich side of the phase diagram? If so, what type of lattice imperfections might be present?

(b) Is the spinel nonstoichiometric on the Al_2O_3 side of the phase diagram? If so, what type of lattice imperfections might be present?

14.12 The specific gravity of Al_2O_3 is 3.96 $Mg.m^{-3}$. A ceramic part is produced by sintering alumina powder. It weighs 80 g when dry, 92 g after it has soaked in water, and 58 g when suspended in water. Calculate the apparent porosity, the true porosity, and the closed pores.

14.13 Silicon carbide (SiC) has a specific gravity of 3.1 $Mg.m^{-3}$. A sintered SiC part is produced, occupying a volume of 500 cm^3 and weighing 1 200 g. After soaking in water, the part weighs 1 250 g. Calculate the bulk density, the true porosity, and the volume fraction of the total porosity that consists of closed pores.

14.14 Calculate the O : Si ratio when 20 wt% Na_2O is added to SiO_2. Explain whether this material will provide good glass-forming tendencies. Above what temperature must the ceramic be heated to be all liquid?

14.15 How many grams of BaO can be added to 1 kg of SiO_2 before the O : Si ratio exceeds 2.5 and glass-forming tendencies are poor? Compare this with the case when Li_2O is added to SiO_2.

14.16 Calculate the O : Si ratio when 30 wt% Y_2O_3 is added to SiO_2. Will this material provide good glass-forming tendencies?

14.17 Lead can be introduced into a glass either as PbO (where the Pb has a valence of +2) or as PbO_2 (where the Pb has a valence of +4). Draw a sketch (similar to Figure 14.17) showing the effect of each of these oxides on the silicate network. Which oxide is a modifier and which is an intermediate?

14.18 A glass composed of 65 mol% SiO_2, 20 Mol% CaO, and 15 mol% Na_2O is prepared. Calculate the O : Si ratio and determine whether the material has good glass-forming tendencies. Estimate the liquidus temperature of the material using Figure 14.31.

14.19 Hot-pressed Si_3N_4 has a tensile strength of 550 $MN.m^{-2}$. Flaws are present due to porosity remaining in the part; the radius of curvature of these flaws is determined to be 0.005 cm. The part should be capable of withstanding an applied stress of 200 $MN.m^{-2}$. What is the maximum length of flaws that can be tolerated?

14.20 A sialon ceramic typically has a flexural strength of 825 $MN.m^{-2}$. In a three-point bend test (*see* Chapter 6), a sialon bar 12 mm thick and 25 mm wide is supported at two points 230 mm apart. The part is known to contain flaws 0.025 mm long with a radius at the tip of 50 nm. At what load during the test is the bar expected to fail? (1 nm = 10^{-9} m.)

14.21 A large ceramic part produced from partially stabilised ZrO_2 has a fracture toughness of 11 $MN.m^{-3/2}$ and an expected yield strength of 450 $MN.m^{-2}$. If the part is to withstand an applied stress of half of its expected yield strength, determine the maximum size of flaws that can be present in the structure. Assume that $f = 1$.

14.22 An alumina part with a square cross-section has a fracture toughness of 5.5 $MN.m^{-3/2}$ and is to be subjected to a tensile force of 89 kN. Nondestructive testing has shown that there are no flaws greater than 0.25 mm in length. Determine the minimum size of the part. Assume that $f = 1.1$.

14.23 You would like a transformation-toughened ZrO_2 part 2 cm thick and 3 cm wide to withstand a force of 20 000 N. The part is supported by blocks 10 cm apart. The fracture toughness of the ceramic is 9 $MN.m^{-3/2}$, and flaws 0.05 cm long are known to be present at the surface of the part. Can this load be sustained without the part failing? Assume that $f = 1$.

14.24 A set of ceramic parts are subjected to bend tests, and the stress required for failure is measured; the results are as follows: 55.5 $MN.m^{-2}$, 54.5 $MN.m^{-2}$, 48.3 $MN.m^{-2}$, 52.4 $MN.m^{-2}$, 56.5 $MN.m^{-2}$, 50.3 $MN.m^{-2}$, 53.4 $MN.m^{-2}$, and 51.7 $MN.m^{-2}$. Calculate the Weibull modulus for the material. Discuss whether the material would be a good choice if we wished to design a reasonably reliable part.

14.25 A set of ceramic parts are tested and the stress required for failure is measured; the result are as follows: 152 $MN.m^{-2}$, 260 $MN.m^{-2}$, 500 $MN.m^{-2}$, 1150 $MN.m^{-2}$, 700 $MN.m^{-2}$, 640 $MN.m^{-2}$, 370 $MN.m^{-2}$, 1020 $MN.m^{-2}$, and 1590 $MN.m^{-2}$. Calculate the Weibull modulus for the material. Discuss whether the material would be a good choice if we wished to design a reasonably reliable part.

14.26 Ceramic moulds used in producing castings contain alcohol, which is burned prior to casting. As a result, a network of microcracks is produced that allow gases to escape through the mould when the liquid metal is poured into the mould. What effect might the microcracks have on the properties of the mould?

14.27 A shear stress of 20 $MN.m^{-2}$ is to be used to deform the surface of a 1 cm-thick plate of soda-lime glass, providing a surface velocity of 1 $cm.s^{-1}$. The velocity at the opposite surface is to be zero. What viscosity is required, assuming a linear velocity gradient in the glass? To what temperature must the glass be heated to do this?

14.28 Calculate and compare the activation energies for viscous flow in fused silica and soda-lime glass. Explain how this influences the processing of the glass.

14.29 A stress of 25 $MN.m^{-2}$ is applied to a 150 mm-long rod of borosilicate glass held at a constant temperature. If the rod is to elongate no more than 1 mm in one year, what is the maximum temperature that can be used? The viscosity of the glass is 10^7 $Ns.m^{-2}$ at 1 042 K and 10^3 $Ns.m^{-2}$ at 1 471 K.

14.30 At 70 $MN.m^{-2}$, the creep rate ($d\varepsilon/dt$) of a silicon nitride ceramic is 0.0635 $mm.mm^{-1}.h^{-1}$ at 1 400°C; at 1 250°C, the creep rate is 0.0002 $mm.mm^{-1}.h^{-1}$. Estimate the activation energy for viscous flow, using these data.

14.31 From Figure 14.27, determine the relationship between creep rate and grain size for MgO. Explain why the grain size influences creep rate. If you were designing an MgO refractory for optimum resistance to creep, would you use a large or a small grain size?

14.32 Suppose you combine 6 mol of SiO_2 with 1 mol of Na_2O and 1 mol of CaO. Determine the liquidus temperature for the ceramic.

14.33 A SiO_2-Al_2O_3 fireclay brick can perform satisfactorily at 1 700°C if no more than 20% liquid surrounds the mullite present in the microstructure. What is the minimum percent alumina that must be in the refractory?

14.34 How much kaolinite clay must be added to 100 g of quartz to produce a SiO_2-30% Al_2O_3 fireclay brick after firing?

14.35 Suppose we combine 60 kg of $Al_2O_3 \cdot 4SiO_2 \cdot H_2O$ with 120 kg of $2CaO \cdot Al_2O_3 \cdot SiO_2$ to produce a clay body. The ceramic is dried and fired at 1 600°C. Determine the weight and composition of the clay body after firing.

Design problems

14.36 Using the data in Table 14.4, design a ceramic part containing flaws that are too small to propagate on their own but may prevent other major cracks from easily propagating. During use, the ceramic part will be loaded at 44 500 N in a manner similar to a three-point bend test, where the supports are 150 mm. apart. To fit into the rest of the assembly, the part must be 30 mm wide.

14.37 Design a silica soda-lime glass that can be cast at 1 300°C and that will have a O : Si ratio of less then 2.3 to ensure good glass-forming tendencies. To ensure adequate viscosity, the casting temperature should be at least 100°C above the liquidus temperature.

14.38 Design a ceramic structure that will not fail when a tensile load of 25 000 N is applied, assuming that our processing can only ensure that cracks are shorter than 0.7 mm.

14.39 Design a 5 cm-long glass rod so that when a load of 10 000 N is applied at 1 000°C, the rod will still be no longer than 5.002 cm after one year.

14.40 We wish to produce a complex silicon nitride impeller; the strength of the part is relatively unimportant, but the dimensional accuracy must be very good. Design a method for producing this part.

CHAPTER 15

Polymers

15.1 Introduction

Polymers – which include such diverse materials as plastics, rubbers, and adhesives – are giant organic, chain-like molecules having molecular weights from 10 000 to more than 1 000 000 $g.mol^{-1}$. *Polymerisation* is the process by which smaller molecules are joined to create these giant molecules. Polymers are used in an amazing number of applications, including toys, home appliances, structural and decorative items, coatings, paints, adhesives, automobile tyres, foams, and packaging. Polymers are often used in composites, both as fibres and as a matrix.

Commercial – or standard – polymers are lightweight, corrosion-resistant materials with low strength and stiffness, and they are not suitable for use at high temperatures. These polymers are, however, relatively inexpensive and are readily formed into a variety of shapes, ranging from plastic bags to mechanical gears to bathtubs. *Engineering* polymers are designed to give improved strength or better performance at elevated temperatures. These materials are produced in relatively small quantities and often are expensive. Some of the engineering polymers can perform at temperatures as high as 350°C; others – usually in a fibre form – have strengths that are greater than that of steel.

Polymers also have many useful physical properties. Some, such as Perspex (PMMA), are transparent and can substitute for ceramic glasses. Although most polymers are electrical insulators, special polymers (such as the acetals) and polymer-based composites possess useful electrical conductivity. Teflon has a low coefficient of friction and is the coating for nonstick cookware. Polymers also resist corrosion and chemical attack. Many of these properties will be discussed in later chapters.

15.2 Classification of Polymers

Polymers are classified in several ways: by how the molecules are synthesised, by their molecular structure, or by their chemical family. However, the most commonly used method to describe polymers is in terms of their mechanical and thermal behaviour. Table 15.1 compares the three major polymer categories.

Table 15.1 Comparison of the three polymer categories.

Behaviour	General Structure	Diagram
Thermoplastic	Flexible linear chains	
Thermosetting	Rigid three-dimensional network	Cross-link
Elastomers	Linear cross-linked chains	Cross-link

Thermoplastic polymers are composed of long chains produced by joining together small molecules, or *monomers*; they typically behave in a plastic, flexible manner. These polymers soften and are formed by viscous flow when heated to elevated temperatures. Thermoplastic polymers may be recycled.

Thermosetting polymers are composed of long chains of molecules that are strongly cross-linked to one another to form three-dimensional network structures. These polymers are generally more rigid, stronger, but more brittle, than thermoplastics. Thermosets do not have a fixed melting temperature and cannot easily be reprocessed after the cross-linking reaction has occurred.

Elastomers, including rubbers, have an intermediate structure in which some cross-linking of the chains is allowed to occur. Elastomers have the ability to elastically deform by enormous amounts without being permanently changed in shape.

Polymerisation of all three polymer types normally begins with the production of long chains in which the atoms are strongly joined by covalent bonding. The amount and strength of cross-linking gives each type its special properties. We should note, however, that the distinctions between these three types often become blurred. For example, there is a continuum of change between the simple structure of polyethylene (a thermoplastic) and the more complex structure of epoxy (a thermoset).

Representative Structures All of the polymers have a complex three-dimensional structure that is difficult to describe pictorially. Figure 15.1 shows three ways we could represent a segment of polyethylene, the simplest of the thermoplastic polymers. The polymer chain consists of a backbone of carbon atoms; two hydrogen atoms are bonded to each carbon atom in the chain. The chain twists and turns throughout space. The simple two-dimensional model in Figure 15.1(c) includes the essential elements of the polymer structure and will be used to describe the various polymers. The single lines (——) between the carbon atoms and between the carbon and hydrogen atoms represent a single covalent bond. Two parallel lines (═══) represent a double covalent bond between atoms.

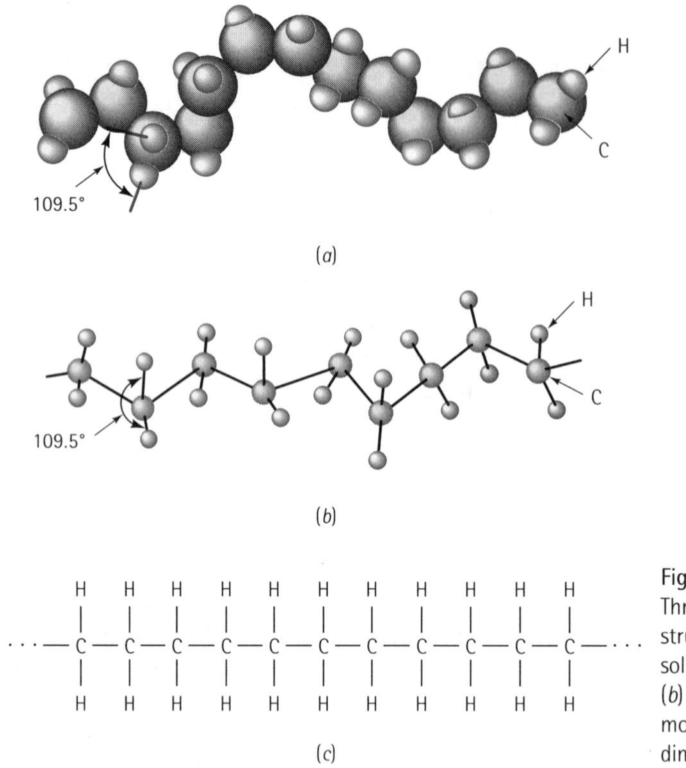

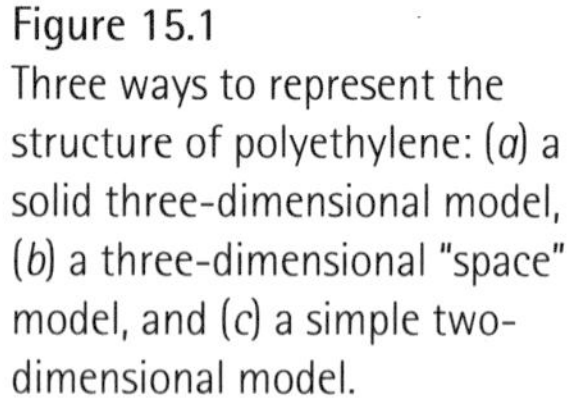
Figure 15.1
Three ways to represent the structure of polyethylene: (*a*) a solid three-dimensional model, (*b*) a three-dimensional "space" model, and (*c*) a simple two-dimensional model.

A number of polymers include ring structures, such as the benzene ring found in styrene and phenol molecules (Figure 15.2). These *aromatic* rings contain six carbon atoms joined with alternating single and double covalent bonds. Rather than showing all of the atoms in the benzene ring, we use a hexagon containing a circle to depict this ring structure.

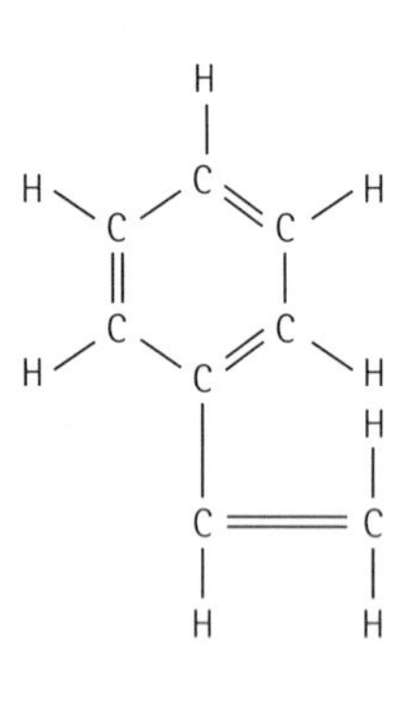

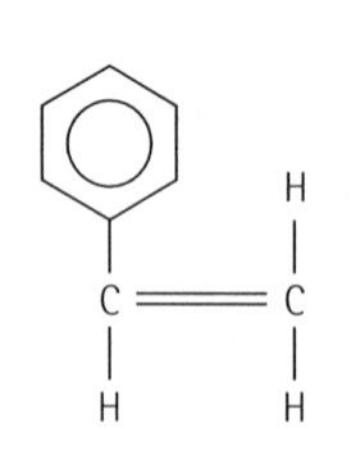

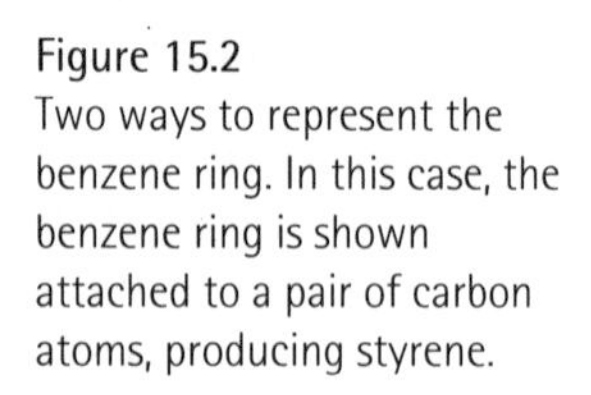
Figure 15.2
Two ways to represent the benzene ring. In this case, the benzene ring is shown attached to a pair of carbon atoms, producing styrene.

EXAMPLE 15.1 Materials Selection for Polymer Components

What type of polymer material might you select for the following applications: a surgeon's glove, a beverage container, and a pulley.

SOLUTION

The glove must be capable of stretching a great deal in order to slip onto the surgeon's hand, yet it must conform tightly to the hand to permit the maximum sensation of touch during surgery. A material that undergoes a large amount of elastic strain – particularly with relatively little applied stress – might be appropriate; this requirement describes an elastomer.

The beverage container should be easily and economically produced. It should have some flexibility and toughness so that it does not accidentally shatter and leak the contents. A thermoplastic will have the necessary formability and flexibility needed for this application.

The pulley will be subjected to some stress and wear as a belt passes over it. A relatively strong, rigid, hard material is required to prevent wear, so a thermosetting polymer might be most appropriate.

15.3 Chain Formation by the Addition Mechanism

The formation of the most common polymer, polyethylene (PE), from ethylene molecules is an example of addition (or chain-growth) polymerisation. Ethylene, a gas, has the formula C_2H_4. The two carbon atoms are joined by a double covalent bond. Each carbon atom shares two of its electrons with the second carbon atom, and two hydrogen atoms are bonded to each of the carbon atoms (Figure 15.3).The ethylene molecule is a monomer

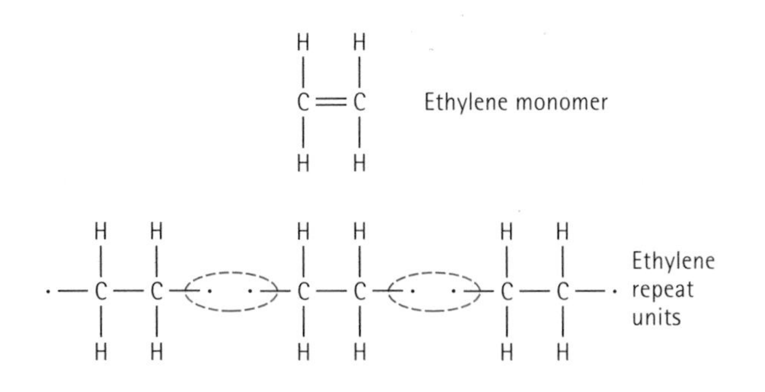

Figure 15.3
The addition reaction for producing polyethylene from ethylene molecules. The unsaturated double bond in the monomer is broken to produce active sites, which then attract additional repeat units to either end to produce a chain.

In the presence of an appropriate combination of heat, pressure, and catalysts, the double bond between the carbon atoms is broken and replaced with a single covalent bond. The ends of the monomer are now *free radicals*; each carbon atom has an unpaired electron that it may share with other free radicals. This reactive molecule – the basic building block for the polymer – is sometimes called a *mer* or, more appropriately, the repeat unit.

Unsaturated Bonds Addition polymerisation occurs because the original monomer contains a double covalent bond between the carbon atoms. The double bond is an unsaturated bond. After changing to a single bond, the carbon atoms are still joined, but they become active; other repeat units can be added to produce the polymer chain.

Functionality The functionality is the number of sites at which new molecules can be attached to the repeat unit of the polymer. In ethylene, there are two locations (each carbon atom) at which molecules can be attached. Thus, ethylene is *bifunctional,* and only chains form. If there are three or more sites at which molecules can be attached, a three-dimensional network forms.

EXAMPLE 15.2

Phenol molecules have the structure shown below. The phenol molecules can be joined to one another when a hydrogen atom is removed from the ring and participates in a condensation reaction. What is the functionality of the phenol? Will a chain or a network structure be produced?

SOLUTION

Hydrogen atoms are available from any of the five corners containing only hydrogen atoms. The hydrogen atom in the OH group is bonded too tightly to the ring. The three sites that are circled are the most reactive and they are preferred sites for a condensation reaction. Thus the effective functionality is three. Two of the reactive sites are used to produce a chain; the third permits cross-linking and formation of a network, or thermosetting, polymer.

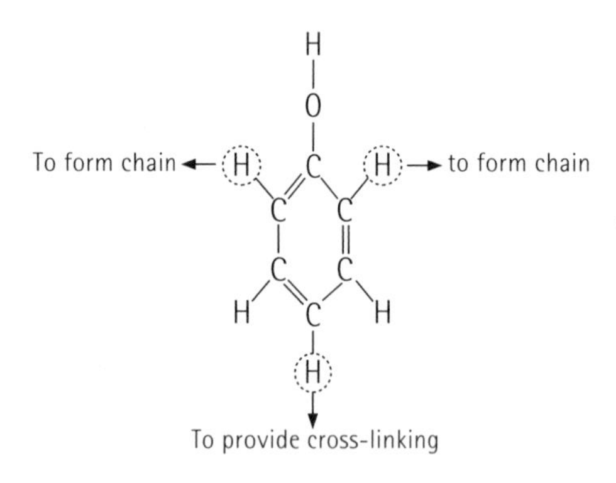

Initiation of Addition Polymerisation To begin the addition, or chain-growth, polymerisation process, an initiator is added to the molecule (Figure 15.4). The initiator forms free radicals with a reactive site that attracts one of the carbon atoms of an ethylene monomer. When this reaction occurs, the reactive site is transferred to the other carbon atom in the monomer and a chain begins to form. A second repeat unit of ethylene can be attached at this new site, and the chain begins to lengthen. This process continues until a long polyethylene chain – an addition polymer – is formed.

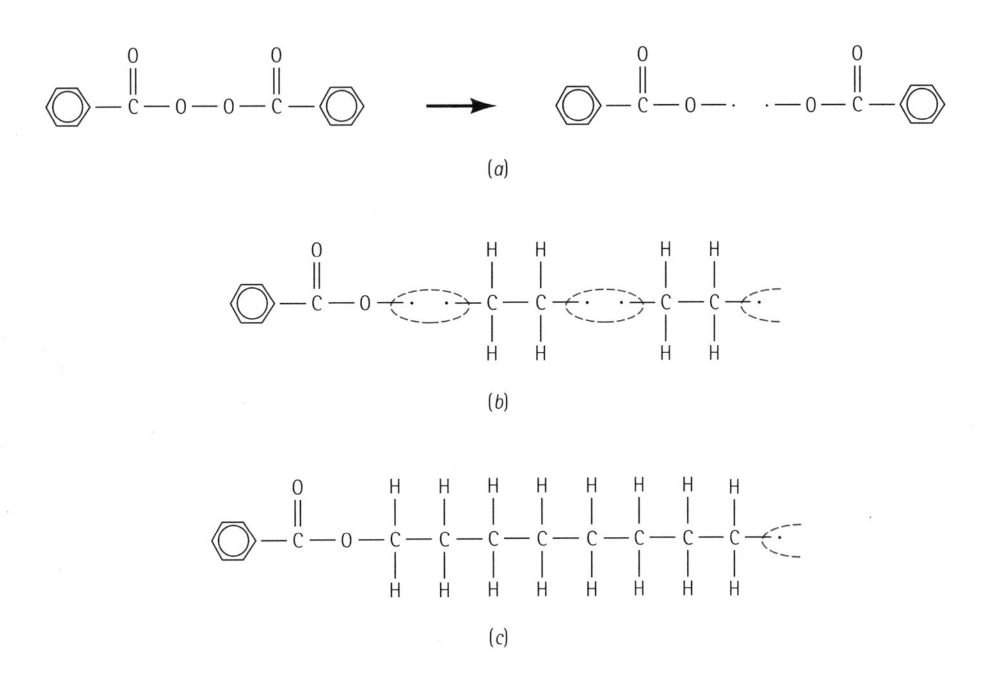

Figure 15.4
Initiation of a polyethylene chain by chain-growth may involve (*a*) producing free radicals from initiators such as benzoyl peroxide, (*b*) attachment of a polyethylene repeat unit to one of the initiator radicals, and (*c*) attachment of additional repeat units to propagate the chain.

Because the initiators – which are often peroxides – react with one another as well as with the monomer, their lifetimes are relatively short. A common initiator is benzoyl peroxide (Figure 15.4).

Growth of the Addition Chain Once the chain is initiated, repeat units are added onto each chain at a high rate, with perhaps several thousand additions each second (Figure 15.5). When polymerisation is nearly complete, the few remaining monomers must diffuse a long distance before reaching an active site at the end of a chain; consequently, the growth rate decreases.

Termination of Addition Polymerisation The chains may be terminated by two mechanisms (Figure 15.6). First, the ends of two growing chains may be joined. This process, called *combination*, creates a single large chain from two smaller chains. Second, the active end of one chain may remove a hydrogen atom from a second chain by a process known as *disproportionation*; this reaction terminates two chains, rather than combining two chains into one larger chain.

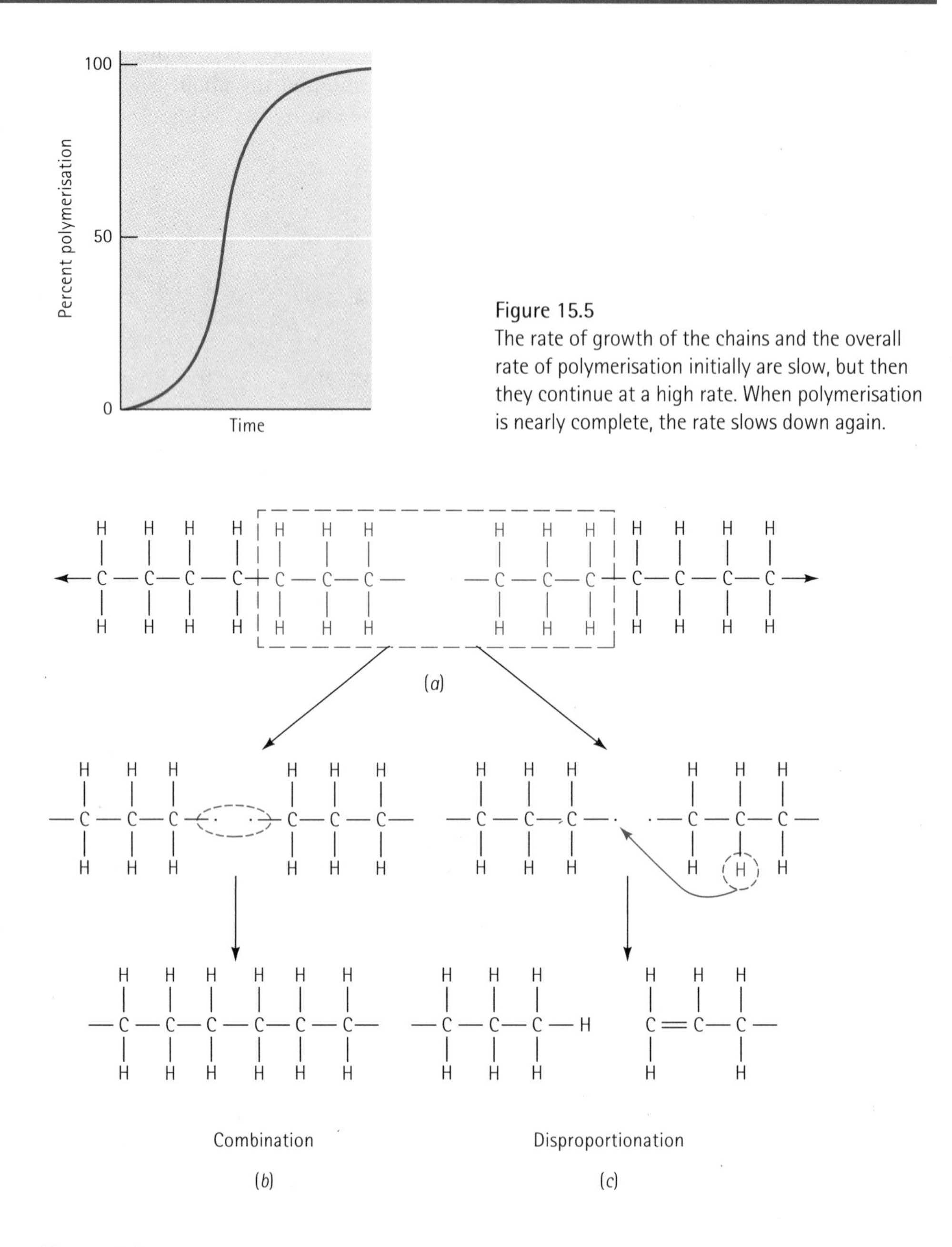

Figure 15.5
The rate of growth of the chains and the overall rate of polymerisation initially are slow, but then they continue at a high rate. When polymerisation is nearly complete, the rate slows down again.

Figure 15.6
Termination of polyethylene chain growth: (*a*) The active ends of two chains come into close proximity, (*b*) the two chains undergo combination and become one large chain, and (*c*) rearrangement of a hydrogen atom and creation of a double covalent bond disproportionation cause termination of two shorter chains.

Tetrahedral Structure of Carbon The structure of addition polymer chains is based on the nature of the covalent bonding in carbon. Carbon, like silicon, has a valence of four. The carbon atom shares its valence electrons with four surrounding atoms, producing a tetrahedral structure [Figure 15.7(a)]. In diamond, all of the atoms in the tetrahedron are carbon and the diamond cubic structure is produced.

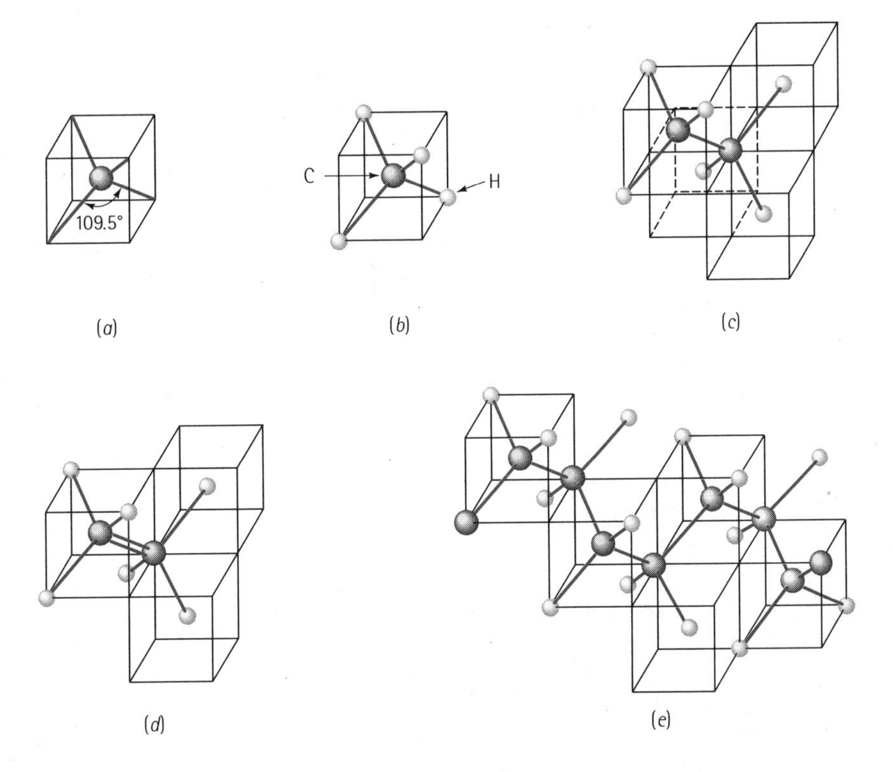

Figure 15.7
The tetrahedral structure of carbon can be combined in a variety of ways to produce solid crystals, nonpolymerisable gas molecules, and polymers: (*a*) carbon tetrahedron; (*b*) methane, with no unsaturated bonds; (*c*) ethane, with no unsaturated bonds; (*d*) ethylene, with no unsaturated bond; and (*e*) polyethylene.

In organic molecules, however, some of the positions in the tetrahedron are occupied by hydrogen, chlorine, fluorine, or even groups of atoms. Since the hydrogen atom has only one electron to share, the tetrahedron cannot be further extended. The structure in Figure 15.7(b) shows an organic molecule (methane) that cannot undergo a simple addition polymerisation process, because all four bonds are satisfied by hydrogen atoms. The initial carbon atom could be joined with one covalent bond to a second carbon atom, with all of the other bonds involving hydrogen, as in ethane [Figure 15.7(c)]. But the bond between the carbon atoms is saturated, and, again, polymerisation cannot occur.

However, in ethylene, the carbon atoms are joined by an unsaturated double bond; the other sites are occupied by hydrogen atoms [Figure 15.7(d)]. During polymerisation, the double bond is broken and each carbon atom can be attached to another ethylene repeat unit, eventually giving polyethylene [Figure 15.7(e)].

Chain Shape The polymer chains are able to twist and turn because of the tetrahedral nature of the covalent bond. Figure 15.8 illustrates two possible geometries in which a chain might grow. The third atom in Figure 15.8(a) can be located at any position on the circle and still preserve the directionality of the covalent bond. A straight chain, as in Figure 15.8(b), could be produced, but it is more likely that the chain will be highly twisted, as in Figure 15.8(c).

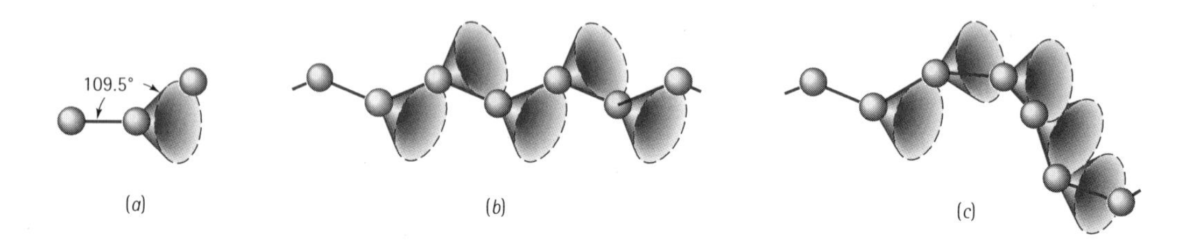

Figure 15.8
The angular relationship between the bonds in the carbon chain can be satisfied when the third carbon atom is placed anywhere on the circle in (*a*). Depending on how the atoms are placed, the chain may either be straight (*b*) or highly twisted (*c*).

The chains twist and turn in response to external factors such as temperature or the location of the next repeat unit to be added to the chain. Eventually, the chains become intertwined with other chains growing simultaneously. The appearance of the polymer chains may resemble that of a bucket of earthworms or a plate of spaghetti. The entanglement of the polymer chains is an important mechanism providing strength to the polymer. As when one grabs a handful of earthworms from a bucket, the entire mass tends to hold together due to the entanglement, even though one is touching only a few of the earthworms. The entanglement of long polymer chains, along with the Van der Waals bonding between the chains, likewise provides strength to the linear polymer.

EXAMPLE 15.3

Calculate the amount of benzoyl peroxide initiator required to produce 1 kg of polyethylene with molecular weight of 200 000 $g.mol^{-1}$. Assume that 20% of the initiator is actually effective and that all termination occurs by the combination mechanism.

SOLUTION

For 100% efficiency, we need one molecule of benzoyl peroxide for each polyethylene chain. (One of the free radicals would initiate one chain, the other free radical a second chain; then, the two chains combine into one larger one.) Since the molecular weight of ethylene = (2 C)(12) + (4H)(1) = 28 g.mol^{-1}:

$$\frac{200\,000 \text{ g.mol}^{-1}}{28 \text{ g.mol}^{-1}} = 7143 \text{ ethylene molecules per chain}$$

$$\frac{(1\,000 \text{ g polyethylene})(6.02\times10^{23})}{28 \text{ g.mol}^{-1}} = 215\times10^{23} \text{ monomers}$$

The combination mechanism requires the number of benzoyl peroxide molecules to be:

$$\frac{215\times10^{23} \text{ ethylene molecules}}{7143 \text{ ethylenes / chain}} = 0.03\times10^{23}$$

The molecular weight of benzoyl peroxide is (14 C)(12) + (10 H)(1) + (4 O)(16) = 242 g.mol^{-1}. Therefore, the amount of initiator needed to form the ends of the chains is:

$$\frac{(0.03\times10^{23})(242 \text{ g.mol}^{-1})}{6.02\times10^{23}} = 1.206 \text{ g}$$

However, only 20% of the initiator actually is effective; the rest recombines or combines with other molecules and does not cause initiation of a chain. Thus, we need five times this amount, or 6.03 g of benzoyl peroxide per kilogram of polyethylene.

15.4 Chain Formation by the Condensation Mechanism

Linear polymers also form by condensation reactions, or *step-growth* polymerisation, producing structures and properties that resemble those of linear addition polymers. However, the mechanism of step polymerisation requires that at least two different monomers participate in the reaction. The polymerisation of dimethyl terephthalate and ethylene glycol to produce *polyester* is an important example (Figure 15.9).

During polymerisation, a hydrogen atom on the end of the ethylene glycol monomer combines with an OCH_3 group from the dimethyl terephthalate. A by-product, methyl alcohol, is driven off and the two monomers combine to produce a larger molecule. Each of the monomers in this example are bifunctional, and step polymerisation can continue by the same reaction. Eventually, a long polymer chain – a polyester – is produced. The repeat unit for this polyester consists of two original monomers: one ethylene glycol and one dimethyl terephthalate.

The length of the polymer chain depends on the ease with which the monomers can diffuse to the ends and undergo the condensation reaction. Chain growth ceases when no more monomers reach the end of the chain to continue the reaction.

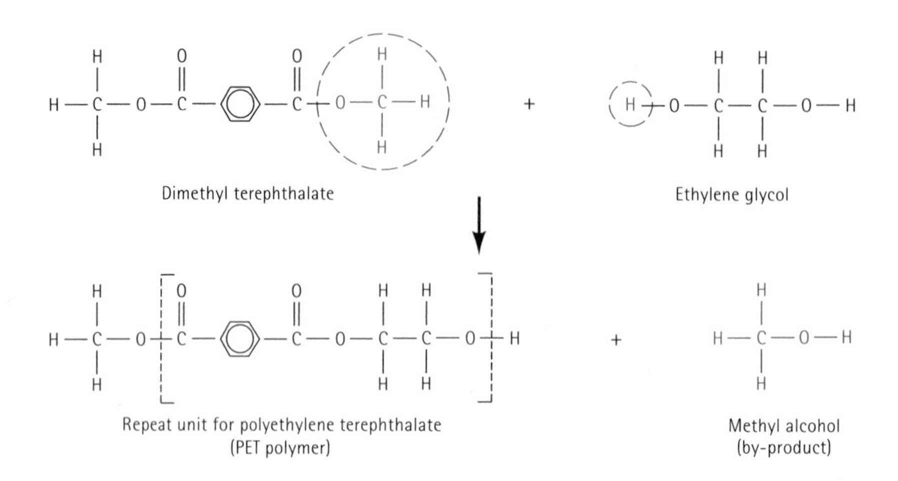

Figure 15.9
The condensation reaction for polyethylene terephthalene (PET), a common polyester. The OCH_3 group and a hydrogen atom are removed from the monomers, permitting the two monomers to join and producing methyl alcohol as a by-product.

EXAMPLE 15.4

The linear polymer 6,6-nylon is to be produced by combining 1000 g of hexamethylene diamine with adipic acid. A condensation reaction then produces the polymer. Show how this reaction occurs and determine the by-product that forms. How many grams of adipic acid are needed, and how much 6,6-nylon is produced, assuming 100% efficiency?

SOLUTION

The molecular structures of the monomers are shown below. The linear nylon chain is produced when a hydrogen atom from the hexamethylene diamine combines with an OH group from adipic to form a water molecule.

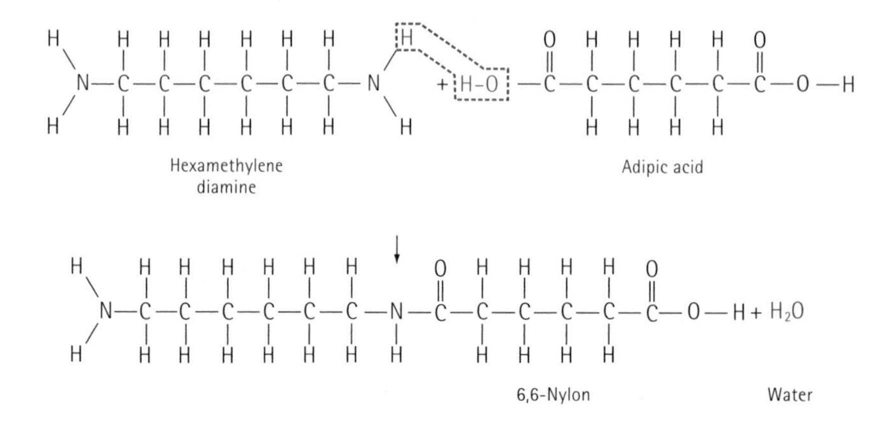

Note that the reaction can continue at both ends of the new molecule; consequently, long chains may form. This polymer is called 6,6-nylon because both monomers contain six carbon atoms.

We can determine that the molecular weight of hexamethylene diamine is 116 $g.mol^{-1}$, of adipic acid is 146 $g.mol^{-1}$, and of water is 18 $g.mol^{-1}$. The number of the moles of hexamethylene diamine added (calculated below) is equal to the number of moles of adipic acid:

$$\frac{1\,000\ g}{116\ g.mol^{-1}} = 8.621\ moles = \frac{x\ g}{146\ g.mol^{-1}}$$

$x = 1\,259$ g of adipic acid required

The number of moles of water lost is also 8.621:

$$y = (8.621\ moles)(18\ g.mol^{-1}) = 155.2\ g\ H_2O$$

But each time one more monomer is attached, another H_2O is released. Therefore, the total amount of nylon produced is 1 000 g + 1 259 g – 2(155.2 g) = 1 948.6 g.

15.5 Degree of Polymerisation

The average length of a linear polymer is represented by the degree of polymerisation or the number of repeat units in the chain. The degree of polymerisation can also be defined as:

$$\text{Degree of polymerisation} = \frac{\text{molecular weight of polymer}}{\text{molecular weight of repeat unit}} \quad (15.1)$$

If the polymer contains only one type of monomer, the molecular weight of the repeat unit is that of the monomer. If the polymer contains more than one type of monomer, the molecular weight of the repeat unit is the sum of the molecular weights of the monomers, less the molecular weight of the by-product.

The length of the chains in a linear polymer varies considerably. Some may be quite short due to early termination; others may be exceptionally long. We can define an average molecular weight in two ways.

The *weight average molecular weight* is obtained by dividing the chains into size ranges and determining the fraction of chains having molecular weights within that range. The weight average molecular weight $\bar{M}_w$ is

$$\bar{M}_w = \Sigma f_i M_i \quad (15.2)$$

where M_i is the mean molecular weight of each range and f_i is the weight fraction of the polymer having chains within that range.

The *number average molecular weight* $\bar{M}_n$ is based on the number fraction, rather than weight fraction, of the chains within each size range. It is always smaller than the weight average molecular weight,

$$\bar{M}_n = \Sigma x_i M_i \tag{15.3}$$

where M_i is again the mean molecular weight of each size range, but x_i is the fraction of the total number of chains within each range. Either $\bar{M}_w$ or $\bar{M}_n$ can be used to calculate the degree of polymerisation.

EXAMPLE 15.5

Calculate the degree of polymerisation if 6,6-nylon has a molecular weight of 120 000 g.mol^{-1}.

SOLUTION

The reaction by which 6,6-nylon is produced was described in Example 15.4. Hexamethylene diamine and adipic acid combine and release a molecule of water. When a long chain forms, there is, on average, one water molecule released for each reacting molecule. The molecular weights are 116 g.mol^{-1} for hexamethylene diamine, 146 g.mol^{-1} for adipic acid, and 18 g.mol^{-1} for water. The repeat unit for 6,6-nylon is:

```
   H   H   H   H   H   H   H   H   O   H   H   H   H   O
   |   |   |   |   |   |   |   |   ||  |   |   |   |   ||
 —N — C — C — C — C — C — C — N — C — C — C — C — C — C—
       |   |   |   |   |   |           |   |   |   |
       H   H   H   H   H   H           H   H   H   H
```

The molecular weight of the repeat unit is the sum of the molecular weights of the two monomers, minus that of the two water molecules that are evolved:

$$M_{\text{repeat unit}} = 116 + 146 - 2(18) = 226 \text{ g.mol}^{-1}$$

$$\text{Degree of polymerisation} = \frac{120\,000}{226} = 531$$

The degree of polymerisation refers to the total number of repeat units in the chain. The chain contains 531 hexamethylene diamine and 531 adipic acid molecules.

EXAMPLE 15.6

We have a polyethylene sample containing 4 000 chains with molecular weights between 0 and 5 000 g.mol^{-1}, 8 000 chains with molecular weights between 5 000 and 10 000 g.mol^{-1}, 7 000 chains with molecular weights between 10 000 and 15 000 g.mol^{-1}, and 2 000 chains with molecular weights between 15 000 and 20 000 g.mol^{-1}. Determine both the number and weight average molecular weights.

SOLUTION

First we need to determine the number fraction x_i and weight fraction f_i for each of the four ranges. For x_i, we simply divide the number in each range by 21 000, the total number of chains. To find f_i, we first multiply the number of chains by the mean molecular weight of the chains in each range, giving the 'weight' of each group, then find f_i by dividing by the total weight of 192.5×10^6. We can then use Equations 15.2 and 15.3 to find the molecular weights.

Number of Chains	Mean M per Chain	x_i	x_iM_i	Weight	f_i	f_iM_i
4 000	2 500	0.191	477.5	10×10^6	0.0519	129.75
8 000	7 500	0.381	2 857.5	60×10^6	0.3118	2 338.50
7 000	12 500	0.333	4 162.5	87.5×10^6	0.4545	5 681.25
2 000	17 500	0.095	1 662.5	35×10^6	0.1818	3 181.50
$\Sigma = 21\,000$		$\Sigma = 1.00$	$\Sigma = 9\,160$	$\Sigma = 192.5 \times 10^6$	$\Sigma = 1$	$\Sigma = 11\,331$

$$\bar{M}_n = \Sigma x_i M_i = 9\,160 \text{ g.mol}^{-1}$$

$$\bar{M}_w = \Sigma f_i M_i = 11\,331 \text{ g.mol}^{-1}$$

The weight average molecular weight is larger than the number average molecular weight.

15.6 Arrangement of Polymer Chains in Thermoplastics

In typical thermoplastic polymers, bonding within the chains is covalent, but the long coiled chains are held to one another by weak secondary bonds and by entanglement (Figure 15.10). When stress is applied to the thermoplastic, the weak bonding between the chains can be overcome and the chains can rotate and slide relative to one another. The ease with which the chains slide depends on both temperature and the polymer structure. Several critical temperatures, summarised in Figures 15.11 and 15.12, may be observed.

Degradation Temperature At very high temperatures, the covalent bonds between the atoms in the linear chain may be destroyed, and the polymer may burn or char. This temperature T_d is the degradation (or decomposition) temperature. Exposure to oxygen, ultraviolet radiation, and attack by bacteria also cause a polymer to degrade, even at low temperatures.

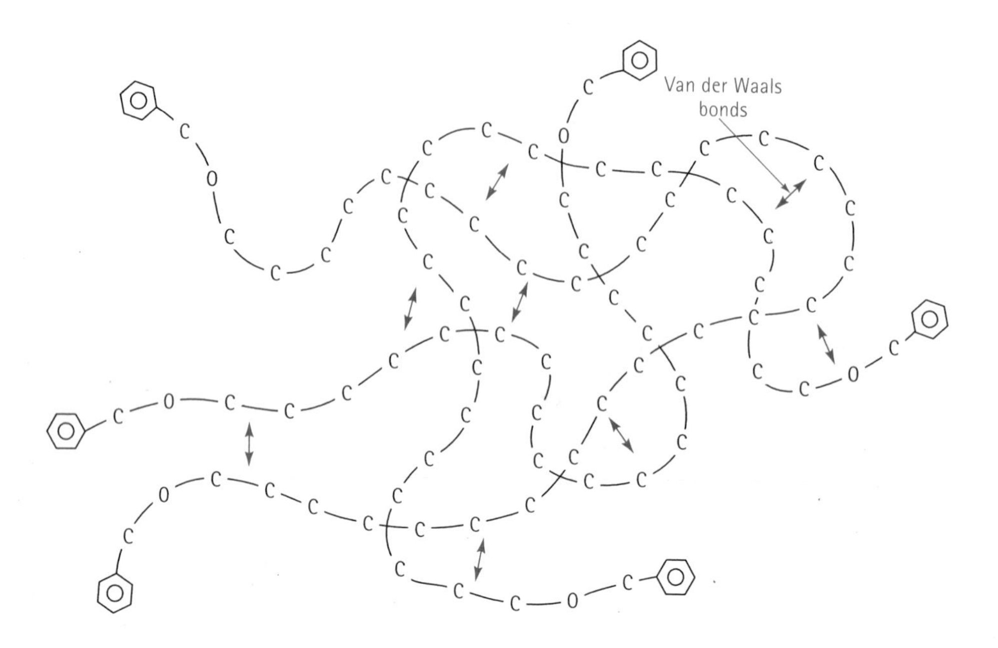

Figure 15.10
The chains are held loosely together by Van der Waals bonds and mechanical entanglement.

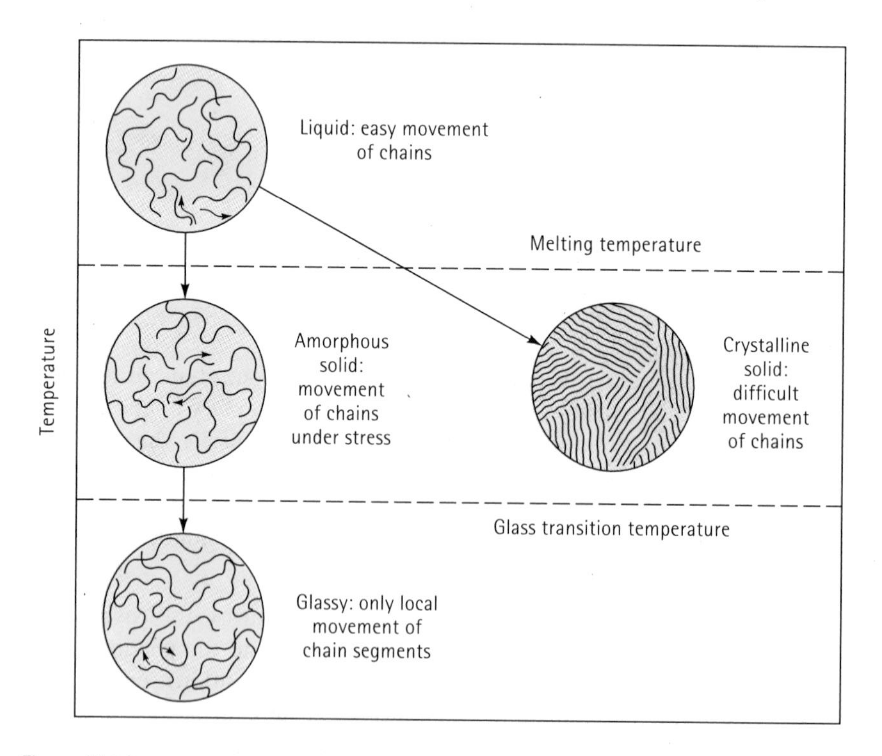

Figure 15.11
The effect of temperature on the structure and behaviour of thermoplastic polymers.

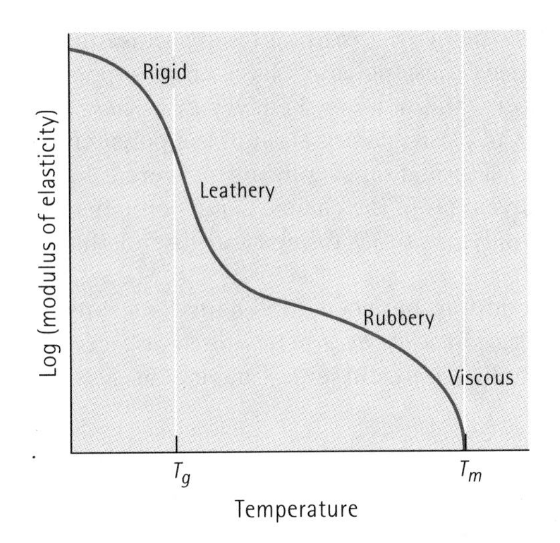

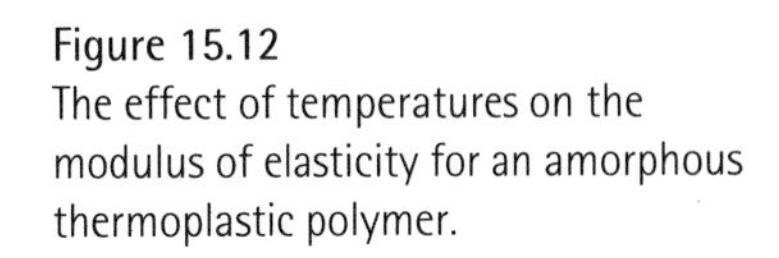
Figure 15.12
The effect of temperatures on the modulus of elasticity for an amorphous thermoplastic polymer.

Liquid Polymers At or above the melting temperature T_m, bonding between the twisted and intertwined chains is weak. If a force is applied, the chains slide past one another and the polymer flows with virtually no elastic strain. The strength and modulus of elasticity are nearly zero and the polymer is suitable for casting and many forming processes. The melting points of typical polymers are included in Table 15.2.

Table 15.2 Melting and glass transition temperatures for selected thermoplastics and elastomers.

Polymer	T_m (°C)	T_g (°C)
Addition Polymers		
Low-density (LD) polyethylene	115	−120
High-density (HD) polyethylene	137	−120
Polyvinyl chloride	175–212	87
Polypropylene	168–176	−16
Polystyrene	240	85–125
Polyacrylonitrile	320	107
Polytetrafluoroethylene (Teflon)	327	
Polychlorotrifluoroethylene	220	
Polymethyl methacrylate (acrylic)		90–105
ABS		88–125
Condensation polymers		
Acetal	181	−85
Polyamide (6,6-nylon)	265	50
Cellulose acetate	230	
Polycarbonate	230	145
Polyester	255	75
Elastomers		
Silicone		−123
Polybutadiene	120	−90
Polychloroprene	80	−50
Polyisoprene	30	−73

Rubbery or Leathery Polymers Below the melting temperature, the polymer chains are still twisted and intertwined. These polymers have an amorphous structure. Just below the melting temperature, the polymer behaves in a *rubbery* manner; when stress is applied, both elastic and plastic deformation of the polymer occur. When the stress is removed, the elastic deformation is quickly recovered, but the polymer is permanently deformed by movement of the chains. Large permanent elongations can be achieved, permitting the polymer to be formed into useful shapes by moulding and extrusion.

At lower temperatures, bonding between the chains is stronger, the polymer becomes stiffer and stronger, and a *leathery* behaviour is observed. Many of the commercial polymers – including polyethylene – have a useable strength in this condition.

Glassy Polymers Below the glass transition temperature T_g, the linear polymer becomes hard, brittle, and glasslike. The arrangement of the polymer chains is still amorphous. When the polymer cools below the glass transition temperature, certain properties – such as density or modulus of elasticity – change at a different rate (Figure 15.13).

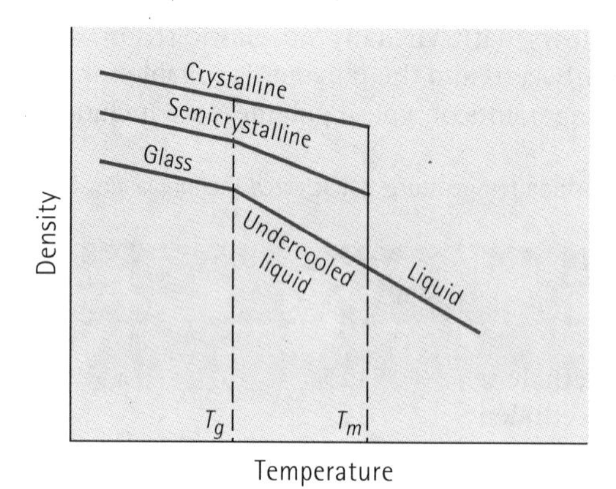

Figure 15.13
The relationship between the density and the temperature of the polymer shows the melting and glass transition temperatures.

Although glassy polymers have poor ductility and formability, they do have good strength, stiffness, and creep resistance. A number of important polymers, including polystyrene and polyvinyl chloride, have glass transition temperatures above room temperature (Table 15.2).

The glass transition temperature is typically about 0.5 to 0.75 times the absolute melting temperature T_m. Polymers such as polyethylene, which have no complicated side groups attached to the carbon backbone, have low glass transition temperatures (even below room temperature) compared with polymers such as polystyrene, which have more complicated side groups.

Crystalline Polymers Many thermoplastics partially crystallise when cooled below the melting temperature, with the chains becoming closely aligned over appreciable distances. A sharp increase in the density occurs as the coiled and intertwined chains in the liquid are rearranged into a more orderly, close-packed structure (Figure 15.13).

One model describing the arrangement of the chains in a crystalline polymer is shown in Figure 15.14. In this *folded chain* model, the chains loop back on themselves, with each loop being approximately 100 carbon atoms long. The folded chain extends

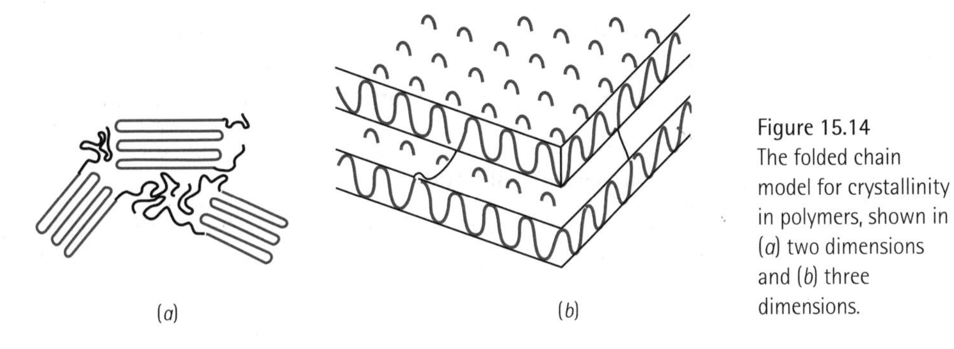

Figure 15.14
The folded chain model for crystallinity in polymers, shown in (*a*) two dimensions and (*b*) three dimensions.

Table 15.3 Crystal structures of several polymers.

Polymer	Crystal Structure	Lattice Parameters (nm)		
Polyethylene	Orthorhombic	$a_0 = 0.742$	$b_0 = 0.495$	$c_0 = 0.255$
Polypropylene	Orthorhombic	$a_0 = 1.450$	$b_0 = 0.569$	$c_0 = 0.740$
Polyvinyl chloride	Orthorhombic	$a_0 = 1.040$	$b_0 = 0.530$	$c_0 = 0.510$
Polyisoprene (cis)	Orthorhombic	$a_0 = 1.246$	$b_0 = 0.886$	$c_0 = 0.810$

in three dimensions, producing thin plates or lamellae. The crystals can take various forms, with the spherulitic shape shown in Figures 8.12 and 15.15 being particularly common. The crystals have a unit cell that describes the regular packing of the chains. The crystal structure for polyethylene, shown in Figure 3.30, describes one such unit cell. Structures for several polymers are described in Table 15.3. Some polymers are polymorphic, having more than one crystal structure.

However, there are always thin regions between the lamellae, as well as between spherulites, that are amorphous transition zones. The weight percentage of the structure that is crystalline can be calculated from the density of the polymer:

$$\% \text{ Crystalline } = \frac{\rho_c}{\rho} \frac{(\rho - \rho_a)}{(\rho_c - \rho_a)} \times 100 \tag{15.4}$$

where ρ is the measured density of the polymer, ρ_a is the density of amorphous polymer, and ρ_c is the density of completely crystalline polymer.

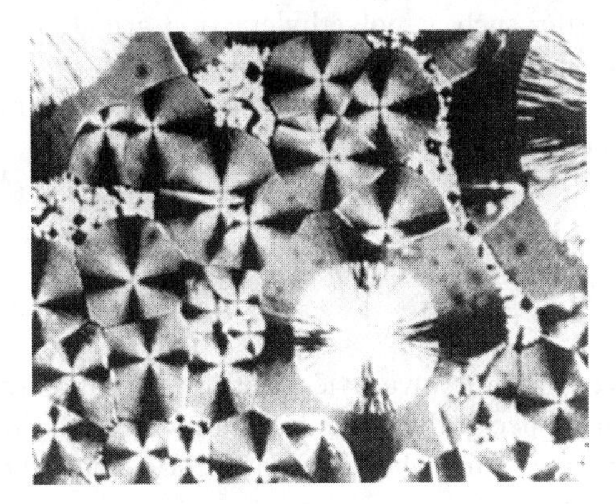

Figure 15.15
Photograph of spherulitic crystals in an amorphous matrix of polyamide (× 200). (*From R. Brick, A. Pense and R. Gordon,* Structure and Properties of Engineering Materials, *4th Ed., McGraw-Hill, 1977.*)

Several factors influence the ability of a polymer to crystallise:

1. *Complexity*. Crystallisation is easiest for simple addition polymers, such as polyethylene, in which no bulky molecules or atom groups that might interfere with the close packing of the chains are attached to the carbon chain.
2. *Cooling rate*. Slow cooling, which permits more time for the chains to become aligned, encourages crystallisation.
3. *Annealing*. Heating an amorphous structure just below the melting temperature provides the thermal activation that permits crystals to nucleate and grow.
4. *Degree of Polymerisation*. Polymers containing long chains are more difficult to crystallise.
5. *Deformation*. Slow deformation of the polymer between the melting and glass transition temperatures may promote crystallisation by straightening the chains, thus permitting them to move closer together.

A completely crystalline polymer would not display a glass transition temperature; however, the amorphous regions in semicrystalline polymers do change to a glass below the glass transition temperature (Figure 15.13).

EXAMPLE 15.7 Specification of a Polymer Insulation Material

A storage tank for liquid hydrogen will be made of metal, but we wish to coat the metal with a 3-mm thickness of polymer as an intermediate layer between metal and additional insulation layers. The temperature of the intermediate layer may drop to –80°C. Specify a material for this layer.

SOLUTION

We want the material to have reasonable ductility. As the temperature of the tank changes, stresses develop in the coating due to differences in thermal expansion, and we do not want the polymer to fail due to these stresses. A material that has good flexibility and/or can undergo large elastic strains is needed. We therefore would prefer either a thermoplastic that has a glass transition temperature below –80°C or an elastomer, also with a glass transition temperature below –80°C. Of the polymers listed in Table 15.2, thermoplastics such as polyethylene and acetal are satisfactory. Suitable elastomers include silicone and polybutadiene.

We might prefer one of the elastomers, for they can accommodate thermal stress by elastic, rather than plastic, deformation.

EXAMPLE 15.8

A new grade of flexible, impact-resistant polyethylene for use as a thin film requires a density of 0.88 to 0.915 $Mg.m^{-3}$. Specify the polyethylene required to produce these properties. (The density of completely amorphous polyethylene is about 0.87 $Mg.m^{-3}$.)

SOLUTION

To produce the required properties and density, we must control the percent crystallinity of the polyethylene. We can use Equation 15.4 to determine the crystallinity that corresponds to the required density range. To do so, however, we must know the density of completely crystalline polyethylene. We can use the data in Table 15.3 to calculate this density if we recognise that there are two polyethylene repeat units in each unit cell (see Example 3.16):

$$\rho_c = \frac{(4C)(12)+(8H)(1)}{(7.42)(4.95)(2.55)(10^{-24})(6.02\times10^{23})} = 0.9932\ \text{Mg.m}^{-3}$$

We know that ρ_a = 0.87 Mg.m^{-3} and that ρ varies from 0.88 to 0.915 Mg.m^{-3}. The required crystallinity then varies from:

$$\%\ \text{crystalline} = \frac{(0.9932)(0.88-0.87)}{(0.88)(0.9932-0.87)}\times100 = 9.2$$

$$\%\ \text{crystalline} = \frac{(0.9932)(0.915-0.87)}{(0.915)(0.9932-0.87)}\times100 = 39.6$$

Therefore, we must be able to process the polyethylene to produce a range of crystallinity between 9.2 and 39.6 percent.

15.7 Deformation and Failure of Thermoplastic Polymers

When an external force is applied to a thermoplastic polymer, both elastic and plastic deformation occur. The mechanical behaviour is closely tied to the manner in which the polymer chains move relative to one another under load. Deformation is more complicated in thermoplastic polymers than it is in most metals and ceramics, because the deformation process depends on both time and the rate at which the load is applied. Figure 15.16 shows a stress-strain curve for a typical thermoplastic polymer obtained under normal loading conditions.

Elastic Behaviour Elastic deformation in these polymers is the result of two mechanisms. An applied stress causes the covalent bonds within the chain to stretch and distort, allowing the chains to elongate elastically. When the stress is removed, recovery from this distortion is almost instantaneous. This behaviour is similar to that in metals and ceramics, which also deform elastically by stretching of metallic, ionic, or covalent bonds.

But in addition, entire segments of the polymer chains may be distorted; when the stress is removed, the segments move back to their original positions only over a period of time – often, hours or even months. This time-dependent, or viscoelastic behaviour may contribute to some nonlinear elastic behaviour.

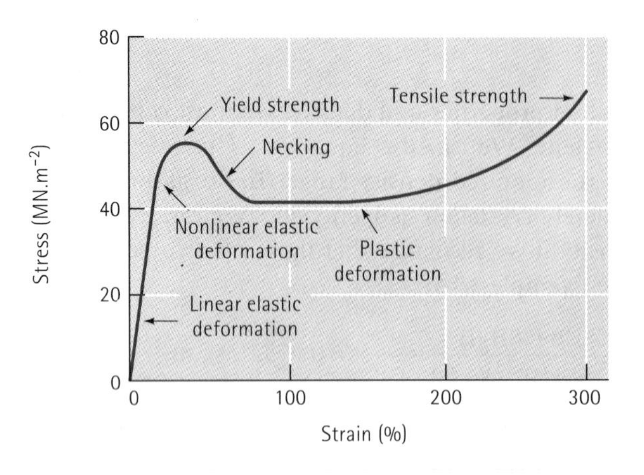

Figure 15.16
The stress-strain curve for 6,6-nylon (polyamide), a typical thermoplastic polymer.

Plastic Behaviour of Amorphous Thermoplastics These polymers deform plastically when the stress exceeds the yield strength. Unlike deformation in the case of metals, however, plastic deformation is not a consequence of dislocation movement. Instead, chains stretch, rotate, slide, and disentangle under load to cause permanent deformation. The drop in the stress beyond the yield point can be explained by this phenomenon. Initially, the chains may be highly tangled and intertwined. When the stress is sufficiently high, the chains begin to untangle and straighten. Necking also occurs, permitting continued sliding of the chains at a lesser stress. Eventually, however, the chains become almost parallel and close together; stronger Van der Waals bonding between the more closely aligned chains requires higher stresses to complete the deformation and fracture process (Figure 15.16).

EXAMPLE 15.9

An amorphous polymer is pulled in a tensile test. After a sufficient stress is applied, necking is observed to begin on the gauge length. However, the neck disappears as the stress continues to increase. Explain this behaviour.

SOLUTION

Normally, when necking begins, the smaller cross-sectional area increases the stress at the neck and necking is accelerated. However, during the tensile test, the chains in the amorphous structure are straightened out and the polymer becomes more crystalline (Figure 15.17). When necking begins, the chains at the neck align and the polymer is locally strengthened sufficiently to resist further deformation at that location. Consequently, the remainder of the polymer, rather than the necked region, continues to deform until the neck disappears.

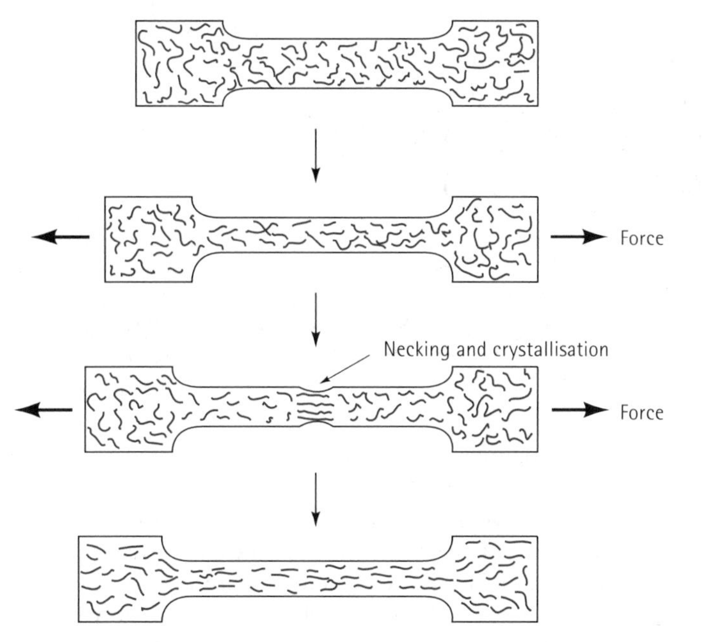

Figure 15.17
Necks are not stable in amorphous polymers, because local alignment strengthens the necked region and reduces its rate of deformation.

Viscoelasticity The ability of a stress to cause chain slippage and plastic deformation is related to time and strain rate. If stress is applied slowly (a low strain rate), the chains slide easily past one another; if it is applied rapidly, sliding does not occur and the polymer behaves in a brittle manner.

The time dependency of elastic and plastic deformation of thermoplastics is explained by the viscoelastic behaviour of the materials. At low temperatures or high rates of loading, the polymer behaves like other solid materials, such as metals or ceramics. In the elastic region, the stress and strain are directly related. However, at high temperatures or low rates, the material behaves as a viscous liquid. This viscoelastic behaviour helps to explain how the polymer deforms under load and also permits us to shape the polymer into useful products.

The viscosity of the polymer describes the ease with which the chains move and cause deformation. The viscosity η, as described in Chapter 14 and again in Figure 15.18, is

$$\eta = \frac{\tau}{dv / dz} \tag{15.5}$$

where τ is the shear stress causing adjacent chains to slide and dv/dz is the velocity gradient, which is related to how rapidly the chains are displaced relative to one another. The temperature effect on the viscosity, as described in ceramic glasses, is

$$\eta_0 \exp\left(\frac{Q_\eta}{RT}\right) \tag{15.6}$$

where η_0 is a constant and Q_η is the activation energy, which is related to the ease with which the chains slide past one another. As the temperature increases, the polymer is less viscous and deforms more easily.

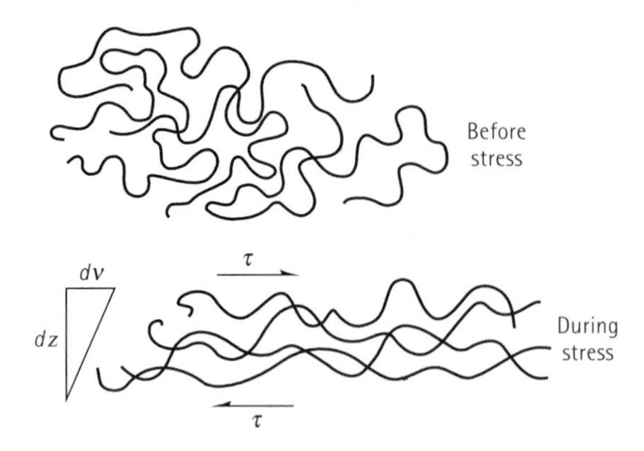

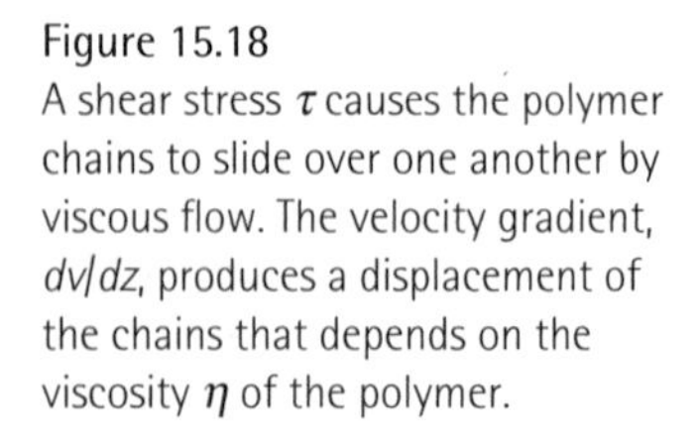

Figure 15.18
A shear stress τ causes the polymer chains to slide over one another by viscous flow. The velocity gradient, dv/dz, produces a displacement of the chains that depends on the viscosity η of the polymer.

Creep In amorphous polymers, the activation energy and viscosity are low, and the polymer deforms at low stresses. When a constant stress is applied to the polymer, the polymer quickly undergoes a strain as chain segments deform. Unlike metals or ceramics, the strain does not reach a constant value (Figure 15.19). Instead, due to the low viscosity, the strain continues to increase with time as the chains slowly slide relative to one another. This condition describes *creep* of the polymer, and it occurs in some polymers even at room temperature. The rate of creep increases with higher stresses and temperature (decreasing viscosity).

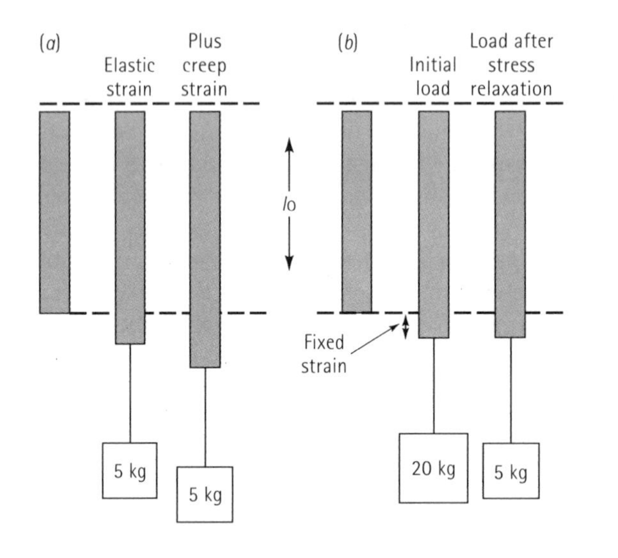

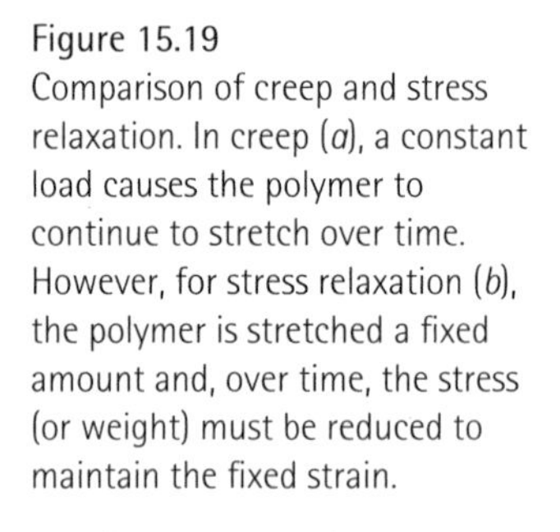

Figure 15.19
Comparison of creep and stress relaxation. In creep (*a*), a constant load causes the polymer to continue to stretch over time. However, for stress relaxation (*b*), the polymer is stretched a fixed amount and, over time, the stress (or weight) must be reduced to maintain the fixed strain.

Several techniques can be used to design a component from creep data. Stress-rupture curves similar to those described in Chapter 6 may be obtained for polymers (Figure 15.20). For known applied stress and operating temperature, the length of time required before the part fails can be determined.

Another method of representing creep data is to measure the strain as a function of time and applied stress (Figure 15.21). The effect of time on the creep curves can often be represented by

$$\varepsilon(t) = at^n \qquad (15.7)$$

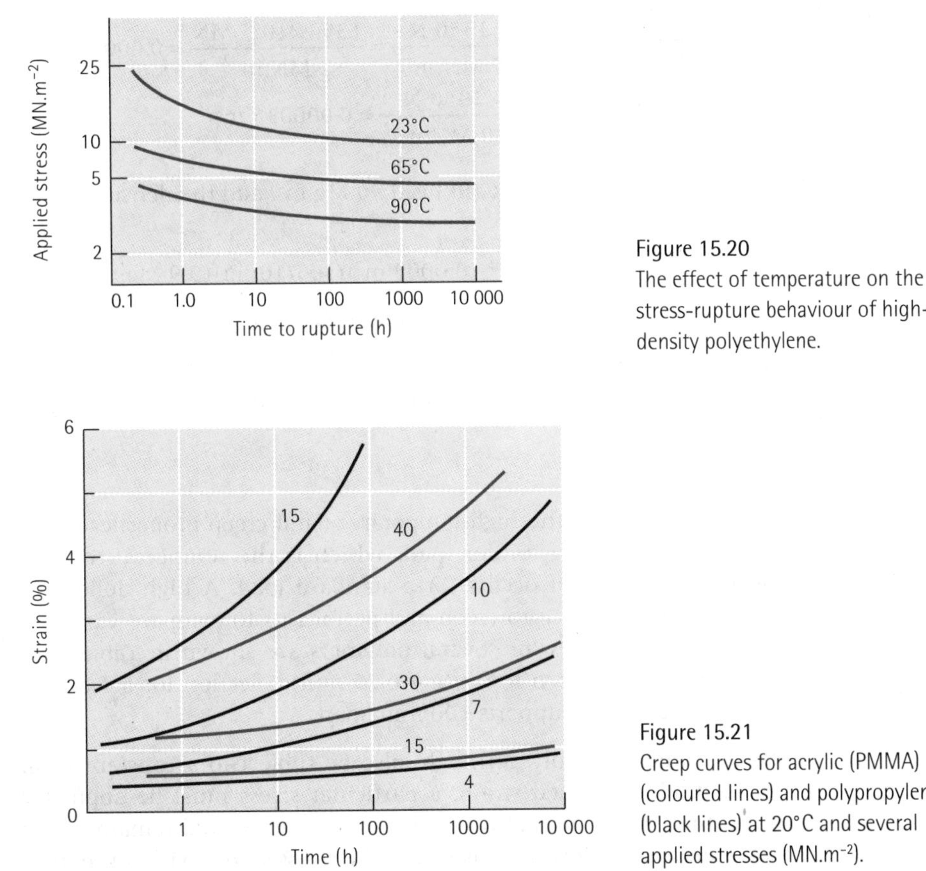

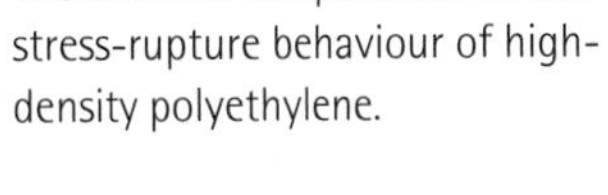

Figure 15.20
The effect of temperature on the stress-rupture behaviour of high-density polyethylene.

Figure 15.21
Creep curves for acrylic (PMMA) (coloured lines) and polypropylene (black lines) at 20°C and several applied stresses ($MN.m^{-2}$).

where $\varepsilon(t)$ is the time-dependent strain while a and n are constants for a given stress and temperature. By predetermining the maximum allowable strain that may occur during the polymer's expected lifetime, we can calculate the maximum applied stress and design the component.

EXAMPLE 15.10 Specification of a Creep-Resistant Polymer Part

A 305 mm-long polymer part is to withstand a tensile force of 1 350 N at 20°C and must be able to survive for 1 000 hours with no more than 2% change in length. Specify the polymer part.

SOLUTION

We can use creep curves in Figure 15.21 to help us design the part. Let's compare polypropylene (which costs about £730.Mg^{-1}) to the acrylic (which costs about £1 900.Mg^{-1}). At 1 000 h, we find that the maximum allowable stress for the polypropylene is about 7 $MN.m^{-2}$ and that for the acrylic is about 30 $MN.m^{-2}$. To withstand a tensile force of 1 350 N, the cross-sectional area of the parts must be:

$$\text{Polypropylene: } A = \frac{F}{\sigma} = \frac{1350 \text{ N}}{7 \text{ MN.m}^{-2}} = \frac{1350 \times 10^{-6} \text{ MN}}{7 \text{ MN.m}^{-2}} = 0.0002 \text{ m}^2$$

$$\text{Acrylic : } A = \frac{F}{\sigma} = \frac{1350 \text{ N}}{30 \text{ MN.m}^{-2}} = 0.000045 \text{ m}^2$$

The density of polypropylene (PP) is 0.90 $Mg.m^{-3}$ and the density of the acrylic is 1.22 $Mg.m^{-3}$.

For PP, cost = (£7.30.Mg^{-1})(0.0002 m^2)($305/10^3$ m)(0.9 $Mg.m^{-3}$) = £0.040

For acrylic, cost = (£1900.Mg^{-1})(0.000045 m^2)($305/10^3$ m)(1.22 $Mg.m^{-3}$) = £0.032

Because the acrylic is more creep-resistant, the part may be smaller and less expensive, even though acrylic is three times more costly than polyethylene.

A third measure of the high temperature and creep properties of a polymer is the deflection temperature under load, which is the temperature at which a given deformation of a beam occurs for a standard load. A high deflection temperature indicates good resistance to creep and permits us to compare various polymers. The deflection temperatures for several polymers are shown in Table 15.4, which gives the temperature required to cause a 0.25 mm deflection for a 1.82 $MN.m^{-2}$ stress at the centre of a bar on supports 100 mm apart.

Stress Relaxation A polymer might also be subject to a constant strain. Initially, in order to produce a fixed strain, a particular stress must be applied. In a metal or ceramic, the stress required to maintain this strain would remain constant. However, the polymer chains flow viscously, and the stress stored within the material decreases (Figure 15.19). This is **stress relaxation** and, like creep, it is a consequence of the viscoelastic behaviour of the polymer. Perhaps the most familiar example of this behaviour is a rubber band (an elastomer) stretched around a pile of books. Initially, the tension in the rubber band is high, and the rubber band is taut. After several weeks, the strain in the rubber band is unchanged (it still completely encircles the books), but the stress will have decreased – that is, the band is no longer taut.

Table 15.4 Deflection temperatures for selected polymers for a 1.82 $MN.m^{-2}$ stress.

Polymer	Deflection Temperature (°C)
Polyester	40
Polyethylene (ultra-high density)	40
Polypropylene	60
Phenolic	80
Polyamide (6,6-nylon)	90
Polystyrene	100
Polyoxymethylene (acetal)	130
Polyamide-imide	280
Epoxy	290

The rate at which stress relaxation occurs is related to the relaxation time λ, which is a property of the polymer. The stress after time t is given by

$$\sigma = \sigma_0 \exp(-t/\lambda) \tag{15.8}$$

where σ_0 is the original stress. The relaxation time, in turn, depends on the viscosity and, thus, the temperature;

$$\lambda = \lambda_0 \exp(Q_n/RT) \tag{15.9}$$

where λ_0 is a constant. Relaxation of the stress occurs more rapidly at higher temperatures and for polymers with a low viscosity.

EXAMPLE 15.11 Design of Initial Stress in a Polymer

A band of polyisoprene is to hold together a bundle of steel rods for up to one year. If the stress on the band is less than 10 MN.m^{-2} the rods will not be held tightly by the band. Design the initial stress that must be applied to a polyisoprene band when it is slipped over the steel. A series of tests showed that an initial stress of 7 MN.m^{-2} decreased to 6.85 MN.m^{-2} after six weeks.

SOLUTION

Although the strain of the elastomer band may be constant, the stress will decrease over time due to stress relaxation. We can use Equation 15.8 and our initial tests to determine the relaxation time for the polymer:

$$\sigma = \sigma_0 \exp\left(-\frac{t}{\lambda}\right)$$

$$6.85 = 7.0 \exp\left(-\frac{6}{\lambda}\right)$$

$$-\frac{6}{\lambda} = \ln\left(\frac{6.87}{7.0}\right) = \ln(0.98) = -0.0202$$

$$\lambda = \frac{6}{0.0202} = 297 \text{ weeks}$$

Now that we know the relaxation time, we can determine the stress that must be initially placed onto the band in order that it still be stressed to 10 MN.m^{-2} after 1 year (52 weeks).

$$10 = \sigma_0 \exp(-52/297) = \sigma_0 \exp(-0.175) = 0.839\sigma_0$$

$$\sigma_0 = \frac{10}{0.839} = 11.9 \text{ MN.m}^{-2}$$

The polyisoprene band must be made significantly undersized so it can slip over the materials it is holding together with a tension of 11.9 MN.m^{-2}. After one year, the stress will still be 10 MN.m^{-2}.

Impact Viscoelastic behaviour also helps us understand the impact properties of polymers. At very high rates of strain, as in an impact test, there is insufficient time for the chains to slide and cause plastic deformation. For these conditions, the thermoplastics behave in a brittle manner and have poor impact values. As Figure 6.13 demonstrated, polymers may have a transition temperature. At low temperatures, brittle behaviour is observed in an impact test, whereas more ductile behaviour is observed at high temperatures, where the chains move more easily.

Deformation of Crystalline Polymers A number of polymers are used in the crystalline state. As we discussed earlier, however, the polymers are never completely crystalline. Instead, small regions – between crystalline lamellae and between crystalline spherulites – are amorphous transition regions. Polymer chains in the crystalline region extend into these amorphous regions as tie chains.

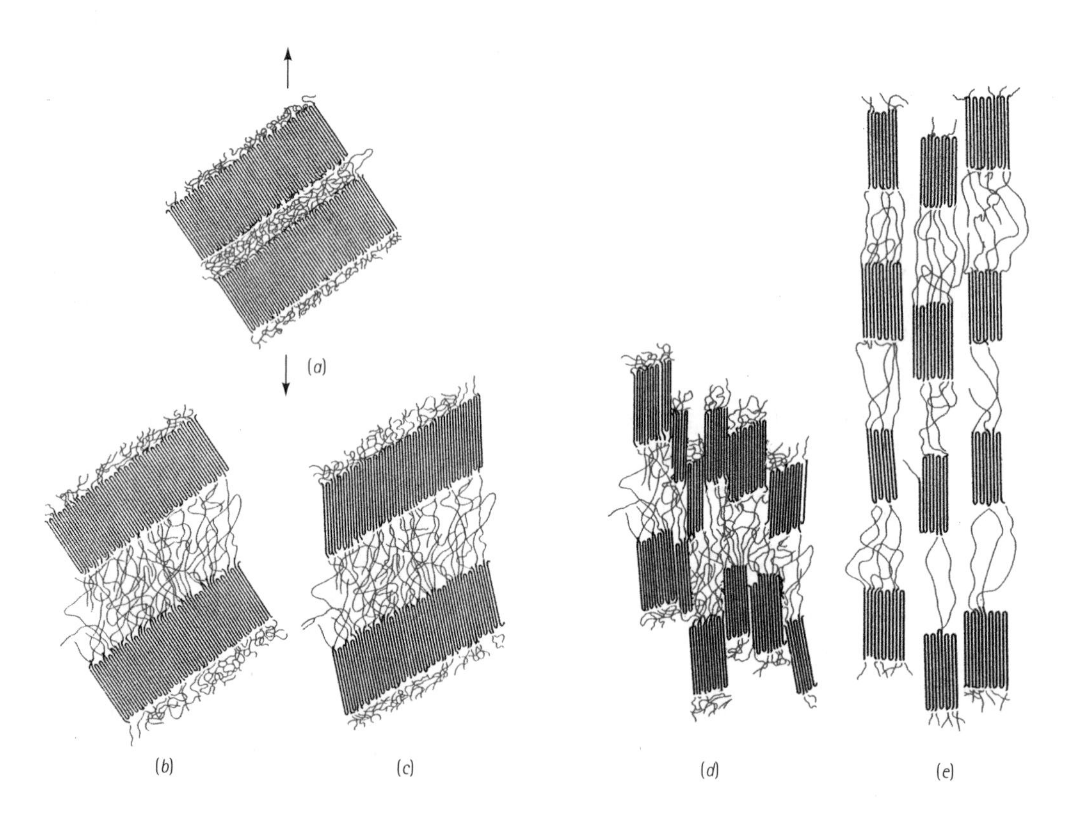

Figure 15.22
Deformation of a semicrystalline polymer, whose original structure (*a*) includes crystalline lamellae separated by amorphous tie chains. (*b*) When a stress is applied, the lamellae slide and the tie chains extend. (*c*) Further deformation tilts the folded chains in the lamellae and further extends the tie chains. (*d*) Lamellae break into smaller blocks. (*e*) Smaller crystalline blocks become aligned, developing a preferred orientation.

When a tensile load is applied to the polymer, the crystalline lamellae within the spherulites slide past one another and begin to separate as the tie chains are stretched (Figure 15.22). The folds in the lamellae tilt and become aligned with the direction of the tensile load. The crystalline lamellae break into smaller units and slide past one another, until eventually the polymer is composed of small aligned crystals joined by tie chains and oriented parallel to the tensile load. The spherulites also change shape and become elongated in the direction of the applied stress. With continued stress, the tie chains disentangle or break, causing the polymer to fail.

Crazing Crazing occurs in thermoplastics when localised regions of plastic deformation occur in a direction perpendicular to that of the applied stress. In transparent thermoplastics, such as some of the glassy polymers, the craze produces a translucent or opaque region that looks like a crack. The craze can grow until it extends across the entire cross-section of the polymer part. But the craze is not a crack, and, in fact, it can continue to support an applied stress.

Figure 15.23 describes how a craze might form in semicrystalline polyethylene. The process is similar to that for the plastic deformation of the polymer, but the process can proceed even at a low stress over an extended period of time. Due to the stress, the tie chains between the crystalline lamellae stretch and disentangle.

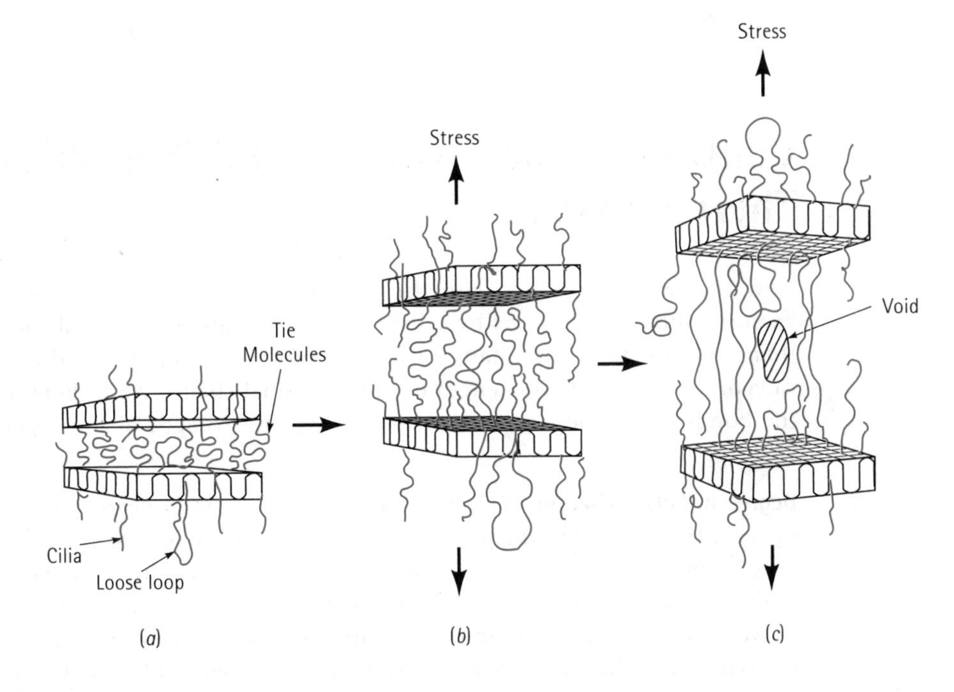

Figure 15.23
(*a*) Crazing in a semicrystalline polymer. When stress is applied, the tie chains between adjacent lamellae stretch (*b*) and eventually begin to disentangle (*c*). Voids may be created along the most highly deformed plane, leading to crazing.

Additives in the polymer or environmental factors that reduce the glass transition temperature of the amorphous regions encourage disentanglement. As disentanglement occurs, the blocks of crystalline polymer separate and voids form between the blocks or between aligned fibres. A similar behaviour occurs in glassy polymers, except no crystalline blocks are present to become aligned.

Crazing can lead to brittle fracture of the polymer (Figure 15.24). As voids grow, they become separated by only a thin fibril of highly strained polymer; continued growth and elongation of the voids caused by the applied stress eventually extends and reduces the diameter of the fibrils to the point where they fail, creating a true crack. The thickening of the craze increases the stress at its tip, nucleating new voids and extending both the craze and the crack. Finally, the crack reaches a size that allows it to rapidly propagate and cause fracture.

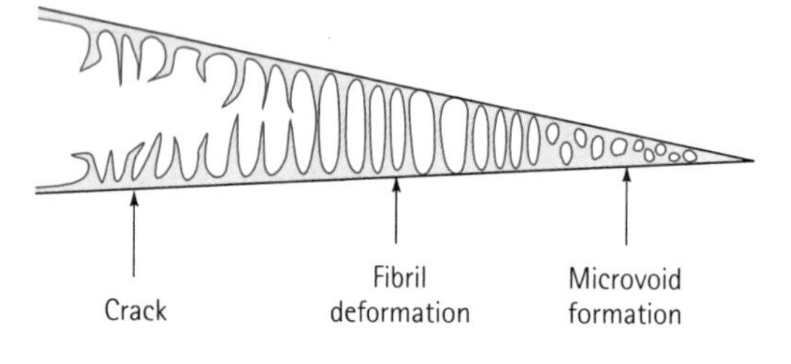

Figure 15.24
Cracks may form from a craze. As the voids elongate, the polymer fibrils stretch, thin, and eventually fail as the cracks widens. A higher stress develops at the tip of the craze, forming more microvoids and continuing the growth of the craze.

15.8 Controlling the Structure and Properties of Thermoplastics

Now that we are familiar with the effects of deformation and temperature, let's examine some of the ways by which we can modify and control the properties of thermoplastic polymers. Many of these methods can be grouped into three main categories: controlling the length of the individual chains, controlling the strength of the bonds *within* the chains, and controlling the strength of the bonds *between* the chains.

Degree of Polymerisation Longer chains – that is, a larger degree of polymerisation – increase the strength of the polymer – at least, up to a point. As the chains increase in length, they become more tangled and the polymer has a higher melting temperature and improved strength, including resistance to creep. The ethylene monomer illustrates this effect. Commercial polyethylene typically has a degree of polymerisation less than about 7 000 (or a molecular weight of less than 200 000 $g.mol^{-1}$). High-performance, high-density polyethylene has a degree of polymerisation of up to 18 000. Ultrahigh-molecular-weight polyethylene may have a degree of polymerisation of 150 000, providing impact properties that exceed those of all other polymers, plus good strength and ductility.

Effect of Monomer on Bonding Between Chains In this section, we consider only homopolymers. These polymers contain identical repeat units. In homopolymers, the type of monomer influences the bonding between chains and the ability of the chains to rotate or slide past one another when a stress is applied.

Let's examine monomers containing just two carbon atoms in the backbone:

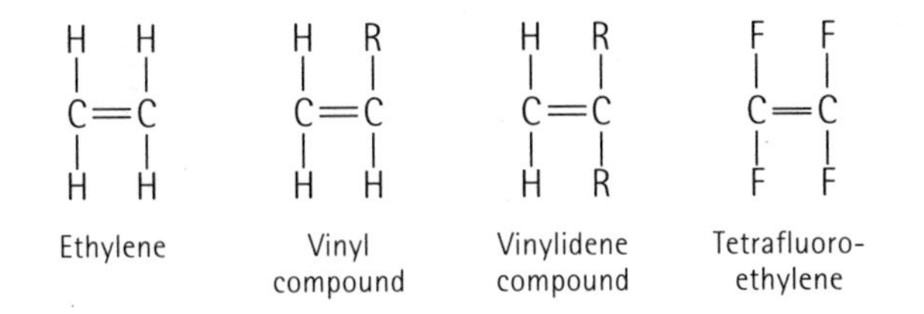

In the above, R can be one or more types of atoms or groups of atoms. Table 15.5 shows the repeat units and typical applications for a number of these polymers. In polyethylene, the linear chains easily rotate and slide when stress is applied, and no strong polar bonds are formed between the chains; thus, polyethylene has a low strength.

Vinyl compounds have one of the hydrogen atoms replaced with a different atom or atom group. When R is chlorine, we produce polyvinyl chloride (PVC); when R is CH_3, we produce polypropylene (PP); addition of a benzene ring gives polystyrene (PS); and a CN group produces polyacrylonitrile (PAN). Generally, a head-to-tail arrangement of the repeat units in the polymers is obtained (Figure 15.25). When two of the hydrogen atoms are replaced, the monomer is a *vinylidene compound*, important examples of which include polyvinylidene chloride (the basis for Saran Wrap) and polymethyl methacrylate (acrylics such as Perspex and Plexiglas).

The effects of adding other atoms or atom groups to the carbon backbone in place of hydrogen atoms are illustrated by the typical properties given in Table 15.6. Larger atoms such as chloride or groups of atoms such as methyl (CH_3) and benzene groups make it more difficult for the chains to rotate, uncoil, disentangle, and deform by viscous flow when a stress is applied or when the temperature is increased. This condition leads to higher strength, stiffness, and melting temperature than in polyethylene. In addition, some of the more polar atoms or atom groups provide stronger Van der Waals bonds between chains.

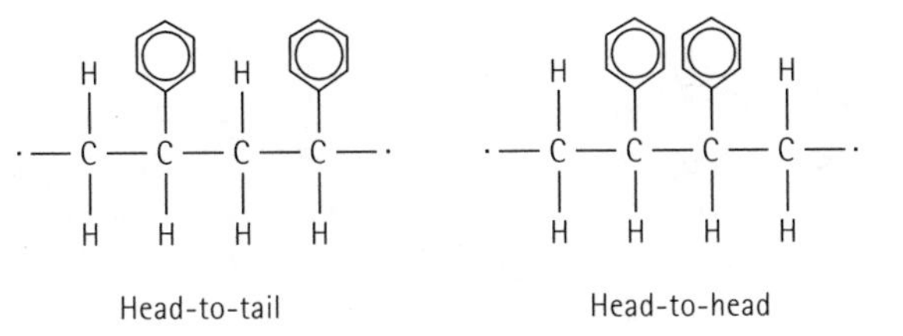

Figure 15.25
Head-to-tail versus head-to-head arrangement of repeat units. The head-to-tail arrangement is most typical.

The chlorine atom in PVC and the carbon-nitrogen group in PAN are strongly attracted by hydrogen bonding to hydrogen atoms on adjacent chains.

In polytetrafluoroethylene (PTFE or Teflon), all four hydrogen atoms are replaced by fluorine. The monomer again is symmetrical, and the strength of the polymer is not much greater than that of polyethylene. However, the C – F bond permits PTFE to have a high melting point, with the added benefit of low-friction, nonstick characteristics that make the polymer useful for bearings and cookware.

Table 15.5 Repeat units and applications for selected addition thermoplastics.

Polymer	Repeat Unit	Applications
Polyethylene (PE)	$\cdots-CH_2-CH_2-\cdots$	Packing films, wire insulation, squeeze bottles, tubing, household items
Polyvinyl chloride (PVC)	$\cdots-CH_2-CHCl-\cdots$	Pipe, valves, fittings, floor tile, wire insulation, vinyl automobile roofs
Polypropylene (PP)	$\cdots-CH_2-CH(CH_3)-\cdots$	Tanks, carpet fibres, rope, packaging
Polystyrene (PS)	$\cdots-CH_2-CH(C_6H_5)-\cdots$	Packaging and insulation foams, lighting panels, appliance components, egg boxes
Polyacrylonitrile (PAN)	$\cdots-CH_2-CH(C\equiv N)-\cdots$	Textile fibres, precursor for carbon fibres, food containers
Polymethyl methacrylate (PMMA) (acrylic-Plexiglas)	$\cdots-CH_2-C(CH_3)(C{=}O-O-CH_3)-\cdots$	Windows, windscreens, coatings, hard contact lenses, internally lighted signs
Polychlorotrifluoro-ethylene	$\cdots-CF_2-CFCl-\cdots$	Valve components, gaskets, tubing, electrical insulation
Polytetrafluoro-ethylene (PTFE) (Teflon)	$\cdots-CF_2-CF_2-\cdots$	Seals, valves, nonstick coatings

Table 15.6 Properties of selected thermoplastics.

	Tensile Strength ($MN.m^{-2}$)	% Elongation	Elastic Modulus ($MN.m^{-2}$)	Density ($Mg.m^{-3}$)	Izod Impact ($J.m^{-1}$)
Polyethylene (PE):					
Low-density	20	800	270	0.92	480
High-density	37	130	1200	0.96	214
Ultrahigh molecular weight	47	350	660	0.934	1600
Polyvinyl chloride (PVC)	60	100	4000	1.40	
Polypropylene (PP)	40	700	1460	0.90	53
Polystyrene (PS)	53	60	3000	1.06	21
Polyacrylonitrile (PAN)	60	4	3860	1.15	256
Polymethyl methacrylate (PMMA) (acrylic, Perspex)	80	5	3000	1.22	27
Polychlorotrifluoroethylene	40	250	2000	2.15	139
Polytetrafluoroethylene (PTFE, Teflon)	47	400	540	2.17	160
Polyoxymethylene (POM) (acetal)	80	75	3460	1.42	123
Polyamide (PA) (nylon)	80	300	3330	1.14	112
Polyester (PET)	70	300	4000	1.36	32
Polycarbonate (PC)	73	130	2660	1.20	83
Polyimide (PI)	113	10	2000	1.39	80
Polyetheretherketone (PEEK)	68	150	3660	1.31	85
Polyphenylene sulphide (PPS)	63	2	3200	1.30	27
Polyether sulphone (PES)	81	80	2330	1.37	83
Polyamide-imide (PAI)	180	15	4860	1.39	214

As illustrated in Figure 15.26, simple polymers such as polyethylene, polyvinyl chloride, polystyrene and polypropylene make up the greatest percentage of polymers by use, and they typically are available at a low cost (Table 15.7).

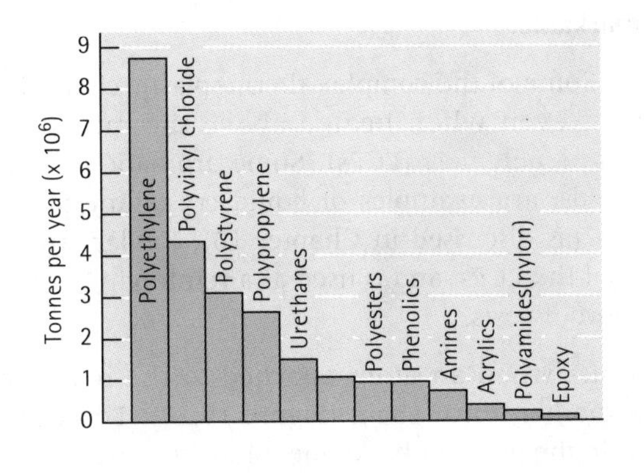

Figure 15.26
Typical annual consumption of various polymers in the United States. All others total less than about 3 million tonnes.

Table 15.7 Approximate cost of polymers in bulk form (1988).

Polymer	$/tonne^{-1}	£/tonne^{-1}
Polyethylene (low-density)	880	590
(ultra-high molecular weight)	2430	1620
Polystyrene	660	440
Polyvinyl chloride	2750	1840
Polymethyl methacrylate	2870	1910
Polypropylene	1100	730
Acetal	4080	2720
Polyester (PET)	3750	2500
Nylon-6,6 (polyamide)	4850	3230
Polycarbonate	4630	3090
PEEK	47400	31600
Polyetherimide	10580	7050
ABS	2650	1760
Liquid crystalline polymer	35270	23510
Polybutylene	3200	2130
Thermoplastic elastomer	3970	2650
Amine	2200	1470
Phenolic	1320	880

Effect of Monomers on Bonding within Chains A large number of polymers, which typically are used for special applications and in relatively small quantities, are formed from complex monomers, often by the condensation mechanism. Oxygen, nitrogen, sulphur, and benzene rings (or aromatic groups) may be incorporated into the chain. Table 15.8 shows the repeat units and typical applications for a number of these complex polymers. Polyoxymethylene, or acetal, is a simple example in which the backbone of the polymer chain contains alternating carbon and oxygen atoms. A number of these polymers, including polyimides and polyetheretherketone (PEEK), are important aerospace materials.

Because bonding within the chains is stronger, rotation and sliding of the chains is more difficult, leading to higher strengths, higher stiffnesses, and higher melting points than the simpler addition polymers (Table 15.2). In some cases, good impact properties (Table 15.6) can be gained from these complex chains, with polycarbonates being particularly remarkable.

Liquid Crystalline Polymers Some of the complex thermoplastic chains become so stiff that they act as rigid rods, even when heated above the melting point. These materials are liquid crystalline polymers (LCPs). Some aromatic polyesters and aromatic polyamides (or aramids) are examples of liquid crystalline polymers and are used as high-strength fibres (as discussed in Chapter 16). Kevlar, an aromatic polyamide, is the most familiar of the LCPs, and is used as a reinforcing fibre for aerospace applications and for bullet-proof vests.

Branching Branching occurs when an atom attached to the main linear chain is removed and replaced by another linear chain (Figure 15.27). This can occur several times per 100 carbon atoms in the polymer backbone. Branching prevents dense packing

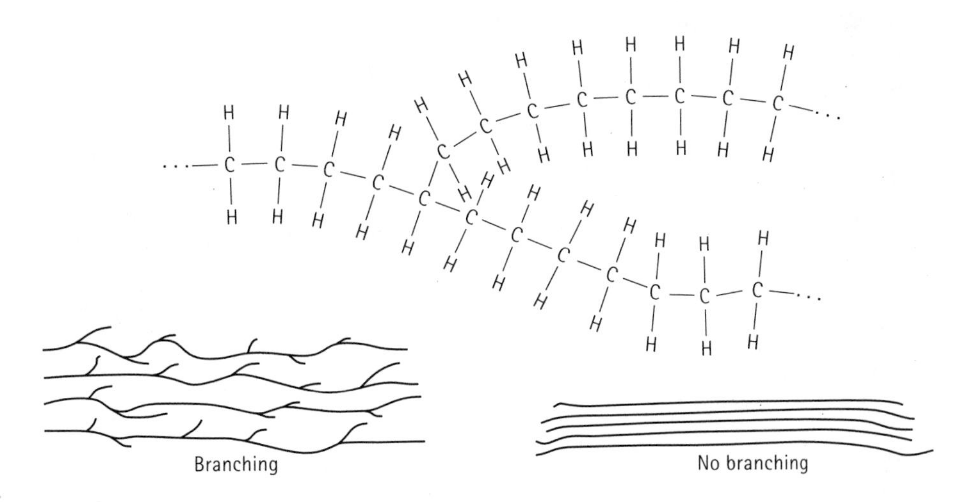

Figure 15.27
Branching can occur in linear polymers. Branching makes crystallisation more difficult.

and crystallisation of the chains, thereby reducing the density, stiffness, and strength of the polymer. Low-density (LD) polyethylene, which has many branches, is weaker than high-density (HD) polyethylene, which has virtually no branching (Table 15.6).

Copolymers Copolymers are linear addition chains composed of two or more types of molecules. ABS, composed of acrylonitrile, butadiene (a synthetic elastomer), and styrene, is one of the most common polymer materials (Figure 15.28). Styrene and acrylonitrile form a linear copolymer (SAN) that serves as a matrix. Styrene and butadiene also form a linear copolymer, BS rubber, that acts as the filler material. The combination of the two copolymers gives ABS an excellent combination of strength, rigidity, and toughness.

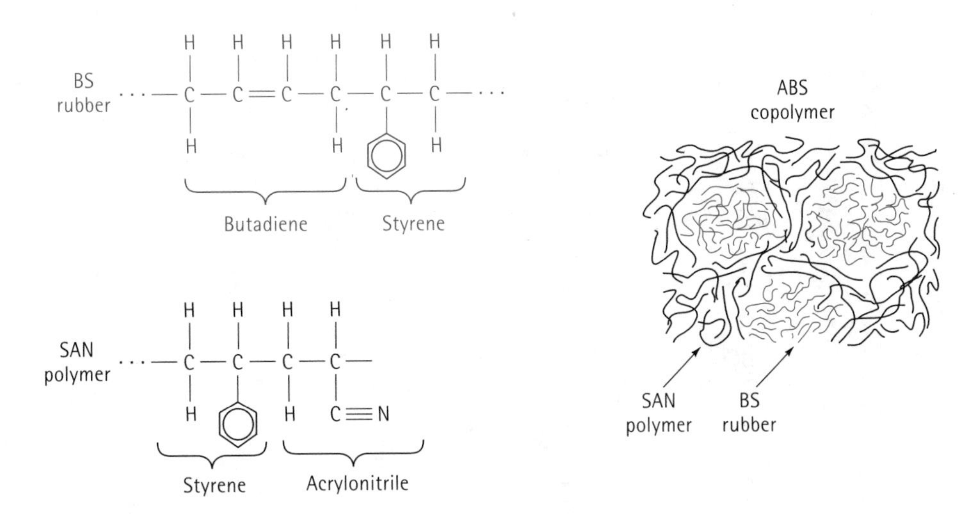

Figure 15.28
Copolymerisation produces the polymer ABS, which is really made up of two copolymers, SAN and BS, grafted together.

Table 15.8 Repeat units and applications for complex thermoplastics.

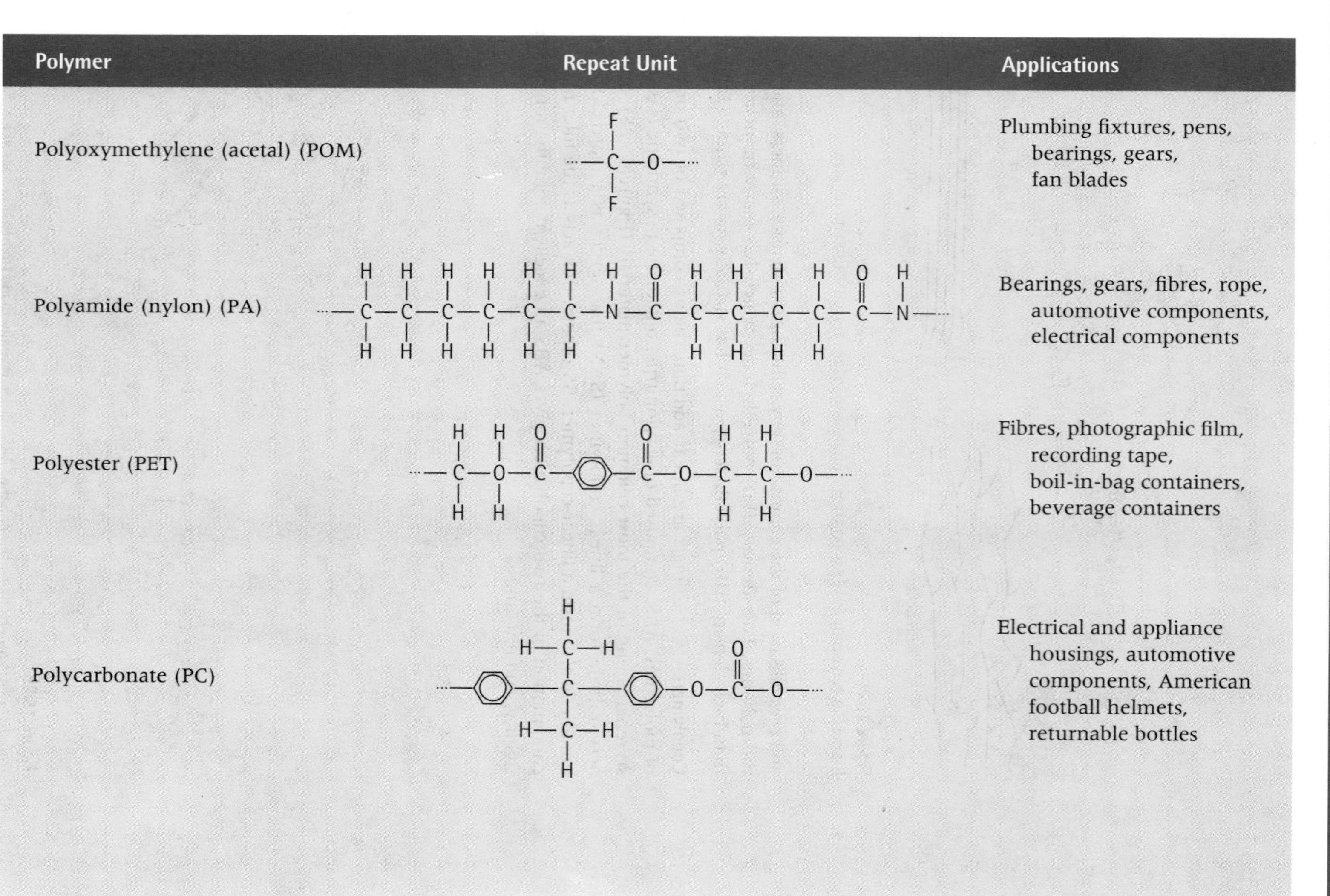

Polymer	Repeat Unit	Applications
Polyoxymethylene (acetal) (POM)	$\cdots-CF_2-O-\cdots$	Plumbing fixtures, pens, bearings, gears, fan blades
Polyamide (nylon) (PA)	$\cdots-CH_2-CH_2-CH_2-CH_2-CH_2-CH_2-NH-C(=O)-CH_2-CH_2-CH_2-CH_2-C(=O)-NH-\cdots$	Bearings, gears, fibres, rope, automotive components, electrical components
Polyester (PET)	$\cdots-CH_2-OH_2-C(=O)-C_6H_4-C(=O)-O-CH_2-CH_2-O-\cdots$	Fibres, photographic film, recording tape, boil-in-bag containers, beverage containers
Polycarbonate (PC)	$\cdots-C_6H_4-C(CH_3)_2-C_6H_4-O-C(=O)-O-\cdots$	Electrical and appliance housings, automotive components, American football helmets, returnable bottles

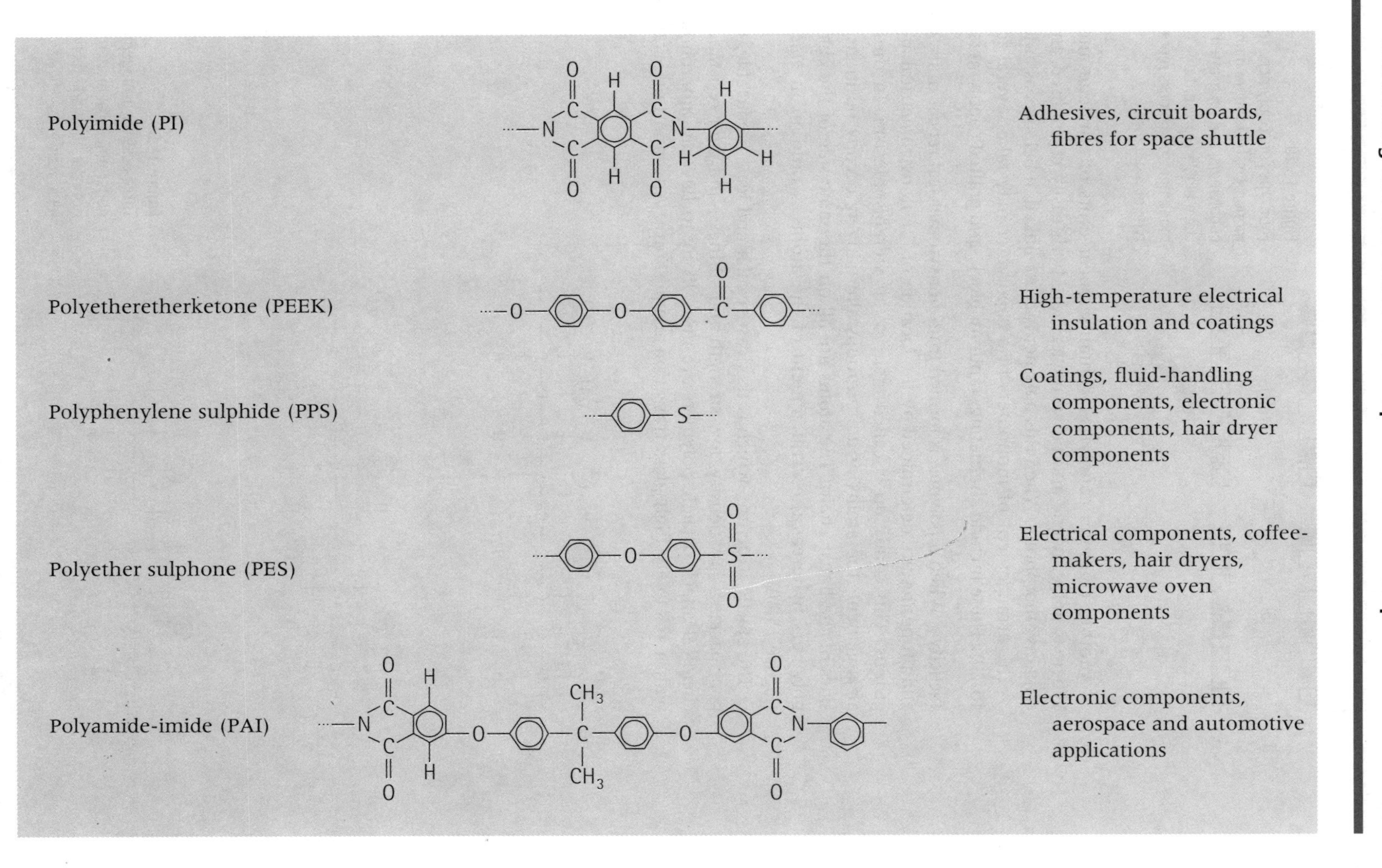

Polymer	Applications
Polyimide (PI)	Adhesives, circuit boards, fibres for space shuttle
Polyetheretherketone (PEEK)	High-temperature electrical insulation and coatings
Polyphenylene sulphide (PPS)	Coatings, fluid-handling components, electronic components, hair dryer components
Polyether sulphone (PES)	Electrical components, coffee-makers, hair dryers, microwave oven components
Polyamide-imide (PAI)	Electronic components, aerospace and automotive applications

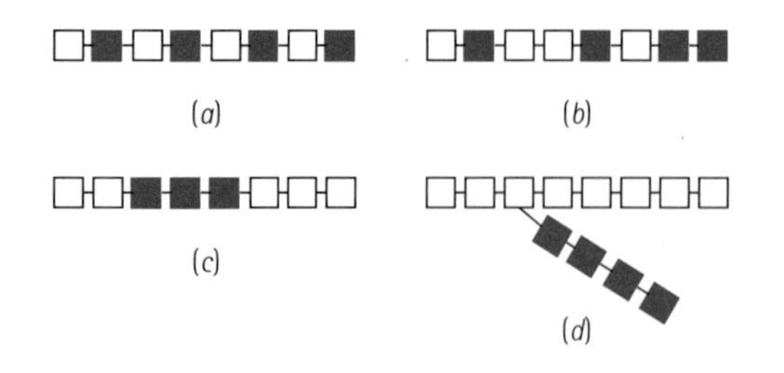

Figure 15.29
Four types of copolymers: (*a*) alternating monomers, (*b*) random monomers, (*c*) block copolymers, and (*d*) grafted copolymers. Open blocks represent one type of monomer; solid blocks represent a second type of monomer.

Another common copolymer contains repeat units of ethylene and propylene. Whereas polyethylene and polypropylene are both easily crystallised, the copolymer remains amorphous. When this polymer is cross-linked, it behaves as an elastomer.

The arrangement of monomers in a copolymer may take several forms (Figure 15.29). These include alternating, random, block, and grafted copolymers.

Tacticity When a polymer is formed from nonsymmetrical repeat units, the structure and properties are determined by the location of the nonsymmetrical atoms or atom groups. This condition is called tacticity, or stereoisomerism. In the syndiotactic arrangement, the atoms or atom groups alternatively occupy positions on opposite sides of the linear chain. The atoms are all on the same side of the chain in *isotactic* polymers, whereas the arrangement of the atoms is random in *atactic* polymers (Figure 15.30).

The atactic structure, which is the least regular and least predictable, tends to give poor packing, low density, low strength and stiffness, and poor resistance to heat or chemical attack. Atactic polymers are more likely to have an amorphous structure with a relatively high glass transition temperature. An important example of the

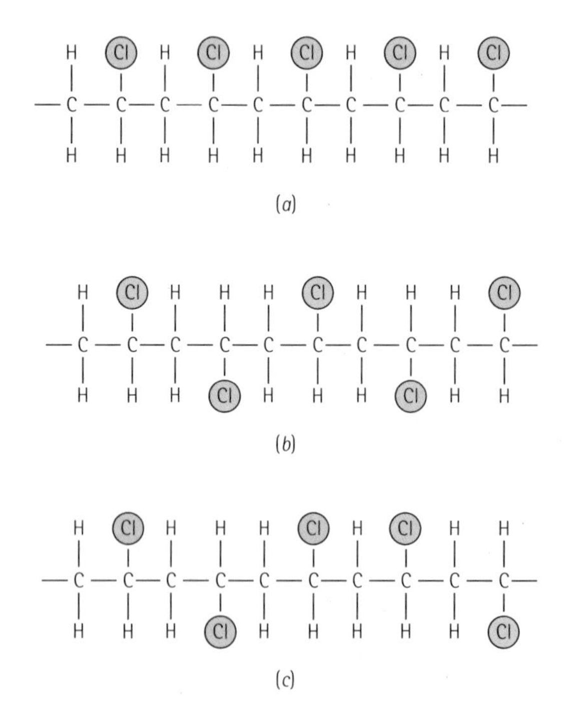

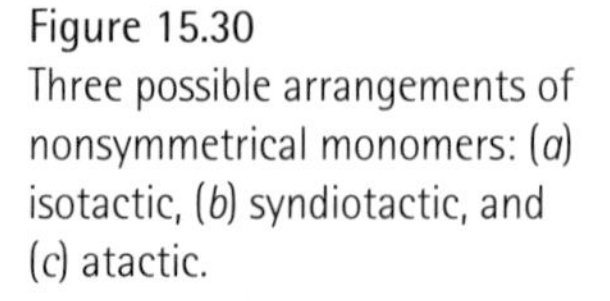

Figure 15.30
Three possible arrangements of nonsymmetrical monomers: (*a*) isotactic, (*b*) syndiotactic, and (*c*) atactic.

importance of tacticity occurs in polypropylene. Atactic polypropylene is an amorphous waxlike polymer with poor mechanical properties, whereas isotactic polypropylene may crystallise and is one of the most widely used commercial polymers.

Crystallisation and Deformation As we have discussed previously, encouraging crystallisation of the polymer also helps to increase density, resistance to chemical attack, and mechanical properties – even at higher temperatures – because of the stronger bonding between the chains. In addition, deformation straightens and aligns the chains, producing a preferred orientation. Deformation of a polymer is often used in producing fibres having mechanical properties in the direction of the fibre that exceed those of many metals and ceramics.

Blending and Alloying We can improve the mechanical properties of many of the thermoplastics by blending or alloying. By mixing an immiscible elastomer with the thermoplastic, we produce a two-phase polymer, as we found in ABS. The elastomer does not enter the structure as a copolymer but, instead, helps to absorb energy and improve toughness. Polycarbonates used to produce transparent aircraft canopies are toughened by elastomers in this manner.

EXAMPLE 15.12

Compare the mechanical properties of LD polyethylene, HD polyethylene, polyvinyl chloride, polypropylene, and polystyrene and explain their differences in terms of their structures.

SOLUTION

Let's look at the maximum tensile strength and modulus of elasticity for each polymer.

Polymer	Tensile Strength ($MN.m^{-2}$)	Modulus of Elasticity ($MN.m^{-2}$)	Structure
LD polyethylene	20	270	Highly branched, amorphous structure with symmetrical monomers
HD polyethylene	37	1 200	Amorphous structure with symmetrical monomers but little branching
Polypropylene	40	1 460	Amorphous structure with small methyl side groups
Polystyrene	53	3 000	Amorphous structure with benzene side groups
Polyvinyl chloride	60	4 000	Amorphous structure with large chlorine atoms as side groups

We can conclude that:

1. Branching, which reduces the density and close packing of chains, reduces the mechanical properties of polyethylene.
2. Adding atoms or atom groups other than hydrogen to the chain increases strength and stiffness. The methyl group in polypropylene provides some improvement, the benzene ring of styrene provides higher properties, and the chloride atom in polyvinyl chloride provides a large increase in properties.

15.9 Elastomers (Rubbers)

A number of natural and synthetic linear polymers called *elastomers* display a large amount of elastic deformation when a force is applied. Rubber bands, automobile tyres, O-rings, hoses, and insulation for electrical wires are common uses for these materials.

Geometric Isomers Some monomers that have different structures, even though they have the same composition, are called geometric isomers. Isoprene, or natural rubber, is an important example (Figure 15.31). The monomer includes two double bonds between carbon atoms; this type of monomer is called a diene. Polymerisation occurs by breaking both double bonds, creating a new double bond at the centre of the molecule and active sites at both ends.

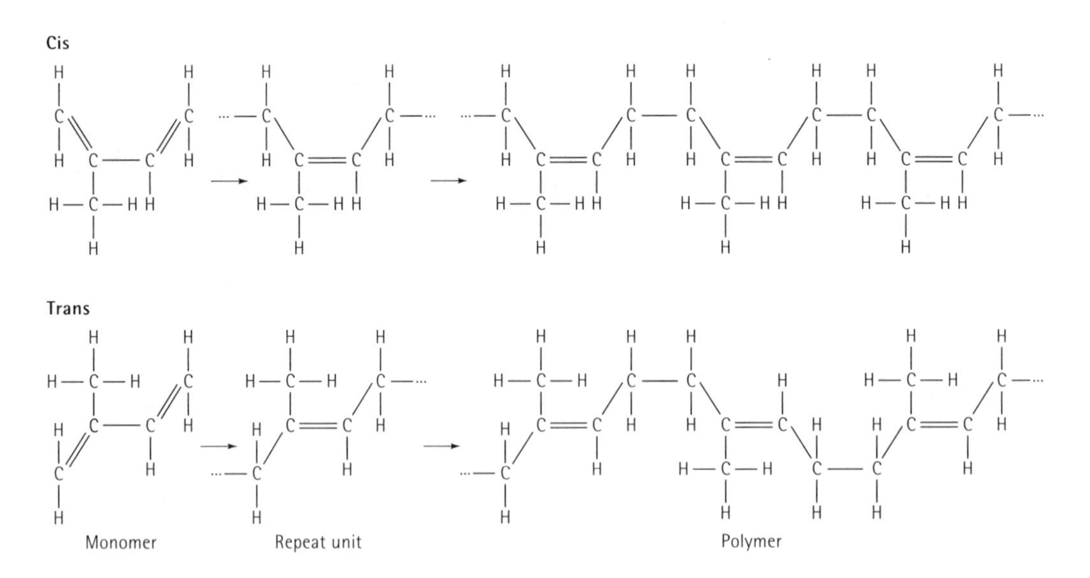

Figure 15.31 The *cis* and *trans* structures of isoprene.

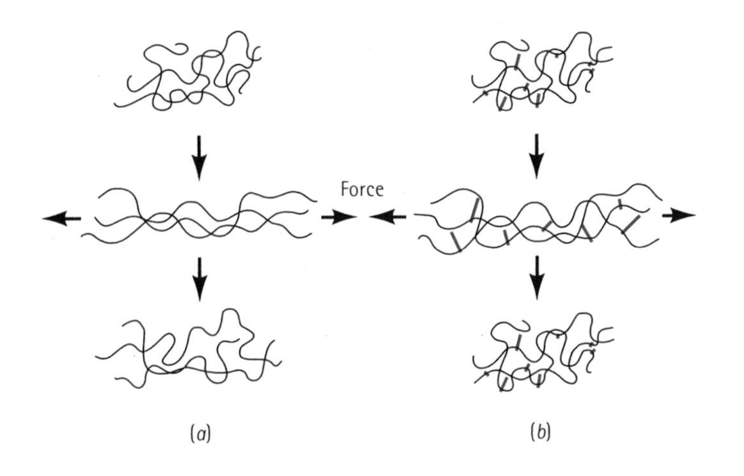

Figure 15.32
(a) When the elastomer contains no cross-links, the application of a force causes both elastic and plastic deformation; after the load is removed, the elastomer is permanently deformed. (b) When cross-linking occurs, the elastomer still may undergo large elastic deformation; however, when the load is removed, the elastomer returns to its original shape.

In the *trans* form of isoprene, the hydrogen atom and the methyl group at the centre of the repeat unit are located on opposite sides of the newly formed double bond. This arrangement leads to relatively straight chains; the polymer crystallises and forms a hard rigid polymer called *gutta percha*.

In the *cis* form, however, the hydrogen atom and the methyl group are located on the same side of the double bond. This different geometry causes the polymer chains to develop a highly coiled structure, preventing close packing and leading to an amorphous, rubbery polymer. If a stress is applied to the cisisoprene, the polymer behaves in a viscoelastic manner. The chains uncoil and bonds stretch, producing elastic deformation, but the chains also slide past one another, producing non-recoverable plastic deformation. The polymer behaves as a thermoplastic rather than an elastomer (Figure 15.32).

Cross-Linking We prevent viscous plastic deformation while retaining large elastic deformation by cross-linking the chains. Vulcanisation, which uses sulphur atoms, is a common method for cross-linking. Figure 15.33 describes how polymer chains can be linked by strands of sulphur atoms as the polymer is processed and shaped at temperatures of about 120 to 180°C. The cross-linking steps may include rearranging a hydrogen atom and replacing one or more of the double bonds with single bonds. The cross-linking process is not reversible; consequently, the elastomer cannot easily be recycled.

The stress-strain curve for an elastomer is shown in Figure 15.34. Virtually all of the curve represents elastic deformation; thus, elastomers display a nonlinear elastic behaviour. Initially, the modulus of elasticity decreases because of the uncoiling of the chains. However, after the chains have been extended, further elastic deformation occurs by stretching of the bonds, leading to a higher modulus of elasticity.

The elasticity of the rubber is determined by the number of cross-links, or the amount of sulphur added to the material. Low sulphur additions leave the rubber soft

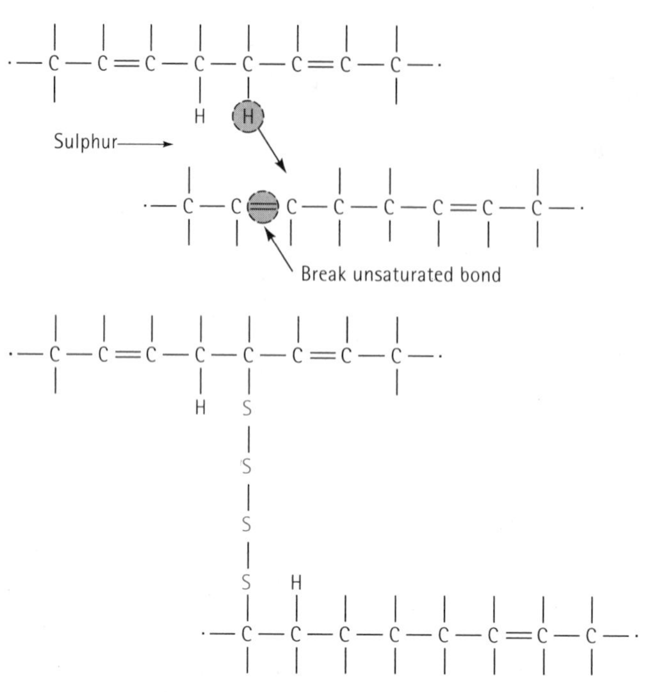

Figure 15.33
Cross-linking of polyisoprene chains may occur by introducing strands of sulphur atoms. Sites for attachment of the sulphur strands occur by rearrangement or loss of a hydrogen atom and breaking of an unsaturated bond.

and flexible, as in elastic bands or rubber gloves. Increasing the sulphur content restricts the uncoiling of the chains and the rubber becomes harder, more rigid, and brittle, as in rubber used for motor mounts. Typically 0.5 to 5% sulphur is added to provide cross-linking in elastomers.

Typical Elastomers Elastomers, which are amorphous polymers, do not easily crystallise during processing. They have a low glass transition temperature, and chains can easily be deformed elastically when a force is applied. The typical elastomers (Tables 15.9 and 15.10) meet these requirements.

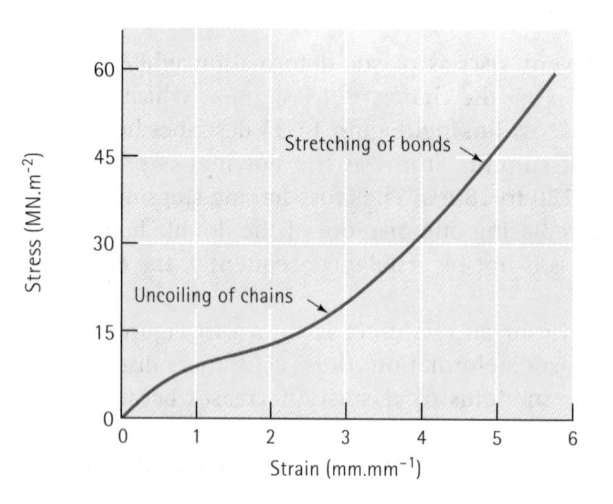

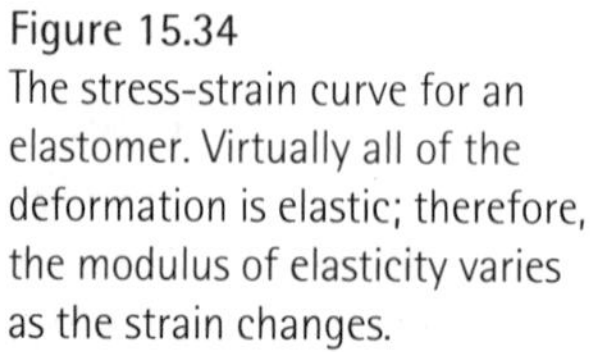
Figure 15.34
The stress-strain curve for an elastomer. Virtually all of the deformation is elastic; therefore, the modulus of elasticity varies as the strain changes.

Table 15.9 Repeat units and applications for selected elastomers.

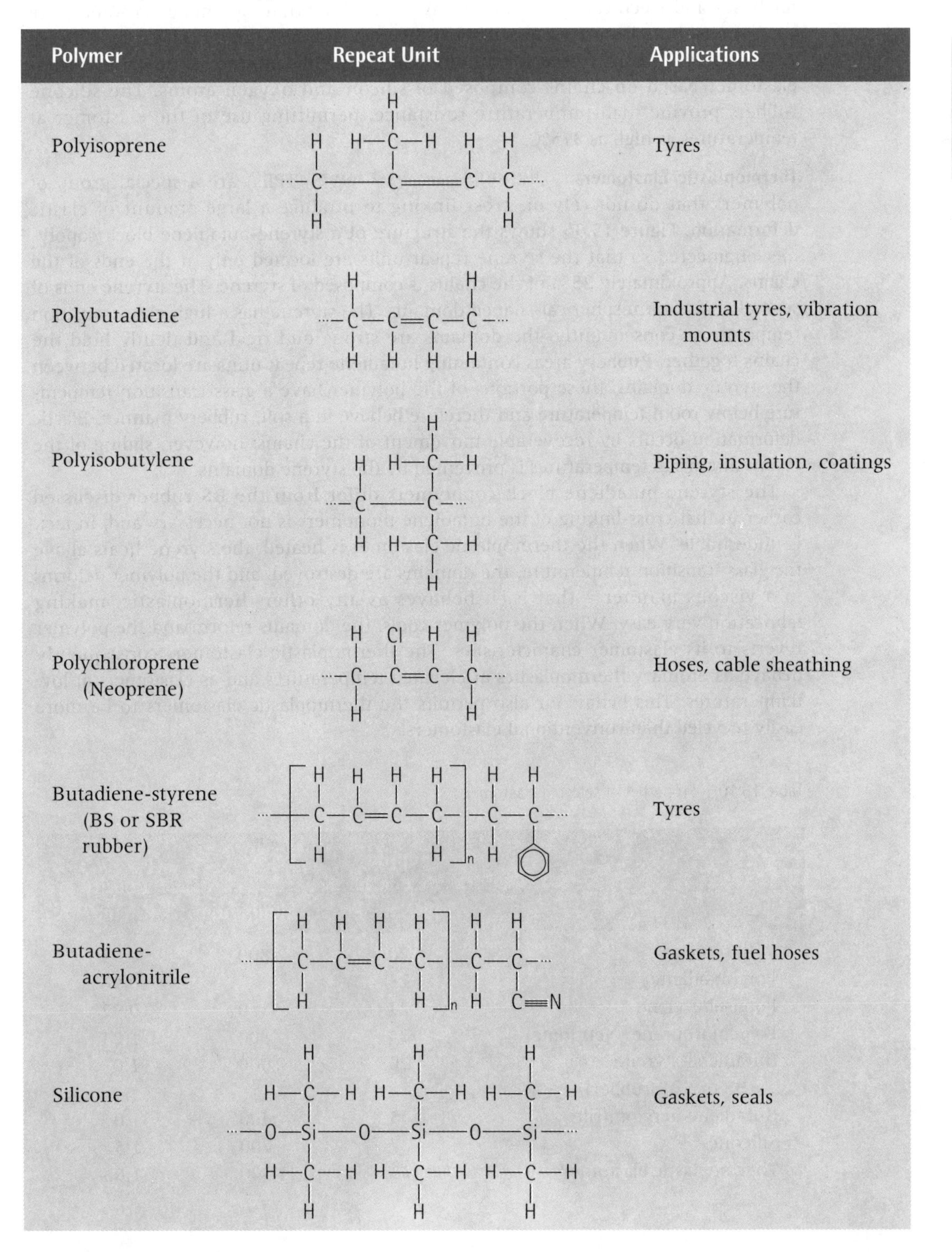

Polymer	Repeat Unit	Applications
Polyisoprene	$-CH_2-C(CH_3)=CH-CH_2-$	Tyres
Polybutadiene	$-CH_2-CH=CH-CH_2-$	Industrial tyres, vibration mounts
Polyisobutylene	$-CH_2-C(CH_3)_2-$	Piping, insulation, coatings
Polychloroprene (Neoprene)	$-CH_2-CCl=CH-CH_2-$	Hoses, cable sheathing
Butadiene-styrene (BS or SBR rubber)	$-[CH_2-CH=CH-CH_2]_n-CH_2-CH(C_6H_5)-$	Tyres
Butadiene-acrylonitrile	$-[CH_2-CH=CH-CH_2]_n-CH_2-CH(C\equiv N)-$	Gaskets, fuel hoses
Silicone	$-O-Si(CH_3)_2-O-Si(CH_3)_2-O-Si(CH_3)_2-$	Gaskets, seals

Polyisoprene is natural rubber. Polychloroprene, or Neoprene, is a common material for hoses and electrical insulation. Many of the important synthetic elastomers are copolymers. Butadiene-styrene rubber (BSR), which is also one of the components of ABS (Figure 15.28), is used for automobile tyres. Silicones are another important elastomer based on chains composed of silicon and oxygen atoms. The silicone rubbers provide high-temperature resistance, permitting use of the elastomer at temperatures as high as 315°C.

Thermoplastic Elastomers Thermoplastic elastomers (TPEs) are a special group of polymers that do not rely on cross-linking to produce a large amount of elastic deformation. Figure 15.35 shows the structure of a styrene-butadiene block copolymer engineered so that the styrene repeat units are located only at the ends of the chains. Approximately 25% of the chains is composed of styrene. The styrene ends of several chains form spherical-shaped domains. The styrene has a high glass transition temperature; consequently, the domains are strong and rigid and tightly hold the chains together. Rubbery areas containing butadiene repeat units are located between the styrene domains; these portions of the polymer have a glass transition temperature below room temperature and therefore behave in a soft, rubbery manner. Elastic deformation occurs by recoverable movement of the chains; however, sliding of the chains at normal temperatures is prevented by the styrene domains.

The styrene-butadiene block copolymers differ from the BS rubber discussed earlier in that cross-linking of the butadiene monomers is not necessary and, in fact, is undesirable. When the thermoplastic elastomer is heated, the styrene heats above the glass transition temperature, the domains are destroyed, and the polymer deforms in a viscous manner – that is, it behaves as any other thermoplastic, making fabrication very easy. When the polymer cools, the domains reform and the polymer reverts to its elastomer characteristics. The thermoplastic elastomers consequently behave as ordinary thermoplastics at elevated temperatures and as elastomers at low temperatures. This behaviour also permits the thermoplastic elastomers to be more easily recycled than conventional elastomers.

Table 15.10 Properties of selected elastomers.

	Tensile Strength ($MN.m^{-2}$)	% Elongation	Density ($Mg.m^{-3}$)
Polyisoprene	20	800	0.93
Polybutadiene	23		0.94
Polyisobutylene	27	350	0.92
Polychloroprene (Neoprene)	23	800	1.24
Butadiene-styrene (BS or SBR rubber)	20	2 000	1.0
Butadiene-acrylonitrile	5	400	1.0
Silicone	7	700	1.5
Thermoplastic elastomer	33	1 300	1.06

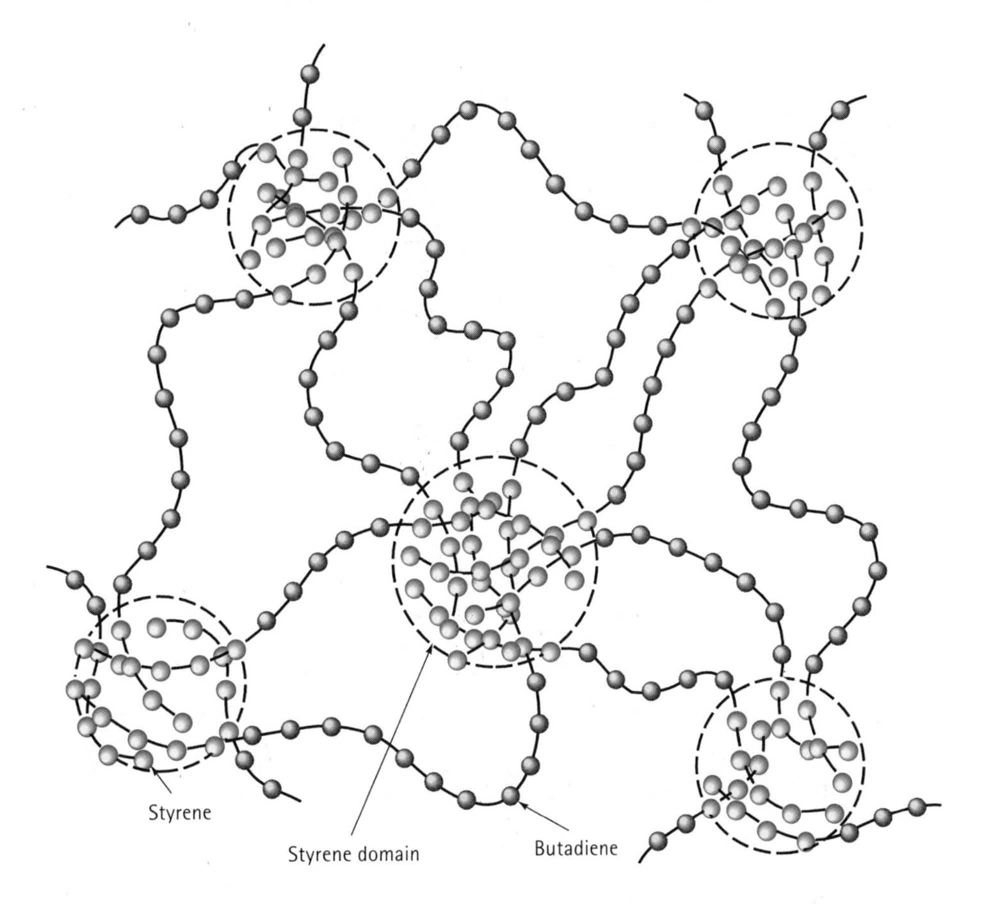

Figure 15.35
The structure of the SB copolymer in a thermoplastic elastomer. The glassy nature of the styrene domains provides elastic behaviour without cross-linking of the butadiene.

15.10 Thermosetting Polymers

Thermosets are highly cross-linked polymer chains that form a three-dimensional network structure. Because the chains cannot rotate or slide, these polymers possess good strength, stiffness, and hardness. However, thermosets also have poor ductility and impact properties and a high glass transition temperature. In a tensile test, thermosetting polymers display the same behaviour as a brittle metal or ceramic.

Thermosetting polymers often begin as linear chains. Depending on the type of repeat units and the degree of polymerisation, the initial polymer may be either a solid or a liquid resin; in some cases, a two- or three-part liquid resin is used (as in the case of two tubes of epoxy glue that we often use). Heat, pressure, mixing of the various resins, or other methods initiate the cross-linking process. Cross-linking is not reversible; once formed, the thermoset cannot be reused or conveniently recycled.

The functional groups for a number of common thermosetting polymers are summarised in Table 15.11, and representative properties are given in Table 15.12.

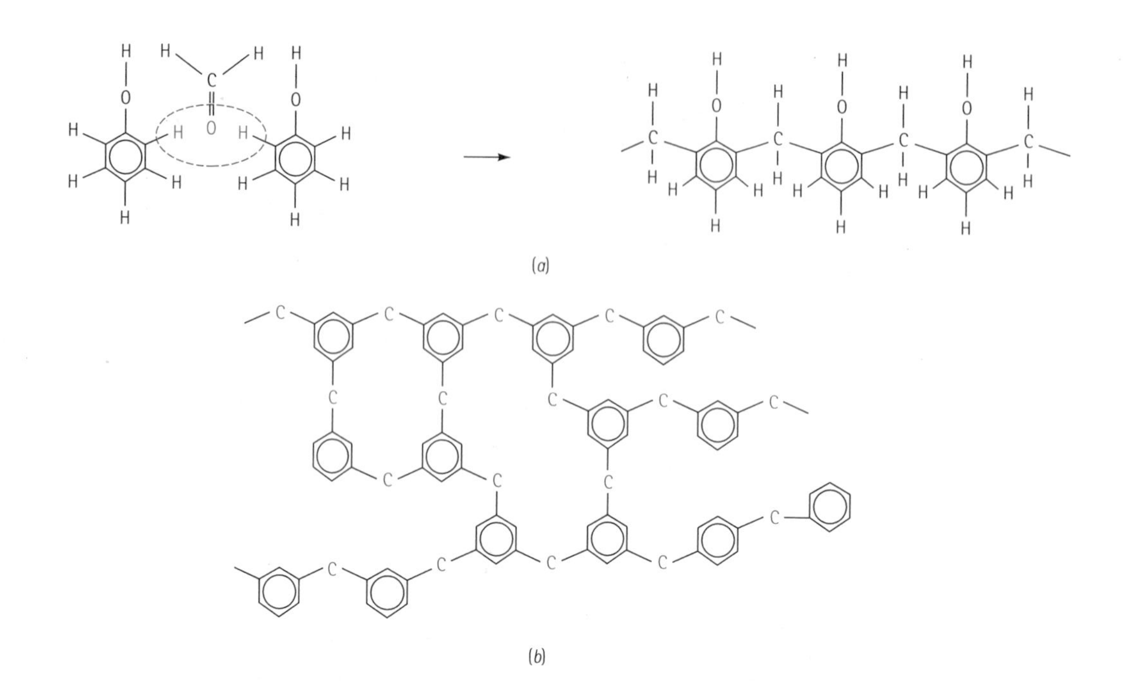

Figure 15.36
Structure of a phenolic. In (*a*), two phenol rings are joined by a condensation reaction through a formaldehyde molecule. Eventually, a linear chain forms. In (*b*), excess formaldehyde serves as the cross-linking agent, producing a network, thermosetting polymer.

Phenolics Phenolics, the most commonly used thermosets, are often used as adhesives, coatings, laminates, and moulded components for electrical or motor applications. Bakelite is one of the common phenolic thermosets.

A condensation reaction joining phenol and formaldehyde molecules produces the initial linear phenolic resin (Figure 15.36). The oxygen atom in the formaldehyde molecule reacts with a hydrogen atom on each of two phenol molecules, and water is evolved as the by-product. The two phenol molecules are then joined by the carbon atom remaining in the formaldehyde.

This process continues until a linear phenol-formaldehyde chain is formed. However, the phenol is trifunctional; after the chain has formed, there is a third location on each phenol ring that provides a site for cross-linking with adjacent chains.

Amines Amino resins, produced by combining urea or melamine monomers with formaldehyde, are similar to the phenolics. The monomers are joined by a formaldehyde link to produce linear chains. Excess formaldehyde provides the cross-linking needed to give strong, rigid polymers suitable for adhesives, laminates, moulding materials for cookware, and electrical hardware such as circuit breakers, switches, outlets, and wall plates.

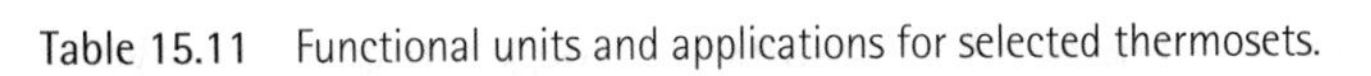

Table 15.11 Functional units and applications for selected thermosets.

Polymer	Functional Units	Typical Applications
Phenolics	C_6H_5OH (phenol: benzene ring with –O–H and five H)	Adhesives, coatings, laminates
Amines	$H_2N{-}C(=O){-}NH_2$ Urea	Adhesives, cookware, electrical mouldings
Polyesters	$\cdots{-}O{-}C(=O){-}CH{-}CH{-}C(=O){-}O{-}\cdots$	Electrical mouldings, decorative laminates, matrix in fibreglass
Epoxies	$H_2C(-O-)C{-}R{-}C(-O-)CH_2$	Adhesives, electrical mouldings, matrix for composites
Urethanes	$\cdots{-}O{-}C(=O){-}NH{-}R{-}NH{-}C(=O){-}O{-}\cdots$	Fibres, coatings, foams, insulation
Silicone	$\cdots{-}Si(O)(CH_3){-}O{-}Si(CH_3){-}O{-}\cdots$	Adhesives, gaskets, sealants

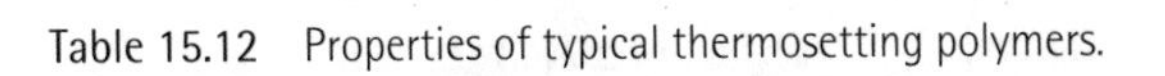

Table 15.12 Properties of typical thermosetting polymers.

	Tensile Strength ($MN.m^{-2}$)	% Elongation	Elastic Modulus ($MN.m^{-2}$)	Density ($Mg.m^{-3}$)
Phenolics	60	2	9	1.27
Amines	67	1	11	1.50
Polyesters	87	3	4	1.28
Epoxies	100	6	3	1.25
Urethanes	67	6		1.30
Silicone	27	0	8	1.55

Urethanes Depending on the degree of cross-linking, the urethanes behave as thermosetting polymers, thermoplastic polymers, or elastomers. These polymers find application as fibres, coatings, and foams for furniture, mattresses, and insulation.

Polyesters Polyesters form chains from acid and alcohol molecules by a condensation reaction, giving water as a by-product. When these chains contain unsaturated bonds, a styrene molecule may provide cross-linking. Polyesters are used as moulding or casting materials for a variety of electrical applications, decorative laminates, boats and other marine equipment, and as a matrix for composites such as fibreglass.

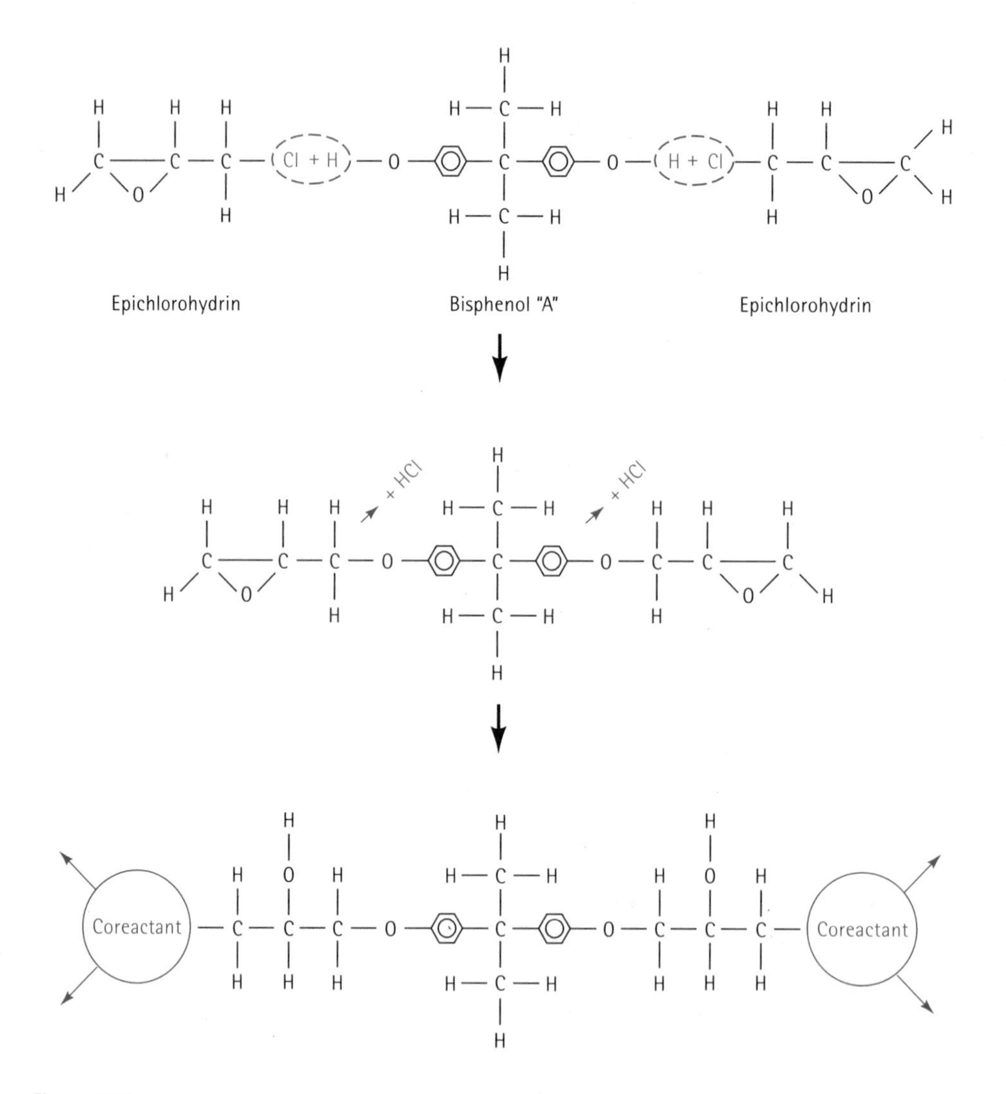

Figure 15.37
One type of epoxy is prepared by combining bisphenol A with epichlorohydrin (resulting in the formation of HCl as a by-product) to produce the epoxy resin. In the presence of a trifunctional coreactant, the rings are opened and the polymer is extended in two directions.

Epoxies Epoxies are thermosetting polymers formed from molecules containing a tight C–O–C ring. During polymerisation, the C–O–C rings are opened and the bonds are rearranged to join the molecules (Figure 15.37). The most common of the commercial epoxies is based on bisphenol A, to which have been added two epoxide units. These molecules are polymerised to produce chains and then coreacted with curing agents that provide cross-linking.

Epoxies are used as adhesives, as rigid moulded parts for electrical applications, automotive components, circuit boards, sporting goods and as a matrix for high-performance fibre-reinforced composite materials for aerospace.

Polyimides Polyimides display a ring structure that contains a nitrogen atom. One special group, the bismaleimides (BMI), are important in the aircraft and aerospace industry. They can operate continuously at temperatures of 175°C and do not decompose until reaching 460°C.

Interpenetrating Polymer Networks Some special polymer materials can be produced when linear thermoplastic chains are intertwined through a thermosetting framework, forming interpenetrating polymer networks. For example, nylon, acetal, and polypropylene chains can penetrate into a cross-linked silicone thermoset. In more advanced systems, two interpenetrating thermosetting framework structures can be produced.

15.11 Adhesives

Adhesives are polymers used to join other polymers, metals, ceramics, composites, or combinations of these materials. The adhesives are used for a variety of applications. The most critical of these are the 'structural adhesives,' which find use in the automotive, aerospace, appliance, electronics, construction, and sporting equipment areas.

Chemically Reactive Adhesives These adhesives include polyurethane, epoxy, silicone, phenolics, anaerobics, and polyimides. One-component systems consist of a single polymer resin cured by exposure to moisture, heat, or – in the case of anaerobics – the absence of oxygen. Two-component systems (such as epoxies) cure when two resins are combined.

Evaporation or Diffusion Adhesives The adhesive is dissolved in either an organic solvent or water and is applied to the surfaces to be joined. When the carrier evaporates, the remaining polymer provides the bond. Water-base adhesives are preferred from the standpoint of environmental and safety considerations. The polymer may be completely dissolved in water or may consist of a latex, or a stable dispersion of polymer in water. A number of elastomers, vinyls, and acrylics are used.

Hot-Melt Adhesives These thermoplastic polymers and thermoplastic elastomers melt when heated. On cooling, the polymer solidifies and joins the materials. Typical melting temperatures of commercial hot-melts are about 80°C to 110°C, which limits the elevated-temperature use of these adhesives. High-performance hot-melts, such as polyamides and polyesters, can be used up to 200°C.

Pressure-Sensitive Adhesives These adhesives are primarily elastomers or elastomer copolymers produced as films or coatings. Pressure is required to cause the polymer to stick to the substrate. They are used to produce electrical and packaging tapes, labels, floor tiles, and wall coverings, and wood-grained textured films.

Conductive Adhesives A polymer adhesive may contain a filler material such as silver, copper, or aluminium flakes or powders to provide electrical and thermal conductivity. In some cases, thermal conductivity is desired but electrical conductivity is not wanted; alumina, beryllia, boron nitride, and silica may be used as fillers to provide this combination of properties.

15.12 Polymer Additives

Most polymers contain additives that impart special characteristics to the materials.

Fillers Fillers are added for a variety of purposes. One of the best known examples is the addition of carbon black to rubber to improve the strength and wear resistance of tyres. Some fillers, such as short fibres or flakes of inorganic materials, improve the mechanical properties of the polymer. Others, called extenders, permit a large volume of a polymer material to be produced with relatively little actual resin, thus reducing cost. Calcium carbonate, silica, talc, and clay are frequently used extenders.

Pigments Used to produce colours in polymers and paints, pigments are finely ground particles, such as TiO_2, that are uniformly dispersed in the polymer.

Stabilisers Stabilisers prevent deterioration of the polymer due to environmental effects. Heat stabilisers are required in processing polyvinyl chloride; otherwise, hydrogen and chloride atoms may be removed as hydrochloric acid, causing the polymer to be embrittled. Stabilisers also prevent deterioration of polymers due to ultraviolet radiation.

Antistatic Agents Most polymers, because they are poor conductors, build up a charge of static electricity. Antistatic agents attract moisture from the air to the polymer surface, improving the surface conductivity of the polymer and reducing the likelihood of a spark or discharge.

Flame Retardants Because they are organic materials, most polymers are flammable. Additives that contain chlorine, bromine, phosphorus, or metallic salts reduce the likelihood that combustion will occur or spread.

Plasticisers Low-molecular-weight molecules or chains called plasticisers reduce the glass transition temperate and provide internal lubrication, thereby improving the forming characteristics of the polymer. Plasticisers are particularly important for polyvinyl chloride, which has a glass transition temperature well above room temperature.

Reinforcements The strength and rigidity of polymers are improved by introducing glass, polymer, or carbon filaments as reinforcements. For example, fibreglass consists of short filaments of glass in a polymer matrix.

15.13 Forming of Polymers

There are a number of methods for producing polymer shapes, including moulding, extrusion, and manufacture of films and fibres. The techniques used to form the polymers depend to a large extent on the nature of the polymer – in particular, whether it is thermoplastic or thermosetting. Typical processes are shown in Figures 15.38 to 15.40.

The greatest variety of techniques are used to form the thermoplastic polymers. The polymer is heated to near or above the melting temperature so that it becomes rubbery or liquid. The polymer is then formed in a mould or die to produce the required shape. Thermoplastic elastomers can be formed in the same manner. In these processes, scrap can be easily recycled and waste is minimised.

Fewer forming techniques are used for the thermosetting polymers because, once cross-linking has occurred, the thermosetting polymers are no longer capable of being formed. After vulcanisation, elastomers also can be formed no further. In these cases, scrap cannot be recycled.

Extrusion A screw mechanism forces heated thermoplastic through a die opening to produce solid shapes, films, sheets, tubes, pipes, and even plastic bags. One special extrusion process for producing films is illustrated in Figure 15.39. Extrusion can be used to coat wires and cables with either thermoplastics or elastomers.

Blow moulding A hollow preform of a thermoplastic called a parison is introduced into a die and, by gas pressure, expanded against the walls of the die. This process is used to produce plastic bottles, containers, automotive fuel tanks, and other hollow shapes.

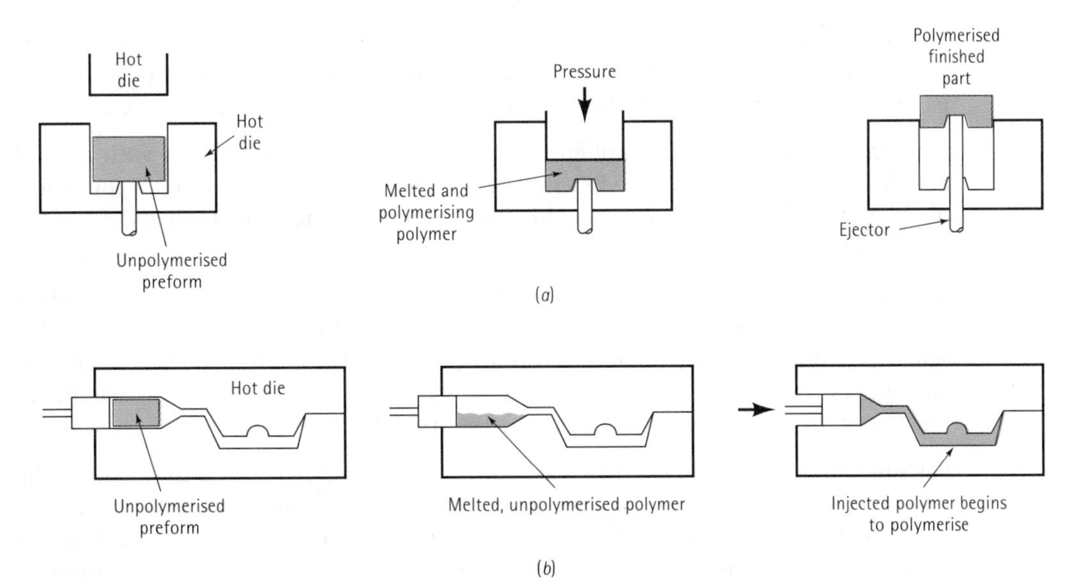

Figure 15.38
Typical forming processes for thermosetting polymers: (*a*) compression moulding and (*b*) transfer moulding.

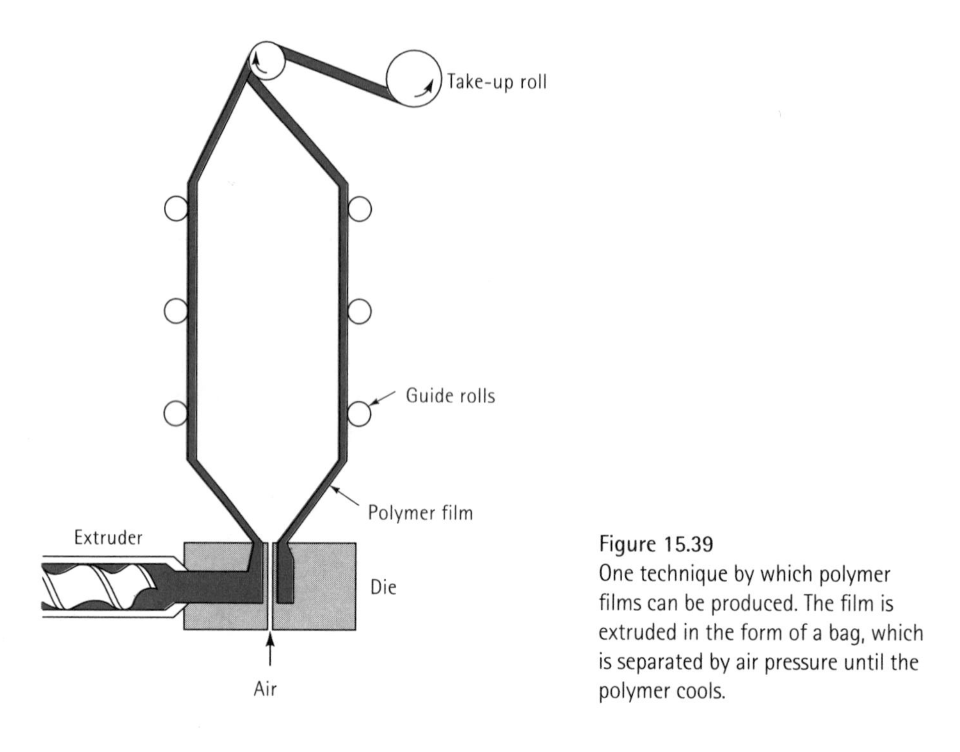

Figure 15.39
One technique by which polymer films can be produced. The film is extruded in the form of a bag, which is separated by air pressure until the polymer cools.

Injection Moulding Thermoplastics heated above the melting temperature are forced into a closed die to produce a moulding. This process is similar to die casting of molten metals. A plunger or special screw mechanism applies pressure to force the hot polymer into the die. A wide variety of products, ranging from cups, combs, and gears to garbage cans, can be produced in this manner.

Thermoforming Thermoplastic polymer sheets heated into the plastic region can be formed over a die to produce such diverse products as egg boxes and decorative panels. The forming can be done using matching dies, a vacuum, or air pressure.

Calendaring In a calendar, molten plastic is poured into a set of rolls with a small opening. The rolls, which may be embossed with a pattern, squeeze out a thin sheet of the polymer – often, polyvinyl chloride. Typical products include vinyl floor tile and shower curtains.

Spinning Filaments, fibres, and yarns may be produced by spinning. The molten thermoplastic polymer is forced through a die containing many tiny holes. The die, called a spinnerette, can rotate and produce a yarn. For some materials, including polyamides, the fibres may subsequently be stretched to align the chains parallel to the axis of the fibre; this process increases the strength of the fibres.

Casting Many polymers can be cast into moulds and permitted to solidify. The moulds may be plate glass for producing individual thick plastic sheets or moving stainless steel belts for continuous casting of thinner sheets. *Rotational moulding* is a special casting process in which molten polymer is poured into a mould rotating about two axes. Centrifugal action forces the polymer against the walls of the mould, producing a thin shape such as a camper top.

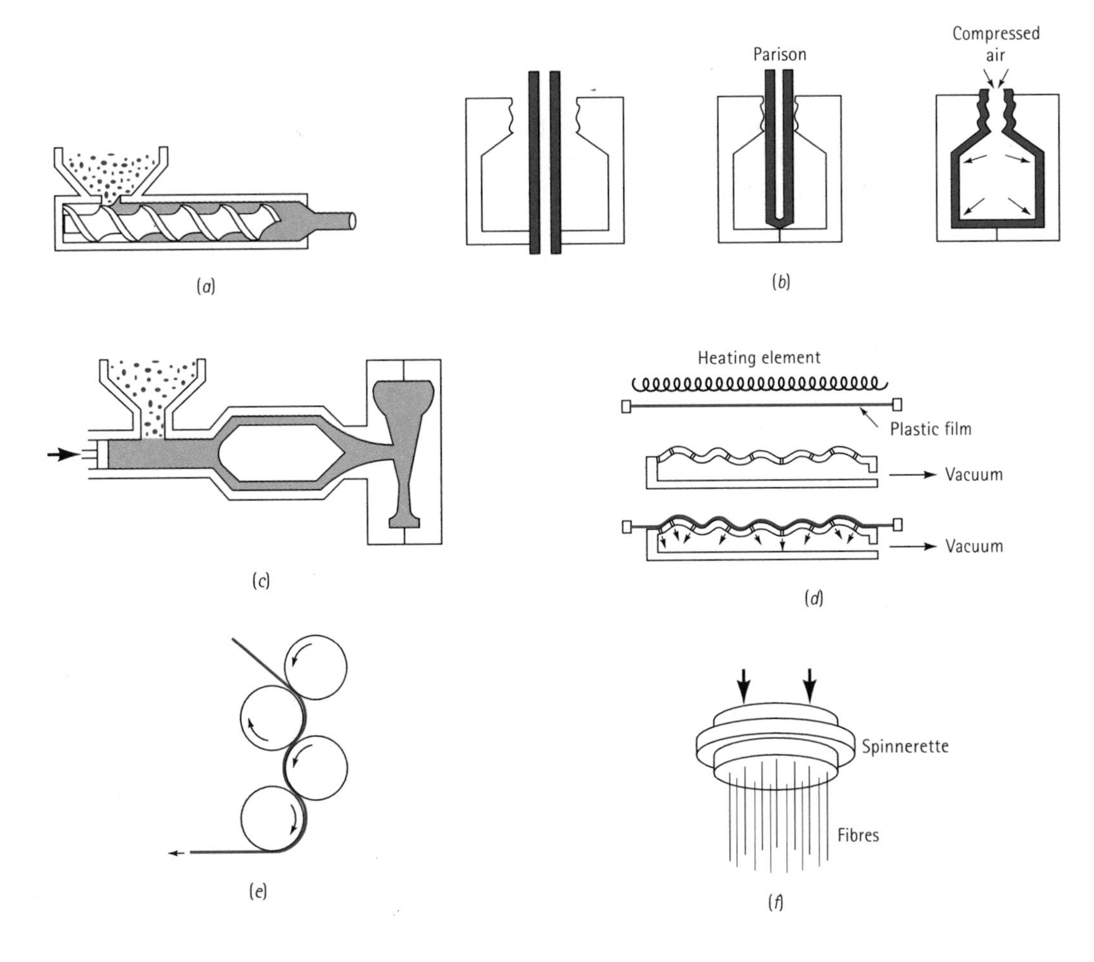

Figure 15.40
Typical forming processes for thermoplastic polymers: (*a*) extrusion, (*b*) blow moulding, (*c*) injection moulding, (*d*) thermoforming, (*e*) calendaring, and (*f*) spinning.

Compression Moulding Thermoset moulding are most often formed by placing the solid material before cross-linking into a heated die. Application of high pressure and temperature causes the polymer to melt, fill the die, and immediately begin to harden. Small electrical housings as well as bumpers, bonnets, and side panels for automobiles can be produced by this process.

Transfer Moulding A double chamber is used in transfer moulding of thermosetting polymers. The polymer is heated under pressure in one chamber. After melting, the polymer is injected into the adjoining die cavity. This process permits some of the advantages of injection moulding to be used for thermosetting polymers.

Reaction Injection Moulding (RIM) Thermosetting polymers in the form of liquid resins are first injected into a mixer and then directly into a heated mould to produce a shape. Forming and curing occur simultaneously in the mould. In reinforced reaction injection moulding (RRIM), a reinforcing material consisting of particles or short fibres is introduced into the mould cavity and is impregnated by the liquid resins to produce a composite material.

Foams Foamed products can be produced in polystyrene, urethanes, polymethyl methacrylate, and a number of other polymers. The polymer is produced in the form of tiny beads, often containing a foaming agent that will decompose to nitrogen, carbon dioxide, pentane, or another gas when heated. During this pre-expansion process, the bead increases in diameter by as many as 50 times and becomes hollow. The pre-expanded beads are then injected into a die, with the individual beads fusing together, to form exceptionally lightweight products with densities of perhaps only 0.02 $Mg.m^{-3}$. Expanded polystyrene (including Styrofoam) cups, packaging, and insulation are some of the applications for foams.

SUMMARY

Polymers are large, high-molecular-weight molecules produced by joining smaller molecules called monomers. Compared with most metals and ceramics, polymers have low strength, stiffness and melting temperatures; however, they also have a low density and good chemical resistance.

- Thermoplastic polymers are linear chains that permit the material to be easily formed into useful shapes, to have good ductility, and to be economically recycled. Thermoplastics can have an amorphous structure, which provides low strength and good ductility, when the ambient temperature is above the glass transition temperature. The polymers are more rigid and brittle when the temperature falls below their glass transition. Many thermoplastics can also partially crystallise, thereby increasing their strength.

  The thermoplastic chains can be made more rigid and stronger by using nonsymmetrical monomers that increase the bonding strength between the chains and make it more difficult for the chains to disentangle when stress is applied. In addition, many monomers produce more rigid chains containing atoms or groups of atoms other than carbon; this structure also produces high-strength thermoplastics.

- Elastomers are linear polymer chains that are lightly cross-linked. The cross-linking makes it possible to obtain very large elastic deformations without permanent plastic deformation. Increasing the number of cross-links increases the stiffness and reduces the amount of elastic deformation of the elastomers.

- Thermoplastic elastomers combine features of both thermoplastics and elastomers. At high temperatures, these polymers behave as thermoplastics and are plastically formed into shapes; at low temperatures, they behave as elastomers.

- Thermosetting polymers are highly cross-linked into a three-dimensional network structure. Typically, high glass transition temperatures, good strength, and brittle behaviour are found. Once cross-linking occurs, these polymers cannot easily be recycled.

- Manufacturing processes depend on the behaviour of the polymers. Processes such as extrusion, injection moulding, thermoforming, casting, drawing, and spinning are made possible by the viscoelastic behaviour of the thermoplastics. The nonreversible behaviour of bonding in thermosetting polymers limits their processing to fewer techniques, such as compression moulding and transfer moulding.

GLOSSARY

Addition polymerisation
Process by which chains are built up by adding monomers together without creating a by-product.

Aramids
Polyamide polymers containing aromatic groups of atoms in the linear chain.

Branching
When a separate polymer side chain is attached to another chain.

Condensation reaction
Process by which polymer chains are built up by a chemical reaction between two or more molecules, producing a by-product.

Copolymer
An addition polymer produced by joining more than one type of monomer.

Crazing
Localised plastic deformation in a polymer. A craze may lead to the formation of cracks in the material.

Cross-linking
Attaching chains of polymers together to produce a three-dimensional network polymer.

Deflection temperature
The temperature at which a polymer will deform a given amount under a standard load.

Degradation temperature
The temperature above which a polymer burns, chars, or decomposes.

Degree of polymerisation
The number of monomers in a polymer.

Diene
A group of monomers that contain two double covalent bonds. These monomers are often used in producing elastomers.

Elastomers
Polymers possessing a highly coiled and partly cross-linked chain structure, permitting the polymer to have exceptional elastic deformation.

Extenders
Additives or fillers to polymers that provide bulk at a low cost.

Functionality
The number of sites on a monomer at which polymerisation can occur.

Geometric isomer
A molecule that has the same composition as, but a structure different from, a second molecule.

Glass transition temperature
The temperature below which the amorphous polymer assumes a rigid glassy structure.

Homopolymers
An addition polymer containing only one type of monomer.

Interpenetrating polymer networks
Polymer structures produced by intertwining two separate polymer structures or networks.

Liquid-crystalline polymers
Exceptionally stiff polymer chains that act as rigid rods, even above their melting point.

Monomer
The molecule from which a polymer is produced.

Parison
A hot glob of soft or molten polymer that is blown or formed into a useful shape.

Plasticiser
An additive that, by decreasing the glass transition temperature, improves the formability of a polymer.

Reinforcement
Additives to polymers designed to provide significant improvement in strength. Fibres are typical reinforcements.

Relaxation time
A property of a polymer that is related to the rate at which stress relaxation occurs.

Repeat unit
The repeating structural unit from which a polymer is built. Also called a *mer*.

Spinnerette
An extrusion die containing many small openings through which hot or molten polymer is forced to produce filaments. Rotation of the spinnerette twists the filaments into a yarn.

Stress relaxation
A reduction of the stress acting on a material over a period of time at a constant strain due to viscoelastic deformation.

Tacticity
Describes the location in the polymer chain of atoms or atom groups in nonsymmetrical monomers.

Thermoplastic elastomers
Polymers that behave as thermoplastics at high temperatures but elastomers at lower temperatures.

Thermoplastic polymers
Polymers that can be reheated and remelted numerous times.

Thermosetting polymers
Polymers that are heavily cross-linked to produce a strong network structure.

Unsaturated bond
The double or even triple covalent bond joining two atoms together in an organic molecule. When a single covalent bond replaces the unsaturated bond, polymerisation can occur.

Viscoelasticity
The deformation of a polymer by viscous flow of the chains or segments of the chains when stress is applied.

Vulcanisation
Cross-linking elastomer chains by introducing sulphur atoms at elevated temperatures and pressures.

PROBLEMS

15.1 The molecular weight of polymethyl methacrylate is 250 000 $g.mol^{-1}$. If all of the polymer chains are the same length, calculate

(a) the degree of polymerisation and

(b) the number of chains in 1 g of the polymer.

15.2 The degree of polymerisation of polytetrafluoroethylene is 7 500. If all of the polymer chains are the same length, calculate

(a) the molecular weight of the chains and

(b) the total number of chains in 1 000 g of the polymer.

15.3 The distance between the centres of two adjacent carbon atoms in linear polymers is approximately 0.15 nm. Calculate the length of an ultrahigh-molecular-weight polyethylene chain that has a molecular weight of 1 000 000 $g.mol^{-1}$.

15.4 A polyethylene rope weighs 370 g per metre. If each chain contains 7 000 repeat units, calculate

(a) the number of polyethylene chains in a 3 m length of rope and

(b) the total length of chains in the rope, assuming that carbon atoms in each chain are approximately 0.15 nm apart.

15.5 Suppose that 20 g of benzoyl peroxide are introduced to 5 kg of propylene monomer. If 30% of the initiator groups are effective, calculate the expected degree of polymerisation and the molecular weight of the polypropylene polymer if

(a) all of the termination of the chains occurs by combination and

(b) all of the termination occurs by disproportionation.

15.6 Suppose hydrogen peroxide (H_2O_2) is used as the initiator for 10 kg of vinyl chloride monomer. Show schematically how the hydrogen peroxide will initiate the polymer chains. Calculate the required amount of hydrogen peroxide (assuming that it is 10% effective) required to produce a degree of polymerisation of 4 000 if

(a) termination of the chains occurs by combination

(b) termination occurs by disproportionation.

15.7 A common copolymer is produced by including both ethylene and propylene monomers in the same chain. Calculate the molecular weight of the polymer produced using 1 kg of ethylene and 3 kg of propylene, giving a degree of polymerisation of 5 000.

15.8 The formula for formaldehyde is HCHO.

(a) Draw the structure of the formaldehyde molecule and repeat unit.

(b) Does formaldehyde polymerise to produce an acetal polymer (see Table 15.8) by the addition mechanism or the condensation mechanism? Try to draw a sketch of the reaction and the acetal polymer by both mechanisms.

15.9 You would like to combine 5 kg of dimethyl terephthalate with ethylene glycol to produce polyester (PET). Calculate

(a) the amount of ethylene glycol required,

(b) the amount of by-product evolved, and

(c) the amount of polyester produced.

15.10 Would you expect polyethylene to polymerise at a faster or slower rate than polymethylmethacrylate? Explain. Would you expect polyethylene to polymerise at a faster or slower rate than a polyester? Explain.

15.11 You would like to combine 10 kg of ethylene glycol with terephthalic acid to produce a polyester. The monomer for terephthalic acid is shown below.

(a) Determine the by-product of the condensation reaction and

(b) calculate the amount of terephthalic acid required, the amount of by-product evolved, and the amount of polyester produced.

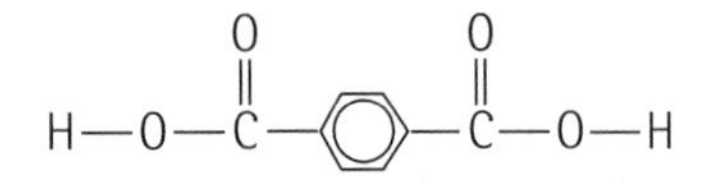

15.12 The data below were obtained for polyethylene. Determine

(a) the weight average molecular weight and degree of polymerisation and

(b) the number average molecular weight and degree of polymerisation.

Molecular Weight Range ($g.mol^{-1}$)	f_i	x_i
0–3 000	0.01	0.03
3 000–6 000	0.08	0.10
6 000–9 000	0.19	0.22
9 000–12 000	0.27	0.36
12 000–15 000	0.23	0.19
15 000–18 000	0.11	0.07
18 000–21 000	0.06	0.02
21 000–24 000	0.05	0.01

15.13 Analysis of a sample of polyacrylonitrile (PAN) (*see* Table 15.6) shows that there are six lengths of chains, with the following number of chains of each length. Determine

(a) the weight average molecular weight and degree of polymerisation and

(b) the number average molecular weight and degree of polymerisation.

Number of Chains	Mean Molecular Weight of Chains ($g.mol^{-1}$)
10 000	3 000
18 000	6 000
17 000	9 000
15 000	12 000
9 000	15 000
4 000	18 000

15.14 Explain why you would prefer that the number average molecular weight of a polymer be as close as possible to the weight average molecular weight.

15.15 Using Table 15.2, plot the relationship between the glass transition temperatures and the melting temperatures of the addition thermoplastic polymers. What is the approximate relationship between these two critical temperatures? Do the condensation thermoplastic polymers and the elastomers also follow the same relationship?

15.16 List the addition polymers in Table 15.2 that might be good candidates for making the bracket that holds the rear view mirror onto the outside of an automobile, assuming that temperatures frequently fall below zero degrees Celsius. Explain your choices.

15.17 Based on Table 15.2, which of the elastomers might be suited for use as a gasket in a pump for liquid CO_2 at –78°C? Explain.

15.18 How do the glass transition temperatures of polyethylene, polypropylene, and polymethyl methacrylate compare? Explain their differences, based on the structure of the monomer.

15.19 Which of the addition polymers in Table 15.2 are used in their leathery condition at room temperature? How is this condition expected to affect their mechanical properties compared with those of addition polymers?

15.20 The density of polypropylene is approximately 0.89 $Mg.m^{-3}$. Determine the number of propylene repeat units in each unit cell of crystalline polypropylene.

15.21 The density of polyvinyl chloride is approximately 1.4 $Mg.m^{-3}$. Determine the number of vinyl chloride repeats units, hydrogen atoms, chlorine atoms, and carbon atoms in each unit cell of crystalline PVC.

15.22 A polyethylene sample is reported to have a density of 0.97 $Mg.m^{-3}$. Calculate the percent crystallinity in the sample. Would you expect that the structure of this sample has a large or small amount of branching? Explain.

15.23 Amorphous polyvinyl chloride is expected to have a density of 1.38 $Mg.m^{-3}$. Calculate the % crystallisation in PVC that has a density of 1.45 $Mg.m^{-3}$. (*Hint*: Find the density of completely crystallised PVC from its lattice parameters, assuming four repeat units per unit cell.)

15.24 Describe the relative tendencies of the following polymers to crystallise. Explain your answer.

(a) branched polyethylene versus linear polyethylene

(b) polyethylene versus polyethylene-polypropylene copolymer

(c) isotactic polypropylene versus atactic polypropylene

(d) polymethyl methacrylate versus acetal (polyoxymethylene)

15.25 At room temperature, a polymer is found to have a creep rate of 0.007 $mm.mm^{-1}.h^{-1}$ when the applied stress is 18 $MN.m^{-2}$, a creep rate of 0.002 $mm.mm^{-1}.h^{-1}$ when the applied stress is 15.5 $MN.m^{-2}$, and a creep rate of 0.0009 $mm.mm^{-1}.h^{-1}$ when the applied stress is 14 $MN.m^{-2}$. The creep rate is found to depend on $a\sigma^n$, and where a and n are constants. Find the constants a and n and determine the maximum stress that will ensure that the polymer will deform no more than 2% in one year.

15.26 A stress of 17 $MN.m^{-2}$ is applied to a polymer serving as a fastener in a complex assembly. At a constant strain, the stress drops to 16.5 $MN.m^{-2}$ after 100 h. If the stress on the part must remain above 14.5 $MN.m^{-2}$ in order for the part to function properly, determine the life of the assembly.

15.27 A stress of 7 $MN.m^{-2}$ is applied to a polymer that operates at a constant strain; after six months, the stress drops to 5.9 $MN.m^{-2}$. For a particular application, a part made of the same polymer must maintain a stress of 6.2 $MN.m^{-2}$ after 12 months. What should be the original stress applied to the polymer for this application?

15.28 Data for the rupture time of polyethylene are shown in Figure 15.20. At an applied stress of 4.8 $MN.m^{-2}$, the figure indicates that the polymer ruptures in 0.2 h at 90°C but survives 10 000 h at 65°C. Assuming that the rupture time is related to the viscosity, calculate the activation energy for the viscosity of the polyethylene and estimate the rupture time at 23°C.

15.29 Figure 15.21 shows the effect of stress and time on the strain in polypropylene at 20°C.

(a) Determine from this data the constants a and n in Equation 15.7 for each applied stress.

(a) Determine the percent strain in the polypropylene if a stress of 8.6 $MN.m^{-2}$ is applied at 20°C for one year.

15.30 A polymer in the shape of a 100 mm-long rod is placed into service under a constant tensile stress. The creep rate, which is measured as a function of temperature, is shown below. Determine the time required for the rod to stretch to 130 mm at 85°C.

T(°C)	$d\varepsilon/dt$ ($mm.mm^{-1}.h^{-1}$)
25	0.0011
50	0.0147
75	0.1375

15.31 For each of the following pairs, recommend the one that will most likely have the better impact properties at 25°C. Explain each of your choices.

(a) polyethylene versus polystyrene

(b) low-density polyethylene versus high-density polyethylene

(c) polymethyl methacrylate versus polytetrafluoroethylene.

15.32 The polymer ABS can be produced with varying amounts of styrene, butadiene, and acrylonitrile monomers, which are present in the form of two copolymers: BS rubber and SAN.

(a) How would you adjust the composition of ABS if you wanted to obtain good impact properties?

(b) How would you adjust the composition if you wanted to obtain good ductility at room temperature?

(c) How would you adjust the composition if you wanted to obtain good strength at room temperature?

15.33 Figure 15.34 shows the stress-strain curve for an elastomer. From the curve, calculate and plot the modulus of elasticity versus strain and explain the results.

15.34 The maximum number of cross-linking sites in polyisoprene is the number of unsaturated bonds in the polymer chain. If three sulphur atoms are in each cross-linking sulphur strand, calculate the amount of sulphur required to provide cross-links at every available site in 5 kg of polymer and the wt% S that would be present in the elastomer. Is this typical?

15.35 Suppose we vulcanise polychloroprene, obtaining the desired properties by adding 1.5% sulphur by weight to the polymer. If each cross-linking strand contains an average of four sulphur atoms, calculate the fraction of the unsaturated bonds that must be broken.

15.36 The monomers for adipic acid, ethylene glycol, and maleic acid are shown below. These monomers can be joined into chains by condensation reactions, then cross-linked by breaking unsaturated bonds and inserting a styrene molecule as the cross-linking agent.

(a) Show how a linear chain composed of these three monomers can be produced.

(b) Explain why a thermosetting polymer cannot be produced using just adipic acid and ethylene glycol.

(c) Show how styrene provides cross-linking between the linear chains.

(d) If 50 g of adipic acid, 100 g of maleic acid, and 50 g of ethylene glycol are combined, calculate the amount of styrene required to completely cross-link the polymer.

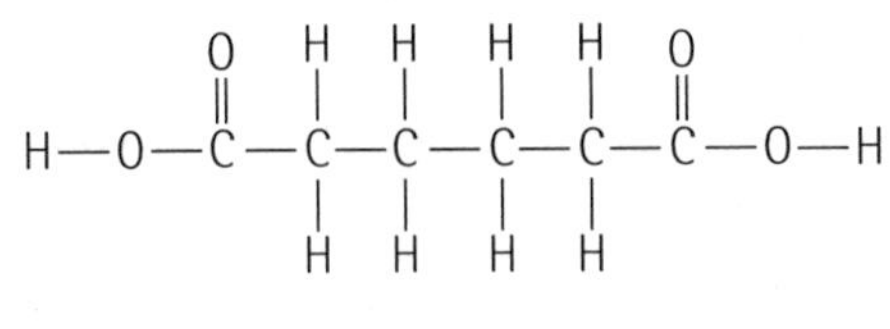

Adipic acid

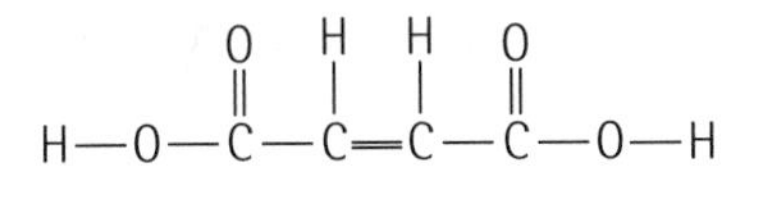

Maleic acid

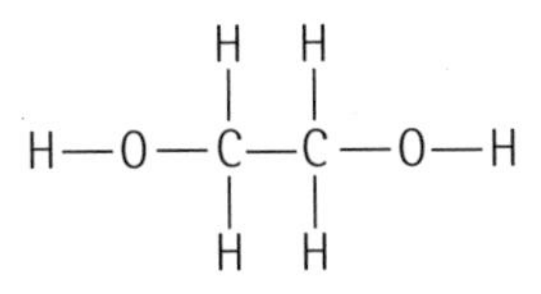

Ethylene glycol

15.37 How much formaldehyde is required to completely cross-link 10 kg of phenol to produce a thermosetting phenolic polymer? How much by-product is evolved?

15.38 Explain why the degree of polymerisation is not usually used to characterise thermosetting polymers.

15.39 Defend or contradict the choice to use the following materials as hot-melt adhesives for an application in which the assembled part is subjected to impact-type blows:

(a) polyethylene

(b) polystyrene

(c) styrene-butadiene thermoplastic elastomer

(d) polyacrylonitrile

(e) polybutadiene

15.40 Many paints are polymeric materials. Explain why plasticisers are added to paints. What must happen to the plasticisers after the paint is applied?

15.41 You want to extrude a complex component from an elastomer. Should you vulcanise the rubber before or after the extrusion operation? Explain.

15.42 Suppose a thermoplastic polymer can be produced in sheet form either by rolling (deformation) or by continuous casting (with a rapid cooling rate). In which case would you expect to obtain the higher strength? Explain.

Design problems

15.43 Figure 15.41 shows the behaviour of polypropylene, polyethylene, and acetal at two temperatures. You would like to produce a 300 mm-long rod of a polymer that will operate at 40°C for 6 months under a constant load of 2 225 N. Design the material and size of the rod such that no more than 5% elongation will occur by creep.

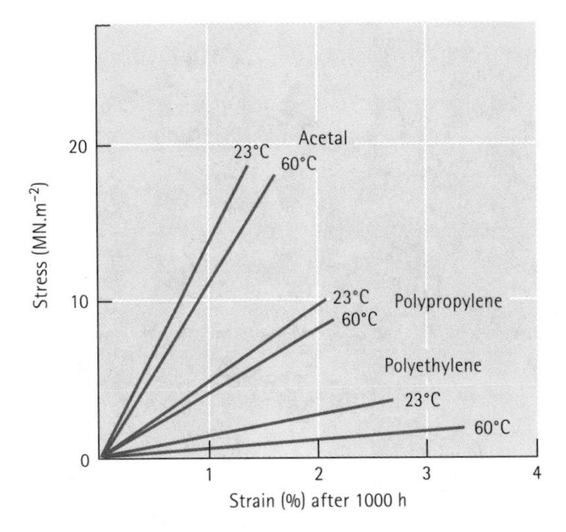

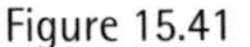

Figure 15.41
The effect of applied stress on the percent creep strain for three polymers (*for* Problem 15.43).

15.44 Design a polymer material that might be used to produce a 75 mm-diameter gear to be used to transfer energy from a low-power electric motor. What are the design requirements? What class of polymers (thermoplastics, thermosets, elastomers) might be most appropriate? What particular polymer might you first consider? What additional information concerning the application and polymer properties do you need to know to complete your design?

15.45 Design a polymer material and a forming process to produce the case for a personal computer.

(a) What are the design and forming requirements?

(b) What class of polymers might be most appropriate?

(c) What particular polymer might you first consider?

(d) What additional information do you need to know?

15.46 Design a polymer part for which, under an applied stress of 7 $MN.m^{-2}$, no more than 0.025 $mm.mm^{-1}$ creep strain will occur at room temperature in one year. Be sure to consider cost in your design.

CHAPTER 16

Composite Materials

16.1 Introduction

Composites are produced when two materials are joined to give a combination of properties that cannot be attained in the original materials. Composite materials may be selected to give unusual combinations of stiffness, strength, weight, high-temperature performance, corrosion resistance, hardness, or conductivity.

Composites can be placed into three categories – particulate, fibre, and laminar – based on the shapes of the materials (Figure 16.1). Concrete, a mixture of cement and gravel, is a particulate composite; fibreglass, containing glass fibres embedded in a polymer, is a fibre-reinforced composite; and plywood, having alternating layers of wood veneer, is a laminar composite. If the reinforcing particles are uniformly distributed, particulate composites have isotropic properties; fibre composites may be either isotropic or anisotropic; laminar composites always display anisotropic behaviour.

16.2 Dispersion-Strengthened Composites

By stretching slightly our definition of a composite, we can consider a special group of dispersion-strengthened materials containing particles 10 nm to 250 nm in diameter as particulate composites. These dispersoids, usually a metallic oxide, are introduced into the matrix by means other than traditional phase transformations. Even though the small particles are not coherent with the matrix, they block the movement of dislocations and produce a pronounced strengthening effect.

At room temperature, the dispersion-strengthened composites may be weaker than traditional age-hardened alloys, which contain a coherent precipitate. However, because the composites do not catastrophically soften by overageing, overtempering, grain growth, or coarsening of the dispersed phase, the strength of the composite decreases only gradually with increasing temperature (Figure 16.2). Furthermore, their creep resistance is superior to that of metals and alloys.

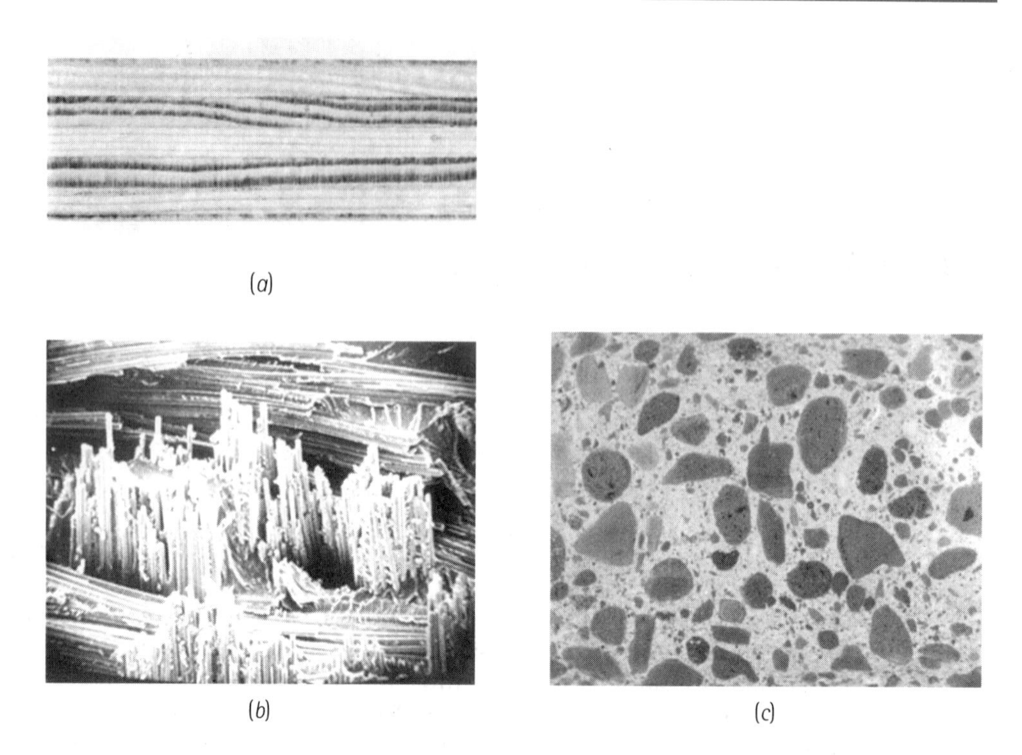

(a)

(b)

(c)

Figure 16.1
Some examples of composite materials: (a) Plywood is a laminar composite of layers of wood veneer. (b) Fibreglass is a fibre-reinforced composite containing stiff, strong glass fibres in a softer polymer matrix (×175). (c) Concrete is a particulate composite containing coarse sand or gravel in a cement matrix (reduced 50%).

The dispersant must have a low solubility in the matrix and must not chemically react with the matrix, but a small amount of solubility may help improve the bonding between the dispersant and the matrix. Copper oxide (Cu_2O) dissolves in copper at high temperatures; thus, the Cu_2O-Cu system would not be effective. However, Al_2O_3 does not dissolve in aluminium; the Al_2O_3-Al system does give an effective dispersion-strengthened material.

Examples of Dispersion-Strengthened Composites Table 16.1 lists some materials of interest. Perhaps the classic example is the sintered aluminium powder (SAP) composite. SAP has an aluminium matrix strengthened by up to 14% Al_2O_3. The composite is formed by powder metallurgy. In one method, aluminium and alumina powders are blended, compacted at high pressures, and sintered. In a second technique, the aluminium powder is treated to add a continuous oxide film on each particle. When the powder is compacted, the oxide film fractures into tiny flakes that are surrounded by the aluminium metal during sintering.

Another important group of dispersion-strengthened composites includes thoria-dispersed metals such as TD-nickel (Figure 16.3). TD-nickel can be produced by internal oxidation. Thorium is present in nickel as an alloying element. After a powder metallurgy compact is made, oxygen is allowed to diffuse into the metal, react with the thorium, and produce thoria (ThO_2).

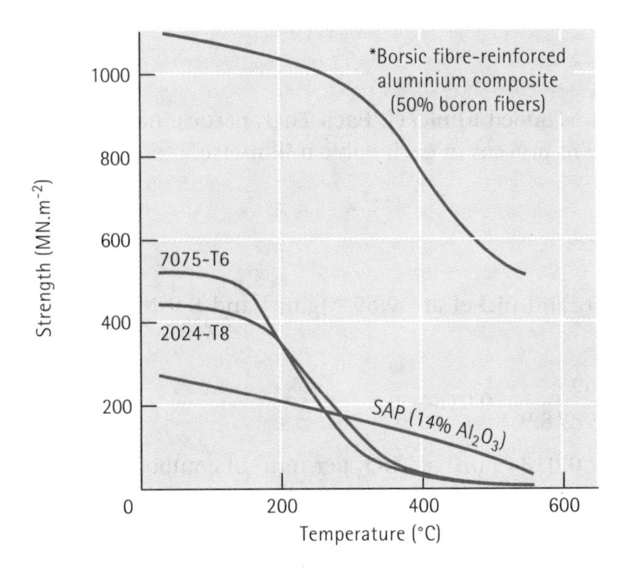

Figure 16.2
Comparison of the yield strength of dispersion-strengthened sintered aluminium powder (SAP) composite with that of two conventional two-phase high-strength aluminium alloys. The composite has benefits above about 300°C. A fibre-reinforced aluminium composite is shown for comparison.

**Borsic is the term used to describe boron fibres coated with silicon carbide. The SiC coating prevents a reaction between the boron and the aluminium during processing.*

Table 16.1 Examples and applications of selected dispersion-strengthened composites.

System	Applications
Ag-CdO	Electrical contact materials
$Al-Al_2O_3$	Possible use in nuclear reactors
Be-BeO	Aerospace and nuclear reactors
$Co-ThO_2$, Y_2O_3	Possible creep-resistant magnetic materials
Ni-20% $Cr-ThO_2$	Turbine engine components
Pb-PbO	Battery grids
$Pt-ThO_2$	Filaments, electrical components
$W-ThO_2$, ZrO_2	Filaments, heaters

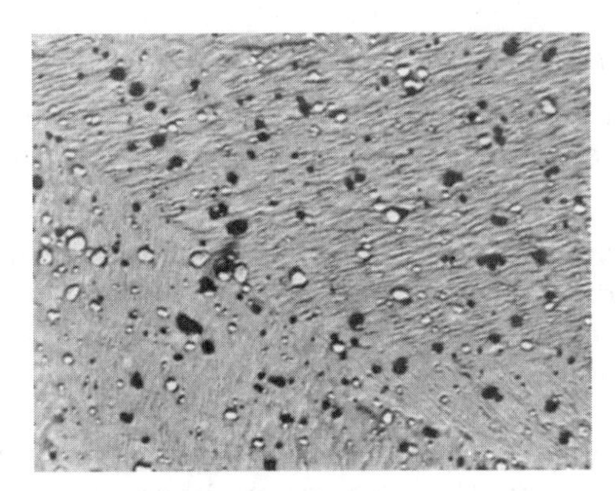

Figure 16.3
Electron micrograph of TD-nickel.
The dispersed ThO_2 particles have a diameter of 300 nm or less (× 2 000). (*From* Oxide Dispersion Strengthening, *p. 714, Gordon and Breach, 1968. © AIME.*)

EXAMPLE 16.1

Suppose 2 wt% ThO_2 is added to nickel. Each ThO_2 particle has a diameter of 100 nm. How many particles are present in each cubic millimetre?

SOLUTION

The densities of ThO_2 and nickel are 9.69 $Mg.m^{-3}$ and 8.9 $Mg.m^{-3}$, respectively. The volume fraction is:

$$f_{ThO_2} = \frac{2/9.69}{2/9.69 + 98/8.9} = 0.0184$$

Therefore, there is 0.0184 mm^3 of ThO_2 per mm^3 of composite. The volume of each ThO_2 sphere is:

$$V_{ThO_2} = \frac{4}{3}\pi r^3 = \frac{4}{3}\pi(50 \times 10^{-6}\ \text{mm})^3 = 5.24 \times 10^{-13}\ \text{mm}^3$$

$$\text{Number of ThO}_2 = \frac{0.0184}{5.24 \times 10^{-13}} = 35.1 \times 10^9\ \text{particles / mm}^3$$

16.3 True Particulate Composites

The true particulate composites contain large amounts of coarse particles that do not effectively block slip. The particulate composites are designed to produce unusual combinations of properties rather than to improve strength.

Rule of Mixtures Certain properties of a particle composite depend only on the relative amounts and properties of the individual constituents. The rule of mixtures can accurately predict these properties. The density of a particulate composite, for example, is

$$\rho_c = \Sigma f_i \rho_i = f_1 \rho_1 + f_2 \rho_2 + \ldots + f_n \rho_n \tag{16.1}$$

where ρ_c is the density of the composite, ρ_1, ρ_2,..., ρ_n are the densities of each constituent in the composite, and f_1, f_2,..., f_n are the volume fractions of each constituent.

Cemented Carbides Cemented carbides, or cermets, contain hard ceramic particles dispersed in a metallic matrix. Tungsten carbide inserts used for cutting tools in machining operations are typical of this group. Tungsten carbide (WC) is a hard, stiff, high-melting-temperature ceramic. Unfortunately, tools constructed from tungsten carbide are extremely brittle.

To improve toughness, tungsten carbide particles are combined with cobalt powder and pressed into powder compacts. The compacts are heated above the melting temperature of the cobalt. The liquid cobalt surrounds each of the solid tungsten carbide particles (Figure 16.4). After solidification, the cobalt serves as the binder for tungsten carbide and provides good impact resistance. Other carbides, such as TaC and TiC, may also be included in the cermet.

Figure 16.4
Microstructure of tungsten carbide – 20% cobalt-cemented carbide (× 1300). (*From* Metals Handbook, *Vol. 7, 8th Ed., American Society for Metals, 1972.*)

EXAMPLE 16-2

A cemented carbide cutting tool used for machining contains 75 wt% WC, 15 wt% TiC, 5 wt% TaC, and 5 wt% Co. Estimate the density of the composite.

SOLUTION

First, we must convert the weight percentages to volume fractions. The densities of the components of the composite are:

$$\rho_{\text{WC}} = 15.77 \text{ Mg.m}^{-3} \qquad \rho_{\text{TiC}} = 4.94 \text{ Mg.m}^{-3}$$

$$\rho_{\text{TaC}} = 14.5 \text{ Mg.m}^{-3} \qquad \rho_{\text{Co}} = 8.90 \text{ Mg.m}^{-3}$$

$$f_{\text{WC}} = \frac{75/15.77}{75/15.77 + 15/4.94 + 5/14.5 + 5/8.9} = \frac{4.76}{8.70} = 0.547$$

$$f_{\text{TiC}} = \frac{15/4.94}{8.70} = 0.349$$

$$f_{\text{TaC}} = \frac{5/14.5}{8.70} = 0.040$$

$$f_{\text{Co}} = \frac{5/8.90}{8.70} = 0.064$$

From the rule of mixtures, the density of the composite is

$$\rho_c = \Sigma f_i \rho_i = (0.547)(15.77) + (0.349)(4.94) + (0.040)(14.5) + (0.064)(8.9)$$

$$= 11.50 \text{ Mg.m}^{-3}$$

Abrasives Grinding and cutting wheels are formed from alumina (Al_2O_3), silicon carbide (SiC), and cubic boron nitride (BN). To provide toughness, the abrasive particles are bonded by a glass or polymer matrix. Diamond abrasives are typically bonded with a metal matrix. As the hard particles wear, they fracture or pull out of the matrix, exposing new cutting surfaces.

Electrical Contacts Materials used for electrical contacts in switches and relays must have a good combination of wear resistance and electrical conductivity. Otherwise, the contacts erode, causing poor contact and arcing. Tungsten-reinforced silver provides this combination of characteristics. A tungsten powder compact is made using conventional powder metallurgical processes (Figure 16.5) to produce high interconnected porosity. Liquid silver is then vacuum infiltrated to fill the interconnected voids. Both the silver and the tungsten are continuous. Thus, the pure silver efficiently conducts current while the hard tungsten provides wear resistance.

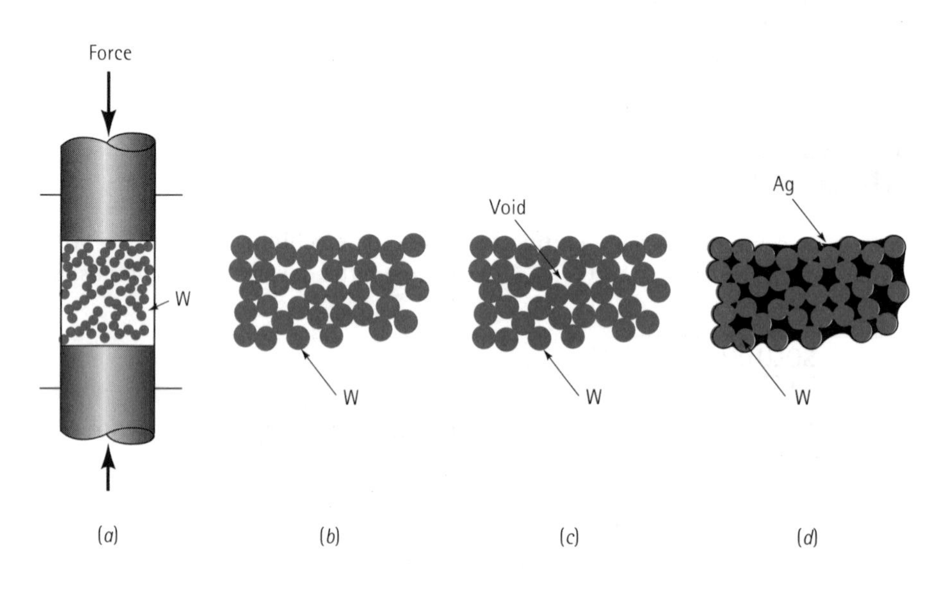

Figure 16.5
The steps in producing a silver-tungsten electrical composite: (*a*) Tungsten powders are pressed, (*b*) a low-density compact is produced, (*c*) sintering joins the tungsten powders, and (*d*) liquid silver is infiltrated into the pores between the particles.

EXAMPLE 16.3

A silver-tungsten composite for an electrical contact is produced by first making a porous tungsten powder metallurgy compact, then infiltrating pure silver into the pores. The density of the tungsten compact before infiltration is 14.5 $Mg.m^{-3}$. Calculate the volume fraction of porosity and the final weight percent of silver in the compact after infiltration.

SOLUTION

The densities of pure tungsten and pure silver are 19.3 Mg.m^{-3} and 10.49 Mg.m^{-3}. We can assume that the density of a pore is zero, so from the rule of mixtures:

$$\rho_c = f_w \rho_w + f_{\text{pore}} \rho_{\text{pore}}$$

$$14.5 = f_w(19.3) + f_{\text{pore}}(0)$$

$$f_w = 0.75$$

$$f_{\text{pore}} = 1 - 0.75 = 0.25$$

After infiltration, the volume fraction of silver equals the volume fraction of pores:

$$f_{\text{Ag}} = f_{\text{pore}} = 0.25$$

$$\text{wt\% Ag} = \frac{(0.25)(10.49)}{(0.25)(10.49) + (0.75)(19.3)} \times 100 = 15.3\%$$

This solution assumes that all of the pores are open, or interconnected.

Polymers Many engineering polymers that contain fillers and extenders are particulate composites. A classic example is carbon black in vulcanised rubber. Carbon black consists of tiny carbon spheroids only 5 nm to 500 nm in diameter. The carbon black improves the strength, stiffness, hardness, wear resistance, and heat resistance of the rubber.

Extenders, such as calcium carbonate, solid glass spheres, and various clays, are added so that a smaller amount of the more expensive polymer is required. The extenders may stiffen the polymer, increase the hardness and wear resistance, increase thermal conductivity, or improve resistance to creep; however, strength and ductility normally decrease (Figure 16.6). Introducing hollow glass spheres may impart the same changes in properties while significantly reducing the weight of the composite.

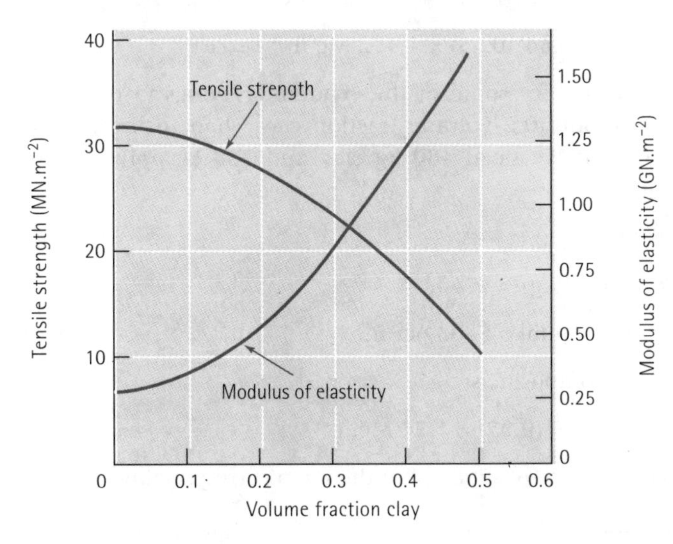

Figure 16.6
The effect of clay on the properties of polyethylene.

Other special properties can be obtained. Elastomer particles are introduced into polymers to improve toughness. Polyethylene may contain metallic powders, such as lead, to improve absorption of fission products in nuclear applications.

EXAMPLE 16.4 Particulate Polymer Composite Formulation

Establish a formulation for a clay-filled polyethylene composite suitable for injection moulding of inexpensive components. The final part must have a tensile strength of at least 20 $MN.m^{-2}$ and a modulus of elasticity of at least 0.5 $GN.m^{-2}$. Polyethylene costs approximately \$1.00 per kilogram and clay costs approximately 10 cents per kilogram. The density of polyethylene is 0.95 $Mg.m^{-3}$ and that of clay is 2.4 $Mg.m^{-3}$.

SOLUTION

From Figure 16.6, a volume fraction of clay below 0.36 is required to maintain a tensile strength greater than 20 $Mn.m^{-2}$, whereas a volume fraction of at least 0.2 is needed for the minimum modulus of elasticity. For lowest cost, we use the maximum allowable clay, or a volume fraction of 0.36 clay.

In 1 m^3 of composite parts, there are 0.36 m^3 of clay and 0.64 m^3 of polyethylene in the composite, or:

$$(0.36\ m^3)(2.4\ Mg.m^{-3}) = 0.864\ Mg = 864\ kg\ clay$$

$$(0.64\ m^3)(0.95\ Mg.m^{-3}) = 0.608\ Mg = 608\ kg\ polyethylene$$

The cost of materials is:

$$(864\ kg\ clay)(\$0.10/kg) = \$86.40$$

$$(608\ kg\ PE)(\$1.00/kg) = \$608$$

$$total = \$694.40\ per\ m^3$$

Suppose that weight is critical. The composite's density is:

$$\rho_c = (0.36)(2.4) + (0.64)(0.95) = 1.472\ Mg.m^{-3}$$

We may wish to sacrifice some of the economic savings in order to obtain lighter weight. If we use only 0.2 volume fraction clay, then (using the same method as above) we find that we need 480 kg clay and 760 kg polyethylene. The cost of materials is now:

$$(480\ kg)(\$0.10/kg) = \$48$$

$$(760\ kg)(\$1.00/kg) = \$760$$

$$total = \$808\ per\ m^3$$

The density of the composite is:

$$\rho_c = (0.2)(2.4) + (0.8)(0.95) = 1.24\ Mg.m^{-3}$$

The material costs about 16% more, but there is a corresponding weight saving of 16%.

Cast Metal Particulate Composites Aluminium castings containing dispersed SiC particles for automotive applications, including pistons and connecting rods, represent an important commercial application for particulate composites (Figure 16.7). With special processing, the SiC particles can be wet by the liquid, helping to keep the ceramic particles from sinking during freezing.

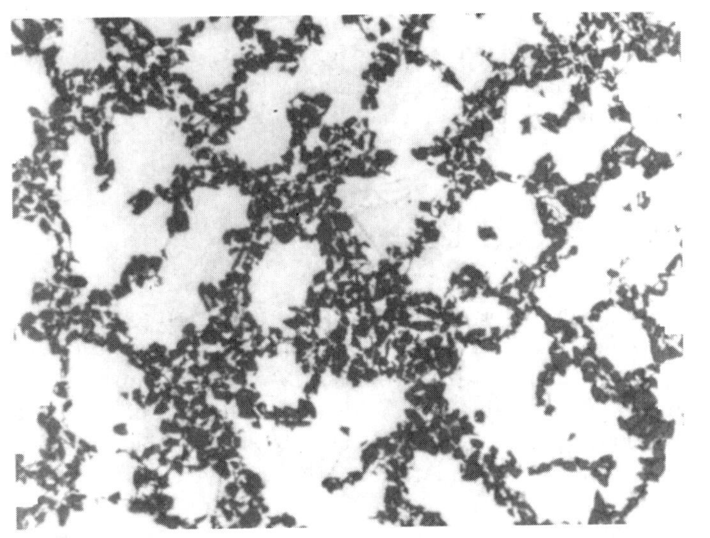

Figure 16.7
Microstructure of an aluminium casting alloy reinforced with silicon carbide particles. In this case, the reinforcing particles have segregated to interdendritic regions of the casting (× 125). (*Courtesy of David Kennedy, Lester B. Knight Cast Metals Inc.*)

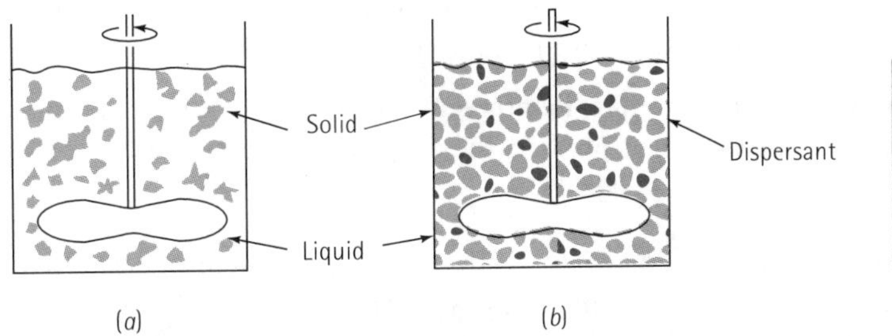

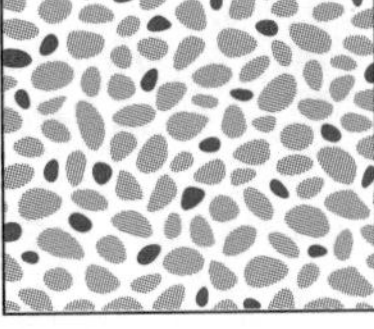

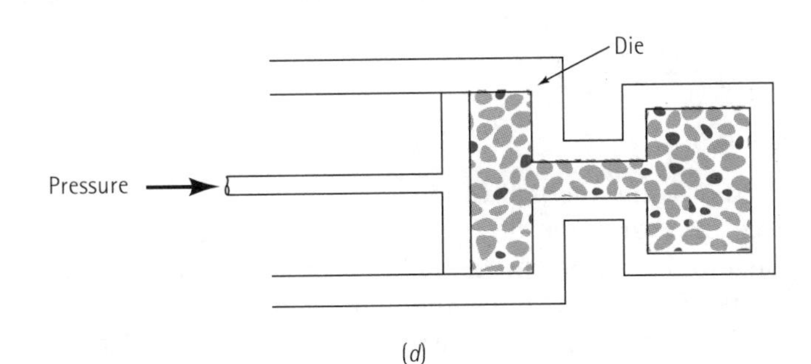

Figure 16.8
In compocasting, (*a*) a solidifying alloy is stirred to break up the dendritic network, (*b*) a reinforcement is introduced into the slurry, (*c*) when no force is applied, the solid-liquid mixture does not flow, and (*d*) high pressures cause the solid-liquid mixture to flow into a die.

An unusual technique for producing particulate-reinforced castings is based on the thixotropic behaviour of partly liquid–partly solid melts. A liquid alloy is allowed to cool until about 40% solids have formed; during solidification, the solid-liquid mixture is vigorously stirred to break up the dendritic structure (Figure 16.8). A particulate material is introduced during stirring. The resulting solid-liquid slurry displays thixotropic behaviour – the slurry behaves as a solid when no stress is applied, but flows like a liquid when pressure is exerted. Consequently, the thixotropic slurry can be injected into a die under pressure, a process called compocasting A variety of ceramic particles and glass beads have been incorporated into aluminium and magnesium alloys by this technique.

16.4 Fibre-Reinforced Composites

Most fibre-reinforced composites provide improved strength, fatigue resistance, stiffness, and strength-to-weight ratio by incorporating strong, stiff, but brittle fibres into a softer, more ductile matrix. The matrix material transmits the force to the fibres, which carry most of the applied force. The strength of the composite may be high at both room temperature and elevated temperatures (Figure 16.2).

Many types of reinforcing materials are employed. Straw has been used to strengthen mud bricks for centuries. Steel reinforcing bars are introduced into concrete structures. Glass fibres in a polymer matrix produce fibreglass for transportation and aerospace applications. Fibres made of boron, carbon, polymers, and ceramics provide exceptional reinforcement in advanced composites based on matrices of polymers, metals, ceramics, and even intermetallic compounds.

The Rule of Mixtures in Fibre-Reinforced Composites As for particulate composites, the rule of mixtures always predicts the density of fibre-reinforced composites:

$$\rho_c = f_m \rho_m + f_f \rho_f \tag{16.2}$$

where the subscripts m and f refer to the matrix and the fibre. Note that $f_m = 1 - f_f$.

In addition, the rule of mixtures accurately predicts the electrical and thermal conductivity of fibre-reinforced composites along the fibre direction if the fibres are *continuous* and *unidirectional*:

$$K_c = f_m K_m + f_f K_f \tag{16.3}$$

$$\sigma_c = f_m \sigma_m + f_f \sigma_f \tag{16.4}$$

where K is the thermal conductivity and σ is the electrical conductivity. Thermal or electrical energy can be transferred through the composite at a rate that is proportional to the volume fraction of the conductive material. In a composite with a metal matrix and ceramic fibres, the bulk of the energy would be transferred through the matrix; in a composite consisting of a polymer matrix containing metallic fibres, energy would be transferred through the fibres.

When the fibres are not continuous or unidirectional, the simple rule of mixtures may not apply. For example, in a metal fibre–polymer matrix composite, electrical conductivity would be low and would depend on the length of the fibres, the volume fraction of the fibres, and how often the fibres touch one another.

Modulus of Elasticity The rule of mixtures is used to predict the modulus of elasticity when the fibres are continuous and unidirectional. Parallel to the fibres, the modulus of elasticity may be as high as:

$$E_c = f_m E_m + f_f E_f \tag{16.5}$$

However, when the applied stress is very large, the matrix begins to deform and the stress-strain curve is no longer linear (Figure 16.9). Since the matrix now contributes little to the stiffness of the composite, the modulus can be approximated by:

$$E_c = f_f E_f \tag{16.6}$$

When the load is applied perpendicular to the fibres, each component of the composite acts independently of the other. The modulus of the composite is now:

$$\frac{1}{E_c} = \frac{f_m}{E_m} + \frac{f_f}{E_f} \tag{16.7}$$

Again, if the fibres are not continuous and unidirectional, the rule of mixtures does not apply.

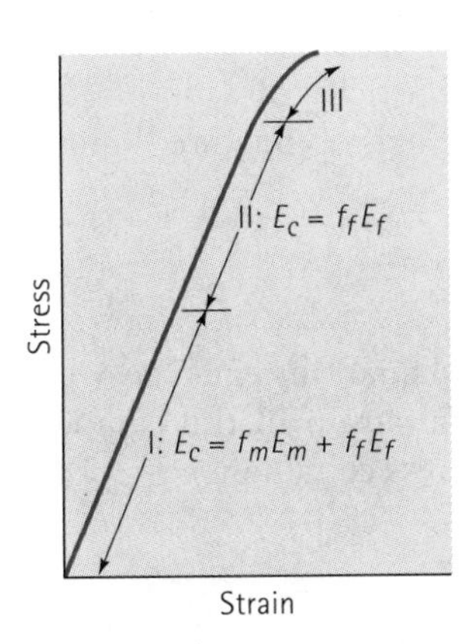

Figure 16.9
The stress-strain curve for a fibre-reinforced composite. At low stresses, the modulus of elasticity is given by the rule of mixtures. At higher stresses, the matrix deforms and the rule of mixtures is no longer obeyed.

EXAMPLE 16.5

Derive the rule of mixtures (Equation 16.5) for the modulus of elasticity of a fibre-reinforced composite when a stress is applied along the axis of the fibres.

SOLUTION

The total force acting on the composite is the sum of the forces carried by each constituent:

$$F_c = F_m + F_f$$

Since $F = \sigma A$:

$$\sigma_c A_c = \sigma_m A_m + \sigma_f A_f$$

$$\sigma_c = \sigma_m\left(\frac{A_m}{A_c}\right) + \sigma_f\left(\frac{A_f}{A_c}\right)$$

If the fibres have a uniform cross-section, the area fraction equals the volume fraction f:

$$\sigma_c = \sigma_m f_m + \sigma_f f_f$$

From Hooke's law, $\sigma = \varepsilon E$. Therefore:

$$E_c\varepsilon_c = E_m\varepsilon_m f_m + E_f \varepsilon_f f_f$$

If the fibres are rigidly bonded to the matrix, both the fibres and the matrix must stretch equal amounts (iso-strain conditions):

$$\varepsilon_c = \varepsilon_m\varepsilon_f$$

$$E_c = f_m E_m + f_f E_f$$

EXAMPLE 16.6

Derive the equation for the modulus of elasticity of a fibre-reinforced composite when a stress is applied perpendicular to the axis of the fibre (Equation 16.7).

SOLUTION

In this example, the strains are no longer equal; instead, the weighted sum of the strains in each component equals the total strain in the composite, whereas the stresses in each component are equal (iso-stress conditions):

$$\varepsilon_c = f_m\,\varepsilon_m + f_f\varepsilon_f$$

$$\frac{\sigma_c}{E_c} = f_m\left(\frac{\sigma_m}{E_m}\right) + f_f\left(\frac{\sigma_f}{E_f}\right)$$

Since $\sigma_c = \sigma_m = \sigma_f$:

$$\frac{1}{E_c} = \frac{f_m}{E_m} + \frac{f_f}{E_f}$$

Strength of Composites The strength of a fibre-reinforced composite depends on both the strength of the raw fibre and the bonding between the fibres and the matrix. However, the rule of mixtures is sometimes used to approximate the tensile strength of a composite containing continuous, parallel fibres:

$$\sigma_c = f_f\,\sigma_f + f_m\sigma_m \qquad (16.8)$$

where σ_f is the tensile strength of the fibre and σ_m is the stress acting on the matrix when the composite is strained to the point where the fibre fractures. Thus, σ_m is *not* the actual tensile strength of the matrix. Other properties, such as ductility, impact properties, fatigue properties, and creep properties, are difficult to predict even for unidirectionally aligned fibres.

EXAMPLE 16.7

Borsic-reinforced aluminium containing 40 vol% fibres is an important high-temperature, lightweight composite material. Estimate the density, modulus of elasticity, and tensile strength parallel to the fibre axis. Also estimate the modulus of elasticity perpendicular to the fibres.

SOLUTION

The properties of the individual components are shown below.

Material	Density ($Mg.m^{-3}$)	Modulus of Elasticity ($GN.m^{-2}$)	Tensile Strength ($MN.m^{-2}$)
Fibres	2.36	380	2760
Aluminium	2.70	70	35

From the rule of mixtures:

$$\rho_c = (0.6)(2.7) + (0.4)(2.36) = 2.56 \text{ Mg.m}^{-3}$$

$$E_c = (0.6)(70) + (0.4)(380) = 194 \text{ GN.m}^{-2}$$

Using the strength of the matrix in the absence of the true σ_m value:

$$\sigma_c = (0.6)(35) + (0.4)(2760) = 1\,125 \text{ MN.m}^{-2}$$

Perpendicular to the fibres:

$$\frac{1}{E_c} = \frac{0.6}{70} + \frac{0.4}{380} = 9.624 \times 10^{-3}$$

$$E_c = 103.9 \text{ GN.m}^{-2}$$

The actual modulus and strength parallel to the fibres are shown in Figure 16.10. The calculated modulus of elasticity (194 $GN.m^{-2}$) is exactly the same as the measured modulus. However, the estimated strength (1125 $MN.m^{-2}$) is substantially higher than the actual strength (about 830 $MN.m^{-2}$). We also note that the modulus of elasticity is very anisotropic, with the modulus perpendicular to the fibre being only half the modulus parallel to the fibres.

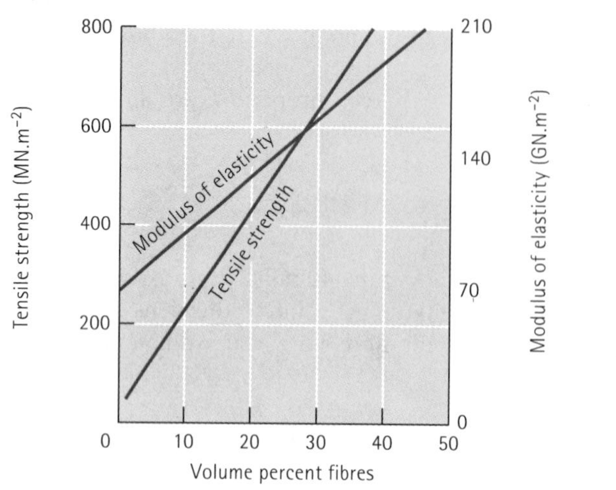

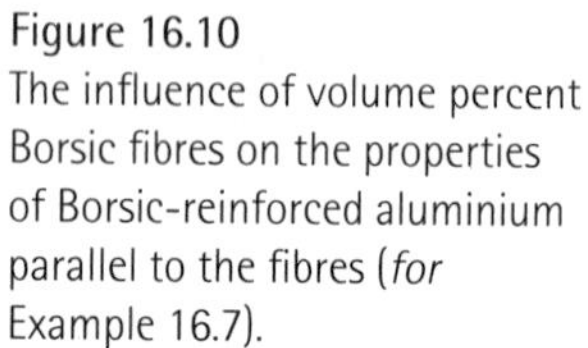
Figure 16.10
The influence of volume percent Borsic fibres on the properties of Borsic-reinforced aluminium parallel to the fibres (*for* Example 16.7).

EXAMPLE 16.8

Continuous glass fibres in nylon provide reinforcement. If the nylon contains 30 vol% E-glass, what fraction of the applied force is carried by the glass fibres?

SOLUTION

The modulus of elasticity for each component of the composite is:

$E_{\text{glass}} = 72\ \text{GN.m}^{-2}$ $\qquad E_{\text{nylon}} = 2.8\ \text{GN.m}^{-2}$

Both the nylon and the glass fibres have equal strain if bonding is good, so:

$$\varepsilon_c = \varepsilon_m = \varepsilon_f$$

$$\varepsilon_m = \frac{\sigma_m}{E_m} = \varepsilon_f = \frac{\sigma_f}{E_f}$$

$$\frac{\sigma_f}{\sigma_m} = \frac{E_f}{E_m} = \frac{72}{2.8} = 25.71$$

$$\text{Fraction} = \frac{F_f}{F_f + F_m} = \frac{\sigma_f A_f}{\sigma_f A_f + \sigma_m A_m} = \frac{\sigma_f(0.3)}{\sigma_f(0.3) + \sigma_m(0.7)}$$

$$= \frac{0.3}{0.3 + 0.7(\sigma_m / \sigma_f)} = \frac{0.3}{0.3 + 0.7(1/25.71)} = 0.92$$

Almost all of the load is carried by the glass fibres.

16.5 Characteristics of Fibre-Reinforced Composites

Many factors must be considered when designing a fibre-reinforced composite, including the length, diameter, orientation, amount, and properties of the fibres; the properties of the matrix; and the bonding between the fibres and the matrix.

Fibre Length and Diameter Fibres can be short, long, or even continuous. Their dimensions are often characterised by the aspect ratio l/d, where l is the fibre length and d is the diameter. Typical fibres have diameters varying from 10 microns (10×10^{-3} mm) to 150 microns (150×10^{-3} mm).

The strength of a composite improves when the aspect ratio is large. Fibres often fracture because of surface imperfections. Making the diameter as small as possible gives the fibre less surface area and, consequently, fewer flaws that might propagate during processing or under load. We also prefer long fibres. The ends of a fibre carry less of the load than the remainder of the fibre; consequently, the fewer the ends, the higher the load-carrying ability of the fibres (Figure 16.11).

In many fibre-reinforced systems, discontinuous fibres with an aspect ratio greater than some critical value are used to provide an acceptable compromise between processing ease and properties. A critical fibre length l_c, for any given fibre diameter d, can be determined:

$$l_c = \frac{\sigma_f d}{2\tau_i} \tag{16.9}$$

where σ_f is the strength of the fibre and τ_i is related to the strength of the bond between the fibre and the matrix, or the stress at which the matrix begins to deform. If the fibre length l is smaller than l_c, little reinforcing effect is observed; if l is greater than about $15l_c$, the fibre behaves almost as if it were continuous. The strength of the composite can be estimated from

$$\sigma_c = f_f \sigma_f \left(1 - \frac{l_c}{2l}\right) + f_m \sigma_m \tag{16.10}$$

where σ_m is the stress on the matrix when the fibres break.

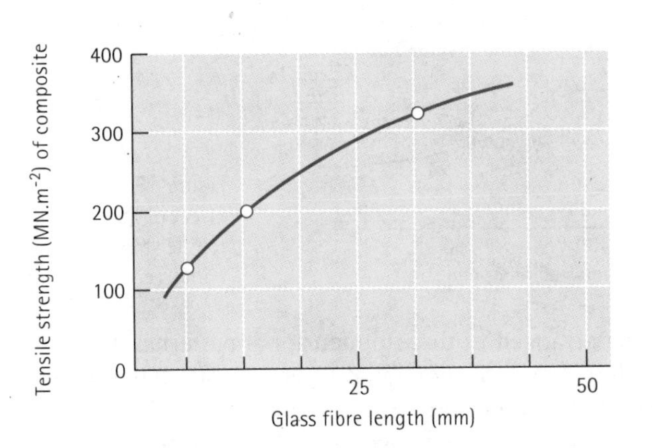

Figure 16.11
Increasing the length of chopped E-glass fibres in an epoxy matrix increases the strength of the composite. In this example, the volume fraction of glass fibres is about 0.5.

Amount of Fibre A greater volume fraction of fibres increases the strength and stiffness of the composite, as we would expect from the rule of mixtures. However, the maximum volume fraction is about 80%, beyond which fibres can no longer be completely surrounded by the matrix.

Orientation of Fibres The reinforcing fibres may be introduced into the matrix in a number of orientations. Short, randomly oriented fibres having a small aspect ratio – typical of fibreglass – are easily introduced into the matrix and give relatively isotropic behaviour in the composite.

Long, or even continuous, unidirectional arrangements of fibres produce anisotropic properties, with particularly good strength and stiffness parallel to the fibres. These fibres are often designated as 0° plies, indicating that all of the fibres are aligned with the direction of the applied stress. However, unidirectional orientations provide poor properties if the load is perpendicular to the fibres (Figure 16.12).

One of the unique characteristics of fibre-reinforced composites is that their properties can be tailored to meet different types of loading conditions. Long, continuous fibres can be introduced in several directions within the matrix (Figure 16.13); in orthogonal arrangements (0°/90° plies), good strength is obtained in two perpendicular directions. More complicated arrangements (such as 0°/–45°/90° plies) provide reinforcement in multiple directions.

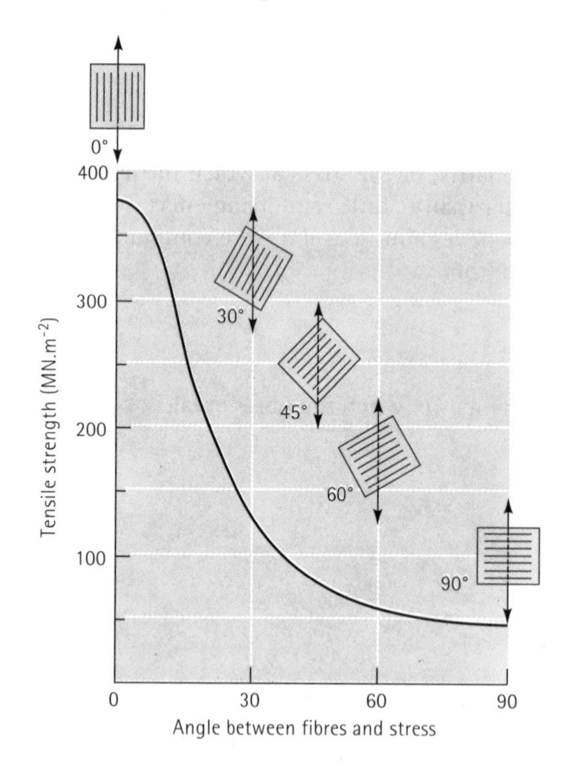

Figure 16.12
Effect of fibre orientation on the tensile strength of E-glass fibre-reinforced epoxy composites.

Fibres can also be arranged in three-dimensional patterns. In even the simplest of fabric weaves, the fibres in each individual layer of fabric have some small degree of orientation in a third direction. Better three-dimensional reinforcement occurs when fabric layers are knitted or stitched together. More complicated three-dimensional weaves (Figure 16.14) can also be used.

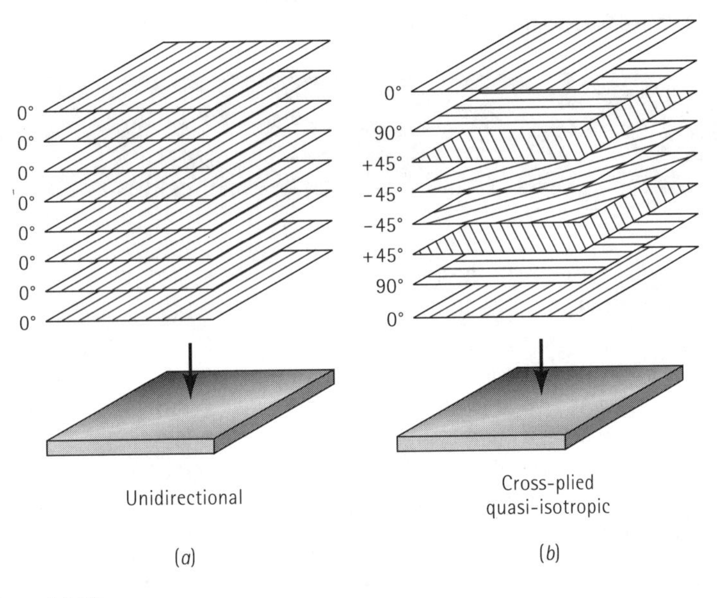

Figure 16.13
(*a*) Tapes containing aligned fibres can be joined to produce a multilayered unidirectional composite structure. (*b*) Tapes containing aligned fibres can be joined with different orientations to produce a quasi-isotropic composite. In this case, a 0°/±45°/90° composite is formed.

Fibre Properties In most fibre-reinforced composites, the fibres are strong, stiff, and lightweight. If the composite is to be used at elevated temperatures, the fibre should also have a high melting temperature. Thus the specific strength and specific modulus of the fibre are important characteristics:

$$\text{Specific strength} = \frac{\sigma}{\rho} \quad (16.11)$$

$$\text{Specific modulus} = \frac{E}{\rho} \quad (16.12)$$

where σ is the yield or tensile strength, ρ is the density, and E is the modulus of elasticity. Properties of typical fibres are shown in Table 16.2 and Figure 16.15. The highest specific modulus is usually found in materials having a low atomic number and covalent bonding, such as carbon and boron. These two elements also have a high strength and melting temperature.

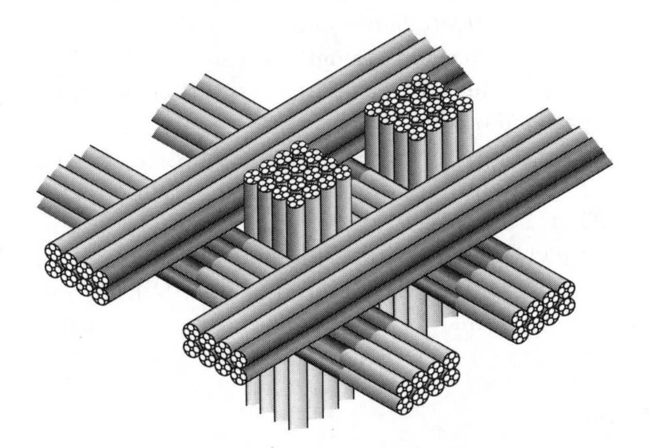

Figure 16.14
A three-dimensional weave for fibre-reinforced composites.

Table 16.2 Properties of selected fibre-reinforcing materials.

Material	Density ($Mg.m^{-3}$)	Tensile Strength ($MN.m^{-2}$)	Modulus of Elasticity ($GN.m^{-2}$)	Melting Temperature (°C)	Specific Modulus ($\times 10^6\ m^2.s^{-2}$)	Specific Strength ($\times 10^6\ m^2.s^{-2}$)
Polymers:						
Kevlar	1.44	4480	124	500	86	3.11
Nylon	1.14	825	2.8	249	2.46	0.72
Polyethylene	0.97	3300	172	147	177	3.40
Metals:						
Be	1.83	1275	303	1277	166	0.70
Boron	2.36	3450	379	2030	161	1.46
W	19.40	4000	407	3410	21	0.21
Glass:						
E-glass	2.55	3450	72.4	<1725	28	1.35
S-glass	2.50	4480	86.9	<1725	35	1.79
Carbon:						
HS (high strength)	1.75	5650	276	3700	158	3.23
HM (high modulus)	1.90	1860	531	3700	279	0.98
Ceramics:						
Al_2O_3	3.95	2070	379	2015	96	0.52
B_4C	2.36	2275	483	2450	205	0.96
SiC	3.00	3930	483	2700	161	1.31
ZrO_2	4.84	2070	345	2677	71	0.43
Whiskers:						
Al_2O_3	3.96	20700	427	1982	108	5.23
Cr	7.20	8890	241	1890	33	1.23
Graphite	1.66	20700	703	3700	423	12.47
SiC	3.18	20700	483	2700	152	12.47
Si_3N_4	3.18	13790	379		119	4.34

Aramid fibres, of which Kevlar is the best known example, are aromatic polyamide polymers strengthened by a backbone containing benzene rings (Figure 16.16) and are examples of liquid-crystalline polymers in that the polymer chains are rodlike and very stiff. Specially prepared polyethylene fibres are also available. Both the aramid and polyethylene fibres have excellent strength and stiffness but are limited to low-temperature use. Because of their lower density, polyethylene fibres have superior specific strength and specific modulus.

Ceramic fibres and whiskers, including alumina, glass, and silicon carbide, are strong and stiff. Glass fibres, which are the most commonly used, include pure silica, S-glass (SiO_2-25% Al_2O_3-10% MgO), and E-glass (SiO_2-18% CaO-15% Al_2O_3). Although they are considerably denser than the polymer fibres, the ceramics can be used at much higher temperatures. Beryllium and tungsten, although metallically bonded, have a high modulus that makes them attractive fibre materials for certain applications.

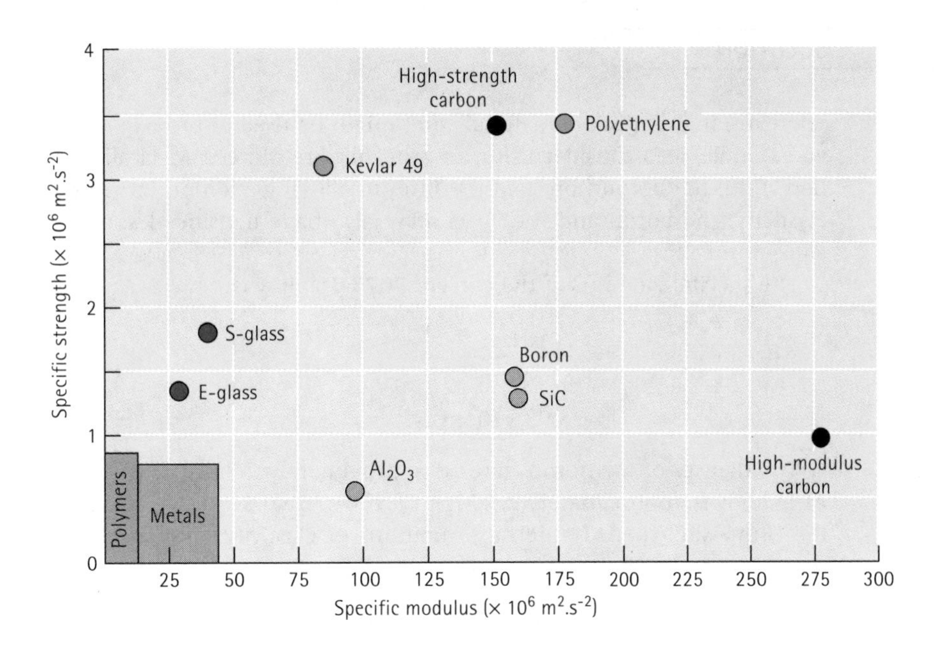

Figure 16.15
Comparison of the specific strength and specific modulus of fibres versus metals and polymers.

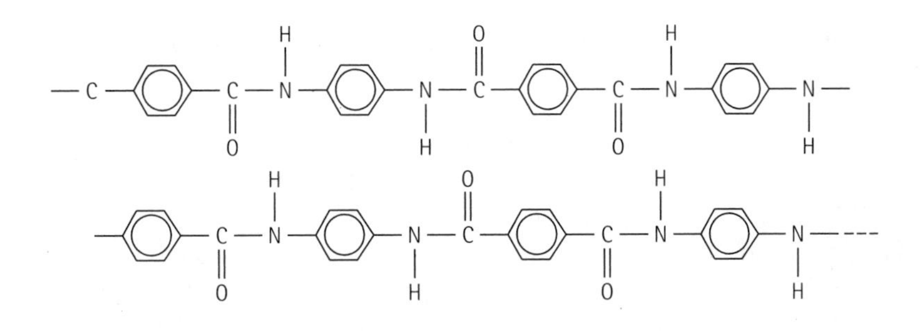

Figure 16.16
The structure of Kevlar. The fibres are joined by secondary bonds between oxygen and hydrogen on adjoining chains.

EXAMPLE 16.9

We are now using a 7075-T6 aluminium alloy (modulus of elasticity of 70 $GN.m^{-2}$) to make a 225 kg panel on a commercial aircraft. Experience has shown that each kilogram reduction in weight on the aircraft reduces the fuel consumption by 5000 litres each year. Select a material for the panel that will reduce weight, yet maintain the same specific modulus, and will be economical over a 10-year lifetime of the aircraft.

SOLUTION

There are many possible materials that might be used to provide a weight savings. As an example, let's consider using a boron fibre-reinforced Al-Li alloy in the T6 condition. Both the boron fibre and the lithium alloying addition increase the modulus of elasticity; the boron and the Al-Li alloy also have densities less than that of typical aluminium alloys.

The specific modulus of the current 7075-T6 alloy is:

$$\text{Specific modulus} = \frac{70\,\text{GN.m}^{-2}}{2.7\,\text{Mg.m}^{-3}}$$
$$= 25.93 \times 10^{6}\,\text{m}^{2}.\text{s}^{-2}$$

The density of the boron fibres is approximately 2.36 Mg.m^{-3} and that of a typical Al-Li alloy is approximately 2.5 Mg.m^{-3}. If we use 0.6 volume fraction boron fibres in the composite, then the density, modulus of elasticity, and specific modulus of the composite are:

$$\rho_c = (0.6)(2.36) + (0.4)(2.5) = 2.416\ \text{Mg.m}^{-3}$$

$$E_c = (0.6)(379) + (0.4)(75) = 257.4\ \text{GN.m}^{-2}$$

$$\text{Specific modulus} = \frac{257.4}{2.416} = 106.54 \times 10^{6}\,\text{m}^{6}.\text{s}^{-2}$$

If the specific modulus is the only factor influencing the design of the component, the thickness of the part might be reduced by 75%, giving a component weight of 56 kg rather than 225 kg. The weight savings would then be 169 kg, or (5 000 l/kg) (169 kg) = 845 000 l per year. At $0.50 per litre, $422 500 in fuel savings could be realised each year, or $4.225 million over the 10-year aircraft lifetime.

This is certainly an optimistic comparison, since strength or fabrication factors may not permit the part to be made as thin as suggested. In addition, the high cost of boron fibres (around $700/kg) and higher manufacturing costs of the composite compared with those of 7075 aluminium would reduce cost savings.

Matrix Properties The matrix supports the fibres and keeps them in the proper position, transfers the load to the strong fibres, protects the fibres from damage during manufacture and use of the composite, and prevents cracks in the fibre from propagating throughout the entire composite. The matrix usually provides the major control over electrical properties, chemical behaviour, and elevated-temperature use of the composite.

Polymer matrices are particularly common. Most polymer materials – both thermoplastics and thermosets – are available in short glass fibre-reinforced grades. These composites are formed into useful shapes by the processes described in Chapter 15. Sheet-moulding compounds (SMCs) and bulk-moulding compounds (BMCs) are typical of this type of composite. Thermosetting aromatic polyimides are used for somewhat higher temperature applications.

Metal matrix composites include aluminium, magnesium, copper, nickel, and intermetallic compound alloys reinforced with ceramic and metal fibres. A variety of aerospace and automotive applications are satisfied by the MMCs. The metal matrix permits the composite to operate at high temperatures, but producing the composite is often more difficult and expensive than producing the polymer matrix materials.

Amazingly, brittle ceramics may be used as a matrix in composites. The ceramic matrix composites have good properties at elevated temperatures and are lighter in weight than the high-temperature metal matrix composites. In a later section, we discuss how to develop toughness in CMCs.

Bonding and Failure Particularly in polymer and metal matrix composites, good bonding must be obtained between the various constituents. The fibres must be firmly bonded to the matrix material if the load is to be properly transmitted from the matrix to the fibres. In addition, the fibres may pull out of the matrix during loading, reducing the strength and fracture resistance of the composite, if bonding is poor. Figure 16.17 illustrates poor bonding of carbon fibres in a copper matrix. In some

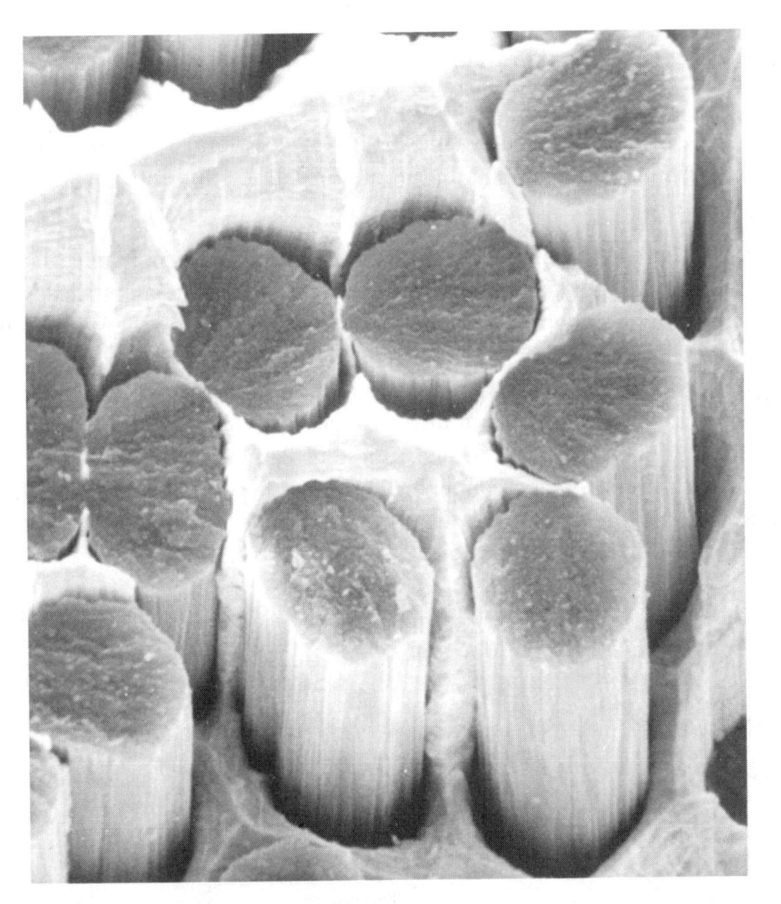

Figure 16.17
Scanning electron micrograph of the fracture surface of a silver-copper alloy reinforced with carbon fibres. Poor bonding causes much of the fracture surface to follow the interface between the metal matrix and the carbon tows (× 3000). (*From* Metals Handbook, *American Society for Metals, Vol. 9, 9th Ed., 1985.*)

cases, special coatings may be used to improve bonding. Glass fibres may be coated with a silane 'keying' agent (called sizing) to improve bonding and moisture resistance in fibreglass composites. Carbon fibres are similarly coated with an organic material to improve bonding. Boron fibres have been coated with silicon carbide or boron nitride to improve bonding with an aluminium matrix; in fact, these fibres have been called Borsic fibres to reflect the presence of the silicon carbide (SiC) coating.

Another property that must be considered when combining fibres into a matrix is the similarity between the coefficients of thermal expansion for the two materials. If the fibre expands and contracts at a rate much different from that of the matrix, fibres may break or bonding can be disrupted, causing premature failure.

In many composites, individual plies or layers of fabric are joined. Bonding between these layers must also be good or another problem – delamination – may occur. The layers may tear apart under load and cause failure. Using composites with a three-dimensional weave will help prevent delamination.

16.6 Manufacturing Fibres and Composites

Producing a fibre-reinforced composite involves several steps, including producing the fibres, arranging the fibres into bundles or fabrics, and introducing the fibres into the matrix.

Making the Fibre Metallic fibres, glass fibres, and many polymer fibres (including nylon, aramid, and polyacrylonitrile) can be formed by drawing processes, as described in Chapter 7 (wire drawing of metal) and Chapter 15 (using the spinnerette for polymer fibres).

Boron, carbon, and ceramics are too brittle and reactive to be worked by conventional drawing processes. Boron fibre is produced by chemical vapour deposition (CVD) [Figure 16.18(a)]. A very fine heated tungsten filament is used as a substrate, passing through a seal into a heated chamber. Vaporised boron compounds such as BCl_3 are introduced into the chamber, decompose, and permit boron to precipitate onto

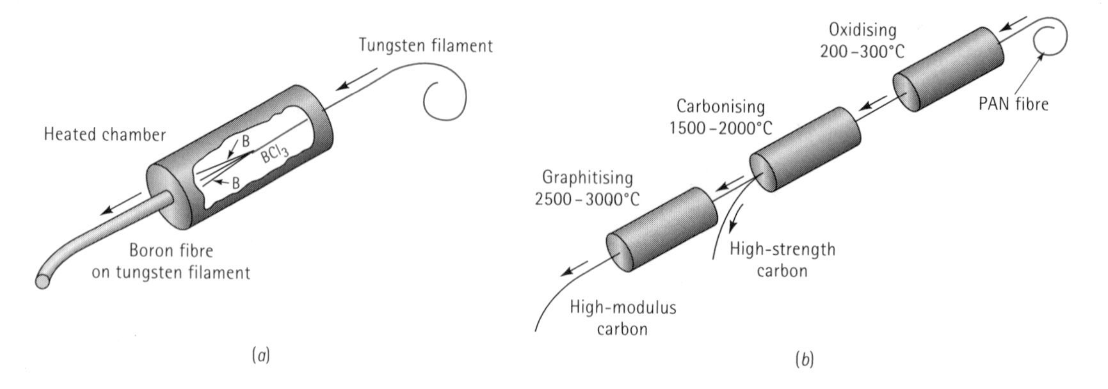

Figure 16.18 Methods for producing (*a*) boron and (*b*) carbon fibres.

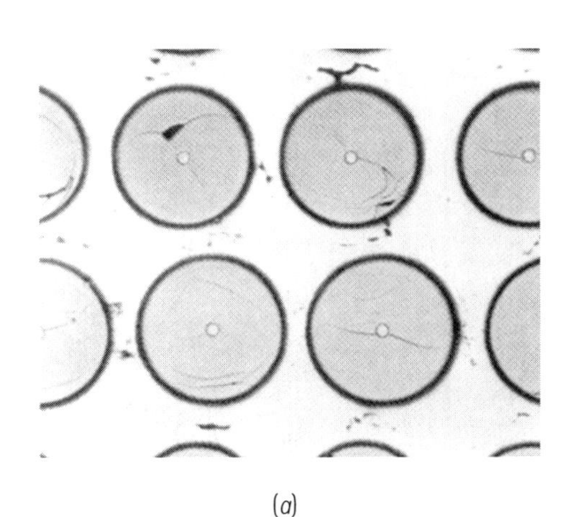

(a)

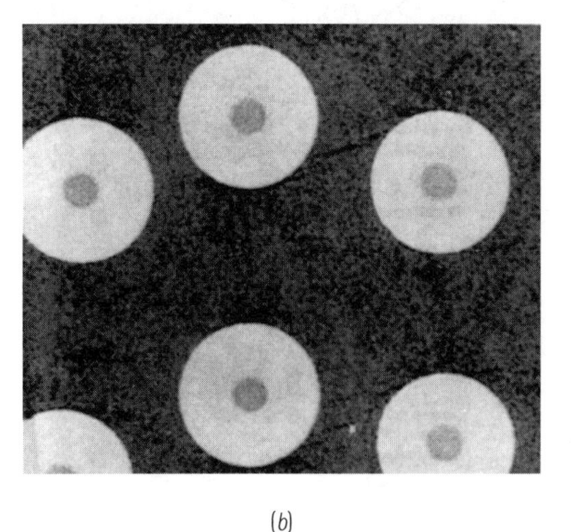

(b)

Figure 16.19
Photomicrographs of two fibre-reinforced composites: (*a*) In Borsic fibre-reinforced aluminium, the fibres are composed of a thick layer of boron deposited on a small-diameter tungsten filament (× 1000). (*From* Metals Handbook, *American Society for Metals, Vol. 9, 9th Ed., 1985.*) (*b*) In this microstructure of a ceramic fibre-ceramic matrix composite, silicon carbide fibres are used to reinforce a silicon nitride matrix. The SiC fibre is vapour-deposited on a small carbon precursor filament (× 125). (*Courtesy of Dr. R. T. Bhatt, NASA Lewis Research Center.*)

the tungsten wire (Figure 16.19). SiC fibres are made in a similar manner, with carbon fibres as the substrate for the vapour deposition of silicon carbide.

Carbon fibres are made by carbonising, or pyrolising, an organic filament, which is more easily drawn or spun into thin, continuous lengths [Figure 16.18(b)]. The organic filament, known as a precursor, is often rayon (a cellulosic polymer), polyacrylonitrile (PAN), or pitch (various aromatic organic compounds). High temperatures decompose the organic polymer, driving off all of the elements but carbon. As the carbonising temperature increases from 1 000°C to 3 000°C, the tensile strength decreases while the modulus of elasticity increases (Figure 16.20). Drawing the carbon filaments at critical times during carbonising may produce desirable preferred orientations in the final carbon filament.

Whiskers are single crystals with aspect ratios of 20 to 1 000. Because the whiskers contain no mobile dislocations, slip cannot occur and they have exceptionally high strengths.

Because of the complex processing required to produce fibres, their cost may be quite high. Table 16.3 gives approximate costs of common fibre reinforcements.

Arranging the Fibres Exceptionally fine filaments are bundled together as rovings, yarns, or tows. In yarns, as many as 10 000 filaments are twisted together to produce the fibre. A tow contains a few hundred to more than 100 000 untwisted filaments (Figure 16.21). Rovings are untwisted bundles of filaments, yarns, or tows.

Often, fibres are chopped into short lengths of 10 mm or less. These fibres, also called staples, are easily incorporated into the matrix and are typical of the sheet moulding and bulk moulding compounds for polymer matrix composites. The fibres often are present in the composite in a random orientation.

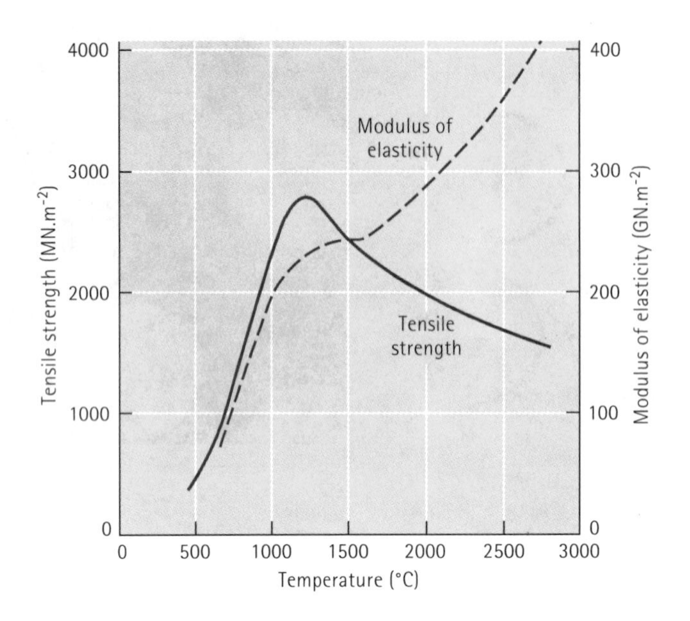

Figure 16.20
The effect of heat treatment temperature on the strength and modulus of elasticity of carbon fibres.

Table 16.3 Approximate costs of fibres.

Fibre	Cost (US $/kg)
Boron	700
SiC	220
Al_2O_3	66
Carbon	66
Aramid (Kevlar)	44
E-glass	7

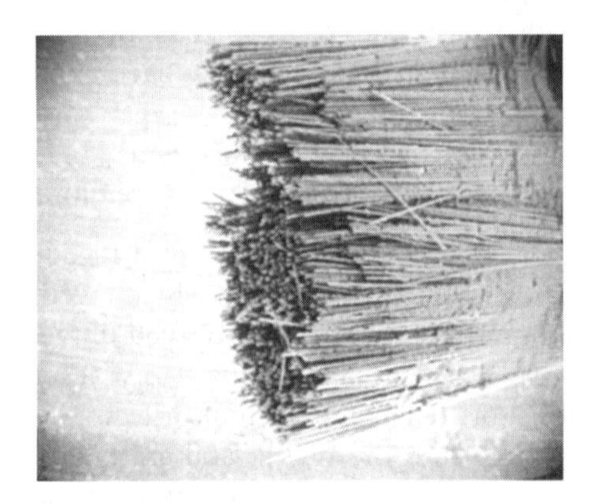

Figure 16.21
A scanning electron micrograph of a carbon tow containing many individual carbon filaments (× 200).

Long or continuous fibres for polymer matrix composites can be processed into mats or fabrics. Mats contain nonwoven, randomly oriented fibres loosely held together by a polymer resin. The fibres can also be woven, braided, or knitted into two-dimensional or three-dimensional fabrics. The fabrics are then impregnated with a polymer resin. The resins at this point in the processing have not yet been completely polymerised; these mats or fabrics are called prepregs.

When unidirectionally aligned fibres are to be introduced into a polymer matrix, tapes may be produced. Tows of fibres can be unwoven from spools onto a mandrel, which determines the spacing of the fibres, and prepregged with a polymer resin. These tapes, only a fraction of a millimetre thick, may be up to 1.2 m wide. Figure 16.22 illustrates that tapes can also be produced by covering the fibres with upper and lower layers of metal foil that are then joined by diffusion bonding.

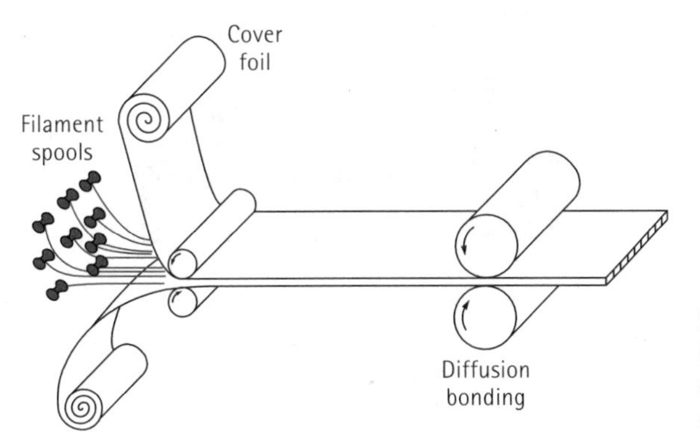

Figure 16.22
Production of fibre tapes by encasing fibres between metal cover sheets by diffusion bonding.

Producing the Composite A variety of methods for producing composite parts are used, depending on the application and materials. Short-fibre-reinforced composites are normally formed by mixing the fibres with a liquid or plastic matrix, then using relatively conventional techniques such as injection moulding for polymer-base composites or casting for metal matrix composites. Polymer matrix composites can also be produced by a spray-up method, in which short fibres mixed with a resin are sprayed against a form and cured.

Special techniques, however, have been devised for producing composites using continuous fibres, either in unidirectionally aligned, mat, or fabric form (Figure 16.23). In hand lay-up techniques, the tapes, mats, or fabrics are placed against a form, saturated with a polymer resin, rolled to ensure good contact and freedom from porosity, and finally cured. Fibreglass car and truck bodies might be made in this manner, which is generally slow and labour intensive.

Tapes and fabrics can also be placed in a die and formed by bag moulding. High-pressure gases or a vacuum are introduced to force the individual plies together so that good bonding is achieved during curing. Large polymer matrix components for the skins of military aircraft have been produced by these techniques. In matched die moulding, short fibres or mats are placed into a two-part die; when the die is closed, the composite shape is formed.

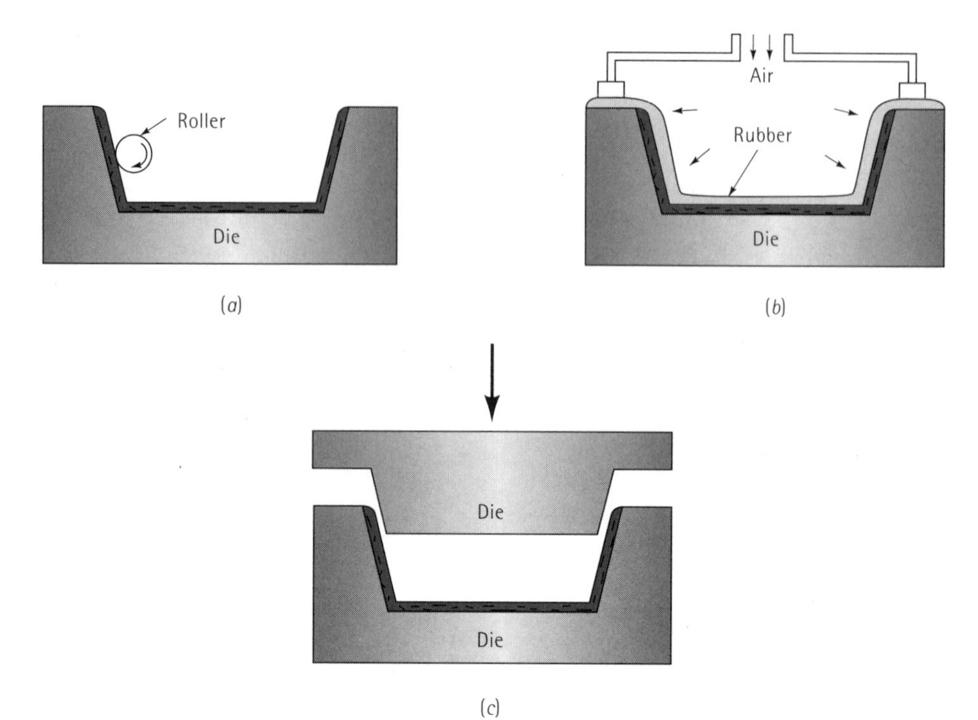

Figure 16.23
Producing composite shapes in dies by (*a*) hand lay-up, (*b*) pressure bag moulding, and (*c*) matched die moulding.

Filament winding is used to produce products such as pressure tanks and rocket motor casings (Figure 16.24). Fibres are wrapped around a form or mandrel to gradually build up a solid shape that may be up to a metre in thickness. The filament can be dipped in the polymer matrix resin prior to winding, or the resin can be impregnated around the fibre during or after winding. Curing completes the production of the composite part.

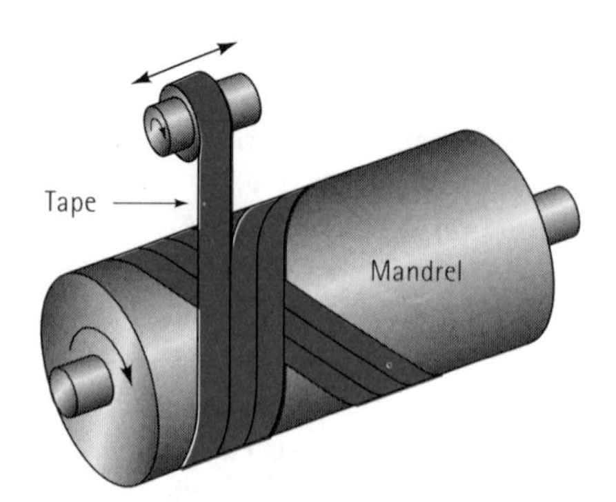

Figure 16.24
Producing composite shapes by filament winding.

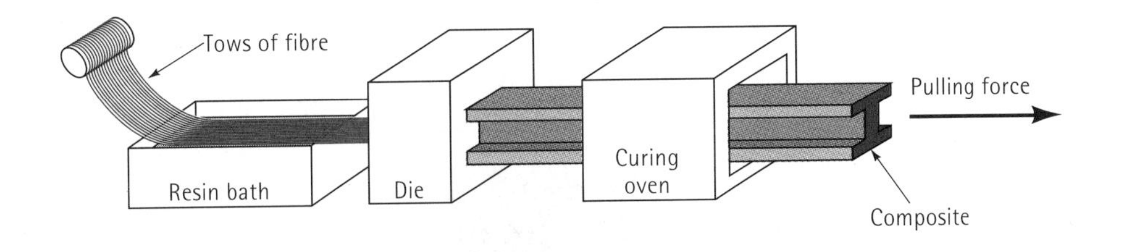

Figure 16.25
Producing composite shapes by pultrusion.

Pultrusion is used to form a simple-shaped product with a constant cross section, such as round, rectangular, pipe, plate, or sheet shapes (Figure 16.25). Fibres or mats are drawn from spools, passed through a polymer resin bath for impregnation, and gathered together to produce a particular shape before entering a heated die for curing. Curing of the resin is accomplished almost immediately, so a continuous product is produced. The pultruded stock can subsequently be formed into somewhat more complicated shapes, such as fishing rods, golf club shafts, and ski poles.

Metal matrix composites with continuous fibres are more difficult to produce than are the polymer matrix composites. Casting processes that force liquid around the fibres using capillary rise, pressure casting, vacuum infiltration, or continuous casting are illustrated in Figure 16.26. Various solid-state compaction processes can also be used. Figure 16.27 illustrates how several tapes can be placed into a closed die and deformed into shape; interdiffusion between the individual tapes produces a solid form.

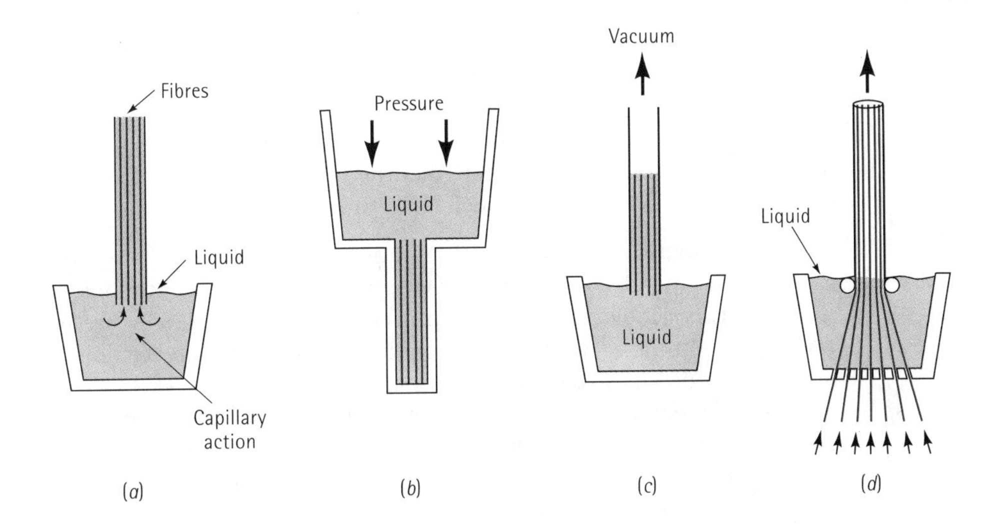

Figure 16.26
Casting techniques for producing composite materials: (*a*) capillary rise, (*b*) pressure casting, (*c*) vacuum infiltration, and (*d*) continuous casting.

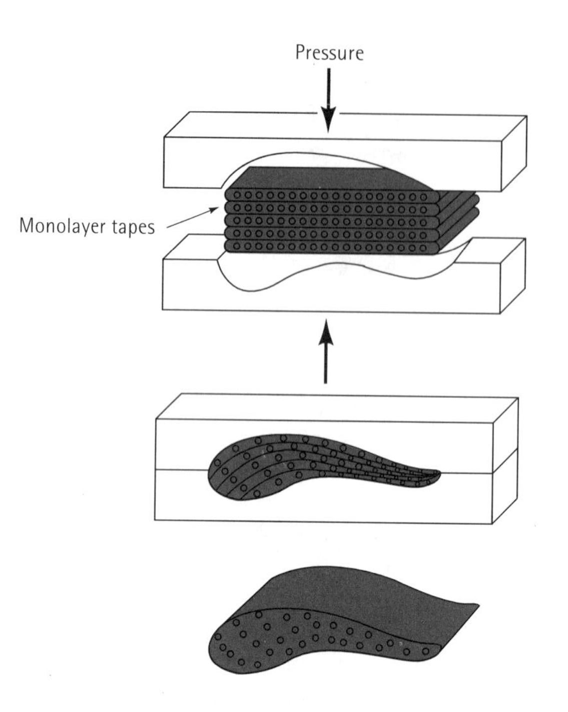

Figure 16.27
Closed die deformation and bonding of a composite composed of several tapes.

16.7 Fibre-Reinforced Systems and Applications

Before completing our discussion of fibre-reinforced composites, let's look at the behaviour and applications of some of the most common of these materials. Figure 16.28 compares the specific modulus and specific strength of several composites with those of metals and polymers. Note that the values in this figure are lower than those in Figure 16.15, since we are now looking at the composite, not just the fibre.

Advanced Composites The term advanced composites is often used when the composite is intended to provide service in very critical applications, as in the aerospace industry (Table 16.4). The advanced composites normally are polymer matrix composites reinforced with high-strength polymer, metal, or ceramic fibres. Carbon fibres are used extensively where particularly good stiffness is required; aramid – and, to an even greater extent, polyethylene – fibres are better suited to high-strength applications in which toughness and damage resistance are also important. Unfortunately, the polymer fibres lose their strength at relatively low temperatures, as do all of the polymer matrices (Figure 16.29).

The advanced composites are also frequently used for sporting goods. Tennis rackets, golf clubs, skis, ski poles, and fishing rods often contain carbon or aramid fibres because the higher stiffness provides better performance. In the case of golf clubs, carbon fibres allow less weight in the shaft and therefore more weight in the head. Fabric reinforced with polyethylene fibres is used for lightweight sails for racing yachts.

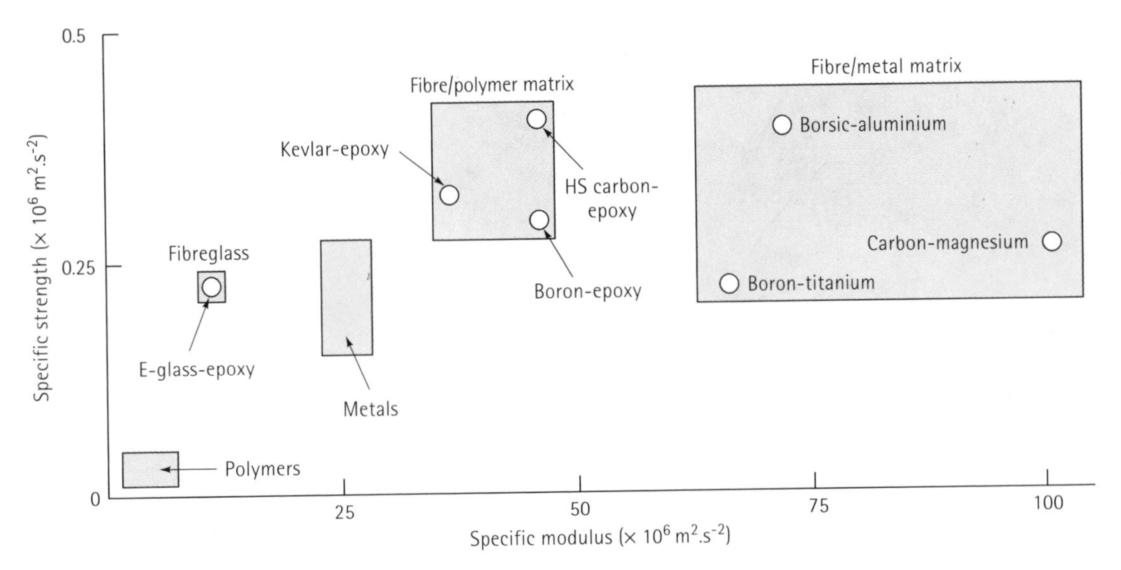

Figure 16.28
A comparison of the specific modulus and specific strength of several composite materials with those of metals and polymers.

A unique application for aramid fibre composites is armour. Tough Kevlar composites provide better ballistic protection than do other materials, making them suitable for lightweight, flexible bulletproof clothing.

Hybrid composites are composed of two or more types of fibres. For instance, Kevlar fibres may be mixed with carbon fibres to improve the toughness of a stiff composite, or Kevlar may be mixed with glass fibres to improve stiffness. Particularly good tailoring of the composite to meet specific applications can be achieved by controlling the amounts and orientations of each fibre type.

Tough composites can also be produced if careful attention is paid to the choice of materials and processing techniques. Better fracture toughness in the usually rather brittle composites can be obtained by using long fibres, amorphous (such as PEEK and PPS) rather then crystalline or cross-linked matrices, thermoplastic elastomer matrices, or interpenetrating network polymers.

Table 16.4 Examples of fibre-reinforced materials and applications.

Material	Applications
Borsic aluminium	Fan blades in engines, other aircraft and aerospace applications
Kevlar-epoxy and Kevlar-polyester	Aircraft, aerospace applications (including space shuttle), boat hulls, sporting goods (including tennis rackets, golf club shafts, fishing rods), flak jackets
Graphite-polymer	Aerospace and automotive applications, sporting goods
Glass-polymer	Lightweight automotive applications, water and marine applications, corrosion-resistant applications, sporting goods equipment, aircraft and aerospace components

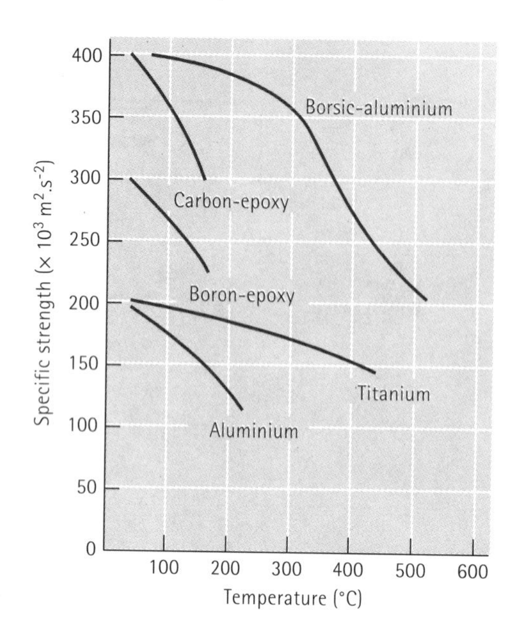

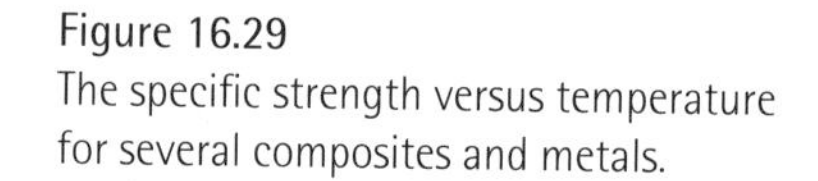
Figure 16.29
The specific strength versus temperature for several composites and metals.

Metal Matrix Composites These materials, strengthened by metal or ceramic fibres, provide high-temperature resistance. Aluminium reinforced with borsic fibres has been used extensively in aerospace applications, including struts for the space shuttle. Copper-base alloys have been reinforced with SiC fibres producing high-strength propellers for ships.

Aluminium is commonly used in metal matrix composites. Al_2O_3 fibres reinforce the pistons for some diesel engines; SiC fibres and whiskers are used in aerospace applications, including stiffeners and missile fins; and carbon fibres provide reinforcement for the aluminium antenna mast of the Hubble telescope. Polymer fibres, because of their low melting or degradation temperatures, are not normally used in a metallic matrix. *Polymets*, however, are produced by hot-extruding aluminium powder and high-melting-temperature liquid-crystalline polymers. A reduction of 1 000 to 1 during the extrusion process elongates the polymer into aligned filaments and bonds the aluminium powder particles into a solid matrix.

Metal matrix composites may find important applications in components for rocket or aircraft engines. Superalloys reinforced with metal fibres (such as tungsten) or ceramic fibres (such as SiC or B_4N) maintain their strength at higher temperatures, permitting jet engines to operate more efficiently. Similarly, titanium and titanium aluminides reinforced with SiC fibres are considered for turbine blades and disks.

A unique application for metal matrix composites is in the superconducting wire required for fusion reactors. The intermetallic compound Nb_3Sn has good superconducting properties but is very brittle. To produce Nb_3Sn wire, pure niobium wire is surrounded by copper as the two metals are formed into a wire composite (Figure 16.30). The niobium-copper composite wire is then coated with tin. The tin diffuses through the copper and reacts with the niobium to produce the intermetallic compound. Niobium-titanium systems are also used.

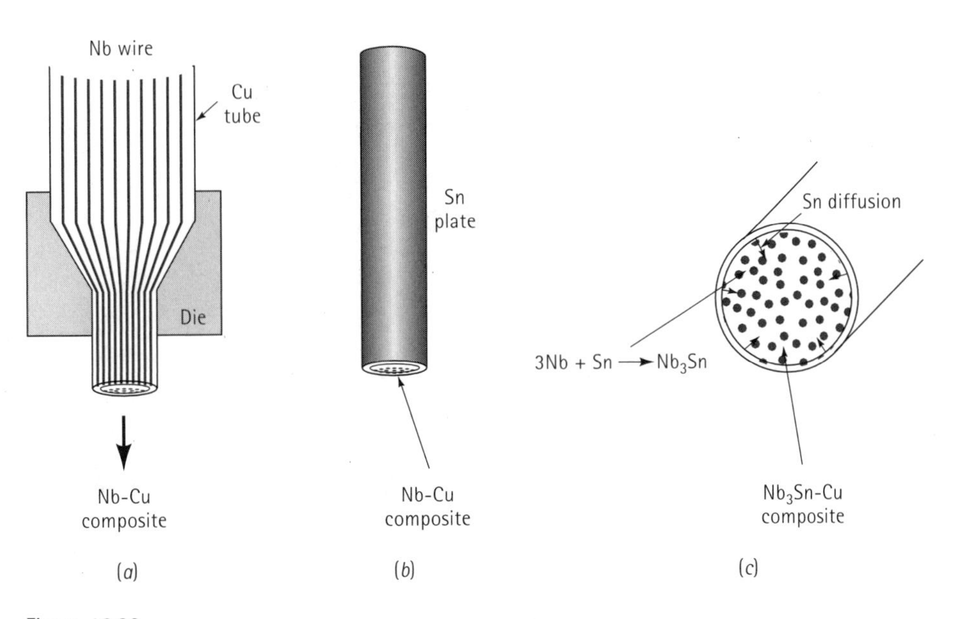

Figure 16.30
The manufacture of composite superconductor wires: (*a*) Niobium wire is surrounded with copper during forming. (*b*) Tin is placed onto Nb-Cu composite wire. (*c*) Tin diffuses to niobium to produce the Nb_3Sn-Cu composite.

Ceramic Matrix Composites Composites containing ceramic fibres in a ceramic matrix are also finding applications. Two important uses will be discussed to illustrate the unique properties that can be obtained with these materials.

Carbon-carbon composites are used for extraordinary temperature resistance in aerospace applications. Carbon-carbon composites can operate at temperatures of up to 3 000°C and, in fact, are stronger at high temperatures than at low temperatures (Figure 16.31). Carbon-carbon composites are made by forming a polyacrylonitrile or carbon fibre fabric into a mould, then impregnating the fabric with an organic resin, such as a phenolic. The part is pyrolised to convert the phenolic resin to carbon. The composite, which is still soft and porous, is impregnated and pyrolised several more times, continually increasing the density, strength, and stiffness. Finally the part is coated with silicon carbide to protect the carbon-carbon composite from oxidation. Strengths of 2 070 $MN.m^{-2}$ and stiffnesses of 345 $GN.m^{-2}$ can be obtained. Carbon-carbon composites have been used as nose cones and leading edges of high-performance aerospace vehicles such as the space shuttle, and as brake discs on racing cars and commercial jet aircraft.

Ceramic fibre-ceramic matrix composites provide improved strength and fracture toughness compared with conventional ceramics (Table 16.5). Fibre reinforcements improve the toughness of the ceramic matrix in several ways. First, a crack moving through the matrix encounters a fibre; if the bonding between the matrix and the fibre is poor, the crack is forced to propagate around the fibre in order to continue the fracture process. In addition, poor bonding allows the fibre to begin to pull out of

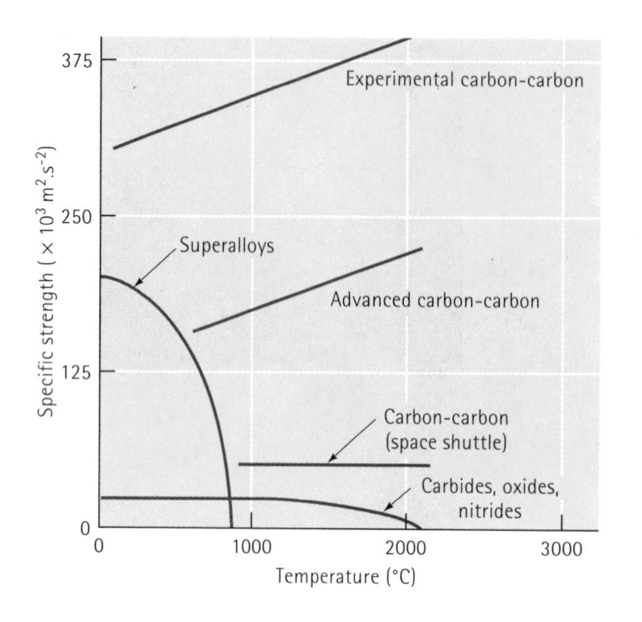

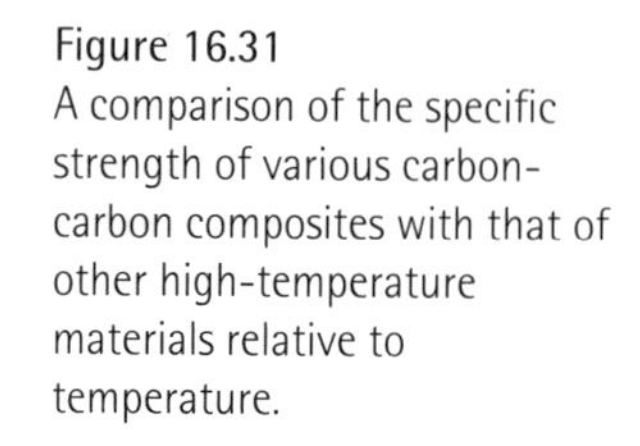
Figure 16.31
A comparison of the specific strength of various carbon-carbon composites with that of other high-temperature materials relative to temperature.

the matrix [Figure 16.32(a)]. Both processes consume energy, thereby increasing fracture toughness. Finally, as a crack in the matrix begins, unbroken fibres may bridge the crack, providing a compressive stress that helps keep the crack from opening [Figure 16.32(b)].

Unlike polymer and metal matrix composites, poor bonding – rather than good bonding – is required! Consequently, control of the interface structure is crucial. In a glass ceramic (based on $Al_2O_3 \cdot SiO_2 \cdot Li_2O$) reinforced with SiC fibres, an interface layer containing carbon and NbC is produced that makes debonding of the fibre from the matrix easy. If, however, the composite is heated to a high temperature, the interface is oxidised; the oxide occupies a large volume, exerts a clamping force on the fibre, and prevents easy pull-out. Fracture toughness is then decreased.

Table 16.5 Effect of SiC reinforcement fibres on the properties of selected ceramic materials.

Material	Flexural strength ($MN.m^{-2}$)	Fracture toughness ($MN.m^{-3/2}$)
Al_2O_3	550	5.5
Al_2O_3/SiC	790	8.8
SiC	495	4.4
SiC/SiC	760	25.3
ZrO_2	205	5.5
ZrO_2/SiC	450	22.2
Si_3N_4	470	4.4
Si_3N_4/SiC	790	56.0
Glass	62	1.1
Glass/SiC	825	18.7
Glass ceramic	205	2.2
Glass ceramic/SiC	825	17.6

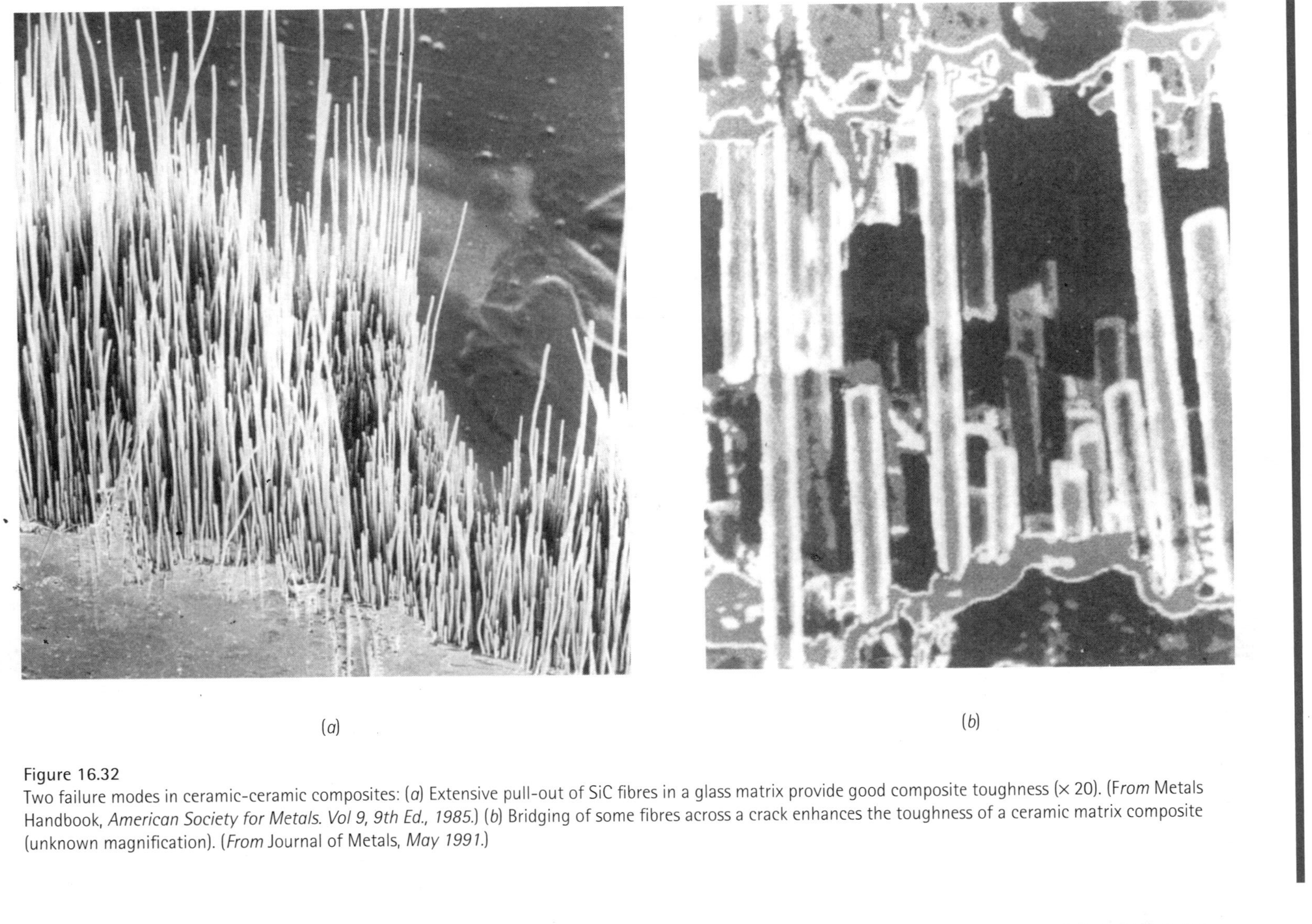

Figure 16.32
Two failure modes in ceramic-ceramic composites: (*a*) Extensive pull-out of SiC fibres in a glass matrix provide good composite toughness (× 20). (*From* Metals Handbook, *American Society for Metals. Vol 9, 9th Ed., 1985.*) (*b*) Bridging of some fibres across a crack enhances the toughness of a ceramic matrix composite (unknown magnification). (*From* Journal of Metals, *May 1991.*)

EXAMPLE 16.10 **Design of a Composite Strut**

Design a unidirectional fibre-reinforced epoxy matrix strut having a round cross-section. The strut is 3 m long and, when a force of 2.2 kN is applied, it should stretch no more than 2.5 mm. We want to ensure that the stress acting on the strut is less than the yield strength of the epoxy matrix, 80 MN.m^{-2}. If the fibres should happen to break, the strut will stretch an extra amount but may not catastrophically fracture. Epoxy costs about \$1.75/kg and has a modulus of elasticity of 3.5 GN.m^{-2}.

SOLUTION

Suppose that the strut were made entirely of epoxy (that is, no fibres):

$$\varepsilon_{\max} = \frac{2.5\text{ mm}}{3000\text{ mm}} = 0.83\times10^{-3}$$

$$\sigma_{\max} = E\varepsilon = (3.5\times10^{3})(0.83\times10^{-3}) = 2.92\text{ MN.m}^{-2}$$

$$A_{\text{strut}} = \frac{F}{\sigma} = \frac{2200}{2.92} = 753.4\text{ mm}^2 \text{ or } d = 31\text{ mm}$$

Since ρ_{epoxy} = 1.25 Mg.m^{-3}:

$$\text{Weight}_{\text{strut}} = (1.25)(753.4\times10^{-6})(3) = 2.83\times10^{-3}\text{ Mg} = 2.83\text{ kg}$$

$$\text{Cost}_{\text{strut}} = (2.83\text{ kg})(\$1.75/\text{kg}) = \$4.95$$

With no reinforcement, the strut is large and heavy; the materials cost is high due to the large amount of epoxy needed.

In a composite, the maximum strain is still 0.83 × 10^{-3}. If we make the strut as small as possible – that is, it operates at 80 MN.m^{-2} – then the minimum modulus of elasticity E_c of the composite is:

$$E_c > \frac{\sigma}{\varepsilon_{\max}} = \frac{80}{0.83\times10^{-3}} = 96.4\text{ GN.m}^{-2}$$

Let's look at several possible composite systems. The modulus of glass fibres is less than 96.4 GN.m^{-2}; therefore, glass reinforcement is not a possible choice.

For high modulus carbon fibres, E = 531 GN.m^{-2}; the density is 1.9 Mg.m^{-3}, and the cost is about \$66/kg. The minimum volume fraction of carbon fibres needed to give a composite modulus of 96.4 GN.m^{-2} is:

$$E_c = f_c(531) + (1 - f_c)(3.5) > 96.4$$

$$f_c = 0.176$$

The volume fraction of epoxy remaining is 0.824. An area of 0.824 times the total cross-sectional area of the strut must support a 2.2 kN load with no more than 80 MN.m^{-2} if all of the fibres should fail:

$$A_{\text{epoxy}} = 0.824 A_{\text{total}} = \frac{F}{\sigma} = \frac{2200}{80} = 27.5 \text{ mm}^2$$

$$A_{\text{total}} = \frac{27.5}{0.824} = 33.4 \text{ mm}^2 \quad \text{or} \quad d = 6.5 \text{ mm}$$

$$\text{Volume}_{\text{strut}} = (33.4 \times 10^{-6} \text{m}^2)(3 \text{ m}) = 0.1 \times 10^{-3} \text{ m}^3$$

$$\text{Weight}_{\text{strut}} = \rho V = [(1.9)(0.176) + (1.25)(0.824)](0.1 \times 10^{-3}) = 0.137 \times 10^{-3} \text{ Mg} = 0.137 \text{ kg}$$

$$\text{Weight fraction carbon} = \frac{(0.176)(1.9)}{(0.176)(1.9) + (0.824)(1.25)} = 0.245$$

$$\text{Weight carbon} = (0.245)(0.137) = 0.034 \text{ kg}$$

$$\text{Weight epoxy} = (0.755)(0.137) = 0.103 \text{ kg}$$

$$\text{Cost}_{\text{strut}} = (0.034 \text{ kg})(\$66/\text{kg}) + (0.103 \text{ kg})(\$1.75/\text{kg}) = \$2.42$$

The carbon-fibre reinforced strut is less than one-quarter the diameter of an all-epoxy structure, with only 5% of the weight and half of the cost.

We might also repeat these calculations using Kevlar fibres, with a modulus of 124 GN.m^{-2}, a density of 1.44 Mg.m^{-3}, and a cost of about $44/kg. By doing so, we would find that a volume fraction of 0.8 fibres is required. Note that 0.8 volume fraction is at the maximum of fibre volume that can be incorporated into a matrix. We would also find that the required diameter of the strut is 13.1 mm and that the strut weighs 0.57 kg and costs $20.26. The modulus of the Kevlar is not high enough to offset its high cost.

Although the carbon fibres are the most expensive, they permit the lightest weight and the lowest material cost strut. (This calculation does not, however, take into consideration the costs of manufacturing the strut.) Our design, therefore, is to use a 6.5 mm-diameter strut containing 0.176 volume fraction high modulus carbon fibre.

16.8 Laminar Composite Materials

Laminar composites include very thin coatings, thicker protective surfaces, claddings, bimetallics, laminates, and a host of other applications. In addition, the fibre-reinforced composites produced from tapes or fabrics can be considered as partly laminar. Many laminar composites are designed to improve corrosion resistance while retaining low cost, high strength, or light weight. Other important characteristics include superior wear or abrasion resistance, improved appearance, and unusual thermal expansion characteristics.

Rule of Mixtures Some properties of the laminar composite materials parallel to the lamellae are estimated from the rule of mixtures. Density, electrical and thermal conductivity, and modulus of elasticity can be calculated with little error:

$$\begin{aligned} \text{Density} &= \rho_c = \Sigma f_i \rho_i \\ \text{Electrical conductivity} &= \sigma_c = \Sigma f_i \sigma_i \\ \text{Thermal conductivity} &= K_c = \Sigma f_i K_i \\ \text{Modulus of elasticity} &= E_c = \Sigma f_i E_i \end{aligned} \tag{16.13}$$

The laminar composites are very anisotropic. The properties perpendicular to the lamellae are:

$$\begin{aligned} \text{Electrical conductivity} &= \frac{1}{\sigma_c} = \Sigma \frac{f_i}{\sigma_i} \\ \text{Thermal conductivity} &= \frac{1}{K_c} = \Sigma \frac{f_i}{K_i} \\ \text{Modulus of elasticity} &= \frac{1}{E_c} = \Sigma \frac{f_i}{E_i} \end{aligned} \tag{16.14}$$

However, many of the really important properties, such as corrosion and wear resistance, depend primarily on only one of the components of the composite, so the rule of mixtures is not applicable.

Producing Laminar Composites Several methods are used to produce laminar composites, including a variety of deformation and joining techniques (Figure 16.33).

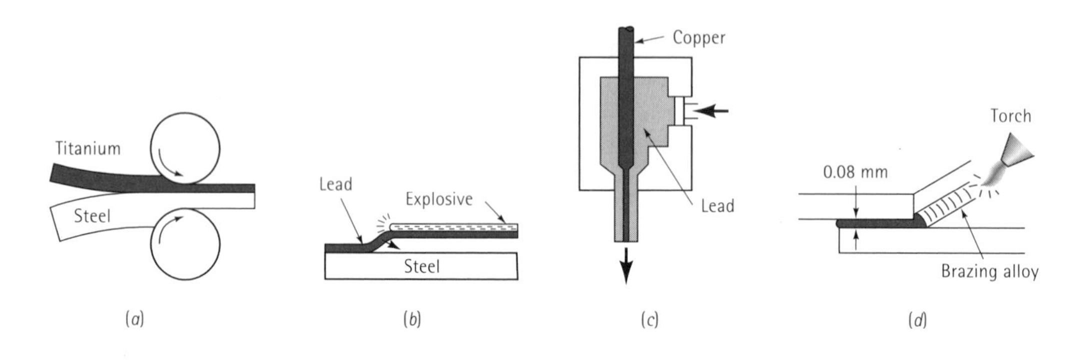

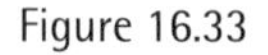
Figure 16.33
Techniques for producing laminar composites: (*a*) roll bonding, (*b*) explosive bonding, (*c*) coextrusion, and (*d*) brazing.

EXAMPLE 16.11

Capacitors used to store electrical charge are essentially laminar composites built up from alternating layers of a conductor and an insulator (Figure 16.34). Suppose we construct a capacitor from 10 sheets of mica, each 0.1 mm thick, and 11 sheets of aluminium, each 0.006 mm thick. The electrical conductivity of aluminium is 38×10^6 ohm^{-1}.m^{-1} and the conductivity of the mica is 10^{-11} ohm^{-1}.m^{-1}. Determine the electrical conductivity of the capacitor parallel and perpendicular to the sheets.

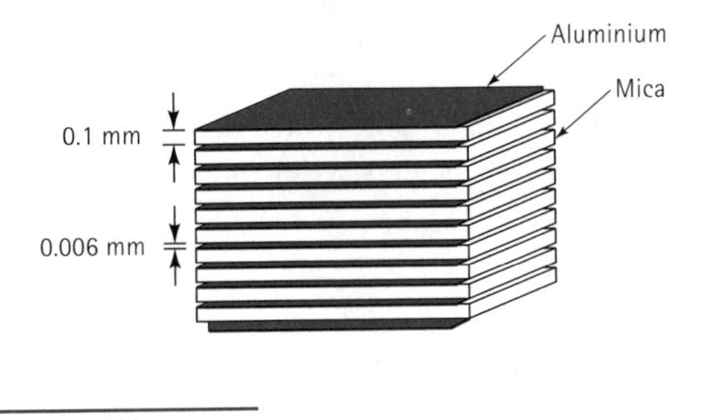

Figure 16.34
A capacitor, composed of alternating layers of aluminium and mica, is one example of a laminar composite (*for* Example 16.11).

SOLUTION

Suppose the capacitor plates are 100 mm². Then the volume fractions are:

$$V_{\text{Al}} = (11\ \text{sheets})(0.006)(100) = 6.6\ \text{mm}^3$$

$$V_{\text{mica}} = (10\ \text{sheets})(0.1)(100) = 100\ \text{mm}^3$$

$$f_{\text{Al}} = \frac{6.6}{6.6+100} = 0.062 \qquad f_{\text{mica}} = \frac{100}{6.6+100} = 0.938$$

Parallel:

$$\sigma = (0.062)(38\times10^6) + (0.938)(10^{-11}) = 2.4\times10^6\ \text{ohm}^{-1}.\text{m}^{-1}$$

Perpendicular :

$$\frac{1}{\sigma} = \frac{0.062}{38\times10^6} + \frac{0.938}{10^{-11}} = 93.8\times10^9$$

$$\sigma = \frac{1}{93.8\times10^9} = 10.66\times10^{-12}\ \text{ohm}^{-1}.\text{m}^{-1}$$

The composite, or capacitor, has high conductivity parallel to the plates, but acts as an insulator perpendicular to the plates.

Individual plies are often joined by *adhesive bonding*, as is the case in producing plywood. Polymer matrix composites built up from several layers of fabric are also joined by adhesive bonding; a film of unpolymerised polymer is placed between each layer of fabric. When the layers are pressed at an elevated temperature, polymerisation is completed and the layers of fibres are joined to produce composites that may be dozens of layers thick.

Most of the metallic laminar composites, such as claddings and bimetallics, are produced by *deformation bonding*, such as hot or cold roll bonding. The pressure exerted by the rolls breaks up the oxide film at the surface, brings the surfaces into atom-to-atom contact, and permits the two surfaces to be joined. Explosive bonding can also be used. An explosive charge provides the pressure required to join metals. This process is particularly well suited for joining very large plates that will not fit into a rolling mill.

Very simple laminar composites, such as coaxial cable, are produced by coextruding two materials through a die in such a way that the soft material surrounds the harder material. Metal conductor wire can be coated with an insulating thermoplastic polymer in this manner.

Brazing can join composite plates. The metallic sheets are separated by a very small clearance – preferably, about 0.08 mm – and heated above the melting temperature of the brazing alloy. The molten brazing alloy is drawn into the thin joint by capillary action.

16.9 Examples and Applications of Laminar Composites

The number of laminar composites is so varied and their applications and intentions are so numerous that we cannot make generalisations concerning their behaviour. Instead we will examine the characteristics of a few commonly used examples.

Laminates Laminates are layers of materials joined by an organic adhesive. In safety glass, a plastic adhesive, such as polyvinyl butyral, joins two pieces of glass; the adhesive prevents fragments of glass from flying about when the glass is broken. Laminates are used for insulation in motors, for gears, for printed circuit boards, and for decorative items such as Formica® worktops and furniture.

Microlaminates include composites composed of alternating layers of aluminium sheet and fibre-reinforced polymer. *Arall* (aramid aluminium laminate) and *Glare* (glass aluminium laminate) have been developed as possible skin materials for aircraft. In Arall, an aramid fibre such as Kevlar is prepared as a fabric or unidirectional tape, impregnated with adhesive, and laminated between layers of aluminium alloy (Figure 16.35). The composite laminate has an unusual combination of strength, stiffness, corrosion resistance, and light weight. Fatigue resistance is improved, since the interface between the layers may block cracks. Compared with polymer matrix composites, the microlaminates have good resistance to lightning strike damage (which is important in aerospace applications), are formable and machinable, and are easily repaired.

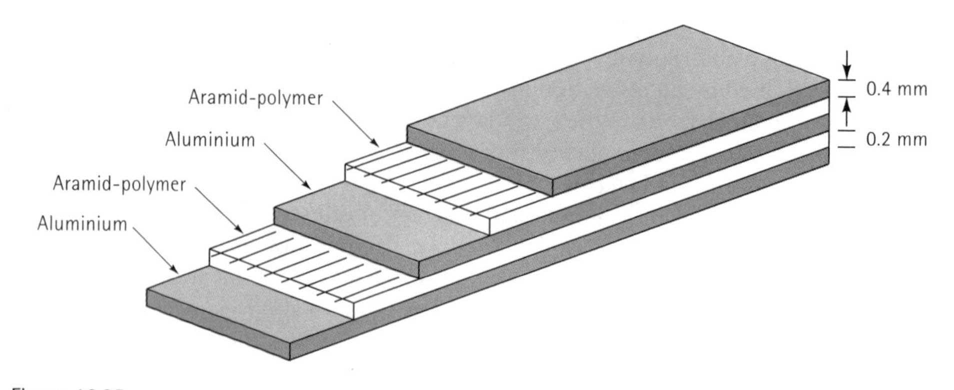

Figure 16.35
Schematic diagram of an aramid aluminium laminate, Arall, which has potential for aerospace applications.

Clad Metals Clad materials are metal-metal composites. A common example of cladding is United States silver coinage. A Cu-80% Ni alloy is bonded to both sides of a Cu-20% Ni alloy. The ratio of thicknesses is about 1/6 : 2/3 : 1/6. The high-nickel alloy is a silver colour, while the predominantly copper core provides relatively low cost.

Clad materials provide a combination of good corrosion resistance with high strength. *Alclad* is a clad composite in which commercially pure aluminium is bonded to higher-strength aluminium alloys. The pure aluminium protects the higher-strength alloy from corrosion. The thickness of the pure aluminium layer is about 1% to 15% of the total thickness. Alclad is used in aircraft construction, heat exchangers, building construction, and storage tanks, where combinations of corrosion resistance, strength, and light weight are desired.

Bimetallics Temperature indicators and controllers take advantage of the different coefficients of thermal expansion of the two metals in laminar composite. If two pieces of metal are heated, the metal with the higher coefficient of thermal expansion becomes longer (Figure 16.36). If two pieces of metal are rigidly bonded together, the difference in their coefficients causes the strip to bend and produce a curved surface. The amount of movement depends on the temperature; by measuring the curvature or deflection of the strip, we can determine the temperature. Likewise, if the free end of the strip activates a relay, the strip can turn on or off a furnace or air conditioner to regulate temperature.

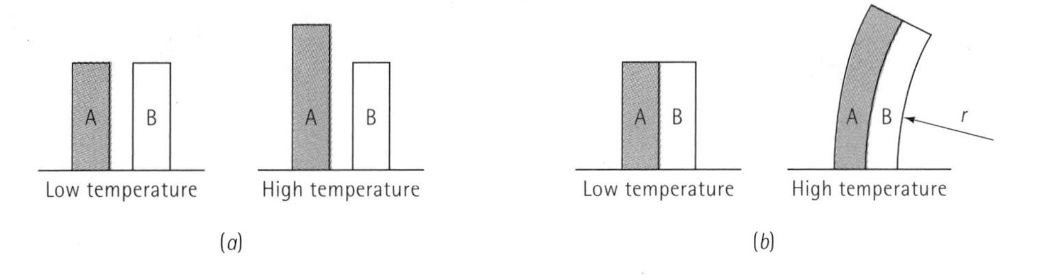

Figure 16.36
The effects of the thermal expansion coefficient on the behaviour of bimetallics: (*a*) increasing the temperature increases the length of one metal more than the other. (*b*) If the two metals are joined, the difference in expansion causes a radius of curvature to be produced.

Metals selected for bimetallics must have (a) very different coefficients of thermal expansion, (b) expansion characteristics that are reversible and repeatable, and (c) a high modulus of elasticity, so that the bimetallic device can do work. Often the low-expansion strip is made from Invar, an iron-nickel alloy, whereas the high-expansion strip may be brass, Monel, or pure nickel.

Bimetallics can act as circuit breakers as well as thermostats; if a current passing through the strip becomes too high, heating causes the bimetallic to deflect and break the circuit.

16.10 Sandwich Structures

Sandwich materials have thin layers of a facing material joined to a lightweight filler material, such as a polymer foam. Neither the filler nor the facing material is strong or rigid, but the composite possesses both properties. A familiar example is corrugated cardboard. A corrugated core of paper is bonded on either side to flat, thick paper. Neither the corrugated core nor the facing paper is rigid, but the combination is.

Another important example is the honeycomb structure used in aircraft applications. A honeycomb is produced by bonding thin aluminium strip at selected locations. The honeycomb material is then expanded to produce a very low density cellular panel that, by itself, is unstable (Figure 16.37). When an aluminium facing sheet is adhesively bonded to either side of the honeycomb, however, a very stiff, rigid, strong, and exceptionally lightweight sandwich with a density as low as 0.04 $Mg.m^{-3}$ is obtained.

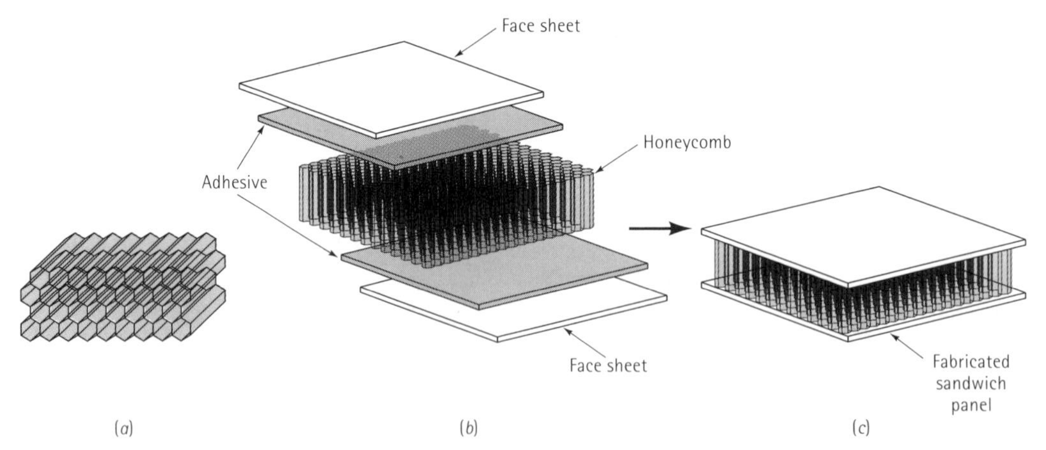

Figure 16.37
A hexagonal cell honeycomb core (*a*) can be joined to two face sheets by means of adhesive sheets (*b*), producing an exceptionally lightweight yet stiff, strong honeycomb sandwich structure (*c*).

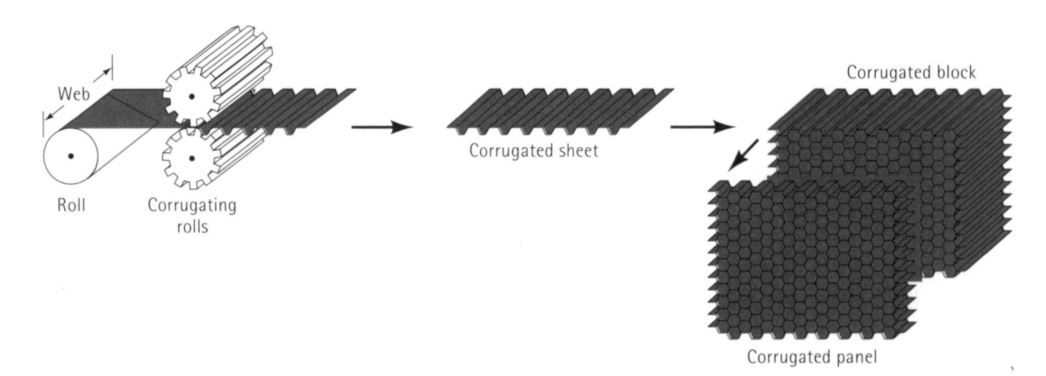

Figure 16.38
In the corrugation method for producing a honeycomb core, a roll of the material (such as aluminium) is corrugated between two rolls. The corrugated sheets are joined together with adhesive and then cut to the desired thickness.

The honeycomb cells can have a variety of shapes, including hexagonal, square, rectangular, and sinusoidal, and they can be made from aluminium, fibreglass, paper, aramid polymers, and other materials. The honeycomb cells can be filled with foam or fibreglass to provide excellent sound and vibration absorption. Figure 16.38 describes one method by which the honeycomb can be fabricated.

SUMMARY

Composites are composed of two or more materials joined to give a combination of properties that cannot be attained in any of the original materials. Virtually any combination of metals, polymers, and ceramics is possible. In many cases, the rule of mixtures can be used to estimate the properties of the composite.

- Dispersion strengthened materials, which are not true composites, contain exceptionally small oxide particles in a metal matrix. The small stable dispersoids interfere with slip, providing good mechanical properties at elevated temperatures.
- True particulate composites contain particles that impart combinations of properties to the composite. Metal matrix composites contain ceramic or metallic particles that provide improved strength and wear resistance and ensure good electrical conductivity, toughness, or corrosion resistance. Polymer matrix composites contain particles that enhance stiffness, heat resistance, or electrical conductivity while maintaining light weight, ease of fabrication, or low cost.
- Fibre-reinforced composites provide improvements in strength, stiffness, or high-temperature performance in metals and polymers and impart toughness to ceramics:
    - Fibres typically have low densities, giving high specific strength and specific modulus, but they often are very brittle.
    - Fibres can be continuous or discontinuous. Discontinuous fibres with a high aspect ratio l/d produce better reinforcement.
    - Fibres are introduced into a matrix in a variety of orientations. Random orientations and isotropic behaviour are obtained using discontinuous fibres; unidirectionally aligned fibres produce composites with anisotropic behaviour, with large improvements in strength and stiffness parallel to the fibre direction. Properties can be tailored to meet the imposed loads by orienting the fibres in multiple directions.
- Laminar composites are built of layers of different materials. These layers may be sheets of different metals, with one metal providing strength and the other providing hardness or corrosion resistance. Layers may also include sheets of fibre-reinforced material bonded to metal or polymer sheets or even to fibre-reinforced sheets having different fibre orientations. The laminar composites are always anisotropic.

Sandwich materials, including honeycombs, are exceptionally lightweight laminar composites, with solid facings joined to an almost hollow core.

GLOSSARY

Aramid fibres

Polymer fibres, such as Kevlar, formed from polyamides, which contain the benzene ring in the backbone of the polymer.

Aspect ratio

The length of a fibre divided by its diameter.

Bimetallic

A laminar composite material produced by joining two strips of metal with different thermal expansion coefficients, making the material sensitive to temperature changes.

Brazing

A process in which a liquid filler metal is introduced by capillary action between two solid base materials that are to be joined. Solidification of the brazing alloy provides the bond.

Carbonising

Driving off the noncarbon atoms from a polymer fibre, leaving behind a carbon fibre of high strength. Also known as pyrolising.

Cemented carbides

Particulate composites containing hard ceramic particles bonded with a soft metallic matrix. The composite combines high hardness and cutting ability, yet still has good shock resistance.

Chemical vapour deposition

Method for manufacturing materials by condensing the material from a vapour onto a solid substrate.

Cladding

The good corrosion-resistant or high-hardness layer of a laminar composite formed onto a less expensive or higher-strength backing.

Compocasting

Injection of a thixotropic mixture of an alloy and a filler material into a die at high pressure to form a composite.

Delamination

Separation of individual plies of a fibre-reinforced composite.

Dispersoids

Tiny oxide particles formed in a metal matrix that interfere with dislocation movement and provide strengthening, even at elevated temperatures.

Filament winding

Process for producing fibre-reinforced composites in which continuous fibres are wrapped around a form or mandrel. The fibres may be prepregged or the filament-wound structure may be impregnated to complete the production of the composite.

Honeycomb
A lightweight but stiff assembly of aluminium strip joined and expanded to form the core of a sandwich structure.

Precursor
The polymer fibre that is carbonised to produce carbon fibres.

Prepregs
Layers of fibres in unpolymerised resins. After the prepregs are stacked to form a desired structure, polymerisation joins the layers together.

Pultrusion
A method for producing composites containing mats or continuous fibres.

Rovings
Bundles of less than 10 000 filaments.

Rule of mixtures
The statement that the properties of a composite material are a function of the volume fraction of each material in the composite.

Sandwich
A composite material constructed of a lightweight, low-density material surrounded by dense, solid layers. The sandwich combines overall light weight with excellent stiffness.

Sizing
Coating glass fibres with an organic material to improve bonding and moisture resistance in fibreglass.

Specific modulus
The modulus of elasticity divided by the density.

Specific strength
The strength of a material divided by the density.

Staples
Fibres chopped into short lengths.

Tapes
Single-filament-thick strips of prepregs, with the filaments either unidirectional or woven fibres. Several layers of tapes can be joined to produce composite structures.

Thixotropic
The ability of a partly liquid–partly solid material to maintain its shape until a stress is applied, when it then flows like a liquid.

Tow
A bundle of more than 10 000 filaments.

Whiskers
Very fine fibres grown in a manner that produces single crystals with no mobile dislocations, thus giving nearly theoretical strengths.

Yarns
Continuous fibres produced from a group of twisted filaments.

PROBLEMS

16.1 Nickel containing 2 wt% thorium is produced in powder form, consolidated into a part, and sintered in the presence of oxygen, causing all of the thorium to produce ThO_2 spheres 80 nm in diameter. Calculate the number of spheres per mm^3. The density of ThO_2 is 9.86 $Mg.m^{-3}$.

16.2 Spherical aluminium powder 0.002 mm in diameter is treated to create a thin oxide layer and is then used to produce a SAP dispersion-strengthened material containing 10 vol% Al_2O_3. Calculate the average thickness of the oxide film prior to compaction and sintering of the powder into the part.

16.3 Yttria (Y_2O_3) particles 75 nm in diameter are introduced into tungsten by internal oxidation. Measurements using an electron microscope show that there are 5×10^{11} oxide particles per mm^3. Calculate the wt% Y originally in the alloy. The density of Y_2O_3 is 5.01 $Mg.m^{-3}$.

16.4 With no special treatment, aluminium is typically found to have an Al_2O_3 layer that is 3 nm thick. If spherical aluminium powder prepared with a total diameter of 0.01 mm is used to produce the SAP dispersion-strengthened material, calculate the volume percent Al_2O_3 in the material and the number of oxide particles per mm^3. Assume that the oxide breaks into disk-shaped flakes 3 nm thick and 3×10^{-4} mm in diameter. Compare the number of oxide particles per mm^3 with the number of solid solution atoms per mm^3 when 3 at % of an alloying element is added to aluminium.

16.5 Calculate the density of a cemented carbide, or cermet, based on a titanium matrix if the composite contains 50 wt% WC, 22 wt% TaC, and 14 wt% TiC. (*See* Example 16.2 for densities of the carbides.)

16.6 A typical grinding wheel is 230 mm in diameter, 25 mm thick, and weighs 2.7 kg. The wheel contains SiC (density of 3.2 $Mg.m^{-3}$) bonded by silica glass (density of 2.5 $Mg.m^{-3}$); 5 vol% of the wheel is porosity. The SiC is in the form of 0.4 mm cubes. Calculate

(a) the volume fraction of SiC particles in the wheel and

(b) the number of SiC particles lost from the wheel after it is worn to a diameter of 200 mm.

16.7 An electrical contact material is produced by infiltrating copper into a porous tungsten carbide (WC) compact. The density of the final composite is 12.3 $Mg.m^{-3}$. Assuming that all of the pores are filled with copper, calculate

(a) the volume fraction of copper in the composite,

(b) the volume fraction of pores in the WC compact prior to infiltration, and

(c) the original density of the WC compact before infiltration.

16.8 An electrical contact material is produced by first making a porous tungsten compact that weighs 125 g. Liquid silver is introduced into the compact; careful measurement indicates that 105 g of silver is infiltrated. The final density of the composite is 13.8 $Mg.m^{-3}$. Calculate the volume fraction of the original compact that is interconnected porosity and the volume fraction that is closed porosity (no silver infiltration).

16.9 How much clay must be added to 10 kg of polyethylene to produce a low-cost composite having a modulus of elasticity greater than 0.8 $GN.m^{-2}$ and a tensile strength greater than 14 $MN.m^{-2}$? The density of the clay is 2.4 $Mg.m^{-3}$ and that of polyethylene is 0.92 $Mg.m^{-3}$.

16.10 We would like to produce a lightweight epoxy part to provide thermal insulation. We have available hollow glass beads for which the outside diameter is 1.6 mm and the wall thickness is 0.25 mm. Determine the weight and number of beads that must be added to the epoxy to produce a one-kilogram composite with a density of 0.65 $Mg.m^{-3}$. The density of the glass is 2.5 $Mg.m^{-3}$ and that of the epoxy is 1.25 $Mg.m^{-3}$.

16.11 Five kg of continuous boron fibres are introduced in a unidirectional orientation into eight kg of an aluminium matrix. Calculate

(a) the density of the composite,

(b) the modulus of elasticity parallel to the fibres, and

(c) the modulus of elasticity perpendicular to the fibres.

16.12 We want to produce 5 kg of a continuous unidirectional fibre-reinforced composite of HS carbon in a polyimide matrix that has a modulus of elasticity of at least 170 GN.m^{-2} parallel to the fibres. How many kilograms of fibres are required? *See* Chapter 15 for properties of polyimide.

16.13 We produce a continuous unidirectionally reinforced composite containing 60 vol% HM carbon fibres in an epoxy matrix. The epoxy has a tensile strength of 100 MN.m^{-2}. What fraction of the applied force is carried by the fibres?

16.14 A polyester matrix with a tensile strength of 90 MN.m^{-2} is reinforced with Al_2O_3 fibres. What vol% fibres must be added to ensure that the fibres carry 75% of the applied load?

16.15 An epoxy matrix is reinforced with 40 vol% E-glass fibres to produce a 20 mm diameter composite that is to withstand a load of 25 000 N. Calculate the stress acting on each fibre. $E_m = 3$ GN.m^{-2}, $E_f = 72.4$ GN.m^{-2}.

16.16 A titanium alloy with a modulus of elasticity of 110 GN.m^{-2} is used to make a 450 kg part for a manned space vehicle. Determine the weight of a part having the same modulus of elasticity parallel to the fibres, if the part is made of

(a) aluminium reinforced with boron fibres and

(b) polyester (with a modulus of 4.5 GN.m^{-2}) reinforced with high-modulus carbon fibres.

(c) Compare the specific modulus for all three materials.

16.17 Short but aligned Al_2O_3 fibres with a diameter of 20 μm are introduced into a 6,6-nylon matrix. The strength of the bond between the fibres and the matrix is estimated to be 7 MN.m^{-2}. Calculate the critical fibre length and compare with the case when 1 μm alumina whiskers are used instead of the coarser fibres. What is the minimum aspect ratio in each case?

16.18 We prepare several epoxy matrix composites using different lengths of 3-μm-diameter ZrO_2 fibres and find that the strength of the composite increases with increasing fibre length up to 5 mm. For longer fibres, the strength is virtually unchanged. Estimate the strength of the bond between the fibres and the matrix.

16.19 In one polymer matrix composite, as produced discontinuous glass fibres are introduced directly into the matrix; in a second case, the fibres are first 'sized'. Discuss the effect this difference might have on the critical fibre length and the strength of the composite.

16.20 A borsic fibre-reinforced aluminium composite is shown in Figure 16.19. Estimate the volume fractions of tungsten, boron, and the matrix for this composite. Calculate the modulus of elasticity parallel to the fibres for this composite. What would the modulus be if the same size boron fibre could be produced without the tungsten precursor?

16.21 A silicon nitride matrix reinforced with silicon carbide fibres containing a HS carbon precursor is shown in Figure 16.19. Estimate the volume fractions of the SiC, Si_3N_4, and carbon in this composite. Calculate the modulus of elasticity parallel to the fibres for this composite. What would the modulus be if the same size SiC fibre could be produced without the carbon precursor?

16.22 Explain why bonding between carbon fibres and an epoxy matrix should be excellent, whereas bonding between silicon nitride fibres and a silicon carbide matrix should be poor.

16.23 A polyimide matrix is to be reinforced with 70 vol% carbon fibres to give a minimum modulus of elasticity of 275 GN.m^{-2}. Recommend a process for producing the carbon fibres required. Estimate the tensile strength of the fibres that are produced.

16.24 An electrical capacitor is produced by sandwiching 19 layers of 0.02 mm-thick Teflon between 20 layers of 0.08 mm-thick silver. Determine the electrical conductivity of the capacitor

(a) parallel to the sheets and

(b) perpendicular to the sheets.

The electrical conductivity of the silver is 68×10^6 $ohm^{-1}.m^{-1}$ and that of Teflon is 10^{-18} $ohm^{-1}.m^{-1}$.

16.25 A microlaminate, Arall, is produced using 5 sheets of 0.4 mm-thick aluminium and 4 sheets of 0.2 mm-thick epoxy reinforced with unidirectionally aligned Kevlar fibres. The volume fraction of Kevlar fibres in these intermediate sheets is 55%. Calculate the modulus of elasticity of the microlaminate parallel and perpendicular to the unidirectionally aligned Kevlar fibres. What are the principal advantages of the Arall material compared with those of unreinforced aluminium?

16.26 A laminate composed of 0.1 mm-thick aluminium sandwiched around a 20 mm-thick layer of polystyrene foam (Styrofoam) is produced as an insulation material. Calculate the thermal conductivity of the laminate parallel and perpendicular to the layers. The thermal conductivity of aluminium is 238.6 $W.m^{-1}.K^{-1}$ and that of the foam is 0.032 $W.m^{-1}.K^{-1}$.

16.27 A 0.1 mm-thick sheet of a polymer with a modulus of elasticity of 5 $GN.m^{-2}$ is sandwiched between two 4 mm-thick sheets of glass with a modulus of elasticity of 83 $GN.m^{-2}$. Calculate the modulus of elasticity of the composite parallel and perpendicular to the sheets.

16.28 A U.S. quarter is 23 mm in diameter and is about 2 mm thick. Copper costs about $2.40 per kg and nickel costs about $9.00 per kg. Compare the material cost in a composite quarter versus a quarter made entirely of nickel.

16.29 Calculate the density of a honeycomb structure composed of the following elements: The two 2 mm-thick cover sheets are produced using an epoxy matrix prepreg containing 55 vol% E-glass fibres. The aluminium honeycomb is 20 mm thick; the cells are in the shape of 5 mm squares and the walls are 0.1 mm thick. Estimate the density of the structure. Compare the weight of a 1 m × 2 m panel of the honeycomb compared with a solid aluminium panel of the same dimensions.

16.30 Select and calculate the constituents of a composite material for an aeroplane wing that has an electrical conductivity of at least 200×10^3 $ohm^{-1}.m^{-1}$ (*see* Table 18.1 for typical values for metals and polymers), a modulus of elasticity of at least 83 $GN.m^{-2}$, and reasonable corrosion resistance.

16.31 Consider a Cu-15 wt% Sn alloy [Figure 13.10(b)]. Design a compocasting process that will permit the introduction of 40 vol% SiC. The density of the alloy is 8.5 $Mg.m^{-3}$ and that of SiC is 3.0 $Mg.m^{-3}$.

16.32 Select the materials and processing required to produce a discontinuous but aligned fibre-reinforced fibreglass composite that will form the bonnet of a sports car. The composite should provide a density of less than 1.6 $Mg.m^{-3}$ and a strength of 140 $MN.m^{-3}$. Be sure to list all of the assumptions you make in selecting your material.

16.33 A 900 mm inside-diameter spherical tank is to be designed to store liquid Cl. The tank must have a modulus of elasticity in the tangential direction of at least 100 $GN.m^{-2}$, it should have a thermal conductivity in the radial direction of no more than 2.5 $W.m^{-1}.K^{-1}$, and it should weigh no more than 75 kg. Using only the materials listed in Table 21.3, formulate a composite material and determine the tank thickness that will be suitable. Estimate the cost of materials in your tank to ensure that it is not prohibitively expensive.

16.34 Select the constituents of an electrical contact material and describe a method for producing the material which will result in a density of no more than 6 $Mg.m^{-3}$, yet at least 50 vol% of the material will be conductive.

CHAPTER 17

Construction Materials

17.1 Introduction

A number of important materials are used primarily for construction of buildings, highways, bridges, and much of the infrastructure of any country. In this chapter, we look at three of the most important materials: wood, concrete, and asphalt. These materials are, in fact, composite materials, and at least part of their characteristics can be explained in terms of our discussion in the previous chapter.

17.2 The Structure of Wood

Wood is one of our most familiar materials. Although it is not a 'high-tech' material, we are literally surrounded by it in our homes and value it for its beauty. In addition, wood is a strong, lightweight material that still dominates much of the construction industry.

We can consider wood to be a complex fibre-reinforced composite composed of long, unidirectionally aligned tubular polymer cells in a polymer matrix. Furthermore, the polymer tubes are composed of bundles of partially crystalline cellulose fibres aligned at various angles to the axes of the tubes. This arrangement provides excellent tensile properties in the longitudinal direction.

Wood consists of four main constituents. Cellulose fibres make up about 40% to 50% of wood. Cellulose is a naturally occurring thermoplastic polymer with a degree of polymerisation of about 10 000. The structure of cellulose is shown in Figure 17.1. About 25% to 35% of a tree is hemicellulose, a polymer having a degree of polymerisation of about 200. Another 20% to 30% of a tree is lignin, a low-molecular-weight organic cement that bonds the various constituents of the wood. Finally, extractives are organic impurities such as oils, which provide colour to the wood or act as preservatives against the environment and insects, and inorganic minerals such as silica, which help dull saw blades during cutting of the wood. As much as 10% of the wood may be extractives.

There are three important levels in the structure of wood: the fibre structure, the cell structure, and the macrostructure (Figure 17.2).

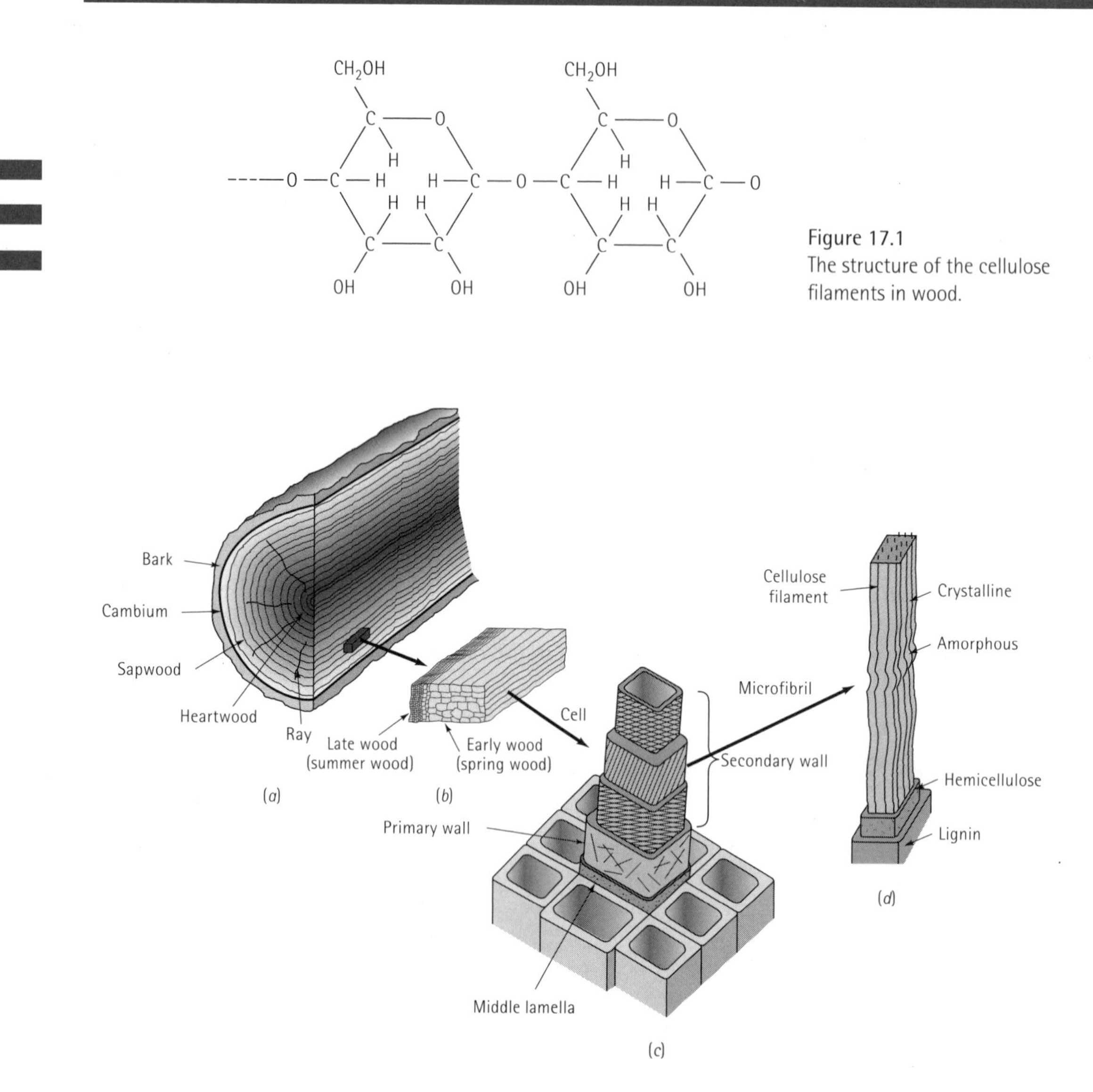

Figure 17.1
The structure of the cellulose filaments in wood.

Figure 17.2
The structure of wood: (*a*) the macrostructure, including a layer structure outlined by the annual growth rings, (*b*) detail of the cell structure within one annual growth ring, (*c*) the structure of a cell, including several layers composed of microfibrils of cellulose fibres, hemicellulose fibres, and lignin, and (*d*) the microfibril's aligned, partly crystalline cellulose chains.

Fibre Structure The basic component of wood is cellulose, $C_6H_{10}O_5$, arranged in polymer chains that form long fibres. Much of the fibre length is crystalline, with the crystalline regions separated by small lengths of amorphous cellulose. A bundle of cellulose chains is encased in a layer of randomly oriented, amorphous hemicellulose chains. Finally, the hemicellulose is covered with lignin. The entire bundle, consisting of cellulose chains, hemicellulose chains, and lignin, is called a microfibril; it can have a virtually infinite length.

Cell Structure The tree is composed of elongated cells, often having an aspect ratio of 100 or more, that constitute about 95% of the solid material in wood. The hollow cells are composed of several layers built up from the microfibrils. The first, or primary, wall of the cell contains randomly oriented microfibrils. As the cell walls thicken, three more distinct layers are formed. The outer and inner walls contain microfibrils oriented in two directions that are not parallel to the cell. The middle wall, which is the thickest, contains microfibrils that are unidirectionally aligned, usually at an angle not quite parallel to the axis of the cell.

Macrostructure A tree is composed of several layers. The outer layer, or *bark*, protects the tree. The cambium, just beneath the bark, contains new growing cells. The sapwood contains a few hollow living cells that store nutrients and serve as the conduit for water. And, finally, the heartwood, which contains only dead cells, provides most of the mechanical support for the tree.

The tree grows when new elongated cells develop in the cambium. Early in the growing season, the cells are large; later they have a smaller diameter, thicker walls, and a higher density. This difference between the early (or *spring*) wood and the late (or *summer*) wood permits us to observe annual growth rings. In addition, some cells grow in a radial direction; these cells, called *rays*, provide storage and transport of food.

Hardwood Versus Softwood The hardwoods are deciduous trees such as oak, elm, beech, birch, walnut, and maple. In these trees, the elongated cells are relatively short, with a diameter of less than 0.1 mm and a length of less than 1 mm. Contained within the wood are longitudinal pores, or vessels, which carry water through the tree (Figure 17.3).

The softwoods, evergreens such as pine, fir, spruce, and cedar, have similar structures. In softwoods, the cells tend to be somewhat longer than in the hardwoods. The hollow centre of the cells is responsible for transporting water. In general, the density of softwoods tends to be lower than that of hardwoods because of a greater percentage of void space.

17.3 Moisture Content and Density of Wood

The material making up the individual cells in virtually all woods has essentially the same density – about 1.45 $Mg.m^{-3}$. However, wood contains void space that causes the actual density to be much lower.

The density of wood depends primarily on the species of the tree (or the amount of void space peculiar to that species) and the percentage of water in wood (which depends on the amount of drying and on the relative humidity to which the wood is exposed during use). Completely dry wood varies in density from about 0.3 to 0.8 $Mg.m^{-3}$, with hardwoods having higher densities than softwoods. But the measured density is normally higher due to the water contained in the wood. The percentage water is given by:

$$\%\text{Water} = \frac{\text{weight of water}}{\text{weight of dry wood}} \times 100 \quad (17.1)$$

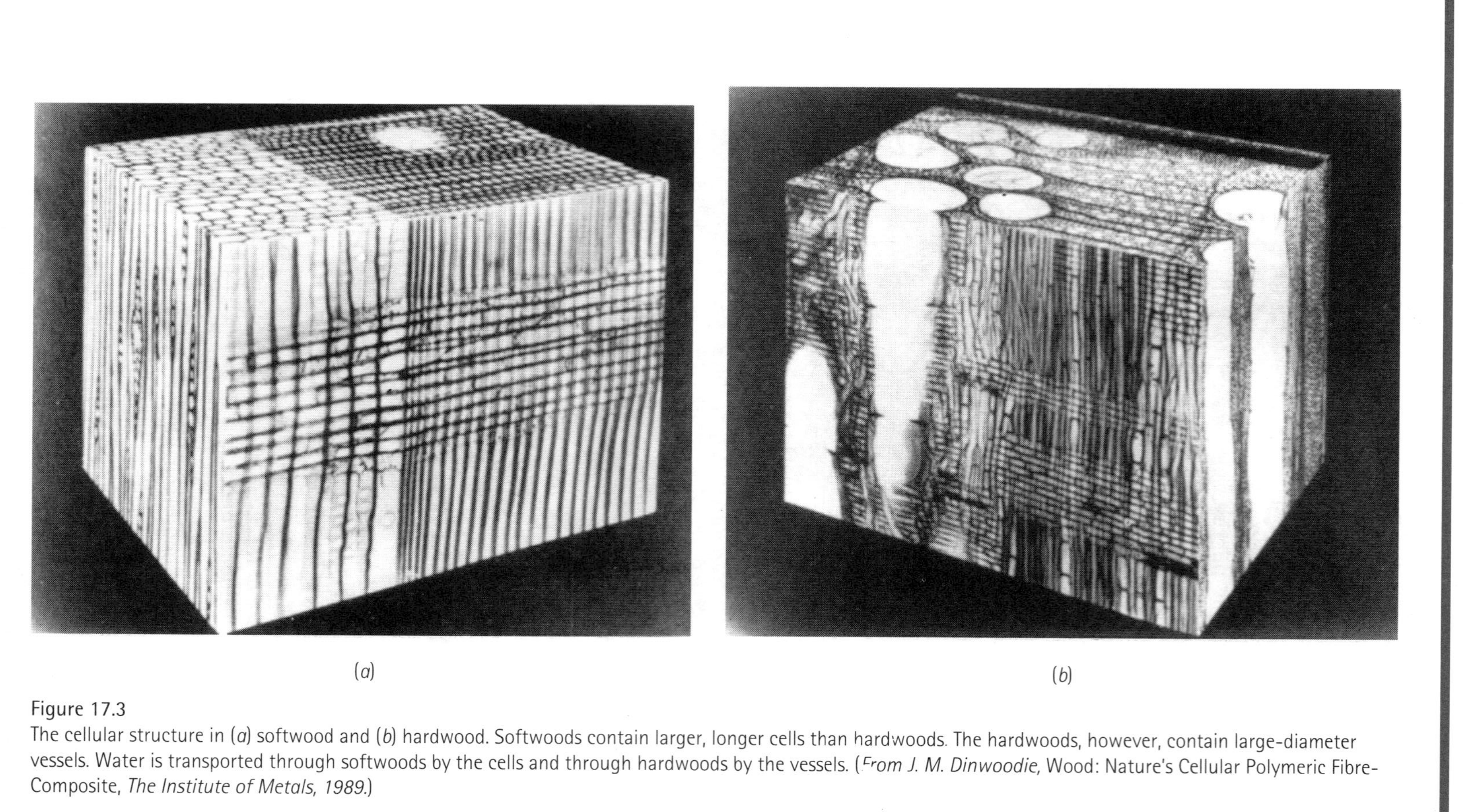

Figure 17.3
The cellular structure in (*a*) softwood and (*b*) hardwood. Softwoods contain larger, longer cells than hardwoods. The hardwoods, however, contain large-diameter vessels. Water is transported through softwoods by the cells and through hardwoods by the vessels. (*From J. M. Dinwoodie*, Wood: Nature's Cellular Polymeric Fibre-Composite, *The Institute of Metals, 1989.*)

On the basis of this definition, it is possible to describe a wood as containing more than 100% water. The water is contained both in the hollow cells or vessels, where it is not tightly held, and in the cellulose structure in the cell walls, where it is more tightly bonded to the cellulose fibres.

While a large amount of water is stored in a live tree, the amount of water in the wood after the tree is harvested depends eventually on the humidity to which the wood is exposed; higher humidity increases the amount of water held in the cell walls. The density of a wood is usually given at a moisture content of 12%, which corresponds to 65% humidity. The density and modulus of elasticity parallel to the grain of several common woods are included in Table 17.1 for this typical water content.

Table 17.1 Properties of typical woods.

Wood	Density (for 12% water) ($Mg.m^{-3}$)	Modulus of Elasticity ($MN.m^{-2} \times 10^3$)
Cedar	0.32	7.6
Pine	0.35	8.3
Fir	0.48	13.8
Maple	0.48	10.3
Birch	0.62	13.8
Oak	0.68	12.4

EXAMPLE 17.1

A green wood has a density of 0.86 $Mg.m^{-3}$ and contains 175% water. Calculate the density of the wood after it has completely dried.

SOLUTION

A 1 m^3 sample of the wood would weigh 0.86 Mg. From Equation 17.1, we can calculate the weight of the dry wood to be:

$$\% \text{ Water} = \frac{\text{weight of water}}{\text{weight of dry wood}} \times 100 = 175$$

$$= \frac{\text{green weight} - \text{dry weight}}{\text{dry weight}} \times 100 = 175$$

$$\text{Dry weight of wood} = \frac{(100)(\text{green weight})}{275} = \frac{(100)(0.86)}{275} = 0.313 \text{ Mg}$$

$$\text{Density of dry wood} = \frac{0.313 \text{ Mg}}{1 \text{ m}^3} = 0.313 \text{ Mg.m}^{-3}$$

17.4 Mechanical Properties of Wood

The strength of a wood depends on its density, which in turn depends on both the water content and the type of wood. As a wood dries, water is eliminated first from the vessels and later from the cell walls. As water is removed from the vessels, practically no change in the strength or stiffness of the wood is observed (Figure 17.4). But on continued drying to less than about 30% water, there is water loss from the actual cellulose fibres. This loss permits the individual fibres to come closer together, increasing the bonding between the fibres and the density of the wood and, thereby, increasing the strength and stiffness of the wood.

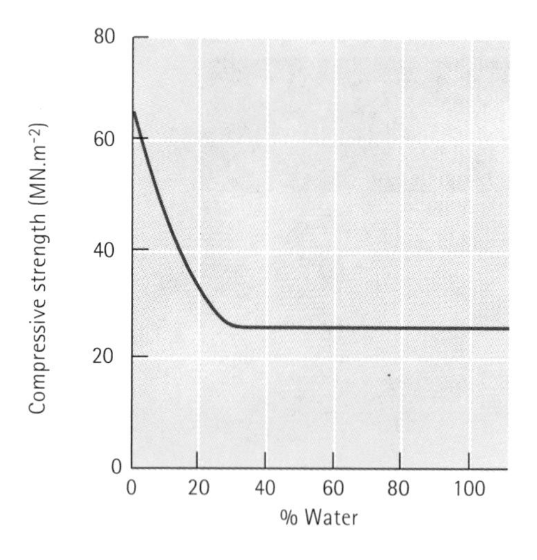

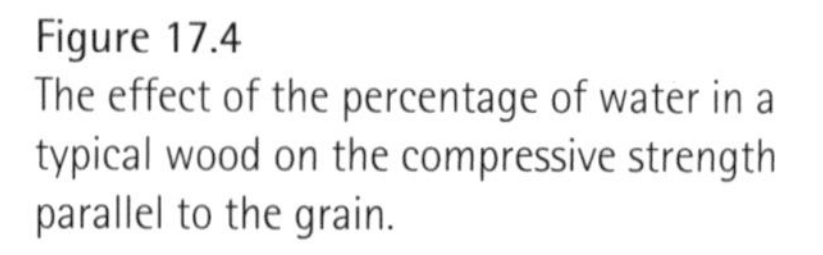
Figure 17.4
The effect of the percentage of water in a typical wood on the compressive strength parallel to the grain.

The type of wood also effects the density. Because they contain less of the higher-density late wood, softwoods typically are less dense and therefore have lower strengths than hardwoods. In addition, the cells in softwoods are larger, longer, and more open than those in hardwoods, leading to lower density.

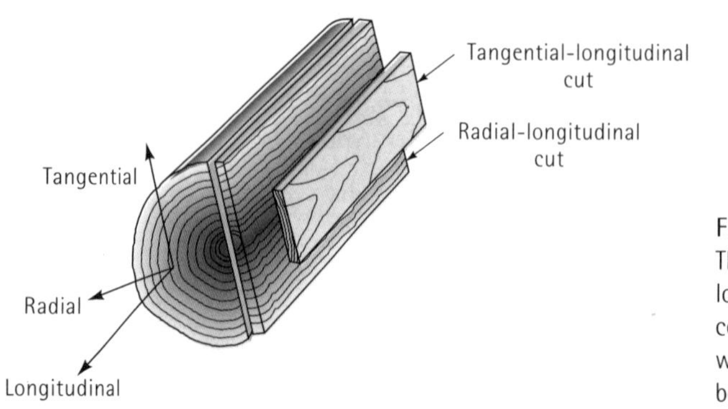

Figure 17.5
The different directions in a log. Because of differences in cell orientation and the grain, wood displays anisotropic behaviour.

The mechanical properties of wood are highly anisotropic. In the longitudinal direction (Figure 17.5), an applied tensile load acts parallel to the microfibrils and cellulose chains in the middle section of the secondary wall. These chains are strong – mostly crystalline – and are able to carry a relatively high load.

However, in the radial and tangential directions, the weaker bonds between the microfibrils and cellulose fibres may break, resulting in very low tensile properties. Similar behaviour is observed in compression and bending loads. Because of the anisotropic behaviour, most timber is cut in a tangential-longitudinal or radial-longitudinal manner. These cuts maximise the longitudinal behaviour of the wood.

Wood has poor properties in compression and bending (which produces a combination of compressive and tensile forces). In compression, the fibres in the cells tend to buckle, causing the wood to deform and break at low stresses. Unfortunately, most applications for wood place the component in compression or bending and therefore do not take full advantage of the engineering properties of the material. Similarly, the modulus of elasticity is highly anisotropic; the modulus perpendicular to the grain is about 1/20th of that given in Table 17.1 parallel to the grain. Table 17.2 compares the tensile and compressive strengths parallel and perpendicular to the cells for several woods.

Table 17.2 Anisotropic behaviour of several woods (at 12% moisture).

	Tensile Strength Longitudinal ($MN.m^{-2}$)	Tensile Strength Radial ($MN.m^{-2}$)	Compressive Strength Longitudinal ($MN.m^{-2}$)	Compressive Strength Radial ($MN.m^{-2}$)
Beech	86	7	50	7
Elm	120	5	38	5
Maple	108	8	54	10
Oak	78	6	43	6
Cedar	46	2	42	6
Fir	78	3	38	4
Pine	73	2	33	3
Spruce	59	3	39	4

Clear wood has a specific strength and specific modulus that compare well with those of other common construction materials (Table 17.3). Wood also has good toughness, largely due to the slight misorientation of the cellulose fibres in the middle layer of the secondary wall. Under load, the fibres straighten, permitting some ductility and energy absorption.

The mechanical properties of wood also depend on imperfections in the wood. Clear wood, free of imperfections such as knots, may have a longitudinal tensile strength of 70 to 140 $MN.m^{-2}$. Less expensive construction, timber which usually contains many imperfections, may have a tensile strength below 35 $MN.m^{-2}$. The knots also disrupt the grain of the wood in the vicinity of the knot, causing the cells to be aligned perpendicular to the tensile load.

Table 17.3 Comparison of the specific strength and specific modulus of wood with those of other common construction materials.

Material	Specific Strength ($kg.m^2.s^{-2}$)	Specific Modulus ($kg.m^2.s^{-2}$)
Clear wood	178	2.4×10^4
Aluminium	127	2.7×10^4
1020 steel	50	2.7×10^4
Copper	38	1.4×10^4
Concrete	15	0.9×10^4

After F. F. Wangaard, 'Wood: Its Structure and Properties,' J. Educ. Models for Mat. Sci. and Engr., *Vol. 3, No. 3, 1979.*

Note: Specific strength definition is given in Section 13.1.

17.5 Expansion and Contraction of Wood

Like other materials, wood changes dimension when heated or cooled. Dimensional changes in the longitudinal direction are very small in comparison with those in metals, polymers, and ceramics. However, the dimensional changes in the radial and tangential directions are greater than those for most other materials.

In addition to dimensional changes caused by temperature fluctuations, the moisture content of the wood causes significant changes in dimension. Again, the greatest changes occur in the radial and tangential directions, where the moisture content affects the spacing between the cellulose chains in the microfibrils. The change in dimensions Δx in wood in the radial and tangential directions is approximated by

$$\Delta x = x_0[c(M_f - M_i)] \tag{17.2}$$

where x_0 is the initial dimension, M_i is the initial water content, M_f is the final water content, and c is a coefficient that describes the dimensional change and can be measured in either the radial or the tangential direction. Table 17.4 includes the dimensional coefficients for several woods. In the longitudinal direction, no more than 0.1 to 0.2% change is observed.

Table 17.4 Dimensional coefficient c ($mm.mm^{-1}.\%H_2O$) for several woods.

Wood	Radial	Tangential	Wood	Radial	Tangential
Beech	0.00190	0.00431	Cedar	0.00111	0.00234
Elm	0.00144	0.00338	Fir	0.00155	0.00278
Maple	0.00165	0.00353	Pine	0.00141	0.00259
Oak	0.00183	0.00462	Spruce	0.00148	0.00263

During the initial drying of wood, the large dimensional changes perpendicular to the cells may cause warping and even cracking. In addition, when the wood is used, its water content may change, depending on the relative humidity in the environment. As the wood gains or loses water during use, shrinkage or swelling continues to occur. If a wood construction does not allow movement caused by moisture fluctuations, warping and cracking can occur – a particularly severe condition in large expanses of wood, such as the floor of a large room. Excessive expansion may cause large bulges in the floor; excessive shrinkage may cause large gaps between individual planks of the flooring.

17.6 Plywood

The anisotropic behaviour of wood can be reduced and wood products can be made in larger sizes by producing plywood. Thin layers of wood called plies are cut from logs – normally, softwoods. The plies are stacked together with the grains between adjacent plies oriented at 90° angles; usually an odd number of plies are used. Assuring that these angles are as precise as possible is important in assuring that the plywood does not warp or twist when the moisture content in the material changes. The individual plies are generally bonded to one another using a thermosetting phenolic resin. The resin is introduced between the plies, which are then pressed together while hot to cause the resin to polymerise.

Similar wood products are also produced as 'laminar' composite materials. The facing (visible) plies may be of a more expensive hardwood, with the centre plies of a less expensive softwood. Wood particles can be compacted into sheets and laminated between two wood plies, producing particle board. Wood plies can be used as the facings for honeycomb materials.

17.7 Concrete Materials

Concrete, another common construction material, is a particulate composite in which both the particulate and the matrix are ceramic materials. In concrete, sand and a coarse aggregate are bonded in a matrix of Portland cement. A cementation reaction between water and the minerals in the cement provides a strong matrix that holds the aggregate in place and provides good compressive strength to the concrete.

Cements The cement binder, which is very fine in size, is composed of various ratios of $3CaO \cdot Al_2O_3$, $2CaO \cdot SiO_2$, $3CaO \cdot SiO_2$, $4CaO \cdot Al_2O_3 \cdot Fe_2O_3$, and other minerals. When water is added to the cement, a hydration reaction occurs, producing a solid gel that bonds the aggregate particles. Possible reactions include

$$3CaO \cdot Al_2O_3 + 6H_2O \rightarrow Ca_3Al_2(OH)_{12} + \text{heat}$$

$$2CaO \cdot SiO_2 + xH_2O \rightarrow Ca_2SiO_4 \cdot xH_2O + \text{heat}$$

$$3CaO + SiO_2 + (x + 1)H_2O \rightarrow Ca_2SiO_4 \cdot xH_2O + Ca(OH)_2 + \text{heat}$$

After hydration, the cement provides the bond for the aggregate particles. Consequently, enough cement must be added to coat all of the aggregate particles. The cement typically constitutes on the order 15 vol% of the solids in the concrete.

The composition of the cement helps determine the rate of curing and the final properties of the concrete. For example, $3CaO \cdot Al_2O_3$ and $3CaO \cdot SiO_2$ produce rapid setting but low strengths. The $2CaO \cdot Al_2O_3$ reacts more slowly during hydration, but eventually produces higher strengths (Figure 17.6). Nearly complete curing of the concrete is normally expected within 28 days (Figure 17.7), although some additional curing may continue for years.

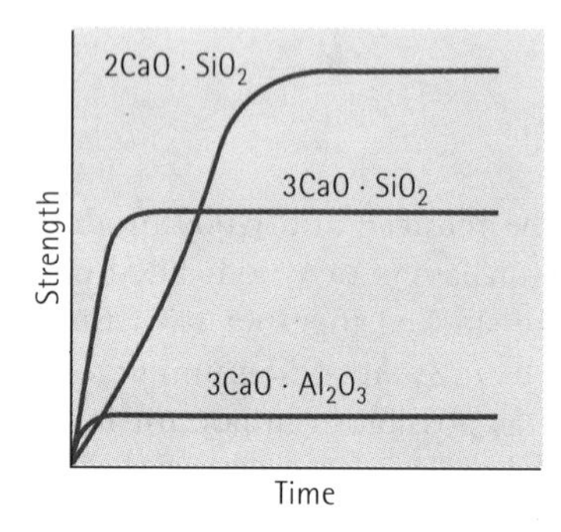

Figure 17.6
The rate of hydration of the minerals in Portland cement.

Several general types of cements are used (Table 17.5). In large structures such as dams, curing should be slow in order to avoid excessive heating caused by the hydration reaction. These cements typically contain low percentages of $3CaO \cdot SiO_2$, such as in Types II and IV. Some construction jobs, however, require that concrete forms be removed and reused as quickly as possible; cements for these purposes may contain large amounts of $3CaO \cdot SiO_2$, as in Type III.

The composition of the cement also affects the resistance of the concrete to the environment. For example, sulphates in the soil may attack the concrete; using higher proportions of $4CaO \cdot Al_2O_3 \cdot Fe_2O_3$ and $2CaO \cdot SiO_2$ helps produce concretes more resistant to sulphates, as in type V.

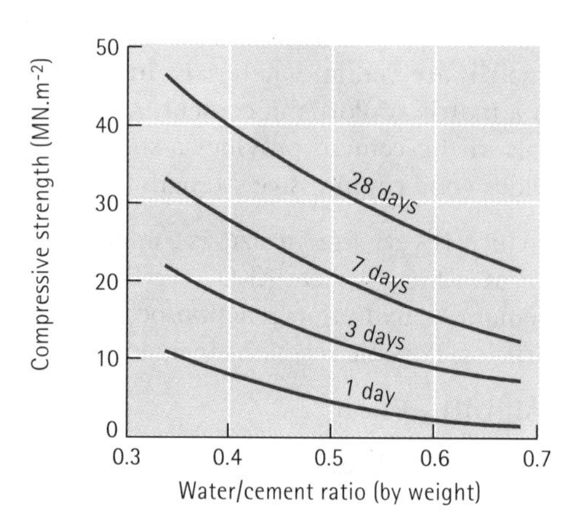

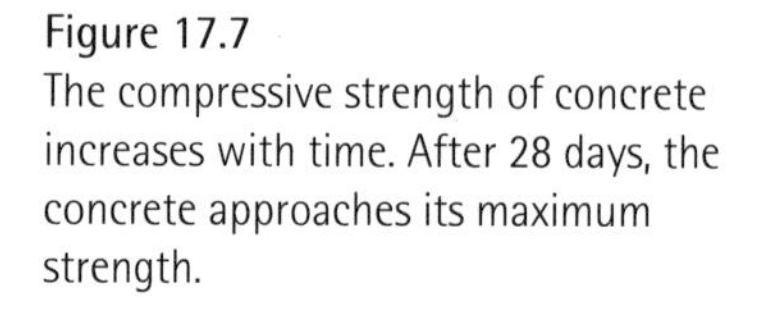
Figure 17.7
The compressive strength of concrete increases with time. After 28 days, the concrete approaches its maximum strength.

Table 17.5 Types of cements.

	Approximate Composition				Characteristics
	3C · S	2C · S	3C · A	4C · A · F	
Type I	55	20	12	9	General purpose
Type II	45	30	7	12	Low rate of heat generation, moderate resistance to sulphates
Type III	65	10	12	8	Rapid setting
Type IV	25	50	5	13	Very low rate of heat generation
Type V	40	35	3	14	Good sulphate resistance

Sand Sands are fine minerals, typically on the order of 0.01 mm in diameter. They often contain at least some adsorbed water, which should be taken into account when preparing a concrete mix. The sand helps fill voids between the coarser aggregate, giving a high-packing factor, reducing the amount of open (or interconnected) porosity in the finished concrete, and reducing problems with disintegration of the concrete due to repeated freezing and thawing during service.

Aggregate Coarse aggregate is composed of gravel and rock. Aggregate must be clean, strong, and durable. Aggregate particles that have an angular rather than a round shape provide strength due to mechanical interlocking between particles, but angular particles also provide more surface on which voids or cracks may form. It is normally preferred that the aggregate size be large; this condition also minimises the surface area at which cracks or voids form. The size of the aggregate must, of course, be matched to the size of the structure being produced; aggregate particles should not be any larger than about 20% of the thickness of the structure.

In some cases, special aggregates may be used. Lightweight concretes can be produced by using mineral slags produced during steel-making operations; these concretes have improved thermal insulation. Particularly heavy concretes can be produced using dense minerals or even metal shot; these heavy concretes can be used in building nuclear reactors to absorb radiation better. The densities of several aggregates are included in Table 17.6.

Table 17.6 Characteristics of concrete materials.

Material	True Density ($Mg.m^{-3}$)	
Cement	1.75	where 1 sack = 50 kg
Sand	2.56	
Aggregate	2.72	Normal
	1.28	Lightweight slag
	0.48	Lightweight vermiculite
	4.49	Heavy Fe_3O_4
	6.25	Heavy ferrophosphorous
Water	1.00	

17.8 Properties of Concrete

Many factors influence the properties of concrete. Some of the most important are the water-cement ratio, the amount of air entrainment, and the type of aggregate.

Water-Cement Ratio The ratio of water to cement affects the behaviour of concrete in several ways:

1. A minimum amount of water must be added to the cement to ensure that all of it undergoes the hydration reaction. Too little water therefore causes low strength. Normally, however, other factors such as workability place the lower limit on the water-cement ratio.
2. A high water-cement ratio improves the workability of concrete – that is, how easily the concrete slurry can fill all of the space in the form. Air pockets or interconnected porosity caused by poor workability reduce the strength and durability of the concrete structure. Workability can be measured by the *slump* test. For example, a wet concrete shape 300 mm tall is produced (Figure 17.8) and is permitted to stand under its own weight. After some period of time, the shape deforms. The reduction in height of the form is the slump. A minimum water-cement ratio of about 0.4 (by weight) is usually required for workability. A larger slump, caused by a higher water-cement ratio, indicates greater workability. Slumps of 25 to 150 mm are typical; high slumps are needed for pouring narrow or complex forms, while low slumps may be satisfactory for large structures such as dams.

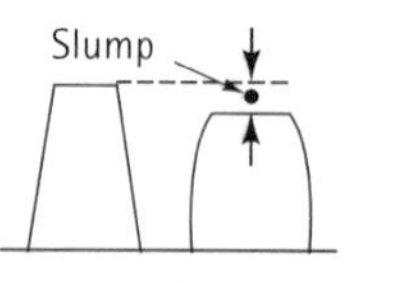

Figure 17.8
The slump test, in which deformation of a concrete shape under its own weight is measured, is used to describe the workability of concrete mix.

3. Increasing the water-cement ratio beyond the minimum required for workability decreases the compressive strength of the concrete. This strength is usually measured by determining the stress required to crush a concrete cylinder 150 mm in diameter and 300 mm tall. Figure 17.9 shows the effect of water-cement ratio on concrete's strength.

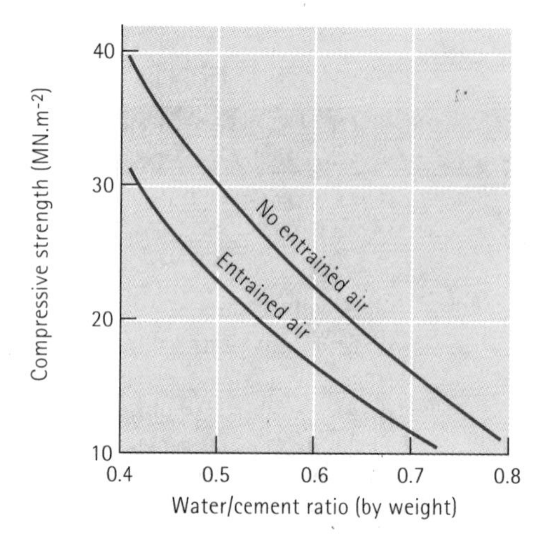

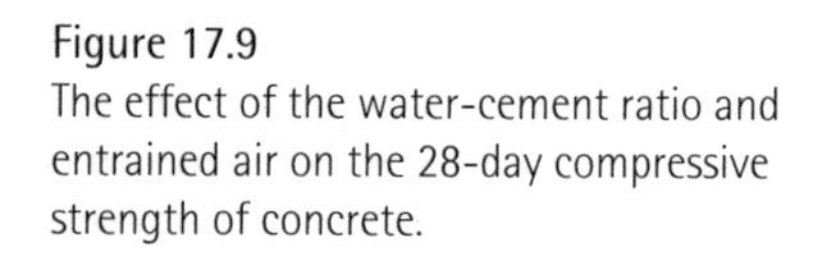

Figure 17.9
The effect of the water-cement ratio and entrained air on the 28-day compressive strength of concrete.

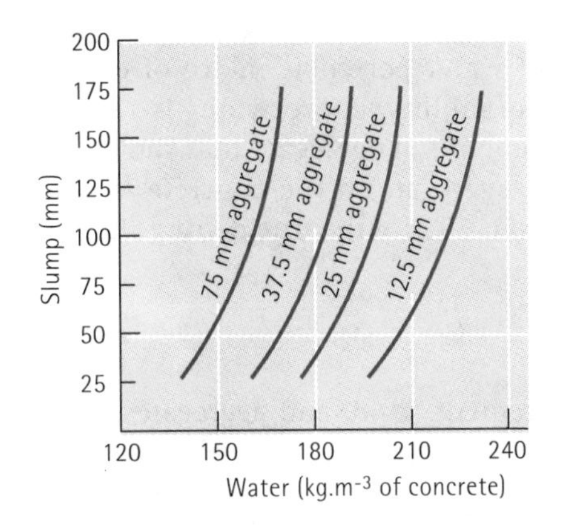

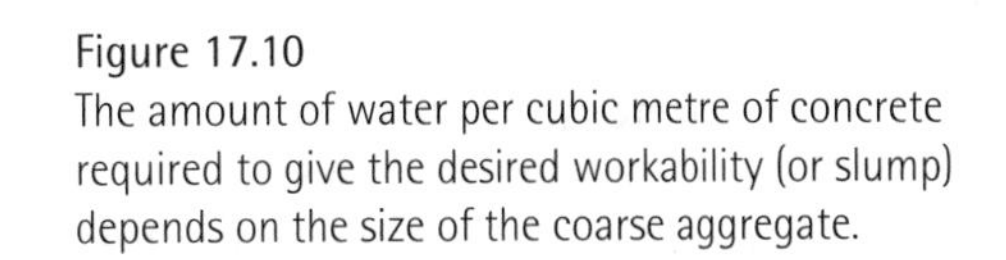
Figure 17.10
The amount of water per cubic metre of concrete required to give the desired workability (or slump) depends on the size of the coarse aggregate.

4. High water-cement ratios increase the shrinkage of concrete during curing, creating a danger of cracking.

Because of the different effects of the water-cement ratio, a compromise between strength, workability, and shrinkage may be necessary. A weight ratio of 0.45 to 0.55 is typical. To maintain good workability, organic plasticisers may be added to the mix with little effect on strength.

Air-Entrained Concrete Almost always, a small amount of air is entrained into the concrete during pouring. For coarse aggregate, such as 40 mm rock, 1% by volume of the concrete may be air. For finer aggregate, such as 12 mm gravel, 2.5% air may be trapped.

We sometimes intentionally entrain air into concrete – sometimes as much as 8% for fine gravel. The entrained air improves workability of the concrete and helps minimise problems with shrinkage and freeze-thaw conditions. However, air-entrained concrete has a lower strength (Figure 17.9).

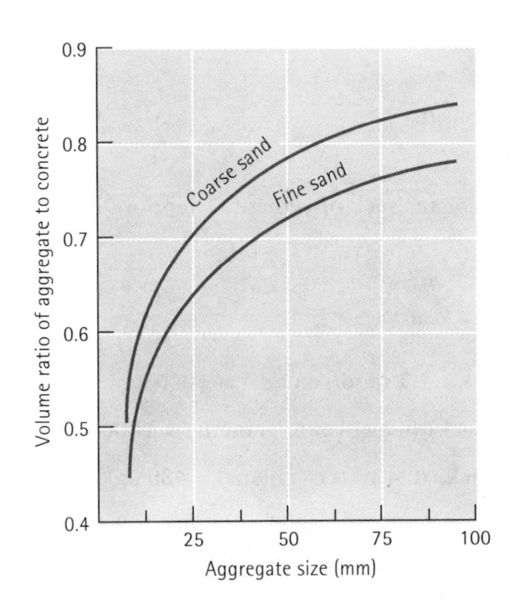

Figure 17.11
The volume ratio of aggregate to concrete depends on the sand and aggregate sizes. Note that the volume ratio uses the bulk density of the aggregate – about 60% of the true density.

Type and Amount of Aggregate The size of the aggregate affects the concrete mix. Figure 17.10 shows the amount of water per cubic metre of concrete required to produce the desired slump, or workability; more water is required for smaller aggregate. Figure 17.11 shows the amount of aggregate that should be present in the concrete mix. The volume ratio of aggregate in the concrete is based on the bulk density of the aggregate, which is about 60% of the true density shown in Table 17.6.

EXAMPLE 17.2

Determine the amounts of water, cement, sand, and aggregate in 4 cubic metres of concrete, assuming that we want to obtain a water/cement ratio of 0.4 (by weight) and that the cement/sand/aggregate ratio is 1 : 2.5 : 4 (by weight). A 'normal' aggregate will be used, containing 1% water, and the sand contains 4% water. Assume that no air is entrained into the concrete, and the water : cement ratio is 0.4.

SOLUTION

One method by which we can calculate the concrete mix is to first determine the volume of each constituent based on one sack (50 kg) of cement. We should remember that after the concrete is poured, there are no void spaces between the various constituents; therefore, we need to consider the true density – not the bulk density – of the constituents in our calculations.

For each sack of cement we use, the volume of materials required is:

$$\text{Cement} = \frac{(50\text{kg / sack})}{(10^3\ \text{Mg / sack})(1.75\ \text{Mg.m}^{-3})} = 0.029\ \text{m}^3$$

$$\text{Sand} = \frac{(2.550)}{(10^3)(2.56)} = 0.049\ \text{m}^3$$

$$\text{Aggregate} = \frac{(4\times 50)}{(10^3)(2.72)} = 0.074\ \text{m}^3$$

$$\text{Water} = \frac{(0.4\times 50)}{(10^3)(1.00)} = 0.020\ \text{m}^3$$

Total volume of concrete = 0.172 m^3/sack of cement. Therefore in 4 cubic metres we need:

$$\text{Cement} = \frac{4\ \text{m}^3}{0.172\ \text{m}^3\text{ / sack}} = 24\ \text{sacks}$$

Sand = (24 sacks) (50 kg/sack) (2.5 sand/cement) = 3 000 kg

Aggregate = (24 sacks) (50/sack) (4 aggregate/cement) = 4 800 kg

Water = (24 sacks) (50 kg/sack) (0.4 water/cement) = 480 kg

But the sand contains 4% water and the gravel contains 1% water. To obtain the weight of wet sand and gravel, we must adjust for the water content of each:

Sand = (3 000 kg dry) (1.04) = 3120 kg, and water = 120 kg

Aggregate = (4 800 kg dry) (1.01) = 4848 kg, and water = 48 kg

Therefore we only need to add water = 480 kg – 120 – 48 = 312 kg and with density of water being 1.00 $Mg.m^{-3}$

$$\text{Volume of water} = \frac{312}{10^3 \times 1.00} = 0.312 \text{ m}^3$$

Accordingly we recommend that 24 sacks of cement, 3120 kg of sand and 4848 kg of aggregate be combined with 0.312 m^3 of water.

EXAMPLE 17.3 Design of a Concrete Mix for a Retaining Wall

Design a concrete mix that will provide a 28-day compressive strength of 25 $MN.m^{-2}$ in a concrete intended for producing a 125 mm thick retaining wall 2 m high. We expect to have about 2% air entrained in the concrete, although we will not intentionally entrain air. Our aggregate contains 1% moisture, and we have only coarse sand containing 5% moisture available.

SOLUTION

Some workability of the concrete is needed to ensure that the form will fill properly with concrete. A slump of 75 mm might be appropriate for such an application.

The wall thickness is 125 mm. To help minimise cost, we would use a large aggregate: A 25 mm diameter aggregate size would be appropriate (about 1/5 of the wall thickness).

To obtain the desired workability of the concrete using 25 mm aggregate, we should use about 190 kg of water per cubic metre (Figure 17.10).

To obtain the 25 $MN.m^{-2}$ compressive strength after 28 days (assuming no intentional entrained air), we need a water/cement weight ratio of 0.60 (Figure 17.9).

Consequently, the weight of cement required per cubic metre of concrete is 190 kg water/0.60 water-cement) = 317 kg cement.

Because our aggregate size is 25 mm and we have only coarse sand available, the volume ratio of the aggregate to the concrete is 0.7 (Figure 17.11). Thus, the amount of aggregate required per metre of concrete is 0.7 m^3; however, this amount is in terms of the bulk density of the aggregate. Because the bulk density is about 60% of the true density, the actual volume occupied by the aggregate in the concrete is 0.7 $m^3 \times 0.6 = 0.42$ m^3

Let's determine the volume of each constituent per cubic metre of concrete in order to calculate the amount of sand required:

$$\text{Water} = \frac{(190 \text{ Mg})(1)}{(10^3)(1.00 \text{ Mg.m}^{-3})} = 0.19 \text{ m}^3$$

$$\text{Cement} = \frac{317}{10^3 \times 1.75} = 0.18 \text{ m}^3$$

Aggregate = 0.42 m^3

Air = 0.02 × 1 m^3 = 0.02 m^3

Sand = 1– 0.19 – 0.18 – 0.42 – 0.02 = 0.19 m^3

Converting these to other units and assuming the sand and aggregate are dry:

Weight of aggregate/m^3 concrete = 0.42 m^3 × 2.72 $Mg.m^{-3}$ = 1.14 Mg

Weight of sand/m^3 concrete = 0.19 × 2.56 = 0.49 Mg

However the aggregate and sand are wet so the actual needed are:

Aggregate = 1.14 × 1.01 = 1.15 Mg, with 0.01 Mg water

Sand = 0.49 × 1.05 = 0.51 Mg, with 0.02 Mg water

The actual amount of water needed = 0.19 – 0.01 – 0.02 = 0.16 m^3

Thus for each cubic metre of concrete, we will combine 180 kg of cement, 1150 kg of aggregate, 510 kg of sand and 0.16 m^3 water. This concrete should give us a slump of 75 mm (the desired workability) and have a compressive strength of 25 $MN.m^{-2}$ after 28 days.

17.9 Reinforced and Prestressed Concrete

Concrete, like other ceramic-based materials, develops good compressive strength. Due to the porosity and interfaces present in the brittle structure, however, it has very poor tensile properties. Several methods are used to improve the load-bearing ability of concrete in tension.

Reinforced Concrete Steel rods, wires, or mesh are frequently introduced into concrete to provide improvement in resisting tensile and bending forces. The tensile stresses are transferred by the concrete to the steel, which has good tensile properties. Polymer fibres are less likely to corrode, can also be used as reinforcement.

Prestressed Concrete Instead of simply being laid as reinforcement rods in a form, the steel can initially be pulled in tension between an anchor and a jack, thus remaining under tension during pouring and curing of the concrete. After the concrete sets, the tension on the steel is released. The steel then tries to relax from its stretched condition, but the restraint caused by the surrounding concrete places the concrete in compression. Now higher tensile and bending stresses can be applied to the concrete because of the compressive residual stresses introduced by the pretensioned steel. In order to permit the external tension to be removed in a timely manner, the early-setting Type III cements are often used for these applications.

Poststressed Concrete An alternative method of placing concrete under compression is to place hollow tubes in the concrete prior to pouring. After the concrete cures, steel rods running through the tubes can then be pulled in tension, acting against the concrete. As the rods are placed in tension, the concrete is placed in compression. The rods are then secured permanently in their stretched condition.

17.10 Asphalt

Asphalt is a composite of aggregate and bitumen, which is a thermoplastic polymer most frequently obtained from petroleum. Asphalt is an important material for paving roads. The properties of the asphalt are determined by the characteristics of the aggregate and binder, their relative amounts, and additives.

The aggregate, as in concrete, should be clean and angular and should have a distribution of grain sizes to provide a high packing factor and good mechanical interlocking between the aggregate grains (Figure 17.12). The binder, composed of thermoplastic chains, bonds the aggregate particles. The binder has a relatively narrow useful temperature range, being brittle at sub-zero temperatures and beginning to melt at relatively low temperatures. Additives such as gasoline or kerosene can be used to modify the binder, permitting it to liquefy more easily during mixing and causing the asphalt to cure more rapidly after application.

The ratio of binder to aggregate is important. Just enough binder should be added so that the aggregate particles touch, but voids are minimised. Excess binder permits viscous deformation of the asphalt under load. Approximately 5% to 10% bitumen is present in a typical asphalt. Some void space is also required – usually, about 2 to 5%. When the asphalt is compressed, the binder can squeeze into voids, rather than be squeezed from the surface of the asphalt and lost. Too much void space, however, permits water to enter the structure; this increases the rate of deterioration of the asphalt and may also embrittle the binder.

The aggregate for asphalt is typically sand and fine gravel. However, there is some interest in using recycled glass products as the aggregate. Glasphalt provides a useful application for crushed glass.

(*a*)

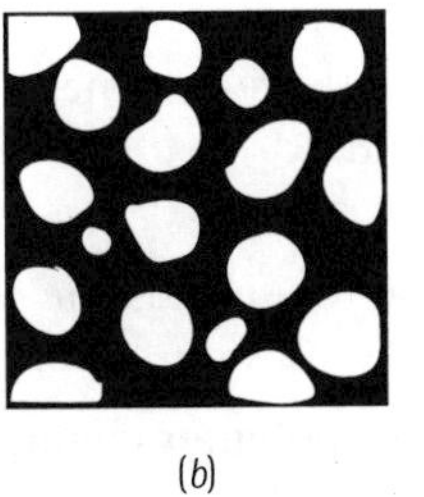
(*b*)

Figure 17.12
The ideal structure of asphalt (*a*) compared with the undesirable structure (*b*) in which round grains, a narrow distribution of grains, and excess binder all reduce the strength of the final material.

SUMMARY

Construction materials are composite materials occurring in nature or produced from natural-occurring materials.

- Wood is a natural fibre-reinforced polymer composite material. Cellulose fibres constitute aligned cells that provide excellent reinforcement in longitudinal directions in wood, but give poor strength and stiffness in directions perpendicular to the cells and fibres. The properties of wood therefore are highly anisotropic and depend on the species of the tree and the amount of moisture present in the wood.

- Concrete is a particulate composite. In concrete, ceramic particles such as sand and aggregate are used as a filler in a ceramic cement matrix. The water-cement ratio is a particularly important factor governing the behaviour of the concrete. This behaviour can be modified by entraining air and by varying the composition of the cement and aggregate materials.
- Asphalt also is a particulate composite, using the same type of aggregates as in concrete, but an organic, polymer binder.

GLOSSARY

Bitumen

The organic binder, composed of low melting point polymers and oils, for asphalt.

Cambium

The layer of growing cells in wood.

Cellulose

With a high degree of polymerisation, a naturally occurring polymer fibre that is the major constituent of wood.

Extractives

Impurities in wood.

Glasphalt

Asphalt in which the aggregate includes recycled glass.

Heartwood

The centre of a tree, comprised of dead cells, which provides mechanical support to a tree.

Hemicellulose

With a low degree of polymerisation, a naturally occurring polymer fibre that is an important constituent of wood.

Lignin

The polymer cement in wood that bonds the cellulose fibres in the wood cells.

Microfibril

Bundles of cellulose and other polymer chains that serve as the fibre reinforcement in wood.

Plies

The individual sheet of wood veneer from which plywood is constructed.

Sapwood

Hollow, living cells in wood that store nutrients and conduct water.

Slump

The decrease in height of a standard concrete from when the concrete settles under its own weight.

Workability

The ease with which a concrete slurry fills all of the space in a form.

PROBLEMS

17.1 A sample of wood with dimensions 75 mm × 100 mm × 300 mm has a dry density of 0.35 $Mg.m^{-3}$.

(a) Calculate the number of gallons of water that must be absorbed by the sample to contain 120% water.

(b) Calculate the density after the wood absorbs this amount of water.

17.2 The density of a sample of oak is 0.90 $Mg.m^{-3}$. Calculate

(a) the density of completely dry oak and

(b) the percent water in the original sample.

17.3 Boards of maple 25 mm thick, 150 mm wide, and 5 m long are used as the flooring for a 20 m × 20 m hall. The boards were cut from logs with a tangential-longitudinal cut. The floor is laid when the boards have a moisture content of 12%. After some particularly humid days, the moisture content in the boards increases to 45%. Determine the dimensional change in the flooring parallel to the boards and perpendicular to the boards. What will happen to the floor? How can this problem be corrected?

17.4 A wall 10 m long is built using radial-longitudinal cuts of 125 mm-wide pine, with the boards arranged in a vertical fashion. The wood contains a moisture content of 55% when the wall is built; however, the humidity level in the room is maintained to give 45% moisture in the wood. Determine the dimensional changes in the wood boards and estimate the size of the gaps that will be produced as a consequence of these changes.

17.5 We have been asked to prepare 100 m^3 of normal concrete using a volume ratio of cement-sand-coarse aggregate of 1 : 2 : 4. The water-cement ratio (by weight) is to be 0.5. The sand contains 6 wt% water and the coarse aggregate contains 3 wt% water. No entrained air is expected.

(a) Determine the number of 50 kg sacks of cement that must be ordered, the kg of sand and aggregate required, and the amount of water needed.

(b) Calculate the total weight of the concrete per cubic metre.

(c) What is the weight ratio of cement-sand-coarse aggregate?

17.6 We plan to prepare 10 m^3 of concrete using a 1 : 2.5 : 4.5 weight ratio of cement-sand-coarse aggregate. The water-cement ratio (by weight) is 0.45. The sand contains 3 wt% water, the coarse aggregate contains 2 wt% water, and 5% entrained air is expected. Determine the number of 50 kg sacks of cement, kg of sand and coarse aggregate, and cubic metres of water required.

Design Problems

17.7 A wooden structure is functioning in an environment controlled at 65% humidity. Design a wood support column that is to hold a compressive load of 90 kN. The distance from the top to the bottom of the column should be 2400 – 6 mm when the load is applied.

17.8 Design a wood floor that will be 15250 × 15250 mm and will be in an environment in which humidity changes will cause a fluctuation of plus or minus 5% water in the wood. We want to minimise any buckling or gap-formation in the floor.

17.9 We would like to produce a concrete that is suitable for use in building a large structure in a sulphate environment. For these situations, the maximum water-cement ratio should be 0.45 (by weight). The compressive strength of the concrete after 28 days should be at least 25 $MN.m^{-2}$. We have available coarse aggregate containing 2% moisture in a variety of sizes, and both fine and coarse sand containing 4% moisture. Design a concrete that will be suitable for this application.

17.10 We would like to produce a concrete sculpture. The sculpture will be as thin as 75 mm in some areas and should be light in weight, but it must have a 28-day compressive strength of at least 12 $MN.m^{-2}$. Our available aggregate contains 1% moisture and our sands contain 5% moisture. Design a concrete that will be suitable for this application.

17.11 The binder used in producing asphalt has a density of about 1.3 $Mg.m^{-3}$. Design an asphalt, including the weight and volumes of each constituent, that might be suitable for use as pavement. Assume that the sands and aggregates are the same as those for a normal concrete.

The physical behaviour of materials is described by a variety of electrical, magnetic, optical, and thermal properties. Most of these properties are determined by the atomic structure, atomic arrangement, and crystal structure of the material. In Chapter 18, we find that the atomic structure – in particular, the energy gap between the electrons in the valence and conduction bands – helps us to divide materials into conductors, semiconductors, and insulators. Atomic structure is responsible for the ferromagnetic behaviour discussed in Chapter 19 and explains many optical (Chapter 20) and thermal (Chapter 21) properties.

Physical properties can be altered to a significant degree by changing the short- and long-range order of the atoms as well as by introducing and controlling imperfections in the atomic structure and arrangement. Strengthening mechanisms and metal processing techniques, for example, have a significant effect on electrical conductivity of metals. Improved magnets are obtained by introducing lattice defects or by controlling grain size. In this section, we again demonstrate the importance of the structure-property-processing relationship.

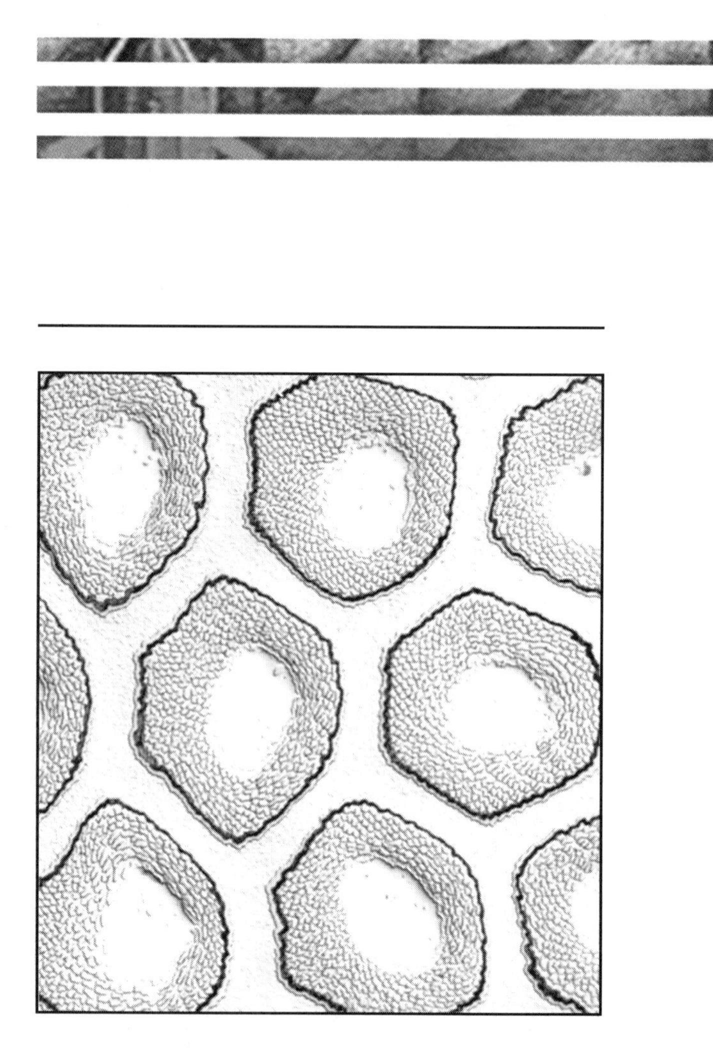

The structure of a Nb_3Sn superconductor wire for high-field magnetic applications includes many complex fibres within the wire. In this photomicrograph taken at a magnification of 1 000 ×, fibres are embedded in a continuous copper matrix. The white core of each fibre is tin, and the 'pebbly' layer around the tin core contains over 100 tiny niobium filaments. The dark ring around each fibre is vanadium; the light ring around the vanadium is niobium. These two rings keep the copper from diffusing into the fibre. During later heat treatment, the niobium filaments react with the tin, converting the niobium filaments to Nb_3Sn filaments (*From* Metals Handbook, *Vol. 2, 10th Ed., ASM International, 1990.*)

Part 4

Physical Properties of Engineering Materials

CHAPTER 18

Electrical Behaviour of Materials

18.1 Introduction

In many applications, the electrical behaviour of a material is more critical than its mechanical behaviour. Metal wire used to transfer current over long distances must have a high electrical conductivity so that little power is lost by heating of the wire. Ceramic and polymer insulators must possess dielectric properties that prevent breakdown of the materials and arcing between conductors. Semiconductor devices used to convert solar energy to electrical power must be as efficient as possible to make solar cells a practical alternative energy source.

To select and use materials for electrical and electronic applications, we must understand how properties such as electrical conductivity are controlled. We must also realise that electrical behaviour is influenced by the structure of a material, the processing of a material, and the environment to which a material is exposed.

18.2 Ohm's Law and Electrical Conductivity

Most of us are familiar with the common form of Ohm's law,

$$V = IR \tag{18.1}$$

where V is the voltage (volts, V), I is the current (amperes or amps, A), and R is the resistance (ohms, Ω) to the current flow. The resistance R is a characteristic of the size, shape, and properties of the materials that compose the circuit,

$$R = \rho\frac{l}{A} = \frac{l}{\sigma A} \tag{18.2}$$

where l is the length (m) of the conductor, A is the cross-sectional area (m^2) of the conductor, ρ is the electrical resistivity (Ω.m), and σ, which is the reciprocal of ρ, is the electrical conductivity ($\Omega^{-1}.m^{-1}$). We can use this equation to design resistors, because we can vary the length or cross-sectional area of the device.

A quantity of electrical charge is measured in coulombs, C. One coulomb is the amount of electrical charge that has been transferred when a current of one amp has flowed for one second. So,

$$1\ \text{C} = 1\ \text{A.s}$$

In components designed to conduct electrical energy, minimising power losses is important not only to conserve energy but also to minimise heating. The electrical power, P (in watts, W) lost when a current flows through a resistance is given by:

$$P = VI = I^2R \qquad (18.3)$$

A high resistance R results in larger power losses.

A second form of Ohm's law is obtained if we combine Equations 18.1 and 18.2 to give:

$$\frac{I}{A} = \sigma \frac{V}{l}$$

If we define I/A as the current density J (A.m^{-2}) and V/l as the electric field strength ξ (V.m^{-1}), then:

$$J = \sigma\xi \qquad (18.4)$$

We can also determine that the current density J is

$$J = nq\bar{v}$$

where n is the number of charge carriers (carriers/m^3), q is the charge on each carrier (1.6×10^{-19} C is the charge on one electron), and $\bar{v}$ is the average drift velocity (m.s^{-1}) at which the charge carriers move (Figure 18.1). Thus:

$$\sigma\xi = nq\bar{v} \qquad \text{or} \qquad \sigma = nq\frac{\bar{v}}{\xi}$$

The term $\bar{v}/\xi$ is called the mobility μ (m^2.V^{-1}.s^{-1}):

$$\mu = \frac{\bar{v}}{\xi}$$

Finally:

$$\sigma = nq\mu \qquad (18.5)$$

The charge q is a constant; from inspection of Equation 18.5, we find that we can control the electrical conductivity of materials by (1) controlling the number of charge carriers in the material or (2) controlling the mobility – or ease of movement – of the charge carriers. The mobility is particularly important in metals, whereas the number of carriers is more important in semiconductors and insulators.

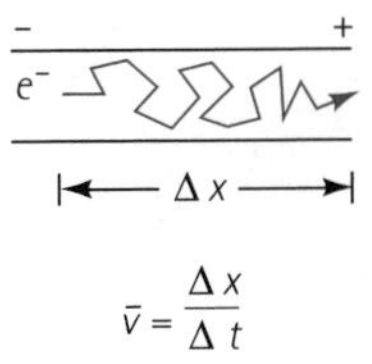

Figure 18.1
Charge carriers, such as electrons, are deflected by atoms or lattice defects and take an irregular path through a conductor. The average rate at which the carriers move is the drift velocity $\bar{v}$.

Electrons are the charge carriers in conductors (such as metals), semiconductors, and many insulators, whereas ions carry most of the charge in ionic compounds (Figure 18.2). The mobility depends on atomic bonding, lattice imperfections, microstructure, and, in ionic compounds, diffusion rates. Because of these effects, the electrical conductivity of materials varies tremendously, as illustrated in Table 18.1. Table 18.2 includes some useful units and relationships.

Table 18.1 Electrical conductivity of selected materials.

Material	Electronic Structure	Conductivity ($\Omega^{-1}.m^{-1}$)
Alkali metals:		
Na	$1s^22s^22p^63s^1$	2.13×10^7
K	.... $3s^23p^64s^1$	1.64×10^7
Alkali earth metals:		
Mg	$1s^22s^22p^63s^2$	2.25×10^7
Ca	.... $3s^23p^64s^2$	3.16×10^7
Group IIIA metals:		
Al	$1s^22s^22p^63s^23p^1$	3.77×10^7
Ga	... $3s^23p^63d^{10}4s^24p^1$	0.66×10^7
Transition metals:		
Fe	..... $3d^64s^2$	1.00×10^7
Ni	..... $3d^64s^2$	1.46×10^7
Group IB metals:		
Cu	..... $3d^{10}4s^1$	5.98×10^7
Ag	..... $4d^{10}5s^1$	6.80×10^7
Au	..... $5d^{10}6s^1$	4.26×10^7
Group IV materials		
C (diamond)	$1s^22s^22p^2$	$<10^{-16}$
Si	...... $3s^23p^2$	5×10^{-4}
Ge	...... $4s^24p^2$	2.0
Sn	...... $5s^25p^2$	9.0×10^6
Polymers:		
Polyethylene		10^{-13}
Polytetrafluoroethylene		10^{-16}
Polystyrene		10^{-15} to 10^{-17}
Epoxy		10^{-10} to 10^{-15}
Ceramics:		
Alumina (Al_2O_3)		10^{-12}
Silica glass		10^{-15}
Boron nitride(BN)		10^{-11}
Silicon carbide (SiC)		1 to 10
Boron carbide (B_4C)		100 to 200

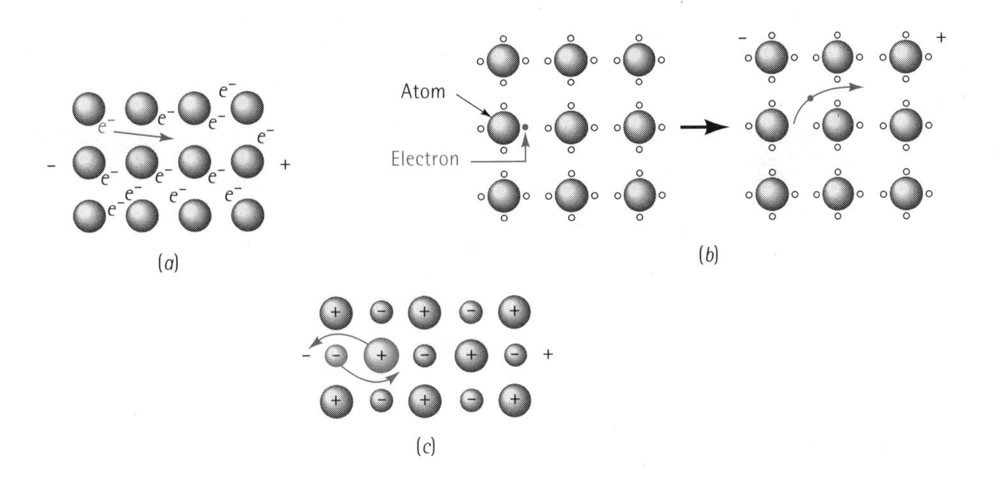

Figure 18.2
Charge carriers in different materials: (*a*) Valence electrons in the metallic bond move easily, (*b*) covalent bonds must be broken in semiconductors and insulators for an electron to be able to move, and (*c*) entire ions must diffuse to carry charge in many ionically bonded materials.

Table 18.2 Some useful relationships and units.

Electron volt = 1 eV = 1.6×10^{-19} joule (J)
1 amp = 1 coulomb/second (1 A = 1 $C.s^{-1}$)
1 volt = 1 amp · ohm (1 V = 1 A.Ω)
k = Boltzmann's constant = 8.63×10^{-5} eV/K = 1.38×10^{-23} $J.K^{-1}$
kT at room temperature = 0.025 eV = 4.1×10^{-21} J

EXAMPLE 18.1 Materials Selection for a Transmission Line

Choose a suitable material for an electrical transmission line 1 500 m long that will carry a current of 50 A with no more than 5×10^5 W loss in power. The electrical conductivity of several materials is included in Table 18.1.

SOLUTION

Electrical power is given by the product of the voltage and current, or:

$$P = VI = I^2R = (50)^2R = 5 \times 10^5 \text{ W}$$

So $R = 200\ \Omega$

From Equation 18.2:

$$A = l/R\sigma = \frac{(1\,500 \text{ m})}{(200\ \Omega)\sigma} = \frac{7.5}{\sigma}$$

Let's consider three metals – aluminium, copper, and silver – that have excellent electrical conductivity. The following table includes appropriate data and some characteristics of the transmission line on each metal (based on the above relationship).

	$\sigma(\Omega^{-1}.m^{-1})$	A (mm²)	Diameter (mm)
Aluminium	3.77×10^7	0.199	0.50
Copper	5.98×10^7	0.125	0.40
Silver	6.80×10^7	0.110	0.37

Any of the three metals will work but cost may be a factor. The weight of the 1500 m line is the volume (area times length) times the density, where the density is given in Appendix A.

$$\text{Al: weight} = A.l.\rho = (1.99 \times 10^{-7}\ \text{m}^2)(1\,500\ \text{m})(2.7\ \text{Mg.m}^{-3})$$
$$= 8.06 \times 10^{-4}\ \text{Mg} = 8.06 \times 10^{-4}\ \text{tonnes}$$
$$\text{cost} = (8.06 \times 10^{-4}\ \text{tonnes})(\$1\,760/\text{tonne}) = \$1.42$$

$$\text{Cu: weight} = A.l.\rho = (1.25 \times 10^{-7}\ \text{m}^2)(1\,500\ \text{m})(8.93\ \text{Mg.m}^{-3})$$
$$= 1.67 \times 10^{-3}\ \text{Mg} = 1.67 \times 10^{-3}\ \text{tonnes}$$
$$\text{cost} = (1.67 \times 10^{-3}\ \text{tonnes})(\$2\,430/\text{tonne}) = \$4.06$$

$$\text{Ag: weight} = A.l.\rho = (1.10 \times 10^{-7}\ \text{m}^2)(1\,500\ \text{m})(10.49\ \text{Mg.m}^{-3})$$
$$= 1.73 \times 10^{-3}\ \text{Mg} = 1.73 \times 10^{-3}\ \text{tonnes}$$
$$\text{cost} = (1.73 \times 10^{-3}\ \text{tonnes})(\$141\,100/\text{tonne}) = \$244.10$$

Based on this analysis, aluminium is the most economical choice, even though the wire has the largest diameter. However, other factors, such as whether the wire can support itself between transmission poles, also contribute to the final choice.

EXAMPLE 18.2

Assuming that all of the valence electrons contribute to current flow, (a) calculate the mobility of an electron in copper and (b) calculate the average drift velocity for electrons in a 1 m copper wire when 10 V are applied.

SOLUTION

1. The valence of copper is one: therefore the number of valence electrons equals the number of copper atoms in the material. The lattice parameter of copper is 3.6151×10^{-10} m and, since copper is FCC, there are 4 atoms/unit cell. From Table 18.1, the resistivity $= 1/\sigma = 1/5.98 \times 10^7 = 1.67 \times 10^{-8}\ \Omega$.m:

$$n = \frac{(4 \text{ atoms / cell})(1 \text{ electron / atom})}{(3.6151 \times 10^{-10} \text{ m})^3} = 8.466 \times 10^{28} \text{ electrons / m}^3$$

$$q = 1.6 \times 10^{-19} \text{C}$$

$$\mu = \frac{\sigma}{nq} = \frac{1}{\rho nq} = \frac{1}{(1.67 \times 10^{-8})(8.466 \times 10^{28})(1.6 \times 10^{-19})}$$

$$= 4.42 \times 10^{-3} \text{ m}^2.\Omega^{-1}.\text{C}^{-1}$$

$$= 4.42 \times 10^{-3} \text{ m}^2.\text{V}^{-1}.\text{s}^{-1}$$

2. The electric field is:

$$\xi = \frac{V}{l} = \frac{10}{1} = 10 \text{ V.m}^{-1}$$

The mobility is 4.42×10^{-3} m^2.V^{-1}.s^{-1}; therefore:

$\bar{v} = \mu\xi = (4.42 \times 10^{-3})(10) = 0.0442$ m.s^{-1}

18.3 Band Theory

In Chapter 2 we found that the electrons in a single atom occupy discrete energy levels. The Pauli exclusion principle permits each energy level to contain only two electrons. For example, the 2*s* level of a single atom contains one energy level and two electrons. The 2*p* level contains three energy levels and a total of six electrons.

When *N* atoms come together to produce a solid, the Pauli principle still requires that only two electrons in the entire solid have the same energy. Each energy level broadens into a band (Figure 18.3). Consequently, the 2*s* band in a solid contains *N* discrete energy levels and 2*N* electrons, two in each energy level. Each of the 2*p* levels contains *N* energy levels and 2*N* electrons. Since the three 2*p* bands actually overlap, we could alternately describe a single, broad 2*p* band containing 3*N* energy levels and 6*N* electrons.

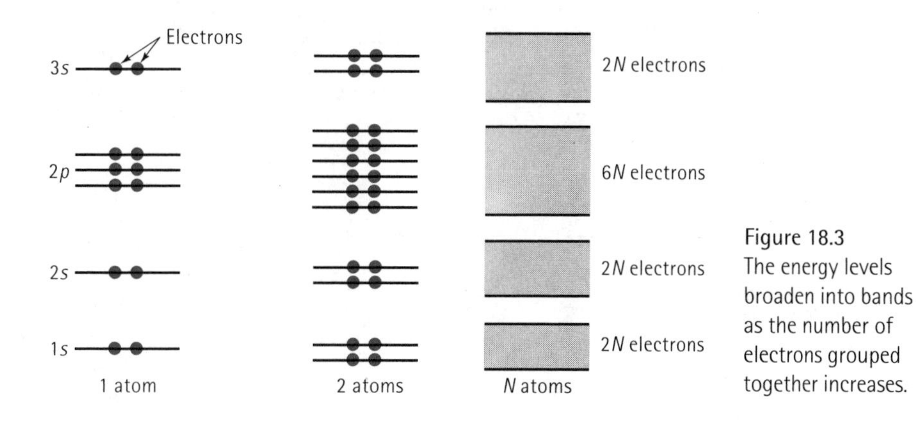

Figure 18.3
The energy levels broaden into bands as the number of electrons grouped together increases.

Band Structure of Sodium Figure 18.4 shows an idealised picture of the band structure in sodium, which has an electronic structure of $1s^2 2s^2 2p^6 3s^1$. The energies within the bands depend on the spacing between the atoms; the vertical line represents the equilibrium interatomic spacing between the atoms in solid sodium. The $3s$ energy levels are the valence band. The empty $3p$ energy levels, which are separated from the $3s$ band by an energy gap, form the conduction band.

Sodium and other alkali metals in column IA of the periodic table have only one electron in the outermost s level. The $3s$ valence band in sodium is half filled and, at absolute zero, only the lowest energy levels are occupied. The *Fermi energy* (E_f) is the energy at which half of the possible energy levels in the band are occupied by electrons.

But when the temperature of the metal increases, some electrons gain energy and are excited into the empty energy levels in the valence band (Figure 18.5). This condition creates an equal number of empty energy levels, or holes, vacated by the excited electrons. Only a small increase in energy is required. An electrical charge can then be carried by both the excited electrons and the newly created holes, by their movement through the solid.

Band Structure of Magnesium and Other Metals The simplified band structure for magnesium ($1s^2 2s^2 2p^6 3s^2$) is shown in Figure 18.6. Magnesium and other metals in column IIA of the periodic table have two electrons in their outermost s band. These

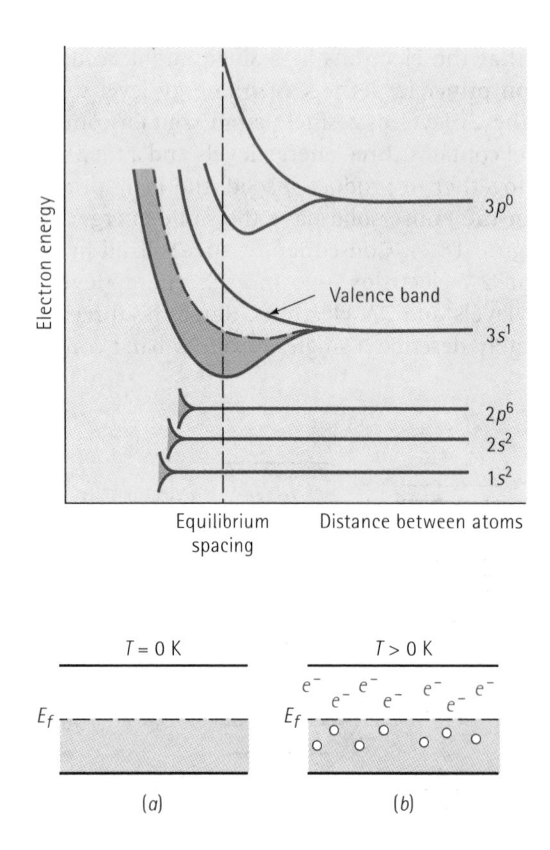

Figure 18.4
The simplified band structure for sodium. The energy levels broaden into bands. The $3s$ band, which is only half filled with electrons, is responsible for conduction in sodium.

Figure 18.5
(*a*) At absolute zero, all of the electrons in the outer energy level have the lowest possible energy. (*b*) When the temperature is increased, some electrons are excited into unfilled levels. Note that the Fermi energy is unchanged.

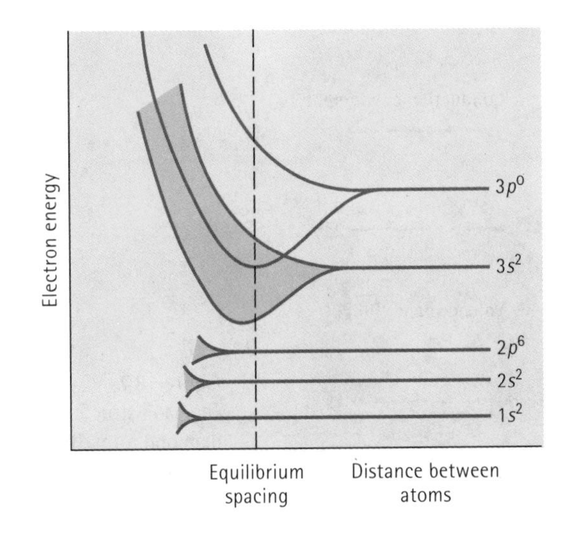

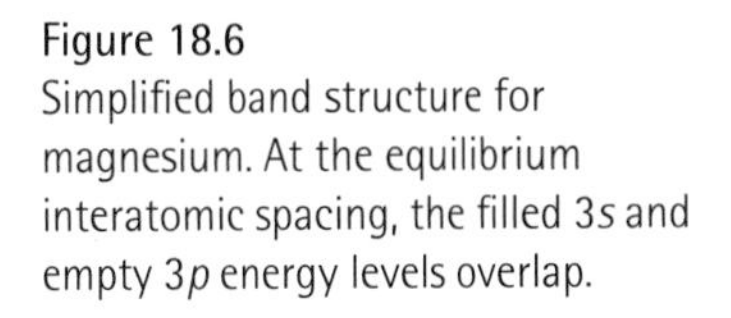
Figure 18.6
Simplified band structure for magnesium. At the equilibrium interatomic spacing, the filled $3s$ and empty $3p$ energy levels overlap.

metals have a high conductivity because the p band overlaps the s band at the equilibrium interatomic spacing. This overlap permits electrons to be excited into the large number of unoccupied energy levels in the combined $3s$ and $3p$ band. Overlapping $3s$ and $3p$ bands in aluminium and other metals in column IIIA provide a similar effect.

In the transition metals, including scandium through to nickel, an unfilled $3d$ band overlaps the $4s$ band. This overlap provides energy levels into which electrons can be excited; however, complex interactions between the bands prevent the conductivity from being as high as in some of the better conductors. But in copper, the inner $3d$ band is full, and these electrons are tightly held by the atom core. Consequently there is little interaction between the electrons in the (unfilled) $4s$ and (full) $3d$ bands, and copper has a high conductivity. A similar situation is found for silver and gold.

Band Structure of Semiconductors and Insulators The elements in Group IVA – carbon (diamond), silicon, germanium, and tin – contain two electrons in their outer p shell and have a valence of four. Based on our discussion in the previous section, we might expect these elements to have a high conductivity due to the unfilled p band, but this behaviour is not observed!

These elements are covalently bonded; consequently, the electrons in the outer s and p bands are rigidly bound to the atoms. The covalent bonding produces a complex change in the band structure, or hybridisation. The $2s$ and $2p$ levels of the carbon atoms in diamond can contain up to eight electrons, but there are only four valence electrons available. When carbon atoms are brought together to form solid diamond, the $2s$ and $2p$ levels interact and produce two bands (Figure 18.7). Each hybrid band can contain $4N$ electrons. Since there are only $4N$ electrons available, the lower (or valence) band is completely filled, whereas the upper (or conduction) band is empty.

A large energy gap, E_g, separates the electrons from the conduction band in diamond. Few electrons possess sufficient energy to jump the forbidden zone to the conduction band. Consequently, diamond has an electrical conductivity of less than $10^{-16}\ \Omega^{-1}.\text{m}^{-1}$. Other covalently and ionically bonded materials have a similar band structure and, like diamond, behave as electrical insulators. Table 18.1 shows the electrical conductivity of several polymer and ceramic materials at room temperature.

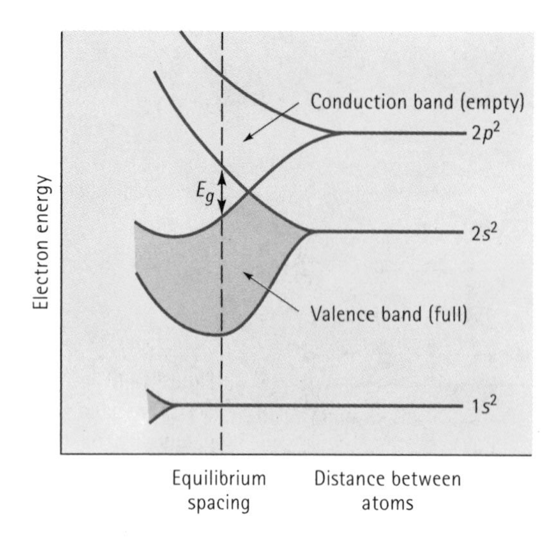

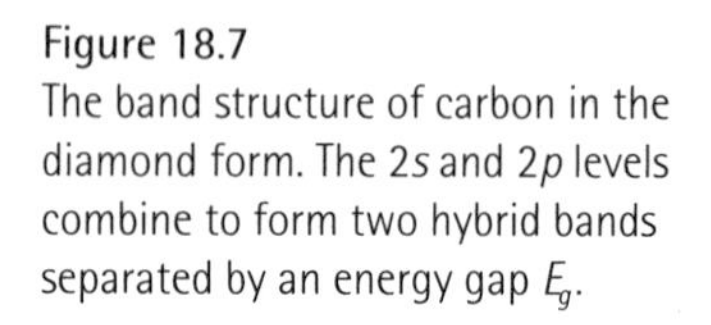
Figure 18.7
The band structure of carbon in the diamond form. The $2s$ and $2p$ levels combine to form two hybrid bands separated by an energy gap E_g.

Increasing the temperature or the applied voltage supplies the energy required for electrons to overcome the energy gap. For example, the electrical conductivity of boron nitride increases from about 10^{-11} at room temperature to 10^{-2} at 800°C.

Although germanium, silicon, and tin have the same crystal and band structure as diamond, the energy gap is smaller. In fact, the energy gap in tin is so small that tin behaves as a metal. The energy gap is somewhat larger in silicon and germanium – these elements behave as semiconductors. Table 18.1 includes the electrical conductivity of these four elements.

18.4 Controlling the Conductivity of Metals

The conductivity of a pure, defect-free metal is determined by the electronic structure of the atoms. But we can change the conductivity by influencing the mobility, μ, of the carriers (refer to Equation 18.5). The mobility is proportional to the drift velocity, $\bar{v}$, which is low if the electrons collide with imperfections in the lattice. The mean free path is the average distance between collisions; a long mean free path permits high mobilities and high conductivities.

Temperature Effect When the temperature of a metal increases, thermal energy causes the atoms to vibrate (Figure 18.8). At any instant, the atom may not be in its equilibrium position, and it therefore interacts with and scatters electrons. The mean free path decreases, the mobility of electrons is reduced, and the resistivity increases. The change in resistivity with temperature can be estimated from the equation:

$$\rho_T = \rho_r(1 + a\Delta T) \tag{18.6}$$

where ρ_T is the resistivity due only to thermal vibration, ρ_r is the resistivity at room temperature (25°C), ΔT is the difference between the temperature of interest and room temperature, and a is the *temperature resistivity coefficient*. The relationship between resistivity and temperature is linear over a wide temperature range (Figure 18.9). Examples of the temperature resistivity coefficient are given in Table 18.3.

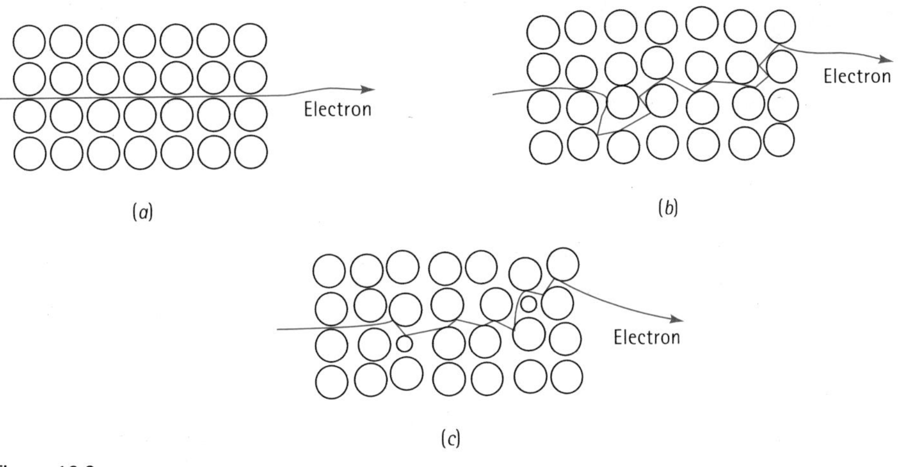

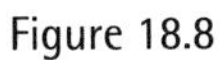

Figure 18.8
Movement of an electron through (*a*) a perfect crystal, (*b*) a crystal heated to a high temperature, and (*c*) a crystal containing lattice defects. Scattering of the electrons reduces the mobility and conductivity.

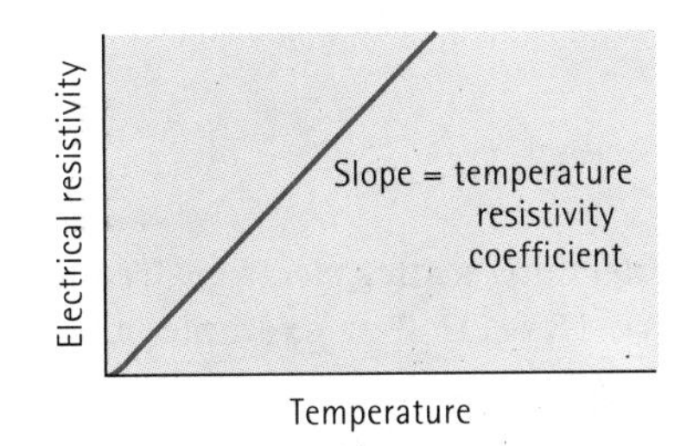

Figure 18.9
The effect of temperature on the electrical resistivity of a metal with a perfect lattice. The slope of the curve is the temperature resistivity coefficient.

Table 18.3 The temperature resistivity coefficient for selected metals.

Metal	Room Temperature Resistivity (Ω.m)	Temperature Resistivity Coefficient (°C^{-1})
Be	4.0×10^{-8}	0.0250
Mg	4.45×10^{-8}	0.0165
Ca	3.91×10^{-8}	0.0042
Al	2.65×10^{-8}	0.0043
Cr	1.29×10^{-7}	0.0030
Fe	9.71×10^{-8}	0.0065
Co	6.24×10^{-8}	0.0060
Ni	6.84×10^{-8}	0.0069
Cu	1.67×10^{-8}	0.0068
Ag	1.59×10^{-8}	0.0041
Au	2.35×10^{-8}	0.0040

From ASM Metals Handbook, *Vol. 2, 9th Ed., 1979.*

EXAMPLE 18.3

Calculate the electrical conductivity of pure copper at (a) 400°C and (b) –100°C.

SOLUTION

The resistivity of copper at room temperature is 1.67×10^{-8} Ω.m and the temperature resistivity coefficient is 0.0068 °C^{-1}.

1. At 400°C:

   $\rho = \rho_r(1 + a\Delta T) = (1.67 \times 10^{-8})[1 + 0.0068(400 - 25)]$

   $\rho = 5.929 \times 10^{-8}$ Ω.m

   $\sigma = 1/\rho = 1.69 \times 10^{7}$ Ω^{-1}.m^{-1}

2. At –100°C:

   $\rho = (1.67 \times 10^{-8})[1 + 0.0068(-100 - 25)] = 0.251 \times 10^{-8}$ Ω.m

   $\sigma = 39.8 \times 10^{7}$ Ω^{-1}.m^{-1}

Effect of Lattice Defects Lattice imperfections scatter electrons, reducing the mobility and conductivity of the metal [Figure 18.8(c)]. For example, the increase in the resistivity due to solid solution atoms is

$$\rho_d = b(1 - x)x \tag{18.7}$$

where ρ_d is the increase in resistivity due to the defects, x is the atomic fraction of the impurity or solid solution atoms present, and b is the defect resistivity coefficient. In a similar manner, vacancies, dislocations, and grain boundaries reduce the conductivity of the metal. Each defect contributes to an increase in the resistivity of the metal. Thus, the overall resistivity is

$$\rho = \rho_T + \rho_d \tag{18.8}$$

where ρ_d equals the contributions from all of the imperfections. The effect of the defects is independent of temperature (Figure 18.10).

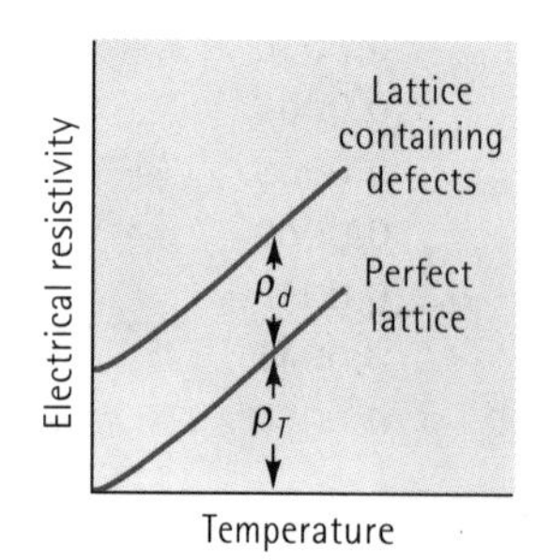

Figure 18.10
The electrical resistivity of a metal is composed of a constant defect contribution ρ_d and a variable temperature contribution ρ_T.

Table 18.4 The effect of alloying, strengthening, and processing on the electrical conductivity of copper and its alloys.

Alloy	$\frac{\sigma_{alloy}}{\sigma_{Cu}} \times 100$ *	Remarks
Commercially pure annealed copper	101	Few lattice defects to scatter electrons; the mean free path is long.
Commercially pure copper deformed 80%	98	Many dislocations, but because of the tangled nature of the dislocation networks, the mean free path is still long.
Dispersion-strengthened Cu-0.7% Al_2O_3	85	The dispersed phase is not as closely spaced as solid solution atoms, nor is it coherent, as in age hardening. Thus, the effect on conductivity is small.
Solution-treated Cu-2% Be	18	The alloy is single phase; however, the small amount of solid solution strengthening from the supersaturated beryllium greatly decreases conductivity.
Aged Cu-2% Be	23	During ageing, the beryllium leaves the copper lattice to produce a coherent precipitate. The precipitate does not interfere with conductivity as much as the solid solution atoms.
Cu-35% Zn	28	This alloy is solid solution strengthened by zinc, which has an atomic radius near that of copper. The conductivity is low, but not as low as when beryllium is present.

* where $\sigma_{Cu} = 5.8 \times 10^7\ \Omega^{-1}.m^{-1}$, the conductivity of a standard grade of copper, otherwise known as 100% IACS (International Annealed Copper Standard).

Effect of Processing and Strengthening Strengthening mechanisms and metal processing techniques affect the electrical properties of a metal in different ways (Table 18.4). Solid solution strengthening is a poor way to obtain high strength in metals intended to have high conductivities. The mean free paths are very short due to the random distribution of the interstitial or substitutional atoms. Figure 18.11 shows the effect of zinc and other alloying elements on the conductivity of copper; as the amount of alloying element increases, the conductivity decreases substantially.

Age hardening and dispersion strengthening reduce the conductivity less than solid solution strengthening, since there is a longer mean free path between precipitates, as compared with the path between point defects. Strain hardening and grain size control have even less effect on conductivity (Figure 18.11 and Table 18.4). Since dislocations and grain boundaries are further apart than solid solution atoms, there are large volumes of metal that have a long mean free path. Consequently, cold working is an effective way to increase the strength of a metallic conductor without seriously impairing the electrical properties of that material. In addition, the effects of cold working on conductivity can be eliminated by the low temperature recovery heat treatment, in which good conductivity is restored while the strength is retained.

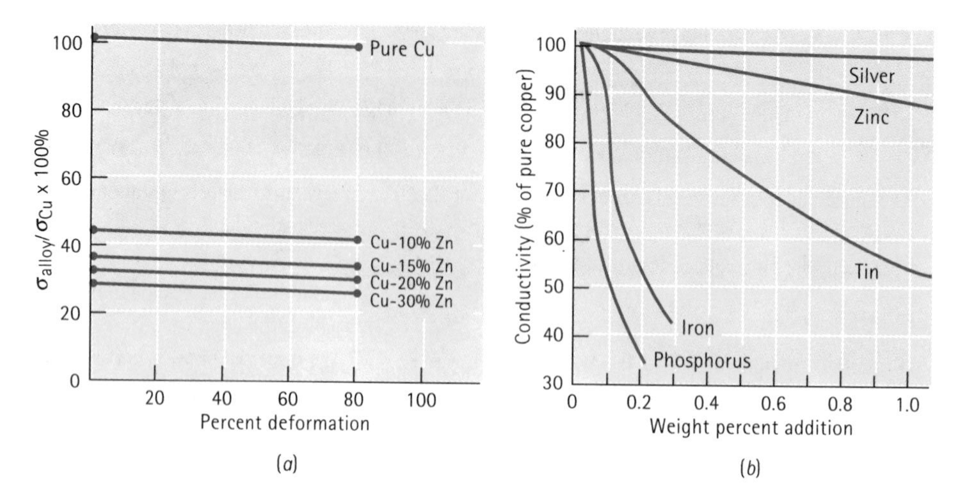

Figure 18.11
(*a*) The effect of solid solution strengthening and cold working on the electrical conductivity of copper and (*b*) the effect of selected elements on the electrical conductivity of copper.

18.5 Superconductivity

Some crystals cooled to absolute zero behave as superconductors, since the electrical resistivity becomes almost zero and a current flows indefinitely in the material. Unfortunately, obtaining absolute zero is not practical. However, some materials display superconductive behaviour above absolute zero, even when the crystal contains defects. The change from normal conduction to superconduction, which occurs abruptly at a critical temperature T_c (Figure 18.12), is a complex phenomenon; electrons having the same energy but opposite spin combine to form pairs. When the frequency of atom vibrations (phonons, *see* Section 21.2) within the lattice is synchronised with the frequency or wavelength of the electron pairs, superconductivity can occur. A variety of metals and intermetallic compounds display this effect (Table 18.5), with critical temperatures up to about 20 K. This effect permits materials to be made superconductive by cooling with liquid helium at 4 K. A number of medical diagnostic tools, including magnetic resonance imaging, use these superconductors.

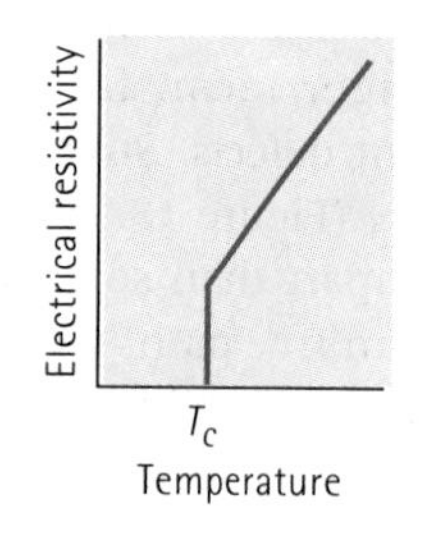

Figure 18.12
The electrical resistivity of a superconductor becomes zero below some critical temperature T_c.

Table 18.5 The critical temperature, T_c, and magnetic field, H_0, for superconductivity of selected materials.

Material	T_c (K)	H_0 (A.m^{-1})
Type I superconductors:		
W	0.015	91.5
Al	1.180	8 360
Sn	3.720	24 300
Type II superconductors:		
Nb	9.25	156 800
Nb_3Sn	18.05	2.0×10^7
GaV_3	16.80	2.8×10^7
Ceramic superconductors		
$(La, Sr)_2CuO_4$	40.0	
$YBa_2Cu_3O_{7-x}$	93.0	
$TlBa_2Ca_3Cu_4O_{11}$	122.0	

In order for a material to be a superconductor, the magnetic field must be excluded from the conductor. (This explains the ability of superconductors to levitate above a magnet.) For type I superconductors, which include tungsten and tin, increasing the magnetic field decreases the temperature required for superconduction. For a magnetic field greater than H_c, lines of magnetic flux enter the conductor and completely suppress superconduction. The effect of the field on the critical temperature is given by:

$$H_c = H_0\left[1-\left(\frac{T}{T_c}\right)^2\right] \tag{18.9}$$

where H_0 is the critical magnetic field at 0 K [Figure 18.13(a)]. Combinations of temperatures and magnetic fields within the envelope created by H_0 and T_c provide superconductivity in Type I materials.

Most materials, including Nb_3Sn, are Type II or high-field superconductors. These materials go through three stages as the magnetic field increases, changing from completely superconductive (1) to a mixed state (2), to normal conduction (3) over a range of fields. At one point in this range, the surface of the material is superconductive, while the centre provides normal conduction. The Type II superconductors typically have higher critical temperatures and magnetic fields than Type I materials.

In addition, the current density (A.m^{-2}) flowing through the conductor affects superconductivity; if the current density J is too high, superconductivity is lost [Figure 18.13(b)]. Superconducting materials should have high critical temperatures, magnetic fields, and current densities.

Until about 1986, liquid helium (at 4 K) was required to cool the materials below the critical temperature and, consequently, the applications for superconductors were limited. However, a group of ceramic materials have been discovered whose critical temperatures exceed 77 K, permitting cooling by relatively inexpensive liquid nitrogen.

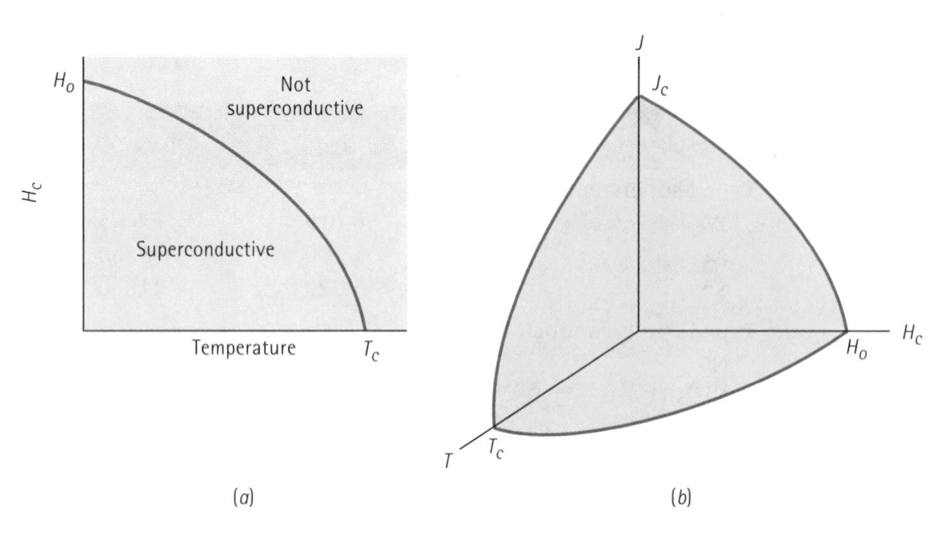

Figure 18.13
(a) The effect of a magnetic field on the temperature below which superconductivity occurs and (b) the superconducting envelope, showing the combined effects of temperature, magnetic field, and current density. Conditions within the envelope produce superconductivity.

These materials include the so-called 1-2-3 compounds of the form $YBa_2Cu_3O_{7-x}$, where x indicates that some oxygen ions are missing from the complicated perovskite crystal structure (Figure 18.14). Similar behaviour has been found for a variety of other ceramics, including some (such as $TlBa_2Ca_3Cu_4O_{11}$) that have critical temperatures above 100 K.

Current densities of 10^{10} $A.m^{-2}$ are obtained using conventional Type II superconductors, but current densities of less than 10^7 $A.m^{-2}$ are obtained in ceramics

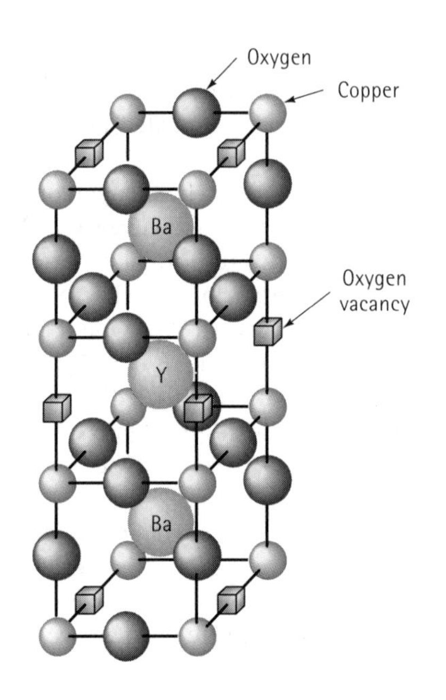

Figure 18.14
The crystal structure of the $YBa_2Cu_3O_{7-x}$ superconducting compound is composed of three perovskite cells in a layer form, giving an orthorhombic structure.

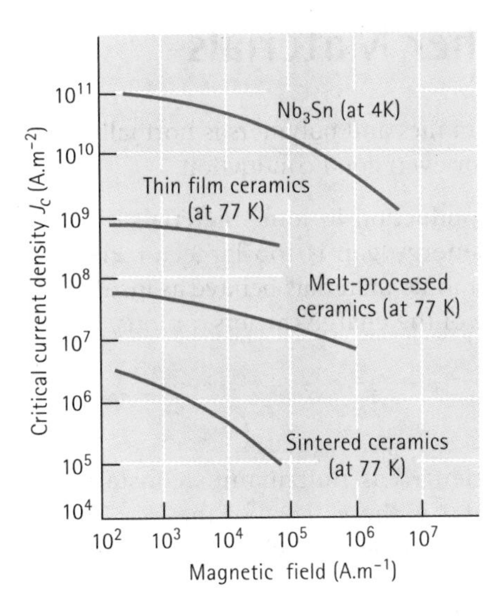

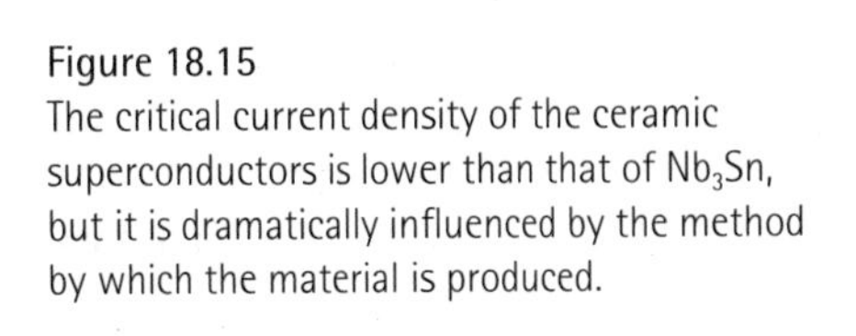

Figure 18.15
The critical current density of the ceramic superconductors is lower than that of Nb_3Sn, but it is dramatically influenced by the method by which the material is produced.

produced by conventional sintering processes. Higher current densities can be obtained in ceramics with a preferred orientation, requiring sophisticated directional solidification processes. Use of thin films rather than bulk ceramics reduces these processing problems and permits even higher current densities (Figure 18.15). The ceramics also are brittle and difficult to fabricate. One approach is to incorporate the superconductor into a metal tube to permit wires to be drawn.

In addition to medical applications, potential uses for superconductive materials include electronic switches, more efficient power transmission lines, high-speed computers, magnetically levitated trains, improved batteries for non-polluting electric road vehicles, fusion power plants and the superconducting supercollider for nuclear physics.

EXAMPLE 18.4 **Specification of a Superconductor System**

Specify the limiting magnetic field that will permit niobium to serve as a superconductor at liquid helium temperatures.

SOLUTION

From Table 18.5, T_c = 9.25 K and H_0 = 156 800 A.m^{-1}. Since the operating temperature will be 4 K, we need to determine the maximum permissible magnetic field, using Equation 18.9:

$$H_c = H_0[1 - (T/T_c)^2]$$

$$H_c = 156\,800[1 - (4/9.25)^2] = (156\,800)(0.813) = 127\,500 \text{ A.m}^{-1}$$

Therefore, the magnetic field must remain below 127 500 A.m^{-1} in order for the niobium to remain superconductive.

18.6 Conductivity in Other Materials

Electrical conductivity in most ceramics and polymers is normally very low. However, special materials provide limited or even good conduction.

Conduction in Ionic Materials Conduction in ionic materials often occurs by movement of entire ions, since the energy gap is too large for electrons to enter the conduction band. Therefore, most ionic materials behave as insulators.

In ionic materials, the mobility of the charge carriers, or ions, is

$$\mu = \frac{ZqD}{kT} \tag{18.10}$$

where D is the diffusion coefficient, k is Boltzmann's constant, T is the absolute temperature, q is the charge, and Z is the valence of the ion. The mobility is many orders of magnitude lower than the mobility of electrons; hence, the conductivity is very small:

$$\sigma = nZq\mu \tag{18.11}$$

Impurities and vacancies increase conductivity; vacancies are necessary for diffusion in substitutional types of crystal structures, and impurities can also diffuse and help carry the current. High temperatures increase conductivity because the rate of diffusion increases.

EXAMPLE 18.5

Suppose that the electrical conductivity of MgO is determined primarily by the diffusion of the Mg^{2+} ions. Estimate the mobility of the Mg^{2+} ions and calculate the electrical conductivity of MgO at 1 800°C.

SOLUTION

From Figure 5.9, the diffusion coefficient of Mg^{2+} ions in MgO at 1 800°C is 10^{-14} $m^2.s^{-1}$. For MgO, $Z = 2$/ion, $q = 1.6 \times 10^{-19}$ C, $k = 1.38 \times 10^{-23}$ $J.K^{-1}$, and $T = 2073$ K:

$$\mu = \frac{ZqD}{kT} = \frac{(2)(1.6\times10^{-19})(10^{-14})}{(1.38\times10^{-23})(2073)} = 1.12\times10^{-13}\ \mathrm{C.m^2.J^{-1}.s^{-1}}$$

Since $C = \mathrm{A \cdot s}$ and $\mathrm{J = A \cdot V \cdot s}$:

$$\mu = 1.12 \times 10^{-13}\ \mathrm{m^2.V^{-1}.s^{-1}}$$

MgO has the NaCl structure, with four magnesium ions per unit cell. The lattice parameter is 3.96×10^{-10} m, so the number of Mg^{2+} ions per cubic centimetre is:

$$n = \frac{4\mathrm{Mg^{2+}\ ions/cell}}{(3.96\times10^{-10}\mathrm{m})^3} = 6.4\times10^{28}\ \mathrm{ions/m^3}$$

$$\sigma = nZq\mu = (64 \times 10^{28})(2)(1.6 \times 10^{-19})(1.12 \times 10^{-13})$$
$$= 2.29 \times 10^{-3}\,\text{C.m}^{-1}.\text{V}^{-1}.\text{s}^{-1}$$

Since $C = A \cdot s$ and $V = A \cdot \Omega$:

$$\sigma = 2.29 \times 10^{-3}\ \Omega^{-1}.\text{m}^{-1}$$

Conduction in Polymers Because their electrons are involved in covalent bonding, polymers have a band structure with a large energy gap, leading to low electrical conductivity. Polymers are frequently used in applications that require electrical insulation to prevent short circuits, arcing, and safety hazards. Table 18.1 includes the conductivity of four common polymers.

In some cases, however, the low conductivity is a hindrance. For example, static electricity can accumulate on housings for electronic equipment, making the polymer transparent to electromagnetic radiation that damages the internal solid-state devices. If lightning strikes the polymer matrix composite wing of an aircraft, severe damage can occur. We can solve these problems by two approaches: (1) introducing an additive to the polymer to improve the conductivity or (2) creating polymers that inherently have good conductivity.

Resistivity can be reduced by adding ionic compounds to the polymer. The ions migrate to the polymer surface and attract moisture, which in turn dissipates static charges. Static charge can also be dissipated by introducing conductive filler materials such as carbon black. Polymer matrix composites containing carbon or nickel-plated carbon fibres combine high stiffness with improved conductivity; hybrid composites containing metal fibres, along with normal carbon, glass, or aramid fibres, also produce lightning-safe aircraft skins. Figure 18.16 shows that when enough carbon fibres are introduced to nylon to ensure fibre-to-fibre contact, the resistivity is reduced by nearly thirteen orders of magnitude. Conductive fillers and fibres are also used to produce polymers that shield against electromagnetic radiation.

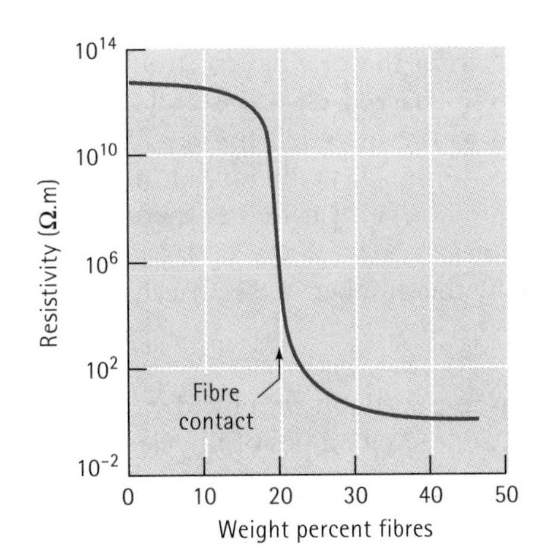

Figure 18.16
Effect of carbon fibres on the electrical resistivity of nylon.

Table 18.6 Electronic structure and electrical conductivity of the Group IVA elements at 25°C.

Metal	Electronic Structure	Electrical Conductivity ($\Omega^{-1}.m^{-1}$)	Energy Gap (E_g) (eV)	Electron Mobility (μ_e) ($m^2.V^{-1}.s^{-1}$)	Hole Mobility (μ_h) ($m^2.V^{-1}.s^{-1}$)
C (diamond)	$1s^2 2s^2 2p^2$	$<10^{-16}$	5.4	0.18	0.14
Si	$1s^2 2s^2 2p^6 3s^2 3p^2$	5×10^{-4}	1.11	0.19	0.05
Ge	.... $4s^2 4p^2$	2.0	0.67	0.38	0.18
Sn	.... $5s^2 5p^2$	9.0×10^6	0.8	0.25	0.24

Some polymers inherently have good conductivity as a result of doping or processing techniques. When acetal polymers are doped with agents such as arsenic pentafluoride, electrons or holes are able to jump freely from one atom to another along the backbone of the chain, increasing the conductivity to near that of metals. Some polymers, such as polyphthalocyanine, can be cross-linked by special curing processes to raise the conductivity to as high as $10^4\ \Omega^{-1}.m^{-1}$, a process that permits the polymer to behave as a semiconductor. Because of the cross-linking, electrons can move more easily from one chain to another.

18.7 Intrinsic Semiconductors

Semiconductor materials, including silicon and germanium, provide the building blocks for many of our electronic devices. These materials have an easily controlled electrical conductivity and, when properly combined, can act as switches, amplifiers, or storage devices.

Pure silicon and germanium behave as intrinsic semiconductors. The energy gap E_g between the valence and conduction bands in the semiconductors is small (Table 18.6) and, as a consequence, some electrons possess enough thermal energy to exceed the gap and enter the conduction band. The excited electrons leave behind unoccupied energy levels, or holes, in the valence band. When an electron moves to fill a hole, another hole is created from the original electron source; consequently, the holes appear to act as positively charged electrons and also carry an electrical charge. When a voltage is applied to the material, the electrons in the conduction band accelerate toward the positive terminal, while holes in the valence band move toward the negative terminal (Figure 18.17). Current is therefore conducted by the movement of both electrons and holes.

The conductivity is determined by the number of electron-hole pairs,

$$\sigma = n_e q \mu_e + n_h q \mu_h \quad (18.12)$$

where n_e is the number of electrons in the conduction band, n_h is the number of holes in the valence band, and μ_e and μ_h are the mobilities of the electrons and holes (Table 18.6). In intrinsic semiconductors:

$$n = n_e = n_h$$

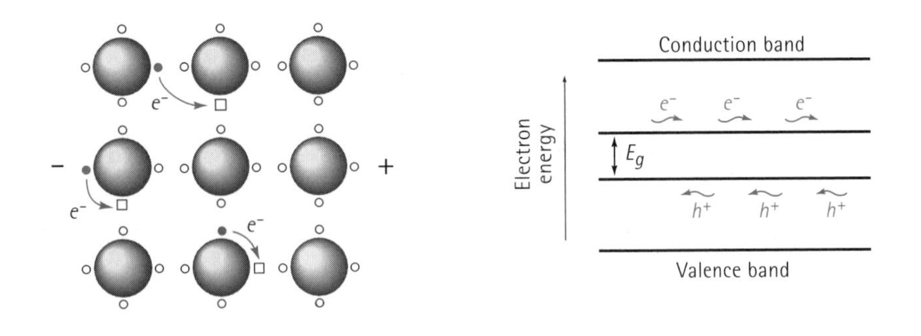

Figure 18.17
When a voltage is applied to a semiconductor, the electrons move through the conduction band, while the electron holes move through the valence band in the opposite direction.

Therefore, the conductivity is:

$$\sigma = nq(\mu_e + \mu_h) \tag{18.13}$$

In intrinsic semiconductors, we control the number of charge carriers and, hence, the electrical conductivity by controlling the temperature. At absolute zero, all of the electrons are in the valence band, whereas all of the levels in the conduction band are unoccupied [Figure 18.18(a)]. As the temperature increases, there is a greater probability that an energy level in the conduction band is occupied (and an equal probability that a level in the valence band is unoccupied, or that a hole is present) [Figure 18.18(b)]. The number of electrons in the conduction band, which is equal to the number of holes in the valence band, is given by

$$n = n_e = n_h = n_o \exp\left(\frac{-E_g}{2kT}\right) \tag{18.14}$$

where n_o can be considered a constant, although it, too, actually depends on temperature. Higher temperatures permit more electrons to cross the forbidden zone and, hence, the conductivity increases according to the following relationship:

$$\sigma = n_o q(\mu_e + \mu_h)\exp\left(\frac{-E_g}{2kT}\right) \tag{18.15}$$

Note that both n and σ are related to temperature by an Arrhenius equation, rate = A exp $(-Q/RT)$. The behaviour of the semiconductor is opposite to that of metals (Figure 18.19). As the temperature increases, the conductivity of a semiconductor increases because more charge carriers are present, whereas the conductivity of a metal decreases due to lower mobility of the charge carriers.

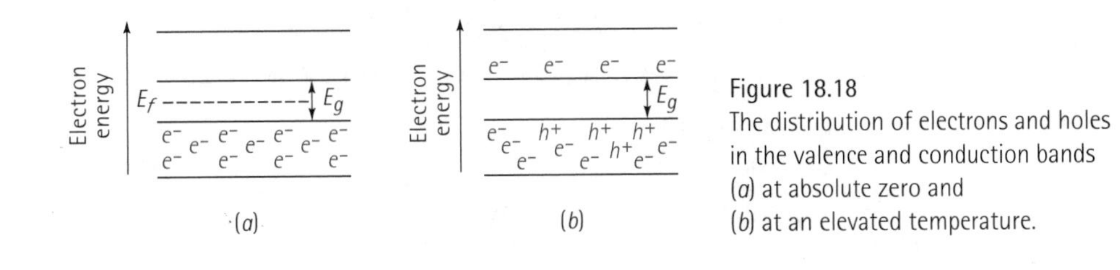

Figure 18.18
The distribution of electrons and holes in the valence and conduction bands
(a) at absolute zero and
(b) at an elevated temperature.

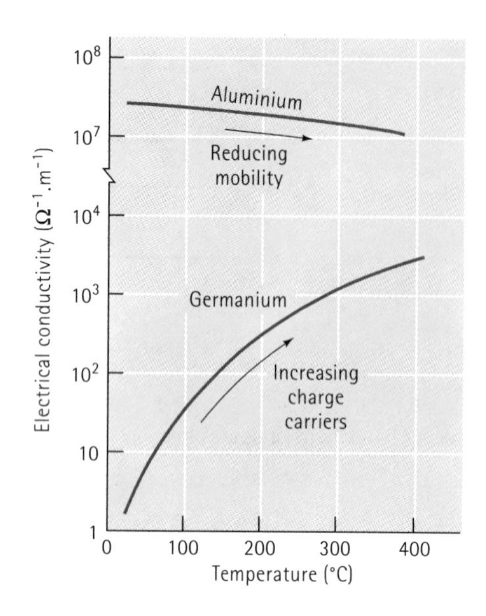

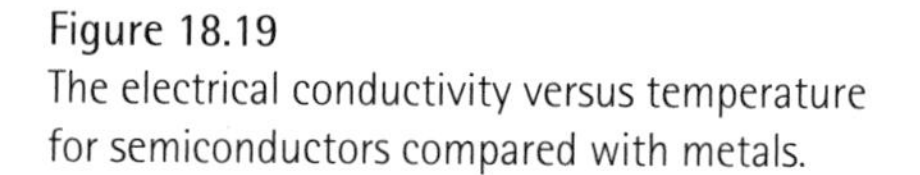
Figure 18.19
The electrical conductivity versus temperature for semiconductors compared with metals.

If the source of the exciting energy or voltage is removed, the holes and electrons recombine, but only over a period of time. The number of electrons in the conduction band decreases at a rate given by

$$n = n_o \exp\left(\frac{-t}{\tau}\right) \tag{18.16}$$

where t is the time after a field is removed, n_o is a constant, and τ is a constant called the recombination time. This characteristic is important in the operation of a number of semiconductor devices.

EXAMPLE 18.6

For germanium at 25°C, estimate (a) the number of charge carriers, (b) the fraction of the total electrons in the valence band that are excited into the conduction band, and (c) the constant n_o.

SOLUTION

From Tables 18.2 and 18.6:

$\sigma = 2.0\ \Omega^{-1}.m^{-1}$ $\qquad E_g = 0.67$ eV

$\mu_e = 0.38\ m^2.V^{-1}.s^{-1}$ $\qquad \mu_h = 0.18\ m^2.V^{-1}.s^{-1}$

$2kT = (2)(8.63 \times 10^{-5}$ eV/K$)(273 + 25) = 0.0514$ eV at $T = 25$°C.

1. From Equation 18.13:

$$n = \frac{\sigma}{q(\mu_e + \mu_h)} = \frac{2.0}{(1.6\times10^{-19})(0.38+0.18)} = 22\times10^{19}$$

There are 2.2×10^{19} electrons/m^3 and 2.2×10^{19} holes/m^3 helping to conduct a charge in germanium at room temperature.

2. The lattice parameter of diamond cubic germanium is 5.6575×10^{-10} m. The total number of electrons in the valence band of germanium is:

$$\text{Total electrons} = \frac{(8 \text{ atoms/cell})(4 \text{ electrons/atom})}{(5.6575 \times 10^{-10}\,\text{m})^3} = 1.77 \times 10^{29}$$

$$\text{Fraction excited} = \frac{2.2 \times 10^{19}}{1.77 \times 10^{29}} = 1.24 \times 10^{-10}$$

3. From Equation 18.14:

$$n_o = \frac{n}{\exp(-E_g/2kT)} = \frac{2.2 \times 10^{19}}{\exp(-0.67/0.0514)}$$

$$= 1.01 \times 10^{25}\,\text{carriers/m}^3$$

18.8 Extrinsic Semiconductors

We cannot accurately control the behaviour of an intrinsic semiconductor because slight variations in temperature change the conductivity. However, by intentionally adding a small number of impurity atoms to the material (called doping), we can produce an extrinsic semiconductor. The conductivity of the extrinsic semiconductor depends primarily on the number of impurity, or dopant, atoms and, within a certain temperature range, may even be independent of temperature.

***n*-Type Semiconductors** Suppose we add an impurity atom such as antimony, which has a valence of five, to silicon or germanium. Four of the electrons from the antimony atom participate in the covalent bonding process, while the extra electron enters an energy level in a donor state just below the conduction band (Figure 18.20).

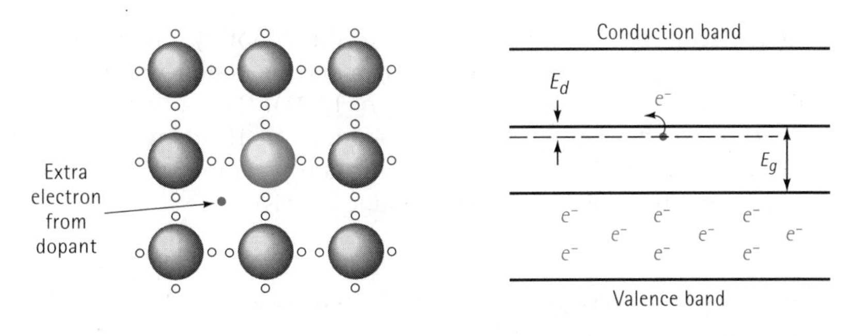

Figure 18.20
When a dopant atom with a valence greater than four is added to silicon, an extra electron is introduced and a donor energy state is created. Now electrons are more easily excited into the conduction band.

Table 18.7 The donor and acceptor energy gaps in electron volts (eV) when silicon and germanium semiconductors are doped.

Dopant	Silicon (E_g = 1.11)		Germanium (E_g = 0.67)	
	E_d	E_a	E_d	E_a
P	0.045		0.0120	
As	0.049		0.0127	
Sb	0.039		0.0096	
B		0.045		0.0104
Al		0.057		0.0102
Ga		0.065		0.0108
In		0.160		0.0112

Since the extra electron is not tightly bound to the atoms, only a small increase in energy, E_d, is required for the electron to enter the conduction band. (Alternatively, E_d is sometimes defined as the energy difference between the top of the valence band and the donor band. In this case, the energy increase required would be defined as $E_g - E_d$.) The energy gap controlling conductivity is now E_d rather than E_g (Table 18.7.) No corresponding holes are created when the donor electrons enter the conduction band.

Some intrinsic semiconduction still occurs, with a few electrons gaining enough energy to jump the large E_g gap. The total number of charge carriers is:

$$n_{\text{total}} = n_e(\text{dopant}) + n_e(\text{intrinsic}) + n_h(\text{intrinsic})$$

or

$$n_{\text{total}} = n_{0d}\exp\left(\frac{-E_d}{kT}\right) + 2n_0\exp\left(\frac{-E_g}{2kT}\right) \quad (18.17)$$

where n_{0d}, n_0 are approximately constant. At low temperatures, few intrinsic electrons and holes are produced and the number of electrons is about:

$$n_{\text{total}} \approx n_{0d}\exp\left(\frac{-E_d}{kT}\right) \quad (18.18)$$

As the temperature increases, more of the donor electrons jump the E_d gap until, eventually, all of the donor electrons enter the conduction band. At this point, we have reached donor exhaustion (Figure 18.21). Over a restricted temperature range, the conductivity is now virtually constant; no more donor electrons are available and the temperature is still too low to produce many intrinsic electrons and holes, particularly when E_g is large. The conductivity is

$$\sigma = n_d q \mu_e \quad (18.19)$$

where n_d is the maximum number of donor electrons, determined by the number of impurity atoms that are added.

At high temperatures, the term exp $(-E_g/2kT)$ becomes significant, and the conductivity increases again according to:

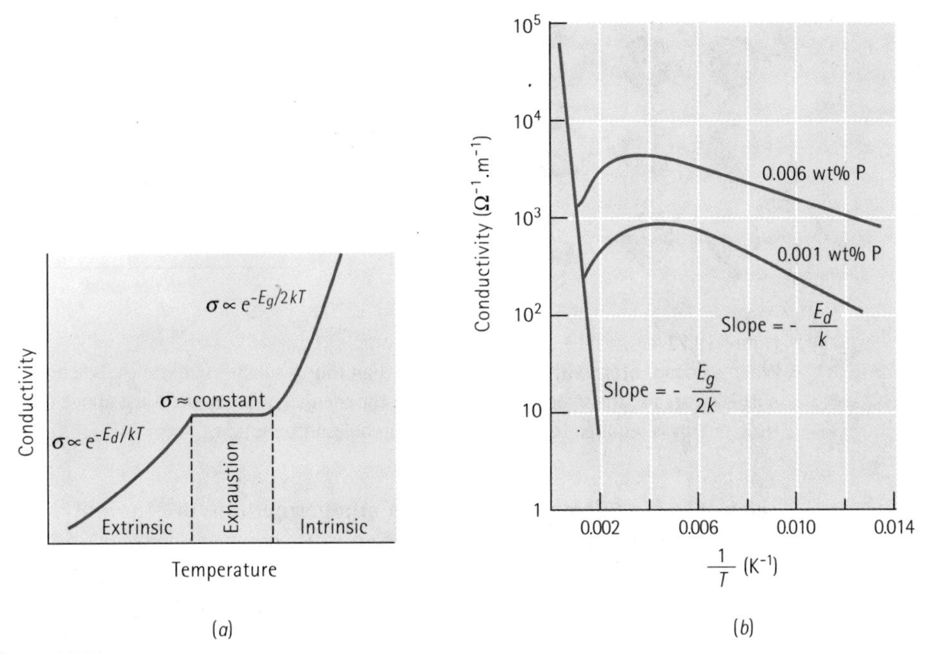

Figure 18.21
The effect of temperature on the conductivity of phosphorus-doped silicon. At low temperatures, the conductivity increases as more donor electrons enter the conduction band. At moderate temperatures, donor exhaustion occurs. At high temperatures, intrinsic semiconduction becomes important.
(a) Schematic diagram and (b) Arrhenius plot.

$$\sigma = qn_d\mu_e + q(\mu_e + \mu_h)n_0 \exp\left(\frac{-E_g}{2kT}\right) \tag{18.20}$$

If we plot σ versus $1/T$, we obtain an Arrhenius relationship from which E_g and E_d can be calculated [Figure 18.21(b)].

p-Type Semiconductors When we add an impurity such as gallium, which has a valence of three, to a semiconductor, there are not enough electrons to complete the covalent bonding process. An electron hole is created in the valence band that can be filled by electrons from other locations in the band (Figure 18.22). The holes act as acceptors of electrons. These hole sites have a somewhat higher than normal energy and create an acceptor level of possible electron energies just above the valence band (Table 18.7). An electron must gain an energy of only E_a in order to create a hole in the valence band. The hole then moves and carries the charge. Now we have a p-type semiconductor.

As in n-type semiconductors, the temperature eventually becomes high enough to cause acceptor saturation and

$$\sigma = n_a q \mu_h \tag{18.21}$$

where n_a is the maximum number of acceptor levels, or holes, introduced by the dopant.

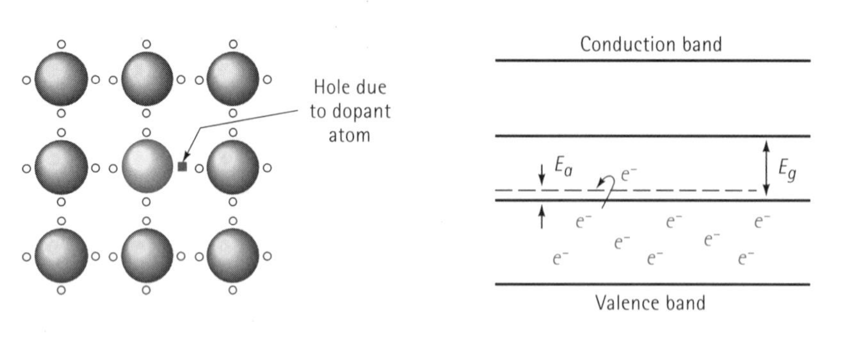

Figure 18.22
When a dopant atom with a valence of less than four is substituted into the silicon lattice, an electron hole is created in the structure and an acceptor energy level is created just above the valence band. Little energy is required to excite the electron holes into motion.

Semiconducting Compounds Silicon and germanium are the only elements that have practical applications as semiconductors. However, a large number of ceramic and intermetallic compounds display the same effect. Examples are given in Table 18.8.

The stoichiometric semiconductors, usually intermetallic compounds, have crystal structures and band structures similar to silicon and germanium. Elements from Group III and Group V of the periodic table are classic examples. Gallium from Group III and arsenic from Group V combine to form a compound GaAs, with an average of four valence electrons per atom. The $4s^24p^1$ levels of gallium and the $4s^24p^3$ levels of arsenic produce two hybrid bands, each capable of containing $4N$ electrons. An energy gap of 1.35 eV separates the valence and conduction bands. The GaAs compound can be doped to produce either an n-type semiconductor or a p-type semiconductor. The large energy gap E_g leads to a broad exhaustion plateau (i.e. over a wide temperature range), and high mobilities of charge carriers in the compound lead to high conductivities.

Table 18.8 Energy gaps and mobilities for semiconducting compounds.

Compound	Energy Gap (E_g) (eV)	Electron Mobility (μ_e) ($m^2.V^{-1}.s^{-1}$)	Hole Mobility (μ_h) ($m^2.V^{-1}.s^{-1}$)
ZnS	3.54	0.018	0.0005
GaP	2.24	0.030	0.010
GaAs	1.35	0.880	0.040
GaSb	0.67	0.400	0.140
InSb	0.165	7.80	0.075
InAs	0.36	3.30	0.046
ZnO	3.2	0.018	
CdS	2.42	0.040	
PbS	0.37	0.060	0.060

From Handbook of Chemistry and Physics, *56th Ed., CRC Press, 1975.*

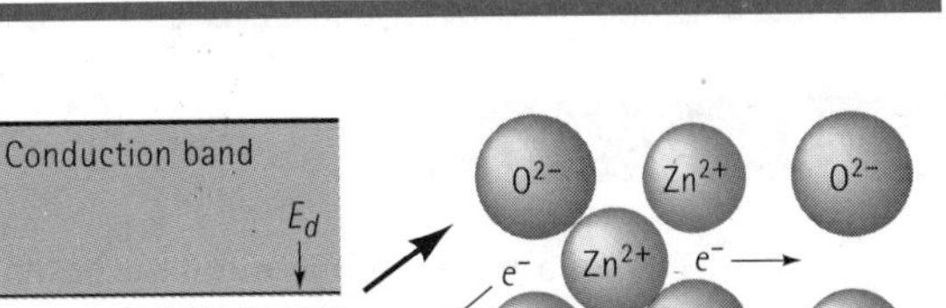

Figure 18.23
Interstitial zinc atoms can ionise and introduce extra electrons, creating an *n*-type defect semiconductor in ZnO.

The nonstoichiometric, or defect semiconductors are ionic compounds containing an excess of either anions (producing a *p*-type semiconductor) or cations (producing an *n*-type semiconductor). A number of oxides and sulphides have this behaviour, including ZnO, which has the zinc blende crystal structure. For example, if an extra zinc atom is added to ZnO, the zinc atom enters the structure as an ion, Zn^{2+}, giving up two electrons that contribute to the number of charge carriers. These electrons can be excited by a small increase in energy to carry current (Figure 18.23). The ZnO now behaves like an *n*-type semiconductor.

Another defect semiconductor is created when two Fe^{3+} ions are substituted for three Fe^{2+} ions in FeO, thereby creating a vacancy (Figure 18.24). The Fe^{3+} ions act as electron acceptors, and a *p*-type semiconductor is produced. This structure was discussed more thoroughly in Chapter 14.

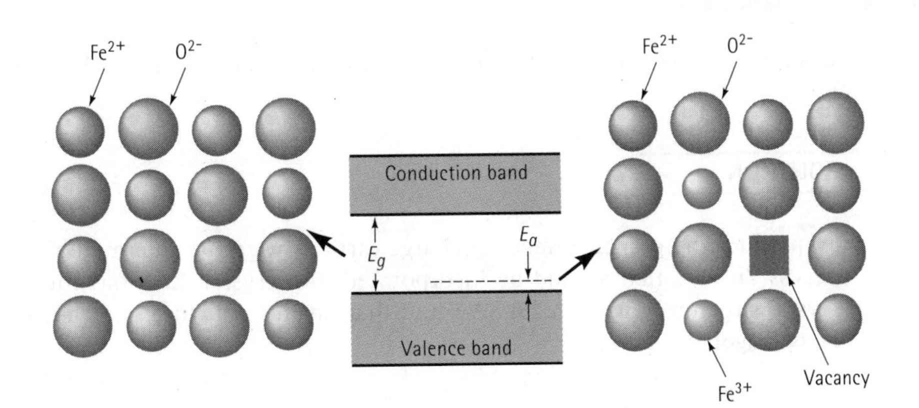

Figure 18.24
Two Fe^{3+} ions and a vacancy substitute for three Fe^{2+} ions, maintaining overall charge balance but creating an acceptor level. The result is a *p*-type defect semiconductor.

EXAMPLE 18.7 **Design of a Semiconductor**

Design a *p*-type semiconductor based on silicon, which provides a constant conductivity of $10^4\ \Omega^{-1}.m^{-1}$ over a range of temperatures.

SOLUTION

In order to obtain the desired conductivity, we must dope the silicon with atoms having a valence of +3, adding enough dopant to provide the required number of charge carriers. If we assume that the number of intrinsic carriers is small, then

$$\sigma = n_d q \mu_h$$

where $\sigma = 10^4\ \Omega^{-1}.m^{-1}$ and $\mu_h = 0.05\ m^2.V^{-1}.s^{-1}$. If we remember that C = A · s and V = A · Ω, the number of charge carriers required is:

$$n_d = \frac{\sigma}{q\mu_h} = \frac{10^4}{(1.6\times10^{-19})(0.05)} = 1.25\times10^{24}\ \text{electrons}/\text{m}^3$$

In addition, n_d is given by:

$$n_d = \frac{(1\ \text{electron}/\text{dopant atom})(x\ \text{dopant atom}/\text{Si atom})(8\ \text{Si atoms}/\text{unit cell})}{(5.4307\times10^{-10}\ \text{m})^3}$$

$$x = (1.25\times10^{24})(5.4307\times10^{-10})^3/8 = 2.5\times10^{-5}\ \text{dopant atom}/\text{Si atom}$$

or 25 dopant atoms per 10^6 Si atoms

Possible dopants include boron, aluminium, gallium, and indium.

EXAMPLE 18.8 **Design of a Defect Semiconductor**

Design an *n*-type ZnO defect semiconductor that will provide 20×10^{26} charge carriers per m^3.

SOLUTION

To produce the proper number of charge carriers, we must determine the number of excess Zn ions that should be incorporated into crystal. ZnO has the zinc blende crystal structure (Figure 3.25). We find that, in this structure, ions touch along the body diagonal, where:

$$4r_{Zn} + 4r_O = \sqrt{3}a_0$$

The ionic radii are $r_{Zn} = 0.074$ nm and $r_O = 0.132$ nm. Thus:

$$4(0.074) + 4(0.132) = 0.824 = \sqrt{3}a_0$$

$$a_0 = 0.476\ \text{nm} = 4.76\times10^{-10}\ \text{m}$$

The number of Zn ions in one cubic metre of stoichiometric ZnO is:

$$\frac{4 \text{ Zn ions / cell}}{(4.7 \times 10^{-10} \text{ m})^3} = 3.72 \times 10^{28}$$

The charge carriers in the nonstoichiometric ZnO are electrons introduced by the excess Zn^{2+} ions. Two electrons are introduced for each excess ion. If we want 20×10^{26} charge carriers per m^3, we must add 10×10^{26} excess ions/m^3, or the number of excess ions per regular ion is:

$$\frac{10 \times 10^{26}}{3.72 \times 10^{28}} = 0.027 \text{ excess / normal, or:}$$

$$\frac{2.7 \text{ excess ions}}{100 \text{ normal ions}}$$

The atomic percent Zn therefore is

$$\frac{102.7 \text{ Zn}}{102.7 \text{ Zn} + 100 \text{ O}} \times 100 = 50.67 \text{ at \% Zn}$$

or the weight percent Zn is:

$$\frac{50.67(65.38)}{50.67(65.38) + 49.33(16)} \times 100 = 80.8 \text{ wt\% Zn}$$

We now know the correct percentages of elements required to build the ZnO crystal to produce the desired number of charge carriers.

18.9 Applications of Semiconductors to Electrical Devices

Many electronic devices have been developed using the characteristics of semiconduction. A few of these are described here; others – particularly those that interact with light – are discussed in Chapter 20.

Thermistors Electrical conductivity depends on temperature. Thermistors use this relationship to measure temperature. Thermistors are also used in other devices, including fire alarms. When the thermistor heats, it passes a larger current through a circuit and activates the alarm.

Pressure Transducers The band structure and the energy gap are a function of the spacing between the atoms in a material. When pressure is applied to the semiconductor, atoms are forced closer together, the energy gap decreases, and conductivity increases. If we measure conductivity, we can, in turn, calculate the pressure acting on the material.

Rectifiers (*p* - *n* Junction Devices) Rectifiers are produced by joining an *n*-type semiconductor to a *p*-type semiconductor, forming a *p*-*n* junction (Figure 18.25). Electrons are concentrated in the *n*-type junction; holes are concentrated in the *p*-type

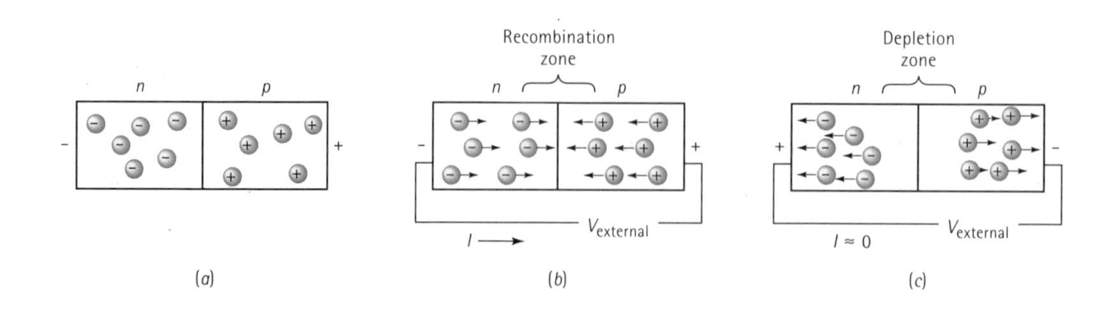

Figure 18.25
Behaviour of a *p-n* junction device: (*a*) Equilibrium is caused by electrons concentrating in the *n*-side and holes in the *p*-side, (*b*) forward bias causes a current to flow, and (*c*) reverse bias does not permit a current to flow.

junction. The resulting electrical imbalance creates a voltage, or contact potential, across the junction.

If we place an external voltage on the *p-n* junction so that the negative terminal is at the *n*-type side, both the electrons and holes move toward the junction and eventually recombine. This movement of the electrons and holes causes a net current to be produced. This is called forward bias [Figure 18.25(b)]. By increasing the forward bias, the current passing through the junction increases [Figure 18.26(a)].

However, if the applied voltage is reversed, creating a reverse bias, both the holes and electrons move away from the junction [Figure 18.25(c)]. With no charge carriers in the depletion zone, the junction behaves as an insulator, and virtually no current flows [Figure 18.26(a)].

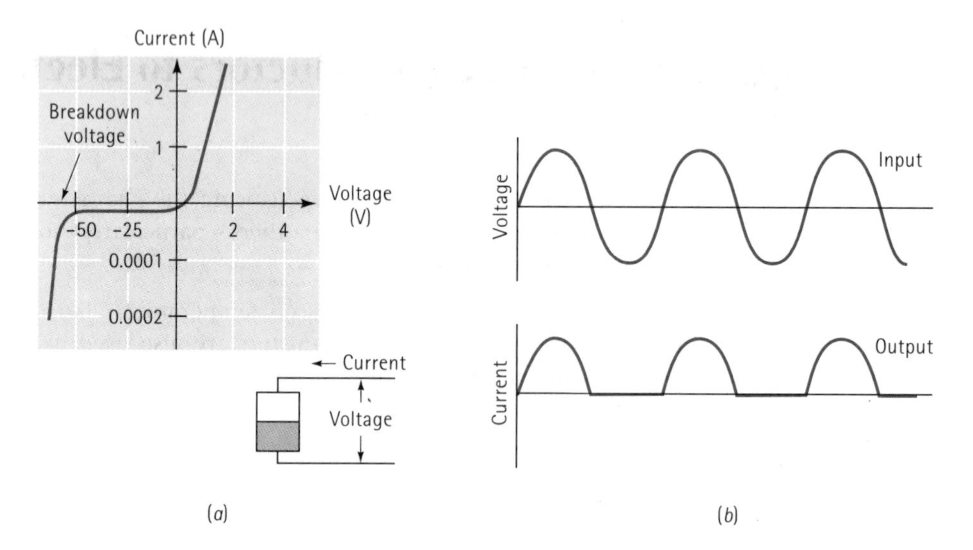

Figure 18.26
(*a*) The current-voltage characteristics for a p-n junction. Note the different scales in the first and third quadrants. (*b*) If an alternating signal is applied, rectification occurs and only half of the input signal passes the rectifier.

Because the *p-n* junction permits current flow in only one direction, it passes only half of an alternating current, therefore converting the alternating current to direct current [Figure 18.26(b)]. These junctions are called *rectifier diodes.*

Typically a small leakage current is produced for reverse bias due to the movement of thermally activated electrons and holes. When the reverse bias becomes too large, however, any carriers that do leak through the insulating barrier of the junction are highly accelerated, excite other charge carriers, and cause a high current in the reverse direction [Figure 18.26(a)]. We can use this phenomenon to design voltage-limiting devices. By proper doping and construction of the *p-n* junction, the breakdown or avalanche voltage can be preselected. When the voltage in the current exceeds the breakdown voltage, a high current flows through the junction and is diverted from the rest of the circuit. These devices, called *Zener diodes,* are used to protect circuitry from accidental high voltages.

Bipolar Junction Transistors A transistor can be used as a switch or an amplifier. One example is the *bipolar junction transistor* (BJT), which is often used in the central processing units of computers because of their rapid switching response. A bipolar junction transistor is a sandwich of either *n-p-n* or *p-n-p* semiconductor materials. There are three zones in the transistor: the emitter, the base, and the collector. As in the *p-n* junction, electrons are initially concentrated in the *n*-type material and holes are concentrated in the *p*-type material.

Figure 18.27 shows an *n-p-n* transistor and its electrical circuit, both schematically and as it might appear when implanted in a silicon chip. The electrical signal to be amplified is connected between the base and the emitter, with a small voltage between these two zones. The output from the transistor, or the amplified signal, is connected between the emitter and the collector and operates at a higher voltage. The circuit is connected so that a forward bias is produced between the emitter and the base (the positive voltage is at the *p*-type base), while a reverse bias is produced between the base and the collector (with the positive voltage at the *n*-type collector). The forward bias causes electrons to leave the emitter and enter the base.

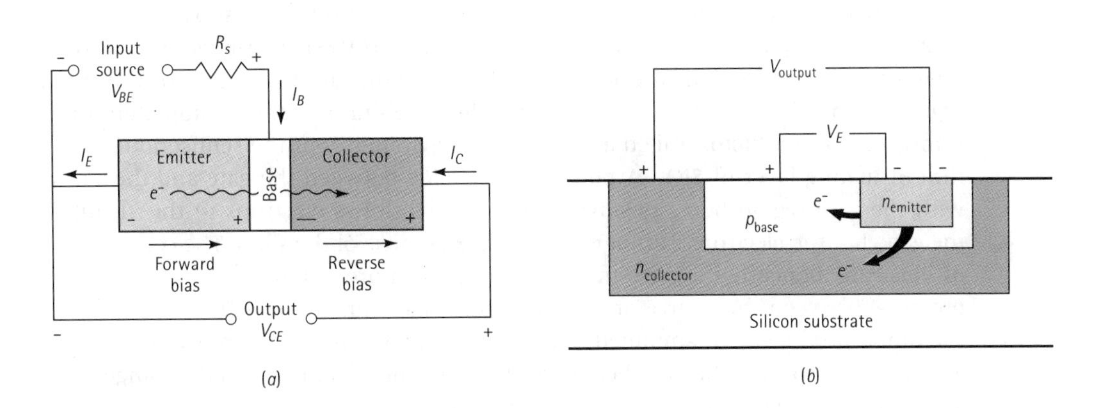

Figure 18.27
(*a*) A circuit for an *n-p-n* bipolar junction transistor. The input creates a forward and reverse bias that causes electrons to move from the emitter, through the base, and into the collector, creating an amplified output. (*b*) Sketch of the cross-section of the transistor.

Electrons and holes attempt to recombine in the base; however, if the base is exceptionally thin and lightly doped, or if the recombination time τ is long, almost all of the electrons pass through the base and enter the collector. The reverse bias between the base and collector accelerates the electrons through the collector, the circuit is completed, and an output signal is produced. The current through the collector is given by

$$I_c = I_0 \exp\left(\frac{V_E}{B}\right) \tag{18.22}$$

where I_0 and B are constants and V_E is the voltage between the emitter and the base. If the input voltage V_E is increased, a very large current I_c is produced.

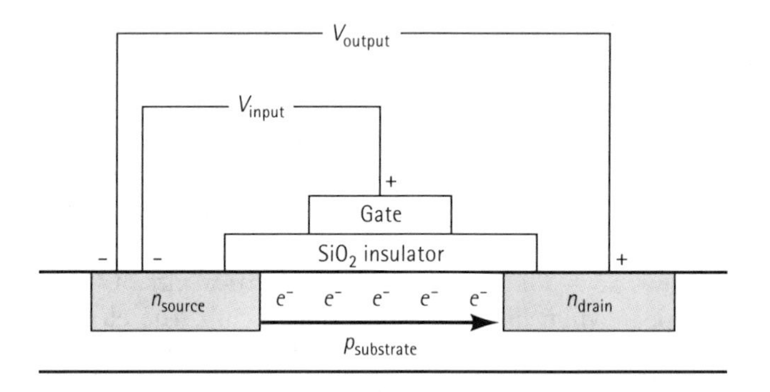

Figure 8.28
An *n-p-n* metal oxide semiconductor field effect transistor.

Field Effect Transistors A second type of transistor, which is more often used for strong data in computer memories, is the *field effect transistor* (FET), which behaves in a somewhat different manner than the bipolar junction transistor. Figure 18.28 shows an example of a metal oxide semiconductor (MOS) field effect transistor, in which two *n*-type regions are formed within a *p*-type substrate. One of the *n*-type regions is called the source; the second is called the drain. A third component of the transistor is a conductor, called a gate, that is separated from the semiconductor by a thin insulating layer of SiO_2. A potential is applied between the gate and the source, with the gate region being positive. The potential draws electrons to the vicinity of the gate, but the electrons cannot enter the gate because of the silica. The concentration of electrons beneath the gate makes this region more conductive, so that a large potential between the source and drain permits electrons to flow from the source to the drain, producing an amplified signal. By changing the input voltage between the gate and the source, the number of electrons in the conductive path changes, thus also changing the output signal.

The field effect transistors are generally less expensive to produce than the bipolar junction transistors. Because FETs occupy less space, they are preferred in microelectronic integrated circuits, where perhaps 100 000 transistors are present in a single silicon chip.

18.10 Manufacture and Fabrication of Semiconductor Devices

In order to produce electronic components that require little power, operate very rapidly, yet are inexpensive, microelectronic integrated circuits formed on silicon chips may contain as many as 1 000 000 transistors or other devices, each having dimensions of as little as 10^{-6} m (or 1 μm). In order to produce these circuits, special technologies are required. The starting point for the most common devices is pure, single-crystal silicon. The steps described in Figure 18.29 summarise the production of an FET transistor.

While many exotic materials and technologies are used to produce integrated circuits, the next generation of semiconducting devices must be even faster. Faster response times can be achieved by making the individual devices smaller by using electron or X-ray beams in place of ultraviolet radiation for photolithography; by producing better insulators for the gate regions in FETs; by producing three-dimensional chips with alternating layers of silicon and silica; or by using semiconductors such as GaAs, in which the electrons move two to five times faster than in silicon-based devices. As explained in Chapter 20, semiconducting devices using light rather than electricity may provide still faster computers and communication systems.

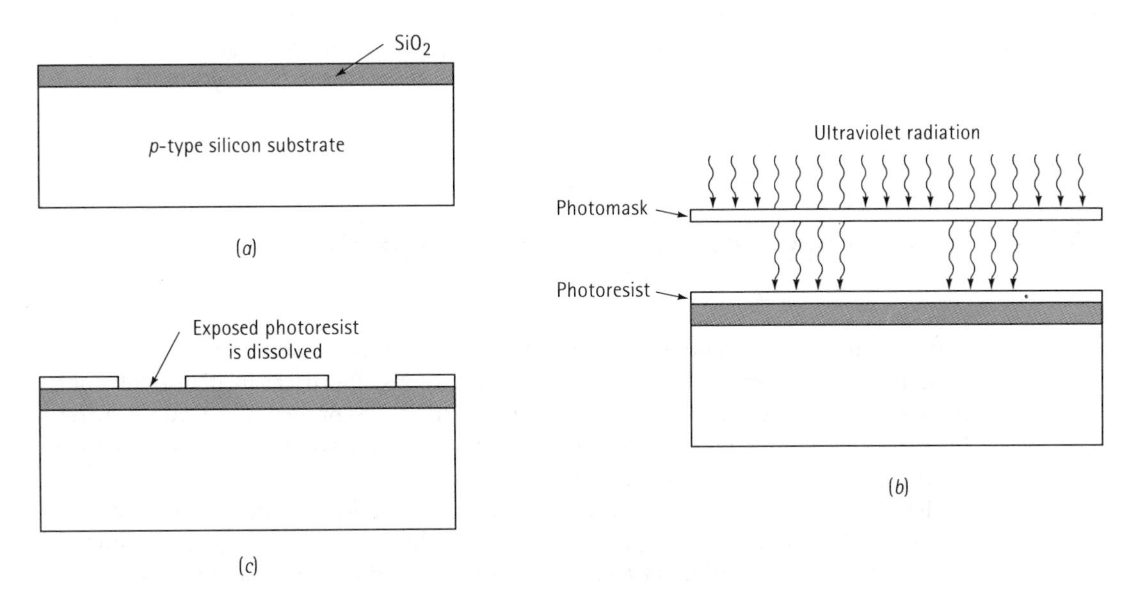

Figure 18.29
Production of an FET semiconductor device: (*a*) A *p*-type silicon semiconductor is sliced into wafers about 0.25 mm thick, polished and then heated in the presence of oxygen to form a thin layer of SiO_2 on its surface. (*b*) An ultraviolet-sensitive polymer, called a photoresist, is coated onto the silica layer. Ultraviolet radiation is then passed through a photomask (which contains the details of the integrated circuit), exposing a portion of the photoresist layer. This process is known as photolithography. (*c*) The exposed photoresist areas are dissolved away by solvents, leaving the unexposed photoresist untouched.

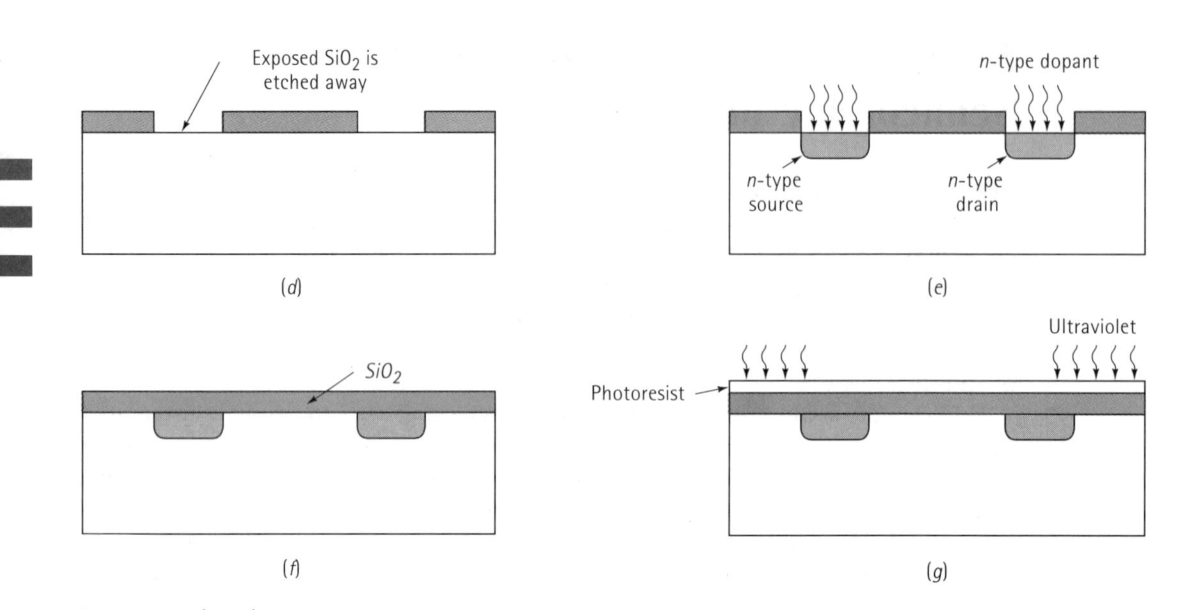

Figure 18.29 (*cont.*)
(*d*) The device is immersed in acid, which removes the silica from exposed areas and leaves a clean silicon surface. The remainder of the photoresist is then removed chemically. (*e*) An *n*-type dopant is introduced onto the exposed silicon areas, by either diffusion or ion implantation techniques. This produces the source and drain for the device. (*f*) The wafer is again oxidised, to form a layer of SiO_2 over the doped surface. A portion of this silica layer forms the insulating layer beneath the gate of the FET. (*g*) Photolithography is repeated to introduce other components for the device, including electrical leads.

18.11 Insulators and Dielectric Properties

Materials used to insulate an electrical field from its surroundings are required in a large number of electrical and electronic applications. Electrical insulators obviously must have a very low conductivity, or high resistivity, to prevent the flow of current. Insulators are produced from ceramic and polymer materials, in which there is a large energy gap between the valence and conduction bands.

However, the high electrical resistivity of these materials is not always sufficient. At high voltages, a catastrophic breakdown of the insulator may occur, similar to what happens in *p-n* diodes at a large reverse bias, and current may flow. In order to properly select an insulating material we must understand how the material stores, as well as conducts, electrical charge. In order to do this, we must examine the dielectric behaviour of these materials. In doing so, we will find that these materials possess special properties beyond merely providing electrical insulation; in fact, they can be used as capacitors to store electrical charge or as transducers to create or receive information.

18.12 Dipoles and Polarisation

The application of an electric field causes the formation and movement of dipoles contained within materials. (*Dipoles* are atoms or groups of atoms that have an unbalanced charge.) In an imposed electric field, the dipoles become aligned in the material, causing *polarisation*.

Dipoles When an electric field is applied to a material, dipoles are induced within the atomic or molecular structure and become aligned with the direction of the field. In addition, any permanent dipoles already present in the material are aligned with the field, and the material is polarised. The polarisation P (C.m^{-2}) is

$$P = Zqd \quad (18.23)$$

where Z is the number of charge centres that are displaced per cubic metre, q is the electronic charge, and d is the displacement between the positive and negative ends of the dipole. Four mechanisms cause polarisation: (1) electronic polarisation, (2) ionic polarisation, (3) molecular polarisation, and (4) space charges (Figure 18.30).

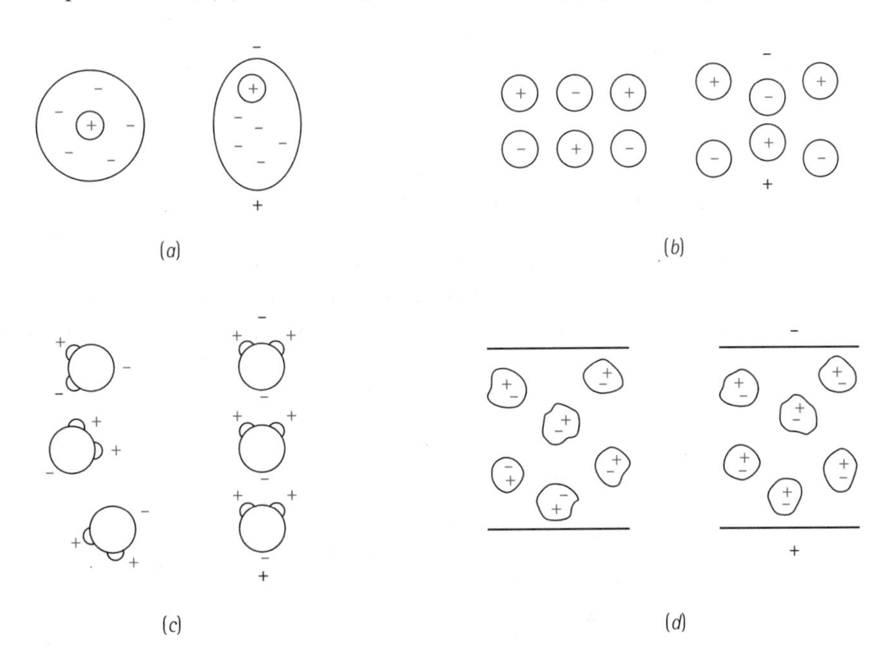

Figure 18.30
Polarisation mechanisms in materials: (*a*) electronic polarisation, (*b*) ionic polarisation, (*c*) molecular polarisation, and (*d*) space charges.

Electronic Polarisation When an electric field is applied to an atom, the electronic arrangement is distorted, with electrons concentrating on the side of the nucleus near the positive end of the field. The atom acts as a temporary, induced dipole. This effect, which occurs in all materials, is small and temporary.

EXAMPLE 18.9

Suppose that the average displacement of the electrons relative to the nucleus in a copper atom is 1×10^{-18} m when an electric field is imposed on a copper plate. Calculate the polarisation.

SOLUTION

The atomic number of copper is 29, so there are 29 electrons in each copper atom. The lattice parameter of copper is 0.3615 nm. Thus

$$Z = \frac{(4 \text{ atoms/cell})(29 \text{ electrons/atom})}{(3.615 \times 10^{-10} \text{ m})^3} = 2.46 \times 10^{30} \text{ electrons/m}^3$$

$$P = Zqd = \left(2.46 \times 10^{30} \frac{\text{electrons}}{\text{m}^3}\right)$$

$$\times \left(1.6 \times 10^{-19} \frac{\text{C}}{\text{electron}}\right)(10^{-18} \text{ m})$$

$$= 3.94 \times 10^{-7} \text{ C.m}^{-2}$$

Ionic Polarisation When an ionically bonded material is placed in an electric field, the bonds between the ions are elastically deformed. Consequently, the charge is minutely redistributed within the material. Depending on the direction of the field, cations and anions move either closer together or further apart. These temporarily induced dipoles provide polarisation and may also change the overall dimensions of the material.

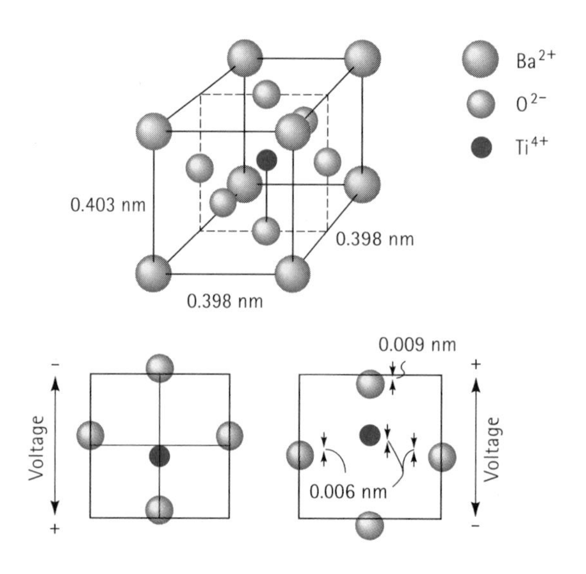

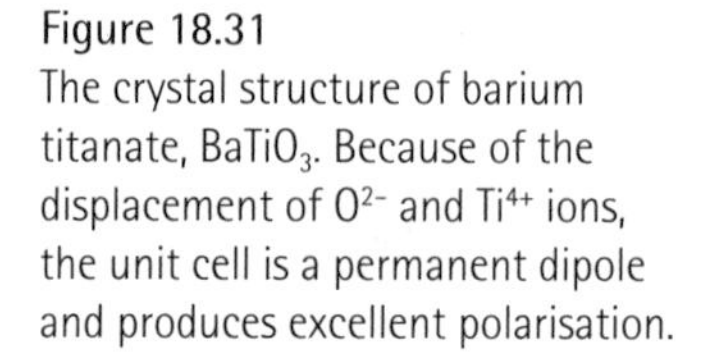
Figure 18.31
The crystal structure of barium titanate, $BaTiO_3$. Because of the displacement of O^{2-} and Ti^{4+} ions, the unit cell is a permanent dipole and produces excellent polarisation.

Molecular Polarisation Some materials contain natural dipoles. When a field is applied, the dipoles rotate to line up with the imposed field. Water molecules, shown schematically in Figure 18.30(c), represent a material that possesses molecular polarisation. Many organic molecules behave in a similar manner, as do a variety of organic oils and waxes.

In a number of materials, the dipoles remain in alignment when the electric field is removed, causing permanent polarisation. Barium titanate ($BaTiO_3$), a crystalline ceramic, has an asymmetrical structure at room temperature (Figure 18.31). The titanium ion is displaced slightly from the centre of the unit cell and the oxygen ions are displaced slightly in the opposite directions from their face-centred positions, causing the crystal to be tetragonal and permanently polarised. When an alternating current is applied to barium titanate, the titanium ion moves back and forth between its two allowable positions to ensure that polarisation is aligned with the field. In this particular material, polarisation is highly anisotropic and the crystal must be properly aligned with respect to the applied field.

EXAMPLE 18.10

Calculate the maximum polarisation per cubic millimetre and the maximum charge that can be stored per square millimetre for barium titanate.

SOLUTION

The strength of the dipoles is given by the product of the charge and the distance between the charges. In $BaTiO_3$, the separations are the distances that the Ti^{4+} and O^{2-} ions are displaced from the normal lattice points (Figure 18.31). The charge on each ion is the product of q and the number of excess or missing electrons. Thus, the dipole moments are:

$$Ti^{4+} : (1.6 \times 10^{-19})(4 \text{ electrons/ion})(6 \times 10^{-12} \text{ m}) = 3.84 \times 10^{-30} \text{ C.m/ion}$$

$$O^{2-}_{(top)} : (1.6 \times 10^{-19})(2 \text{ electrons/ion})(9 \times 10^{-12} \text{ m}) = 2.88 \times 10^{-30} \text{ C.m/ion}$$

$$O^{2-}_{(side)} : (1.6 \times 10^{-19})(2 \text{ electrons/ion})(6 \times 10^{-12} \text{ m}) = 1.92 \times 10^{-30} \text{ C.m/ion}$$

Each oxygen ion is shared with another unit cell, so the total dipole moment in the unit cell is:

$$\begin{aligned}\text{Dipole moment} &= (1\ Ti^{4+}/\text{cell})(3.84 \times 10^{-30}) \\ &+ (1\ O^{2-} \text{ from top and bottom of cell})(2.88 \times 10^{-30}) \\ &+ (2\ O^{2-} \text{ from four sides of cell})(1.92 \times 10^{-30}) \\ &= 1.056 \times 10^{-29} \text{ C.m/cell}\end{aligned}$$

The polarisation per cubic millimetre is:

$$P = \frac{1.056 \times 10^{29} \text{ C.m / cell}}{(3.98 \times 10^{-10} \text{ m})^2 (4.03 \times 10^{-10} \text{ m})}$$

$$= 0.165 \text{C.m}^{-2} = 1.65^{-10} \times 10^{-7} \text{C.mm}^{-2}$$

The total charge on a $BaTiO_3$ crystal 1 mm × 1 mm is:

$$Q = PA = (1.65 \times 10^{-7}\ \text{C.mm}^{-2})(1\ \text{mm})^2$$

$$= 1.65 \times 10^{-7}\ \text{C}$$

Space Charges A charge may develop at interfaces between phases within a material, normally as a result of the presence of impurities. The charge moves on the surface when the material is placed in an electric field. This type of polarisation is not an important factor in most common dielectrics.

18.13 Dielectric Properties and Their Control

We will examine a number of applications for dielectrics in the following sections. First, we should define some of the important dielectric characteristics and examine how these characteristics are affected by service conditions.

Dielectric Constant When a voltage is imposed on two conductive materials that are separated from one another by a vacuum [Figure 18.32(a)], we would expect no current flow. Instead, the electrical charge produced by the voltage remains stored in the circuit. The magnitude of the charge that can be stored between the conductors is called the *capacitance* C and is related to the imposed voltage by

$$Q = CV \qquad (18.24)$$

where V is the voltage across the conductors and Q is the stored charge in coulombs. The units for capacitance are coulombs/volt (C.V^{-1}), or farads (F).

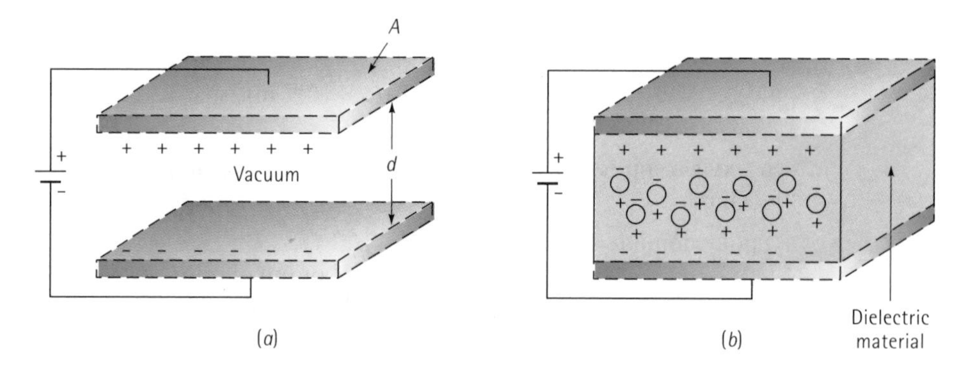

Figure 18.32
A charge can be stored at the conductor plates in a vacuum (a). However, when a dielectric is placed between the plates (b), the dielectric polarises and additional charge is stored. (*a*) Shows a total of 12 units of charge and (*b*) shows a total of 22 units.

The capacitance depends on the material between the conductors, the size and geometry of the conductors, and the separation between the conductors. For the geometry shown in Figure 18.32(a), the capacitance in a vacuum is given by

$$C = \varepsilon_0 \frac{A}{d} \tag{18.25}$$

where A is the area of each conductor and d is the distance between the plates. The constant ε_0 is the permittivity of a vacuum and is 8.85×10^{-12} F.m^{-1}.

When a dielectric material replaces the vacuum between conductors, polarisation can occur in the dielectric and permit additional charges to be stored [Figure 18.32(b)]. The ability of the dipoles in the dielectric to polarise and store charge is reflected by the permittivity ε, which is a property of the dielectric material. Now the capacitance is given by:

$$C = \varepsilon \frac{A}{d} \tag{18.26}$$

Normally we describe the ability of a material to polarise and store electrical charge by the relative permittivity or dielectric constant κ, which is simply the ratio of the permittivity of the material to the permittivity of a vacuum:

$$\kappa = \frac{\varepsilon}{\varepsilon_0} \tag{18.27}$$

The dielectric constant κ is the normal way to describe the ability of a material to store a charge.

The dielectric constant, as expected, is related to the polarisation that can be achieved in the material:

$$P = (\kappa - 1)\, \varepsilon_0 \xi \tag{18.28}$$

where ξ is the strength of the electrical field (V.m^{-1}). For materials that polarise easily, both the dielectric constant and the capacitance are large and, in turn, a large quantity of charge can be stored. In addition, Equation 18.28 suggests that polarisation increases, at least until all of the dipoles are aligned, as the voltage (expressed by the strength of the electric field) increases.

Dielectric Strength Unfortunately, we find that if the applied voltage is too high or the separation between the two conductors is too small, the dielectric device breaks down and discharges and the electrical charge is lost. The dielectric strength is the maximum electric field ξ that the dielectric material can maintain between the conductors. The dielectric strength therefore places an upper limit on both C and Q:

$$\text{Dielectric strength} = \xi_{max} = \left(\frac{V}{d}\right)_{max} \tag{18.29}$$

In order to construct a small device capable of storing large charges in an intense field, we must select materials with both a high dielectric strength and a high dielectric constant. Dielectric strengths and dielectric constants for typical materials are shown in Table 18.9.

Table 18.9 Properties of selected dielectric materials.

Material	Dielectric constant (κ) (at 60 Hz)	(at 10^6 Hz)	Dielectric Strength (E_{max}) ($V.m^{-1}$)	Dissipation Factor, tan δ (at 10^6 Hz)	Resistivity (Ω.m)
Polyethylene	2.3	2.3	2.0×10^7	0.0001	$>10^{14}$
Teflon (PTFE)	2.1	2.1	2.0×10^7	0.00007	10^{16}
Polystyrene	2.5	2.5	2.0×10^7	0.0002	10^{16}
PVC	3.5	3.2	4.0×10^7	0.05	10^{10}
Nylon (Polyamide)	4.0	3.6	2.0×10^7	0.04	10^{13}
Rubber	4.0	3.2	2.4×10^7		
Phenolic	7.0	4.9	1.2×10^7	0.05	10^{10}
Epoxy	4.0	3.6	1.8×10^7		10^{13}
Paraffin wax		2.3	1.0×10^7		10^{11}–10^{17}
Fused silica	3.8	3.8	1.0×10^7	0.0004	10^{9}–10^{10}
Soda-lime glass	7.0	7.0	1.0×10^7	0.009	10^{13}
Al_2O_3	9.0	6.5	6.0×10^6	0.001	10^{9}–10^{11}
TiO_2		14–110	8.0×10^6	0.0002	10^{11}–10^{16}
Mica		7.0	4.0×10^7		10^{11}
$BaTiO_3$		3 000.0	1.2×10^7		10^{6}–10^{13}
Water		78.3			10^{12}

Electrical Conductivity In order for the dielectric to store energy, charge carriers such as electrons or ions must be prevented from moving through the material from one conductor to the other. As a consequence, dielectric materials always have a very high electrical resistivity, as shown in Table 18.9. Ceramic and polymer materials, which normally have electrical resistivities in excess of 10^9 Ω.m, are used as dielectric materials.

Effect of Material Structure Polarisation, and therefore the ability of the material to store charge, are closely related to the structure of the material. The material should possess permanent dipoles that move easily in an electric field and still produce high dielectric constants. In water, organic liquids, oils, and waxes, molecular polarisation is accomplished easily, because the molecules making up the liquid or wax are mobile and respond quickly to the application of the electric field.

Segments of the chains in amorphous polymers have sufficient mobility to polarise and, because they are solid, are easily made into electrical devices. Chains in more rigid structures, such as glassy or crystalline polymers, are less mobile and have lower dielectric constants and dielectric strengths than their amorphous counterparts. Amorphous polymers with asymmetrical chains have a higher dielectric constant, even though the chains may not easily align, because the strength of each molecular dipole is greater. Thus, polyvinyl chloride and polystyrene have dielectric constants greater than polyethylene.

Ceramic glasses, also amorphous structures, permit some movement of segments of the glassy structure. Electronic and ionic polarisation in crystalline ceramics also provide dielectric constants of the same order of magnitude as the polymer materials.

However, certain ceramics, such as barium titanate ($BaTiO_3$), provide exceptionally large dielectric constants because of the molecular polarisation caused by the asymmetrical structure of the unit cell, as described earlier (*see* Figure 18.31).

Imperfections in the structure are also critical. Often, breakdown of a dielectric is a result of a current flow through the material following cracks, grain boundary impurities, or moisture.

Dissipation and Dielectric Loss Factors When an alternating current is applied to a perfect dielectric, the current will lead the voltage by 90°. However, due to losses, the current leads the voltage by only 90° − δ, where δ is called the dielectric loss angle. When the current and voltage are out of phase by the dielectric loss angle, electrical energy or power is lost, often in the form of heat. The dissipation factor is given by:

$$\text{Dissipation factor} = \tan \delta \qquad (18.30)$$

The dielectric loss factor is:

$$\text{Dielectric loss factor} = \kappa \tan \delta \qquad (18.31)$$

The total power lost, P_L, is related to the dissipation factor, the dielectric constant, the electric field, the frequency, and the volume of the dielectric material,

$$P_L = 5.556 \times 10^{-11}\ \kappa \tan \delta \xi^2 f v \qquad (18.32)$$

where the electric field is given in volts per metre, the frequency f in hertz, the volume v in cubic metres, and the power loss in watts. We can minimise heating, even with a large dielectric constant, if a material with a small loss angle is selected.

Frequency Dielectric materials are often used in alternating-current circuits, and the dipoles must therefore switch directions – often at a high frequency. *Dipole friction* occurs when reorientation of the dipoles is difficult, causing energy losses. The greatest loss occurs at frequencies at which the dipoles can almost – but not quite – be reoriented (Figure 18.33).

Because of the dielectric loss, the dielectric constant and the polarisation are frequency dependent (Figure 18.33). At frequencies greater than about 10^{16} Hz, none of the dipoles moves and no polarisation occurs. Below 10^{16} Hz, electronic polarisation

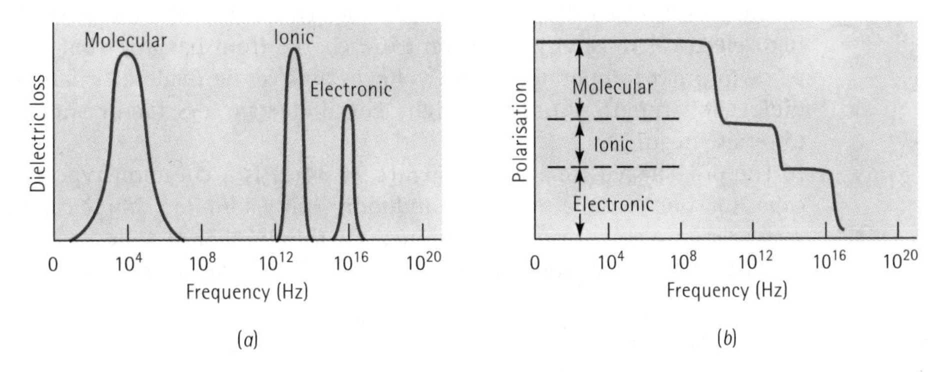

Figure 18.33
The effect of frequency on dielectric properties: (*a*) the dielectric loss is greatest at frequencies at which one of the contributions to polarisation is lost. (*b*) The total polarisation depends on the number of mechanisms that are active. At low frequencies, all types of polarisation may occur.

occurs, since no rearrangement of atoms is necessary. For frequencies less than about 10^{13} Hz, ionic polarisation also occurs; only a simple elastic distortion of the bonds between the ions is required. However, molecular polarisation occurs only at low frequencies, since entire atoms or groups of atoms must be rearranged. The maximum polarisation occurs at low frequencies, where all three types of polarisation are possible.

The structure also influences the frequency effect. Gases and liquids polarise at higher frequencies than solids. Amorphous polymers and ceramics polarise at higher frequencies than their crystalline counterparts. Polymers with bulky asymmetrical groups attached to the chain polarise only at low frequencies. Therefore, polyethylene and polytetrafluoroethylene (Teflon) have the same dielectric constant at almost all frequencies, but the dielectric constant for polyvinyl chloride decreases as the frequency increases.

We can intentionally select a frequency so that materials with permanent dipoles have a high dielectric loss and materials that polarise only by electronic or ionic contributions have a low dielectric loss. Consequently, the permanent dipole materials heat, but the other materials remain cool. Microwave ovens are used to cure many polymer adhesives. The materials to be joined, including metals, have a low loss factor, while the adhesive has a high loss factor. The heat produced in the adhesive due to dielectric losses initiates the thermosetting reaction.

18.14 Dielectric Properties and Capacitors

A capacitor is used to store charge received from a circuit. The capacitor may smooth out fluctuations in the signal, accumulate charge to prevent damage to the rest of the circuit, store charge for later distribution, or even change the frequency of the electric signal. Capacitors are designed so that a charge is stored in a polarised material between two conductors, as described in Figure 18.32. As we found earlier, the charge that can be stored depends on the capacitance, which in turn depends on the design of the capacitor and the dielectric material that is used. The material between the conductors must easily polarise, so the dielectric constant should be high yet have a high electrical resistivity to prevent the charge from passing from one plate to the next. In order to operate at high voltages and yet be made as small as possible, the dielectric strength should be high. The dielectric loss factor should be small to minimise heating.

The disc-shaped capacitor in Figure 18.34(a) is a common type of parallel plate capacitor, but – from a practical standpoint – only a limited charge can be stored. One way to improve performance is to increase the number of plates, as shown in Figure 18.34(b). For a capacitor containing n parallel conductor plates, the capacitance is:

$$C = \varepsilon_0 \kappa (n-1)\frac{A}{d} \tag{18.33}$$

The use of many large plates, a small separation between the plates, a high dielectric constant, and a high dielectric strength improve the ability of the capacitor to store a charge.

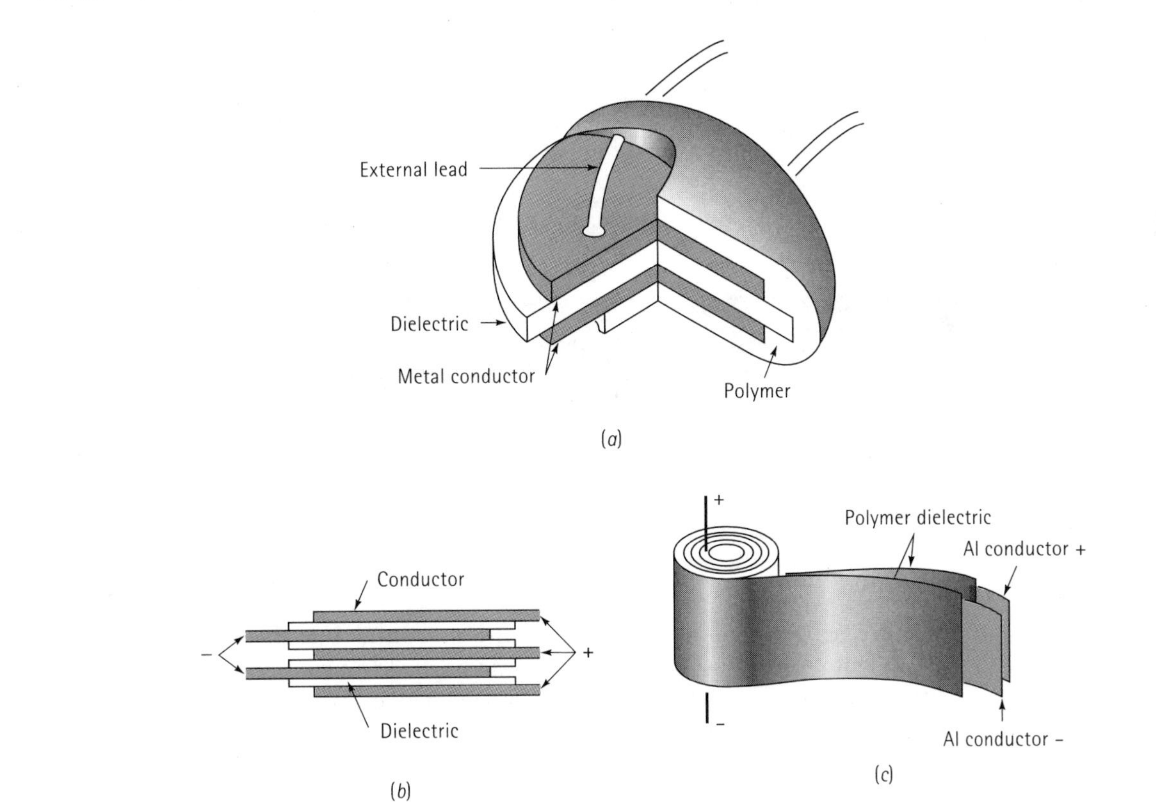

Figure 18.34
Examples of capacitors: (*a*) A disc-shaped capacitor, (*b*) a multiple conductor parallel plate capacitor, and (*c*) a tube capacitor.

EXAMPLE 18.11 Design of a Parallel Plate Capacitor

Design a parallel capacitor using mica that will have a capacitance of 0.0252 μF. We are able to obtain the mica only in a thickness of 2.5 μm.

SOLUTION

From Table 18.9, typical properties for mica include $\kappa = 7$ and a dielectric strength of 4.0×10^7 V.m^{-1}. The thickness of the mica is 2.5 μm (2.5×10^{-6} m). We must specify the size of the capacitor plates and the number of plates to use. From Equation 18.33:

$$C = \varepsilon_0 \kappa (A/d)(n-1) = 0.0252 \times 10^{-6}\,\text{F}$$

$$A(n-1) = \frac{Cd}{\varepsilon_0 \kappa} = \frac{(0.0252 \times 10^{-6}\,\text{F})(2.5 \times 10^{-6}\,\text{m})}{(8.85 \times 10^{-12}\,\text{F.m}^{-1})(7)}$$

$$A(n-1) = 1.017 \times 10^{-3}\,\text{m}^2$$

If we use a single layer of dielectric, the number of conductor plates n is 2 and the area of the plates must be:

$$A = \frac{1.017 \times 10^{-3}}{(2-1)} = 1.017 \times 10^{-3}\,\text{m}^2 = 1017\,\text{mm}^2$$

Other combinations, assuming a square conductor plate, include those given in the table below.

Dielectric Layers	Conductor Plates	Area (mm^2)	Dimensions (mm)
1	2	1017	31.9 × 31.9
2	3	515	22.7 × 22.7
3	4	343	18.5 × 18.5
4	5	258	16.0 × 16.0
5	6	206	14.4 × 14.4

Other geometries or more plates could also be considered. By using five layers of dielectric rather than just one, the device can be made much more compact.

We must still limit the voltage that is applied between any set of conductors. Since the dielectric strength is 4×10^7 V.m^{-1}:

$$\frac{V}{d} = \frac{V}{(2.5 \times 10^{-6}\,\text{m})} = 4 \times 10^7\,\text{V.m}^{-1}$$

$$V_{max} = (4 \times 10^7)(2.5 \times 10^{-6}) = 100\,\text{V}$$

18.15 Dielectric Properties and Electrical Insulators

Materials used to insulate an electric field from its surroundings must also be dielectric. Electrical insulators possess a high electrical resistivity, a high dielectric strength, and a low loss factor. However, a high dielectric constant is not a necessary requirement for insulators and, in fact may even be undesirable. Most polymer and ceramic materials, including glass, satisfy some or all of these requirements.

High electrical resistivity prevents current leakage; high dielectric strength prevents catastrophic breakdown of the insulator at high voltages. Internal failure of the insulator occurs if impurities provide donor or acceptor levels that permit electrons to be excited into the conduction band. External failure is caused by arcing along the surface of the insulator or through interconnected porosity within the insulator body. In particular, adsorbed moisture on the surface of ceramic insulators presents a problem. Glazes on ceramic insulators seal off porosity and reduce the effect of surface contaminants.

The small dielectric constant prevents polarisation, so charge is not stored locally at the insulator. Low dielectric constants are desirable for insulators, but high constants are required for capacitors.

18.16 Piezoelectricity and Electrostriction

When an electric field is applied, polarisation may change the dimensions of the material, an effect called electrostriction. This might occur as a result of atoms acting as egg-shaped particles rather than spheres, or the bonds between ions changing in length, or by distortion to the orientation of the permanent dipoles in the material.

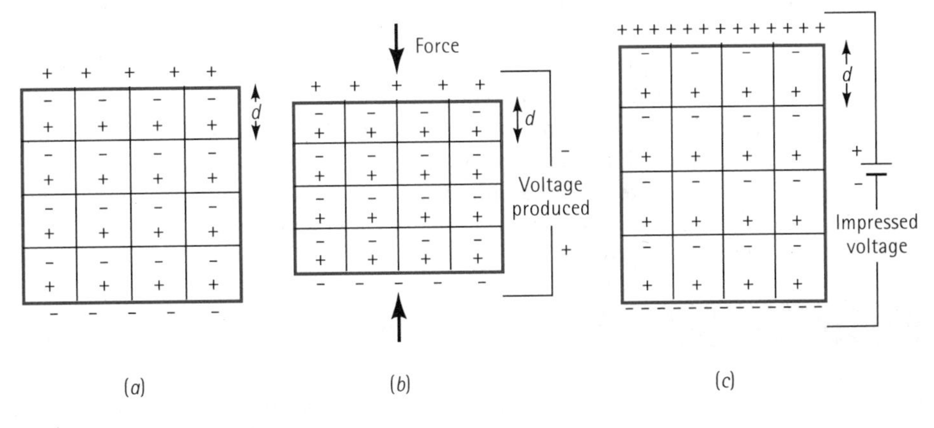

Figure 18.35
The piezoelectric effect: (*a*) Piezoelectric crystals have a charge difference because of permanent dipoles. (*b*) A compressive force reduces the distance between charge centres, changes the polarisation, and introduces a voltage. (*c*) A voltage changes the distance between charge centres, causing a change in dimensions.

However, certain dielectric materials display a further property. When a dimensional change is imposed on the dielectric, polarisation occurs and a voltage or field is created (Figure 18.35). Dielectric materials that display this reversible behaviour are piezoelectric. Quartz, barium titanate, a solid solution of $PbZrO_3$-$PbTiO_3$ (or PZT), and more complicated ceramics such as (Pb, La)-(Ti, Zr)O_3 (or PLTZ) all are permanently polarised and display this behaviour. We can write the two reactions that occur in piezoelectrics as:

$$\text{Field produced by stress} = \xi = g\sigma \qquad (18.34)$$

$$\text{Strain produced by field} = \varepsilon = d\xi \qquad (18.35)$$

where ξ is the electric field strength (V.m^{-1}), σ is the applied stress (N.m^{-2}), ε is the strain, and g and d are constants. Typical values for d are given in Table 18.10. The constant g is related to d through the modulus of elasticity E:

$$E=\frac{1}{gd} \qquad (18.36)$$

Table 18.10 The piezoelectric constant *d* for selected materials.

Material	Piezoelectric Constant d ($C.N^{-1}$ or $m.V^{-1}$)
Quartz	2.3×10^{-12}
$BaTiO_3$	100×10^{-12}
$PbZrTiO_6$	250×10^{-12}
$PbNb_2O_6$	80×10^{-12}

The piezoelectric effect is used in transducers, which convert acoustical waves (sound) into electric fields, or electric fields into acoustical waves. Sound of a particular frequency produces a strain in a piezoelectric material. The dimensional changes polarise the crystal, creating an electric field. In turn, the electric field is transmitted to a second piezoelectric crystal. There the electric field produces dimensional changes in the second crystal; these changes produce an acoustical wave that is amplified. This description depicts the telephone. Similar electromechanical transducers are used for stereo record players and other audio devices.

In some materials, changing the temperature also produces a distortion of the unit cell, causing polarisation and creating a voltage. These pyroelectric materials can be used in heat-sensing devices.

EXAMPLE 18.12 Design of a Pressure-Limiting Device

You would like to mount a 0.25 mm thick barium titanate wafer on the end of a 2.5 mm-diameter probe. When a force of more than 200 N acts on the wafer, an electrical circuit is to be activated to stop the application of the force. Design this system. Assume that the modulus of elasticity of barium titanate is 69 $GN.m^{-2}$.

SOLUTION

One method would be to take advantage of the piezoelectric behaviour of the barium titanate. The applied force strains the crystal and produces a voltage. Our electrical circuit can be designed to be activated when a critical voltage (corresponding to a force of 200 N) is reached. The maximum stress is:

$$\sigma = \frac{F}{A} = \frac{200\,\text{N}}{(\pi/4)(2.5 \times 10^{-3})^2} = 40.7\,\text{MN.m}^{-2}$$

Since $E = 69$ $GN.m^{-2}$, the strain is:

$$\varepsilon = \frac{\sigma}{E} = \frac{40.7 \times 10^6}{69 \times 10^9} = 5.90 \times 10^{-4}\,\text{m/m}$$

From Table 18.10, $d = 100 \times 10^{-12}$ $m.V^{-1}$:

$$\xi = \frac{\varepsilon}{d} = \frac{5.90\times10^{-4}}{100\times10^{-12}} = 5.90\times10^{6}\ \text{V.m}^{-1}$$

$$V = (\xi)(\text{thickness}) = (5.90\times10^{6}\ \text{V.m}^{-1})(0.25\times10^{-3}\ \text{m})$$

$$= 1475\ \text{V}$$

Therefore, we must design an electrical circuit that is activated when the voltage increases to 1 475 V.

18.17 Ferroelectricity

The presence of polarisation in a material after the electric field is removed can be explained in terms of a residual alignment of permanent dipoles. Barium titanate is again an excellent example. Materials that retain a net polarisation when the field is removed are called ferroelectric.

In ferroelectric materials, the orientation of one dipole influences the surrounding dipoles to have an identical alignment. We can examine this behaviour by describing the effect of an electric field on polarisation (Figure 18.36).

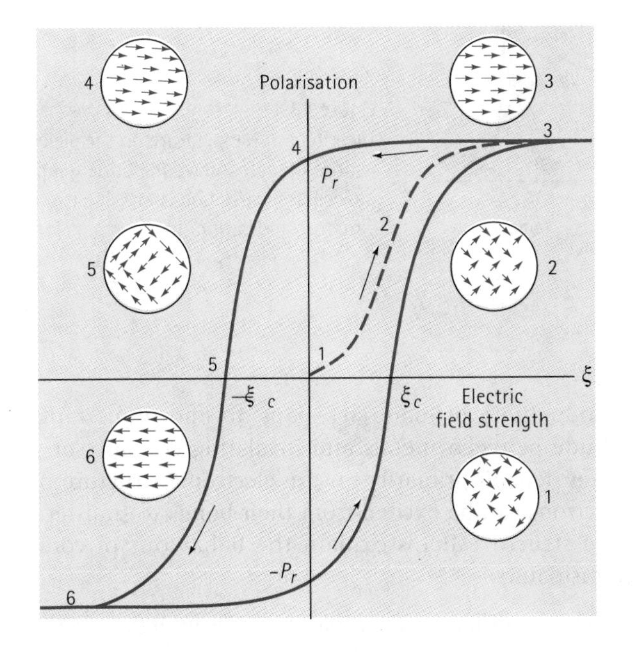

Figure 18.36
The ferroelectric hysteresis loop, showing the influence of the electric field on polarisation and the alignment of the dipoles.

Let's begin with a crystal whose dipoles are randomly oriented, so that there is no net polarisation. When a field is applied, the dipoles begin to line up with the field (points 1 to 3 in Figure 18.36). Eventually the field aligns all of the dipoles and the maximum, or saturation, polarisation P_s is obtained (point 3). When the field is subsequently removed, a remanent polarisation P_r remains (point 4) due to the coupling between the dipoles; the material is permanently polarised. The ability to retain polarisation permits the ferroelectric material to retain information, making the material useful in computer circuitry.

When a field is applied in the opposite direction, the dipoles must be reversed. A coercive field ξ_c must be applied to remove the polarisation and randomise the dipoles (point 5). If the reverse field is increased further, saturation occurs with the opposite polarisation (point 6). As the field continues to alternate, a hysteresis loop is described showing how the polarisation of the ferroelectric varies with the field. The area contained within the hysteresis loop is related to the energy required to cause polarisation to switch from one direction to the other.

Above the critical Curie temperature, dielectric and, consequently, ferroelectric behaviour are lost (Figure 18.37). In some materials, such as barium titanate, the Curie temperature corresponds to a change in crystal structure from the distorted tetragonal structure in Figure 18.31 to a normal cubic perovskite unit cell [Figure 14.1(a)]. Consequently, the permanent dipoles in each unit cell no longer exist.

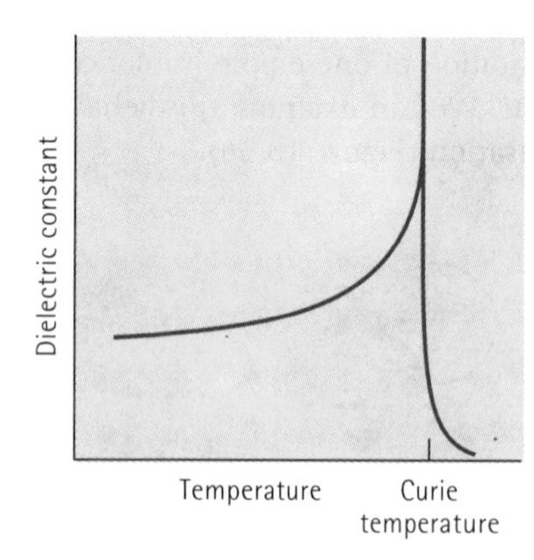

Figure 18.37
The effect of temperature on the dielectric constant of barium titanate. Above the Curie temperature, the molecular polarisation is lost due to a change in crystal structure and barium titanate is no longer ferroelectric.

SUMMARY

- The electrical conductivity of materials spans an enormous range – nearly 25 orders of magnitude between metals and insulating ceramics or polymers. The electrical properties depend primarily on the electronic structure of the material, or how easily electrons can be excited from their bonds to transfer their electrical charge. The band structure helps explain the behaviour of conductors, semiconductors, and insulators.

- Because metals have unfilled valence bands, little energy is required to excite their electrons; consequently, electrical conductivities are high. Electrical conductivity is reduced when the temperature or number of lattice defects increases. Solid solution atoms and coherent precipitates cause the largest decrease; non-coherent precipitates, grain boundaries, and dislocations have a less pronounced effect.

- In ionic materials, electrical charge is carried by ions; high diffusion rates, caused by high temperatures, provide improved conductivity.
- Although the electrical conductivity of polymers is normally very low, a few polymers have inherently good conductivity; others can contain fillers or fibres that provide conductivity.
- Below a critical temperature, superconductive materials (including certain metals, intermetallic compounds, and ceramics) transfer electrical charge with virtually no resistance. Liquid helium (4 K) is required to cool most materials below the critical temperature, although some ceramics need only be cooled to liquid nitrogen (77 K) temperatures.
- The energy gap for intrinsic semiconductors is small, causing moderate conductivities if sufficient energy is introduced. By introducing doping elements to form extrinsic semiconductors, the required energy is reduced and electrical conductivity can be made nearly constant over a range of temperatures, allowing semiconductors to be used for electronic devices such as diodes, transistors, and integrated circuits. In semiconductors, increasing the temperature or number of lattice defects increases electrical conductivity.
- Electrical insulators, which include most ceramic and polymer materials, have a large energy gap; their conductivity is very low. The performance of insulators also depends on their dielectric properties, including a high dielectric strength and a low dielectric constant. The dielectric properties of these materials, including their low ability to polarise, also lead to special behaviour – for example, capacitors that store electrical energy and piezoelectric materials whose dimensions are related to the applied voltage.

GLOSSARY

Acceptor saturation
When all of the extrinsic acceptor levels in a p-type semiconductor are filled.

Avalanche voltage
The reverse bias that causes a large current flow in a *p-n* junction.

Capacitor
An electrical device, constructed from alternating layers of a dielectric and a conductor, that is capable of storing a charge.

Conduction band
The unfilled energy levels into which electrons can be excited to provide conductivity.

Curie temperature
The temperature above which ferroelectric behaviour is lost.

Current density
The current flowing through a given cross-sectional area.

Defect semiconductors

Compounds, such as ZnO and FeO, that contain lattice defects that provide semiconduction.

Dielectric constant

The ratio of the permittivity of a material to the permittivity of a vacuum, thus describing the relative ability of a material to polarise and store a charge.

Dielectric loss

The fraction of energy lost each time an electric field in a material is reversed.

Dielectric strength

The maximum electric field that can be maintained between two conductor plates.

Donor exhaustion

When all of the extrinsic donor levels in an n-type semiconductor are filled.

Doping

Addition of controlled amounts of impurities to increase the number of charge carriers in a semiconductor.

Drift velocity

The average rate at which electrons or other charge carriers move through a material.

Electric field strength

The voltage gradient, or volts per unit length.

Electrostriction

The dimensional change that occurs in a material when an electric field is acting on it.

Energy gap

The energy between the top of the valence band and the bottom of the conduction band that a charge carrier must obtain before it can transfer a charge.

Extrinsic semiconductor

A semiconductor prepared by adding impurities or dopants which determine the number of charge carriers.

Ferroelectricity

Alignment of domains so that a net polarisation remains after the electric field is removed.

Forward bias

Connecting a junction device so that holes and electrons flow toward the junction to produce a net current flow.

Holes

Unfilled energy levels in the valence band. Because electrons move to fill these holes, the holes move and produce a current.

Hybridisation

When valence and conduction bands are separated by an energy gap, leading to the semiconductive behaviour of silicon and germanium.

Hysteresis loop

The loop traced out by the polarisation as the electric field is cycled. A similar loop occurs in magnetic materials.

Intrinsic semiconductor
A semiconductor in which the temperature determines the conductivity.

Mean free path
The average distance that electrons can move without being scattered by the lattice.

Mobility
The ease with which a charge carrier moves through a material.

Permittivity
The ability of a material to polarise and store a charge within it.

Piezoelectricity
The ability in some materials for a change in electric field to change the dimensions of the material, whereas a change in dimensions produces an electric field.

Polarisation
Alignment of dipoles so that a charge can be permanently stored.

Pyroelectricity
The ability of a material to polarise and produce a voltage due to changes in temperature.

Recombination time
A constant related to the time required for electrons and holes to recombine when an electric field is removed.

Rectifiers
p-n junction devices that permit current to flow in only one direction in a circuit.

Reverse bias
Connecting a junction device so that holes and electrons flow away from the junction, preventing a net current flow.

Stoichiometric semiconductors
Intermetallic compounds, such as GaAs, that can be doped to provide semiconduction.

Superconductivity
Flow of current through a material that has virtually no resistance to that flow.

Thermistor
A semiconductor device that is particularly sensitive to changes in temperature, permitting it to serve as an accurate measure of temperature.

Transducer
A device that receives one type of input (such as strain or light) and provides an output that may be of a different type (such as an electrical signal).

Transistor
A semiconductor device that can be used to amplify electrical signals.

Valence band
The energy levels filled by electrons in their lowest energy states.

Zener diode
A *p-n* junction device which, with a very large reverse bias, causes a current to flow.

PROBLEMS

18.1 A current of 10 A is passed through a 1 mm-diameter wire 1 km long. Calculate the power loss if the wire is made of

(a) aluminium,

(b) silicon, and

(c) silicon carbide (*see* Table 18.1.).

18.2 The power lost in a 2 mm-diameter copper wire is to be less than 250 W when a 5-A current is flowing in the circuit. What is the maximum length of the wire?

18.3 A current density of 10^9 A.m^{-2} is applied to a gold wire 50 m in length. The resistance of the wire is found to be 2 Ω. Calculate the diameter of the wire and the voltage applied to the wire.

18.4 We would like to produce a 5 000 Ω resistor from boron carbide fibres having a diameter of 0.1 mm. What is the required length of the fibres?

18.5 Suppose we estimate that the mobility of electrons in silver is 7.5×10^{-3} m^2.V^{-1}.s^{-1}. Estimate the fraction of the valence electrons that are carrying an electrical charge.

18.6 A current density of 5×10^7 A.m^{-2} is applied to a magnesium wire. If half of the valence electrons serve as charge carriers, calculate the average drift velocity of the electrons.

18.7 We apply a voltage of 10 V to an aluminium wire 2 mm in diameter and 20 m long. If 10% of the valence electrons carry the electrical charge, calculate the average drift velocity of the electrons in km.h^{-1}.

18.8 In a welding process, a current of 400 A flows through the arc when the voltage is 35 V. The length of the arc is about 2.5 mm and the average diameter of the arc is about 4.5 mm. Calculate the current density in the arc, the electric field across the arc, and the electrical conductivity of the hot gases in the arc during welding.

18.9 Calculate the electrical conductivity of nickel at −50°C and at +500°C.

18.10 The electrical resistivity of pure chromium is found to be 18×10^{-8} Ω.m. Estimate the temperature at which the resistivity measurement was made.

18.11 After finding the electrical conductivity of cobalt at 0°C, we decide we would like to double that conductivity. To what temperature must we cool the metal?

18.12 From Figure 18.11(b), estimate the defect resistivity coefficient for tin in copper.

18.13 The electrical resistivity of a beryllium alloy containing 5 at% of an alloying element is found to be 50×10^{-8} Ω.m at 400°C. Determine the contributions to resistivity due to temperature and due to impurities by finding the expected resistivity of pure beryllium at 400°C, the resistivity due to impurities, and the defect resistivity coefficient. What would be the electrical resistivity if the beryllium contained 10 at% of the alloying element at 200°C?

18.14 Is Equation 18.7 valid for the copper-zinc system? If so, calculate the defect resistivity coefficient for zinc in copper.

18.15 GaV_3 is to operate as a superconductor in liquid helium (at 4 K). What is the maximum magnetic field that can be applied to the material?

18.16 Nb_3Sn and GaV_3 are candidates for a superconductive application when the magnetic field is 1.2×10^7 A.m^{-1}. Which would require the lower temperature in order for the material to be superconductive?

18.17 A filament of Nb_3Sn 0.05 mm in diameter operates in a magnetic field of 80 000 A.m^{-1} at 4 K. What is the maximum current that can be applied to the filament in order for the material to behave as a superconductor?

18.18 Assume that most of the electrical charge transferred in MgO is caused by the diffusion of Mg^{2+} ions. Determine the mobility and electrical conductivity of MgO at 25°C and at 1500°C. (*See* Table 5.1.)

18.19 Assume that most of the electrical charge transferred in Al_2O_3 is caused by the diffusion of Al^{3+} ions. Determine the mobility and electrical conductivity of Al_2O_3 at 500°C and at 1 500°C. (*See* Table 5.1 and Example 14.1.)

18.20 Calculate the electrical conductivity of a fibre-reinforced polyethylene part that is reinforced with 20 vol% of continuous, aligned nickel fibres.

18.21 For germanium, silicon, and tin, compare, at 25°C.

(a) the number of charge carriers per cubic millimetre,

(b) the fraction of the total electrons in the valence band that are excited into the conduction band, and

(n) the constant n_0.

18.22 For germanium, silicon, and tin, compare the temperature required to double electrical conductivity from the room temperature value.

18.23 When an electric field is applied to a semiconductor, 5×10^6 electrons/mm^3 serve as charge carriers. When the field is removed, 30 electrons/mm^3 remain after 10^{-6} s. Calculate

(a) the recombination time and

(b) the time required for 99.9% of the electrons and holes to recombine.

18.24 Calculate the number of extrinsic charge carriers per cubic metre in an n-type semiconductor when one of every 1 000 000 atoms in silicon is replaced by an antimony atom. Estimate the conductivity of the semiconductor in the exhaustion zone.

18.25 Determine the electrical conductivity of silicon when 0.0001 at% antimony is added as a dopant and compare it to the electrical conductivity when 0.0001 at% indium is added.

18.26 We would like to produce an extrinsic germanium semiconductor having an electrical conductivity of $2 \times 10^5\ \Omega^{-1}.m^{-1}$ in the exhaustion or saturation region. Determine the amount of phosphorous and the amount of gallium required.

18.27 Estimate the electrical conductivity of silicon doped with 0.0002 at% arsenic at 600°C, which is above the exhaustion plateau.

18.28 Determine the amount of arsenic that must be combined with 1 kg of gallium to produce a p-type semiconductor with an electrical conductivity of $5 \times 10^4\ \Omega^{-1}.m^{-1}$ at 25°C. The lattice parameter of GaAs is about 0.565 nm, and GaAs has the zinc blende structure.

18.29 A ZnO crystal is produced in which one interstitial Zn atom is introduced for every 500 Zn lattice sites. Estimate

(a) the number of charge carriers per cubic millimetre and

(b) the electrical conductivity at 25°C.

18.30 Each Fe^{3+} ion in FeO serves as an acceptor site for an electron. If there is one vacancy per 750 unit cells of the FeO crystal (with the sodium chloride structure), determine the number of possible charge carriers per cubic millimetre. The lattice parameter of FeO is 0.429 nm.

18.31 When a voltage of 5 mV is applied to the emitter of a transistor, a current of 2 mA is produced. When the voltage is increased to 8 mV, the current through the collector rises to 6 mA. By what percentage will the collector current increase when the emitter voltage is doubled from 9 mV to 18 mV?

18.32 A 0.5-mm diameter fibre 10 mm in length made of boron nitride is placed into a 120-V circuit.

Using Table 18.1, calculate

(a) the current flowing in the circuit and

(b) the number of electrons passing through the boron nitride fibre per second.

(c) What would the current and number of electrons be if the fibre were made of magnesium instead of boron nitride?

18.33 Calculate the displacement of the electrons or ions for the following conditions:

(a) electronic polarisation in nickel of 2×10^{-7} C.m^{-2}.

(c) electronic polarisation in aluminium of 2×10^{-8} C.m^{-2}.

(c) ionic polarisation in NaCl of 4.3×10^{-8} C.m^{-2}.

(d) ionic polarisation in ZnS of 5×10^{-8} C.m^{-2}.

18.34 A 2 mm-thick alumina dielectric is used in a 60-Hz circuit. Calculate the voltage required to produce a polarisation of 5×10^{-7} C.m^{-2}.

18.35 Suppose we are able to produce a polarisation of 5×10^{-5} C.m^{-2} in a 5 mm cube of barium titanate. What voltage is produced?

18.36 Calculate the thickness of polyethylene required to store the maximum charge in a 24 000 V circuit without breakdown.

18.37 Calculate the maximum voltage that can be applied to a 1 mm-thick barium titanate dielectric without causing breakdown.

18.38 A 120-V circuit is to operate at 10^6 Hz. A Teflon film 10 mm in diameter is a part of the circuit.

(a) What thickness of Teflon is required to ensure a power loss of no more than 1 W?

(b) What thickness would be required if the dielectric was made of polyvinyl chloride?

18.39 Calculate the capacitance of a parallel plate capacitor containing 5 layers of mica, where each mica sheet is 10 mm × 20 mm × 0.05 m.

18.40 Determine the number of Al_2O_3 sheets, each 15 mm × 15 mm × 0.01 mm, required to obtain a capacitance of 0.0142 μF in a 10^6 Hz parallel plate capacitor.

18.41 We would like to construct a barium titanate device with a 2.5 mm diameter that will produce a voltage of 250 V when a 20 N force is applied. How thick should the device be?

18.42 A force of 90 N is applied to the face of a 5 mm × 5 mm × 1 mm thick quartz crystal. Determine the voltage produced by the force. The modulus of elasticity of quartz is 71.7 GN.m^{-2}.

18.43 Determine the strain produced when a 300 V signal is applied to a barium titanate wafer 2 mm × 2 mm × 0.1 mm thick.

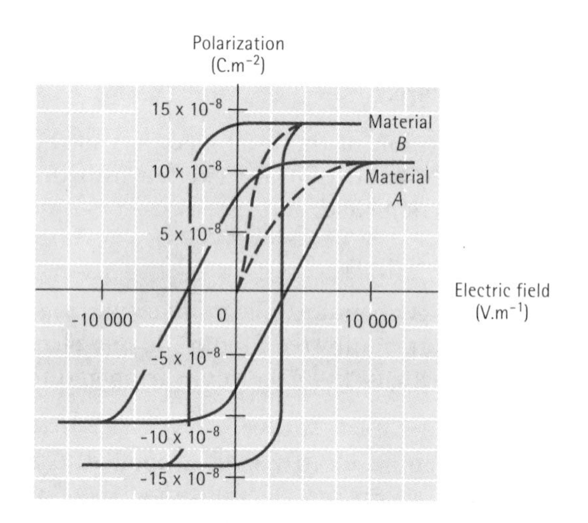

Figure 18.38
Ferroelectric hysteresis loops (*for* Problem 18.44).

18.44 Figure 18.38 shows the hysteresis loops for two ferroelectric materials:

(a) Determine the voltage required to eliminate polarisation in a 1 mm-thick dielectric made from material *A*.

(b) Determine the thickness of a dielectric made from material *B* if 10 V is required to eliminate polarisation.

(c) What electric field is required to produce a polarisation of 8×10^{-8} C.m^{-2} in material

A, and what is the dielectric constant at this polarisation?

(d) An electric field of 2 500 $V.m^{-1}$ is applied to material *B*. Determine the polarisation and the dielectric constant at this electric field.

Design Problems

18.45 We would like to produce a 100 Ω resistor using a thin wire of a material. Design such a device.

18.46 Design a capacitor that is capable of storing 1 μF when 100 V is applied.

18.47 Design an epoxy-matrix composite that has a modulus of elasticity of at least 24.1 $GN.m^{-2}$ and an electrical conductivity of at least $1 \times 10^7\ \Omega^{-1}.m^{-1}$.

18.48 Design a semiconductor thermistor that will activate a cooling system when the ambient temperature reaches 500°C.

18.49 Design a dielectric device that will detect whether sand is at a particular level in a sand storage tank.

18.50 Design a piezoelectric part that will produce 25 000 V when subjected to a stress of 35 $kN.m^{-2}$.

CHAPTER 19

Magnetic Behaviour of Materials

19.1 Introduction

Most materials interact with a magnetic field, just as dielectrics interact with an electric field. Magnetic materials are used to operate electrical motors, generators, and transformers, to store and retrieve information on magnetic tape or in computers, to serve as actuators and sensors, to focus electron beams, to assist in medical diagnostic devices, and for a host of other applications. The most widely used magnetic materials are based on ferromagnetic metals and alloys such as iron, nickel, and cobalt or ferrimagnetic ceramics, including various ferrites and garnets.

Magnetic behaviour is determined primarily by the electronic structure of a material, which provides magnetic dipoles. Interactions between these dipoles determine the type of magnetic behaviour that is observed. Magnetic behaviour can be modified by composition, microstructure, and processing of these basic materials.

19.2 Magnetic Dipoles and Magnetic Moments

Magnetisation occurs when included or permanent magnetic dipoles are oriented by an interaction between the magnetic material and a magnetic field. Magnetisation enhances the influence of the magnetic field, permitting larger magnetic energies to be stored than if the material were absent. This energy can be stored permanently or temporarily and can be used to do work.

Each electron in an atom has two magnetic moments. A magnetic moment is simply the strength of the magnetic field associated with the electron. This moment, called the Bohr magneton, is

$$\text{Bohr magneton} = \frac{qh}{4\pi m_e} = 9.27 \times 10^{24}\ \text{A·m}^2 \qquad (19.1)$$

where q is the charge on the electron, h is Planck's constant, and m_e is the mass of the electron. The magnetic moments are caused by the orbital motion of the electron around the nucleus and the spin of the electron about its own axis (Figure 19.1).

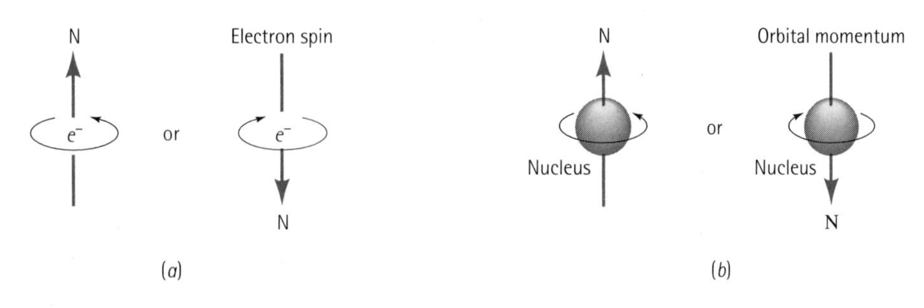

Figure 19.1
Origin of magnetic dipoles: (*a*) The spin of the electron produces a magnetic field with a direction dependant on the quantum number m_s. (*b*) Electrons orbiting around the nucleus create a magnetic field around the atom.

When we discussed electronic structure and quantum numbers in Chapter 2, we pointed out that each discrete energy level could contain two electrons, each having an opposite spin. The magnetic moments of each electron pair in an energy level are opposed. Consequently, whenever an energy level is completely full, there is no net magnetic moment.

Based on this reasoning, we expect any atom of an element with an odd atomic number to have a net magnetic moment from the unpaired electron, but this is not the case. In most of these elements, the unpaired electron is a valence electron. Because the valence electrons from each atom interact, the magnetic moments, on average, cancel each other out and no net magnetic moment is associated with the material.

However, certain elements, such as the transition metals, have an inner energy level that is not completely filled. The elements scandium through to copper, whose electronic structures are shown in Table 19.1, are typical. Except for chromium and copper, the valence electrons in the 4*s* level are paired; the unpaired electrons in chromium and copper are cancelled by interactions with other atoms. Copper also has a completely filled 3*d* shell and thus does not display a net moment.

Table 19.1 The electron spins in the 3*d* level in transition metals, with arrows indicating the direction of spin.

Metal	3*d*					4*s*
Sc	↑					↑↓
Ti	↑	↑				↑↓
V	↑	↑	↑			↑↓
Cr	↑	↑	↑	↑	↑	↑
Mn	↑	↑	↑	↑	↑	↑↓
Fe	↑↓	↑	↑	↑	↑	↑↓
Co	↑↓	↑↓	↑	↑	↑	↑↓
Ni	↑↓	↑↓	↑↓	↑	↑	↑↓
Cu	↑↓	↑↓	↑↓	↑↓	↑↓	↑

The electrons in the 3*d* level of the remaining transition elements do not enter the shells in pairs. Instead, as in manganese, the first five electrons have the same spin. Only after half of the 3*d* level is filled do pairs with opposing spins form. Therefore, each atom in a transition metal has a permanent magnetic moment, which is related to the number of unpaired electrons. Each atom behaves as a magnetic dipole.

The response of the atom to an applied magnetic field depends on how the magnetic dipoles represented by each atom react to the field. Most of the transition elements react in such a way that the sum of the individual atoms' magnetic moments is zero. However, the atoms in nickel, iron, and cobalt undergo an exchange interaction, whereby the orientation of the dipole in one atom influences the surrounding atoms to have the same dipole orientation, producing a desirable amplification of the effect of the magnetic field.

19.3 Magnetisation, Permeability, and the Magnetic Field

Let's look at the relationship between the magnetic field and magnetisation. Figure 19.2 depicts a coil having n turns. When an electric current is passed through the coil, a magnetic field H is produced, with the strength of the field given by

$$H = \frac{nI}{l} \tag{19.2}$$

where n is the number of turns, l is the length of the coil (m), and I is the current (A). The units of H are therefore ampere · turn/m, or simply A.m^{-1}. An alternate unit for magnetic field is the oersted, obtained by multiplying A.m^{-1} by $4\pi \times 10^{-3}$ (*see* Table 19.2).

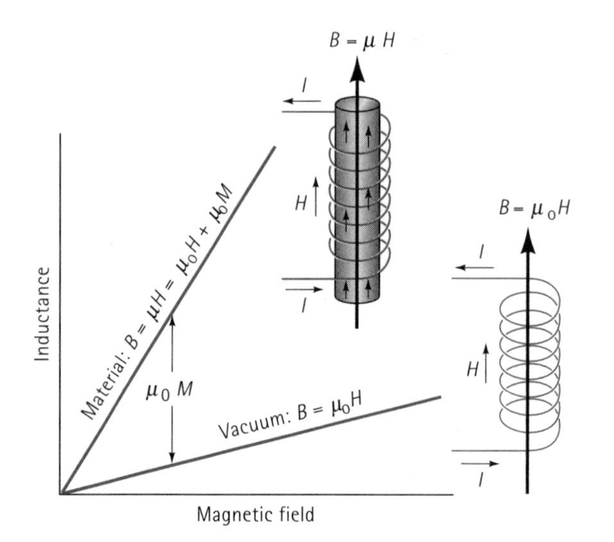

Figure 19.2
A current passing through a coil sets up a magnetic field H with a flux density B. The flux density is higher when a magnetic core is placed within the coil.

Table 19.2 Units for magnetic behaviour.

	SI Units	cgs Units	Conversion
Inductance B	tesla, T (or weber/m^2, Wb.m^{-2})	gauss	$1\ T = 1 \times 10^4$ gauss
Magnetic Field H	A.m^{-1}	oersted	$1\ A.m^{-1} = 4\pi \times 10^{-3}$ oersted
Magnetisation M	A.m^{-1}	oersted	
Permeability μ	T.m.A^{-1} (or Wb.A^{-1}.m^{-1})	gauss/oersted	$4\pi \times 10^{-7}$ T.m.A^{-1} = 1 gauss/oersted

When a magnetic field is applied in a vacuum, lines of magnetic flux are induced. The number of lines of flux, called the flux density, or *inductance B*, is related to the applied field by

$$B = \mu_0 H \tag{19.3}$$

where B is the inductance (tesla, T or weber/metre^2, Wb.m^{-2}), H is the magnetic field (A.m^{-1}) and μ_0 is a constant called the magnetic permeability of a vacuum ($4\pi \times 10^{-7}$ T.m.A^{-1}, or 1 gauss/oersted).

When we place a material within the magnetic field, the magnetic inductance is determined by the manner in which induced and permanent magnetic dipoles interact with the field. The inductance now is

$$B = \mu H \tag{19.4}$$

where μ is the permeability of the material in the field. If the magnetic moments reinforce the applied field, then $\mu > \mu_0$, a greater number of lines of flux that can accomplish work are created, and the magnetic field is magnified. If the magnetic moments oppose the field, however, $\mu < \mu_0$.

We can describe the influence of the magnetic material by the relative permeability μ_r, where:

$$\mu_r = \frac{\mu}{\mu_0} \tag{19.5}$$

A large relative permeability means that the material amplifies the effect of the magnetic field. Thus, the relative permeability has the same importance that the dielectric constant, or relative permittivity, has in dielectrics.

The magnetisation M represents the increase in the inductance due to the core material, so we can rewrite the equation for inductance as:

$$B = \mu_0 H + \mu_0 M \tag{19.6}$$

The magnetic susceptibility χ, which is the ratio between magnetisation and the applied field, gives the amplification produced by the material:

$$\chi = \frac{M}{H} \tag{19.7}$$

Both μ_r and χ refer to the degree to which the material enhances the magnetic field and are therefore related, by:

$$\mu_r = 1 + \chi \tag{19.8}$$

For important magnetic materials, the term $\mu_0 M$ is much greater than $\mu_0 H$. Thus, for these materials:

$$B \cong \mu_0 M \tag{19.9}$$

We sometimes interchangeably refer to either inductance or magnetisation. Normally, we are interested in producing a high inductance B or magnetisation M. This is accomplished by selecting materials that have a high relative permeability or magnetic susceptibility.

EXAMPLE 19.1

Calculate the maximum, or saturation, magnetisation that we expect in iron. The lattice parameter of BCC iron is 0.2866 nm.

SOLUTION

Based on the unpaired electronic spins, we expect each iron atom to have four electrons that act as magnetic dipoles. The number of atoms per m³ in iron is:

$$\text{Number of atoms/m}^3 = \frac{2\text{ atoms/cell}}{(2.866 \times 10^{-10}\text{ m})^3} = 0.085 \times 10^{30}$$

The magnetisation is then:

$$M = (0.085 \times 10^{30}\text{ atoms/m}^3)(4\text{ magnetons/atom}) \times (9.27 \times 10^{-24}\text{ A.m}^2\text{/magneton})$$

$$M = 3.15 \times 10^6\text{ A.m}^{-1}\text{ in one cubic metre}$$

19.4 Interactions Between Magnetic Dipoles and the Magnetic Field

When a magnetic field is applied to a collection of atoms, several types of behaviour are observed (Figure 19.3).

Diamagnetic Behaviour A magnetic field acting on any atom induces a magnetic dipole for the entire atom by influencing the magnetic moment caused by the orbiting electrons. These dipoles can oppose the magnetic field, causing the magnetisation to be less then zero. This behaviour, called diamagnetism, gives a relative permeability of about 0.99995 (or a negative susceptibility). Materials such as

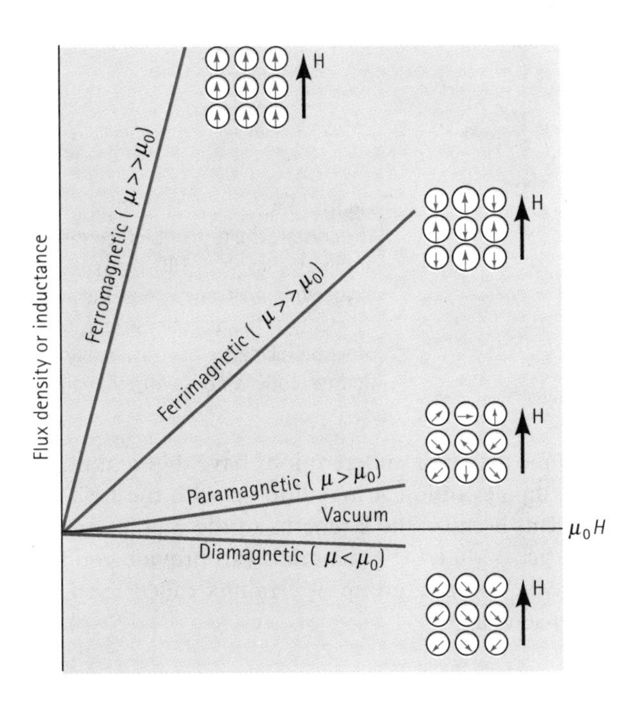

Figure 19.3
The effect of the core material on the flux density. The magnetic moment opposes the field in diamagnetic materials. Progressively stronger moments are present in paramagnetic, ferrimagnetic, and ferromagnetic materials for the same applied field.

copper, silver, gold and alumina are diamagnetic at room temperature. Superconductors must be diamagnetic; they lose their superconductivity when other magnetic effects, such as paramagnetism, become active and permit the field to enter the material.

Paramagnetism When materials have unpaired electrons, a net magnetic moment due to electronic spin is associated with each atom. When a magnetic field is applied, the dipoles line up with the field, causing a positive magnetisation. However, because the dipoles do not interact, extremely large magnetic fields are required to align all of the dipoles. In addition, the effect is lost as soon as the magnetic field is removed. This effect, called paramagnetism, is found in metals such as aluminium, titanium, and alloys of copper. The relative permeability of paramagnetic materials lies between 1 and 1.01.

Ferromagnetism Ferromagnetic behaviour is caused by the unfilled energy levels in the 3*d* level of iron, nickel, and cobalt. Similar behaviour is found in a few other materials, including gadolinium. In ferromagnetic materials, the permanent unpaired dipoles easily line up with the imposed magnetic field due to the exchange interaction, or mutual reinforcement of the dipoles. Large magnetisations are obtained even for small magnetic fields, giving relative permeabilities as high as 10^6.

Antiferromagnetism In materials such as manganese, chromium, MnO, and NiO, the magnetic moments produced in neighbouring dipoles line up in opposition to one another in the magnetic field, even though the strength of each dipole is very high. This effect is illustrated for MnO in Figure 19.4. These materials are antiferromagnetic and have zero magnetisation.

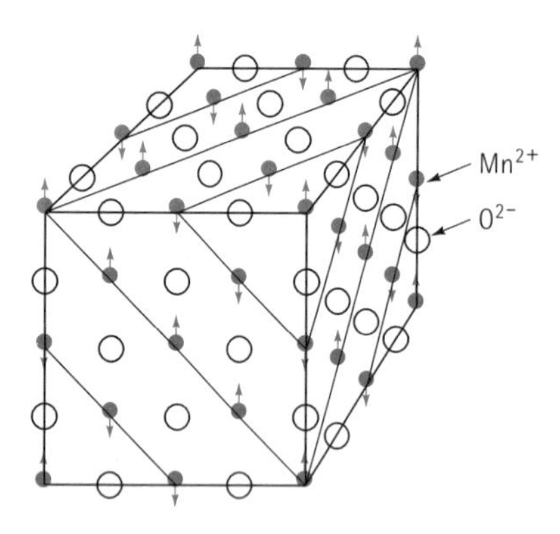

Figure 19.4
The crystal structure of MnO consists of alternating layers of {III} type planes of oxygen and manganese ions. The magnetic moments of the manganese ions in every other (III) plane are oppositely aligned. Consequently, MnO is antiferromagnetic.

Ferrimagnetism In ceramic materials, different ions have different magnetic moments. In a magnetic field, the dipoles of ion *A* may line up with the field, while dipoles of ion *B* oppose the field. But because the strengths of the dipoles are not equal, a net magnetisation results. The ferrimagnetic materials can provide good amplification of the imposed field. We will look at a group of ceramics called ferrites, which display this behaviour, in a later section.

EXAMPLE 19.2 Materials Selection for a Solenoid

We want to produce a solenoid coil that produces an inductance of at least 0.2 tesla when a 10 mA current flows through the conductor. Due to space limitations, the coil should be composed of 10 turns over a 10 mm length. Select a core material for the coil.

SOLUTION

First, we can determine the magnetic field *H* produced by the coil. From Equation 19.2:

$$H = \frac{nI}{l} = \frac{(10)(0.01\,\text{A})}{0.01\,\text{m}} = 10\,\text{A.m}^{-1}$$

If the inductance *B* must be at least 0.2 T, then the permeability of the core material must be:

$$\mu = \frac{B}{H} = \frac{0.2}{10} = 0.02\ \text{T.m.A}^{-1}$$

The relative permeability of the core material must be at least:

$$\mu_r = \frac{\mu}{\mu_0} = \frac{0.02}{4\pi \times 10^{-7}} = 15\,915$$

If we examine the electrical magnetic materials listed in Table 19.3. we find that 45 Permalloy has a maximum relative permeability of 25 000 and might be a good selection for the core material.

19.5 Domain Structure and the Hysteresis Loop

Ferromagnetic materials have their powerful influence on magnetisation because of the positive interaction between the dipoles of neighbouring atoms. Within the grain structure of a ferromagnetic material, a substructure composed of magnetic domains is produced, even in the absence of an external field. Domains are regions in the material in which all of the dipoles are aligned. In a material that has never been exposed to a magnetic field, the individual domains have a random orientation. The net magnetisation in the material as a whole is zero.

Boundaries, called Bloch walls, separate the individual domains. The Bloch walls are narrow zones in which the direction of the magnetic moment gradually and continuously changes from that of one domain to that of the next (Figure 19.5). The domains are typically very small, about 0.05 mm or less, while the Bloch walls are about 100 nm thick.

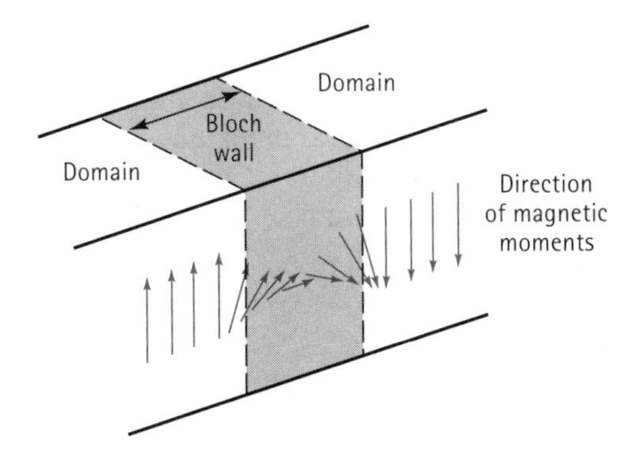

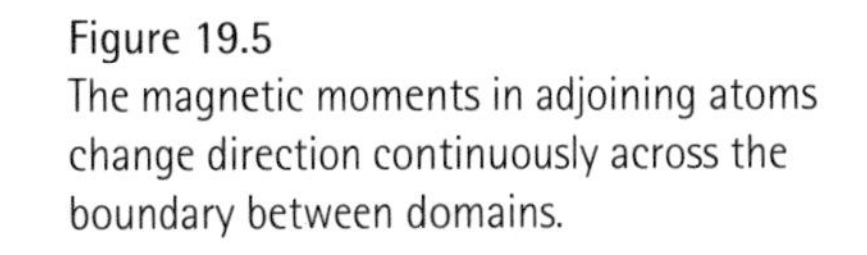

Figure 19.5
The magnetic moments in adjoining atoms change direction continuously across the boundary between domains.

Movement of Domains in a Magnetic Field When a magnetic field is imposed on the material, domains that are nearly lined up with the field grow at the expense of unaligned domains. In order for the domains to grow, the Bloch walls must move; the field provides the force required for this movement. Initially the domains grow with difficulty, and relatively large increases in the field are required to produce even a little magnetisation. This condition is indicated in Figure 19.6 by a shallow slope, which is the initial permeability of the material, μ_i. As the field increases in strength, favourably oriented domains grow more easily, with permeability increasing as well. A maximum permeability, μ_{max}, can be calculated, as shown in the figure. Eventually, the unfavourably oriented domains disappear and rotation completes the alignment of the domains with the field. The saturation magnetisation, produced when all of the domains are properly oriented, is the greatest amount of magnetisation that the material can obtain.

Effect of Removing the Field When the field is removed, the resistance offered by the domain walls prevents regrowth of the domains into random orientations. As a result, many of the domains remain oriented near the direction of the original field and a residual magnetisation or inductance, known as the remanence, (B_r), is present in the material. The material acts as a permanent magnet. Figure 19.7 shows this effect in the magnetisation-field curve.

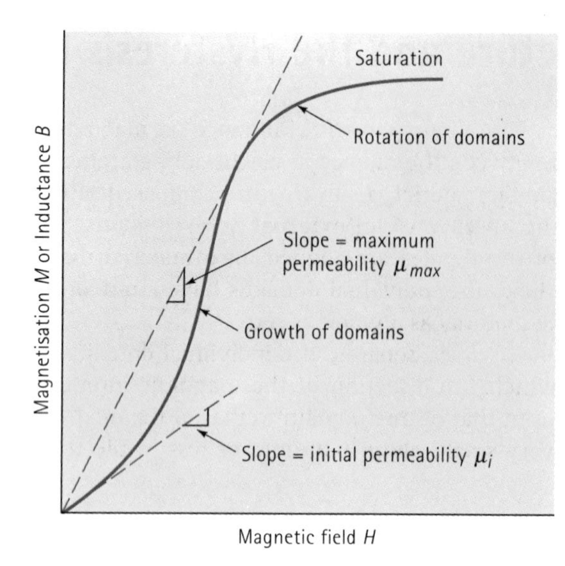

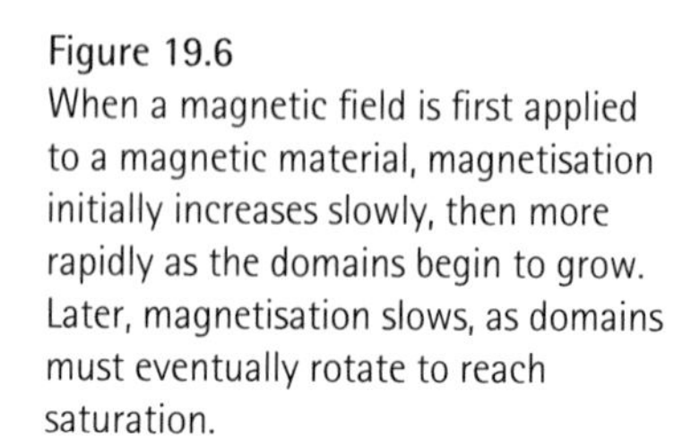
Figure 19.6
When a magnetic field is first applied to a magnetic material, magnetisation initially increases slowly, then more rapidly as the domains begin to grow. Later, magnetisation slows, as domains must eventually rotate to reach saturation.

Effect of an Alternating Field If we now apply a field in the reverse direction, the domains grow with an alignment in the opposite direction. A coercive field H_c (or coercivity) is required to force the domains to be randomly oriented and cancel one another's effect. Further increases in the strength of the field eventually align the domains to saturation in the opposite direction.

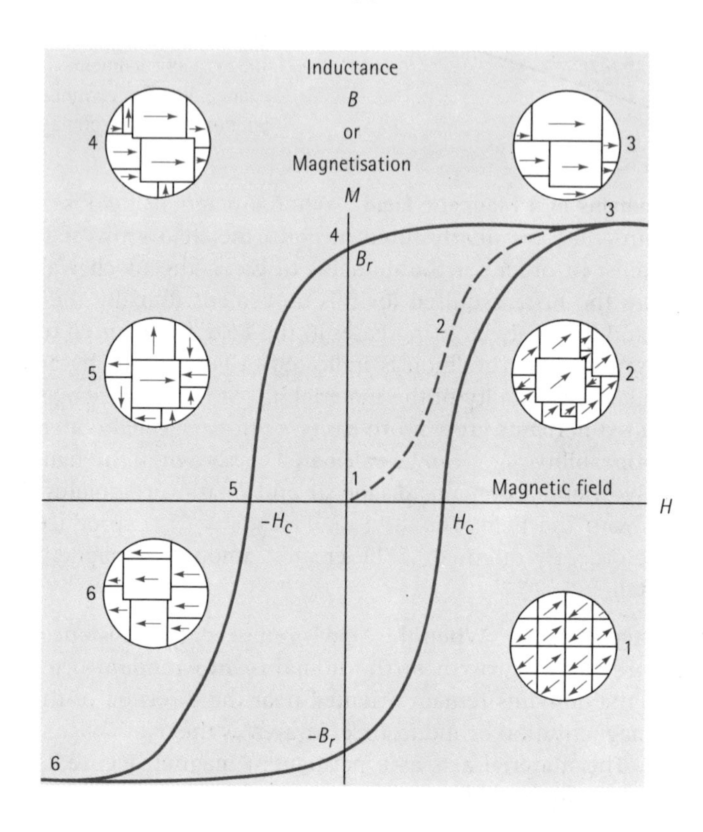

Figure 19.7
The ferromagnetic hysteresis loop, showing the effect of the magnetic field on inductance or magnetisation. The dipole alignment leads to saturation magnetisation (point 3), a remanence (point 4), and a coercive field (point 5).

As the field continually alternates, the magnetisation versus field relationship traces out a hysteresis loop. The area contained within the hysteresis loop is related to the energy consumed during one cycle of the alternating field.

19.6 Application of the Magnetisation–Field Curve

The behaviour of a material in a magnetic field is related to the size and shape of the hysteresis loop (Figure 19.8). Let's look at three applications for magnetic materials.

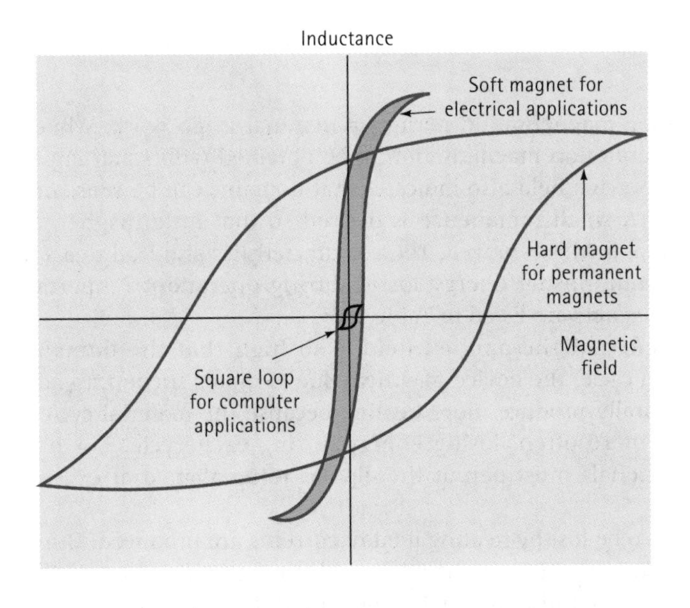

Figure 19.8
Comparison of the hysteresis loops for three applications of ferromagnetic materials: electrical applications, computer applications, and permanent magnets.

Magnetic Materials for Electrical Applications Ferromagnetic materials are used to enhance the magnetic field produced when an electric current is passed through the material. The magnetic field is then expected to do work. Applications include cores for electromagnets, electric motors, transformers, generators, and other electrical equipment. Because these devices utilise an alternating field, the core material is continually cycled through the hysteresis loop. Electrical magnetic materials, often called soft magnets, have several characteristics:

1. High saturation magnetisation.
2. High permeability.
3. Small coercive field.
4. Small remanence.
5. Small hysteresis loop.
6. Rapid response to high-frequency magnetic fields.
7. High electrical resistivity.

Table 19.3 Properties of selected soft, or electrical, magnetic materials.

Material	Maximum Relative Permeability	Saturation Inductance (tesla)	Coercive Field ($A.m^{-1}$)
99.95% iron	5000	2.14	71.60
Fe-3% Si (oriented)	50000	2.01	7.16
Fe-3% Si (not oriented)	8000	2.01	55.70
45 Permalloy (55% Fe-45% Ni)	25000	1.60	19.89
Supermalloy (79% Ni-16% Fe-5% Mo)	800000	0.80	0.48
A6 Ferroxcube (Mn, Zn)Fe_2O_4		0.40	
B2 Ferroxcube (Ni, Zn)Fe_2O_4		0.30	

High saturation magnetisation permits a material to do work, while high permeability permits saturation magnetisation to be obtained with small imposed magnetic fields. A small coercive field also indicates that domains can be reoriented with small magnetic fields. A small remanence is desired so that little magnetisation remains when the external field is removed. These characteristics also lead to a small hysteresis loop, therefore minimising energy losses during operation. Properties for several important soft magnets are listed in Table 19.3.

If the frequency of the applied field is so high that the domains cannot be realigned in each cycle, the device may heat due to dipole friction. In addition, higher frequencies naturally produce more heating because the material cycles through the hysteresis loop more often, losing energy during each cycle. For high-frequency applications, materials must permit the dipoles to be aligned at exceptionally rapid rates.

Energy can also be lost by heating if eddy currents are produced. During operation, electrical currents can be introduced into the magnetic material. These currents produce power losses and joule, or I^2R, heating. Eddy current losses are particularly severe when the material operates at high frequencies. If the electrical resistivity is high, eddy current losses can be held to a minimum. Soft magnets produced from ceramic materials have a high resistivity and therefore are less likely to heat than metallic magnets.

Magnetic Materials for Computer Memories Magnetic materials are used to store bits of information in computers. Memory is stored by magnetising the material in a certain direction. For example, if the 'north' pole is up, the bit of information stored is 1. If the 'north' pole is down, then a 0 is stored.

For this application, materials with a square hysteresis loop, a low remanence, a low saturation magnetisation, and a low coercive field are preferable. Ferrites containing manganese, magnesium, or cobalt may satisfy these requirements. The square loop ensures that a bit of information placed in the material by a field remains stored; a steep and abrupt change in magnetisation is required to remove the information from storage in the ferromagnet. Furthermore, the magnetisation is produced by small external fields, so the coercive field, saturation magnetisation, and remanence should be low.

Table 19.4 Selected properties of hard, or permanent, magnetic materials.

	Remanence (tesla)	Coercive Field ($A.m^{-1}$)	$(BH)_{max.}$ ($T.A.m^{-1}$)
Steel (0.9% C, 1.0% Mn)	1.00	4000	1600
Alnico 1 (21% Ni, 12% Al, 5% Co, bal Fe)	0.71	35000	11100
Alnico 5 (24% Co, 14% Ni, 8% Al, 3% Cu, bal Fe)	1.31	50900	47700
Alnico 12 (35% Co, 18% Ni, 8% Ti, 6% Al, bal Fe)	0.58	75600	12700
Cunife (60% Cu, 20% Fe, 20% Ni)	0.54	43800	11900
Co_5Sm	0.95	756000	200000
$BaO \cdot 6Fe_2O_3$	0.40	191000	20000
$SrO \cdot 6Fe_2O_3$	0.34	263000	29000
Neodymium-iron-boron ($Nd_2Fe_{12}B$)	1.20	875000	360000

Magnetic Materials for Permanent Magnets Finally, magnetic materials are used to make strong permanent magnets (Table 19.4). Strong permanent magnets, often called hard magnets, require the following:

1. High remanence (stable domains).
2. High permeability.
3. High coercive field.
4. Large hysteresis loop.
5. High power (or *BH* product).

The power of the magnet is related to the size of the hysteresis loop, or the maximum product of *B* and *H*. The area of the largest rectangle that can be drawn in the second or fourth quadrants of the *B-H* curve is related to the energy required to demagnetise the magnet (Figure 19.9). For the product to be large, both the remanence and the coercive field should be large.

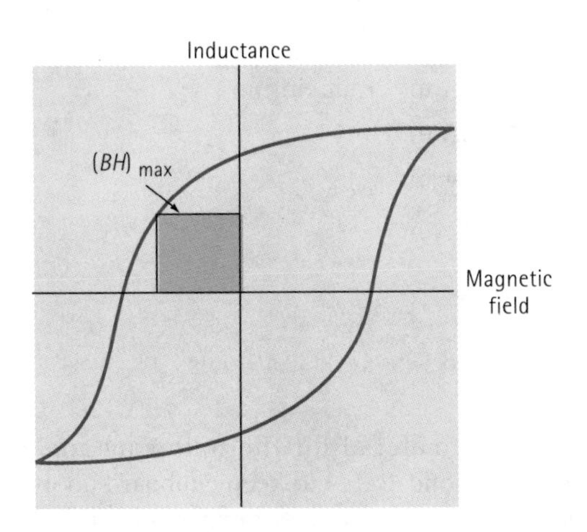

Figure 19.9
The largest rectangle drawn in the second or fourth quadrant of the *B-H* curve gives the maximum *BH* product. $(BH)_{max.}$ is related to the power, or energy, required to demagnetise the permanent magnet.

EXAMPLE 19.3

Determine the power, or *BH* product, for the magnetic material whose properties are shown in Figure 19.10.

SOLUTION

Several rectangles have been drawn in the fourth quadrant of the *B-H* curve. The *BH* product in each is:

$$BH_1 = (1.2)(22\,300) = 26\,760 \text{ T.A.m}^{-1}$$

$$BH_2 = (1.1)(28\,600) = 31\,460 \text{ T.A.m}^{-1}$$

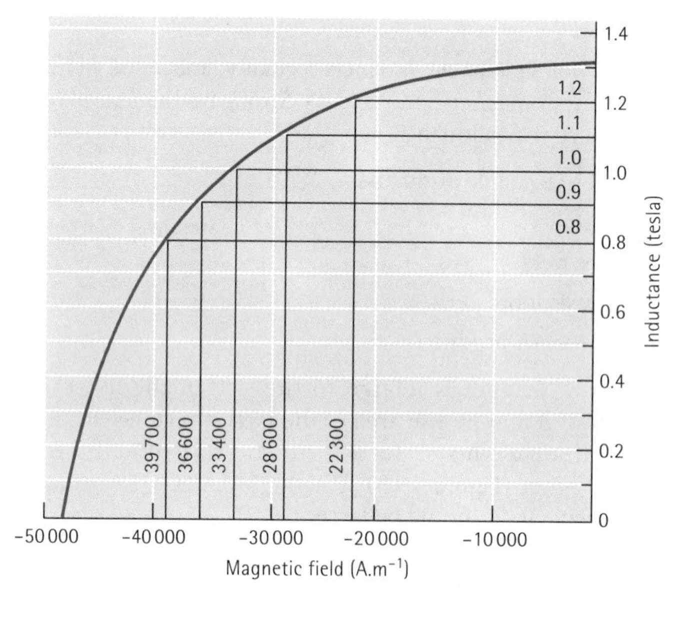

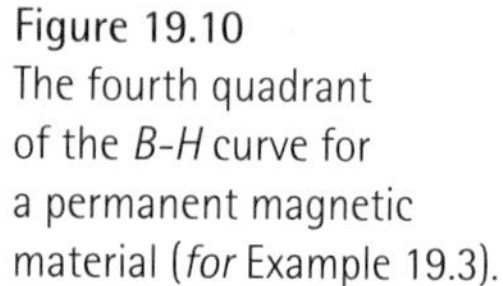

Figure 19.10
The fourth quadrant of the *B-H* curve for a permanent magnetic material (*for* Example 19.3).

$$BH_3 = (1.0)(33\,400) = 33\,400 \text{ T.A.m}^{-1} = \text{maximum}$$

$$BH_4 = (0.9)(36\,600) = 32\,940 \text{ T.A.m}^{-1}$$

$$BH_5 = (0.8)(39\,700) = 31\,760 \text{ T.A.m}^{-1}$$

Thus, the power is about 33 400 T.A.m^{-1}

EXAMPLE 19.4 Selection of Magnetic Materials

Select an appropriate magnetic material for the following applications: a high-electrical-efficiency motor; a magnetic device to keep cupboard doors closed; a magnet used in a ammeter or voltmeter; and magnetic resonance imaging.

SOLUTION

High-electrical-efficiency motor: To minimise hysteresis losses, we might use an oriented silicon iron, taking advantage of its anisotropic behaviour and its small hysteresis loop. Since the iron-silicon alloy is electrically conductive, we would produce a laminated structure, with thin sheets of the silicon iron sandwiched between a nonconducting dielectric material. Sheets thinner than about 0.5 mm might be recommended.

Magnet for cupboard doors: The magnetic latches used to fasten cupboard doors must be permanent magnets; however, low cost is a more important design feature than high power. An inexpensive ferritic steel or a low-cost $BaO \cdot 6Fe_2O_3$ ferrite would be recommended.

Magnets for an ammeter or voltmeter: For these applications, Alnico alloys are particularly effective. We find that these alloys are among the least sensitive to changes in temperature, assuring accurate current or voltage readings over a range of temperatures.

Magnetic resonance imaging: One of the applications for MRI is in medical diagnostics. In this case, we want a very powerful permanent magnet. A $Nd_2Fe_{12}B$ magnetic material, which has an exceptionally high *BH* product, might be recommended for this application.

19.7 The Curie Temperature

When the temperature of a ferromagnetic material is increased, the added thermal energy increases the mobility of the domains, making it easier for them to become aligned but also preventing them from remaining aligned when the field is removed. Consequently, saturation magnetisation, remanence, and the coercive field are all reduced at high temperatures (Figure 19.11). If the temperature exceeds the Curie temperature, ferromagnetic behaviour is no longer observed. The Curie temperature (Table 19.5), which depends on the material, can be changed by alloying elements.

The dipoles can be still be aligned in a magnetic field above the Curie temperature, but they become randomly aligned when the field is removed. Above the Curie temperature, the material displays paramagnetic behaviour.

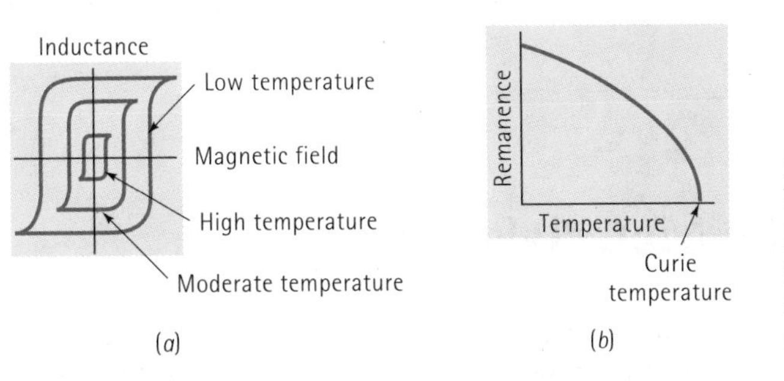

Figure 19.11
The effect of temperature on (*a*) the hysteresis loop and (*b*) the remanence. Ferromagnetic behaviour disappears above the Curie temperature.

Table 19.5 Curie temperatures for selected materials.

Material	Curie Temperature (°C)
Gadolinium	16
$Nd_2Fe_{12}B$	310
Nickel	358
$BaO \cdot 6Fe_2O_3$	450
Co_5Sm	725
Iron	770
Alnico 1	780
Cunico	855
Alnico 5	900
Cobalt	1131

EXAMPLE 19.5 Materials Selection for a High-Temperature Magnet

Select a permanent magnet for an application in an aerospace vehicle that must re-enter Earth's atmosphere. During re-entry, the magnet may be exposed to magnetic fields as high as 50 000 $A.m^{-1}$ and may briefly reach temperatures as high as 500°C. We want the material to have the highest power possible and to maintain its magnetisation after re-entry.

SOLUTION

It is first necessary to select potential materials having sufficient coercive field H_c and Curie temperature that re-entry will not demagnetise them. From Table 19.5, we can eliminate materials such as gadolinium, nickel, $Nd_2Fe_{12}B$, and the ceramic ferrites, since their Curie temperatures are below 500°C. From Table 19.4, other materials, such as steel, Cunife, and Alnico 1, can be eliminated because their coercive fields are below 50 000 $A.m^{-1}$.

Of the permanent magnetic materials remaining in Table 19.4, Alnico 12 has the lowest power and can be eliminated. Thus, our choice is between Alnico 5 and Co_5Sm. The Co_5Sm has four times the power of the Alnico 5 and, based on performance, might be our best choice. Some of the benefits of Co_5Sm, however, are offset by its higher cost.

19.8 Magnetic Materials

Let's look at typical metallic alloys and ceramic materials used in magnetic applications and discuss how their properties and behaviour can be enhanced.

Magnetic Metals Pure iron, nickel, and cobalt are not usually used for electrical applications because they have electrical conductivities and relatively large hysteresis loops, leading to excessive power loss. Furthermore, they are relatively poor permanent magnets; the domains are easily reoriented and both the remanence and the *BH* product are small compared with those of more complex alloys. Some change in the magnetic properties is obtained by introducing defects into the structure. Dislocations, grain boundaries, boundaries between multiple phases, and point defects help pin the domain boundaries, therefore keeping the domains aligned when the original magnetising field is removed.

Iron-Nickel Alloys Some iron-nickel alloys, such as Permalloy, have high permeabilities, making them useful as soft magnets. One example of an application for these magnets is the 'head' that stores or reads information on a computer disk (Figure 19.12). As the disk rotates beneath the head, a current produces a magnetic field in the head. The magnetic field in the head, in turn, magnetises a portion of the disk. The direction of the field produced in the head determines the orientation of the magnetic particles embedded in the disk and, consequently, stores information. The information can be retrieved by again spinning the disk beneath the head. The magnetised region in the disk induces a current in the head; the direction of the current depends on the direction of the magnetic field in the disk.

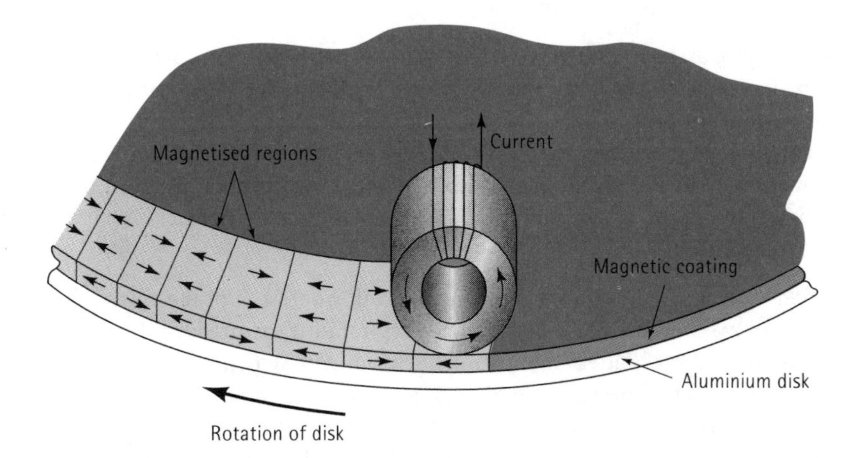

Figure 19.12
Information can be stored or retrieved from a magnetic disk by use of an electromagnetic head. A current in the head magnetises domains in the disk during storage; the domains in the disk induce a current in the head during retrieval.

Silicon Iron Introduction of 3% to 5% Si into iron produces an alloy that, after proper processing, is useful in electrical applications such as motors and generators. We take advantage of the anisotropic magnetic behaviour of silicon iron to obtain the best performance (Figure 19.13). As a result of rolling and subsequent annealing, a sheet texture is formed in which the <100> directions in each grain are aligned. Because the silicon iron is most easily magnetised in <100> directions, the field required to give saturation magnetisation is very small, and both a small hysteresis loop and a small remanence are observed.

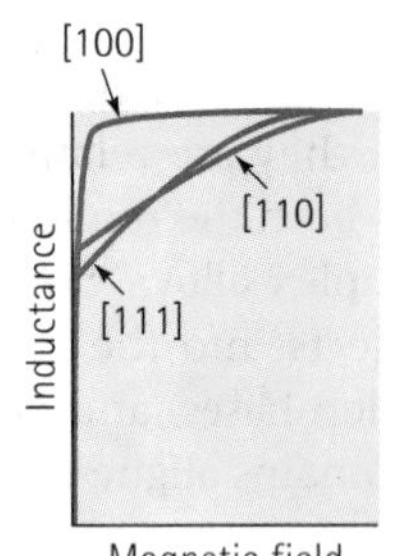

Figure 19.13
The initial magnetisation curve for silicon iron is highly anisotropic; magnetisation is easiest when the <100> directions are aligned with the field.

Composite Magnets Composite materials are used to reduce eddy current losses. Thin sheets of silicon iron are laminated with sheets of a dielectric material. The laminated layers are then built up to the desired overall thickness. The laminant increases the resistivity of the composite magnets and makes them successful at low and intermediate frequencies.

At very high frequencies, losses are even more significant because the domains do not have time to realign. In this case, a composite material containing domain-sized magnetic particles in a polymer matrix may be used. The particles, or domains, rotate easily in the soft polymer, while eddy current losses are minimised because of the high resistivity of the polymer.

Metallic glasses Amorphous metallic glasses, often complex iron-boron alloys, are produced by employing extraordinarily high cooling rates during solidification (rapid solidification processing). The metallic glasses are produced in the form of thin tapes, which are stacked together to produce larger materials. These materials behave as soft magnets with a high magnetic permeability; the absence of grain boundaries permits easy movement of the domains, while a high electrical resistivity minimises eddy current losses.

Magnetic Tape Magnetic materials for information storage must have a square loop and a low coercive field, permitting very rapid transmission of information. Magnetic tape for audio or video applications is produced by evaporating, sputtering, or plating particles of a magnetic material such as Fe_2O_3 onto a polyester tape.

Both floppy disks and hard disks for computer data storage are produced in a similar manner. In a hard disk, magnetic particles are embedded in a polymer film on a flat aluminium substrate. Because of the polymer matrix and the small particles, the domains can rotate quickly in response to a magnetic field.

Complex Metallic Alloys for Permanent Magnets Improved permanent magnets are produced by making the grain size so small that only one domain is present in each grain. Now the boundaries between domains are grain boundaries rather then Bloch walls. The domains can change their orientation only by rotating, which requires greater energy than domain growth.

Two techniques are used to produce these magnetic materials: phase transformations and powder metallurgy. Alnico, one of the most common of the complex metallic alloys, has a single-phase BCC structure at high temperatures. But when Alnico slowly cools below 800°C, a second BCC phase rich in iron and cobalt precipitates. This second phase is so fine that each precipitate particle is a single domain, producing a very high remanence, coercive field, and power. Often the alloys are

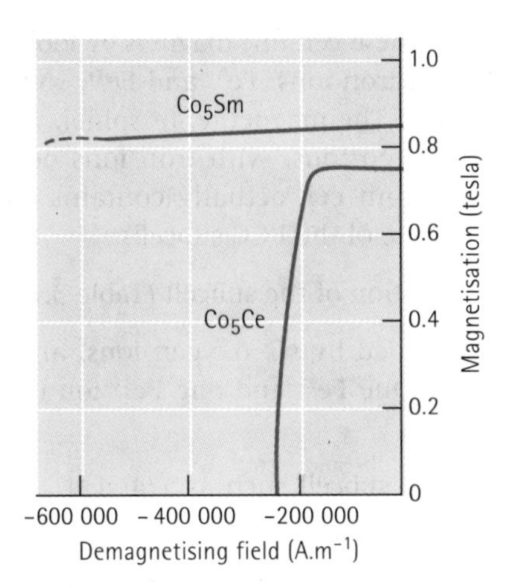

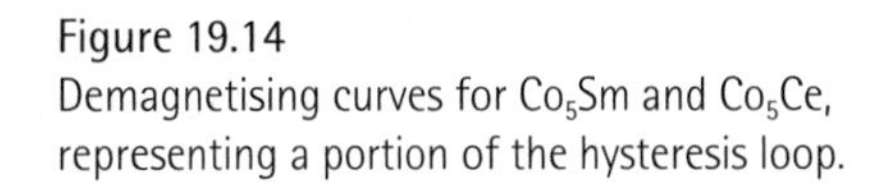

Figure 19.14
Demagnetising curves for Co_5Sm and Co_5Ce, representing a portion of the hysteresis loop.

permitted to cool and transform while in a magnetic field to align the domains as they form.

A second technique – powder metallurgy – is used for a group of rare earth metal alloys, including samarium-cobalt. A composition giving Co_5Sm, an intermetallic compound, has a high *BH* product (Figure 19.14) due to unpaired magnetic spins in the 4*f* electrons of samarium. The brittle intermetallic is crushed and ground to produce a fine powder in which each particle is a domain. The powder is then compacted while in an imposed magnetic field to align the powder domains. Careful sintering to avoid growth of the particles produces a solid powder metallurgy magnet. Another rare earth magnet based on neodymium, iron, and boron has a *BH* product of 360 000 T.A.m^{-1}. In these materials, a fine-grained intermetallic compound, $Nd_2Fe_{14}B$, provides the domains, and a fine HfB_2 precipitate prevents movement of the domain walls.

Ferrimagnetic Ceramic Materials Common magnetic ceramics are the ferrites, which have a spinel crystal structure [Figure 19.15 and Figure 14.1(c)]. Each metallic ion in the crystal structure behaves as a dipole. Although the dipole moments of each type of ion may oppose one another, the strengths of the dipoles are different, a net magnetisation develops, and ferrimagnetic behaviour is observed.

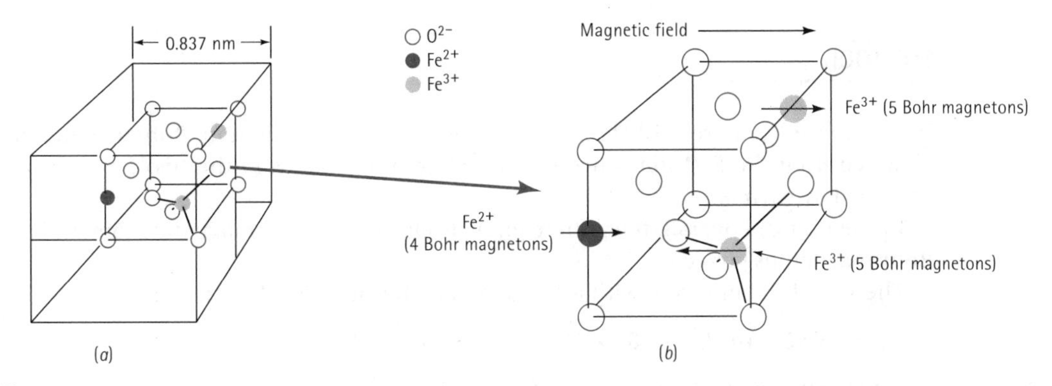

Figure 19.15
(*a*) The structure of magnetite, Fe_3O_4. (*b*) The subcell of magnetite. The ions in the octahedral sites line up with the magnetic field, but ions in tetrahedral sites oppose the field. A net magnetic moment is produced by this ionic arrangement.

We can understand the behaviour of these ceramic magnets by looking at magnetite, Fe_3O_4. Magnetite contains two different iron ions, Fe^{2+} and Fe^{3+}, so we could rewrite the formula for magnetite as $Fe^{2+}Fe_2^{3+}O_4^{2-}$. The magnetite, or spinel, crystal structure is based on an FCC arrangement of oxygen ions, with iron ions occupying selected interstitial sites. Although the spinel unit cell actually contains eight of the FCC arrangements, we need examine only one of the FCC subcells:

1. Four oxygen ions are in the FCC position of the subcell (Table 3.2).
2. Octahedral sites, which are surrounded by six oxygen ions, are present at each edge and the centre of the subcell. One Fe^{2+} and one Fe^{3+} ion occupy octahedral sites.
3. Tetrahedral sites have indices in the subcell such as 1/4, 1/4, 1/4. One Fe^{3+} ion occupies one of the tetrahedral sites.
4. When Fe^{2+} ions form, the two 4*s* electrons of iron are removed, but all of the 3*d* electrons remain. Because there are four unpaired electrons in the 3*d* level of iron, the magnetic strength of the Fe^{2+} dipole is four Bohr magnetons. However, when Fe^{3+} forms, both 4*s* electrons and one of the 3*d* electrons are removed. The Fe^{3+} ion has five unpaired electrons in the 3*d* level and, thus, has a strength of five Bohr magnetons.
5. The ions in the tetrahedral sites of the magnetite line up so that their magnetic moments oppose the applied magnetic field, but the ions in the octahedral sites reinforce the field [Figure 19.15(b)]. Consequently, the Fe^{3+} ion in the tetrahedral sites neutralises the Fe^{3+} ion in the octahedral site (the Fe^{3+} ions have antiferromagnetic behaviour). However, the Fe^{2+} ion in the octahedral site is not opposed by any other ion, and it therefore reinforces the magnetic field. In a reversing magnetic field, magnetite displays a hysteresis loop.

EXAMPLE 19.6

Calculate the total magnetic moment per cubic millimetre in magnetite.

SOLUTION

In the subcell [Figure 15(b)], the total magnetic moment is four Bohr magnetons, obtained from the Fe^{2+} ion, since the magnetic moments from the two Fe^{3+} ions are cancelled by each other.

In the unit cell overall, there are eight subcells, so the total magnetic moment is 32 Bohr magnetons per cell.

The size of the unit cell, with a lattice parameter of 8.37×10^{-10} m, is:

$$V_{cell} = (8.37 \times 10^{-10})^3 = 5.86 \times 10^{-28} \text{ m}^3 = 5.86 \times 10^{-19} \text{ mm}^3$$

The magnetic moment per cubic millimetre is:

$$\text{Total moment} = \frac{32 \text{ Bohr magnetons / cell}}{5.86 \times 10^{-19} \text{ mm}^3 \text{ / cell}} = 5.46 \times 10^{19} \text{ magnetons / mm}^3$$

$$= (5.46 \times 10^{19})(9.27 \times 10^{-24} \text{ A} \cdot \text{m}^2 \text{ / magneton})$$

$$= 5.1 \times 10^{-4} \text{ A} \cdot \text{m}^2 \text{ per mm}^3$$

$$(= 5.1 \times 10^5 \text{ A} \cdot \text{m}^2 \text{ per m}^3, \text{ or } 5.1 \times 10^5 \text{ A} \cdot \text{m}^{-1})$$

When ions are substituted for Fe^{2+} ions in the spinel structure, the magnetic behaviour may be changed. Ions that may not produce ferromagnetism in a pure metal may contribute to ferrimagnetism in the spinels, as shown by the magnetic moments in Table 19.6. Soft electrical magnets are obtained when the Fe^{2+} ion is replaced by various mixtures of manganese, zinc, nickel, and copper. The nickel and manganese ions have magnetic moments that partly cancel the effect of the two iron ions, but a net ferrimagnetic behaviour, with a small hysteresis loop, is obtained. The high electrical resistivity of these ceramic compounds helps minimise eddy currents and permits the materials to operate at high frequencies. Ferrites used in computer applications may contain additions of manganese, magnesium, or cobalt to produce a square loop hysteresis behaviour.

Table 19.6 Magnetic moments for ions in the spinel structure.

Ion	Bohr Magnetons
Fe^{3+}	5
Mn^{2+}	5
Fe^{2+}	4
Co^{2+}	3
Ni^{2+}	2
Cu^{2+}	1
Zn^{2+}	0

Another group of soft ceramic magnets is based on garnets, which include yttria iron garnet, $Y_3Fe_5O_{12}$ (YIG). These complex oxides, which may be modified by substituting aluminium or chromium for iron or by replacing yttrium with lanthanum or praesydium, behave much like the ferrites. Another garnet, based on gadolinium and gallium, can be produced in the form of a thin film. Tiny magnetic domains can be produced in the garnet film; these domains, or *magnetic bubbles*, can then serve as storage units for computers. Once magnetised, the domains do not lose their memory in case of a sudden power loss.

Hard ceramic magnets selected as permanent magnets include another complex oxide family, the hexagonal ferrites. The hexagonal ferrites include $SrFe_{12}O_{19}$, $BaFe_{12}O_{19}$, and $PbFe_{12}O_{19}$.

EXAMPLE 19.7 Materials Selection for a Ceramic Magnet

Design a cubic ferrite magnet that has a total magnetic moment per cubic metre of 5.5×10^5 A.m^{-1}.

SOLUTION

We found in Example 19.6 that the magnetic moment per cubic metre for Fe_3O_4 is 5.1×10^5 A.m^{-1}. To obtain a higher saturation magnetisation, we must replace Fe^{2+} ions with ions having more Bohr magnetons per atom. One such possibility (Table 19.6) is Mn^{2+}, which has five Bohr magnetons.

Assuming that the addition of Mn ions does not appreciably affect the size of the unit cell, we find from Example 19.6 that:

$$V_{cell} = 5.86 \times 10^{-28} \text{ m}^3$$

Let x be the fraction of Mn^{2+} ions that have replaced the Fe^{2+} ions, which have now been reduced to $1 - x$. Then, the total magnetic moment is:

$$\text{Total moment} = \frac{(8 \text{ subcells})[(x)(5 \text{ magnetons}) + (1-x)(4 \text{ magnetons})](9.27 \times 10^{-24} \text{ A.m}^2)}{5.86 \times 10^{-28} \text{ m}^3}$$

$$= \frac{(8)(5x + 4 - 4x)(9.27 \times 10^{-24})}{5.86 \times 10^{-28}} = 5.5 \times 10^5$$

$$x = -4 + 4.346 = 0.346$$

Therefore we need to replace 34.6 at% of the Fe^{2+} ions with Mn^{2+} ions to obtain the desired magnetisation.

SUMMARY

- The magnetic properties of materials are related to the interaction of magnetic dipoles with a magnetic field. The magnetic dipoles originate with the electronic structure of the atom, causing several types of behaviour:
  - In diamagnetic materials, the magnetic dipoles weakly oppose the applied magnetic field.
  - In paramagnetic materials, the magnetic dipoles weakly reinforce the applied magnetic field, increasing the net magnetisation or inductance.
  - In ferromagnetic materials (such as iron, nickel, and cobalt), the magnetic dipoles strongly reinforce the applied magnetic field, producing large net magnetisation or inductance. Magnetisation may remain even after the magnetic field is removed. Increasing the temperature above the Curie temperature destroys the ferromagnetic behaviour.

- In ferrimagnetic ceramics, some magnetic dipoles reinforce the field, whereas others oppose the field. However, a net increase in magnetisation or inductance occurs. Ferrimagnetic materials also display a Curie temperature.

- The structure of ferromagnetic and ferrimagnetic materials includes domains, within which all of the magnetic dipoles are aligned. When a magnetic field is applied, the dipoles become aligned with the field, increasing the magnetisation to its maximum, or saturation, value. When the field is removed, some alignment of the domains may remain:
  - For soft, or electrical, magnetic materials, little remanence remains, only a small coercive field is required to remove any alignment of the domains, and little energy is consumed in reorienting the domains when an alternating magnetic field is applied.
  - For hard, or permanent, magnetic materials, the domains remain almost completely aligned when the field is removed, large coercive fields are required to randomise the domains, and a large hysteresis loop is observed. This condition provides the material with a high power.

GLOSSARY

Antiferromagnetism
Opposition of adjacent magnetic dipoles, causing zero net magnetisation.

Bloch walls
The boundaries between magnetic domains.

Bohr magneton
The strength of a magnetic moment.

Curie temperature
The temperature above which ferromagnetic behaviour is lost.

Diamagnetism
The effect caused by the magnetic moment due to the orbiting electrons, which produces a slight opposition to the imposed magnetic field.

Domains
Small regions within a material in which all of the dipoles are aligned.

Ferrimagnetism
Magnetic behaviour obtained when two types of dipoles having different strengths oppose one another, but a net magnetisation remains.

Ferromagnetism
Alignment of domains so that a net magnetisation remains after the magnetic field is removed.

Hard magnet
Ferromagnetic material that has a large hysteresis loop and remanence.

Hysteresis loop
The loop traced out by magnetisation as the magnetic field is cycled.

Magnetic moment
The strength of the magnetic field associated with an electron.

Magnetic permeability
The ratio between the magnetic field and the inductance or magnetisation.

Magnetic susceptibility
The ratio between magnetisation and the applied field.

Magnetisation
The sum of all magnetic moments per unit volume.

Paramagnetism
The net magnetic moment caused by alignment of the electron spins when a magnetic field is applied.

Power
The strength of a permanent magnet as expressed by the maximum product of the inductance and magnetic field.

Remanence
The polarisation or magnetisation that remains in a material after it has been removed from the field due to permanent alignment of the dipoles.

Saturation magnetisation
When all of the dipoles have been aligned by the field, producing the maximum magnetisation.

Soft magnet
Ferromagnetic material that has a small hysteresis loop and little energy loss in an alternating field.

PROBLEMS

19.1 Calculate and compare the maximum magnetisation we would expect in iron, nickel, cobalt, and gadolinium. There are seven electrons in the 4*f* level of gadolinium.

19.2 An alloy of nickel and cobalt is to be produced to give a magnetisation of 2×10^6 A.m^{-1}. The crystal structure of the alloy is FCC with a lattice parameter of 0.3544 nm. Determine the atomic percent cobalt required, assuming no interaction between the nickel and cobalt.

19.3 Estimate the magnetisation that might be produced in an alloy containing nickel and 70 at% copper, assuming that no interaction occurs.

19.4 An Fe-80% Ni alloy has a maximum relative permeability of 300 000 when an inductance of 0.35 tesla is obtained. The alloy is placed in a 20-turn coil that is 20 mm in length. What current must flow through the conductor coil to obtain this field?

19.5 An Fe-49% Ni alloy has a maximum relative permeability of 64 000 when a magnetic field of 10 A.m^{-1} is applied. What inductance is obtained and what current is needed to obtain this inductance in a 200-turn, 30-mm long coil?

19.6 The following data describe the effect of the magnetic field on the inductance in a silicon steel. Calculate

(a) the initial permeability and

(b) the maximum permeability for the material.

H (A.m^{-1})	B (tesla)
0	0
20	0.08
40	0.30
60	0.65
80	0.85
100	0.95
150	1.10
250	1.25

19.7 A magnetic material has a coercive field of 167 A.m^{-1}, a saturation magnetisation of 0.616 tesla, and a residual inductance of 0.3 tesla. Sketch the hysteresis loop for the material.

19.8 A magnetic material has a coercive field of 10.74 A.m^{-1}, a saturation magnetisation of 2.158 tesla, and a remanence induction of 1.183 tesla. Sketch the hysteresis loop for the material.

19.9 Using Figure 19.16, determine the following properties of the magnetic material:

(a) remanence,

(b) saturation magnetisation,

(c) coercive field,

(d) initial permeability,

(e) maximum permeability,

(f) power (maximum BH product).

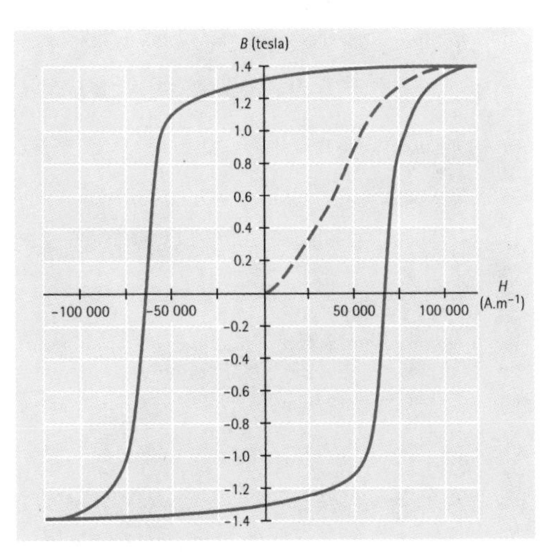

Figure 19.16
Hysteresis curve for a hard magnetic material (*for* Problem 19.9).

19.10 Using Figure 19.17, determine the following properties of the magnetic material:

(a) remanence,

(b) saturation magnetisation,

(c) coercive field,

(d) initial permeability,

(e) maximum permeability,

(f) power (maximum BH product).

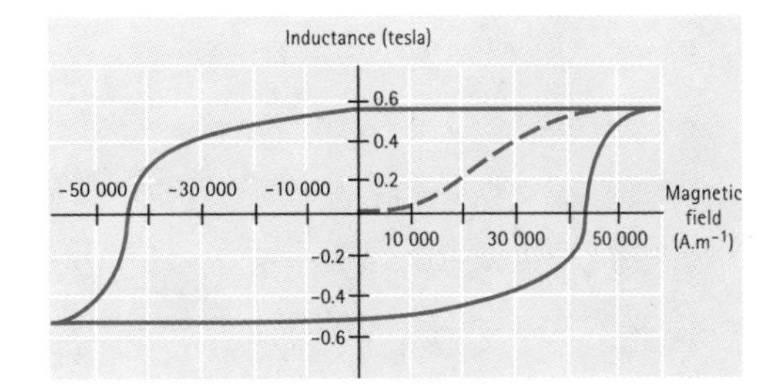

Figure 19.17
Hysteresis curve for a hard magnetic material (*for* Problem 19.10).

19.11 Estimate the power of the Co_5Ce material shown in Figure 19.14.

19.12 Why are eddy current losses important design factors in ferromagnetic materials but less important in ferrimagnetic materials?

19.13 What advantage does the Fe-3% Si material have compared with Supermalloy for use in electric motors?

19.14 The coercive field for pure iron is related to the grain size of the iron by the relationship $H_c = 1.83 + 4.14/\sqrt{A}$, where A is the area of the grain in two dimensions (mm^2) and H_c is in $A.m^{-1}$. If only the grain size influences the 99.95% iron given in Table 19.3, estimate the size of the grains in the material. What happens when the iron is annealed to increase the grain size?

19.15 Suppose we replace 10% of the Fe^{2+} ions in magnetite with Cu^{2+} ions. Determine the total magnetic moment per cubic metre.

19.16 Suppose that the total magnetic moment per cubic metre in a spinel structure in which Ni^{2+} ions have replaced a portion of the Fe^{2+} ion is 4.6×10^5 $A.m^{-1}$. Calculate the fraction of the Fe^{2+} ions that have been replaced and the wt% Ni present in the spinel.

Design Problems

19.17 Design a solenoid no longer than 10 mm that will produce an inductance of 0.3 tesla.

19.18 Design a permanent magnet that will have a remanence of at least 0.5 tesla, that will not be demagnetised if exposed to a temperature of 400°C or to a magnetic field of 80 000 $A.m^{-1}$, and that has good magnetic power.

19.19 Design a spinel structure that will produce a total magnetic moment per cubic metre of 5.6×10^5 $A.m^{-1}$.

19.20 Design a spinel structure that will produce a total magnetic moment per cubic metre of 4.1×10^5 $A.m^{-1}$.

CHAPTER 20

Optical Behaviour of Materials

20.1 Introduction

Optical properties are related to the interaction of a material with electromagnetic radiation in the form of waves or particles of energy called photons. The radiation may have characteristics that fall in the visible light spectrum, or it may be invisible to the human eye. In this chapter, we explore two avenues by which we can utilise the optical properties of materials: emission of photons from materials and interaction of photons with materials.

Several sources cause the emissions of photons having a certain frequency, wavelength, and energy. For example, gamma rays are produced by changes in the structure of the nucleus of the atom; X-rays, ultraviolet radiation, and the visible spectrum are produced by changes in the electronic structure of the atom. Microwaves and radio waves are low-energy, long-wavelength radiation caused by vibration of atoms or the crystal structure.

When photons interact with a material, a variety of optical effects are produced, including absorption, transmission, reflection, refraction, and electronic behaviour. Examining these phenomena enables us not only to better understand the behaviour of the materials but also to use them to produce aircraft that cannot be detected by radar; lasers for medical use, communications, or manufacturing; fibre optic devices; light emitting diodes; solar cells; analytical instruments for determining crystal structure of material composition; and many more critical devices.

20.2 The Electromagnetic Spectrum

Energy, or radiation in the form of waves or particles called photons, can be emitted from a material. The important characteristics of the photons – their energy E, wavelength λ, and frequency ν – are related by the equation

$$E = h\nu = \frac{hc}{\lambda} \tag{20.1}$$

where c is the speed of light (3×10^8 m.s^{-1}) and h is Planck's constant (6.62×10^{-34} J · s). Since there are 1.6×10^{-19} J per electron volt (eV), h also is given by 4.14×10^{-15} eV · s.

This equation permits us to consider the photon either as a particle of energy E or as a wave with a characteristic wavelength and frequency.

Earth is constantly bathed by a stream of photons produced by the sun, some of which lie in the visible spectrum. We can use these photons as the energy source to make solar cells function, providing us with the potential for a renewable energy source. However, we can also deliberately produce photons from a variety of sources within a material. Depending on the source of the photons, we can produce radiation over an enormous range of wavelengths; the entire spectrum of electromagnetic radiation is shown in Figure 20.1. Gamma and X-rays have a very short wavelength, or a high frequency, and possess very high energies; microwaves and radio waves possess very low energies; and visible light represents only a very narrow portion of the electromagnetic spectrum.

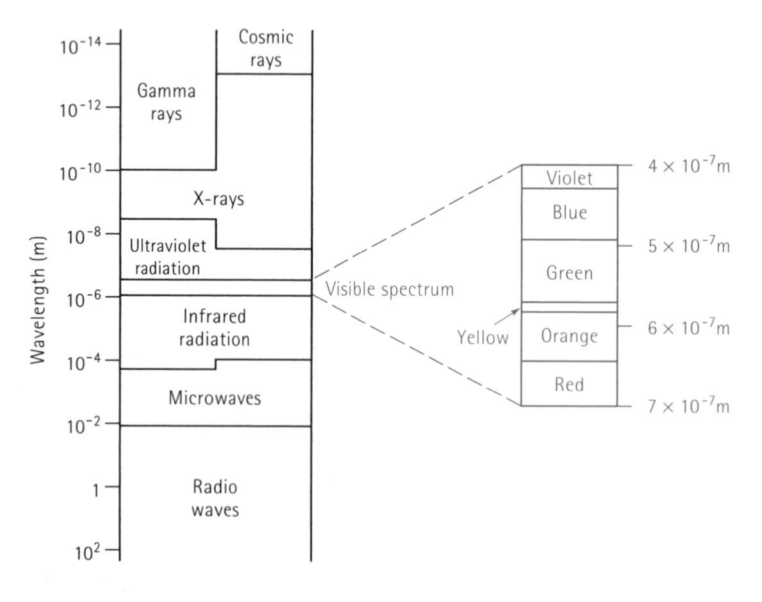

Figure 20.1
The electromagnetic spectrum of radiation.

20.3 Examples and Use of Emission Phenomena

Let's look at some particular examples of emission phenomena which, by themselves, provide some familiar and important functions.

Gamma Rays – Nuclear Interactions Gamma rays, which are very high-energy photons, are emitted during radioactive decay of unstable nuclei of certain atoms. The energy of the gamma rays therefore depends on the structure of the atom nucleus and varies for different materials. The gamma rays produced from a material have fixed wavelengths. When cobalt 60 decays, for example, gamma rays having energies of 1.87×10^{-13} J and 1.60×10^{-13} J (or wavelengths of 1.06×10^{-12} m and 0.93×10^{-12} m) are emitted. The gamma rays can be used as the radiation source to detect defects lying within a material.

X-rays – Inner Electron Shell Interactions X-rays, which have somewhat lower energy than gamma rays, are produced when electron in the inner shells of the atom are stimulated. The stimulus could be high-energy electrons or other x-rays. When stimulation occurs, X-rays of a wide range of energies are emitted. Both a continuous and a characteristic spectrum of X-ray are produced.

Suppose that a high-energy electron strikes a material. As the electron decelerates, energy is given up and emitted as photons. Each time the electron strikes an atom, more of its energy is given up. Each interaction, however, may be more or less severe, so the electron gives up a different fraction of its energy each time and produces photons of different wavelengths (Figure 20.2). A continuous spectrum is produced (the smooth portion of the curves in Figure 20.3). If the electron were to lose all of its energy in one impact, the minimum wavelength of the emitted photons would be equivalent to the original energy of the stimulus. The minimum wavelength of X-rays produced is called the short wavelength limit λ_{swl}. The short wavelength limit decreases, and the number and energy of the emitted photons increase, when the energy of the stimulus increases.

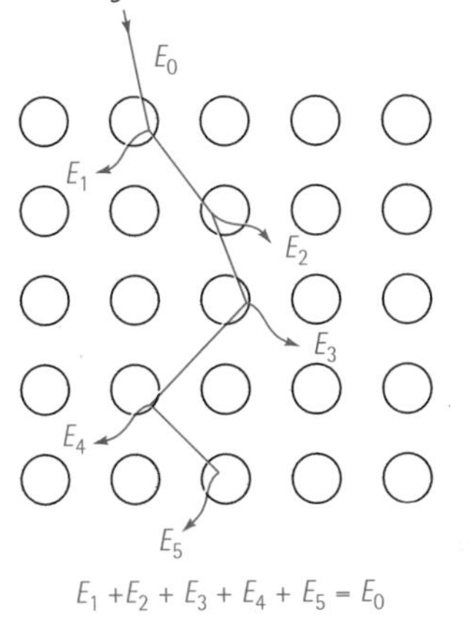

Figure 20.2
When an accelerated electron strikes and interacts with a material, its energy may be reduced in a series of steps. In the process, several photons of different energies E_1 to E_5 are emitted, each with a unique wavelength.

The incoming stimulus may have also sufficient energy to excite an electron from an inner energy level into an outer energy level. The excited electron is not stable and, to restore equilibrium, the empty inner level is filled by electrons from a higher level. This process leads to the emission of a characteristic spectrum of X-rays that is different for each type of atom.

The characteristic spectrum is produced because there are discrete energy differences between any two energy levels. When an electron drops from one level to a second level, a photon having that particular energy and wavelength is emitted. This effect is illustrated in Figure 20.4. We typically refer to the energy levels by the *K, L, M,...* designation, as described in Chapter 2. If an electron is exited from the *K* shell, electrons may fill that vacancy from any outer shell. Normally, electrons in the closest shells fill the vacancies. Thus, photons with energy $\Delta E = E_K - E_L$ (K_α X-rays) or $\Delta E = E_K - E_M$ (K_β X-rays) are emitted. When an electron from the *M* shell fills the *L* shell, a photon with energy $\Delta E = E_L - E_M$ (L_α X-rays) is emitted; it has a long wavelength, or low energy. Note that we need a more energetic stimulus to produce K_α X-rays than that required for L_α X-rays.

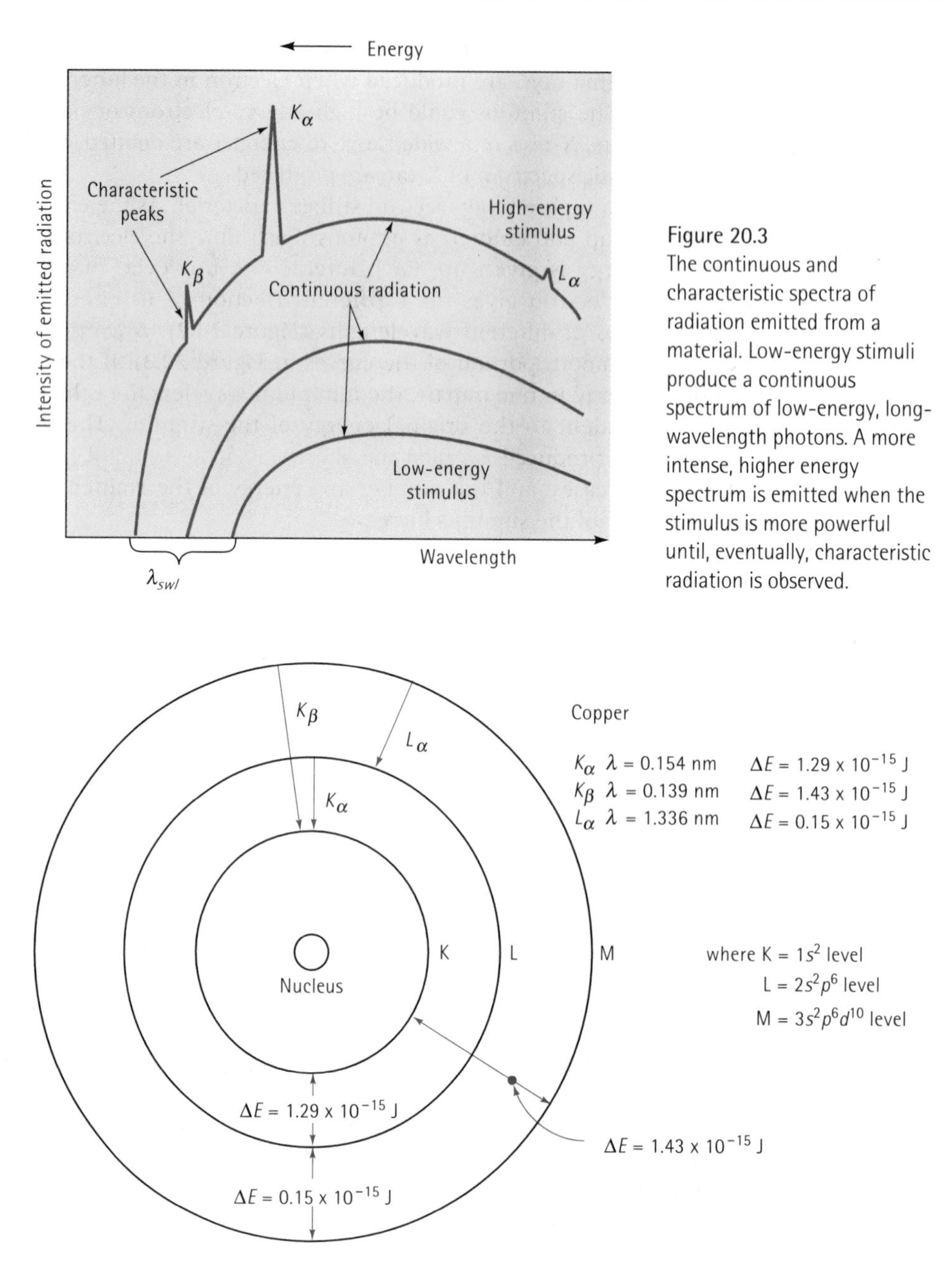

Figure 20.3
The continuous and characteristic spectra of radiation emitted from a material. Low-energy stimuli produce a continuous spectrum of low-energy, long-wavelength photons. A more intense, higher energy spectrum is emitted when the stimulus is more powerful until, eventually, characteristic radiation is observed.

Figure 20.4
Characteristic X-rays are produced when electrons change from one energy level to a lower energy level, as illustrated here for copper. The energy and wavelength of the X-rays are fixed by the energy differences between the energy levels.

As a consequence of the emission of photons having a characteristic wavelength, a series of peaks is superimposed on the continuous spectrum (Figure 20.3). The wavelengths at which these peaks occur are peculiar to the type of atom. Thus, each element produces a different characteristic spectrum, which serves as a 'fingerprint'

Table 20.1 Characteristic emission lines and absorption edges for selected elements.

Metal	K_α (nm)	K_β (nm)	L_α (nm)	Absorption Edge (nm)
Al	0.8337	0.7981	—	0.7951
Si	0.7125	0.6768	—	0.6745
S	0.5372	0.5032	—	0.5018
Cr	0.2291	0.2084	—	0.2070
Mn	0.2104	0.1910	—	0.1896
Fe	0.1937	0.1757	—	0.1743
Co	0.1790	0.1621	—	0.1608
Ni	0.1660	0.1500	—	0.1488
Cu	0.1542	0.1392	1.3357	0.1380
Mo	0.0711	0.0632	0.5724	0.0620
W	0.0211	0.0184	0.1476	0.0178

Note: 1nm = 10^{-9} m

for that type of atom. If we match the emitted characteristic wavelengths with those expected for various elements, the identity of the material can be determined. We can also measure the intensity of the characteristic peaks. By comparing measured intensities with standard intensities, we can estimate the percentage of each type of atom in the material and, hence, we can estimate the composition of the material. We can perform this test on large samples of the material using X-ray fluorescent analysis. Or, on a microscopic scale, we can use the scanning electron microscope (SEM) to permit us to identify individual phases or even inclusions in the microstructure.

Examples of a portion of the characteristic spectra for several elements are included in Table 20.1. The absorption edge in the table will be explained in a later section.

EXAMPLE 20.1

Suppose an electron accelerated at 5 000 V strikes a copper target. Will K_α, K_β, or L_α X-rays be emitted from the copper target?

SOLUTION

The electron must possess enough energy to excite an electron to a higher level, or its wavelength must be less than that corresponding to the energy difference between the shells:

$$E = (5000 \text{ eV})(1.6\times10^{-19} \text{ J/eV}) = 8\times10^{-16} \text{ J}$$

$$\lambda = \frac{hc}{E} = \frac{(6.62\times10^{-34})(3\times10^{8})}{8\times10^{-16}}$$

$$= 2.48\times10^{-10} \text{ m} = 0.248 \text{ nm}$$

For copper, K_α is 0.1542 nm, K_β is 0.1392 nm and L_α is 1.3357 nm. Therefore, the L_α peak may be produced but K_α and K_β will not.

EXAMPLE 20.2

The photograph in Figure 20.5 was obtained using a scanning electron microscope at a magnification of 1 000. The beam of electrons in the SEM was directed at the three different phases, creating X-rays and producing the characteristic peaks. From the energy spectra, determine the probable identity of each phase.

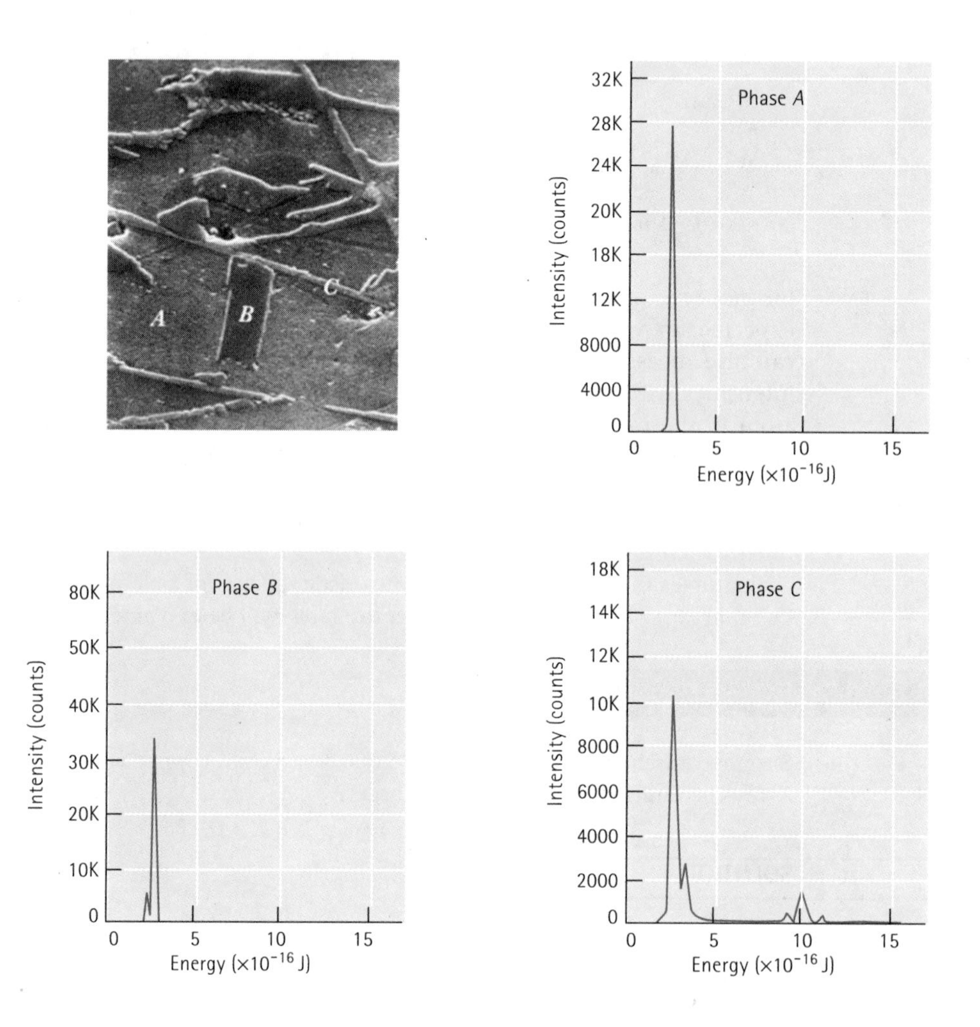

Figure 20.5
Scanning electron micrograph of a multiple-phase material. The energy distribution of emitted radiation from the three phases marked *A*, *B*, and *C* is shown. The identity of each phase is determined in Example 20.2.

SOLUTION

All three phases have an energy peak of about 2.4 × 10^{-16} J, which corresponds to a wavelength of:

$$\lambda = \frac{hc}{E} = \frac{(6.62\times10^{-34}\,\mathrm{J\cdot s})(3\times10^{8}\,\mathrm{m.s^{-1}})}{(2.4\times10^{-16}\,\mathrm{J})(10^{-9}\,\mathrm{m/nm})} = 0.8275\ \mathrm{nm}$$

In a similar manner, energies and wavelengths can be found for the other peaks. These wavelengths are compared with those in Table 20.1, and the identity of the elements in each phase can be found, as summarised in the table.

Phase	Peak Energy (J)	λ (nm)	λ (Table 20.1) (nm)	Element
A	2.4 × 10^{-16}	0.8275	0.8337	K_α Al
B	2.4 × 10^{-16}	0.8275	0.8337	K_α Al
	2.72 × 10^{-16}	0.730	0.7125	K_α Si
C	2.4 × 10^{-16}	0.8275	0.8337	K_α Al
	2.72 × 10^{-16}	0.730	0.7125	K_α Si
	9.28 × 10^{-16}	0.214	0.2104	K_α Mn
	1.024 × 10^{-15}	0.194	0.1937	K_α Fe
	1.136 × 10^{-15}	0.175	0.1757	K_β Fe

Thus, phase *A* appears to be an aluminium matrix, phase *B* appears to be a silicon needle (perhaps containing some aluminium), and phase *C* appears to be an Al-Si-Mn-Fe compound. Actually, this is an aluminium-silicon alloy. The stable phases are aluminium and silicon, with inclusions forming when manganese and iron are present as impurities.

Luminescence – Outer Electron Shell Interactions Whereas X-rays are produced by electron transitions in the inner energy levels of an atom, luminescence is the conversion of radiation or other forms of energy to visible light. Luminescence occurs when the incident radiation excites electrons from the valence band, through the energy gap, and into the conduction band. The excited electrons remain in the higher energy levels only briefly. When the electrons drop back to the valence band, photons are emitted. If the wavelength of those photons is in the visible light range, luminescence occurs.

Luminescence does not occur in metals. Electrons are merely excited into higher energy levels within the unfilled valance band and, when the excited electron returns to the lower energy level, the photon that is produced has a very small energy and a wavelength longer than that of visible light [Figure 20.6(a)].

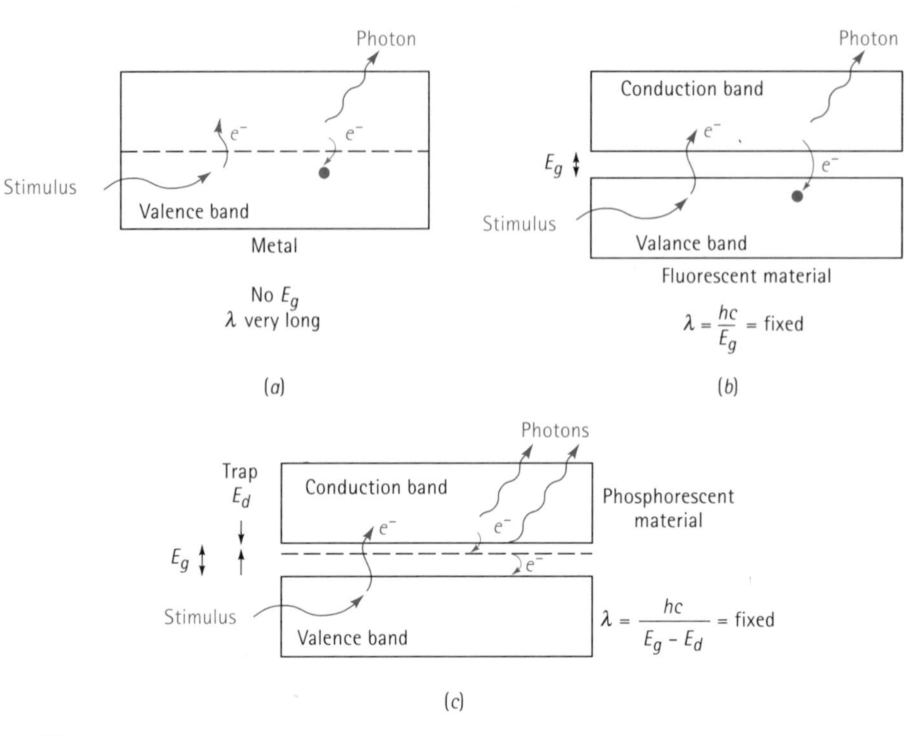

Figure 20.6
Luminescence occurs when photons have a wavelength in the visible spectrum. (*a*) In metals, there is no energy gap, so luminescence does not occur. (*b*) Fluorescence occurs when there is an energy gap. (*c*) Phosphorescence occurs when the photons are emitted over a period of time due to donor traps in the energy gap.

In certain ceramics and semiconductors, however, the energy gap between the valence and conduction bands is such that an electron dropping through this gap produces a photon in the visible range. Two different effects are observed in these luminescent materials: fluorescence and phosphorescence. In fluorescence, all of the excited electrons drop back to the valence band and the corresponding photons are emitted within a fraction of a second after the stimulus is removed [Figure 20.6(b)]. One wavelength, corresponding to the energy gap E_g, predominates. Fluorescent lamps, which are coated with a fluorescent material, behave in this manner. The lamp immediately ceases to emit light when the power is turned off.

In contrast, phosphorescent materials have impurities that introduce a donor level within the energy gap [Figure 20.6(c)]. The stimulated electrons first drop into the donor level and are trapped. The electrons must then escape the trap before returning to the valence band. There is a delay before the photons are emitted. When the source is removed, electrons in the traps gradually escape and emit light over some additional period of time. The intensity of the luminescence is given by

$$\ln\left(\frac{I}{I_0}\right) = -\frac{t}{\tau} \tag{20.2}$$

where τ is the relaxation time, a constant for the material. After time t following removal of the source, the intensity of the luminescence is reduced from I_0 to I.

Phosphorescent materials are very important in the operation of television screens. In this case, the relaxation time must not be too long or the images begin to overlap. In colour television, three types of phosphorescent materials are used; the energy gaps are engineered so that red, green, and blue colours are produced. Oscilloscope and radar screens rely on the same principle.

EXAMPLE 20.3 **Materials Selection for a Television Screen**

Select a material that will produce a blue image on a television screen.

SOLUTION

Photons having energies that correspond to the colour blue have wavelengths of about 4.5×10^{-5} cm (Figure 20.1). The energy of the emitted photons therefore is:

$$E = \frac{hc}{\lambda} = \frac{(6.62\times10^{-34}\text{ J.s})(3\times10^{8}\text{ m.s}^{-1})}{4.5\times10^{-7}\text{ m}}$$
$$= 4.422\times10^{-19}\text{ J}$$

Table 18.8 includes energy gaps for a variety of materials. None of the materials listed has an E_g of 4.422×10^{-19} J, but ZnS has an E_g of 5.671×10^{-19} J. If a suitable dopant were introduced to provide a trap $(5.671 - 4.422)\ 10^{-19} = 1.249 \times 10^{-19}$ J below the conduction band, phosphorescence would occur.

We would also need information concerning the relaxation time to ensure that phosphorescence would not persist long enough to distort the image. Typical phosphorescent materials for television screens might include $CaWO_4$, which produces photons with a wavelength of 4.3×10^{-7} m (blue). This material has a relaxation time of 4×10^{-6} s. ZnO doped with excess zinc produces photons with a wavelength of 5.1×10^{-7} m (green), whereas $Zn_3(PO_4)_2$ doped with manganese gives photons with a wavelength of 6.45×10^{-7} m (red).

Light-Emitting Diodes – Electroluminescence Luminescence can be used to advantage in creating light-emitting diodes (LEDs). LEDs are used to provide the display for watches, clocks, calculators and other electronic devices. The stimulus for these devices is an externally applied voltage, which causes electron transitions and electroluminescence. LEDs are *p-n* junction devices engineered so that the E_g is in the visible spectrum (often red). A voltage applied to the diode in the forward-bias direction causes holes and electrons to recombine at the junction and emit photons (Figure 20.7). GaAs, GaP, GaAlAs, and GaAsP are typical materials for LEDs.

Lasers – Amplification of Luminescence The laser (*l*ight *a*mplification by *s*timulated *e*mission of *r*adiation) is another example of a special application of luminescence. In certain materials, electrons excited by a stimulus (such as the flash tube shown in Figure 20.8) produce photons which, in turn, excite additional photons of identical wavelength. Consequently, a large amplification of the photons emitted in the material,

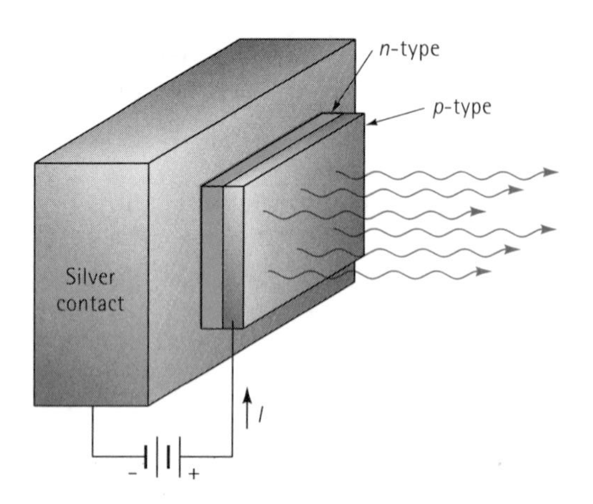

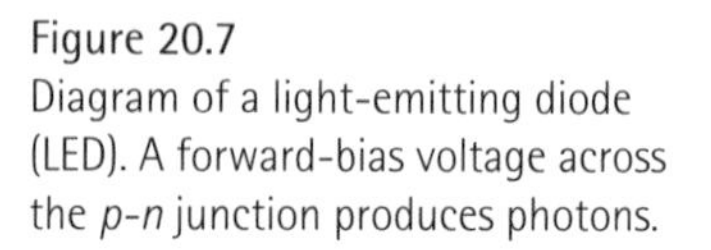
Figure 20.7
Diagram of a light-emitting diode (LED). A forward-bias voltage across the *p-n* junction produces photons.

occurs. By proper choice of stimulant and material, the wavelength of the photons can be in the visible range. The output of the laser is a beam of photons that are parallel, of the same wavelength, and coherent. In a *coherent* beam, the wavelike nature of the photons is in phase, so that destructive interference does not occur. Lasers are useful in heat treating and melting of metals, welding, surgery, mapping, transmission and processing of information, and a variety of other applications, including reading of the compact disks used to produce noise-free stereo recordings.

A variety of materials are used to produce lasers. Ruby, which is Al_2O_3 doped with a small amount of Cr_2O_3, and yttrium aluminium garnet (YAG) doped with neodymium are two common solid-state lasers. Other lasers are based on Co_2 gas.

Semiconductor lasers such as GaAs, which have an energy gap corresponding to a wavelength in the visible range, are also used. Figure 20.9 illustrates how a semiconductor laser might operate. When the semiconductor is excited by a voltage applied to the device, electrons jump from the valence band to the conduction band, leaving behind holes in the valence band. When an electron collapses back to the valence band and recombines with a hole, a photon having an energy and wavelength equivalent to the energy gap produced. This photon stimulates another electron to drop from the conduction band to the valence band, creating a second photon having an identical wavelength and a frequency that is in phase with the first photon. A mirror at one end of the laser crystal completely reflects the photons, trapping them within the semiconductor. The reflected photons stimulate even more recombinations, until an intense wave of photons is produced. The photons then reach the other end of the crystal, which

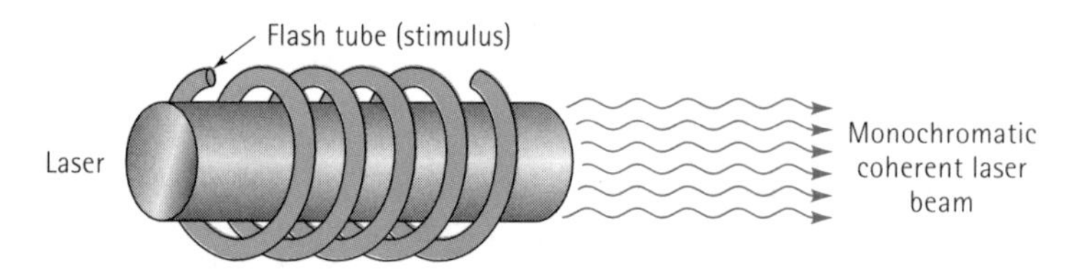

Figure 20.8
The laser converts a stimulus into a beam of coherent photons.

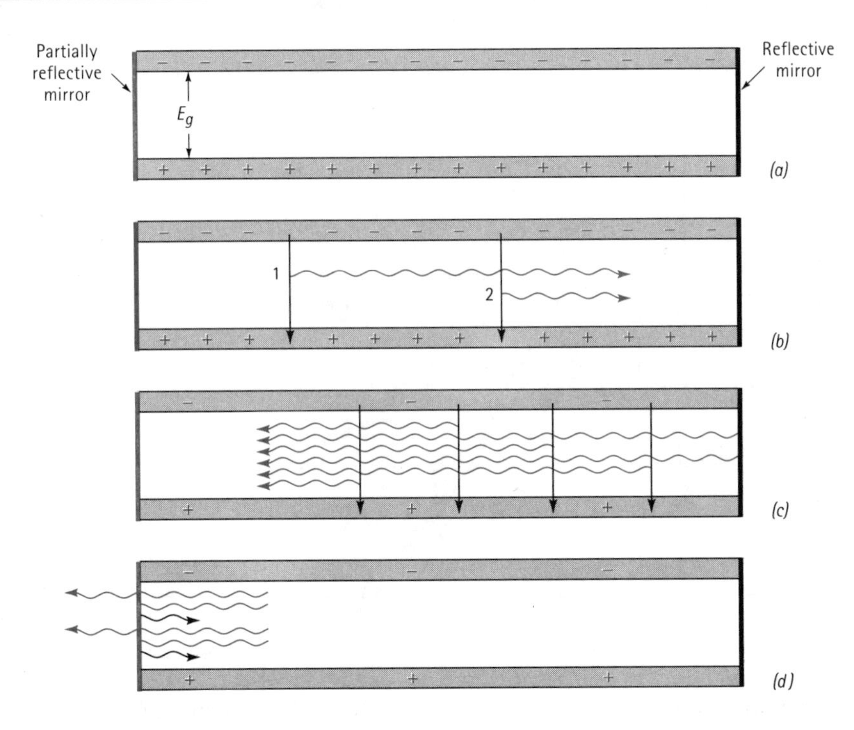

Figure 20.9
Creation of a laser beam from a semiconductor: (*a*) Electrons are excited into the conduction band by an applied voltage. (*b*) Electron I recombines with a hole to produce a photon. The photon stimulates the emission of photon 2 by a second recombination. (*c*) Photons reflected from the mirrored end stimulate even more photons. (*d*) A fraction of the photons are emitted as a laser beam, while the rest are reflected to stimulate more recombinations.

is only partly mirrored. A fraction of the photons emerge from the crystal as a monochromatic, coherent laser beam, while the rest of the photons remain in the crystal to stimulate further recombinations. The applied voltage ensures that a steady source of excited electrons is available to produce additional photons; a continuous laser beam is therefore produced. Figure 20.10 schematically depicts one design for a semiconductor laser.

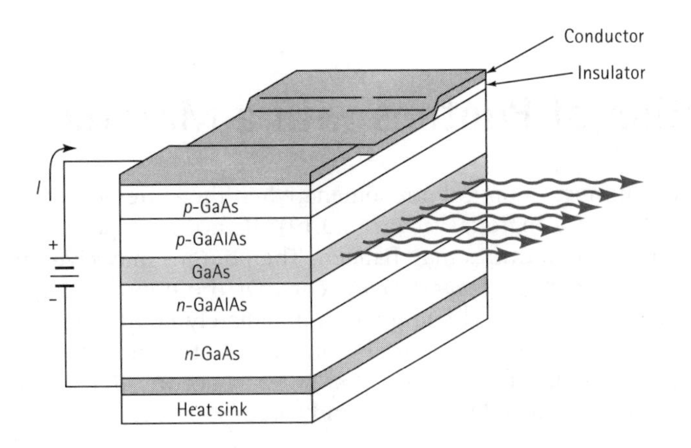

Figure 20.10
Schematic cross-section of a GaAs laser. Because the surrounding *p*- and *n*-type GaAlAs layers have a higher energy gap and a lower index of refraction than GaAs, the photons are trapped in the active GaAs layer.

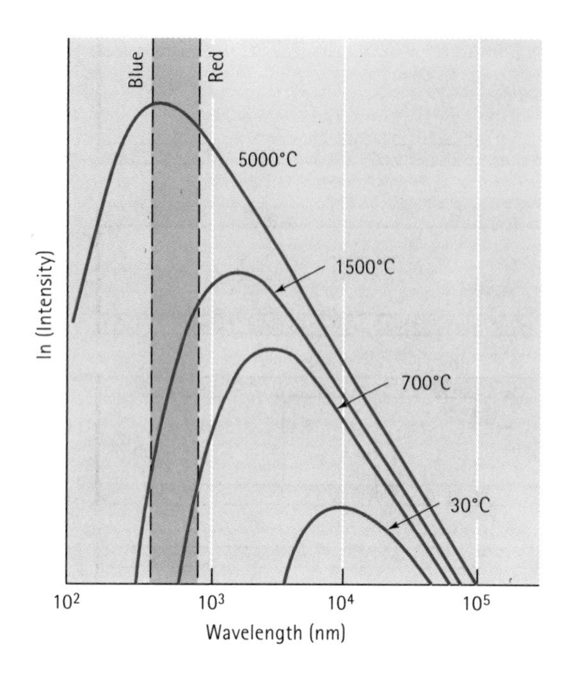

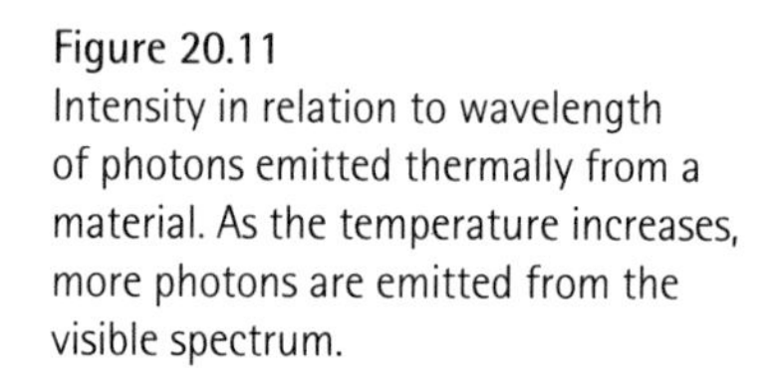

Figure 20.11
Intensity in relation to wavelength of photons emitted thermally from a material. As the temperature increases, more photons are emitted from the visible spectrum.

Thermal Emission When a material is heated, electrons are thermally excited to higher energy levels, particularly in the outer energy levels where the electrons are less tightly bound to the nucleus. The electrons immediately drop back to their normal levels and release photons, an event known as thermal emission.

As the temperature increases, thermal agitation increases and the maximum energy of the emitted photons increases. A continuous spectrum of radiation is emitted, with a minimum wavelength and an intensity distribution dependent on the temperature. The photons may include wavelengths in the visible spectrum; consequently the colour of the material changes with temperature. At low temperatures, the wavelength of the radiation is too long to be visible. As the temperature increases, emitted photons have shorter wavelengths. At 700°C we begin to see a reddish tint; at 1500°C the orange and red wavelengths are emitted (Figure 20.11). Higher temperatures produce all wavelengths in the visible range, and the emitted spectrum is white light. By measuring the intensity of a narrow band of the emitted wavelengths with a pyrometer, we can estimate the temperature of the material.

20.4 Interaction of Photons with a Material

Photons cause a number of optical phenomena when they interact with the electronic or crystal structure of a material (Figure 20.12). If incoming photons interact with valence electrons, several things may happen. The photons may give up their energy to the material, in which case *absorption* occurs. Or the photons may give up their energy, but photons of identical energy are immediately emitted by the material; in this case, *reflection* occurs. Finally, the photons may not interact with the electronic structure of the material; in this case, *transmission* occurs. Even in transmission, however, photons are changed in velocity and *refraction* occurs.

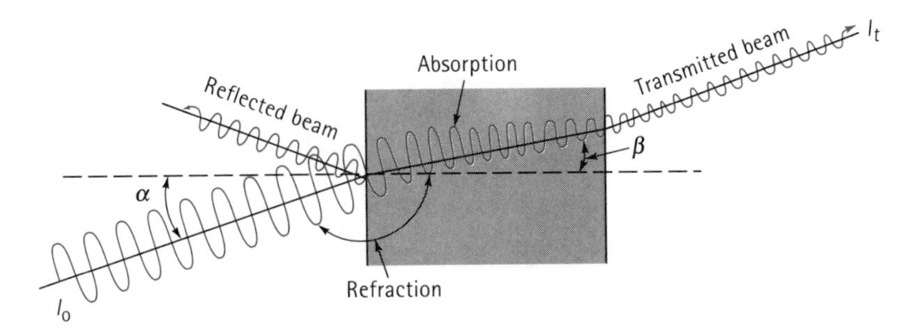

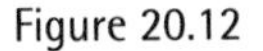

Figure 20.12
Interaction of photons with a material. In addition to reflection, absorption, and transmission, the beam changes direction, or is refracted. The change in direction is given by the index of refraction *n*.

As Figure 20.12 illustrates, an incident beam of intensity I_0 may be partly reflected, partly absorbed, and partly transmitted. The intensity of the incident beam can therefore be expressed as

$$I_0 = I_r + I_a + I_t \tag{20.3}$$

where I_r is the portion of the beam that is reflected, I_a is the portion that is absorbed, and I_t is the portion finally transmitted through the material. Reflection may occur at both the front and back surfaces of the material. Several factors are important in determining the behaviour of the photons in the material, with the energy required to excite an electron to a higher energy state being of particular importance.

Let's examine each of these four phenomena. We begin with refraction, since it is related to reflection and transmission.

Refraction Even when a photon is transmitted, the photon causes polarisation of the electrons in the material and, by interacting with the polarised material, loses some of its energy. The speed of light in a material (v) can be related to the ease with which a material polarises both electrically (permittivity) and magnetically (permeability), or

$$v = \frac{1}{\sqrt{\mu\varepsilon}} \tag{20.4}$$

Generally optical materials are not magnetic and permeability can be neglected.

Because the speed of the photons decreases, the beam of photons changes direction when it enters the material (Figure 20.12). Suppose photons travelling in a vacuum impinge on a material. If α and β, respectively, are the angles that the incident and refracted beams make with the normal to the surface of the material, then:

$$n = \frac{c}{v} = \frac{\lambda_{\text{vacuum}}}{\lambda} = \frac{\sin\alpha}{\sin\beta} \tag{20.5}$$

The ratio n is the index of refraction, c is the speed of light in a vacuum, and v is the speed of light in the material. Typical values of the index of refraction for several materials are listed in Table 20.2.

Table 20.2 Index of refraction of selected materials for photons of wavelength 589 nm.

Material	Index of Refraction
Air	1.00
Ice	1.309
Water	1.333
SiO_2 (glass)	1.46
SiO_2 (quartz)	1.55
Typical glasses	1.50
Leaded glass	2.50
TiO_2	1.74
Silicon	3.49
Diamond	2.417
Polytetrafluoroethylene (PTFE)	1.35
Polyethylene	1.52
Epoxy	1.58
Polystyrene	1.60

If the photons are travelling in Material 1, instead of in a vacuum, and then pass into Material 2, the velocities of the incident and refracted beams depend on the ratio between their indices of refraction, again causing the beam to change direction:

$$\frac{v_1}{v_2} = \frac{n_2}{n_1} = \frac{\sin \alpha}{\sin \beta} \tag{20.6}$$

We can use the latter expression to determine whether the beam will be transmitted as a refracted beam or be reflected. A beam travelling through Material 1 is reflected rather than transmitted if the angle β becomes 90°.

More interaction of the photons with the electronic structure of the materials occurs when the material is easily polarised. Consequently, we expect to find a relationship between the index of refraction and the dielectric constant κ of the material. From Equation 20.5, and for nonmagnetic materials:

$$n = \frac{c}{v} = \sqrt{\frac{\mu\varepsilon}{\mu_0\varepsilon_0}} \cong \sqrt{\frac{\varepsilon}{\varepsilon_0}} = \sqrt{\kappa} \tag{20.7}$$

The index of refraction is also larger for denser materials (such as glasses containing PbO), leading to the exceptional appearance of fine crystal glassware. However, n is not a constant for a particular material; the frequency, or wavelength, of the photons affects the index of refraction.

EXAMPLE 20.4 **Design of a Fibre Optic System**

In designing a fibre optic transmission system, we plan to introduce a beam of photons from a laser into a glass fibre having an index of refraction of 1.5. Design a system to introduce the beam with a minimum of leakage of the beam from the fibre.

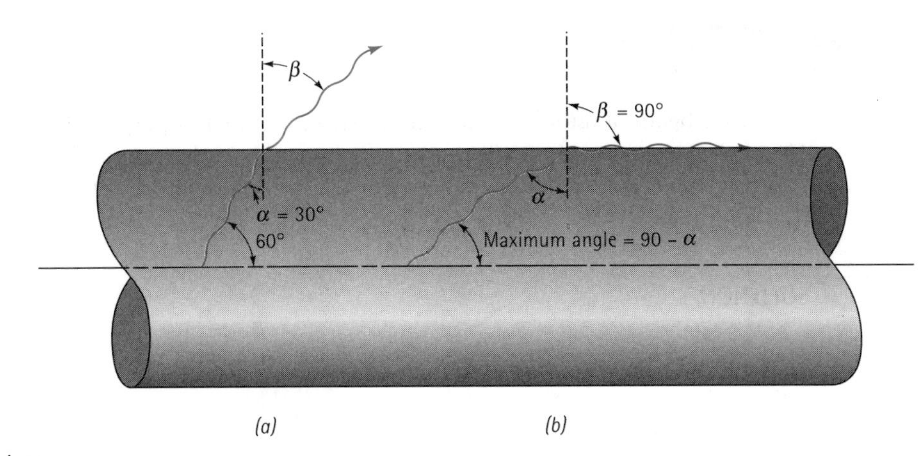

Figure 20.13
Diagram of light beam in glass fibre for Example 20.5: (*a*) calculation of angle β and (*b*) calculation of maximum angle α.

SOLUTION

To prevent leakage of the beam, the angle β must be at least 90°. Suppose that the photons enter at a 60° angle to the axis of the fibre. From Figure 20.13, we find that $\alpha = 90 - 60 = 30°$. If we let the glass be Material 1 and if the glass fibre is in air ($n = 1.0$), then from Equation 20.6:

$$\frac{n_2}{n_1} = \frac{\sin\alpha}{\sin\beta} \quad \text{or} \quad \frac{1}{1.5} = \frac{\sin 30}{\sin\beta}$$

$$\sin\beta = 1.5\sin 30 = 1.5(0.50) = 0.75 \quad \text{or} \quad \beta = 48.6°$$

Because β must be less than 90°, photons escape from the fibre. To prevent transmission, we must introduce the photons at a shallower angle, giving $\beta = 90°$.

$$\frac{1}{1.5} = \frac{\sin\alpha}{\sin\beta} = \frac{\sin\alpha}{\sin 90°} = \sin\alpha$$

$$\sin\alpha = 0.6667 \quad \text{or} \quad \alpha = 41.8°$$

If the angle between the beam and the axis of the fibre is 90 – 41.8 = 48.2° or less, the beam is reflected.

If the fibre were immersed in water ($n = 1.333$), then:

$$\frac{1.333}{1.5} = \frac{\sin\alpha}{\sin\beta} = \frac{\sin\alpha}{\sin 90°} = \sin\alpha$$

$$\sin\alpha = 0.8887 \quad \text{or} \quad \alpha = 62.7°$$

In water, the photons would have to be introduced at an angle of less than 90 – 62.7 = 27.3° in order to prevent transmission.

EXAMPLE 20.5

Suppose a beam of photons in a vacuum strikes a sheet of polyethylene at an angle of 10° to the normal of the surface of the polymer. Calculate the index of refraction of polyethylene and find the angle between the incident beam and the beam as it passes through the polymer.

SOLUTION

The index of refraction is related to the dielectric constant. From Table 18.9, $\kappa = 2.3$:

$$n = \sqrt{\kappa} = \sqrt{2.3} = 1.52$$

The angle β is:

$$n = \frac{\sin \alpha}{\sin \beta}$$

$$\sin \beta = \frac{\sin \alpha}{n} = \frac{\sin 10^\circ}{1.52} = \frac{0.174}{1.52} = 0.114$$

$$\beta = 6.56^\circ$$

Reflection When a beam of photons strikes a material, the photons interact with the valence electrons and give up their energy. In metals, the valence bands are unfilled and radiation of almost any wavelength excites the electrons into higher energy levels. One might expect that, if the photons are totally absorbed, no light would be reflected and the metal would appear to be black. In aluminium or silver, however, photons of almost identical wavelength are immediately re-emitted as the excited electrons return to their lower energy levels – that is, reflection occurs. Since virtually all of the visible spectrum is reflected, these metals have a white, or silvery, colour.

The reflectivity R gives the fraction of the incident beam that is reflected and is related to the index of refraction. If the material is a vacuum or in air:

$$R = \left(\frac{n-1}{n+1}\right)^2 \tag{20.8}$$

If the material is in some other medium with an index of refraction of n_i, then:

$$R = \left(\frac{n-n_i}{n+n_i}\right)^2 \tag{20.9}$$

Materials with a high index of refraction have a higher reflectivity than materials with a low index. Because the index of refraction varies with the wavelength of the photons, so too does the reflectivity.

In metals, the reflectivity is typically on the order of 0.9 to 0.95, whereas the reflectivity of typical glasses is nearer to 0.05. The high reflectivity of metals is one reason that they are *opaque*.

Absorption That portion of the incident beam that is not reflected by the material is either absorbed or transmitted through the material. The fraction of the beam that is absorbed is related to the thickness of the material and the manner in which the photons interact with the material's structure. The intensity of the beam after passing through the material is given by

$$I = I_0 \exp(-\mu x) \tag{20.10}$$

where x is the path through which the photons move (usually the thickness of the material), μ is the linear absorption coefficient of the material for the photons, I_0 is the intensity if the beam after reflection at the front surface, and I is the intensity of the beam when it reaches the back surface.

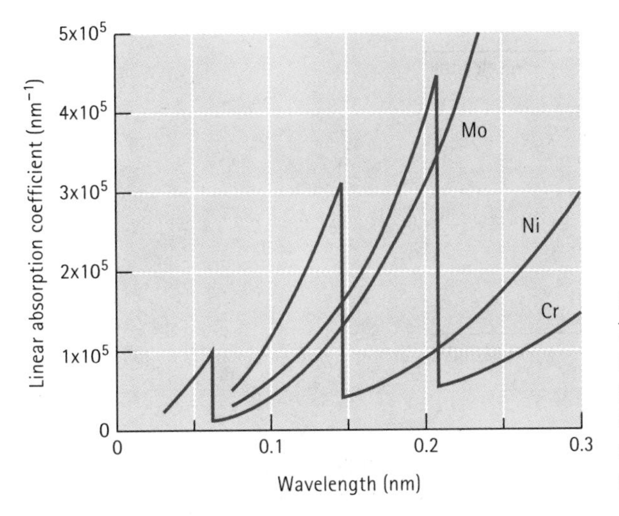

Figure 20.14
The linear absorption coefficient relative to wavelength for several metals. Note the sudden decrease in absorption coefficient for wavelengths greater than the absorption edge.

Absorption occurs by several mechanisms. In *Raleigh scattering,* the photon interacts with the electrons orbiting an atom and is deflected without any change in photon energy; this outcome is more significant for high atomic number atoms and low photon energies. *Compton scattering* is caused by an interaction between the photon and orbiting electrons, causing the electron to be ejected from the atom and, consequently, consuming some of the energy of the photon. Again, higher atomic number atoms and low photon energies cause more scattering. The *photoelectric effect* occurs when the energy of the photon is consumed by breaking the bond between an electron and its nucleus. As the energy of the photon increases (or the wavelength decreases, *see* Figure 20.14), less absorption occurs until the photon has an energy equal to that of the binding energy. At this energy, the absorption coefficient increases significantly. The energy, or wavelength, at which this occurs is called the absorption edge. In Figure 20.14, the abrupt change in absorption coefficient corresponds to the energy required to remove an electron for the K shell of the atom; this absorption edge is important in certain X-ray analytical techniques, as illustrated in Example 20.6.

EXAMPLE 20.6 Materials Selection for an X-ray Filter

Select a filter that preferentially absorbs K_β X-rays from the nickel spectrum but permits K_α X-rays to pass with little absorption.

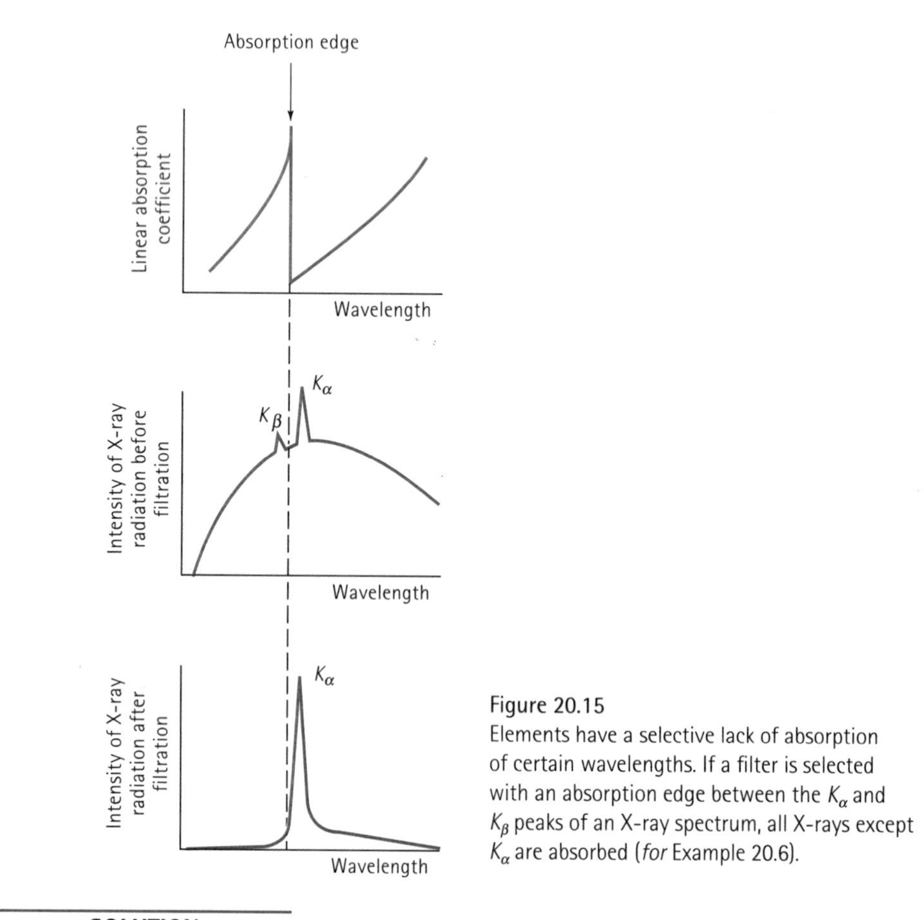

Figure 20.15
Elements have a selective lack of absorption of certain wavelengths. If a filter is selected with an absorption edge between the K_α and K_β peaks of an X-ray spectrum, all X-rays except K_α are absorbed (*for* Example 20.6).

SOLUTION

When determining crystal structure or identifying unknown materials using various X-ray diffraction techniques, we prefer to use X-rays of a single wavelength. If both K_α and K_β characteristics peaks are present and interact with the material, analysis becomes much more difficult.

However, we can use the selective absorption, or the existence of the absorption edge, to isolate the K_α peak. Table 20.1 includes the information that we need. If a filter material is selected such that the absorption edge lies between the K_α and K_β wavelengths, then the K_β is almost completely absorbed, whereas the K_α is almost completely transmitted. In nickel, $K_\alpha = 0.1660$ nm and $K_\beta = 0.1500$ nm. A filter with an absorption edge between these characteristic peaks will work. Cobalt, with an absorption edge of 0.1608 nm, would be our choice. Figure 20.15 shows how this filtering process occurs.

In metals, the absorption coefficient tends to be large [Figure 20.16(a)], particularly in the visible light spectrum. Because there is no energy gap in metals, virtually any photon has sufficient energy to excite an electron into a higher energy level, thus absorbing the energy of the excited photon. Insulators, on the contrary, possess a large energy gap between the valence and conduction bands. If the energy of the incident photons is less than the energy gap, no electrons gain enough energy to escape the valence band, and therefore absorption does not occur [Figure 20.16(b)]. When the photons do not interact with imperfections in the material, the material is said to be *transparent*. This is the case for glass, many high-purity crystalline ceramics, and amorphous polymers such as acrylics, polycarbonates, and polysulphones.

In semiconductors, the energy gap is smaller that in insulators, particularly in extrinsic semiconductors that contain donor or acceptor energy levels. In intrinsic semiconductors, absorption occurs when the photons have energies exceeding the energy gap E_g, whereas transmission occurs for less energetic photons [Figure 20.16 (b)]. In extrinsic semiconductors, absorption occurs when photons have energies greater than E_a or E_d [Figure 20.16(c)]. Semiconductors are therefore opaque to short-wavelength radiation but transparent to long-wavelength photons. For example, silicon and germanium appear opaque to visible light, but they are transparent to the longer wavelength infrared radiation.

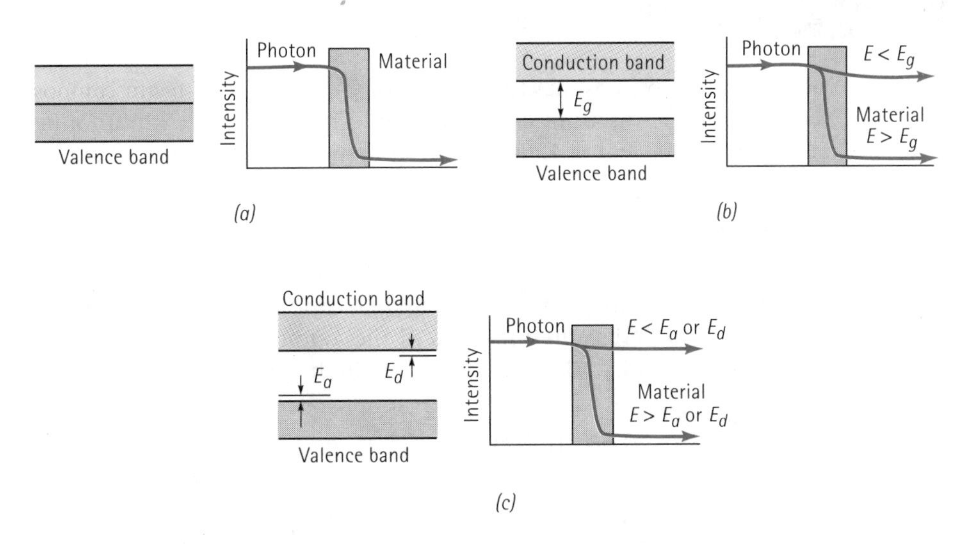

Figure 20.16
Relationships between absorption and the energy gap: (*a*) metals, (*b*) insulators and intrinsic semiconductors, and (*c*) extrinsic semiconductors.

EXAMPLE 20.7

Determine the critical energy gaps that provide complete transmission and complete absorption of photons in the visible spectrum.

SOLUTION

The visible light spectrum varies from 4×10^{-7} m to 7×10^{-7} m. The E_g required to ensure that no photons in the visible spectrum are absorbed is:

$$E_g = \frac{hc}{\lambda} = \frac{(6.62 \times 10^{-34}\,\mathrm{J \cdot s})(3 \times 10^8\,\mathrm{m.s^{-1}})}{(4 \times 10^{-7}\,\mathrm{m})}$$

$$= 4.965 \times 10^{-19}\,\mathrm{J}$$

The E_g below which all the photons in the visible spectrum are absorbed is:

$$E_g = \frac{hc}{\lambda} = \frac{(6.62 \times 10^{-34}\,\mathrm{J \cdot s})(3 \times 10^8\,\mathrm{m.s^{-1}})}{(7 \times 10^{-7}\,\mathrm{m})}$$

$$= 2.837 \times 10^{-19}\,\mathrm{J}$$

For materials with an intermediate E_g, a portion of the photons in the visible spectrum will be absorbed.

EXAMPLE 20.8 Design of an X-ray Filter

Design a filter to transmit at least 95% of the energy of a beam composed of zinc K_α X-rays, using aluminium as the shielding material. (The aluminium has a linear absorption coefficient of 1.08×10^4 m^{-1}.) Assume no loss to reflection.

SOLUTION

Assuming that no losses are caused by reflection of X-rays from the aluminium, we need simply to design the thickness of the aluminium required to transmit 95% of the incident intensity. The final intensity will therefore be $0.95I_0$. Thus:

$$\ln\left(\frac{0.95 I_0}{I_0}\right) = -(1.08 \times 10^4)(x)$$

$$\ln(0.95) = -0.051 = -1.08 \times 10^4 x$$

$$x = \frac{-0.051}{(-1.08 \times 10^4)} = 4.7 \times 10^{-6}\,\mathrm{m} = 0.0047\,\mathrm{mm}$$

We would like to roll the aluminium to a thickness of 0.0047 mm or less. The filter could be thicker if a material were selected that has a lower linear absorption coefficient for zinc K_α X-rays.

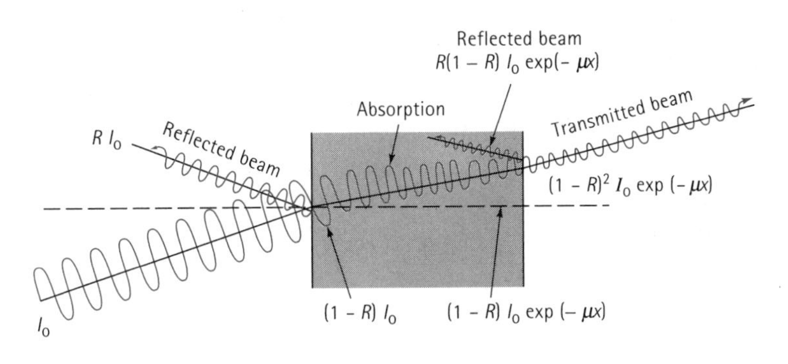

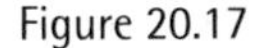
Figure 20.17
Fractions of the original beam that are reflected, absorbed, and transmitted.

Transmission The fraction of the beam that is not reflected or absorbed is transmitted through the material. Using the following steps, we can determine the fraction of the beam that is transmitted (see Figure 20.17):

1. If the incident intensity is I_0, then the loss due to reflection at the front face of the material is RI_0. The fraction of the incident beam that actually enters the materials is $I_0 - RI_0 = (1 - R)I_0$:

$$I_{\text{reflected at front surface}} = RI_0$$

$$I_{\text{after reflection}} = (1 - R)I_0$$

2. A portion of the beam that enters the material is lost by absorption. The intensity of the beam after passing through a material having a thickness x is:

$$I_{\text{after absorption}} = (1 - R)I_0 \exp(-\mu x)$$

3. Before the partially absorbed beam exits the material, reflection occurs at the back surface. The fraction of the beam that reaches the back surface and is reflected is:

$$I_{\text{reflected at back surface}} = R(1 - R)I_0 \exp(-\mu x)$$

4. Consequently, the fraction of the beam that is completely transmitted through the material is:

$$\begin{aligned} I_{\text{transmitted}} &= I_{\text{after absorption}} - I_{\text{reflected at back}} \\ &= (1 - R)I_0 \exp(-\mu x) - R(1 - R)I_0 \exp(-\mu x) \\ &= (1 - R)(1 - R)I_0 \exp(-\mu x) \end{aligned}$$

$$I_t = I_0(1 - R)^2 \exp(-\mu x) \tag{20.11}$$

Again, however, the intensity of the transmitted beam depends on the wavelength of the photons in the beam. Suppose that a beam of white light (containing photons of all wavelengths in the visible spectrum) impinges on a material. If the same fraction of the photons having different wavelength is absorbed, reflected, and therefore transmitted, the transmitted beam would also be white light, or colourless. This is what we find in materials such as diamond. If, however, longer wavelength photons (red, orange, etc.) are absorbed more than the shorter wavelength photons, we expect the transmitted light to appear blue or green.

The intensity of the transmitted beam also depends on microstructural features. Porosity in glasses scatters photons; even a small amount of porosity (less than 1 volume percent) may make the glass opaque. Crystalline precipitates, particularly those that have a much different index of refraction than the matrix material, also cause scattering. These crystalline *opacifiers* cause a glass, which normally may have excellent transparency, to become translucent or even opaque. Typically smaller pores or precipitates cause a greater reduction in transmission of the photons.

EXAMPLE 20.9 Design of a Radiation Shield

A material has a reflectivity of 0.15 and an absorption coefficient of 1×10^4 m^{-1}. Design a shield that will permit only 1% of the incident radiation from being transmitted through the material.

SOLUTION

From Equation 20.11, the fraction of the incident intensity that will be transmitted is:

$$\frac{I_t}{I_0} = (1-R)^2 \exp(-\mu x)$$

$$0.01 = (1-0.15)^2 \exp(-1\times 10^4 x)$$

$$\frac{0.01}{(0.85)^2} = 0.01384 = \exp(-10^4 x)$$

$$\ln(0.01384) = -4.28 = -10^4 x$$

$$x = 4.28 \times 10^{-4} \text{ m} = 0.428 \text{ mm}$$

The material should have a thickness of 0.428 mm in order to transmit 1% of the incident radiation.

If we wished, we could determine the amount of radiation lost in each step:

Reflection at the front face: $I_r = RI_0 = 0.15\, I_0$

Energy after reflection: $I = I_0 - 0.15\, I_0 = 0.85\, I_0$

Energy after absorption: $I_a = (1 - R)I_0 \exp\,[(-1 \times 10^4)(4.28 \times 10^{-4})] = 0.0118\, I_0$

Energy due to absorption: $0.85\, I_0 - 0.0118\, I_0 = 0.838\, I_0$

Reflection at the back face: $I_r = R(1 - R)I_0 \exp\,(-\mu x)$

$$=(0.15)(1 - 0.15)I_0 \exp\,[-(1 \times 10^4)(4.28 \times 10^{-4})]$$

$$= 0.0018\, I_0$$

Selective Absorption, Transmission, or Reflection Unusual optical behaviour is observed when photons are selectively absorbed, transmitted, or reflected. We have already found that semiconductors transmit long-wavelength photons but absorb short-wavelength radiation. There are a variety of other cases in which similar selectivity produces unusual optical properties.

Table 20.3 Effect of ions on colours produced in glasses.

Ion	Colour
Cr^{2+}	Blue
Cr^{3+}	Green
Cu^{2+}	Blue-green
Mn^{2+}	Orange
Fe^{2+}	Blue-green
U^{6+}	Yellow

In certain materials, replacement of normal ions by transition or rare earth elements produces a *crystal field*, which creates new energy levels within the structure. This phenomenon occurs when Cr^{3+} ions replace Al^{3+} ions in Al_2O_3. The new energy levels absorb visible light in the violet and green-yellow portions of the spectrum. Red wavelengths are transmitted, giving the reddish colour in ruby. In addition, the chromium ion replacement creates an energy level that permits luminescence to occur when the electrons are excited by a stimulus. Lasers produced from chromium-doped ruby produce a characteristic red beam because of this.

Glasses can also be doped with ions that produce selective absorption and transmission (Table 20.3). For example, polychromic glass, used for sunglasses, contains silver atoms. The glass darkens in sunlight but becomes transparent in darkness. In bright light, the silver ions in the glass gain an electron through excitation by the photons and are reduced from Ag^+ to metallic silver atoms. Thus, absorption of photons occurs. When the incoming light diminishes in intensity, the silver reverses to silver ions and no absorption occurs.

Electron or hole traps, called *F-centres*, can also be present in crystals. When fluorite (CaF_2) is produced so that there is excess calcium, a fluoride ion vacancy is produced. To maintain electrical neutrality, an electron is trapped in the vacancy, producing energy levels that absorb all visible photons, with exception of purple.

Polymers – particularly those containing an aromatic ring in the backbone – can have complex covalent bonds that produce an energy level structure that causes selective absorption. For this reason, chlorophyll in plants appears green and haemoglobin in blood appears red.

EXAMPLE 20.10 Design of a 'Stealthy' Aircraft

Suggest design guidelines that would create an aircraft that cannot be detected by radar.

SOLUTION

Radar is electromagnetic radiation, typically in the microwave portion of the spectrum. To detect an aircraft by radar, some portion of the microwave must be reflected from the aircraft and returned to a radar receiver. Several approaches can be considered to reduce the radar signature from an aircraft:

1. We might make the aircraft from materials that are transparent to radar. Many polymers, polymer-matrix composites, and ceramics satisfy this requirement.
2. We might design the aircraft so that the radar signal is reflected at severe angles from the source. This could be done with flat surfaces at more than about a 30° angle from the horizontal, or by assuring that all surfaces are curved to help reflect the radar at a variety of angles.
3. The internal structure of the aircraft also can be made to absorb the radar. For example, use of a honeycomb material in the wings may cause the radar waves to be repeatedly reflected within the material. The honeycomb can also be filled with an absorbing material to hasten the dissipation of the radar energy.
4. We might make the aircraft less visible by selecting materials that have electronic transitions of the same energy as the radar. Carbon fibres in a polymer matrix and carbon-carbon composites at high-temperature locations in the aircraft accomplish this end. Coatings containing ferrimagnetic ceramic ferrites on an aircraft skin absorb radar waves and convert the energy to heat, but these materials are very heavy. Dielectrics, which typically are lighter in weight, can be engineered to absorb radiation of the appropriate frequency.

Photoconduction Photoconduction occurs in semiconducting materials if the semiconductor is part of an electrical circuit. In this case, the stimulated electrons produce a current rather than an emission (Figure 20.18). If the energy of an incoming photon is sufficient, an electron is excited into the conduction band or a hole is created in the valence band and the electron or hole then carries a charge through the circuit. The maximum wavelength of the incoming photon required to produce photoconduction is related to the energy gap in the semiconductive material.

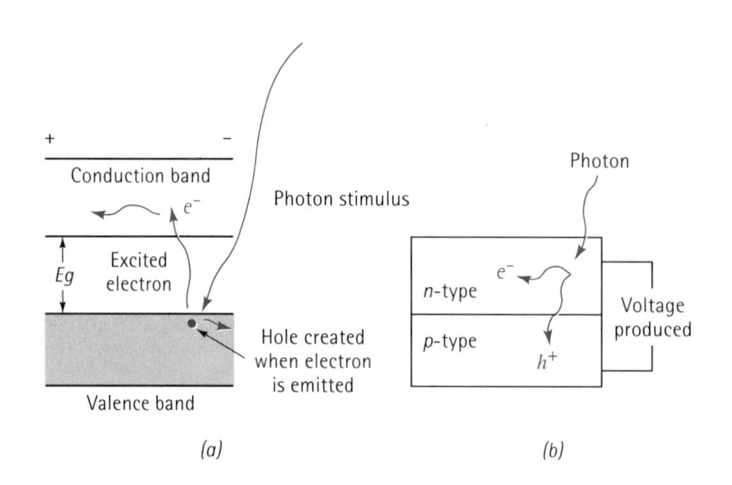

Figure 20.18
(*a*) Photoconduction in semiconductors involves absorption of a stimulus by exciting electrons from the valence band to the conduction band. Rather than dropping back to the valence band to cause emission, the excited electrons carry a charge through an electrical circuit. (*b*) A solar cell takes advantage of this effect.

$$\lambda_{\max} = \frac{hc}{E_g} \tag{20.12}$$

We can use this principle for 'electric eyes' that open or close doors or switches when a beam of light focused on a semiconductive material is interrupted. Note that photoconduction is the inverse of luminescence and LEDs. In the present case, photons produce a voltage and current, whereas voltage in an LED produces photons and light.

Solar cells are *p-n* junctions designed so that photons excite electrons into the conduction band. The electrons move to the *n*-side of the junction, while holes move to the *p*-side of the junction. This movement produces a contact voltage due to the charge imbalance. If the junction device is connected to an electric circuit, the junction acts as a battery to power the circuit.

20.5 Photonic Systems and Materials

Photonic systems use light to transmit information. Telephone communication systems, for instance, use fibre optics as a means for transmitting larger numbers of messages than can be transmitted using conventional techniques. Supercomputers may rely on photonic systems rather than the present electronic systems based largely on transport of electrons through silicon semiconducting devices.

A photonic system must generate a light signal from some other source, such as an electrical signal, transmit the light to a receiver, process the data received, and convert the data to a usable form (Figure 20.19). Photonic materials are required for this purpose. Most of the principles and materials presently utilised in photonic systems have already been introduced in the previous sections. Let us now review these materials in the context of an actual system, pointing out some of the special requirements that are needed.

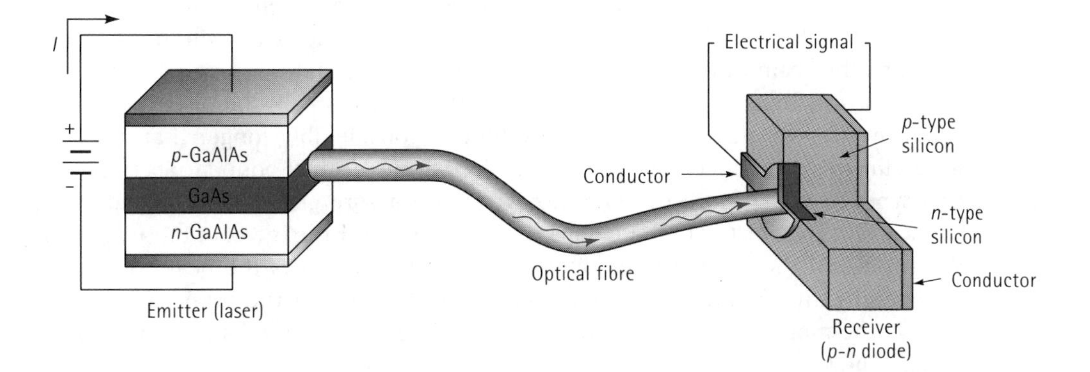

Figure 20.19

A photonic system for transmitting information involves a laser or LED to generate photons from an electrical signal, optical fibres to transmit the beam of photons efficiently, and an LED receiver to convert the photons back into an electrical signal.

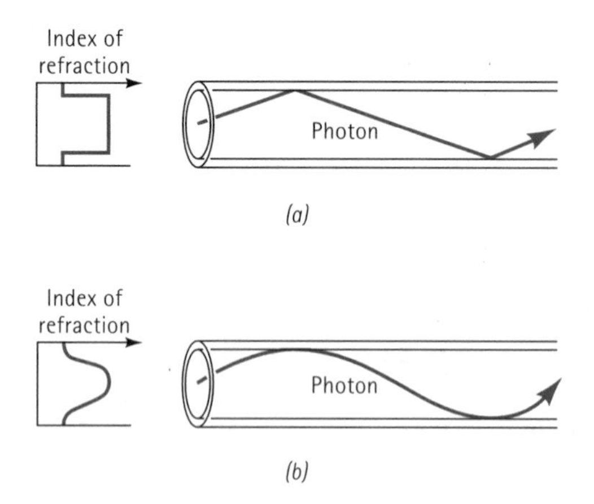

Figure 20.20
Two methods for controlling leakage from an optical fibre: (*a*) A composite glass fibre, in which the index of refraction is slightly different in each glass, and (*b*) a glass fibre doped at the surface to lower the index of refraction.

Generating the Signal In order to best transmit and process information, the light should be coherent and monochromatic. Thus, a laser is an ideal method for generating the photons. The Group III-V semiconductors such as GaAs, GaAlAs, and InGaAsP have energy gaps that provide emitted photons in the visible spectrum. Lasers built from these materials can be energised by a voltage, thereby generating a laser beam. Light-emitting diodes are a second device that can be used to produce the photons.

By varying the voltages applied to these devices, the intensity if the photon beam can be varied. The intensity of the beam can then be exploited to convey information.

Transmitting the Beam Waveguides composed of glass fibres transmit the light from the source to the receiver. In order for the optical fibres to transmit light long distances efficiently, the glass must have exceptional transparency and must not leak any light.

The index of refraction is important when considering losses due to leakage. In previous section, we found that light will be reflected if the index of refraction of the surroundings is less than that of the fibre, provided that the angle between the beam and the length of the fibre is sufficiently small. Optical fibres are often a composite glass material, with the central core having a higher index of refraction than the glass sheath. The difference in index of refraction helps keep the beam within the central core, and the beam travels the path shown in Figure 20.20(a).

Even if the angle between the beam and the fibre is small, the distance that the beam must travel through a composite fibre is considerably longer than the fibre itself, due to sharp reflections that are involved at the distinct boundary between the two types of glass. More complex fibres contain a core glass that is doped at the surface with B_2O_3 or GeO_2; these dopants gradually lower the index of refraction near the surface of the fibre. The gradual change in index allows the beam of light to change direction gradually, rather than sharply, reducing the total path it must follow. The more complex fibres help to transmit a less distorted signal, particularly in long fibres.

Receiving the Signal Either the light-emitting diodes that might be used to generate a signal or more conventional silicon semiconductor diodes can receive the signal. When a photon reaches the *p-n* diode, an electron is excited into the conduction band, leaving behind a hole. If voltage is applied to the diode, the electron-hole pair creates a current that is amplified and further processed.

Processing the Signal Normally the received signal is immediately converted into an electric signal and then processed using conventional silicon-based semiconductor devices. However, certain materials, such as $LiNbO_3$, have a non-linear optical response. When a beam of photons is received by such a material, the material can act as a transistor and amplify the signal or act as a switch (or a logic gate in a computer) and control the path of the beam. Development of photonic transistors could lead to an optically based computer.

SUMMARY

- The optical properties of materials are determined by their emission and interaction with electromagnetic radiation, or photons. Emission of photons occurs by transactions in the atom structure of the material:
    - Gamma rays are produced by the decay of unstable nuclei in some materials. The energy of the gamma rays from a particular type of atom is fixed.
    - X-rays are produced by electronic transistors in the inner energy levels of atoms. Atoms of each element produce a characteristic spectrum that permits the use of X-ray emission to identify materials.
    - Photons in or near the visible portion of the electromagnetic spectrum are produced by electronic transitions in the outer energy levels of atoms. This behaviour, called luminescence, requires an energy gap between the valence and conductive bands, with photons of energy equal to the energy gap being emitted. Fluorescence, phosphorescence, electroluminescence (used in light-emitting diodes), and lasers are examples of luminescence.
    - Photons are emitted by thermal excitation, with photons in the visible portion of the spectrum produced when the temperature is sufficiently high.
- Interactions between a material and the emitted photons provide many optical properties:
    - Refraction, or changing the direction of photons, occur as a result of polarisation of a material. The index of refraction describes the distortion of the beam of photons.
    - Reflection occurs when photons are absorbed by the material and immediately re-emitted from a material.
    - Absorption occurs when photons lose their energy to atoms in a material, either by exciting electrons or by scattering. The linear absorption coefficient helps describe the degree of absorption.
    - Photoconduction occurs when photons excite electrons in a material, creating a voltage.

GLOSSARY

Absorption edge
The wavelength at which the absorption characteristics of a material abruptly change.

Characteristic spectrum
The spectrum of radiation emitted from a material that occurs at fixed wavelengths corresponding to particular energy level differences within the atomic structure of the material.

Continuous spectrum
Radiation emitted from a material having all wavelengths longer than a critical short wavelength limit.

Electroluminescence
Use of an applied electrical signal to stimulate photons from a material.

Fluorescence
Emission of radiation from a material only when the material only when the material is actually being stimulated.

Index of refraction
Relates the change in velocity and direction of radiation as it passes through a transparent medium.

Laser
A beam of monochromatic coherent radiation produced by the controlled emission of photons.

Light-emitting diodes (LEDs)
Electronic *p-n* junction devices that convert an electrical signal into visible light.

Linear absorption coefficient
Describes the ability of a material to absorb radiation (m^{-1}).

Luminescence
Conversion of radiation to visible light.

Phosphorescence
Emission of radiation from a material after the stimulus is removed.

Photoconduction
Production of a voltage due to stimulation of electrons into the conduction band by light radiation.

Photons
Energy or radiation produced from atomic, electronic, or nuclear sources that can be treated as particles or waves.

Reflectivity
The percentage of incident radiation that is reflected.

Relaxation time
The time required for $1/e$ of the electrons to drop from the conduction band to the valence band in luminescence.

Short wavelength limit
The shortest wavelength or highest energy radiation emitted from a material under particular conditions.

Solar cell
p-n junction devices that create a voltage due to excitation by photons.

Thermal emission
Emission of photons from a material due to excitation of the material by heat.

X-rays
Electromagnetic radiation produced by changes in the electronic structure of atoms.

PROBLEMS

20.1 What voltage must be applied to a tungsten filament to produce a continuous spectrum of X-rays having a minimum wavelength of 0.09 nm?

20.2 A tungsten filament is heated with a 12 400 V power supply. What is

(a) the wavelength and

(b) frequency of the highest-energy X-rays that are produced?

20.3 What is the minimum voltage required to produce K_α X-rays in nickel?

20.4 Based on the characteristic X-rays that are emitted, determine the difference in energy between electrons in

(a) the *K* and *L* shells,

(b) the *K* and *M* shells, and

(c) the *L* and *M* shells of tungsten.

20.5 Figure 20.21 shows the results of an X-ray fluorescent analysis, in which the energy of X-rays emitted from a material are plotted relative to the wavelength of the X-rays. Determine

(a) the accelerating voltage used to produce the exciting X-rays and

(b) the identity of the elements in the sample.

20.6 Figure 20.22 shows the energies of X-rays produced from an energy-dispersive analysis of radiation emitted from a specimen in a scanning electron microscope. Determine the identity of the elements in the sample.

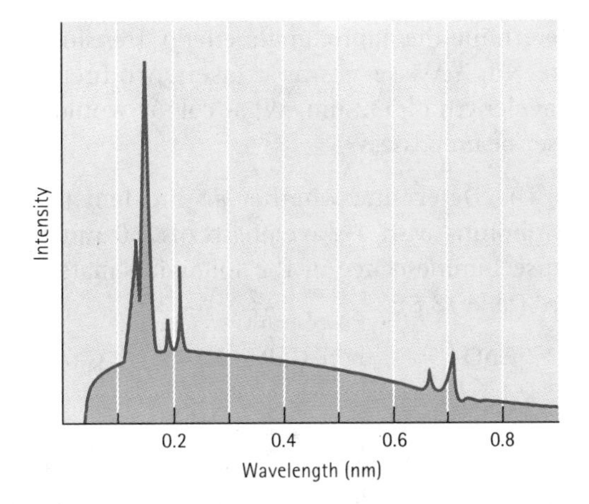

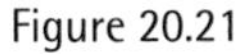
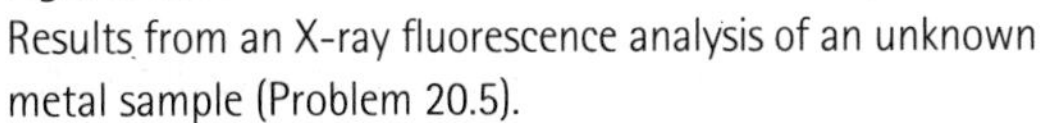
Figure 20.21
Results from an X-ray fluorescence analysis of an unknown metal sample (Problem 20.5).

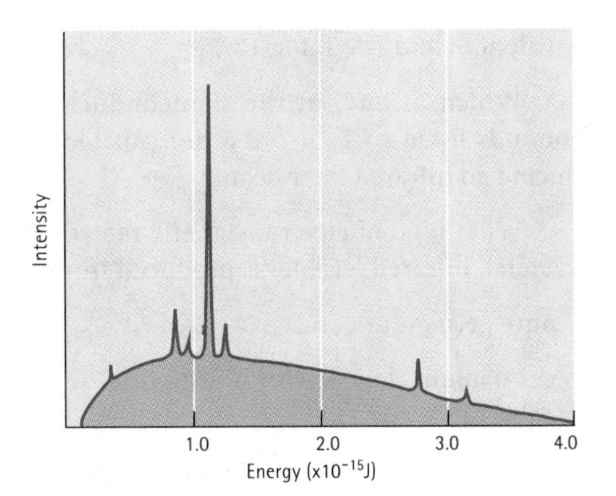

Figure 20.22
X-ray emission spectrum (*for* Problem 20.6).

20.7 $CaWO_4$ has a relaxation time of 4×10^{-6} s. Determine the time required for the intensity of this phosphorescent material to decrease to 1% of the original intensity after the stimulus is removed.

20.8 The intensity of a phosphorescent material is reduced to 90% of its original intensity after 1.95×10^{-7} s. Determine the time required for the intensity to decrease to 1% of its original intensity.

20.9 By appropriately doping yttrium aluminium garnet with neodymium, electrons are excited within the $4f$ energy shell of the Nd atoms. Determine the approximate energy transition if the Nd: YAG serves as a laser, producing a wavelength of 532 nm. What colour would the laser beam possess?

20.10 Determine whether an incident beam of photons with a wavelength of 750 nm will cause luminescence in the following materials (*see* Table 18.8):

(a) ZnO (b) GaP (c) GaAs

(d) GaSb (e) PbS

20.11 Determine the wavelength of photons produced when electrons excited into the conduction band of indium-doped silicon

(a) drop from the conduction band to the acceptor band and

(b) then drop from the acceptor band to the valence band (*see* Table 18.7).

20.12 Which, if any, of the semiconducting compounds listed in Table 18.8 are capable of producing an infrared laser beam?

20.13 What type of electromagnetic radiation (ultraviolet, infrared, visible) is produced from

(a) pure germanium and

(a) germanium doped with phosphorus? (*See* Tables 18.6 *and* 18.7.)

20.14 Which, if any, of the dielectric materials listed in Table 18.9 would reduce the speed of light in air from 3×10^8 m.s^{-1} to less than 0.5×10^8 m.s^{-1}?

20.15 A beam of photons strikes a material at an angle of 25° to the normal of the surface. Which, if any, of the materials listed in Table 20.2 could cause the beam of photons to continue at an angle of 18 to 20° from the normal of the material's surface?

20.16 A laser beam passing through air strikes a 50 mm-thick polystyrene block at a 20° angle to the normal of the block. By what distance is the beam displaced from its original path when the beam reaches the opposite side of the block?

20.17 A beam of photons in air strikes a composite material consisting of a 10-mm thick sheet of polyethylene and a 20-mm thick sheet of soda-lime glass. The incident beam is 10° from the normal of the composite. Determine the angle of the beam with respect to the normal as the beam

(a) passes through the polyethylene,

(b) passes through the glass, and

(c) passes through air on the opposite side of the composite.

(d) By what distance is the beam displaced from its original path when it emerges from the composite?

20.18 A glass fibre (n = 1.5) is coated with PTFE. Calculate the maximum angle that a beam of light can deviate from the axis of the fibre without escaping from the inner portion of the fibre.

20.19 A material has a linear absorption coefficient of 5.91×10^4 m^{-1} for photons of a particular wavelength. Determine the thickness of the material required to absorb 99.9% of the photons.

20.20 What filter material would you use to isolate the K_α peak of the following X-rays: iron, manganese, nickel? Explain your answer.

20.21 A beam of photons passes through air and strikes a soda-lime glass that is part of an aquarium containing water. What fraction of the beam is reflected by the front face of the glass? What fraction of the remaining beam is reflected by the back face of the glass?

20.22 We find that 20% of the original intensity of a beam of photons is transmitted from air through a 10 mm-thick material having a dielectric constant of 2.3 and back into air. Determine the fraction of the beam that is

(a) reflected at the front surface,

(b) absorbed in the material, and

(c) reflected at the back surface.

(d) Determine the linear absorption coefficient of the photons in the material.

20.23 Figure 20.23 shows the intensity of the radiation obtained from a copper X-ray generating tube as a function of wavelength. The accompanying table shows the linear absorption coefficient for a nickel filter for several wavelengths. If the Ni filter is 0.05 mm thick, calculate and plot the intensity of the transmitted X-ray beam versus wavelength.

Wavelength (nm)	Linear Absorption Coefficient (m^{-1})
0.0711	4.22×10^4
0.1436	2.90×10^5
0.1542	4.40×10^4
0.1659	5.43×10^4
0.1790	6.70×10^4
0.1937	8.30×10^4
0.2103	1.03×10^5
0.2291	1.30×10^5

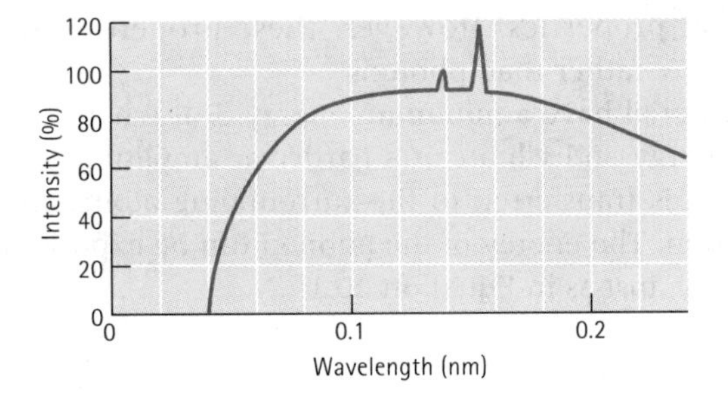

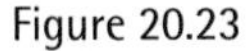
Figure 20.23
Intensity of the initial spectrum from a copper X-ray source before filtering (*for* Problem 20.23).

Design Problems

20.24 Nickel X-rays are to be generated inside a container, with the X-rays being emitted from the container through only a small slot. Design a container that will ensure that no more than 0.01% of the K_α nickel X-rays escape through the rest of the container walls, yet 95% of the K_α nickel X-rays pass through a thin window covering the slot. The following data give the mass absorption coefficients of several metals for nickel K_α X-rays. The mass absorption coefficient μ_m is μ/ρ, where μ is the linear mass absorption coefficient and ρ is the density of the filter material.

Material	μ_m ($m^2 . Mg^{-1}$)
Be	1.8×10^2
Al	5.84×10^3
Ti	2.47×10^4
Fe	3.54×10^4
Co	5.44×10^3
Cu	6.50×10^3
Sn	3.22×10^4
Ta	2.00×10^4
Pb	2.94×10^4

20.25 Design a method by which a photoconductive material be used to measure the temperature of a material from the material's thermal emission.

20.26 Design a method, based on a material's refractive characteristics, that will cause a beam of photons originally at a 2° angle to the normal to the material to be displaced from its original path by 20 mm at a distance of 500 mm from the material.

20.27 Design a 900 mm diameter satellite housing an infrared detector that can be placed into a low Earth orbit and that will not be detected by radar.

CHAPTER 21

Thermal Properties of Materials

21.1 Introduction

In previous chapters, we described how a material's properties change with temperature. We found in many cases that mechanical and physical properties depend on the temperature at which the material serves or on the temperature to which the material may be subjected during processing. An appreciation of the thermal properties of materials is helpful in understanding mechanical failure of materials such as ceramics, coatings, and fibres when the temperature changes; in designing processes in which materials must be heated; or in selecting materials to transfer heat rapidly – as for electronic devices. In this chapter, we discuss the heat capacity, thermal expansion properties, and thermal conductivity of materials in order to gain this appreciation.

21.2 Heat Capacity and Specific Heat

In Chapter 20, we found that optical behaviour depends on how photons are produced and interact with a material. The photon is treated as a particle with a particular energy or as electromagnetic radiation having a particular wavelength or frequency. Some of the thermal properties of materials can also be characterised in the same dual manner as are optical properties. However, these properties are determined by the behaviour of phonons, rather than photons.

At absolute zero, the atoms in a material have a minimum energy. But when heat is supplied, the atoms gain thermal energy and vibrate at a particular amplitude and frequency. The vibration of each atom is transferred to the surrounding atoms and produces an elastic wave called a phonon. The energy of the phonon can be expressed in terms of the wavelength or frequency, just as in Equation 20.1:

$$E = \frac{hc}{\lambda} = h\nu \tag{21.1}$$

A material gains or loses heat by gaining or losing phonons. The energy, or number of phonons, required to change the temperature of the material one degree is of interest. We can express this energy as the heat capacity, or the specific heat.

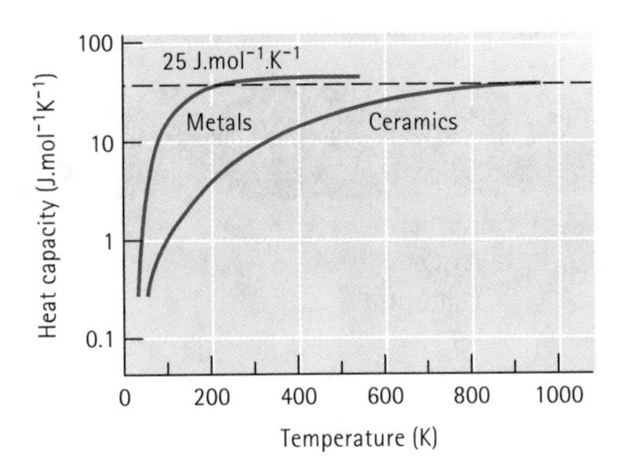

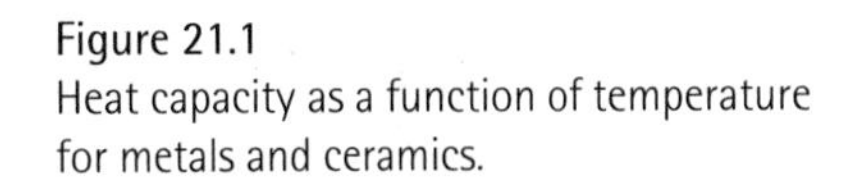
Figure 21.1
Heat capacity as a function of temperature for metals and ceramics.

The heat capacity is the energy required to raise the temperature of one mole of a material by one Kelvin (K) or one degree Celsius (°C). The heat capacity can be expressed either at constant pressure, C_p, or at a constant volume, C_v. At high temperatures, the heat capacity for a given volume of material approaches

$$C_p = 3R = 25 \text{ J.mol}^{-1}.\text{K}^{-1} \tag{21.2}$$

where R is the gas constant (8.314 J.mol^{-1}.K^{-1}). However, as shown in Figure 21.1, heat capacity is not a constant. The heat capacity of metals approaches 25 J.mol^{-1}.K^{-1} near room temperature, but this value is not reached in ceramics until about 700°C (973 K).

Specific heat is the energy required to raise the temperature of a particular weight or mass of a material one degree. The relationship between specific heat and heat capacity is:

$$\text{Specific heat} = c = \frac{\text{heat capacity}}{\text{atomic weight}} \tag{21.3}$$

In most engineering calculations, specific heat is more conveniently used than heat capacity. The specific heat of typical materials is given in Table 21.1. Neither the heat

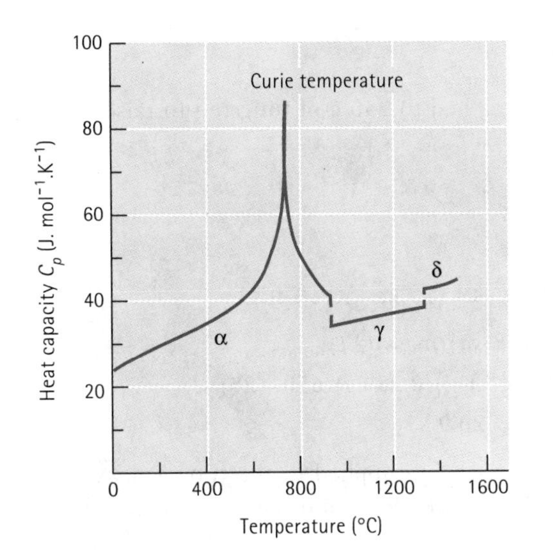

Figure 21.2
The effect of temperature on the heat capacity of iron. Both the change in crystal structure and the change from ferromagnetic to paramagnetic behaviour are indicated.

Table 21.1 The specific heat of selected materials at 27°C.

Material	Specific Heat ($J.Kg^{-1}.K^{-1}$)	Material	Specific Heat ($J.Kg^{-1}.K^{-1}$)
Metals:		Ceramics:	
Al	900	Al_2O_3	837
Cu	385	Diamond	519
B	1 026	SiC	1 047
Fe	444	Si_3N_4	712
Pb	159	SiO_2 (silica)	1 109
Mg	1 017	Polymers:	
Ni	444	High-density polyethylene	1 842
Si	703	Low-density polyethylene	2 302
Ti	523	Nylon-6,6 (polyamide)	1 674
W	134	Polystyrene	1 172
Zn	389	Other:	
		Water	4 186
		Nitrogen	1 042

capacity nor the specific heat depends significantly on the structure of the material; thus, changes in dislocation density, grain size, or vacancies have little effect.

The most important factor affecting specific heat is the lattice vibration, or phonons. However, other factors affect the heat capacity, one striking example occurs in ferromagnetic materials such as iron (Figure 21.2). An abnormally high heat capacity is observed in iron at the Curie temperature, where the normally aligned magnetic moments of the iron atoms are randomised and the iron becomes paramagnetic. Heat capacity also depends on the crystal structure, shown as well in Figure 21.2 for iron.

EXAMPLE 21.1

How much heat must be supplied to 250 g of tungsten to raise its temperature from 25°C to 650°C?

SOLUTION

The specific heat of tungsten is 134 $J.kg^{-1}.K^{-1}$. Thus:

$$\begin{aligned}\text{Heat required} &= (\text{specific heat})(\text{mass})(\Delta T)\\ &= (134\ J.kg^{-1}.K^{-1})(0.25\ kg)(650-25)\\ &= 20938\ J = 20.9\ kJ\end{aligned}$$

If no losses occur, 20.9 kJ must be supplied to the tungsten. A variety of processes might be used to heat the metal. We could use a gas torch, we could place the tungsten in an induction coil to induce eddy currents, we could pass an electrical current through the metal, or we could place the metal into an oven heated by SiC resistors.

EXAMPLE 21.2

Suppose the temperature of 50 g of niobium increases 75°C when heated for a period of time. Estimate the specific heat and determine the heat in joules required.

SOLUTION

The atomic weight of niobium is 92.91 $g.mol^{-1}$. We can use Equation 21.3 to estimate the heat required to raise the temperature of one gram by one °C:

$$c \approx \frac{25}{92.91} = 0.269 \text{ J.g}^{-1}\text{.K}^{-1} = 269 \text{ J.Kg}^{-1}\text{.K}^{-1}$$

Thus the total heat required is:

$$\text{Heat} = (0.269 \text{ J.g}^{-1}\text{.K}^{-1})(50 \text{ g})(75°\text{C}) = 1009 \text{ J}$$

21.3 Thermal Expansion

An atom that gains thermal energy and begins to vibrate behaves as though it has a larger atomic radius. The average distance between the atoms and the overall dimensions of the material increase. The change in the dimensions of the material Δl per unit length is given by the linear coefficient of thermal expansion, α:

$$\alpha = \frac{l_f - l_0}{l_0(T_f - T_0)} = \frac{\Delta l}{l_0 \Delta T} \tag{21.4}$$

where T_0 and T_f are the initial and final temperatures and l_0 and l_f are the initial and final dimension of the material. A *volume* coefficient of thermal expansion (α_υ) can also be defined to describe the change in volume when the temperature of the material is changed. If the material is isotropic, $\alpha_\upsilon = 3\alpha$. Coefficients of thermal expansion for several materials are included in Table 21.2.

The coefficient of thermal expansion is related to the strength of the atomic bonds (Figure 2.15). In order for the atoms to move from their equilibrium positions, energy must be introduced into the material. If a very deep energy trough caused by strong atomic bonding is characteristic of the material, the atoms separate to a lesser degree and the material has a low linear coefficient of expansion. This relationship also suggests that materials having a high melting temperature – also due to strong

Table 21.2 The linear coefficient of thermal expansion at room temperature for selected materials.

Material	Linear Coefficient of Thermal Expansion (K^{-1})
Al	2.50×10^{-5}
Cu	1.66×10^{-5}
Fe	1.20×10^{-5}
Ni	1.30×10^{-5}
Pb	2.90×10^{-5}
Si	3.0×10^{-6}
W	4.5×10^{-6}
0.2% C steel	1.20×10^{-5}
Al-1.2% Mn alloy	2.32×10^{-5}
Grey cast iron	1.20×10^{-5}
Invar (Fe-36% Ni)	1.54×10^{-6}
Stainless steel	1.73×10^{-5}
Yellow brass	1.89×10^{-5}
Epoxy	5.50×10^{-5}
6,6-nylon	8.00×10^{-5}
6,6-nylon-33% glass fibre	2.00×10^{-5}
Polyethylene	1.00×10^{-4}
Polyethylene-30% glass fibre	4.80×10^{-5}
Polystyrene	7.00×10^{-5}
Al_2O_3	6.7×10^{-6}
Fused silica	5.5×10^{-7}
Partially stabilised ZrO_2	1.06×10^{-5}
SiC	4.3×10^{-6}
Si_3N_4	3.3×10^{-6}
Soda-lime glass	9.0×10^{-6}

atomic attractions – have low coefficients of thermal expansion (Figure 21.3). Consequently, lead (Pb) has a much larger coefficient than high-melting-point metals such as tungsten (W). Most ceramics, which have strong ionic or covalent bonds, have low coefficients compared with metals. Certain glasses, such as fused silica, also have a poor packing factor, which helps accommodate thermal energy with little dimensional change. Although bonding within the chains of polymers is covalent, the secondary bonds holding the chains together are weak, leading to high coefficients. Polymers that contain strong cross-linking, such as epoxies, typically have lower coefficients than linear polymers such as polyethylene.

Several precautions must be taken when calculating dimensional changes in materials:

1. The expansion characteristics of some materials, particularly single crystals or materials having a preferred orientation, are anisotropic.

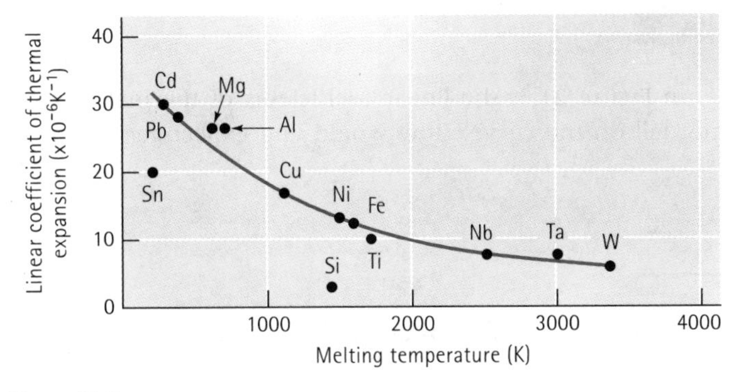

Figure 21.3
The relationship between the linear coefficient of thermal expansion (at 25°C) and the melting temperature in metals. Higher-melting-point metals tend to expand to a lesser degree.

2. Allotropic materials have abrupt changes in their dimensions when phase transformation occurs [Figure 21.4(a)]. These abrupt changes contribute to cracking of refractories on heating or cooling and quench cracks in steels.
3. The linear coefficient of expansion continually changes with temperature. Normally, α either is listed in handbooks as a complicated temperature-dependent function or is given as a constant for only a particular temperature range.
4. Interaction of the material with electric or magnetic fields produced by magnetic domains may prevent normal expansion until temperatures above the Curie temperature are reached. This is the case for Invar, an Fe-36% Ni alloy, which undergoes practically no dimensional changes at temperatures below the Curie temperature (about 200°C). This makes Invar attractive as a material for bimetallics [Figure 21.4(b)].

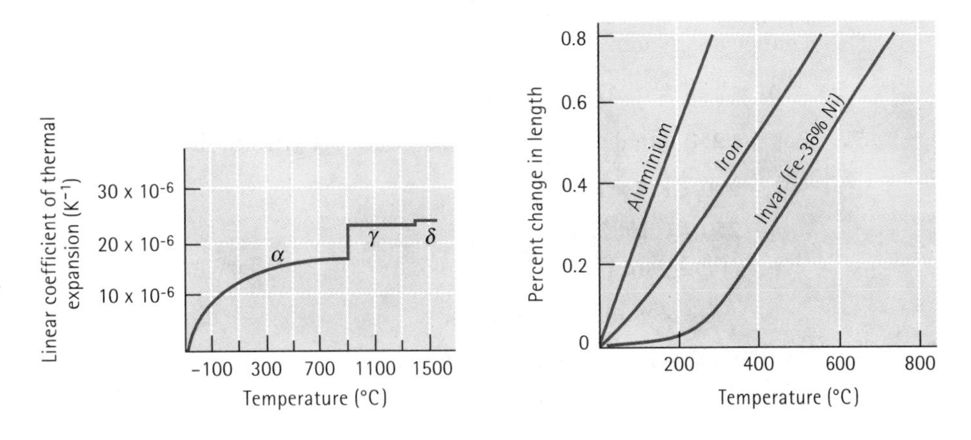

Figure 21.4
(*a*) The linear coefficient of thermal expansion of iron changes abruptly at temperatures where an allotropic transformation occurs. (*b*) The expansion of Invar is very low due to the magnetic properties of the material at low temperatures.

EXAMPLE 21.3

Explain why, in Figure 21.3, the linear coefficients of thermal expansion for silicon and tin do not fall on the curve. How would you expect germanium to fit into this figure?

SOLUTION

Both silicon and tin are covalently bonded. The strong covalent bonds are more difficult to stretch than the metallic bonds (a deeper trough in the energy-separation curve), so these elements have a lower coefficient. Since germanium also is covalently bonded, its thermal expansion should be less than that predicted by Figure 21.3.

EXAMPLE 21.4 **Design of a Pattern for a Casting Process**

Specify the dimensions for a pattern that will be used to produce a rectangular-shaped aluminium casting having dimensions at 25°C of 250 mm × 250 mm × 30 mm.

SOLUTION

To produce a casting having particular final dimensions, the mould cavity into which the liquid aluminium is to be poured must be made oversize. After the liquid solidifies, which occurs at 660°C for pure aluminium, the solid casting contracts as it cools to room temperature. If we calculate the amount of contraction expected, we can make the original pattern used to produce the mould cavity that much larger.

The linear coefficient of thermal expansion for aluminium is 25×10^{-6} K^{-1}. The temperature change from the freezing temperature to 25°C is 660 – 25 = 635°C. The change in any dimension is given by:

$$\Delta l = l_0 - l_f = \alpha l_0 \Delta T$$

For the 250 mm dimensions, $l_f = 250$ mm. We wish to find l_0 :

$$l_0 - 250 = (25 \times 10^{-6})(l_0)(635)$$

$$l_0 - 250 = 0.15875\ l_0$$

$$0.984\ l_0 = 250$$

$$l_0 = 254 \text{ mm}$$

For the 30 mm dimensions, $l_f = 30$ mm

$$l_0 - 30 = (25 \times 10^{-6})(l_0)(635)$$

$$l_0 - 30 = 0.015875\ l_0$$

$$0.984\ l_0 = 30$$

$$l_0 = 30.5 \text{ mm}$$

If we design the pattern to the dimensions 254 mm × 254 mm × 30.5 mm, the casting should contract to the required dimensions.

When an isotropic material is slowly and uniformly heated, the material expands uniformly without creating any residual stress. If, however, the material is restrained from moving, the dimensional changes may not be possible and, instead, stresses develop. These thermal stresses are related to the coefficient of thermal expansion, the modulus of elasticity E of the material, and the temperature change ΔT:

$$\sigma_{\text{thermal}} = \alpha E \Delta T \tag{21.5}$$

Thermal stresses can arise from a variety of sources. In large rigid structures such as bridges, restraint may develop as a result of the design. Some bridges are designed in sections, so that the sections move relative to one another during seasonal temperature changes.

When materials are joined – for example, coating cast iron bathtubs with a ceramic enamel or coating superalloy turbine blades with a zirconia thermal barrier – changes in temperature cause different amounts of contraction or expansion in the different materials. This disparity leads to thermal stresses that may cause the brittle coating to fail. Careful matching of the thermal properties of the coating to those of the substrate material is necessary to prevent the coating from cracking (if the coefficient of the coating is less than that of the underlying substrate) or spalling (flaking of the coating due to its higher expansion coefficient).

A similar situation may occur in composite materials. Brittle fibres that have a lower coefficient than the matrix may be stretched to the breaking point when the temperature of the composite increases.

Thermal stresses may even develop in a nonrigid, isotopic material if the temperature is not uniform. In producing tempered glass (Chapter 14), the surface is cooled more rapidly than the centre, permitting the surface to initially contract. When the centre cools later, its contraction is restrained by the rigid surface, placing compressive residual stresses on the surface.

EXAMPLE 21.5 Design of a Protective Coating

A ceramic enamel is to be applied to a 0.2%C steel plate. The ceramic has a fracture strength of 27.5 $MN.m^{-2}$, a modulus of elasticity of 103 $GN.m^{-2}$, and a coefficient of thermal expansion of 10×10^{-6} K^{-1}. Calculate the maximum temperature change that can be allowed without cracking the ceramic.

SOLUTION

Because the enamel is bonded to the 0.2%C steel, it is essentially restrained. If only the enamel heated (and the steel remained at a constant temperature), the maximum temperature change would be:

$$\sigma_{\text{thermal}} = \alpha E \Delta T = \sigma_{\text{fracture}}$$

$(10 \times 10^{-6}\ K^{-1})(103\,000\ MN.m^{-2})\ \Delta T = 27.5\ MN.m^{-2}$

$\Delta T = 26.7°C$

However, the steel also expands. Its coefficient of thermal expansion (Table 21.2) is $12 \times 10^{-6}\ K^{-1}$ and its modulus of elasticity is 207 $GN.m^{-2}$. Since the steel expands more than the enamel, a stress is still introduced into the enamel. The net coefficient of expansion is $\Delta\alpha = 12 \times 10^{-6} - 10 \times 10^{-6} = 2 \times 10^{-6}\ K^{-1}$:

$\sigma = (2 \times 10^{-6})(103\,000)\ \Delta T = 27.5$

$\Delta T = 133°C$

In order to permit greater temperature variations, we might select an enamel that has a coefficient of thermal expansion closer to that of steel, an enamel that has a lower modulus of elasticity (so that greater strains can be permitted before the stress reaches the fracture stress), or an enamel that has a higher strength.

21.4 Thermal Conductivity

The thermal conductivity k is a measure of the rate at which heat is transferred through a material. The conductivity relates the heat Q transferred across a given plane of area A per second when a temperature gradient $\Delta T/\Delta x$ exists (Figure 21.5):

$$\frac{Q}{A} = k\frac{\Delta T}{\Delta x} \tag{21.6}$$

Note that the thermal conductivity k plays the same role in heat transfer that the diffusion coefficient D does in mass transfer. Values for the thermal conductivity of typical materials are included in Table 21.3.

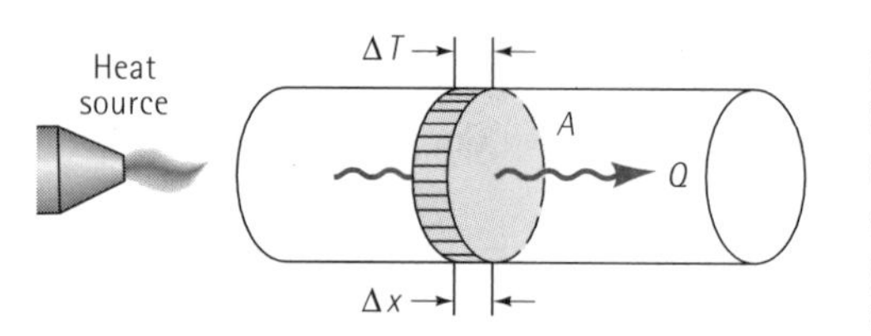

Figure 21.5
When one end of a bar is heated, a heat flux Q/A flows toward the cold end at a rate determined by the temperature gradient produced in the bar and the thermal conductivity of the material.

Thermal energy is transferred by two important mechanisms: transfer of free electrons and lattice vibrations (or phonons). Valence electrons gain energy, move toward the colder areas of the material, and transfer their energy to other atoms. The amount of energy transferred depends on the number of excited electrons and their mobility; these, in turn, depend on the type of material, lattice imperfections, and temperature. In addition, thermally induced vibrations of the atoms transfer energy through the material.

Table 21.3 Thermal conductivity of selected materials.

Material	Thermal Conductivity ($W.m^{-1}.K^{-1}$)	Material	Thermal Conductivity ($W.m^{-1}.K^{-1}$)
Pure metals:		Ceramics:	
Al	238	Al_2O_3	16
Cu	402	Carbon (diamond)	2 320
Fe	80	Carbon (graphite)	335
Mg	100	Fireclay	0.27
Pb	35	Silicon carbide	88
Si	151	Si_3N_4	15
Ti	22	Soda-lime glass	0.96
W	172	Vitreous silica	1.34
Zn	117	Vycor glass	1.26
Zr	23	ZrO_2	5.0
Alloys:		Polymers:	
0.2% C steel	100	6,6-nylon (polyamide)	0.25
Al-1.2% Mn alloy	280	Polyethylene	0.33
304 stainless steel (18/8)	30	Polyimide	0.21
Cementite	50	Polystyrene foam	0.03
Cu-30% Ni	50		
Ferrite	75		
Grey cast iron	80		
Yellow brass	222		

Note: 1 watt (W) = $1.J.s^{-1}$

Metals Because the valence band is not completely filled in metals, electrons require little thermal excitation in order to move and contribute to the transfer of heat. Since the thermal conductivity of metals is due primarily to the electronic contribution, we expect a relationship between thermal and electrical conductivity:

$$\frac{k}{\sigma T} = L = 2.30 \times 10^{-8} \text{ W.}\Omega\text{.K}^{-2} \quad (21.7)$$

where L is the Lorentz constant. This relationship is followed to a limited extent in many metals.

When the temperature of the material increases, two off-setting factors affect thermal conductivity. Higher temperatures are expected to increase the energy of the electrons, creating more 'carriers' and also increasing the contribution from lattice vibration; these effects increase the thermal conductivity. However, the greater lattice vibration scatters the electrons, reducing their mobility, and therefore decreases the thermal conductivity. The combined effect of these factors leads to very different behaviour for different metals. For iron, the thermal conductivity initially decreases with increasing temperature (due to the lower mobility of the electrons), then

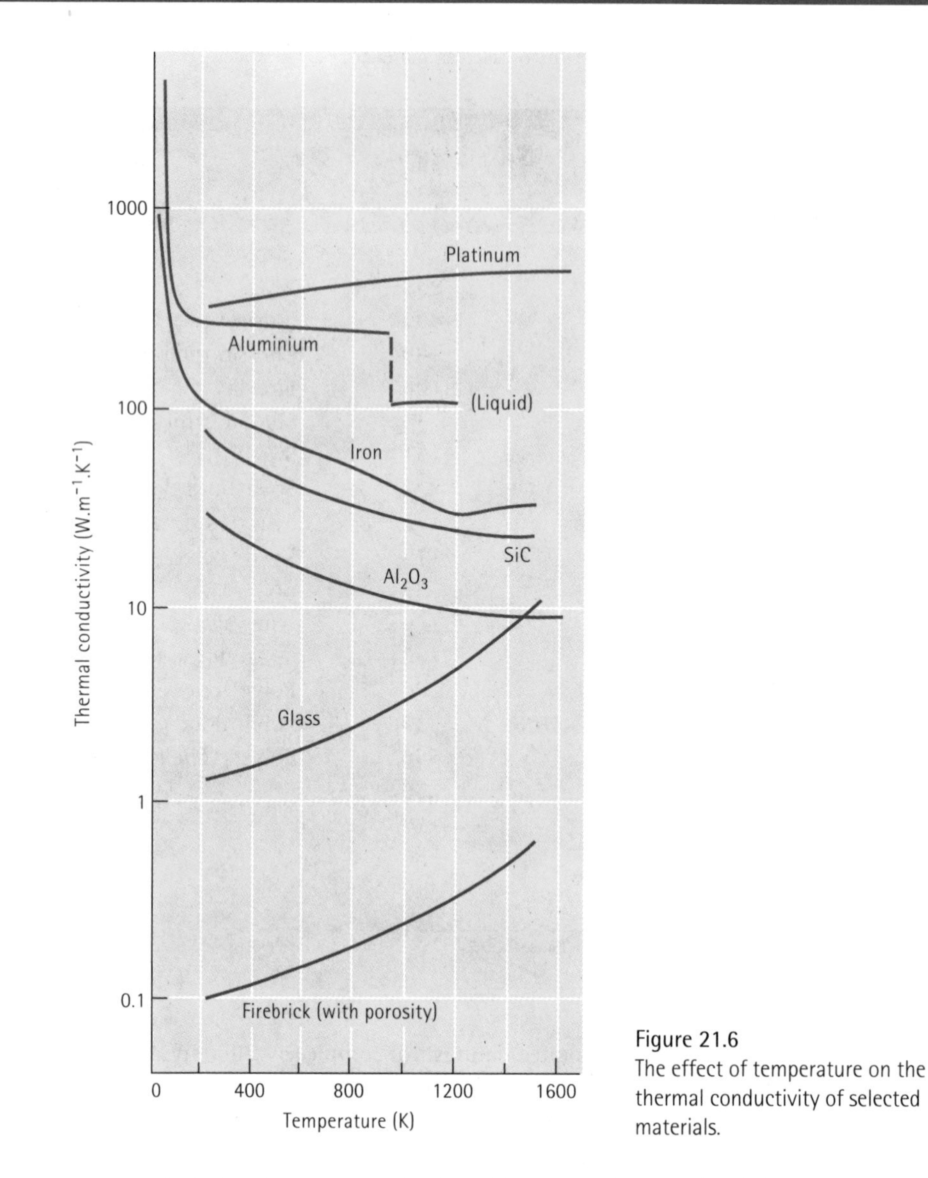

Figure 21.6
The effect of temperature on the thermal conductivity of selected materials.

increases slightly (due to more lattice vibration). The conductivity *decreases* continuously when aluminium is heated but *increases* continuously when platinum is heated (Figure 21.6).

Thermal conductivity in metals also depends on lattice defects, microstructure, and processing. Thus cold-worked metals, solid-solution-strengthened metals, and two-phase alloys might display lower conductivities compared with defect-free counterparts.

Ceramics The energy gap in ceramics is too large for many electrons to be excited into the conduction band except at very high temperatures. Consequently, transfer of heat in ceramics is caused primarily by lattice vibrations (or phonons). Since the electronic contribution is absent, the thermal conductivity of most ceramics is much lower than that of metals.

Glasses have low thermal conductivity. The amorphous loosely packed structure minimises the points at which silicate chains contact one another, making it more difficult for the phonons to be transferred. However, the thermal conductivity increases as the temperature increases; higher temperatures produce more energetic phonons and more rapid transfer of heat.

The more ordered structure of the crystalline ceramics, as well as glass-ceramics that contain large amounts of crystalline precipitates, cause less scattering of phonons. Compared with glasses, these materials have a higher thermal conductivity. As the temperature increases, however, scattering becomes more pronounced and the thermal conductivity decreases (as shown for alumina and silicon carbide in Figure 21.6). At still higher temperatures heat transfer by radiation becomes significant, and the conductivity may increase.

Other factors also influence the thermal conductivity of ceramics. Materials with a close-packed structure and high modulus of elasticity produce high-energy phonons that encourage high thermal conductivities. Lattice defects and porosity increase scattering and reduce conductivity. The best insulating brick, for example, contains a large fraction of porosity.

Some ceramics have thermal conductivities approaching that of metals. Although advanced ceramics such as AlN and SiC are good thermal conductors, they are also electrical insulators; therefore, these materials are good candidates for use in electronic applications where heat dissipation is needed.

Semiconductors Heat is conducted in semiconductors by both phonons and electrons. At low temperatures, phonons are the principal carriers of energy, but at higher temperatures electrons are excited through the small energy gap into the conduction band and thermal conductivity increases significantly.

Polymers The thermal conductivity of polymers is very low – even in comparison with ceramic glasses. Energy is transferred by vibration and movement of the molecular polymer chains. Increasing the degree of polymerisation, increasing the crystallinity, minimising branching, and providing extensive cross-linking all provide for higher thermal conductivity.

Unusually good thermal insulation is obtained using polymer foams, often produced from polystyrene or polyurethane. Polystyrene foam coffee cups are a typical product.

EXAMPLE 21.6 Design of a Window Glass

Design a glass window 1.2 m × 1.2 m square that separates a room at 25°C from the outside at 40°C and allows no more than 20 MJ of heat to enter the room each day.

SOLUTION

Table 21.3 shows that the thermal conductivity of a soda-lime glass – typical of windows – is 0.96 $W.m^{-1}.K^{-1}$. From Equation 21.6:

$$\frac{Q}{A} = k\frac{\Delta T}{\Delta x}$$

where Q/A is the heat transferred per second through the window.

$$1 \text{ day} = (24 \text{ h/day})(3\,600 \text{ s/h}) = 8.64 \times 10^4 \text{ s}$$

$$A = (1.2 \text{ m})^2 = 1.44 \text{ m}^2$$

$$Q = \frac{(2 \times 10^7 \text{J/day})}{8.64 \times 10^4 \text{s/day}} = 231.5 \text{ J.s}^{-1}, \text{or } 231.5 \text{ W}$$

$$\frac{Q}{A} = \frac{231.5 \text{ J.s}^{-1}}{1.44 \text{ m}^2} = 160.8 \text{ J.s}^{-1}.\text{m}^{-2}$$

$$\frac{Q}{A} = 160.8 \text{ J.s}^{-1}.\text{m}^{-2} = (0.96 \text{ W.m}^{-1}.\text{K}^{-1})(40 - 25°\text{C}) / \Delta x$$

$$\Delta x = 0.09 \text{ m} = 90 \text{ mm} = \text{thickness}$$

The glass would have to be exceptionally thick to prevent the desired maximum heat flux. We might do several things to reduce the heat flux. Although all of the ceramic glasses have similar thermal conductivities, we might use instead a transparent polymer material (such as polymethyl methacrylate). The polymers have thermal conductivities approximately one order of magnitude smaller than the ceramic glasses. We could also use a double-paned glass, with the glass panels separated either by an air gap (where air has a very low thermal conductivity) or a sheet of transparent polymer.

21.5 Thermal Shock

Stresses leading to fracture of brittle materials can be introduced thermally as well as mechanically. When a piece is cooled quickly, a temperature gradient is produced. This gradient can lead to different amounts of contraction in different areas. If residual tensile stresses become high enough, flaws may propagate and cause failure. Similar behaviour can occur if a material is heated rapidly. Thermal shock is affected by several factors:

1. *Coefficient of thermal expansion*: A low coefficient minimises dimensional changes and reduces thermal shock.
2. *Thermal conductivity*: The magnitude of the temperature gradient is determined partly by the thermal conductivity of the material. A high thermal conductivity helps to transfer heat and reduce temperature differences quickly in the material.
3. *Modulus of elasticity*: A low modulus of elasticity permits large amounts of strain before the stress reaches the critical level required to cause fracture.
4. *Fracture stress*: A high stress required for fracture permits larger strains. The fracture stress for a particular material is high if flaws are small and few in number.
5. *Phase transformations*: Additional dimensional changes can be caused by phase transformations. Transformation of silica from quartz to cristobalite, for example, introduces residual stresses and increases problems with thermal shock.

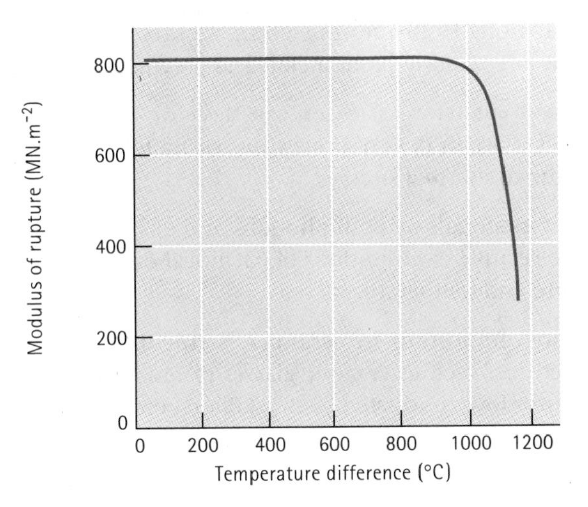

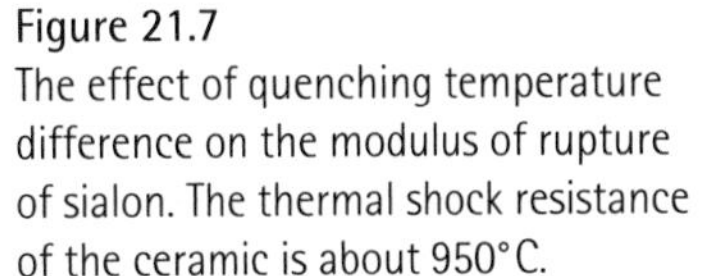
Figure 21.7
The effect of quenching temperature difference on the modulus of rupture of sialon. The thermal shock resistance of the ceramic is about 950°C.

One method for measuring resistance to thermal shock is to determine the maximum temperature difference that can be tolerated during a quench without affecting the mechanical properties of the material. Pure (fused) silica glass has a thermal shock resistance of about 3 000°C. Figure 21.7 shows the effect of quenching temperature difference on the modulus of rupture in sialon ($Si_3Al_3O_3N_5$) after quenching; no cracks and therefore no change in the properties of the ceramic are evident until the quenching temperature difference approaches 950°C. Other ceramics have poorer resistance. Shock resistance for partially stabilised zirconia (PSZ) and Si_3N_4 is about 500°C; for SiC 350°C; and for Al_2O_3 and ordinary glass, about 200°C.

Another way to evaluate the resistance of a material to thermal shock is by the thermal shock parameter:

$$\text{Thermal shock parameter} = \frac{\sigma_f k}{E\alpha} \tag{21.8}$$

where σ_f is the fracture stress of the material, k is the thermal conductivity, E is the modulus of elasticity, and α is the linear coefficient of thermal expansion.

Thermal shock is not a problem in most metals, because metals normally have sufficient ductility to permit deformation rather than fracture.

SUMMARY

- The thermal properties of materials can be at least partly explained by the movement of electrons and phonons, which are discrete elastic waves caused by vibration of the material's structure.
- Heat capacity and specific heat represent the quantity of energy required to raise the temperature of a given amount of material by one degree Celsius and they are influenced by temperature, crystal structure and bonding.
- The coefficient of thermal expansion describes the dimensional changes that occur in a material when its temperature changes. Strong bonding leads to a low

coefficient of expansion. High-melting-point metals and ceramics have low coefficients, whereas low-melting-point metals and polymers have high coefficients.

- Because of thermal expansion, stresses can develop in a material when the temperature changes. Care in design, processing, or materials selection is required to prevent failure due to thermal stresses.
- Heat is transferred in materials by both phonons and electrons. Thermal conductivity depends on the relative contributions of each of these mechanisms, as well as the effect of structure and temperature:
  - Phonons are most important in ceramics, semiconductors, and polymers. Disordered structures, such as ceramic glasses or amorphous polymers, scatter phonons and cause low conductivity. Crystalline ceramics and polymers have higher conductivities than their glassy or amorphous counterparts. Second phases, including porosity, have a significant effect on thermal conductivity in nonmetallic materials.
  - Electronic contributions are most important in metals; consequently, lattice imperfections that scatter electrons reduce conductivity. The effect of temperature is less easy to characterise, since increasing temperature increases phonon energy but also increases scattering of both phonons and electrons.

GLOSSARY

Heat capacity
The energy required to raise the temperature of one mole of a material by one degree Celsius.

Linear coefficient of thermal expansion
Describes the amount by which each unit length of a material changes when the temperature of the material changes by one degree Celsius.

Lorentz constant
The constant that relates electrical and thermal conductivity.

Phonon
An elastic wave, characterised by its energy, wavelength, or frequency, which transfers energy through a material.

Specific heat
The energy required to raise the temperature of one gram of a material by one degree Celsius.

Thermal conductivity
Measures the rate at which heat is transferred through a material.

Thermal shock
Failure of a material caused by stresses introduced by sudden changes in temperature.

Thermal stresses
Stresses introduced into a material due to differences in the amount of expansion or contraction that occur because of a temperature change.

PROBLEMS

21.1 Calculate the heat required to raise the temperature of 1 kg of the following materials by 50°C:

(a) lead (b) nickel

(c) Si_3N_4 (d) 6,6-nylon

21.2 Calculate the temperature of a 100 g sample of the following materials (originally at 25°C) when 12 500 J are introduced:

(a) tungsten (b) titanium

(c) Al_2O_3 (d) low-density polyethylene

21.3 An alumina insulator for an electrical device is also to serve as a heat sink. A 10°C temperature rise in an alumina insulator 10 mm × 10 mm × 0.2 mm is observed during use. Determine the thickness of a high-density polyethylene insulator that would be needed to provide the same performance as a heat sink. The density of alumina is 3.96 $Mg.m^{-3}$.

21.4 A 200 g sample of aluminium is heated to 400°C and is then quenched into 2 litres of water at 20°C. Calculate the temperature of the water after the aluminium and water reach equilibrium. Assume no temperature loss from the system.

21.5 A 2 m-long soda-lime glass sheet is produced at 1400°C. Determine its length after it cools to 25°C.

21.6 A copper casting is to be produced having the final dimensions of 25 mm × 300 mm × 600 mm. Determine the size of the pattern that must be used to make the mould into which the liquid copper is poured during the manufacturing process.

21.7 An aluminium casting is made by the permanent mould process. In this process, the liquid aluminium is poured into a grey cast iron mould that is heated to 350°C. We wish to produce an aluminium casting that is 375 mm long at 25°C. Calculate the length of the cavity that must be machined into the grey cast iron mould.

21.8 We coat a 1 m-long, 2-mm diameter copper wire with a 0.5 mm-thick epoxy insulation coating. Determine the length of the copper and the coating when their temperature increases from 25°C to 250°C. What is likely to happen to the epoxy coating as a result of this heating?

21.9 We produce a 250 mm-long bimetallic composite material composed of a strip of yellow brass bonded to a strip of Invar. Determine the length to which each material would like to expand when the temperature increases from 20°C to 150°C. Draw a sketch showing what will happen to the shape of the bimetallic strip.

21.10 A nickel engine part is coated with SiC to provide corrosion resistance at high temperatures. If no residual stresses are present in the part at 20°C, determine the thermal stresses that develop when the part is heated to 1 000°C during use. (*See* Table 14.4.)

21.11 Alumina fibres 20 mm long are incorporated into an aluminium matrix. Assuming good bonding between the ceramic fibres and the aluminium, estimate the thermal stresses acting on the fibre when the temperature of the composite increases 250°C. Are the stresses on the fibre tensile or compressive? (*See* Table 14.4.)

21.12 A 600 mm-long copper bar with a yield strength of 210 $MN.m^{-2}$ is heated to 120°C and immediately fastened securely to a rigid framework. Will the copper deform plastically during cooling to 25°C? How much will the bar deform if it is released from the framework after cooling?

21.13 Repeat Problem 21.12, but using a silicon carbide rod rather than a copper rod. (*See* Table 14.4.)

21.14 A 30 mm-thick plate of silicon carbide separates liquid aluminium (held at 700°C) from a water-cooled steel shell maintained at 20°C. Calculate the heat Q transferred to the steel per mm^2 of silicon carbide each second.

21.15 A sheet of 0.25 mm-thick polyethylene is sandwiched between two 900 mm × 900 mm × 3 mm sheets of soda-lime glass to produce a window. Calculate

(a) the heat lost through the window each day when room temperature is 25°C and the outside air is 0°C and

(b) the heat entering through the window each day when room temperature is 25°C and the outside air is 40°C.

21.16 We would like to build a heat deflection plate that permits heat to be transferred rapidly parallel to the sheet but very slowly perpendicular to the sheet. Consequently, we incorporate 1 kg of copper wires, each 1 mm in diameter, into 5 kg of a polyimide polymer matrix. Estimate the thermal conductivity parallel and perpendicular to the sheet.

21.17 Suppose we just dip a 10 mm-diameter, 10 mm-long rod of aluminium into one litre of water at 20°C. The other end of the rod is in contact with a heat source operating at 400°C. Determine the length of time required to heat the water to 25°C if 75% of the heat is lost by radiation from the bar.

21.18 Determine the thermal shock parameter for silicon nitride, hot pressed silicon carbide, and alumina and compare it with the thermal shock resistance as defined by the maximum quenching temperature difference. (*See* Table 14.4.)

21.19 Grey cast iron has a higher thermal conductivity than ductile or malleable cast iron. Review Chapter 12 and explain why this difference in conductivity might be expected.

Design Problems

21.20 A chemical reaction vessel contains liquids at a temperature of 680°C. The wall of the vessel must be constructed so that the outside wall operates at a temperature of 35°C or less. Design the vessel wall and appropriate materials if $Q_{maximum} = 25\,000$ W.

21.21 Liquid copper is held in a silicon nitride vessel. The inside diameter of the vessel is 75 mm. The outside of the vessel is in contact with copper that contains cooling channels through which water at 20°C flows at the rate of 50 litres per minute. The copper is to remain at a temperature below 25°C. Design a system that will accomplish this end.

21.22 Design a metal panel coated with a glass enamel capable of thermal cycling between 20°C and 150°C. The glasses generally available are expected to have a tensile strength of 35 $MN.m^{-2}$ and a compressive strength of 350 $MN.m^{-2}$.

21.23 Design a turbine blade for a jet engine that may be capable of operating at higher temperatures.

The failure of materials by corrosion, wear, and fracture is discussed in this part. We again find that deterioration or failure is related to the structure, properties, and processing of materials.

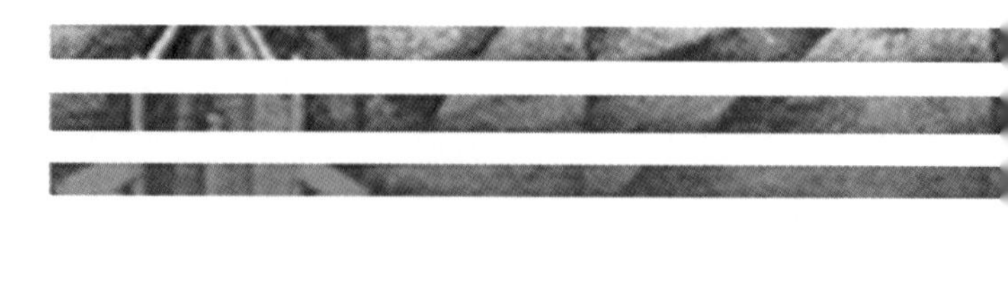

In Chapter 22, corrosion and wear are examined; electrochemical corrosion is found to be particularly important. In addition to examining the mechanism for corrosion, we look at techniques for controlling and preventing damage to a material by these processes.

In Chapter 23, we review the mechanical failure of materials and pick up some hints on how to identify the cause of fracture. We also examine a number of techniques by which we can nondestructively test a material to determine if it is subject to fracture.

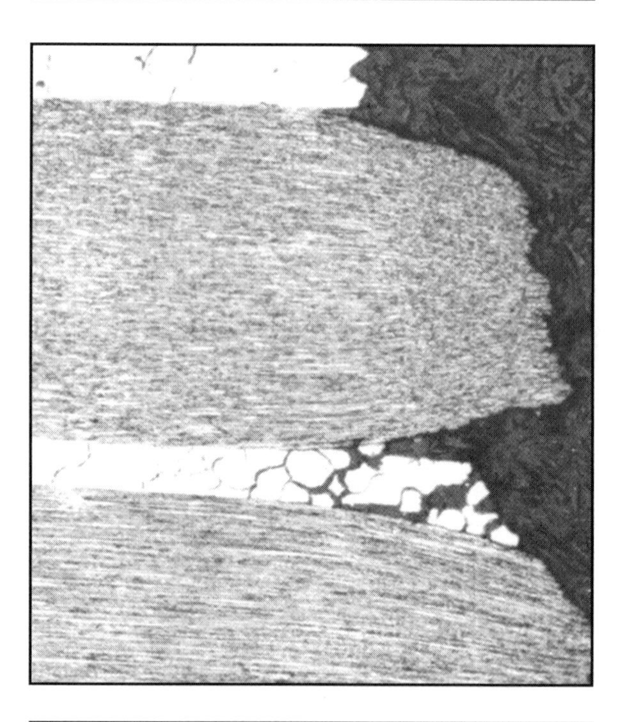

Fracture mechanisms differ in the two components of a nickel-base superalloy reinforced with tungsten fibres. The tungsten fibres (dark) failed in a ductile manner, with necking of the fibres at the fracture surface. The superalloy matrix failed in a brittle manner at the boundaries between the original powder particles used to produce the composite. (Magnification × 100) *(Courtesy of* Metals Handbook, *8th Ed., Vol. 12,* 'Fractography,' *ASM International, 1987.)*

Part

Protection Against Deterioration and Failure of Materials

CHAPTER 22

Corrosion and Wear

22.1 Introduction

The composition and physical integrity of a solid material are altered in a corrosive environment. In chemical corrosion, the material is dissolved by a corrosive liquid. In electrochemical corrosion, metal atoms are removed from the solid material as the result of an electric circuit that is produced. Metals and certain ceramics react with a gaseous environment, usually at elevated temperatures, and the material may be destroyed by formation of oxides or other compounds. Polymers degrade when exposed to oxygen at elevated temperatures. Materials may be altered when exposed to oxygen at elevated temperatures. Materials may be altered when exposed to radiation or even bacteria. Finally, a variety of wear and wear-corrosion mechanisms alter the shape of materials. Billions of dollars are required to repair the damage done by corrosion each year.

22.2 Chemical Corrosion

In chemical corrosion, or direct solution, a material dissolves in a corrosive liquid medium. The material continues to dissolve until either it is consumed or the liquid is saturated. A simple example is salt dissolving in water.

Liquid Metal Attack Liquid metals first attack a solid at high-energy locations, such as grain boundaries. If these regions continue to be attacked preferentially, cracks eventually grow (Figure 22.1). Often this form of corrosion is complicated by the presence of fluxes that accelerate the attack or by electrochemical corrosion. Ceramics can also be attacked by aggressive metals such as liquid lithium.

Selective Leaching One particular element in an alloy may be selectively dissolved, or leached, from the solid. Dezincification occurs in brass containing more than 15% Zn. Both copper and zinc are dissolved by aqueous solutions at high temperatures; the zinc irons remain in solution while the copper irons are replated onto the brass (Figure 22.2). Eventually, the brass becomes porous and weak.

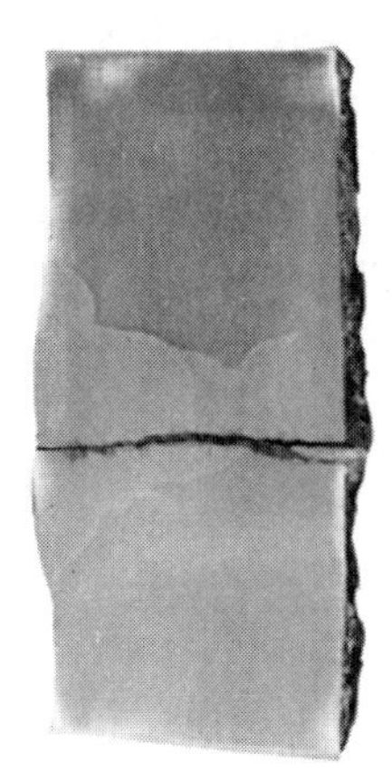

Figure 22.1
Molten lead is held in thick steel pots during refining. In this case, the molten lead has attacked a weld in a steel plate and cracks have developed. Eventually, the cracks propagate through the steel, and molten lead leaks from the pot.

Graphitic corrosion of grey cast iron occurs when iron is selectively dissolved in water or soil, leaving behind interconnected graphite flakes and a corrosion product. Localised graphitic corrosion often causes leakage or failure of buried grey cast iron gas lines, leading to explosions.

Solution of Ceramics Ceramic refractories used to contain molten metal during melting or refining may be may be dissolved by the slags that are produced on the metal surface; an acid (high SiO_2) refractory is rapidly attacked by a basic (high CaO or MgO) slag. A glass produced from SiO_2 and Na_2O is rapidly attacked by water; CaO must be added to the glass to minimise this attack. Nitric acid may selectively leach iron or silica from some ceramics, reducing their strength and density.

Chemical Attack of Polymers Solvents may diffuse into low-molecular-weight thermoplastic polymers heated above the glass transition temperature. As the solvent is incorporated into the polymer, the smaller solvent molecules force apart the chains, causing swelling. The strength of the bonds between the chains decreases, leading to softer, lower-strength polymers with a low glass transition temperature. In extreme cases, the swelling leads to stress cracking. Solution of water into nylon is one example of this phenomenon.

Thermoplastic polymers may also be dissolved into a solvent. Prolonged exposure causes loss of material and weakening of the polymer part. This process occurs most

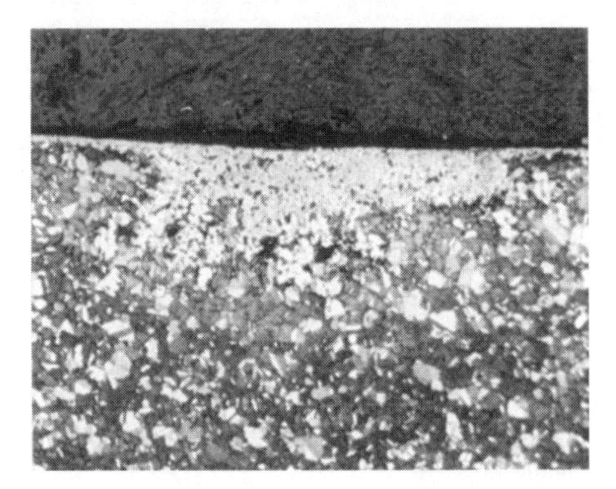

Figure 22.2
Photomicrograph of a copper deposit in brass, showing the effect of dezincification (× 50).

easily when the temperature is high and when the polymer has a low molecular weight, is highly branched and amorphous, and is not cross-linked. The structure of the monomer is also important; the CH_3 groups on the polymer chain in polypropylene are more easily removed from the chain than the chloride or fluoride ions in polyvinyl chloride (PVC) or polytetrafluoroethylene (Teflon). Teflon has exceptional resistance to chemical attack by almost all solvents.

22.3 Electrochemical Corrosion

Electrochemical corrosion, the most common form of attack of metals, occurs when metal atoms lose electrons and become ions, going into solution. As the metal is gradually consumed by this process, a by-product of the corrosion process is typically formed. Electrochemical corrosion occurs most frequently in an aqueous medium, in which ions are present in water or moist air. In this process, an electric circuit is created and the system is called an electrochemical cell. Corrosion of a steel pipe or a steel automobile panel, creating holes in the steel and rust as the by-product, are examples of this reaction.

Although responsible for corrosion, electrochemical cells may also be useful. By deliberately creating an electric circuit, we can *electroplate* protective or decorative coatings onto materials. In some cases, electrochemical corrosion is even desired. For example, in etching a polished metal surface with an appropriate acid, various features in the microstructure are selectively attacked, permitting them to be observed. In fact, most of the photomicrographs in this text were obtained in this way, thus enabling for example, the observation of pearlite in steel or grain boundaries in copper.

Components of an Electrochemical Cell There are four components of an electrochemical cell (Figure 22.3):

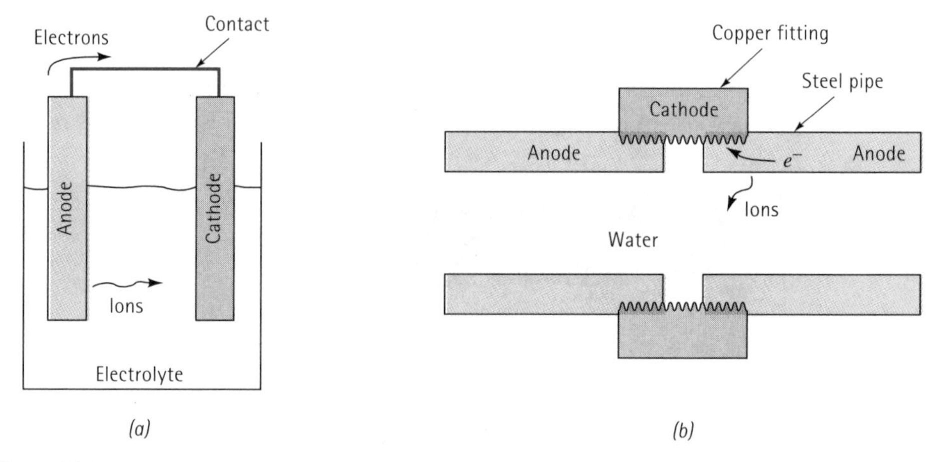

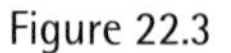

(a) *(b)*

Figure 22.3
The components in an electrochemical cell: *(a)* a simple electrochemical cell and *(b)* a corrosion cell between a steel water pipe and a copper fitting.

1. The anode gives up electrons to the circuit and corrodes.
2. The cathode receives electrons from the circuit by means of a chemical, or cathode, reaction. Ions that combine with the electrons produce a by-product at the cathode.
3. The anode and cathode must be electrically connected, usually by physical contact, to permit the electrons to flow from the anode to the cathode and continue the reaction.
4. A liquid electrolyte must be in contact with the anode and the cathode. The electrolyte is conductive, thus completing the circuit. It provides the means by which metallic ions leave the anode surface and ensures that ions move to the cathode to accept the electrons.

This description of an electrochemical cell defines either electrochemical corrosion or electroplating.

Anode Reaction The anode which is a metal, undergoes an oxidation reaction by which metal atoms are ionised. The metal ions enter the electrolytic solution, while the electrons leave the anode through the electrical connection:

$$M \rightarrow M^{n+} + ne^- \quad (22.1)$$

Because metal ions leave the anode, the anode corrodes.

Cathode Reaction in Electroplating In electroplating, a reduction reaction, which is the reverse of the anode reaction, occurs at the cathode:

$$M^{n+} + ne^- \rightarrow M \quad (22.2)$$

The metal ions, either intentionally added to the electrolyte or formed by the anode reaction, combine with the electrons at the cathode. The metal then plates out and covers the cathode surface.

Cathode Reactions in Corrosion Except in unusual conditions, plating of a metal does not occur during electrochemical corrosion. Instead, the reduction reaction forms a gas, solid or liquid by-product at the cathode (Figure 22.4).

1. *The hydrogen electrode*: In oxygen-free liquids, such as hydrochloric acid (HCl) or stagnant water, hydrogen gas may be evolved at the cathode:

   $$2H^+ + 2e^- \rightarrow H_2\uparrow \quad (22.3)$$

   If zinc were placed in such an environment, we would find that the overall reaction is:

   $$Zn \rightarrow Zn^{2+} + 2e^- \text{ (anode reaction)}$$
   $$2H^+ + 2e^- \rightarrow H_2\uparrow \text{ (cathode reaction)}$$
   $$Zn + 2H^+ \rightarrow Zn^{2+} + H_2\uparrow \text{ (overall reaction)} \quad (22.4)$$

   The zinc anode gradually dissolves, and hydrogen bubbles continue to evolve at the cathode.

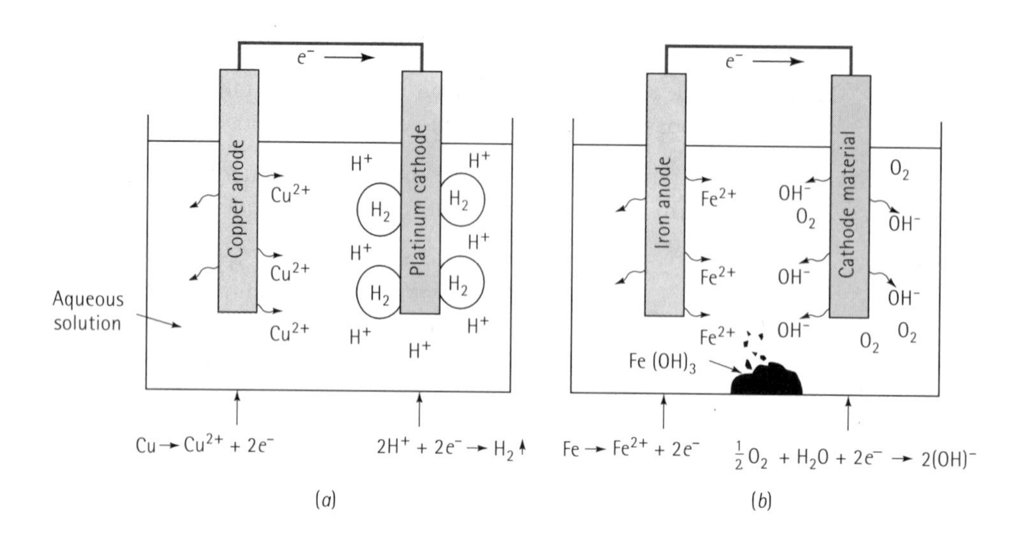

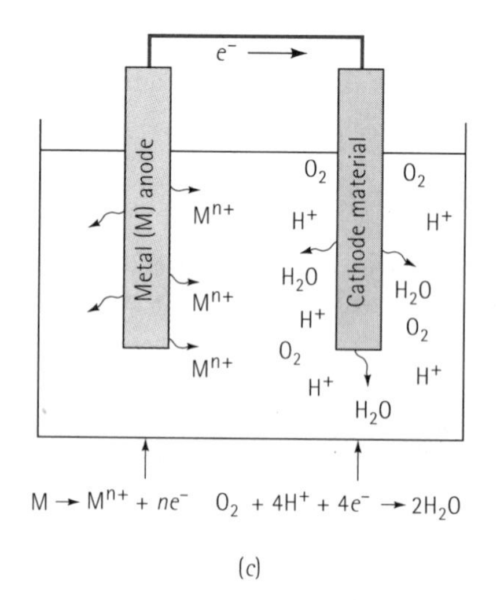

Figure 22.4
The anode and cathode reactions in typical electrolytic corrosion cells: (*a*) the hydrogen electrode, (*b*) the oxygen electrode, and (*c*) the water electrode.

2. *The oxygen electrode*: In aerated water, oxygen is available to the cathode, and hydroxyl, or $(OH)^-$, ions form:

$$\frac{1}{2}O_2 + H_2O + 2e^- \rightarrow 2(OH)^- \tag{22.5}$$

The oxygen electrode enriches the electrolyte in $(OH)^-$ ions. These ions react with positively charged metallic ions and produce a solid product. In the case of rusting of iron:

$$\begin{aligned}
&Fe \rightarrow Fe^{2+} + 2e^- \text{ (anode reaction)}\\
&\left.\begin{aligned}&\tfrac{1}{2}O_2 + H_2O + 2e^- \rightarrow 2(OH)^-\\ &Fe^{2+} + 2(OH)^- \rightarrow Fe(OH)_2\end{aligned}\right\} \text{(cathode reaction)}\\
&Fe + \tfrac{1}{2}O_2 + H_2O \rightarrow Fe(OH)_2 \text{ (overall reaction)}
\end{aligned} \tag{22.6}$$

The reaction continues as the $Fe(OH)_2$ reacts with more oxygen and water:

$$2Fe(OH)_2 + \frac{1}{2}O_2 + H_2O \rightarrow 2Fe(OH)_3 \tag{22.7}$$

$Fe(OH)_3$ is commonly known as *rust.*

3. *The water electrode*: In oxidising acids, the cathode reaction produces water as a by-product:

$$O_2 + 4H^+ + 4e^- \rightarrow 2H_2O \tag{22.8}$$

If a continuous supply of both oxygen and hydrogen is available, the water electrode produces neither a buildup of solid rust nor a high concentration or dilution of ions at the cathode.

22.4 The Electrode Potential in Electrochemical Cells

In electroplating, an imposed voltage is required to cause a current to flow in the cell. But in corrosion, a potential naturally develops when a material is placed in a solution. Let's see how the potential required to drive the corrosion reaction develops.

Electrode Potential When a perfect ideal metal (not an ordinary engineering metal) is placed in an electrolyte, an electrode potential is developed that is related to the tendency of the material to give up its electrons. However, the driving force for the oxidation reaction is offset by an equal but opposite driving force for the reduction reaction. No net corrosion occurs. Consequently, we cannot measure the electrode potential for a single electrode material.

Electromotive Force Series To determine the tendency of a metal to give up its electrons, we measure the potential difference between the metal and a standard electrode using a half-cell (Figure 22.5). The metal electrode to be tested is placed in a 1-M solution of its ions. A standard reference electrode is also placed in a 1-M solution of its ions. The two electrolytes are in electrical contact but are not permitted to mix with one another. Each electrode establishes its own electrode potential. By measuring the voltage between the two electrodes when the circuit is open, we obtain the potential difference. The potential of the hydrogen electrode which is taken as our standard reference electrode, is arbitrarily set equal to zero volts. If the metal has a greater tendency to give up electrons than hydrogen, then the potential of the metal is negative – the metal is anodic with respect to the hydrogen electrode.

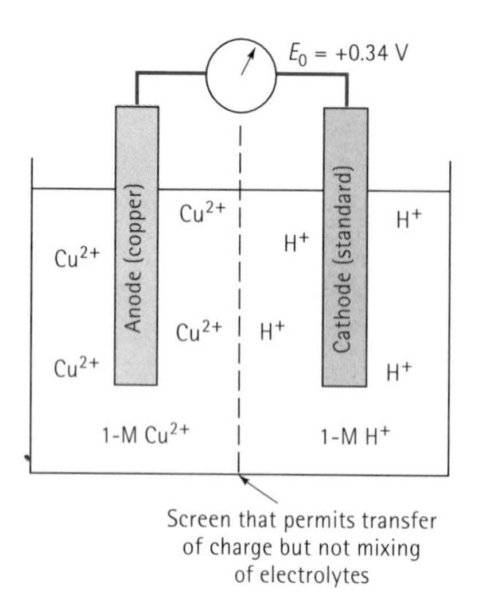

Figure 22.5
The half-cell used to measure the electrode potential of copper under standard conditions. The electrode potential of copper is the potential difference between it and the standard hydrogen electrode in an open circuit. Since E_0 is greater than zero, copper is cathodic compared with the hydrogen electrode.

The electromotive force (or emf) series shown in Table 22.1 compares the electrode potential E_0 for each metal with that of the hydrogen electrode under standard conditions of 25°C and 1-M solution of ions in the electrolyte. Note that the measurement of the potential is made when the electric circuit is open. The voltage difference begins to change as soon as the circuit is closed.

Table 22.1 The electromotive force (emf) series for selected elements.

	Metal	Electrode Potential E_0 (V)
Anodic	$Li \rightarrow Li^+ + e^-$	−3.05
	$Mg \rightarrow Mg^{2+} + 2e^-$	−2.37
	$Al \rightarrow Al^{3+} + 3e^-$	−1.66
	$Ti \rightarrow Ti^{2}+ + 2e^-$	−1.63
	$Mn \rightarrow Mn^{2+} + 2e^-$	−1.63
	$Zn \rightarrow Zn^{2+} + 2e^-$	−0.76
	$Cr \rightarrow Cr^{3+} + 3e^-$	−0.74
	$Fe \rightarrow Fe^{2+} + 2e^-$	−0.44
	$Ni \rightarrow Ni^{2+} + 2e^-$	−0.25
	$Sn \rightarrow Sn^{2+} + 2e^-$	−0.14
	$Pb \rightarrow Pb^{2+} + 2e^-$	−0.13
	$H_2 \rightarrow 2H^+ + 2e^-$	0.00
	$Cu \rightarrow Cu^{2+} + 2e^-$	+0.34
	$4(OH)^- \rightarrow O_2 + 2H_2O + 4e^-$	+0.40
	$Ag \rightarrow Ag^+ + e^-$	+0.80
	$Pt \rightarrow Pt^{4+} + 4e^-$	+1.20
	$2H_2O \rightarrow O_2 + 4H^+ + 4e^-$	+1.23
Cathodic	$Au \rightarrow Au^{3+} + 3e^-$	+1.50

Effect of Concentration on the Electrode Potential The electrode potential depends on the concentration of the electrolyte. At 25°C, the Nernst equation gives the electrode potential in nonstandard solutions:

$$E = E_0 + \frac{0.0592}{n}\log(C_{ion}) \tag{22.9}$$

where E is the electrode potential in a solution containing a concentration C_{ion} of the metal in molar units, n is the valance of the metallic ion, and E_0 is the standard electrode potential in a 1-M solution. Note that when $C_{ion} = 1$, $E = E_0$, because log 1=0.

EXAMPLE 22.1

Suppose 1 g of copper as Cu^{2+} is dissolved in 1 kg of water to produce an electrolyte. Calculate the electrode potential of the copper half-cell in this electrolyte.

SOLUTION

From chemistry, we know that a standard 1-M solution of Cu^{2+} is obtained when we add 1 mol of Cu^{2+} (an amount equal to the atomic mass of copper) to 1 kg of water. The atomic mass of copper is 63.54 g.mol^{-1}. The concentration of the solution when only 1 g of copper is added must be:

$$C_{ion} = \frac{1}{63.54} = 0.0157\text{M}$$

From the Nernst equation, with $n = 2$ and $E_0 = +0.34$ V:

$$E = E_0 + \frac{0.0592}{n} = \log(C_{ion}) = 0.34 + \frac{0.0592}{2}\log\ (0.0157)$$

$$= 0.34 + (0.0296)(-1.8) = 0.29\text{V}$$

Rate of Corrosion or Plating The amount of metal plated on the cathode in electroplating, or removed from the metal by corrosion, can be determined from Faraday's equation

$$w = \frac{ItM}{nF} \tag{22.10}$$

where w is the weight plated or corroded (g), I is the current (A), M is the atomic mass of the metal, n is the valence of the metal ion, t is the time (s), and F is Faraday's constant (96 500 C). Often the current is expressed in terms of current density, $i = I/A$, so Equation 22.10 becomes

$$w = \frac{iAtM}{nF} \tag{22.11}$$

where the area $A(\text{m}^2)$ is the surface area of the anode or cathode.

EXAMPLE 22.2 Design of a Copper Plating Process

Design a process to electroplate a 1 mm-thick layer of copper onto a 10 mm × 10 mm cathode surface.

SOLUTION

In order for us to produce a 1 mm-thick layer on a 100 mm^2 surface area, the weight of copper must be:

ρ_{Cu} = 8.96 Mg.m^{-3} A = 100 / 10^6 = 10^{-4} m^2

Volume of copper = (1 × 10^{-3} m)(10^{-4} m) = 10^{-7} m^3

Weight of copper = (8.96 Mg.m^{-3})(10^{-7} m^3) = 0.896 g

From Faraday's equation, where M_{Cu} = 63.54 g.mol^{-1} and n = 2:

$$It = \frac{wnF}{M} = \frac{(0.896)(2)(96\ 500)}{63.54} = 2\ 772\ \text{A}\cdot\text{s}$$

Therefore, we might use several different combinations of current and time to produce the copper plate:

Current	Time
0.1 A	27 220 s = 7.6 h
1.0 A	2 722 s = 45.4 min
10.0 A	272.2 s = 4.5 min
100.0 A	27.22 s = 0.45 min

Our choice of the exact combination of current and time might be made on the basis of rate of production and quality of the copper plate. Low currents require very long plating times, perhaps making the process economically unsound. High currents, however, may reduce plating efficiencies. The plating effectiveness may depend on the composition of the electrolyte containing the copper ions, as well as on any impurities or additives that are present. Additional background or experimentation may be needed to obtain the most economical and efficient plating process.

EXAMPLE 22.3

An iron container 100 mm × 100 mm at its base is filled to a height of 200 mm with a corrosive liquid. A current is produced as a result of an electrolytic cell, and after 4 weeks, the container has decreased in weight by 70 g. Calculate (1) the current and (2) the current density involved in the corrosion of the iron.

SOLUTION

1. The total exposure time is:

   $t = (4 \text{ wk})(7 \text{ d/wk})(24 \text{ h/d})(3600 \text{ s/h}) = 2.42 \times 10^6 \text{ s}$

   From Faraday's equation, using $n = 2$ and $M = 55.847 \text{ g.mol}^{-1}$

$$I = \frac{wnF}{tM} = \frac{(70)(2)(96\,500)}{(2.42 \times 10^6)(55.847)}$$
$$= 0.1 \text{ A}$$

2. The total surface area of iron in contact with the corrosive liquid and the current density are:

   $A = (4 \text{ sides})(100 \times 200) + (1 \text{ bottom})(100 \times 100) = 90\,000 \text{ mm}^2 = 0.09 \text{ m}^2$

$$i = \frac{I}{A} = \frac{0.1}{0.09} = 1.11 \text{ A.m}^{-2}$$

EXAMPLE 22.4

Suppose that, in a corrosion cell composed of copper and zinc, the current density at the copper cathode is 500 A.m^{-2}. The area of both the copper and zinc electrodes is 0.01 m^2. Calculate (1) the corrosion current, (2) the current density at the zinc anode, and (3) the zinc loss per hour.

SOLUTION

1. The corrosion current is:

   $I = i_{Cu}A_{Cu} = (500 \text{ A.m}^{-2})(0.01 \text{ m}^2) = 5 \text{ A}$

2. The current in the cell is the same everywhere. Thus:

$$i_{Zn} = \frac{I}{A_{Zn}} = \frac{5}{0.01} = 500 \text{ A.m}^{-2}$$

3. The atomic mass of zinc is 65.38 g.mol^{-1}. From Faraday's equation:

$$w_{\text{zinc loss}} = \frac{ItM}{nF} = \frac{(5)(3\,600 \text{ s/h})(65.38)}{(2)(96\,500)}$$
$$= 6.1 \text{ g.h}^{-1}$$

22.5 The Corrosion Current and Polarisation

To protect metals from corrosion, we wish to make the current as small as possible. Unfortunately, the corrosion current is very difficult to measure, control, or predict. Part of this difficulty can be attributed to various changes that occur during operation of the corrosion cell. A change in the potential of an anode or cathode, which in turn affects the current in the cell, is called polarisation. There are three important kinds of polarisation: (1) activation, (2) concentration, and (3) resistance polarisation.

Activation Polarisation This kind of polarisation is related to the energy required to cause the anode or cathode reactions to occur. If we can increase the degree of polarisation, these reactions occur with greater difficulty and the rate of corrosion is reduced. Small differences in composition and structure in the anode and cathode materials dramatically change the activation polarisation. Segregation effects in the electrodes cause the activation polarisation to vary from one location to another. These factors make it difficult to predict the corrosion current.

Concentration Polarisation After corrosion begins, the concentration of ions at the anode or cathode surface may change. For example, a higher concentration of metal ions may be produced at the anode if the ions are unable to diffuse rapidly into the electrolyte. Hydrogen ions may be depleted at the cathode in a hydrogen electrode, or a high $(OH)^-$ concentration may develop at the cathode in an oxygen electrode. When this situation occurs, either the anode or cathode reaction is stifled, because fewer electrons are released at the anode or accepted at the cathode.

In any of these examples, the current density, and thus the rate of corrosion, decreases because of concentration polarisation. Normally, the polarisation is less pronounced when the electrolyte is highly concentrated, the temperature is increased, or the electrolyte is vigorously agitated. Each of these factors increases the current density and encourages electrochemical corrosion.

Resistance Polarisation This type of polarisation is caused by the electrical resistivity of the electrolyte. If a greater resistance to the flow of the current is offered, the rate of corrosion is reduced. Again the degree of resistance polarisation may change as the composition of the electrolyte changes during the corrosion process.

22.6 Types of Electrochemical Corrosion

In this section we will look at some of the more common forms taken by electrochemical corrosion. First, there is *uniform attack*. When an engineering metal (note this is not the same as the ideal metal referred to in 22.4) is placed in an electrolyte, some regions are anodic to other regions. However, the location of these regions moves and even reverses from time to time. Since the anode and cathode regions continually shift, the metal corrodes uniformly.

Galvanic attack occurs when certain areas always act as anodes, whereas other areas always act as cathodes. These electrochemical cells are called galvanic cells and can be separated into three types: composition cells stress cells, and concentration cells

Table 22.2 The galvanic series in seawater.

Anodic	Magnesium and Mg alloys	Anodic	Lead
	Zinc		Tin
	Galvanised steel		Cu-40% Zn brass
	5052 aluminium		Nickel-base alloys (active)
	3003 aluminium		Copper
	1100 aluminium		Cu-30% Ni alloy
	Alclad		Nickel-base alloys (passive)
	Cadmium		Stainless steels (passive)
	2024 aluminium		Silver
	Low-carbon steel		Titanium
	Cast iron		Graphite
	50% Pb-50% Sn solder		Gold
	Stainless steels (active)	Cathodic	Platinum

After ASM Metals Handbook, *Vol. 10, 8th Ed., 1975.*

Composition Cells Composition cells, or dissimilar metal corrosion, develop when two metals or alloys, such as copper and iron, form an electrolytic cell. Because of the effect of alloying elements or electrolyte concentration on polarisation, the emf series may not tell us which regions corrode and which are protected. Instead, we use a galvanic series, in which the different alloys are ranked according to their anodic or cathodic tendencies in a particular environment (Table 22.2). We may find a different galvanic series for seawater, freshwater, and industrial atmospheres.

EXAMPLE 22.5

A brass fitting used in a marine application is joined by soldering with lead-tin solder. Will the brass or the solder corrode?

SOLUTION

From the galvanic series, we find that all of the copper-base alloys are more cathodic than a 50% Pb-50% Sn solder. Thus, the solder is the anode and corrodes. In a similar manner, corrosion of solder can contaminate water in freshwater plumbing systems with lead.

Composition cells also develop in two-phase alloys, where one phase is more anodic than the other. Since ferrite is anodic to cementite in steel, small microcells cause steel to galvanically corrode (Figure 22.6). Almost always, a two-phase alloy has less resistance to corrosion than a single-phase alloy of similar composition.

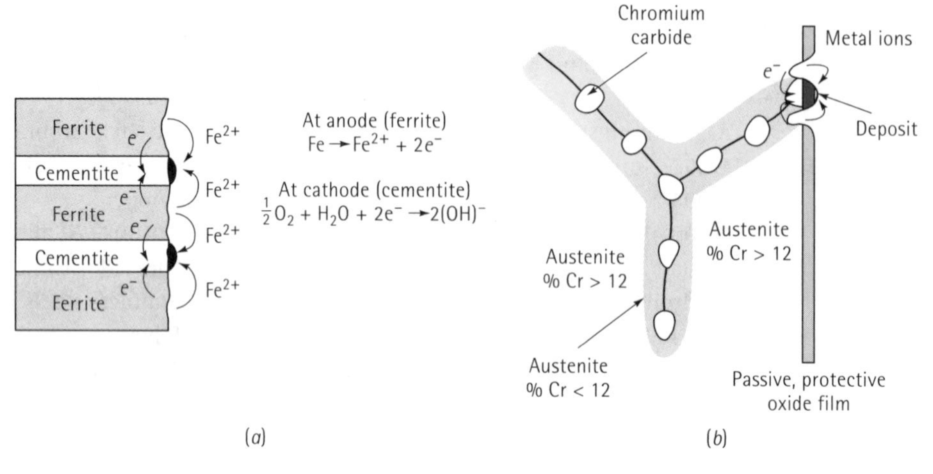

Figure 22.6
Example of microgalvanic cells in two-phase alloys: (*a*) In steel, ferrite is anodic to cementite. (*b*) In austenitic stainless steel, precipitation of chromium carbide makes the austenite in the grain boundaries anodic.

Intergranular corrosion occurs when precipitation of a second phase or segregation at grain boundaries produces a galvanic cell. In zinc alloys, for example impurities such as cadmium, tin, and lead segregate at the grain boundaries during solidification. The grain boundaries are anodic compared with the remainder of the grains, and corrosion of the grain boundary metal occurs (Figure 22.7). In austenitic stainless steels, chromium carbides can precipitate at grain boundaries [Figure 22.6(b)]. The formation of the carbides removes chromium from the austenite adjacent to the boundaries. The low-chromium austenite at the grain boundaries is anodic to the remainder of the grain and corrosion occurs at the grain boundary areas.

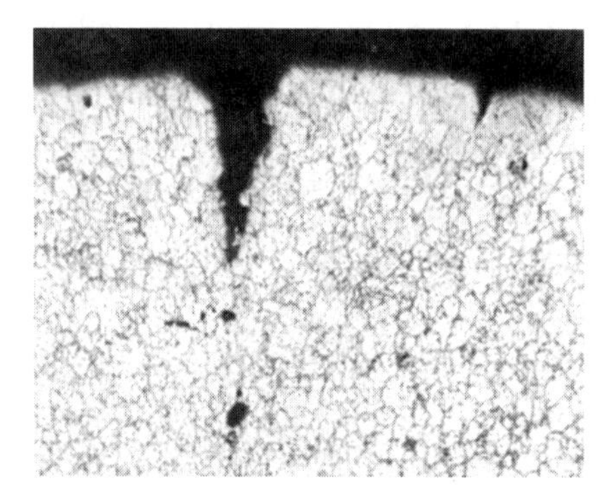

Figure 22.7
Photomicrograph of intergranular corrosion in a zinc die casting. Segregation of impurities to the grain boundaries produces microgalvanic corrosion cells ($\times$ 50).

Stress Cells Stress cells develop when a metal contains regions with different local stresses. The most highly stressed, or high-energy, regions act as anodes to the less stressed cathodic areas (Figure 22.8). Regions with a finer grain size, or a higher density of grain boundaries, are anodic to coarse-grained regions of the same material. Highly cold-worked areas are anodic to less cold-worked areas.

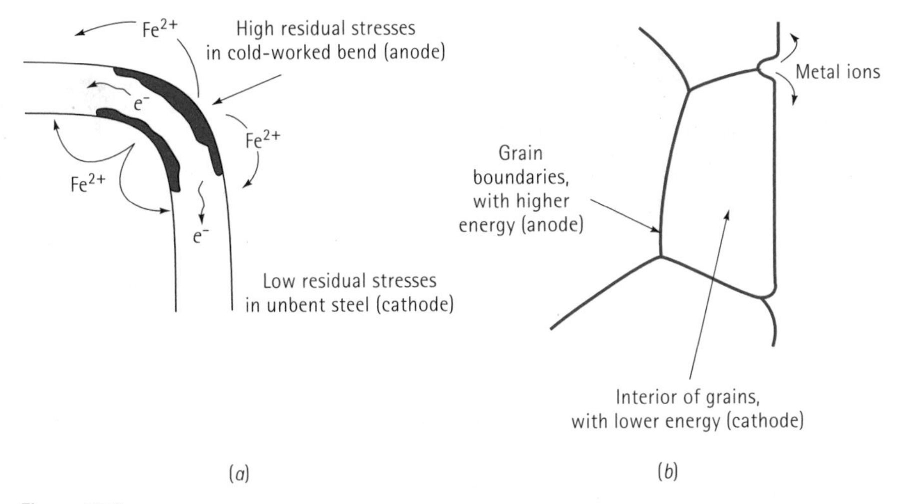

Figure 22.8
Examples of stress cells. (*a*) Cold work required to bend a steel bar introduces high residual stresses at the bend, which then is anodic and corrodes. (*b*) Because grain boundaries have a high energy, they are anodic and corrode.

Stress corrosion occurs by galvanic action, but other mechanisms, such as adsorption of impurities at the tip of an existing crack, may also occur. Failure occurs as a result of corrosion and an applied stress. Higher applied stresses reduce the time required for failure.

Fatigue failures are also initiated or accelerated when corrosion occurs. Corrosion fatigue can reduce fatigue properties by initiating cracks, perhaps by producing pits or crevices, and by increasing the rate at which the cracks propagate.

EXAMPLE 22.6

A cold-drawn steel wire is formed into a nail by additional deformation, producing the point at one end and the head at the other. Where will the most severe corrosion of the nail occur?

SOLUTION

Since the head and point have been cold-worked an additional amount compared with the shank of the nail, the head and point serve as anodes and corrode most rapidly.

Concentration Cells Concentration cells develop due to differences in the electrolyte (Figure 22.9). A difference in metal ion concentration causes a difference in electrode potential, according to the Nernst equation. The metal in contact with the most concentrated solution is the cathode; the metal in contact with the dilute solution is the anode.

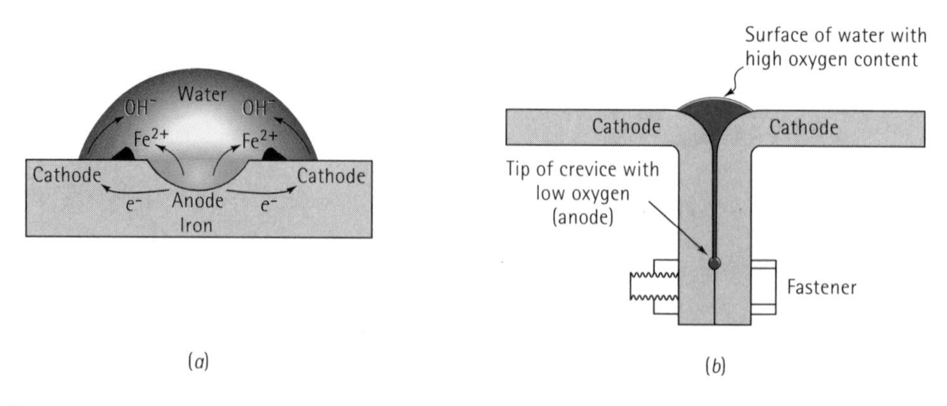

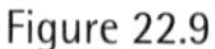

Figure 22.9
Concentration cells: (*a*) Corrosion occurs beneath a water droplet on a steel plate due to low oxygen concentration in the water at the centre of the droplet. (*b*) Corrosion occurs at the tip of a crevice because of limited access to oxygen in the water drawn into this area.

The oxygen concentration cell (often referred to as oxygen starvation) occurs when the cathode reaction is the oxygen electrode, $H_2O + (\frac{1}{2})O_2 + 4e^- \rightarrow 4(OH)^-$. Electrons flow from the low-oxygen region, which serves as the anode, to the high-oxygen region, which serves as the cathode.

Deposits, such as rust or water droplets, shield the underlying metal from oxygen. Consequently, the metal under the deposit is the anode and corrodes. This causes one form of pitting corrosion. Waterline corrosion is similar. Metal at the waterline is exposed to oxygen, while metal beneath the waterline is deprived of oxygen; hence, the metal underwater corrodes. Normally, the metal far below the surface corrodes more slowly than metal just below the waterline, due to differences in the distance that electrons must travel. Because water that has got drawn into cracks and crevices has a lower oxygen concentration than the water at their entrance [Figure 22.9(b)], the tip of a crack or crevice is the anode, causing crevice corrosion.

A pipe buried in soil may corrode because of differences in the composition of the soil. Velocity differences may cause concentration differences. Stagnant water contains low oxygen concentrations whereas fast-moving, aerated water contains higher oxygen concentrations. Metal near stagnant water is anodic and corrodes.

EXAMPLE 22.7

Two pieces of steel are joined mechanically by crimping the edges. Why would this be a bad idea if the steel is then exposed to water? If the water contains salt, would corrosion be affected?

SOLUTION

By crimping the steel edges, we produce a crevice. The region in the crevice is exposed to less air and moisture, so it behaves as the anode in a concentration cell. The steel in the crevice corrodes.

Salt in the water increases the conductivity of the water, permitting electrical charge to be transferred at a more rapid rate. This causes a higher current density and, thus, faster corrosion due to less resistance polarisation.

Microbiological Corrosion Various microbes, such as fungi and bacteria, create conditions that encourage electrochemical corrosion. Particularly in aqueous environments, these organisms grow on metallic surfaces. The organisms typically form colonies that are not continuous. The presence of the colonies and the by-products of the growth of the organisms produce changes in the environment and, hence, in the type of corrosion and the rate at which it occurs.

Some bacteria reduce sulphates in the environment, producing sulphuric acid which, in turn, attacks metal. The bacteria may be either aerobic (which thrive when oxygen is available) or anaerobic (which do not need oxygen to grow). Such bacteria cause attack of a variety of metals, including steels, stainless steels, aluminium, and copper, and some ceramics and concrete. A common example occurs in in aluminium fuel tanks for aircraft. When the fuel, typically kerosene, is contaminated with water, bacteria grow and excrete acids. The acids attack the aluminium, eventually causing the fuel tank to leak.

The growth of colonies of organisms on a metal surface leads to the development of oxygen concentration cells (Figure 22.10). Areas beneath the colonies are anodic, whereas unaffected areas are cathodic. In addition, the colonies of organisms reduce the rate of diffusion of oxygen to the metal, and – even if the oxygen does diffuse into the colony – the organisms tend to consume the oxygen. The concentration cell produces pitting beneath the regions covered with the organisms. Growth of the organisms, which may include products of the corrosion of the metal, produces accumulations (called tubercules) that may plug pipes, clogging water cooling systems in nuclear reactors, submarines, or chemical reactors, for instance.

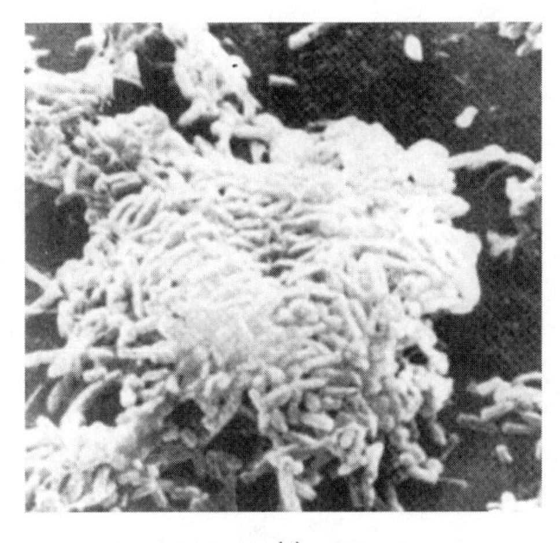

(*a*)

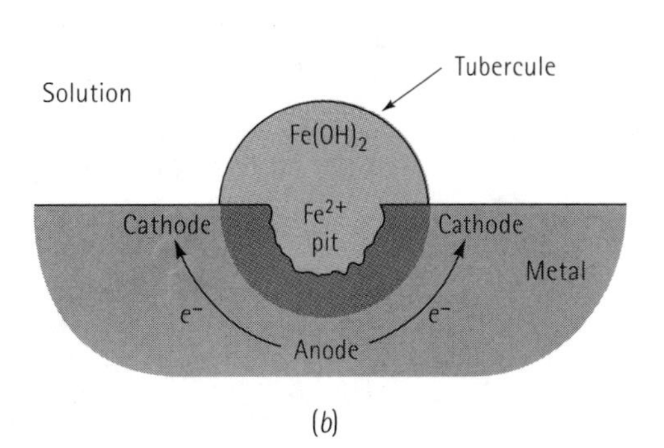

(*b*)

Figure 22.10
(*a*) Bacterial cells growing in a colony (× 2 700). (*b*) Formation of a tubercule and a pit under a biological colony.

22.7 Protection Against Electrochemical Corrosion

A number of techniques are used to combat corrosion, including design, coatings, inhibitors, cathodic protection, passivation, and materials selection.

Design Proper design of metal structures can slow or even avoid corrosion. Some of the steps that should be taken are as follows:

1. Prevent the formation of galvanic cells. For example, steel pipe is frequently connected to brass plumbing fixtures, producing a galvanic cell that causes the steel to corrode. By using intermediate plastic fittings to electrically insulate the steel and brass, this problem can be minimised.
2. Make the anode area much larger than the cathode area. For example, copper rivets can be used to fasten steel sheet. Because of the small area of the copper rivets, a limited cathode reaction occurs. The copper accepts few electrons, and the steel anode reaction proceeds slowly. If, on the other hand, steel rivets are used for joining copper sheet, the small steel anode area gives up many electrons, which are accepted by the large copper cathode area; corrosion of the steel rivets is then very rapid.

EXAMPLE 22.8

Consider a copper-zinc corrosion couple. If the current density at the copper cathode is 500 $A.m^{-2}$, calculate the weight loss per hour if (1) the copper cathode area is 10 000 mm^2 and the zinc anode area is 100 mm^2, and (2) the copper cathode area is 100 mm^2 and the zinc anode area is 10 000 mm^2.

SOLUTION

1. For the small zinc anode area:

$$I = i_{Cu}A_{Cu} = (500\ \text{A.m}^{-2})(0.01\ \text{m}^2) = 5\ \text{A}$$

$$w_{Zn} = \frac{ItM}{nF} = \frac{(5)(3\,600)(65.38)}{(2)(96\,500)} = 6.1\ \text{g.h}^{-1}$$

2. For the large zinc anode area:

$$I = i_{Cu}A_{Cu} = (500\ \text{A.m}^{-2})(0.0001\ \text{m}^2) = 0.05\ \text{A}$$

$$w_{Zn} = \frac{ItM}{nF} = \frac{(0.05)(3\,600)(65.38)}{(2)(96\,500)} = 0.061\ \text{g.h}^{-1}$$

The rate of corrosion of the zinc is reduced significantly when the zinc anode is much larger than the cathode.

3. Design components so that fluid systems are closed rather than open and so that stagnant pools do not collect. Partly filled tanks undergo waterline corrosion. Open systems continuously dissolve gas, providing ions that participate in the cathode reaction and encourage concentration cells.

4. Avoid crevices between assembled or joined materials (Figure 22.11). Welding may be a better joining technique than brazing, soldering or mechanical fastening. Galvanic cells develop in brazing or soldering, since the filler metals have a different composition from the metal being joined. Mechanical fasteners produce crevices that lead to concentration cells. However, if the filler metal is closely matched to the base metal, welding may prevent these cells from developing.

5. In some cases, the rate of corrosion cannot be reduced to a level that will not interfere with the expected lifetime of the component. In such cases, the assembly should be designed in such a manner that the corroded part can easily and economically be replaced.

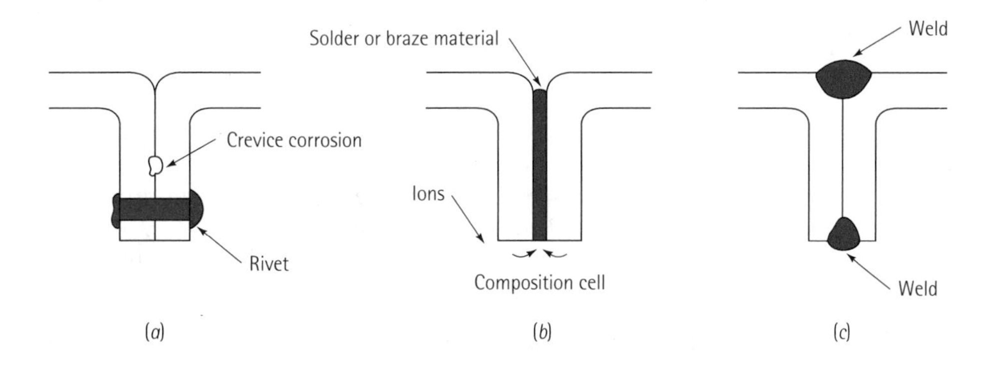

Figure 22.11
Alternative methods for joining two pieces of steel: (*a*) Fasteners may produce a concentration cell, (*b*) brazing or soldering may produce a composition cell, and (*c*) welding with a filler metal that matches the base metal may avoid galvanic cells.

Coatings Coatings are used to isolate the anode and cathode regions. Temporary coatings, such as grease or oil, provide some protection but are easily disrupted. Organic coatings, such as paint, or ceramic coatings, such as enamel or glass, provide better protection. However, if the coating is disrupted, a small anodic site is exposed that undergoes rapid, localised corrosion.

Metallic coatings include tin-plated and galvanised (zinc-plated) steel (Figure 22.12). A continuous coating of either metal isolates the steel from the electrolyte. However, when the coating is scratched, exposing the underlying steel, the two coatings behave differently. The zinc continues to be effective, because zinc is anodic to steel. Since the area of the exposed steel cathode is small, the zinc coating corrodes at a very slow rate and the steel remains protected. However, steel is anodic to tin, so a tiny steel anode is created when tin plate is scratched and rapid corrosion of the steel subsequently occurs.

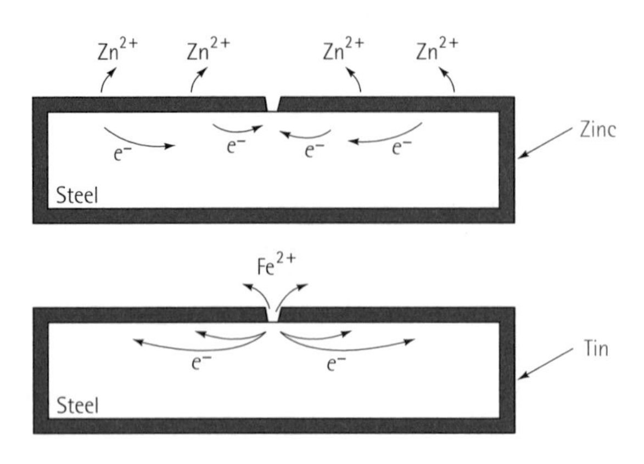

Figure 22.12
Zinc-plated steel and tin-plated steel are protected differently. Zinc protects steel even when the coating is scratched, since zinc is anodic to steel. Tin does not protect steel when the coating is disrupted, since steel is anodic with respect to tin.

Chemical conversion coatings are produced by a chemical reaction with the surface. Liquids such as zinc acid orthophosphate solutions form an adherent phosphate layer on the metal surface. The phosphate layer is, however, rather porous and is more often used to improve paint adherence. Stable, adherent, nonporous, nonconducting oxide layers form on the surface of aluminium, chromium, and stainless steel. These oxides exclude the electrolyte and prevent the formation of galvanic cells.

Inhibitors When added to the electrolyte solution, some chemicals migrate preferentially to the anode or cathode surface and produce concentration or resistance polarisation – that is, they are inhibitors. Chromate salts perform this function in automobile radiators. A variety of chromates, phosphates, molybdates, and nitrites produce protective films on anodes or cathodes in power plants and heat exchangers, thus stifling the electrochemical cell.

Cathodic Protection We can protect against corrosion by supplying the metal with electrons and forcing the metal to be a cathode (Figure 22.13). Cathodic protection can use a sacrificial anode or an impressed voltage.

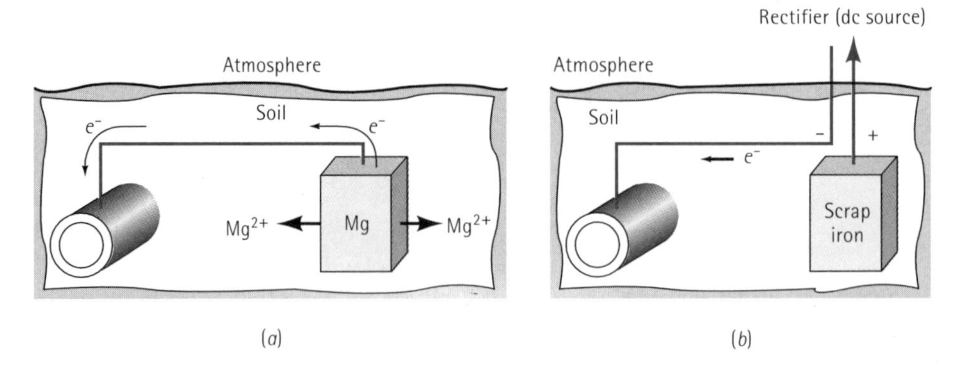

Figure 22.13
Cathodic protection of a buried steel pipeline: (*a*) A sacrificial magnesium anode ensures that the galvanic cell makes the pipeline the cathode. (*b*) An impressed voltage between a scrap iron auxiliary anode and the pipeline ensures that the pipeline is the cathode.

A sacrificial anode is attached to the material to be protected, forming an electrochemical circuit. The sacrificial anode corrodes, supplies electrons to the metal, and thereby prevents an anode reaction at the metal. The sacrificial anode, typically zinc or magnesium is consumed and must eventually be replaced. Applications include preventing corrosion of buried pipelines, ships, off-shore drilling platforms, and water heaters.

An impressed voltage is obtained from a direct current source connected between an auxiliary anode and the metal to be protected. Essentially, we have connected a battery so that electrons flow to the metal, causing the metal to be the cathode. The auxiliary anode, such as scrap iron, corrodes. Alternatively, inert anodes such as platinum can be used, in which case the anodic reactions occur in the electrolyte surrounding them.

Passivation or Anodic Protection Metals near the anodic end of the galvanic series are active and serve as anodes in most electrolytic cells. However, if these metals are made passive or more cathodic, they corrode at slower rates than normal. Passivation is accomplished by producing strong anodic polarisation, preventing the normal anode reaction; thus the term anodic protection.

We cause passivation by exposing the metal to highly concentrated oxidising solutions. If iron is dipped in very concentrated nitric acid, the iron rapidly and uniformly corrodes to form a thin, protective hydroxide coating. The coating protects the iron from subsequent corrosion in nitric acid.

We can also cause passivation by increasing the potential on the anode above a critical level. A passive film forms on the metal surface, causing strong anodic polarisation, and the current decreases to a very low level. Passivation of aluminium is called anodising, and a thick oxide coating is produced. This oxide layer can be dyed to produce attractive colours.

Materials Selection and Treatment Corrosion can be prevented or minimised by selecting appropriate materials and heat treatments. In castings, for example, segregation causes tiny, localised galvanic cells that accelerate corrosion. We can improve corrosion resistance with a homogenisation heat treatment. When metals are formed into finished shapes by bending, differences in the amount of cold work and residual stresses cause local stress cells. These may be minimised by a stress relief anneal or a full recrystallisation anneal.

The heat treatment is particularly important in austenitic stainless steels (Figure 22.14). When the steel cools slowly from 870°C to 425°C, chromium carbides precipitate at the grain boundaries. The austenite at the grain boundaries may contain less than 12% chromium, which is the minimum required to produce a passive oxide layer. The steel is sensitised. Because the grain boundary regions are small and highly anodic, rapid corrosion of the austenite at the grain boundaries occurs. There are several techniques by which we can minimise the problem.

1. If the steel contains less than 0.03% C, the chromium carbides do not form.
2. If the percent chromium is very high, the austenite may not be depleted to below 12% Cr, even if the chromium carbides form.
3. Addition of titanium or niobium ties up the carbon as TiC or NbC, preventing the formation of chromium carbide. The steel is said to be stabilised

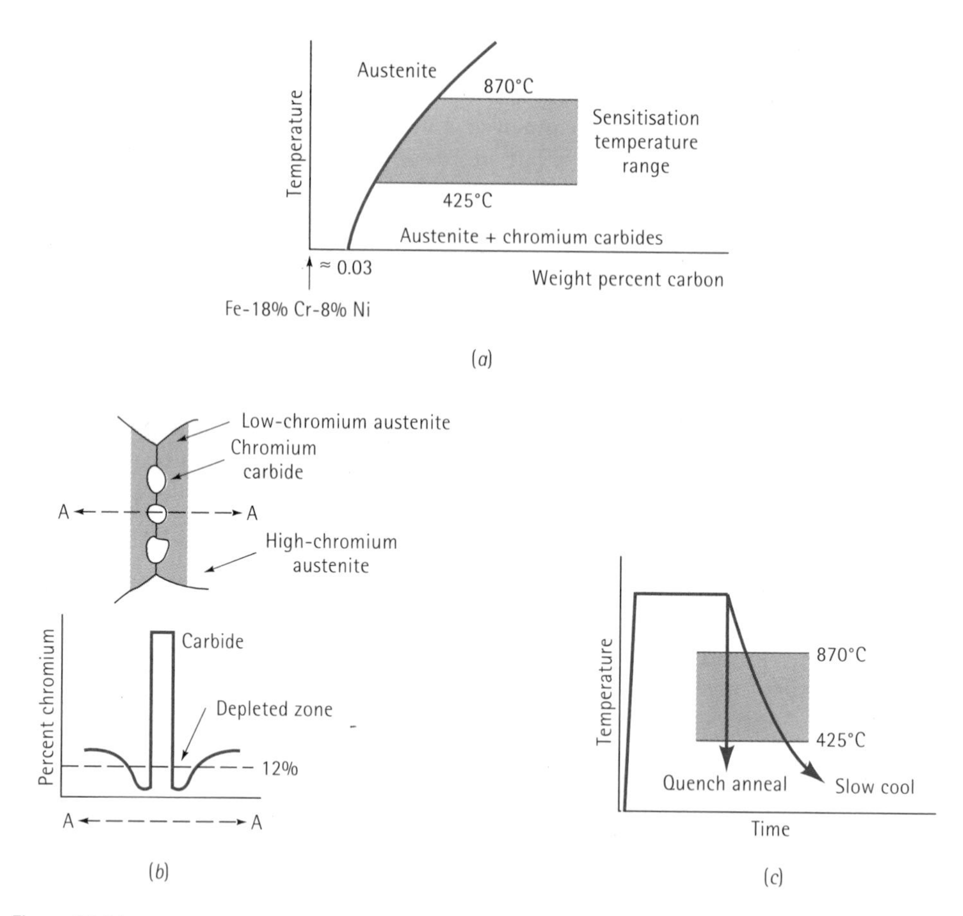

Figure 22.14
(*a*) Intergranular corrosion takes place in austenitic stainless steel. (*b*) Slow cooling permits chromium carbides to precipitate at grain boundaries. (*c*) A quench anneal to dissolve the carbides may prevent intergranular corrosion.

4. The sensitisation temperature range – 425°C to 870°C – should be avoided during manufacture and service.

5. In a quench anneal heat treatment, the stainless steel is heated above 870°C, causing the chromium carbides to dissolve. The structure, now containing 100% austenite, is rapidly quenched to prevent formation of carbides.

EXAMPLE 22.9 Design of a Corrosion Protection System

Steel troughs are located in a field to provide drinking water for a herd of cattle. The troughs frequently rust though and must be replaced. Design a system to prevent or delay this problem.

SOLUTION

The troughs are likely a low-carbon, unalloyed steel containing ferrite and cementite, producing a composition cell. The waterline in the trough, which is partially filled with water, provides a concentration cell. The trough is also exposed to the environment and the water is contaminated with impurities. Consequently, corrosion of the unprotected steel tank is to be expected.

Several approaches might be used to prevent or delay corrosion. We might, for example fabricate the trough using stainless steel or aluminium. Either would provide better corrosion resistance than the plain carbon steel, but both are considerably more expensive than the current material.

We might suggest using cathodic protection; a small magnesium anode could be attached to the inside of the trough. The anode corrodes sacrificially and prevents corrosion of the steel. This would require that the farm operator regularly check the tank to be sure that the anode is not completely consumed. We also want to be sure that magnesium ions introduced into the water are not a health hazard.

Another approach would be to protect the steel trough using a suitable coating. Painting the steel (that is, introducing a protective polymer coating) and using a tin-plated steel to provide protection as long as the coating is not disrupted.

The most likely approach is to use a galvanised steel, taking advantage of the protective coating and the sacrificial behaviour of the zinc. Corrosion is very slow, due to the large anode area, even if the coating is disrupted. Furthermore, the galvanised steel is relatively inexpensive, readily available, and does not require frequent inspection.

EXAMPLE 22.10 Design of a Stainless Steel Weldment

A piping system used to transport a corrosive liquid is fabricated from 304 stainless steel BS304S15. Welding of the pipes is required to assemble the system. Unfortunately, corrosion occurs and the corrosive liquid leaks from the pipes near the weld. Identify the problem and design a system to prevent corrosion in the future.

SOLUTION

Table 12.4 shows that the 304 stainless steel contains 0.08% C, causing the steel to be sensitised if it is improperly heated or cooled during welding. Figure 22.15 shows the maximum temperatures reached in the fusion and heat-affected zones during welding. A portion of the pipe in the HAZ heats into the sensitisation temperature range, permitting chromium carbides to precipitate. If the cooling rate of the weld is very slow, the fusion zone and other areas of the heat-affected zone may also be affected. Sensitisation of the weld area, therefore, is the likely reason for corrosion of the pipe in the region of the weld.

Several solutions to the problem may be considered. We might use a welding process that provides very rapid rates of heat input, causing the weld to heat and cool very quickly. If the steel is exposed to the sensitisation temperature range for only a

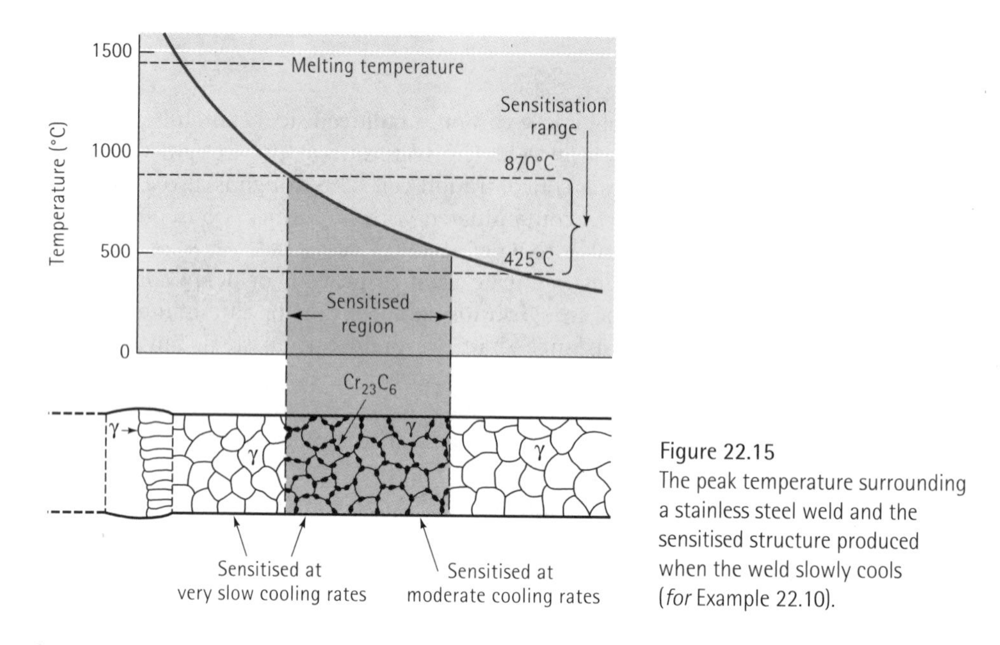

Figure 22.15
The peak temperature surrounding a stainless steel weld and the sensitised structure produced when the weld slowly cools (*for* Example 22.10).

brief time, chromium carbides may not precipitate. Joining processes such as laser welding or electron beam welding are high-rate-of-heat-input processes, but they are expensive. In addition, electron beam welding requires the use of a vacuum, and it may not be feasible to assemble the piping system in a vacuum chamber.

We might heat treat the assembly after the weld is made. By performing a quench anneal, any precipitated carbides are redissolved during the anneal and do not re-form during quenching. However, it may be impossible to perform this treatment on a large assembly.

We might check the original welding procedure to determine if the pipe was preheated prior to joining in order to minimise the development of stresses due to the welding process. If the pipe was preheated, sensitisation would be more likely to occur. We would recommend that any preheat procedure be suspended.

Perhaps our best solution is to use a stainless steel that is not subject to sensitisation. For example, carbides do not precipitate in a 304L stainless steel, BS 304S12, which contains less than 0.03% C. The low-carbon stainless steels are more expensive than the normal 304 steel; however, the extra cost does prevent the corrosion process and still permits us to use conventional joining techniques.

22.8 Microbiological Degradation and Biodegradable Polymers

Attack by a variety of insects and microbes is one form of 'corrosion' of polymers. Relatively simple polymers (such as polyethylene, polypropylene, and polystyrene), high-molecular-weight polymers, crystalline polymers, and thermosets are relatively immune to attack.

However, certain polymers – including polyesters, polyurethanes, cellulosics, and plasticised polyvinyl chloride (which contains additives that reduce the degree of polymerisation) – are particularly vulnerable to microbial degradation. These polymers can be broken into low-molecular-weight molecules by radiation or chemical attack until they are small enough to be ingested by the microbes.

We take advantage of microbial attack by producing *biodegradable* polymers, thus helping to remove the material from the waste stream. Biodegradation requires the complete conversion of the polymer to carbon dioxide, water, inorganic salts, and other small by-products produced by the ingestion of the material by bacteria. Polymers such as cellulosics can easily be broken into molecules with low molecular weights and are therefore biodegradable. In addition, special polymers are produced to degrade rapidly; a copolymer of polyethylene and starch is one example. Bacteria attack the starch portion of the polymer and reduce the molecular weight of the remaining polyethylene. Polymers produced from bacteria and which have mechanical properties similar to those of polypropylene are also rapidly degraded by bacteria when the polymer is returned to the environment.

22.9 Oxidation and Other Gas Reactions

Materials of all types may react with oxygen and other gases. These reactions can, like corrosion, alter the composition, properties, or integrity of a material.

Oxidation of Metals Metals may react with oxygen to produce an oxide at the surface. We are interested in three aspects of this reaction: the ease with which the metal oxidises, the nature of the oxide film that forms, and the rate at which oxidation occurs.

The ease with which oxidation occurs is given by the free energy of formation for the oxide (Figure 22.16). There is a large driving force for the oxidation of magnesium and aluminium, but there is little tendency for the oxidation of nickel or copper.

EXAMPLE 22.11

Explain why we should not add alloying elements such as chromium to pig iron before the pig iron is converted to steel in a basic oxygen furnace at 1700°C.

SOLUTION

In a basic oxygen furnace, we lower the carbon content of the metal from about 4% to much less than 1% by blowing pure oxygen through the molten metal. If chromium were already present before the steel-making began, chromium would oxidise before the carbon (Figure 22.16), since chromium oxide has a lower free energy of formation (or is more stable) than carbon dioxide (CO_2). Thus, any expensive chromium added would be lost before the carbon was removed from the pig iron.

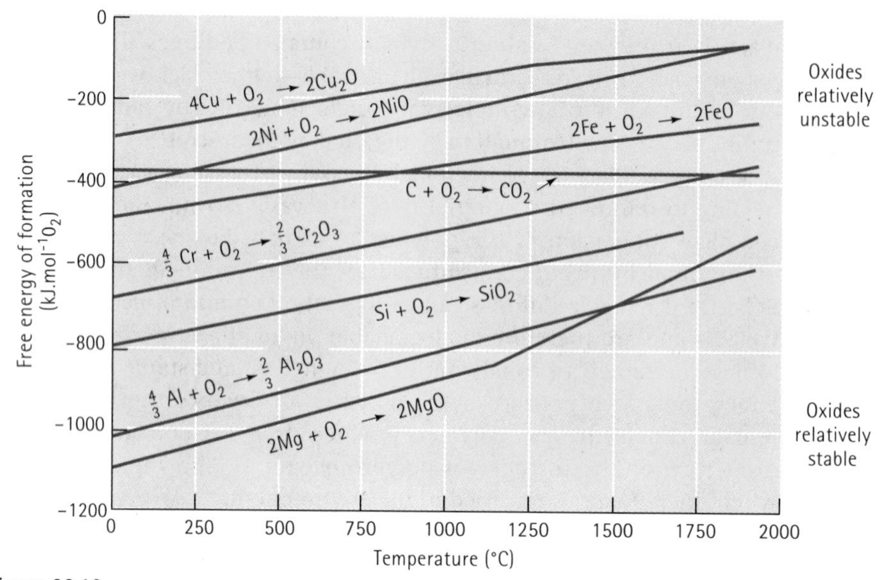

Figure 22.16
The free energy of formation of selected oxides as a function of temperature. A large negative free energy indicates a more stable oxide.

The type of oxide film influences the rate at which oxidation occurs (Figure 22.17). For the oxidation reaction

$$nM + mO_2 \rightarrow M_nO_{2m} \tag{22.12}$$

the Pilling-Bedworth (P-B) ratio is:

$$\text{P}-\text{B ratio} = \frac{\text{oxide volume per metal atom}}{\text{metal volume per metal atom}} = \frac{(M_{\text{oxide}})(\rho_{\text{metal}})}{n(M_{\text{metal}})(\rho_{\text{oxide}})} \tag{22.13}$$

where M is the atom or molecular mass, ρ is the density, and n is the number of metal atoms in the oxide, as defined in Equation 22.12.

If the Pilling-Bedworth ratio is less than one, the oxide occupies a smaller volume than the metal from which it formed; the coating is therefore porous and oxidation continues rapidly – typical of metals such as magnesium. If the ratio is one or two, the volumes of the oxide and metal are similar, and an adherent, nonporous, protective film forms – typical of aluminium and titanium. If the ratio exceeds two, the oxide occupies a large volume and may flake from the surface, exposing fresh metal that continues to oxidise – typical of iron. Although the Pilling-Bedworth equation historically has been used to characterise oxide behaviour many, exceptions to this behaviour are observed.

EXAMPLE 22.12

The density of aluminium is 2.7 Mg.m^{-3} and that of Al_2O^3 is about 4 Mg.m^{-3}. Describe the characteristics of the aluminium oxide film. Compare with the oxide film that forms on tungsten. The density of tungsten is 19.254 Mg.m^{-3} and that of WO_3 is 7.3 Mg.m^{-3}.

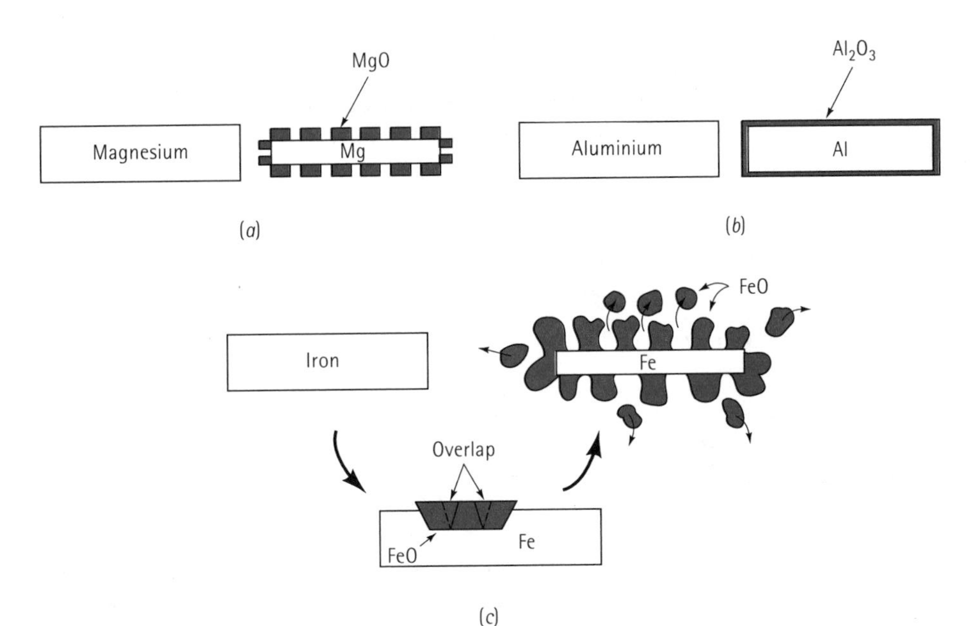

Figure 22.17
Three types of oxides may form, depending on the volume ratio between the metal and the oxide: (*a*) magnesium produces a porous oxide film, (*b*) aluminium forms a protective, adherent, nonporous oxide film, and (*c*) iron forms an oxide film that spalls off the surface and provides poor protection.

SOLUTION

For $2Al + 3/2O_2 \rightarrow Al_2O_3$, the molecular weight of Al_2O_3 is 101.96 and that of aluminium is 26.981.

$$\text{P-B} = \frac{M_{Al_2O_3}\rho_{Al}}{nM_{Al}\,\rho_{Al_2O_3}} = \frac{(101.96)(2.7)}{(2)(26.981)(4)} = 1.28$$

For tungsten, $W + 3/2O_2 \rightarrow WO_3$. The molecular weight of WO_3 is 231.85 and that of tungsten is 183.85:

$$\text{P-B} = \frac{M_{WO_3}\rho_W}{nM_W\rho_{WO_3}} = \frac{(231.85)(19.254)}{(1)(183.85)(7.3)} = 3.33$$

Since P-B $\simeq$ 1 for aluminium, the Al_2O_3 film is nonporous and adherent, providing protection to the underlying aluminium. However, P-B > 2 for tungsten, so the WO_3 should be nonadherent and nonprotective.

The rate at which oxidation occurs depends on the access of oxygen to the metal atoms. A linear rate of oxidation occurs when the oxide is porous (as in magnesium) and oxygen has continued access to the metal surface:

$$y = kt \tag{22.14}$$

where y is the thickness of the oxide, t is the time, and k is a constant that depends on the metal and temperature.

A parabolic relationship is observed when diffusion of ions or electrons through a nonporous oxide layer is the controlling factor. This relationship is observed in iron, copper, and nickel:

$$y = \sqrt{kt} \tag{22.15}$$

Finally, a logarithmic relationship is observed for the growth of thin oxide films that are particularly protective, as for aluminium and possibly chromium:

$$y = k \ln (ct + 1) \tag{22.16}$$

where k and c are constants for a particular temperature, environment, and composition.

EXAMPLE 22.13

At 1 000°C, pure nickel follows a parabolic oxidation curve given by the constant $k = 3.9 \times 10^{-16}$ m^2.s^{-1} in an oxygen atmosphere. If this relationship is not affected by the thickness of the oxide film, calculate the time required for a 1 mm-thick nickel sheet to oxidise completely.

SOLUTION

Assuming that the sheet oxidises from both sides:

$$y = \sqrt{kt} = \sqrt{(3.9\times10^{-16})(t)} = \frac{(1\times10^{-3}\ \text{mm/m})}{2\ \text{sides}} = 0.5\times10^{-3}\ \text{m}$$

$$t = \frac{(0.5\times10^{-3})^2}{3.9\times10^{-16}} = 6.4\times10^{8}\ \text{s} = 20.3\ \text{years}$$

Temperature also affects the rate of oxidation. In many metals, the rate of oxidation is controlled by the rate of diffusion of oxygen or metal ions through the oxide. If oxygen diffusion is more rapid, oxidation occurs between the oxide and the metal; if metal ion diffusion is more rapid, oxidation occurs at the oxide-atmosphere interface. Consequently, we would expect oxidation rates to follow an Arrhenius relationship, increasing exponentially as the temperature increases.

Oxidation and Thermal Degradation of Polymers Polymers degrade when heated and/or exposed to oxygen. A polymer chain may be ruptured, producing two macroradicals. In rigid thermosets, the macroradicals may instantly recombine (a process called the *cage* effect), resulting in no net change in the polymer. However, in the more flexible thermoplastics – particularly for amorphous rather than crystalline polymers – recombination does not occur and the result is a decrease in the molecular weight, the viscosity, and the mechanical properties of the polymer. Depolymerisation continues as the polymer is exposed to the high temperature. Polymer chains can also *unzip*. In this case, individual monomers are removed one after another from the ends of the chain, gradually reducing the molecular weight of the remaining chains. As the degree of polymerisation decreases, the remaining chains become more heavily branched or cyclisation may occur. In *cyclisation*, the two ends of the same chain may be bonded together to form a ring.

Polymers also degrade by the loss of side groups on the chain. Chloride ions (in polyvinyl chloride) and benzene rings (in polystyrene) are lost from the chain, forming by-products. For example, as polyvinyl chlorine is degraded, hydrochloric acid (HCl) is produced. Hydrogen atoms are bonded more strongly to the chains; thus, polyethylene does not degrade as easily as PVC or PS. Fluoride ions (in PTFE) are more difficult to remove than hydrogen atoms, providing PTFE with its high-temperature resistance.

22.10 Wear and Erosion

Wear and erosion remove material from a component by mechanical attack of solids or liquids. Corrosion and mechanical failure also contribute to this type of attack.

Adhesive Wear Adhesive wear – also known as scoring, galling, or seizing – occurs when two solid surfaces slide over one another under pressure. Surface projections, or asperities, are plastically deformed and eventually welded together by the high local pressures (Figure 22.18). As sliding continues, these bonds are broken, producing cavities on one surface, projections on the second surface, and frequently tiny, abrasive particles – all of which contribute to further wear of the surfaces.

Many factors may be considered in trying to improve the wear resistance of materials. Designing components so that loads are small, surfaces are smooth, and continual lubrication is possible helps prevent adhesions that cause the loss of material.

The properties and microstructure of the material are also important. Normally, if both surfaces have high hardnesses, the wear rate is low. High strength, to help resist the applied loads, and good toughness and ductility, which prevent tearing of material from the surface, may be beneficial. Ceramic materials, with their exceptional hardness, are expected to provide good adhesive wear resistance.

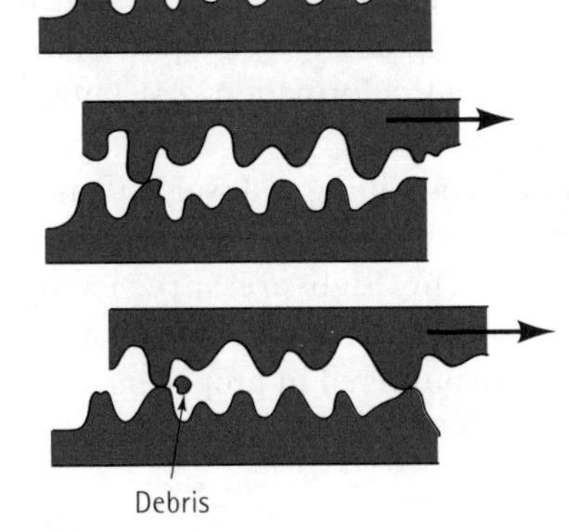

Figure 22.18
The asperities on two rough surfaces may initially be bonded. A sufficient force breaks the bonds and the surfaces slide. As they slide, asperities may be fractured, wearing away the surfaces and producing debris.

Wear resistance of polymers can be improved if the coefficient of friction is reduced by the addition of polytetrafluoroethylene (PTFE, Teflon) or if the polymer is strengthened by the introduction of reinforcing fibres such as glass, carbon, or aramid.

Abrasive Wear When material is removed from a surface by contact with hard particles, abrasive wear occurs. The particles either may be present at the surface of a second material or may exist as loose particles between two surfaces (Figure 22.19). This type of wear is common in machinery such as ploughs, scraper blades, crushers, and grinders used to handle abrasive materials and may also occur when hard particles are unintentionally introduced into moving parts of machinery. Abrasive wear is also used for grinding operations to remove material intentionally.

Materials with a high hardness, good toughness, and high hot strength are most resistant to abrasive wear. Typical materials used for abrasive wear applications include quenched and tempered steels; carburised or surface-hardened steels; cobalt alloys such as Stellite; composite materials, including tungsten carbide cermets; white cast irons; and hard surfaces produced by welding. Most ceramic materials also resist wear effectively because of their high hardness; however, their brittleness may sometimes limit their usefulness in abrasive wear conditions.

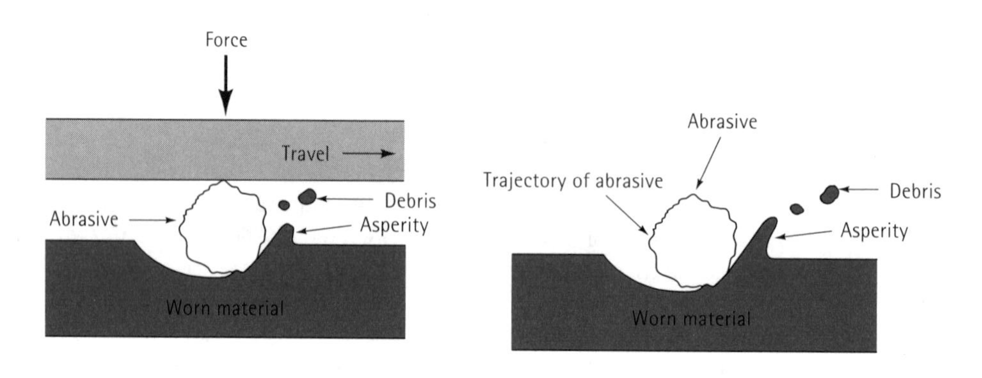

Figure 22.19
Abrasive wear, caused by either trapped or free-flying abrasives, produces troughs in the material, piling up asperities that may fracture into debris.

Liquid Erosion The integrity of a material may be destroyed by erosion caused by high pressures associated with a moving liquid. The liquid causes strain hardening of the metal surface, leading to localised deformation, cracking, and loss of material. Two types of liquid erosion deserve mention.

Cavitation occurs when a liquid containing a dissolved gas enters a low-pressure region. Gas bubbles, which precipitate and grow in the liquid, collapse when the pressure subsequently increases. The high-pressure, local shock wave that is produced may exert a pressure of hundreds of $MN.m^{-2}$ against the surrounding material. Cavitation is frequently encountered in propellers, dams, and spillways, and hydraulic pumps.

Liquid impingement occurs when liquid droplets carried in a rapidly moving gas strike a metal surface. High localised pressures develop because of the initial impact and the rapid lateral movement of the droplets from the impact point along the metal surface. Water droplets carried by steam may erode turbine blades in steam generators and nuclear power plants.

Liquid erosion can be minimised by proper materials selection and design. Minimising the liquid velocity, assuring that the liquid is deaerated, selection of hard, tough materials to absorb the impact of the droplets, and coating the material with an energy-absorbing elastomer all may help minimise erosion.

SUMMARY

- Corrosion causes deterioration of all types of materials. Designers and engineers must know how corrosion occurs in order to consider suitable designs, materials selection, or protective measures.
- In chemical corrosion, a material is dissolved in a solvent, resulting in the loss of material. All materials – metals, ceramics, polymers and composites – are subject to this form of attack. Choice of appropriate materials having a low solubility in a given solvent or use of inert coatings on materials helps avoid or reduce chemical corrosion.
- Electrochemical corrosion requires that a complete electric circuit develop. The anode corrodes and a by-product such as rust forms at the cathode. Because an electric circuit is required, this form of corrosion is most serious in metals and alloys:
  - Composition cells are formed by the presence of two different metals, two different phases within a single alloy, or even segregation within a single phase.
  - Stress cells form when the level of residual or applied stresses vary within the metal; the regions subjected to the highest stress are the anode and, consequently, they corrode. Stress corrosion cracking and corrosion fatigue are examples of stress cells.
  - Concentration cells form when a metal is exposed to a nonuniform electrolyte; for example, the portion of a metal exposed to the lowest oxygen content corrodes. Pitting corrosion, waterline corrosion, and crevice corrosion are examples of these cells. Microbiological corrosion, in which colonies of organisms such as bacteria grow on the metal surface, is another example of a concentration cell.
- Electrochemical corrosion can be minimised or prevented by using electrical insulators to break the electric circuit, designing and manufacturing assemblies without crevices, designing assemblies so that the anode area is unusually large compared with that of the cathode, using protective and even sacrificial coatings, inhibiting the action of the electrolyte, supplying electrons to the metal by means

of an impressed voltage, using heat treatments that reduce residual stresses or segregation, and a host of other actions.

- Oxidation degrades most materials. While an oxide coating provides protection for some metals such as aluminium, most materials are attacked by oxygen. Diffusion of oxygen and metallic atoms is often important; therefore oxidation occurs most rapidly at elevated temperatures.
- Other factors, such as attack by microbes, damage caused by radiation, and wear or erosion of a material may also cause a material to deteriorate.

GLOSSARY

Abrasive wear
Removal of material from surfaces by the cutting action of particles.

Adhesive wear
Removal of material from surfaces of moving equipment by momentary local bonding, then bond fracture, at the surfaces.

Anode
The location at which corrosion occurs as electrons and ions are given up in an electrochemical cell.

Anodising
An anodic protection technique in which a thick oxide layer is deliberately produced on a metal surface.

Cathode
The location at which electrons are accepted and a by-product is produced during corrosion.

Cavitation
Erosion of a material surface by the pressures produced when a gas bubble collapses within a moving liquid.

Chemical corrosion
Removal of atoms from a material by virtue of the solubility or chemical reaction between the material and the surrounding liquid.

Composition cells
Electrochemical corrosion cells produced between two materials having a different composition. Also known as galvanic cells.

Concentration cells
Electrochemical corrosion cells produced between two locations on a material at which the composition of the electrolyte is different.

Crevice corrosion
A special concentration cell in which corrosion occurs in crevices because of the low concentration of oxygen.

Dezincification

A special chemical corrosion process by which both zinc and copper atoms are removed from brass, but the copper is replated back onto the metal.

Electrochemical cell

A cell in which electrons and ions can flow by separate paths between two materials, producing a current which, in turn, leads to corrosion or plating.

Electrochemical corrosion

Corrosion produced by the development of a current in an electrochemical cell that removes ions from the material.

Electrode potential

Related to the tendency of a material to corrode. The potential is the voltage produced between the material and a standard electrode.

Electrolyte

The conductive medium through which ions move to carry current in an electrochemical cell.

Emf series

The arrangement of elements according to their electrode potential, or their tendency to corrode.

Faraday's equation

The relationship that describes the rate at which corrosion or plating occurs in an electrochemical cell.

Galvanic series

The arrangement of alloys according to their tendency to corrode in a particular environment.

Graphitic corrosion

A special chemical corrosion process by which iron is leached from cast iron, leaving behind a weak, spongy mass of graphite.

Impressed voltage

A cathodic protection technique by which a direct current is introduced into the material to be protected, thus preventing the anode reaction.

Inhibitors

Additions to the electrolyte that preferentially migrate to the anode or cathode, cause polarisation, and reduce the rate of corrosion.

Intergranular corrosion

Corrosion at grain boundaries because grain boundary segregation or precipitation produces local galvanic cells.

Liquid impingement

Erosion of a material caused by the impact of liquid droplets carried by a gas stream.

Nernst equation

The relationship that describes the effect of electrolyte concentration on the electrode potential in an electrochemical cell.

Oxidation
Reaction of a metal with oxygen to produce a metallic oxide. This normally occurs most rapidly at high temperatures.

Oxidation reaction
The anode reaction by which electrons are given up to the electrochemical cell.

Oxygen starvation
The concentration cell in low-oxygen regions of the electrolyte cause the underlying material to behave as the anode and to corrode.

Passivation
Producing strong anodic polarisation by causing a protective coating to form on the anode surface and to thereby interrupt the electric circuit.

Pilling-Bedworth ratio
Describes the type of oxide film that forms on a metal surface during oxidation.

Polarisation
Changing the voltage between the anode and cathode to reduce the rate of corrosion. *Activation* polarisation is related to the energy required to cause the anode or cathode reaction; *concentration* polarisation is related to change in the composition of the electrolyte; and *resistance* polarisation is related to the electrical resistivity of the electrolyte.

Quench anneal
The heat treatment used to dissolve carbides and prevent intergranular corrosion in stainless steels.

Reduction reaction
The cathode reaction by which electrons are accepted from the electrochemical cell.

Sacrificial anode
Cathodic protection by which a more anodic material is connected electrically to the material to be protected. The anode corrodes to protect the desired material.

Sensitisation
Precipitation of chromium carbides at the grain boundaries in stainless steels, making the steel sensitive to intergranular corrosion.

Stabilisation
Addition of titanium or niobium to a stainless steel to prevent intergranular corrosion.

Stress cells
Electrochemical corrosion cells produced by differences in imposed or residual stresses at different locations in the material.

Stress corrosion
Deterioration of a material in which an applied stress accelerates the rate of corrosion.

Tubercule
Accumulations of microbial organisms and corrosion by-products on the surface of a material.

PROBLEMS

22.1 A grey cast iron pipe is used in the natural gas distribution system for a city. The pipe fails and leaks, even though no corrosion noticeable to the naked eye has occurred. Offer an explanation for why the pipe failed.

22.2 A brass plumbing fitting produced from a Cu-30% Zn alloy operates in the hot water system of a large office building. After some period of use, cracking and leaking occur. On visual examination, no metal appears to have been corroded. Offer an explanation for why the fitting failed.

22.3 Suppose 10 g of Sn^{2+} are dissolved in 1 000 ml of water to produce an electrolyte. Calculate the electrode potential of the tin half-cell.

22.4 A half-cell produced by dissolving copper in water produces an electrode potential of +0.32 V. Calculate the amount of copper that must have been added to 1 000 ml of water to produce this potential.

22.5 An electrode potential in a platinum half-cell is 1.10 V. Determine the concentration of Pt^{4+} ions in the electrolyte.

22.6 A current density of 500 $A.m^{-2}$ is applied to a 0.015 m^2 cathode. What period of time is required to plate out a 1 mm-thick coating of silver onto the cathode?

22.7 We wish to produce 100 g of platinum per hour on a 0.1 m^2 cathode by electroplating. What plating current density is required? Determine the current required.

22.8 A 1 m-square steel plate is coated on both sides with a 0.05 mm-thick layer of zinc. A current density of 200 $A.m^{-2}$ is applied to the plate in an aqueous solution. Assuming that the zinc corrodes uniformly, determine the length of time required before the steel is exposed.

22.9 A 50 mm-inside-diameter, 4 m-long copper distribution pipe in a plumbing system is accidentally connected to the power system of a manufacturing plant, causing a current of 0.05 A to flow through the pipe. If the wall thickness of the pipe is 3 mm, estimate the time required before the pipe begins to leak, assuming a uniform rate of corrosion.

22.10 A steel surface 0.1 m × 1 m is coated with a 0.02 mm-thick layer of chromium. After one year of exposure to an electrolytic cell, the chromium layer is completely removed. Calculate the current density required to accomplish this removal.

22.11 A corrosion cell is composed of a 0.03 m^2 copper sheet and a 0.002 m^2 iron sheet, with a current density of 6 000 $A.m^{-2}$ applied to the copper. Which material is the anode? What is the rate of loss of metal from the anode per hour?

22.12 A corrosion cell is composed of a 0.002 m^2 copper sheet and a 0.04 m^2 iron sheet, with a current density of 7 000 $A.m^{-2}$ applied to the copper. Which material is the anode? What is the rate of loss of metal from the anode per hour?

22.13 Alclad is a laminar composite composed of two sheets of commercially pure aluminium (alloy 1100) sandwiched around a core of 2024 aluminium alloy. Discuss the corrosion resistance of the composite. Suppose that a portion of one of the 1 100 layers was machined off, exposing a small patch of the 2024 alloy. How would this affect the corrosion resistance? Explain. Would there be a difference in behaviour if the core material were 3003 aluminium? Explain.

22.14 The leaf springs for an automobile are formed from a high-carbon steel. For best corrosion resistance, should the springs be formed by hot working or cold working? Explain. Would corrosion still occur even if you use the most desirable forming process? Explain.

22.15 Several types of metallic coatings are used to protect steel, including zinc, lead, tin, cadmium, aluminium, and nickel. In which of these cases will the coating provide protection even when the coating is locally disrupted? Explain.

22.16 An austenitic stainless steel corrodes in all of the heat-affected zone surrounding the fusion zone of a weld. Explain why corrosion occurs and discuss the type of welding process or procedure that might have been used. What might you do to prevent corrosion in this region?

22.17 A steel is securely tightened onto a bolt in an industrial environment. After several months, the nut is found to contain numerous cracks, even though no externally applied load acts on the nut. Explain why cracking might have occurred.

22.18 The shaft for a propeller on a ship is carefully designed so that the applied stresses are well below the endurance limit for the material. Yet after several months, the shaft cracks and fails. Offer an explanation for why failure might have occurred under these conditions.

22.19 An aircraft wing composed of carbon fibre reinforced epoxy is connected to a titanium forging on the fuselage. Will the anode for a corrosion cell be the carbon fibre, the titanium, or the epoxy? Which will most likely be the cathode? Explain.

22.20 The inside surface of a cast iron pipe is covered with tar, which provides a protective coating. Acetone in a chemical laboratory is drained through the pipe on a regular basis. Explain why, after several weeks, the pipe begins to corrode.

22.21 A cold-worked copper tube is soldered, using a lead-tin alloy, into a steel connector. What types of electrochemical cells might develop due to this connection? Which of the materials would you expect to serve as the anode and suffer the most extensive damage due to corrosion? Explain.

22.22 Pure tin is used to provide a solder connection for copper in many electrical uses. Which metal will most likely act as the anode?

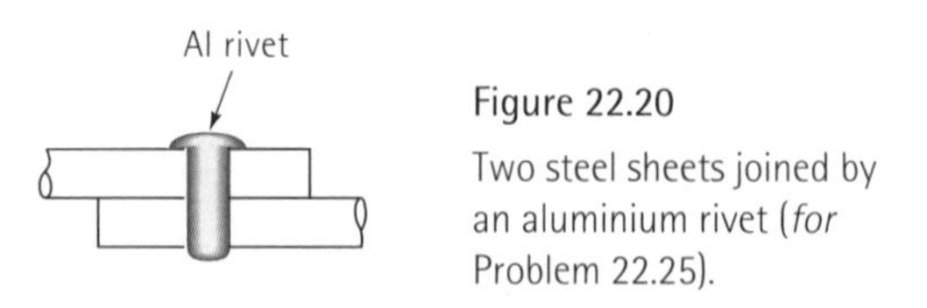

Figure 22.20

Two steel sheets joined by an aluminium rivet (*for* Problem 22.25).

22.23 Sheets of annealed nickel, cold-worked nickel, and recrystallised nickel are placed into an electrolyte. Which would be most likely to corrode? Which would be least likely to corrode? Explain.

22.24 A pipeline carrying liquid fertiliser crosses a small creek. A large tree washes down the creek and is wedged against the steel pipe. After some time, a hole is produced in the pipe at the point where the tree touches the pipe, with the diameter of the hole larger on the outside of the pipe than on the inside of the pipe. The pipe then leaks fertiliser into the creek. Offer an explanation of why the pipe corroded.

22.25 Two sheets of a 0.4%C steel (BS080A40) are joined together with an aluminium rivet (Figure 22.20). Discuss the possible corrosion cells that might be created as a result of this joining process. Recommend a joining process that might minimise some of these cells.

22.26 Figure 22.21 shows a cross-section through an epoxy-encapsulated integrated circuit, including a small microgap between the copper lead frame and the epoxy polymer. Suppose chloride ions from the manufacturing process penetrate the package. What types of corrosion cells might develop? What portions of the integrated circuit are most likely to corrode?

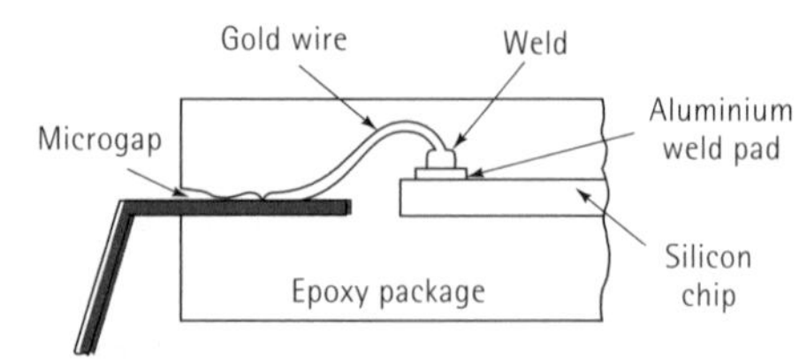

Figure 22.21

Cross-section through an integrated circuit showing the external lead connection to the chip (*for* Problem 22.26).

22.27 A current density of 1 000 A.m^{-2} is applied to the iron in an iron-zinc corrosion cell. Calculate the weight loss of zinc per hour

(a) if the zinc has a surface area of 0.001 m^2 and the iron has a surface area of 0.1 m^2 and

(b) if the zinc has a surface area of 0.1 m^2 and the iron has a surface area of 0.01 m^2.

22.28 Determine the Pilling-Bedworth ratio for the following metals and predict the behaviour of the oxide that forms on the surface. Is the oxide protective, does it flake off the metal, or is it permeable? (*See* Appendix A for the metal density.)

	Oxide Density (Mg.m^{-3})
Mg-MgO	3.60
Na-Na_2O	2.27
Ti-TiO_2	5.10
Fe-Fe_2O_3	5.30
Ce-Ce_2O_3	6.86
Nb-Nb_2O_5	4.47
W-WO_3	7.30

22.29 Oxidation of most ceramics is not considered to be a problem. Explain.

22.30 A sheet of copper is exposed to oxygen at 1 000°C. After 100 h, 24.6 g of copper are lost per mm^2 of surface area; after 250 h, 38.8 g.mm^{-2} are lost; and after 500 h, 55.0 g.mm^{-2} are lost. Determine whether oxidation is parabolic, linear, or logarithmic, then determine the time required for a 7.5 mm sheet of copper to be completely oxidised. The sheet of copper is oxidised from both sides.

22.31 At 800°C, iron oxidises at a rate of 1.4 g.mm^{-2} per hour; at 1 000°C, iron oxidises at a rate of 6.56 g.mm^{-2} per hour. Assuming a parabolic oxidation rate, determine the maximum temperature at which iron can be held if the oxidation rate is to be less than 0.50 g.mm^{-2} per hour.

Design Problems

22.32 A cylindrical steel tank 1 m in diameter and 2.5 m long is filled with water. We find that a current density of 150 A.m^{-2} acting on the steel is required to prevent corrosion. Design a sacrificial anode system hat will protect the tank.

22.33 The drilling platforms for offshore oil rigs are supported on large steel columns resting on the bottom of the ocean. Design an approach to ensure that corrosion of the supporting steel columns does not occur.

22.34 A storage building is to be produced using steel sheet for the siding and roof. Design a corrosion-protection system for the steel.

22.35 Design the materials for the scraper blade for a piece of earthmoving equipment.

CHAPTER 23

Failure – Origin, Detection, and Prevention

23.1 Introduction

Despite our understanding of the behaviour of materials, failures frequently occur. The sources of these failures include improper design, materials selection, and materials processing, as well as abuse. The engineer must anticipate potential failures and consequently, exercise good design, materials and processing selection, quality control, and testing to prevent failures. When failures do occur, the engineer must determine the cause so that failures can be prevented in the future.

The topic of failure analysis is far too complex to cover in one chapter, but we can discuss a few general principles. First, we identify the fracture mechanism by which failure occurs. Then, we discuss some measures – including nondestructive testing techniques – that may help us prevent failures.

23.2 Determining the Fracture Mechanism in Metal Failures

Failure analysis requires a combination of technical understanding, careful observation, detective work, and common sense. An understanding of fracture mechanisms is essential for us to determine the cause of failure. This section concentrates on identifying the mechanism by which a metal fails when subjected to stress. We consider six common fracture mechanisms: ductile, brittle, fatigue, creep and stress rupture, and stress corrosion.

Ductile Fracture Ductile fracture normally occurs in a transgranular manner (through the grains) in metals that have good ductility and toughness. Often, a considerable amount of deformation – including necking – is observed in the failed component. The deformation occurs before the final fracture. Ductile fractures are usually caused by simple overloads, or by applying too high a stress to the material.

In a simple tensile test, ductile fracture begins with the nucleation, growth, and coalescence of microvoids at the centre of the test bar (Figure 23.1). Microvoids form

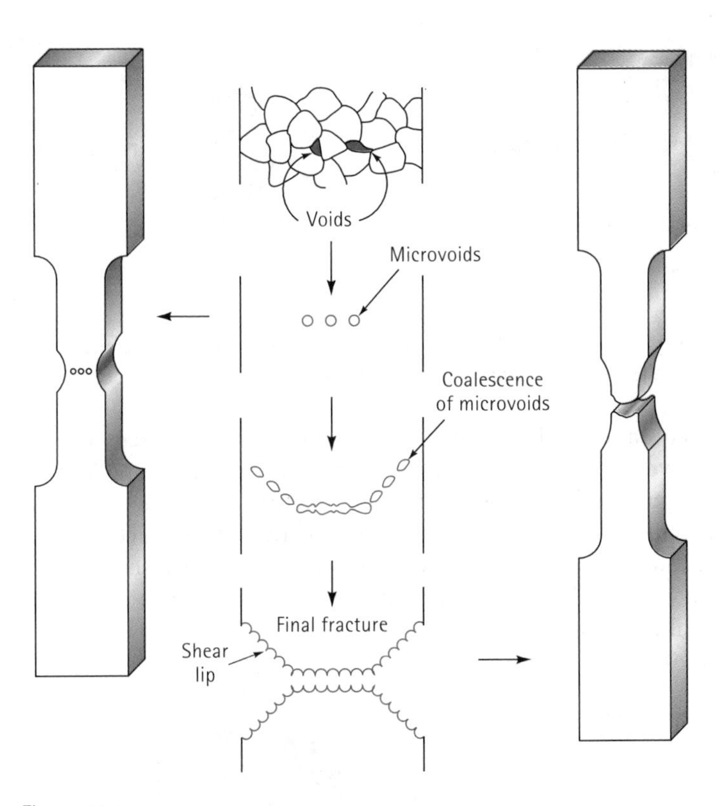

Figure 23.1
When a ductile material is pulled in a tensile test, necking begins and voids form – starting near the centre of the bar – by nucleation at grain boundaries or inclusions. As deformation continues, a 45° shear lip may form, producing a final cup and cone fracture.

when a high stress causes separation of the metal at grain boundaries or interfaces between the metal and inclusions. As the local stress increases, the microvoids grow and coalesce into larger cavities. Eventually, the metal-to-metal contact area is too small to support the load and fracture occurs.

Deformation by slip also contributes to the ductile fracture of a metal. We know that slip occurs when the resolved shear stress reaches the critical resolved shear stress and that the resolved shear stresses are highest at a 45° angle to the applied tensile stress (Schmid's law).

These two aspects of ductile fracture give the failed surface characteristic features. In thick metal sections, we expect to find evidence of necking, with a significant portion of the fracture surface having a flat face where microvoids first nucleated and coalesced, and a small shear lip, where the fracture surface is at a 45° angle to the applied stress. The shear lip, indicating that slip occurred, gives the fracture a cup and cone appearance (Figure 23.2). Simple macroscopic observation of this fracture may be sufficient to identify the ductile fracture mode.

Examination of the fracture surface at a high magnification – perhaps using a scanning electron microscope – reveals a dimpled surface (Figure 23.3). The dimples are traces of the microvoids produced during fracture. Normally, these microvoids are

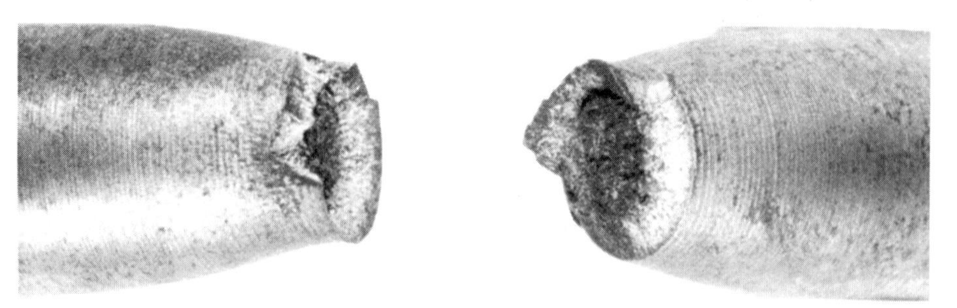

Figure 23.2
The cup and cone fracture observed when a ductile material (in this case, an annealed low carbon steel) fractures in a tensile test. The original diameter of the test bar was 12.8 mm.

round, or equiaxed, when a normal tensile stress produces the failure [Figure 23.4(a)]. However, on the shear lip, the dimples are oval-shaped, or elongated, with the ovals pointing towards the origin of the fracture [Figure 23.4(b)].

In thin plate, less necking is observed and the entire fracture surface may be a shear face. Microscopic examination of the fracture surface shows elongated dimples rather than equiaxed dimples, indicating a greater proportion of 45° slip than in thicker metals.

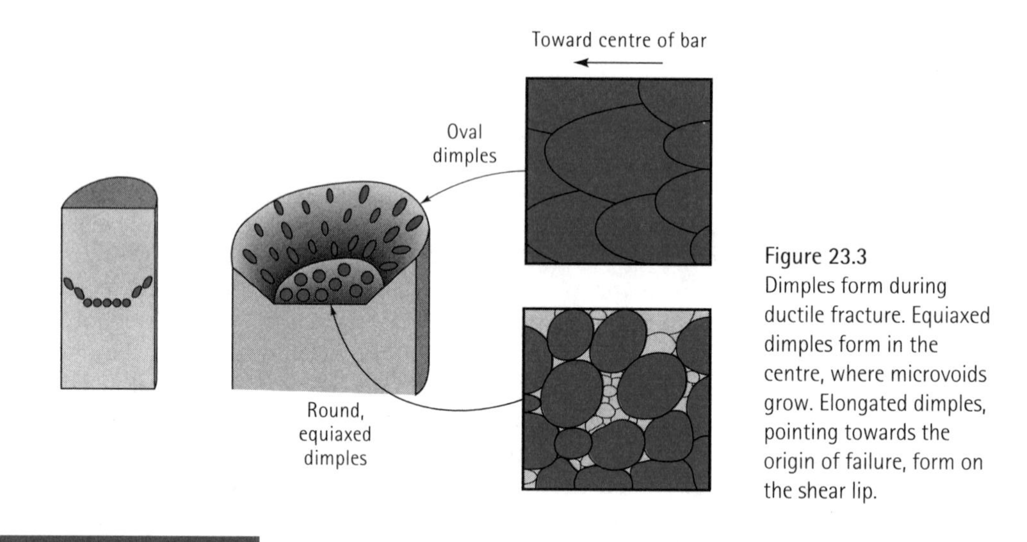

Figure 23.3
Dimples form during ductile fracture. Equiaxed dimples form in the centre, where microvoids grow. Elongated dimples, pointing towards the origin of failure, form on the shear lip.

EXAMPLE 23.1 Failure Analysis of a Hoist Chain

A chain used to hoist heavy loads fails. Examination of the failed link indicates considerable deformation and necking prior to failure. List some of the possible reasons for failure.

SOLUTION

This description suggests that the chain failed in a ductile manner by a simple tensile overload. Two factors could be responsible for this failure:

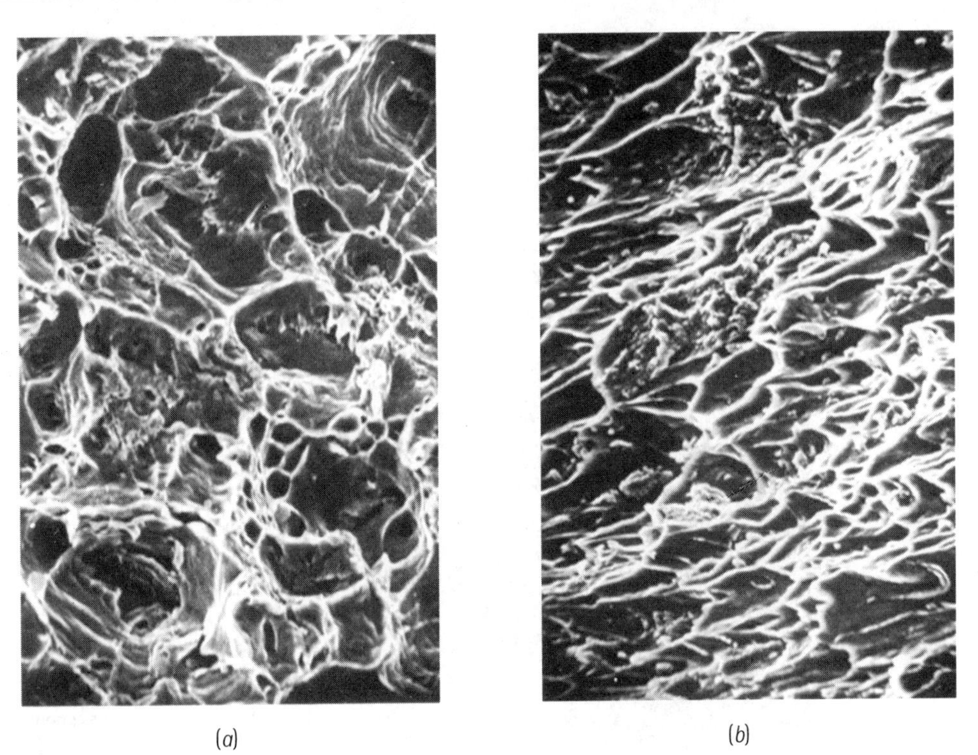

(a) (b)

Figure 23.4
Scanning electron micrographs of an annealed low carbon steel exhibiting ductile fracture in a tensile test. (a) Equiaxed dimples at the flat centre of the cup and cone, and (b) elongated dimples at the shear lip (× 1 250).

1. The load exceeded the hoisting capacity of the chain. Thus, the stress due to the load exceeded the yield strength of the chain, permitting failure. Comparison of the load to the manufacturer's specifications will indicate that the chain was not intended for such a heavy load. This is the fault of the user!
2. The chain was of the wrong composition or was improperly heat-treated. Consequently, the yield strength was lower than intended by the manufacturer and could not support the load. This is the fault of the manufacturer!

Brittle Fracture Brittle fracture occurs in high-strength metals or metals with poor ductility and toughness. Furthermore, even metals that are normally ductile may fail in a brittle manner at low temperatures, in thick sections, at high strain rates (such as impact), or when flaws play an important role. Brittle fractures are frequently observed when impact rather than overload causes failure.

In brittle fracture, little or no plastic deformation is required. Initiation of the crack normally occurs at small flaws, which cause a concentration of stress. The crack may move at a rate approaching the velocity of sound in the metal. Normally, the crack propagates most easily along specific crystallographic planes, often the {100} planes, by cleavage. In some cases, however, the crack may take an intergranular (along the grain boundaries) path, particularly when segregation or inclusions weaken the grain boundaries.

Brittle fracture can be identified by observing the features on the failed surface. Normally, the fracture surface is flat and perpendicular to the applied stress in a tensile test. If failure occurs by cleavage, each fractured grain is flat and differently oriented, giving a crystalline or 'rock candy' appearance to the fracture surface (Figure 23.5). Often the layman claims that the metal failed because it crystallised. Of course, we know that the metal was crystalline to begin with and the surface appearance is due to the cleavage faces.

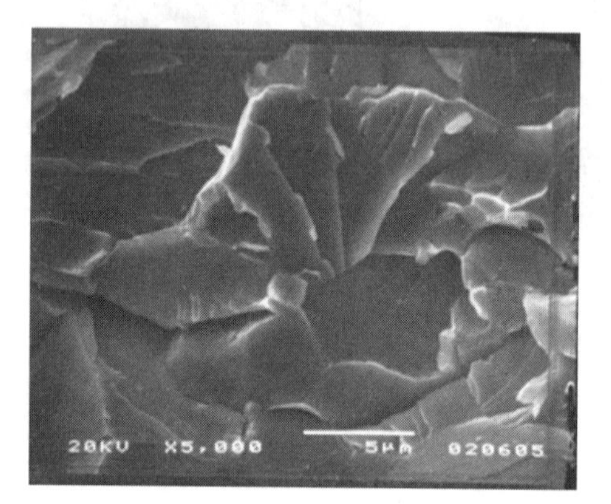

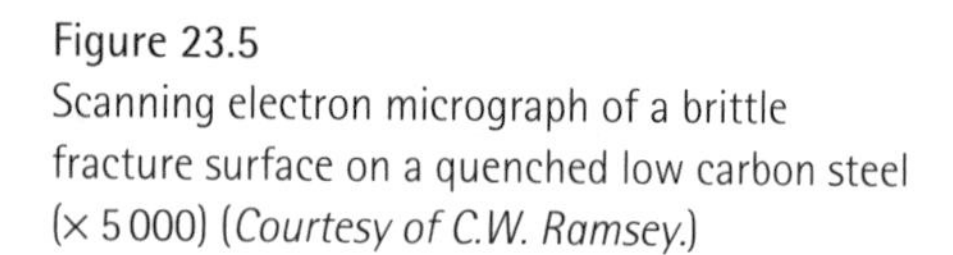
Figure 23.5
Scanning electron micrograph of a brittle fracture surface on a quenched low carbon steel (× 5 000) (*Courtesy of C.W. Ramsey.*)

Another common fracture feature is the chevron pattern (Figure 23.6), produced by separate crack fronts propagating at different levels in the material. A radiating pattern of surface markings, or ridges, fans away from the origin of the crack (Figure 23.7). The chevron pattern is visible with the naked eye or a magnifying glass and helps us identify both the brittle nature of the failure process as well as the origin of the failure.

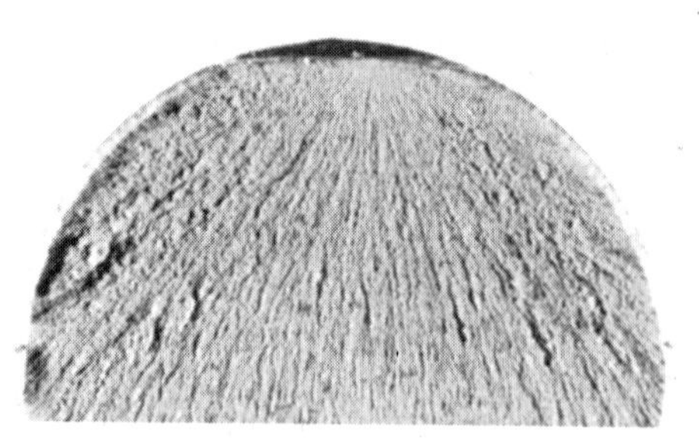

Figure 23.6
The chevron pattern in a 12 mm-diameter quenched Ni-Cr-Mo steel. The steel failed in a brittle manner by an impact blow.

EXAMPLE 23.2 **Failure Analysis of an Automobile Axle**

An engineer investigating the cause of an automobile accident finds that the right rear wheel has broken off at the axle. The axle is bent. The fracture surface reveals a chevron pattern pointing towards the surface of the axle. Suggest a possible cause for the fracture.

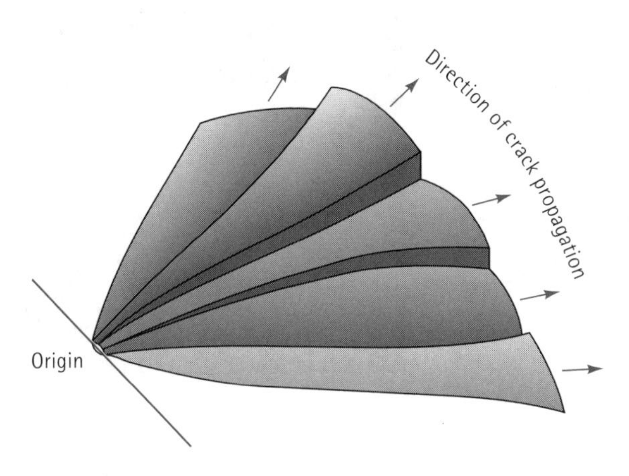

Figure 23.7
The chevron pattern forms as the crack propagates from the origin at different levels. The pattern points back to the origin.

SOLUTION

The evidence suggests that the axle did not break prior to the accident. The deformed axle means that the wheel was still attached when the load was applied. The chevron pattern indicates that the wheel was subjected to an intense impact blow, which was transmitted to the axle, causing failure. The preliminary evidence suggests that the driver lost control and crashed and the force of the crash caused the axle to break. Further examination of the fracture surface, microstructure, composition, and properties may verify that the axle was properly manufactured.

Fatigue Fracture A metal fails by fatigue when an alternating stress greater than the endurance limit is applied. Fracture occurs by a three-step process involving (1) nucleation of a crack, (2) slow, cyclic propagation of the crack, and (3) catastrophic failure of the metal. Normally, nucleation sites are at or near the surface, where the stress is at a maximum, and include surface defects such as scratches or pits, sharp corners due to poor design or manufacture, inclusions, grain boundaries, or dislocation concentrations.

Once nucleated, the crack grows towards lower stress regions. Because of the stress concentration at the tip, the crack propagates a little bit further during each cycle until the load-carrying capacity of the remaining metal is approached. The crack then grows spontaneously, usually in a brittle manner.

Fatigue failures are often easy to identify. The fracture surface – particularly near the origin – is typically smooth. The surface becomes rougher as the original crack increases in size and may be fibrous during final crack propagation.

Microscopic and macroscopic examination reveal a fracture surface including a beach mark pattern and striations (Figure 23.8). Beach marks are normally formed when the load is changed during service or when the loading is intermittent, perhaps permitting time for oxidation inside the crack. Striations, which are on a much finer scale, show the position of the crack tip after each cycle. Beach marks always suggest a fatigue failure, but – unfortunately – the absence of beach marks does not rule out fatigue failure.

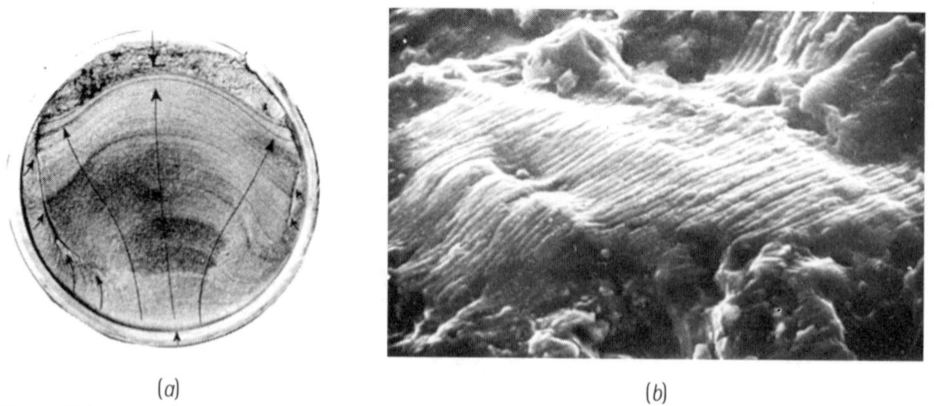

(a) (b)

Figure 23.8
Fatigue fracture surface: (*a*) at low magnifications, the beach mark pattern indicates fatigue as the fracture mechanism. The arrows show the direction of growth of the crack front, whose origin is at the bottom of the photograph. (*b*) At very high magnifications, closely spaced striations formed during fatigue are observed (× 1 000). (*Image (a) is from C.A. Cottell, 'Fatigue Failures with Special Reference to Fracture Characteristics,'* Failure Analysis: The British Engine Technical Reports, *American Society for Metals, 1981, p. 318.*)

EXAMPLE 23.3 **Failure Analysis of a Crankshaft**

A crankshaft in a diesel engine fails. Examination of the crankshaft reveals no plastic deformation. The fracture surface is smooth. In addition, several other cracks appear at other locations in the crankshaft. What type of failure mechanism would you expect?

SOLUTION

Since the crankshaft is a rotating part, the surface experiences cyclical loading. We should immediately suspect fatigue. The absence of plastic deformation supports our suspicion. Furthermore, the presence of other cracks is consistent with fatigue; the other cracks didn't have time to grow to the size that produced catastrophic failure. Examination of the fracture surface will probably reveal beach marks or fatigue striations.

Creep and Stress Rupture At elevated temperatures, a metal undergoes thermally induced plastic deformation, even though the applied stress is below the nominal yield strength. Often fracture is accompanied by necking, void nucleation and coalescence, or grain boundary sliding (Figure 23.9). Creep failures are defined as excessive deformation or distortion of the metal part, even if fracture has not occurred. Stress-rupture failures are defined as the actual fracture of the metal part.

Normally, ductile stress-rupture fractures include necking and the presence of many cracks that did not have an opportunity to produce final fracture. Furthermore, grains near the fracture surface tend to be elongated. Ductile stress-rupture failures

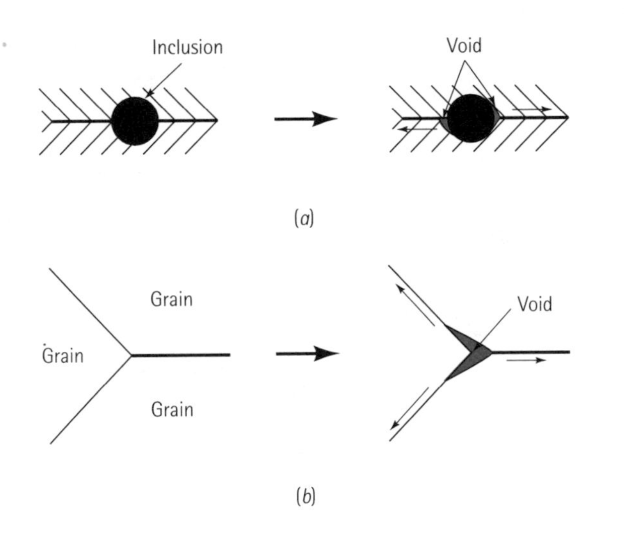

Figure 23.9
Grain boundary sliding during creep causes (a) the creation of voids at an inclusion trapped at the grain boundary and (b) the creation of a void at a triple point where three grains are in contact.

generally occur at high creep rates and relatively low exposure temperatures and have short rupture times.

Brittle stress-rupture fractures usually show little necking and occur more often at slow creep rates and high temperatures. Equiaxed grains are observed near the fracture surface. Brittle failure typically occurs by formation of voids at the intersection of three grain boundaries and precipitation of additional voids along grain boundaries by diffusion processes (Figure 23.10).

Stress-Corrosion Fractures Stress-corrosion fractures occur at stresses well below the yield strength of the metal due to attack accompanied by a corrosive medium. Deep, fine corrosion cracks are produced, even though the metal as a whole shows little uniform attack. The stresses can be either externally applied or stored residual stresses. Stress-corrosion failures are often identified by microstructural examination of the nearby metal. Ordinarily, extensive branching of the cracks along grain boundaries is observed (Figure 23.11). The location at which cracks initiated may be identified by the presence of a corrosion product.

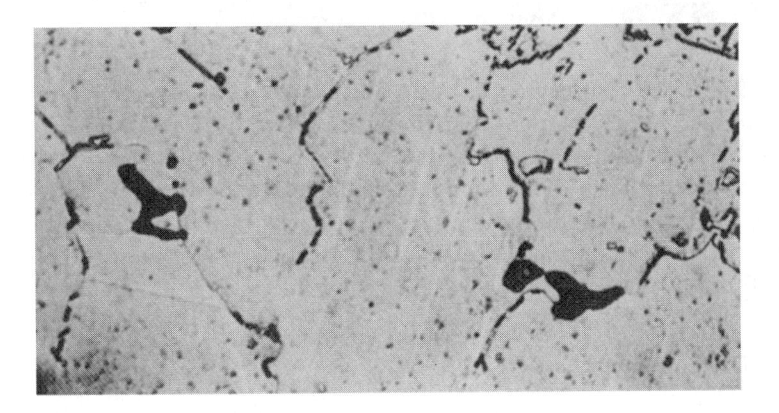

Figure 23.10
Creep cavities formed at grain boundaries in an austenitic stainless steel (× 500).
(*From* Metals Handbook, *Vol. 7, 8th Ed., American Society for Metals, 1972.*)

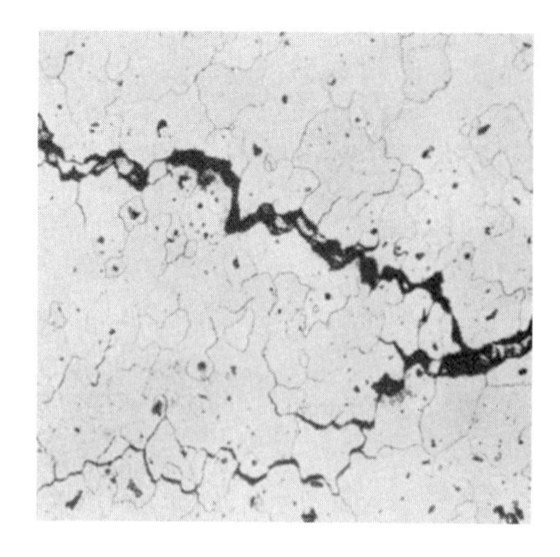

Figure 23.11
Photomicrograph of a metal near a stress-corrosion fracture, showing the many intergranular cracks formed as a result of the corrosion process (× 200). (*From* Metals Handbook, *Vol. 7, 8th Ed., American Society for Metals, 1972.*)

EXAMPLE 23.4 **Failure Analysis of a Pipe**

A titanium pipe used to transport a corrosive material at 400°C is found to fail after several months. How would you determine the cause for the failure?

SOLUTION

Since a period of time at a high temperature was required before failure occurred, we might first suspect a creep or stress-corrosion mechanism for failure. Microscopic examination of the material near the fracture surface would be advisable.

If many tiny, branched cracks leading away from the surface are noted, stress-corrosion is a strong possibility. However, if the grains near the fracture surface are elongated, with many voids between the grains, creep is a more likely culprit.

EXAMPLE 23.5 **Design of a Bridge Component**

Figure 23.12 shows a diagram of a pin-and-hanger-strap assembly used on a bridge. The connection is designed in such a manner that as the ambient temperature changes, the resulting expansion or contraction of the bridge occurs by rotation of the pin. In one such case, the pin cracked, as indicated, and eventually failed. A cross-section of the pin, including grooves and cracks introduced into the pin, is also shown. Determine a probable cause of the fracture and suggest methods to protect against future failures.

SOLUTION

Inspection of the overall assembly shows that the bridge accommodates expansion or contraction when the two halves of the girder move sideways, opening or closing the gap at their junction. The pin-and-strap assembly holds the two halves of the girder

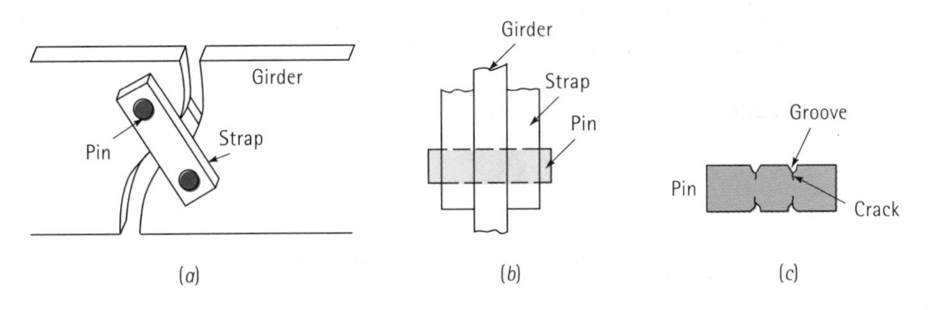

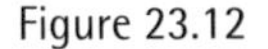

Figure 23.12
Sketches of a strap-pin-girder connection for a highway bridge: (*a*) overall picture of the assembly, (*b*) section through the connection, and (*c*) cross-section of pin showing grooves and cracks just prior to fracture.

together. Consequently, the girder and strap rotate around the pin as the bridge heats or cools. To avoid a torsional stress acting on the pins, the assembly should be lubricated to ensure that the girder and strap rotate freely on the pin. Our inspection of the bridge reveals that the connection is exposed to the environment. In particular, water on the surface of the bridge can trickle through open expansion plates and collect at the girder-strap-pin interface.

On the basis of our chemical and microscopic examinations, we find that the pin is produced from a relatively low-carbon steel and contains a mixture of ferrite and pearlite. Slip lines – a series of parallel lines caused by movement of dislocations – are observed in the ferrite near the cracked surface. A corrosion product is observed in branches of the crack that did not completely propagate.

Visual examination reveals that grooves formed in the pin at the junction between the girder, strap, and pin; the cracks that led to final failure initiated at the base of these grooves.

Based on these observations, we may decide that the pin was unable to rotate freely, causing a cyclical torsional stress to be placed on the pin. The pin was unable to rotate freely due to corrosion. As moisture collected at the joint, galvanic corrosion (perhaps a consequence of a concentration cell) occurred, removing metallic atoms from the pin and creating the groove. In addition, corrosion within the crevice produced by-products that 'froze' the pin and prevented it from rotating. Subsequent movement of the bridge introduced stresses at the pin, creating stress cells and cracking due to stress-corrosion or corrosion-fatigue. Growth of the crack continued to reduce the effective cross-section of the pin, increasing the stress and eventually leading to failure. The overall process was accelerated by the use of salts to keep ice from forming on the bridge during the winter; the salts produce an electrolyte that encourages galvanic corrosion.

Several changes in the the design of the bridge connection might be considered. The pin and other portions of the assembly could be produced from more corrosion-resistant or higher-strength materials – for example, stainless steel. However, the cost of building bridges from such materials would be prohibitive. The assembly could be designed to permit easier maintenance and lubrication, perhaps reducing the rate of corrosion by excluding moisture and minimising stresses by preventing the pin from freezing in the strap or girder. The assembly might be redesigned to include rubber gaskets or other devices to help exclude the environment.

In addition, nondestructive techniques might be used to assist in detecting corrosion and cracking early in the process so that the pins could be replaced before they fail. Use of ultrasonic testing, described in a later section, might be one process by which these problems could be detected.

23.3 Fracture in Nonmetallic Materials

In ceramic materials, the ionic or covalent bonds permit little or no slip. Consequently, failure is a result of brittle fracture. Most crystalline ceramics fail by cleavage along widely spaced, closely packed planes. The fracture surface typically is smooth, and frequently no characteristic surface features point to the origin of the fracture [Figure 23.13(a)].

Glasses also fracture in a brittle manner. Frequently, a conchoidal fracture surface is observed. This surface contains a very smooth mirror zone near the origin of the fracture, with tear lines comprising the remainder of the surface [Figure 23.13(b)]. The tear lines point back to the mirror zone and the origin of the crack, much like the chevron pattern in metals.

Polymers can fail by either a ductile or a brittle mechanism. Below the glass transition temperature, thermoplastic polymers fail in a brittle manner – much like a ceramic glass. Likewise, the hard thermosetting polymers fail by a brittle mechanism. Thermoplastics, however, fail in a ductile manner above the glass transition temperature, giving evidence of extensive deformation and even necking prior to failure. The ductile behaviour is a result of sliding of the polymer chains, which is not possible in glassy or thermosetting polymers.

Fracture in fibre-reinforced composite materials is more complex. Typically, these composites contain strong, brittle fibres surrounded by a soft, ductile matrix, as in

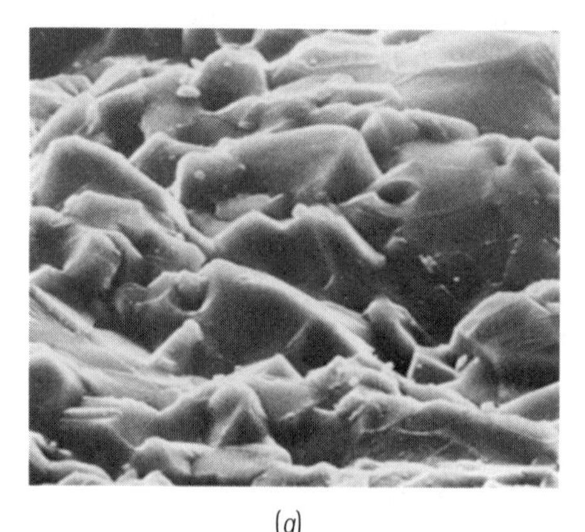

(*a*)

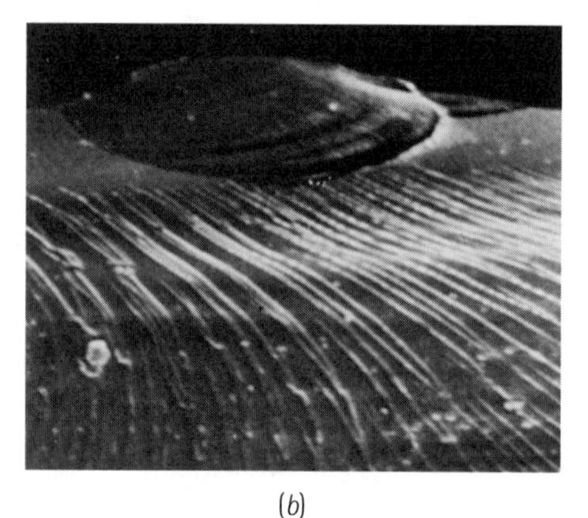

(*b*)

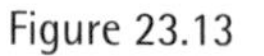

Figure 23.13

Scanning electron micrographs of fracture surfaces in ceramics: (*a*) the fracture surface Al_2O_3, showing the cleavage faces (× 1 250); and (*b*) the fracture surface of glass, showing the mirror zone (*top*) and tear lines characteristic of conchoidal fracture (× 300).

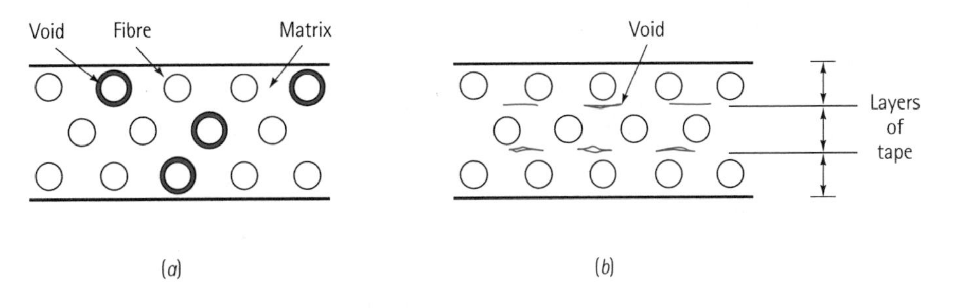

Figure 23.14
Fibre-reinforced composites can fail by several mechanisms: (*a*) Due to weak bonding between the matrix and fibres, fibres can pull out of the matrix, creating voids. (*b*) If the individual layers of the matrix are poorly bonded, the matrix may delaminate, creating voids.

boron-reinforced aluminium. When a tensile stress is applied along the fibres, the soft aluminium deforms in a ductile manner, with void formation and coalescence eventually producing a dimpled fracture surface. As the aluminium deforms, the load is no longer transmitted effectively to the fibres; the fibres break in a brittle manner until there are too few of them left intact to support the final load.

Fracture becomes easier if bonding between the fibres is poor. Voids can form between the fibres and the matrix, causing pull-out. Voids can also form between layers of the matrix if composite tapes or sheets are not properly bonded, causing delamination (Figure 23.14).

EXAMPLE 23.6

Describe the difference in fracture mechanism between a boron-reinforced aluminium composite and a glass fibre-reinforced composite.

SOLUTION

In the boron-aluminium composite, the aluminium matrix is soft and ductile; thus we expect the matrix to fail in a ductile manner.

Boron fibres, in contrast, fail in a brittle manner. Both glass fibres and epoxy are brittle; thus the composite as a whole should display little evidence of ductile fracture.

23.4 Source and Prevention of Failures in Metals

We can prevent metal failures by several approaches: design of components, selection of appropriate materials and processing techniques, and consideration of the service conditions.

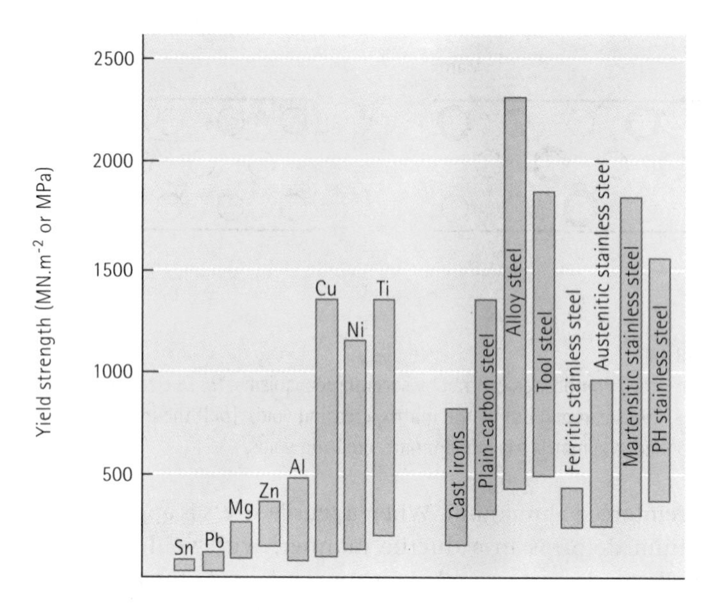

Figure 23.15
Comparison of the range of properties available for many important metals and alloys. A wide range of properties is possible for each alloy system, depending on composition and treatment.

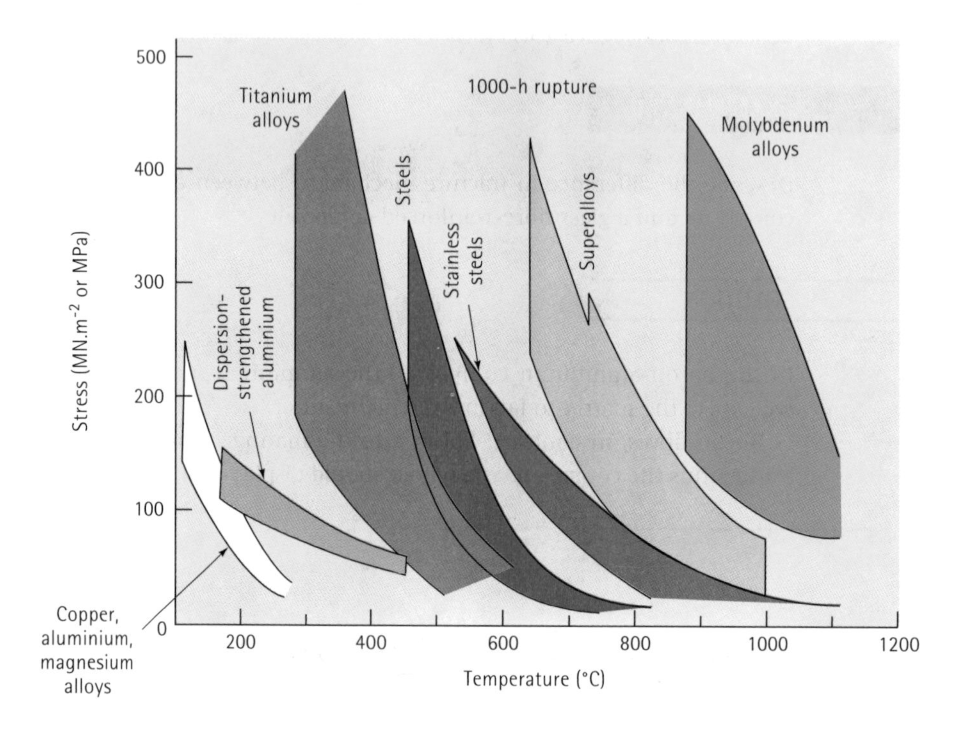

Figure 23.16
Stress-rupture plots for a 1000-h lifetime, showing the range of suitable temperatures for several alloys.

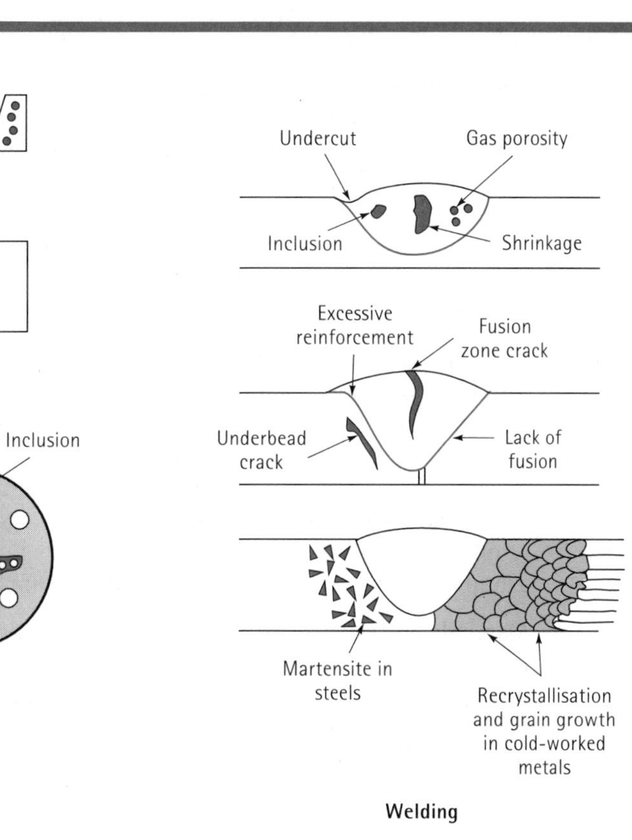

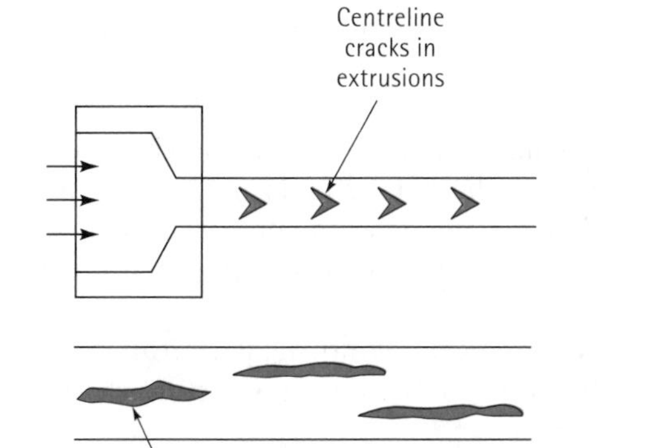

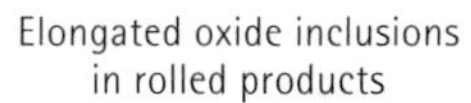

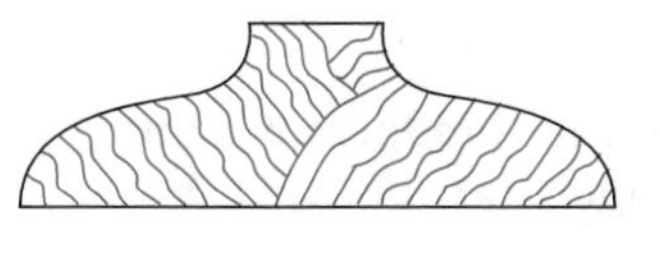

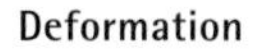

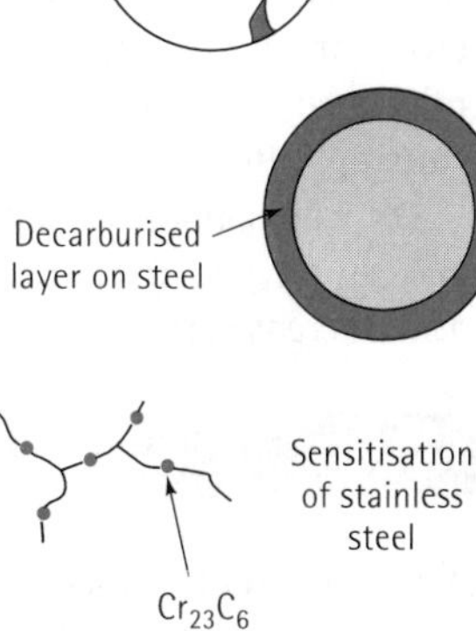

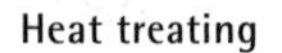

Figure 23.17
Typical defects introduced into a metal during processing.

Design Components must be designed to (a) permit the material to withstand the maximum stress that is expected to be applied during service, (b) avoid stress raisers that cause the metal to fail at lower than expected loads, and (c) ensure that deterioration of the material during service does not cause failure at lower than expected loads.

Creep, fatigue, and stress-corrosion failures occur at stresses well below the yield strength. The design of the component must be based on the appropriate creep, fatigue, and stress-corrosion data, not yield strength.

Materials Selection A tremendous variety of materials are available to the engineer for any application, many of which are capable of withstanding high applied stresses (Figure 23.15). Selection of a material is based both on the ability of the material to serve and on the cost of the material and its processing.

The engineer must consider the condition of the material. For example, age hardened, cold-worked, or quenched and tempered alloys lose their strength at high temperatures. Figure 23.16 shows the temperature ranges over which a number of groups of alloys can operate as a function of applied stress.

Materials Processing All finished components are at one time passed through some type of processing – casting, forming, machining, joining, or heat treatment – to produce the appropriate shape, size, and properties. However, a variety of flaws can be introduced. The engineer must design to compensate for the flaws or must detect their presence and either reject the material or correct the flaw. Detection of flaws will be discussed in the next section. Figure 23.17 illustrates some of the typical flaws that might be introduced in metals.

Service Conditions The performance of a material is influenced by service conditions, including the type of loading, the environment, and the temperature to which an assembly is exposed.

Another source of failure is abuse of the material in service. Such abuse includes overloading a material – for example, using a chain to lift a tank whose weight exceeds the capacity of the chain. An ordinary carpenter's hammer should not be used as a crowbar by striking the hammer with another metallic instrument. Flakes of metal may spall off the face of the hammer (which is heat-treated to a high hardness), causing injury.

Improper maintenance, such as inadequate lubrication of moving parts, can lead to adhesive wear, overheating and oxidation. If overheated, the microstructure changes and decreases the strength or ductility of the metal.

EXAMPLE 23.7 Failure Analysis of a Weld

An alloy steel, which is welded using an electrode that produces a high-hydrogen atmosphere, is found to fail in the heat-affected zone near the weld. What factors may have contributed to the failure?

SOLUTION

If we assume that the joint is designed so that the maximum stresses can be withstood, failure probably occurred as a result of flaws introduced during welding.

Since the steel failed in the heat-affected zone rather than the fusion zone, possible problems include the following:

1. Severe grain growth may have weakened the metal near the fusion zone.
2. Since an alloy steel has good hardenability, martensite may have formed during cooling. A large grain size also increases the likelihood of martensite formation.
3. Hydrogen dissolved in the fusion zone may have diffused to the heat-affected zone and caused hydrogen embrittlement, or underbead cracking.
4. Because of temperature differences in the steel during welding, residual stresses may build up in the weld. Tensile residual stresses reduce the local load-bearing ability of the weld.
5. Elongated inclusions in the steel could act as stress raisers and initiate cracks.

Additional information obtained concerning the initial state of the steel, the welding parameters, and the microstructure would permit us to pin down the cause for failure.

EXAMPLE 23.8 Failure Analysis of a Steel Rope

A steel rope, composed of many tiny strands of steel, passes over a 50 mm-diameter pulley. After several months, the steel rope fails, dropping its heavy load. Suggest a possible reason for failure.

SOLUTION

The long period of time required before failure occurred suggests that fatigue might be the culprit. Each time the steel rope passes over the pulley, the outer strands of the steel experience a high stress. This stress may exceed the endurance limit. After passing over the pulley enough times, the strands begin to fail by fatigue. This failure increases the stress on the remaining strands, accelerating their failure, until the rope is overloaded and breaks. A larger-diameter pulley will reduce the stress on the fibres to below the endurance limit, so that the rope will not fail.

EXAMPLE 23.9 Failure Analysis of a Gear

After a helicopter crash, the teeth on a gear in the transmission are found to have worn away. The gear, which is a carburised alloy steel, is intended to have a surface hardness of HRC 60. However, the hardness measured on a portion of one tooth that is still intact reveals a hardness of HRC 30. Suggest possible causes for the failure.

SOLUTION

We know that the rate of wear increases when the hardness decreases. Thus, we confidently blame the failure on the soft gear. However, we still need to determine why the gear was soft. One explanation would be that the gear was not carburised or heat-treated. Microscopic examination would reveal this fact. Suppose that examination shows the presence of a proper case depth and an overtempered martensitic structure. This condition would suggest that the gear – originally properly manufactured – overheated during use, softened, and then began to wear. In this case, failure may have been caused by loss of oil from the gear box, permitting the gear to overheat.

23.5 Nondestructive Testing Methods

Care in design, materials selection, materials processing, and service conditions helps prevent failure of materials. But how do we determine whether our engineering process has been successful? Often a simple visual examination will reveal defects, such as an undercut in a weld or a hot tear in a casting. Destructive tests can also be used to determine the properties of the finished product. For example, chain manufacturers routinely pull samples of their product to failure to determine the maximum rating for the product.

Obviously, however, we cannot destructively test all of the product! Instead, a number of techniques are used to determine the quality of a product and to detect the presence, location, and size of any flaws without causing any damage to the material. These *nondestructive tests* can be coupled with the concepts of fracture toughness described in Chapter 6 to predict whether flaws will cause fracture for a given applied stress. If flaws are sufficiently small, they may not propagate under the expected loads and, consequently, the part may continue in service. By periodically monitoring flaw size and charting flaw growth, we can determine when a part must be scrapped or repaired. This approach is often used for aerospace components.

In this section, a number of such tests will be discussed, including hardness testing, proof testing, radiography, ultrasonic inspection, eddy current inspection, magnetic particle inspection, liquid penetrant inspection, thermography, and acoustic-emission inspection.

Hardness Test In some situations, a hardness test can be used to ensure proper heat treatment. We may determine whether an aluminium alloy has been properly aged, or a steel adequately tempered, or a grey cast iron fully annealed. However, the hardness test does not tell us the microstructure, or whether a crack is present at the surface, or whether shrinkage cavity lies within a casting.

Proof Test In many instances, a proof test can be designed. We load a part to its rated capacity and see if the part remains intact. If the part is never abused or used to support loads higher than the proof test, we can be confident that the part will perform properly.

Radiography In radiographic testing, the transmission and absorption characteristics of a material are utilised to produce a visual image of flaws in a material. There are several requirements for using a radiographic technique:

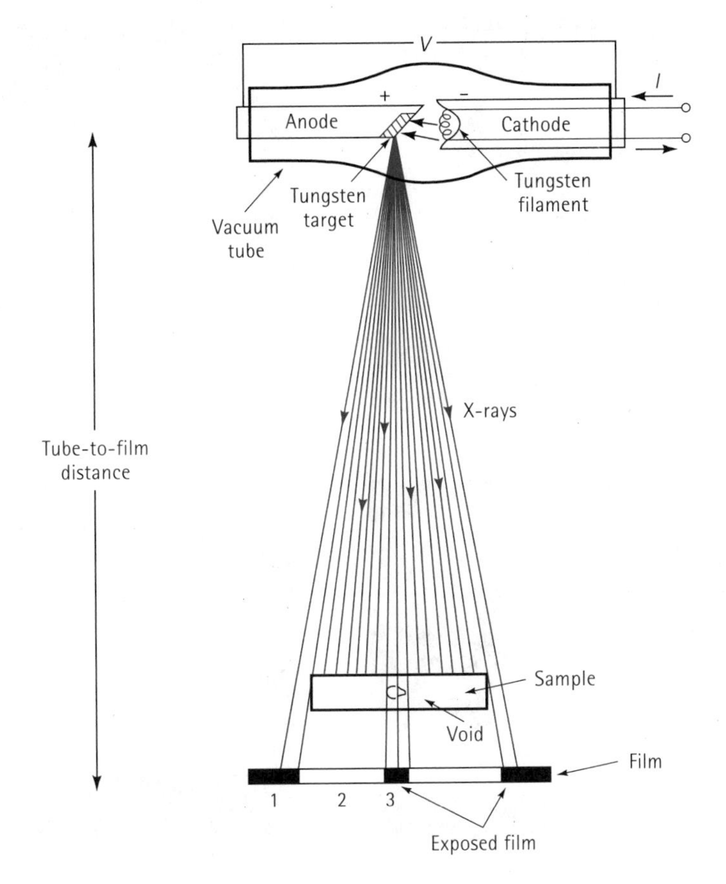

Figure 23.18
The elements in an X-ray radiographic nondestructive testing set-up.

1. A source of penetrating radiation is required. X-rays emitted from a tungsten target are most common, but occasionally gamma rays or neutrons are required.
2. A detection system is needed. Normally, special film is used to detect the amount of radiation transmitted through the material. Other detectors include fluorescent screens, Geiger counters, and Xerographs.
3. The flaw – or discontinuity – in the material must have a different absorption characteristic than the material itself.

In X-ray radiography (Figure 23.18), an X-ray tube provides the source of the radiation. Electrons are emitted from a tungsten filament cathode and accelerated at a high voltage onto an anode material, also tungsten. The beam excites electrons in the inner shells of the tungsten target and a continuous spectrum of X-rays are emitted when the electrons fall back to their equilibrium states. The emitted X-rays are directed to the material to be inspected. A small fraction of the X-rays are transmitted through the material to expose the film.

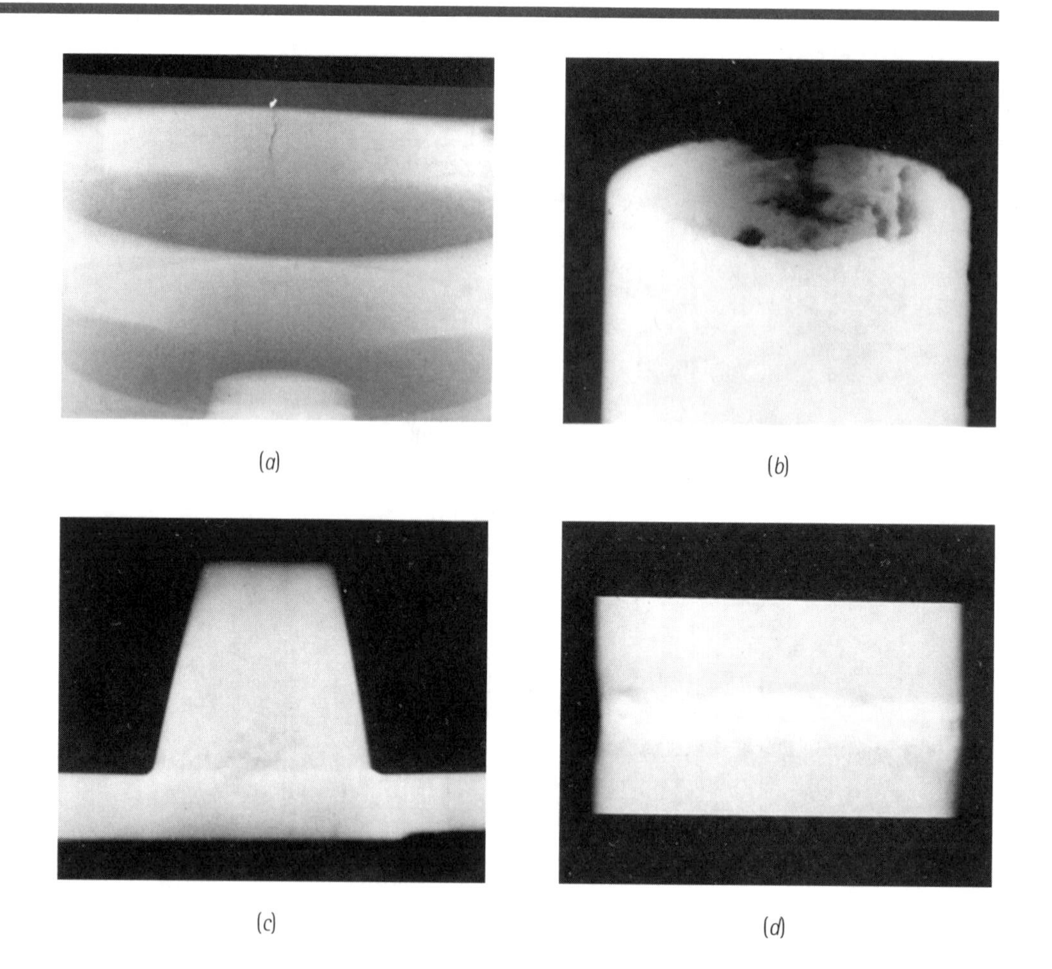

Figure 23.19
Radiographs of (*a*) crack and hydrogen porosity in a magnesium casting, (*b*) graphitic corrosion of a cast iron pipe, (*c*) solidification shrinkage in an aluminium casting, and (*d*) incomplete penetration of a steel weld.

The intensity I of the transmitted X-ray beam depends on the absorption coefficient and the thickness of the material:

$$I = I_0 \exp(-\mu x) = I_0 \exp(-\mu_m \rho x) \tag{23.1}$$

where I_0 is the intensity of the incoming beam, μ is the linear absorption coefficient (m^{-1}), μ_m is the mass absorption coefficient ($m^2.Mg^{-1}$), ρ is the density ($Mg.m^{-3}$), and x is the thickness of the material (m). The mass absorption coefficients for selected materials are included in Table 23.1. If a large shrinkage cavity is present inside a casting, the absorption of the X-rays by the cavity is lower than in the solid metal and, consequently, the intensity of the transmitted beam is greater. The film is exposed to more radiation and, after developing, is much darker (Figure 23.19).

Table 23.1 Absorption coefficients of selected elements for tungsten X-rays and for neutrons.

Element	Density ($Mg.m^{-3}$)	μ_m (X-rays) λ = 0.0098 nm ($m^2.Mg^{-1}$)	μ_m (neutrons) λ = 0.108 nm ($m^2.Mg^{-1}$)
H		28	11
Be	1.85	13.1	0.03
B	2.3	13.8	2 400
C	2.2	14.2	0.015
N	0.00116	14.3	4.8
O	0.00133	14.4	0.002
Mg	1.74	15.2	0.1
Al	2.7	15.6	0.3
Si	2.33	15.9	0.1
Ti	4.54	21.7	0.1
Fe	7.87	26.5	1.5
Ni	8.9	31	2.8
Cu	8.96	32.5	2.1
Zn	7.133	35	0.55
Mo	10.2	79	0.9
Sn	7.3	117	0.2
W	19.3	288	3.6
Pb	11.34	350	0.03

After W. McGonnagle, Nondestructive Testing, *McGraw-Hill, 1961.*

EXAMPLE 23.10

In Figure 23.18, let's assume that the material being inspected is a copper plate 25 mm thick and that a discontinuity 6 mm thick is present at point 3. The discontinuity contains air. Estimate the intensity I at locations 1, 2, and 3 if the copper plate is placed on the film 760 mm from the source. The average wavelength of the tungsten X-rays is 0.0098 nm.

SOLUTION

Point 1: The beam must pass through air only. If we assume that air is 80% N_2 and 20% O_2, then using the rule of mixtures:

$$\begin{aligned}\mu_{m,\text{air}} &= f_{O_2}\,\mu_{m,O_2} + f_{N_2}\,\mu_{m,N_2}\\ &= (0.2)(14.4) + (0.8)(14.3)\\ &= 14.3\ m^2.Mg^{-1}\end{aligned}$$

$$\begin{aligned}\rho_{air} &= f_{O_2}\rho_{O_2} + f_{N_2}\rho_{N_2} \\ &= (0.2)(1.33 \times 10^{-3}) + (0.8)(1.16 \times 10^{-3}) \\ &= 1.19 \times 10^{-3}\ \text{Mg.m}^{-3}\end{aligned}$$

$$\begin{aligned}\frac{I}{I_0} &= \exp(-\mu_m \rho x) \\ &= \exp(-14.3)(1.19 \times 10^{-3})(760)(10^{-3}\ \text{mm/m}) = 0.987\end{aligned}$$

Virtually no absorption occurs in air.

Point 2: Since virtually no absorption occurs in air, let's neglect absorption of the beam before it reaches the copper. Thus $x = 25$ mm. From Table 23.1, $\mu_{m,\,Cu} = 32.5$ $\text{m}^2.\text{Mg}^{-1}$

$$\frac{I}{I_0} = \exp(-\mu_m \rho x) = \exp(32.5)(8.96)(25)(10^{-3}\text{mm/m}) = 0.00061$$

Practically none of the beam is transmitted.

Point 3: Since the 6 mm cavity absorbs no X-rays compared with the copper, we can ignore it and consider the thickness to be 25 – 6 = 19 mm.

$$\frac{I}{I_0} = \exp(-\mu_m \rho x) = \exp(-32.5)(8.96)(19 \times 10^{-3}) = 0.00395$$

The X-ray beam intensity is about one order of magnitude greater when it passes through the discontinuity. Thus, the film will be darker beneath the discontinuity.

The geometry and physical properties of the discontinuity also affect how easily it can be detected (Figure 23.20):

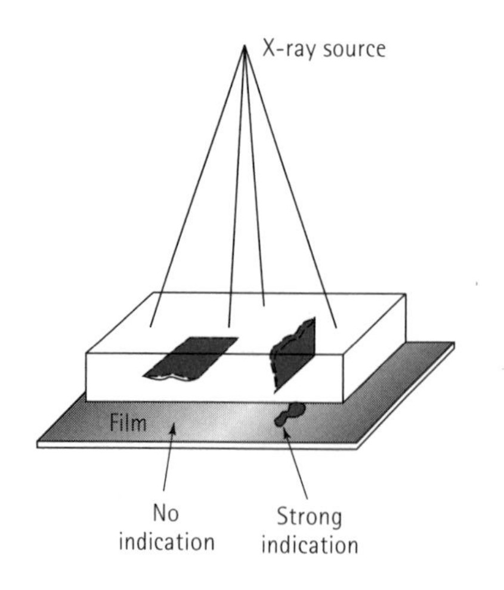

Figure 23.20
The importance of orientation on the detection of discontinuities in a material by radiography.

1. Shrinkage voids or gas porosity in castings can be detected if their size falls within the sensitivity of the radiographic technique. The voids or pores are normally round, so the direction of the X-rays compared with the orientation of the casting is not critical. The voids or pores are filled with either a gas or a vacuum, which have very different absorption coefficients than the metal.
2. Nonmetallic inclusions in rolled products may be flattened during deformation processing. If the X-rays are perpendicular to the inclusion, the thin inclusions may not cause a noticeable change in intensity. Either the geometry between the X-ray source and the flaw must be changed or some other nondestructive testing technique may be required.
3. Detection of cracks caused by casting, welding, incipient fatigue, or stress-corrosion is dependent on the crack geometry. If the crack is perpendicular to the incident radiation, it may not be detected, but the crack may be easily observed if it is parallel to the radiation.

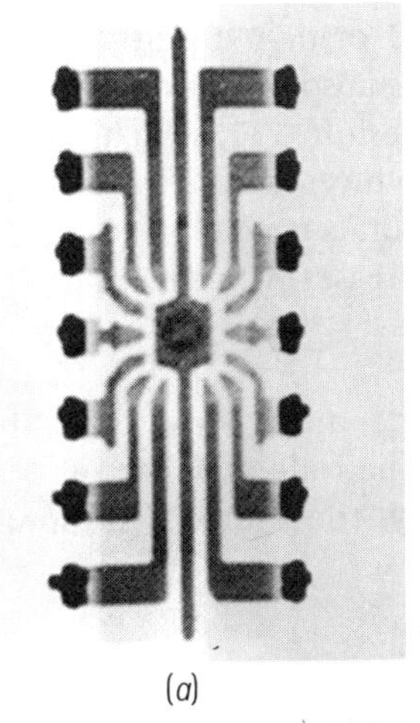

(*a*)

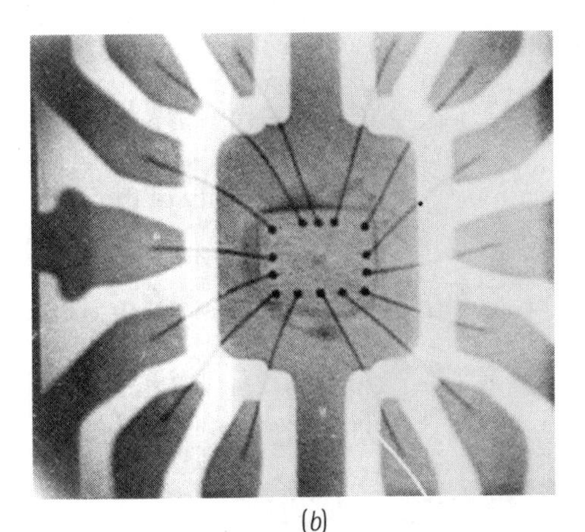

(*b*)

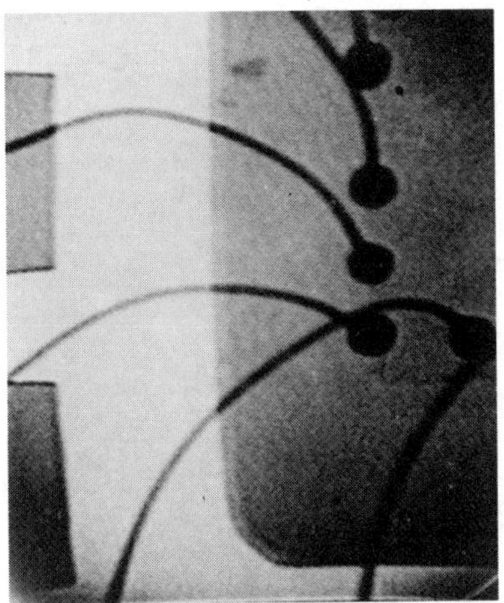

(c)

Figure 23.21
X-ray radiographs of an integrated circuit at three magnifications obtained by electronic reception of the transmitted beam. The radiograph shows the 14 copper lead pads, gold wire and weld balls, and an aluminium weld pad. [(a) × 5, (b) × 20, (c) × 50.] (*From* Nondestructive Testing Handbook, *Vol. 3*: 'Radiography and Radiation Testing,' *2nd Ed., American Society or Nondestructive Testing, 1985.*)

Improvements in radiographic techniques have made possible better resolutions an even three-dimensional representations of flaws. Semiconductor materials such as *p-n* diodes receive the transmitted X-ray beam and convert the beam to a current pulse. The current pulses are then electronically and computer enhanced to provide improved sensitivity and sharpness of the radiograph. Inspection of very small parts, including integrated circuits (Figure 23.21), can be accomplished. By using computerised axial tomography (the CAT scan), either the X-ray source and detector or the sample is rotated during radiography. Several exposures at different locations can be made. Again, the transmitted beam is received by electronic means and analysed by computer to provide information in three dimensions. Two-dimensional radiographs can then be generated by the computer for any plane within the sample. This technique is also used frequently for medical purposes.

Gamma ray radiography employs an intense radiation of a single wavelength produced by nuclear disintegration of radioactive materials. The intensity of the radiation depends on the type and size of the radioactive source. Cobalt 60 produces gamma rays having an average energy of about 2.13×10^{-13} J (or a wavelength of about 0.0009 nm). The gamma rays are much more energetic than the X-rays obtained from tungsten (about 2.0×10^{-14} J or 0.0098 nm). Consequently, Co^{60} is used for thick, absorptive materials. The mass absorption coefficient of most elements for Co gamma rays is about 5.5 $m^2.Mg^{-1}$. Caesium 137, which produces gamma rays of 1.06×10^{-13} J, and indium 192, which produces gamma rays of 0.5×10^{-13} J to 1.0×10^{-13} J, are used when less energetic radiation is needed.

The intensity of the gamma ray source decreases with time:

$$I = I_0 \exp(-\lambda t) \tag{23.2}$$

where λ is the decay constant for the material and t is the time. The half-life is the time required for I to decrease to $0.5I_0$. The half-life of cobalt 60 is about 5.27 years and that for iridium 192 is 74 days. Usually, the source is no longer suitable after about two half-lives have elapsed.

EXAMPLE 23.11

Determine the decay constant for cobalt 60, which has a half-life of 5.27 years.

SOLUTION

In equation 23.2, $I = 0.5I_0$ when $t = 5.27$ years:

$$0.5I_0 = I_0 \exp(-5.27\lambda)$$

$$0.5 = \exp(-5.27\lambda)$$

$$\ln(0.5) = -5.27\lambda$$

$$\lambda = \frac{-0.693}{-5.27} = 0.131 \text{ years}^{-1} = 4.15 \times 10^{-9} \text{s}^{-1}$$

Table 23.2 Ultrasonic velocities for selected materials.

Material	Bulk Velocity ($m.s^{-1} \times 10^3$)	Modulus of Elasticity ($GN.m^{-2}$)	Density ($Mg.m^{-3}$)
Al	6.25	70	2.7
Cu	4.62	127	8.96
Pb	1.96	14	11.34
Mg	5.77	45.5	1.74
Ni	6.02	209	8.9
60% Ni-40%Cu	5.33	175	8.9
Ag	3.63	72	10.49
Stainless steel	5.74	210	7.91
Sn	3.38	53.8	7.3
W	5.18	414	19.25
Air	0.33		0.0013
Glass	5.64	70	2.32
Lucite	2.67	3	1.18
Polyethylene (h.d.)	1.96	1.2	0.9
Quartz	5.74	67	2.65
Water	1.50		1.00

From W.McGonnagle, Nondestructive Testing, *McGraw-Hill, 1961.*

Neutron radiography is occasionally used because the neutrons are absorbed by nuclear rather than electronic interactions. Materials that have similar absorption coefficients for X-rays or gamma rays may be easily distinguished by neutron radiography (Table 23.1) However, the source for the neutrons is a nuclear reactor, which is not widely available.

Ultrasonic Testing A material may both transmit and reflect elastic waves. An ultrasonic transducer composed of quartz, barium titanate, or lithium sulphate uses the piezoelectric effect to introduce a series of elastic pulses into the material at a high frequency – usually greater than 100 000 Hz. The pulses create a compressive strain wave that propagates through the material. The elastic wave is transmitted through the material at a rate that depends on the modulus of elasticity and the density of the material. For a thin rod,

$$v = \sqrt{\frac{Eg}{\rho}} \tag{23.3}$$

where E is the modulus of elasticity, g is the acceleration due to gravity (9.81 $m.s^{-2}$), and ρ is the density of the material. More complicated expressions are required for pulses propagating in bulkier material. Examples of the bulk ultrasonic velocity v for several materials are given in Table 23.2.

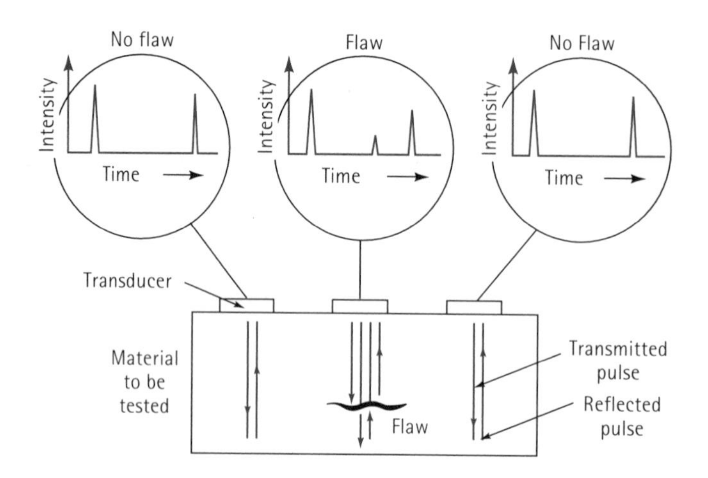

Figure 23.22
The pulse-echo ultrasonic test. The time required for a pulse to travel through the metal, reflect off a discontinuity on the opposite side, and return to the transducer is measured with an oscilloscope.

Three common techniques are used to inspect a material ultrasonically. First, in the pulse-echo or reflection method, an ultrasonic pulse is generated and transmitted through the material (Figure 23.22). When the elastic wave strikes an interface, a portion is reflected and returns to the transducer. Both the initial and the reflected pulse can be displayed on an oscilloscope.

From the oscilloscope display, we measure the time required for the pulse to travel from the transducer, to the reflecting interface, and back to the transducer. If we know the velocity at which the pulse travels in the material, we can determine how far the elastic wave travelled and can calculate the distance below the surface at which the reflecting interface is located. If there are no flaws in the material, the beam reflects from the opposite side of the material and our measured distance corresponds to twice the wall thickness.

If a discontinuity is present and properly oriented beneath the transducer, at least a portion of the pulse reflects from the discontinuity and registers at the transducer in a shorter period of time. Now our calculations show that a discontinuity lies within the material and even tell us the depth of the discontinuity below the surface. By moving the transducer we can estimate the size of the discontinuity. Automatic scanning techniques can move the transducer over the surface and display the results of the entire scan, showing the exact location of flaws. If we also combine holography with ultrasonics, we can obtain a three-dimensional picture of the discontinuities.

EXAMPLE 23.12

In a pulse-echo ultrasonic inspection of an aluminium rod, the oscilloscope shows three peaks. The first, at time zero, is the initial transmitted pulse; the second, at time 1.63×10^{-5} s, is the reflection from an internal discontinuity; and the third peak, at 2.44×10^{-5} s, is the reflection from the opposite surface of the material. Calculate the thickness of the material and the depth below the surface at which the flaw exists.

SOLUTION

From Table 23.2, we expect the ultrasonic velocity in aluminium to be 6.25×10^3 m.s^{-1}. Thus, the total distance that the pulse travelled in each case is:

Discontinuity: Distance $= (6.25 \times 10^3)(1.63 \times 10^{-5})$
$= 0.1019$ m $= 101.9$ mm

Back surface: Distance $= (6.25 \times 10^3)(2.44 \times 10^{-5})$
$= 0.1525$ m $= 152.5$ mm

Since the total distance includes the return path, the actual depth of the discontinuity is 101.9 / 2 = 50.95 mm and the actual thickness of the rod is 152.5 / 2 = 76.25 mm.

Second, in the through-transmission method, an ultrasonic pulse is generated at one transducer and detected at the opposite surface by a second transducer (Figure 23.23). The initial and transmitted pulses are displayed on an oscilloscope. The loss of energy from the initial to the transmitted pulse depends on whether or not a discontinuity is present in the material.

Third, in the resonance method, we utilise the wavelike nature of the ultrasonic wave. A continuum of pulses is generated and travels as an elastic wave through the material (Figure 23.24). By selecting a wavelength or frequency so that the thickness of the material is a whole number of half-wavelengths, a stationary elastic wave is produced and reinforced in the material. A discontinuity in the material prevents resonance from occurring. However, this technique is used more frequently to determine the thickness of the material.

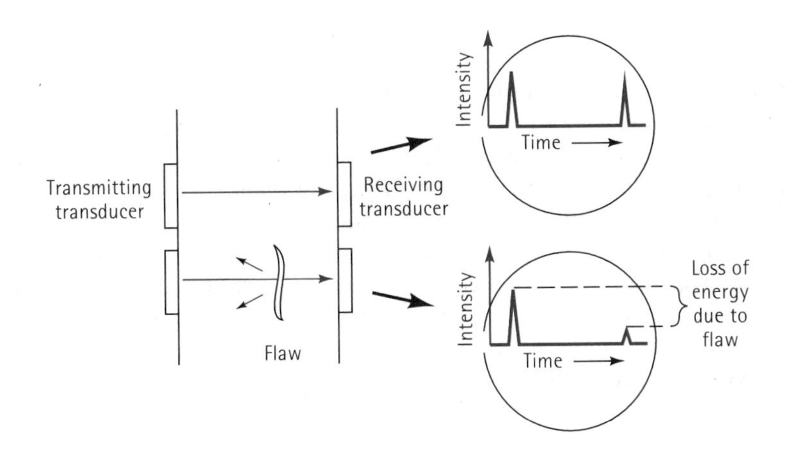

Figure 23.23
The through-transmission ultrasonic test. The presence of a discontinuity reflects a portion of the transmitted beam, thus reducing the intensity of the pulse at the receiving transducer.

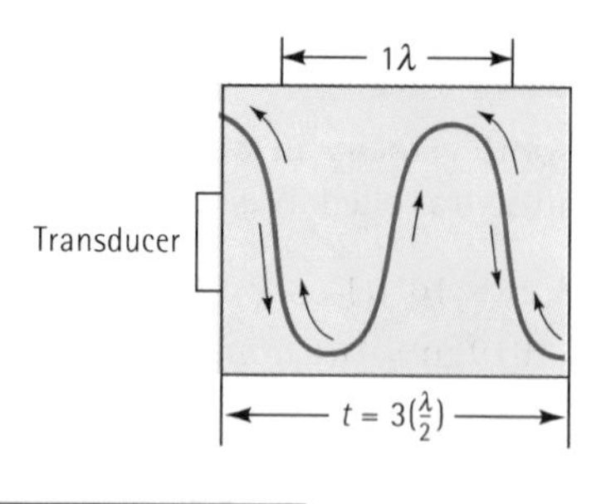

Figure 23.24
In the resonance method of ultrasonic testing, a stationary elastic wave is produced by changing the wavelength or frequency of the ultrasonic pulse until a whole number of half-wavelengths matches the thickness of the material.

EXAMPLE 23.13

The thickness of copper is to be determined by the resonance method. A frequency v of 1 213 000 Hz is required to produce 12 half-wavelengths at resonance. Determine the thickness of the copper.

SOLUTION

From Table 23.2, the ultrasonic velocity in copper is 4.62×10^3 m.s^{-1}:

$$\lambda = \frac{\upsilon}{v} = \frac{4.62\times10^3}{1213\,000} = 0.00381 \text{ m} = 3.81 \text{ mm}$$

$$\frac{\lambda}{2} = 1.905 \text{ mm}$$

$$\frac{\lambda}{2}(12) = (1.905)(12) = 22.86 \text{ mm} = \text{thickness of the copper}$$

Magnetic Particle Inspection Discontinuities near the surface of ferromagnetic materials are detected by magnetic particle testing. A magnetic field is induced in the material to be tested (Figure 23.25), producing lines of flux. If a discontinuity is present in the material, the reduction in the magnetic permeability of the material due to the discontinuity alters the flux density of the magnetic field. Leakage of the lines of flux into the surrounding atmosphere creates local north and south poles that attract magnetic powder particles. The particles may be added dry or in a fluid such as water or light oil for better movement; they also may be dyed or coated with a fluorescent material to aid in their detection.

Several requirements must be satisfied in order to detect discontinuities by magnetic particle inspection:

1. The discontinuity must be perpendicular to the lines of flux. Thus, different methods of introducing the magnetic field detect differently oriented discontinuities.
2. The discontinuity must be near the surface or the flux lines will merely crowd together rather than leak from the material. Magnetic particle testing is well

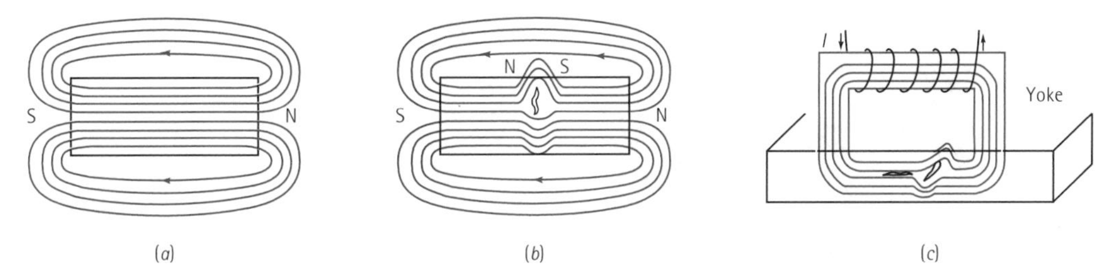

Figure 23.25
(*a*) Normal lines of magnetic flux in a defect-free material. (*b*) A flaw in a ferromagnetic material that causes a disruption of the normal lines of magnetic flux. Magnetic particles are attracted to the flux leakage and indicate the location of the flaw. (*c*) Effect of flaw orientation on disruption of the lines of flux.

suited for locating quench cracks, fatigue cracks, or cracks induced by grinding, all of which occur at the surface.

3. The discontinuity must have a lower magnetic permeability than the metal.
4. Only ferromagnetic materials can be tested.

EXAMPLE 23.14

Two cylindrical bars of a steel are joined by friction welding. The ends of the bars are rotated against one another, heating the surfaces, and are then forced together under high pressure to produce a bond. Describe a magnetic particle test that will determine if the bars are properly joined.

SOLUTION

The potential weld defect is perpendicular to the cylindrical bars. Thus, we want to produce longitudinal magnetisation to detect any flaw. The yoke technique [Figure 23.25(c)] produces the proper orientation of the magnetic flux lines with the weld.

Eddy Current Testing In eddy current testing we rely on the interaction between the material and an electromagnetic field. An alternating current flowing in a conductive coil produces an electromagnetic field. If a conductive material is placed near or within the coil (Figure 23.26), the field of the coil will induce eddy currents and additional electromagnetic fields in the sample that, in turn, interact with the original field of the coil. By measuring the effect of the sample on the coil, we can detect changes in electrical conductivity or magnetic permeability of the sample caused by differences in composition, microstructure, and properties. Because discontinuities in the sample will alter the electromagnetic fields, potentially harmful defects can be detected. Even changes in the size of the sample or the thickness of the plating on a sample might be detected by the test.

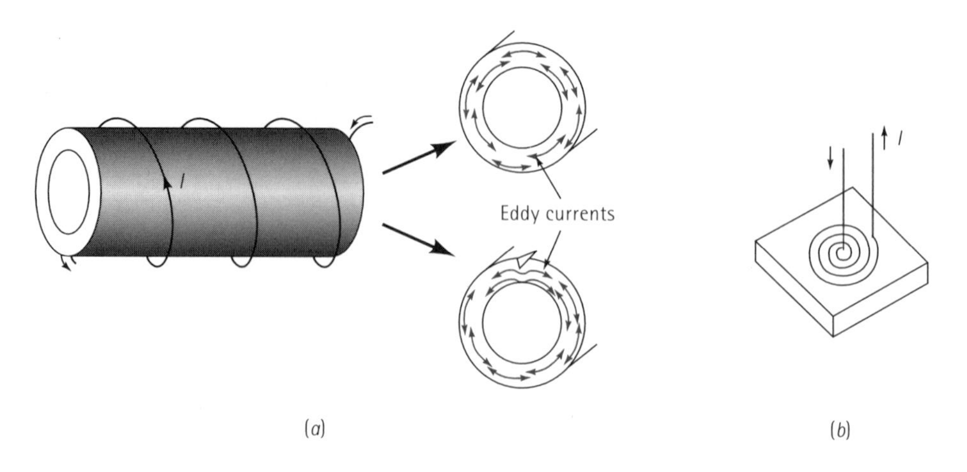

Figure 23.26
(*a*) The through-coil method and (*b*) the probe method for eddy current inspection.

Eddy current testing, like magnetic particle inspection, is best suited for detecting flaws near the surface of a sample. Particularly at high frequencies, the eddy currents do not penetrate deeply below the surface.

The eddy current test is particularly rapid compared with most other non-destructive testing techniques. Therefore, large numbers of parts can be tested quickly and economically. Often the eddy current test is set up as a 'go or no go' test that is standardised on good parts. If the interaction between the coil and the part is the same when other parts are tested, the parts may be assumed to be good.

Liquid Penetrant Inspection Discontinuities such as cracks that penetrate to the surface can be detected by liquid penetrant inspection, or the dye penetrant technique. A liquid dye is drawn by capillary action into a thin crack that might otherwise be invisible. A four-step process is involved (Figure 23.27). The surface is first thoroughly cleaned. A liquid dye is sprayed onto the surface and permitted to stand for a period of time, during which the dye is drawn into any surface discontinuities. Excess dye is then cleaned from the metal surface. Finally, a developing solution is sprayed onto the surface. The developer reacts with any dye that remains, drawing the dye from the cracks. The dye can then be observed, either because it changes the colour of the developer or because it fluoresces under an ultraviolet lamp.

Thermography Often imperfections in a material alter the rate of heat flow in the vicinity, leading to high-temperature gradients or hot spots. In thermography, a temperature-sensitive coating is applied to the surface of a material, then the material is uniformly heated and cooled. The temperature is higher in the vicinity of an imperfection than at other locations; therefore, the colour of the coating at this location is different and easily detected.

A variety of coatings can be used. Heat-sensitive paints, heat-sensitive papers, organic compounds or phosphors that produce visible light when excited by infrared radiation, and crystalline organic materials known as liquid crystals are commonly used.

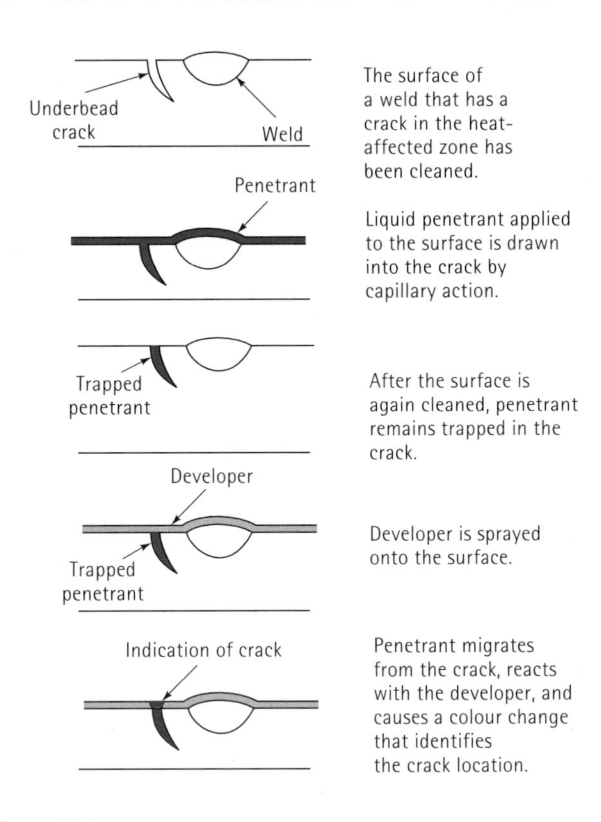

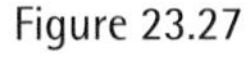

Figure 23.27
The elements of the dye penetrant test, used to detect cracks that penetrate to the surface.

One important use of thermography is in detecting lack of bonding between or delamination of the individual monolayers or tapes that make up many fibre-reinforced composite material structures, particularly in the aerospace industry.

Acoustic-Emission Inspection Accompanying many microscopic phenomena, such as crack growth or phase transformations, is the release of strain energy in the form of high-frequency elastic stress waves, much like those produced during an earthquake. In the acoustic-emission test, a stress below the nominal yield strength of the material is applied. Due to stress concentrations at the tip of a pre-existing crack, the crack may enlarge, releasing the strain energy surrounding the crack tip. The elastic stress wave associated with the movement of the crack can be detected by a piezoelectric sensor, amplified, and analysed. Cracks as small as 10 nm long can be detected by this technique. By using several sensors simultaneously, the location of the crack can also be determined.

Acoustic emission testing can be used for all materials. The technique is used to detect microcracks in aluminium aircraft components even before the micro-cracks are large enough to imperil the safety of the aircraft. Cracks in polymers and ceramics can be detected. The test will detect breakage of fibres in fibre-reinforced composite materials as well as debonding between the fibres and the matrix.

SUMMARY

- Flaws are present in materials, and failures do occur. To learn from our mistakes, we analyse a fracture to determine the origin and cause of the failure. Details of the fracture surface and the microstructure near the fracture may reveal its origin and cause. Certain key observations for failures of metals include:
  - Ductile overload, which occurs when an applied stress exceeds the yield strength, is normally accompanied by overall deformation of the part, necking, a cup-and-cone fracture, shear lips, and dimpled fractured surfaces.
  - Impact failures often include a chevron pattern pointing to the fracture origin, and flat cleavage planes observed on the fracture surface.
  - Fatigue failures include beach marks, striations, smooth fracture surfaces, and the absence of overall plastic deformation. Beach marks point to the fracture origin.
  - Creep and stress-rupture failures often include deformation of a part, along with the formation of voids within the material.
  - Stress-corrosion cracks typically are highly branched and contain the by-products of the corrosion process.
- A number of nondestructive methods, including hardness testing and proof testing, are used to help ensure quality parts. Particularly valuable, however, are nondestructive tests that enable us to determine the presence, location, and size of potentially dangerous flaws in material:
  - Radiography relies on differences in the absorption of X-rays or gamma rays between flawed material, on the one hand, and sound material, on the other hand. If the absorption coefficients are different and if the dimension of the flaw parallel to the beam is large, radiography reveals the presence of the flaw.
  - In ultrasonic inspection, the flaw reflects or attenuates a high-frequency elastic wave, permitting the size and location of a flaw to be determined.
  - In magnetic particle inspection, properly oriented flaws at or near the surface of a part disrupt the magnetic field, whereas in eddy current inspection, flaws at or near the surface interact with an electromagnetic field. For magnetic particle inspection, the part must be composed of a ferromagnetic material; in eddy current testing, the material must be an electrical conductor.
  - Dye penetrant inspection can locate imperfections only at the surface of a part; thermography relies on temperature gradients introduced into the material due to differences in thermal properties; acoustic emission detects the growth of existing flaws by means of sensitive microphones.

GLOSSARY

Acoustic emission testing
Detection of elastic stress waves produced when a small applied stress causes enlargement of a flaw in a material.

Beach marks
Marks on the surface of a fatigue fracture representing the position of the crack front at various times during failure.

Chevron pattern
Markings caused by the merging of crack fronts in brittle fracture. The markings form arrows that point back towards the origin of the brittle fracture.

Cleavage
Brittle fracture along particular crystallographic planes in the grains of the material.

Conchoidal fracture
A characteristic fracture surface in glass, with a mirror zone near the origin of the fracture and tear lines pointing towards the origin.

Eddy current testing
A nondestructive testing technique to detect flaws or evaluate structure and properties by determining the interaction between the material and an electric field.

Intergranular fracture
Fracture of a material along the grain boundaries.

Liquid penetrant inspection
A nondestructive testing technique in which a liquid, drawn into a surface imperfection by capillary action, is exposed by a dye or ultraviolet light.

Magnetic particle inspection
A nondestructive testing technique that relies on the interruption of lines of magnetic flux by imperfections near the surface of a ferromagnetic material.

Mass absorption coefficient
A measure of a material's ability to absorb X-rays or other radiation.

Microvoids
Tiny voids at the fracture surface formed by separation of the material at grain boundaries or other interfaces during ductile failure.

Proof test
Loading a material to its designed capacity to determine if it is capable of proper service.

Radiography
A nondestructive testing technique that relies on a difference in the absorption of radiation by the material, on the one hand, and flaws in the material, on the other hand.

Shear lip
The surface formed by ductile fracture that is at a 45° angle to the direction of the applied stress.

Stress-corrosion fracture

Fracture caused by a combination of corrosion and a stress below the yield strength.

Striations

Microscopic traces of the location of a fatigue crack.

Thermography

Detection of discontinuities in a material by the change in colour of a coating caused by temperature differences induced in a material.

Transgranular fracture

Fracture of a material through the grains rather than along the grain boundaries.

PROBLEMS

23.1 Investigation of an automobile accident revealed that an axle for a rear wheel was broken. Expert witnesses disagree on whether the axle failed by fatigue, thereby causing the accident, or whether the axle failed in impact as a consequence of the accident. What features on the axle and the fracture surface would you look for in an attempt to settle this dispute?

23.2 A turbine blade in a jet engine is found to be the cause of an aeroplane crash. Expert witnesses disagree on whether the turbine blade failed because a large bird was ingested into the engine, because the engine overheated and creep failure occurred, or because of stress-corrosion cracking due to the presence of sulphides in the engine. What features of the turbine blade, fracture surface, and microstructure would you look for in an attempt to settle this dispute?

23.3 A wire rope made up of many strands of small-diameter wire passes over a pulley before being used to lift heavy loads. After several months of use, the wire rope fails while lifting a load. Expert witnesses disagree on whether the strands in the rope failed due to fatigue (when a rope passes over a pulley whose diameter is too small, the stresses are unusually high) or because the load being lifted exceeded the limit of the rope. What features would you look for in order to determine the cause of failure?

23.4 A cast iron lever is used to tighten ropes that hold a heavy load on a truck. After several years of use, the lever breaks during a routine tightening operation. What type of cast iron would you recommend for such an application? What types of fracture mechanisms would be possible for such a case? What features of the fractured lever would you look for to determine the cause of the fracture?

23.5 A complex-shaped clamp made of copper alloy is made by a cold-forming process. The clamp is used to fasten a heavy electrical transformer to a pole outdoors. Shortly after installation, cracks are found at a location where the clamp had been bent into a U-shape during forming. What are possible fracture mechanisms for this device? What features in the material or on the fracture surface would you look for to determine the cause of the crack?

23.6 Two thick steel plates are joined by an arc welding process as part of an assembly for a missile-carrying transport vehicle. After the vehicle is placed in service, the assembly fails. Inspection of the failure indicates that the crack propagated through the heat-affected area of the weld, right next to the actual fusion zone. Suggest possible causes of the failure, including the role of the welding process and the microstructure that the welding process may have produced in the heat-affected zone. What recommendations might you have to avoid such failures in the future?

23.7 The titanium tubes in a heat exchanger operating at 500°C are found to crack and leak after several months of use. Suggest possible causes for the failure and describe the features of the heat exchanger and its microstructure that you would look for in order to confirm the actual cause.

23.8 Determine the mass absorption coefficient and the linear absorption coefficient for tungsten X-rays in a copper alloy containing 5 wt% Sn and 15 wt% Pb.

23.9 We would like to determine whether there is a large amount of segregation in an Al-15 wt% Si alloy. To ensure the best chance for determining this radiographically, determine whether we should use X-ray radiography or neutron radiography by calculating the mass absorption coefficients of the alloy and comparing them with the coefficients of pure aluminium and silicon.

23.10 A hydrogen gas bubble 1 mm in diameter is present in a 20 mm-thick magnesium casting. Compare the intensity of an X-ray beam transmitted through the section containing the bubble with that of a beam transmitted through the section containing no bubble. If the difference in intensity must be larger than 5% in order for the bubble to detected, will X-ray radiography be successful? Would the bubble be detected if the casting is made of zinc rather than magnesium?

23.11 The gas tungsten arc-welding process is used to join 6 mm-thick titanium alloy sheet (Ti-7 wt% Al-4 wt% Mo). A 0.76 mm-diameter tungsten inclusion is melted from the electrode and is lodged in the fusion zone. Compare the intensity of an X-ray beam transmitted through the fusion zone containing the inclusion with that of a beam transmitted through the unaffected titanium alloy.

23.12 On the basis of data in Table 23.1, determine the relationship between mass absorption coefficient and atomic number Z for X-rays with $\lambda = 0.0098$ nm. Is a similar relationship obtained for the absorption of neutrons?

23.13 We wish to determine whether there is a crack perpendicular to the surface of a 25 mm-thick aluminium plate. We are able to see the crack if there is at least a 2% difference in the intensities of a transmitted X-ray beam. What is the length of the smallest crack that we can detect?

23.14 A 4 mm-thick composite material is obtained when magnesium is reinforced with 60 vol% carbon fibres, each fibre having a diameter of 0.08 mm. Determine the ratio of intensities of an X-ray beam transmitted through a portion of the composite containing a cracked fibre to that transmitted through a portion of the composite containing no cracked fibres. Would you expect a broken fibre to be easily detected by X-ray radiography?

23.15 The intensity of a radiation source and the absorption capability of a material are often related by the *half value layer* (HVL) – the thickness of the material that will reduce the intensity of a radiation beam by half. Calculate the HVL for magnesium and copper for

(a) tungsten X-rays,

(b) cobalt 60 gamma rays, and

(c) neutrons.

23.16 Strontium has a half-life of 28 years. Calculate

(a) the decay constant for Sr^{90} and

(b) the percentage of the intensity of the strontium source after 500 years.

23.17 Polonium has a half-life of 138 days. Determine the number of years before the intensity of polonium is reduced to 10% of its original value.

23.18 Calculate the thin-rod ultrasonic velocity in magnesium, silver and tin and compare with the values given in Table 23.3.

23.19 An ultrasonic pulse introduced into the wall of a 100 mm-thick nickel pressure vessel returns to the transducer in 2.4×10^{-5} s. Is this time lapse caused by the opposite wall of the vessel or by a discontinuity within the wall? If caused by a discontinuity, how far beneath the surface of the wall is the discontinuity located?

23.30 An ultrasonic pulse is introduced into an 80 mm-thick material, and a pulse is received by the transducer after 8×10^{-5} s. Calculate the ultrasonic velocity in the material. If this velocity were the result of a test using a thin rod, determine the specific modulus of the material.

23.21 Figure 23.28 shows the results on an oscilloscope screen of an ultrasonic inspection of a composite material produced by centrifugally casting aluminium on the inside of a stainless steel cylinder. The transducer in this inspection is located on the outside of the cylinder, in contact with the steel. Determine the thickness of both the stainless steel cylinder and the cast aluminium.

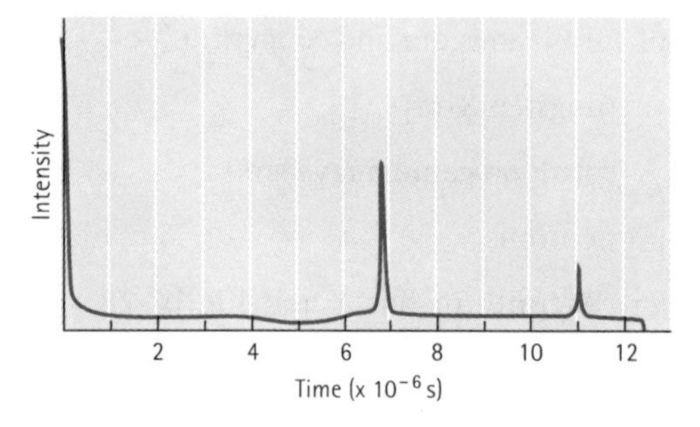

Figure 23.28
Oscilloscope trace for a pulse-echo ultrasonic test (*for* Problem 23.21).

23.22 A laminar composite is produced by gluing a 6 mm-thick sheet of glass onto a 50 mm-thick plate of polyethylene. Determine the length of time required for an ultrasonic pulse to be received from

(a) the interface between the glass and the polymer, and

(b) from the back surface of the polymer. The transducer is located on the glass side of the composite.

23.23 Sketch the signal you expect to receive when an ultrasonic pulse passes through a sandwich structure composed of 4 mm-thick aluminium surrounded by two layers of 2.5 mm-thick nickel.

23.24 We would like to monitor the thickness of nominally 0.01 mm-thick aluminium foil using a resonance ultrasonic test. What frequency must we select if we want to produce 5 half-wavelengths in the foil?

23.25 A resonance ultrasonic test is to be used to determine the thickness of a polyethylene coating on copper. A frequency of 31×10^6 Hz is required to produce 3 half-wavelengths in the coating. Calculate the thickness of the coating.

23.26 What nondestructive testing method(s) might be helpful in detecting the amount of ferrite produced in an austenitic stainless steel as a result of a welding process?

23.27 What nondestructive testing method(s) might be helpful in detecting delamination between the plies of a fibre-reinforced composite material?

23.28 What nondestructive testing method(s) might be helpful in detecting a crack that is parallel to the surface of a steel plate?

23.29 What nondestructive testing method(s) might be helpful in detecting a crack that is perpendicular to and intersects the surface of an aluminium casting?

Design Problems

23.30 Select a nondestructive method to continuously monitor the thickness of lead foil during a cold-rolling process. The lead foil is intended to be 0.0025 mm thick.

23.31 Select a nondestructive ultrasonic inspection method that will allow us to separate white cast iron, grey cast iron, and ductile cast iron from one another. You may wish to review the section on cast irons in Chapter 12.

23.32 Select the shielding around a Co^{60} gamma ray source to ensure that no more than 0.001% of the radiation escapes into the surrounding environment.

23.33 The boom of a crane is subjected to large stresses when it lifts heavy loads. As a result, cracks sometimes initiate and, if not detected, cause the boom to fail, with catastrophic results. Select an inspection method to detect and locate cracks before they become large enough to cause fracture.

Answers to selected problems

Chapter 1

1 Protective covering and sacrificial corrosion; need good bonding; zinc that boils off must be collected.

3 No; brittle ceramic.

5 Lightweight polymer matrix composites and stiff reinforcing fibres.

7 Hardness, impact properties; forging and heat treatment.

9 Hardness and wear resistance plus good electrical conductivity; metal matrix composite, including ceramic particles might work; No, Al_2O_3 is an insulator.

11 Density, conductivity, hardness, magnetic behaviour, etc.

13 Good heat transfer, ductility, and processing from the aluminium; high temperature strength, hardness, and wear resistance from SiC.

Chapter 2

1 1.12×10^{25} atoms.

3 1.078×10^{28} atoms.

5 (a) 5.933×10^{23} atoms.

(b) 0.9856 mol.

7 3.

9 142×10^{23} carriers.

11 1.036×10^{-19}.

13 0.086.

15 Similar electronegativities.

21 Al_2O_3, with strong ionic bonding, has lower coefficient.

23 Weak secondary bonds between chains.

Chapter 3

1 (a) 1.426×10^{-10} m.

(b) 1.4447×10^{-10} m.

3 (a) 0.5336 nm. (b) 0.2310 nm.

5 FCC ($x = 4.025$).

7 BCT.

9 (a) 8 atoms/cell. (b) 0.387.

11 0.6% contraction.

13 (a) 3.185×10^{21} cells.

(b) 6.37×10^{21} iron atoms.

15 A:$[00\bar{1}]$. B:$[1\bar{2}0]$. C:$[\bar{1}11]$. D:$[2\bar{1}\bar{1}]$.

17 A:$(1\bar{1}1)$. B:(030). C:$(10\bar{2})$.

19 A:$[1\bar{1}0]$ or $[1\bar{1}00]$.
B:$[11\bar{1}]$ or $[11\bar{2}\bar{3}]$.
C:$[011]$ or $[\bar{1}2\bar{1}3]$.

21 A:$(1\bar{1}01)$. B:(0003). C:$(1\bar{1}00)$.

27 $[\bar{1}10]$, $[1\bar{1}0]$, $[101]$, $[\bar{1}0\bar{1}]$, $[011]$, $[0\bar{1}\bar{1}]$.

29 Tetragonal–4; orthorhombic–2; cubic–12.

31 (a) (111). (c) $(0\bar{1}2)$.

33 [100]: 0.35089 nm, 2.85 nm^{-1}, 0.866.
[110]: 0.496 nm, 2.015 nm^{-1}, 0.612.
[111]: 0.3039 nm, 3.291 nm^{-1}, 1.

35 (100): $1.617 \times 10^{-13}/m^2$, 0.7854.
(110): 1.144×10^{-13} m^2, 0.555.
(111): $1.867 \times 10^{-13}/m^2$, 0.907.

37 4563000.

39 (a) 0.02797 nm. (b) 0.0629 nm.

41 (a) 6. (c) 8. (e) 4. (g).6.

43 Fluorite.
(a) 0.52885 nm. (b) 12.13 $Mg.m^{-3}$.
(c) 0.624.

45 Caesium chloride.
(a) 0.41916 nm. (b) 4.8 $Mg.m^{-3}$.
(c) 0.693.

47 (111): 0.202 (Mg).
(222): 0.806 (O).

49 8 SiO_2, 8 Si ions, 16 O ions.

51 0.40497 nm.

53 (a) BCC. (c) 0.23 nm.

Chapter 4

1 (a) $[0\bar{1}1]$, $[01\bar{1}]$, $[\bar{1}10]$, $[1\bar{1}0]$, $[\bar{1}01]$, $[10\bar{1}]$.

3 $(0\bar{1}1)$, $(01\bar{1})$, $(1\bar{1}0)$, $(\bar{1}10)$, $(10\bar{1})$, $(\bar{1}01)$.

5 Expected:
$\boldsymbol{b}$ = 0.2863 nm, $\boldsymbol{d}$ = 0.2338 nm.
$(110)[1\bar{1}1]$:
$\boldsymbol{b}$ = 0.7014 nm, $\boldsymbol{d}$ = 0.2863 nm.
Ratio = 0.44.

7 0.13 g.

9 Below, where atoms are less closely spaced.

11 $\tau[110] = 0$; $\tau[011] = 14.29$ $MN.m^{-2}$;
$\tau[101] = 14.29$ $MN.m^{-2}$.

13 5.1×10^{25} vacancies/m^3.

15 (a) 0.002375.
(b) 1.61×10^{26} vacancies/m^3.

17 (a) 1.157×10^{26} vacancies/m^3.
(b) 0.532 $Mg.m^{-3}$.

19 0.345.

21 8.265 $Mg.m^{-3}$.

23 (a) 0.0081.
(b) one H atom per 123 unit cells.

25 (a) 0.0534 defects/unit cell.
(b) 2.52×10^{26} defects/m^3.

27 (a) $K = 0.131$ $MN.m^{-3/2}$. $\sigma_0 = 418$ $MN.m^{-2}$.
(b) 712 $MN.m^{-2}$.

29 2048 grains/mm^2.

31 6.29.

35 28.4 nm.

Chapter 5

1 1.08×10^9 jumps/s.

3 (a) 2.48558×10^5 $J.mol^{-1}$.
(b) 5.7616×10^{-6} $m^2.s^{-1}$.

5 (a) −2.495 at% Sb/m.
(b) -1.246×10^{27} Sb/m^3.m.

7 (a) -1.969×10^{19} H atoms/m^3.m.
(b) 3.3×10^{11} H atoms m^2.s.

9 0.001245 g/hour.

11 −198°C.

13 $D_O = 3.47 \times 10^{-20}$ $m^2.s^{-1}$
versus $D_{Al} = 2.48 \times 10^{-17}$ $m^2.s^{-1}$.

15 $D_H = 1.07 \times 10^{-8}$ $m^2.s^{-1}$
versus $D_N = 3.9 \times 10^{-13}$ $m^2.s^{-1}$.

17 0.1 mm: 0.87% C.
0.5 mm: 0.43% C.
1.0 mm: 0.18% C.

19 907°C.

21 0.53% C.

23 2.9 min.

25 12.8 min.

27 667°C.

29 213486 $J.mol^{-1}$; yes.

48 11 mm.

50 12.5 kN.

Chapter 6

1 (a) Deforms. (b) Does not neck.

3 4891 N.

5 92105 N.

7 49.89 mm.

9 15.265 m.

11 (a) 274 $MN.m^{-2}$. (b) 417 $MN.m^{-2}$.
(c) 172 $GN.m^{-2}$. (d) 18.55%.
(e) 15.8%. (f) 397.9 $MN.m^{-2}$.
(g) 473 $MN.m^{-2}$. (h) 0.17 $MN.m^{-2}$.

13 l = 305.08 mm. d = 9.9992 mm.

15 (a) 618.06 $MN.m^{-2}$.
(b) 183 $GN.m^{-2}$.

17 41 mm; will not fracture.

19 HB = 29.8.

22 No transition temperature.

24 Not notch-sensitive; poor toughness.

26 No; stress required to propagate the crack is 20 times greater than the tensile strength.

28 0.99 $MN.m^{-3/2}$.

30 67.15 N.

32 d = 41.1 mm.

34 22 $MN.m^{-2}$;
max = +22 $MN.m^{-2}$,
min = –22 $MN.m^{-2}$,
mean = 0 $MN.m^{-2}$;
reduce fatigue strength due to heating.

36 (a) 2.5 mm. (b) 0.0039 mm.

38 $C = 2.047 \times 10^{-3}$. n = 3.01.

42 11.97 years.

44 n = 6.86. m = –6.9.

46 3500 hours.

Chapter 7

1 n = 0.12; BCC.

3 n = 0.18.

5 Total = 0.0109 mm,
% increase = 32630%.

7 3.65 mm.

9 165 $MN.m^{-2}$ tensile,
135 $MN.m^{-2}$ yield, 5.5% elongation.

11 First step:
36% CW giving 165 $MN.m^{-2}$ tensile,
135 $MN.m^{-2}$ yield, 6% elongation.
Second step:
64% CW giving 188 $MN.m^{-2}$ tensile,
170 $MN.m^{-2}$ yield, 3% elongation.
Third step:
84% CW giving 200 $MN.m^{-2}$ tensile,
185 $MN.m^{-2}$ yield, 2% elongation.

13 20 to 25 mm.

15 47% CW: 175 $MN.m^{-2}$ tensile,
150 $MN.m^{-2}$ yield, 4% elongation.

17 (a) 6185 N. (b) Will not break.

19 (a) 550°C, 750°C, 950°C. (b) 700°C.
(c) 900°C. (d) 2285°C.

21 Increase grain growth temperature and keep grain size small.

23 Slope = 0.4.

25 CW 75% from 50 to 25 mm, anneal.
CW 75% from 25 to 12.5 mm, anneal.
CW 71.3% from 12.5 to 6.7 mm, anneal.
CW 44.3% from 6.7 to 5 mm or
hot work 98.2% from 50 to 6.7 mm, then
CW 44.3% from 6.7 to 5 mm.

Chapter 8

1 (a) 0.665 nm. (b) 109 atoms.

3 1.136×10^6 atoms.

5 (a) 0.0333. (b) 0.333. (c) All.

7 1265°C.

9 31.15 s.

11 $B = 7.6$ s.mm^{-2}, $n = 1.58$.

13 (a) 0.0416 mm. (b) 90 s.

15 $K = 0.032$ s, $m = 0.34$.

17 0.033 mm.

19 (a) 900°C. (b) 420°C. (c) 480°C.
(d) 312°C/min. (e) 9.7 min.
(f) 8.1 min. (g) 60°C. (h) Zinc.
(i) 8.38 s.mm^{-2}.

21 $D = 139.4$ mm,
$H = 209.1$ mm,
$V = 3.19 \times 10^6$ mm^3.

23 V/A (riser) = 18.75,
V/A (middle) = 28.28,
V/A (end) = 22.22; not effective.

25 $D_{Cu} = 37.08$ mm. $D_{Fe} = 32.40$ mm.

27 (a) 1970 mm^3. (b) 4.1%.

29 230.4 mm.

31 0.46 mm^3/g Al.

Chapter 9

3 (a) Yes. (c) No. (e) No. (g) No.

5 *Cd* should give smallest decrease in conductivity; none should give unlimited solid solubility.

7 (a) 2330°C, 2150°C, 180°C,
(c) 2570°C, 2380°C, 190°C.

9 (a) 100% L containing 30% MgO.
(b) 70.8% L containing 38% MgO,
29.2% S containing 62% MgO.
(c) 8.3% L containing 38% MgO, 91.7% S containing 62% MgO.
(d) 100% S containing 85% MgO.

11 44.1 at% Cu – 55.9 at% Al.

13 (a) *L*:15 mol% MgO or 8.69 wt% MgO.
S: 38 mol% MgO or 24.85 wt% MgO.
(b) 78.26 mol% L or 80.1 wt% L;
21.74 mol% S or 19.9 wt% MgO.
(c) 78.1 vol% L, 21.9 vol% S.

15 750 g Ni, Ni/Cu = 1.62.

17 332 g MgO.

19 64.1 wt% FeO.

21 (a) 49 wt% W in L, 70 wt% W in α.
(b) Not possible.

23 88.2 kg W; 500 kg W.

25 Ni dissolves. When liquid reaches 10% Ni, the bath begins to freeze.

27 (a) 2900°C, 2690°C, 210°C.
(b) 60% L containing 49% W,
40% α containing 70% W.

29 (a) 55% W. (b) 18% W.

31 (a) 2000°C. (b) 1450°C. (c) 550°C.
(d) 40% FeO. (e) 92% FeO.
(f) 65.5% L containing 75% FeO,
34.5% S containing 46% FeO.
(g) 30.3% L containing 88% FeO,
69.7% S containing 55% FeO.

33 (a) 3100°C. (b) 2720°C. (c) 380°C.
(d) 90% W. (e) 40% W.
(f) 44.4% L containing 70% W,
55.6% α containing 88% W.
(g) 9.1% L containing 50% W,
90.9% α containing 83% W.

35 (a) 2900°C. (b) 2710°C. (c) 190°C.
(d) 2990°C. (e) 90°C. (f) 300 s.
(g) 340 s. (h) 60% W.

Chapter 10

1 (a) θ. (b) $\alpha, \beta, \gamma, \eta$.

(c) 1 100°C: peritectic.
900°C: monotectic.
680°C: eutectic.
600°C: peritectoid.
300°C: eutectoid.

3 (a) $CuAl_2$.

(b) 548°C, eutectic, $L \rightarrow \alpha + \theta$,
33.2% Cu in L, 5.65% Cu in α,
52.5% Cu in θ.

5 $SnCu_3$.

7 3 solid phases.

9 (a) 2.5% Mg.

(b) 600°C, 470°C, 400°C, 130°C.

(c) 74% α containing 7% Mg,
26% L containing 26% Mg.

(d) 100% α containing 12% Mg.

(e) 67% α containing 1% Mg,
33% β containing 34% Mg.

11 (a) Hypereutectic.

(b) 98% Sn.

(c) 22.8% β containing 97.5% Sn,
77.2% L containing 61.9% Sn.

(d) 35% α containing 19% Sn,
65% β containing 97.5% Sn.

(e) 22.8% primary β containing 97.5% Sn,
77.2% eutectic containing 61.9% Sn.

(f) 30% α containing 2% Sn,
70% β containing 100% Sn.

13 (a) Hypoeutectic.

(b) 1% Si.

(c) 78.5% α containing 1.65% Si,
21.5% L containing 12.6% Si.

(d) 97.6% α containing 1.65% Si,
2.4% β containing 99.83% Si.

(e) 78.5 primary α containing 1.65% Si,
21.5% eutectic containing 12.6% Si.

(f) 96% α containing 0% Si,
4% β containing 100% Si.

15 Hypoeutectic.

17 52% Sn.

19 (a) 4% Li. (c) 3% Cu.

21 Hypereutectic.

(b) 64% α, 36% β.

23 0.54.

25 (a) 1 150°C. (b) 150°C. (c) 1 000°C.
(d) 577°C. (e) 423°C. (f) 10.5 min.
(g) 11.5 min. (h) 45% Si.

29 (a) Yes, $T_m = 2040°C > 1900°C$.

(b) No, forms 5% L.

31 (a) 390°C, γ, $\gamma + \alpha$.

(b) 330°C, β, $\alpha + \beta$.

(c) 290°C, β, $\alpha + \beta + \gamma$.

Chapter 11

1 $c = 6.47 \times 10^{-6}$, $n = 2.89$.

3 252049 $J.mol^{-1}$.

5 For Al-4% Mg: solution treat between 210 and 451°C, quench, age below 210°C.
For Al-12% Mg: solution treat between 390 and 451°C, quench, age below 390°C.

7 (a) Solution treat between 290 and 400°C, quench, age below 290°C.

(c) Not good candidate.

(e) Not good candidate.

9 (a) 795°C.

(b) Primary ferrite.

(c) 56.1% ferrite containing 0.0218% C and 43.9% austenite containing 0.77% C.

(d) 95.1% ferrite containing 0.0218% C and 4.9% cementite containing 6.67% C.

(e) 56.1% primary ferrite containing 0.0218% C and 43.9% pearlite containing 0.77% C.

11 0.53% C, hypoeutectoid.

13 0.156% C, hypoeutectoid.

15 0.281% C.

17 760°C, 0.212% C.

19 (a) 900°C: 12% CaO in tetragonal, 3% CaO in monoclinic, 14% CaO in cubic; 18% monoclinic and 82% cubic.

(c) 250°C; 47% Zn in β', 36% Zn in α, 59% Zn in γ; 52.2% α, 47.8% γ.

21 (a) 615°C. (b) 0.167×10^{-3} mm.

23 Bainite with HRC 47.

25 Martensite with HRC 66.

27 37.2% martensite with 0.77% C and HRC 65.

(c) 84.8% martensite with 0.35% C and HRC 58.

29 (a) 750°C. (b) 0.455% C.

31 3.06% expansion.

33 Austenitise at 750°C, quench, temper above 330°C.

Chapter 12

1 (a) 97.2% ferrite, 2.2% cementite, 82.9% primary ferrite, 17.1% pearlite.

(c) 85.8% ferrite, 14.2% cementite, 3.1% primary cementite, 96.9% pearlite.

3 For 1035: A_1 = 727°C; A_3 = 790°C; anneal = 820°C; normalise = 845°C; process anneal = 557–647°C; not usually spheroidised.

5 (a) Ferrite and pearlite.

(c) Martensite.

(e) Ferrite and bainite.

(g) Tempered martensite.

7 Austenitise at 820°C, hold at 600°C for 10 s, cool.

(c) Austenitise at 780°C, hold at 600°C for 10 s, cool.

(e) Austenitise at 900°C, hold at 320°C for 5 000 s, cool.

9 (a) Austenitise at 820°C, quench, temper between 420 and 480°C; 1 030 to 1 240 MN.m^{-2} tensile, 970 to 1 100 MN.m^{-2} yield.

(b) 1 210 to 1 240 MN.m^{-2} tensile, 900 to 930 MN.m^{-2} yield.

(c) 690 MN.m^{-2} tensile, 450 yield, 20% elongation.

11 0.48% C in martensite; austenitised at 770°C; should austenitise at 860°C.

13 1080: fine pearlite. 4340: martensite.

15 May become hypereutectoid, with grain boundary cementite.

17 Not applicable.

(c) 8 to 10°C/s. (e) 32 to 36°C/s.

19 (a) 16°C/s. (b) Pearlite with HRC 38.

21 (a) Pearlite with HRC 36.

(c) Pearlite and martensite with HRC 46.

23 (a) 33 mm. (c) 48 mm.

(e) greater than 64 mm.

25 0.26 hours.

27 0.05 mm: pearlite and martensite with HRC 53. 0.15 mm: medium pearlite with HRC 38.

29 δ-ferrite; nonequilibrium freezing; quench anneal.

31 2.4% Si.

33 (a) Less ferrite. (b) Cooling rate effect.

35 Ductile iron is most hardenable; steel is least hardenable.

Chapter 13

3 Eutectic microconstituent contains 97.6% β, so is brittle.

5 Sand: 0.071 mm, 200 s.
Permanent: 0.02 mm, 20 s.
Die: 0.01 mm, 2.0 s.
Die casting has highest strength.

7 200 $MN.m^{-2}$ tensile, 180 $MN.m^{-2}$ yield, 3% elongation.

9 27% β for Al-7% Li versus 2.2% β for 2090.

11 Al-10% Mg.

13 (a) 2.89 mm, 6.8 g, $0.020.
(b) 2.89 mm, 10.6 g, $0.014.

15 $\alpha + \varepsilon$ (equilibrium conditions).

17 Al: 440%. Mg: 130%. Cu: 1100%.

19 Lead may melt during hot working.

21 γ' more at low temperature.

23 Large formed first at higher temperature; solubility decreases as temperature decreases.

25 Ti-15% V: 100% β transforms to 100% α', which then transforms to 24% β precipitate in an α matrix.
Ti-35% V: 100% β transforms to 100% β', which then transforms to 27% α precipitate in a β matrix.

27 Spalls off; cracks.

29 Al: 192 000 $m^2.s^{-2}$.
Cu: 135 000 $m^2.s^{-2}$.
Ni: 85 000 $m^2.s^{-2}$.

Chapter 14

1 0.4 nm; 0.63; 6.053 $Mg.m^{-3}$.

3 (a) 3. (b) 0.52.

5 (a) Metasilicate.
(c) Metasilicate.
(e) Pyrosilicate.

7 12.8 g.

9 4.40×10^{21} vacancies/cm^3.

11 (b) Magnesium vacancies.

13 B = 2.4; true = 22.58%; fraction = 0.044.

15 1.257 kg BaO; 0.245 kg Li_2O.

17 PbO_2 is modifier; PbO is intermediate.

19 0.095 mm.

21 0.76 mm.

23 No.

25 m = 1.33.

27 2×10^7 poise; 597°C.

29 327°C.

31 $d\varepsilon/dt = 0.049/d^2$.

33 60.8% Al_2O_3.

35 177 kg;
34.8 wt% Al_2O_3-37.5 wt% SiO_2-27.7 wt% CaO.

Chapter 15

1 (a) 2500. (b) 2.4×10^{18}.

3 8.748×10^{-6} m.

5 (a) 4798. (b) 9597.

7 186.69 $g.mol^{-1}$.

9 (a) 1.598 kg. (b) 1.649 kg. (c) 4.948 kg.

11 (a) H_2O. (b) 26.77 kg, 5.81 kg, 30.96 kg.

13 (a) 211. (b) 175.

17 Polybutadiene and silicone.

19 Polyethylene and polypropylene.

21 4 repeat units;
8 C atoms, 12 H atoms, 4 Cl atoms.

23 74.2%.

25 $a = 4 \times 10^{-13}$, $n = 8.16$, $\sigma = 6.7$ MPa.

27 8.73 $MN.m^{-2}$.

29 (a) at 4 MN.m $a = 0.004$,
$n = 0.09$
at 10 $MN.m^{-2}$,
$a = 0.0132$, $n = 0.16$.
(b) 3.9%.

31 (a) PE. (b) LDPE. (c) PTFE.

33 At $\varepsilon = 1$, $E = 5.7$ MN.m^{-2}; at $\varepsilon = 4$, $E = 13.9$ MN.m^{-2}.

35 0.0105.

37 6.383 kg; 3.83 kg.

41 After.

Chapter 16

1 7.65×10^{10} per mm^3.

3 2.47%.

5 9.408 Mg.m^{-3}.

7 (a) 0.507.
(b) 0.507.
(c) 7.775 Mg.m^{-3}.

9 11.18 to 22.2 kg.

11 (a) 2.53 Mg.m^{-3}. (b) 200.7 GN.m^{-2}.
(c) 106.8 GN.m^{-2}.

13 0.965.

15 187.3 MN.m^{-2}.

17 For $d = 20$ μm, $l_c = 2.96$ mm, $l_c/d = 148$.
For $d = 1\mu$m, $l_c = 1.48$ mm, $l_c/d = 1480$.

19 Sizing improves strength and reduces l_c.

23 Pyrolise at 2000°C; 1900 MN.m^{-2}.

25 $E_{\text{parallel}} = 69.84$ GN.m^{-2};
$E_{\text{perpendicular}} = 18.36$ GN.m^{-2}.

27 $E_{\text{parallel}} = 82.04$ GN.m^{-2};
$E_{\text{perpendicular}} = 69.60$ GN.m^{-2}.

29 0.417 Mg.m^{-3}; 20.0 kg versus 129.6 kg.

Chapter 17

1 (a) 0.945 litres.
(b) 0.77 Mg.m^{-3}.

3 Expands 2330 mm perpendicular to boards and 40 mm parallel to boards.

5 (a) 3500 sacks of cement, 1088 Mg aggregate, 24.14 m^3 water, 512 Mg sand.
(b) 18 Mg.m^{-3}. (c) 1 : 2.93 : 6.22.

Chapter 18

1 (a) 3380 W,
(b) 2.546×10^{14} W.
(c) 1.273×10^{10} to 1.273×10^{11} W.

3 $d = 0.865$ mm; 1174 V.

5 0.968.

7 0.234 km.h^{-1}.

9 3.03×10^7 Ω^{-1}.m^{-1} at –50°C;
0.34×10^7 Ω^{-1}.m^{-1} at 500°C.

11 –70.8°C.

13 At 400°C, $\rho = 41.5 \times 10^{-8}$ Ω.m;
$\rho_d = 8.5 \times 10^{-8}$ Ω.m;
$b = 178.9 \times 10^{-8}$ Ω.m.
At 200°C, $\rho = 37.6 \times 10^{-8}$ Ω.m.

15 2.64×10^7 A.m^{-1}.

17 39.3 A.

19 $\mu_{500} = 7.3 \times 10^{-34}$ m^2.V^{-1}.s^{-1};
$\sigma_{500} = 1.66 \times 10^{-23}$ Ω^{-1}.m^{-1}.

21 (a) n(Ge) = 1.767×10^{20} per mm^3.
(b) f(Ge) = 1.259×10^{-10}.
(c) n_0(Ge) = 1.017×10^{16} per mm^3.

23 (a) 8.32×10^{-8} s. (b) 5.75×10^{-7} s.

25 Sb: 1520 Ω^{-1}.m^{-1}. In: 399 Ω^{-1}.m^{-1}.

27 3754 Ω^{-1}.m^{-1}.

29 (a) 1.485×10^{17} per mm^3.
(b) 428000 Ω^{-1}.m^{-1}.

31 2600%.

33 (a) 4.85×10^{-19} m.
(c) 1.12×10^{-17} m.

35 9.4 V.

37 12000 V.

39 0.001238 μF.

41 0.42 mm.

43 0.0003 m.m^{-1}.

Chapter 19

1 Fe: 3.15×10^6 A.m^{-1}.
Co: 2.51×10^6 A.m^{-1}.

3 5.1×10^5 A.m^{-1}.

5 0.8 T; 1.49 mA.

9 (a) 1.3 T.
(b) 1.4 T.
(c) 65 000 A.m^{-1}.
(d) 7.3×10^{-6} T.m.A^{-1}.
(e) 2.0×10^{-5} T.m.A^{-1}.
(f) 54 000 T.A. m^{-1}.

11 12 000 T.A.m^{-1}.

13 High saturation inductance.

15 4.68×10^5 A.m^2 per m^3.

Chapter 20

1 13 790 V.

3 7 477 V.

5 (a) 24 825 V.
(b) Cu, Mn, Si.

7 1.84×10^{-5} s.

9 3.733×10^{-19} J; green.

11 (a) 1.311×10^{-6} m.
(b) 7.758×10^{-6} m.

13 (a) 1.853×10^{-6} m.
(b) 1.034×10^{-4} m.

15 Ice, water, Teflon.

17 (a) 6.60°. (b) 6.69°. (c) 10°.
(d) 1.78 mm.

19 0.117 mm.

21 4%; 0.36%.

23 At 0.0711 nm, $I = 8.7$;
at 0.1436 nm, $I = 45 \times 10^{-6}$.

Chapter 21

1 (a) 7950 J. (c) 35.564 J.

3 0.375 mm.

5 1.975 m.

7 379.6 mm.

9 Brass: 250.614 mm, Invar; 250.050 mm.

11 1 766 MN.m^{-2}; tensile.

13 No; 169 MN.m^{-2};
0.24 mm decrease in length.

15 (a) 2.44×10^8 J/day.
(b) 1.46×10^8 J/day.

17 19.6 min.

19 Interconnected graphite flakes in grey iron.

Chapter 22

1 Graphitic corrosion.

3 −0.172 V.

5 0.000034 g/1000 ml.

7 55A.

9 34 years.

11 187.5 g Fe lost/hour.

13 1100 alloy is anode and continues to protect 2024; 1100 alloy is cathode and the 3003 will corrode.

15 Al, Zn, Cd.

17 Stress corrosion cracking.

19 Ti is the anode, carbon will be the cathode.

23 Cold worked will corrode most rapidly, annealed most slowly.

27 (a) 12.2 g/hour. (b) 1.22 g/hour.

29 Most ceramics are already oxides.

31 698 °C.

Chapter 23

9 For X-radiography, the μ_m for the alloy, pure Si and pure Al is very similar (15.64, 15.9 and 15.6 $m^2.Mg^{-1}$, respectively). For neutron radiography there is a large percentage difference between the alloy ($\mu_m = 0.1$ $m^2.Mg^{-1}$) and the pure Al ($\mu_m =$ 0.3 $m^2.Mg^{-1}$). Hence neutron radiography can determine segregation in the Al-Si alloy.

11 I/I_0 solid = 0.557;
I/I_0 for inclusion = 0.0088.

13 0.41 mm.

15 (a) For Mg, HVL = 26.2 mm.
(b) For Cu, HVL = 14.1 mm.
(c) For Mg, HVL = 3 984 mm.

17 1.26 years.

19 72.24 mm.

21 Stainless steel: 18.94 mm.
Aluminium: 12.81 mm.

23 Signals expected after 8.31×10^{-7} s, 2.11×10^{-6} s, and 2.94×10^{-6} s.

25 0.095 mm.

27 Thermography, ultrasonic.

29 Dye penetrant, radiography, eddy current.

Metal		Atomic Number	Crystal Structure	Lattice Parameter (nm)	Atomic Mass ($g.mol^{-1}$)	Density ($Mg.m^{-3}$)	Melting Temperature (°C)
Aluminium	Al	13	FCC	0.404958	26.981	2.699	660.4
Antimony	Sb	51	hex	a = 0.4307	121.75	6.697	630.7
				c = 1.1273			
Arsenic	As	33	hex	a = 0.3760	74.9216	5.778	816
				c = 1.0548			
Barium	Ba	56	BCC	0.5025	137.3	3.5	729
Beryllium	Be	4	hex	a = 0.22858	9.01	1.848	1290
				c = 0.35842			
Bismuth	Bi	83	hex	a = 0.4546	208.98	9.808	271.4
				c = 1.186			
Boron	B	5	rhomb	a = 1.012	10.81	2.3	2300
				α = 65.5°			
Cadmium	Cd	48	HCP	a = 0.29793	112.4	8.642	321.1
				c = 0.56181			
Caesium	Cs	55	BCC	0.613	132.91	1.892	28.6
Calcium	Ca	20	FCC	0.5588	40.08	1.55	839
Cerium	Ce	58	HCP	a = 0.3681	140.12	6.6893	798
				c = 1.1857			
Chromium	Cr	24	BCC	0.28844	51.996	7.19	1875
Cobalt	Co	27	HCP	a = 0.25071	58.93	8.832	1495
				c = 0.40686			
Copper	Cu	29	FCC	0.36151	63.54	8.93	1084.9
Gadolinium	Gd	64	HCP	a = 0.36336	157.25	7.901	1313
				c = 0.57810			
Gallium	Ga	31	ortho	a = 0.45258	69.72	5.904	29.8
				b = 0.45186			
				c = 0.76570			
Germanium	Ge	32	FCC	0.56575	72.59	5.324	937.4
Gold	Au	79	FCC	0.40786	196.97	19.302	1064.4
Hafnium	Hf	72	HCP	a = 0.31883	178.49	13.31	2227
				c = 0.50422			
Indium	In	49	tetra	a = 0.32517	114.82	7.286	156.6
				c = 0.49459			
Iridium	Ir	77	FCC	0.384	192.9	22.65	2447
Iron	Fe	26	BCC	0.2866	55.847	7.87	1538
			FCC	0.3589	(>912°C)		
			BCC		(>1394°C)		
Lanthanum	La	57	HCP	a = 0.3774	138.91	6.146	918
				c = 1.217			
Lead	Pb	82	FCC	0.49489	207.19	11.36	327.4
Lithium	Li	3	BCC	0.35089	6.94	0.534	180.7
Magnesium	Mg	12	HCP	a = 0.32087	24.312	1.738	650
				c = 0.5209			
Manganese	Mn	25	cubic	0.8931	54.938	7.47	1244

Metal		Atomic Number	Crystal Structure	Lattice Parameter (nm)	Atomic Mass (g.mol^{-1})	Density (Mg.m^{-3})	Melting Temperature (°C)
Mercury	Hg	80	rhomb		200.59	13.546	−38.9
Molybdenum	Mo	42	BCC	0.31468	95.94	10.22	2610
Nickel	Ni	28	FCC	0.35167	58.71	8.902	1453
Niobium	Nb	41	BCC	0.3294	92.91	8.57	2468
Osmium	Os	76	HCP	$a = 0.27341$ $c = 0.43197$	190.2	22.57	2700
Palladium	Pd	46	FCC	0.38902	106.4	12.02	1552
Platinum	Pt	78	FCC	0.39231	195.09	21.45	1769
Potassium	K	19	BCC	0.5344	39.09	0.855	63.2
Rhenium	Re	75	HCP	$a = 0.2760$ $c = 0.4458$	186.21	21.04	3180
Rhodium	Rh	45	FCC	0.3796	102.99	12.41	1963
Rubidium	Rb	37	BCC	0.57	85.467	1.532	38.9
Ruthenium	Ru	44	HCP	$a = 0.26987$ $c = 0.42728$	101.07	12.37	2310
Selenium	Se	34	hex	$a = 0.43640$ $c = 0.49594$	78.96	4.809	217
Silicon	Si	14	FCC	0.54307	28.08	2.33	1410
Silver	Ag	47	FCC	0.40862	107.868	10.49	961.9
Sodium	Na	11	BCC	0.42906	22.99	0.967	97.8
Strontium	Sr	38	FCC BCC	0.60849 0.484	87.62 (>557°C)	2.6	768
Tantalum	Ta	73	BCC	0.33026	180.95	16.6	2996
Technetium	Tc	43	HCP	$a = 0.2735$ $c = 0.4388$	98.9062	11.5	2200
Tellurium	Te	52	hex	$a = 0.44565$ $c = 0.59268$	127.6	6.24	449.5
Thorium	Th	90	FCC	0.5086	232	11.72	17.75
Tin	Sn	50	FCC	0.64912	118.69	5.765	231.9
Titanium	Ti	22	HCP BCC	$a = 0.29503$ $c = 0.46831$ 0.332	47.9 (>882°C)	4.507	1668
Tungsten	W	74	BCC	0.31652	183.85	19.254	3410
Uranium	U	92	ortho	$a = 0.2854$ $b = 0.5869$ $c = 0.4955$	238.03	19.05	1133
Vanadium	V	23	BCC	0.30278	50.941	6.1	1900
Yttrium	Y	39	HCP	$a = 0.3648$ $c = 0.5732$	88.91	4.469	1522
Zinc	Zn	30	HCP	$a = 0.26648$ $c = 0.49470$	65.38	7.133	420
Zirconium	Zr	40	HCP BCC	$a = 0.32312$ $c = 0.51477$ 0.36090	91.22 (>862°C)	6.505	1852

A

Element	Atomic Radius (nm)	Valence	Ionic Radius (nm)
Aluminium	0.1432	+3	0.051
Antimony		+5	0.062
Arsenic		+5	0.222
Barium	0.2176	+2	0.134
Beryllium	0.1143	+2	0.035
Bismuth		+5	0.074
Boron	0.046	+3	0.023
Bromine	0.119	–1	0.196
Cadmium	0.149	+2	0.097
Caesium	0.265	+1	0.167
Calcium	0.1976	+2	0.099
Carbon	0.077	+4	0.016
Cerium	0.184	+3	0.1034
Chlorine	0.0905	–1	0.181
Chromium	0.1249	+3	0.063
Cobalt	0.1253	+2	0.072
Copper	0.1278	+1	0.096
Fluorine	0.06	–1	0.133
Gallium	0.1218	+3	0.062
Germanium	0.1225	+4	0.053
Gold	0.1442	+1	0.137
Hafnium		+4	0.078
Hydrogen	0.046	+1	0.154
Indium	0.1570	+3	0.081
Iodine	0.135	–1	0.220
Iron	0.1241 (BCC)	+2	0.074
	0.1269 (FCC)	+3	0.064
Lanthanum	0.1887	+3	0.1016
Lead	0.175	+4	0.084
Lithium	0.1519	+1	0.068
Magnesium	0.1604	+2	0.066
Manganese	0.112	+2	0.080
		+3	0.066
Mercury	0.155	+2	0.110
Molybdenum	0.1363	+4	0.070
Nickel	0.1243	+2	0.069
Niobium	0.1426	+4	0.074
Nitrogen	0.071	+5	0.015
Oxygen	0.060	–2	0.132

Element	Atomic Radius (nm)	Valence	Ionic Radius (nm)
Palladium	0.1375	+4	0.065
Phosphorus	0.110	+5	0.035
Platinum	0.1387	+2	0.080
Potassium	0.2314	+1	0.133
Rubidium	0.2468	+1	0.070
Selenium		−2	0.191
Silicon	0.1176	+4	0.042
Silver	0.1445	+1	0.126
Sodium	0.1858	+1	0.097
Strontium	0.2151	+2	0.112
Sulphur	0.106	−2	0.184
Tantalum	0.143	+5	0.068
Tellurium		−2	0.211
Thorium	0.1798	+4	0.102
Tin	0.1405	+4	0.071
Titanium	0.1475	+4	0.068
Tungsten	0.1371	+4	0.070
Uranium	0.138	+4	0.097
Vanadium	0.1311	+3	0.074
Yttrium	0.1824	+3	0.089
Zinc	0.1332	+2	0.074
Zirconium	0.1616	+4	0.079

Atomic		K	L		M			N				O			P	
Number	Element	1s	2s	2p	3s	3p	3d	4s	4p	4d	4f	5s	5p	5d	6s	6p
1	Hydrogen	1														
2	Helium	2														
3	Lithium	2	1													
4	Beryllium	2	2													
5	Boron	2	2	1												
6	Carbon	2	2	2												
7	Nitrogen	2	2	3												
8	Oxygen	2	2	4												
9	Fluorine	2	2	5												
10	Neon	2	2	6												
11	Sodium	2	2	6	1											
12	Magnesium	2	2	6	2											
13	Aluminium	2	2	6	2	1										
14	Silicon	2	2	6	2	2										
15	Phosphorus	2	2	6	2	3										
16	Sulphur	2	2	6	2	4										
17	Chlorine	2	2	6	2	5										
18	Argon	2	2	6	2	6										
19	Potassium	2	2	6	2	6		1								
20	Calcium	2	2	6	2	6		2								
21	Scandium	2	2	6	2	6	1	2								
22	Titanium	2	2	6	2	6	2	2								
23	Vanadium	2	2	6	2	6	3	2								
24	Chromium	2	2	6	2	6	5	1								
25	Manganese	2	2	6	2	6	5	2								
26	Iron	2	2	6	2	6	6	2								
27	Cobalt	2	2	6	2	6	7	2								
28	Nickel	2	2	6	2	6	8	2								
29	Copper	2	2	6	2	6	10	1								
30	Zinc	2	2	6	2	6	10	2								
31	Gallium	2	2	6	2	6	10	2	1							
32	Germanium	2	2	6	2	6	10	2	2							
33	Arsenic	2	2	6	2	6	10	2	3							
34	Selenium	2	2	6	2	6	10	2	4							
35	Bromine	2	2	6	2	6	10	2	5							
36	Krypton	2	2	6	2	6	10	2	6							
37	Rubidium	2	2	6	2	6	10	2	6			1				
38	Strontium	2	2	6	2	6	10	2	6			2				
39	Yttrium	2	2	6	2	6	10	2	6	1		2				
40	Zirconium	2	2	6	2	6	10	2	6	2		2				
41	Niobium	2	2	6	2	6	10	2	6	4		1				
42	Molybdenum	2	2	6	2	6	10	2	6	5		1				
43	Technetium	2	2	6	2	6	10	2	6	6		1				
44	Ruthenium	2	2	6	2	6	10	2	6	7		1				

Atomic		K	L			M			N				O		P	
Number	Element	1s	2s	2p	3s	3p	3d	4s	4p	4d	4f	5s	5p	5d	6s	6p
45	Rhodium	2	2	6	2	6	10	2	6	8		1				
46	Palladium	2	2	6	2	6	10	2	6	10						
47	Silver	2	2	6	2	6	10	2	6	10		1				
48	Cadmium	2	2	6	2	6	10	2	6	10		2				
49	Indium	2	2	6	2	6	10	2	6	10		2	1			
50	Tin	2	2	6	2	6	10	2	6	10		2	2			
51	Antimony	2	2	6	2	6	10	2	6	10		2	3			
52	Tellurium	2	2	6	2	6	10	2	6	10		2	4			
53	Iodine	2	2	6	2	6	10	2	6	10		2	5			
54	Xenon	2	2	6	2	6	10	2	6	10		2	6			
55	Caesium	2	2	6	2	6	10	2	6	10		2	6		1	
56	Barium	2	2	6	2	6	10	2	6	10		2	6		2	
57	Lanthanum	2	2	6	2	6	10	2	6	10	1	2	6		2	
	.	.	.	.	.	.	.	.	.	.	.	.	.		.	
	.	.	.	.	.	.	.	.	.	.	.	.	.		.	
	.	.	.	.	.	.	.	.	.	.	.	.	.		.	
71	Lutetium	2	2	6	2	6	10	2	6	10	14	2	6	1	2	
72	Hafnium	2	2	6	2	6	10	2	6	10	14	2	6	2	2	
73	Tantalum	2	2	6	2	6	10	2	6	10	14	2	6	3	2	
74	Tungsten	2	2	6	2	6	10	2	6	10	14	2	6	4	2	
75	Rhenium	2	2	6	2	6	10	2	6	10	14	2	6	5		
76	Osmium	2	2	6	2	6	10	2	6	10	14	2	6	6		
77	Iridium	2	2	6	2	6	10	2	6	10	14	2	6	9		
78	Platinum	2	2	6	2	6	10	2	6	10	14	2	6	9	1	
79	Gold	2	2	6	2	6	10	2	6	10	14	2	6	10	1	
80	Mercury	2	2	6	2	6	10	2	6	10	14	2	6	10	2	
81	Thallium	2	2	6	2	6	10	2	6	10	14	2	6	10	2	1
82	Lead	2	2	6	2	6	10	2	6	10	14	2	6	10	2	2
83	Bismuth	2	2	6	2	6	10	2	6	10	14	2	6	10	2	3
84	Polonium	2	2	6	2	6	10	2	6	10	14	2	6	10	2	4
85	Astatine	2	2	6	2	6	10	2	6	10	14	2	6	10	2	5
86	Radon	2	2	6	2	6	10	2	6	10	14	2	6	10	2	6

Index

Italic page numbers are references to Glossary

A

B

C

D

E

F

G

H

I

J

K

L

M

N

O

P

Q

R

S

T

U

V

W

X

Y

Z

Periodic Table of Elements

IA		IIIB	IVB	VB	VIB	VIIB	VIIIB			IB	IIB	IIIA	IVA	VA	VIA	VIIA	
1 H 1.00797																	2 He 4.003
3 Li 6.939	4 Be 9.012											5 B 10.81	6 C 12.011	7 N 14.007	8 O 15.9994	9 F 19.00	10 Ne 20.183
[illegible]	lg .31											13 Al 26.98	14 Si 28.09	15 P 30.974	16 S 32.064	17 Cl 35.453	18 Ar 39.948
[illegible]	a .08	21 Sc 44.96	22 Ti 47.90	23 V 50.94	24 Cr 52.00	25 Mn 54.94	26 Fe 55.85	27 Co 58.93	28 Ni 58.71	29 Cu 63.54	30 Zn 65.37	31 Ga 69.72	32 Ge 72.59	33 As 74.92	34 Se 78.96	35 Br 79.909	36 Kr 83.80
[illegible]	r .62	39 Y 88.905	40 Zr 91.22	41 Nb 92.91	42 Mo 95.94	43 Tc 98	44 Ru 101.1	45 Rh 102.90	46 Pd 106.4	47 Ag 107.87	48 Cd 112.4	49 In 114.82	50 Sn 118.69	51 Sb 121.75	52 Te 127.60	53 I 126.90	54 Xe 131.30
[illegible]	a 7.34		72 Hf 178.49	73 Ta 180.95	74 W 183.85	75 Re 186.2	76 Os 190.2	77 Ir 192.2	78 Pt 195.09	79 Au 196.97	80 Hg 200.59	81 Tl 204.37	82 Pb 207.19	83 Bi 208.98	84 Po 210	85 At 210	86 Rn 222
[illegible]	a 26																

Lanthanide series

Actinide series

57 La 138.91	58 Ce 140.12	59 Pr 140.91	60 Nd 144.24	61 Rm 147	62 Sm 150.35	63 Eu 152	64 Gd 157.25	65 Tb 158.92	66 Dy 162.50	67 Ho 164.93	68 Er 167.26	69 Tm 168.93	70 Yb 173.04	71 Lu 174.97
89 Ac 227	90 Th 232.04	91 Pa 231	92 U 238.03	93 Np 237	94 Pu 242	95 Am 243	96 Cm 247	97 Bk 247	98 Cf 251	99 Es 254	100 Fm 253	101 Md 256	102 No 254	103 Lw 257